T0298028

Monopoles and Three-Manifolds

Originating with Andreas Floer in the 1980s, Floer homology has proved to be an effective tool in tackling many important problems in 3- and 4-dimensional geometry and topology. This book provides a comprehensive treatment of Floer homology, based on the Seiberg–Witten monopole equations. After first providing an overview of the results, the authors develop the analytic properties of the Seiberg–Witten monopole equations, assuming only a basic grounding in differential geometry and analysis. The Floer groups of a general 3-manifold are then defined, and their properties studied in detail. Two final chapters are devoted to the calculation of Floer groups, and to applications of the theory in topology.

Suitable for beginning graduate students and researchers in the field, this book provides the first full discussion of a central part of the study of the topology of manifolds since the mid 1990s.

NEW MATHEMATICAL MONOGRAPHS

Editorial Board
Béla Bollobás
William Fulton
Anatole Katok
Frances Kirwan
Peter Sarnak
Barry Simon
Burt Totaro

All the titles listed below can be obtained from good booksellers or from Cambridge University Press. For a complete series listing visit http://www.cambridge.org/uk/series/sSeries.asp?code=NMM

1 M. Cabanes and M. Enguehard *Representation Theory of Finite Reductive Groups*
2 J. B. Garnett and D. E. Marshall *Harmonic Measure*
3 P. Cohn *Free Ideal Rings and Localization in General Rings*
4 E. Bombieri and W. Gubler *Heights in Diophantine Geometry*
5 Y. J. Ionin and M. S. Shrikhande *Combinatorics of Symmetric Designs*
6 S. Berhanu, P. D. Cordaro and J. Hounie *An Introduction to Involutive Structures*
7 A. Shlapentokh *Hilbert's Tenth Problem*
8 G. Michler *Theory of Finite Simple Groups I*
9 A. Baker and G. Wüstholz *Logarithmic Forms and Diophantine Geometry*
10 P. Kronheimer and T. Mrowka *Monopoles and Three-Manifolds*
11 B. Bekka, P. de la Harpe and A. Valette *Kazhdan's Property (T)*
12 J. Neisendorfer *Algebraic Methods in Unstable Homotopy Theory*
13 M. Grandis *Directed Algebraic Topology*
14 G. Michler *Theory of Finite Simple Groups II*
15 R. Schertz *Complex Multiplication*
16 S. Bloch *Lectures on Algebraic Cycles (2nd Edition)*
17 B. Conrad, O. Gabber and G. Prasad *Pseudo-reductive Groups*
18 T. Downarowicz *Entropy in Dynamical Systems*

Monopoles and Three-Manifolds

PETER KRONHEIMER
Harvard University
TOMASZ MROWKA
Massachusetts Institute of Technology

CAMBRIDGE
UNIVERSITY PRESS

CAMBRIDGE UNIVERSITY PRESS
Cambridge, New York, Melbourne, Madrid, Cape Town,
Singapore, São Paulo, Delhi, Tokyo, Mexico City

Cambridge University Press
The Edinburgh Building, Cambridge CB2 8RU, UK

Published in the United States of America by Cambridge University Press, New York

www.cambridge.org
Information on this title: www.cambridge.org/9780521880220

© P. Kronheimer and T. Mrowka 2007

This publication is in copyright. Subject to statutory exception
and to the provisions of relevant collective licensing agreements,
no reproduction of any part may take place without the written
permission of Cambridge University Press.

First published 2007
Paperback edition 2011

A catalogue record for this publication is available from the British Library

ISBN 978-0-521-88022-0 Hardback
ISBN 978-0-521-18476-2 Paperback

Cambridge University Press has no responsibility for the persistence or
accuracy of URLs for external or third-party internet websites referred to in
this publication, and does not guarantee that any content on such websites is,
or will remain, accurate or appropriate. Information regarding prices, travel
timetables, and other factual information given in this work is correct at
the time of first printing but Cambridge University Press does not guarantee
the accuracy of such information thereafter.

To Jenny and Gigliola

Contents

Preface

Gauge theory and related areas of geometry have been an important tool for the study of 4-dimensional manifolds since the early 1980s, when Donaldson introduced ideas from Yang–Mills theory to solve long-standing problems in topology. In dimension 3, the same techniques formed the basis of Floer's construction of his "instanton homology" groups of 3-manifolds [32]. Today, Floer homology is an active area, and there are several varieties of Floer homology theory, all with closely related structures. While Floer's construction used the anti-self-dual Yang–Mills (or *instanton*) equations, the theory presented in this book is based instead on the Seiberg–Witten equations (or *monopole* equations).

We have aimed to lay a secure foundation for the study of the Seiberg–Witten equations on a general 3-manifold, and for the construction of the associated Floer groups. Our goal has been to write a book that is complete in its coverage of several aspects of the theory that are hard to find in the existing literature, providing at the same time an introduction to the techniques from analysis and geometry that are used. We have omitted some background topics that are now well treated in several good sources: in particular, the Seiberg–Witten invariants of closed 4-manifolds and related topics in gauge theory are given only a brief exposition here. The main results of this book – the formal properties of the Floer groups that we construct – can be summarized without delving too far into the techniques which lie beneath; so we present such a summary in Chapter I. The final chapter of the book touches on some further topics and describes how the theory has been applied to questions in topology.

The definition of the Floer groups that we present here is new in some aspects. We believe that our approach to the Morse homology of a manifold with circle action has not appeared before. It is described in Section 2, along with a closely related approach to Morse theory on a manifold with boundary. Our definition of the groups that we call $\widecheck{HM}(Y)$ and $\widehat{HM}(Y)$ has roots in lectures given by Donaldson in Oxford in 1993. For the case that the first Betti number of Y is zero, a similar construction is described in [22] for the case of the instanton Floer theory, and there is related material due to Frøyshov in [40]. Another approach to the Seiberg–Witten version of Floer homology is presented by Marcolli and Wang in [71].

During the course of this work, a completely different approach to Floer homology was introduced by Ozsváth and Szabó in [93]. The construction of their "Heegaard homology" of 3-manifolds is not based on gauge theory, but appears to be entirely equivalent to the Seiberg–Witten version. Ozsváth and Szabó's theory has influenced the development of this book, most particularly because of the way in which it has clarified the formal structure of Floer homology. We have sometimes tailored our account to emphasize the similarities between the two versions. Heegaard homology has spurred tremendous activity in the topological applications of Floer theory. Chapter X provides a small sample of results from this rapidly moving field.

Acknowledgements. Gauge theory is now a mature subject, and the analysis on which it rests has deep roots. Much of the material that we present is therefore not original. When a particular argument is taken directly from a unique source, we have tried to cite the source at the relevant point in the text. More often, however, pointers to the earlier literature are to be found in the remarks collected at the end of each chapter. Among the many mathematicians who have contributed to this field, we would like to acknowledge particularly our debt to Simon Donaldson, Kim Frøyshov, Peter Ozsváth, Zoltán Szabó and Cliff Taubes.

This work was supported in part by the Institute for Advanced Study and the National Science Foundation through grants DMS-9531964, DMS-9803166, DMS-0100771, DMS-011129, DMS-0244663, DMS-0206485 and DMS-0405271. The authors would also like to thank Ron Fintushel, Larry Guth, Yi-Jen Lee, Yanki Lekili, Max Lipyanski, Tim Perutz, Yann Rollin, Jake Rasmussen, Peter Ozsváth, Zoltán Szabó and Fangyun Yang for many comments and corrections.

I
Outlines

The three parts of this chapter provide outline accounts of three different topics. While all three are central to the subject of this book, the outlines serve three different purposes. In Section 1, we give a brief account of the Seiberg–Witten invariants, or monopole invariants, of smooth, closed 4-manifolds. These invariants, discovered by Seiberg and Witten and originally described in Witten's paper [125], are now the subject of several expository papers, published lecture notes and books. Our purpose here is to review the definition and main properties of these invariants, while establishing our notation and conventions.

Section 2 covers Morse theory, and specifically the manner in which one can recover the ordinary homology of a manifold *with boundary* from a "Morse complex", constructed from the data provided by the critical points and gradient-flow lines of a suitable Morse function. There are no proofs in this section. In the main part of this book, the Floer homology of a 3-manifold will be constructed by taking these constructions of Morse theory and repeating them in an infinite-dimensional setting. Proofs of the main propositions are presented only in the more difficult context of Floer homology; the finite-dimensional constructions are presented here for motivation, to provide a framework that explains the origin of many arguments. Although some notation is introduced, no essential use is made of this material in the later chapters.

Finally in this chapter, Section 3 provides an outline of the main results of this book. We describe the principal features and properties of the monopole Floer homology groups of 3-manifolds; we explain how their construction is related to the Morse theory of Section 2, and we explain the role of Floer homology in computing the monopole invariants of closed 4-manifolds.

1 Monopole invariants of four-manifolds

1.1 Spin^c structures

Spin^c structures can be considered on manifolds of any dimension, but we will focus here on dimensions 3 and 4, the two cases we will need. We begin with 3-manifolds.

Let Y be a closed, oriented, Riemannian 3-manifold. A spin^c structure on Y consists of a unitary rank-2 vector bundle $S \to Y$ with a Clifford multiplication

$$\rho : TY \to \mathrm{Hom}(S, S).$$

Clifford multiplication is a bundle map that identifies TY isometrically with the subbundle $\mathfrak{su}(S)$ of traceless, skew-adjoint endomorphisms equipped with the inner product $\frac{1}{2}\,\mathrm{tr}(a^*b)$. It also respects orientation, which by convention means that

$$\rho(e_1)\rho(e_2)\rho(e_3) = 1$$

when the e_i are an oriented frame. Given any oriented frame at a point y in Y, these conditions mean that we can choose a basis for the fiber S_y such that the matrices of the linear transformations $\rho(e_i)$ are the three Pauli matrices σ_i:

$$\sigma_1 = \begin{bmatrix} i & 0 \\ 0 & -i \end{bmatrix}, \quad \sigma_2 = \begin{bmatrix} 0 & -1 \\ 1 & 0 \end{bmatrix}, \quad \sigma_3 = \begin{bmatrix} 0 & i \\ i & 0 \end{bmatrix}. \tag{1.1}$$

The action of ρ is extended to cotangent vectors using the metric, and then to forms using the rule

$$\rho(\alpha \wedge \beta) = \frac{1}{2}\big(\rho(\alpha)\rho(\beta) + (-1)^{\deg(\alpha)\deg(\beta)}\rho(\beta)\rho(\alpha)\big).$$

We also extend ρ to complex forms, so that it gives, for example, an isomorphism

$$\rho : T^*Y \otimes \mathbb{C} \to \mathfrak{sl}(S).$$

Our orientation convention means that $\rho(*\alpha) = -\rho(\alpha)$ for 1-forms α.

Because the tangent bundle of an oriented 3-manifold is always trivial, a spin^c structure always exists: we can simply take S to be the product bundle $\mathbb{C}^2 \times Y$ and then define Clifford multiplication globally by the matrices (1.1), using any trivialization of TY. To understand the classification of spin^c structures in general, the important observation is that if we are given one spin^c structure, say (S_0, ρ_0) on Y, then we can construct a new spin^c structure (S, ρ) as follows. Choose any hermitian line bundle $L \to Y$, and define

$$\begin{aligned} S &= S_0 \otimes L \\ \rho(e) &= \rho_0(e) \otimes 1_L. \end{aligned} \tag{1.2}$$

The following proposition tells us that any (S, ρ) can be obtained from (S_0, ρ_0) in this way, for a uniquely determined L, up to isomorphism.

Proposition 1.1.1. *Given a single spin^c structure (S_0, ρ_0), the construction* (1.2) *establishes a one-to-one correspondence between:*

(i) *the isomorphism classes of spin^c structures (S, ρ) on Y; and*
(ii) *the isomorphism classes of complex line bundles $L \to Y$.*

Because line bundles L are classified by their first Chern class $c_1(L) \in H^2(Y;\mathbb{Z})$, we can equivalently replace (ii) *here by:*

(iii) *the elements of $H^2(Y;\mathbb{Z})$.*

Proof. Let us show that any (S, ρ) can be obtained from (S_0, ρ_0) by tensoring with a suitable line bundle.

Given spin^c structures (S', ρ') and (S, ρ) on Y, we can define a vector bundle L on Y as the subbundle of $\mathrm{Hom}(S', S)$ consisting of homomorphisms that intertwine ρ' and ρ. This L has rank 1 (it is a line bundle): this is a manifestation of Schur's lemma and reflects the fact that only the scalar endomorphisms of S commute with the image of $\rho : TX \to \mathrm{End}(S)$. We call L the difference line bundle. If the difference line bundle is trivial, then a global section of unit length provides an isomorphism between the spin^c structures.

To apply this construction, let (S, ρ) be a spin^c structure and consider the difference line bundle L between (S_0, ρ_0) and (S, ρ). Set $S' = S_0 \otimes L$, and let ρ' be the Clifford multiplication $\rho \otimes 1_L$. Then the difference line bundle between S' and S is the trivial bundle $L^{-1} \otimes L$. So (S', ρ') and (S, ρ) are isomorphic spin^c structures. □

We will usually use $\mathfrak{s}$ to denote a typical spin^c structure (S, ρ). If $\mathfrak{s}_0$ is a chosen spin^c structure and L has first Chern class $l \in H^2(Y;\mathbb{Z})$, then we write

$$\mathfrak{s} = \mathfrak{s}_0 + l$$

for the spin^c structure defined by (1.2). The way we have defined it, a spin^c structure depends on a prior choice of Riemannian metric. However, if g_0 and

g_1 are two metrics on Y, and $\mathfrak{s}_0$, $\mathfrak{s}_1$ are corresponding spin^c structures, we can still compare the two: we can ask if there is a path g_t in the (contractible) space of metrics, joining g_0 to g_1, and a corresponding family (S_t, ρ_t), forming a continuous family over $[0, 1]$. We can therefore think of the set of isomorphism classes of spin^c structures as being associated to a smooth oriented manifold Y.

On an oriented 4-dimensional Riemannian manifold X, a spin^c structure again provides a hermitian vector bundle $S_X \to X$, this time of rank 4, with a Clifford multiplication

$$\rho : TX \to \mathrm{Hom}(S_X, S_X),$$

such that at each $x \in X$ we can find an oriented orthonormal frame $e_0, \ldots, e_3$ with

$$\rho(e_0) = \begin{bmatrix} 0 & -I_2 \\ I_2 & 0 \end{bmatrix}, \quad \rho(e_i) = \begin{bmatrix} 0 & -\sigma_i^* \\ \sigma_i & 0 \end{bmatrix} \quad (i = 1, 2, 3) \tag{1.3}$$

in some orthonormal basis of the fiber S_x. Here I_2 is the 2-by-2 identity matrix and σ_i is as above. If we extend Clifford multiplication to (complex) forms as before, then in the same basis for S_x we have

$$\rho(\mathrm{vol}_x) = \begin{bmatrix} -I_2 & 0 \\ 0 & I_2 \end{bmatrix}$$

where $\mathrm{vol} = e_0 \wedge e_1 \wedge e_2 \wedge e_3$ is the oriented volume form. So the eigenspaces of $\rho(\mathrm{vol})$ give a decomposition of S_X into two orthogonal rank-2 bundles. We define S^+ to be the -1 eigenspace, and write

$$S_X = S^+ \oplus S^-.$$

Clifford multiplication by a tangent vector is an *odd* linear transformation: it interchanges the two summands, and we can write

$$\rho(e) : S^+ \to S^-.$$

If ν is a 2-form, then $\rho(\nu)$ preserves the two summands. In dimension 4, the bundle of 2-forms $\Lambda^2 X$ decomposes as a sum of the self-dual and anti-self-dual forms,

$$\Lambda^2 X = \Lambda^+ \oplus \Lambda^-,$$

the $+1$ and -1 eigenbundles of the Hodge $*$ operator. A short calculation with the matrices above shows that, if $\nu \in \Lambda^+$, then $\rho(\nu)$ restricts to zero on S^-, and vice versa. We have maps

$$\begin{aligned} \rho &: \Lambda^+ \to \mathfrak{su}(S^+) \\ \rho &: \Lambda^- \to \mathfrak{su}(S^-) \end{aligned} \tag{1.4}$$

which are bundle isometries. For $e \in T_xX$ a unit vector, the determinant of $\rho(e) : S_x^+ \to S_x^-$ is a map

$$\det \rho(e) : \Lambda^2 S_x^+ \to \Lambda^2 S_x^-$$

that is independent of e. So the complex line bundles $\Lambda^2 S^+$ and $\Lambda^2 S^-$ are canonically identified.

Proposition 1.1.1 continues to hold in dimension 4, but the existence of at least one spin^c structure is a slightly more subtle question than in dimension 3. The isomorphisms (1.4) mean that $w_2(\Lambda^+)$ is equal to the mod 2 reduction of $c_1(S^+)$; so the existence of a spin^c structure implies the existence of an integral lift of $w_2(\Lambda^+)$, or equivalently of $w_2(X)$ since these two are equal. This condition is also sufficient: the existence of a spin^c structure is equivalent to the existence of an integral lift of $w_2(X)$. On an orientable 4-manifold, $w_2(X)$ always has an integral lift, see [52], so spin^c structures always exist.

In any dimension, an *automorphism* of a spin^c structure (S, ρ) means a unitary bundle automorphism of S which commutes with Clifford multiplication. The group of automorphisms can be identified with the group of $\mathcal{G}$ of S^1-valued functions $u : X \to S^1$, acting by scalar multiplication. We call $\mathcal{G}$ the *gauge group* and we call its elements *gauge transformations*. The gauge group acts on sections Φ of S by

$$\Phi \mapsto u\Phi.$$

1.2 Dirac operators

Let $\mathfrak{s} = (S_X, \rho)$ be a spin^c structure on an oriented Riemannian 4-manifold X. A unitary connection A on S_X is a *spin^c connection* if ρ is parallel. This implies, in particular, that parallel transport preserves the decomposition of S_X as $S^+ \oplus S^-$. Given such a connection A, one defines the Dirac operator $D_A : \Gamma(S_X) \to \Gamma(S_X)$ as the composite

$$\Gamma(S_X) \xrightarrow{\nabla_A} \Gamma(T^*X \otimes S_X) \longrightarrow \Gamma(S_X),$$

in which the second map is constructed from the Clifford multiplication. The difference between two spin^c connections A and $\tilde{A}$, regarded as a 1-form with values in the endomorphisms of S_X, has the form

$$\tilde{A} - A = a \otimes 1_{S_X} \tag{1.5}$$

for some $a \in \Omega^1(X; i\mathbb{R})$. Conversely, if A is a spin^c connection and $a \in \Omega^1(X; i\mathbb{R})$, then $\tilde{A} = A + a \otimes 1_{S_X}$ is a spin^c connection. In this way, the spin^c connections on S_X form an affine space, with underlying vector space $\Omega^1(X; i\mathbb{R})$. If $\tilde{A}$ and A are related as above, then the corresponding Dirac operators are related by

$$D_{\tilde{A}} - D_A = \rho(a).$$

Because Clifford multiplication by 1-forms interchanges S^+ and S^-, we can write

$$D_A = D_A^+ + D_A^-,$$

where

$$D_A^+ : \Gamma(S^+) \to \Gamma(S^-)$$
$$D_A^- : \Gamma(S^-) \to \Gamma(S^+).$$

If we are given a spin^c connection A, then the associated line bundles $\Lambda^2 S^+$, $\Lambda^2 S^-$ inherit connections too. The canonical isomorphism between these line bundles respects the connections. We give this connection a name:

Notation 1.2.1. If A is a spin^c connection on the spin bundle $S_X = S^+ \oplus S^-$ on X, we write A^{t} for the associated connection in the line bundle $\Lambda^2 S^+ = \Lambda^2 S^-$. So if $\tilde{A}$ and A are related by (1.5), then

$$\tilde{A}^{\mathrm{t}} = A^{\mathrm{t}} + 2a.$$

◊

In dimension 3, we define a spin^c connection B for the spin^c bundle $S \to Y$ in the same way. The spin^c connections are again an affine space, now over $\Omega^1(Y; i\mathbb{R})$, for we can write

$$\tilde{B} = B + b \otimes 1_S, \tag{1.6}$$

just as in (1.5). For each spin^c connection B, we have a Dirac operator

$$D_B : \Gamma(S) \to \Gamma(S).$$

We write B^{t} for the associated connection on the line bundle $\Lambda^2 S$. There is no decomposition of this operator as there is in dimension 4.

In any dimension, the full Dirac operator is *self-adjoint*. In dimension 4, this means that D_A^- is the adjoint of D_A^+. The Dirac operator is elliptic, so if the underlying manifold is compact, then the operator is *Fredholm*: it has finite-dimensional kernel and cokernel. In dimension 3, because it is self-adjoint, the Dirac operator has index zero. On a compact 4-manifold, the complex index of the operator D_A^+ (the difference in the complex dimensions of the kernel and cokernel) is given by the Atiyah–Singer index theorem,

$$\text{index } D_A^+ = \frac{1}{8}\big(c_1(S^+)^2[X] - \sigma(X)\big), \tag{1.7}$$

where $\sigma(X)$ is the signature of X. (We write $\alpha[X]$, typically, for the evaluation of a cohomology class α on the fundamental class.)

The gauge group $\mathcal{G}$ acts on the space of spin^c connections A on S, by pull-back. If $u : X \to S^1 \subset \mathbb{C}$ is a gauge transformation, we write the action as

$$\begin{aligned} A &\mapsto u(A) \\ &= A - u^{-1}du. \end{aligned} \tag{1.8}$$

1.3 The Seiberg–Witten equations

On an oriented Riemannian 4-manifold X with spin^c structure $\mathfrak{s}_X$, the Seiberg–Witten equations, or monopole equations, are equations for a pair (A, Φ) consisting of a spin^c connection A and a section Φ of the associated spin bundle S^+. The equations are the following:

$$\begin{aligned} \frac{1}{2}\rho(F_{A^{\mathrm{t}}}^+) - (\Phi\Phi^*)_0 &= 0 \\ D_A^+\Phi &= 0. \end{aligned} \tag{1.9}$$

Here $F_{A^{\mathrm{t}}}^+$ is the self-dual part of the curvature 2-form $F_{A^{\mathrm{t}}}$ of the connection A^{t},

$$\begin{aligned} F_{A^{\mathrm{t}}} &= F_{A^{\mathrm{t}}}^+ + F_{A^{\mathrm{t}}}^- \\ &\in \Omega^+(X; i\mathbb{R}) \oplus \Omega^-(X; i\mathbb{R}), \end{aligned}$$

and $(\Phi\Phi^*)_0$ denotes the trace-free part of the hermitian endomorphism $\Phi\Phi^*$ of the bundle S^+,

$$\begin{aligned}(\Phi\Phi^*)_0 &= \Phi\Phi^* - \frac{1}{2}\operatorname{tr}(\Phi\Phi^*)1_{S^+} \\ &= \Phi\Phi^* - \frac{1}{2}|\Phi|^2 1_{S^+}.\end{aligned}$$

Note that F_{A^t} is an imaginary-valued 2-form, so $\rho(F_{A^t})$ is hermitian: the map ρ in (1.4) carries real self-dual forms to skew-adjoint endomorphisms of S^+.

If ω is a smooth imaginary-valued 2-form and ω^+ its self-dual part, we can also consider the monopole equations *perturbed by* ω. These are the equations

$$\begin{aligned}\frac{1}{2}\rho(F^+_{A^t} - 4\omega^+) - (\Phi\Phi^*)_0 &= 0 \\ D^+_A\Phi &= 0.\end{aligned} \tag{1.10}$$

The left-hand sides of the two equations in (1.9) define a map

$$\mathfrak{F} : \mathcal{A} \times \Gamma(S^+) \to \Gamma(i\,\mathfrak{su}(S^+) \oplus S^-), \tag{1.11}$$

where $\mathcal{A}$ denotes the affine space of all spin^c connections A, and $i\,\mathfrak{su}(S^+)$ is the bundle of hermitian endomorphisms of S^+. We can then write the monopole equations as $\mathfrak{F}(A, \Phi) = 0$. We write the perturbed equations similarly, as

$$\mathfrak{F}_\omega(A, \Phi) = 0. \tag{1.12}$$

The set of solutions (A, Φ) of the perturbed equations is invariant under the action of the gauge group $\mathcal{G}$. We will write $[A, \Phi]$ to denote the gauge-equivalence class of a pair (A, Φ): the orbit of (A, Φ) under the action of $\mathcal{G}$.

Definition 1.3.1. If X is an oriented Riemannian 4-manifold with spin^c structure $\mathfrak{s}_X = (S_X, \rho)$, and ω is an imaginary-valued 2-form, we write $N(X, \mathfrak{s}_X)$ for the quotient space of the set of solutions of the equations (1.12) by the action of $\mathcal{G}$:

$$N(X, \mathfrak{s}_X) = \{\, [A, \Phi] \mid \mathfrak{F}_\omega(A, \Phi) = 0 \,\}.$$

This is the *monopole moduli space* for $(X, \mathfrak{s}_X)$ with perturbing 2-form ω. It is a subset of the *configuration space*

$$\mathcal{B}(X, \mathfrak{s}_X) = \big(\mathcal{A} \times \Gamma(S^+)\big) \,/\, \mathcal{G}.$$

◇

The configuration space $\mathcal{B}(X, \mathfrak{s}_X)$ is Hausdorff, so the moduli space is Hausdorff also. The following result reflects the very special nature of the monopole equations. (It would not be true, for example, if the sign of the second term in (1.9) were changed.)

Theorem 1.3.2. *If the 4-manifold X is compact (without boundary), then the moduli space $N(X, \mathfrak{s}_X) \subset \mathcal{B}(X, \mathfrak{s}_X)$ is compact.* □

1.4 Regularity

From this point on, we will always assume that our 4-manifold X is *connected*. Let (A, Φ) be a solution of the equations $\mathfrak{F}_\omega(A, \Phi) = 0$ on X, as above. We can take the derivative of the map

$$\mathfrak{F}_\omega : \mathcal{A} \times \Gamma(S^+) \to i\,\mathfrak{su}(S^+) \oplus \Gamma(S^-),$$

at the point (A, Φ) in the affine space $\mathcal{A} \times \Gamma(S^+)$, to obtain a linear map

$$\mathcal{D}_{(A,\Phi)}\mathfrak{F}_\omega : \Omega^1(X; i\mathbb{R}) \times \Gamma(S^+) \to \Gamma(i\,\mathfrak{su}(S^+) \oplus S^-),$$

given by

$$(a, \phi) \mapsto \left(\frac{1}{2}\rho(d^+a) - (\phi\Phi^* + \Phi\phi^*)_0, D_A^+\phi + \rho(a)\Phi\right). \tag{1.13}$$

Definition 1.4.1. A solution (A, Φ) to the perturbed monopole equations $\mathfrak{F}_\omega = 0$ is *regular* if the linearization (1.13) is a *surjective* linear operator. We say that the moduli space $N(X, \mathfrak{s}_X)$ is regular if all solutions are regular. ◊

Proposition 1.4.2. *Suppose that the oriented Riemannian 4-manifold X is compact (without boundary), and let $\mathfrak{s}_X$ be a given* spin^c *structure. Then there is an open and dense subset of the space of imaginary-valued 2-forms ω for which the corresponding moduli space $N(X, \mathfrak{s}_X)$ is regular.* □

The action of the gauge group $\mathcal{G}$ on $\mathcal{A} \times \Gamma(S^+)$ is free on the set of pairs (A, Φ) with Φ non-zero. We call such a pair *irreducible*. For a *reducible* configuration $(A, 0)$, the equations (1.9) reduce to the equations

$$F_{A^t}^+ = 4\omega^+.$$

Suppose κ is a 2-form on X that is both closed and self-dual. Then if A satisfies the equation above, we have

$$\begin{aligned}\int_X \omega \wedge \kappa &= \int_X \omega^+ \wedge \kappa \\ &= \int_X F_{A^t} \wedge \kappa \\ &= (2\pi/i)\big(c_1(S^+) \cup [\kappa]\big)[X]. \qquad (1.14)\end{aligned}$$

If κ is non-zero, this is a non-trivial linear constraint on ω, which must be satisfied if a reducible solution is to exist. The closed self-dual (real) 2-forms κ form a subspace $\mathcal{H}^+$ of the space $\mathcal{H}^2$ of harmonic 2-forms, and determine a metric-dependent subspace

$$\mathcal{H}^+ \subset H^2(X;\mathbb{R}).$$

This is a maximal positive-definite subspace for the quadratic form

$$\begin{aligned}Q : H^2(X;\mathbb{R}) &\to \mathbb{R} \\ Q(\alpha) &= \alpha^2[X].\end{aligned}$$

We write b^+ for the dimension of $\mathcal{H}^+$. Defining b^- similarly, we have $b^+ + b^- = b^2$ and $b^+ - b^- = \sigma(X)$. From the calculation (1.14), we deduce:

Lemma 1.4.3. *If X is a compact manifold with $b^+ \geq 1$, then for all ω in the complement of a proper linear subspace, the corresponding moduli space $N(X, \mathfrak{s}_X)$ contains no reducible solutions.* □

When the moduli space is regular and contains no reducibles, it is a smooth manifold whose dimension can be computed as the index of a certain operator (essentially the sum of the two operators in (27.3); see Lemma 27.1.1):

Theorem 1.4.4. *Let X be a closed, connected, oriented Riemannian manifold with a spin^c structure $\mathfrak{s}_X$. Suppose ω is chosen so that the moduli space $N(X, \mathfrak{s}_X)$ is regular and contains no reducible solutions, as we can always do when $b^+(X) \geq 1$. Then the moduli space $N(X, \mathfrak{s}_X)$ is a smooth, compact manifold, whose dimension d is given by the formula*

$$\begin{aligned}d &= \big(b^1(X) - b^+(X) - 1\big) + 2\,\text{index}\, D_A^+ \\ &= \frac{1}{4}\big(c_1(S^+)^2[X] - 2\chi(X) - 3\sigma(X)\big). \qquad (1.15)\end{aligned}$$

□

Remark. In the formula for d, the quantity $\chi(X)$ is the Euler number of X, which can be expressed as $2-2b^1+b^2$. The equality between the two formulae follows from the index theorem (1.7) for the Dirac operator. From the first line, we see that the parity of d is independent of the choice of $\mathfrak{s}_X$: the dimension d is even if and only if $b^1 - b^+$ is odd, and vice versa.

1.5 Monopole invariants

Suppose ω satisfies the conditions of Theorem 1.4.4, so that $N(X,\mathfrak{s}_X)$ is a smooth manifold contained in the *irreducible* part of the configurations space

$$\begin{aligned}\mathcal{B}^*(X,\mathfrak{s}_X) &= \{\,[A,\Phi] \mid \Phi \neq 0\,\}\\ &\subset \mathcal{B}(X,\mathfrak{s}_X).\end{aligned} \tag{1.16}$$

The monopole invariants are defined using the *fundamental class* $[N(X,\mathfrak{s}_X)]$ of the moduli space in the ambient space $\mathcal{B}^*(X,\mathfrak{s}_X)$. The moduli space is always orientable, but does not have a preferred orientation. An orientation for $N(X,\mathfrak{s}_X)$ can be fixed if we are given some extra data: a *homology orientation* for the 4-manifold X.

Definition 1.5.1. A *homology orientation* μ_X for a closed oriented 4-manifold X is an orientation of the line

$$\Lambda^{\max} H^1(X;\mathbb{R}) \otimes \Lambda^{\max} H^+(X),$$

where $H^+(X) \subset H^2(X;\mathbb{R})$ is any chosen maximal positive-definite subspace for the quadratic form Q. (The set of all maximal positive-definite subspaces H^+ is a contractible subset of the Grassmannian, so orienting $\Lambda^{\max} H^+$ for any one of these defines orientations for all others.) ◊

When X has been given a homology orientation, we therefore have a well-defined class

$$[N(X,\mathfrak{s}_X)] \in H_d(\mathcal{B}^*(X,\mathfrak{s})),$$

where d is given by the formula (1.15) for the dimension of the moduli space. To obtain numerical quantities from this homology class, we pair it with fixed cohomology classes.

By construction, $\mathcal{B}^*(X,\mathfrak{s}_X)$ is the quotient of $\mathcal{A} \times (\Gamma(S^+)\setminus 0)$ by a *free* action of the group $\mathcal{G}$. The quotient map

$$\mathcal{A} \times \big(\Gamma(S^+) \setminus 0\big) \to \mathcal{B}^*(X,\mathfrak{s}_X)$$

is a principal $\mathcal{G}$-bundle. If we choose a basepoint $x_0 \in X$, then we obtain a homomorphism $\mathcal{G} \to S^1$ by evaluation at x_0, and there is an associated principal S^1 bundle

$$P \to \mathcal{B}^*(X, \mathfrak{s}_X),$$

whose isomorphism class is independent of the choice of x_0 (because X is connected). There is therefore a well-defined 2-dimensional cohomology class

$$u_2 \in H^2(\mathcal{B}^*(X, \mathfrak{s}_X); \mathbb{Z})$$

given by $u_2 = c_1(P)$.

Theorem 1.5.2. *Let X be a closed, connected, Riemannian manifold with a spin^c structure $\mathfrak{s}_X$, and let ω be an imaginary-valued 2-form, chosen so that the moduli space $N = N(X, \mathfrak{s}_X)$ is regular and contains no reducibles. Choose a homology orientation μ_X, such that the moduli space is a smooth, compact oriented manifold, of dimension d, and suppose that d is even. Then, if $b^+(X) \geq 2$, the quantity*

$$\langle u_2^{d/2}, [N] \rangle \in \mathbb{Z} \tag{1.17}$$

is independent of the choice of Riemannian metric and the choice of ω. □

A short digression will explain the condition $b^+(X) \geq 2$ in this theorem. When $b^+(X) = 1$, we can still choose a metric and ω so that the moduli spaces are regular and contain no reducibles, but the pairing (1.17) is not independent of the choices made. To understand this, fix a metric on X and let κ_1 be a generator for the 1-dimensional space $\mathcal{H}^+$. The moduli space $N(X, \mathfrak{s}_X)$ with perturbing 2-form ω contains reducible solutions when

$$\int_X \omega \wedge \kappa_1 = (2\pi/i)\left(c_1(S^+) \smile [\kappa_1]\right)[X], \tag{1.18}$$

as in (1.14). This condition defines a codimension-1 affine subspace of the space of imaginary-valued 2-forms ω, and divides $\Omega^2(X; i\mathbb{R})$ into two connected components. We can choose an ω in either one of these two connected components, so that $N(X, \mathfrak{s}_X)$ is regular; but the value of the pairing (1.17) will depend on which component we choose. We will return to this case, which is special to manifolds with $b^+ = 1$, in Subsection 27.5.

Remarks. In the case that $d = 0$, each component of $N(X, \mathfrak{s}_X)$ is a point. The orientation of each component, determined by the choice of μ_X, may or may not agree with the canonical orientation of the 1-point space, and by comparing orientations we may assign each component a sign. The pairing $\langle 1, [N] \rangle$ "counts" the points of N, with signs.

Because it is independent of the choice of metric and perturbation, the quantity (1.17) is an invariant of the smooth manifold X, the choice of spin^c structure and the homology orientation. The last only affects the overall sign of the invariant. We make a definition, first for the case $d = 0$.

Definition 1.5.3. Let X be a closed, connected, oriented 4-manifold with a homology orientation μ_X and $b^+(X) \geq 2$. Let $\mathfrak{s}_X$ be a spin^c structure on X, and choose a metric and perturbing 2-form ω so that the moduli space $N(X, \mathfrak{s}_X)$ is regular and contains no reducible solutions. We define an invariant $\mathfrak{m}(X, \mathfrak{s}_X) \in \mathbb{Z}$ by the formula

$$\mathfrak{m}(X, \mathfrak{s}_X) = \big\langle 1, [N(X, \mathfrak{s}_X]\big\rangle$$

if the dimension d of the moduli space $N(X, \mathfrak{s}_X)$ is zero. We define $\mathfrak{m}(X, \mathfrak{s}_X)$ to be zero otherwise. ◊

For the case of moduli spaces of higher dimension, we extend our notation:

Definition 1.5.4. Let X be a manifold with $b^+(X) \geq 2$ as in the previous definition, and let $N(X, \mathfrak{s}_X)$ be as before. For each integer $e \geq 0$, we define an invariant $\mathfrak{m}(u_2^e \mid X, \mathfrak{s}_X) \in \mathbb{Z}$ by the formula

$$\mathfrak{m}(u_2^e \mid X, \mathfrak{s}_X) = \big\langle u_2^e, [N(X, \mathfrak{s}_X)]\big\rangle.$$

We understand the pairing to be zero if the dimension of the moduli space is not equal to $2e$. ◊

We refer to the invariants $\mathfrak{m}(X, \mathfrak{s}_X)$ and $\mathfrak{m}(u_2^e \mid X, \mathfrak{s}_X)$ as the *monopole invariants* of X. As functions on the set of spin^c structures, these functions have common support a finite set:

Proposition 1.5.5. *Let ω be a perturbing 2-form, chosen so that all the moduli spaces $N(X, \mathfrak{s}_X)$ on the compact Riemannian manifold X are regular. (Such a form exists, by the Baire category theorem, because the space of 2-forms has the topology of a complete metric space.) Then the moduli space $N(X, \mathfrak{s}_X)$ is empty for all but finitely many isomorphism classes of spin^c structures $\mathfrak{s}_X$. Consequently, $\mathfrak{m}(u_2^e \mid X, \mathfrak{s}_X)$ is non-zero for only finitely many pairs $(\mathfrak{s}_X, e)$.* □

On a manifold X with $b^+(X) > 1$, the term *basic class* is generally used to refer to a class $c \in H^2(X;\mathbb{Z})$ arising as $c_1(\mathfrak{s}_X)$ for some spin^c structure $\mathfrak{s}_X$ with $\mathfrak{m}(X, \mathfrak{s}_X)$ non-zero.

1.6 Simple type

Having introduced the monopole invariants $\mathfrak{m}(u_2^e \mid X, \mathfrak{s}_X)$, we should point out that there are no known examples in which these invariants are non-zero for $e > 0$. Another way to say this is that when $b^+(X) \geq 2$ and when the the moduli space $N(X, \mathfrak{s}_X)$ is regular and has strictly positive dimension $2e$, then the pairing of the fundamental class $[N(X, \mathfrak{s}_X)]$ with the class $u_2^e = c_1(P)^e$ is zero in all known cases: it is only the zero-dimensional moduli spaces $N(X, \mathfrak{s}_X)$ which contribute. Accordingly, we make the following definition:

Definition 1.6.1. A closed oriented 4-manifold X with $b^+(X) \geq 2$ is said to have *simple type* if the monopole invariants $\mathfrak{m}(u_2^e \mid X, \mathfrak{s}_X)$ are zero for all $e > 0$ and all spin^c structures $\mathfrak{s}_X$ on X. ◊

As long as $b^+(X) \geq 2$, the simple type condition is known to hold for many classes of 4-manifolds, including all symplectic 4-manifolds and all 4-manifolds containing a "tight" surface in the sense of [61] (an embedded surface Σ of positive genus g representing a non-trivial homology class and satisfying $2g - 2 = \Sigma \cdot \Sigma$). It is very unclear to what extent the familiar examples of 4-manifolds are representative of the general case, so there is little reason to extrapolate with any confidence. Nevertheless, we state what is usually called the "simple type conjecture":

Conjecture 1.6.2. *All closed oriented 4-manifolds with $b^+ \geq 2$ have simple type.*

2 Morse theory

In this section, we first review how the ordinary homology of a compact manifold B can be recovered as the homology of the "Morse complex", associated with a suitable smooth function f on B and a Riemannian metric. We then present a non-standard variant of the Morse complex that can be used to compute both the relative and absolute homologies of a manifold with boundary, $(B, \partial B)$. The latter construction can be adapted to study a manifold with circle action, as we explain in Subsection 2.5.

2.1 The Morse complex

Let B be a smooth, closed Riemannian manifold, and $f : B \to \mathbb{R}$ a smooth function. Let $V = \operatorname{grad} f$ be the gradient vector field, and $\phi_t : B \to B$ be the flow generated by $-V$, the downward gradient flow:

$$\frac{d}{dt}\phi_t(x) = -V(\phi_t(x)) \quad (t \in \mathbb{R}).$$

We shall suppose that f is a Morse function. This means that at each critical point a of f, the self-adjoint endomorphism ∇V of T_aB has no kernel. We can then decompose the tangent space T_aB at each critical point as a direct sum,

$$T_aB = K_a^+ \oplus K_a^-,$$

of the eigenspaces belonging to positive and negative eigenvalues of ∇V at a. The *index* of a is the dimension of K_a^-:

$$\text{index}\,(a) = \dim K_a^-.$$

The *stable* and *unstable manifolds* of a critical point a are the smooth submanifolds S_a and U_a defined by

$$S_a = \left\{ x \,\middle|\, \lim_{t\to\infty} \phi_t(x) = a \right\}$$

$$U_a = \left\{ x \,\middle|\, \lim_{t\to-\infty} \phi_t(x) = a \right\}.$$

These submanifolds are tangent at a to the linear spaces K_a^+ and K_a^- respectively.

If a and b are two critical points, we define $M(a,b)$ to be the set of points $x \in B$ that flow to b under the downward gradient flow and to a under the upward flow:

$$M(a,b) = U_a \cap S_b.$$

We can alternatively regard $M(a,b)$ as parametrizing a family of trajectories of the flow, the trajectories

$$\gamma(t) = \phi_t(x) \quad (x \in M(a,b)).$$

We will sometimes switch between these two viewpoints on $M(a,b)$ – regarding it as a subset of B or as a family of trajectories. The latter viewpoint is often more useful.

The gradient flow of f satisfies the *Morse–Smale* condition if the intersection $U_a \cap S_b$ is transverse for all critical points a and b. In this case, $M(a, b)$ is a smooth submanifold of B. If it is non-empty, the space $M(a, b)$ has dimension given by the difference of the indices:

$$\dim M(a, b) = \mathrm{index}(a) - \mathrm{index}(b).$$

If a and b are distinct, the flow ϕ_t is a free $\mathbb{R}$-action on $M(a, b)$. The quotient is Hausdorff: it is a manifold of 1 smaller dimension that we denote by $\breve{M}(a, b)$:

$$\breve{M}(a, b) = M(a, b)/\mathbb{R}.$$

We refer to elements of $\breve{M}(a, b)$ as *unparametrized* trajectories.

Let $\mathfrak{c}$ be the critical set of f, and $\mathfrak{c}_k \subset \mathfrak{c}$ the set of critical points of index k. Under our hypothesis that f is a Morse function, the set $\mathfrak{c}$ is finite. Let R be an abelian group. Until we deal with questions of orientation later in this section, we suppose R has characteristic 2: that is, $r + r = 0$ for all $r \in R$. Define C_* as the tensor product of R with a free abelian group on a set of generators e_a, indexed by the elements a of $\mathfrak{c}$:

$$C_* = \left(\bigoplus_{a \in \mathfrak{c}} \mathbb{Z} e_a \right) \otimes R.$$

This is a graded abelian group: we can write $C_* = \bigoplus C_k$, where

$$C_k = \left(\bigoplus_{a \in \mathfrak{c}_k} \mathbb{Z} e_a \right) \otimes R.$$

If a and b are critical points of index k and $k - 1$ respectively, the space $\breve{M}(a, b)$ has dimension zero. The Morse–Smale hypothesis implies that it is compact, and therefore a finite set. Define $n(a, b)$ to be its cardinality:

$$n(a, b) = \left| \breve{M}(a, b) \right|.$$

The integer $n(a, b)$ is the number of *unparametrized trajectories* of the flow, running from a to b. Define an operator $\partial : C_k \to C_{k-1}$ for all k, by

$$\partial e_a = \sum_{b \in \mathfrak{c}_{k-1}} n(a, b) e_b. \tag{2.1}$$

(It is understood here, and in similar formulae below, that we are writing a formula for an operator on a free abelian group with generators e_a; and the actual operator ∂ is obtained by taking the tensor product with 1_R.)

The main result is now the following:

Theorem 2.1.1. *The operator $\partial : C_* \to C_*$ satisfies $\partial\partial = 0$, so (C_*, ∂) is a complex. The homology of this complex is isomorphic to the ordinary homology of the manifold B:*

$$H_k(C_*, \partial) = H_k(B; R).$$

□

The complex (C_*, ∂) is the *Morse complex* of the function f on the Riemannian manifold B.

We will not present a proof of this theorem. However, the proof of the first part (the proof that $\partial\partial = 0$) will appear again in the context of Floer homology, in Section 22. So we outline the argument here. The proof is based on two points. The first is a simple reinterpretation of the matrix entries of the operator $\partial\partial$. For a a critical point of index k, let us write

$$\partial\partial e_a = \sum_{c \in \mathfrak{C}_{k-2}} p(a, c) e_c,$$

so that

$$p(a, c) = \sum_{b \in \mathfrak{C}_{k-1}} n(a, b) n(b, c).$$

This is the matrix entry of $\partial\partial$ corresponding to the pair of basis elements e_a, e_c. We can reinterpret $p(a, c)$ as the number of unparametrized *broken trajectories* running from a to c. Here, in general, a broken trajectory running from a critical point a to a' means a sequence of non-constant trajectories with matching limit points: that is, it consists of an l-tuple

$$(\breve{x}_1, \ldots, \breve{x}_l) \in \breve{M}(a_0, a_1) \times \cdots \times \breve{M}(a_{l-1}, a_l),$$

where the a_i are distinct critical points with $a_0 = a$ and $a_l = a'$, and $\breve{x}_i \in \breve{M}(a_{i-1}, a_i)$ is the orbit of $x_i \in M(a_{i-1}, a_i)$ under the flow. The closure of the union of the l orbits in such a broken trajectory consists of the orbits themselves together with the $l + 1$ limiting critical points:

$$X = \mathrm{Cl}_B\left(\bigcup_i \breve{x}_i\right). \tag{2.2}$$

This space $X \subset B$ is topologically an embedded arc joining a to a'. In the case that a and c differ in index by 2, any strictly broken trajectory between them must have exactly two components, each with index drop 1.

To complete the proof that $\partial\partial = 0$ then, one must show that the number of strictly broken trajectories from a to c is even, when the index difference is 2. (Recall that we have assumed R to have characteristic 2.) This brings us to the second point. Our hypotheses imply that the space $\breve{M}(a,c)$ is 1-dimensional, but it will not in general be compact. To form a compact space, one must take also the broken trajectories. The topology on the space of broken trajectories can be described, for now, as coming from a metric defined using the Hausdorff distance between the corresponding arcs $X \subset B$. We write $\breve{M}^+(a,c)$ for this space of (unparametrized) trajectories. The main part of the proof is now to show that $\breve{M}^+(a,c)$ is compact and is a 1-manifold with boundary: its boundary consists of the strictly broken trajectories, while its interior is $\breve{M}(a,c)$. The fact that $\partial\partial = 0$ then follows from the fact that a compact 1-manifold with boundary has an even number of endpoints.

2.2 Orientations and integer coefficients

If we wish to use $\mathbb{Z}$ as the coefficients in the Morse complex, or any group R not of characteristic 2, then we must deal with orientations of the trajectory spaces $M(a,b)$. Recall that if a is a critical point of the Morse function f, then the unstable manifold U_a is tangent at a to the linear space $K_a^- \subset T_aB$. The unstable manifold is contractible, so an orientation of the vector space K_a^- provides an orientation of U_a. The same vector space K_a^- is the fiber of the normal bundle to the *stable* manifold at a. So an orientation of K_a^- orients the normal bundle of the (contractible) stable manifold $S_a \subset B$: thus the stable manifold is *cooriented.*

Suppose now that we choose an orientation μ_a for K_a^- at every critical point a of f. Then all unstable manifolds are oriented and all stable manifolds are cooriented. If the flow is Morse–Smale, we can then orient the space $M(a,b)$ as the intersection of U_a and S_b. To spell out a recipe, observe that at each point $x \in M(a,b)$ we have a short exact sequence:

$$0 \to T_xM(a,b) \xrightarrow{i} T_xU_a \xrightarrow{\pi} N_xS_b \to 0. \tag{2.3}$$

We orient $T_xM(a,b)$ as the fiber of the second map, using the following "fiber-first" convention. Let $w_1, \ldots, w_l$ in T_xU_a have the property that their images under π are an oriented basis of N_xS_b and let $v_1, \ldots, v_d$ be any basis of

$T_xM(a,b)$. Then we declare that the v_j are an oriented basis if

$$i(v_1), \dots, i(v_d), w_1, \dots, w_l$$

is an oriented basis of T_xU_a.

Having so oriented $M(a,b)$, we can similarly orient the space of unparametrized trajectories $\breve{M}(a,b) = M(a,b)/\mathbb{R}$ using the sequence

$$0 \to \mathbb{R} \to T_xM(a,b) \to T_{\breve{x}}\breve{M}(a,b) \to 0,$$

as long as $a \neq b$. (The $\mathbb{R}$ action is the *downward* gradient flow ϕ_t.) In the special case that $\breve{M}(a,b)$ is zero-dimensional, we can compare the orientation of $\breve{M}(a,b)$ at an unparametrized trajectory $\breve{x}$ with the canonical orientation of the 1-point space, and in this way obtain a sign

$$\epsilon(\breve{x}) = \pm 1$$

for each $\breve{x}$ in $\breve{M}(a,b)$. The sign is still dependent on the choice of μ_a and μ_b. We can then modify the definition of the boundary map ∂ on the complex C_*: the new formula for ∂ still takes the form (2.1), but instead of defining the integer $n(a,b)$ as the cardinality of $\breve{M}(a,b)$, we redefine

$$n(a,b) = \sum_{\breve{x} \in \breve{M}(a,b)} \epsilon(\breve{x}).$$

With this signed version of ∂, the proof that $\partial\partial = 0$ continues to work, whether or not R has characteristic 2. What needs to be proved is that for all $a \in \mathfrak{c}_k$ and $c \in \mathfrak{c}_{k-2}$, the integer

$$p(a,c) = \sum_{b \in \mathfrak{c}_{k-1}} \sum_{\breve{x}_1 \in \breve{M}(a,b)} \sum_{\breve{x}_2 \in \breve{M}(b,c)} \epsilon(\breve{x}_1)\epsilon(\breve{x}_2)$$

is zero, not just zero modulo 2. The proof is to interpret the above sum as (minus) the signed count of the endpoints of the compact oriented 1-manifold $\breve{M}^+(a,c)$. The isomorphism

$$H_*(C_*, \partial) = H_*(B; R)$$

continues to hold, now with arbitrary coefficient group.

As we have described it, the complex C_* now depends on the choice of orientations of all the unstable manifolds, though it is easy to check that different choices lead to isomorphic chain complexes. If we prefer, we can slightly modify the construction of (C_*, ∂) so that no explicit a priori choice of orientations of the spaces K_a^- is required. For each critical point a, let $\Lambda(a) = \{\mu, \mu'\}$ denote the 2-element set consisting of the possible orientations of K_a^-. Define $\mathbb{Z}\Lambda(a)$ to be the free abelian group of rank 1, given by a presentation with generators μ, μ' and a single relation $\mu = -\mu'$. We can then define

$$C_* = \sum_{a \in \mathfrak{C}} \mathbb{Z}\Lambda(a) \otimes R$$

and we can reinterpret $\epsilon(\breve{x})$ above as an isomorphism $\epsilon(\breve{x}) : \mathbb{Z}\Lambda(a) \to \mathbb{Z}\Lambda(b)$. This choice of viewpoints concerning orientations is familiar also when setting up the homology of simplicial complexes.

2.3 Linear flows on projective space

Let L be a hermitian matrix, regarded as defining a linear transformation of $\mathbb{C}^n$, and let Λ be the function on $\mathbb{C}^n \setminus \{0\}$ defined by

$$\Lambda(z) = \langle z, Lz \rangle \Big/ \|z\|^2.$$

This function is invariant under the action of the scalars $\mathbb{C}^*$, and so descends to a function

$$\Lambda^* : \mathbb{CP}^{n-1} \to \mathbb{R}.$$

We equip $\mathbb{CP}^{n-1}$ with its standard metric as the quotient of the unit sphere S^{2n-1} by S^1, and we examine the downward gradient flow of Λ^* on the projective space. To understand the flow lines, we may consider first another function on $\mathbb{C}^n$ defined in terms of L, namely the function

$$f(z) = \frac{1}{2}\langle z, Lz \rangle.$$

The downward gradient flow for the function f is easy to write down: it is described by the linear equation

$$dz/dt = -Lz \tag{2.4}$$

for a path $z(t)$ in $\mathbb{C}^n$.

Lemma 2.3.1. *The trajectories of the gradient flow of the function* $\frac{1}{2}\Lambda^*$ *on* $\mathbb{CP}^{n-1}$ *are the images under the quotient map*

$$\pi : \mathbb{C}^n \setminus 0 \to \mathbb{CP}^{n-1}$$

of the non-zero solutions $z : \mathbb{R} \to \mathbb{C}^n$ *of the linear equation* (2.4).

Proof. At a point w in the unit sphere S^{2n-1}, we may decompose $\mathrm{grad}(f)$ into parts tangent and normal to the sphere. The part tangent to the sphere is

$$\mathrm{grad}(f|_{S^{2n-1}})(w) = Lw - \Lambda(w)w.$$

The right-hand side is homogeneous. The image in S^{2n-1} of a trajectory $z(t)$ of the linear equation (2.4) is therefore a solution $w : \mathbb{R} \to S^{2n-1}$ of the equation

$$dw/dt = -Lw + \Lambda(w)w, \tag{2.5}$$

which is the downward gradient-flow equation for the restriction of f to the sphere. The function f is invariant under the action of S^1 on S^{2n-1}, and the corresponding function on the quotient $\mathbb{CP}^{n-1}$ is $\frac{1}{2}\Lambda^*$. So the image of a solution $w(t)$ in $\mathbb{CP}^{n-1}$ is a flow line for $\frac{1}{2}\Lambda^*$. □

From the lemma it follows that the gradient trajectories of Λ^* in $\mathbb{CP}^{n-1}$ are the paths $t \mapsto [z(2t)]$, where $z(t)$ is a solution of the linear equation. (We write $[z]$ for the image of z in $\mathbb{CP}^{n-1}$.) The next lemma is a simple consequence:

Lemma 2.3.2. *The critical points of the function* Λ^* *on* $\mathbb{CP}^{n-1}$ *are the images* $[z]$ *of the eigenvectors* $z \in \mathbb{C}^n$ *of the hermitian operator* L. □

The critical points are therefore isolated in $\mathbb{CP}^{n-1}$ if and only if the non-trivial eigenspaces of L are 1-dimensional: that is, if and only if the spectrum of L is *simple*. Let us suppose that this condition holds, and let us write the eigenvalues of L as

$$\lambda_1 < \lambda_2 < \cdots < \lambda_n.$$

Let $w_1,\ldots,w_n$ be corresponding unit eigenvectors.

Lemma 2.3.3. *When the spectrum of* L *is simple, each critical point* $[w_i]$ *of* Λ^* *is non-degenerate. The index of* $[w_i]$ *is* $2(i-1)$. *The closures of the unstable and stable manifolds of* $[w_i]$ *in* $\mathbb{CP}^{n-1}$ *are the projective subspaces spanned by*

$$[w_1], \ldots, [w_i]$$

and

$$[w_i], \ldots, [w_n]$$

respectively. The unstable and stable manifolds themselves are the affine subspaces of these projective spaces where the component of w_i is non-zero.

Proof. We can write down the Hessian of $\frac{1}{2}\Lambda^*$ at $[w_i]$ using the description (2.5) of the flow. We can identify $T_{w_i}\mathbb{CP}^{n-1}$ with the complex subspace W of $\mathbb{C}^n$ orthogonal to w_i, and from (2.5) we obtain the formula

$$w \mapsto Lw - \lambda_i w$$

for the Hessian, as an operator on W. There is no zero eigenvalue, and the eigenvectors of this operator belonging to negative and positive eigenvalues are the vectors w_j with $j < i$ and $j > i$ respectively. The statements about non-degeneracy and the index both follow. The description of the unstable and stable manifolds follows from Lemma 2.3.1. □

We can elaborate on the description of the unstable and stable manifolds. Suppose we look at $M(i, j)$, the space of trajectories from $[w_i]$ to $[w_j]$, viewed as the intersection $U_i \cap S_j$ of the unstable and stable manifolds of the two critical points in $\mathbb{CP}^{n-1}$. The lemma tells us that this is empty unless $i > j$, in which case it is the image in $\mathbb{CP}^{n-1}$ of the subspace of $\mathbb{C}^n$ consisting of vectors of the form

$$z = \alpha_j w_j + \alpha_{j+1} w_{j+1} + \cdots + \alpha_i w_i,$$

where the coefficients α are complex numbers and the first and last are non-zero. The corresponding solution of the linear equation $dz/dt = -Lz$ is given by

$$z(t) = \sum_{m=j}^{i} \alpha_m e^{-\lambda_m t} w_m.$$

As t goes to $-\infty$ and $+\infty$, the dominant terms in this sum are the first and the last respectively. This gives us the following result:

Proposition 2.3.4. *Under the above hypothesis on the spectrum of L, the space $M(i, j)$ can be identified with the quotient by $\mathbb{C}^*$ of the set of non-zero solutions* *z to the linear equation $dz/dt = -Lz$ which have the asymptotics*

$$z(t) \sim c_0 e^{-\lambda_i t} w_i, \quad \text{as } t \to -\infty$$

$$z(t) \sim c_1 e^{-\lambda_j t} w_i, \quad \text{as } t \to +\infty,$$

where c_0, c_1 are non-zero complex constants. This can be identified with the subset of $\mathbb{CP}^{i-j}$ on which both the first and the last homogeneous coordinate are non-zero: the complement of two hyperplanes. □

Because the critical points are only in even index, there are no non-zero differentials in the Morse complex. The Morse complex computes the ordinary homology of $\mathbb{CP}^{n-1}$ as expected.

2.4 Manifolds with boundary

Now let $(B, \partial B)$ be a Riemannian manifold with boundary. There is a standard variant of the Morse complex that computes the relative homology $H_k(B, \partial B; R)$. One takes a Morse function that is constant on ∂B and achieves its minimum there. The Morse complex formed using the interior critical points of f computes the relative homology. If instead f achieves its maximum on the boundary, one obtains in this way the absolute homology group, $H_k(B; R)$. We wish to deal rather differently with $(B, \partial B)$. Rather than take a function that is constant on the boundary, we instead choose a smooth Morse function f whose gradient vector field V is everywhere *tangent* to the boundary ∂B. To be more precise, we envision the following situation.

We suppose that a smooth manifold $\tilde{B}$ without boundary is given, and is equipped with an involution $i : \tilde{B} \to \tilde{B}$ whose fixed-point set is of codimension 1. The quotient $\tilde{B}/i$ is then a manifold with boundary, and we suppose that this is identified with B. (Thus $\tilde{B}$ is a *double* of B.) The fixed-point set of the involution is ∂B. We suppose that the Riemannian metric on B arises as the restriction of a metric on $\tilde{B}$ that is invariant under the involution. Similarly, we suppose that $f : B \to \mathbb{R}$ is the restriction to B of an i-invariant Morse function $\tilde{f}$ on $\tilde{B}$.

If f is a Morse function of this sort, we can divide its set of critical points first into two sets: the critical points in the interior and those on the boundary. The boundary critical points can be further divided into two types. If a is a critical point on ∂B, then the normal vector ν to ∂B at a is an eigenvector of ∇V, and belongs to either K_a^+ or K_a^-.

Definition 2.4.1. Let a be a critical point of f on ∂B. We say a is *boundary-stable* if the normal vector ν belongs to K_a^+. Otherwise, we say a is *boundary-unstable*. ◊

We write $\mathfrak{c}^s$ and $\mathfrak{c}^u$ for the sets of boundary-stable and boundary-unstable critical points, and $\mathfrak{c}^o$ for the interior critical points. So

$$\mathfrak{c} = \mathfrak{c}^o \cup \mathfrak{c}^s \cup \mathfrak{c}^u. \tag{2.6}$$

If a is boundary-stable, then the stable manifold $S_a \subset B$ meets the interior of B: it is a manifold with (possibly empty) boundary, and the boundary ∂S_a is the intersection $S_a \cap \partial B$. If a is boundary-unstable on the other hand, the stable manifold S_a is contained entirely in ∂B. For the unstable manifold U_a the situation is reversed: it is a manifold with boundary if a is boundary-unstable, and is contained in ∂B if a is boundary-stable.

In this situation, we should not ask that the Morse–Smale condition hold without modification. The reason can be explained as follows. Suppose that a and b are critical points on the boundary, and suppose a is boundary-stable while b is boundary-unstable. Then the unstable manifold U_a and the stable manifold S_b are both contained in ∂B. The intersection $U_a \cap S_b$ cannot be a transverse intersection in B: the most we can expect is that the intersection is transverse in ∂B. This leads us to the following definition.

Definition 2.4.2. We say that f is *regular* if the following conditions hold. We require that the intersection

$$M(a, b) = U_a \cap S_b$$

be transverse in B, except in the special case that a is boundary-stable and b is boundary-unstable. In this special case, both U_a and S_b are contained in ∂B, and we require that their intersection be transverse as an intersection in ∂B, so that

$$T_x U_a + T_x S_b = T_x \partial B$$

for all x in $M(a, b)$. We refer to this special case as the *boundary-obstructed* case: we may refer either to $M(a, b)$, or to the corresponding trajectories $\gamma(t)$ of the flow, as being boundary-obstructed . ◊

These conditions imply that each space $M(a, b)$ is either a manifold, or, in the case that $M(a, b)$ meets both ∂B and the interior, a manifold with boundary. The latter can happen only if a and b are on the boundary, a is boundary-unstable and b is boundary-stable. The dimension of $M(a, b)$ is given by

$$\dim M(a, b) = \begin{cases} \text{index } (a) - \text{index } (b) + 1, & \text{if boundary-obstructed} \\ \text{index } (a) - \text{index } (b), & \text{otherwise.} \end{cases}$$

If we orient the vector spaces K_a^- for all the critical points a, then the spaces $M(a, b)$ again become oriented; but we need an extra word about orienting $M(a, b)$ in the boundary-obstructed case, because of the lack of transversality. In the sequence (2.3), the second map π is no longer surjective in the boundary-obstructed case: instead we have an exact sequence

$$0 \to T_x M(a, b) \xrightarrow{i} T_x U_a \xrightarrow{\pi} N_x S_b \xrightarrow{q} \mathbb{R} \to 0,$$

where we identify the last $\mathbb{R}$ with the outward normal to ∂B at x. We choose an oriented basis $\bar{w}_1, \dots, \bar{w}_{j+1}$ of $N_x S_b$, so that the first j vectors are the images under π of vectors $w_1, \dots, w_j$ and $q(\bar{w}_{j+1}) = 1$. Then, as before, we declare a basis $v_1, \dots, v_d$ of $T_x M(a, b)$ to be an oriented basis if

$$i(v_1), \dots, i(v_d), w_1, \dots, w_j$$

is an oriented basis of $T_x U_a$.

Having described the raw ingredients, we now form the Morse complexes. Because we have three types of critical points, we can form three graded groups, C_*^o, C_*^s and C_*^u: we set

$$\left. \begin{aligned} C_k^o &= \left(\bigoplus_{a \in \mathfrak{c}^o} \mathbb{Z} e_a \right) \otimes R \\ C_k^s &= \left(\bigoplus_{a \in \mathfrak{c}_k^s} \mathbb{Z} e_a \right) \otimes R \\ C_k^u &= \left(\bigoplus_{a \in \mathfrak{c}_k^u} \mathbb{Z} e_a \right) \otimes R. \end{aligned} \right\} \tag{2.7}$$

If $M(a,b)$ is non-empty and of dimension 1, so that $\breve{M}(a,b)$ is a finite set, we again write

$$n(a,b) = \sum_{\breve{x} \in \breve{M}(a,b)} \epsilon(\breve{x}),$$

using the just-described orientation convention for the boundary-obstructed case. If $M(a,b)$ is not 1-dimensional, we set $n(a,b) = 0$.

When a and b are both boundary critical points, there is a possible variant of $n(a,b)$. Let $M^{\partial}(a,b)$ be the intersection of $M(a,b)$ with ∂M. This manifold parametrizes trajectories along the boundary, joining a to b. We write $\breve{M}^{\partial}(a,b)$ for its quotient by $\mathbb{R}$. Recall that $M(a,b)$ can meet the interior of B only when a is boundary-unstable and b is boundary-stable; so only in this one case is $M^{\partial}(a,b)$ any different from $M(a,b)$. The zero-dimensional spaces among the $\breve{M}^{\partial}(a,b)$ are the ingredients of the boundary map for the usual Morse complex of $f|_{\partial B}$. But to write down the Morse complex, we need orientations of the unstable manifolds in the boundary. For a critical point a on the boundary, let us write

$$\bar{K}_a^- = K_a^- \cap T_a(\partial B).$$

If a is boundary-stable, then $\bar{K}_a^- = K_a^-$ and we orient the former using our chosen orientation of the latter. If a is boundary-unstable, we orient $\bar{K}_a^-$ by regarding it as the tangent space at a to ∂U_a, and using the outward-normal-first convention for the orientation of the boundary. After orienting all $\bar{K}_a^-$ in this way, we again obtain signs

$$\bar{\epsilon}(\breve{x}) = \pm 1$$

for all trajectories belonging to zero-dimensional trajectory spaces $\breve{M}^{\partial}(a,b)$. We write

$$\bar{n}(a,b) = \sum_{\breve{x} \in \breve{M}(a,b)} \epsilon(\breve{x}).$$

We consider now how the integers $n(a,b)$ and $\bar{n}(a,b)$ can be used to define operators on the groups (2.7). The first thing we can do is form the Morse complex of the boundary manifold ∂B, using the Morse flow on the boundary and the integers $\bar{n}(a,b)$. This is straightforward, but there is a slight twist concerning the grading. Suppose a is a critical point on the boundary that is boundary-unstable, and index $(a) = k$, so that U_a is k-dimensional. The dimension of $U_a \cap \partial M$ is then $k-1$. For the flow on the boundary therefore, a has index $k-1$. To form the Morse complex $\bar{C}_*$ of ∂B, we must set

$$\bar{C}_k = C^s_k \oplus C^u_{k+1}.$$

The differential $\bar{\partial} : \bar{C}_k \to \bar{C}_{k-1}$ is defined by

$$\bar{\partial} e_a = \sum_b \bar{n}(a,b) e_b.$$

As an instance of Theorem 2.1.1, we have $\bar{\partial}\bar{\partial} = 0$, and

$$H_k(\bar{C}, \bar{\partial}) = H_k(\partial B; R).$$

We now describe Morse complexes that incorporate the interior critical points also, and compute $H_k(B; R)$ and $H_k(B, \partial B; R)$. To motivate the definition, let us look back at the proof outlined above, that $\partial\partial = 0$ in the case of a manifold B without boundary. We can still define the space of unparametrized broken trajectories, $\breve{M}^+(a, a')$, and it is still the case that this is compact. But two things change.

The first change to notice when B has boundary is that not every strictly broken trajectory will be a limit point of a sequence of unbroken trajectories. Suppose for example that a is an interior critical point, b is a boundary-stable critical point, and c is boundary-unstable. Suppose that $(\breve{x}_1, \breve{x}_2)$ is a broken trajectory with two components, joining a to c via b. This broken trajectory is not the limit of unbroken trajectories, simply because there can be no trajectory from a to c: the stable manifold of c is contained in ∂B, while a is in the interior.

The second change is that, in the case that a and c have index difference 2 (the case relevant to the earlier proof that $\partial\partial$ is zero), there can be broken trajectories from a to c with more than two components. Consider the case, for example, that a and c are both in the interior:

Lemma 2.4.3. *Suppose a and c are interior critical points with indices k and $k-2$. Then a strictly broken trajectory in $\breve{M}^+(a, c)$ has either two or three components, so takes one of the two forms*

$$(\breve{x}_1, \breve{x}_2) \in \breve{M}(a, b) \times \breve{M}(b, c)$$

or

$$(\breve{x}_1, \breve{x}_2, \breve{x}_3) \in \breve{M}(a, b_1) \times \breve{M}(b_1, b_2) \times \breve{M}(b_2, c).$$

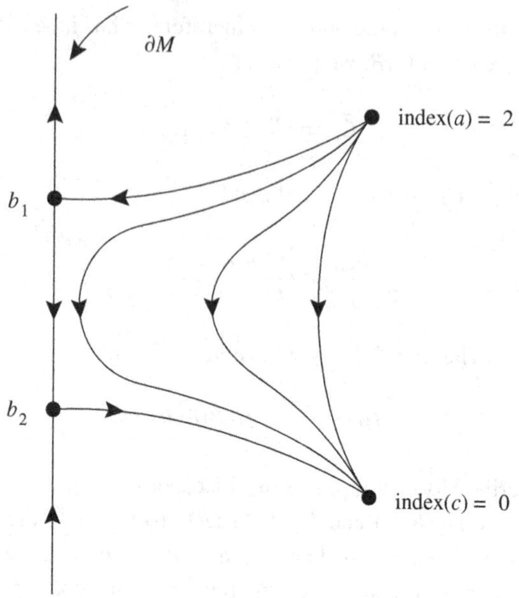

Fig. 1. A family of index-2 trajectories limiting to a 3-component broken trajectory.

In the first case, b must be an interior critical point of index $k-1$. In the second case b_1 and b_2 must both be boundary critical points of index $k-1$ and the space of trajectories $\breve{M}(b_1, b_2)$ is boundary-obstructed.

Proof. If there are two components, the intermediate critical point must be in the interior, because no boundary critical point is both a forward and a backward limit point of an interior trajectory. Its index must be $k-1$ if $M(a,b)$ and $M(b,c)$ are both to be non-empty.

If there are more than two components, then some component is boundary-obstructed: only in the boundary-obstructed case can there be a trajectory joining critical points of equal index. There cannot be more than three components, because two adjacent components of a broken trajectory cannot both be boundary-obstructed. Figure 1 illustrates the picture in dimension 2. □

In the situation described in this lemma, the structure of $\breve{M}^+(a,c)$ near the second type of broken trajectory is a little more complicated to analyze than the previous case. But it is still true that $\breve{M}^+(a,c)$ behaves as if it were a 1-manifold with boundary, to the extent that the number of strictly broken trajectories in

$\breve{M}^+(a,c)$ is even. Thus, for any such a and c, the sum

$$\sum_{b\in\mathfrak{C}_{k-1}} n(a,b)n(b,c) + \sum_{b_1\in\mathfrak{C}^s_{k-1}} \sum_{b_2\in\mathfrak{C}^u_{k-1}} n(a,b_1)\bar{n}(b_1,b_2)n(b_2,c)$$

is even. If we take more care with the orientations, and consider the *oriented* boundary of $\breve{M}^+(a,c)$, we establish

$$-\sum_{b\in\mathfrak{C}_{k-1}} n(a,b)n(b,c) + \sum_{b_1\in\mathfrak{C}^s_{k-1}} \sum_{b_2\in\mathfrak{C}^u_{k-1}} n(a,b_1)\bar{n}(b_1,b_2)n(b_2,c) = 0. \quad (2.8)$$

Let us see how to interpret the left side of (2.8) as the matrix entry of an operator on C^o_*. First, using the decomposition of $\bar{C}$ as a sum of C^s and C^u, we write $\bar{\partial}$ as a 2-by-2 matrix,

$$\bar{\partial} = \begin{bmatrix} \bar{\partial}^s_s & \bar{\partial}^u_s \\ \bar{\partial}^u_s & \bar{\partial}^u_u \end{bmatrix},$$

so that $\bar{\partial}^u_s$, for example, is a homomorphism $\bar{\partial}^u_s : C^u \to C^s$. We use the integers $n(a,b)$, which count interior trajectories, to define similar operators

$$\begin{aligned}
\partial^o_o &: C^o_k \to C^o_{k-1} \\
\partial^o_s &: C^o_k \to C^s_{k-1} \\
\partial^u_o &: C^u_k \to C^o_{k-1} \\
\partial^u_s &: C^u_k \to C^s_{k-1}
\end{aligned}$$

by

$$\begin{aligned}
\partial^o_o e_a &= \sum_{b\in\mathfrak{C}^o} n(a,b)e_b \qquad (a \in \mathfrak{C}^o) \\
\partial^o_s e_a &= \sum_{b\in\mathfrak{C}^s} n(a,b)e_b \qquad (a \in \mathfrak{C}^o) \\
\partial^u_o e_a &= \sum_{b\in\mathfrak{C}^o} n(a,b)e_b \qquad (a \in \mathfrak{C}^u) \\
\partial^u_s e_a &= \sum_{b\in\mathfrak{C}^s} n(a,b)e_b \qquad (a \in \mathfrak{C}^u).
\end{aligned}$$

(Note that there are two different maps $C^u \to C^s$, namely $\bar{\partial}^u_s$ and ∂^u_s: the first is defined by counting isolated trajectories in ∂B, and the second counts isolated trajectories in the interior of B.)

Now the quantity on the left of (2.8) is the matrix entry between e_a and e_c of the operator

$$-\partial^o_o \partial^o_o + \partial^u_o \bar{\partial}^s_u \partial^o_s : C^o_k \to C^o_{k-2}.$$

Thus the identity (2.8) can be written:

$$-\partial^o_o \partial^o_o + \partial^u_o \bar{\partial}^s_u \partial^o_s = 0. \tag{2.9}$$

There are similar identities corresponding to the other possibilities for the types of the critical points a and c. (See Lemma 22.1.5 later in the book.) With this as motivation, we come to the following definition of two chain complexes, $\left(\check{C}_*, \check{\partial}\right)$ and $\left(\hat{C}_*, \hat{\partial}\right)$.

Definition 2.4.4. We define $\check{C}_k = C^o_k \oplus C^s_k$, and we define an operator

$$\check{\partial} : \check{C}_k \to \check{C}_{k-1}$$

$$\check{\partial} = \begin{bmatrix} \partial^o_o & -\partial^u_o \bar{\partial}^s_u \\ \partial^o_s & \bar{\partial}^s_s - \partial^u_s \bar{\partial}^s_u \end{bmatrix}.$$

We similarly define $\hat{C}_k = C^o_k \oplus C^u_k$, we define

$$\hat{\partial} : \hat{C} \to \hat{C}$$

$$\hat{\partial} = \begin{bmatrix} \partial^o_o & \partial^u_o \\ -\bar{\partial}^s_u \partial^o_s & -\bar{\partial}^u_u - \bar{\partial}^s_u \partial^u_s \end{bmatrix}.$$

◊

We have the following result.

Theorem 2.4.5. *The operators $\check{\partial}$ and $\hat{\partial}$ both have square zero. The complexes $(\check{C}_*, \check{\partial})$ and $(\hat{C}_*, \hat{\partial})$ compute the absolute and relative homology groups: there are isomorphisms*

$$H_k(\check{C}_*, \check{\partial}) = H_k(B; R)$$

$$H_k(\hat{C}_*, \hat{\partial}) = H_k(B, \partial B; R).$$

Proof. We do not present a complete proof here, because much of it would be repeating material that appears later, in the context of Floer homology. There are three steps. First one must show that the operators have square zero; then one shows that the resulting homology groups are independent of f; and finally one identifies the homology groups.

The proof that these two operators have square zero is a consequence of the identity (2.9) and its relatives. (See the proof of Lemma 22.1.5 and Proposition 22.1.4.) From the material of Section 22, the reader can also extract a proof that the homology of the two complexes is independent of the choice of a suitable f. To identify the homology groups as the absolute and relative groups of the pair, one can choose f so that all the critical points on ∂B are boundary-unstable, and some collar W of ∂B has the property that no trajectory enters W from $B \setminus W$. The complex $\check{C}$ is now just the Morse complex of the interior critical points, and the situation is essentially no different from the more standard picture described at the start of this subsection, in which f achieves its maximum on ∂B and the Morse complex computes the absolute homology. One can treat the complex $\hat{C}$ similarly using Poincaré duality. □

2.5 Circle actions and blowing up

We now turn to a slightly different situation. Let P be a compact, Riemannian manifold, without boundary, and suppose that S^1 acts on P by isometries. Let $Q = P^{S^1}$ be the set of fixed points, and suppose that S^1 acts freely elsewhere. The fixed-point set Q will be a smooth submanifold with even codimension. The S^1 action on the normal bundle $N \to Q$ equips the normal bundle with the structure of a complex vector bundle. We write B for the quotient,

$$B = P/S^1,$$

which is a smooth manifold away from the image of the fixed-point set.

The singularity in B along the image of Q can be resolved by blowing up. To describe this, we begin by recalling how one constructs the real oriented blow-up of a manifold such as P along a compact submanifold Q. Let $S(N)$ be the unit sphere bundle in the normal bundle to Q, and let

$$p : S(N) \times [0, \epsilon) \to P \tag{2.10}$$

be the map which assigns to (v, t) the point $\gamma_v(t)$, where γ_v is the geodesic in P with initial tangent vector v. The tubular neighborhood theorem tells us that if we choose ϵ sufficiently small, then the image of p is an open neighborhood W of Q, and that the restriction of p to the subset $t \neq 0$ is a diffeomorphism,

$$p^o : S(N) \times (0, \epsilon) \to W \setminus Q.$$

The oriented blow-up of P along Q is defined to be the manifold with boundary P^σ obtained as the union of $P \setminus Q$ and $S(N) \times [0, \epsilon)$, patched together using p^o:

$$P^\sigma = \big((S(N) \times [0, \epsilon)) \cup (P \setminus Q)\big)/p^o.$$

There is a smooth map $\pi : P^\sigma \to P$, the blow-down map, which is a diffeomorphism over $P \setminus Q$, and has fibers the spheres $S(N_q)$ over points q in Q.

Although we used the Riemannian metric as a tool, the set of points that comprise P^σ can be described without the metric by regarding $S(N_q)$ as a quotient space of $T_qP \setminus T_qQ$; and the smooth structure on this set P^σ is then independent of the choice of metric. Indeed, we can construct the tubular neighborhood coordinates in any way we choose (not necessarily from geodesics), and the result is the same. The proof of this assertion is easily reduced to the case that a neighborhood of Q is a product, when the essential point is contained in the following lemma (which is applied in the case that M and N have the same dimension).

Lemma 2.5.1. *Let M and N be Euclidean spaces, and let M^σ, N^σ be obtained from these by blowing up the origin. Let $h : M \to N$ be a smooth map with $h(0) = 0$, and suppose that dh is injective at 0. Then there is a smooth map $h^\sigma : M^\sigma \to N^\sigma$ such that the following diagram commutes:*

$$\begin{array}{ccc} M^\sigma & \xrightarrow{h^\sigma} & N^\sigma \\ \pi \big\downarrow & & \big\downarrow \pi \\ M & \xrightarrow[h]{} & N. \end{array}$$

The map h^σ is an embedding in a neighborhood of $\pi^{-1}(0) \subset M^\sigma$. If we have a family of smooth maps h_t depending smoothly on a parameter $t \in T$, then the maps h_t^σ depend smoothly on t also.

Proof. We prove the first part. On the boundary $\mathbb{P}(M)$ of M, define h^σ to be the projectivization of the linear map $dh|_0 : M \to N$. This produces a map h^σ on M^σ so that the diagram commutes. To see that h^σ is smooth, let $x_1, \ldots, x_m$ and $y_1, \ldots, y_n$ be coordinates on the two vector spaces, and write $\xi_i = x_i/x_m$ and $\eta_i = y_i/y_n$. Then

$$\xi_1, \ldots, \xi_{m-1}, x_m$$

$$\eta_1, \ldots, \eta_{n-1}, y_n$$

are coordinates on open subsets U and V of the manifolds with boundary, M^σ and N^σ: the coordinates map U and V to $\mathbb{R}^{m-1} \times [0, \infty)$ and $\mathbb{R}^{n-1} \times [0, \infty)$. The boundaries of U and V are where x_m and y_n are zero. Because dh is injective at 0, we can suppose that $h^*(dy_n)|_0 = dx_m|_0$. With this assumption, the functions $h^*(\eta_i)$ and $h^*(y_n)$ are defined in a neighborhood of ∂U; and $h^*(y_n)$ has the same derivative as x_m at ∂U. The functions

$$\xi_1, \ldots, \xi_{m-1}, h^*(y_n)$$

therefore define a coordinate system on U near the boundary, and the boundary is defined by $h^*(y_n) = 0$. The main assertion of the lemma is that $h^*(\eta_i)$ is smooth. We can write

$$h^*(\eta_i) = h^*(y_i)/h^*(y_n).$$

The numerator is a smooth function that vanishes on ∂U, while the denominator vanishes with non-zero derivative at ∂U. The ratio is therefore a smooth function. □

Let us return to the case that P has an isometric circle action with Q as fixed-point set and no other non-trivial stabilizer. The circle action lifts to a circle action on P^σ, because the circle acts naturally on $S(N)$ and the map p^o is equivariant. We note that on P^σ the circle action is free, because the action on $S(N)$ is. We define

$$B^\sigma = P^\sigma / S^1.$$

This is again a smooth manifold with boundary, and we have a blow-down map

$$\pi : B^\sigma \to B.$$

This is a diffeomorphism except over the set of fixed points. Restricted to the boundary of B^σ, the map π is a fiber bundle, $\pi : \partial B^\sigma \to Q$, with fibers the complex projective spaces $S(N_q)/S^1$, so we can write

$$\partial B^\sigma = \mathbb{P}(N). \tag{2.11}$$

Now we again introduce a smooth function $\tilde{f} : P \to \mathbb{R}$ and its gradient vector field $\tilde{V}$ on P. We suppose that $\tilde{f}$ is invariant under the circle action, so that it descends to a function $f : B \to \mathbb{R}$ which is smooth away from the singular

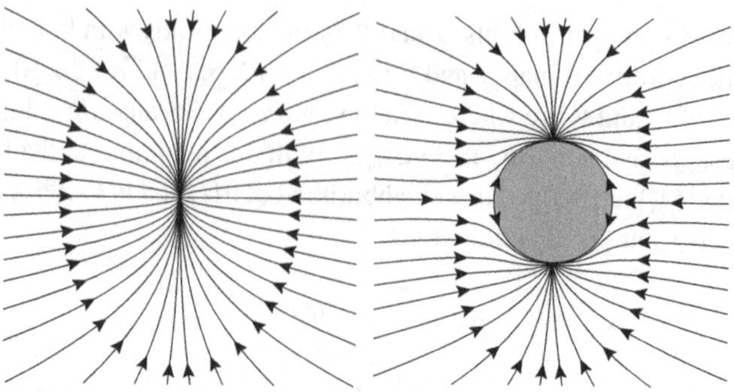

Fig. 2. A linear flow on $\mathbb{R}^2$ and the corresponding flow on the blow-up, $S^1 \times \mathbb{R}^{\geq}$.

locus in the quotient space. Because the blow-down map $\pi : P^\sigma \to P$ is a diffeomorphism except over Q, we can regard $\tilde{V}$ as giving a vector field also on the interior of P^σ.

Lemma 2.5.2. *The vector field $\tilde{V}$ on $P^\sigma \setminus \partial P^\sigma$ extends to a smooth vector field $\tilde{V}^\sigma$ on P^σ, everywhere tangent to the boundary.* □

We will not prove the lemma here, but it can be deduced from Lemma 2.5.1: one can consider the flow ϕ_t generated by the vector field on P, and use the lemma to deduce the existence of a smooth flow ϕ_t^σ on the blow-up. Figure 2 shows the picture of a linear flow on $\mathbb{R}^2$ (without an S^1 symmetry) and the corresponding flow on the manifold with boundary obtained by blowing up at the origin, namely the space $S^1 \times \mathbb{R}^{\geq}$: there is no circle symmetry in this picture.

Because the circle action is free on P^σ, the vector field $\tilde{V}^\sigma$ on P^σ passes to the quotient space B^σ: we obtain a vector field V^σ on B^σ. The vector field V^σ is also tangent to the boundary of B^σ.

Remark. The vector field V^σ on B^σ is not a gradient vector field in any natural way. Nevertheless, it shares with gradient vector fields the important property that, at each zero a of V^σ, the derivative ∇V^σ is an operator with real eigenvalues. We shall say that a zero of V^σ is non-degenerate if 0 is not an eigenvalue of ∇V^σ, in which case we can decompose $T_a B^\sigma$ as $K^- \oplus K^+$ as usual. These are the tangent spaces to the unstable and stable manifolds at a of the downward flow. We still refer to the dimension of K^- as the *index* of a.

The model case is when P is a complex vector space, say $\mathbb{C}^n$ with the standard metric, the circle action is the standard one, Q is the origin, and $\tilde{f}$ is given in terms of a hermitian operator L by the formula

$$\tilde{f}(p) = \frac{1}{2}\langle p, Lp \rangle, \qquad p \in P.$$

As in (2.4), the downward gradient-flow equation on P is the linear equation

$$dp/dt = -Lp. \tag{2.12}$$

In this case, we can identify the blow-up P^σ as

$$\begin{aligned} P^\sigma &= S(P) \times [0, \infty) \\ &\cong S^{2n-1} \times [0, \infty). \end{aligned} \tag{2.13}$$

The blow-down map is given by $\pi(\phi, s) = s\phi$ for ϕ in S^{2n-1} and $s \geq 0$. To describe the vector field $-\tilde{V}$ on the interior of P^σ, we write the downward gradient-flow equations for the function $\tilde{f}$ on P in the "polar coordinates" (ϕ, s). Writing the equation $dp/dt = -Lp$ in polar coordinates is essentially the same exercise as Lemma 2.3.1. Let us again write Λ for the function

$$\Lambda(p) = \langle p, Lp \rangle / \|p\|^2 \tag{2.14}$$

on $P \setminus 0$. In the coordinates (ϕ, s), this function is independent of s and so defines a function on P^σ:

$$\Lambda : P^\sigma \to \mathbb{R}.$$

The gradient-flow equations (2.12) can now be written in polar coordinates as

$$\begin{aligned} \dot{\phi} &= -L\phi + \Lambda(\phi)\phi \\ \dot{s} &= -\Lambda(\phi)s. \end{aligned} \tag{2.15}$$

The first of these two equations is the gradient-flow equation for the restriction of $\frac{1}{2}\Lambda$ to the unit sphere $S(P)$, just as in (2.5). In this form, it is clear that the flow extends to a smooth flow on the blow-up P^σ, preserving the boundary ∂P^σ. This means that $\tilde{V}$ extends to a smooth vector field on P^σ, tangent to the boundary, as the lemma claimed.

Let us look at the quotient $B^\sigma = P^\sigma / S^1$ and the zeros of the vector field V^σ in this example. We write

$$B^\sigma = \mathbb{CP}^{n-1} \times [0, \infty), \tag{2.16}$$

so that a point of B^σ is a pair $([\phi], s)$.

Proposition 2.5.3. *In the model case* (2.16)*, suppose that the spectrum of L is simple and that* 0 *is not an eigenvalue. Let* $\phi_1, \ldots, \phi_n$ *be orthogonal unit eigenvectors of L in P, belonging to eigenvalues*

$$\lambda_1 < \cdots < \lambda_n.$$

Then the zeros of the vector field V^σ *on* B^σ *are precisely the points* $([\phi_i], 0)$*, all of which belong to the boundary of* B^σ*. The index of* $([\phi_i], 0)$ *is given by*

$$\operatorname{index}([\phi_i], 0) = \begin{cases} 2(i-1) & (\lambda_i > 0) \\ 2(i-1)+1 & (\lambda_i < 0). \end{cases}$$

Proof. Because the first equation of (2.15) is the same equation that appeared in (2.5), we see that $([\phi], s)$ can be a stationary point of the flow only if ϕ is an eigenvector of L, just as in Lemma 2.3.2. From the second equation in (2.15), we see that it is also necessary that $\Lambda(\phi)s$ is zero. At an eigenvector ϕ, the value of $\Lambda(\phi)$ is the corresponding eigenvalue. So if zero is not in the spectrum of L, then s must be zero at the critical points.

At $([\phi_i], 0)$, we can identify the tangent space to B^σ with $W \oplus \mathbb{R}$, where W is the orthogonal complement of ϕ_i, as in the proof of Lemma 2.3.3. The operator ∇V^σ on $W \oplus \mathbb{R}$ is the sum of the operator $w \mapsto Lw - \lambda_i w$ that appeared in the proof of that lemma and the operator $s \mapsto \lambda_i s$. The index of this operator is as stated. □

We turn from the model, linear case back to the general case of a manifold P with circle action. The normal bundle $N \to Q$ carries a bundle endomorphism L: at each point q in the Q, we have an operator L_q on the normal space N_q, given by the Hessian of $\tilde{f}$ in the normal directions (the restriction of $\nabla\tilde{V}$ to N_q). Because it commutes with the circle action, this real symmetric operator is a complex-linear, hermitian operator on N_q. We impose the following condition, which reduces to the hypothesis of the proposition above if Q is just a point:

Hypothesis 2.5.4. We suppose that the restriction of $\tilde{f}$ to Q is a Morse function on Q. At each critical point $q \in Q$ of the $\tilde{f}$, we suppose that the spectrum of the hermitian operator L_q is simple, and that zero is not an eigenvalue. We write the eigenvalues of L_q as

$$\lambda_1(q) < \lambda_2(q) < \cdots < \lambda_n(q),$$

and choose corresponding eigenvectors

$$\phi_1(q) < \phi_2(q) < \cdots < \phi_n(q).$$

◊

The next lemma describes those zeros of V^σ which lie on the boundary.

Lemma 2.5.5. *Let a in ∂B^σ be written as a pair $(q, [\phi])$, where $q \in Q$ and $[\phi]$ is in the projectivization of N_q. Then a is a stationary point for the vector field V^σ if and only if q is a critical point for the restriction $\tilde{f}|_Q$ and ϕ is an eigenvector of L_q. Under the above hypothesis, the zeros $(q, [\phi_i(q)])$ are non-degenerate. In this case, the index of the critical point is given by*

$$\text{index} = \begin{cases} \text{index}_Q(q) + 2(i-1) & (\lambda_i(q) > 0) \\ \text{index}_Q(q) + 2(i-1) + 1 & (\lambda_i(q) < 0), \end{cases}$$

where $\text{index}_Q(q)$ *is the index of q as a critical point of $\tilde{f}|_Q$.*

Proof. Let $(q, [\phi])$ be a point of $\partial B^\sigma = \mathbb{P}(N)$. We can decompose the tangent space of B^σ at this point as an orthogonal direct sum

$$T_q Q \oplus W \oplus \mathbb{R}. \tag{2.17}$$

Here, the first summand is the horizontal part of the tangent space to the fiber bundle $\mathbb{P}(N)$, the summand W denotes the complex-orthogonal complement of ϕ, which we identify with the tangent space $T_{[\phi]}\mathbb{P}(N_q)$ (the vertical part of the tangent space to the fiber bundle $\mathbb{P}(N)$), and $\mathbb{R}$ is the normal direction to the boundary. In terms of this decomposition, we can write the vector $V^\sigma_{(q,[\phi])}$ as $(v_1, v_2, 0)$. The fact that the third component is zero is the statement that V^σ is tangent to the boundary. The vector $v_1 \in T_q Q$ is equal to $\text{grad}\tilde{f}|_Q$; so $(q, [\phi])$ cannot be a critical point of the flow unless q is a critical point of $\tilde{f}$. When v_1 is zero the vector v_2 is

$$v_2 = L_q\phi - \Lambda_q(\phi)\phi,$$

where

$$\Lambda_q(\phi) = \langle \phi, L_q\phi \rangle / \|\phi\|^2 \tag{2.18}$$

as in the model case. This can be zero only when ϕ is one of the eigenvectors $\phi_i(q)$ of L_q.

This completes the identification of the critical points. To verify the remaining statements, we must compute the derivative ∇V^σ at $(q, [\phi_i(q)])$ as an endomorphism of the tangent space (2.17). Let us write this operator H in 3-by-3 block form as

$$H = \begin{bmatrix} h_{11} & h_{12} & h_{13} \\ h_{21} & h_{22} & h_{23} \\ h_{31} & h_{32} & h_{33} \end{bmatrix}.$$

Because V^σ is tangent to the boundary, the subspace $T_qQ \oplus W$ of (2.17) is invariant under H, so h_{31} and h_{32} are zero.

The $\mathbb{R}$ summand of (2.17) is also invariant under H. This is most easily seen as follows. The map (2.10) which we used to form the blow-up can be extended to negative t, by the same formula, to give a map $S(N) \times (-\epsilon, \epsilon) \to P$. We can use this extension to form a natural *double* of the manifold with boundary, P^σ, by taking two copies $(P \setminus Q)^\pm$ of $P \setminus Q$ and attaching them to $S(N) \times (0, \epsilon)$ and $S(N) \times (-\epsilon, 0)$. The vector field $\tilde{V}$ gives rise to a vector field on $S(N) \times (-\epsilon, \epsilon)$, and in this way, the vector field V^σ can be extended smoothly to a natural double of the manifold with boundary, B^σ. Because the original vector field on P is invariant under S^1, the extension of V^σ to the doubled manifold is invariant under the involution corresponding to the map $t \mapsto -t$ on $(-\epsilon, \epsilon)$. It follows that H commutes with the involution of the tangent space (2.17) that changes the sign on the $\mathbb{R}$ summand. This means that h_{13} and h_{23} are zero.

The block h_{12} is zero because the horizontal part of V^σ at ∂B^σ is equal to $\operatorname{grad} \tilde{f}|_Q$, which is independent of $[\phi]$. The remaining blocks are easily identified, and we are left with

$$H = \begin{bmatrix} h_q & 0 & 0 \\ x & L_q - \lambda_i & 0 \\ 0 & 0 & \lambda_i \end{bmatrix} \tag{2.19}$$

where h_q is the Hessian of $\tilde{f}|_Q$ at q, and x is the operator

$$x(v) = (\partial_v L_q)\phi_i(q) \quad (v \in T_qQ).$$

We see that this zero of V^σ is non-degenerate provided that h_q is non-degenerate, λ_i is a simple eigenvalue of L_q and λ_i is non-zero. We can also read off the index and obtain the result of the proposition. □

In the case that all of the critical points of V^σ on B^σ are non-degenerate, we can summarize the situation as follows. The critical points will be of two sorts.

First, there is the set of critical points in the interior, which we call $\mathfrak{c}^o$ as in the previous subsection. Then there are the critical points on the boundary, which are the points $(q, [\phi_i(q)])$, where q is a critical point of $\tilde{f}|_Q$. As in the previous subsection, the critical points on the boundary can be further divided into the boundary-stable and boundary-unstable critical points, $\mathfrak{c}^s$ and $\mathfrak{c}^u$. The critical point corresponding to $(q, [\phi_i(q)])$ is boundary-stable if and only if $\lambda_i(q) > 0$.

We can now study the Morse theory of the flow generated by the vector field $-V^\sigma$ on the manifold with boundary, B^σ. Although the vector field is not a gradient vector field, it behaves sufficiently like a gradient vector field that we can still construct the Morse complex. For example, $\tilde{f}$ is always non-increasing along any flow line. It is possible to choose $\tilde{f}$ so that the flow of $-V^\sigma$ on B^σ satisfies the regularity condition in Definition 2.4.2, and the same compactness theorem holds for the spaces of broken trajectories. Thus we can form complexes $\check{C}_*$, $\hat{C}_*$ and $\bar{C}_*$, as before: see Definition 2.4.4 and Theorem 2.4.5 above. These complexes still compute the corresponding ordinary homology groups:

Theorem 2.5.6. *Suppose that the flow generated by the vector field V^σ is regular, in that the stable and unstable manifolds of the critical points satisfy the condition in Definition 2.4.2. Then the operators $\check{\partial}$, $\hat{\partial}$ and $\bar{\partial}$ on $\check{C}_*$, $\hat{C}_*$ and $\bar{C}_*$ have square zero. The complexes compute the absolute and relative homology for the pair $(B^\sigma, \partial B^\sigma)$, and the homology of ∂B^σ, respectively.* □

Let us take a closer look at the complex $\bar{C}_*$. According to the theorem just stated, the complex $\bar{C}_*$ computes the ordinary homology (say with coefficients $\mathbb{Z}$) of ∂B^σ, which is a bundle over Q with fiber $\mathbb{CP}^{n-1}$, the projectivization of the normal bundle (2.11). According to the Leray–Hirsch theorem, the homology is isomorphic to n copies of the homology $H_*(Q)$. The normal bundle N carries the self-adjoint bundle map $L : N \to N$, and a connection ∇ arising from the Levi-Cività connection on P. We write

$$e_{q,i} \in \bar{C}_*$$

for the generator of $\bar{C}_*$ corresponding to the stationary point $(q, [\phi_i(q)])$ in ∂B^σ. In the complex $\bar{C}_*$, this generator has grading index $\varrho(q) + 2(i-1)$.

The boundary map in the complex $\bar{C}_*$ is defined using the flow lines of the vector field $-V^\sigma$ on $\mathbb{P}(N)$. We can write the equations for a flow line in two slightly different ways. These correspond to two different ways of viewing a linear flow on $\mathbb{CP}^{n-1}$, as either the flow (2.5) on the unit sphere or the linear flow (2.4) on $\mathbb{C}^n$. In the present setting, the first way is to lift the vector field V^σ on $\mathbb{P}(N)$ to a vector field $\tilde{V}^\sigma$ on the unit sphere bundle $S(N)$. If $(q(t), [\phi(t)])$ is a trajectory of $-V^\sigma$, we can lift it to a trajectory $(q(t), \phi(t))$ of $-\tilde{V}^\sigma$ on $S(N)$.

Thus $|\phi(t)| = 1$ for all t. The equations can then be written

$$\frac{d}{dt}q + (\operatorname{grad}\tilde{f}|_Q)_{q(t)} = 0 \tag{2.20a}$$

$$q^*(\nabla)\phi + \big((L_{q(t)} - \Lambda_q(\phi))\phi\big)dt = 0. \tag{2.20b}$$

The first term in the second equation denotes the derivative of ϕ "along the path", and the term $\Lambda_q(\phi)$ is defined as before.

The second way to write the equations is to consider pairs $(q(t), \phi(t))$, where $q : \mathbb{R} \to Q$ is a smooth path in Q and $\phi(t) \in N_{q(t)}$ is a nowhere-zero section of N along the path. For such a pair, we write down the pair of equations

$$\frac{d}{dt}q + (\operatorname{grad} f)_{q(t)} = 0 \tag{2.21a}$$

$$q^*(\nabla)\phi + (L_{q(t)}\phi)dt = 0. \tag{2.21b}$$

Note that the second equation is linear in ϕ. This provides the following counterpart to Lemma 2.3.1.

Lemma 2.5.7. *The trajectories of $-V^\sigma$ on the boundary of B^σ are the images under the quotient map*

$$\pi : N \setminus 0 \to \mathbb{P}(N) = \partial B^\sigma$$

of the solutions $(q(t), \phi(t))$ to the equations (2.21) *with ϕ non-zero. Each flow-line of $-V^\sigma$ lifts to a solution $(q(t), \phi(t))$ of the above equations which is unique up to the action of the scalars $\mathbb{C}^*$ on ϕ.* □

In a similar spirit, we can characterize the solutions of the flow asymptotic to particular stationary points, the counterpart of Proposition 2.3.4.

Proposition 2.5.8. *Let $\mathfrak{a}_0 = (q_0, [\phi_{i_0}(q_0)])$ and $\mathfrak{a}_1 = (q_1, [\phi_{i_1}(q_1)])$ be two zeros of V^σ on ∂B^σ, and let λ_0, λ_1 be the corresponding eigenvalues. Then the space of trajectories $M^\partial(\mathfrak{a}_0, \mathfrak{a}_1)$ of the flow $-V^\sigma$ on ∂B^σ can be identified with the quotient by $\mathbb{C}^*$ of the set of solutions $(q(t), \phi(t))$ to the equations* (2.21) *for which ϕ has the asymptotics*

$$\phi(t) \sim c_0 e^{-\lambda_0 t}\phi_{i_0}(q_0), \quad \text{as } t \to -\infty$$

$$\phi(t) \sim c_1 e^{-\lambda_1 t}\phi_{i_1}(q_1), \quad \text{as } t \to +\infty,$$

where c_0, c_1 are non-zero complex constants. □

Having examined the trajectories on the boundary, let us now examine trajectories $x(t)$ contained in the interior of B^σ that are asymptotic to a boundary critical point (say as t goes to $+\infty$). The boundary critical point $(q, [\phi_i(q)])$ must be boundary-stable. Because the vector field V^σ is pulled back from $B = P/S^1$ except at the boundary, the image of this trajectory under the blow-down map $\pi : B^\sigma \to B$ is a trajectory $y(t)$ for the gradient flow of the function f on $B \setminus Q$, asymptotic to the critical point q of $\tilde{f}|_Q$. From the form (2.19) of the linearization of V^σ, we see that the distance to the boundary, $d(x(t), \partial B^\sigma)$, has the form

$$d(x(t), \partial B^\sigma) \sim ce^{-\lambda_i t}$$

as $t \to +\infty$. This means that, after blowing down, we have

$$d(y(t), \partial Q) \sim ce^{-\lambda_i t}.$$

Thus we can characterize these flow lines as follows:

Proposition 2.5.9. *Trajectories $x(t)$ of the vector field $-V^\sigma$ in the interior of B^σ that approach the boundary critical point $(q, [\phi_i(q)])$ as $t \to +\infty$ are in one-to-one correspondence with trajectories $y(t)$ to the gradient flow of f on $B \setminus Q$ which approach q and for which the distance from $y(t)$ to Q has the asymptotics $ce^{-\lambda_i t}$.* □

2.6 Comparison with the homotopy quotient

We continue to consider a manifold P with circle action, with fixed-point set Q as above, and the associated manifold with boundary, B^σ. The theorem above states that the resulting complex $\check{C}_*$ computes the ordinary homology of B^σ, which is the same as the homology of $(P \setminus Q)/S^1$:

$$H_*(\check{C}_*) \cong H_*((P \setminus Q)/S^1).$$

If the (real) codimension of Q is $2n$, then the pair $(P, P \setminus Q)$ is $(2n-1)$-connected, and so too therefore is the pair obtained by forming the *homotopy quotients* by S^1. In the case of $P \setminus Q$, the group S^1 acts freely, and the homotopy quotient is equivalent to the ordinary quotient. It follows that

$$H_j(\check{C}_*) \cong H_j(P /\!/ S^1)$$

for $j \leq 2n - 2$, where $P /\!/ S^1$ denotes the homotopy quotient.

There is a simple way to increase the codimension of the fixed-point set: we can replace P by $P \times \mathbb{C}^k$, and make S^1 act in the standard way on $\mathbb{C}^k$. The fixed-point set is now $Q \times \{0\}$, which has codimension $2(n+k)$. Let $B^\sigma_{(k)}$ be the associated manifold with boundary. Of course, we have now lost the compactness of our manifold, but this does not matter if we choose our Morse function carefully. Let $\tilde{f}$ be the original Morse function on P, and on $P \times \mathbb{C}^k$, take the Morse function

$$\tilde{f}_{(k)} = \tilde{f} + \sum_{r=1}^{k} \mu_r |z_r|^2.$$

We suppose that the μ_r are positive and increasing, and that μ_1 is greater than $\lambda_n(q)$ for all critical points $q \in Q$. Let $V^\sigma_{(k)}$ denote the vector field which we obtain on $B^\sigma_{(k)}$. Although the total space is non-compact, there is a compact subset which contains all the critical points and within which every trajectory between critical points is contained: this compact subset is the union of $\partial B^\sigma_{(k)}$ and the proper transform of $B \times \{0\}$. This means that we can still construct the complex $\check{C}_*$ for the flow of the vector field $-V^\sigma_{(k)}$. Let us denote this complex by $\check{C}^{(k)}_*$

The homology of the complex $\check{C}^{(k)}_*$ is isomorphic to the ordinary homology of $B^\sigma_{(k)}$, just as in the compact case. So, arguing as before and using the fact that $P \times \mathbb{C}^k$ has the same homotopy type as P, we conclude that there is an isomorphism

$$H_j(\check{C}^{(k)}_*) \cong H_j(P//S^1)$$

for $j \leq 2(n+k-1)$. We can compare the complex $\check{C}^{(k)}_*$ with the original $\check{C}_*$. The extra generators all arise from critical points on the boundary, one for each pair (q, μ_r), where q is a critical point of $\tilde{f}|_Q$ and μ_r is one of the k eigenvalues of the Hessian of $\tilde{f}_{(k)}$ in the $\mathbb{C}^k$ directions. Because the μs are all bigger than the λs, Lemma 2.5.5 tells us that the new generator corresponding to (q, μ_r) is in degree

$$\text{index}_Q(q) + 2(n+r-1).$$

As we increase k therefore, the complexes $\check{C}_*(V^\sigma_{(k)})$ stabilize, and there is a well-defined limiting complex $\check{C}^{(\infty)}_*$, defined by

$$\check{C}^{(\infty)}_* = \check{C}^{(k)}_j, \quad j < 2(n+k-1).$$

The homology of the limiting complex is isomorphic to the homology of the homotopy quotient in all dimensions:

$$H_i(\check{C}_*^{(\infty)}) \cong H_i(P/\!/S^1), \quad \text{for all } i.$$

We can also examine $H_*(\hat{C}_*^{(k)})$ in this situation. Unlike $H_*(\check{C}_*^{(k)})$, this group is independent of k, essentially because the new critical points introduced by increasing k are boundary-stable, not boundary-unstable, when the μ_i are positive. So for all k, this group is the relative homology of the pair $(B^\sigma, \partial B^\sigma)$, or equivalently of the pair $(P/S^1, Q)$. We can also write

$$H_i(\hat{C}_*^{(\infty)}) \cong (P/\!/S^1, Q/\!/S^1).$$

Finally, the group $H_i(\bar{C}_*^{(\infty)})$ in this context is the ordinary homology of $Q/\!/S^1 \cong Q \times \mathbb{CP}^\infty$.

2.7 Morse theory and local coefficients

Ordinary homology has a variant, homology with *local coefficients*, which we can also obtain from a Morse complex. We consider a local system, Γ, of abelian groups on a compact manifold B (a manifold without boundary, for simplicity). This means that for each point $a \in B$, we have an abelian group Γ_a, and for each relative homotopy class of paths z from a to b we have an isomorphism

$$\Gamma(z) : \Gamma_a \to \Gamma_b,$$

satisfying the obvious composition law for composite paths. We adopt the (perhaps non-standard) convention that if z and w are homotopy classes of paths from a to b and b to c respectively, then the homotopy class of the composite paths ("first z then w") is written

$$w \circ z. \tag{2.22}$$

With this convention, the isomorphisms $\Gamma(z)$ satisfy

$$\Gamma(w \circ z) = \Gamma(w) \circ \Gamma(z).$$

Suppose now that f is a Morse function on B and that the gradient flow of f, for a suitable choice of Riemannian metric on B, satisfies the Morse–Smale transversality condition. We define a graded abelian group $C_*(f,\Gamma)$, the Morse complex of f with coefficients in Γ, as

$$C_*(f,\Gamma) = \bigoplus_{a\in\mathfrak{c}} \Gamma_a.$$

(The sum is over the set of critical points of f, as it was earlier.) As usual, we set

$$C_k(f,\Gamma) = \bigoplus_{a\in\mathfrak{c}_k} \Gamma_a.$$

Unless the groups Γ_a have characteristic 2, we need also to choose orientations for each K_a^- at this stage.

Let a and b be chosen so that $\breve{M}(a,b)$ is zero-dimensional. To each $\breve{x}$ in $\breve{M}(a,b)$, we have an associated sign $\epsilon(\breve{x}) = \pm 1$, as before. The closure of the orbit $\breve{x}$ of the flow also provides a path from a to b, and there is an associated isomorphism $\Gamma(\breve{x})$, which we combine with the sign ϵ to obtain an isomorphism

$$\epsilon(\breve{x})\Gamma(\breve{x}) : \Gamma_a \to \Gamma_b.$$

We use these to define the boundary map

$$\partial : C_k(f,\Gamma) \to C_{k-1}(f,\Gamma)$$

by the formula

$$\partial = \sum_{a\in\mathfrak{c}_k} \sum_{b\in\mathfrak{c}_{k-1}} \sum_{\breve{x}\in\breve{M}(a,b)} \epsilon(\breve{x})\Gamma(\breve{x}).$$

The proof that $\partial\partial = 0$ is not much altered. The homology of the resulting complex is indeed $H_*(B;\Gamma)$, the ordinary homology of B with coefficients in the local system Γ.

We can also construct the *cohomology* with local coefficients. We define a cochain complex $C^*(f,\Gamma)$ by

$$C^k(f,\Gamma) = \sum_{a\in\mathfrak{c}_k} \Gamma_a,$$

and define a coboundary map $\delta : C^{k-1}(f,\Gamma) \to C^k(f,\Gamma)$ by the formula

$$\partial = \sum_{a\in\mathfrak{c}_k} \sum_{b\in\mathfrak{c}_{k-1}} \sum_{\breve{x}\in\breve{M}(a,b)} \epsilon(\breve{x})\Gamma(\breve{x})^{-1}.$$

2.8 Functoriality in Morse theory

A continuous map $h : B_1 \to B_2$ gives rise, of course, to a homomorphism $h_* : H_*(B_1; R) \to H_*(B_2; R)$. When B_1, B_2 are smooth, compact manifolds (without boundary, at present), there is a generalization. Rather than a smooth map $h : B_1 \to B_2$, we can consider more generally a compact manifold Z and a smooth map

$$r : Z \to B_1 \times B_2.$$

We could take r to be the inclusion of the graph of a smooth map h, to recover the special case. Composing r with the projections, we obtain two maps

$$r_1 : Z \to B_1$$
$$r_2 : Z \to B_2.$$

We shall need the first of these two maps to be *oriented*. That is, we shall suppose that there is an isomorphism of real line bundles

$$\mu : \Lambda^{\dim Z}(Z) \to r_1^*(\Lambda^{\dim B_1}(B_1)). \tag{2.23}$$

When μ is given, there is a well-defined map

$$(r_1)^* : H_*(B_1; R) \to H_*(Z; R)$$

defined by

$$(r_1)^* = \omega_Z^{-1}(r_1)_* \omega_{B_1}.$$

Here $(r_1)_*$ is the usual induced map on homology, and the two ωs denote the two Poincaré duality isomorphisms

$$\omega_{B_1} : H_*(B_1; R) \to H^*(B_1; \Lambda \otimes R)$$
$$\omega_Z : H_*(Z; R) \to H^*(Z; \Lambda \otimes R),$$

where Λ denotes the orientation bundle in both cases, regarded as a local system. In this way, the pair (Z, r) gives rise to a map, by "pull-up and push-down":

$$(r_2)_*(r_1)^* : H_*(B_1; R) \to H_*(B_2; R). \tag{2.24}$$

In the special case that (Z, r) is the graph of h, we recover the usual map $h_* : H_*(B_1; R) \to H_*(B_2; R)$. This construction has an extension that we will not pursue here, but which we shall take up again in the context of Floer homology. Given a cohomology class u on Z, one can construct a map $H_*(B_1; R) \to H_*(B_2; R)$ by

$$x \mapsto (r_2)_*(u \frown (r_1)^*(x)).$$

In the special case that u is Poincaré dual to an embedded submanifold $Z_u \subset Z$, this map is the "pull-up and push-down" map for the pair $(Z_u, r|_{Z_u})$.

We now describe how to obtain the map (2.24) from a chain map on Morse complexes. Equip both manifolds with Riemannian metrics, and let f_1, f_2 be Morse functions giving rise to flows $\phi_{1,t}$, $\phi_{2,t}$ on B_1 and B_2, satisfying the Morse–Smale transversality conditions. Choose orientations for the unstable manifolds of all critical points on both manifolds, so that we have Morse complexes

$$C_*(f_1), C_*(f_2),$$

which compute the ordinary homology groups $H_*(B_1; R)$ and $H_*(B_2; R)$. Given critical points

$$a \in B_1, \quad b \in B_2$$

for the Morse functions f_1, f_2, we define a space

$$Z(a, b) \subset Z$$

by

$$Z(a, b) = \{ w \in Z \mid r(w) \in U_a \times S_b \}.$$

Here U_a and S_b are the unstable and stable manifolds, regarded as submanifolds of B_1 and B_2. The spaces $Z(a, b)$ need not be smooth, unless we impose an additional transversality condition (which can be achieved by adjusting r):

Hypothesis 2.8.1. The map $r : Z \to B_1 \times B_2$ is transverse to $U_a \times S_b$, for all critical points a and b, of f_1 and f_2 respectively. ◊

When the transversality hypothesis holds, $Z(a, b)$ is a smooth submanifold of Z. For each $w \in Z(a, b)$, we have an exact sequence

$$0 \to T_w Z(a, b) \to T_w Z \to N_{r_1(w)} U_a \oplus N_{r_2(w)} S_b \to 0,$$

where NU_a is the normal bundle of $U_a \subset B_1$ and NS_b is the normal bundle of $S_b \subset B_2$. From the sequence, we can read off

$$\begin{aligned}\dim Z(a,b) &= \dim Z - \dim S_a - \dim U_b \\ &= \dim Z - \dim B_1 + \text{index}\,(a) - \text{index}\,(b).\end{aligned}$$

We can also use the sequence and our isomorphism μ to orient $Z(a,b)$, using our fiber-first convention. In this way, to each point w belonging to a zero-dimensional space $Z(a,b)$, we can associate a sign

$$\epsilon(w) = \pm 1.$$

We then define an integer

$$m(a,b) = \sum_{w \in Z(a,b)} \epsilon(w).$$

When $\dim Z(a,b)$ is non-zero, we set $m(a,b) = 0$. The integers $m(a,b)$ are the matrix entries of a homomorphism

$$m : C_*(f_1) \to C_*(f_2)$$

defined by the formula

$$m(e_a) = \sum_{b \in \mathfrak{c}(f_2)} m(a,b) e_b.$$

The following proposition states that we have achieved our objective, of describing the "pull-up push-down" map in terms of the Morse complexes:

Proposition 2.8.2. *The map $m : C_*(f_1) \to C_*(f_2)$ is a chain map between the Morse complexes. The resulting map $m_* : H_*(B_1; R) \to H_*(B_2; R)$ coincides with the map $(r_2)_*(r_1)^*$ described at* (2.24). □

The construction of the map associated to the pair (Z, r) can also be extended to the case of local coefficients. Suppose Γ_1 and Γ_2 are local systems of abelian groups on B_1 and B_2 respectively. Via r_1 and r_2, we can pull these back to Z, and we suppose that an isomorphism of local systems, Γ_Z, is given between these pull-backs:

$$\Gamma_Z : r_1^* \Gamma_1 \to r_2^* \Gamma_2.$$

This means, in particular, that for each w in Z, we have an isomorphism of groups,

$$\Gamma_Z(w) : \Gamma_{1,r_1(w)} \to \Gamma_{2,r_2(w)}.$$

Let $C_*(f_1, \Gamma_1)$ and $C_*(f_2, \Gamma_2)$ be the Morse complexes with local coefficients. Then we can define a map

$$m : C_*(f_1, \Gamma_1) \to C_*(f_2, \Gamma_2)$$

by the formula

$$m = \sum_{a \in \mathfrak{c}(f_1)} \sum_{b \in \mathfrak{c}(f_2)} \sum_{w \in Z(a,b)} \epsilon(w) \Gamma_Z(w),$$

where the sum is over all zero-dimensional spaces $Z(a, b)$ and the sign $\epsilon(w)$ is as before.

One can also carry over the construction of the homomorphisms $C_*(f_1) \to C_*(f_2)$ to the situation considered in Subsection 2.5. Let P_1 and P_2 be two compact manifolds with circle action, and suppose again that the action is free outside the respective fixed-point sets Q_1, Q_2. Let $\tilde{Z}$ be another manifold with an action of S^1, and

$$\tilde{r} = (\tilde{r}_1, \tilde{r}_2) : \tilde{Z} \to P_1 \times P_2$$

an S^1-equivariant map. Let $W \subset \tilde{Z}$ be the fixed-point set, let $\tilde{Z}^\sigma$ be the blow-up along W, and let Z^σ be the quotient by S^1. If we impose the condition

$$(\tilde{r}_i)^{-1}(Q_i) = W$$

for $i = 1$ and 2, and ask also that the derivative of $\tilde{r}_i$ gives an injective map on the normal direction,

$$d\tilde{r}_i N_w W \to N_{\tilde{r}_i(w)}(Q_i),$$

then we obtain well-defined maps on the blow-up, $\tilde{r}_i^\sigma : \tilde{Z}^\sigma \to P_i^\sigma$ (see Lemma 2.5.1), and hence on the quotient spaces also:

$$r^\sigma = (r_1^\sigma, r_2^\sigma) : Z^\sigma \to B_1^\sigma \times B_2^\sigma.$$

These maps have the property that $(r_i^\sigma)^{-1}(\partial B_i^\sigma) = \partial Z^\sigma$, for $i = 1$ and 2. Given an orientation for the map r_1^σ, one can then use this data to construct chain maps between the associated chain complexes $\check{C}_*$, $\hat{C}_*$, $\bar{C}_*$. We do not describe the details here, but they can be extracted from our treatment of the Floer homology case, in Subsection 25.3.

3 Monopole Floer homology for three-manifolds

We present here a summary of most of the formal properties of monopole Floer homology as developed in this book, stating the results usually as propositions without proof. There are some further refinements that are best explained when we have more machinery in place, and these will be added later.

3.1 Three flavors of monopole Floer homology

Throughout this section, we will be considering 3-manifolds Y that are non-empty, closed, connected and oriented. We will use the notation $-Y$ to denote the same manifold with the opposite orientation. The monopole Floer homology theory that we consider associates to each such oriented manifold Y three abelian groups,

$$\widecheck{HM}_\bullet(Y), \quad \widehat{HM}_\bullet(Y), \quad \overline{HM}_\bullet(Y), \tag{3.1}$$

pronounced "*H-M-to*", "*H-M-from*" and "*H-M-bar*" respectively, which we will define in Section 22.

In outline, the construction of these groups is as follows. After some auxiliary choices, including a choice of Riemannian metric, we associate to Y a Hilbert manifold $\mathcal{B}^o(Y)$ with a circle action, and its quotient $\mathcal{B}(Y) = \mathcal{B}^o(Y)/S^1$, which is singular along the fixed-point set. As we did in Subsection 2.5 above, we blow up along the singular set to obtain a Hilbert manifold with boundary, $\mathcal{B}^\sigma(Y)$. On $\mathcal{B}^o(Y)$ there is a smooth function, the Chern–Simons–Dirac functional $\mathcal{L}$, whose downward gradient gives a (partially defined) flow on the blow-up $\mathcal{B}^\sigma(Y)$, just as we defined the vector field V^σ above. After perturbing the functional to achieve the necessary transversality, we mimic the construction of the Morse complexes $\check{C}$, $\hat{C}$ and $\bar{C}$ from Subsection 2.4, now in this infinite-dimensional setting. Neither the stable nor the unstable manifolds of the critical points of the flow are finite-dimensional, so the critical points do not have an "index" in the usual sense of Morse theory. In the light of Theorem 2.4.5, the Floer groups (3.1) can be thought of as the "middle-dimensional" homology groups of $\mathcal{B}^\sigma(Y)$, of the pair $(\mathcal{B}^\sigma(Y), \partial\mathcal{B}^\sigma(Y))$, and of $\partial\mathcal{B}^\sigma(Y)$ respectively. It should be noted, however, that one cannot expect to give meaning to the middle-dimensional homology groups of these Hilbert manifolds without reference to the functional $\mathcal{L}$ or some related data.

The space we have referred to as $\mathcal{B}^{\sigma}(Y)$ is a disjoint union of components, one for each isomorphism class of spin^c structure $\mathfrak{s}$ on Y. We write the components as $\mathcal{B}^{\sigma}(Y,\mathfrak{s})$. Corresponding to this decomposition, there is a direct sum decomposition of $\widecheck{HM}_{\bullet}(Y)$ and its companions into summands $\widecheck{HM}_{\bullet}(Y,\mathfrak{s})$,

$$\widecheck{HM}_{\bullet}(Y) = \bigoplus_{\mathfrak{s}} \widecheck{HM}_{\bullet}(Y,\mathfrak{s}). \tag{3.2}$$

This is a finite decomposition, a fact that we will prove in Section 23 and which we record as a proposition:

Proposition 3.1.1. *For a given Y, the groups $\widecheck{HM}_{\bullet}(Y,\mathfrak{s})$, $\widehat{HM}_{\bullet}(Y,\mathfrak{s})$ and $\overline{HM}_{\bullet}(Y,\mathfrak{s})$ are non-zero for only finitely many spin^c structures $\mathfrak{s}$ on Y.* □

Like the ordinary homology groups of a space, the Floer groups are graded abelian groups, but there are two caveats. First, the "grading" is not by the integers. Second, there is a completion involved. We explain these two points. In general, let $\mathbb{J}$ be any set with an action of $\mathbb{Z}$ (not necessarily transitive), and let the action of $n \in \mathbb{Z}$ on $\mathbb{J}$ be written as $j \mapsto j+n$. We can consider abelian groups graded by $\mathbb{J}$:

$$G_{*} = \bigoplus_{j \in \mathbb{J}} G_j.$$

A group homomorphism $h : G_{*} \to G'_{*}$ has degree n if h maps G_j to G'_{j+n} for all j.

In this sense, the relevant grading set for the Floer groups is the set of homotopy classes of oriented 2-plane fields, a set which has a natural action of $\mathbb{Z}$, as we shall explain. The simplest case to understand is the case of S^3. The classification, up to homotopy, of oriented 2-plane fields on S^3 is the same as the classification of oriented line fields; and because the tangent bundle is trivial, these are classified by $\pi_3(S^2) = \mathbb{Z}$. The classification by $\pi_3(S^2)$ depends on our choice of trivialization of the tangent bundle; so the more natural statement is simply that the homotopy classes of 2-plane fields are a set with a free, transitive action of $\mathbb{Z}$. If a 3-manifold has non-trivial homology, then a 2-plane field also has a 2-dimensional characteristic class, the Euler class. There is a slight refinement of this last statement: a 2-plane field determines a spin^c structure on Y. The general situation is explained in detail later in this book, in Section 28, where we explain that the homotopy classes of 2-plane fields admit an action of $\mathbb{Z}$ whose orbits correspond to the different spin^c structures. The orbit corresponding to $\mathfrak{s}$ is a free orbit of $\mathbb{Z}$ if and only if the corresponding Euler class is torsion. For now, we make precise how the action of $\mathbb{Z}$ is constructed:

Definition 3.1.2. We write $\Xi(Y)$ for the space of all oriented 2-plane fields ξ on Y: oriented, rank-2 subbundles of the tangent bundle. We write $\pi_0(\Xi(Y))$ for its set of connected components, the homotopy classes of 2-plane fields, and we define an action of $\mathbb{Z}$ on this set as follows. We specify that $[\xi] + n$ is the homotopy class $[\tilde{\xi}]$ of a 2-plane field $\tilde{\xi}$ which is equal to ξ outside a standard ball $B^3 \subset Y$. Inside the ball, trivialize the tangent bundle, take a map $\rho : (B^3, \partial B^3) \to (SO(3), 1)$ of degree $-2n$, and define

$$\tilde{\xi}(y) = \rho(y)\xi(y).$$

◊

Remark. The degree of a map from $(B^3, \partial B^3)$ to $(SO(3), 1)$ is necessarily even. Our convention on signs is fixed in the following way. Let $SU(2)_+$ and $SU(2)_-$ be the subgroups of $SO(4)$ which act trivially on the bundles of anti-self-dual and self-dual 2-forms respectively. Let ξ_+ and ξ_- be 2-plane fields on S^3 (the oriented boundary of B^4) which are invariant under $SU(2)_-$ and $SU(2)_+$ respectively. Then $[\xi_+] = [\xi_-] - 1$. Note that if the orientation of Y is reversed, then the action of $\mathbb{Z}$ changes sign also.

We now come to the second point mentioned above: there is a completion involved in the construction of the Floer homology groups. Each $j \in \pi_0(\Xi(Y))$ defines subgroups of the Floer groups,

$$\begin{aligned} \widecheck{HM}_j(Y) &\subset \widecheck{HM}_{\bullet}(Y) \\ \widehat{HM}_j(Y) &\subset \widehat{HM}_{\bullet}(Y) \\ \overline{HM}_j(Y) &\subset \overline{HM}_{\bullet}(Y); \end{aligned}$$

but these do not provide direct sum decompositions. Instead, the direct sums

$$\left.\begin{aligned} \widecheck{HM}_*(Y) &= \bigoplus_j \widecheck{HM}_j(Y) \\ \widehat{HM}_*(Y) &= \bigoplus_j \widehat{HM}_j(Y) \\ \overline{HM}_*(Y) &= \bigoplus_j \overline{HM}_j(Y) \end{aligned}\right\} \tag{3.3}$$

are subgroups of $\widecheck{HM}_\bullet(Y)$, $\widehat{HM}_\bullet(Y)$ and $\overline{HM}_\bullet(Y)$; and the larger groups are obtained from the $*$ versions by the following type of completion:

Definition 3.1.3. Let G_* be a graded abelian group graded by a set $\mathbb{J}$. Let O_α ($\alpha \in A$) be the free $\mathbb{Z}$-orbits in $\mathbb{J}$, and let an element $j_\alpha \in O_\alpha$ be given for each α. Let $G_*[n] \subset G_*$ be the subgroup

$$G_*[n] = \bigoplus_\alpha \bigoplus_{m \geq n} G_{j_\alpha - m}.$$

This is a decreasing filtration of G_*. We define the *negative completion* of G_* to be the topological group $G_\bullet \supset G_*$ obtained by completing with respect to this filtration. ◊

(The simplest example would be the case that G is a ring of finite Laurent series, with the obvious grading, in which case the negative completion is the ring of Laurent series that extend infinitely in the direction of the negative powers.) Our groups $\widecheck{HM}_\bullet(Y)$, $\widehat{HM}_\bullet(Y)$ and $\overline{HM}_\bullet(Y)$ are obtained from the graded $*$ versions by negative completion.

Remark. As well as being graded by the set $\pi_0(\Xi(Y))$, the Floer groups also have a canonical mod 2 grading: a decomposition into "odd" and "even" pieces. This decomposition arises because we can define a natural map

$$\pi_0(\Xi(Y)) \to \mathbb{Z}/2,$$

which respects the action of $\mathbb{Z}$. (Thus we can talk about odd and even 2-plane fields on an oriented 3-manifold.) This mod 2 grading is discussed in Subsection 22.4.

For each grading j, the abelian group $\widecheck{HM}_j(Y)$ and its companions are finitely generated. Indeed, they arise as the homology groups of complexes, graded by the same set, that are also finitely generated in each grading. Formally, they behave like the ordinary homology groups $H_*(Z)$, $H_*(Z,A)$ and $H_*(A)$ for a pair of spaces (Z,A), in that they are related by a long exact sequence,

$$\cdots \xrightarrow{i_*} \widecheck{HM}_\bullet(Y) \xrightarrow{j_*} \widehat{HM}_\bullet(Y) \xrightarrow{p_*} \overline{HM}_\bullet(Y) \xrightarrow{i_*} \widecheck{HM}_\bullet(Y) \xrightarrow{j_*} \cdots, \tag{3.4}$$

where the maps i_* and j_* have degree 0, and p_* has degree -1. (The last statement, for example, means that p_* maps $\widehat{HM}_j$ to $\overline{HM}_{j-1}$; so in particular, these maps are defined also on the $*$ versions.) These maps are constructed in Subsection 22.2. Because they are defined as the (completions of) homology groups of complexes, the Floer homology groups come with companion *co*homology groups, related by a sequence

$$\cdots \xleftarrow{i^*} \widecheck{HM}^\bullet(Y) \xleftarrow{j^*} \widehat{HM}^\bullet(Y) \xleftarrow{p^*} \overline{HM}^\bullet(Y) \xleftarrow{i^*} \widecheck{HM}^\bullet(Y) \xleftarrow{j^*} \cdots.$$

These Floer cohomology groups are obtained from graded abelian groups $\widecheck{HM}^*(Y)$, $\widehat{HM}^*(Y)$ and $\overline{HM}^*(Y)$ by *positive* completion. There is the expected $\mathbb{Z}$-valued pairing between $\widecheck{HM}_j(Y)$ and $\widecheck{HM}^j(Y)$, with similar pairings for the other two flavors.

There is a duality relating the Floer homology and cohomology groups that formally resembles Poincaré duality for an oriented manifold with boundary: we can think of it as Poincaré duality for the Hilbert manifold $\mathcal{B}^\sigma(Y)$. The duality isomorphism depends on a choice of orientation, in this case an orientation for the vector space $H^1(Y;\mathbb{R})$. We call such an orientation a *homology orientation* of Y. (See Definition 22.5.2.) If we choose a homology orientation μ, we have isomorphisms:

$$\begin{aligned}
\check{\omega}_\mu &: \widecheck{HM}_\bullet(-Y) \to \widehat{HM}^\bullet(Y)\\
\hat{\omega}_\mu &: \widehat{HM}_\bullet(-Y) \to \widecheck{HM}^\bullet(Y)\\
\bar{\omega}_\mu &: \overline{HM}_\bullet(-Y) \to \overline{HM}^\bullet(Y).
\end{aligned}$$

These isomorphisms are straightforward consequences of the definitions, just as in the finite-dimensional Morse-theory setting: for example, if f is changed to $-f$ in the constructions of Subsection 2.4, then the complex $\check{C}_*$ becomes the complex $\hat{C}^*$, and so on. On a *compact* manifold, the Morse homology is independent of the choice of Morse function, so one deduces an isomorphism between $H_*(B)$ and $H^*(B, \partial B)$. In the Floer context, $\mathcal{B}^\sigma(Y)$ and $\mathcal{B}^\sigma(-Y)$ are the same: only $\mathcal{L}$ changes sign. But one cannot replace the functional $\mathcal{L}$ by $-\mathcal{L}$ without changing the homology of the Morse complex, so we cannot replace Y by $-Y$ in the above isomorphisms.

The grading sets $\pi_0(\Xi(Y))$ and $\pi_0(\Xi(-Y))$ are the same object, because an oriented 2-plane field on Y is also an oriented 2-plane field on $-Y$. But the actions of $\mathbb{Z}$, as we mentioned, are opposite. So we give the canonical identification of these two sets a name,

$$o : \pi_0(\Xi(-Y)) \to \pi_0(\Xi(Y)),$$

and record the relation

$$o(j+n) = o(j) - n$$

for $j \in \pi_0(\Xi(-Y))$. With this understood, we can state how the duality isomorphisms affect the grading:

$$\begin{aligned}\check{\omega}_\mu &: \widecheck{HM}_j(-Y) \to \widehat{HM}^{o(j)}(Y)\\ \hat{\omega}_\mu &: \widehat{HM}_j(-Y) \to \widecheck{HM}^{o(j)}(Y)\\ \bar{\omega}_\mu &: \overline{HM}_j(-Y) \to \overline{HM}^{o(j)-1}(Y).\end{aligned}$$

We shall see that, just as for the duality maps on a manifold with boundary, we have a diagram

$$\begin{array}{ccccccc} \cdots\ \overline{HM}_j(-Y) & \xrightarrow{i_*} & \widecheck{HM}_j(-Y) & \xrightarrow{j_*} & \widehat{HM}_j(-Y) & \xrightarrow{p_*} & \overline{HM}_{j-1}(-Y)\ \cdots \\ \downarrow \bar{\omega}_\mu & & \downarrow \check{\omega}_\mu & & \downarrow \hat{\omega}_\mu & & \downarrow \bar{\omega}_\mu \\ \cdots\ \overline{HM}^{o(j)-1}(Y) & \xrightarrow{p^*} & \widehat{HM}^{o(j)}(Y) & \xrightarrow{j^*} & \widecheck{HM}^{o(j)}(Y) & \xrightarrow{i^*} & \overline{HM}^{o(k)}(Y)\ \cdots \end{array} \tag{3.5}$$

in which the first two squares commute and the third square commutes up to sign. For details of the sign, see Corollary 22.5.10.

3.2 Module structure

The ordinary cohomology groups of a space are provided with a ring structure by the cup product, and the homology groups are a module over the cohomology ring, via the cap product. The Floer homology groups bear a resemblance to middle-dimensional homology groups, or the groups within finite distance of the middle dimension, in an infinite-dimensional setting. Based on this loose analogy, one does not expect the Floer cohomology groups, for example, to be rings, but one does expect them to be modules for the *ordinary* cohomology of the ambient space $\mathcal{B}^\sigma(Y)$. This is indeed the case: we can construct "cap" and "cup" products,

$$\begin{aligned}\frown &: H^*(\mathcal{B}^\sigma(Y)) \times \widecheck{HM}_\bullet(Y) \to \widecheck{HM}_\bullet(Y)\\ \smile &: H^*(\mathcal{B}^\sigma(Y)) \times \widecheck{HM}^\bullet(Y) \to \widecheck{HM}^\bullet(Y),\end{aligned}$$

with similar maps on $\widehat{HM}_\bullet$ and $\overline{HM}_\bullet$. As expected, if u is a class of degree d in the ordinary cohomology ring, then the cap and cup products by u are endomorphisms of the Floer homology and cohomology groups of degrees $-d$ and d respectively. These operations satisfy the "associativity" laws

$$(u \cup v) \cup x = u \cup (v \cup x)$$
$$(u \cup v) \cap \xi = v \cap (u \cap \xi).$$

Note the order of u and v in the second formula: these operations make the Floer cohomology groups a module over the ordinary cohomology ring $H^*(\mathcal{B}^\sigma(Y))$; and – as is natural for the dual – they make the homology groups a module for the opposite ring, $H^*(\mathcal{B}^\sigma(Y))^{\mathrm{opp}}$.

It is not hard to identify the ordinary cohomology ring of $\mathcal{B}^\sigma(Y)$. First of all, we recall that this space is a union of components $\mathcal{B}^\sigma(Y, \mathfrak{s})$, one for each isomorphism class of spin^c structure $\mathfrak{s}$ on Y. Each component $\mathcal{B}^\sigma(Y, \mathfrak{s})$ has the homotopy type

$$\mathcal{B}^\sigma(Y, \mathfrak{s}) \simeq \mathbb{T}(Y) \times \mathbb{CP}^\infty, \tag{3.6}$$

where $\mathbb{T}(Y)$ is the torus $H^1(Y; \mathbb{R})/H^1(Y; \mathbb{Z})$, and there is a canonical identification of the cohomology rings of the two sides. There is an isomorphism

$$H_1(Y; \mathbb{Z})/\text{torsion} \to H^1(\mathbb{T}(Y); \mathbb{Z})$$

arising from the pairing $H^1(Y)$ and $H_1(Y)$, and the cohomology of the torus $\mathbb{T}(Y)$ is an exterior algebra on the image of this map:

$$H^*(\mathbb{T}(Y); \mathbb{Z}) \cong \Lambda\big(H_1(Y; \mathbb{Z})/\text{torsion}\big). \tag{3.7}$$

Using this map on each component $\mathcal{B}^\sigma(Y, \mathfrak{s})$, we obtain a map

$$\nu : H_1(Y; \mathbb{Z})/\text{torsion} \to H^1(\mathcal{B}^\sigma(Y)).$$

We denote by u_2 the 2-dimensional generator of the cohomology of $\mathcal{B}^\sigma(Y)$ arising from the standard generator of $H^2(\mathbb{CP}^\infty)$ on each component $\mathcal{B}^\sigma(Y, \mathfrak{s})$. (The sign of u_2 is fixed by stating that it evaluates to 1 on a copy of $\mathbb{CP}^1$ with its complex orientation.)

Combining the Floer cap product with ν and the class u_2, the Floer cohomology groups $\widecheck{HM}^\bullet(Y)$ etc. become modules for the graded ring

$$\begin{aligned} \mathbb{A}(Y) &= \Lambda\,(H_1(Y)/\text{torsion}) \otimes \mathbb{Z}[U] \\ &\cong H^*(\mathbb{T}(Y); \mathbb{Z}) \otimes \mathbb{Z}[U]. \end{aligned} \tag{3.8}$$

The elements $\gamma \in H_1(Y)$ and the 2-dimensional generator U act by

$$\gamma x = \nu(\gamma) \cup x$$
$$Ux = u_2 \cup x$$

respectively. There are similar actions on the Floer homology using the Floer cap product, which make it a module over the graded ring $\mathbb{A}_\dagger(Y)$ – essentially the same ring, but with the negative grading and *opposite* multiplication. To keep $\mathbb{A}(Y)$ and $\mathbb{A}_\dagger(Y)$ separate, we will sometimes denote the generators of $\mathbb{A}_\dagger(Y)$ by $\gamma_\dagger$ and $U_\dagger$. So $U_\dagger$ has degree -2, and

$$\gamma_\dagger x = \nu(\gamma) \cap x$$
$$U_\dagger x = u_2 \cap x.$$

The maps i_*, j_*, in the long exact sequence of the pair (3.4) are all module homomorphisms, while p_* respects the module structure up to sign, as is detailed in Corollary 25.5.2: we have, for example $p_*(\gamma x) = -\gamma p_*(x)$, for γ in $H_1(Y)$, as one would expect for a map p_* of odd degree. The duality isomorphisms $\breve{\omega}_\mu$, $\hat{\omega}_\mu$ and $\bar{\omega}_\mu$ are isomorphisms of $\mathbb{A}(Y)$-modules, up to sign. See Corollary 25.5.4.

Having well-defined degrees, the maps U and $U_\dagger$ are endomorphisms of the graded subgroups (3.3). After completion, the action of $\mathbb{Z}[U]$ on the "$\bullet$" versions extends continuously to an action of the ring of formal power series $\mathbb{Z}[[U]]$.

3.3 The simplest example

Take Y to be S^3. We have already mentioned the classes of 2-plane fields $[\xi_-]$ and $[\xi_+]$, invariant under the actions of $SU(2)_+$ and $SU(2)_-$ respectively. (See the remark following Definition 3.1.2.) Either of these is a generator for $\pi_0(\Xi(S^3))$ as a $\mathbb{Z}$-set, and we have $[\xi_-] = [\xi_+] + 1$ when S^3 is oriented as the boundary of B^4. We now state without proof a description of the Floer homology and cohomology groups of the sphere. See also Figure 3.

Proposition 3.3.1. *Orient S^3 as the boundary of B^4, and let $\xi_\pm$ be the 2-plane fields described above.*

(i) *The Floer homology groups $\widecheck{HM}(S^3)$ are given by*

$$\widecheck{HM}_{[\xi_-]+n}(S^3) = \begin{cases} \mathbb{Z}, & n \text{ even}, n \geq 0 \\ 0, & \text{otherwise.} \end{cases}$$

(ii) *The groups* $\widehat{HM}_*(S^3)$ *are given by*

$$\widehat{HM}_{[\xi_-]+n}(S^3) = \begin{cases} \mathbb{Z}, & n \text{ odd, } n < 0 \\ 0, & \text{otherwise.} \end{cases}$$

(iii) *The groups* $\overline{HM}_*(S^3)$ *are given by*

$$\overline{HM}_{[\xi_-]+n}(S^3) = \begin{cases} \mathbb{Z}, & n \text{ even} \\ 0, & \text{otherwise.} \end{cases}$$

By the universal coefficient theorem, the same result holds for the Floer cohomology groups. □

The Floer homology groups are graded modules over $\mathbb{A}(S^3)_\dagger = \mathbb{Z}[U_\dagger]$ respectively, with $U_\dagger$ of degree -2. Referring to Figure 3, we can describe the action of $U_\dagger$ as a left shift in all three cases. In the case of $\widecheck{HM}_\bullet(S^3)$, multiplication by $U_\dagger$ has kernel equal to the copy of $\mathbb{Z}$ which sits at grading $[\xi_-]$. In the other two cases, $U_\dagger$ is injective. Similarly, the Floer cohomology groups are modules over $\mathbb{Z}[U]$, with U having degree 2 and acting by a right shift.

The map j_* is zero for S^3, and there is therefore a short exact sequence

$$0 \longrightarrow \widehat{HM}_\bullet(S^3) \xrightarrow{p_*} \overline{HM}_\bullet(S^3) \xrightarrow{i_*} \widecheck{HM}_\bullet(S^3) \longrightarrow 0, \tag{3.9}$$

with p_* having degree -1. As a sequence of graded topological $\mathbb{Z}[[U_\dagger]]$-modules, this sequence is isomorphic to the sequence

$$0 \longrightarrow \mathbb{Z}[[U_\dagger]] \xrightarrow{p_*} \mathbb{Z}[U_\dagger^{-1}, U_\dagger]] \xrightarrow{i_*} \mathbb{Z}[U_\dagger^{-1}, U_\dagger]/\mathbb{Z}[U_\dagger] \longrightarrow 0, \tag{3.10}$$

where $\mathbb{Z}[U_\dagger^{-1}, U_\dagger]]$ denotes the formal Laurent series that are finite in the direction of negative powers of $U_\dagger$. (Although these series extend infinitely in the direction of positive powers of $U_\dagger$, this is still – somewhat confusingly – a *negative* completion, because $U_\dagger$ itself has negative degree.)

n	$\cdots$	-4	-3	-2	-1	0	$+1$	$+2$	$+3$	$+4$	$\cdots$
$\widecheck{HM}_{[\xi_-]+n}$	$\cdots$	0	0	0	0	$\mathbb{Z}$	0	$\mathbb{Z}$	0	$\mathbb{Z}$	$\cdots$
$\widehat{HM}_{[\xi_-]+n}$	$\cdots$	0	$\mathbb{Z}$	0	$\mathbb{Z}$	0	0	0	0	0	$\cdots$
$\overline{HM}_{[\xi_-]+n}$	$\cdots$	$\mathbb{Z}$	0	$\mathbb{Z}$	0	$\mathbb{Z}$	0	$\mathbb{Z}$	0	$\mathbb{Z}$	$\cdots$

Fig. 3. The Floer groups of S^3.

Some features of this example are reflections of more general properties. For a general Y, the image of j_* need not be zero but is always *finitely generated* as an abelian group. In particular, it is non-zero in only finitely many degrees.

Second, the calculation of $\overline{HM}_\bullet(S^3)$ can be extended: essentially the same result holds for any homology 3-sphere. More generally, we have the following result.

Proposition 3.3.2. *Suppose Y and Y' are two oriented 3-manifolds with isomorphic cohomology rings. Then there is an identification $\pi_0(\Xi(Y)) \to \pi_0(\Xi(Y')$ such that the groups $\overline{HM}_*(Y)$ and $\overline{HM}_*(Y')$ are isomorphic as $\mathbb{A}(Y)_\dagger$ modules graded by $\pi_0(\Xi(Y))$. In all cases, the endomorphism $U_\dagger$ is invertible.*

We shall describe $\overline{HM}_\bullet(Y)$ in some detail in Chapter IX, where the above proposition is also proved.

3.4 Cobordisms and functoriality

If Y_0 and Y_1 are compact, oriented 3-manifolds, a *cobordism* from Y_0 to Y_1 is a compact, oriented 4-manifold W with boundary, together with a diffeomorphism of oriented manifolds

$$\alpha = (\alpha_0 \amalg \alpha_1) : \partial W \to (-Y_0 \amalg Y_1).$$

We will generally use the letter W, rather than X, to denote a 4-manifold that is being considered as a cobordism in this way. If (W, α) and (W', α') are two cobordisms from Y_0 to Y_1, an *isomorphism* between them is an orientation-preserving diffeomorphism intertwining α with α'. The trivial cobordism from Y to itself is the oriented cylinder $W = [0, 1] \times Y$, with α being the restriction to the boundary of the projection to Y. If (W_1, α_1) and (W_2, α_2) are cobordisms from Y_0 to Y_1 and from Y_1 to Y_2 respectively, there is a composite cobordism (W, α) from Y_0 to Y_2: the manifold W is formed from $W_1 \amalg W_2$ using $\beta = (\alpha_2)^{-1}\alpha_1$ to identify the two copies of Y_1 in the boundary. We will write the composite cobordism as

$$\begin{aligned} W &= W_2 \circ W_1 \\ &= (W_1 \amalg W_2)/\beta. \end{aligned} \tag{3.11}$$

Note that we order the factors as we would for composing maps. Usually, we will omit the diffeomorphism α from our notation, and regard α as if it were the identity. The reason for introducing α formally is that we now have a category in which the morphisms are the isomorphism classes of cobordisms: the isomorphism class of the trivial cobordism from Y to Y is the identity morphism from Y to Y in this category.

A key feature of the monopole Floer homology groups, and one which motivates several aspects of their definition, is that a connected, oriented cobordism W from Y_0 to Y_1 gives rise to maps

$$\left.\begin{aligned} \widecheck{HM}_\bullet(Y_0) &\to \widecheck{HM}_\bullet(Y_1) \\ \widehat{HM}_\bullet(Y_0) &\to \widehat{HM}_\bullet(Y_1) \\ \overline{HM}_\bullet(Y_0) &\to \overline{HM}_\bullet(Y_1), \end{aligned}\right\} \tag{3.12}$$

well-defined up to sign. The map associated to a composite cobordism is the composite of the corresponding maps. In terms of the analogy with finite-dimensional Morse theory, the existence of such maps can be understood as an example of the construction described in Subsection 2.8 above. Associated to the manifold W is a space $\mathcal{B}^\sigma(W)$, the configuration space for the Seiberg–Witten equations on the compact manifold with boundary, blown up along the fixed-point set of the circle action. Inside $\mathcal{B}^\sigma(W)$ is the locus of solutions to the Seiberg–Witten equations, $M(W)$. (We will actually define $M(W)$ using something like the proper transform of the equations, on the blow-up.) The space $M(W)$ is still infinite-dimensional, unlike the moduli space associated to a closed 4-manifold. Restriction to the boundary provides a map

$$r : M(W) \to \mathcal{B}^\sigma(Y_0) \times \mathcal{B}^\sigma(Y_1). \tag{3.13}$$

By the construction of Subsection 2.8, the space $M(W)$ and the map r give rise to chain maps on the Morse complexes defining the Floer groups, and so give rise to the maps (3.12).

We stated that the maps (3.12) are defined only up to overall sign. To fix the ambiguity, an auxiliary choice must be made: a *homology orientation* for the cobordism. This is analogous to the isomorphism μ in (2.23) that was needed in the finite-dimensional case. Recall from Definition 1.5.1 that a homology orientation of a *closed* oriented 4-manifold X is an orientation of the line $\Lambda^{\max} H^1(X;\mathbb{R}) \otimes \Lambda^{\max} H^+(X)$, where $H^+(X) \subset H^2(X;\mathbb{R})$ was any chosen maximal positive-definite subspace for the quadratic form Q. We shall extend this definition first to 4-manifolds with boundary, and then to cobordisms, regarded as 4-manifolds whose boundary is partitioned into "incoming" and "outgoing" components. On an oriented 4-manifold with boundary, there is a non-degenerate pairing

$$H^2(X,\partial X;\mathbb{R}) \times H^2(X;\mathbb{R}) \to H^4(X,\partial X;\mathbb{R}) \\ \to \mathbb{R}.$$

Using the restriction map $j^* : H^2(X,\partial X;\mathbb{R}) \to H^2(X;\mathbb{R})$, we obtain a pairing on the relative group having null space the kernel of j^*, and hence a non-degenerate quadratic form Q on

$$I^2(X) = \mathrm{im}\{H^2(X,\partial X;\mathbb{R}) \to H^2(X;\mathbb{R})\}.$$

Definition 3.4.1. Let X be a compact, connected, oriented 4-manifold with boundary. Let $I^+(X)$ be a chosen maximal non-negative subspace for the quadratic form Q on $I^2(X)$. A *homology orientation* of X is an orientation of the line

$$\Lambda^{\max} H^1(X;\mathbb{R}) \otimes \Lambda^{\max} I^+(X) \otimes \Lambda^{\max} H^1(Y;\mathbb{R})$$

(with the understanding that Y may be disconnected). Let $W : Y_0 \to Y_1$ be an oriented cobordism between 3-manifolds. We define a *homology orientation* of W to be an orientation of the line

$$\Lambda^{\max} H^1(W;\mathbb{R}) \otimes \Lambda^{\max} I^+(W) \otimes \Lambda^{\max} H^1(Y_1;\mathbb{R}).$$

◊

Note the asymmetric treatment of the two ends in the second definition. The first definition is consistent with the second, if we choose to regard a manifold X with boundary as a cobordism with the empty manifold as "incoming" end.

As motivation for this definition, we cite the fact that there is a natural composition law for homology orientations of cobordisms, as we have defined them. That is, if $W = W_2 \circ W_1$, then a homology orientation for each of the W_i determines a homology orientation for W. The explanation for this is best understood by reinterpreting the definition as an orientation for the determinant line of Fredholm operator. We will return to this, and fix our convention for the composition law, in Subsection 25.2, but we note in passing that the cylinder $W = [0,1] \times Y$ has a canonical homology orientation when viewed as a cobordism, because $H^1(Y)$ and $H^1(W)$ are the same, while $I^2(W)$ is trivial.

Definition 3.4.2. We write COB for the category whose objects are compact, connected, oriented 3-manifolds and whose morphisms are isomorphism classes of *connected* cobordisms equipped with homology orientations. The identity morphism from Y to Y is the cylindrical cobordism with its canonical homology orientation. ◊

We are now in a position to state:

Theorem 3.4.3. *The Floer homology groups define covariant functors from the cobordism category* COB *to the category* $\text{MOD}_\dagger$, *of topological modules for the ring* $\mathbb{Z}[[U_\dagger]]$*:*

$$\widecheck{HM}_\bullet : \text{COB} \to \text{MOD}_\dagger$$
$$\widehat{HM}_\bullet : \text{COB} \to \text{MOD}_\dagger$$
$$\overline{HM}_\bullet : \text{COB} \to \text{MOD}_\dagger .$$

Similarly, the Floer cohomology groups define contravariant functors

$$\widecheck{HM}^\bullet : \text{COB} \to \text{MOD}$$
$$\widehat{HM}^\bullet : \text{COB} \to \text{MOD}$$
$$\overline{HM}^\bullet : \text{COB} \to \text{MOD}$$

to the category of topological modules over the ring $\mathbb{Z}[[U]]$*. The maps* i_*, j_*, p_*, *and* i^*, j^*, p^* *are all natural transformations between these functors.*

We usually use a notation such as $\widecheck{HM}_\bullet(W) : \widecheck{HM}_\bullet(Y_0) \to \widecheck{HM}_\bullet(Y_1)$ to denote the homomorphism arising from a cobordism. In particular, we will omit mention of the homology orientation.

We have not referred to the homomorphisms $\widecheck{HM}_\bullet(W)$ as homomorphisms of *graded* modules. Indeed, the modules $\widecheck{HM}_\bullet(Y_0)$ and $\widecheck{HM}_\bullet(Y_1)$ are graded by two different sets, $\mathbb{J}_0 = \pi_0(\Xi(Y_0))$ and $\mathbb{J}_1 = \pi_0(\Xi(Y_1))$, and we cannot talk of a homomorphism as having an integer degree in such a setting. There is a weaker notion that is applicable in this situation however. A cobordism $W : Y_0 \to Y_1$ determines a *relation* $\sim_W$ between the grading sets $\mathbb{J}_0$ and $\mathbb{J}_1$, compatible with the $\mathbb{Z}$ action. To define the relation, let j_r be an element of $\mathbb{J}_r$ $(r = 0, 1)$, and let ξ_r be corresponding 2-plane fields. Then $j_0 \sim_W j_1$ if there is an almost complex structure J on W such that the 2-plane fields ξ_0 and ξ_1

are invariant under J and have the complex orientation. With this notation, we can state that the map $\widecheck{HM}_\bullet(W)$ has a non-zero component from $\widecheck{HM}_{j_0}(Y_0)$ to $\widecheck{HM}_{j_1}(Y_1)$ only if $j_0 \sim_W j_1$. The same applies to $\widehat{HM}_\bullet$ and $\overline{HM}_\bullet$.

A slightly more informative version of the theorem would state that a cobordism W gives rise to chain map between the chain complexes that define these groups; and $\widecheck{HM}_\bullet(W)$ and $\widecheck{HM}^\bullet(W)$, for example, are the maps induced on the completions of their homology and cohomology. It follows that these two maps have the expected adjoint relationship with respect to the pairing between Floer homology and cohomology.

There is a different sort of duality that comes into play when one observes that a cobordism $W : Y_0 \to Y_1$ gives rise to a cobordism $W^\dagger : -Y_1 \to -Y_0$. The manifold $W^\dagger$ has the *same* orientation as W: we have only changed our mind about which is the "incoming" and which is the "outgoing" end. Note that a homology orientation for W does not determine one for $W^\dagger$: we need additional data, in the form of homology orientations μ_0 and μ_1 for Y_0 and Y_1, to translate between the two. We can then state the naturality of the duality isomorphisms in the form

$$\begin{aligned}
\check{\omega}_{\mu_0} \circ \widecheck{HM}_\bullet(W^\dagger)(\check{x}) &= \pm \widehat{HM}^\bullet(W) \circ \check{\omega}_{\mu_1}(\check{x}) \\
\hat{\omega}_{\mu_0} \circ \widehat{HM}_\bullet(W^\dagger)(\hat{x}) &= \pm \widecheck{HM}^\bullet(W) \circ \hat{\omega}_{\mu_1}(\hat{x}) \\
\bar{\omega}_{\mu_0} \circ \overline{HM}_\bullet(W^\dagger)(\bar{x}) &= \pm \overline{HM}^\bullet(W) \circ \bar{\omega}_{\mu_1}(\bar{x}),
\end{aligned}$$

for $\check{x}, \hat{x}, \bar{x}$ in $\widecheck{HM}_j(-Y_1)$, $\widehat{HM}_j(-Y_1)$ and $\overline{HM}_j(-Y_1)$ respectively. The signs here depend on j and on the characteristic numbers of W. See Proposition 25.5.3.

Although Theorem 3.4.3 tells us that the maps arising from cobordisms respect multiplication by the 2-dimensional class U or $U_\dagger$, it does not yet say anything about the structure that $\widecheck{HM}_\bullet(Y)$ carries as an $\mathbb{A}(Y)_\dagger$-module. Now that we have introduced cobordisms into the story, we can explain this module multiplication as arising from a special case of a more general algebraic structure. Define $\mathbb{A}(W)_\dagger$ (like $\mathbb{A}(Y)_\dagger$) as the algebra $\Lambda^*(H_1(W)/\text{torsion}) \otimes \mathbb{Z}[U]$, with the generators of $H_1(W)$ lying in degree -1, and with the opposite multiplication. Then we can augment the definition of the category COB by taking as morphisms the pairs (a, W), where W is a cobordism with homology orientation as before, and a is an element of $\mathbb{A}(W)_\dagger$. The composition of morphisms $(a_2, W_2) \circ (a_1, W_1)$ is defined to be the morphism (a, W), where W is the composite cobordism W and

$$a = (i_1)_*(a_1) \smile (i_2)_*(a_2),$$

by which we mean the product in $\mathbb{A}(W)_\dagger$ of the images of a_2 and a_1 under the maps $H_1(W_r) \to H_1(W)$ induced by the inclusions i_r. We can then state:

Theorem 3.4.4. *To each pair (a, W), where $W : Y_0 \to Y_1$ is a morphism in* COB *and $a \in \mathbb{A}(W)_\dagger$ as above, there are associated homomorphisms*

$$\begin{aligned}
\widecheck{HM}_\bullet(a \mid W) &: \widecheck{HM}_\bullet(Y_0) \to \widecheck{HM}_\bullet(Y_1) \\
\widehat{HM}_\bullet(a \mid W) &: \widehat{HM}_\bullet(Y_0) \to \widehat{HM}_\bullet(Y_1) \\
\overline{HM}_\bullet(a \mid W) &: \overline{HM}_\bullet(Y_0) \to \overline{HM}_\bullet(Y_1),
\end{aligned}$$

respecting the composition law just defined. In the special case that W is a cylindrical cobordism $W = [0,1] \times Y$, the maps $\widecheck{HM}_\bullet(a \mid W)$ etc. coincide with the module multiplication by the element a in $\mathbb{A}(W)_\dagger \cong \mathbb{A}(Y)_\dagger$. There is a similar statement for the Floer cohomology groups.

To illustrate this result, one can consider the situation that there are 1-cycles γ^0 and γ^1 in Y_0 and Y_1 whose images in $H_1(W)$ coincide. Then the theorem implies the relation

$$\gamma^1_\dagger\big(\widecheck{HM}_\bullet(W)(x)\big) = \widecheck{HM}_\bullet(W)(\gamma^0_\dagger x)$$

for x in $\widecheck{HM}_\bullet(Y_0)$.

3.5 Cobordisms with b^+ positive

Although the map j_* defines a natural transformation

$$j_* : \widecheck{HM}_\bullet \to \widehat{HM}_\bullet,$$

there is no non-trivial natural transformation in the opposite direction. However, a map $\widehat{HM}_\bullet(Y_0) \to \widecheck{HM}_\bullet(Y_1)$ does arise from a cobordism if the underlying 4-manifold W satisfies an additional topological constraint. We begin with a definition.

Definition 3.5.1. For a compact, oriented 4-manifold X with boundary, we define $b^+(X)$ to be the dimension of a maximal positive-definite subspace for the quadratic form Q on $I^2(X)$. ◊

The relevance of $b^+(W)$ for a cobordism $W : Y_0 \to Y_1$ is seen first in the following result.

Proposition 3.5.2. *If W is a morphism in* COB *with $b^+(W) \neq 0$, then the maps $\overline{HM}_\bullet(W)$ and $\overline{HM}^\bullet(W)$ are zero, as are their generalizations $\overline{HM}_\bullet(a \mid W)$ and $\overline{HM}^\bullet(a \mid W)$ for $a \in \mathbb{A}(W)$.*

Consider now the commutative diagram resulting from a cobordism W from Y_0 to Y_1.

$$\begin{array}{ccccccccc}
\longrightarrow & \overline{HM}_\bullet(Y_0) & \xrightarrow{i_*} & \widecheck{HM}_\bullet(Y_0) & \xrightarrow{j_*} & \widehat{HM}_\bullet(Y_0) & \xrightarrow{p_*} & \overline{HM}_\bullet(Y_0) & \longrightarrow \\
& \downarrow{\scriptstyle \overline{HM}_\bullet(W)} & & \downarrow{\scriptstyle \widecheck{HM}_\bullet(W)} & & \downarrow{\scriptstyle \widehat{HM}_\bullet(W)} & & \downarrow{\scriptstyle \overline{HM}_\bullet(W)} & \\
\longrightarrow & \overline{HM}_\bullet(Y_1) & \xrightarrow{i_*} & \widecheck{HM}_\bullet(Y_1) & \xrightarrow{j_*} & \widehat{HM}_\bullet(Y_1) & \xrightarrow{p_*} & \overline{HM}_\bullet(Y_1) & \longrightarrow .
\end{array}$$

The fact that $\overline{HM}_\bullet(W)$ is zero when $b^+(W) \neq 0$ means that for $x \in \widehat{HM}_\bullet(Y_0)$ the element $\widehat{HM}_\bullet(W)(x)$ is in the image of j_*. At the level of sets at least, the map $\widehat{HM}_\bullet(W)$ therefore factors through a map

$$\widehat{HM}_\bullet(Y_0) \to \widecheck{HM}_\bullet(Y_1)$$

when $b^+(W) \neq 0$. This map is not canonical, but when $b^+(W) \geq 2$, there *is* a canonical choice. This is the content of the next theorem:

Theorem 3.5.3. *For each morphism $W : Y_0 \to Y_1$ in* COB *with $b^+(W) \geq 2$ there is a map*

$$\overrightarrow{HM}_\bullet(W) : \widehat{HM}_\bullet(Y_0) \to \widecheck{HM}_\bullet(Y_1)$$

such that the following diagram commutes:

$$\begin{array}{ccccccccc}
\longrightarrow & \overline{HM}_\bullet(Y_0) & \xrightarrow{i_*} & \widecheck{HM}_\bullet(Y_0) & \xrightarrow{j_*} & \widehat{HM}_\bullet(Y_0) & \xrightarrow{p_*} & \overline{HM}_\bullet(Y_0) & \longrightarrow \\
& \downarrow & & \downarrow & \swarrow{\scriptstyle \overrightarrow{HM}_\bullet(W)} & \downarrow & & \downarrow & \\
\longrightarrow & \overline{HM}_\bullet(Y_1) & \xrightarrow[i_*]{} & \widecheck{HM}_\bullet(Y_1) & \xrightarrow[j_*]{} & \widehat{HM}_\bullet(Y_1) & \xrightarrow[p_*]{} & \overline{HM}_\bullet(Y_1) & \longrightarrow .
\end{array}$$

This map satisfies the following two composition laws for a composite cobordism $W = W_2 \circ W_1$:

$$\begin{aligned}
\overrightarrow{HM}_\bullet(W) &= \widecheck{HM}_\bullet(W_2) \circ \overrightarrow{HM}_\bullet(W_1), \quad \text{if } b^+(W_1) \geq 2 \\
\overrightarrow{HM}_\bullet(W) &= \overrightarrow{HM}_\bullet(W_2) \circ \widehat{HM}_\bullet(W_1), \quad \text{if } b^+(W_2) \geq 2.
\end{aligned} \tag{3.14}$$

For $a \in \mathbb{A}_\dagger(W)$, there are also maps $\overrightarrow{HM}_\bullet(a \mid W)$ with parallel properties.

Although $b^+(W)$ is not in general additive for composite cobordisms, we can get some understanding of this last theorem by considering a cobordism W that is the composite $W_2 \circ W_1$ with $b^+(W_i) \geq 1$ for $i = 1, 2$. We have

$$b^+(W) \geq b^+(W_1) + b^+(W_2) \geq 2,$$

so the theorem applies to W. On the other hand, we can look at the commutative diagram

$$\begin{array}{ccccccccc}
\longrightarrow & \overline{HM}_\bullet(Y_0) & \xrightarrow{i_*} & \widecheck{HM}_\bullet(Y_0) & \xrightarrow{j_*} & \widehat{HM}_\bullet(Y_0) & \xrightarrow{p_*} & \overline{HM}_\bullet(Y_0) & \longrightarrow \\
& \Big\downarrow {\scriptstyle \overline{HM}_\bullet(W_1)} & & \Big\downarrow {\scriptstyle \widecheck{HM}_\bullet(W_1)} & & \Big\downarrow {\scriptstyle \widehat{HM}_\bullet(W_1)} & & \Big\downarrow {\scriptstyle \overline{HM}_\bullet(W_1)} & \\
\longrightarrow & \overline{HM}_\bullet(Y_1) & \xrightarrow{i_*} & \widecheck{HM}_\bullet(Y_1) & \xrightarrow{j_*} & \widehat{HM}_\bullet(Y_1) & \xrightarrow{p_*} & \overline{HM}_\bullet(Y_1) & \longrightarrow \\
& \Big\downarrow {\scriptstyle \overline{HM}_\bullet(W_2)} & & \Big\downarrow {\scriptstyle \widecheck{HM}_\bullet(W_2)} & & \Big\downarrow {\scriptstyle \widehat{HM}_\bullet(W_2)} & & \Big\downarrow {\scriptstyle \overline{HM}_\bullet(W_2)} & \\
\longrightarrow & \overline{HM}_\bullet(Y_2) & \xrightarrow{i_*} & \widecheck{HM}_\bullet(Y_2) & \xrightarrow{j_*} & \widehat{HM}_\bullet(Y_2) & \xrightarrow{p_*} & \overline{HM}_\bullet(Y_2) & \longrightarrow .
\end{array}$$

The maps $\overline{HM}_\bullet(W_i)$ are both zero, by the proposition above. Given $x \in \widehat{HM}_\bullet(Y_0)$, we can therefore choose y in $\widecheck{HM}_\bullet(Y_1)$ with

$$j_*(y) = \widehat{HM}_\bullet(W_1)(x).$$

If y' is another choice, then $y - y' = i_*(z)$ for some $z \in \overline{HM}_\bullet(Y_1)$; and because $\overline{HM}_\bullet(W_2)(z) = 0$, we see that

$$\widecheck{HM}_\bullet(W_2)(y') = \widecheck{HM}_\bullet(W_2)(y).$$

This diagram chase constructs a homomorphism, independent of the intermediate choices,

$$\begin{aligned} Z(W_1, W_2) : \widehat{HM}_\bullet(Y_0) &\to \widecheck{HM}_\bullet(Y_2) \\ x &\mapsto \widecheck{HM}_\bullet(W_2)(y). \end{aligned} \tag{3.15}$$

The map $Z(W_1, W_2)$ defined by this diagram chase coincides with the map $\overrightarrow{HM}_\bullet(W)$, where $W = W_2 \circ W_1$. We cannot use $Z(W_1, W_2)$ as a *definition* of $\overrightarrow{HM}_\bullet(W)$, because we cannot show directly that the result of the diagram chase does not depend on the choice of factorization of W as a composite cobordism.

3.6 Closed four-manifolds

As a special case of a cobordism, we can consider the complement W of two disjoint standard balls in a closed, oriented 4-manifold X, viewed as a cobordism

$$W : S^3 \to S^3.$$

Note that $I^2(W) = H^2(X)$ and $b^+(W) = b^+(X)$, and a homology orientation of the cobordism W, according to Definition 3.4.1, is the same thing as a homology orientation of the closed manifold X. We described the Floer homology of S^3 in Subsection 3.3, where we exhibited isomorphisms

$$\begin{aligned}\widecheck{HM}_\bullet(S^3) &= \mathbb{Z}[U_\dagger^{-1}, U_\dagger]/\mathbb{Z}[U_\dagger]\\ \widehat{HM}_\bullet(S^3) &= \mathbb{Z}[[U_\dagger]].\end{aligned}$$

Suppose that $b^+(X) \geq 2$. As in the previous subsection, we then have a homomorphism of $\mathbb{Z}[[U_\dagger]]$-modules

$$\overrightarrow{HM}_\bullet(W) : \mathbb{Z}[[U_\dagger]] \to \mathbb{Z}[U_\dagger^{-1}, U_\dagger]/\mathbb{Z}[U_\dagger]. \tag{3.16}$$

Because of the module structure, the map $\overrightarrow{HM}_\bullet(W)$ is entirely determined by the image of the generator 1 in the domain: the map is given by multiplication by a fixed Laurent series (finite in the direction of negative powers of $U_\dagger$), followed by projection to the quotient. If we take the coefficient of $U_\dagger^{-1}$, we obtain an integer,

$$n = \mathrm{coeff}\left(\overrightarrow{HM}_\bullet(W)(1); U_\dagger^{-1}\right) \in \mathbb{Z}. \tag{3.17}$$

Our construction tells us that this integer n depends only on the original closed 4-manifold X and its homology orientation; and the only additional property of X that we required was that $b^+(X) \geq 2$. If we write $\check{1} \in \widecheck{HM}^\bullet(S^3) \cong \mathbb{Z}[[U]]$ for the generator of cohomology, we can rewrite this as

$$n = \left\langle \overrightarrow{HM}_\bullet(W)(1), \check{1} \right\rangle, \tag{3.18}$$

where the angle brackets denote the pairing between Floer homology and cohomology of S^3. What we have recovered here is the sum of the Seiberg–Witten monopole invariants of the 4-manifold:

Proposition 3.6.1. *The invariant n just defined is equal to the sum over all spin^c structures $\mathfrak{s}_X$ of the invariants $\mathfrak{m}(X, \mathfrak{s}_X)$, defined in Definition 1.5.3: that is,*

$$\langle \overrightarrow{HM}_\bullet(W)(1), \check{1} \rangle = \sum_{\mathfrak{s}_X} \mathfrak{m}(X, \mathfrak{s}_X)$$

when W is obtained from X by removing two balls.

We give this invariant a name:

Definition 3.6.2. If X is a closed, oriented 4-manifold with $b^+(X) \geq 2$, we write

$$\mathfrak{m}(X) = \sum_{\mathfrak{s}_X} \mathfrak{m}(X, \mathfrak{s}_X)$$

for the sum of the monopole invariants of X taken over all isomorphism classes of spin^c structures. ◊

Thus we can write the result of the proposition above as:

$$\mathfrak{m}(X) = \langle \overrightarrow{HM}_\bullet(W)(1), \check{1} \rangle. \tag{3.19}$$

This discussion of invariants of closed 4-manifolds can be extended by considering the situation in which a 4-manifold X is divided into two components X_1, X_2 by a 3-manifold Y, with $\partial X_1 = Y$ and $\partial X_2 = -Y$. If we remove a single ball from each, we obtain cobordisms

$$W_1 : S^3 \to Y$$
$$W_2 : Y \to S^3.$$

The composite cobordism $W = W_2 \circ W_1$ is a cobordism from S^3 to S^3 obtained by removing two balls from X. If we choose homology orientations for the two cobordisms, then their composite determines a homology orientation for W and hence for X. Then, if $b^+(X_1) \geq 2$, we can write:

$$\begin{aligned}\mathfrak{m}(X) &= \langle \overrightarrow{HM}_\bullet(W)(1), \check{1} \rangle \\ &= \langle \widetilde{HM}_\bullet(W_2) \circ \overrightarrow{HM}_\bullet(W_1)(1), \check{1} \rangle \\ &= \langle \overrightarrow{HM}_\bullet(W_1)(1), \widetilde{HM}^\bullet(W_2)(\check{1}) \rangle.\end{aligned}$$

Here the angle brackets denote pairings between a Floer homology class on the left, and a Floer cohomology class on the right. In this way, the invariant of X is expressed as a pairing between an element of $\widecheck{HM}_\bullet(Y)$ depending only on X_1, and an element of $\widecheck{HM}^\bullet(Y)$ depending only on X_2. Note that if we choose a homology orientation μ for Y, then we can regard the invariant of X_2 as living in a homology group, using the duality isomorphism

$$\hat{\omega}_\mu : \widehat{HM}_\bullet(-Y) \cong \widecheck{HM}^\bullet(Y).$$

This allows us to write the formula as

$$\mathfrak{m}(X) = \langle \overrightarrow{HM}_\bullet(W_1)(1), \widehat{HM}_\bullet(W_2^\dagger)(1)\rangle_{\omega_\mu},$$

where the pairing

$$\langle -, - \rangle_{\omega_\mu} : \widecheck{HM}_\bullet(Y) \times \widehat{HM}_\bullet(-Y) \to \mathbb{Z} \tag{3.20}$$

is defined using $\hat{\omega}_\mu$. There is a similar pairing formula when $b^+(W_2) \geq 2$: we can write

$$\mathfrak{m}(X) = \langle \widehat{HM}_\bullet(W_1)(1), \overrightarrow{HM}^\bullet(W_2)(\check{1})\rangle,$$

which is a pairing between an element of $\widehat{HM}_\bullet(Y)$ and an element of $\widehat{HM}^\bullet(Y)$.

In the case that $b^+(X_1)$ and $b^+(X_2)$ are *both* at least 1, there is a symmetrical way of writing this formula. We introduce:

Definition 3.6.3. We define $HM_\bullet(Y)$ as the image of $j_* : \widecheck{HM}_\bullet(Y) \to \widehat{HM}_\bullet(Y)$ and we call this the *reduced* Floer homology group of Y. ◊

If $b^+(X_1) \geq 1$, then we have seen that the element $\widehat{HM}_\bullet(W_1)(1)$ in $\widehat{HM}_\bullet(Y)$ is in the image of j_*: this followed from Proposition 3.5.2 and a diagram chase. So we have a well-defined invariant:

Definition 3.6.4. Let X be an oriented 4-manifold with connected boundary Y, and suppose $b^+(X) \geq 1$. Given a homology orientation of X, we define

$$\psi_X \in HM_\bullet(Y)$$

to be the element $\widehat{HM}_\bullet(W)(1)$ in $HM_\bullet(Y) \subset \widehat{HM}_\bullet(Y)$, where W is the complement of a ball in X, regarded as a cobordism from S^3 to Y. ◊

In the situation that $\partial X_1 = Y$ and $\partial X_2 = -Y$ with $b^+ \geq 1$ on both sides, we have elements

$$\psi_{X_1} \in HM_\bullet(Y), \quad \psi_{X_2} \in HM_\bullet(-Y).$$

Because of the diagram (3.5), the pairing (3.20) gives rise to a pairing of the reduced groups,

$$\langle -, - \rangle_{\omega_\mu} : HM_\bullet(Y) \times HM_\bullet(-Y) \to \mathbb{Z}, \tag{3.21}$$

and we have the pairing formula

$$\mathfrak{m}(X) = \langle \psi_{X_1}, \psi_{X_2} \rangle_{\omega_\mu}. \tag{3.22}$$

In this formula, the homology orientation of X is constructed using the given homology orientations of the two pieces and the homology orientation μ of Y.

The invariant $\mathfrak{m}(X)$ has contributions only from the spin^c structures $\mathfrak{s}_X$ for which the corresponding moduli space on X is zero-dimensional. We can also form the similar quantity

$$\mathfrak{m}(u_2^e \,|\, X) = \sum_{\mathfrak{s}_X} \mathfrak{m}(u_2^e \,|\, X, \mathfrak{s}_X)$$

involving the moduli spaces of dimension $2e$: see Definition 1.5.4. There are similar pairing formulae for this quantity. For example, in the situation of (3.22), we can also write

$$\mathfrak{m}(u_2^e \,|\, X) = \langle U_\dagger^e \psi_{X_1}, \psi_{X_2} \rangle_\mu. \tag{3.23}$$

Remark. The reduced Floer group $HM_\bullet(Y)$ is always of finite rank, unlike $\widecheck{HM}_\bullet(Y)$, for example. This is stated and proved as Proposition 22.2.3, in Subsection 22.2.

3.7 Floer homology with local coefficients

When a closed 4-manifold X is decomposed along Y into two pieces X_1 and X_2, the formula (3.22) expresses the invariant $\mathfrak{m}(X)$ (equal to the sum of the invariants $\mathfrak{m}(X, \mathfrak{s}_X)$, by Proposition 3.6.1) in terms of invariants ψ_{X_1}, ψ_{X_2} of the two pieces, each taking values in the reduced Floer groups of their boundary manifold. This formula has two shortcomings. First, it is a formula only for the *sum* $\mathfrak{m}(X)$, and does not by itself allow us to recover the monopole invariant

$\mathfrak{m}(X, \mathfrak{s}_X)$ for a *particular* spin^c structure. Second, the formula can be applied only when $b^+(X_1)$ and $b^+(X_2)$ are both positive. There are alternative versions when just one $b^+(X_i)$ is at least 2 (see above), but these versions are still insufficient to deal with all cases. The issue is that b^+ is not additive. For the invariant $\mathfrak{m}(X)$ to be defined, we need $b^+(X) \geq 2$; but such an X may be decomposed along Y into two pieces with $b^+(X_1) = b^+(X_2) = 0$. In such a case, we have not yet provided a formula for $\mathfrak{m}(X)$ in terms of invariants of the two pieces. A particularly simple illustration of the non-additivity is the decomposition of the manifold $X = \#_k(S^2 \times S^2)$ into pieces X_1, X_2 both of which are boundary-connected sums $\natural_k(D^2 \times S^2)$. The manifold X has $b^+(X) = k$, but the two pieces each have $I^2(X_i) = 0$.

These two shortcomings, though apparently of rather different origin, are both remedied by a single device. Recall that the Floer groups $\widecheck{HM}_\bullet(Y)$ and their companions arise from the Morse complex of a partially defined flow on a space $\mathcal{B}^\sigma(Y)$. If we introduce a local system of groups on this space, say Γ, we can introduce the Morse complex with coefficients in Γ (see Subsection 2.7 for the finite-dimensional analog), and so define Floer homology groups $\widecheck{HM}_\bullet(Y;\Gamma)$, $\widehat{HM}_\bullet(Y;\Gamma)$ and $\overline{HM}_\bullet(Y;\Gamma)$.

The components of the space $\mathcal{B}^\sigma(Y)$ have the homotopy type described by (3.6), and in particular their fundamental groups are isomorphic to $H^1(Y;\mathbb{Z})$. Up to isomorphism, we can specify a local system of groups (on one or all components) by specifying the fiber Γ_0 at a point and describing a homomorphism from $H^1(Y;\mathbb{Z})$ to the group of automorphisms of Γ_0, to specify the monodromy of the local system around loops in $\mathcal{B}^\sigma(Y)$. As a particular case, we can take the fiber to be $\mathbb{R}$, so that the monodromy data is a homomorphism $\chi : H^1(Y;\mathbb{Z}) \to \mathbb{R}^\times$. In terms of de Rham representatives, the sort of homomorphism ξ that we deal with can be expressed as

$$\chi : [\alpha] \mapsto \exp \int_\eta \alpha,$$

where η is a smooth singular 1-cycle in Y with real coefficients. (Our choice to express this map in terms of a singular 1-cycle, rather than, say, a smooth closed 2-form dual to $[\eta]$, is a matter of convenience.) In Subsection 22.6, we will carry out this program, and show (amongst other things) how such a 1-cycle η gives rise canonically to a local system Γ_η on $\mathcal{B}^\sigma(Y)$ (rather than just an isomorphism class of local systems), and we shall construct Floer groups $\widecheck{HM}(Y;\Gamma_\eta)$ and so on. At present, we are content to describe the formal properties of the resulting groups, as we have done for the Floer groups with $\mathbb{Z}$ coefficients. This we do in the present subsection. In the following two subsections, we will show how local coefficients can be used to overcome the weaknesses of the pairing formula (3.22). We use $\Omega_i(Y;\mathbb{R})$ for the space of C^∞ singular i-chains on Y with real coefficients, and $Z_i(Y;\mathbb{R})$ for the subgroup of chains. We also write, for example, $Z_i(X,Y;\mathbb{R})$ for the relative cycles: the chains η in X with $\partial\eta \in Z_{i-1}(Y;\mathbb{R})$.

To begin the summary, we can state that for each cycle η in $Z_1(Y;\mathbb{R})$, there is a long exact sequence of real vector spaces

$$\cdots \xrightarrow{i_*} \widecheck{HM}_\bullet(Y;\Gamma_\eta) \xrightarrow{j_*} \widehat{HM}_\bullet(Y;\Gamma_\eta) \xrightarrow{p_*} \overline{HM}_\bullet(Y;\Gamma_\eta) \xrightarrow{i_*} \cdots,$$

with a similar sequence for cohomology. The grading set is $\pi_0(\Xi(Y))$ as before. In the case that η is zero, these groups are the tensor products $\widecheck{HM}_\bullet(Y)\otimes\mathbb{R}$ etc. If η and η' are homologous, then the corresponding Floer groups are isomorphic. But the isomorphism is not canonical. If we are given $\theta \in \Omega_2(Y;\mathbb{R})$ with $\partial\theta = \eta - \eta'$, then we can use θ to determine a canonical isomorphism. This is why the notation for the group involves the cycle η and not just its homology class. There is a pairing between Floer homology and cohomology which provides isomorphisms of real vector spaces

$$\left.\begin{aligned} \widecheck{HM}^\bullet(Y;\Gamma_{-\eta}) &\cong \widecheck{HM}_\bullet(Y;\Gamma_\eta)^* \\ \widehat{HM}^\bullet(Y;\Gamma_{-\eta}) &\cong \widehat{HM}_\bullet(Y;\Gamma_\eta)^* \\ \overline{HM}^\bullet(Y;\Gamma_{-\eta}) &\cong \overline{HM}_\bullet(Y;\Gamma_\eta)^*. \end{aligned}\right\} \tag{3.24}$$

Most of the previous structure carries over with little change. For example, a homology orientation for Y allows the definition of duality isomorphisms forming a diagram like (3.5) (in which the third square anti-commutes); and there are cup and cap products, so that the Floer cohomology groups with coefficients Γ_η, for example, are modules for the ring $\mathbb{A}(Y)\otimes\mathbb{R}$.

Consider next an oriented cobordism W from Y_0 to Y_1, equipped with a homology orientation. Let η_0 and η_1 be C^∞ singular 1-cycles in Y_0 and Y_1. In order to obtain a homomorphism between the corresponding Floer groups, we need to specify an additional piece of data. When the coefficients were simply $\mathbb{Z}$, we explained that the map $\widecheck{HM}_\bullet(W)$ associated to a cobordism arises formally from the restriction map r described in (3.13). In line with the discussion for the finite-dimensional case from Subsection 2.8, we would expect that to extend this construction to local coefficients, we would need to specify an isomorphism on $M(W)$,

$$r^*(\Gamma_{\eta_0}) \to r^*(\Gamma_{\eta_1}),$$

between the pull-backs of the two local systems. We shall see in Subsection 23.3 how we can write down such an isomorphism, given a C^∞ singular 2-chain ν on W which provides a homology between η_0 and η_1:

$$\begin{aligned} \nu &\in Z_2(W, \partial W; \mathbb{R}), \\ \partial\nu &= \eta_1 - \eta_0. \end{aligned}$$

We shall write the resulting isomorphism between the pull-backs as

$$\Gamma_\nu : r^*(\Gamma_{\eta_0}) \to r^*(\Gamma_{\eta_1}).$$

The existence of such a ν constrains the homology classes $[\eta_i]$. When a ν exists, it determines (as our discussion is meant to explain) homomorphisms of $\mathbb{R}[[U_\dagger]]$-modules

$$\begin{aligned} \widecheck{HM}_\bullet(W;\Gamma_\nu) &: \widecheck{HM}_\bullet(Y_0;\Gamma_{\eta_0}) \to \widecheck{HM}_\bullet(Y_1;\Gamma_{\eta_1}) \\ \widehat{HM}_\bullet(W;\Gamma_\nu) &: \widehat{HM}_\bullet(Y_0;\Gamma_{\eta_0}) \to \widehat{HM}_\bullet(Y_1;\Gamma_{\eta_1}) \\ \overline{HM}_\bullet(W;\Gamma_\nu) &: \overline{HM}_\bullet(Y_0;\Gamma_{\eta_0}) \to \overline{HM}_\bullet(Y_1;\Gamma_{\eta_1}) \end{aligned}$$

commuting with i_*, j_* and p_*, with dual maps on the cohomology groups. The construction of $\overrightarrow{HM}_\bullet(W)$ carries over also, so if $b^+(W) \geq 2$ we have maps

$$\begin{aligned} \overrightarrow{HM}_\bullet(W;\Gamma_\nu) &: \widehat{HM}_\bullet(Y_0;\Gamma_{\eta_0}) \to \widecheck{HM}_\bullet(Y_1;\Gamma_{\eta_1}) \\ \overrightarrow{HM}^\bullet(W;\Gamma_\nu) &: \widecheck{HM}^\bullet(Y_1;\Gamma_{\eta_0}) \to \widehat{HM}^\bullet(Y_0;\Gamma_{\eta_1}). \end{aligned}$$

We state a proposition with a few important properties.

Proposition 3.7.1.

(i) *Let $W : Y_0 \to Y_1$ be a cobordism and ν, ν' be elements of $Z_2(W, \partial W; \mathbb{R})$ whose difference is a boundary:*

$$\begin{aligned} \partial\nu = \partial\nu' &= \eta_1 - \eta_0 \\ \nu' &= \nu + \partial\theta. \end{aligned}$$

Then

$$\widecheck{HM}_\bullet(W;\Gamma_\nu) = \widecheck{HM}_\bullet(W;\Gamma_{\nu'})$$

as homomorphisms from $\widecheck{HM}_\bullet(Y_0;\Gamma_{\eta_0})$ to $\widecheck{HM}_\bullet(Y_1;\Gamma_{\eta_1})$. The same holds for $\widehat{HM}_\bullet$, $\overline{HM}_\bullet$ and $\overrightarrow{HM}_\bullet$ when defined.

(ii) *Suppose W is the composite of two cobordisms $W_1 : Y_0 \to Y_1$ and $W_2 : Y_1 \to Y_2$, and let W be equipped with the composite homology orientation. Let ν_i belong to $Z_2(W_i, \partial W_i; \mathbb{R})$, and suppose*

$$\begin{aligned} \partial \nu_1 &= \eta_1 - \eta_0 \\ \partial \nu_2 &= \eta_2 - \eta_1 \\ \eta_i &\in \Omega_1(Y_i; \mathbb{R}), \quad i = 0, 1, 2. \end{aligned}$$

Let $\nu = \nu_1 + \nu_2 \in Z_2(W, \partial W; \mathbb{R})$. Then we have

$$\widecheck{HM}_\bullet(W; \Gamma_\nu) = \widecheck{HM}_\bullet(W_2; \Gamma_{\nu_2}) \circ \widecheck{HM}_\bullet(W_1; \Gamma_{\nu_1})$$

as homomorphisms $\widecheck{HM}_\bullet(Y_0; \Gamma_{\eta_0}) \to \widecheck{HM}_\bullet(Y_2; \Gamma_{\eta_2})$. The same holds with $\widehat{HM}_\bullet$ or $\overline{HM}_\bullet$ in place of $\widecheck{HM}_\bullet$; and the composition laws for $\overrightarrow{HM}_\bullet(W, \nu)$ hold also, when they make sense.

(iii) *Let $W^\dagger : -Y_1 \to -Y_0$ be the cobordism obtained from W by switching ends. Let W be given a homology orientation, and let μ_0, μ_1 be homology orientations of the two ends, used to determine a homology orientation of $W^\dagger$. Then*

$$\begin{aligned} \widecheck{HM}_\bullet(W^\dagger; \Gamma_\nu) &= \check{\omega}_{\mu_0}^{-1} \widehat{HM}^\bullet(W; \Gamma_\nu) \check{\omega}_{\mu_1} \\ \widehat{HM}_\bullet(W^\dagger; \Gamma_\nu) &= \hat{\omega}_{\mu_0}^{-1} \widecheck{HM}^\bullet(W; \Gamma_\nu) \hat{\omega}_{\mu_1} \\ \overline{HM}_\bullet(W^\dagger; \Gamma_\nu) &= \bar{\omega}_{\mu_0}^{-1} \overline{HM}^\bullet(W; \Gamma_\nu) \bar{\omega}_{\mu_1}. \end{aligned}$$

Local coefficients can also be combined with the generalized cap and cup products described in Theorem 3.4.4. Thus for a in $\mathbb{A}_\dagger(W)$ we have homomorphisms $\widecheck{HM}_\bullet(a \mid W; \Gamma_\nu)$ and so on. We formulate this eventually, as Theorem 23.3.4, later in the book.

3.8 Local coefficients and invariants of closed four-manifolds

Let X be a closed, oriented 4-manifold with $b^+(X) \geq 2$. We can combine the invariants $\mathfrak{m}(X, \mathfrak{s}_X)$ to form a function on $H_2(X; \mathbb{R})$, by writing

$$\mathfrak{m}(X, h) = \sum_{\mathfrak{s}_X} \mathfrak{m}(X, \mathfrak{s}_X) \exp\langle c_1(\mathfrak{s}_X), h \rangle, \quad h \in H_2(X; \mathbb{R}), \tag{3.25}$$

where the angle brackets denote pairing between cohomology and homology. When h is zero, this recovers the sum $\mathfrak{m}(X)$. If $H_2(X; \mathbb{Z})$ is torsion-free, so that a spin^c structure $\mathfrak{s}_X$ is entirely determined by the image of $c_1(\mathfrak{s}_X)$ in real cohomology, then knowing $\mathfrak{m}(X, h)$ as a function on $H_2(X;\mathbb{R})$ determines the individual monopole invariants $\mathfrak{m}(X, \mathfrak{s}_X)$.

Now let $W : S^3 \to S^3$ be the cobordism obtained by removing two balls from X, and let ν be a cycle in $Z_2(W;\mathbb{R})$. If $b^+(X) \geq 2$, we have a map

$$\overrightarrow{HM}_\bullet(W;\Gamma_\nu) : \widehat{HM}_\bullet(S^3) \otimes \mathbb{R} \to \widecheck{HM}_\bullet(S^3) \otimes \mathbb{R}.$$

By Proposition 3.7.1, this map depends only on the homology class $[\nu]$ in $H_2(W;\mathbb{R}) \cong H^2(X;\mathbb{R})$. Using the pairing between Floer homology and Floer cohomology, we can use this map to form the quantity

$$\left\langle \overrightarrow{HM}_\bullet(W;\Gamma_\nu)(1), \check{1} \right\rangle \in \mathbb{R} \tag{3.26}$$

as in (3.18). The analog of Proposition 3.6.1 is now the following proposition.

Proposition 3.8.1. *We have*

$$\left\langle \overrightarrow{HM}_\bullet(W;\Gamma_\nu)(1), \check{1} \right\rangle = \mathfrak{m}(X, [\nu]),$$

where the right-hand side is defined at (3.25).

Suppose now that X is decomposed as $X_1 \cup X_2$ with $\partial X_1 = -\partial X_2 = Y$, as before. Remove balls from X_1 and X_2 to obtain a composite cobordism $W = W_2 \circ W_1$ from S^3 to S^3, and let $\nu \in Z_2(W;\mathbb{R})$ be a 2-cycle on W, and suppose that $\nu = \nu_1 + \nu_2$, with

$$\nu_i \in Z_1(W_i, Y;\mathbb{R}).$$

Let $\eta = \partial\nu_1 = -\partial\nu_2$, regarded as a cycle in $Z_1(Y;\mathbb{R})$. As in Subsection 3.6, Proposition 3.8.1 and the composition laws of Proposition 3.7.1 lead to formulae

$$\mathfrak{m}(X, [\nu]) = \left\langle \overrightarrow{HM}_\bullet(W_1;\Gamma_{\nu_1})(1), \widecheck{HM}^\bullet(W_2;\Gamma_{\nu_2})(\check{1}) \right\rangle, \quad \text{when } b^+(X_1) \geq 2,$$

and

$$\mathfrak{m}(X, [\nu]) = \left\langle \widehat{HM}_\bullet(W_1;\Gamma_{\nu_1})(1), \overrightarrow{HM}^\bullet(W_2;\Gamma_{\nu_2})(\check{1}) \right\rangle, \quad \text{when } b^+(X_2) \geq 2.$$

In the first case, the pairing refers to the map

$$\widecheck{HM}_\bullet(Y;\Gamma_\eta) \times \widecheck{HM}^\bullet(Y;\Gamma_{-\eta}) \to \mathbb{R},$$

and in the second case to the map

$$\widehat{HM}_\bullet(Y;\Gamma_\eta) \times \widehat{HM}^\bullet(Y;\Gamma_{-\eta}) \to \mathbb{R}.$$

Finally, there is the symmetrical form, the local coefficient version of (3.22). Write

$$HM_\bullet(Y;\Gamma_\eta) = \operatorname{im}\bigl(j_* : \widetilde{HM}_\bullet(Y;\Gamma_\eta) \to \widehat{HM}_\bullet(Y;\Gamma_\eta)\bigr)$$

as before, the *reduced Floer homology* groups with local coefficients. Given a homology orientation of Y, we can use the duality isomorphisms ω_μ to obtain a pairing

$$\langle -, -\rangle_{\omega_\mu} : HM_\bullet(Y;\Gamma_\eta) \times HM_\bullet(-Y;\Gamma_{-\eta}) \to \mathbb{R}.$$

If both $b^+(X_1)$ and $b^+(X_2)$ are positive, we can write

$$\mathfrak{m}(X,[\nu]) = \bigl\langle \psi_{(X_1,\nu_1)}, \psi_{(X_2,\nu_2)} \bigr\rangle_{\omega_\mu}. \tag{3.27}$$

Here, for a general manifold X with connected boundary Y, satisfying $b^+(X) \geq 1$, we write

$$\psi_{(X,\nu)} \in HM_\bullet\bigl(Y;\Gamma_\eta\bigr) \quad (\nu \in Z_2(X,Y;\mathbb{R})),$$

for the element $\widehat{HM}_\bullet(W;\Gamma_\nu)(1)$ in $HM_\bullet(Y;\Gamma_{\partial\nu}) \subset \widehat{HM}_\bullet(Y;\Gamma_{\partial\nu})$, just as we did in Definition 3.6.4 when the coefficients were in $\mathbb{Z}$. We can also introduce the 2-dimensional class u_2, and write

$$\mathfrak{m}(u_2^e \,|\, X, h) = \sum_{\mathfrak{s}_X} \mathfrak{m}(u_2^e \,|\, X, \mathfrak{s}_X) \exp\langle c_1(\mathfrak{s}_X), h\rangle, \quad h \in H_2(X;\mathbb{R}). \tag{3.28}$$

Then (3.27) has the generalization

$$\mathfrak{m}(u_2^e \,|\, X, [\nu]) = \bigl\langle U_\dagger^e \psi_{(X_1,\nu_1)}, \psi_{(X_2,\nu_2)} \bigr\rangle_{\omega_\mu}. \tag{3.29}$$

3.9 Non-additivity of b^+

We now turn to the resolution of the second of the two shortcomings of the pairing formula (3.22): the fact that a closed manifold X with $b^+(X) = 2$ can be decomposed into two pieces X_i, each with $b^+ = 0$, so that the invariants ψ_{X_i} are undefined. The key point is the following proposition, proved in Subsection 35.2.

Proposition 3.9.1. *If the homology class* $[\eta] \in H_2(Y;\mathbb{R})$ *is non-zero, then* $\overline{HM}_\bullet(Y;\Gamma_\eta) = 0$.

Thus when $[\eta]$ is non-zero, the theory simplifies considerably. The groups $\widecheck{HM}_\bullet(Y;\Gamma_\eta)$, $\widehat{HM}_\bullet(Y;\Gamma_\eta)$ and $HM_\bullet(Y;\Gamma_\eta)$ are all isomorphic via j_*. We can identify them all with $HM_\bullet(Y;\Gamma_\eta)$, a vector space whose dual space is $HM^\bullet(Y;\Gamma_{-\eta})$. The negative completion (Definition 3.1.3) has no effect, as these vector spaces are all finite-dimensional. A homology orientation μ identifies $HM^\bullet(Y;\Gamma_{-\eta})$ with $HM_\bullet(-Y;\Gamma_{-\eta})$. In particular, if X is a manifold with oriented boundary Y, and if ν is a relative 2-cycle on X with $[\eta] = [\partial\nu]$ a non-zero class in $H_1(Y;\mathbb{R})$, then we can always construct an invariant

$$\psi_{(X,\nu)} \in HM_\bullet(Y;\Gamma_\eta),$$

defining it as $HM_\bullet(W;\Gamma_\nu)(1)$, where $W = X \setminus B^4$.

To see how this resolves the issue arising from non-additivity of b^+, we need the following lemma, whose proof we omit. (It is a straightforward application of the Mayer–Vietoris sequence, and is equivalent to Novikov's additivity for the signature.) We state the lemma for a composite cobordism.

Lemma 3.9.2. *Let* W *be an oriented 4-dimensional cobordism formed as the composite of* $W_1 : Y_0 \to Y_1$ *and* $W_2 : Y_1 \to Y_2$. *Then*

$$b^+(W) = b^+(W_1) + b^+(W_2) + i,$$

where i *is the rank of the restriction map*

$$H^2(W;\mathbb{R}) \to H^2(Y_1;\mathbb{R}),$$

or equivalently (via Poincaré duality) the map

$$\partial_1 : H_2(W;\mathbb{R}) \to H_1(Y_1;\mathbb{R})$$

arising from the Mayer–Vietoris sequence of $W_1 \cup W_2$. □

The lemma tells us that if a closed manifold X is decomposed along Y into pieces X_1 and X_2, then either b^+ is additive or we can find a 2-dimensional class $[\nu]$ on X such that $\partial_1[\nu]$ is in a non-zero homology class in Y. In the latter case, we can find a representative cycle $\nu = \nu_1 + \nu_2$, where

$$\nu_i \in Z_2(X_i, Y;\mathbb{R})$$
$$\partial\nu_1 = -\partial\nu_2 = \eta;$$

and whatever the values of $b^+(X_i)$, we have invariants

$$\begin{aligned}\psi_{(X_1,\nu_1)} &\in HM_\bullet(Y;\Gamma_\eta)\\ \psi_{(X_2,\nu_2)} &\in HM_\bullet(-Y;\Gamma_{-\eta})\end{aligned}$$

because of the vanishing of $\overline{HM}_\bullet(Y;\Gamma_{\pm\eta})$. We then have the following result:

Proposition 3.9.3. *If $b^+(X) \geq 2$ and the restriction $\partial_1[\nu]$ is non-zero in $H_1(Y;\mathbb{R})$, then we have*

$$\mathfrak{m}(X,[\nu]) = \big\langle \psi_{(X_1,\nu_1)}, \psi_{(X_2,\nu_2)} \big\rangle_{\omega_\mu},$$

where $\nu = \nu_1 + \nu_2$ as above, whether or not either of the X_i has $b^+ = 0$. Under the same hypotheses, we also have the generalization

$$\mathfrak{m}(u_2^e \mid X,[\nu]) = \big\langle U_\dagger^e \psi_{(X_1,\nu_1)}, \psi_{(X_2,\nu_2)} \big\rangle_{\omega_\mu}.$$

Whenever the rank i of the restriction map $H^2(X;\mathbb{R}) \to H^2(Y;\mathbb{R})$ is non-zero (that is, whenever b^+ is non-additive), this proposition provides a pairing formula for $\mathfrak{m}(X,[\nu])$ for *generic* $[\nu]$ (all $[\nu]$ outside a proper linear subspace of the space of $H_2(X;\mathbb{R})$); and this determines the function $\mathfrak{m}(X,h)$ on $H_2(X;\mathbb{R})$ entirely. The above proposition is proved in Section 32.

Remarks. In the situation described in the proposition, the fact that the rank i of the restriction map is non-zero implies that $b^+(X)$ is at least 1, but does not imply the hypothesis $b^+(X) \geq 2$. There remains the case that $b^+(X) = 1$, the rank i is 1 and $b^+(X_1) = b^+(X_2) = 0$. The right-hand side of the equality in the proposition is then well-defined, and is an invariant of X, Y and ν. It is natural to ask whether we can interpret it. There is an answer to this question, which can be formulated by introducing the monopole invariants $\mathfrak{m}$ for manifolds X with $b^+ = 1$: these invariants are dependent on the choice of metric and perturbing 2-form on X, as we briefly mentioned in our discussion after the statement of Theorem 1.5.2. We will return to this matter in Subsection 27.5.

A second remark to make at this point is that, although the function $\mathfrak{m}(X,-)$ on $H^2(X;\mathbb{R})$ determines the individual invariants $\mathfrak{m}(X,\mathfrak{s}_X)$ when $H^2(X;\mathbb{Z})$ is torsion-free, we cannot in general isolate the invariants belonging to spin^c structures whose difference element is torsion. (It is possible to resolve this issue with a more elaborate version of Floer homology with local coefficients.)

3.10 The manifolds T^3 and $S^1 \times S^2$

We now describe the Floer groups of the next two simplest manifolds after the 3-sphere. We begin with the 3-torus. On T^3, let ξ_0 be an oriented 2-plane field that is invariant under translation. The homotopy class of ξ_0 in $\pi_0(\Xi(T^3))$ is independent of the choice, and the orbit of this class under the action of $\mathbb{Z}$ on $\pi_0(\Xi(T^3))$ is free.

Proposition 3.10.1. *The group $\widecheck{HM}_\bullet(T^3)$ is isomorphic to $\mathbb{Z}^3$ in grading $[\xi_0]+j$ for all $j \geq 0$, and is zero otherwise. The group $\widehat{HM}_\bullet(T^3)$ is isomorphic to $\mathbb{Z}^3$ in grading $[\xi_0]-j$ for all $j \geq 0$, and is zero otherwise. The map j_* is zero, so $HM_\bullet(T^3) = 0$, and $\overline{HM}_\bullet(T^3)$ is isomorphic to $\mathbb{Z}^3$ in grading $[\xi_0]+j$ for all j.*

The group of orientation-preserving diffeomorphisms of T^3 acts on the Floer groups, and we have, for example

$$\widecheck{HM}_{[\xi_0]+j}(T^3) \cong \begin{cases} H^1(T^3), & j \text{ even}, j \geq 0 \\ H^2(T^3), & j \text{ odd}, j \geq 0 \end{cases}$$

as representations of the diffeomorphism group. For $\widehat{HM}_\bullet(T^3)$, the roles of j even and odd are reversed.

We can also describe the module structure for these groups. In each case, the map $U_\dagger$ is simply a shift; so we can write

$$\begin{aligned}
\widecheck{HM}_\bullet(T^3) &\cong \big(\mathbb{Z}[U_\dagger^{-1}, U_\dagger]/\mathbb{Z}[U_\dagger]\big) \otimes \big(H^1(T^3) \oplus H^2(T^3)\big) \\
\widehat{HM}_\bullet(T^3) &\cong \mathbb{Z}[[U_\dagger]] \otimes \big(H^1(T^3) \oplus H^2(T^3)\big) \\
\overline{HM}_\bullet(T^3) &\cong \mathbb{Z}[U_\dagger^{-1}, U_\dagger]] \otimes \big(H^1(T^3) \oplus H^2(T^3)\big).
\end{aligned}$$

The action of $\mathbb{A}(T^3)_\dagger$ can be described by saying, for example, that for $\gamma \in H_1(T^3)$, the map

$$\gamma_\dagger : \widecheck{HM}_{[\xi_0]+1}(T^3) \to \widecheck{HM}_{[\xi_0]}(T^3)$$

coincides under the above isomorphisms with the map

$$H^2(T^3) \to H^1(T^3) = \mathrm{Hom}(H_1(T^3), \mathbb{Z})$$

given by identifying $H^2(T^3)$ with skew bilinear forms on $H_1(T^3)$ and contracting with γ.

When we introduce local coefficients, the situation changes. Recall from Proposition 3.9.1 that $\overline{HM}_\bullet(T^3;\Gamma_\eta)$ vanishes if $[\eta]$ is non-zero, so j_* is an isomorphism and we only have $HM_\bullet(T^3;\Gamma_\eta)$ to calculate.

Proposition 3.10.2. *For any* 1*-cycle* η *with* $[\eta] \neq 0$ *in* $H_1(T^3;\mathbb{R})$*, the Floer groups with local coefficients* Γ_η *are*

$$HM_{[\xi]}(T^3;\Gamma_\eta) \cong \begin{cases} \mathbb{R}, & [\xi] = [\xi_0] \\ 0, & \textit{otherwise.} \end{cases}$$

The diffeomorphisms of T^3 *which preserve orientation and preserve* η *act trivially on* $HM_\bullet(T^3;\Gamma_\eta)$.

The next manifold to consider is $S^1 \times S^2$. Let ξ_0 be an oriented 2-plane field on this manifold that has Euler class zero and is invariant under translations in the S^1 coordinate. The homotopy class $[\xi_0]$ is uniquely determined by these conditions, and the orbit of this class under the $\mathbb{Z}$ action is again free. With $\mathbb{Z}$ coefficients, the description of the Floer groups of $S^1 \times S^2$ is very similar to the case of T^3.

Proposition 3.10.3. *The group* $\widecheck{HM}_\bullet(S^1 \times S^2)$ *is isomorphic to* $\mathbb{Z}$ *in grading* $[\xi_0] + j$ *for all* $j \geq 0$*, and is zero otherwise. The group* $\widehat{HM}_\bullet(S^1 \times S^2)$ *is isomorphic to* $\mathbb{Z}$ *in grading* $[\xi_0] - j$ *for all* $j \geq 0$*, and is zero otherwise. The map* j_* *is zero, so* $HM_\bullet(S^1 \times S^2) = 0$*, and* $\overline{HM}_\bullet(S^1 \times S^2)$ *is isomorphic to* $\mathbb{Z}$ *in grading* $[\xi_0] + j$ *for all* j.

Again, the action of $U_\dagger$ is by a shift in each case. If γ is a generator of $H_1(S^1 \times S^2)$, then the map

$$\gamma_\dagger : \widecheck{HM}_{[\xi_0]+1}(S^1 \times S^2) \to \widecheck{HM}_{[\xi_0]}(S^1 \times S^2)$$

is an isomorphism, as is

$$\gamma_\dagger : \widehat{HM}_{[\xi_0]}(S^1 \times S^2) \to \widehat{HM}_{[\xi_0]-1}(S^1 \times S^2).$$

When we twist the coefficients, all the Floer groups vanish:

Proposition 3.10.4. *If* $[\eta] \neq 0$*, then* $HM_\bullet(S^1 \times S^2;\Gamma_\eta) = 0$.

3.11 Simple vanishing theorems

The most straightforward application of the pairing formulae is to prove the vanishing of the monopole invariants of a closed manifold X when it is decomposed along a 3-manifold Y for which an appropriate Floer group is zero. We give a proposition with three variations on this theme, all of which follow immediately from our pairing formulae.

Proposition 3.11.1. *Let X be a closed, oriented 4-manifold with $b^+(X) \geq 2$, and suppose that $X = X_1 \cup X_2$ with $\partial X_1 = -\partial X_2 = Y$, a connected 3-manifold.*

(i) *If $HM_\bullet(Y)$ is zero, and $b^+(X_1)$ and $b^+(X_2)$ are both positive, then $\mathfrak{m}(X) = 0$.*
(ii) *Suppose the Mayer–Vietoris map $H_2(X;\mathbb{R}) \to H_1(Y;\mathbb{R})$ is trivial. If $HM_\bullet(Y)$ is zero, and $b^+(X_1)$ and $b^+(X_2)$ are both positive, then the function $\mathfrak{m}(X,h)$ is identically zero on $H^2(X;R)$*
(iii) *Suppose the Mayer–Vietoris map $\partial_1 : H_2(X;\mathbb{R}) \to H_1(Y;\mathbb{R})$ is non-trivial, and suppose that $HM_\bullet(Y;\Gamma_\eta)$ is zero for at least one non-zero $[\eta]$ in the image of ∂_1. Then the function $\mathfrak{m}(X,h)$ on $H_2(X;\mathbb{R})$ is identically zero, without any hypothesis on $b^+(X_i)$.*

Proof. Only the third part needs further comment. For η belonging to a non-zero class, the vanishing of $HM_\bullet(Y;\Gamma_\eta)$ is the same as the vanishing of $\widetilde{HM}_\bullet(Y;\Gamma_\eta)$, which is equivalent to saying that the complex defining this group is exact. As η varies, only the differentials vary, not the generators; so this vanishing is an open condition. So $HM_\bullet(Y;\Gamma_\eta)$ will vanish for all $[\eta]$ in an open subset U of $H_1(Y;\mathbb{R})$. The analytic function $\mathfrak{m}(X,h)$ will therefore vanish for all h such that $\partial_1 h \in U$. Since this is an open set, the analytic function vanishes identically. □

Remark. The vanishing of $\mathfrak{m}(X,h)$ on $H_2(X;\mathbb{R})$ is equivalent to the statement

$$\sum_{l \in T} \mathfrak{m}(X, \mathfrak{s}_X + l) = 0$$

for all spin^c structures $\mathfrak{s}_X$ on X, where $T \subset H^2(X;\mathbb{Z})$ is the torsion subgroup.

The three 3-manifolds for which we have described the Floer group provide illustrations of the three parts of this proposition. We state these illustrations as three corollaries.

Corollary 3.11.2. *If X is the union of X_1 and X_2 with common boundary T^3, and $b^+(X_i) \geq 1$ for $i = 1, 2$, then $\mathfrak{m}(X) = 0$.* □

Corollary 3.11.3. *If X is a connected sum of manifolds X_1 and X_2 with $b^+(X_i) \geq 1$ for $i = 1, 2$, then $\mathfrak{m}(X, h)$ is identically zero.* □

Corollary 3.11.4. *If X is the union of X_1 and X_2 with common boundary $S^1 \times S^2$, and the homology class of the S^2 is non-zero in $H_2(X; \mathbb{R})$, then $\mathfrak{m}(X, h)$ is identically zero.* □

The last two of these three corollaries have long histories, going back to Donaldson's vanishing theorem for the instanton invariants of a connected sum [20]. Corollary 3.11.2 is of a different nature from the other two: it reflects the cancellation of terms in the sum defining $\mathfrak{m}(X)$. The invariants $\mathfrak{m}(X, \mathfrak{s}_X)$ belonging to individual spin^c structures may well be non-zero under the hypotheses of the corollary. Elliptic surfaces provide an example of this phenomenon: see [118] and Section 38.

3.12 Other variants

One of the ways that the Morse theory of the Chern–Simons–Dirac function $\mathcal{L}$ on the manifold $\mathcal{B}^\sigma(Y)$ is different from the usual setup for Morse theory is that $\mathcal{L}$ is not in general a single-valued real function on $\mathcal{B}^\sigma(Y)$ if $b_1(Y)$ is non-zero: we must instead define it as a circle-valued function. The derivative of the functional is formally a closed 1-form with non-trivial periods. Finite-dimensional Morse theory for circle-valued functions was developed by Novikov in [87]. With appropriate adaptations, the Morse complex still produces topological invariants, but it is clear that one should expect these invariants to depend on the periods of the closed 1-form.

The simplest example already illustrates this phenomenon. One can take the space S^1, with the identity map as a circle-valued Morse function. There are no critical points, and it is clear that the Morse complex will not compute the homology groups of the circle in the same way that they are obtained from an ordinary Morse function.

With monopole Floer homology, it is similarly possible to consider perturbations of the functional $\mathcal{L}$ which change the periods. We call these *non-exact* perturbations of the functional. In Section 29 we will investigate under what circumstances we can still construct a Morse complex.

A particularly interesting case occurs when one perturbs $\mathcal{L}$ so as to make the periods zero, to obtain a single-valued function on $\mathcal{B}^\sigma(Y)$. We call such a perturbation *balanced*. For a balanced perturbation, there is a Morse complex obtained from $\mathcal{B}^\sigma(Y)$ with coefficients $\mathbb{Z}$. We call the resulting Floer groups the *balanced* Floer groups, and denote them by $\widecheck{HM}_\bullet(Y, c_b)$, $\widehat{HM}_\bullet(Y, c_b)$ and $\overline{HM}_\bullet(Y, c_b)$.

(The notation c_b stands for the period class of the balanced perturbation.) These groups share some of the formal properties of the ordinary Floer groups, though not the duality isomorphisms ω_μ. On the basis of the simplest calculations, it is natural to guess that these groups are the same as the *Heegaard homology groups* introduced by Ozsváth and Szabó in [93]. More specifically, it appears likely that there are isomorphisms

$$\begin{aligned} \widecheck{HM}_\bullet(Y, c_b) &\cong HF^+(Y) \\ \widehat{HM}_\bullet(Y, c_b) &\cong HF^-(Y) \\ \overline{HM}_\bullet(Y, c_b) &\cong HF^\infty(Y), \end{aligned}$$

where the groups on the right are the Heegaard homology groups, in the notation used by Ozsváth and Szabó. When $b_1(Y)$ is zero, the balanced Floer groups coincide with the ordinary Floer groups that we have been considering up to now.

Notes and references for Chapter I

The Seiberg–Witten equations were introduced to mathematicians in Witten's paper [125], and since then there have been several good expositions of the material outlined in Section 1: see [70, 76, 77, 86] and [117]. The fact that $w_2(X)$ always has an integer lift if X is of dimension 4 appears to be due to Whitney, with an alternative argument having been given by Hirzebruch and Hopf in [52].

The term "basic class" was introduced in [61], in connection with a structure theorem for Donaldson's polynomial invariants. The phrase has since been adapted to the Seiberg–Witten invariants.

Floer's adoption of Morse theory as the starting point for what is now called Floer homology was motivated by Witten's paper [124], which contains the first written account of how the homology of a manifold can be recovered from the "Morse complex". This formulation appears to have been known earlier to topologists familiar with the more traditional exposition of the relationship between the gradient flow of a Morse function and the topology of a manifold, namely the cell decomposition which arises from the unstable manifolds of the critical points, as described in [75]. See [13] for an interesting survey. The variant of Morse theory for manifolds with boundary that we discuss here seems not to have appeared in print before, but is also probably not new. For example, Bott writes in [13]:

> ... *Thom does his mathematics with his fingers and hands, and I still recall the motions of his hands as he taught me that for manifolds with boundary, only half of the critical points on the boundary really "counted".*

This suggests that the role of what we have called boundary-stable and boundary-unstable critical points was understood, at least at the level of the Morse inequalities.

The description of the formal structure of Seiberg–Witten Floer homology in Section 3 has many antecedents. Floer introduced his "instanton" homology groups $I_*(Y)$ in [32], and showed that cobordisms between 3-manifolds gave rise to homomorphisms between their instanton homology groups. Floer restricted his attention initially to 3-manifolds Y with $H_1(Y) = 0$; and this allowed him to define a topological invariant $I_*(Y)$ using only the irreducible critical points. It was soon realized that a complete theory would require a construction which took account of the reducibles, but a complete theory is still lacking in the instanton case, due to the difficulties which arise from the phenomenon of bubbling in the instanton moduli spaces.

In the case of instanton homology, an understanding emerged that, when incorporating the reducibles, one should naturally construct two, dual variants of the Floer groups: such constructions appear in the work of Donaldson, Frøyshov and Taubes [22, 40, 113]. This is the origin of our Seiberg–Witten Floer groups $\widecheck{HM}_\bullet(Y)$ and $\widehat{HM}_\bullet(Y)$. For the case of rational homology spheres, a related Seiberg–Witten Floer theory is developed in [71]. The interesting special case of gluing along 3-tori was treated in detail in [118].

The importance of a third group (which appears as $\overline{HM}_\bullet(Y)$ in our version) became apparent with the work of Ozsváth and Szabó [93], when they developed their Heegaard Floer homology groups and successfully exploited the long exact sequence relating the three variants.

II

The Seiberg–Witten equations and compactness

In this chapter we introduce the Chern–Simons–Dirac functional and discuss some of its basic properties. In particular, we relate the gradient flow of the Chern–Simons–Dirac functional to the 4-dimensional Seiberg–Witten equations on a cylinder $I \times Y$. For a pair (A, Φ) consisting of a spin^c connection and a spinor on a 4-manifold with boundary, we introduce two notions of *energy*, the analytic energy and the topological energy, related by an inequality

$$\mathcal{E}^{\mathrm{an}}(A, \Phi) \geq \mathcal{E}^{\mathrm{top}}(A, \Phi).$$

We will see that solutions to the Seiberg–Witten equations are characterized by equality here. In the case that the 4-manifold is a cylinder, the topological energy is simply the difference in the value of the Chern–Simons–Dirac functional between the two ends. In Subection 5.1, we prove a compactness theorem for solutions to the equations on a general 4-manifold with boundary, which states that a sequence of solutions with bounded energy has a subsequence that converges on interior domains after gauge transformation.

Motivated by the ideas from finite-dimensional Morse theory which we described in Subsection 2.5, Section 6 introduces the blown-up configuration space in both the 3- and 4-dimensional contexts. We will examine how the formal gradient flow of the Chern–Simons–Dirac functional gives rise to flow also on the blown-up configuration space. After a discussion of unique-continuation results for our equations in Section 7, the final section of this chapter proves a refinement of the compactness theorem in the blown-up setting.

4 Basic terms

4.1 The Chern–Simons–Dirac functional

Let Y be a closed, connected, oriented Riemannian 3-manifold. For each isomorphism class of spin^c structure $\mathfrak{s}$ on Y, we choose a reference spin^c connection $B_0 = B_0(\mathfrak{s})$ on a spin bundle $S = S(\mathfrak{s})$.

Definition 4.1.1. The Chern–Simons–Dirac functional is a function of a spin^c connection B and a section Ψ of the corresponding spin bundle. It is defined by

$$\mathcal{L}(B,\Psi) = -\frac{1}{8}\int_Y (B^{\mathrm{t}} - B_0^{\mathrm{t}}) \wedge (F_{B^{\mathrm{t}}} + F_{B_0^{\mathrm{t}}}) + \frac{1}{2}\int_Y \langle D_B\Psi, \Psi\rangle \, d\mathrm{vol}.$$

Here B^{t} is the connection in $\Lambda^2 S$, as in Notation 1.2.1, and $F_{B^{\mathrm{t}}}$ is the curvature of B^{t}, as an imaginary-valued 2-form. We denote by $\mathcal{C}$, or $\mathcal{C}(Y,\mathfrak{s})$, the space on which $\mathcal{L}$ is defined:

$$\mathcal{C}(Y,\mathfrak{s}) = \{\, (B,\Psi) \mid B \text{ is a spin}^c \text{ connection and } \Psi \in C^\infty(Y;S)\,\}. \tag{4.1}$$

Thus we have a map $\mathcal{L} : \mathcal{C}(Y,\mathfrak{s}) \to \mathbb{R}$. ◊

Using (1.6), we can regard $\mathcal{C}(Y,\mathfrak{s})$ as an affine space for which the underlying vector space (or the tangent space at any point) is

$$T_{(B,\Psi)}\mathcal{C}(Y,\mathfrak{s}) = C^\infty(Y; iT^*Y \oplus S).$$

It is sometimes useful to have a notation for the space of spin^c connections also: we write

$$\begin{aligned}\mathcal{A}(Y,\mathfrak{s}) &= \{\, B \mid B \text{ is a spin}^c \text{ connection}\,\}\\ &= B_0 + C^\infty(Y; iT^*Y) \otimes 1_S.\end{aligned} \tag{4.2}$$

On the affine space $\mathcal{C}(Y,\mathfrak{s})$ we can compute the formal gradient of $\mathcal{L}$ with respect to the L^2 inner product. Before doing so, we should note that, with our convention that the inner product on $\mathfrak{gl}(S)$ is given by $\frac{1}{2}\operatorname{tr}(a^*b)$ (see Subsection 1.1), we have

$$\|\tilde{B} - B\|_{L^2} = \|b\|_{L^2}$$

when $\tilde{B} = B + b \otimes 1_S$. So the appropriate L^2 norm on $C^\infty(Y; iT^*Y \oplus S)$ is just $\|b\|^2 + \|\psi\|^2$, without any factor of 2. With this remark out of the way, let $(b \otimes 1_S, \psi)$ be a small change in (B, Ψ). Then the change in B^{t} is $2b$, and to first order the change in $\mathcal{L}$ is

$$-\frac{1}{8}\int_Y \big(2b \wedge (F_{B^{\mathrm{t}}} + F_{B_0^{\mathrm{t}}}) + (B^{\mathrm{t}} - B_0^{\mathrm{t}}) \wedge (2\,db)\big)$$
$$+ \frac{1}{2}\int_Y \langle \rho(b)\Psi, \Psi\rangle \, d\mathrm{vol} + \int_Y \mathrm{Re}\langle \psi, D_B \Psi\rangle \, d\mathrm{vol},$$

which we can rewrite using Stokes' theorem as

$$-\frac{1}{2}\int_Y b \wedge F_{B^{\mathrm{t}}} + \int_Y \langle b, \rho^{-1}(\Psi\Psi^*)_0\rangle \, d\mathrm{vol} + \int_Y \mathrm{Re}\langle \psi, D_B\Psi\rangle \, d\mathrm{vol}.$$

Here $(\Psi\Psi^*)_0$ denotes the traceless part of the hermitian endomorphism $\Psi\Psi^*$ of S, and $\rho^{-1}(\Psi\Psi^*)_0$ is an imaginary-valued 1-form. Finally we rewrite this as

$$\frac{1}{2}\int_Y \langle b, *F_{B^{\mathrm{t}}}\rangle \, d\mathrm{vol} + \int_Y \langle b, \rho^{-1}(\Psi\Psi^*)_0\rangle \, d\mathrm{vol} + \int_Y \mathrm{Re}\langle \psi, D_B\Psi\rangle \, d\mathrm{vol}.$$

So we have a formula for the gradient

$$\operatorname{grad} \mathcal{L} = \big((\tfrac{1}{2} * F_{B^{\mathrm{t}}} + \rho^{-1}(\Psi\Psi^*)_0) \otimes 1_S,\ D_B\Psi\big) \tag{4.3}$$

as a section

$$\operatorname{grad}\mathcal{L} : \mathcal{C}(Y, \mathfrak{s}) \to T\mathcal{C}(Y, \mathfrak{s}).$$

(Note that the chosen base connection B_0 does not appear in the formula for the gradient, reflecting the fact that a change in our choice of B_0 changes $\mathcal{L}$ only by the addition of a constant.) Remembering that $\rho(*\alpha) = -\rho(\alpha)$, we can write the equations for a stationary point (where the right-hand side of (4.3) is zero) in the following form.

Definition 4.1.2. The equations

$$\begin{aligned} \tfrac{1}{2}\rho(F_{B^{\mathrm{t}}}) - (\Psi\Psi^*)_0 &= 0 \\ D_B\Psi &= 0 \end{aligned} \tag{4.4}$$

for a pair $(B, \Psi) \in \mathcal{C}(Y, \mathfrak{s})$ are the *monopole equations*, or the *Seiberg–Witten equations*, on Y for the spin^c structure $\mathfrak{s}$. ◇

The space $\mathcal{C}(Y, \mathfrak{s})$ of pairs (B, Ψ) is acted on naturally by the automorphism group of the spin^c structure, the *gauge group*

$$\mathcal{G}(Y) = \mathrm{Map}(Y, S^1).$$

(See Subsection 1.1 and (1.8).) The action is given by

$$u : B \mapsto B - u^{-1}du \otimes 1_S, \qquad \Psi \mapsto u\Psi. \tag{4.5}$$

The group $\mathcal{G}(Y)$ is not necessarily connected: the group of components of $\mathrm{Map}(Y, S^1)$ is $H^1(Y; \mathbb{Z})$, and we write $[u] \in H^1(Y; \mathbb{Z})$ for the cohomology class corresponding to u. In terms of de Rham cohomology, regarding S^1 as the circle in $\mathbb{C}$, we can represent $[u]$ by the real 1-form $(1/2\pi i)u^{-1}du$. Note that, as u runs through $\mathcal{G}(Y)$, the forms $u^{-1}du$ which appear in the formula (4.5) run through the set of all closed imaginary-valued 1-forms whose de Rham classes lie in $2\pi i H^1(Y; \mathbb{Z})$.

The functional $\mathcal{L}$ is not invariant under the action of $\mathcal{G}(Y)$ in general:

Lemma 4.1.3. *For $(B, \Psi) \in \mathcal{C}(Y, \mathfrak{s})$ and $u : Y \to S^1$ an element of $\mathcal{G}(Y)$, we have*

$$\mathcal{L}(u(B, \Psi)) - \mathcal{L}(B, \Psi) = 2\pi^2\big([u] \smile c_1(S)\big)[Y],$$

where $[u]$ denotes the homotopy class of $u : Y \to S^1$ as an element of $H^1(Y, \mathbb{Z})$.

Proof. We compute the change in $\mathcal{L}$:

$$\begin{aligned}
\mathcal{L}(u(B, \Psi)) - \mathcal{L}(B, \Psi) &= -\frac{1}{8}\int_Y (-2u^{-1}du) \wedge (F_{B^t} + F_{B_0^t}) \\
&= \frac{1}{2}\int_Y u^{-1}du \wedge F_{B_0^t} \\
&= \frac{1}{2}\big((2\pi i[u]) \smile (-2\pi i c_1(S))\big)[Y] \\
&= 2\pi^2\big([u] \smile c_1(S)\big)[Y].
\end{aligned}$$

This is the assertion in the lemma. □

The formula shows that $\mathcal{L}$ is invariant under the action of the identity component of the group $\mathcal{G}(Y)$, and furthermore defines a map

$$\bar{\mathcal{L}} : \mathcal{C}(Y, \mathfrak{s}) \to \mathbb{R}/(2\pi^2\mathbb{Z})$$

which is invariant under the full group. Note that in the case that $b_1(Y) = 0$, the function $\mathcal{L}$ on $\mathcal{C}(Y, \mathfrak{s})$ is invariant under all of $\mathcal{G}(Y)$. More generally, this is the case whenever $\mathfrak{s}$ is a spin^c structure for which $c_1(S)$ is torsion.

4.2 Reducible solutions

Following the notation that we introduced in the 4-dimensional case, we shall say that a pair (B, Ψ) in $\mathcal{C}(Y, \mathfrak{s})$ is *reducible* if Ψ is zero. If such a reducible configuration is also a solution of the 3-dimensional Seiberg–Witten equations (4.4), then the connection B^{t} in $\Lambda^2 S$ is flat, which in turn means that $c_1(S)$ is a torsion class in $H^2(Y; \mathbb{Z})$. Conversely, if $\mathfrak{s}$ is a spin^c structure with $c_1(S)$ torsion, then there are reducible solutions:

Proposition 4.2.1. *If $\mathfrak{s}$ is a spin^c structure on Y with $c_1(S)$ torsion, then there exist reducible solutions $(B, 0)$ to the Seiberg–Witten equations in $\mathcal{C}(Y, \mathfrak{s})$. If $(B_1, 0)$ is one such solution, then all others are of the form $(B, 0)$ with $B = B_1 + b \otimes 1_S$, with b a closed imaginary-valued 1-form.*

Proof. We are simply seeking spin^c connections B for which $F_{B^{\mathrm{t}}}$ is zero. If B_0 is a base connection and we seek B in the form $B_0 + b \otimes 1_S$, then the equation to be satisfied by b is $2db = -F_{B_0^{\mathrm{t}}}$. The hypothesis that $c_1(S)$ is torsion means that $F_{B_0^{\mathrm{t}}}$ is exact, so the set of solutions is non-empty and consists of a single coset of the space of closed forms. □

As pointed out in the previous subsection, a gauge transformation u in $\mathcal{G}(Y)$ changes B by the addition of a closed form whose de Rham class lies in $2\pi i H^1(Y; \mathbb{Z})$. Combining this observation with the proposition above, we can state:

Corollary 4.2.2. *If $\mathfrak{s}$ is a spin^c structure with $c_1(S)$ torsion, then the quotient of the set of reducible solutions by the action of the gauge group $\mathcal{G}(Y)$ can be identified with the torus $H^1(Y; i\mathbb{R})/(2\pi i H^1(Y; \mathbb{Z}))$. In particular, if $b_1(Y) = 0$, then there is exactly one gauge orbit of reducible solutions to the Seiberg–Witten equations in $\mathcal{C}(Y, \mathfrak{s})$.* □

4.3 The gradient flow and the four-dimensional equations

Having computed the gradient of $\mathcal{L}$ with respect to an L^2 inner product at (4.3) above, we now examine the formal, downward gradient flow, namely the equations

$$\begin{aligned}\frac{d}{dt}B &= -(\tfrac{1}{2} * F_{B^{\mathrm{t}}} + \rho^{-1}(\Psi\Psi^*)_0) \otimes 1_S \\ \frac{d}{dt}\Psi &= -D_B\Psi.\end{aligned} \tag{4.6}$$

The first of these two equations can be rewritten in terms of B^{t} rather than B; their derivatives differ by a factor of 2:

$$\begin{aligned}\frac{d}{dt}B^{\mathrm{t}} &= - * F_{B^{\mathrm{t}}} - 2\rho^{-1}(\Psi\Psi^*)_0 \\ \frac{d}{dt}\Psi &= -D_B\Psi.\end{aligned} \tag{4.7}$$

We can define a spin^c structure on the cylinder

$$Z = \mathbb{R} \times Y$$

using the spin^c structure defined by S and ρ on Y. For the spin bundle $S_Z = S^+ \oplus S^-$ on the cylinder, we take $S \oplus S$. (Of course, we omit from our notation the pull-back or product that is really involved.) For the Clifford multiplication $\rho_Z : TZ \to \mathrm{Hom}(S_Z, S_Z)$ we take

$$\rho_Z(\partial/\partial t) = \begin{pmatrix} 0 & -1 \\ 1 & 0 \end{pmatrix}$$

$$\rho_Z(v) = \begin{pmatrix} 0 & -\rho(v)^* \\ \rho(v) & 0 \end{pmatrix} \qquad \text{for } v \in TY.$$

(Compare this definition with the formulae (1.3).) As before, ρ_Z is extended to forms, and gives for example an isomorphism

$$\rho_Z : \Lambda^+ \otimes \mathbb{C} \to \mathfrak{sl}(S^+),$$

where $\Lambda^+ \subset \Lambda^2$ is the bundle of self-dual 2-forms on the 4-manifold.

A time-dependent spin^c connection B on S gives a spin^c connection A on S_Z whose t component is ordinary differentiation:

$$\nabla_A = \frac{d}{dt} + \nabla_B. \tag{4.8}$$

We say that a 4-dimensional spin^c connection A on the cylinder Z is in *temporal gauge* if it has this restricted form. With A as above, we have the Dirac operator

$D_A^+ : \Gamma(S^+) \to \Gamma(S^-)$, and its formal adjoint D_A^-. As $S^\pm$ are both identified with (the pull-back of) S, we can write

$$D_A^+ = \frac{d}{dt} + D_B.$$

We can also reinterpret the time-dependent spinor Ψ on Y as giving rise to a section Φ of S^+ on the cylinder.

We can now reinterpret the gradient-flow equations in terms of these 4-dimensional objects. If we write A^{t} for the 4-dimensional connection in $\Lambda^2 S^+$, then the curvature of this connection is

$$F_{A^{\mathrm{t}}} = dt \wedge \left(\frac{d}{dt} B^{\mathrm{t}}\right) + F_{B^{\mathrm{t}}}.$$

The 4-dimensional Hodge star of this 2-form, written out in terms of the star operator $*$ on Y, is

$$*_4 F_{A^{\mathrm{t}}} = * \left(\frac{d}{dt} B^{\mathrm{t}}\right) + dt \wedge *F_{B^{\mathrm{t}}},$$

and so the self-dual part of $F_{A^{\mathrm{t}}}$ is

$$F_{A^{\mathrm{t}}}^+ = \frac{1}{2}\left(* \left(\frac{d}{dt} B^{\mathrm{t}}\right) + F_{B^{\mathrm{t}}} + dt \wedge \left(\frac{d}{dt} B^{\mathrm{t}} + *F_{B^{\mathrm{t}}}\right)\right).$$

The Clifford actions of these forms are related by

$$\rho_Z(F_{A^{\mathrm{t}}}^+) = -\rho\left(\frac{d}{dt} B^{\mathrm{t}} + *F_{B^{\mathrm{t}}}\right)$$

in $\mathfrak{sl}(S^+) \cong \mathfrak{sl}(S)$. (The ρ on the right side is still the Clifford multiplication on Y.) Thus we can write the gradient-flow equations (4.7) as the following equations on the cylinder Z:

$$\begin{aligned} \frac{1}{2}\rho_Z(F_{A^{\mathrm{t}}}^+) - (\Phi\Phi^*)_0 &= 0 \\ D_A^+ \Phi &= 0. \end{aligned} \tag{4.9}$$

The equations (4.9) are precisely the (4-dimensional) Seiberg–Witten equations (1.9) for the pair (A, Φ), that we introduced in Subsection 1.3.

In this way, a solution of the downward gradient-flow equation for the functional $\mathcal{L}$ can be reinterpreted as a solution (A, Φ) to the 4-dimensional Seiberg–Witten equations, with the additional property that the 4-dimensional connection A is in temporal gauge on the cylinder. If A is a general connection, not in temporal gauge, we can write

$$A = B + (c\,dt) \otimes 1_S, \tag{4.10}$$

where B is a time-dependent connection in $S \to Y$ and c is a time-dependent element of $C^\infty(Y; i\mathbb{R})$, the dt component of the connection. If we write out the 4-dimensional Seiberg–Witten equations for a pair (A, Φ) in a form parallel to (4.6), we obtain:

$$\begin{aligned} \frac{d}{dt}B - dc &= -\Big(\tfrac{1}{2} * F_{B^t} + \rho^{-1}(\Psi\Psi^*)_0\Big) \otimes 1_S \\ \frac{d}{dt}\Psi + c\Psi &= -D_B\Psi. \end{aligned} \tag{4.11}$$

For a Riemannian 4-manifold X, not necessarily cylindrical, equipped with a spin^c structure $\mathfrak{s}_X$, we write $\mathcal{C}(X, \mathfrak{s}_X)$ for the space on which the 4-dimensional Seiberg–Witten equations are defined (*cf.* (4.1)):

$$\mathcal{C}(X, \mathfrak{s}_X) = \{\, (A, \Phi) \mid A \text{ is a spin}^c \text{ connection and } \Phi \in C^\infty(X; S^+) \,\}.$$

The equations are invariant under the action of the 4-dimensional gauge group $\mathcal{G}(X) = \mathrm{Map}(X, S^1)$. The left-hand sides of the equations (4.9) take their values in

$$C^\infty(X; i\,\mathfrak{su}(S^+) \oplus S^-).$$

As in Subsection 1.3, we write the equations as

$$\mathfrak{F}(A, \Phi) = 0,$$

where

$$\mathfrak{F}(A, \Phi) = \left(\frac{1}{2}\rho\big(F^+_{A^t}\big) - (\Phi\Phi^*)_0,\ D^+_A\Phi\right). \tag{4.12}$$

4.4 Some notation

We have just noted the fact that, when a 4-manifold is a cylinder, say $Z = I \times Y$, a time-dependent spin^c connection $B(t)$ ($t \in I$) on a spin bundle $S \to Y$ gives rise to a spin^c connection A on Z in temporal gauge on the pull-back spin bundle (see (4.8)). Thus configurations $(A, \Phi) \in \mathcal{C}(Z, \mathfrak{s}_Z)$, with A in temporal gauge, correspond to smooth paths $I \to \mathcal{C}(Y, \mathfrak{s})$.

It is useful to have a notation for the opposite construction:

Definition 4.4.1. If $(A, \Phi) \in \mathcal{C}(Z, \mathfrak{s}_Z)$, where $Z = I \times Y$, we write $(\check{A}, \check{\Phi})$ for the corresponding path

$$(\check{A}, \check{\Phi}) : I \to \mathcal{C}(Y, \mathfrak{s}).$$

That is, $\check{A}(t)$ is the connection on Y obtained by restricting A to $\{t\} \times Y$. Note that the t component of A is lost, and cannot be recovered from $\check{A}$. ◊

With this definition we can write a general connection A on Z as

$$A = \check{A} + c \otimes 1_{S_X} dt$$

where $c \in C^\infty(Z; i\mathbb{R})$. The distinction between Φ and $\check{\Phi}$ is, of course, only a change of viewpoint, and we will not always maintain a distinction.

We shall often write

$$\gamma = (A, \Phi)$$

for a typical configuration on the cylinder, and we may then write $\check{\gamma} = (\check{A}, \check{\Phi})$ for the corresponding path. If γ satisfies the 4-dimensional Seiberg–Witten equations on the cylinder, $\mathfrak{F}(\gamma) = 0$, we shall refer to γ as a *trajectory* (whether or not γ is in temporal gauge).

4.5 Integration by parts

For the moment, we consider a compact, oriented Riemannian 4-manifold X with boundary Y. If we start with a spin^c structure $\mathfrak{s}_X = (S_X, \rho_X)$ on X, we can use the outward normal $\vec{n}$ to identify S^+ and S^- at the boundary:

$$\rho_X(\vec{n}) : S^+|_Y \to S^-|_Y.$$

We therefore recover a spin^c structure $\mathfrak{s} = (S, \rho)$ on Y with $S = S^+|_Y \cong S^-|_Y$ and Clifford multiplication ρ defined by

$$\rho(v) = \rho_X(\vec{n})^{-1} \rho_X(v). \qquad (4.13)$$

Let A be a spin^c connection on X, and let B be its restriction to the boundary Y. We have the Lichnerowicz–Weitzenböck formula [66] for the Dirac operator on S^+, which tells us:

$$D_A^- D_A^+ \Phi = \nabla_A^* \nabla_A \Phi + \frac{1}{2}\rho_X(F_{A^t}^+)\Phi + \frac{1}{4}s\Phi, \tag{4.14}$$

where s is the scalar curvature. In the absence of a boundary, we can take the inner product with Φ and integrate by parts to obtain

$$\int_X |D_A^+\Phi|^2 = \int_X |\nabla_A\Phi|^2 + \frac{1}{2}\int_X \langle \Phi, \rho_X(F_{A^t}^+)\Phi\rangle + \frac{1}{4}\int_X s|\Phi|^2.$$

When the boundary is non-empty, we pick up two boundary terms, coming from the formulae

$$\int_X \langle \Phi, \nabla_A^*\nabla_A\Phi\rangle = \int_X |\nabla_A\Phi|^2 - \int_Y \langle \Phi, \nabla_{A,\vec{n}}\Phi\rangle,$$
$$\int_X \langle \Phi, D_A^- D_A^+\Phi\rangle = \int_X |D_A^+\Phi|^2 - \int_Y \langle \rho_X(\vec{n})\Phi, D_A^+\Phi\rangle.$$

With our present conventions, the difference of the boundary terms can be expressed using the boundary Dirac operator D_B. In the case that the Riemannian metric on X is cylindrical in a neighborhood of the boundary Y, the formula is

$$D_B\Phi|_Y = \rho_X(\vec{n})^{-1}D_A^+\Phi - \nabla_{A,\vec{n}}\Phi,$$

as follows easily from the definition of the Dirac operator and (4.13). In the general case, we have the following.

Lemma 4.5.1. *Let X be a Riemannian 4-manifold with boundary Y, let A be a spin^c connection, D_A^+ the 4-dimensional Dirac operator on sections of S^+, and let D_B be the boundary Dirac operator, for the spin^c connection B obtained on the boundary. Then for any section Φ of S^+, we have*

$$D_B(\Phi|_Y) = \rho_X(\vec{n})^{-1}(D_A^+\Phi)|_Y - (\nabla_{A,\vec{n}}\Phi)|_Y + (H/2)\Phi|_Y,$$

where H is the mean curvature of the boundary. (The orientation convention for mean curvature makes H negative in the case that X is the ball.)

Proof. Let e_1, e_2, e_3 be an oriented orthonormal frame for TY whose covariant derivative vanishes at some $y \in Y$. Then $\vec{n}, e_1, e_2, e_3$ is an oriented frame for

TX along Y. Let ν, e^1, e^2, e^3 be the dual coframe. Along Y, the Levi-Cività derivatives ∇_X and ∇_Y are related by

$$\nabla_{X,e_i} V = \nabla_{Y,e_i} V + \sum_j N(e_i, e_j) e^j(V) \otimes \vec{n},$$

where V is a section of TY and N is the second fundamental form. The corresponding spin^c connections ∇_A and ∇_B, acting on sections of $S^+|_Y$, are therefore similarly related by

$$\nabla_{A,e_i} \Phi = \nabla_{B,e_i} \Phi + \frac{1}{2} \sum_j N(e_i, e_j) \rho_X(e^j) \rho_X(\nu) \Phi,$$

because the element of $\mathfrak{so}(4)$ given by

$$e^j \otimes \vec{n} - \nu \otimes e_j$$

acts in the spin representation by $(1/2)\rho_X(e^j)\rho_X(\nu)$. We now compute

$$\begin{aligned}
D_A^+ \Phi &= \rho_X(\nu) \nabla_{A,\vec{n}} \Phi + \sum_i \rho_X(e^i) \nabla_{A,e_i} \Phi \\
&= \rho_X(\nu) \nabla_{A,\vec{n}} \Phi + \sum_i \rho_X(e^i) \nabla_{B,e_i} \Phi \\
&\quad + \frac{1}{2} \sum_{i,j} \rho_X(e^i) N(e_i, e_j) \rho_X(e^j) \rho_X(\nu) \Phi \\
&= \rho_X(\nu) \Big(\nabla_{A,\vec{n}} \Phi + D_B \Phi - \frac{1}{2} \sum_i N(e_i, e_i) \Phi \Big) \\
&= \rho_X(\nu) \Big(\nabla_{A,\vec{n}} \Phi + D_B \Phi - \frac{H}{2} \Phi \Big).
\end{aligned}$$

This is the formula claimed. □

So in the presence of boundary, the Lichnerowicz–Weitzenböck formula (4.14) gives us

$$\begin{aligned}
\int_X |D_A^+ \Phi|^2 &= \int_X |\nabla_A \Phi|^2 + \frac{1}{2} \int_X \langle \Phi, \rho_X(F_{A^t}^+) \Phi \rangle + \frac{1}{4} \int_X s |\Phi|^2 \\
&\quad + \int_Y \langle \Phi, D_B \Phi \rangle - \int_Y (H/2) |\Phi|^2,
\end{aligned} \tag{4.15}$$

for all sections Φ of S^+. The second term on the right-hand side appears again, with the opposite sign, when we expand the L^2 norm of the expression appearing in the first of the two equations (4.9):

$$\begin{aligned}\int_X \left|\tfrac{1}{2}\rho_X(F^+_{A^t}) - (\Phi\Phi^*)_0\right|^2 &= \frac{1}{4}\int_X |\rho_X(F^+_{A^t})|^2 + \int_X |(\Phi\Phi^*)_0|^2 - \frac{1}{2}\int \operatorname{tr}\big(\rho_X(F^+_{A^t})(\Phi\Phi^*)_0\big) \\ &= \frac{1}{2}\int_X |F^+_{A^t}|^2 + \frac{1}{4}\int_X |\Phi|^4 - \frac{1}{2}\int_X \langle \Phi, \rho_X(F^+_{A^t})\Phi\rangle \\ &= \frac{1}{4}\int_X |F_{A^t}|^2 - \frac{1}{4}\int_X F_{A^t}\wedge F_{A^t} + \frac{1}{4}\int_X |\Phi|^4 - \frac{1}{2}\int_X \langle \Phi, \rho_X(F^+_{A^t})\Phi\rangle.\end{aligned}$$

(Again, our norm on $\mathfrak{sl}(S^+)$ is $\frac{1}{2}\operatorname{tr}(a^*b)$.) Combining this formula with the previous one, and completing the square, we have the following version of the formula.

Proposition 4.5.2. *Let A be a spin^c connection on a compact manifold X with boundary Y, let B be the boundary spin^c connection and D_B the boundary Dirac operator. Then for all sections Φ of S^+ we have:*

$$\begin{aligned}\int_X \left|\tfrac{1}{2}\rho_X(F^+_{A^t}) - (\Phi\Phi^*)_0\right|^2 + \int_X |D^+_A\Phi|^2 &= \frac{1}{4}\int_X |F_{A^t}|^2 + \int_X |\nabla_A\Phi|^2 + \frac{1}{4}\int_X \big(|\Phi|^2 + (s/2)\big)^2 - \int_X \frac{s^2}{16} \\ &\quad - \frac{1}{4}\int_X F_{A^t}\wedge F_{A^t} + \int_Y \langle \Phi|_Y, D_B\Phi|_Y\rangle - \int_Y (H/2)|\Phi|^2.\end{aligned}$$

□

(The quantity on the left in this formula is the L^2 norm of $\mathfrak{F}(A,\Phi)$.)

Corollary 4.5.3. *Solutions of the equations* (4.9) *on X are characterized by the equality*

$$\begin{aligned}0 &= \frac{1}{4}\int_X |F_{A^t}|^2 + \int_X |\nabla_A\Phi|^2 + \frac{1}{4}\int_X \big(|\Phi|^2 + (s/2)\big)^2 - \int_X \frac{s^2}{16} \\ &\quad - \frac{1}{4}\int_X F_{A^t}\wedge F_{A^t} + \int_Y \langle \Phi|_Y, D_B\Phi|_Y\rangle - \int_Y (H/2)|\Phi|^2.\end{aligned}$$

□

We can gain some understanding of this formula by returning to the case that the 4-manifold is a cylinder. Before doing so, we put the terms in the above corollary in two groups, and make the following definition.

Definition 4.5.4. Let X be a compact Riemannian 4-manifold with oriented boundary Y. For a configuration $(A, \Phi) \in \mathcal{C}(X, \mathfrak{s}_X)$ we define the *topological energy* and the *analytic energy* by the formulae:

$$\mathcal{E}^{\text{top}}(A, \Phi) = \frac{1}{4}\int_X F_{A^t} \wedge F_{A^t} - \int_Y \langle \Phi|_Y, D_B \Phi|_Y \rangle + \int_Y (H/2)|\Phi|^2$$

and

$$\mathcal{E}^{\text{an}}(A, \Phi) = \frac{1}{4}\int_X |F_{A^t}|^2 + \int_X |\nabla_A \Phi|^2 + \frac{1}{4}\int_X \left(|\Phi|^2 + (s/2)\right)^2 - \int_X \frac{s^2}{16}.$$

Here B is the boundary spin^c connection as usual, and H is the mean curvature of the boundary, which is zero in the cylindrical case. ◊

With these definitions, we can rephrase the above corollary as saying that solutions are characterized by the equality $\mathcal{E}^{\text{an}}(A, \Phi) = \mathcal{E}^{\text{top}}(A, \Phi)$, while Proposition 4.5.2 states that, in general,

$$\mathcal{E}^{\text{an}}(A, \Phi) = \mathcal{E}^{\text{top}}(A, \Phi) + \|\mathfrak{F}(A, \Phi)\|^2. \tag{4.16}$$

Now let us examine the meaning of the topological and analytic energy in the cylindrical case, for the manifold $Z = [t_1, t_2] \times Y$. The oriented boundary is $\{t_2\} \times Y \cup \{t_1\} \times (-Y)$, and the mean curvature H of the boundary is zero. The term involving $F_{A^t} \wedge F_{A^t}$ in $\mathcal{E}^{\text{top}}(A, \Phi)$ can be written as a boundary integral:

$$\frac{1}{4}\int_Z F_{A^t} \wedge F_{A^t} = \frac{1}{4}\left(\int_{t_2 \times Y} - \int_{t_1 \times Y}\right)(A^t - A_0^t) \wedge (F_{A^t} + F_{A_0^t}). \tag{4.17}$$

The topological energy is therefore twice the change in $\mathcal{L}$:

$$\mathcal{E}^{\text{top}}(A, \Phi) = 2(\mathcal{L}(t_1) - \mathcal{L}(t_2)), \tag{4.18}$$

where $\mathcal{L}(t_i)$ stands for $\mathcal{L}(\check{A}(t_i), \check{\Phi}(t_i))$. To understand the analytic energy in the cylindrical case, we begin with the case that (A, Φ) is in temporal gauge and so corresponds to a path $\check{\gamma}$ in $\mathcal{C}(Y)$. Using (4.18), we can rewrite (4.16) in this case as

$$
\begin{aligned}
\mathcal{E}^{\mathrm{an}}(A,\Phi) &= 2(\mathcal{L}(t_1)-\mathcal{L}(t_2)) + \int_{t_1}^{t_2} \|\tfrac{d}{dt}\check{\gamma}(t) + \operatorname{grad}\mathcal{L}\|^2\,dt \\
&= 2(\mathcal{L}(t_1)-\mathcal{L}(t_2)) + \int_{t_1}^{t_2} \left(\|\tfrac{d}{dt}\check{\gamma}(t)\|^2 + \|\operatorname{grad}\mathcal{L}\|^2\right) dt \\
&\quad + 2\int_{t_1}^{t_2} \langle \tfrac{d}{dt}\check{\gamma}(t), \operatorname{grad}\mathcal{L}\rangle\,dt \\
&= \int_{t_1}^{t_2} \left(\|\tfrac{d}{dt}\check{\gamma}(t)\|^2 + \|\operatorname{grad}\mathcal{L}\|^2\right) dt. \qquad (4.19)
\end{aligned}
$$

What we have recovered here is the quite formal fact that solutions of the downward gradient-flow equation are characterized by the equality

$$2(\mathcal{L}(t_1)-\mathcal{L}(t_2)) = \int_{t_1}^{t_2} \left(\|\tfrac{d}{dt}\check{\gamma}(t)\|^2 + \|\operatorname{grad}\mathcal{L}\|^2\right) dt.$$

Although the right-hand side is equal to the analytic energy for solutions in temporal gauge, it is not a gauge-invariant quantity. To derive a formula for $\mathcal{E}^{\mathrm{an}}$ for general configurations on the cylinder, we need to replace the integral of $\|\frac{d}{dt}\check{\gamma}(t)\|^2$ by a gauge-invariant quantity. For general A, let $\check{A}$ denote the corresponding path of 3-dimensional connections, and let $c : Z \to i\mathbb{R}$ be the dt component of the connection as in (4.10):

$$\nabla_{A,\frac{d}{dt}} = \frac{d}{dt} + c \otimes 1_S.$$

Then the quantity

$$\int_{t_1}^{t_2} \left\| \frac{d}{dt}\check{A} - dc \otimes 1_S \right\|^2 dt + \int_{t_1}^{t_2} \left\| \frac{d}{dt}\check{\Phi} + c\check{\Phi} \right\|^2 dt \qquad (4.20)$$

is (as the reader may verify) gauge-invariant; and it is evidently equal to the integral of $\|\frac{d}{dt}\check{\gamma}(t)\|^2$ when c is zero. We have therefore verified:

Lemma 4.5.5. *For configurations (A,Φ) on the cylinder $Z = [t_1,t_2] \times Y$, the analytic energy is given by the formula*

$$
\begin{aligned}
&\mathcal{E}^{\mathrm{an}}(A,\Phi) \\
&= \int_{t_1}^{t_2} \left\| \frac{d}{dt}\check{A} - dc \otimes 1_S \right\|^2 dt + \int_{t_1}^{t_2} \left\| \frac{d}{dt}\check{\Phi} + c\check{\Phi} \right\|^2 dt + \int_{t_1}^{t_2} \|\operatorname{grad}\mathcal{L}\|^2 dt.
\end{aligned}
$$

□

The fact that the topological energy on the cylinder is equal to the change in $\mathcal{L}$ means in particular that it only depends on the endpoints of the corresponding path in $\mathcal{C}(Y, \mathfrak{s})$. For a general 4-manifold with boundary, it is again true that $\mathcal{E}^{\mathrm{top}}(A, \Phi)$ depends only on the restriction of (A, Φ) to $Y = \partial X$. If (B, Ψ) denotes the restriction, then we have

$$\mathcal{E}^{\mathrm{top}}(A, \Phi) = -2\mathcal{L}(B, \Psi) + C, \tag{4.21}$$

where the constant C depends on the base connection B_0 used to define $\mathcal{L}$ and the topology of the spin bundle S_X.

When we are considering solutions γ of the equations $\mathfrak{F}(\gamma) = 0$, we may just talk unambiguously of the *energy* of γ, written $\mathcal{E}(\gamma)$, without distinguishing between the topological and analytic energies.

4.6 Positive scalar curvature

In the case that the manifold X is closed, we can make further use of the formula in Corollary 4.5.3. The two boundary terms in the formula are absent; and we can combine the two expressions involving F_{A^t} using

$$\int_X |F_{A^t}|^2 - \int_X F_{A^t} \wedge F_{A^t} = 2 \int_X |F^+_{A^t}|^2.$$

Thus solutions on a closed 4-manifold can be characterized by the identity

$$0 = \frac{1}{2}\int_X |F^+_{A^t}|^2 + \int_X |\nabla_A \Phi|^2 + \frac{1}{4}\int_X \left(|\Phi|^2 + (s/2)\right)^2 - \int_X \frac{s^2}{16},$$

or

$$0 = \frac{1}{2}\int_X |F^+_{A^t}|^2 + \int_X |\nabla_A \Phi|^2 + \frac{1}{4}\int_X |\Phi|^4 + \frac{1}{4}\int_X s|\Phi|^2.$$

From this, we deduce:

Proposition 4.6.1. *On a closed 4-manifold X with non-negative scalar curvature s, all solutions (A, Φ) to the equations (4.9) have $\Phi = 0$.* □

There is a variation of this argument, which uses the same basic Lichnerowicz–Weitzenböck formula (4.14) in a different way, to reach the same conclusion. For any configuration (A, Φ) on a 4-manifold X, we can calculate, using (4.14),

$$\begin{aligned}\Delta|\Phi|^2 &= 2\langle\Phi, \nabla_A^*\nabla_A\Phi\rangle - 2\langle\nabla_A\Phi, \nabla_A\Phi\rangle \\ &\leq 2\langle\Phi, \nabla_A^*\nabla_A\Phi\rangle \\ &= 2\langle\Phi, D_A^- D_A^+\Phi\rangle - \langle\Phi, \rho_X(F_{A^t}^+)\Phi\rangle - \left\langle\Phi, \frac{1}{2}s\Phi\right\rangle.\end{aligned}$$

If (A, Φ) is a solution of the equations, then the first term on the last line is zero, and we can substitute for $\rho_X(F_{A^t}^+)$ in the second term to obtain

$$\begin{aligned}\Delta|\Phi|^2 &\leq -\langle\Phi, 2(\Phi\Phi^*)_0\Phi\rangle - \left\langle\Phi, \frac{1}{2}s\Phi\right\rangle \\ &= -|\Phi|^4 - \left\langle\Phi, \frac{1}{2}s\Phi\right\rangle.\end{aligned}$$

Previously, we used an integrated version of essentially the same calculation. We can, instead, use the maximum principle at this point. For example, if the scalar curvature s is everywhere positive and $s_0 > 0$ is its infimum, then we have

$$\Delta|\Phi|^2 \leq -\frac{s_0}{2}|\Phi|^2 \tag{4.22}$$

from which it follows that Φ is identically zero (for otherwise, $|\Phi|^2$ achieves a local maximum, where the left-hand side is non-negative and the right-hand side is strictly negative). So we recover the result of Proposition 4.6.1 by this route.

5 Compactness and properness

5.1 A compactness theorem

A basic compactness result for the Seiberg–Witten equations, $\mathfrak{F}(A, \Phi) = 0$, on a closed 4-manifold X, was stated without proof as Theorem 1.3.2: if (A_n, Φ_n) is any sequence of solutions, then there exist gauge transformations $u_n \in \mathcal{G}(X)$ such that, after passing to a subsequence, the transformed solutions $u_n(A_n, \Phi_n)$ converge in the C^∞ topology. Here we state and prove a version of this basic result for solutions on a compact manifold with boundary, Theorem 5.1.1. When the boundary is non-empty, an additional hypothesis is required: a bound on the energy (see (5.1) below). The conclusion is also weaker: we have C^∞ convergence only on interior domains.

In the course of the proof of this theorem, and indeed for the sharpest statement of the result itself, we need to introduce the *Sobolev spaces* $L^p_k(X)$, defined for $1 \leq p < \infty$ and $k \geq 0$ and any compact X as the completion of $C^\infty(X)$ in the L^p_k norm

$$\|f\|^p_{L^p_k(X)} = \int_X \left(|f|^p + |\nabla f|^p + \cdots + |\nabla^k f|^p\right) d\text{vol}.$$

If f is a section of a bundle E with an inner product, and if a connection ∇_A on the bundle is given, then the L^p_k norm is defined similarly for sections f of E; the space $L^p_k(X; E)$ is then defined as the completion of $C^\infty(X; E)$. We have given the definition in the case that k is a non-negative integer: we shall also have need of the Sobolev spaces L^2_k for fractional (and in particular half-integer) values of k. In the case of a closed manifold, these fractional Sobolev norms can be defined using the pseudo-differential operator $(1 + \Delta)^{k/2}$:

$$\|f\|_{L^2_k} = \left\|(1 + \Delta)^{k/2} f\right\|_{L^2}.$$

(In the case of integer k, this defines a norm equivalent to the previous formula.) We can now state the compactness theorem.

Theorem 5.1.1. *Let X be a compact Riemannian 4-manifold with boundary. Then the following hold.*

(i) *For given C, there are only finitely many spin^c structures $\mathfrak{s}_X$ on X for which there exist solutions (A, Φ) satisfying the bound*

$$\mathcal{E}^{\text{top}}(A, \Phi) \leq C. \tag{5.1}$$

(ii) *Suppose (A_n, Φ_n) is a sequence of smooth solutions to the Seiberg–Witten equations (4.9) on X satisfying the same upper bound $\mathcal{E}^{\text{top}}(A_n, \Phi_n) \leq C$. Then there is a sequence of (smooth) 4-dimensional gauge transformations $u_n : X \to S^1$ with the following properties:*

(a) *after passing to a subsequence, the transformed solutions $u_n(A_n, \Phi_n)$ converge weakly in L^2_1 to a (possibly only L^2_1) configuration (A, Φ) on X;*

(b) *if*

$$\limsup_{n \to \infty} \mathcal{E}^{\text{top}}(A_n, \Phi_n) = \mathcal{E}^{\text{top}}(A, \Phi)$$

then the weakly convergent subsequence in the previous statement is convergent in the strong L^2_1 topology on X;

(c) *the weakly convergent subsequence of* (ii)(a) *converges strongly in* C^∞ *on every interior domain* $X' \Subset X$.

Remarks. (a) Note that, by Equation (4.21), in the case that $H = 0$ the hypothesis of the theorem is a lower bound on $\mathcal{L}$ on the boundary. (b) In the case that X is a *closed* 4-manifold, the formula for the topological energy becomes the topological quantity

$$\mathcal{E}^{\text{top}}(A, \Phi) = -\pi^2 c_1^2(S^+)[X],$$

so the first part of the theorem above tells us that, for any C, there are only finitely many spin^c structures $\mathfrak{s}_X$ with

$$c_1^2(S^+)[X] \geq C$$

for which there exist solutions to the Seiberg–Witten equations. The topological quantity on the left appears also in the formula (1.15) for the dimension d of the moduli space, when transversality holds. From the formula for d, we also see that there are only finitely many spin^c structures $\mathfrak{s}_X$ for which the moduli space is non-empty and $d \geq 0$. This basic finiteness result lies behind Proposition 1.5.5 which we stated in Subsection 1.5, concerning the finiteness of the Seiberg–Witten invariants of a closed 4-manifold.

Proof of Theorem 5.1.1. Corollary 4.5.3 tells us that the topological energy of a solution is equal to the analytic energy, so the hypothesis of the theorem provides a uniform upper bound on $\mathcal{E}^{\text{an}}(A_n, \Phi_n)$, and hence uniform upper bounds

$$\int_X |F_{A_n^{\text{t}}}|^2 \leq C_1, \qquad \int_X |\Phi_n|^4 \leq C_2, \qquad \int_X |\nabla_{A_n}\Phi_n|^2 \leq C_3.$$

The L^2 bound on the curvature $F_{A^{\text{t}}}$ gives an upper bound on the pairing of $c_1(\mathfrak{s}_X)$ with any closed form on X, and therefore leaves only finitely many possibilities for $c_1(\mathfrak{s}_X)$. There are therefore only finitely many possibilities for $\mathfrak{s}_X$, as stated in the rider to the theorem. We can therefore restrict our attention to a fixed spin^c structure $\mathfrak{s}_X$.

Pick a smooth spin^c connection A_0 on X, and choose gauge transformations $u_n' : X \to S^1$ so that

$$\begin{aligned} d^*(A_n^{\text{t}} - A_0^{\text{t}} - 2(u_n')^{-1}du_n') &= 0, \quad \text{in } X \\ \langle A_n^{\text{t}} - A_0^{\text{t}} - 2(u_n')^{-1}du_n', \vec{n}\rangle &= 0, \quad \text{at } \partial X. \end{aligned} \tag{5.2}$$

Such a gauge transformation can be found in the form $u'_n = e^{\xi_n}$ by solving the Neumann boundary-value problem for a function $\xi_n : X \to i\mathbb{R}$:

$$2\Delta\xi_n = d^*(A^{\mathrm{t}}_n - A^{\mathrm{t}}_0), \quad \text{in } X$$
$$2\langle d\xi_n, \vec{n}\rangle = \langle A^{\mathrm{t}}_n - A^{\mathrm{t}}_0, \vec{n}\rangle, \quad \text{at } \partial X.$$

If $(A^{\mathrm{t}}_n - A^{\mathrm{t}}_0)$ is smooth the solution, ξ_n, will be smooth as well. In this context, the condition $d * a = 0$ on a 1-form a is referred to as a *Coulomb* condition: when the two conditions (5.2) both hold, we say that the connection $A_n - (u'_n)^{-1}du'_n$ is in *Coulomb–Neumann gauge*, with respect to the chosen base connection A_0. If only the first of the two conditions (5.2) holds, we say that the connection is in *Coulomb gauge* with respect to A_0.

Up to multiplication by a constant, $u'_n = e^{\xi_n}$ is the unique solution to the constraints (5.2) in the trivial homotopy class. In each non-trivial homotopy class, there is a map $v : X \to S^1$ satisfying the homogeneous equation

$$d^*(v^{-1}dv) = 0, \quad \text{in } X$$
$$\langle v^{-1}dv, \vec{n}\rangle = 0, \quad \text{at } \partial X,$$

so a solution u'_n can be found for the equations (5.2) in each homotopy class. If we pick a basis $\{\gamma_1, \ldots, \gamma_b\}$ for $H_1(X;\mathbb{Z})/\text{torsion}$ and pick smooth 3-forms $\{\beta_1, \ldots, \beta_b\}$ representing their Poincaré duals, we can therefore find gauge transformations u_n satisfying (5.2) and in addition the constraint:

$$i\int \beta_r \wedge (A^{\mathrm{t}}_n - A^{\mathrm{t}}_0 - 2u_n^{-1}du_n) \in [0, 2\pi) \qquad (r = 1, \ldots, b). \tag{5.3}$$

The gauge transformation u_n is then unique up to the addition of a constant.

Lemma 5.1.2. *For the class of* 1*-forms* a *on* X *satisfying the boundary condition* $\langle a, \vec{n}\rangle = 0$ *and the constraints* $i\int \beta_r \wedge a \in [0, 2\pi)$ $(r = 1, \ldots, b)$, *there are constants* K_1 *and* K_2 *such that*

$$\int_X (|\nabla a|^2 + |a|^2)d\mathrm{vol} \le K_1 \int_X (|d^*a|^2 + |da|^2)d\mathrm{vol} + K_2.$$

Proof. For a general 1-form a satisfying the boundary conditions, the Weitzenböck formula for the Laplace–Beltrami operator gives

$$\int_X (|\nabla a|^2 + Ricci(a,a))d\mathrm{vol} + \int_{\partial X} N(a,a)d\mathrm{vol} = \int_X (|d^*a|^2 + |da|^2)d\mathrm{vol},$$

because the boundary terms coming from the two integrations by parts cancel except for a term involving N, the second fundamental form. In particular if the boundary is totally geodesic this term vanishes. If not, the following lemma allows us to proceed.

Lemma 5.1.3. *For any $\epsilon > 0$ there is a constant $K_3 > 0$ such that for all 1-forms a we have the following estimate:*

$$\int_{\partial X} |a|^2 d\mathrm{vol} \leq \int_X (\epsilon|\nabla a|^2 + K_3|a|^2) d\mathrm{vol}.$$

Proof. The key point is that the restriction map $C^\infty(X) \to C^\infty(\partial X)$ extends to a bounded linear map

$$L^2_1(X) \to L^2(\partial X),$$

and that this last map is a *compact* linear operator: that is, if f_i is a bounded sequence in $L^2_1(X)$, then the restrictions have a convergent subsequence in $L^2(\partial X)$. This compactness property can be deduced by combining the Sobolev restriction theorem (see Subsection 17.1), which tells us that the restriction map extends to a continuous map from $L^2_1(X)$ to the fractional Sobolev space $L^2_{1/2}(\partial X)$, with the Rellich lemma, which tells us that the inclusion of $L^2_{1/2}(\partial X)$ in $L^2(\partial X)$ is compact.

With this as background, the result can now be proved by arguing by contradiction. Fixing $\epsilon > 0$, let us suppose that no such K_3 exists. This means that we can find a_i contradicting the inequality for $K_3 = i$: that is, we have

$$\int_{\partial X} |a_i|^2 d\mathrm{vol} \geq \int_X (\epsilon|\nabla a_i|^2 + i|a_i|^2) d\mathrm{vol}.$$

Rescale and rename this sequence so that $\|a_i\|_{L^2_1(X)} = 1$. It cannot be the case that some subsequence has $\|a_i\|_{L^2(X)}$ bounded away from 0 for this would imply that $\int_{\partial X} |a_i|^2 d\mathrm{vol} \to \infty$ contradicting continuity of the restriction map. So it must be the case that $\|a_i\|_{L^2} \to 0$ and $\|\nabla a_i\|_{L^2} \to 1$ as $i \to \infty$. Thus we must have that $\int_{\partial X} |a_i|^2 d\mathrm{vol} \geq \epsilon$. We can pass to a subsequence converging weakly in L^2_1 on X. The weak limit must be zero since it is converging strongly in $L^2(X)$ to 0. So by the compactness of the restriction map we can pass to a subsequence where the sequence of restrictions converges strongly in $L^2(\partial X)$ to 0, contradicting the uniform bound $\int_{\partial X} |a_i|^2 d\mathrm{vol} \geq \epsilon$. □

Thus we only have to bound the L^2 norm of a to complete the proof of Lemma 5.1.2. If the lemma is false, there is a sequence $\{a_i\}$ satisfying the hypotheses of the lemma, with unbounded L^2 norm, and with

$$\int_X (|d^* a_i|^2 + |da|^2) \Big/ \int_X (|a_i|^2) \to 0.$$

Then a rescaled sequence a_i' has L^2 norm 1, bounded L_1^2 norm and has

$$\int \beta_r \wedge a_i' \to 0,$$

$$\int_X (|d^* a_i|^2 + |da|^2) \to 0.$$

A subsequence would converge strongly in L^2 and weakly in L_1^2 to a 1-form a_∞ satisfying the homogeneous equations $da = 0$ and $d^* a = 0$, satisfying the boundary conditions, and having zero periods on all the loops γ_r. By considering a primitive for a_∞, which would be a harmonic function satisfying Neumann boundary conditions, one sees that a_∞ would be zero, in contradiction to it being a strong limit of forms with L^2 norm 1. □

We can now prove Assertion (ii)(a) of the theorem. We write

$$\begin{aligned}(\tilde{A}_n, \tilde{\Phi}_n) &= u_n(A_n, \Phi_n) \\ &= (A_n - u_n^{-1} du_n, u_n \Phi_n),\end{aligned}$$

where u_n are the unique gauge transformations satisfying the conditions (5.2) and (5.3). Lemma 5.1.2 provides an L_1^2 bound on $\tilde{A}_n^{\mathrm{t}} - A_0^{\mathrm{t}}$, and hence by the Sobolev embedding theorem we obtain an L^4 bound on $\tilde{A}_n^{\mathrm{t}} - A_0^{\mathrm{t}}$. We also have an L^4 bound on $\tilde{\Phi}_n$ and an L^2 bound on $\nabla_{\tilde{A}_n} \tilde{\Phi}_n$. These last two give us an L^2 bound on $\nabla_{A_0} \tilde{\Phi}_n$, because we have

$$\nabla_{A_0} \tilde{\Phi}_n = \nabla_{\tilde{A}_n} \tilde{\Phi}_n + (\tilde{A}_n - A_0) \tilde{\Phi}_n$$

and the L^2 norm of the last term can be bounded using the inequality

$$\|fg\|_{L^2} \le \|f\|_{L^4} \|g\|_{L^4}.$$

So we have L_1^2 bounds on both $\tilde{\Phi}_n$ and $\tilde{A}_n - A_0$. It follows that we can pass to a subsequence converging weakly in L_1^2 to some L_1^2 configuration (A, Φ).

Suppose the hypothesis of Part (ii)(b) of the theorem holds:

$$\limsup_{n \to \infty} \mathcal{E}^{\mathrm{top}}(A_n, \Phi_n) = \mathcal{E}^{\mathrm{top}}(A, \Phi).$$

Since $(\tilde{A}_n, \tilde{\Phi}_n)$ are solutions we have that

$$\mathcal{E}^{\text{top}}(\tilde{A}_n, \tilde{\Phi}_n) = \mathcal{E}^{\text{an}}(\tilde{A}_n, \tilde{\Phi}_n).$$

We can pass to a further subsequence where $F^+_{\tilde{A}_n}$, $\nabla_{\tilde{A}_n}\tilde{\Phi}_n$ and $(\tilde{\Phi}_n\tilde{\Phi}_n^*)_0$ converge weakly in L^2 while $(\tilde{A}_n, \tilde{\Phi}_n)$ converges strongly in L^2. Since weak limits are preserved under linear maps, the weak limit of $F^+_{\tilde{A}_n}$ is F^+_A. We shall show that the weak limit of $\nabla_{\tilde{A}_n}\tilde{\Phi}_n$ is $\nabla_A\Phi$ and that of $(\tilde{\Phi}_n\tilde{\Phi}_n^*)_0$ is $(\Phi\Phi^*)_0$.

To show the first of these, write $\tilde{A}_n = A_0 + a_n$ for some basepoint A_0. Then we know that $\nabla_{A_0}\tilde{\Phi}_n$ converges weakly to $\nabla_{A_0}\Phi$ and that $a_n\tilde{\Phi}_n$ converges weakly to some limit. Since a_n and $\tilde{\Phi}_n$ converge strongly in L^2 it follows that $a_n\tilde{\Phi}_n$ converges strongly in L^1 to $a\Phi$ giving the desired conclusion. Similar reasoning gives the result for $(\tilde{\Phi}_n\tilde{\Phi}_n^*)_0$. Notice that we can conclude that the weak limit is a solution to the Seiberg–Witten equations at this point, since the weak limit of $0 = D_{\tilde{A}_n}\tilde{\Phi}_n$ is $D_A\Phi$ and the weak limit of $0 = \rho(F^+_{\tilde{A}_n}) - (\tilde{\Phi}_n\tilde{\Phi}_n^*)_0$ is $\rho(F^+_A) - (\Phi\Phi^*)_0$.

If $\{x_n\}$ is a weakly convergent sequence in a Banach space, then the norm of the limit x can only be smaller than or equal to the limit inferior of the norms of the elements of the sequence:

$$\limsup_{n\to\infty} \|x_n\| \geq \liminf_{n\to\infty} \|x_n\| \geq \|x\|;$$

and furthermore, in the case of a Hilbert space or L^p space with $1 < p < \infty$, if all three of these are equal, then the convergence is strong. In our situation, this means that in any event

$$\limsup_{n\to\infty} \mathcal{E}^{\text{an}}(\tilde{A}_n, \tilde{\Phi}_n) \geq \mathcal{E}^{\text{an}}(A, \Phi);$$

and our assumption in Part (ii)(b) implies equality here. So it must be that $F^+_{\tilde{A}_n}$, $\nabla_{\tilde{A}_n}\tilde{\Phi}_n$ and $(\tilde{\Phi}_n\tilde{\Phi}_n^*)_0$ all converge strongly in L^2. The first of these implies that $\tilde{A}_n$ converges strongly in L^2_1. The third implies that

$$\lim_{n\to\infty} \|\tilde{\Phi}_n\|_{L^4(X)} = \|\Phi\|_{L^4(X)}$$

and hence that $\tilde{\Phi}_n$ converges strongly in L^4 to Φ. Once we know these two facts, the L^2 convergence of $\nabla_{\tilde{A}_n}\tilde{\Phi}_n$ to $\nabla_A\Phi$ implies the L^2 convergence of $\nabla_{A_0}\tilde{\Phi}_n$ to $\nabla_{A_0}\Phi$, so that $\tilde{\Phi}_n$ converges strongly in L^2_{1,A_0}.

It remains to prove Part (ii)(c). We will prove this statement in two steps: first we prove that a strongly L^2_1-convergent sequence in Coulomb gauge converges in the C^∞ topology on interior domains; then we will prove that on any interior domain, the condition in Assertion (ii)(b) holds, so we can deduce L^2_1 and hence C^∞ convergence on interior domains.

For the first of these two steps, we obtain bounds on higher Sobolev norms by a process called bootstrapping. To keep the notation to a minimum we shall write the monopole equations together with the Coulomb gauge condition as

$$D\gamma + \gamma \sharp \gamma = b. \tag{5.4}$$

Here D is the differential operator

$$D : C^\infty(X; iT^*X \oplus S^+) \to C^\infty(X; i\mathbb{R} \oplus i\,\mathfrak{su}(S^+) \oplus S^-)$$

given by $D(a, \phi) = (d^*a, \rho(d^+a), D_{A_0}\phi)$ and $\sharp$ denotes a bilinear operator involving only pointwise multiplication. The key point is that the operator is *elliptic*, which allows us to appeal to the Gårding inequality (see [53]):

Theorem 5.1.4 (Gårding inequality). *If D is a first-order elliptic operator with smooth coefficients on a manifold X, possibly non-compact, and $X^{(1)}$ is an open subset with compact closure, then there is a constant $C > 0$ such that for any smooth γ we have*

$$\|\gamma\|_{L^p_{k+1}(X^{(1)})} \le C\big(\|D\gamma\|_{L^p_k(X)} + \|\gamma\|_{L^p(X)}\big).$$

□

We apply the Gårding inequality repeatedly to prove the following lemma.

Lemma 5.1.5. *Suppose that $\{\gamma_n\}$ is a sequence of smooth solutions to the equation (5.4) on X, a compact 4-manifold with boundary, and suppose that γ_n converges strongly in L^2_1 to γ. Then $\{\gamma_n\}$ converges in the C^∞ topology on every open subset compactly contained in the interior of X. In particular, γ is smooth in the interior of X.*

Proof. Choose a cut-off function β, with $\beta = 1$ on an interior domain X', and with the support of β compactly contained in the interior of X. Fix $\epsilon > 0$ and choose i_0 so that for all $i \ge i_0$ we have $\|\gamma_i - \gamma_{i_0}\|_{L^2_1} \le \epsilon$. Then the Gårding inequality (applied to the support of β) tells us that we have

$$\begin{aligned}
&\|\beta(\gamma_i - \gamma_j)\|_{L^3_1(X)} \\
&\quad \le C\big(\|D(\beta(\gamma_i - \gamma_j))\|_{L^3(X)} + \|\beta(\gamma_i - \gamma_j)\|_{L^3(X)}\big) \\
&\quad \le C\big(\|\beta(\gamma_i - \gamma_j) \sharp (\gamma_i + \gamma_j)\|_{L^3(X)} + \|\sigma(D, d\beta)(\gamma_i - \gamma_j)\|_{L^3(X)} \\
&\qquad + \|\beta(\gamma_i - \gamma_j)\|_{L^3(X)}\big) \\
&\quad \le CC_{\text{mult}}\|\beta(\gamma_i - \gamma_j)\|_{L^3_1}\|\gamma_i + \gamma_j - 2\gamma_{i_0}\|_{L^2_1(X)} \\
&\qquad + C\|2\gamma_{i_0}\|_{C^0}\|\beta(\gamma_i - \gamma_j)\|_{L^3(X)} + C_1\|\gamma_i - \gamma_j\|_{L^3(X)}.
\end{aligned}$$

In the above formulae, σ is the symbol of the operator D. We have used the continuity of the multiplication $L^2_1 \times L^3_1 \to L^3$ and C_{mult} is the norm of this multiplication. We can now choose ϵ (and hence i_0) so that for $i, j \ge i_0$ we have $CC_{\text{mult}}\epsilon \le \frac{1}{4}$. Then we can rearrange this inequality to read

$$\begin{aligned}
\|\beta(\gamma_i - \gamma_j)\|_{L^3_1(X)} &\le 2\big(C_1 + C\|2\gamma_{i_0}\|_{C^0}\big)\|\gamma_i - \gamma_j\|_{L^3(X)} \\
&\le C_2\|\gamma_i - \gamma_j\|_{L^2_1(X)}.
\end{aligned}$$

Since β is arbitrary we conclude that $\{\gamma_i\}$ converges to γ in $L^3_{1,\text{loc}}(\text{int}X)$: that is, γ converges in L^3_1 on any compact subset of the interior. We can run through this argument twice more: first to prove convergence in L^2_2 norm, using the continuity of the multiplication $L^2_2 \times L^3_1 \to L^2_1$ in place of $L^2_1 \times L^3_1 \to L^3$, and a C^1 bound on γ_{i_0} in place of a C^0 bound; and second to prove convergence in L^2_3, using the continuity of the multiplication $L^2_3 \times L^2_2 \to L^2_2$ and a C^2 bound on γ_{i_0}. In each of these two iterations, we pass to a smaller subdomain (with appropriate change in β). Once we know that γ_i converges in $L^2_3(X')$ for each $X' \Subset X$ we can exploit the fact that L^2_k is an algebra for $k \ge 3$: with β supported in the interior of X' and with $\beta = 1$ on some $X'' \Subset X'$, we have

$$\begin{aligned}
&\|\beta(\gamma_i - \gamma_j)\|_{L^2_{k+1}(X')} \\
&\quad \le C\big(\|D(\beta(\gamma_i - \gamma_j))\|_{L^2_k(X')} + \|\beta(\gamma_i - \gamma_j)\|_{L^2(X)}\big) \\
&\quad \le C\big(\|\beta(\gamma_i - \gamma_j) \sharp (\gamma_i + \gamma_j)\|_{L^2_k(X')} + \|\sigma(D, d\beta)(\gamma_i - \gamma_j)\|_{L^2_k(X')} \\
&\qquad + \|\beta(\gamma_i - \gamma_j)\|_{L^2(X')}\big) \\
&\quad \le C\big(C_1\|\beta(\gamma_i - \gamma_j)\|_{L^2_k(X')}\|\gamma_i + \gamma_j\|_{L^2_k(X')} \\
&\qquad + \|\sigma(D, d\beta)(\gamma_i - \gamma_j)\|_{L^2_k(X')} + \|\beta(\gamma_i - \gamma_j)\|_{L^2(X')}\big),
\end{aligned}$$

which shows that $\{\gamma_i\}$ is Cauchy in $L^2_{k+1}(X'')$. Thus for all k and all $X' \Subset X$, the sequence is Cauchy in $L^2_k(X')$. This proves the lemma. □

To finish the proof of Part (ii)(c), we need to show that for every $X' \Subset X$, where X' is the interior of a manifold with boundary, we can pass to a subsequence where

$$\lim_{n\to\infty} \mathcal{E}^{\mathrm{an}}_{X'}(\tilde{A}_n, \tilde{\Phi}_n) = \mathcal{E}^{\mathrm{an}}_{X'}(A, \Phi). \tag{5.5}$$

We will give the proof only in the case that the boundary of X is a metric cylinder: that is, there is an open subset X'' of X and an isometry of $X \setminus X''$ with $[-\epsilon, 0] \times Y$, so the boundary of X is identified with $\{0\} \times Y$. (The general case is a slight modification.) Let $X_s \subset X$ be the part of X corresponding to the union of X'' and $[-\epsilon, s] \times Y$.

Since the configurations (A_n, Φ_n) are smooth, the functions

$$f_n(s) = \mathcal{E}^{\mathrm{an}}_{X_s}(A_n, \Phi_n)$$

on $[-\epsilon, 0]$ are smooth. They are uniformly bounded above and below, and the sequence of derivatives

$$f_n'(s)$$

are non-negative functions with uniformly bounded integrals. It follows that for M large enough, the measure of the compact set $f_n' \le M$ is bounded away from zero, by a constant δ independent of n.

Lemma 5.1.6. *Let $\{S_\alpha\}_{\alpha\in A}$ be a collection of measurable subsets of an interval $[a, b]$ with measure $\mu(S_\alpha) \ge \delta$ for some $\delta > 0$ and all α. Then there is an infinite subset B of A such that the intersection*

$$\bigcap_{\alpha\in B} S_\alpha$$

is non-empty.

Proof. This is a consequence of the monotone convergence theorem, applied to the characteristic functions of the decreasing family of sets $T_n = \bigcup_{n\ge k} S_{\alpha_k}$, where α_k is any infinite sequence taken from A. If the conclusion of the lemma does not hold, then these characteristic functions converge pointwise to zero, while their integrals remain bounded below by δ. □

Applying this lemma to the subsets $f_n' \leq M$ we can find $s_0 \in [-\epsilon, 0]$ such that for all n in a subsequence we have

$$f_n'(s_0) \leq M\,.$$

On the other hand, as we have seen,

$$\begin{aligned} f_n'(s_0) &= -\frac{d}{ds}\mathcal{L}(\gamma_n|_{\{s\}\times Y})(s_0) \\ &= \|\operatorname{grad}_{\gamma_n(s_0)}\mathcal{L}\|^2. \end{aligned}$$

To continue we need the following lemma.

Lemma 5.1.7. *If (B_n, Ψ_n) is a sequence of configurations on Y with*

$$\|\operatorname{grad}_{(B,\Psi)}\mathcal{L}\|^2_{L^2(Y)} \leq M\,,$$

then there is a sequence of gauge transformations v_n such that $v_n(B_n, \Psi_n)$ converges in the $L^2_{1/2}$ topology, to an L^2_1 configuration.

Proof. Put (B, Ψ) into Coulomb gauge on Y with respect to a basepoint B_0 and require that the 3-dimensional version of the period condition (5.3) holds as well. Then $\|\operatorname{grad}_{(B,\Psi)}\mathcal{L}\|^2_{L^2(Y)}$ controls the L^2_1 norm, so we can pass to a weakly convergent subsequence. The compactness of the inclusion $L^2_1 \subset L^2_{1/2}$ implies that we can pass to a further subsequence converging strongly in the $L^2_{1/2}$ topology. □

By this lemma we find gauge transformations v_n on $s_0 \times Y$ such that $v_n(A_n, \Phi_n)$ is uniformly bounded in L^2_1 norm on $s_0 \times Y$ and such that a subsequence converges strongly in the $L^2_{1/2}$ topology. If the first Chern class of the spin^c bundle on Y is torsion then $\mathcal{L}$ is fully gauge-invariant, and is continuous in the $L^2_{1/2}$ topology; so we can conclude that $f_n(s_0)$ is convergent in this subsequence, and (5.5) holds, which completes the argument and proves the theorem. In general, if the first Chern class is not torsion, then we argue as follows. Write B_0 for the restriction of A_0 to ∂X_{s_0} and write $B_0 + b_n \otimes 1_S$ for the restriction of $\tilde{A}_n$ (our weakly L^2_1-converging sequence of connections on X.) Writing the Chern–Simons–Dirac functional on our sequence in these terms we have:

$$\mathcal{L}(B_n, \tilde{\Phi}_n) = -\frac{1}{8}\int_Y (2b_n) \wedge (2F_{B_0^t} + 2db_n) + \frac{1}{2}\int_Y \langle D_{B_n}\tilde{\Phi}_n, \tilde{\Phi}_n\rangle\, d\mathrm{vol}.$$

The strong $L^2_{1/2}$ convergence up to gauge transformation implies the convergence of

$$-\frac{1}{8}\int_Y (2b_n \wedge 2db_n) + \frac{1}{2}\int_Y \langle D_{B_n}\tilde{\Phi}_n, \tilde{\Phi}_n\rangle \, d\text{vol}.$$

The only term above that is not fully gauge-invariant is

$$-\frac{1}{8}\int_Y (2b_n) \wedge (2F_{B_0^t}).$$

Since the sequence b_n is weakly $L^2_{1/2}$-convergent, we can pass to a further subsequence where the b_n converge strongly in L^2, in which case this term converges as well. This completes the proof of Theorem 5.1.1. □

We can apply this theorem to the case that the 4-manifold is a cylinder, to deduce:

Corollary 5.1.8. *Let $\gamma_n \in \mathcal{C}(Z, \mathfrak{s}_Z)$ be a sequence of solutions of the 4-dimensional Seiberg–Witten equations, $\mathfrak{F}(\gamma) = 0$, on a cylinder $Z = [t_1, t_2] \times Y$. Suppose that the drop in the Chern–Simons–Dirac functional is uniformly bounded:*

$$\mathcal{L}(\check{\gamma}_n(t_1)) - \mathcal{L}(\check{\gamma}_n(t_2)) \le C.$$

Then there is a sequence of gauge transformations, $u_n \in \mathcal{G}(Z)$, such that, after passing to a subsequence, the transformed solutions $u_n(\gamma_n)$ converge in the C^∞ topology on $[t_1', t_2'] \times Y$, for any interval $[t_1', t_2']$ in the interior of $[t_1, t_2]$. Solutions satisfying this bound can exist only for finitely many spin^c structures on Y. □

Remark. The corollary provides 4-dimensional gauge transformations $u_n \in \mathcal{G}(Z)$, such that the transformed 4-dimensional connections and spinors, $\tilde{\gamma}_n = (\tilde{A}_n, \tilde{\Phi}_n)$ converge. If $\gamma_n = (A_n, \Phi_n)$ is in temporal gauge for each n (so that $\check{\gamma}_n$ is a solution to the gradient-flow equations), then in fact we can take the gauge transformations to be time-independent. Indeed, let v_n be the time-independent gauge transformation obtained from evaluating u_n at $\bar{t} = (t_1 + t_2)/2$ (or any other intermediate point). To show that $v_n(B_n, \Psi_n)$ also converges, it is enough to show that the quotient $v_n u_n^{-1}$ converges in C^∞ on interior domains. This quotient is equal to 1 at time $\bar{t}$ and transforms $\tilde{A}_n^t$ to a connection with no $\partial/\partial t$ component. It is therefore e^{ξ_n}, where ξ_n is the unique solution on $[t_1, t_2] \times Y$ to the equation

$$2d\xi_n = (\tilde{A}_n^{\mathrm{t}} - A_0^{\mathrm{t}})_{\partial/\partial t}$$
$$\xi_n(\bar{t}) = 0,$$

where A_0 is any connection with no $\partial/\partial t$ component. The ξ_n converge in C^∞ on interior intervals because the connections $\tilde{A}_n^{\mathrm{t}}$ do.

5.2 Properness properties of the Seiberg–Witten map

The ideas that go into proving the compactness of the moduli space can be extended to show that on a closed manifold the Seiberg–Witten map (together with gauge fixing) is a proper map (for example when the domain and range have the C^∞ topology. In the case of a manifold with boundary if we include the behavior of the Chern–Simons–Dirac functional then we can prove a subset in the configuration space with compact image under the Seiberg–Witten map in the C^∞ topology and bounded Chern–Simons–Dirac functional has compact image under the restrictions to interior domains $X' \Subset X$. These results are important for proving compactness of perturbations of the Seiberg–Witten equations.

Theorem 5.2.1. *Let X be a compact Riemannian 4-manifold with boundary. Then the following hold.*

(i) *For given C, there are only finitely many spin^c structures $\mathfrak{s}_X$ on X for which there exist configurations (A, Φ) satisfying the bounds*

$$\begin{aligned} \mathcal{E}^{\mathrm{top}}(A, \Phi) &\leq C \\ \|\mathfrak{F}(A, \Phi)\|_{L^2(X)} &\leq C. \end{aligned} \tag{5.6}$$

(ii) *Furthermore, if (A_n, Φ_n) is a sequence of configurations satisfying the topological energy bound $\mathcal{E}^{\mathrm{top}}(A_n, \Phi_n) \leq C$, the gauge-fixing conditions* (5.2) *and period conditions* (5.3)*, then the following holds.*
   (a) *If the sequence $\mathfrak{F}(A_n, \Phi_n)$ converges in the L^2 topology on X , then a subsequence of the sequence (A_n, Φ_n) converges weakly in the L^2_1 topology, to an L^2_1 configuration (A, Φ) on X, still satisfying the topological energy bound, the gauge-fixing condition and the period conditions (but possibly with the periods in* $[0, 2\pi]$*).*
   (b) *If in addition*

$$\limsup_{n\to\infty} \mathcal{E}^{\mathrm{top}}(A_n, \Phi_n) = \mathcal{E}^{\mathrm{top}}(A, \Phi),$$

then the subsequence converges strongly in L^2_1.

(c) *Finally, if $\mathfrak{F}(A_n, \Phi_n)$ converges in the C^∞ topology then the subsequence (A_n, Φ_n) converges in the C^∞ topology on every interior domain $X' \Subset X$. Indeed, for any $j \geq 1$, if $\mathfrak{F}(A_n, \Phi_n)$ converges in the L^2_{j-1} topology then (A_n, Φ_n) converges in the L^2_j topology on every interior domain $X' \Subset X$*

Proof. The finiteness statement is again simply the fact the two estimates control the analytic energy and hence the L^2 norm of the curvature form.

To prove (ii)(a), we first pass to a subsequence in a fixed spin^c structure. The convergence of the values of the Seiberg–Witten map implies again a uniform L^2_1 bound on the configuration in the sequence, and we can pass to a weakly convergent subsequence. Let (A, Φ) denote the weak limit. We can assume, as above, strong L^2 convergence of the sequence, as well as weak L^2 convergence of F_{A_n}, $(\Phi_n\Phi_n^*)$ and $\nabla_{A_n}\Phi_n$ to the corresponding quantities F_A, $(\Phi\Phi^*)$ and $\nabla_A\Phi$ for the weak limit. The Coulomb–Neumann gauge condition will be preserved under the weak limit; the period condition may not quite be preserved, because of the half-open interval condition, but will satisfy the closure interval conditions listed in the theorem.

If the hypotheses of both (ii)(a) and (ii)(b) hold then from (4.16) we have that

$$\liminf_{n\mapsto\infty} \mathcal{E}^{\mathrm{an}}(A_n, \Phi_n) = \mathcal{E}^{\mathrm{an}}(A, \Phi).$$

We can now conclude as we did in the proof of Theorem 5.1.1 that after passing to a subsequence the L^2 norms of $F_{A_n^t}$, $\nabla_{A_n}\Phi_n$, and $D_{A_n}\Phi_n$ do not drop in the limit so these sequences converge strongly in L^2. Together with the gauge-fixing conditions, this implies the strong L^2_1 convergence of (A_n, Φ_n) also as in the proof of Theorem 5.1.1. The proof of (ii)(c) similarly follows the same line of argument as the corresponding statement in Theorem 5.1.1. □

6 The blown-up configuration space

6.1 Blowing up

In Subsection 2.5 we emphasized the role of the blow-up of a manifold as an important tool in dealing with Morse theory on a manifold with a circle action. In this section we will begin setting this up in the infinite-dimensional setting of the Seiberg–Witten equations. Recall that a configuration (A, Φ) in $\mathcal{C}(Y, \mathfrak{s})$

is *reducible* if Φ is zero: we write

$$\mathcal{C}^*(Y,\mathfrak{s}) \subset \mathcal{C}(Y,\mathfrak{s}) \tag{6.1}$$

for the space of *irreducible* configurations (see (1.16) in Subsection 1.5). The blown-up configuration space will be constructed by blowing up $\mathcal{C}(Y,\mathfrak{s})$ along the locus of reducibles, and we shall make a similar definition for the case of a 4-manifold.

For any topological vector space V equipped with a continuous norm (not necessarily defining the same topology), we write $\mathbb{S}(V)$ for the associated sphere

$$\mathbb{S}(V) = \{\, v \in V \mid \|v\| = 1 \,\}.$$

The continuity of the norm ensures that this sphere is naturally identified with the quotient space

$$V \setminus \{0\}/\mathbb{R}^+.$$

Thus we can define the (oriented real) blow-up of a normed topological vector space as

$$V^\sigma = \mathbb{R}^{\geq} \times \mathbb{S}(V)$$

as we did for a finite-dimensional vector space (2.13).

For a 4-manifold X, the blown-up Seiberg–Witten configuration space is defined to be:

$$\begin{aligned} \mathcal{C}^\sigma(X,\mathfrak{s}_X) &= \mathcal{A}(X,\mathfrak{s}_X) \times \mathbb{R}^{\geq} \times \mathbb{S}(C^\infty(X;S^+)) \\ &\subset \mathcal{A}(X,\mathfrak{s}_X) \times \mathbb{R}^{\geq} \times C^\infty(X;S^+); \end{aligned} \tag{6.2}$$

or more explicitly, it is the space of triples

$$\mathcal{C}^\sigma(X,\mathfrak{s}_X) = \{\, (A,s,\phi) \mid \|\phi\|_{L^2} = 1,\ s \geq 0 \,\}.$$

The "blow-down" map $\pi : \mathcal{C}^\sigma(X,\mathfrak{s}_X) \to \mathcal{C}(X,\mathfrak{s}_X)$ is given by

$$\pi(A,s,\phi) = (A,s\phi)$$

and is a bijection over the open set $\mathcal{C}^*(X,\mathfrak{s}_X) \subset \mathcal{C}(X,\mathfrak{s}_X)$ of configurations with Φ non-zero. The fiber over $(A,0)$ is the sphere $\mathbb{S}(C^\infty(X;S^+))$. The notation

$\mathcal{C}^\sigma(X, \mathfrak{s}_X)$ is meant to suggest the term "σ-process", as it is used in algebraic geometry.

Recall that we have regarded the Seiberg–Witten equations on X as being expressed by the vanishing of a map

$$\mathfrak{F} : \mathcal{C}(X, \mathfrak{s}_X) \to C^\infty(X; i\,\mathfrak{su}(S^+) \oplus S^-).$$

With a slight change of viewpoint, we now define $\mathcal{V}(X, \mathfrak{s}_X)$ as a trivial bundle over $\mathcal{C}(X, \mathfrak{s}_X)$ with fiber the right-hand side above, so that we may regard $\mathfrak{F}$ as a *section* of this bundle:

$$\mathfrak{F} : \mathcal{C}(X, \mathfrak{s}_X) \to \mathcal{V}(X, \mathfrak{s}_X).$$

The bundle $\mathcal{V}(X, \mathfrak{s}_X)$ is a direct sum,

$$\mathcal{V}(X, \mathfrak{s}_X) = \mathcal{V}^0 \oplus \mathcal{V}^1,$$

with the fibers of the summands being $C^\infty(X, i\,\mathfrak{su}(S^+))$ and $C^\infty(X, S^-)$ respectively. Let

$$\mathcal{V}^\sigma(X, \mathfrak{s}_X) \to \mathcal{C}^\sigma(X, \mathfrak{s}_X) \tag{6.3}$$

be defined as the pull-back under π of $\mathcal{V}(X, \mathfrak{s}_X)$. Finally, we define a section

$$\mathfrak{F}^\sigma : \mathcal{C}^\sigma(X, \mathfrak{s}_X) \to \mathcal{V}^\sigma(X, \mathfrak{s}_X) \tag{6.4}$$

by the formula

$$\mathfrak{F}^\sigma(A, s, \phi) = \left(\frac{1}{2}\rho_X(F^+_{A^t}) - s^2(\phi\phi^*)_0,\ D^+_A\phi \right). \tag{6.5}$$

Notice that if s is not zero, the vanishing of $\mathfrak{F}^\sigma(A, s, \phi)$ is equivalent to the vanishing of $\mathfrak{F}(A, s\phi)$, while if s is zero it is equivalent to the two conditions

$$\mathfrak{F}(A, 0) = 0$$
$$D^+_A\phi = 0.$$

Notice also that $\mathfrak{F}^\sigma$ is a continuous section. The gauge group $\mathcal{G}(X)$ acts on $\mathcal{C}^\sigma(X, \mathfrak{s}_X)$ and on the vector bundle $\mathcal{V}^\sigma(X, \mathfrak{s}_X)$. The section $\mathfrak{F}^\sigma$ is equivariant.

Remarks. (i) The construction of the section $\mathfrak{F}^\sigma$ on the blow-up from the section $\mathfrak{F}$ on the space $\mathcal{C}(X, \mathfrak{s}_X)$ is a natural one that is somewhat disguised by our choice of description of the blow-up, which made use of the L^2 norm. Although our chosen description is well-suited to the calculations we need to make, we could alternatively adopt the viewpoint suggested above, that $\mathbb{S}(V)$ is the quotient of $V \setminus \{0\}$ by $\mathbb{R}^+$, and so regard the real blow-up of $C^\infty(X, S^+)$ as the tautological space of pairs

$$\{ (\mathbb{R}^+\phi, \Phi) \in \mathbb{S}(C^\infty(X, S^+)) \times C^\infty(X, S^+) \mid \Phi \in \mathbb{R}^{\geq}\phi \}.$$

With this notation, $\mathcal{C}^\sigma(X, \mathfrak{s}_X)$ is a space of triples $(A, \mathbb{R}^+\phi, \Phi)$. On the blow-up there is a tautological complex line bundle $\mathcal{O}(-1)$, whose fiber at $(A, \mathbb{R}^+\phi, \Phi)$ is $\mathbb{C}\phi$. The natural construction for the section $\mathfrak{F}^\sigma$ then makes it a section of the bundle

$$\mathcal{V}^\sigma = \mathcal{O}(-1)^* \otimes \pi^*(\mathcal{V}) \tag{6.6}$$

given by

$$\mathfrak{F}^\sigma : \mathcal{O}(-1) \to \pi^*(\mathcal{V})$$
$$\mathfrak{F}^\sigma(A, \mathbb{R}^+\phi, \Phi)(\psi) = \left(\frac{1}{2}\rho(F^+_{A^t}) - (\Phi\Phi^*)_0, D^+_A \psi\right).$$

The use of the L^2 norm effectively trivializes the bundle $\mathcal{O}(-1)$, and thus disguises the relation between $\mathcal{V}$ and $\mathcal{V}^\sigma$. Note, in particular, that $\mathfrak{F}^\sigma$ is not the pull-back section.

(ii) Later, we will replace $\mathcal{C}(X, \mathfrak{s}_X)$ etc. with a suitable Banach space completion. At that time, we will also introduce a version of $\mathcal{C}^\sigma(X, \mathfrak{s}_X)$ that is a Banach manifold with boundary, and we will consider the differentiability of (a perturbation of) $\mathfrak{F}^\sigma$ as a section of a Banach vector bundle. At present, our discussion is simply topological.

In a similar vein, we introduce the space

$$\mathcal{C}^\sigma(Y, \mathfrak{s}) = \mathcal{A}(Y, \mathfrak{s}) \times \mathbb{R}^{\geq} \times \mathbb{S}(C^\infty(Y, S))$$

as the (oriented, real) blow-up of $\mathcal{C}(Y, \mathfrak{s})$ along the locus $\Psi = 0$.

If $X' \subset X$ is an interior domain, there is a partially defined restriction map

$$r : \mathcal{C}^\sigma(X, \mathfrak{s}_X) \dashrightarrow \mathcal{C}^\sigma(X', \mathfrak{s}_X).$$

Its domain of definition is the set of configurations $\gamma^\sigma = (A, s, \phi)$ where $\phi|_{X'} \neq 0$, when we are using the unit sphere model for the blow-up the restriction map is given by

$$r(A, s, \phi) = (A, s\|\phi\|_{L^2(X')}, \phi/\|\phi\|_{L^2(X')}).$$

(We will sometimes use this notation, γ^σ in place of γ, for a typical point of the blown-up configuration space, when it is useful to make the distinction. Typically we will drop the σ.) We will prove a unique continuation result, Proposition 7.1.4, which implies that any solution of $\mathfrak{F}^\sigma = 0$ is in this domain: the point is that, for a solution of the equations, ϕ cannot vanish on an open set without vanishing identically. In a similar spirit, in the case that the 4-manifold is a cylinder $Z = I \times Y$, there is no restriction map to the slices $\{t\} \times Y$ *in general*; but if $\gamma^\sigma \in \mathcal{C}^\sigma(Z, \mathfrak{s}_Z)$ is a solution of $\mathfrak{F}^\sigma(\gamma^\sigma) = 0$, then by a unique continuation result, the restriction to each slice *is* defined (see Proposition 7.1.2 below). We write $\check{\gamma}^\sigma$ for the corresponding path in $\mathcal{C}^\sigma(Y, \mathfrak{s})$.

6.2 The blown-up equations as a flow

Suppose $\gamma^\sigma = (A, s, \phi) \in \mathcal{C}^\sigma(Z, \mathfrak{s}_Z)$ is a solution of $\mathfrak{F}^\sigma(\gamma^\sigma) = 0$ on the cylinder $Z = I \times Y$, and let $\check{\gamma}^\sigma(t)$ be the corresponding path in $\mathcal{C}^\sigma(Y, \mathfrak{s})$. We wish to write down an equation satisfied by this path. Using the L^2 unit sphere model for $\mathbb{S}$ we have

$$\begin{aligned}\mathcal{C}^\sigma(Z, \mathfrak{s}_Z) &= \mathcal{A}(Z, \mathfrak{s}_Z) \times [0, \infty) \times \mathbb{S}(C^\infty(Z; S^+)) \\ &= \{\, (A, s, \phi) \mid s \geq 0, \|\phi\|_{L^2} = 1 \,\}.\end{aligned}$$

Similarly

$$\begin{aligned}\mathcal{C}^\sigma(Y, \mathfrak{s}) &= \mathcal{A}(Y, \mathfrak{s}) \times [0, \infty) \times \mathbb{S}(C^\infty(Y; S)) \\ &= \{\, (B, r, \psi) \mid r \geq 0, \|\psi\|_{L^2} = 1 \,\}.\end{aligned}$$

In these coordinates, the path $\check{\gamma}^\sigma$ is given by $(B(t), r(t), \psi(t))$, where

$$\begin{aligned}B(t) &= \check{A}(t) \\ r(t) &= s\|\check{\phi}(t)\|_{L^2(Y)} \\ \psi(t) &= \check{\phi}(t)/\|\check{\phi}(t)\|_{L^2(Y)}.\end{aligned}$$

Note that the function $\mathcal{L}$ is everywhere defined on $\mathcal{C}^\sigma(Y,\mathfrak{s})$. Also of importance in the discussion is the function

$$\Lambda(B,r,\psi) = \langle \psi, D_B\psi \rangle_{L^2(Y)}. \tag{6.7}$$

This function generalizes the function $\Lambda_q(\phi)$ of Equation (2.18).

Lemma 6.2.1. *An element γ^σ in $\mathcal{C}^\sigma(Z,\mathfrak{s}_Z)$, in temporal gauge, is a solution of $\mathfrak{F}^\sigma(\gamma^\sigma)=0$ on the cylinder $Z = I \times Y$ if and only if the corresponding path $\check{\gamma}^\sigma(t) = (B(t), r(t), \psi(t))$ in $\mathcal{C}^\sigma(Y,\mathfrak{s})$ satisfies the equations*

$$\left.\begin{aligned} \frac{1}{2}\frac{d}{dt}B^{\mathrm{t}} &= -\frac{1}{2} * F_{B^{\mathrm{t}}} - r^2\rho^{-1}(\psi\psi^*)_0 \\ \frac{d}{dt}r &= -\Lambda(B,r,\psi)r \\ \frac{d}{dt}\psi &= -D_B\psi + \Lambda(B,r,\psi)\psi. \end{aligned}\right\} \tag{6.8}$$

Proof. This is a straightforward verification. □

Definition 6.2.2. The right-hand sides of the equations (6.8) define a vector field on $\mathcal{C}^\sigma(Y,\mathfrak{s})$. We denote this vector field by $(\operatorname{grad}\mathcal{L})^\sigma$. ◊

On the locus $r \neq 0$, the vector field $(\operatorname{grad}\mathcal{L})^\sigma$ coincides with the vector field $\operatorname{grad}\mathcal{L}$ on $\mathcal{C}^*(Y,\mathfrak{s})$; and it is tangent to the locus $r=0$ (the boundary of $\mathcal{C}^\sigma(Y,\mathfrak{s})$). The construction of this vector field on the blow-up is exactly parallel to the finite-dimensional construction described in the introduction, Subsection 2.5.

As in the finite-dimensional case, the critical points of $(\operatorname{grad}\mathcal{L})^\sigma$ (i.e. the stationary points of the flow) can be easily characterized in terms of the critical points of $\operatorname{grad}\mathcal{L}$:

Proposition 6.2.3. *Let (B,r,ψ) represent a point of $\mathcal{C}^\sigma(Y,\mathfrak{s})$, so ψ is of unit L^2 norm and $r \geq 0$ and let $(B, r\psi)$ be the corresponding point in $\mathcal{C}(Y,\mathfrak{s})$. Then (B,r,ψ) is a critical point of* $(\operatorname{grad}\mathcal{L})^\sigma$ *if and only if either:*

(i) *$r \neq 0$ and $(B, r\psi)$ is a critical point of* $\operatorname{grad}\mathcal{L}$*; or*
(ii) *$r=0$, the point $(B,0)$ is a critical point of* $\operatorname{grad}\mathcal{L}$*, and ψ is an eigenvector of D_B.*

Proof. Only the second case needs comment. The stationary equation for r is always satisfied at $r=0$. The equation for B is the equation $F_{B^{\mathrm{t}}}=0$, which means that $(B, 0)$ is a critical point of grad $\mathcal{L}$. The equation for ψ is

$$D_B\psi - \Lambda(B, 0, \psi)\psi = 0,$$

which states that ψ is an eigenvector of the Dirac operator. (Compare the statement and proof here with the first statement in Lemma 2.5.5, which deals with the finite-dimensional model.) □

6.3 The τ model for the blow-up

In the case that Z is a cylinder $I \times Y$, we saw that an element (A, s, ϕ) of $\mathcal{C}^\sigma(Z, \mathfrak{s}_Z)$ gives rise to a path in $\mathcal{C}^\sigma(Y, \mathfrak{s})$ *provided* that the non-zero spinor ϕ is also non-zero on each slice $\{t\} \times Y$. There is another version of the blown-up configuration space on the cylinder that is useful and is more closely tied to the set of paths in $\mathcal{C}^\sigma(Y, \mathfrak{s})$.

Definition 6.3.1. For a cylinder $Z = I \times Y$, we define

$$\mathcal{C}^\tau(Z, \mathfrak{s}_Z) \subset \mathcal{A}(Z, \mathfrak{s}_Z) \times C^\infty(I; \mathbb{R}) \times C^\infty(Z; S^+)$$

to be the space consisting of triples (A, s, ϕ), with

$$s(t) \geq 0$$
$$\|\phi(t)\|_{L^2(Y)} = 1$$

for all $t \in I$. ◇

Thus the difference between the σ version of the blown-up configuration space and the τ model is that, in the latter, s is a non-negative real *function of* t, rather than a single constant, and ϕ is normalized to have unit L^2 norm *on each slice*, rather than satisfy a single normalization condition on the cylinder. Note that the configuration space $\mathcal{C}^\sigma(X, \mathfrak{s})$ makes sense for any 4-manifold X (possibly with boundary); but $\mathcal{C}^\tau$ can only be defined in the special case of a cylindrical 4-manifold Z.

An element $\gamma = (A, s, \phi)$ in $\mathcal{C}^\tau(Z, \mathfrak{s}_Z)$ determines a path $\check{\gamma}$ in $\mathcal{C}^\sigma(Y, \mathfrak{s})$; and there is a one-to-one correspondence between paths $\check{\gamma}$ and elements $\gamma = (A, s, \phi)$ with A in temporal gauge.

We can regard the equations (6.8) as defining equations for an element $\gamma \in \mathcal{C}^\tau(Z, \mathfrak{s}_Z)$ in temporal gauge. If $\gamma = (A, s, \phi)$, we write the equations in terms of the corresponding path $\check{\gamma}$ as:

$$\left.\begin{aligned} \frac{1}{2}\frac{d}{dt}\check{A}^{\mathrm{t}} &= -\tfrac{1}{2} * F_{\check{A}^{\mathrm{t}}} - s^2\rho^{-1}(\check{\phi}\check{\phi}^*)_0 \\ \frac{d}{dt}s &= -\Lambda(\check{A}, s, \check{\phi})s \\ \frac{d}{dt}\check{\phi} &= -D_{\check{A}}\check{\phi} + \Lambda(\check{A}, s, \check{\phi})\check{\phi}. \end{aligned}\right\} \tag{6.9}$$

For a general configuration that is not necessarily in temporal gauge, we write again

$$\nabla_{A,\frac{d}{dt}} = \frac{d}{dt} + c \otimes 1_S.$$

For such a configuration, we have the following gauge-invariant equations, which reduce to (6.9) when c is zero:

$$\left.\begin{aligned} \frac{1}{2}\frac{d}{dt}\check{A}^{\mathrm{t}} &= -\tfrac{1}{2} * F_{\check{A}^{\mathrm{t}}} + dc - s^2\rho^{-1}(\check{\phi}\check{\phi}^*)_0 \\ \frac{d}{dt}s &= -\Lambda(\check{A}, s, \check{\phi})s \\ \frac{d}{dt}\check{\phi} &= -D_{\check{A}}\check{\phi} - c\check{\phi} + \Lambda(\check{A}, s, \check{\phi})\check{\phi}. \end{aligned}\right\} \tag{6.10}$$

Their 4-dimensional gauge invariance is seen when one writes the equations as

$$\left.\begin{aligned} \frac{1}{2}\rho_Z(F^+_{A^{\mathrm{t}}}) - s^2(\phi\phi^*)_0 &= 0 \\ \frac{d}{dt}s + \mathrm{Re}\langle D^+_A\phi, \rho_Z(dt)^{-1}\phi\rangle_{L^2(Y)}s &= 0 \\ D^+_A\phi - \mathrm{Re}\langle D^+_A\phi, \rho_Z(dt)^{-1}\phi\rangle_{L^2(Y)}\phi &= 0. \end{aligned}\right\} \tag{6.11}$$

(In these equations, the terms that are inner products on Y are to be interpreted as functions on Z that are dependent on t only.)

To understand solutions (A, s, ϕ) to the equations (6.11) in $\mathcal{C}^\tau(Z, \mathfrak{s}_Z)$, note first that s is either identically zero or everywhere positive. In the latter case, the equations express the fact that $(A, s\phi)$ is a solution of the Seiberg–Witten equations, $\mathfrak{F}(A, s\phi) = 0$. In the former case ($s = 0$), let s_0 be a positive solution of the equation

$$\frac{d}{dt}s_0 + \mathrm{Re}\langle D^+_A\phi, \rho_Z(dt)^{-1}\phi\rangle_{L^2(Y)}s_0 = 0. \tag{6.12}$$

Then $s_0\phi$ satisfies the Dirac equation $D_A^+(s_0\phi) = 0$. In other words,

$$(A, 0, s_0\phi/\|s_0\phi\|_{L^2(Z)})$$

satisfies $\mathfrak{F}^\sigma = 0$, when we regard it as defining an element of $\mathcal{C}^\sigma(Z, \mathfrak{s}_Z)$ using the original coordinates (6.2). We summarize the above equations with a definition.

Definition 6.3.2. We define $\mathcal{V}^\tau \to \mathcal{C}^\tau(Z, \mathfrak{s}_Z)$ to be the vector bundle whose fiber over (A, s, ϕ) is the vector space

$$\begin{aligned}\mathcal{V}^\tau_{(A,s,\phi)} &= \{\, (\eta, r, \psi) \mid \operatorname{Re}\langle \check{\phi}(t), \check{\psi}(t)\rangle_{L^2(Y)} = 0 \text{ for all } t \,\} \\ &\subset C^\infty(Z; i\,\mathfrak{su}(S^+)) \oplus C^\infty(I; \mathbb{R}) \oplus C^\infty(Z; S^-).\end{aligned}$$

We write $\mathfrak{F}^\tau$ for the section of $\mathcal{V}^\tau$ defined by the left-hand sides of (6.11). ◊

Remarks. Our discussion shows that solutions of $\mathfrak{F}^\tau = 0$ and solutions of $\mathfrak{F}^\sigma = 0$ are in one-to-one correspondence. The choice between these two viewpoints for the blown-up equations on the cylinder is a matter of convenience. The τ model fits much better with the dynamical viewpoint on the equations, which formally describe trajectories of the vector field $-(\operatorname{grad} \mathcal{L})^\sigma$ on $\mathcal{C}^\sigma(Y, \mathfrak{s})$. For example, zeros of the vector field correspond to translation-invariant solutions of the equations $\mathfrak{F}^\tau = 0$ in temporal gauge on the cylinder. The disadvantages of the τ model are first that the constraint on ϕ (that it has L^2 norm 1 on each slice $\{t\} \times Y$) runs somewhat counter to our desire to treat the equations directly as an elliptic system; and second, that the τ model is not available to us when we wish to look at non-cylindrical manifolds.

In this book, we will regularly switch between the two viewpoints, emphasizing the τ model most often when dealing with a cylinder.

7 Unique continuation

In this section we shall prove unique continuation results for the type of equations that we have been dealing with: the (linear) Dirac equation, and the non-linear Seiberg–Witten equations. Our treatment of these topics is based on the approach of Agmon and Nirenberg [1].

7.1 The linear case

Let us begin by recalling from Subsection 2.3 that if L is a hermitian matrix acting on $\mathbb{C}^n$ and z is a non-zero solution of the linear equation

$$dz/dt = -Lz, \tag{7.1}$$

then the function

$$\Lambda(z) = \langle z, Lz\rangle / \|z\|^2$$

is a non-increasing function of t, because the corresponding path $z^*(t)$ in $\mathbb{CP}^{n-1}$ is a trajectory of the downward gradient-flow equation for the corresponding function $\Lambda^* : \mathbb{CP}^{n-1} \to \mathbb{R}$. Since -2Λ is the time derivative of $\log \|z(t)\|^2$, we can reinterpret this monotonicity as a differential inequality

$$\frac{d^2}{dt^2} \log \|z\|^2 \geq 0.$$

For the formula (7.1), of course, we know the uniqueness of the solution to the initial-value problem, because the equation can be solved in a straightforward manner; but this uniqueness can also be deduced from the above differential inequality. The point is that the inequality does not allow $\log \|z\|^2$ to go to $-\infty$ in finite time (either forward or backwards), so a non-zero solution z on an open interval (t_1, t_2) cannot approach zero as t approaches either endpoint.

The first extension of this idea is to consider the same calculation in the case that L is time-dependent (and hermitian for all t):

$$dz/dt = -L(t)z. \tag{7.2}$$

Lemma 7.1.1. *For a non-zero solution $z(t)$ to the equation (7.2), the quantity $\log \|z\|^2$ satisfies the differential inequality*

$$\frac{d^2}{dt^2} \log \|z\| \geq -\|\dot{L}\|_{\mathrm{Op}},$$

where the right-hand side denotes the operator norm of the derivative of L.

Proof. We calculate:

$$\begin{aligned}\frac{d^2}{dt^2}\log\|z\| &= \frac{d}{dt}\left(-\langle Lz, z\rangle/\|z\|^2\right)\\ &= \frac{1}{\|z\|^4}\left(2\|Lz\|^2\|z\|^2 - 2|\langle Lz, z\rangle|^2 - \langle \dot{L}z, z\rangle\|z\|^2\right)\\ &= \frac{2}{\|z\|^2}\left\|Lz - \frac{\langle Lz, z\rangle}{\|z\|^2}z\right\|^2 - \frac{1}{\|z\|^2}\langle \dot{L}z, z\rangle\\ &\geq -1\|\dot{L}\|_{\mathrm{Op}}.\end{aligned} \tag{7.3}$$

□

For future reference, we note that the penultimate line in the above calculation can be rewritten conveniently in terms of the normalized object $z_1(t) = z(t)/\|z(t)\|$. We have an identity:

$$\frac{d^2}{dt^2}\log\|z\| = 2\|\dot{z}_1\|^2 - \langle \dot{L}z_1, z_1\rangle. \tag{7.4}$$

We can apply the same argument to the Dirac operator on a cylinder.

Proposition 7.1.2. *Suppose A is a smooth spin^c connection on $[t_1, t_2] \times Y$, and let Φ be a solution of the 4-dimensional Dirac equation $D_A^+\Phi = 0$ on the same cylinder. If Φ is zero on the slice $\{t\} \times Y$ for some t in the interval $[t_1, t_2]$, then Φ is identically zero.*

Proof. We suppose the contrary. Replacing the interval by a smaller one if necessary, we can arrive at the case that Φ is non-zero on the slice $\{t\} \times Y$ for all t in the open interval (t_1, t_2) but is zero at one of the endpoints. Since the question is gauge-invariant, we may suppose that A is in temporal gauge, so that we may write the Dirac equation as

$$\frac{d}{dt}\check{\Phi}(t) = -D_{\check{A}(t)}\check{\Phi}(t).$$

The proof of the preceding lemma makes equally good sense in this context: the operator $\dot{L}$ is Clifford multiplication by $d\check{A}/dt$, and we therefore have

$$\frac{d^2}{dt^2}\log\left\|\check{\Phi}(t)\right\|_{L^2} \geq -\left\|d\check{A}/dt\right\|_{C^0(Y)}, \tag{7.5}$$

because the C^0 norm of $d\check{A}/dt$ is the operator norm of its action on L^2. Since the right-hand side is bounded on the closed interval, we can integrate twice to see that $\log \|\check{\Phi}(t)\|$ is bounded below on (t_1, t_2): for example, we can write

$$\log \left\|\check{\Phi}(t)\right\| \geq C_0 + C_1(t - t_0) - \frac{C_2}{2}(t - t_0)^2,$$

where t_0 is any interior point of the interval, C_0 and C_1 are the values of $\log \|\check{\Phi}(t)\|$ and its first derivative at $t = t_0$, and C_2 is the supremum of $\left\|d\check{A}/dt\right\|_{C^0(Y)}$. In particular, $\|\check{\Phi}(t)\|$ cannot approach zero as t approaches either endpoint. □

The same approach can be used to prove unique continuation for the Dirac operator on an arbitrary 4-manifold. (Of course, at this point the choice of dimension 4 is immaterial.) Suppose then that A is a smooth spin^c connection on a 4-manifold X, and Φ satisfies the equation $D_A^+\Phi = 0$. Pick a basepoint x_0, and use geodesic coordinates to identify a punctured neighborhood of x_0 with $(0, \epsilon) \times S^3$. In these coordinates, the metric has the form $dt^2 + g_t$, where t is the radial coordinate and g_t is a family of metrics on the 3-sphere.

Let $\tilde{g}$ be the standard round metric on S^3, let $\tilde{S}$ be the standard spin bundle, let $\tilde{\nabla}$ be a spin^c connection, and let $\tilde{\rho}$ be Clifford multiplication for the standard metric. For the family of metrics g_t, we can identify the spin bundles S_t isometrically with $\tilde{S}$: only the Clifford multiplication $\rho_t : T^*S^3 \to C^\infty(S^3; \mathfrak{su}(\tilde{S}))$ changes with t. From this point of view, Φ gives rise to a time-dependent section $\check{\Phi}(t)$ of the bundle $\tilde{S} \to S^3$ satisfying an equation

$$\frac{d}{dt}\check{\Phi}(t) = -L(t)\check{\Phi}(t),$$

where $L(t)$ is given by

$$L(t)\psi = \rho_t \circ \tilde{\nabla}\psi + R_1\psi,$$

and R_1 is an operator of degree zero arising from the difference between the difference between the spin^c connections. (See the definition of the Dirac operator in Subsection 1.2.)

There are two qualitative differences between the family of operators $L(t)$ that arise here and the family of Dirac operators that we saw in the case of a cylinder $I \times Y$. The first difference is that, unlike the true Dirac operator, the operator $L(t)$ is not self-adjoint. Its skew-symmetric part, however, is an operator of degree zero. Indeed, writing the operator as

$$L(t)\psi = \sum_i \rho_t(dy^i)\tilde{\nabla}_i\psi + R_1\psi,$$

we compute the formal adjoint as

$$\begin{aligned} L(t)^*\psi &= \sum_i (\tilde{\nabla}_i)^*(\rho_t(dy^i)^*\psi) + R_1^*\psi \\ &= -\sum_i (\tilde{\nabla}_i)^*(\rho_t(dy^i)\psi) + R_1^*\psi \\ &= \sum_i \tilde{\nabla}_i(\rho_t(dy^i)\psi) + R_2\psi \\ &= \sum_i \rho_t(dy^i)\tilde{\nabla}_i\psi + R_3\psi \\ &= L(t)\psi + R_4\psi, \end{aligned}$$

where each R_i has order zero. The second difference is that the time derivative of $L(t)$ is not of order zero: it has a first-order term $\dot{\rho}_t \circ \tilde{\nabla}$. Nevertheless, because $L(t)$ is elliptic, we can control $\dot{L}(t)$ in terms of $L(t)$ using the Gårding inequality: for any closed interval $[t_1, t_2]$ in $(0, \epsilon)$, we can find constants C_1 and C_2 such that

$$\|\dot{L}(t)\psi\| \le C_1\|L(t)\psi\| + C_2\|\psi\|,$$

where the norms are L^2 norms for the standard metric on S^3.

For the statement of the next lemma, we abstract the situation following [1] as follows. Let us consider a Hilbert space H (the space $L^2(S^3; \tilde{S})$ in our present situation), and a family of unbounded linear operators $L(t)$ with common domain $D \subset H$, parametrized by $t \in [t_1, t_2]$. Let $f : [t_1, t_2] \to H$ be a given continuous path, and consider a solution z to an equation

$$\frac{d}{dt}z + L(t)z = f(t)$$

on the interval, by which we understand that $z : [t_1, t_2] \to H$ is a C^1 path with values in D. We make two suppositions about $L(t)$, encapsulating our two observations about the Dirac-type operators above.

(i) We suppose that we can write

$$L(t) = L_+(t) + L_-(t), \tag{7.6a}$$

where $L_+(t)$ is symmetric, and $L_-(t)$ is bounded and skew-adjoint.

(ii) We suppose that $L(t)$ has a time-derivative, in the sense that the limit

$$\lim_{h\to 0} \frac{L(t+h)x - L(t)x}{h} = \dot{L}(t)x$$

exists for all x in D and all t in $[t_1, t_2]$, and that the derivative satisfies a bound

$$\|\dot{L}(t)x\| \leq C_1\|L(t)x\| + C_2\|x\|, \tag{7.6b}$$

for constants C_i that are independent of t and x.

Lemma 7.1.3. *Let $z : [t_1, t_2] \to H$ be a solution of the equation*

$$\frac{dz}{dt} + L(t)z = f(t), \tag{7.7}$$

where $L(t)$ satisfies the hypotheses above. Suppose in addition that $f(t)$ satisfies

$$\|f(t)\| \leq \delta\|z(t)\|, \qquad \forall t \in [t_1, t_2], \tag{7.8}$$

for some constant δ. Then if $z(t)$ is zero for any t in the interval, it follows that z is identically zero.

Proof. Again, we may as well suppose z is non-zero everywhere on (t_1, t_2). We may also suppose that $L = L_+$ is symmetric; for if L_- is non-zero we can absorb the term $L_-(t)z(t)$ in Equation (7.7) into $f(t)$ on the right-hand side: the hypothesis (7.8) will not be affected, because L_- is bounded.

We repeat the part of the calculation from Lemma 7.1.1, taking account of the weakened hypotheses. (We take the Hilbert space to be real.) Pick t_0 in the open interval, and define

$$l(t) = \log\|z(t)\| - \int_{t_0}^{t} \frac{\langle f(\tau), z(\tau)\rangle}{\|z(\tau)\|^2}\, d\tau.$$

We have

$$\dot{l}(t) = \frac{-\langle Lz, z\rangle}{\|z\|^2},$$

and differentiating a second time we obtain

$$\begin{aligned}
\ddot{l}(t) &= \frac{d}{dt}\Big(-\langle Lz, z\rangle/\|z\|^2\Big) \\
&= \frac{1}{\|z\|^4}\Big(2\|Lz\|^2\|z\|^2 - 2|\langle Lz, z\rangle|^2 - 2\langle Lz, f\rangle\|z\|^2 + 2\langle Lz, z\rangle\langle f, z\rangle \\
&\qquad - \langle \dot{L}z, z\rangle\|z\|^2\Big) \\
&\geq \frac{2}{\|z\|^2}\left\|Lz - \frac{\langle Lz, z\rangle}{\|z\|^2}z\right\|^2 - C_3\frac{\|Lz\|}{\|z\|} - C_4.
\end{aligned}$$

The hypotheses on $\dot{L}$ and $\|f\|$ are both used in the last line. Writing θ for the angle between z and Lz, we can continue:

$$\begin{aligned}
\ddot{l}(t) &\geq 2\frac{\|Lz\|^2}{\|z\|^2}\sin^2\theta - C_3\frac{\|Lz\|}{\|z\|} - C_4 \\
&= 2\frac{\|Lz\|^2}{\|z\|^2}\sin^2\theta - C_3\frac{\|Lz\|}{\|z\|}(\sin^2\theta + \cos^2\theta) - C_4 \\
&\geq 2\frac{\|Lz\|^2}{\|z\|^2}\sin^2\theta - C_3\frac{\|Lz\|}{\|z\|}(\sin^2\theta + |\cos\theta|) - C_4 \\
&= \sin^2\theta\left(\frac{\|Lz\|^2}{\|z\|^2} - C_3\frac{\|Lz\|}{\|z\|}\right) - C_3\frac{|\langle Lz, z\rangle|}{\|z\|^2} - C_4 \\
&\geq -C_3\frac{|\langle Lz, z\rangle|}{\|z\|^2} - C_5 \\
&= -C_3\big|\dot{l}(t)\big| - C_5.
\end{aligned}$$

Thus the quantity l satisfies the differential inequality

$$\ddot{l} + C_3\big|\dot{l}\big| + C_5 \geq 0, \tag{7.9}$$

which means that the quantity $u(t) = e^{C_3 t}\dot{l}(t)$ satisfies

$$\dot{u} + C_5 e^{C_3 t} \geq 0$$

at all points where $u < 0$. This means that the function $u_- = \min(0, u)$ satisfies

$$u_-(t) \geq u_-(t_0) - C_5 e^{C_3(t-t_0)}$$

for all $t \geq t_0$ in the open interval (t_1, t_2). In particular, $u(t)$ is bounded below on (t_0, t_2); and $l(t)$ is therefore bounded below on (t_0, t_2) also. Our hypothesis on f implies

$$\log \|z(t)\| \geq l(t) - \delta|t - t_0|$$

for $t \in (t_1, t_2)$. So $\log \|z\|$ is bounded below also. It follows that $\|z(t)\|$ cannot approach zero as $t \to t_2$. Similarly, $\|z(t)\|$ is bounded away from zero as $t \to t_1$. □

From this lemma we deduce:

Proposition 7.1.4. *Let A be a smooth spin^c connection on a 4-manifold X, and let Φ solve the Dirac equation $D_A^+ \Phi = 0$. If X is connected and Φ vanishes on a non-empty open set, then Φ is identically zero.*

Proof. Let $Z \subset X$ be the zero locus of Φ. We observed above, that the Dirac equation on a ball in X can be written in a form that fits the hypotheses of Lemma 7.1.3. This means that if the solution Φ vanishes on a geodesic ball of radius t_1 about a point x_0, then it vanishes on the larger ball of radius t_2, for all t_2 less than the injectivity radius at x_0. It follows that the interior of Z is a closed set, which must therefore coincide with X. □

7.2 The non-linear case

Lemma 7.1.3 above can also be used to prove a unique continuation result for non-linear equations, and in particular for the Seiberg–Witten equations. We start with the case of a cylinder:

Proposition 7.2.1. *Let γ_1 and γ_2 be two solutions of the Seiberg–Witten equations $\mathfrak{F}(\gamma) = 0$ on a cylinder $Z = [t_1, t_2] \times Y$, and let $\check{\gamma}_i$ be the corresponding paths, $\check{\gamma}_i : [t_1, t_2] \to \mathcal{C}(Y, \mathfrak{s})$. Suppose there exists some t_0 in the interval for which $\check{\gamma}_1(t_0)$ and $\check{\gamma}_2(t_0)$ are gauge-equivalent on Y. Then γ_1 and γ_2 are gauge-equivalent on Z.*

The same result holds for solutions to the equations $\mathfrak{F}^\tau(\gamma^\tau) = 0$ in the blown-up picture.

Proof. After applying a gauge transformation, we may assume that γ_1 and γ_2 are both in temporal gauge on the cylinder, and that $\check{\gamma}_1(t_0)$ and $\check{\gamma}_2(t_0)$ are equal as elements $\mathcal{C}(Y, \mathfrak{s})$, not just gauge-equivalent. For γ in temporal gauge, we can write the equations $\mathfrak{F}(\gamma) = 0$ schematically as

$$\frac{d}{dt}\gamma + L\gamma = -\gamma \sharp \gamma + c,$$

where $\sharp$ is a bilinear bundle map (which we can take to be symmetric), L is an operator on Y, and we are now regarding γ as a path in the space $C^\infty(Y; iT^*Y \oplus S)$, relative to a chosen basepoint in $\mathcal{C}(Y, \mathfrak{s})$. (Compare (5.4).) To be precise, L is the self-adjoint operator

$$L : (b, \psi) \mapsto (*db, D_{B_0}\psi),$$

as we see from the explicit form of the equations in (4.7). If we take $z = \gamma_2 - \gamma_1$, then z satisfies

$$\frac{d}{dt}z + Lz = f(t),$$

where

$$f(t) = -(\gamma_1 + \gamma_2) \sharp z.$$

The hypothesis of Lemma 7.1.3 is satisfied, with

$$\delta = \|\gamma_1 + \gamma_2\|_{C^0(Z)},$$

and it follows from the lemma that z is identically zero, because $z(t_0)$ is zero.

The argument for solutions to $\gamma_i^\tau = (A_i, s_i, \phi_i)$ to the equations $\mathfrak{F}^\tau = 0$ is similar. □

The proof of the proposition above can be adapted to prove a unique continuation result on an arbitrary 4-manifold.

Proposition 7.2.2. *Let $\gamma_1 = (A_1, \Phi_1)$ and $\gamma_2 = (A_2, \Phi_2)$ be two smooth solutions to the Seiberg–Witten equations $\mathfrak{F}(\gamma) = 0$ on a connected 4-manifold X. If the restrictions of the γ_i are gauge equivalent on a non-empty open set, and if Φ_1 is not identically zero, then γ_1 and γ_2 are gauge-equivalent on all of X.*

The same conclusion holds for solutions of the blown-up equations $\mathfrak{F}^\sigma(\gamma^\sigma) = 0$, although in this case the extra hypothesis on the non-vanishing of the spinor can be dropped.

Proof. As in the proof of Proposition 7.1.4, we choose a point x_0 in X and identify a punctured neighborhood of x_0 with $(0, \epsilon) \times S^3$ using geodesic coordinates. Let $t_1 < t_2$ both be smaller than ϵ, and suppose that γ_1 and γ_2 are gauge-equivalent on the geodesic ball $B_{t_1}(x_0)$ of radius t_1. After applying a gauge transformation, we can suppose that both solutions γ_i are in radial gauge on $B_\epsilon(x_0)$, and that they are equal on the ball $B_{t_1}(x_0)$. In this gauge, let z be the difference of γ_1 and γ_2.

We regard z as a time-dependent 1-form and spinor on S^3, which we make into a Hilbert space using the round metric. As in the proof of Proposition 7.2.1, we can cast the equation for z in the form

$$\frac{d}{dt}z + L(t)z = f(t),$$

where $f(t)$ again absorbs the non-linear terms and satisfies the bound (7.8) required for Lemma 7.1.3. We have to examine the operator $L(t)$, as we did when we dealt with the Dirac operator alone in the proof of Proposition 7.1.4. The operator L now has the form

$$L : (b, \psi) \mapsto (*_t db, D_{A(t)}\psi),$$

where $*_t$ is the Hodge star operator for the metric g_t at radius t. With this form of the equation, we cannot apply Lemma 7.1.3 directly: the difficulty is that the operator $*_t d$ is not a symmetric operator with respect to the inner product defined by the round metric, nor is its skew-symmetric part a bounded operator. (Compare this with our earlier discussion of the Dirac operator). To remedy this, we introduce the bundle map

$$\Gamma_t : \Lambda^1 S^3 \to \Lambda^1 S^3$$

which relates the L^2 inner product defined by g_t and the L^2 inner product defined by the round metric $g_\circ$, so that

$$\langle v, w\rangle_{L^2(S^3, g_t)} = \langle v, \Gamma_t w\rangle_{L^2(S^3, g_\circ)}$$

for all v, w in $\Lambda^1(S^3)$. Then $\Gamma_t^{1/2}(*_t d)\Gamma_t^{-1/2}$ is a symmetric operator with respect to the inner product of $L^2(S^3, g_\circ)$. We may now replace $b(t)$ by $\tilde{b}(t) = \Gamma_t^{1/2} b(t)$, and proceed as before.

This argument shows that, if γ_1 and γ_2 are gauge-equivalent in a small geodesic ball about x_0, then the gauge equivalence extends to a larger ball, up to the injectivity radius around x_0. When X is connected, it follows that for every $x \in X$, we can find a neighborhood U_x of x and a gauge transformation u_x with $u_x(\gamma_1) = \gamma_2$ on U_x. If the spinor Φ_1 is not identically zero, then (by unique continuation) it is non-zero on each open set U_x; and this implies that the gauge transformation u_x is unique. So in this case, the gauge transformations u_x patch together to give a single gauge transformation u on all of X.

For the blown-up equations $\mathfrak{F}^\sigma(\gamma^\sigma) = 0$, the argument is very similar. In this case, the gauge group $\mathcal{G}(U_x)$ always acts freely on $\mathcal{C}^\sigma(U_x, \mathfrak{s}_X)$ for each open set U_x, so there is no question that the gauge transformations agree on overlaps. □

8 Compactness in the blown-up configuration space

8.1 Statement of the compactness result

We shall now refine the basic compactness result, Theorem 5.1.1, to the context of the blown-up configuration space. We deal first with the case of a cylinder, and state a result modelled on Corollary 5.1.8. In addition to a hypothesis on the drop in the Chern–Simons–Dirac functional, we need an additional hypothesis to control the values of the function Λ introduced in (6.7).

Theorem 8.1.1. *Let $Z = [t_1, t_2] \times Y$ be a cylinder, and let $Z_\epsilon = [t_1 + \epsilon, t_2 - \epsilon]$ be a smaller cylinder, so $Z_\epsilon \Subset Z$. Let $\gamma_n^\tau \in \mathcal{C}^\tau(Z, \mathfrak{s})$ be a sequence of solutions of the equations $\mathfrak{F}^\tau(\gamma^\tau) = 0$ on Z, and let $\check{\gamma}^\tau$ be the corresponding paths in $\mathcal{C}^\sigma(Y, \mathfrak{s})$. Suppose that the drop in the Chern–Simons–Dirac functional is uniformly bounded on the larger cylinder Z,*

$$\mathcal{L}(\check{\gamma}_n^\tau(t_1)) - \mathcal{L}(\check{\gamma}_n^\tau(t_2)) \leq C_1,$$

and suppose that there are one-sided bounds on the value of Λ at the endpoints of the smaller cylinder Z_ϵ:

$$\Lambda(\check{\gamma}_n^\tau(t_1 + \epsilon)) \leq C_2$$
$$\Lambda(\check{\gamma}_n^\tau(t_2 - \epsilon)) \geq -C_2.$$

Then there is a sequence of gauge transformations, $u_n \in \mathcal{G}(Z)$, such that, after passing to a subsequence, the transformed solutions $u_n(\gamma_n^\tau)$ have the following property: for every interior domain $Z' \Subset Z_\epsilon$, the transformed solutions converge in the C^∞ topology of $\mathcal{C}^\tau(Z, \mathfrak{s})$ to a solution $\gamma^\tau \in \mathcal{C}^\tau(Z', \mathfrak{s})$ of the equations $\mathfrak{F}^\tau(\gamma^\tau) = 0$. □

Proof. Let $\gamma_n \in \mathcal{C}(Z, \mathfrak{s})$ be the image of γ_n^τ under the blow-down map, and let $(\check{A}_n, \check{\Phi}_n)$ be the corresponding path in $\mathcal{C}(Y, \mathfrak{s})$. After passing to a subsequence, we can assume we are in one of the following two cases: either $\check{\Phi}_n$ is identically zero for all n, or it is always non-zero. We treat the latter case first.

By Corollary 5.1.8, we may assume that γ_n is converging in the C^∞ topology on the smaller cylinder Z_ϵ. In particular, $\check{\Phi}_n$ is converging. If the limit of the $\check{\Phi}_n$

is non-zero on Z_ϵ, then the statement of the theorem reduces to the statement of Corollary 5.1.8. So we suppose that $\check{\Phi}_n$ converges to zero.

Recall now from the proof of Proposition 7.1.2 the inequality (7.5), which we can recast as a statement about Λ. The following version has the additional advantage of being formulated in a gauge-invariant manner.

Lemma 8.1.2. *For a solution γ^τ to the equations $\mathfrak{F}^\tau(\gamma^\tau) = 0$ on a cylinder Z, we have the inequality*

$$\frac{d}{dt}\Lambda(\check{\gamma}^\tau(t)) \leq \| \operatorname{grad} \mathcal{L}(\check{\gamma}^\tau(t))\|_{C^0}.$$

Proof. We have

$$\frac{d}{dt}\Lambda = -\frac{d^2}{dt^2}\log \|\check{\Phi}(t)\|;$$

and for a solution of the equations in temporal gauge, the norm of $d\check{A}/dt$, which appears on the right in (7.5), is controlled by the norm of grad $\mathcal{L}$. □

Because the solutions $(\check{A}_n, \check{\Phi}_n)$ are converging on Z_ϵ, we have a uniform bound on $\| \operatorname{grad} \mathcal{L}(\check{\gamma}^\tau(t))\|_{C^0}$; so if we apply the lemma above to the sequence $(\check{A}_n, \check{\Phi}_n)$, we obtain a uniform upper bound on $(d/dt)\Lambda(\check{\gamma}_n^\tau)$ on the interval $[t_1+\epsilon, t_2-\epsilon]$. Combining this with the one-sided bounds on the endpoints, we obtain a two-sided uniform bound

$$|\Lambda(\check{\gamma}_n^\tau(t))| \leq M, \qquad \forall t \in [t_1+\epsilon, t_2-\epsilon], \forall n.$$

Since $-\Lambda$ is the derivative of $\log \|\check{\Phi}(t)\|$, we deduce that there is a constant K, independent of n, such that

$$\|\check{\Phi}_n(a)\|_{L^2(Y)} \leq K\|\check{\Phi}_n(b)\|_{L^2(Y)}$$

for all a, b in $[t_1+\epsilon, t_2-\epsilon]$, from which it follows that, for any t_0 in the same interval,

$$\|\check{\Phi}_n(t_0)\|_{L^2(Y)} \geq K'\|\Phi_n\|_{L^2(Z_\epsilon)}. \tag{8.1}$$

Consider now the normalized spinors

$$\Phi_n^1 = \Phi_n/\|\Phi_n\|_{L^2(Z_\epsilon)}. \tag{8.2}$$

Because their L^2 norm is 1 and they satisfy the elliptic equations $D_{A_n}\Phi_n^1 = 0$, a subsequence of these normalized spinors converges on any interior domain $Z' = [t_1', t_2'] \times Y$. So we may assume that

$$\Phi_n^1 \xrightarrow{C^\infty(Z')} \Phi^1.$$

The inequality (8.1) assures us that Φ^1 is non-zero on each slice $\{t\} \times Y \subset Z'$.

We now return to our original configurations γ_n^τ, which we write as

$$\gamma_n^\tau = (A_n, s_n, \phi_n),$$

where $\|\phi_n\|_{L^2(Y)} = 1$ on each slice. On the cylinder Z', we have already arranged that the A_n are converging in the C^∞ topology, and we know that the real functions s_n are converging to zero, because $s_n(t)$ is the norm of Φ_n on the slice $\{t\} \times Y$. Finally, we have

$$\check{\phi}_n(t) = \check{\Phi}_n^1(t)/\|\check{\Phi}_n^1(t)\|_{L^2(Y)};$$

and since the Φ_n^1 are converging C^∞ to a limit Φ^1 which is non-zero on each slice, it follows that

$$\phi_n \xrightarrow{C^\infty(Z')} \phi,$$

where

$$\check{\phi}(t) = \check{\Phi}^1(t)/\|\check{\Phi}^1(t)\|_{L^2(Y)}.$$

This concludes the proof in the case that Φ_n is non-zero for all n.

The remaining case is that the configurations γ_n^τ all have s_n identically zero. In this case, a similar argument to the one above can be applied to the spinors $s_{0,n}\phi_n$ on the cylinder, where $s_{0,n}$ is defined as a solution of the equations (6.12), so that $s_{0,n}\phi_n$ is a solution of the Dirac equation $D_{A_n}\Phi = 0$. □

Notes and references for Chapter II

Our exposition of the Chern–Simons–Dirac functional and its relation to the 4-dimensional Seiberg–Witten equations is modelled on the similar, well-developed story of the Chern–Simons functional, as it relates to the anti-self-dual Yang–Mills (or "instanton") equation $F_A^+ = 0$, e.g. for an $SU(2)$

connection A. A general account of the latter is given in [22]. In the instanton theory, the role of the topological energy is played by the Chern–Weil integral for the second Chern class, while the analytic energy corresponds to the Yang–Mills functional.

The compactness theorems for solutions of the Seiberg–Witten equations are what makes these equations so special. There are various versions of the compactness argument, all exploiting the Lichnerowicz–Weitzenböck formula in essentially the same way. The original model is in Witten's paper [125]. The proof presented here gives a sharper result with weaker Sobolev norms. For example, Theorem 5.2.1 gives weak L^2_1 convergence of a subsequence of (A_n, Φ_n) after gauge transformation, assuming only an energy bound and the L^2 convergence of $\mathfrak{F}(A_n, \Phi_n)$. The criterion for strong L^2_1 convergence on a manifold with boundary (rather than convergence on an interior domain) is new also.

A unique continuation result for the anti-self-dual Yang–Mills equations was proved in [23] using the same argument as is used in this chapter for the Seiberg–Witten equations. A slightly stronger result than Proposition 7.2.2 is true (see [112, 115] for the case of the anti-self-dual Yang–Mills equation): if γ_1 and γ_2 are gauge-equivalent to infinite order at a point in X, then they are gauge-equivalent in a neighborhood of that point. This result is proved using the same techniques.

III
Hilbert manifolds and perturbations

In finite-dimensional Morse theory, the reconstruction of the ordinary homology of a manifold from a suitable Morse function relies on the familiar notions of transversality in differential topology. Finite-dimensional results such as Sard's theorem have their analogs for infinite-dimensional manifolds modelled on Banach or Hilbert spaces. This chapter introduces suitable Hilbert manifolds, to replace our configuration spaces of smooth pairs (A, Φ). We then examine with care how to introduce perturbations of the Seiberg–Witten equations. In the next chapter, this framework will allow us to carry over the transversality results, from finite-dimensional Morse theory, to the Morse theory of the Chern–Simons–Dirac functional and its perturbations.

9 Completions and Hilbert manifolds

9.1 Completions of the configuration spaces

Although Sobolev norms were introduced in the proof of the compactness theorems, we have so far considered only smooth configurations. We now introduce the Sobolev completions of $\mathcal{C}(X, \mathfrak{s}_X)$ and $\mathcal{C}(Y, \mathfrak{s})$, and of the corresponding gauge groups. In order to be able to deal with X and Y side by side, we will temporarily introduce M to stand for either one. Thus M will be a compact Riemannian manifold (whose dimension will be 3 or 4) with a spin^c structure. The boundary ∂M may be non-empty. We write $\mathfrak{s}$ for a spin^c structure on M, and we write W to stand for either S^+ in the case of a 4-manifold, or S in the case of a 3-manifold. Thus, for example, $\mathcal{C}(M, \mathfrak{s})$ will be an affine space whose underlying vector space is $C^\infty(M; iT^*M \oplus W)$.

For any integer or half-integer $k \geq 0$, we write

$$\begin{aligned} \mathcal{C}_k(M, \mathfrak{s}) &= (A_0, 0) + L^2_k(M; iT^*M \oplus W), \\ &= \mathcal{A}_k(M, \mathfrak{s}) \times L^2_k(M; W) \end{aligned} \tag{9.1}$$

where $(A_0, 0) \in \mathcal{C}(M, \mathfrak{s})$ is a smooth configuration, and we have set $\mathcal{A}_k(M, \mathfrak{s}) = A_0 + L^2_k(M; iT^*M)$ (compare (4.2)). The blown-up configuration space $\mathcal{C}^\sigma_k(M, \mathfrak{s})$ is defined similarly, again using the L^2_k Sobolev spaces (*cf.* (6.2)),

$$\begin{aligned} \mathcal{C}^\sigma_k(M, \mathfrak{s}) &= \{ (A, s, \phi) \mid \|\phi\|_{L^2(M)} = 1, s \geq 0 \} \\ &= \mathcal{A}_k(M, \mathfrak{s}) \times \mathbb{R}^{\geq} \times \mathbb{S}(L^2_k(M; W)). \end{aligned} \tag{9.2}$$

(Note that the space $\mathbb{S}$ is still defined using the L^2 norm.)

Recall that if $2(k+1) > \dim M$ (so that $L^2_{k+1}(M) \subset C^0(M)$), then the Sobolev space $L^2_{k+1}(M)$ is a Banach algebra; indeed, the multiplication

$$L^2_{k+1}(M) \times L^2_j(M) \to L^2_j(M)$$

is continuous for $j \leq k+1$. For $2(k+1) > \dim M$, we define the gauge group $\mathcal{G}_{k+1}(M)$ as a subset of $L^2_k(M; \mathbb{C})$ consisting of functions whose norm is 1 pointwise. The gauge group inherits the topology of the normed space $L^2_{k+1}(M; \mathbb{C})$.

Lemma 9.1.1. *The space $\mathcal{C}^\sigma_k(M, \mathfrak{s})$ is naturally a Hilbert manifold with boundary. When $2(k+1) > \dim M$, the space $\mathcal{G}_{k+1}(M)$ is a Hilbert Lie group, and acts smoothly and freely on $\mathcal{C}^\sigma_k(M, \mathfrak{s})$.*

Proof. First, the sphere $\mathbb{S}(L^2_k(M; W))$ is a Hilbert manifold. Indeed, it is a submanifold of $L^2_k(M; W)$ because the squared L^2 norm is a smooth function with 1 as a regular value. The other two factors in the definition (9.2) are the half-line $\mathbb{R}^{\geq} = [0, \infty)$ and the affine Hilbert manifold $\mathcal{A}_k(M, \mathfrak{s})$, so making $\mathcal{C}^{\sigma_k}$ a manifold with boundary. The Sobolev multiplication theorems just discussed show at once that $\mathcal{G}_{k+1}(M)$ is a Hilbert Lie group acting smoothly on $\mathcal{C}^\sigma_k(M, \mathfrak{s})$.

To see that the action is free, notice first that a connection A is invariant under a gauge transformation u only if u is constant. Then observe that constant gauge transformations (a copy of the circle group) act freely on the sphere $\mathbb{S}(L^2_k(M; W))$. □

Remark. In the borderline case, $2(k+1) = \dim M$, we recover the same result if we replace $L^2_{k+1}(M)$ with $L^2_{k+1}(M) \cap L^\infty(M)$ in our definition of the gauge group. The space $L^2_{k+1}(M) \cap L^\infty(M)$ is a Banach algebra, and the multiplication

$$(L^2_{k+1}(M) \cap L^\infty(M)) \times L^2_j(M) \to L^2_j(M)$$

is continuous for $j \leq k$. With this in mind, for $2(k+1) = \dim M$, we can define the gauge group $\mathcal{G}_{k+1}(M)$ as a subset of $L^2_k(M;\mathbb{C}) \cap L^\infty(M;\mathbb{C})$. The gauge group inherits the topology of the normed space $L^2_{k+1}(M;\mathbb{C}) \cap L^\infty(M;\mathbb{C})$. (Note that this is different from the topology it would inherit as a subset of $L^2_{k+1}(M;\mathbb{C})$ in the borderline case. With the latter topology, it would not be a topological group.) The lemma above then holds without further modification. We will not make further use of the borderline gauge group.

In addition to the tangent bundle $T\mathcal{C}^\sigma_k(M,\mathfrak{s})$ and $T\mathcal{C}_k(M,\mathfrak{s})$, we have the vector bundles which are the fiberwise completions of these bundles in the L^2_j norms for $j \leq k$. To define these, we first realize $T\mathcal{C}^\sigma_k(M,\mathfrak{s})$ explicitly, by regarding $\mathcal{C}^\sigma_k(M,\mathfrak{s})$ as a submanifold of the affine space

$$\mathcal{A}_k \times \mathbb{R} \times L^2_k(M;W),$$

and so regard $T_\gamma \mathcal{C}^\sigma_k$ as a subspace of the corresponding vector space

$$L^2_k(M;iT^*M) \times \mathbb{R} \times L^2_k(M;W).$$

If (A_0, s_0, ϕ_0) belongs to $\mathcal{C}^\sigma_k(M,\mathfrak{s})$ (so that $\|\phi_0\|_{L^2} = 1$), then a tangent vector is a triple (a, s, ϕ) with $\mathrm{Re}\langle\phi_0, \phi\rangle_{L^2} = 0$. The bundle

$$\mathcal{T}^\sigma_j \to \mathcal{C}^\sigma_k(M,\mathfrak{s}) \tag{9.3}$$

is then defined by declaring its fiber at (A_0, ϕ_0, s_0) to be the space

$$\{\, (a,s,\phi) \in L^2_j(M;iT^*M) \times \mathbb{R} \times L^2_j(M;W) \mid \mathrm{Re}\langle\phi_0, \phi\rangle = 0 \,\}.$$

So defined, the total space of the bundle is a smooth subbundle of the trivial bundle

$$\mathcal{C}^\sigma_k(M,\mathfrak{s}) \times L^2_k(M;iT^*M) \times \mathbb{R} \times L^2_k(M;W).$$

In the case of $\mathcal{C}_k(M,\mathfrak{s})$, the bundle is a product bundle,

$$\begin{aligned} &\mathcal{T}_j \to \mathcal{C}_k(M,\mathfrak{s}) \\ &\mathcal{T}_j = L^2_j(M;iT^*M \oplus W) \times \mathcal{C}_k(M,\mathfrak{s}). \end{aligned} \tag{9.4}$$

Note that, when $j = k$, the bundles $\mathcal{T}_j$ and $\mathcal{T}_j^\sigma$ are the ordinary tangent bundles of $\mathcal{C}_k(M, \mathfrak{s})$ and $\mathcal{C}_k^\sigma(M, \mathfrak{s})$.

9.2 Completion for the τ model

In the special case of the cylinder, $Z = I \times Y$, with I a compact interval, we can also introduce a Sobolev completion of $\mathcal{C}^\tau(Z, \mathfrak{s}_Z)$. We define

$$\mathcal{C}_k^\tau(Z, \mathfrak{s}_Z) \subset \mathcal{A}_k(Z, \mathfrak{s}_Z) \times L_k^2(I; \mathbb{R}) \times L_k^2(Z; S^+)$$

to be the subset consisting of triples (A, s, ϕ) with

$$\begin{aligned} s(t) &\geq 0 \\ \|\phi(t)\|_{L^2(Y)} &= 1 \end{aligned}$$

for all $t \in I$ (compare Definition 6.3.1). This space is not a Hilbert manifold, nor even a manifold with boundary, because of the condition $s \geq 0$. However, it is a closed subspace of a Hilbert manifold: we define

$$\tilde{\mathcal{C}}_k^\tau(Z, \mathfrak{s}_Z) \subset \mathcal{A}_k(Z, \mathfrak{s}_Z) \times L_k^2(I; \mathbb{R}) \times L_k^2(Z; S^+) \tag{9.5}$$

to be the subset consisting of triples (A, s, ϕ) with

$$\|\phi(t)\|_{L^2(Y)} = 1 \tag{9.6}$$

for all t, but no condition on s. On $\tilde{\mathcal{C}}_k^\tau(Z, \mathfrak{s}_Z)$ there is an involution

$$\begin{aligned} \mathbf{i} : \tilde{\mathcal{C}}_k^\tau(Z, \mathfrak{s}_Z) &\to \tilde{\mathcal{C}}_k^\tau(Z, \mathfrak{s}_Z) \\ (A, s, \phi) &\mapsto (A, -s, \phi). \end{aligned} \tag{9.7}$$

Certainly $\mathcal{C}_k^\tau(Z, \mathfrak{s}_Z)$ is closed in $\tilde{\mathcal{C}}_k^\tau(Z, \mathfrak{s}_Z)$, and we have:

Lemma 9.2.1. *For $k \geq 1$, the space $\tilde{\mathcal{C}}_k^\tau(Z, \mathfrak{s}_Z)$ is a Hilbert manifold. When $2(k+1) > 4$, it is acted on smoothly and freely by $\mathcal{G}_{k+1}(Z)$.*

Proof. The set of L_k^2 spinors on Z satisfying the condition (9.6) is the inverse image of the constant function 1 under the map

$$\begin{aligned} n : L_k^2(Z; S^+) &\to L_k^2(I; \mathbb{R}) \\ n(\phi) &= \left(t \mapsto \|\phi(t)\|_{L^2(Y)}^2\right). \end{aligned}$$

When $k \geq 1$, this map is smooth, because multiplication in $L^2_k(I;\mathbb{R})$ is continuous. If $n(\phi) = 1$, then the derivative $\mathcal{D}_\phi n$ is surjective, because for any $f \in L^2_k(I;\mathbb{R})$, we have

$$\mathcal{D}_\phi n(f\phi) = 2f.$$

It follows that $n^{-1}(1)$ is a Hilbert submanifold of the space of spinors. The space $\tilde{\mathcal{C}}^\tau_k(Z,\mathfrak{s}_Z)$ is therefore a Hilbert submanifold of the affine space (9.5). The gauge group acts smoothly on the affine space when $2(k+1) > 4$, and preserves the submanifold $\tilde{\mathcal{C}}^\tau_k(Z,\mathfrak{s}_Z)$. □

We write $\mathcal{T}^\tau_j$ for the L^2_j completion of the tangent bundle of $\tilde{\mathcal{C}}^\tau_k(Z,\mathfrak{s}_Z)$. At $\gamma = (A_0, s_0, \phi_0)$, the fiber is

$$\mathcal{T}^\tau_{j,\gamma} = \{\,(a,s,\phi) \mid \operatorname{Re}\langle \phi_0|_t, \phi|_t\rangle_{L^2(Y)} = 0 \text{ for all } t\,\}. \tag{9.8}$$

9.3 The quotient configuration spaces

We continue to use the conventions of Subsection 9.1. In particular, M will denote either a closed 3 manifold Y, or a 4-manifold X, possibly with boundary. For $2(k+1) > \dim M$, we introduce the quotient spaces

$$\mathcal{B}_k(M,\mathfrak{s}) = \mathcal{C}_k(M,\mathfrak{s})/\mathcal{G}_{k+1}(M)$$
$$\mathcal{B}^\sigma_k(M,\mathfrak{s}) = \mathcal{C}^\sigma_k(M,\mathfrak{s})/\mathcal{G}_{k+1}(M).$$

In the latter case, the action of $\mathcal{G}_{k+1}(M)$ is free. In the case of a cylinder $Z = I \times Y$ (with I again a compact interval), we also introduce the quotient spaces

$$\mathcal{B}^\tau_k(Z,\mathfrak{s}_Z) = \mathcal{C}^\tau_k(Z,\mathfrak{s}_Z)/\mathcal{G}_{k+1}(Z)$$
$$\tilde{\mathcal{B}}^\tau_k(Z,\mathfrak{s}_Z) = \tilde{\mathcal{C}}^\tau_k(Z,\mathfrak{s}_Z)/\mathcal{G}_{k+1}(Z).$$

Proposition 9.3.1. *Suppose $2(k+1) > \dim M$. Then the quotient space $\mathcal{B}^\sigma_k(M,\mathfrak{s})$ is Hausdorff. In the case that M is a cylinder $Z = I \times Y$, the same applies to $\tilde{\mathcal{B}}^\tau_k(Z,\mathfrak{s}_Z)$.*

Proof. Suppose we have two sequences in $\mathcal{C}^\sigma_k(M,\mathfrak{s})$, say γ_n converging to γ, and γ'_n converging to γ'. Suppose that $u_n \in \mathcal{G}_{k+1}(M)$ is a sequence of gauge transformations with $u_n\gamma_n = \gamma'_n$. We wish to show that γ and γ' are gauge-equivalent by some $u \in \mathcal{G}_{k+1}(M)$. We shall write $\gamma_n = (A_n, s_n, \phi_n)$, with similar notation for γ, γ'_n and γ'.

First, all the u_n eventually lie in the same connected component of $\mathcal{G}_{k+1}(M)$, because the component in which u_n lies is determined by knowing the integers

$$(1/2\pi i)\int_M \beta \wedge (u_n^{-1}du_n) = (1/2\pi i)\int_M \beta \wedge (A_n - A'_n)$$

for finitely many closed forms β supported in the interior of M, dual to a basis for $H_1(M;\mathbb{Z})$. So we may assume that all the u_n are in the identity component, and write $u = e^{\xi_n}$.

Write $\xi_n = \xi_n^0 + \xi_n^\perp$, where ξ_n^0 is constant on each component of M and the integral of $\xi_n^\perp$ over each component is zero. The relation $\gamma'_n = u_n\gamma_n$ means that

$$d\xi_n^\perp = (A_n - A'_n),$$

and the fact that the right-hand side is Cauchy in L^2_k means that $\xi_n^\perp$ is Cauchy in L^2_{k+1}.

Replacing γ_n with $e^{\xi_n^\perp}\gamma_n$, we are now reduced to the case that $u_n : M \to S^1$ is a constant for each n. We can therefore pass to a subsequence in which the gauge transformations u_n converge to some u; and in this situation the fact that $u\gamma = \gamma'$ follows from the continuity of the gauge group action.

This proof, for the case of $\mathcal{B}^\sigma_k$, applies verbatim to $\tilde{\mathcal{B}}^\tau_k(Z, \mathfrak{s}_Z)$ in the cylindrical case also. □

We wish to show that $\mathcal{B}^\sigma_k(M, \mathfrak{s})$ is a Hilbert manifold with boundary, and we can base the argument on a quite general principle:

Lemma 9.3.2. *Suppose we have a Hilbert Lie group G acting smoothly and freely on a Hilbert manifold C with Hausdorff quotient. Suppose that at each $c \in C$, the map*

$$d_0 : T_eG \to T_cC$$

(obtained from the derivative of the action) has closed range. Then the quotient C/G is also a Hilbert manifold. □

For a proof of this lemma, see [98]. The Hilbert manifold structure on the quotient can be characterized by the following property. If $S \subset C$ is any locally closed submanifold containing c, with the property that

$$T_cC = \operatorname{im}(d_0) \oplus T_cS,$$

then the restriction of the quotient map, $S \to C/G$, is a diffeomorphism from a neighborhood of c in S to a neighborhood of Gc in C/G.

We will verify the necessary closed-range hypothesis to show that $\mathcal{C}^\sigma_k(M,\mathfrak{s})/\mathcal{G}_{k+1}(M)$ is a Hilbert manifold (with boundary), and in so doing construct a particular slice. If the Hilbert manifold structure were our only concern, any slice S would do. For studying solutions of the Seiberg–Witten equations, we will need something more specific.

We begin with the irreducible part, the locus

$$\mathcal{C}^*_k(M,\mathfrak{s}) \subset \mathcal{C}_k(M,\mathfrak{s})$$

consisting of configurations (A, Φ) with Φ not zero. The map $\mathcal{C}^\sigma_k(M,\mathfrak{s}) \to \mathcal{C}_k(M,\mathfrak{s})$ is a diffeomorphism over $\mathcal{C}^*_k(M,\mathfrak{s})$, and is $\mathcal{G}_{k+1}(M)$-equivariant, so we may regard the quotient $\mathcal{B}^*(M,\mathfrak{s})$ as a subspace of $\mathcal{B}^\sigma(M,\mathfrak{s})$ also.

Write $\mathbf{d}$ for the map obtained by linearizing the gauge group action of $\mathcal{G}_{k+1}(M)$ on $\mathcal{C}_k(M,\mathfrak{s})$, possibly extended to Sobolev spaces of lower regularity: at $\gamma = (A_0, \Phi_0) \in \mathcal{C}_k(M,\mathfrak{s})$, the map is given by

$$\begin{aligned} \mathbf{d}_\gamma : L^2_{j+1}(M; i\mathbb{R}) &\to \mathcal{T}_{j,\gamma} \\ \mathbf{d}_\gamma(\xi) &= (-d\xi, \xi\Phi_0). \end{aligned}$$

Define

$$\mathcal{J}_{j,\gamma} \subset \mathcal{T}_{j,\gamma}$$

to be the image of $\mathbf{d}_\gamma$, and let

$$\mathcal{K}_{j,\gamma} \subset \mathcal{T}_{j,\gamma}$$

be the subspace of vectors orthogonal to $\mathcal{J}_{j,\gamma}$ with respect to the standard L^2 inner product:

$$\mathcal{K}_{j,\gamma} = \{\, v \in \mathcal{T}_{j,\gamma} \mid \langle v, w\rangle_{L^2(M)} = 0, \forall w \in \mathcal{J}_{j,\gamma} \,\}. \tag{9.9}$$

Lemma 9.3.3. *The space $\mathcal{K}_{j,\gamma}$ is*

$$\{\, (a,\phi) \mid -d^*a + i\,\mathrm{Re}\langle i\Phi_0, \phi\rangle = 0 \text{ and } \langle a, \vec{n}\rangle = 0 \text{ at } \partial M \,\},$$

where $\vec{n}$ is the outward normal to the boundary of M, and $\gamma = (A_0, \Phi_0)$.

Proof. We recall first that the correctly normalized inner product is

$$\langle (a_1, \phi_1), (a_2, \phi_2) \rangle_{L^2} = \int_M \big(\langle a_1, a_2 \rangle + \mathrm{Re}\langle \phi_1, \phi_2 \rangle \big).$$

Taking the first tangent vector to be $\mathbf{d}_\gamma(\xi)$, and integrating by parts (first with ξ compactly supported in the interior and then with a general ξ), gives the desired result. □

In the statement of the lemma, the condition

$$-d^* a + i\,\mathrm{Re}\langle i\Phi_0, \phi \rangle = 0$$

can be written

$$\mathbf{d}_\gamma^*(a, \phi) = 0,$$

where

$$\mathbf{d}_\gamma^* : \mathcal{T}_{j,\gamma} \to L^2_{j-1}(M; i\mathbb{R})$$

is the formal adjoint of $\mathbf{d}_\gamma$.

Proposition 9.3.4. *As γ varies over $\mathcal{C}_k^*(M, \mathfrak{s})$, the subspaces $\mathcal{J}_{j,\gamma}$ and $\mathcal{K}_{j,\gamma}$ define complementary closed subbundles of the restriction of $\mathcal{T}_j$ to $\mathcal{C}_k^*(M, \mathfrak{s})$, and we have a smooth decomposition*

$$\mathcal{T}_j|_{\mathcal{C}_k^*(M,\mathfrak{s})} = \mathcal{J}_j \oplus \mathcal{K}_j.$$

In particular,

$$T\mathcal{C}_k^*(M, \mathfrak{s}) = \mathcal{J}_k \oplus \mathcal{K}_k.$$

Proof. With $\gamma = (A_0, \Phi_0)$, we show that

$$\mathcal{T}_{j,\gamma} = \mathcal{J}_{j,\gamma} \oplus \mathcal{K}_{j,\gamma}.$$

This means that, given any (a, ϕ), we can solve uniquely for ξ the equations

$$\mathbf{d}_\gamma^*\big((a, \phi) + \mathbf{d}_\gamma(\xi)\big) = 0$$
$$\langle a - d\xi, \vec{n} \rangle = 0 \text{ at } \partial M.$$

The first equation is

$$\Delta\xi + |\Phi_0|^2\xi = \eta$$

where $\eta = \mathbf{d}_\gamma^*(a,\phi)$. Under our hypothesis $2(k+1) > \dim M$ and $j \geq 0$, the map

$$\xi \mapsto \Delta\xi + |\Phi_0|^2\xi$$

is a bounded linear operator from L^2_{j+1} to L^2_{j-1}. Since Φ_0 is non-zero, the boundary-value problem has a unique solution in L^2_{j+1}, by standard theory.

The decomposition is smooth, because $\mathcal{J}_j$ is the image of a smooth, injective bundle map and $\mathcal{K}_j$ is the kernel of a smooth, surjective bundle map. □

When we regard $\mathcal{C}_k^*(M,\mathfrak{s})$ as a subspace of $\mathcal{C}_k^\sigma(M,\mathfrak{s})$ (the interior of the Hilbert manifold with boundary), the subbundle $\mathcal{K}_{j,\gamma}$ has a smooth extension to the boundary:

Proposition 9.3.5. *There is a smooth bundle decomposition, over the whole of $\mathcal{C}_k^\sigma(M,\mathfrak{s})$, extending the decomposition in Proposition* 9.3.4*:*

$$\mathcal{T}_j^\sigma = \mathcal{J}_j^\sigma \oplus \mathcal{K}_j^\sigma.$$

In the above proposition, the construction of $\mathcal{J}_j^\sigma$ is straightforward: adapting the earlier definition of $\mathcal{J}_j$, we can consider the derivative of the gauge group action as defining a map

$$\mathbf{d}_\gamma^\sigma : T_e\mathcal{G}_{k+1}(M) \to T_\gamma\mathcal{C}_k^\sigma(M,\mathfrak{s}) \tag{9.10}$$

for each γ in $\mathcal{C}^\sigma(M,\mathfrak{s})$; we define $\mathcal{J}_{k,\gamma}^\sigma$ to be the image of $\mathbf{d}_\gamma^\sigma$, and we define $\mathcal{J}_{j,\gamma}^\sigma$ for $j \leq k$ by extending the map to Sobolev spaces of lower regularity. If we write γ as (A_0, s_0, ϕ_0) and write a typical tangent vector as (a, s, ϕ), with $\mathrm{Re}\langle i\phi_0, \phi\rangle_{L^2(M)} = 0$, then the map is given by

$$\mathbf{d}_\gamma^\sigma(\xi) = (-d\xi, 0, \xi\phi_0).$$

The definition of the subspace $\mathcal{K}_j^\sigma(M)$ in the proposition is as follows.

Definition 9.3.6. At $\gamma = (A_0, s_0, \phi_0)$ in $\mathcal{C}_j^\sigma(M, \mathfrak{s})$, we define a subspace $\mathcal{K}_{j,\gamma}^\sigma \subset \mathcal{T}_{j,\gamma}^\sigma$ as the set of all (a, s, ϕ) satisfying the conditions

$$\left.\begin{aligned} \langle a, \vec{n}\rangle &= 0 \text{ at } \partial M \\ -d^*a + is_0^2 \operatorname{Re}\langle i\phi_0, \phi\rangle &= 0 \\ \operatorname{Re}\langle i\phi_0, \phi\rangle_{L^2(M)} &= 0. \end{aligned}\right\} \tag{9.11}$$

◊

This is *not* the orthogonal complement of $\mathcal{J}_{j,\gamma}^\sigma$ with respect to any particularly natural inner product on the tangent space. We shall see that it is a complement of $\mathcal{J}_{j,\gamma}^\sigma$ as claimed, and that it coincides with the earlier $\mathcal{K}_{j,\gamma}$ at irreducible configurations γ. It should be pointed out that we can define $\mathcal{J}_{k,\gamma} \subset T\mathcal{C}_k(M, \mathfrak{s})$ also when $\gamma = (A, 0)$ is reducible, and we can define $\mathcal{K}_{k,\gamma}$ to be its L^2-orthogonal complement, given by (9.9); but with these definitions, neither is a subbundle of $T\mathcal{C}_k(M, \mathfrak{s})$ because $\mathbf{d}_\gamma$ acquires a kernel at reducible configurations.

Proof of Proposition 9.3.5. Note first that, if s_0 is not zero, then by integrating the second condition over M we see that the first two conditions in the definition of $\mathcal{K}_j^\sigma$ imply the third, because the integral of d^*a is zero when the boundary condition is satisfied. It follows that, under the map $\pi : \mathcal{C}_k^\sigma(M, \mathfrak{s}) \to \mathcal{C}_k(M, \mathfrak{s})$ given by $(A_0, s_0, \phi_0) \mapsto (A_0, s_0\phi_0)$, the subspaces $\mathcal{K}_j^\sigma$ and $\mathcal{K}_j$ coincide where s_0 is not zero. We also see that the three conditions can be rewritten as two,

$$\begin{aligned} \langle a, \vec{n}\rangle &= 0 \text{ at } \partial M \\ -d^*a + is_0^2 \operatorname{Re}\langle i\phi_0, \phi\rangle + i|\phi_0|^2 \operatorname{Re}\mu_M(\langle i\phi_0, \phi\rangle) &= 0, \end{aligned} \tag{9.12}$$

where in the second equation we write $\mu_M(f)$ for the average value of f:

$$\mu_M(f) = \left(\frac{\int_M f}{\int_M 1}\right).$$

Written this way, $\mathcal{K}_j^\sigma$ is once again the kernel of a smooth family of surjective operators, and is therefore a smooth Hilbert subbundle of $\mathcal{T}_j^\sigma$.

The proof that $\mathcal{J}_{j,\gamma}^\sigma$ and $\mathcal{K}_{j,\gamma}^\sigma$ are complementary at a reducible configuration $\gamma = (A_0, \phi_0, 0)$ is much the same as in the irreducible case. One arrives at the equation

$$\begin{aligned} \Delta\xi + |\phi_0|^2\mu_M(|\phi_0|^2\xi) &= \eta \\ \langle d\xi, \vec{n}\rangle &= \langle a, \vec{n}\rangle \text{ at } \partial M. \end{aligned} \tag{9.13}$$

The operator defined by the right-hand side of the equation is a compact perturbation of the Laplacian and hence it is an index-zero Fredholm operator. Thus it suffices to prove that it has trivial kernel. If ξ is in the kernel then multiplying the homogeneous version of the above equation by ξ and integrating gives

$$\left(\int_M |d\xi|^2\right) + \mathrm{Vol}(M)\left|\mu_M(|\phi_0|^2\xi)\right|^2 = 0$$

and we conclude $\xi = 0$. □

In line with the discussion that follows Lemma 9.3.2, we now describe, for each $\gamma \in \mathcal{C}_k^\sigma(M, \mathfrak{s})$, a closed Hilbert submanifold $\mathcal{S}_{k,\gamma}^\sigma \subset \mathcal{C}_k^\sigma(M, \mathfrak{s})$ which contains γ and whose tangent space at γ is $\mathcal{K}_{k,\gamma}^\sigma \subset T_\gamma \mathcal{C}_k^\sigma(M, \mathfrak{s})$. We begin by describing these in the irreducible part. For any γ in the affine space $\mathcal{C}_k(M, \mathfrak{s})$ (not the blow-up), we define $\mathcal{S}_{k,\gamma}$ to be the affine space with tangent space $\mathcal{K}_{k,\gamma}$ at γ: if $\gamma = (A_0, \Phi_0)$, this can be described as

$$\mathcal{S}_{k,\gamma} = \{\, (A, \Phi) \mid -d^*a + i\,\mathrm{Re}\langle i\Phi_0, \Phi\rangle = 0 \text{ and } \langle a|_{\partial M}, \vec{n}\rangle = 0 \,\}, \qquad (9.14)$$

where $A = A_0 + a \otimes 1$ as usual. (We used the fact that $\mathrm{Re}\langle i\Phi_0, \Phi_0\rangle$ is zero pointwise.) If γ is irreducible, then we can regard γ as a point (A_0, s_0, ϕ_0) in the blow-up, $\mathcal{C}_k^\sigma(M, \mathfrak{s})$, with $s_0 \neq 0$. In this case, we define a submanifold

$$\mathcal{S}_{k,\gamma}^\sigma \subset \mathcal{C}_k^\sigma(M, \mathfrak{s})$$

by taking the closure of $\mathcal{S}_{k,\gamma} \cap \mathcal{C}_k^*(M, \mathfrak{s})$, identifying $\mathcal{C}_k^*(M, \mathfrak{s})$ with an open subset of $\mathcal{C}_k^\sigma(M, \mathfrak{s})$. In other language, $\mathcal{S}_{k,\gamma}^\sigma$ is the proper transform of the affine space $\mathcal{S}_{k,\gamma}$ under the blow-up. In coordinates, this proper transform is

$$\begin{aligned}\{\, (A, s, \phi) \mid -d^*a + iss_0\,\mathrm{Re}\langle i\phi_0, \phi\rangle = 0,&\\ \mathrm{Re}\langle i\phi_0, \phi\rangle_{L^2(M)} = 0 \text{ and } \langle a|_{\partial M}, \vec{n}\rangle = 0 \,\}.&\end{aligned}$$

As before, the vanishing of the L^2 inner product in this definition is a consequence of the other two conditions if s is non-zero. As γ approaches the reducible locus $s_0 = 0$, the equations above have a well-defined limit, which we use as a *definition* of $\mathcal{S}_{k,\gamma}^\sigma$ at a reducible point $\gamma = (A_0, 0, \phi_0)$:

$$\{\, (A, s, \phi) \mid -d^*a = 0,\ \mathrm{Re}\langle i\phi_0, \phi\rangle_{L^2(M)} = 0 \text{ and } \langle a|_{\partial M}, \vec{n}\rangle = 0 \,\}.$$

With this definition, $\mathcal{S}_{k,\gamma}^\sigma$ is again a submanifold of $\mathcal{C}_{k,\gamma}^\sigma$ with tangent space $\mathcal{K}_{k,\gamma}^\sigma$. The following definition summarizes the situation.

Definition 9.3.7. For $\gamma \in \mathcal{C}_k(M, \mathfrak{s})$, we define $\mathcal{S}_{k,\gamma} \subset \mathcal{C}_k(M, \mathfrak{s})$ to be the affine subspace with tangent space $\mathcal{K}_{k,\gamma}$ at γ. For γ a point (A_0, s_0, ϕ_0) in the blow-up $\mathcal{C}_k^\sigma(M, \mathfrak{s})$, we define $\mathcal{S}_{k,\gamma}^\sigma$ to be the closed submanifold of $\mathcal{C}_k^\sigma(M, \mathfrak{s})$ consisting of the triples (A, s, ϕ) satisfying the conditions

$$\begin{aligned} -d^*a + iss_0 \operatorname{Re}\langle i\phi_0, \phi\rangle &= 0 \\ \langle a, \vec{n}\rangle &= 0 \text{ at } \partial M \\ \operatorname{Re}\langle i\phi_0, \phi\rangle_{L^2(M)} &= 0. \end{aligned}$$

The last condition follows from the other two if ss_0 is non-zero. ◊

We write ι for the inclusion of $\mathcal{S}_{k,\gamma}^\sigma$ in $\mathcal{C}_k^\sigma(M, \mathfrak{s})$, and

$$\bar{\iota} : \mathcal{S}_{k,\gamma}^\sigma \to \mathcal{B}_k^\sigma(M, \mathfrak{s})$$

for the composite of the inclusion and the quotient map. In view of Lemma 9.3.2 we have:

Corollary 9.3.8. *When $2(k+1) > \dim M$, the quotient space $\mathcal{B}_k^\sigma(M, \mathfrak{s})$ is a Hilbert manifold with boundary. For each $\gamma \in \mathcal{C}_k^\sigma(M, \mathfrak{s})$, there is an open neighborhood of γ in the slice,*

$$\mathcal{U} \subset \mathcal{S}_{k,\gamma}^\sigma$$

such that $\bar{\iota} : \mathcal{U} \to \mathcal{B}_k^\sigma(M, \mathfrak{s})$ is a diffeomorphism onto its image, which is an open neighborhood of $[\gamma] \in \mathcal{B}_k^\sigma(M, \mathfrak{s})$. □

Definition 9.3.9. In the context of this corollary, we refer to $\bar{\iota} : \mathcal{U} \to \mathcal{B}_k^\sigma(M, \mathfrak{s})$ as a *Coulomb–Neumann chart* at $[\gamma] \in \mathcal{B}_k^\sigma(M, \mathfrak{s})$. The submanifold $\mathcal{S}_{k,\gamma}^\sigma \subset \mathcal{C}_k^\sigma(M, \mathfrak{s})$ is the *Coulomb–Neumann slice* at $\gamma \in \mathcal{C}_k^\sigma(M, \mathfrak{s})$. ◊

The terminology is applicable to both the 3- and 4-dimensional configuration spaces, though in the 3-dimensional case, the boundary will be empty and the second condition of Definition 9.3.7 is vacuous. In the absence of boundary, we refer to the "Coulomb" slice. Note that if $\gamma = (A_0, 0)$ is a reducible configuration in $\mathcal{C}_k(M, \mathfrak{s})$, then the condition that (A, Φ) be in the Coulomb–Neumann slice $\mathcal{S}_{k,\gamma}$ is a condition on $A - A_0$ alone, and coincides with the Coulomb–Neumann gauge condition introduced earlier in Subsection 5.1.

9.4 Slices in the τ model

We now turn to the quotient space $\mathcal{B}_k^\tau(Z, \mathfrak{s}_Z)$ in the case that Z is a cylinder $I \times Y$. Recall that $\mathcal{B}_k^\tau(Z, \mathfrak{s}_Z)$ is a closed subset of $\tilde{\mathcal{B}}_k^\tau(Z, \mathfrak{s}_Z)$. We will show that $\tilde{\mathcal{B}}_k^\tau(Z, \mathfrak{s}_Z)$ is a Hilbert manifold by exhibiting preferred slices, as we did for $\mathcal{B}_k^\sigma(Z, \mathfrak{s}_Z)$ above. Parallel to our previous discussion, for $\gamma \in \tilde{\mathcal{C}}_k^\tau(Z, \mathfrak{s}_Z)$ we write

$$\mathbf{d}_\gamma^\tau : L^2_{j+1}(Z; i\mathbb{R}) \to \mathcal{T}_{j,\gamma}^\tau$$

for the map defined by the derivative of the gauge group action on $\mathcal{C}_k^\tau(Z, \mathfrak{s}_Z)$. If γ is (A_0, s_0, ϕ_0) and we describe $\mathcal{T}_{j,\gamma}^\tau$ as in (9.8), the formula for $\mathbf{d}_\gamma^\tau$ is

$$\mathbf{d}_\gamma^\tau(\xi) = (-d\xi, 0, \xi\phi_0).$$

We write $\mathcal{J}_{j,\gamma}^\tau$ for the image of this map. Our definition of a complement, $\mathcal{K}_{j,\gamma}^\tau$, to the subspace $\mathcal{T}_{j,\gamma}^\tau$ is somewhat more ad hoc than in the previous case, in that, for example, it does not reduce to the standard $\mathcal{K}_{j,\gamma}$ at a configuration with s_0 non-zero. Motivated by the characterization (9.12) of the Coulomb–Neumann condition in $\mathcal{C}_k^\sigma$, we define

$$\mathcal{K}_{j,\gamma}^\tau \subset \mathcal{T}_{j,\gamma}^\tau \tag{9.15}$$

to be the subspace consisting of elements (a, s, ϕ) (in the description (9.8)) satisfying the conditions

$$\begin{aligned} \langle a, \vec{n}\rangle &= 0 \text{ at } \partial Z \\ -d^*a + is_0^2 \operatorname{Re}\langle i\phi_0, \phi\rangle + i|\phi_0|^2 \operatorname{Re}\mu_Y\langle i\phi_0, \phi\rangle &= 0. \end{aligned} \tag{9.16}$$

Here μ_Y is the operator on functions on Z, defined by averaging over each slice $\{t\} \times Y$. That is, $\mu_Y(f)$ is the function on Z which is constant on each slice and has the same integrals on the slices as f. We then have, just as in Proposition 9.3.4, the following result:

Proposition 9.4.1. *The subspaces $\mathcal{J}_{j,\gamma}^\tau$ and $\mathcal{K}_{j,\gamma}^\tau$ define complementary closed subbundles of $\mathcal{T}_j^\tau \to \tilde{\mathcal{C}}_k^\tau(Z, \mathfrak{s}_Z)$, and we have a smooth bundle decomposition*

$$\mathcal{T}_j^\tau = \mathcal{J}_j^\tau \oplus \mathcal{K}_j^\tau.$$

In particular,

$$T\tilde{\mathcal{C}}_k^\tau(Z, \mathfrak{s}_Z) = \mathcal{J}_k^\tau \oplus \mathcal{K}_k^\tau.$$

Proof. Compare the proof of Proposition 9.3.5. In showing that

$$\mathcal{T}^{\tau}_{j,\gamma} = \mathcal{J}^{\tau}_{j,\gamma} \oplus \mathcal{K}^{\tau}_{j,\gamma},$$

we arrive at the following equations for a $\xi \in L^2_{j+1}(Z; i\mathbb{R})$ –

$$\begin{aligned} \Delta\xi + (\mathrm{Vol}(Y))^{-1}|\phi_0|^2\mu_Y(|\phi_0|^2\xi) + s_0^2|\phi_0|^2\xi &= \eta \\ \langle d\xi, \vec{n}\rangle &= \langle a, \vec{n}\rangle \text{ at } \partial Z \end{aligned}$$

– which once again admit a unique non-zero solution. ☐

For each γ in $\tilde{\mathcal{C}}^{\tau}_{k,\gamma}$, we can define a slice $\mathcal{S}^{\tau}_{k,\gamma} \subset \tilde{\mathcal{C}}^{\tau}_{k,\gamma}(Z, \mathfrak{s}_Z)$ which is a Hilbert submanifold near γ with tangent space $\mathcal{K}^{\tau}_{k,\gamma}$, by imitating the definition of $\mathcal{S}^{\sigma}_k$ in Definition 9.3.7:

Definition 9.4.2. For a configuration $\gamma = (A_0, s_0, \phi_0)$ in $\mathcal{C}^{\tau}_k(Z, \mathfrak{s}_Z)$, we define $\mathcal{S}^{\tau}_{k,\gamma}$ to be the set of triples $(A, s, \phi) \in \tilde{\mathcal{C}}^{\tau}_k(Z, \mathfrak{s}_Z)$ satisfying

$$\begin{aligned} \langle a, \vec{n}\rangle &= 0 \text{ at } \partial Z \\ -d^*a + iss_0 \operatorname{Re}\langle i\phi_0, \phi\rangle + i|\phi_0|^2 \operatorname{Re}\mu_Y(\langle i\phi_0, \phi\rangle) &= 0, \end{aligned} \tag{9.17}$$

where $A = A_0 + a \otimes 1$. ◇

The linearizations of the conditions defining $\mathcal{S}^{\tau}_{k,\gamma}$ at γ coincide with the conditions defining $\mathcal{K}^{\tau}_{k,\gamma}$, so $\mathcal{S}^{\tau}_{k,\gamma}$ is a submanifold near γ, as claimed. Although we are only really concerned with the behavior near γ, we take the trouble to check:

Lemma 9.4.3. *For each $\gamma \in \mathcal{C}^{\tau}_k(Z, \mathfrak{s}_Z)$, the space $\mathcal{S}^{\tau}_{k,\gamma}$ is a Hilbert submanifold of $\tilde{\mathcal{C}}^{\tau}_k(Z, \mathfrak{s}_Z)$ at all points of the closed subset $\mathcal{C}^{\tau}_k(Z, \mathfrak{s}_Z) \subset \tilde{\mathcal{C}}^{\tau}_k(Z, \mathfrak{s}_Z)$.*

Proof. In the same notation as in the definition, let $\gamma_1 = (A_1, s_1, \phi_1)$ belong to $\mathcal{S}^{\tau}_{k,\gamma}$. At (A_1, s_1, ϕ_1), we consider the linearization of the second equation in the definition. This gives an operator

$$\begin{aligned} K :&(a, s, \phi) \mapsto \\ &- d^*a + is_1s_0 \operatorname{Re}\langle i\phi_0, \phi\rangle + iss_0 \operatorname{Re}\langle i\phi_0, \phi_1\rangle + i|\phi_0|^2 \operatorname{Re}\mu_Y\langle i\phi_0, \phi\rangle, \end{aligned}$$

acting on triples (a, s, ϕ), where a satisfies the boundary condition and ϕ is orthogonal to ϕ_1 on each $\{t\} \times Y$. We wish to see that this operator is surjective, mapping onto $L^2_{k-1}(Z; i\mathbb{R})$. By considering variations $(a, 0, 0)$, we see that the cokernel is at most 1-dimensional, represented by the constant function on Z. So to establish surjectivity we need only find a variation $(0, s, \phi)$ with the integral of $K(0, s, \phi)$ non-zero.

Because γ and γ_1 are both in $C_k^\tau(Z, \mathfrak{s}_Z)$, the function $s_1 s_0$ is non-negative; so if we consider a variation $(0, 0, \phi)$ where ϕ has positive inner product with $i\phi_0$ on each slice, then both of the two non-zero terms in the formula for K will be positive multiples of i. □

Remark. The argument shows that the only non-manifold points which can occur arise when $s_0 s_1$ is the constant function $-1/\mathrm{Vol}(Y)$.

We again write ι for the inclusion of $\mathcal{S}_{k,\gamma}^\tau$ in $\tilde{C}_k^\tau(Z, \mathfrak{s}_Z)$, and

$$\bar{\iota} : \mathcal{S}_{k,\gamma}^\tau \to \tilde{\mathcal{B}}_k^\tau(Z, \mathfrak{s}_Z)$$

for the quotient map. As in the earlier models, we now have:

Corollary 9.4.4. *The quotient space $\tilde{\mathcal{B}}_k^\tau(Z, \mathfrak{s}_Z)$ is a Hilbert manifold. For each $\gamma \in C_k^\tau(Z, \mathfrak{s}_Z)$, there is an open neighborhood of γ,*

$$\mathcal{U} \subset \tilde{\mathcal{S}}_{k,\gamma}^\tau,$$

such that $\bar{\iota} : \mathcal{U} \to \tilde{\mathcal{B}}_k^\tau(Z, \mathfrak{s}_Z)$ is a diffeomorphism onto its image, which is an open neighborhood of $[\gamma] \in \mathcal{B}_k^\tau(Z, \mathfrak{s}_Z)$. □

Definition 9.4.5. In the context of this corollary, we refer to $\bar{\iota} : \mathcal{U} \to \tilde{\mathcal{B}}_k^\tau(Z, \mathfrak{s}_Z)$ as a *τ-Coulomb–Neumann chart* at $[\gamma] \in \mathcal{B}_k^\tau(Z, \mathfrak{s}_Z)$. The submanifold $\mathcal{S}_{k,\gamma}^\tau \subset C_k^\sigma(M)$ is the τ-Coulomb–Neumann slice at $\gamma \in C_k^\tau(Z, \mathfrak{s}_Z)$. ◊

9.5 The equations on the completion

We go back to the case of a compact 4-manifold X (perhaps with boundary). On the Sobolev completion $C_k^\sigma(X, \mathfrak{s}_X)$ of $C^\sigma(X, \mathfrak{s}_X)$, there is a construction parallel to (6.4). We define

$$\mathcal{V}_j \to \mathcal{C}_k$$

to be the trivial vector bundle with fiber $L_j^2(X; i\,\mathfrak{su}(S^+) \oplus S^-)$; and on $C_k^\sigma(X, \mathfrak{s}_X)$, we write $\mathcal{V}_j^\sigma$ for the pull-back bundle. The section $\mathfrak{F}^\sigma$, defined by (6.5), extends to these Sobolev completions as a section of $\mathcal{V}_{k-1}^\sigma \to C_k^\sigma(X, \mathfrak{s}_X)$, as is clear from the formula, using the Sobolev multiplication theorems. As long as $j \le k+1$, the gauge group $\mathcal{G}_{k+1}(X)$ acts smoothly on the Hilbert bundles $\mathcal{V}_j$ and $\mathcal{V}_j^\sigma$. In the case $j = k-1$, the sections $\mathfrak{F}$ and $\mathfrak{F}^\sigma$ of these bundles are $\mathcal{G}_{k+1}(X)$-equivariant.

There are similar comments to be made about the 3-dimensional configuration space $\mathcal{C}_k(Y, \mathfrak{s})$ and its blow-up $\mathcal{C}_k^\sigma(Y, \mathfrak{s})$. In these terms, $\operatorname{grad} \mathcal{L}$ is a section of

$$\mathcal{T}_{k-1} \to \mathcal{C}_k(Y, \mathfrak{s})$$

and similarly $(\operatorname{grad} \mathcal{L})^\sigma$ is a section of

$$\mathcal{T}_{k-1}^\sigma \to \mathcal{C}_k^\sigma(Y, \mathfrak{s}).$$

Both of these sections are smooth.

Finally, there is the parallel construction for $\mathcal{C}^\tau(Z, \mathfrak{s}_Z)$ on a compact cylinder $Z = I \times Y$. Recalling Definition 6.3.2, we define $\mathcal{V}_j^\tau \to \mathcal{C}^\tau(Z, \mathfrak{s}_Z)$ to be the vector bundle whose fiber over $\gamma = (A, s, \phi)$ is the vector space

$$\begin{aligned} \mathcal{V}_{j,\gamma}^\tau &= \{ (\eta, r, \psi) \mid \operatorname{Re}\langle \check{\phi}(t), \check{\psi}(t) \rangle_{L^2(Y)} = 0 \text{ for all } t \} \\ &\subset L_j^2(Z; i\,\mathfrak{su}(S^+)) \oplus L_j^2(I; \mathbb{R}) \oplus L_j^2(Z; S^-). \end{aligned} \tag{9.18}$$

Then the left-hand sides of (6.11) define a section $\mathfrak{F}^\tau$ of $\mathcal{V}_{k-1}^\tau \to \mathcal{C}_k^\tau(Z, \mathfrak{s}_Z)$:

$$\begin{aligned} \mathfrak{F}^\tau(A, s, \phi) = \Big(&\frac{1}{2}\rho_Z(F_{A^t}^+) - s^2(\phi\phi^*)_0, \frac{d}{dt}s + \operatorname{Re}\langle D_A^+\phi, \phi \rangle_{L^2(Y)} s, \\ &D_A^+\phi - \operatorname{Re}\langle D_A^+\phi, \phi\rangle_{L^2(Y)}\phi \Big). \end{aligned} \tag{9.19}$$

We have omitted the identification $\rho_Z(dt)$ between S^- and S^+ for the sake of compactness in this formula. Note that $\mathcal{V}_j^\tau$ can be extended as a bundle over the larger space $\tilde{\mathcal{C}}_k^\tau(Z, \mathfrak{s}_Z)$, as can the smooth section $\mathfrak{F}^\tau$.

9.6 The global slice

At a reducible configuration $\gamma_0 = (A_0, 0)$ in $\mathcal{C}_k(M, \mathfrak{s})$ (where M is either a 4-manifold, possibly with boundary, or a 3-manifold), the Coulomb–Neumann condition defining $\mathcal{K}_{k,\gamma_0}$ becomes simply

$$\begin{aligned} -d^*a &= 0 \\ \langle a, \vec{n} \rangle &= 0 \quad \text{at } \partial M. \end{aligned}$$

To find a gauge transformation of the form $u = e^{\xi}$ to put a general element $(A, \Phi) = (A_0 + a \otimes 1, \Phi)$ into the slice $\iota(\mathcal{K}_{k,\gamma_0})$, one solves the equations

$$\begin{aligned} \Delta\xi &= d^*a \\ \langle d\xi, \vec{n}\rangle &= \langle a, \vec{n}\rangle \quad \text{at } \partial M. \end{aligned}$$

(These equations appeared in the proof of Theorem 5.1.1.) There is a unique solution $\xi \in L^2_{k+1}(M; i\mathbb{R})$ subject to the additional constraint $\int_M \xi = 0$. We introduce the subgroup

$$\mathcal{G}^{\perp}_{k+1} = \{\, e^{\xi} \mid \int_M \xi = 0 \,\}.$$

Then the action of the gauge group gives a diffeomorphism

$$\begin{aligned} \mathcal{G}^{\perp}_{k+1} \times \mathcal{K}_{k,\gamma_0} &\to \mathcal{C}_k \\ \big(e^{\xi}, (a, \phi)\big) &\mapsto (A_0 + (a - d\xi) \otimes 1, e^{\xi}\phi). \end{aligned}$$

In the case that the boundary is empty, the inverse map can be written

$$(A_0 + a \otimes 1, \phi) \mapsto \big(e^{-Gd^*a}, (A_0 + a - dGd^*a, e^{Gd^*a}\phi)\big),$$

where $G : L^2_{k-1}(M) \to L^2_{k+1}(M)$ is the Green's operator of $\Delta = d^*d$. The quotient space $\mathcal{B}_k(M, \mathfrak{s})$ is the quotient of the Coulomb slice $\iota(\mathcal{K}_k)$ by the action of the quotient group $\mathcal{G}_{k+1}/\mathcal{G}^{\perp}_{k+1}$. The quotient group has a concrete realization, because $\mathcal{G}_{k+1}$ can be written as a product,

$$\mathcal{G}_{k+1} = \mathcal{G}^h \times \mathcal{G}^{\perp}_{k+1},$$

where $\mathcal{G}^h$ consists of the *harmonic* maps $u : M \to S^1$ with (homogeneous) Neumann boundary conditions. So we can identify $\mathcal{B}_k(M, \mathfrak{s})$ with $\iota(\mathcal{K}_k)/\mathcal{G}^h$, or with $\mathcal{K}_k/\mathcal{G}^h$. The group $\mathcal{G}^h$ is isomorphic to $S^1 \times H^1(M; \mathbb{Z})$, but not canonically. What is canonical is the sequence

$$S^1 \to \mathcal{G}^h \to H^1(M; \mathbb{Z}), \tag{9.20}$$

where the first map is the inclusion of the constant maps and the second map assigns to each u its homotopy class as a map to the circle. We have also proved:

Lemma 9.6.1. *The group of components of the gauge group $\mathcal{G}_{k+1}$ is isomorphic to $H^1(M; \mathbb{Z})$.* □

9.7 The homotopy type of configuration spaces

The global slice results of the previous subsection give a very simple description of the homotopy type of the configuration spaces on 3- and 4-dimensional manifolds, using the identification

$$\mathcal{B}_k(M,\mathfrak{s}) = \iota(\mathcal{K}_{k,\gamma_0})/\mathcal{G}^h.$$

The projection map $\mathcal{K}_{k,\gamma_0} \to H^1(M;i\mathbb{R})$ (taking the harmonic part of the 1-form) induces a homotopy equivalence between the quotient space and a torus,

$$\mathcal{B}_k(M,\mathfrak{s}) \simeq H^1(M;i\mathbb{R})/H^1(M;i\mathbb{Z}).$$

For the blown-up configuration spaces, we blow up

$$\mathcal{K}_{k,\gamma_0} = \{\,(a,\Phi) \mid a \text{ is in Coulomb–Neumann gauge}\,\}$$

along the locus $\Phi = 0$ and then take the quotient. Up to homotopy type, we can simply consider the complement

$$\mathcal{K}^*_{k,\gamma_0} = \{\,(a,\Phi) \mid a \text{ is in Coulomb–Neumann gauge}, \Phi \neq 0\,\}$$

and we have a homotopy equivalence,

$$\mathcal{B}^\sigma_{k,\gamma_0} \simeq \iota(\mathcal{K}^*_{k,\gamma_0})/\mathcal{G}^h.$$

This quotient is a fiber bundle

$$\begin{aligned}\iota(\mathcal{K}^*_{k,\gamma_0})/\mathcal{G}^h &\to \{\,a \mid a \text{ is in Coulomb–Neumann gauge}\,\}/H^1(M;i\mathbb{R})\\ &\simeq H^1(M;i\mathbb{R})/H^1(M;i\mathbb{Z})\end{aligned}$$

by projecting away the spinor, with fiber

$$\big(L^2_k(M,S)\setminus\{0\}\big)/S^1.$$

Kuiper's theorem on the contractibility of the group of unitary transformations of a Hilbert space [64] gives this bundle a product structure; so the homotopy type of the quotient space $\mathcal{B}^\sigma_k(M,\mathfrak{s})$ is

$$\mathbb{CP}^\infty \times H^1(M;i\mathbb{R})/H^1(M;i\mathbb{Z}).$$

Without appealing to Kuiper's theorem, we can use the Leray–Hirsch theorem to identify the cohomology:

Proposition 9.7.1. *The cohomology ring of $\mathcal{B}^{\sigma}_k(M, \mathfrak{s})$ is isomorphic to*

$$\Lambda^*(H_1(M;\mathbb{Z})/\text{torsion}) \otimes \mathbb{Z}[u].$$

The isomorphism is natural with respect to the action of the group of orientation-preserving diffeomorphisms of M which preserve the spin^c structure $\mathfrak{s}$. □

10 Abstract perturbations

We need to introduce perturbations in order to obtain various transversality results, by the use of the Sard–Smale theorem. The perturbations must be mild enough to allow us to repeat the proofs of the compactness results above, but sufficiently flexible to achieve transversality. In order to apply the Sard–Smale theorem, we will eventually want a Banach space structure on our space of perturbations.

In this section, Y will continue to be a closed, connected, oriented Riemannian 3-manifold, but in addition *a spin^c structure $\mathfrak{s}$ will be fixed, throughout.* Thus we will just write $\mathcal{C}(Y)$, for example, instead of $\mathcal{C}(Y, \mathfrak{s})$. On the cylinder $Z = I \times Y$, the corresponding spin^c structure $\mathfrak{s}_Z$ will also be understood. We will be working with the Sobolev completions $\mathcal{C}_k(Z, \mathfrak{s})$ introduced in the previous section, and we will again fix an integer $k \geq 2$, so that $L^2_{k+1}(Z) \subset C^0(Z)$

10.1 Abstract perturbations

We consider a function $f : \mathcal{C}(Y) \to \mathbb{R}$ which is invariant under the gauge group $\mathcal{G}$ (not just the identity component), and write $\mathcal{L}\!\!\!-$ for the perturbed Chern–Simons–Dirac functional

$$\not{\mathcal{L}} = \mathcal{L} + f : \mathcal{C}(Y) \to \mathbb{R}.$$

We will be considering the gradient-flow equations for such a perturbed functional, and we need to put some constraints on f to allow, for example, the compactness theorems to be carried over from the unperturbed case. Because of its appearance in the equations, it is really the gradient of f with respect to the L^2 inner product that is the primary object for us. This gradient is, roughly speaking, a vector field on $\mathcal{C}(Y)$. However, it will be necessary to consider, more generally, a section of the completion of the tangent bundle in some Sobolev norm. The most general such section that we might need to consider would be a continuous map

$$\mathfrak{q} : \mathcal{C}(Y) \to L^2(Y; iT^*Y \oplus S),$$

or equivalently a section

$$\mathfrak{q} : \mathcal{C}(Y) \to \mathcal{T}_0, \tag{10.1}$$

where the notation $\mathcal{T}_j$ denotes the L^2_j completion of the tangent bundle, as in (9.4). We will say that such a $\mathfrak{q}$ is the *formal gradient* of a continuous function $f : \mathcal{C}(Y) \to \mathbb{R}$ if for every smooth path $\gamma : [0,1] \to \mathcal{C}(Y)$, we have

$$f \circ \gamma(1) - f \circ \gamma(0) = \int_0^1 \langle \dot{\gamma}, \mathfrak{q} \rangle_{L^2} \, dt.$$

When this relation holds, we may write $\mathfrak{q} = \operatorname{grad} f$, and will write

$$\operatorname{grad} \pounds = \operatorname{grad} \mathcal{L} + \mathfrak{q}.$$

We will require that our perturbation $\mathfrak{q}$ be $\operatorname{grad} f$ for some f that is invariant under $\mathcal{G}(Y)$. It follows from this hypothesis that $\mathfrak{q}$ is a $\mathcal{G}(Y)$-invariant section of $\mathcal{T}_0$ over $\mathcal{C}(Y)$.

In Section 4, we reinterpreted the gradient-flow equations as an equation for a pair $(A, \Phi) \in \mathcal{C}(Z)$, where Z was the cylinder $[t_1, t_2] \times Y$. Any continuous map $\mathfrak{q}$ as in (10.1) determines also a map

$$\hat{\mathfrak{q}} : \mathcal{C}(Z) \to \mathcal{V}_0(Z) \tag{10.2}$$

as follows. (Recall that $\mathcal{V}_j = \mathcal{V}_j(Z)$ is the trivial bundle with fiber $L^2_j(Z; i\,\mathfrak{su}(S^+) \oplus S^-)$.) By restricting to the slices $\{t\} \times Y$, we obtain from the pair (A, Φ) a continuous path $(\check{A}(t), \check{\Phi}(t))$ in $\mathcal{C}(Y)$ (see Definition 4.4.1). Thus we obtain a continuous path $\mathfrak{q}(\check{A}(t), \check{\Phi}(t))$ in $L^2(Y; iT^*Y \oplus S)$. A continuous path in $L^2(Y)$ determines an element of $L^2(Z)$, and the bundle $iT^*Y \oplus S$ on Z is identified with $i\,\mathfrak{su}(S^+) \oplus S^-$, using Clifford multiplication on the first summand.

The first analytic requirement we make of our perturbation $\mathfrak{q}$ is that the associated 4-dimensional perturbation $\hat{\mathfrak{q}}$ should extend to a section

$$\hat{\mathfrak{q}} : \mathcal{C}_k(Z) \to \mathcal{V}_k$$

that is infinitely differentiable. The fact that $\mathfrak{q}$ is invariant under $\mathcal{G}(Y)$ implies that the section $\hat{\mathfrak{q}}$ is invariant under $\mathcal{G}(Z)$, and hence by continuity it is also invariant under the Sobolev gauge group $\mathcal{G}_{k+1}(Z)$.

Using the inclusion $\mathcal{V}_k \subset \mathcal{V}_{k-1}$, we now regard $\mathfrak{F}+\hat{\mathfrak{q}}$ as a section of $\mathcal{V}_{k-1}(Z)$ (also infinitely differentiable). The vanishing of this section defines an equation in $\mathcal{C}_k(Z)$ that is invariant under $\mathcal{G}_{k+1}$:

Definition 10.1.1. Let $\mathfrak{q} : \mathcal{C}(Y) \to \mathcal{T}_0$ be a section that is the formal L^2 gradient of a continuous, $\mathcal{G}(Y)$-invariant function f, and having the property that the corresponding 4-dimensional perturbation $\hat{\mathfrak{q}}$ extends to a C^∞ section

$$\hat{\mathfrak{q}} : \mathcal{C}_k(Z) \to \mathcal{V}_k$$

on the cylinder $Z = [t_1, t_2] \times Y$. The equations

$$\mathfrak{F}(A, \Phi) + \hat{\mathfrak{q}}(A, \Phi) = 0$$

are then the *perturbed Seiberg–Witten equations* on Z for the perturbation $\mathfrak{q}$. As an abbreviation, we write $\mathfrak{F}_{\mathfrak{q}}$ for $\mathfrak{F} + \hat{\mathfrak{q}}$. ◊

To write out the perturbed equations more fully, we introduce notation for the two components of each of $\mathfrak{q}$ and $\hat{\mathfrak{q}}$. We write

$$\mathfrak{q} = (\mathfrak{q}^0, \mathfrak{q}^1)$$
$$\hat{\mathfrak{q}} = (\hat{\mathfrak{q}}^0, \hat{\mathfrak{q}}^1),$$

where

$$\mathfrak{q}^0 \in L^2(Y; iT^*Y)$$
$$\mathfrak{q}^1 \in L^2(Y; S)$$
$$\hat{\mathfrak{q}}^0 \in L^2(Z; i\,\mathfrak{su}(S^+))$$
$$\hat{\mathfrak{q}}^1 \in L^2(Z; S^-).$$

The bundle $\mathcal{V}_k \to \mathcal{C}_k(Z)$ is a sum $\mathcal{V}_k^0 \oplus \mathcal{V}_k^1$, and in these terms,

$$\hat{\mathfrak{q}}^0 : \mathcal{C}_k(Z) \to \mathcal{V}_k^0$$
$$\hat{\mathfrak{q}}^1 : \mathcal{C}_k(Z) \to \mathcal{V}_k^1$$

are smooth sections. The gradient-flow equation for a path $(B(t), \Psi(t))$ is

$$\begin{aligned} \frac{d}{dt}B^{t} &= -*F_{B^{t}} - 2\rho^{-1}(\Psi\Psi^{*})_{0} - 2\mathfrak{q}^{0}(B,\Psi) \\ \frac{d}{dt}\Psi &= -D_{B}\Psi - \mathfrak{q}^{1}(B,\Psi) \end{aligned} \tag{10.3}$$

(*cf.* (4.7)). If (A, Φ) denotes the 4-dimensional configuration on $Z = [t_1, t_2] \times Y$ obtained from the path (B, Ψ) (so that A is a 4-dimensional spin^c connection in temporal gauge), these equations are equivalent to the perturbed Seiberg–Witten equations for (A, Φ), which can be written

$$\begin{aligned} \rho_Z(F^{+}_{A^{t}}) - 2(\Phi\Phi^{*})_0 &= -2\hat{\mathfrak{q}}^{0}(A,\Phi) \\ D^{+}_{A}\Phi &= -\hat{\mathfrak{q}}^{1}(A,\Phi), \end{aligned} \tag{10.4}$$

where ρ_Z is the 4-dimensional Clifford multiplication. Under our assumption, these equations make sense for an arbitrary pair $(A, \Phi) \in \mathcal{C}_k(Z)$, and are invariant under $\mathcal{G}_{k+1}(Z)$. The invariance of $\hat{\mathfrak{q}}$ under $\mathcal{G}_{k+1}(Z)$ means simply that

$$\begin{aligned} \hat{\mathfrak{q}}^{0}(A', \Phi') &= \hat{\mathfrak{q}}^{0}(A,\Phi), \\ \hat{\mathfrak{q}}^{1}(A', \Phi') &= u\hat{\mathfrak{q}}^{1}(A,\Phi) \end{aligned}$$

when $(A', \Phi') = u(A, \Phi)$.

10.2 Abstract perturbations on the blow-up

The section $\hat{\mathfrak{q}}$ gives rise to a section $\hat{\mathfrak{q}}^{\sigma}$ of $\mathcal{V}^{\sigma}_{k} \to \mathcal{C}^{\sigma}_{k}(Z)$. To define this, we use the decomposition $\hat{\mathfrak{q}} = (\hat{\mathfrak{q}}^{0}, \hat{\mathfrak{q}}^{1})$, and we observe that the $\mathcal{G}$-equivariance (in particular the invariance under S^1) implies that $\hat{\mathfrak{q}}^{1}(A, \Phi)$ vanishes at $\Phi = 0$. The section $\hat{\mathfrak{q}}^{\sigma}$ is defined as

$$\hat{\mathfrak{q}}^{\sigma} = (\hat{\mathfrak{q}}^{0}, \hat{\mathfrak{q}}^{1,\sigma})$$

where $\hat{\mathfrak{q}}^{1,\sigma}$ is the section of $\mathcal{V}^{1,\sigma}_{k}$ given by

$$\hat{\mathfrak{q}}^{1,\sigma}(A, s, \phi) = \int_0^1 (\mathcal{D}_{(A, rs\phi)}\hat{\mathfrak{q}}^{1})(\phi)\, dr.$$

This definition should be compared with the definition of $\mathfrak{F}^{\sigma}$ in (6.5). The vanishing of $\hat{\mathfrak{q}}^{1}$ at the reducibles means that, for $\Phi = s\phi \neq 0$, the integral in the above definition can be integrated, to give the following formula:

$$\hat{\mathfrak{q}}^{1,\sigma}(A,s,\phi) = (1/s)\hat{\mathfrak{q}}^1(A,s\phi). \tag{10.5}$$

At $\Phi = 0$ on the other hand, the formula is

$$\hat{\mathfrak{q}}^{1,\sigma}(A,0,\phi) = (\mathcal{D}_{(A,0)}\hat{\mathfrak{q}}^1)(\phi).$$

The definition makes clear the following lemma:

Lemma 10.2.1. *If $\hat{\mathfrak{q}}$ is a C^ℓ section of $\mathcal{V}_k$, then $\hat{\mathfrak{q}}^\sigma$ is a $C^{\ell-1}$ section of $\mathcal{V}_k^\sigma$. Thus $\hat{\mathfrak{q}}^\sigma$ will be C^∞, under our assumption that $\hat{\mathfrak{q}}$ is C^∞.* □

We write $\mathfrak{F}_{\mathfrak{q}}^\sigma$ for the smooth section $\mathfrak{F}^\sigma + \hat{\mathfrak{q}}^\sigma$ of $\mathcal{V}_{k-1}^\sigma \to \mathcal{C}_k^\sigma$. The equations $\mathfrak{F}_{\mathfrak{q}}^\sigma(A,s,\phi) = 0$ are the *perturbed Seiberg–Witten equations* in the blow-up picture. When s is non-zero, the equations for (A,s,ϕ) can be written

$$\begin{aligned} \rho_Z(F_{A^t}^+) - 2s^2(\phi\phi^*)_0 &= -2\hat{\mathfrak{q}}^0(A,s\phi) \\ D_A^+\phi &= -(1/s)\hat{\mathfrak{q}}^1(A,s\phi), \end{aligned} \tag{10.6}$$

while when $s = 0$ we must replace the last term on the right by the derivative $-(\mathcal{D}_{(A,0)}\hat{\mathfrak{q}}^1)(\phi)$.

10.3 The perturbed flow on the blow-up

The discussion of the blow-up also has its 3-dimensional counterpart. The vector field $\mathfrak{q}$ on $\mathcal{C}_k(Y)$ gives rise to a vector field $\mathfrak{q}^\sigma$ on the blow-up $\mathcal{C}_k^\sigma(Y)$. We then have a perturbation $(\operatorname{grad} \text{£})^\sigma$ of $(\operatorname{grad} \mathcal{L})^\sigma$:

$$(\operatorname{grad} \text{£})^\sigma = (\operatorname{grad} \mathcal{L})^\sigma + \mathfrak{q}^\sigma.$$

This will be a smooth section of the vector bundle

$$\mathcal{T}_{k-1}^\sigma \to \mathcal{C}_k^\sigma(Y).$$

We can verify that $\mathfrak{q}^\sigma$ is smooth by writing it down in the coordinates. So we consider a path $(B(t), r(t), \psi(t))$ in $\mathcal{C}_k^\sigma(Y)$, with $\psi(t)$ having L^2 norm 1, and $r(t) \geq 0$. The corresponding path in $\mathcal{C}_k(Y)$ is $(B(t), r(t)\psi(t))$. To write down the equations we need the function that generalizes the function Λ which we defined in the unperturbed case in (6.7). In the perturbed case, we define

$$\Lambda_{\mathfrak{q}}(B,r,\psi) = \operatorname{Re}\langle \psi, D_B\psi + \tilde{\mathfrak{q}}^1(B,r,\psi)\rangle_{L^2} \tag{10.7}$$

where $\tilde{\mathfrak{q}}^1$ is defined in the same way as $\hat{\mathfrak{q}}^{1,\sigma}$ was above:

$$\tilde{\mathfrak{q}}^1(B, r, \psi) = \int_0^1 \mathcal{D}_{(B, sr\psi)}\mathfrak{q}^1(0, \psi)\, ds. \tag{10.8}$$

If $r(t)$ is not zero, we can write the equations (10.3) of the path $(B(t), r(t)\psi(t))$ as

$$\left.\begin{aligned} \frac{1}{2}\frac{d}{dt}B^{\mathrm{t}} &= -\frac{1}{2} * F_{B^{\mathrm{t}}} - r^2\rho^{-1}(\psi\psi^*)_0 - \mathfrak{q}^0(B, r\psi), \\ \frac{d}{dt}r &= -\Lambda_{\mathfrak{q}}(B, r, \psi)r \\ \frac{d}{dt}\psi &= -D_B\psi - \tilde{\mathfrak{q}}^1(B, r, \psi) + \Lambda_{\mathfrak{q}}(B, r, \psi)\psi. \end{aligned}\right\} \tag{10.9}$$

(Compare Equations (6.8).) This means that we can write

$$(\operatorname{grad} \mathcal{L})^\sigma = \big((\operatorname{grad} \mathcal{L})^{\sigma,0}, (\operatorname{grad} \mathcal{L})^{\sigma,r}, (\operatorname{grad} \mathcal{L})^{\sigma,1}\big)$$

in these coordinates as the vector field

$$\begin{bmatrix} (1/2) * F_{B^{\mathrm{t}}} + r^2\rho^{-1}(\psi\psi^*)_0 + \mathfrak{q}^0(B, r\psi) \\ \Lambda_{\mathfrak{q}}(B, r, \psi)r \\ D_B\psi + \tilde{\mathfrak{q}}^1(B, r, \psi) - \Lambda_{\mathfrak{q}}(B, r, \psi)\psi \end{bmatrix}. \tag{10.10}$$

In these coordinates, therefore, we can write $\mathfrak{q}^\sigma$ as

$$\mathfrak{q}^\sigma(B, r, \psi) = \big(\mathfrak{q}^0(B, r\psi),\ \langle \tilde{\mathfrak{q}}^1(B, r, \psi), \psi\rangle_{L^2(Y)} r,\ \tilde{\mathfrak{q}}^1(B, r, \psi)^\perp\big), \tag{10.11}$$

where the superscript $\perp$ denotes the projection to the real-orthogonal complement of ψ (the tangent space to the unit sphere at ψ). For the critical points, the following version of Proposition 6.2.3 holds.

Proposition 10.3.1. *Let (B, r, ψ) represent a point of $\mathcal{C}_k^\sigma(Y)$, so ψ is of unit L^2 norm and $r \geq 0$. Let $(B, r\psi)$ be the corresponding point in $\mathcal{C}_k(Y)$. Then (B, r, ψ) is a critical point of* $(\operatorname{grad} \mathcal{L})^\sigma$ *if and only if either:*

(i) *$r \neq 0$ and $(B, r\psi)$ is a critical point of* grad $\mathcal{L}$*; or*
(ii) *$r = 0$, the point $(B, 0)$ is a (reducible) critical point of* grad $\mathcal{L}$*, and ψ is an eigenvector of $\phi \mapsto D_B\phi + \mathcal{D}_{(B,0)}\mathfrak{q}^1(0, \phi)$.*

□

10.4 The perturbed equations in the τ model

We define a section $\hat{\mathfrak{q}}^\tau$ of $\mathcal{V}_k^\tau \to \mathcal{C}_k^\tau(Z)$ by defining $\hat{\mathfrak{q}}^\tau(\gamma)$ slicewise in terms of $\mathfrak{q}^\sigma$. In more detail, for a smooth element $\gamma = (A, s, \phi)$ in $\mathcal{C}^\tau(Z)$, we let $\check{\gamma} = (\check{A}, \check{s}, \check{\phi})$ be the corresponding smooth path in $\mathcal{C}^\sigma(Y)$, and we then have a continuous path

$$\mathfrak{q}^\sigma(\check{\gamma}(t)) \in L^2(Y; iT^*Y) \oplus \mathbb{R} \oplus L^2(Y; S)$$

given by the formula (10.11). The component in S is orthogonal to $\check{\phi}(t)$, for all t. We can regard this path, in turn, as defining an element

$$\hat{\mathfrak{q}}^\tau(\gamma) \in L^2(Z; i\,\mathfrak{su}(S^+)) \oplus L^2(I; \mathbb{R}) \oplus L^2(Z; S^-)$$

using Clifford multiplication to identify iT^*Y with $i\,\mathfrak{su}(S^+)$ on the cylinder as before. The orthogonality condition means that $\hat{\mathfrak{q}}^\tau(\gamma)$ lies in the subspace $\mathcal{V}_{0,\gamma}^\tau$ (see Definition 6.3.2). Our hypothesis that $\hat{\mathfrak{q}}$ extends to a smooth section of $\mathcal{V}_k \to \mathcal{C}_k$ implies that $\hat{\mathfrak{q}}^\tau$ extends to a smooth section

$$\hat{\mathfrak{q}}^\tau : \mathcal{C}_k^\tau(Z) \to \mathcal{V}_k^\tau,$$

much as in the previous cases. In the τ model, the perturbed Seiberg–Witten equations are the equations $\mathfrak{F}_\mathfrak{q}^\tau(\gamma) = 0$, where

$$\mathfrak{F}_\mathfrak{q}^\tau = \mathfrak{F}^\tau + \hat{\mathfrak{q}}^\tau,$$

regarded as a C^∞ section of $\mathcal{V}_{k-1}^\tau \to \mathcal{C}_k^\tau(Z)$.

10.5 Tame perturbations

We continue to write Z for the cylinder $[t_1, t_2] \times Y$, and consider a perturbation $\mathfrak{q}$. We wish to impose additional conditions on $\mathfrak{q}$ to obtain compactness results for the solutions to the perturbed equations. To define suitable constraints, we need a norm on the fibers of $\mathcal{V}_k$ that is gauge-invariant: on the fiber of $\mathcal{V}_k$ at $\gamma = (A, \Phi)$, we therefore take the Sobolev $L_{k,A}^2$ norm, defined using A as the covariant derivative. (Although the bundle is still trivial, as a normed vector bundle $\mathcal{V}_k$ is no longer trivial, because the norm is varying.) Similarly, on the bundle $T\mathcal{C}_k(Z)$, we have a gauge-invariant $L_{k,A}^2$ norm. Using the affine structure of $\mathcal{C}_k(Z)$, we will regard the mth derivative of a smooth section $\hat{\mathfrak{q}}$ of $\mathcal{V}_k$ at

$\gamma = (A, \Phi) \in \mathcal{C}_k(Z)$ as an element of the space of multi-linear maps

$$\begin{aligned}\mathcal{D}^m_\gamma \hat{\mathfrak{q}} &\in \mathrm{Mult}^m \left(\times_m L^2_{k,A}(Z; iT^*Z \oplus S^+), L^2_{k,A}(Z; i\,\mathfrak{su}(S^+) \oplus S^-) \right) \\ &= \mathrm{Mult}^m \left(\times_m T_\gamma \mathcal{C}_k(Z), \mathcal{V}_k \right).\end{aligned}$$

The norms of such derivatives will always be measured using these gauge-invariant norms. We also mention that, in the definition below, we use L^2 Sobolev spaces with *negative* Sobolev exponent. The space L^2_{-k} on a compact manifold is the dual space of L^2_k.

We are now ready to write down a suitable set of constraints on our perturbations: a catalog of estimates that we will need. Most of these conditions on a perturbation $\mathfrak{q}$ are phrased as conditions on the associated 4-dimensional perturbation $\hat{\mathfrak{q}}$ on the cylinder Z; these 4-dimensional conditions always imply a corresponding constraint for $\mathfrak{q}$ on the 3-manifold Y, because we can consider translation-invariant configurations as a special case.

Definition 10.5.1. Let k be an integer not less than 2. A perturbation $\mathfrak{q}$, given as a section

$$\mathfrak{q} : \mathcal{C}(Y) \to \mathcal{T}_0,$$

will be called *k-tame* if it is the formal gradient of a continuous, $\mathcal{G}(Y)$-invariant function on $\mathcal{C}(Y)$ and satisfies in addition:

(i) the corresponding 4-dimensional perturbation $\hat{\mathfrak{q}}$ defines an element

$$\hat{\mathfrak{q}} \in C^\infty\big(\mathcal{C}_k(Z), \mathcal{V}_k(Z)\big);$$

(ii) for every integer $j \in [1, k]$, the 4-dimensional perturbation $\hat{\mathfrak{q}}$ also defines an element

$$\hat{\mathfrak{q}} \in C^0\big(\mathcal{C}_j(Z), \mathcal{V}_j(Z)\big);$$

(iii) for every integer $j \in [-k, k]$, the first derivative,

$$\mathcal{D}\hat{\mathfrak{q}} \in C^\infty\big(\mathcal{C}_k(Z), \mathrm{Hom}(T\mathcal{C}_k(Z), \mathcal{V}_k(Z))\big),$$

extends to a map

$$\mathcal{D}\hat{\mathfrak{q}} \in C^{\infty}\big(\mathcal{C}_k(Z), \mathrm{Hom}(T\mathcal{C}_j(Z), \mathcal{V}_j(Z))\big);$$

(iv) there is a constant m_2 such that

$$\|\mathfrak{q}(B, \Psi)\|_{L^2} \leq m_2\big(\|\Psi\|_{L^2} + 1\big)$$

holds for all configurations (B, Ψ) in $\mathcal{C}_k(Y)$;

(v) for any choice of smooth connection A_0, there is a real function μ_1 such that the inequality

$$\|\hat{\mathfrak{q}}(A, \Phi)\|_{L^2_{1,A}} \leq \mu_1\left(\|(A - A_0, \Phi)\|_{L^2_{1,A_0}}\right)$$

holds, for all $(A, \Phi) \in \mathcal{C}_k(Z)$;

(vi) the 3-dimensional perturbation $\mathfrak{q}$ defines a C^1 section

$$\mathfrak{q} : \mathcal{C}_1(Y) \to \mathcal{T}_0.$$

We simply say that $\mathfrak{q}$ is *tame* if it is k-tame for all $k \geq 2$. ◊

Remark. In these definitions, the notation C^0 (for example) refers simply to the space of continuous sections, not to the Banach space of bounded continuous sections. The k-tame perturbations do not form a Banach space. Later, however, we will introduce a smaller (Banach) space of perturbations, in order to be able to apply the Sard-Smale theorem.

We briefly mention where the various hypotheses will be used. Condition (i), as indicated above, is a natural condition allowing us to set up the perturbed equations. Condition (iii) is relevant when we study the Fredholm properties of the linearization of the equation $\mathfrak{F}_{\mathfrak{q}} = 0$ on the Coulomb slice. Condition (iv) is used to obtain an inequality that plays the role of the identity leading to Lemma 4.5.5. Conditions (ii), (iv) and (v) are used for a bootstrapping argument in the proof of the compactness theorem for the perturbed equations.

With the exception of Condition (iii), all these matters will be dealt with in the remainder of this section.

10.6 Perturbations and integration by parts

We consider first the issue of the identity in Lemma 4.5.5, and prove a version for the perturbed gradient flow. Let $\mathfrak{q}$ be a k-tame perturbation, the formal gradient of a function f on $\mathcal{C}(Y)$.

For the perturbed equations on the cylinder $Z = [t_1, t_2] \times Y$, we define the (perturbed, analytic) energy using the same formula that appeared in Lemma 4.5.5, but using the perturbed functional $\mathcal{L}\!\!\!- = \mathcal{L} + f$ in place of $\mathcal{L}$:

$$\mathcal{E}^{\mathrm{an}}_{\mathfrak{q}}(A, \Phi) = \int_{t_1}^{t_2} \left\| \frac{d}{dt}\check{A} - dc \right\|^2 dt + \int_{t_1}^{t_2} \left\| \frac{d}{dt}\check{\Phi} + c\check{\Phi} \right\|^2 dt + \int_{t_1}^{t_2} \| \operatorname{grad} \mathcal{L}\!\!\!-\|^2 dt. \tag{10.12}$$

For a solution of the perturbed Seiberg–Witten equations, this is twice the drop in $\mathcal{L}\!\!\!-$:

$$\mathcal{E}^{\mathrm{an}}_{\mathfrak{q}}(\gamma) = 2\big(\mathcal{L}\!\!\!-(\check{\gamma}(t_1)) - \mathcal{L}\!\!\!-(\check{\gamma}(t_2))\big). \tag{10.13}$$

Recall that in the unperturbed case, a formula such as (10.12) was not taken as the primary definition of the analytic energy $\mathcal{E}^{\mathrm{an}}$: the primary definition of $\mathcal{E}^{\mathrm{an}}$ is given in Definition 4.5.4, and a formula of the above sort was derived in Lemma 4.5.5. The importance of the formula in Definition 4.5.4 is that it shows that $\mathcal{E}^{\mathrm{an}}$ controls the L^2 norm of such quantities as F_A and $\nabla_A \Phi$. The next lemma plays the same role for the perturbed energy $\mathcal{E}^{\mathrm{an}}_{\mathfrak{q}}$.

Lemma 10.6.1. *Let $\mathfrak{q}$ be a k-tame perturbation. Let Z be the cylinder $[t_1, t_2] \times Y$. Then for all $(A, \Phi) \in \mathcal{C}(Z)$ we have*

$$\mathcal{E}^{\mathrm{an}}_{\mathfrak{q}}(A, \Phi) \geq \frac{1}{2} \int_Z \left(\tfrac{1}{4}|F_{A^t}|^2 + |\nabla_A \Phi|^2 + \tfrac{1}{4}(|\Phi|^2 - C)^2 \right) - C'(t_2 - t_1) \tag{10.14}$$

for some constants C, C' which do not depend on A and Φ.

Proof. We have

$$\begin{aligned} \| \operatorname{grad} \mathcal{L}\!\!\!-\|^2 &= \| \operatorname{grad} \mathcal{L} + \mathfrak{q} \|^2 \\ &\geq \tfrac{1}{2} \| \operatorname{grad} \mathcal{L} \|^2 - \|\mathfrak{q}\|^2, \end{aligned}$$

by an application of a Peter–Paul inequality to the cross-term. From this, using the formula (10.12) for $\mathcal{E}^{\mathrm{an}}_{\mathfrak{q}}$, we obtain

$$\mathcal{E}^{\mathrm{an}}_{\mathfrak{q}}(A, \Phi) \geq \frac{1}{2} \mathcal{E}^{\mathrm{an}}(A, \Phi) - \int_{t_1}^{t_2} \|\mathfrak{q}\|^2 \, dt,$$

where $\mathcal{E}^{\text{an}}$ on the right is the unperturbed analytic energy. Using the formula for $\mathcal{E}^{\text{an}}$ from Definition 4.5.4, we rewrite this as

$$\mathcal{E}^{\text{an}}_{\mathfrak{q}}(A, \Phi) \geq$$
$$\frac{1}{2}\int_Z \left(\frac{1}{4}|F_{A^t}|^2 + |\nabla_A \Phi|^2 + \frac{1}{4}\left(|\Phi|^2 + (s/2)\right)^2 - s^2/16\right) - \int_{t_1}^{t_2} \|\mathfrak{q}\|^2 \, dt.$$

The last term can be bounded in terms of the squared L^2 norm of Φ on the cylinder, using Condition (iv) of Definition 10.5.1. This results in an extra quadratic term in Φ, which can be absorbed by the quartic term in Φ on the right, to give an inequality

$$\int_{t_1}^{t_2} \left(\|\dot{\gamma}(t)\|^2 + \|\operatorname{grad} \mathcal{L}\|^2\right) dt$$
$$\geq \frac{1}{2}\int_Z \left(\tfrac{1}{4}|F_{A^t}|^2 + |\nabla_A \Phi|^2 + \tfrac{1}{4}(|\Phi|^2 - C_1)^2\right) - C_2(t_2 - t_1)$$

which can be rewritten to give the statement of the lemma. □

Corollary 10.6.2. *With the same hypotheses as Lemma* 10.6.1, *if* $(A, \Phi) \in \mathcal{C}_k(Z)$ *is a solution of the perturbed Seiberg–Witten equations* (10.4), *then*

$$2(\mathcal{L}(t_1) - \mathcal{L}(t_2))$$
$$\geq \frac{1}{2}\int_Z \left(\tfrac{1}{4}|F_{A^t}|^2 + |\nabla_A \Phi|^2 + \tfrac{1}{4}(|\Phi|^2 - C)^2\right) - C'(t_2 - t_1).$$

Proof. The inequality in the lemma holds for $(A, \Phi) \in \mathcal{C}_k(Z)$ (not just for smooth configurations) because the terms that appear are continuous in the L^2_k topology. □

10.7 Compactness for the perturbed Seiberg–Witten equations

The properness results of Theorem 5.2.1 lead in a straightforward way to a proof of a compactness theorem for the perturbed Seiberg–Witten equations (10.4).

Theorem 10.7.1. *Let* $\mathfrak{q}$ *be a* k*-tame perturbation. Let* $\gamma_n \in \mathcal{C}_k(Z)$ *be a sequence of solutions of the perturbed Seiberg–Witten equations,* $\mathfrak{F}_{\mathfrak{q}}(\gamma) = 0$, *on the cylinder* $Z = [t_1, t_2] \times Y$. *Suppose that the drop in the perturbed Chern–Simons–Dirac functional is uniformly bounded:*

$$\mathcal{L}(\check{\gamma}_n(t_1)) - \mathcal{L}(\check{\gamma}_n(t_2)) \leq C.$$

Then there is a sequence of gauge transformations, $u_n \in \mathcal{G}_{k+1}(Z)$, such that, after passing to a subsequence, the transformed solution $u_n(\gamma_n)$ is a convergent sequence in $\mathcal{C}_{k+1}(Z')$, for any interior cylinder $Z' = [t_1', t_2'] \times Y$.

Proof. Write $\gamma_n = (A_n, \Phi_n)$. The bound in the change of $\mathcal{L}$ gives us bounds

$$\int_Z |F_{A_n^t}|^2 \le C_1, \qquad \int_Z |\Phi_n|^4 \le C_2, \qquad \int_Z |\nabla_{A_n}\Phi_n|^2 \le C_3 \tag{10.15}$$

on the cylinder $Z = [t_1, t_2] \times Y$, via Corollary 10.6.2. We can perform gauge transformations $u_n \in \mathcal{G}_{k+1}$ on Z so that with respect to some base connection A_0 on Z the connections A_n are all in Coulomb–Neumann gauge (5.2) and satisfy the period conditions (5.3). The transformed pairs

$$(\tilde{A}_n, \tilde{\Phi}_n) = u_n(A_n, \Phi_n)$$

have $\tilde{\Phi}_n$ and $\tilde{A}_n - A_0$ uniformly bounded in the norm $L^2_1(Z)$. Then by Condition (v) of Definition 10.5.1, the perturbing terms $\hat{\mathfrak{q}}(\tilde{A}_n, \tilde{\Phi}_n)$ are a uniformly bounded sequence in L^2_1, and so we can pass to a subsequence with $\hat{\mathfrak{q}}(\tilde{A}_n, \tilde{\Phi}_n)$ converging in L^2. This means that $\mathfrak{F}(\tilde{A}_n, \tilde{\Phi}_n)$ is converging in L^2 (where $\mathfrak{F}$ is the unperturbed Seiberg–Witten map). By Part (ii)(a) of Theorem 5.2.1, we can therefore pass to a further subsequence where $(\tilde{A}_n, \tilde{\Phi}_n)$ converges on interior domains in the L^2_1 topology. (The hypothesis in Theorem 5.2.1, that the topological energy is uniformly bounded, follows from the bounds (10.15).) By Condition (ii) of Definition 10.5.1, the sequence $\hat{\mathfrak{q}}(\tilde{A}_n, \tilde{\Phi}_n)$ converges in the L^2_1 topology on interior domains. Another application of Theorem 5.2.1 gives convergence of $(\tilde{A}_n, \tilde{\Phi}_n)$ in the L^2_2 topology on interior domains. We can continue in this manner to prove convergence on interior domains in the L^2_{k+1} topology. □

We note in addition a regularity result that comes from the bootstrapping argument at the end of the proof above.

Proposition 10.7.2. *If $\gamma \in \mathcal{C}_k(Z)$ is a solution of $\mathfrak{F}_{\mathfrak{q}}(\gamma) = 0$ for some k-tame $\mathfrak{q}$, then there exists a gauge transformation $u \in \mathcal{G}_{k+1}$ such that the restriction of $u(\gamma)$ to any interior cylinder Z' is in the image of the inclusion $\mathcal{C}_{k+1}(Z') \to \mathcal{C}_k(Z')$.* □

Corollary 10.7.3. *If $\gamma \in \mathcal{C}_k(Z)$ is a solution of $\mathfrak{F}_{\mathfrak{q}}(\gamma) = 0$ for some k-tame $\mathfrak{q}$, then there exists a gauge transformation $u \in \mathcal{G}_{k+1}$ such that if we write $u(\gamma) = (A, \Phi)$ and write $(\check{A}, \check{\Phi})$ for the corresponding path in the 3-dimensional configuration space, then $(\check{A}, \check{\Phi})$ is an $L^2_{1,\mathrm{loc}}$ path, $(t_1, t_2) \to \mathcal{C}_k(Y)$.* □

From Theorem 10.7.1, we can deduce a compactness result for solutions to the perturbed 3-dimensional Seiberg–Witten equations,

$$(\mathrm{grad}\,\mathcal{L})(B, \Psi) = 0, \tag{10.16}$$

for $(B, \Psi) \in \mathcal{C}_k(Y)$.

Corollary 10.7.4. *The image in* $\mathcal{B}_k(Y, \mathfrak{s})$ *of the set of solutions* (B, Ψ) *to the perturbed 3-dimensional Seiberg–Witten equations* (10.16) *is compact.*

Proof. A sequence of solutions to the 3-dimensional equations on Y gives rise to a sequence of translation-invariant solutions γ_i of the 4-dimensional Seiberg–Witten equations on $\mathbb{R} \times Y$. We can then apply Theorem 10.7.1 to the 4-dimensional solutions, restricted to any finite cylinder, say $[-1, 1] \times Y$. The drop in $\mathcal{L}$ along the cylinder is zero, so we obtain, for example,

$$\int_Y \left|F_{B_i^t}\right|^2 \leq C_1, \tag{10.17}$$

with similar estimates for the integrals of $|\Psi_i|^4$ and $|\nabla_B \Psi|^2$. After applying a 4-dimensional gauge-transformation and passing to a subsequence, the solutions γ_i will converge in the L^2_{k+1} topology on any interior cylinder. They will therefore converge in the L^2_k topology on the slice $\{0\} \times Y$. □

In the statement of Corollary 10.7.4, we emphasize that it is only for a fixed spin^c structure $\mathfrak{s}$ that a compactness statement is being made. There is the auxiliary question of whether there may be infinitely many spin^c structures $\mathfrak{s}$ on Y for which the 3-dimensional equations admit solutions. For the unperturbed equations, this cannot occur, as follows easily from Item (i) of Theorem 5.1.1. For the perturbed equations, it is necessary to have chosen perturbations $\mathfrak{q}_\mathfrak{s}$, one for each isomorphism class of spin^c structure, before we are even able to formulate the question; and for a finiteness result to hold, we certainly need some uniformity in these perturbations. The following lemma clarifies the situation.

Lemma 10.7.5. *Suppose* $\{\mathfrak{q}_\mathfrak{s}\}$ *are tame perturbations, one for each spin^c structure* $\mathfrak{s}$ *on the Riemannian manifold* Y*. Suppose that these satisfy a uniform version of the inequality in Item* (iv) *of Definition* 10.5.1:

$$\|\mathfrak{q}_\mathfrak{s}(B, \Psi)\|_{L^2} \leq m_2\big(\|\Psi\|_{L^2} + 1\big),$$

where m_2 *is independent of* $\mathfrak{s}$*. Then the perturbed 3-dimensional equations on* Y *admit solutions for only finitely many* $\mathfrak{s}$.

Proof. The constant C_1 in (10.17) depends only on the Riemannian metric and the constant m_2. So if m_2 is uniform, we obtain a bound on the L^2 norm of the curvature. This means that the real first Chern class of the spin^c structure lies in a bounded subset of $H^2(Y;\mathbb{R})$, leaving only finitely many possibilities for $\mathfrak{s}$. □

10.8 Unique continuation for the perturbed equations

Let $\mathfrak{q}$ be a k-tame perturbation, and consider a solution (A, Φ) of the perturbed Seiberg–Witten equations $\mathfrak{F}_{\mathfrak{q}}(A, \Phi) = 0$ on a cylinder $Z = [t_1, t_2] \times Y$. We wish to extend the unique continuation results of Section 7 to the perturbed setting, starting with a version of Proposition 7.1.2, which says that the spinor Φ is identically zero if it vanishes on a slice.

Proposition 10.8.1. *Suppose $(A, \Phi) \in \mathcal{C}_k(Z, \mathfrak{s}_Z)$ is a solution of the perturbed Seiberg–Witten equations $\mathfrak{F}_{\mathfrak{q}}(A, \Phi) = 0$ on $Z = [t_1, t_2] \times Y$, for some k-tame perturbation $\mathfrak{q}$. If Φ is zero on the slice $\{t\} \times Y$ for some t in the open interval (t_1, t_2), then Φ is identically zero.*

Proof. Recall that our underlying assumption about k is that it is an integer not less than 2, so that L^2_{k+1} is in the range of the Sobolev embedding theorem. According to Proposition 10.7.2, we may assume that A and Φ are of class L^2_{k+1} in the interior. Replacing Z by a slightly smaller cylinder, we may then assume that (A, Φ) belongs to $\mathcal{C}_{k+1}(Z, \mathfrak{s}_Z)$. They are therefore continuous. At this point we may as well consider the case that t is an endpoint of the interval, say t_2.

The equation for Φ can be written in terms of the corresponding path $(\check{A}, \check{\Phi})$ and the dt component c of the connection:

$$\frac{d}{dt}\check{\Phi} + c\check{\Phi} + D_{\check{A}(t)}\check{\Phi} = \mathfrak{q}^1(\check{A}, \check{\Phi}). \tag{10.18}$$

The action of the gauge group includes the action of S^1, acting on Φ by scalar multiplication and trivially on A; so $(A, e^{i\theta}\Phi)$ is a solution of the equations, for all θ; and by differentiating (10.18) with respect to θ we obtain

$$i\left(\frac{d}{dt}\check{\Phi} + c\check{\Phi}\right) = -iD_{\check{A}}\check{\Phi} - (\mathcal{D}_{(\check{A},\check{\Phi})}\mathfrak{q}^1)(0, i\check{\Phi}). \tag{10.19}$$

Let H denote the Hilbert space of L^2 spinors on Y. Corollary 10.7.3 tells us that $(\check{A}, \check{\Phi})$ defines a continuous path in $\mathcal{C}_k(Y)$; and Condition (iii) in

Definition 10.5.1 then tell us that the operator

$$\psi \mapsto \left(\mathcal{D}_{(\check{A}(t),\check{\Phi}(t))}\mathfrak{q}^1\right)(0,\psi) \tag{10.20}$$

is a bounded operator $H \to H$ depending continuously on t. Because c is continuous, multiplication by c defines a bounded operator on H, also depending continuously on t. Because Φ is in $L^2_{k+1}(Z;S)$, it defines an L^2_{k+1} path

$$[t_1,t_2] \to H,$$

where H is the Hilbert space of L^2 spinors on Y. In dimension 1, we have $L^2_{k+1} \hookrightarrow C^k$; so the path is at least C^2. The equation therefore fits into the general framework of Lemma 7.1.3, for we can write it as

$$\frac{d}{dt}\check{\Phi} + D_{\check{A}}\check{\Phi} = f(t),$$

where $\|f(t)\|$ is bounded by $\delta\|\check{\Phi}(t)\|$ by our observations about the operator (10.20) and c. The remaining hypothesis that needs to be checked is the bound (7.6b) on the derivative of the linear operator. Because A is C^0, this bound can be equivalently expressed as

$$\|\rho((d/dt)\check{A})\check{\Phi}\|_{L^2(Y)} \le C\|D\check{\Phi}\|_{L^2(Y)} + C'\|\check{\Phi}\|_{L^2(Y)},$$

or equivalently

$$\|\rho((d/dt)\check{A})\check{\Phi}\|_{L^2(Y)} \le C''\|\check{\Phi}\|_{L^2_1(Y)}.$$

The path $\check{A}$ is C^1 in the space of L^2_1 connections on Y. So the required bound follows from the continuity of multiplication $L^2_1 \times L^2_1 \to L^2$ in dimension 3. Lemma 7.1.3 therefore applies. □

We can also prove a unique continuation result for the full perturbed equations, along the same lines as Proposition 7.2.1. For this version, we need some extra regularity for the perturbation: and we assume that the perturbation $\mathfrak{q}$ is both k-tame and $(k+1)$-tame.

Proposition 10.8.2. *Suppose $k \ge 2$, and let $\mathfrak{q}$ be a perturbation that is k-tame and $(k+1)$-tame. Let $\gamma_1, \gamma_2 \in \mathcal{C}_k(Z, \mathfrak{s}_Z)$ be two solutions of the perturbed Seiberg–Witten equations $\mathfrak{F}_{\mathfrak{q}}(\gamma) = 0$ on a cylinder $Z = [t_1, t_2] \times Y$. Suppose there exists some t_0 in the open interval (t_1, t_2) for which $\check{\gamma}_1(t_0)$ and $\check{\gamma}_2(t_0)$ are gauge-equivalent on Y. Then γ_1 and γ_2 are gauge-equivalent on Z.*

The same result holds for solutions to the equations $\mathfrak{F}^\tau(\gamma^\tau) = 0$ in the blown-up picture.

Proof. As in the proof of Proposition 10.8.1, we can appeal to Proposition 10.7.2 and replace the cylinder by a smaller one, so as to have γ_i in $\mathcal{C}_{k+2}(Z, \mathfrak{s}_Z)$. Next, we can apply a gauge transformation so that γ_1 and γ_2 are both in temporal gauge and are equal on the slice $\{t_0\} \times Y$. The gauge transformation that achieves this may only be of class L^2_{k+2}, so after the gauge transformation, the solutions γ_i will belong only to $\mathcal{C}_{k+1}(Z, \mathfrak{s}_Z)$; but this still means that the spinors and connection forms are continuous on the cylinder, as in the previous proposition.

We now have two continuous paths $\check{\gamma}_1(t)$, $\check{\gamma}_2(t)$ in $\mathcal{C}_k(Y)$. We can join the two paths by a straight-line homotopy $\check{\gamma}(s,t)$, and then use Condition (iii) of Definition 10.5.1 to obtain an inequality

$$\big\|\mathfrak{q}(\check{\gamma}_1(t)) - \mathfrak{q}(\check{\gamma}_2(t))\big\|_{L^2} \le \delta \|\check{\gamma}_1(t) - \check{\gamma}_2(t)\|_{L^2}$$

for some constant δ independent of t. This allows us to apply Lemma 7.1.3, and the remaining points in the proof can be modelled on the argument we used in the unperturbed case, Proposition 7.2.1. The same remarks apply to the blown-up equations. □

10.9 Convergence to reducible solutions

We now turn to the refined compactness result for the blown-up configuration space. Our task is to extend the results of Section 8 to the perturbed situation.

We introduced earlier the gauge-invariant function $\Lambda_\mathfrak{q}$ (see (10.7)) on $\mathcal{C}^\sigma_k(Y)$, a version of the function Λ adapted to the perturbed equation. It is a C^∞ function on the blown-up configuration space. Note that if Z is the cylinder $[t_1, t_2] \times Y$, and $\gamma^\tau = (A, s, \phi)$ belongs to $\mathcal{C}^\tau_k(Z)$, then the function

$$t \mapsto \Lambda_\mathfrak{q}(\check{A}(t), s(t), \check{\phi}(t))$$

is a C^{k-1} function (in particular a C^1 function) of t on the open interval (t_1, t_2): this follows from Proposition 10.7.2 and the definition of $\Lambda_\mathfrak{q}$. The following is a version of Lemma 8.1.2 in the perturbed situation.

Lemma 10.9.1. *There is a continuous function positive ζ on $\mathcal{C}^\sigma_k(Y)$, such that for any solution γ^τ to the equations $\mathfrak{F}^\tau_\mathfrak{q}(\gamma^\tau) = 0$ on a cylinder $Z = [t_1, t_2] \times Y$,* *we have the inequality*

$$\frac{d}{dt}\Lambda_{\mathfrak{q}}(\check{\gamma}^\tau(t)) \le \zeta(\check{\gamma}(t))\|\operatorname{grad}\mathcal{L}(\check{\gamma}(t))\|_{L^2_k(Y)},$$

where γ on the right is the trajectory in $\mathcal{C}_k(Z)$ obtained from γ^τ by blowing down.

Remark. Although the left-hand side is a continuous function of t on the open interval, the right-hand side need not be: the right-hand side is defined for almost all t and defines a locally square-integrable function on the open interval.

Proof of Lemma 10.9.1. Suppose first that $\gamma^\tau = (A, s, \phi)$ is not reducible, so $s(t)$ is everywhere non-zero, and write $\gamma = (A, \Phi)$ for the image of γ^τ under the blow-down. As in the proof of Proposition 10.8.1, we use the fact that $(A, e^{i\theta}\Phi)$ is a solution of the equations, for all θ, and we differentiate (10.18) with respect to θ to obtain

$$i\left(\frac{d}{dt}\check{\Phi} + c\check{\Phi}\right) = -iD_{\check{A}}\check{\Phi} - \left(\mathcal{D}_{(\check{A},\check{\Phi})}\mathfrak{q}^1\right)(0, i\check{\Phi}).$$

Condition (iii) of Definition 10.5.1 (with $j = 0$) implies that $\mathcal{D}_{(\check{A}(t),\check{\Phi}(t))}\mathfrak{q}^1$ is a bounded operator on L^2. As the derivative of a gradient, it is also self-adjoint. (It is not complex-linear, however, for $\check{\Phi} \ne 0$.) We write the identity above as

$$\left(\frac{d}{dt}\check{\Phi} + c\check{\Phi}\right) = -L\check{\phi}$$

where L is a symmetric, time-dependent linear operator:

$$L\psi = D_{\check{A}}\psi - i(\mathcal{D}_{(\check{A},\check{\Phi})}\mathfrak{q}^1)(0, i\psi). \tag{10.21}$$

We now follow the proof of Lemma 7.1.1, to arrive at the inequality

$$\frac{d}{dt}\Lambda_{\mathfrak{q}}(\check{\gamma}^\tau(t)) \le \langle\check{\Phi}(t), L'\check{\Phi}(t)\rangle \;/\; \|\check{\Phi}(t)\|^2. \tag{10.22}$$

Here L' denotes the commutator $L' = [(d/dt) + c, L]$. We compute this gauge-equivariant operator in a temporal gauge ($c = 0$):

$$\begin{aligned} L'\psi &= \rho\left(\frac{d}{dt}\check{A}\right)\psi - i\mathcal{D}^2_{(\check{A},\check{\Phi})}\mathfrak{q}^1\left(\left(\frac{d}{dt}\check{A}, \frac{d}{dt}\check{\Phi}\right), (0, i\psi)\right) \\ &= -\rho((\operatorname{grad}\mathcal{L})^0)\psi - i\mathcal{D}^2_{(\check{A},\check{\Phi})}\mathfrak{q}^1(\operatorname{grad}\mathcal{L}, (0, i\psi)). \end{aligned}$$

(We have written grad $\mathcal{L} = ((\text{grad}\,\mathcal{L})^0, (\text{grad}\,\mathcal{L})^1)$.) The last line is a gauge-invariant formula. Condition (iii) of Definition 10.5.1, with $j = 0$, tells us that the second derivative of $\mathfrak{q}^1$ is a bounded bilinear operator

$$\mathcal{D}^2_{(\check{A}(t),\check{\Phi}(t))}\mathfrak{q}^1 : L^2_{k,\check{A}}(Y; iT^*Y \oplus S) \times L^2(Y; iT^*Y \oplus S) \to L^2(Y; S),$$

and that its norm depends continuously on $\check{\gamma}(t) = (\check{A}(t), \check{\Phi}(t)) \in \mathcal{C}_k(Y)$. We therefore have

$$\|L'\psi\|_{L^2} \leq \left(\|(\text{grad}\,\mathcal{L})^0\|_{C^0} + \zeta_1(\check{\gamma}(t))\|\,\text{grad}\,\mathcal{L}\|_{L^2_{k,\check{A}}} \right) \|\psi\|_{L^2}$$

for some continuous function ζ_1. Because $L^2_k(Y) \hookrightarrow C^0$, the first term in the parenthesis can be absorbed by the second, and this gives the result. □

We are now in a position to prove a version of Theorem 8.1.1 for the perturbed equations.

Theorem 10.9.2. *Let $\mathfrak{q}$ be a k-tame perturbation. Let $Z = [t_1, t_2] \times Y$ be a cylinder, and let $Z_\epsilon = [t_1 + \epsilon, t_2 - \epsilon]$ be a smaller cylinder, so $Z_\epsilon \Subset Z$. Let $\gamma_n^\tau \in \mathcal{C}_k^\tau(Z, \mathfrak{s})$ be a sequence of solutions of the perturbed equations $\mathfrak{F}^\tau_{\mathfrak{q}}(\gamma^\tau) = 0$ on Z, and let $\check{\gamma}^\tau$ be the corresponding paths in $\mathcal{C}^\sigma_k(Y, \mathfrak{s})$. Suppose that the drop in the perturbed Chern–Simons–Dirac functional is uniformly bounded on the larger cylinder Z,*

$$\mathcal{L}(\check{\gamma}_n^\tau(t_1)) - \mathcal{L}(\check{\gamma}_n^\tau(t_2)) \leq C_1,$$

and suppose that there are one-sided bounds on the value of Λ at the endpoints of the smaller cylinder Z_ϵ:

$$\Lambda_{\mathfrak{q}}(\check{\gamma}_n^\tau(t_1 + \epsilon)) \leq C_2$$
$$\Lambda_{\mathfrak{q}}(\check{\gamma}_n^\tau(t_2 - \epsilon)) \geq -C_2.$$

Then there is a sequence of gauge transformations, $u_n \in \mathcal{G}_{k+1}(Z)$, such that, after passing to a subsequence, the transformed solutions $u_n(\gamma_n^\tau)$ have the following property: for every interior domain $Z' \Subset Z_\epsilon$, the transformed solutions belong to $\mathcal{C}^\tau_{k+1}(Z', \mathfrak{s})$ and converge in the topology of $\mathcal{C}^\tau_{k+1}(Z', \mathfrak{s})$ to a solution $\gamma^\tau \in \mathcal{C}^\tau_{k+1}(Z', \mathfrak{s})$ of the equations $\mathfrak{F}^\tau_{\mathfrak{q}}(\gamma^\tau) = 0$.

Proof. The proof follows the same course as the proof of Theorem 8.1.1 with minor changes. To begin, we apply Theorem 10.7.1, which allows us to assume that the trajectories $\gamma_n = \pi\gamma_n^\tau$ define a convergent sequence in $\mathcal{C}_{k+1}(Z_\epsilon)$. Thus

$$t \mapsto \operatorname{grad} \mathcal{L}(\check{\gamma}_n(t))$$

is an L^2 path in the Hilbert space $L^2_k(Y; iT^*Y \oplus S)$, converging in the topology of L^2 paths as $n \to \infty$. The functions

$$t \mapsto \zeta(\check{\gamma}_n(t)) \| \operatorname{grad} \mathcal{L}(\check{\gamma}_n(t)) \|_{L^2_k(Y)}$$

that appear in Lemma 10.9.1 are therefore an L^2-convergent sequence of functions on the interval $[t_1 + \epsilon, t_2 - \epsilon]$, and their integrals are therefore uniformly bounded. Combining Lemma 10.9.1 with the assumed bounds on $\Lambda_\mathfrak{q}$ at the endpoints, we obtain a two-sided uniform bound

$$|\Lambda_\mathfrak{q}(\check{\gamma}_n^\tau(t))| \leq M,$$

as in the unperturbed case, leading to the same inequality (8.1).

As before, the interesting case is now when the Φ_n are non-zero but are converging to zero in $L^2(Z_\epsilon)$. In this case, we once again consider the normalized spinors Φ_n^1 as in (8.2). We must show that these converge in the L^2_{k+1} topology on interior cylinders $Z' \Subset Z_\epsilon$. Although the equation satisfied by Φ_n^1 is a non-linear one in the perturbed setting, we can again use the circle-action, as we did in deriving the equation (10.19); in the 4-dimensional setting, this allows us to regard Φ_n^1 as a solution of the equation

$$D^+_{A_n} \Phi_n^1 = i\big(\mathcal{D}_{(A_n,\Phi_n)} \hat{\mathfrak{q}}^1\big)(0, i\Phi_n^1),$$

which we rewrite as

$$D^+_{A_0} \Phi_n^1 = W_n \Phi_n^1$$

where

$$W_n \Phi_n^1 = \rho_Z(A_0 - A_n)\Phi_n^1 + i\big(\mathcal{D}_{(A_n,\Phi_n)} \hat{\mathfrak{q}}^1\big)(0, i\Phi_n^1).$$

According to Condition (iii), the operator $\mathcal{D}_{(A_n,\Phi_n)} \hat{\mathfrak{q}}^1$ maps $L^2_j \to L^2_j$ for $j \leq k$, and in the corresponding operator norms, it converges to $\mathcal{D}_{(A,0)} \hat{\mathfrak{q}}^1$ as A_n converges to A. The same is true of the other term $\rho(A_0 - A_n)$ in W_n since we are in the range $L^2_{k+1} \subset C^0$. Thus by bootstrapping, using the Gårding inequality, and the uniform $L^2(Z)$ bound on Φ^1_n, we find that Φ^1_n is bounded in L^2_{k+1} on interior domains. Hence after passing to a subsequence and bootstrapping once more, we can assume that $\{\Phi^1_n\}$ is Cauchy in L^2_{k+1}, as desired.

The remainder of the proof is essentially unchanged. □

11 Constructing tame perturbations

We now take up the task of showing that the class of k-tame perturbations, defined by the conditions of Definition 10.5.1, is sufficiently large to meet our needs. Thus far, we have not given any examples of tame perturbations. (Recall that a perturbation is *tame* if it is k-tame for all $k \geq 2$.) As a particularly simple example, it is not hard to check that the function

$$f(A, \Phi) = \|\Phi\|^2$$

is the primitive of a tame perturbation, as is $h(\|\Phi\|^2)$ for any smooth function h with bounded derivative. We will give an explicit construction of a large class of tame perturbations, which we call *cylinder functions*. (The example just given is not a cylinder function.)

11.1 Cylinder functions

Recall from Subsection 9.6 that the gauge group $\mathcal{G}_{k+1}(Y)$ can be decomposed as a product $\mathcal{G}^{\perp}_{k+1}(Y) \times \mathcal{G}^h$, and that if $\gamma_0 = (B_0, 0) \in \mathcal{C}_k(Y)$ is a reducible configuration, then there is a diffeomorphism

$$\begin{aligned} \mathcal{G}^{\perp}_{k+1}(Y) \times \mathcal{K}_{k,\gamma_0} &\to \mathcal{C}_k(Y) \\ (u, (b, \Psi)) &\mapsto u(B_0 + b \otimes 1, \Psi) \\ &= (B_0 + (b - u^{-1}du) \otimes 1, u\Psi), \end{aligned}$$

where $\mathcal{K}_{k,\gamma_0}$ is defined by the Coulomb condition $d^*b = 0$. Constructing functions f on $\mathcal{C}_k(Y)$ invariant under the gauge group $\mathcal{G}_{k+1}(Y)$ is equivalent to constructing functions on $\mathcal{K}_{k,\gamma_0}$ which are invariant under the action of $\mathcal{G}^h$. This idea motivates the definitions below. In the case that $b_1(Y) = 0$ (so that $\mathcal{G}^h = S^1$), the functions f that we will construct correspond to functions on $\mathcal{K}_{k,\gamma_0}$ which are obtained as the composite of a linear map $p : \mathcal{K}_{k,\gamma_0} \to \mathbb{R}^n \times \mathbb{C}^m$ with a smooth, S^1-invariant function g on $\mathbb{R}^n \times \mathbb{C}^m$. The map p is defined by taking the L^2 inner products with a collection of smooth elements of $\mathcal{K}_{\gamma_0}$.

Given a coclosed 1-form $c \in \Omega^1(Y; i\mathbb{R})$ we define a function

$$r_c : \mathcal{C}(Y) \to \mathbb{R}$$

by setting

$$\begin{aligned} r_c(B_0 + b \otimes 1, \Psi) &= \int_Y b \wedge *\bar{c} \\ &= \langle b, c \rangle_Y . \end{aligned}$$

Any gauge transformation in the identity component, $\mathcal{G}^e$, of the gauge group changes b by an exact 1-form, so the condition that c is coclosed ensures that r_c is invariant under $\mathcal{G}^e$. For the general element $u \in \mathcal{G}_{k+1}$, the 1-form $-u^{-1}du$ represents an element $h \in 2\pi i H^1(Y;\mathbb{Z})$, and under the action of such a u, the function r_c transforms by

$$r_c(u(B_0 + b \otimes 1, \Psi)) = r_c(B_0 + b \otimes 1, \Psi) + (h \smile [*\bar{c}])[Y].$$

If c is coexact, then r_c is invariant under all of $\mathcal{G}$.

Let $\mathbb{T}$ denote the torus

$$\mathbb{T} = H^1(Y; i\mathbb{R})/(2\pi i H^1(Y;\mathbb{Z}))$$

which we identify with the space of imaginary-valued harmonic 1-forms on Y modulo those with periods in $2\pi i\mathbb{Z}$. We can define a $\mathcal{G}$-invariant map $\mathcal{C}(Y) \to \mathbb{T}$ by

$$(B_0 + b \otimes 1, \Psi) \mapsto [b_{\text{harm}}]$$

where b_{harm} denotes the harmonic part of b and the square brackets denote the equivalence class under the action of $2\pi i H^1(Y;\mathbb{Z})$. If we choose forms $\omega_1, \ldots, \omega_t$ representing an integral basis for $H^1(Y;\mathbb{R})$, we can identify $\mathbb{T}$ with $\mathbb{R}^t/(2\pi\mathbb{Z}^t)$, and write the map as

$$(B, \Psi) \mapsto (r_{\omega_1}(B, \Psi), \ldots, r_{\omega_t}(B, \Psi)) \pmod{2\pi\mathbb{Z}^t}.$$

It is convenient to choose a splitting

$$S^1 \to \mathcal{G}^h \overset{v}{\leftrightarrows} H^1(Y;\mathbb{Z}), \tag{11.1}$$

and define $\mathcal{G}^{h,o}$ to be the image of v. For example, we can choose v so that $\mathcal{G}^{h,o}$ is the subgroup of $\mathcal{G}^h$ consisting of harmonic gauge transformations u with

$u(y_0) = 1$, where y_0 is a chosen basepoint in Y. We then define

$$\mathcal{G}^o(Y) = \mathcal{G}^{h,o} \times \mathcal{G}^{\perp}(Y) \subset \mathcal{G}(Y).$$

Then we have $\mathcal{G}(Y) = S^1 \times \mathcal{G}^o(Y)$. Note that $\mathcal{G}^o(Y)$ is not the "based gauge group" $\{ u \mid u(y_0) = 0 \}$, however v might be chosen. We can also form the Sobolev version of this group, $\mathcal{G}^o_{k+1}(Y) \subset \mathcal{G}_{k+1}(Y)$. We have

$$\mathcal{G}(Y) = S^1 \times \mathcal{G}^{h,o} \times \mathcal{G}^{\perp}(Y). \tag{11.2}$$

The identity component, $\mathcal{G}^e(Y) \subset \mathcal{G}(Y)$, is the product of the first and last factors. The action of $\mathcal{G}^o(Y)$ on $\mathcal{C}(Y)$ is free, and we write $\mathcal{B}^o_k(Y)$ for the quotient of the Sobolev completions:

$$\mathcal{B}^o_k(Y) = \mathcal{C}_k(Y)/\mathcal{G}^o_{k+1}(Y). \tag{11.3}$$

We refer to $\mathcal{B}^o_k(Y)$ as the *based* configuration space. It is a Hilbert manifold. The space $\mathcal{B}_k(Y)$ is the quotient of $\mathcal{B}^o_k(Y)$ by the remaining circle action, $(B, \Psi) \mapsto (B, e^{i\theta}\Psi)$.

Taking the quotient of $H^1(Y; i\mathbb{R}) \times S$ by the group $\mathcal{G}^{h,o}$ gives us a bundle $\mathbb{S}$ over $\mathbb{T} \times Y$. If Υ is a smooth section of $\mathbb{S}$, let $\tilde{\Upsilon}$ denote its lift to a section of $H^1(Y; i\mathbb{R}) \times S \to H^1(Y; i\mathbb{R}) \times Y$. This $\tilde{\Upsilon}$ has the following quasi-periodicity. For each class $\kappa \in H^1(Y; \mathbb{Z})$, there is a unique $u = v(\kappa) \in \mathcal{G}^{h,o}$, where v is the chosen splitting of the sequence (11.1), and we have

$$\tilde{\Upsilon}_{\alpha+\kappa}(y) = u(y)\tilde{\Upsilon}_{\alpha}(y).$$

(We write $\Upsilon_b(y)$ instead of $\Upsilon(b, y)$.) Note that $-u^{-1}du$ is a harmonic representative for $2\pi i\kappa$. Given a section Υ of $\mathbb{S}$, we can now define a $\mathcal{G}^o(Y)$-equivariant map

$$\Upsilon^{\dagger} : \mathcal{C}(Y) \to C^{\infty}(S)$$

by

$$\Upsilon^{\dagger}(B_0 + b \otimes 1, \Psi) = e^{-Gd^*b}\tilde{\Upsilon}_{b_{\text{harm}}}$$

and a $\mathcal{G}^o(Y)$-invariant map

$$q_{\Upsilon} : \mathcal{C}(Y) \to \mathbb{C}$$

by setting

$$\begin{aligned} q_\Upsilon(B, \Psi) &= \int_Y \langle \Psi, \Upsilon^\dagger(B, \Psi) \rangle \\ &= \langle \Psi, \tilde{\Upsilon}^\dagger \rangle_Y . \end{aligned}$$

Note that this map is in fact $\mathcal{G}$-equivariant when we make $\mathcal{G}$ act on $\mathbb{C}$ via the map to S^1 in the decomposition (11.2).

Choose a finite collection of coclosed 1-forms $c_1, \ldots, c_{n+t}$ with the first n being coexact and the remaining t being our basis ω_ν for the harmonic forms. Also choose a collection of smooth sections $\Upsilon_1, \ldots, \Upsilon_m$ of $\mathbb{S}$. These give rise to a map

$$\begin{aligned} p : \mathcal{C}(Y) &\to \mathbb{R}^n \times \mathbb{T} \times \mathbb{C}^m \\ &= \mathbb{R}^n \times (\mathbb{R}^t / 2\pi \mathbb{Z}^t) \times \mathbb{C}^m \end{aligned} \tag{11.4}$$

by setting

$$\begin{aligned} &p(B, \Psi) = \\ &\quad \big(r_{c_1}(B, \Psi), \ldots, r_{c_{n+t}}(B, \Psi), q_{\Upsilon_1}(B, \Psi), \ldots, q_{\Upsilon_m}(B, \Psi) \big) \quad (\text{mod } 2\pi \mathbb{Z}^t). \end{aligned}$$

Such a p is invariant under the action of $\mathcal{G}^o(Y)$ and equivariant under the remaining S^1 action (with S^1 acting by scalar multiplication on the $\mathbb{C}^m$ factor in $\mathbb{R}^n \times \mathbb{T} \times \mathbb{C}^m$). From an S^1-invariant function

$$g : \mathbb{R}^n \times \mathbb{T} \times \mathbb{C}^m \to \mathbb{R},$$

we get an induced function $f = g \circ p : \mathcal{C}(Y) \to \mathbb{R}$ which is now invariant under the full gauge group.

Definition 11.1.1. We call a function f on $\mathcal{C}(Y)$ a *cylinder function* if it arises as $g \circ p$ where:

- the map $p : \mathcal{C}(Y) \to \mathbb{R}^n \times \mathbb{T} \times \mathbb{C}^m$ is defined as above, using any collection of coexact forms $c_1, \ldots, c_n$ and any collection of sections $\Upsilon_1, \ldots \Upsilon_m$, for any $n, m \geq 0$;
- the function g is an S^1-invariant smooth function on $\mathbb{R}^n \times \mathbb{T} \times \mathbb{C}^m$, with compact support.

◊

In the above construction, the map p (and hence also f) is smooth on the Hilbert manifold $\mathcal{C}_k(Y)$. We will take up later the task of estimating the gradient of f and its higher derivatives.

Theorem 11.1.2. *If f is a cylinder function, then its gradient*

$$\operatorname{grad} f : \mathcal{C}(Y) \to \mathcal{T}_0$$

is a tame perturbation, in the sense of Definition 10.5.1.

11.2 Cylinder functions and embeddings

Before taking up the proof of Theorem 11.1.2, we shall show that the class of cylinder functions on $\mathcal{C}_k(Y)$ is a large one in a sense we now make precise. For the statement of the following proposition, note that the map p constructed above is invariant under the action of $\mathcal{G}^o_{k+1}(Y)$ on $\mathcal{C}_k(Y)$ and therefore descends to a map (also called p here) on the Hilbert manifold $\mathcal{B}^o_k(Y)$.

Proposition 11.2.1. *Given a compact subset K of a finite-dimensional C^1 submanifold $M \subset \mathcal{B}^o_k(Y)$, both invariant under the action of S^1, there exist a collection of coclosed forms c_ν, sections Υ_μ of $\mathbb{S}$ and a neighborhood U of K in M, such that the corresponding map*

$$p : \mathcal{B}^o_k(Y) \to \mathbb{R}^n \times \mathbb{T} \times \mathbb{C}^m$$

gives an embedding of U.

Proof. It is enough to prove two things: first, that given a pair of points $x \neq y$ in K, we can find collections c_i and Υ_μ such that p separates x and y; second, that given any point $x \in K$ and any v in T_xM, we can find a p whose differential at x does not annihilate v.

To begin with the first of these, we first note that if $x = [B_x, \Psi_x]$ and $y = [B_y, \Psi_y]$, then if $B_x - B_y$ has harmonic part which does not represent an element of the lattice $2\pi i H^1(Y, \mathbb{Z})$, then the images of x and y in $\mathbb{T}$ are already distinct. So we may assume that the harmonic part does lie in this lattice, and after choosing a different gauge representative, the harmonic parts are equal. If B_x and B_y do not now lie in the same orbit of $\mathcal{G}^\perp_{k+1}$, then the coexact part of $B_x - B_y$ is non-zero, and there is therefore a coexact c_ν such that r_c separates x and y. We are left with the case that B_x and B_y are gauge-equivalent by an element of $\mathcal{G}^\perp_{k+1}$; we may assume then that $B_x = B_y = B$ and that $B = B_0 + b \otimes 1$, with $d^*b = 0$. Now Ψ_x and Ψ_y must be distinct, and we can choose an Υ such that

$\Upsilon(b_{\text{harm}})$ has non-zero inner product with their difference. Then q_Υ separates x and y.

The second matter is not essentially different. □

Corollary 11.2.2. *Given any* $[B, \Psi]$ *in* $\mathcal{B}_k^*(Y)$ *and any non-zero tangent vector* v *to* $\mathcal{B}_k^*(Y)$ *at* $[B, \Psi]$*, there exists a cylinder function* f *whose differential* $\mathcal{D}_{[B,\Psi]} f(v)$ *is non-zero.* □

11.3 Gradients and Hessians of cylinder functions

Next we take up the task of computing the gradient of such a function. To begin consider the 1-forms dx^ν on $\mathbb{R}^{n+t}$ and dz^μ on $\mathbb{C}^m$. Pulling these back by p gives 1-forms $p^*(dx^\nu)$ and $p^*(dz^\mu)$ on $\mathcal{C}_k(Y)$. The formula for $p^*(dx^\nu)$ is straightforward:

$$p^*(dx^\nu)(\delta b, \delta\Psi) = \langle \delta b, c_\nu \rangle_Y.$$

The formula for $p^*(dz^\mu)$ is a little more complicated. We first write $\partial_{n+\nu}\Upsilon_\mu$ for $\partial\Upsilon_\mu/\partial x^{n+\nu}$, as a section of $\mathbb{S}$. Then

$$\begin{aligned} &p^*(dz^\mu)(\delta b, \delta\Psi) \\ &\quad = \langle (Gd^*\delta b)\Psi, \Upsilon_\mu^\dagger \rangle_Y + \sum_{\nu=1}^{t} \langle \Psi, (\partial_{n+\nu}\Upsilon_\mu)^\dagger \rangle_Y \langle \delta b, c_{n+\nu} \rangle_Y + \langle \delta\Psi, \Upsilon_\mu^\dagger \rangle_Y. \end{aligned} \tag{11.5}$$

Here $\Upsilon^\dagger$ stands as an abbreviation for $\Upsilon^\dagger(B, \Psi)$. Working formally we can compute the vector fields X^ν and Z^μ on $\mathcal{C}$ which are L^2-dual to $p^*(dx^\nu)$ and $p^*(dz^\mu)$. The Z^μ formally belong to the complexified tangent bundle of $\mathcal{C}$. We have

$$X^\nu(B, \Psi) = (c_\nu, 0), \tag{11.6a}$$

while

$$Z^\mu(B, \Psi) = \left(dG\langle \Psi, \Upsilon_\mu^\dagger \rangle + \sum_{\nu=1}^{t} \langle \Psi, (\partial_{n+\nu}\Upsilon_\mu)^\dagger \rangle_Y c_{n+\nu}, \Upsilon_\mu^\dagger \right). \tag{11.6b}$$

Just to be clear, in the above expression the first inner product $\langle \Psi, \Upsilon_\mu^\dagger \rangle$ is a pointwise inner product while the second $\langle \Psi, (\partial_{n+\nu}\Upsilon_\mu)^\dagger \rangle_Y$ is an L^2 inner product on Y. Although the X^ν are constant vector fields and thus determine smooth vector fields of every order on any $\mathcal{C}_k(Y)$ the story for the Z^μs is somewhat more complicated. We estimate the size of Z^μ and its derivatives in the gauge-invariant Sobolev norms $L^2_{k,B}$. We again write $\mathcal{T}_j$ for the L^2_j completion of the tangent bundle of $\mathcal{C}(Y)$; and we regard $\mathcal{T}_j \to \mathcal{C}_k(Y)$ as a Hilbert vector bundle equipped with the gauge-invariant Hilbert norm $L^2_{j,B}$ on the fibers.

Proposition 11.3.1. *For all $k \geq 1$ we have:*

(i) *The Z^μ determine C^∞ vector fields on $\mathcal{C}_k(Y)$. There is a constant C such that the differentials*

$$\mathcal{D}^\ell Z^\mu : \mathcal{C}_k(Y) \to \mathrm{Mult}^\ell \left(\times_\ell \mathcal{T}_k, \mathcal{T}_k \right) \otimes \mathbb{C}$$

have norm satisfying a bound

$$\|\mathcal{D}^\ell_{(B,\Psi)} Z^\mu\| \leq C\big(1 + \|b\|_{L^2_{k-1}}\big)^k \big(1 + \|\Psi\|_{L^2_{k,B}}\big).$$

In addition, for each j with $-k \leq j \leq k$, this ℓ-th derivative extends to a continuous map

$$\mathcal{D}^\ell Z^\mu : \mathcal{C}_k(Y) \to \mathrm{Mult}^\ell \left(\times_{\ell-1} \mathcal{T}_k \times \mathcal{T}_j, \mathcal{T}_j \right) \otimes \mathbb{C}$$

whose norm satisfies the same bound.

(ii) *The $p^*(dz^\mu)$ are C^∞ 1-forms on $\mathcal{C}_k(Y)$. There is a constant C such that the differentials*

$$\mathcal{D}^\ell p^*(dz^\mu) : \mathcal{C}_k(Y) \to \mathrm{Mult}^{\ell+1} \left(\times_{\ell+1} \mathcal{T}_k, \mathbb{C} \right)$$

have norm satisfying a bound

$$\|\mathcal{D}^\ell_{(B,\Psi)} p^*(dz^\mu)\| \leq C(1 + \|\Psi\|_{L^2}).$$

In addition, for all j with $-k \leq j \leq k$, this derivative extends to a continuous map

$$\mathcal{D}^\ell p^*(dz^\mu) : \mathcal{C}_k(Y) \to \mathrm{Mult}^\ell \left(\times_\ell \mathcal{T}_k \times \mathcal{T}_j, \mathbb{C} \right)$$

whose norm satisfies the same bound.

Proof. The essential point is in the following lemma:

Lemma 11.3.2. *For all $k \geq 1$ and all $0 \leq j \leq k+1$ the map $\Upsilon^\dagger$ corresponding to any smooth section Υ of $\mathbb{S}$ is a smooth map*

$$\mathcal{C}_k(Y) \to L^2_j(S).$$

Furthermore for each $\ell \geq 0$ there is a constant C such that the differential

$$\mathcal{D}^\ell_{(B,\Psi)}\Upsilon^\dagger \in \mathrm{Mult}^\ell\left(\times_\ell \mathcal{T}_k, L^2_{j,B}(Y;S)\right)$$

satisfies the bound

$$\|\mathcal{D}^\ell_{(B,\Psi)}\Upsilon^\dagger\| \leq C\left(1 + \|b\|_{L^2_{j-1}}\right)^j, \qquad \forall\,(B,\Psi) \in \mathcal{C}_k(Y).$$

In addition, for all j with $-k-1 \leq j \leq k+1$, this ℓ-th derivative extends to an element of

$$\mathcal{D}^\ell_{(B,\Psi)}\Upsilon^\dagger \in \mathrm{Mult}^\ell\left(\times_{\ell-1}\mathcal{T}_k \times \mathcal{T}_j, L^2_{j,B}(Y;S)\right)$$

which is a continuous function of (B,Ψ) and whose norm in this space satisfies the bound

$$\|\mathcal{D}^\ell_{(B,\Psi)}\Upsilon^\dagger\| \leq C\left(1 + \|b\|_{L^2_{k-1}}\right)^k, \qquad \forall\,(B,\Psi) \in \mathcal{C}_k(Y).$$

Proof. Note that $\Upsilon^\dagger(B,\Psi)$ is a function of B alone, and in calculating its ℓ-th derivative with respect to the variable B,

$$\mathcal{D}^\ell_{(B,\Psi)}\Upsilon^\dagger(\delta b_1,\dots,\delta b_\ell),$$

we see terms such as the symmetrization of

$$(Gd^*\delta b_1)(Gd^*\delta b_2)\dots(Gd^*\delta b_m)e^{-Gd^*b}\mathcal{D}^{\ell-m}\tilde{\Upsilon}_{b_{\mathrm{harm}}}(\delta b_{m+1},\dots,\delta b_\ell); \tag{11.7}$$

and the $L^2_{j,B}$ norm of this term is bounded by

$$C\left(\prod \|\delta b_i\|_{L^2_k}\right)\|e^{-Gd^*b}\|_{L^2_{j,b-b_{\mathrm{harm}}}} \times \|\mathcal{D}^{\ell-m}\tilde{\Upsilon}_{b_{\mathrm{harm}}}(\delta b_{m+1},\dots,\delta b_\ell)\|_{L^2_{k+1,B_0+b_{\mathrm{harm}}}}.$$

Here we have interpreted e^{-Gd^*b} as a section of the trivial line bundle with connection form $b - b_{\mathrm{harm}}$. We also used the multiplication

$$\bigotimes_{m+1} L^2_{k+1} \otimes L^2_j \to L^2_j$$

which is continuous if $(k+1-\frac{3}{2}) > 0$ and $-k-1 \le j \le k+1$, for then L^2_{k+1} is an algebra and L^2_j is a module over L^2_{k+1}. The section $\tilde{\Upsilon}$ corresponds to a smooth section Υ of $\mathbb{S}$ on the compact space $\mathbb{T} \times Y$, and the norm $\|\tilde{\Upsilon}_{b_{\mathrm{harm}}}\|_{L^2_{k+1,B_0+b_{\mathrm{harm}}}}$ is invariant under the deck transformations. The same is true for all the partial derivatives of $\tilde{\Upsilon}$ with respect to the coordinates $x_{n+\nu}$. So the last factor in this expression is bounded in L^2_{k+1} norm (or indeed any Sobolev norm) by

$$\|\mathcal{D}^{\ell-m}\tilde{\Upsilon}_{b_{\mathrm{harm}}}(\delta b_{m+1}, \dots, \delta b_\ell)\|_{L^2_{k+1,B_0+b_{\mathrm{harm}}}} \le C\left(\prod \|(\delta b_i)_{\mathrm{harm}}\|_{L^2}\right).$$

Finally, the factor $\|e^{-Gd^*b}\|_{L^2_{j,b-b_{\mathrm{harm}}}}$ is also gauge-invariant, and so can be calculated under the assumption that $d^*b = 0$, when it reduces to $\|1\|_{L^2_{j,b-b_{\mathrm{harm}}}}$. We have

$$\|1\|_{L^2_{j,b-b_{\mathrm{harm}}}} \le C\big(1 + \|b\|_{L^2_{j-1}}\big)^j$$

by a straightforward calculation.

For the additional clause, we estimate the $L^2_{j,B}$ norm of (11.7) by a constant multiple of

$$\left(\prod_{i=1}^{m-1} \|\delta b_i\|_{L^2_k}\right) \|\delta b_m\|_{L^2_j}$$
$$\times \|e^{-Gd^*b}\|_{L^2_{k,b-b_{\mathrm{harm}}}} \|\mathcal{D}^{\ell-m}\tilde{\Upsilon}_{b_{\mathrm{harm}}}(\delta b_{m+1}, \dots, \delta b_\ell)\|_{L^2_{k+1,B_0+b_{\mathrm{harm}}}},$$

and proceed as above using the same multiplication theorem. □

Returning to the proof of the proposition, and beginning with Part (i), we observe that the first term in (11.6b), namely

$$dG\langle \Psi, \Upsilon^\dagger_\mu \rangle,$$

can be built up as the composite of

$$1 \times \Upsilon^\dagger : L^2_{k,B}(Y;S) \times L^2_k(Y;iT^*Y) \to L^2_{k,B}(Y;S) \times L^2_{k,B}(Y;S),$$

a bilinear multiplication

$$L^2_{k,B}(Y;S) \times L^2_{k,B}(Y;S) \to L^2_{k-1}(\mathbb{C}),$$

and a bounded linear map $dG : L^2_{k-1}(\mathbb{C}) \to L^2_k(T^*_{\mathbb{C}}Y)$. Using Lemma 11.3.2, the chain rule and the Leibniz rule, we see that the norm of this term is bounded as claimed in the main clause. The remaining terms are easier, and satisfy stronger bounds. The additional clause of Part (i) is similar.

For the 1-forms $p^*(dz^\mu)$ in Part (ii), the argument is similar. It is only necessary to estimate the norm of Z^μ and its derivatives as a map to L^2_{-k}, or a fortiori, to L^2. □

Now we can estimate the gradient of a general cylinder function $f = g \circ p$, arising from a smooth map

$$g : \mathbb{R}^n \times \mathbb{T} \times \mathbb{C}^m \to \mathbb{R}$$

with $\nabla^s g$ uniformly bounded on $\mathbb{R}^n \times \mathbb{T} \times \mathbb{C}^m$ for all s. The derivative of f is the 1-form

$$df = \sum_{\nu=1}^{n+t} (\partial_{x^\nu} g \circ p) p^*(dx^\nu) + \sum_{\mu=1}^{m} (\partial_{z^\mu} g \circ p) p^*(dz^\mu) + (\partial_{\bar{z}^\mu} g \circ p) p^*(d\bar{z}^\mu),$$

and from this we see that its formal L^2 gradient, $\mathfrak{q} = \operatorname{grad} f$, is the vector field

$$\mathfrak{q} = \sum_{i=1}^{n+t} (\partial_{x^\nu} g \circ p) X^\nu + \sum_{\mu=1}^{m} (\partial_{z^\mu} g \circ p) Z^\mu + (\partial_{\bar{z}^\mu} g \circ p) \bar{Z}^\mu. \qquad (11.8)$$

Proposition 11.3.3. *For $k \geq 1$, the perturbation $\mathfrak{q}$ determines a smooth vector field on $\mathcal{C}_k(Y)$, and for each $\ell \geq 0$, there is a constant C with*

$$\|\mathcal{D}^\ell_{(B,\Psi)}\mathfrak{q}\| \leq C(1 + \|\Psi\|_{L^2})^\ell \left(1 + \|b\|_{L^2_{k-1}}\right)^k \left(1 + \|\Psi\|_{L^2_{k,B}}\right).$$

Here the norm of $\mathcal{D}^\ell_{(B,\Psi)}\mathfrak{q}$ is as an element of

$$\operatorname{Mult}^\ell \left(\times_\ell \mathcal{T}_k, \mathcal{T}_k\right).$$

In addition, for every $-k-1 \leq j \leq k+1$, the first derivative $\mathcal{D}\mathfrak{q}$ extends to a smooth map

$$\mathcal{D}\mathfrak{q} : \mathcal{C}_k(Y) \to \operatorname{Hom}(\mathcal{T}_j, \mathcal{T}_j)$$

whose $(\ell - 1)$*-th derivative, considered as an element of*

$$\mathrm{Mult}^{\ell}\left(\times_{\ell-1}\mathcal{T}_k \times \mathcal{T}_j, \mathcal{T}_j\right),$$

has norm satisfying the same bound.

Proof. In the derivative $\mathcal{D}^{\ell}\mathfrak{q}$, the terms that appear are symmetrizations of

$$\mathcal{D}^r(\partial_{z^\mu} g \circ p) \otimes \mathcal{D}^{\ell-r} Z^\mu \tag{11.9}$$

for $0 \leq r \leq \ell$, and the similar terms involving X^ν and $\bar{Z}^\mu$. If we expand the first factor of the tensor product, we obtain terms

$$\nabla^s(\partial_{z^\mu} g) \circ \left(\bigotimes_{i=1}^{s} \mathcal{D}^{r_i} p\right)$$

with $r_i \geq 1$ and $\sum r_i = r$. As a corollary of the second part of Proposition 11.3.1, we have, for $r_i \geq 1$,

$$\|\mathcal{D}^{r_i}_{(B,\Psi)} p\| \leq C(1 + \|\Psi\|_{L^2}),$$

and so we obtain

$$\|\mathcal{D}^r(\partial_{z^\mu} g \circ p)\| \leq C(1 + \|\Psi\|_{L^2})^s$$

because we have assumed that the derivatives of g are bounded. Using the first part of Proposition 11.3.1 to estimate $\mathcal{D}^{\ell-r}Z^\mu$, we obtain

$$\|\mathcal{D}^{\ell}_{(B,\Psi)}\mathfrak{q}\| \leq C(1 + \|\Psi\|_{L^2})^{\ell}\left(1 + \|b\|_{L^2_{k-1}}\right)^k\left(1 + \|\Psi\|_{L^2_{k,B}}\right).$$

The additional clause is similar. □

11.4 Estimates for perturbations on the cylinder

A cylinder function f, as above, gives rise to a smooth map

$$\mathfrak{q} = \operatorname{grad} f : \mathcal{C}(Y) \to C^{\infty}(Y; iT^*Y \oplus S).$$

For perturbations of this sort, the corresponding 4-dimensional maps over the cylinder $Z = [t_1, t_2] \times Y$ map smooth sections to smooth sections: that is, we have

$$\hat{\mathfrak{q}} : \mathcal{C}(Z) \to \mathcal{V},$$

where $\mathcal{V} = C^\infty(Z; i\,\mathfrak{su}(S^+) \oplus S^-)$, as in Subsection 6.1. Recall that $\hat{\mathfrak{q}}(A, \Phi)$ is constructed from $\mathfrak{q}$ by restricting (A, Φ) to the slices $\{t\} \times Y$ (restricting A as a connection form, and so losing the dt component), and then identifying the bundles as:

$$iT^*Y \oplus S = \left(i\,\mathfrak{su}(S^+) \oplus S^-\right)|_{\{t\}\times Y}.$$

We claim that the induced 4-dimensional map $\hat{\mathfrak{q}}$ has the following smoothness properties on Sobolev spaces. In these statements, $\mathcal{T}_j(Z)$ denotes the tangent bundle of $\mathcal{C}(Z)$, equipped with the $L^2_{j,A}$ norm.

Proposition 11.4.1. *Let*

$$\hat{\mathfrak{q}} : \mathcal{C}(Z) \to \mathcal{V}$$

be a perturbation on Z arising from a cylinder function, and let $k \geq 2$.

(i) *The map $\hat{\mathfrak{q}}$ on $\mathcal{C}(Z)$ extends to a C^∞ map*

$$\mathcal{C}_k(Z) \to \mathcal{V}_k. \tag{11.10}$$

The ℓ-th derivative of the map (11.10)*,*

$$\mathcal{D}^\ell_{(A,\Phi)}\hat{\mathfrak{q}} \in \mathrm{Mult}^\ell\left(\times_\ell \mathcal{T}_k(Z), \mathcal{V}_k\right),$$

satisfies the estimate

$$\|\mathcal{D}^\ell_{(A,\Phi)}\hat{\mathfrak{q}}\| \leq C(1 + \|a\|_{L^2_k(Z)})^{2k(\ell+1)}(1 + \|\Phi\|_{L^2_{k,A}(Z)})^{\ell+1},$$

where $A = A_0 + a \otimes 1$ as usual.

(ii) *For any j in the range $-k \leq j \leq k$, the first derivative $\mathcal{D}\hat{\mathfrak{q}}$ extends to a smooth map*

$$\mathcal{D}\hat{\mathfrak{q}} : \mathcal{C}_k(Z) \to \mathrm{Hom}(\mathcal{V}_j, \mathcal{V}_j)$$

whose $(\ell - 1)$-th derivative, regarded as an element of

$$\mathrm{Mult}^\ell\left(\times_{\ell-1}\mathcal{T}_k(Z) \times \mathcal{T}_j(Z), \mathcal{V}_j\right)$$

satisfies the same bound.

(iii) *For $i = 0, 1$ the map $\hat{\mathfrak{q}} : \mathcal{C}_k(Z) \to \mathcal{V}_k$ satisfies the estimate*

$$\|\hat{\mathfrak{q}}\|_{L^2_{i,A}} \leq C(1 + \|(a, \Phi)\|_{L^2_{i,A}(Z)})^{(2i+1)}$$

where $A = A_0 + a \otimes 1$ as usual. Furthermore for $i = 1$ the map $\hat{\mathfrak{q}}$ is continuous as a map $\mathcal{C}_1(Z) \to \mathcal{V}_1$.

Proof. This will follow by arguments similar to the above 3-dimensional arguments if we can prove the analog of Lemma 11.3.2. The main source of complication is the appearance of the operators Gd^* and dG in (11.5) and (11.6b). In the 3-dimensional case these operators are smoothing of order 1. In the 4-dimensional case these operators give rise to operators, defined slicewise, which smooth in the Y direction while leaving the order of differentiability fixed in the t direction. (We will continue to write G for this slicewise operator on the cylinder, and d_Y for the 3-dimensional exterior derivative.) This has the effect of losing a derivative in the results obtained. In order to deal with this we need to introduce Sobolev spaces which control different orders of differentiability in different directions, the *anisotropic Sobolev spaces* (see [53, Appendix B]).

Define the $L^2_{k,l}$ norm for functions on the cylinder Z to be

$$\|f\|^2_{L^2_{k,l}} = \sum_{\substack{i+j\leq k,\\ j\leq l}} \int_Z \left| \frac{\partial}{\partial t^j} \nabla^i_Y f \right|^2$$

and define $L^2_{k,l}$ to be the completion of smooth functions (or sections) with respect to this norm. One can characterize $L^2_{k,l}$ as the functions f such that

$$f \in L^2_j([t_1,t_2], L^2_i(Y)), \tag{11.11}$$

for all non-negative integers i and j with $i+j \leq k$ and $j \leq l$. In fact, if we use fractional-derivative Sobolev spaces, we can drop the condition that i and j are integers: for f in $L^2_{k,l}$, the inclusion (11.11) holds for real i and j in the same range.

Using this characterization, one sees:

Lemma 11.4.2. *For all k and j, the slicewise operators $d^*_Y G$ and $G d_Y$ define continuous linear operators from $L^2_{k,j}(Z) \to L^2_{k+1,j}(Z)$.* □

We will also need the following facts regarding these spaces.

Lemma 11.4.3. *The anisotropic Sobolev spaces enjoy the following properties whenever $2k > \dim(Z)$ and $l \geq 1$:*

(i) $L^2_{k,l} \subset C^0$;
(ii) $L^2_{k,l}$ *is an algebra under pointwise multiplication;*

(iii) *if* $|j| \leq k$ *and* $|m| \leq l$*, then* $L^2_{j,m}$ *is a module under pointwise multiplication by* $L^2_{k,l}$.

Proof. For the first item, we can use the fact that an element of $L^2_{k,l}$ lies in $L^2_{1-\epsilon}([t_1,t_2], L^2_{k-1+\epsilon}(Y))$. For ϵ smaller than $1/2$, this is contained in $C^0([t_1,t_2], C^0(Y)) = C^0(Z)$. (This argument uses the fact that k is an integer.) The other two items follow easily, as in the case of standard Sobolev spaces. □

The map $\Upsilon^\dagger$ associated to a section Υ of $\mathbb{S}$ gives rise to a 4-dimensional map

$$\Upsilon^\ddagger : \mathcal{C}(Z) \to C^\infty(Z, S^-)$$

and we now investigate its behavior on Sobolev completions.

Lemma 11.4.4. *For all* $k \geq 2$ *and all* j *in the range* $-l \leq j \leq k$*, the map* $\Upsilon^\ddagger$ *corresponding to any smooth section* Υ *of* $\mathbb{S}$ *is a smooth map*

$$\mathcal{C}_k(Z) \to L^2_{j+1,j,A}(Z; S^-)$$

with the following properties.

(i) *For each* $\ell \geq 0$ *there is a constant* C *such that the differential*

$$\mathcal{D}^\ell_{(A,\Phi)}\Upsilon^\ddagger \in \mathrm{Mult}^\ell\left(\times_\ell \mathcal{T}_k(Z), L^2_{j+1,j,A}(Z;S^-)\right)$$

satisfies the bound

$$\|\mathcal{D}^\ell_{(A,\Phi)}\Upsilon^\ddagger\| \leq C\big(1+\|a\|_{L^2_j}\big)^j\big(1+\|a\|_{L^2_k}\big)^k, \qquad \forall\,(A,\Phi) \in \mathcal{C}_k(Z).$$

(ii) *This* ℓ*-th derivative extends to an element of*

$$\mathcal{D}^\ell_{(A,\Phi)}\Upsilon^\ddagger \in \mathrm{Mult}^\ell\left(\times_{\ell-1}\mathcal{T}_k(Z) \times \mathcal{T}_j(Z), L^2_{j+1,j,A}\right)$$

which is a continuous function of (A,Φ) *and whose norm in this space satisfies the bound*

$$\|\mathcal{D}^\ell_{(A,\Phi)}\Upsilon^\ddagger\| \leq C\big(1+\|a\|_{L^2_k}\big)^{2k}.$$

(iii) *For* $i = 0, 1$ *we have the bound*

$$\|\Upsilon^\ddagger\|L^2_{A,i} \leq C\big(1+\|a\|_{L^2_i}\big)^i \qquad \forall\,(A,\Phi) \in \mathcal{C}_k(Z),$$

and furthermore $\Upsilon^{\ddagger}$ *is continuous as map*

$$\mathcal{C}_1(Z) \to L^2_{1,A}(Z; S^-).$$

Proof. We begin with Item (i). As in the proof of Lemma 11.3.2 we need a bound on the $L^2_{j+1,j,A}$ norm of

$$(Gd_Y^*\delta a_1)(Gd_Y^*\delta a_2)\cdots(Gd_Y^*\delta a_m)e^{-Gd_Y^*a}\mathcal{D}^{\ell-m}\tilde{\Upsilon}_{a_{\mathrm{harm}}}(\delta a_{m+1},\ldots,\delta a_\ell). \tag{11.12}$$

Here a_{harm} is the slicewise harmonic projection of a restricted to the slices as a 1-form. The $L^2_{j,A}$ norm of this term is bounded by the product of three factors:

$$C\left(\prod_{i=1}^{m}\|Gd_Y^*\delta a_i\|_{L^2_{k+1,k}}\right) \tag{11.13a}$$

$$\|e^{-Gd_Y^*a}\|_{L^2_{j+1,j,a-a_{\mathrm{harm}}}} \tag{11.13b}$$

$$\|\mathcal{D}^{\ell-m}\tilde{\Upsilon}_{a_{\mathrm{harm}}}(\delta a_{m+1},\ldots,\delta a_\ell)\|_{L^2_{k+1,k,A_0+a_{\mathrm{harm}}}}. \tag{11.13c}$$

Here we have interpreted $e^{-Gd_Y^*a}$ as a section of the trivial line bundle with connection form $a - a_{\mathrm{harm}}$. We have also used the fact that the multiplication

$$\bigotimes_m L^2_{k+1,k} \otimes L^2_{j+1,j} \to L^2_j$$

is continuous: this is the case if $k > 1$ and $-k \le j \le k$; for then $L^2_{k+1,k}$ consists of continuous functions and is hence an algebra, and $L^2_{j+1,j}$ is a module over this algebra.

The factor (11.13a) is bounded by

$$\prod_{i=1}^{m}\|\delta a_i\|_{L^2_k}.$$

To estimate the factor (11.13b), we can find a gauge transformation in $L^2_{k+1,k}$ which transforms a to $\hat{a} \in L^2_{k,k-1}$ which is in slicewise Coulomb gauge, i.e. $d_Y^*\hat{a} = 0$. (Note that $\hat{a} = a - d_Z G d_Y^* a$.) This term then reduces to

$\|1\|_{L^2_{j+1,j,\hat{a}-a_{\mathrm{harm}}}}$. We then have

$$\begin{aligned}\|1\|_{L^2_{j+1,j,\hat{a}-a_{\mathrm{harm}}}} &\le C\big(1+\|\hat{a}\|_{L^2_{j,j-1}}\big)^j \\ &\le C\big(1+\|a\|_{L^2_j}\big)^j\end{aligned}$$

much as in the 3-dimensional case. To understand the factor (11.13c), note first that an a in L^2_k gives rise to an L^2_k path a_{harm} in $\mathbb{R}^t$, the space of harmonic 1-forms on Y. We begin by estimating this term when $\ell = m$. The connection $A_0 + a_{\mathrm{harm}}$ will be implicit in our notation below, when differentiating in the Y directions. We also introduce ∇_Y and $\nabla_{\mathbb{R}^t}$ for the components of the covariant derivative acting on $\tilde{\Upsilon}$. When expanding those parts of

$$\nabla_Z^{k+1}\tilde{\Upsilon}_{a_{\mathrm{harm}}}$$

which contribute to the $L^2_{k+1,k}$ norm, we obtain terms

$$\nabla_Y^{k'}\left(\frac{d}{dt}\right)^{k''}\tilde{\Upsilon}_{a_{\mathrm{harm}}}\otimes(dt)^{k''}$$

with $k'+k''=k+1$ and $k''\le k$. This in turn expands to give terms

$$\nabla_Y^{k'}\Big(\nabla^n_{\mathbb{R}^t}\tilde{\Upsilon}(\partial_t^{r_1}a_{\mathrm{harm}},\ldots,\partial_t^{r_n}a_{\mathrm{harm}})\Big)$$

with $\sum r_i = k''$ and $r_i \ge 1$. The $L^2(Y)$ norm of such a term is bounded by

$$C\prod_{i=1}^{n}\|\partial_t^{r_n}a_{\mathrm{harm}}\|_{\mathbb{R}^t},$$

because the covariant derivatives of $\tilde{\Upsilon}$ are uniformly bounded on $\mathbb{R}^t\times Y$. The norm of this term in $L^2(Z)$ is then bounded by

$$C\big(1+\|a_{\mathrm{harm}}\|_{L^2_{k''}}\big)^n.$$

This is no larger than

$$C\big(1+\|a_{\mathrm{harm}}\|_{L^2_k}\big)^k.$$

For the functional derivatives $\mathcal{D}^{\ell-m}\tilde{\Upsilon}_{a_{\text{harm}}}$, we have

$$\mathcal{D}^{\ell-m}\tilde{\Upsilon}_{a_{\text{harm}}}(\delta a_{m+1},\ldots,\delta a_\ell) = \sum \partial_{n+i_1}\ldots\partial_{n+i_{\ell-m}}\tilde{\Upsilon}_{a_{\text{harm}}}\prod\langle\delta a_{m+i},c_{n+i}\rangle_Y.$$

This is similarly bounded by

$$C\big(1+\|a_{\text{harm}}\|_{L^2_k}\big)^k\prod_{i=1}^{\ell-m}\|\delta a_{m+i}\|_{L^2_k}.$$

Putting the estimates for all three factors together gives the result of the main clause of the lemma.

For Item (ii), we bound the $L^2_{j+1,j,A}$ norm of (11.12) by the product of the factors:

$$C\left(\prod_{i=1}^{m-1}\|Gd_Y^*\delta a_i\|_{L^2_{k+1,k}}\right) \tag{11.14a}$$

$$\|Gd_Y^*\delta a_m\|_{L^2_{j+1,j}} \tag{11.14b}$$

$$\|e^{-Gd_Y^*a}\|_{L^2_{k+1,k,a-a_{\text{harm}}}} \tag{11.14c}$$

$$\|\mathcal{D}^{\ell-m}\tilde{\Upsilon}_{a_{\text{harm}}}(\delta a_{m+1},\ldots,\delta a_\ell)\|_{L^2_{k+1,k,A_0+a_{\text{harm}}}}. \tag{11.14d}$$

The arguments then proceed as above.

To prove Item (iii) we need the following lemma.

Lemma 11.4.5. *For all $1\le p\le\infty$ the map $f\mapsto e^{if}$ initially defined on $C^\infty(Z;\mathbb{R})$ extends to a continuous map $L^p\to L^p$. For all $f\in L^1$ we have $\|e^{if}\|_{L^\infty}=1$*

Proof. We have the trivial estimate that for all $s,t\in\mathbb{R}$, $|e^{is}-e^{it}|\le|s-t|$ from which the continuity of the induced map on L^p follows. Given $\{f_j\}$ a sequence of smooth functions converging in L^1 norm it follows that in particular e^{if_j} converges to e^{if} except on a set of measure zero. Since $\|e^{if_j}\|_{L^\infty}=1$, it follows that $\|e^{if}\|_{L^\infty}=1$. □

This lemma immediately implies the case $i=0$ of Item (iii), for then

$$\|e^{-Gd_Y^*a}\|_{L^\infty}=1,$$

and so $\tilde{\Upsilon}$ is certainly $L^2(Z)$-bounded as required. The case $i = 1$ requires us to investigate the covariant derivative $\nabla_A \Upsilon^\ddagger$:

$$\begin{aligned}\nabla_A \Upsilon^\ddagger &= -(d_X G d_Y^* a)\Upsilon^\ddagger + e^{-Gd_Y^* a}\nabla_A \tilde{\Upsilon}_{a_{\text{harm}}} \\ &= -\big(\Pi_Y(a) + dt \wedge G d_Y^* \dot{a}\big) e^{-Gd_Y^* a}\tilde{\Upsilon}_{a_{\text{harm}}} + e^{-Gd_Y^* a}\nabla_A \tilde{\Upsilon}_{a_{\text{harm}}}. \quad (11.15)\end{aligned}$$

Here Π_Y denotes the L^2 projection onto the coclosed 1-forms on Y. From Lemma 11.4.5, we see immediately that both terms are L^2-bounded, but now the L^2 bound involves the L^2_1 norm of a. What remains is to establish the continuity of the map $\Upsilon^\ddagger$ in the L^2_1 topology. Lemma 11.4.5 immediately tells us that the map is continuous as a map from L^2 to L^2. To check the continuity as a map to L^2_1 we need to see that Equation (11.15) defines a continuous map from L^2_1 to L^2. We deal this term by term. To deal with the first two terms, note to begin with that $\tilde{\Upsilon}_{a_{\text{harm}}}$ is continuous as a map from $L^2_1(Z)$ to $C^0(Z)$. For the first term, we now use the fact that the map $a \mapsto e^{-Gd_Y^* a}$ is a continuous map from L^2_1 to $L^2([t_1, t_2], L^2_2(Y))$, and therefore also to $L^2([t_1, t_2], C^0(Y))$; and that the map $a \mapsto \Pi_Y a$ is continuous as a map from L^2_1 to $L^2_1([t_1, t_2], L^2(Y))$ and hence to $C^0([t_1, t_2], L^2(Y))$. So the first term has the required continuity property. To deal with the second term, we use the fact that the map $a \mapsto e^{-Gd_Y^* a}$ is a continuous map from L^2_1 to $L^2_{2/3}([t_1, t_2], L^2_{4/3}(Y))$, and hence also to $C^0([t_1, t_2], L^{18}(Y))$, and that the map $a \mapsto G d_Y^* \dot{a}$ is continuous as a map from L^2_1 to $L^2([t_1, t_2], L^2_1(Y))$, and hence also as a map to $L^2([t_1, t_2], L^6(Y))$. These give the continuity of the second term. The last term is continuous from Lemma 11.4.5. □

We now return to the proof of Proposition 11.4.1, following the outline in the 3-dimensional case. We begin with the first item. The analogue of the term Z^μ in (11.6b) is W^μ:

$$W^\mu(A, \Phi) = \left(d_Y G \langle \Phi, \Upsilon_\mu^\ddagger \rangle + \sum_{i=1}^{t} \langle \Phi, (\partial_{n+i}\Upsilon_\mu)^\ddagger \rangle_Y c_{n+i}, \Upsilon_\mu^\ddagger \right). \quad (11.16)$$

Our estimates in Lemma 11.4.4 imply that

$$\mathcal{D}^\ell W^\mu : \mathcal{C}_k(Z) \to \text{Mult}^\ell\left(\times_\ell \mathcal{T}_k(Z), \mathcal{V}_j\right) \otimes \mathbb{C}$$

has norm satisfying a bound

$$\|\mathcal{D}^\ell_{(A,\Phi)} W^\mu\| \le C\big(1 + \|a\|_{L^2_j}\big)^j \big(1 + \|a\|_{L^2_k}\big)^k \big(1 + \|\Phi\|_{L^2_{j,A}}\big).$$

We have

$$\hat{\mathfrak{q}} = \sum_{i=1}^{n+t} (\partial_{x^\nu} g \circ p) X^\nu + \sum_{\mu=1}^{m} (\partial_{z^\mu} g \circ p) W^\mu + (\partial_{\bar{z}^\mu} g \circ p) \bar{W}^\mu \qquad (11.17)$$

where now p is to be interpreted slicewise. Just as in the 3-dimensional case, we can estimate the functional derivative $\mathcal{D}^\ell \hat{\mathfrak{q}}$ if we can estimate

$$\|\mathcal{D}^{r_i}_{(A,\Phi)} p\|$$

where p is now viewed as a map $\mathcal{C}_k(Z) \to L^2_k([t_1,t_2], \mathbb{R}^{n+t} \times \mathbb{C}^m)$. Each $\mathbb{C}^m$ component of p can be viewed as a composition

$$\sigma(dt) \times \Upsilon^\ddagger_\mu \colon L^2_{k,A}(Z; S^+) \times L^2_k(Z; iT^*Z) \to L^2_{k,A}(Z; S^-) \times L^2_{k+1,k,A}(Z; S^-)$$

with the bilinear map

$$(\Psi, \Upsilon^\ddagger) \mapsto \langle \Psi, \Upsilon^\ddagger \rangle_Y$$

regarded as a map

$$L^2_{k,A}(Z; S^-) \times L^2_{k+1,k,A}(Z; S^-) \to L^2_k([t_1,t_2], \mathbb{C}).$$

This gives us that

$$\|\mathcal{D}^{r_i}_{(A,\Phi)} p\| \le C\big(1 + \|a\|_{L^2_k}\big)^{2k} \big(1 + \|\Phi\|_{L^2_{k,A}}\big),$$

and hence (using Formula (11.9)) we have

$$\|\mathcal{D}^{r}_{(A,\Phi)} (\partial_{z^\mu} g \circ p)\| \le C\big(1 + \|a\|_{L^2_k}\big)^{2kr} \big(1 + \|\Phi\|_{L^2_{k,A}}\big)^r.$$

Putting all this together gives us

$$\|\mathcal{D}^{\ell}_{(A,\Phi)} \hat{\mathfrak{q}}\| \le C\big(1 + \|a\|_{L^2_k}\big)^{2k(\ell+1)} \big(1 + \|\Phi\|_{L^2_{k,A}}\big)^{\ell+1},$$

which is the assertion of Part (i) of Proposition 11.4.1.

The second and third parts follow similarly now, using the second and third parts of Lemma 11.4.4 respectively. $\square$

11.5 Completing the proof of Theorem 11.1.2

The above results contain all we need to prove Theorem 11.1.2. Part (i) of Proposition 11.4.1 tells us that the perturbation $\mathfrak{q}$ arising from a cylinder function satisfies Condition (i) of Definition 10.5.1. This same part of Proposition 11.4.1 (applied to smaller values of k) tells us that $\mathfrak{q}$ satisfies Condition (ii) of Definition 10.5.1 for $j \geq 2$; while for the case $j = 1$ we can apply the last clause of Part (iii) of Proposition 11.4.1. Condition (iii) of Definition 10.5.1 is satisfied on account of Part (ii) of the proposition. Condition (iv) of Definition 10.5.1 holds for $\mathfrak{q}$, as one can see using the formula (11.8), which expresses $\mathfrak{q}$ in terms of the vector fields X^{ν} and Z^{μ}, and the formulae (11.6) for those vector fields. Condition (v) of Definition 10.5.1 holds because of Part (iii) of Proposition 11.4.1, with $i = 1$. Finally, Condition (vi) holds on account of the first statement in Proposition 11.3.3. □

11.6 Banach spaces of tame perturbations

We shall now use cylinder functions to construct a Banach space of tame perturbations, large enough to retain the embedding properties discussed in Subsection 11.2.

Theorem 11.6.1. *Let $\mathfrak{q}_i$ ($i \in \mathbb{N}$) be any countable collection of tame perturbations arising as gradients of cylinder functions on $\mathcal{C}(Y)$. Then there exist a separable Banach space $\mathcal{P}$ and a linear map*

$$\mathfrak{Q} : \mathcal{P} \to C^0(\mathcal{C}(Y), \mathcal{T}_0)$$
$$\lambda \mapsto \mathfrak{q}_\lambda$$

with the following properties.

(i) *For each $\lambda \in \mathcal{P}$, the element $\mathfrak{q}_\lambda$ is a tame perturbation in the sense of Definition* 10.5.1.
(ii) *The image of $\mathfrak{Q}$ contains all the perturbations $\mathfrak{q}_i$ from the given countable collection.*
(iii) *If $Z = [t_1, t_2] \times Y$ is a cylinder, then for all $k \geq 2$, the map*

$$\mathcal{P} \times \mathcal{C}_k(Z) \to \mathcal{V}_k(Z)$$

given by $(\lambda, \gamma) \mapsto \hat{\mathfrak{q}}_\lambda(\gamma)$ is a smooth map of Banach manifolds.
(iv) *The map*

$$\mathcal{P} \times \mathcal{C}_1(Y) \to \mathcal{T}_1(Y)$$

given by $(\lambda, \beta) \mapsto \mathfrak{q}_\lambda(\beta)$ *is continuous and satisfies bounds*

$$\|\mathfrak{q}_\lambda(B, \Psi)\|_{L^2} \leq \|\lambda\| m_2(\|\Psi\|_{L^2} + 1)$$

and

$$\|\mathfrak{q}_\lambda(B, \Psi)\|_{L^2_{1,A_0}} \leq \|\lambda\| \mu_1\big(\|(B - B_0, \Psi\|_{L^2_{1,A_0}}\big)$$

for some constant m_2 *and some continuous real function* μ_1 *(cf. Properties* (iv) and (v) *in Definition* 10.5.1).

Proof. Let C_i be any sequence of positive real numbers, and let $\mathcal{P}$ be the Banach space consisting of all real sequences $\lambda = \{\lambda_i\}$ such that the norm

$$\|\lambda\|_{\mathcal{P}} = \sum_i C_i |\lambda_i| \tag{11.18}$$

is finite.

Lemma 11.6.2. *The real numbers* $C_i > 0$ *can be chosen so that for all* λ *in* $\mathcal{P}$ *and all* k, *the series*

$$\sum_i \lambda_i \hat{\mathfrak{q}}_i$$

is convergent in $C^\infty_{\text{loc}}(\mathcal{C}_k(Z), \mathcal{V}_k)$.

Remark. Convergence in $C^\infty_{\text{loc}}(\mathcal{C}_k(Z), \mathcal{V}_k)$ means that for each $l \geq 0$ and each x in $\mathcal{C}_k(Z)$, there is a neighborhood U of x on which the partial sums of the series are Cauchy in $C^l(U, \mathcal{V}_k)$.

Proof of Lemma 11.6.2. Fix a smooth base connection A_0. Part (i) of Proposition 11.4.1 tells us that for each $k \geq 2$, each $l \geq 0$ and each $R > 0$, the perturbation $\hat{\mathfrak{q}}_i$ has finite C^l norm on the ball $B_{R,k}$ of radius R about $(A_0, 0)$ in $\mathcal{C}_k(Z)$. If we set $C(i, k, l, R)$ to be the corresponding norm,

$$C(i, k, l, R) = \|\hat{\mathfrak{q}}_i\|_{C^l(B_{R,k}, \mathcal{V}_k)},$$

then the series

$$\sum_i \lambda_i \hat{\mathfrak{q}}_i$$

converges in $C^l(B_{R,k}, \mathcal{V}_k)$ whenever the λ_i satisfy

$$\sum_i C(i,k,l,R)|\lambda_i| < \infty.$$

We can then define C_i by diagonalization,

$$C_i = \max_{i' \leq i} C(i, i', i, i),$$

and we will ensure convergence in $C^l(B_{R,k}, \mathcal{V}_k)$ for all k, l and R. □

Because of the lemma, we now have a linear map $\mathfrak{Q} : \mathcal{P} \to C^0(\mathcal{C}, \mathcal{T}_0)$ defined by

$$\lambda \mapsto \mathfrak{q}_\lambda = \sum_i \lambda_i \mathfrak{q}_i.$$

Property (iii) follows from the lemma. Replacing the constants C_i with larger ones if necessary, we can achieve the required properties (i) and (iv) by the same type of diagonalization. □

In the remainder of the book, we will make use of a Banach space of perturbations such as the one provided by the above theorem. In order to obtain transversality results, we will want in addition that the initial countable collection of perturbations $\mathfrak{q}_i$ be sufficiently large. To this end, let us consider the choices to be made when constructing a cylinder function f on $\mathcal{C}(Y)$. We must choose the following things, in order:

- a pair of positive integers n and m;
- coexact forms c_1, ..., c_n and sections Υ_1, ..., Υ_m of the bundle $\mathbb{S}$;
- a compact subset K of $\mathbb{R}^n \times \mathbb{T} \times \mathbb{C}^m$;
- a smooth S^1-invariant function g on $\mathbb{R}^n \times \mathbb{T} \times \mathbb{C}^m$, supported in K.

We can specify a countable collection of cylinder functions as follows. For every pair (n, m), we choose a countable collection of $(n + m)$-tuples $(c_1, \dots c_n, \Upsilon_1, \dots, \Upsilon_m)$ which are dense in the C^∞ topology in the space of all such $(n+m)$-tuples. We also choose a countable collection of compact subsets K of $\mathbb{R}^n \times \mathbb{T} \times \mathbb{C}^m$ which are dense in the Hausdorff topology. Finally, for each K, we choose a collection of functions $g_\alpha = g(n, m, K)_\alpha$ ($\alpha \in \mathbb{N}$) with the properties:

- each g_α is an S^1-invariant function on $\mathbb{R}^n \times \mathbb{T} \times \mathbb{C}^m$ supported in K;
- the functions g_α are dense in the C^∞ topology in the space of smooth, S^1-invariant functions with support in K;
- the subset of the g_α comprising those g_α that vanish on the set

$$K_0 = K \cap (\mathbb{R}^n \times \mathbb{T} \times \{0\})$$

are dense in the C^∞ topology in the space of smooth, S^1-invariant functions with support in K and vanishing on K_0.

Combining all these choices, we arrive at a countable collection of cylinder functions and corresponding perturbations $\mathfrak{q}_i$.

Definition 11.6.3. By a *large* Banach space of tame perturbations, we will mean a separable Banach space $\mathcal{P}$ and a linear map $\mathfrak{Q} : \mathcal{P} \to C^0(\mathcal{C}(Y), \mathcal{T}_0)$, satisfying the conditions of Theorem 11.6.1 and containing a countable collection of tame perturbations $\mathfrak{q}_i$ obtained by making choices as prescribed above. $\diamondsuit$

In what follows, we will presume that a large Banach space of tame perturbations has been chosen, once and for all. In our notation, we will usually confuse $\mathcal{P}$ with its image under $\mathfrak{Q}$. So we will consider $\mathcal{P}$ as a linear space of perturbations, $\mathfrak{q} \in \mathcal{P}$. Of course, the Banach space topology is not the topology of its image as a subspace of the set of tame perturbations, for any natural choice of topology on the latter.

In the conclusions of Theorem 11.6.1, the last condition (unlike the first three) needs some motivation. The reason for imposing it is that it allows us to prove the following properness result:

Proposition 11.6.4. *Let $\mathcal{P}$ be a large Banach space of tame perturbations, let $\mathfrak{q}_i$ be a convergent sequence in $\mathcal{P}$, and let $\beta_i \in \mathcal{C}_k(Y)$ be solutions of the equations*

$$(\operatorname{grad} \mathcal{L} + \mathfrak{q}_i)(\beta_i) = 0.$$

Then there is a sequence of gauge transformations u_i such that the transformed solutions $u_i(\beta_i)$ have a convergent subsequence.

Proof. As with Corollary 10.7.4, this can be deduced from the compactness theorem on the cylinder, Theorem 10.7.1. The conditions imposed by the last part of Theorem 11.6.1 provide the uniform bounds needed to start the bootstrapping argument. $\square$

Notes and references for Chapter III

The material in Section 9 is now a very standard part of gauge theory, going back a long way: see [6, 7, 23, 35, 123] for example, for treatment of such Hilbert-manifold quotients and Coulomb slices. The material here has become a little more elaborate because of our need to treat the blown-up configuration space.

The need to introduce a large class of perturbations when constructing Floer homology was a feature of Floer's original exposition, [32]. In the context of $SU(2)$ gauge theory, Floer introduced perturbations based on the holonomy of a connection around a family of loops: similar perturbations appear in [111] and [19]. In the context of a $U(1)$ gauge theory, one can replace the holonomy idea by a simple integration of the connection 1-form against a closed 2-form, and this provides the simplest model for our cylinder functions.

The rather strange space of perturbations $\mathcal{P}$ is necessitated by the need to fulfill two competing requirements: our perturbations need to be smooth vector fields on the configuration space (not just C^ℓ vector fields), in order to apply the Sard-Smale theorem later; and the space of perturbations needs to be a Banach space, or a Banach manifold. Similar non-standard Banach spaces of smooth functions were used by Floer in [32].

IV
Moduli spaces and transversality

Given a closed, oriented, Riemannian 3-manifold Y, with a spin^c structure $\mathfrak{s}$, we have introduced in the previous chapter a Hilbert manifold with boundary, $\mathcal{C}_k^\sigma(Y)$, obtained by blowing up the configuration space of all pairs (B, Ψ) along the locus $\Psi = 0$. In Subsection 10.3, we saw that each choice of tame perturbation $\mathfrak{q}$ gave rise to a smooth section

$$(\operatorname{grad} \not{\mathcal{L}})^\sigma = (\operatorname{grad} \mathcal{L})^\sigma + \mathfrak{q}^\sigma$$

of the L^2_{k-1} completion of the tangent bundle of $\mathcal{C}_k^\sigma(Y)$:

$$\mathcal{T}_{k-1}^\sigma \to \mathcal{C}_k^\sigma(Y).$$

We shall refer to a section of any of the completions $\mathcal{T}_j^\sigma$ $(j \leq k)$ somewhat loosely as a *vector field* on $\mathcal{C}_k^\sigma(Y)$, even though only in the case $j = k$ is this a true vector field on the Hilbert manifold with boundary. This vector field is invariant under the free action of the gauge group $\mathcal{G}_{k+1}(Y)$, and so it descends to a vector field (in the same, weak sense) on the quotient manifold $\mathcal{B}_k^\sigma(Y)$. We have already studied the formal gradient-flow equations for the Chern–Simons–Dirac functional and its perturbation: these give rise to the perturbed Seiberg–Witten equations on the 4-dimensional cylinder. But so far, we have considered solutions to these equations mainly on a finite time interval, or a cylinder $[t_1, t_2] \times Y$.

Our next task is to study the gradient-flow equations more globally, and to set up the analytic underpinnings which will be needed for us to take the constructions of finite-dimensional Morse theory (discussed in Section 2) and adapt them to this infinite-dimensional setting.

In Section 12 below, we discuss the non-degeneracy of critical points of the flow. This discussion has two aspects: a Fredholm theory for the linearization of the flow, and then an examination of our class of perturbations, to see that the class is indeed large enough to achieve non-degeneracy. Two features complicate the discussion: first, the fact that we have blown up the configuration space; and second, the fact that, while the Fredholm theory is best considered in the context of $\mathcal{C}_k^\sigma(Y)$, the non-degeneracy we are hoping to achieve is in the quotient space $\mathcal{B}_k^\sigma(Y)$.

When all critical points of the flow are non-degenerate, we then consider the trajectory spaces: the spaces of solutions to the formal gradient-flow equations, defined for all time on the cylinder $\mathbb{R} \times Y$. Setting up the Fredholm theory for the appropriate operators on the cylinder takes some time, and the work is begun in Section 13.

12 Transversality for the three-dimensional equations

12.1 Non-degeneracy of critical points

We now have a large class of tame perturbations, constructed via cylinder functions (see Theorem 11.1.2). In this section we shall see that these perturbations are sufficiently general that, for a generic choice of $\mathfrak{q}$, all the critical points of the vector field $(\operatorname{grad} \mathcal{L})^\sigma$ are non-degenerate in the directions normal to the gauge orbit. To state the non-degeneracy condition precisely, recall from Proposition 9.3.5 that for $j \leq k$ we have a bundle decomposition

$$\mathcal{T}_j^\sigma = \mathcal{J}_j^\sigma \oplus \mathcal{K}_j^\sigma,$$

in which the first summand is a bundle of tangent spaces to the gauge orbits.

Definition 12.1.1. A critical point $\mathfrak{a} \in \mathcal{C}_k^\sigma(Y)$ of the vector field $(\operatorname{grad} \mathcal{L})^\sigma$ is *non-degenerate* if the smooth section $(\operatorname{grad} \mathcal{L})^\sigma$ of $\mathcal{T}_{k-1}^\sigma$ is transverse to the subbundle $\mathcal{J}_{k-1}^\sigma$. ◊

If $\mathcal{S}_k^\sigma \subset \mathcal{C}_k^\sigma(Y)$ is the Coulomb slice at $\mathfrak{a}$ (Definition 9.3.9), then the non-degeneracy condition is equivalent to saying that the restriction of $(\operatorname{grad} \mathcal{L})^\sigma$ to $\mathcal{S}_k^\sigma$ is transverse to $\mathcal{J}_{k-1}$ at $\mathfrak{a}$. Although we shall be working primarily with $\mathcal{C}_k^\sigma(Y)$, the non-degeneracy condition has a straightforward interpretation in terms of the quotient Hilbert manifold $\mathcal{B}_k^\sigma(Y)$. The quotient space carries a vector bundle

$$\begin{aligned} [\mathcal{T}_j^\sigma] &\to \mathcal{B}_k^\sigma(Y), \\ [\mathcal{T}_j^\sigma] &= (\mathcal{T}_j^\sigma / \mathcal{J}_j^\sigma)/\mathcal{G}_{k+1}, \end{aligned} \tag{12.1}$$

whose fiber at the gauge equivalence class $[\mathfrak{a}]$ is the completion of the tangent space $T_{[\mathfrak{a}]}\mathcal{B}_k^\sigma(Y)$ in the L_j^2 norm ($j \leq k$). The section $(\operatorname{grad} \mathcal{L})^\sigma$ of $\mathcal{T}_{k-1}$ is invariant under the action of the gauge group $\mathcal{G}_{k+1}(Y)$, and therefore descends to a smooth section $[\operatorname{grad} \mathcal{L}]^\sigma$ of $[\mathcal{T}_{k-1}]$ on $\mathcal{B}_k^\sigma(Y)$. Non-degeneracy of all critical points simply means that the section $[\operatorname{grad} \mathcal{L}]^\sigma$ is everywhere transverse to the zero section of $[\mathcal{T}_{k-1}^\sigma]$.

Remark. A cautionary remark is appropriate here. To say that $[\operatorname{grad} \mathcal{L}]^\sigma$ is transverse to the zero section means that, at each zero $[\mathfrak{a}]$, the derivative defines a surjective linear map $T_{[\mathfrak{a}]}\mathcal{B}_k^\sigma(Y) \to \mathcal{T}_{k-1,[\mathfrak{a}]}^\sigma$. Unlike the situation of a vector field on a finite-dimensional manifold, surjectivity of this derivative for a general section does not imply injectivity. In the course of this section, we will see that the derivative of the particular section $[\operatorname{grad} \mathcal{L}]^\sigma$ is a Fredholm operator of index zero, so our transversality condition will imply that the derivative is an isomorphism. It follows also that transverse zeros in $\mathcal{B}_k^\sigma(Y)$ are isolated.

The rest of this section is devoted to proving:

Theorem 12.1.2. *Let $\mathcal{P}$ be a large Banach space of tame perturbations, as in Definition* 11.6.3. *Then there is a residual (and in particular non-empty) subset of $\mathcal{P}$ such that for every $\mathfrak{q}$ in this subset, all the zeros of the section* $(\operatorname{grad} \mathcal{L})^\sigma$ *of $\mathcal{T}_{k-1}^\sigma \to \mathcal{C}_k^\sigma(Y)$ are non-degenerate in the above sense. For such a perturbation, the image of the zeros in $\mathcal{C}_k(Y)$ comprises a finite set of gauge orbits.*

12.2 Non-degeneracy characterized

Recall Proposition 10.3.1, that gives a characterization of the critical points of $(\operatorname{grad} \mathcal{L})^\sigma$. In the same terms, we now characterize non-degeneracy of the critical points.

As an analytic preliminary, we examine the sort of perturbation of the Hessian of $\mathcal{L}$ that arises when we look at the linearization of $(\operatorname{grad} \mathcal{L})^\sigma$. We introduce a definition to handle the general situation.

Definition 12.2.1. An operator L is called *k-almost self-adjoint first-order elliptic* (k-ASAFOE) if it can be written as

$$L = L_0 + h$$

where

- L_0 is a first-order, self-adjoint elliptic differential operator with smooth coefficients, acting on sections of a vector bundle $E \to Y$, and

- h is an operator on sections of E which we suppose to be a map
$$h : C^\infty(Y;E) \to L^2(Y;E)$$
which extends to a bounded map on $L^2_j(Y;E)$ for all j in the range $|j| \le k$.

We simply say that L is ASAFOE if it is k-ASAFOE for all $k \ge 0$. ◊

Note that we do not assume that h is symmetric. The conditions on h are motivated by the properties of the derivative of $\mathfrak{q}^\sigma$.

We have a simple regularity result:

Lemma 12.2.2. *Suppose L is k-ASAFOE, and let $u \in L^2_{-k}$ be a weak solution of*
$$Lu = v,$$
with $v \in L^2_j$ for some j in the range $|j| \le k$. Then $u \in L^2_{j+1}$.

Proof. This is a straightforward bootstrapping argument. Write $L = L_0 + h$ as above. We have $hu \in L^2_{-k}$, by the hypothesis on h. As a weak solution of the elliptic equation $L_0 u = v - hu$, the element u is then in L^2_{-k+1}, as long as $-k \le j$. □

Corollary 12.2.3. *If L is k-ASAFOE, then for j in the range $-k \le j \le k+1$, it is invertible as an operator*
$$L : L^2_j(E) \to L^2_{j-1}(E)$$
if and only if it is injective. Furthermore, if L is invertible for one j in this range, then it is invertible for all j in the range.

Proof. The first statement holds because L is Fredholm of index zero. The second clause follows from the regularity lemma. □

We can therefore talk unambiguously about the spectrum of L: the operator
$$(L - \lambda) : L^2_j \to L^2_{j-1}$$
is invertible if and only if λ is not an eigenvalue, a condition that is independent of j. As usual, by the *generalized* eigenspace belonging to an eigenvalue λ, we mean the union
$$\bigcup_{n \ge 0} (L - \lambda)^{-n}\{0\}.$$

Lemma 12.2.4. *Let $L = L_0 + h$ be a k-ASAFOE operator. Then the following hold.*

(i) *There are only finitely many eigenvalues of the complexification $L \otimes 1_{\mathbb{C}}$ in any compact subset of the complex plane $\mathbb{C}$, and the generalized eigenspaces of the complexification are finite-dimensional. All the generalized eigenvectors belong to L^2_{k+1}.*
(ii) *If h, like L_0, is symmetric, then the eigenvalues are real, and there is a complete orthonormal system of eigenvectors $\{e_n\}$ in $L^2(E)$. The span of the eigenvectors is dense in L^2_{k+1}.*
(iii) *In the non-symmetric case, the imaginary parts of the eigenvalues λ of $L \otimes 1_{\mathbb{C}}$ are bounded by the L^2-operator norm of $(h - h^*)/2$.*

Proof. The first two parts follow from the standard theory of compact operators, because, for any λ' not in the spectrum, $(L-\lambda')^{-1}$ is a compact operator (on L^2) since it factors through the compact inclusion $L^2_1 \hookrightarrow L^2$. The last part follows from the equality

$$(\lambda - \bar{\lambda})\|v\|^2 = \langle (h - h^*)v, v\rangle,$$

which holds for any v in the kernel of $L \otimes 1_{\mathbb{C}} - \lambda$. □

We should also remark at this point that, in the symmetric case, the eigenvalues of L are a doubly infinite sequence, unbounded both above and below. This is a consequence of L_0 being first-order. The essential point is that a first-order differential operator with constant coefficients on $\mathbb{R}^n$ is conjugate to its negative via the map $x \mapsto -x$ on $\mathbb{R}^n$. For example, if a 3-manifold admits an orientation-reversing isometry, then the spectrum of the Dirac operator is symmetric about zero.

To return to the question of non-degeneracy, we again use the coordinates (6.2), so as to write an element in $\mathcal{C}^\sigma_k(Y)$ as a triple (B, r, ψ), with $r \geq 0$ and ψ of unit L^2 norm. In the following proposition, we have used some additional notation. The affine space $\mathcal{A}_k$ is acted on by $\mathcal{G}_{k+1}$, and we write

$$\mathcal{T}^{\mathrm{red}}_j = \mathcal{J}^{\mathrm{red}}_j \oplus \mathcal{K}^{\mathrm{red}}_j \tag{12.2}$$

for the decomposition of its L^2_j tangent bundle. (The superscript stands for "reducible.") That is, we write $\mathcal{T}^{\mathrm{red}}_j = \mathcal{A}_k \times L^2_j(Y; iT^*Y)$, and $\mathcal{J}^{\mathrm{red}}_j$, $\mathcal{K}^{\mathrm{red}}_j$ are the trivial subbundles with fibers the exact and coclosed imaginary-valued L^2_j 1-forms respectively. The first component of grad $\mathcal{L}$, restricted to $\mathcal{A}_k \times \{0\} \subset$

$\mathcal{C}_k(Y)$, defines a section

$$(\operatorname{grad} \mathcal{L})^{\mathrm{red}} : \mathcal{A}_k \to \mathcal{T}^{\mathrm{red}}_{k-1}.$$

For $B \in \mathcal{A}_k$, we introduce the operator

$$\begin{aligned} D_{\mathfrak{q},B} &: L^2_k(Y;S) \to L^2_{k-1}(Y;S) \\ D_{\mathfrak{q},B} &: \phi \mapsto D_B\phi + \mathcal{D}_{(B,0)}\mathfrak{q}^1(0,\phi). \end{aligned} \tag{12.3}$$

This perturbation of the Dirac operator is k-ASAFOE, in the sense defined above: if we write $B = B_0 + b \otimes 1$, we can write $D_{\mathfrak{q},B}$ as $L_0 + h$, where $L_0 = D_{B_0}$ and

$$h\phi = \rho(b)\phi + \mathcal{D}_{(B,0)}\mathfrak{q}^1(0,\phi).$$

The required mapping properties of h follow from the Sobolev multiplication theorems (for the first term) and Condition (iii) of Definition 10.5.1 (for the second). In addition, the operator h is symmetric, as follows from the fact that $\mathfrak{q}$ is a formal gradient. Lemma 12.2.4 is therefore applicable. The S^1 invariance of the perturbation also ensures that $D_{\mathfrak{q},B}$ is complex-linear.

Proposition 12.2.5. *Let $\mathfrak{a} = (B, r, \psi)$ be a critical point of the vector field $(\operatorname{grad} \mathcal{L})^\sigma$. Then the non-degeneracy of $\mathfrak{a}$ can be characterized by one of the following conditions, according to whether r is zero or non-zero.*

(i) *If $r \neq 0$, then $\mathfrak{a}$ is non-degenerate if and only if the corresponding point $(B, r\psi) \in \mathcal{C}^*_k(Y)$ is a non-degenerate zero of $\operatorname{grad} \mathcal{L}$, in the sense that $\operatorname{grad} \mathcal{L}$ is transverse to the subbundle $\mathcal{J}_{k-1}$ of $\mathcal{T}_{k-1}$.*

(ii) *If $r = 0$, then by Proposition 10.3.1, ψ is an eigenvector of $D_{\mathfrak{q},B}$, with eigenvalue λ, say, and in this case $\mathfrak{a}$ is non-degenerate if and only if the following three conditions hold:*
   (a) *$B \in \mathcal{A}_k$ is a non-degenerate zero of $(\operatorname{grad} \mathcal{L})^{\mathrm{red}}$, in the sense that $(\operatorname{grad} \mathcal{L})^{\mathrm{red}}$ is transverse to the subbundle $\mathcal{J}^{\mathrm{red}}_{k-1}$ of $\mathcal{T}^{\mathrm{red}}_{k-1}$;*
   (b) *λ is a simple eigenvalue of $D_{\mathfrak{q},B}$, as a complex operator;*
   (c) *λ is not zero.*

Proof. The first item is immediate, because the map from $\mathcal{C}^\sigma_k(Y)$ to $\mathcal{C}_k(Y)$ is a diffeomorphism over $\mathcal{C}^*_k(Y)$. For the case $r = 0$, we refer to the formula (10.10) for $(\operatorname{grad} \mathcal{L})^\sigma$ in these coordinates. Differentiating the vector field at $(B, 0, \psi)$, we obtain an operator of the form

$$\mathcal{D}_{(B,0,\psi)}(\operatorname{grad} \mathcal{L})^\sigma : \begin{bmatrix} b \\ t \\ \phi \end{bmatrix} \mapsto \begin{bmatrix} \mathcal{D}_B(\operatorname{grad} \mathcal{L})^{\mathrm{red}} & 0 & 0 \\ 0 & \lambda & 0 \\ x & 0 & D_{\mathfrak{q},B} - \lambda \end{bmatrix} \begin{bmatrix} b \\ t \\ \phi \end{bmatrix}$$

acting on triples (b, t, ϕ) with $\operatorname{Re}\langle \psi, \phi \rangle = 0$. Those zero entries in the above matrix that are not an immediate consequence of the shape of (10.10) follow from the S^1 invariance of the perturbation, which tells us, for example, that $\mathfrak{q}^0(B, r\psi)$ is an even function of r, while $\mathfrak{q}^1(B, r\psi)$ is an odd function of r.

Writing $\mathbf{d}^\sigma$ again for the derivative of the gauge group action on $\mathcal{C}^\sigma(Y)$, the non-degeneracy of $\mathfrak{a} = (B, 0, \phi)$ is equivalent to the surjectivity of

$$\mathbf{d}^\sigma_{\mathfrak{a}} \oplus \mathcal{D}_{(B,0,\psi)}(\operatorname{grad} \mathcal{L})^\sigma,$$

which is given by the matrix

$$\begin{bmatrix} -d & \mathcal{D}_B(\operatorname{grad} \mathcal{L})^{\mathrm{red}} & 0 & 0 \\ 0 & 0 & \lambda & 0 \\ \psi\cdot & x & 0 & D_{\mathfrak{q},B} - \lambda \end{bmatrix}.$$

Condition (ii)(a) is equivalent to the surjectivity of the first diagonal block,

$$(-d, \mathcal{D}_B(\operatorname{grad} \mathcal{L})^{\mathrm{red}}).$$

Condition (ii)(c) is equivalent to the surjectivity of the middle diagonal block, multiplication by λ. Condition (ii)(b) means that, acting on the real-orthogonal complement of ψ, the last diagonal block $D_{\mathfrak{q},B} - \lambda$ has cokernel the real span of $i\psi$. However, the vector $(0, 0, i\psi)$ is the image of $(i, 0, 0, 0)$. □

We note a simple corollary of our regularity results and the above characterization of non-degeneracy.

Corollary 12.2.6. *If $\mathfrak{a}$ is a critical point of $\mathcal{L}$ in $\mathcal{C}^\sigma_k(Y)$, then $\mathfrak{a}$ is gauge-equivalent to a smooth configuration; so the set of gauge-equivalence classes of critical points is independent of k. Furthermore, the notion of non-degeneracy is independent of k also.* □

12.3 Hessians

On the irreducible part $\mathcal{C}_k^*(Y) \subset \mathcal{C}_k(Y)$, we have a decomposition (Proposition 9.3.4),

$$\mathcal{T}_j|_{\mathcal{C}_k^*} = \mathcal{J}_j \oplus \mathcal{K}_j.$$

Because this decomposition is orthogonal with respect to the standard L^2 inner product (the same inner product with respect to which grad $\mathcal{L}$ is a gradient), we have that grad $\mathcal{L}$ is a section of $\mathcal{K}_{k-1} \subset \mathcal{T}_{k-1}$ on $\mathcal{C}_k^*(Y)$. We have the Coulomb slice $\mathcal{S}_{k,\alpha} = \alpha + \mathcal{K}_{k,\alpha}$ as a transverse slice to the gauge orbit at $\alpha \in \mathcal{C}^*(Y)$. (See (9.14).) We define an operator

$$\mathrm{Hess}_{\mathfrak{q},\alpha} : \mathcal{K}_{k,\alpha} \to \mathcal{K}_{k-1,\alpha} \tag{12.4}$$

as the restriction of the linear map

$$\Pi_{\mathcal{K}_{k-1}} \circ \mathcal{D}_\alpha \operatorname{grad} \mathcal{L} : T_\alpha \mathcal{C}_k^*(Y) = \mathcal{T}_{k,\alpha} \to \mathcal{K}_{k-1,\alpha}$$

to $\mathcal{K}_{k,\alpha}$. (Here $\Pi_{\mathcal{K}_{k-1}}$ is the L^2 orthogonal projection, which has kernel $\mathcal{J}_{k-1}$.) As α varies, we have a smooth bundle map

$$\mathrm{Hess}_{\mathfrak{q}} : \mathcal{K}_k \to \mathcal{K}_{k-1}$$

that is $\mathcal{G}_{k+1}(Y)$-equivariant. We can identify $\mathcal{K}_k$ with the pull-back of the tangent bundle of $\mathcal{B}_k^*(Y)$, and $\mathcal{K}_{k-1}$ with the pull-back of $[\mathcal{T}_{k-1}]$ (where $[\mathcal{T}_j]$ is defined as in (12.1)). In these terms, the family of operators $\mathrm{Hess}_{\mathfrak{q},\alpha}$ is the pull-back of a family of operators $[\mathcal{T}_k] \to [\mathcal{T}_{k-1}]$ that is formally the covariant Hessian of the circle-valued function $\bar{\mathcal{L}}$ on $\mathcal{B}_k^*(Y)$:

$$\mathrm{Hessian}(\bar{\mathcal{L}}) : [\mathcal{T}_k] \to [\mathcal{T}_{k-1}].$$

(By covariant Hessian, we mean to refer to what is formally the Levi-Cività derivative corresponding to the L^2 inner product on the tangent bundle.) We will refer to $\mathrm{Hess}_{\mathfrak{q},\alpha} : \mathcal{K}_{k,\alpha} \to \mathcal{K}_{k-1,\alpha}$ also as the *Hessian*, without much danger of confusion.

Proposition 12.3.1. *The operator* $\mathrm{Hess}_{\mathfrak{q},\alpha} : \mathcal{K}_{k,\alpha} \to \mathcal{K}_{k-1,\alpha}$ *is symmetric. There is a complete orthonormal system* $\{e_n\}$ *in* $\mathcal{K}_{0,\alpha}$, *with the property that each* e_n *is smooth, and*

$$\mathrm{Hess}_{\mathfrak{q},\alpha} e_n = \lambda_n e_n$$

for some $\lambda_n \in \mathbb{R}$. *The span of the eigenvectors is dense in* $\mathcal{K}_{k,\alpha}$ *for all* k. *The number of eigenvalues* λ_n *in any bounded interval is finite.*

In particular, $\mathrm{Hess}_{\mathfrak{q},\alpha}$ *is Fredholm of index zero, and is therefore surjective if and only if it is injective.*

Proof. The symmetry is a formal consequence of $\mathrm{Hess}_{\mathfrak{q},\alpha}$ being a Hessian under the above identifications. To deduce the remaining properties from Lemma 12.2.4, we introduce the *extended* Hessian. This is the operator

$$\widehat{\mathrm{Hess}}_{\mathfrak{q},\alpha} : \mathcal{T}_{k,\alpha} \oplus L^2_k(Y; i\mathbb{R}) \to \mathcal{T}_{k-1,\alpha} \oplus L^2_{k-1}(Y; i\mathbb{R}) \tag{12.5}$$

given by

$$\widehat{\mathrm{Hess}}_{\mathfrak{q},\alpha} = \begin{bmatrix} \mathcal{D}_\alpha \operatorname{grad} \mathcal{L} & \mathbf{d}_\alpha \\ \mathbf{d}^*_\alpha & 0 \end{bmatrix}. \tag{12.6}$$

This is a symmetric operator, to which Lemma 12.2.4 applies directly. Indeed if we write $\mathcal{T}_{j,\alpha}$ as

$$\mathcal{T}_{j,\alpha} = L^2_j(Y; S) \oplus L^2_j(Y; iT^*Y),$$

then the extended Hessian has the shape

$$\widehat{\mathrm{Hess}}_{\mathfrak{q},\alpha} = \begin{bmatrix} D_{A_0} & 0 & 0 \\ 0 & *d & -d \\ 0 & d^* & 0 \end{bmatrix} + h$$

where the term h is the sum of a zeroth-order operator and terms arising from the perturbation. The first term in this expression is a self-adjoint elliptic operator on Y, acting on sections of $S \oplus iT^*Y \oplus i\mathbb{R}$, and the second term satisfies the conditions laid out in the definition of k-ASAFOE (Definition 12.2.1).

There is another block decomposition of the extended Hessian. If we decompose $\mathcal{T}_j$ as $\mathcal{J}_j \oplus \mathcal{K}_j$ and decompose the operator $\widehat{\mathrm{Hess}}_{\mathfrak{q},\alpha}$ accordingly, we have

$$\widehat{\mathrm{Hess}}_{\mathfrak{q},\alpha} = \begin{bmatrix} 0 & x & \mathbf{d}_\alpha \\ x^* & \mathrm{Hess}_{\mathfrak{q},\alpha} & 0 \\ \mathbf{d}^*_\alpha & 0 & 0 \end{bmatrix}. \tag{12.7}$$

Here x is the operator

$$x = \Pi_{\mathcal{J}_{k-1}} \circ \mathcal{D}_\alpha \operatorname{grad} \mathcal{L}|_{\mathcal{K}_{k,\alpha}},$$

which can be written

$$x = \mathbf{d}(\mathbf{d}^*\mathbf{d})^{-1}\mathbf{d}^*\mathcal{D}_\alpha \operatorname{grad} \pounds|_{\mathcal{K}_{k,\alpha}}.$$

(We have temporarily dropped the subscript from $\mathbf{d}_\alpha$.)

Lemma 12.3.2. *The operator $x : \mathcal{K}_k \to \mathcal{J}_{k-1}$, and its formal adjoint $x^* : \mathcal{J}_k \to \mathcal{K}_{k-1}$, extend to bounded operators $\mathcal{K}_j \to \mathcal{J}_j$ and $\mathcal{J}_j \to \mathcal{K}_j$ for $0 \leq j \leq k$. At a critical point, the operator x is zero.*

Proof of Lemma. We shall examine only x. We write

$$x = \Pi_{\mathcal{J}_{k-1}} \circ \big(\mathcal{D}_\alpha \operatorname{grad} \mathcal{L} + \mathcal{D}_\alpha \mathfrak{q}\big),$$

and the conclusion of the lemma certainly holds for the second term, because of Condition (iii) in Definition 10.5.1. The first term can be written

$$x = \mathbf{d}(\mathbf{d}^*\mathbf{d})^{-1}\mathbf{d}^*\mathcal{D}_\alpha \operatorname{grad} \mathcal{L}|_{\mathcal{K}_{k,\alpha}}.$$

The operator $\mathbf{d}^*\mathcal{D}_\alpha \operatorname{grad} \mathcal{L}|_{\mathcal{K}_{k,\alpha}}$ is a differential operator of first order, as is easily checked, and $\mathbf{d}(\mathbf{d}^*\mathbf{d})^{-1}$ is smoothing of order 1. This verifies the first part.

The last assertion reflects the general fact that the derivative of a section of a subbundle takes values in the subbundle at all points where the section vanishes. □

From this lemma, it follows that the hypotheses of Lemma 12.2.4 are satisfied by the operator obtained from the extended Hessian $\widehat{\mathrm{Hess}}_{\mathfrak{q},\alpha}$ by dropping the x and x^* terms:

$$\begin{pmatrix} 0 & 0 & \mathbf{d}_\alpha \\ 0 & \mathrm{Hess}_{\mathfrak{q},\alpha} & 0 \\ \mathbf{d}_\alpha^* & 0 & 0 \end{pmatrix}.$$

In particular, this operator has a complete orthonormal system of eigenvectors; and the symmetry of the operator means that the eigenvalues are real. At this point, the operator $\mathrm{Hess}_{\mathfrak{q},\alpha}$ appears as a summand, and the conclusion of the proposition follows. □

12.4 Hessians on the blown-up configuration space

Before moving on, we make some remarks about how to adapt the construction of $\mathrm{Hess}_{\mathfrak{q},\alpha}$ to the setting of the blown-up configuration space. The operator

$\mathrm{Hess}_{\mathfrak{q},\alpha}$ at a general $\alpha \in \mathcal{C}_k(Y)$ has a natural interpretation as the Hessian of the perturbed Chern–Simons–Dirac functional, as we have explained. The vector field grad $\mathcal{L}$ gives rise to a vector field $(\mathrm{grad}\,\mathcal{L})^\sigma$ on the blow-up, $\mathcal{C}^\sigma_k(Y)$; but if we take a derivative of $(\mathrm{grad}\,\mathcal{L})^\sigma$ on the blow-up, we are not dealing any more with the second derivative of a function, because this vector field on the blow-up is not a gradient in any natural way. Furthermore, we need to specify what covariant derivative we are going to use if we are to differentiate this vector field.

If we recall that grad $\mathcal{L}$ is a section of the subbundle $\mathcal{K}_{k-1}$ of $\mathcal{T}_{k-1}$ on $\mathcal{C}_k(Y)$, then we see from Proposition 9.3.5 that $(\mathrm{grad}\,\mathcal{L})^\sigma$ is a section of $\mathcal{K}^\sigma_{k-1}$ on the blow-up, by continuity. The standard coordinates (B, r, ψ) embed the blow-up, as usual, in an affine space carrying an L^2 inner product:

$$\mathcal{C}^\sigma_k(Y) \subset \mathcal{A}_k \times \mathbb{R} \times L^2_k(Y; S).$$

The vector field $(\mathrm{grad}\,\mathcal{L})^\sigma$ can therefore be differentiated as a vector field along a submanifold in this affine space to obtain a section $\mathcal{D}(\mathrm{grad}\,\mathcal{L})^\sigma$. This derivative can then be projected back, first to the tangent space $\mathcal{T}^\sigma_{k-1}$ and then to the summand $\mathcal{K}^\sigma_{k-1}$ using projection with kernel $\mathcal{J}^\sigma_{k-1}$. This defines a smooth bundle map

$$\begin{aligned} \mathcal{T}^\sigma_k &\to \mathcal{K}^\sigma_{k-1} \\ x &\mapsto \Pi_{\mathcal{K}^\sigma_{k-1}} \mathcal{D}(\mathrm{grad}\,\mathcal{L})^\sigma(x), \end{aligned}$$

over $\mathcal{C}^\sigma_k(Y)$. Finally, this operator can be restricted to the subbundle $\mathcal{K}^\sigma_k \subset \mathcal{T}^\sigma_k$, and we have a bundle map

$$\mathrm{Hess}^\sigma_{\mathfrak{q}} : \mathcal{K}^\sigma_k \to \mathcal{K}^\sigma_{k-1} \tag{12.8}$$

over $\mathcal{C}^\sigma_k(Y)$.

Because $\mathcal{K}^\sigma_k$ is the pull-back of the tangent bundle of the quotient configuration space $\mathcal{B}^\sigma_k(Y)$, we can indeed regard $\mathrm{Hess}^\sigma_{\mathfrak{q}}$ formally as a covariant derivative of a vector field on $\mathcal{B}^\sigma_k(Y)$. It is not a Levi-Cività derivative, however, because the decomposition of the tangent bundle of $\mathcal{C}^\sigma_k(Y)$ as $\mathcal{J}^\sigma_k \oplus \mathcal{K}^\sigma_k$ is not an orthogonal decomposition. For want of a better term, we may still refer to $\mathrm{Hess}^\sigma_{\mathfrak{q}}$ as the *Hessian* on the blown-up configuration space.

In terms of the decomposition

$$\mathcal{T}^\sigma_{j,\mathfrak{a}} = \mathcal{J}^\sigma_{j,\mathfrak{a}} \oplus \mathcal{K}^\sigma_{j,\mathfrak{a}},$$

the definition of $\mathrm{Hess}^\sigma_{\mathfrak{q},\mathfrak{a}}$ means that the derivative of the vector field has block form

$$\mathcal{D}_{\mathfrak{a}}(\operatorname{grad} \mathcal{L})^\sigma = \begin{bmatrix} 0 & x \\ y & \mathrm{Hess}^\sigma_{\mathfrak{q},\mathfrak{a}} \end{bmatrix}. \tag{12.9}$$

A critical point $\mathfrak{a}$ is a point at which the vector field vanishes. When this happens, the vector field vanishes along the whole gauge orbit, which means that y is zero. The entry x is zero also at a critical point: as before, this is a formal consequence of the fact that $(\operatorname{grad} \mathcal{L})^\sigma$ is a section of the summand $\mathcal{K}^\sigma_{k-1}$. So at a critical point we have

$$\mathcal{D}_{\mathfrak{a}}(\operatorname{grad} \mathcal{L})^\sigma = \begin{bmatrix} 0 & 0 \\ 0 & \mathrm{Hess}^\sigma_{\mathfrak{q},\mathfrak{a}} \end{bmatrix}.$$

The Hessian $\mathrm{Hess}^\sigma_{\mathfrak{q},\mathfrak{a}}$ allows us to rephrase the non-degeneracy condition for a critical point:

Lemma 12.4.1. *A critical point $\mathfrak{a}$ in the blown-up configuration space is non-degenerate if and only if* $\mathrm{Hess}^\sigma_{\mathfrak{q},\mathfrak{a}}$ *is surjective.*

Proof. The definition of non-degeneracy is that $(\operatorname{grad} \mathcal{L})^\sigma$ should be transverse to the subbundle $\mathcal{J}^\sigma_{k-1}$ at $\mathfrak{a}$. This is the same as saying that the image of $\mathcal{D}_{\mathfrak{a}}(\operatorname{grad} \mathcal{L})^\sigma$ is the whole of $\mathcal{K}^\sigma_{k-1}$. Given the block form of the derivative (above), this is the same as saying that $\mathrm{Hess}^\sigma_{\mathfrak{q},\mathfrak{a}}$ is onto. □

There is also a parallel version of the *extended* Hessian (12.5) in the blown-up configuration space. For $\mathfrak{a}$ in $\mathcal{C}^\sigma_k(Y)$, this is an operator

$$\widehat{\mathrm{Hess}}^\sigma_{\mathfrak{q},\mathfrak{a}} : \mathcal{T}^\sigma_{k,\mathfrak{a}} \oplus L^2_k(Y; i\mathbb{R}) \to \mathcal{T}^\sigma_{k-1,\mathfrak{a}} \oplus L^2_{k-1}(Y; i\mathbb{R})$$

defined by a formula similar to (12.6). Rather than the formal adjoint $\mathbf{d}^*$ that appears in (12.6) however, it is appropriate to put an operator whose kernel is $\mathcal{K}^\sigma_{\mathfrak{a}}$. Such an operator is the map

$$\mathbf{d}^{\sigma,\dagger}_{\mathfrak{a}} : \mathcal{T}^\sigma_{k,\mathfrak{a}} \to L^2_{k-1}(Y; i\mathbb{R}) \tag{12.10}$$

given at $\mathfrak{a} = (B_0, s_0, \psi_0)$ by

$$(b, s, \psi) \mapsto -d^*b + is_0^2 \operatorname{Re}\langle i\psi_0, \psi\rangle + i|\psi_0|^2 \operatorname{Re} \mu_Y(\langle i\psi_0, \psi\rangle).$$

(See the definition of $\mathcal{K}^\sigma$ in the more general setting of a manifold M with boundary at (9.12)). Thus we are led to define the extended Hessian in the blown-up context by the matrix

$$\widehat{\mathrm{Hess}}^\sigma_{\mathfrak{q},\mathfrak{a}} = \begin{bmatrix} \mathcal{D}_\mathfrak{a}(\operatorname{grad} \mathcal{L})^\sigma & \mathbf{d}^\sigma_\mathfrak{a} \\ \mathbf{d}^{\sigma,\dagger}_\mathfrak{a} & 0 \end{bmatrix}. \tag{12.11}$$

This operator $\widehat{\mathrm{Hess}}^\sigma_{\mathfrak{q},\mathfrak{a}}$ will reappear later in Subsection 14.4. We can use the direct sum decomposition

$$\mathcal{T}^\sigma_{j,\mathfrak{a}} = \mathcal{J}^\sigma_{j,\mathfrak{a}} \oplus \mathcal{K}^\sigma_{j,\mathfrak{a}}$$

to decompose $\widehat{\mathrm{Hess}}^\sigma_{\mathfrak{q},\mathfrak{a}}$ as we did in (12.9) above. The summand $\mathcal{J}^\sigma_{j,\mathfrak{a}}$ is the image of $\mathbf{d}^\sigma_\mathfrak{a}$ by definition, and $\mathbf{d}^{\sigma,\dagger}_\mathfrak{a}$ has kernel $\mathcal{K}^\sigma_{j,\mathfrak{a}}$. So as an operator on

$$\mathcal{J}^\sigma_{j,\mathfrak{a}} \oplus \mathcal{K}^\sigma_{j,\mathfrak{a}} \oplus L^2_j(Y; i\mathbb{R}),$$

$\widehat{\mathrm{Hess}}^\sigma_{\mathfrak{q},\mathfrak{a}}$ has the shape

$$\widehat{\mathrm{Hess}}^\sigma_{\mathfrak{q},\mathfrak{a}} = \begin{bmatrix} 0 & x & \mathbf{d}^\sigma_\mathfrak{a} \\ y & \mathrm{Hess}^\sigma_{\mathfrak{q},\mathfrak{a}} & 0 \\ \mathbf{d}^{\sigma,\dagger}_\mathfrak{a} & 0 & 0 \end{bmatrix} \tag{12.12}$$

where $\mathrm{Hess}^\sigma_{\mathfrak{q},\mathfrak{a}}$ is the Hessian on $\mathcal{K}^\sigma_{j,\mathfrak{a}}$.

As above, the terms x and y are zero if $\mathfrak{a}$ is a critical point. We record this observation as a lemma:

Lemma 12.4.2. *At a critical point $\mathfrak{a}$ in the blown-up configuration space $\mathcal{C}^\sigma_k(Y)$, the extended Hessian $\widehat{\mathrm{Hess}}^\sigma_{\mathfrak{q},\mathfrak{a}}$ has the block form*

$$\widehat{\mathrm{Hess}}^\sigma_{\mathfrak{q},\mathfrak{a}} = \begin{bmatrix} 0 & 0 & \mathbf{d}^\sigma_\mathfrak{a} \\ 0 & \mathrm{Hess}^\sigma_{\mathfrak{q},\mathfrak{a}} & 0 \\ \mathbf{d}^{\sigma,\dagger}_\mathfrak{a} & 0 & 0 \end{bmatrix}$$

as an operator on $\mathcal{J}^\sigma_j \oplus \mathcal{K}^\sigma_j \oplus L^2_j(Y; i\mathbb{R})$. □

The operator $\widehat{\mathrm{Hess}}^\sigma_{\mathfrak{q},\mathfrak{a}}$ as it stands is not presented as a perturbation of an elliptic operator on Y. The awkwardness arises from the fact that Hilbert space $\mathcal{T}^\sigma_{j,\mathfrak{a}}$ is not the space of L^2_j sections of a fixed vector bundle on Y. If we write $\mathfrak{a}$

as (B_0, r_0, ψ_0) then $\mathcal{T}^{\sigma}_{j,\mathfrak{a}}$ consists of triples (b, ψ, r) and we have the following two points to consider:

- the spinor ψ is not an arbitrary section of S on Y, but is constrained by the condition that it be real-orthogonal to the spinor $\psi_0(t)$;
- r is simply a real number, so cannot be regarded as an unconstrained section of a vector bundle.

Of course, we could deal with these points by extending the framework developed above to include a mild generalization sufficient to incorporate our situation. An alternative (though less elegant) approach is to avoid the issue by combining ψ and r together into a single spinor,

$$\boldsymbol{\psi} = \psi + r\psi_0, \tag{12.13}$$

so that $\boldsymbol{\psi}$ is an unconstrained section of the spin bundle S.

With this modification, we can regard $\widehat{\mathrm{Hess}}^{\sigma}_{\mathfrak{q},\mathfrak{a}}$ as acting on triples

$$(b, \boldsymbol{\psi}, c) \in L^2_j(Y; iT^*Y \oplus S \oplus i\mathbb{R}).$$

In this way it becomes an operator on sections of a vector bundle. To explicitly write this operator as a perturbation of an elliptic operator, we can begin by writing $\widehat{\mathrm{Hess}}^{\sigma}_{\mathfrak{q},\mathfrak{a}}$ in full at $\mathfrak{a} = (B_0, r_0, \psi_0)$. It is given by the following formula, in which α is the image of $\mathfrak{a}$ in $\mathcal{C}_k(Y)$, the projection $\Pi^{\perp}$ is the orthogonal projection onto the real-orthogonal complement of ψ_0, and $\tilde{\mathfrak{q}}^1$ is defined as in (10.8):

$$\begin{aligned}
&(b, r, \psi, c) \mapsto \\
&\Big(-dc + {*db} + 4r_0 r\rho^{-1}(\psi_0\psi_0^*)_0 + 4r_0^2\rho^{-1}(\psi\psi_0^* + \psi_0\psi^*)_0 \\
&\quad + 2\mathcal{D}_\alpha \mathfrak{q}^0(b, r\psi_0 + r_0\psi), \\
&\quad \Lambda_{\mathfrak{q}}(\mathfrak{a})r + \langle D_{B_0}\psi + \rho(b)\psi_0 + c\psi_0 + \mathcal{D}_{\mathfrak{a}}\tilde{\mathfrak{q}}^1(b, r, \psi), \psi_0\rangle_{L^2(Y)} r_0 \\
&\quad + \langle D_{B_0}\psi_0 + \tilde{\mathfrak{q}}^1(\mathfrak{a}), \psi\rangle_{L^2(Y)} r_0, \\
&\quad \Pi^{\perp}\Big[D_{B_0}\psi + \rho(b)\psi_0 + c\psi_0 + \mathcal{D}_{\mathfrak{a}}\tilde{\mathfrak{q}}^1(b, r, \psi) - \Lambda_{\mathfrak{q}}(\mathfrak{a})\psi\Big], \\
&\quad - d^*b + ir_0^2 \operatorname{Re}\langle i\psi_0, \psi\rangle + irr_0 \operatorname{Re}\langle i\psi_0, \psi_0\rangle + i|\psi_0|^2 \operatorname{Re}\mu_Y(\langle i\psi_0, \psi\rangle)\Big).
\end{aligned} \tag{12.14}$$

In terms of (b, ψ, c), the operator becomes

$$
\begin{aligned}
&(b, \psi, c) \mapsto \\
&\Big(*db - dc + 2\rho^{-1}((r_0\psi - (r_0 - 1)\langle \psi, \psi_0\rangle_{L^2(Y)}\psi_0)\psi_0^* \\
&\quad + \psi_0(r_0\psi - (r_0 - 1)\langle \psi, \psi_0\rangle_{L^2(Y)}\psi_0)^*)_0 \\
&\quad + 2\mathcal{D}_{(B_0, r_0\psi_0)}\mathfrak{q}^0(b, r_0\psi - (r_0 - 1)\langle \psi, \psi_0\rangle_{L^2(Y)}\psi_0), \\
&\quad L(b, \psi) + \big((r_0 - 1)\langle L(b, \psi), \psi_0\rangle_{L^2(Y)} + \langle n(B_0, r_0, \psi_0), \Pi^\perp\psi\rangle_{L^2(Y)}\big)\psi_0 \\
&\quad - \Lambda_{\mathfrak{q}}(B_0, r_0, \psi_0)(\psi - 2\langle \psi, \psi_0\rangle_{L^2(Y)}\psi_0), \\
&\quad - d^*b + ir_0^2 \operatorname{Re}\langle i\psi_0, \Pi^\perp\psi\rangle + i \operatorname{Re}\langle \psi, \psi_0\rangle_{L^2(Y)} r_0 \operatorname{Re}\langle i\psi_0, \psi_0\rangle \\
&\quad + i|\psi_0|^2 \operatorname{Re} \mu_Y(\langle i\psi_0, \Pi^\perp\psi\rangle)\Big),
\end{aligned}
$$

where

$$
\begin{aligned}
L(b, \psi) = D_{B_0}\psi - \langle \psi, \psi_0\rangle_{L^2(Y)} D_{B_0}\psi_0 + \rho(b)\psi_0 + c\psi_0 \\
+ \mathcal{D}_{(B_0, r_0, \psi_0)}\tilde{\mathfrak{q}}^1(b, \langle \psi, \psi_0\rangle, \Pi^\perp\psi)
\end{aligned}
\tag{12.15}
$$

and

$$
n(B_0, r_0, \psi_0) = D_{B_0}\psi_0 + \tilde{\mathfrak{q}}^1(B_0, r_0, \psi_0). \tag{12.16}
$$

Separating the above operator into its first- and zeroth-order parts we arrive at an operator of the form

$$
\widehat{\operatorname{Hess}}^{\sigma}_{\mathfrak{q},\mathfrak{a}}(b, \psi, c) = L_0(b, \psi, c) + h_{\mathfrak{a}}(b, \psi, c) \tag{12.17}
$$

where L_0 is (as before) the operator

$$
\begin{bmatrix} *d & 0 & -d \\ 0 & D_{B_0} & 0 \\ -d^* & 0 & 0 \end{bmatrix}.
$$

This L_0 is elliptic and self-adjoint, being the direct sum of D_{B_0} and the signature operator on Y. In writing the operator this way, the only term which needs attention is the term $(r_0 - 1)\langle L(b, \psi), \psi_0\rangle_{L^2(Y)}$: a priori, this term looks as if it involves the first derivatives of $\boldsymbol{\psi}$ through the Dirac operator; but it can be rewritten as a zeroth-order term using the symmetry

$$\langle D_{B_0}\boldsymbol{\psi}, \psi_0\rangle_{L^2(Y)} = \langle \boldsymbol{\psi}, D_{B_0}\psi_0\rangle_{L^2(Y)}.$$

The association of the operator $h_{\mathfrak{a}}$ to the pair $\mathfrak{a}$ may be seen to give rise to a smooth map

$$\mathcal{C}_k^\sigma(Y) \to \operatorname{Hom}(L_j^2(Y; iT^*Y \oplus S \oplus i\mathbb{R}), L_j^2(Y; iT^*Y \oplus S \oplus i\mathbb{R}))$$

for $j \leq k$. In particular, $h_{\mathfrak{a}}$ satisfies the regularity hypotheses in Definition 12.2.1, so that $\widehat{\operatorname{Hess}}^\sigma_{\mathfrak{q},\mathfrak{a}}$ is k-ASAFOE. The conclusions of Lemma 12.2.4 therefore apply.

The extended Hessian $\widehat{\operatorname{Hess}}^\sigma_{\mathfrak{q},\mathfrak{a}}$, unlike its counterpart $\widehat{\operatorname{Hess}}_{\mathfrak{q},\alpha}$ in the un-blown-up setting, is not a symmetric operator. At a critical point $\mathfrak{a}$, however, its spectrum is real. We prove this as part of the next lemma.

Lemma 12.4.3. *If $\mathfrak{b}$ is a non-degenerate critical point in the blown-up configuration space, then the extended Hessian $\widehat{\operatorname{Hess}}^\sigma_{\mathfrak{q},\mathfrak{b}}$ is invertible and has real spectrum. In particular, it is hyperbolic.*

Proof. We have already seen that, at a critical point, the operator has the block form

$$\widehat{\operatorname{Hess}}^\sigma_{\mathfrak{q},\mathfrak{b}} = \begin{bmatrix} 0 & 0 & \mathbf{d}^\sigma_{\mathfrak{b}} \\ 0 & \operatorname{Hess}^\sigma_{\mathfrak{q},\mathfrak{b}} & 0 \\ \mathbf{d}^{\sigma,\dagger}_{\mathfrak{b}} & 0 & 0 \end{bmatrix}$$

(see Lemma 12.4.2). This is the direct sum of two operators: the Hessian $\operatorname{Hess}^\sigma_{\mathfrak{q},\mathfrak{a}}$ acting on $\mathcal{K}^\sigma_{j,\mathfrak{a}}$, and the operator

$$\begin{bmatrix} 0 & \mathbf{d}^\sigma_{\mathfrak{b}} \\ \mathbf{d}^{\sigma,\dagger}_{\mathfrak{b}} & 0 \end{bmatrix}$$

acting on $\mathcal{J}^\sigma_{j,\mathfrak{a}} \oplus L^2_j(Y; i\mathbb{R})$. Despite the notation, the operator $\mathbf{d}^{\sigma,\dagger}_{\mathfrak{b}}$ is not the adjoint of $\mathbf{d}^\sigma_{\mathfrak{b}}$. Nevertheless, $\mathbf{d}^{\sigma,\dagger}_{\mathfrak{b}}\mathbf{d}^\sigma_{\mathfrak{b}}$, which is the operator

$$\xi \mapsto dd^*\xi + r_0^2|\psi_0|^2\xi + |\psi_0|^2\mu_Y(|\psi_0|^2\xi) \tag{12.18}$$

(*cf.* equation (9.13)), is self-adjoint and strictly positive. Hence this 2-by-2 block is invertible with real spectrum (and symmetric about zero.) Note also that the analysis of this 2-by-2 block makes no reference to the assumption that $\mathfrak{b}$ is a critical point.

Because this 2-by-2 block is invertible, the invertibility of the operator $\widehat{\mathrm{Hess}}^{\sigma}_{\mathfrak{q},\mathfrak{b}}$ is equivalent to the invertibility of the summand $\mathrm{Hess}^{\sigma}_{\mathfrak{q},\mathfrak{b}}$. This summand is a Fredholm operator of index zero, so it is invertible if and only if it is surjective; and we have already observed (in Lemma 12.4.1 above) that the non-degeneracy of the critical point is equivalent to this surjectivity.

What remains is to show that the spectrum of the $\mathrm{Hess}^{\sigma}_{\mathfrak{q},\mathfrak{b}}$ is real. Our discussion of this operator brings in the material of Subsection 12.2. At an irreducible critical point, the operator $\mathrm{Hess}^{\sigma}_{\mathfrak{q},\mathfrak{b}}$ on $\mathcal{K}^{\sigma}_{j,\mathfrak{a}}$ is conjugate to the operator $\mathrm{Hess}_{\mathfrak{q},\mathfrak{b}}$ on $\mathcal{K}_{j,\mathfrak{a}}$ via the blow-down map. Because the latter operator is a covariant second derivative, it is symmetric. The spectrum of $\mathrm{Hess}^{\sigma}_{\mathfrak{q},\mathfrak{b}}$ is therefore real in the irreducible case.

To see that the spectrum is real in the reducible case, we examine the extended Hessian in more detail. We use the expression (12.14): on substituting $r_0 = 0$, we obtain

$$
\begin{aligned}
&(b, r, \psi, c) \mapsto \\
&\qquad \Big(- dc + *db + 2\mathcal{D}_{(B_0,0)}\mathfrak{q}^0(b,0), \quad \lambda r, \\
&\qquad \Pi^{\perp}\Big[(D_{\mathfrak{q},B_0} - \lambda)\psi + \rho(b)\psi_0 + c\psi_0 + \mathcal{D}^2\mathfrak{q}^1_{(B_0,0)}((b,0),(0,\psi_0)) \\
&\qquad\qquad + (r/2)\mathcal{D}^2\mathfrak{q}^1_{(B_0,0)}((0,\psi_0),(0,\psi_0))\Big], \\
&\qquad - d^*b + i|\psi_0|^2 \operatorname{Re}\mu_Y(\langle i\psi_0, \psi\rangle)\Big).
\end{aligned}
$$

Here $D_{\mathfrak{q},B_0}$ is the symmetric perturbation of the Dirac operator D_{B_0} defined previously, so that $D_{\mathfrak{q},B_0}\psi_0 = \lambda\psi_0$, where $\lambda = \Lambda_{\mathfrak{q}}(B_0, 0, \psi_0)$. We have used the formula (10.8) for $\tilde{\mathfrak{q}}^1$ in terms of the derivative of $\mathfrak{q}^1$.

To further analyze this operator, we decompose both c and ψ by writing

$$
\begin{aligned}
c &= i\epsilon_1 + \hat{c} \\
\psi &= i\epsilon_2\psi_0 + \hat{\psi}
\end{aligned}
$$

where ϵ_j are real constants, $\hat{c}$ is orthogonal to the constants and $\hat{\psi}$ is orthogonal to $i\psi_0$ in the real L^2 inner product. The S^1 invariance of the equations implies that

$\mathcal{D}^2\mathfrak{q}^1_{(B_0,0)}((b,0),(0,\psi_0))$ is orthogonal to $i\psi_0$, as is $\mathcal{D}^2\mathfrak{q}^1_{(B_0,0)}((0,\psi_0),(0,\psi_0))$, and the operator can then be written

$$\begin{bmatrix} r \\ \epsilon_1 \\ \epsilon_2 \\ \hat{c} \\ b \\ \hat{\psi} \end{bmatrix} \mapsto \begin{bmatrix} \lambda & 0 & 0 & 0 & 0 & 0 \\ 0 & 0 & -1 & 0 & 0 & 0 \\ 0 & -1 & 0 & 0 & 0 & x_6 \\ 0 & 0 & x_1 & 0 & -d^* & 0 \\ 0 & 0 & 0 & -d & *d_{\mathfrak{q}} & 0 \\ x_2 & 0 & x_5 & x_3 & x_4 & \Pi^{\perp}_{\mathbb{C}}(D_{\mathfrak{q},B_0}-\lambda) \end{bmatrix} \begin{bmatrix} r \\ \epsilon_1 \\ \epsilon_2 \\ \hat{c} \\ b \\ \hat{\psi} \end{bmatrix}. \tag{12.19}$$

Here $*d_{\mathfrak{q}}$ is the operator $b \mapsto *db + 2\mathcal{D}_{(B_0,0)}\mathfrak{q}^0(b,0)$, the x_i denote the non-zero zeroth-order entries, and $\Pi^{\perp}_{\mathbb{C}}$ is the projection to $(\mathbb{C}\psi_0)^{\perp}$. Note that $\hat{\psi}$ belongs to the intersection of the real-orthogonal complements of ψ_0 and $i\psi_0$, so it is complex-orthogonal to ψ_0. The term x_6 is the linear map

$$\begin{aligned} x_6 : (\mathbb{C}\psi_0)^{\perp} &\to \mathbb{R} \\ \hat{\psi} &\mapsto \langle \hat{\psi}, iD_{\mathfrak{q},B_0}\psi_0\rangle_{\mathbb{R}}. \end{aligned}$$

This decomposition of the extended Hessian is applicable to any reducible configuration, whether or not it is a critical point. At a critical point, however, ψ_0 is an eigenvector of $D_{\mathfrak{q},B_0}$, and this Dirac operator therefore preserves the real- and complex-orthogonal complements of ψ_0. So in the case of a critical point, the term x_6 is zero and the entry $\Pi^{\perp}(D_{\mathfrak{q},B_0}-\lambda)$ can simply be written as $(D_{\mathfrak{q},B_0}-\lambda)$, acting on $(\mathbb{C}\psi_0)^{\perp}$. In this case, the matrix is block lower triangular, and the diagonal blocks are

$$\begin{bmatrix}\lambda\end{bmatrix}, \quad \begin{bmatrix} 0 & -1 \\ -1 & 0 \end{bmatrix}, \quad \begin{bmatrix} 0 & -d^* \\ -d & *d_{\mathfrak{q}} \end{bmatrix}, \quad \begin{bmatrix}\Pi^{\perp}_{\mathbb{C}}(D_{\mathfrak{q},B_0}-\lambda)\end{bmatrix}. \tag{12.20}$$

The part of this matrix that corresponds to the summand $\mathrm{Hess}^{\sigma}_{\mathfrak{q},\mathfrak{b}}$ acting on $\mathcal{K}^{\sigma}_{j,\mathfrak{b}}$ is the matrix

$$\begin{bmatrix} \lambda & 0 & 0 \\ 0 & *d_{\mathfrak{q}}|_K & 0 \\ x_2 & x_4|_K & \Pi^{\perp}_{\mathbb{C}}(D_{\mathfrak{q},B_0}-\lambda) \end{bmatrix}$$

where K is the kernel of d^* in the space of imaginary-valued 1-forms. This is an operator on

$$\mathbb{R} \oplus K \oplus (\mathbb{C}\psi_0)^{\perp}.$$

The hypothesis that $\mathfrak{b}$ is non-degenerate means, in particular, that λ is non-zero and a simple eigenvalue of the complex-linear, hermitian operator $D_{\mathfrak{q},B_0}$, by Proposition 12.2.5. The operator $D_{\mathfrak{q},B_0} - \lambda$, acting on the complex-orthogonal complement of ψ_0, is therefore invertible; it is a symmetric operator with real eigenvalues. Finally, the operator $*d_{\mathfrak{q}}$ is the operator $\mathcal{D}_{B_0}(\text{grad } \mathcal{L}^{\text{red}})$ from Subsection 12.2, restricted to K. It is symmetric because it arises from the covariant second derivative of $\mathcal{L}$ on the locus of reducible configurations. □

12.5 Proof of transversality: the irreducible case

To begin the proof of Theorem 12.1.2, we establish the non-degeneracy of the irreducible critical points, for a residual subset of the space of perturbations $\mathcal{P}$. Such transversality arguments for Fredholm problems follow a standard model, organized around the following lemma.

Lemma 12.5.1. *Let $\mathcal{E}$ and $\mathcal{F}$ and $\mathcal{P}$ be separable Banach manifolds, and let $\mathcal{S} \subset \mathcal{F}$ be a closed submanifold. Let*

$$F : \mathcal{E} \times \mathcal{P} \to \mathcal{F}$$

be a smooth map, and write $F_p = F(-, p)$. Suppose that F is transverse to $\mathcal{S}$, and that for all (e, p) in $F_p^{-1}(Z)$, the composite

$$T_e\mathcal{E} \xrightarrow{\mathcal{D}_e F_p} T_f\mathcal{F} \xrightarrow{\pi} T_f\mathcal{F}/T_f\mathcal{S}$$

is Fredholm. Then there is a residual set of p in $\mathcal{P}$ for which the map $F_p : \mathcal{E} \to \mathcal{F}$ is transverse to $\mathcal{S}$.

Proof. The Fredholm hypothesis implies that the kernel of $(\pi \circ \mathcal{D}_a F)$ is a complemented subspace for all a in $F^{-1}(\mathcal{S})$. The implicit function theorem then tells us that $\mathcal{Z} = F^{-1}(\mathcal{S})$ is a Banach submanifold. The projection map $Q : \mathcal{Z} \to \mathcal{P}$ is Fredholm, so the Sard–Smale theorem [103] provides a residual set of regular values of Q in $\mathcal{P}$. If $p \in \mathcal{P}$ is a regular value of Q, then F_p is transverse to $\mathcal{S}$. □

We now set up the transversality argument so as to fit it into the framework of the above lemma. Because of the first part of Proposition 12.2.5, we may as well consider the vector field grad $\mathcal{L}$ on $\mathcal{C}_k^*(Y)$.

We introduce the parametrized critical point set as the inverse image of the zero section $\mathcal{C}_k^* \subset \mathcal{T}_{k-1}$ under the map

$$\mathfrak{g} : \mathcal{C}_k^* \times \mathcal{P} \to \mathcal{K}_{k-1}$$

defined by

$$\begin{aligned}\mathfrak{g}(\alpha, \mathfrak{q}) &= \operatorname{grad} \pounds(\alpha) \\ &= \operatorname{grad} \mathcal{L}(\alpha) + \mathfrak{q}(\alpha).\end{aligned}$$

This map is a smooth map of Banach manifolds.

Lemma 12.5.2. *The map $\mathfrak{g}$ is transverse to the zero section of $\mathcal{K}_{k-1}$.*

Proof. Let $(\alpha, \mathfrak{q})$ lie in the inverse image of the zero section, and let $\pounds = \mathcal{L} + f$, where $\operatorname{grad} f = \mathfrak{q}$. What is asserted is the surjectivity of the map

$$\mathcal{T}_{k,\alpha} \times T_{\mathfrak{q}}\mathcal{P} \to \mathcal{K}_{k-1,\alpha}$$

given by

$$((b, \psi), \delta\mathfrak{q}) \mapsto \mathcal{D}_\alpha \operatorname{grad} \pounds(b, \psi) + \delta\mathfrak{q}(\alpha),$$

or equivalently, the map

$$\mathcal{K}_{k,\alpha} \times T_{\mathfrak{q}}\mathcal{P} \to \mathcal{K}_{k-1,\alpha}$$

given by

$$((b, \psi), \delta\mathfrak{q}) \mapsto \operatorname{Hess}_{\mathfrak{q},\alpha}(b, \psi) + \delta\mathfrak{q}(\alpha).$$

(Note that $\mathcal{D}_\alpha \operatorname{grad} \pounds$ takes values in $\mathcal{K}_{k-1}$ as a consequence of the last part of Lemma 12.3.2.) The cokernel of $\operatorname{Hess}_{\mathfrak{q},\alpha}$ is finite-dimensional and is represented by the L^2 orthogonal complement of the range, which is the kernel of $\operatorname{Hess}_{\mathfrak{q},\alpha}$. So it suffices to prove that given any non-zero v in the kernel of $\operatorname{Hess}_{\mathfrak{q},\alpha}$ there exists a $\delta\mathfrak{q} \in \mathcal{P}$ such that the L^2 inner product of $\delta\mathfrak{q}(\alpha)$ with v is non-zero. If $\delta\mathfrak{q}$ is $\operatorname{grad}(\delta f)$, this is equivalent to saying the differential of δf in the direction of v is non-zero; and the existence of such an element of $\mathcal{P}$ now follows from Corollary 11.2.2 and the denseness conditions in Definition 11.6.3, because the image of v in $\mathcal{B}_k^*(Y)$ is non-zero. □

Lemma 12.5.2 allows us to apply the general transversality argument, Lemma 12.5.1. We learn that the parametrized critical set $\mathfrak{g}^{-1}(0)$ is a Banach manifold, as is the quotient

$$\mathcal{Z} = \mathfrak{g}^{-1}(0)/\mathcal{G}_{k+1} \subset \mathcal{B}_k^*(Y) \times \mathcal{P}.$$

The projection $\mathcal{Z} \to \mathcal{P}$ is a smooth Fredholm map of index zero, and by the Sard–Smale theorem the regular values of the projection are a residual set. If $\mathfrak{q}$ is a regular value, then for this perturbation the irreducible critical points are non-degenerate. This proves the first part of Theorem 12.1.2 as it relates to the irreducible part of the critical set.

12.6 Proof of transversality: the reducible case

Let $\mathcal{K}_j^{\text{red}}$ be as in Equation (12.2) above. We introduce the notation $\mathfrak{g}^{\text{red}}$ for the map

$$\mathfrak{g}^{\text{red}} : \mathcal{A}_k \times \mathcal{P} \to \mathcal{K}_{k-1}^{\text{red}}$$

given by

$$\begin{aligned} \mathfrak{g}^{\text{red}}(B, \mathfrak{q}) &= (\text{grad}\, \rlap{-}\mathcal{L})^{\text{red}}(B) \\ &= (\text{grad}\, \mathcal{L})^{\text{red}}(B) + \mathfrak{q}(B). \end{aligned}$$

Lemma 12.6.1. *The map $\mathfrak{g}^{\text{red}}$ is transverse to the zero section of $\mathcal{K}_{k-1}^{\text{red}}$.*

Proof. This is essentially the same as the proof of Lemma 12.5.2. □

This lemma will allow us to find a perturbation $\mathfrak{q}$ which achieves Condition (ii)(a) of Proposition 12.2.5 at all reducible critical points. To achieve the other two conditions in the reducible case, we must alter the Hessian of the perturbation in the directions normal to the reducibles. We introduce

$$\mathcal{P}^{\perp} \subset \mathcal{P}$$

for the perturbations $\mathfrak{q}$ which vanish at the reducible locus in $\mathcal{C}(Y)$. The definition of "large" in Definition 11.6.3 is phrased so as to ensure that $\mathcal{P}^{\perp}$ contains an ample supply of perturbations. (See the third condition on the collection of functions g_α.)

In the next lemma, we write Op^{sa} for the space of self-adjoint Fredholm maps from $L_k^2(Y; S)$ to $L_{k-1}^2(Y; S)$ having the form $D_{B_0} + h$, where B_0 is a smooth spin^c connection and h is a self-adjoint operator that extends to a bounded operator $h : L_j^2 \to L_j^2$ $(j \leq k)$. (Compare Lemma 12.2.4.) This Banach space has a stratification according to the dimension of the kernel. If $L \in \text{Op}^{sa}$ has kernel V, then the tangent space to the corresponding stratum is the kernel of the map from Op^{sa} to the space of self-adjoint operators on V, given by compression.

In the space Op^{sa}, the set of operators whose spectrum is not simple is a countable union of the images of Fredholm maps F_n of negative index. Indeed, we can take the domain of F_n to be $\mathrm{Op}^{sa}_n \times \mathbb{R}$, where $\mathrm{Op}^{sa}_n \subset \mathrm{Op}^{sa}$ is the space of operators having 0 as an eigenvalue of multiplicity exactly n, and the map F_n is $(L, \lambda) \mapsto L + \lambda$. The map F_n is locally an embedding, and the normal bundle to its image at $(L+\lambda)$ is isomorphic to the space of traceless, self-adjoint endomorphisms of $\mathrm{Ker}(L)$.

We write $M : \mathcal{A}_k \times \mathcal{P}^\perp \to \mathrm{Op}^{sa}$ for the map given by

$$M : (B, \mathfrak{q}^\perp) \mapsto D_{\mathfrak{q}^\perp, B}.$$

Lemma 12.6.2. *The map M is transverse to the stratification of self-adjoint operators according to the dimension of the kernel, and transverse also to the Fredholm maps F_n.*

Proof. For the first assertion, let $\mathfrak{q}^\perp = \operatorname{grad} f^\perp$ be any perturbation in $\mathcal{P}^\perp$, and let V be the kernel of $D_{\mathfrak{q},B}$, which we can regard as a subspace of the normal bundle to $\mathcal{A}_k$ in $\mathcal{C}_k(Y)$. By Proposition 11.2.1, we can choose a p whose differential embeds this S^1-invariant linear subspace in $\mathbb{C}^m \subset T_{p(B,0)}(\mathbb{R}^n \times \mathbb{T} \times \mathbb{C}^m)$. We can assume that p is defined by a collection of coclosed forms c_ν and sections Υ^μ belonging to the C^∞-dense collection of choices which we made in constructing the large Banach space $\mathcal{P}$. By choosing an S^1-invariant function δg on $\mathbb{R}^n \times \mathbb{T} \times \mathbb{C}^m$ which vanishes along $\mathbb{R}^n \times \mathbb{T} \times \{0\}$, we can therefore find a $\delta\mathfrak{q}^\perp = \operatorname{grad} \delta f \in \mathcal{P}^\perp$ such that the Hessian of $\delta f|_V$ is any chosen S^1-equivariant (i.e. complex-linear) self-adjoint endomorphism of V. We can take it that δg belongs to the dense collection g_α used in the construction of $\mathcal{P}$. This is equivalent to the desired transversality. The second assertion follows similarly. □

We can now complete the proof of the transversality theorem for the critical points.

Proof of Theorem 12.1.2. We have already seen that, for a residual set of perturbations, all the irreducible critical points in $\mathcal{C}^\sigma_k(Y)$ are non-degenerate. From Lemma 12.6.1 we have that $(\mathfrak{g}^{\mathrm{red}})^{-1}(0) \subset \mathcal{A}_k \times \mathcal{P}$ is a smooth Banach manifold, as is the quotient

$$\mathcal{Z}^{\mathrm{red}} = (\mathfrak{g}^{\mathrm{red}})^{-1}(0)/\mathcal{G}_{k+1} \subset (\mathcal{A}_k/\mathcal{G}_{k+1}) \times \mathcal{P}.$$

The projection from $\mathcal{Z}^{\mathrm{red}}$ to $\mathcal{P}$ is a Fredholm map of index 0. Thus $\mathcal{P}^\perp \times \mathcal{Z}^{\mathrm{red}}$ also has index-zero projection to $\mathcal{P}^\perp \times \mathcal{P}$. From Lemma 12.6.2, we see that the set

$$\mathcal{W} \subset \mathcal{P}^\perp \times (\mathcal{A}_k/\mathcal{G}_{k+1})$$

of pairs $(\mathfrak{q}^\perp, [B])$, where the spectrum of $D_{\mathfrak{q}^\perp,B}$ either is non-simple or contains zero, is a countable union of Banach submanifolds $\mathcal{W}_n$, each of which has finite, positive codimension. The proof of the lemma actually gives a little more: at each point x of each $\mathcal{W}_k$, there is a complement to $T_x\mathcal{W}_k$ which is entirely contained in the $T\mathcal{P}^\perp$ directions. Taking the product with $\mathcal{P}$, we have a similar decomposition of $\mathcal{W} \times \mathcal{P}$ as a union of Banach submanifolds,

$$\mathcal{W}_k \times \mathcal{P} \subset \mathcal{P}^\perp \times (\mathcal{A}_k/\mathcal{G}_{k+1}) \times \mathcal{P}.$$

The additional property of the $\mathcal{W}_k$, just mentioned, shows that the submanifolds $\mathcal{P}^\perp \times \mathcal{Z}$ and $\mathcal{W}_k \times \mathcal{P}$ are transverse so the intersection

$$(\mathcal{P}^\perp \times \mathcal{Z}^{\text{red}}) \cap (\mathcal{W} \times \mathcal{P})$$

is a locally finite union of Banach submanifolds $\mathcal{U}_k \subset \mathcal{P}^\perp \times \mathcal{Z}^{\text{red}}$, of finite, positive codimension in $\mathcal{P}^\perp \times \mathcal{Z}^{\text{red}}$. The projection of each $\mathcal{U}_k$ to $\mathcal{P}^\perp \times \mathcal{P}$ is therefore Fredholm of negative index. Thus the Smale-Sard theorem guarantees a residual set of pairs $(\mathfrak{q}^\perp, \mathfrak{q})$ in $\mathcal{P}^\perp \times \mathcal{P}$ which are regular values of the projection $\mathcal{P}^\perp \times \mathcal{Z}^{\text{red}} \to \mathcal{P}^\perp \times \mathcal{P}$ and not in the image of the projection

$$(\mathcal{P}^\perp \times \mathcal{Z}^{\text{red}}) \cap (\mathcal{W} \times \mathcal{P}^{\text{red}}) \to \mathcal{P}^\perp \times \mathcal{P}.$$

For such a pair $(\mathfrak{q}^\perp, \mathfrak{q})$, the combined perturbation $\mathfrak{q}^\perp + \mathfrak{q}$ satisfies the three conditions (ii)(a)–(ii)(c) of Proposition 12.2.5 at all reducible critical points. The subset of $\mathcal{P}$ arising as such sums $\mathfrak{q}^\perp + \mathfrak{q}$ is residual: indeed, a submersive linear map between Banach spaces carries residual sets to residual sets in general.

This completes the verification of the first part of Theorem 12.1.2. For the finiteness assertion which finishes the theorem, we recall the compactness result for the critical set, Corollary 10.7.4. A non-degenerate critical point is isolated in $\mathcal{B}_k(Y)$, so the compactness result implies finiteness. □

13 Moduli spaces of trajectories

Until this point, we have usually studied the perturbed 4-dimensional Seiberg–Witten equations on a compact cylinder $I \times Y$, for I a closed interval $[t_1, t_2]$. In this section, we shift our focus and study the infinite cylinder

$$Z = \mathbb{R} \times Y.$$

In the case of the compact cylinder $I \times Y$, we introduced in Definition 6.3.1 the configuration space $\mathcal{C}^\tau(I \times Y)$ consisting of triples (A, s, ϕ), where $s : I \to \mathbb{R}$ is non-negative, and ϕ has L^2 norm 1 on each slice $\{t\} \times Y$. The quotient by the 4-dimensional gauge group $\mathcal{G}(I \times Y)$ is a space $\mathcal{B}^\tau(I \times Y)$ which we could identify with a space of smooth paths in $\mathcal{B}^\sigma(Y)$. We also introduced Sobolev completions of these configuration spaces. In the case of the infinite cylinder Z, we must give thought to the appropriate topology and completions. In this section, we will introduce two Sobolev versions of these configuration spaces: these are defined by taking a suitable reference configuration and considering all configurations (A, s, ϕ) whose difference from the reference configuration either is in $L^2_{k,\mathrm{loc}}$ (so that it lies in L^2_k on each compact subcylinder) or is globally in L^2_k on the infinite cylinder. Our first main task is then to show that these two notions of configuration space lead eventually to the same trajectory spaces of solutions. Thus, given critical points $[\mathfrak{a}]$ and $[\mathfrak{b}]$ in $\mathcal{B}^\sigma(Y)$, for the perturbed flow, we will be able to talk of a space of connecting trajectories $M([\mathfrak{a}], [\mathfrak{b}])$, just as in our earlier finite-dimensional discussion.

13.1 Definitions

Let A_0 be any smooth spin^c connection on the cylinder $Z = I \times Y$. For any interval $I \subset \mathbb{R}$ (possibly $\mathbb{R}$ itself), we introduce the configuration space

$$\begin{aligned} \tilde{\mathcal{C}}^\tau_{k,\mathrm{loc}}(I \times Y) \subset \Big(A_0 + L^2_{k,\mathrm{loc}}(I \times Y; iT^*Z)\Big) \times L^2_{k,\mathrm{loc}}(I; \mathbb{R}) \\ \times L^2_{k,\mathrm{loc}}(I \times Y; S^+) \\ = \mathcal{A}_{k,\mathrm{loc}}(I \times Y) \times L^2_{k,\mathrm{loc}}(I; \mathbb{R}) \times L^2_{k,\mathrm{loc}}(I \times Y; S^+) \end{aligned} \tag{13.1}$$

consisting of $L^2_{k,\mathrm{loc}}$ triples (A, s, ϕ), where $\check{\phi}(t)$ is of L^2 norm 1 on $\{t\} \times Y$ for all t in I. If I is compact, the "loc" is irrelevant, and the definition coincides with the previous definition of $\tilde{\mathcal{C}}^\tau_k$. We have a closed subspace

$$\mathcal{C}^\tau_{k,\mathrm{loc}}(I \times Y) \subset \tilde{\mathcal{C}}^\tau_{k,\mathrm{loc}}(I \times Y)$$

consisting of triples (A, s, ϕ) satisfying the additional condition $s \geq 0$. This space carries the topology of L^2_k convergence on compact subsets. As usual, a choice of spin^c structure is understood.

The appropriate gauge group is $\mathcal{G}_{k+1,\mathrm{loc}}(I \times Y)$, the group of $L^2_{k+1,\mathrm{loc}}$ maps with values in the circle, $S^1 \subset \mathbb{C}$. The quotient spaces will be denoted

$$\begin{aligned}\mathcal{B}^{\tau}_{k,\mathrm{loc}}(I \times Y) &= \mathcal{C}^{\tau}_{k,\mathrm{loc}}(I \times Y)/\mathcal{G}_{k+1,\mathrm{loc}}(I \times Y)\\ \tilde{\mathcal{B}}^{\tau}_{k,\mathrm{loc}}(I \times Y) &= \tilde{\mathcal{C}}^{\tau}_{k,\mathrm{loc}}(I \times Y)/\mathcal{G}_{k+1,\mathrm{loc}}(I \times Y).\end{aligned} \tag{13.2}$$

Throughout this section, we suppose that $\mathfrak{q} \in \mathcal{P}$ is chosen so that all the critical points of the vector field $(\mathrm{grad}\,\mathcal{L})^{\sigma}$ are non-degenerate (Theorem 12.1.2). We write $\mathfrak{a}$ for a typical critical point in $\mathcal{C}^{\sigma}_k(Y)$, and $[\mathfrak{a}]$ for the corresponding critical point in the quotient space $\mathcal{B}^{\sigma}_k(Y)$. The choice of $\mathfrak{q}$ leads to the perturbed 4-dimensional Seiberg–Witten equations in the blow-up, defined by a section

$$\mathfrak{F}^{\tau}_{\mathfrak{q}} : \tilde{\mathcal{C}}^{\tau}_{k,\mathrm{loc}}(I \times Y) \to \mathcal{V}^{\tau}_{k-1,\mathrm{loc}}(I \times Y).$$

Here the fiber of $\mathcal{V}^{\tau}_{j,\mathrm{loc}}$ at $\gamma = (A_0, s_0, \phi_0)$ is defined as the subspace

$$\mathcal{V}^{\tau}_{j,\mathrm{loc},\gamma} \subset L^2_{j,\mathrm{loc}}(I \times Y; i\,\mathfrak{su}(S^+)) \oplus L^2_{j,\mathrm{loc}}(I;\mathbb{R}) \oplus L^2_{j,\mathrm{loc}}(I \times Y; S^-)$$

consisting of triples (a, s, ϕ) with $\mathrm{Re}\langle \check{\phi}_0(t), \check{\phi}(t)\rangle_{L^2(Y)} = 0$ for all t, as in Definition 6.3.2. (We are careful not to use the language of vector bundles in this instance, because $\mathcal{V}^{\tau}_{j,\mathrm{loc}}$ is not a locally trivial bundle in any straightforward way.)

We suppose that a perturbation $\mathfrak{q}$ has been chosen so that all the critical points of the perturbed equations in $\mathcal{C}^{\sigma}_k(Y)$ are non-degenerate. If $\mathfrak{b}$ is a critical point, there is a translation-invariant element $\gamma_{\mathfrak{b}}$ in $\mathcal{C}^{\tau}_{k,\mathrm{loc}}(Z)$, which is a solution of the equations, $\mathfrak{F}^{\tau}_{\mathfrak{q}}(\gamma_{\mathfrak{b}}) = 0$. We write $[\gamma_{\mathfrak{b}}]$ for its gauge-equivalence class. We say that a configuration $[\gamma] \in \tilde{\mathcal{B}}^{\tau}_{k,\mathrm{loc}}(Z)$ is *asymptotic* to $[\mathfrak{b}]$ as $t \to \pm\infty$ if

$$[\tau_t^* \gamma] \to [\gamma_{\mathfrak{b}}] \qquad \text{in } \tilde{\mathcal{B}}^{\tau}_{k,\mathrm{loc}}(Z),$$

as $t \to \pm\infty$. Here $\tau_t : Z \to Z$ is the map $(s, y) \mapsto (s + t, y)$. We write

$$\underset{\rightarrow}{\lim}[\gamma] = [\mathfrak{b}] \qquad \text{or} \qquad \underset{\leftarrow}{\lim}[\gamma] = [\mathfrak{b}]$$

if $[\gamma]$ is asymptotic to $[\mathfrak{b}]$ as $t \to +\infty$ or $t \to -\infty$ respectively.

Definition 13.1.1. We write $M([\mathfrak{a}], [\mathfrak{b}])$ for the space of all configurations $[\gamma]$ in $\mathcal{B}^{\tau}_{k,\mathrm{loc}}(Z)$ which are asymptotic to $[\mathfrak{a}]$ as $t \to -\infty$, asymptotic to $[\mathfrak{b}]$ as $t \to +\infty$ and which solve the perturbed Seiberg–Witten equations:

$$M([\mathfrak{a}], [\mathfrak{b}]) = \{\, [\gamma] \in \mathcal{B}^{\tau}_{k,\mathrm{loc}}(Z) \mid \mathfrak{F}^{\tau}_{\mathfrak{q}}(\gamma) = 0, \underset{\leftarrow}{\lim}[\gamma] = [\mathfrak{a}], \underset{\rightarrow}{\lim}[\gamma] = [\mathfrak{b}] \,\}.$$

We refer to $M([\mathfrak{a}],[\mathfrak{b}])$ as a *moduli space* of trajectories on the cylinder $Z = \mathbb{R}\times Y$. We denote by $\tilde{M}([\mathfrak{a}],[\mathfrak{b}])$ the similarly defined subset of the larger space $\tilde{\mathcal{B}}^\tau_{k,\mathrm{loc}}(Z)$. ◊

Our notation for $M([\mathfrak{a}],[\mathfrak{b}])$ does not indicate the value of k used in its definition, because this space is essentially independent of k:

Proposition 13.1.2. *Let $M([\mathfrak{a}],[\mathfrak{b}])_k$ temporarily denote the moduli space $M([\mathfrak{a}],[\mathfrak{b}]) \subset \mathcal{B}^\tau_{k,\mathrm{loc}}(Z)$.*

(i) *If $[\gamma]$ is an element of $M([\mathfrak{a}],[\mathfrak{b}])_k$, then there is a gauge representative $\gamma \in \mathcal{C}^\tau_{k,loc}(Z)$ that is C^∞ on Z. So there are natural bijections $M([\mathfrak{a}],[\mathfrak{b}])_{k_1} \to M([\mathfrak{a}],[\mathfrak{b}])_{k_2}$ for all $k_1, k_2 \geq 2$.*
(ii) *The above bijections are homeomorphisms.*

Proof. Let γ represent an element of $M([\mathfrak{a}],[\mathfrak{b}])_k$. Because the perturbation is tame (which means k-tame for all k, let us recall), Proposition 10.7.2 tells us that, for any bounded subcylinder $I \times Y \subset Z$, there is a gauge transformation u_I on $I \times Y$ such that $u_I(\gamma)$ is smooth on the interior of $I \times Y$. We can also arrange that u_I is in the identity component, because each component contains smooth gauge transformations. Thus we can write $u_I = \exp(\zeta_I)$. On the overlaps $I \cap I'$, the difference $\zeta_I - \zeta_{I'}$ is smooth, and we can now paste together the ζ_I using a partition of unity to obtain a gauge transformation $u = \exp(\zeta)$ on the whole of Z, so that $u(\gamma)$ is smooth.

For the second part, we note that the compactness result, Theorem 10.9.2, tells us that if $[\gamma_i]$ is a convergent sequence of solutions in $M([\mathfrak{a}],[\mathfrak{b}])_k$, and I is any bounded interval, then there are gauge transformations $u_{I,i}$ such that $u_{I,i}(\gamma_i)$ converge in the C^∞ topology on interior cylinders contained in $I \times Y$. We may assume these are in the identity component, and then patch together the $u_{I,i}$ into a gauge transformation u_i as above. □

If $[\gamma] \in M([\mathfrak{a}],[\mathfrak{b}])$, then there is a corresponding (smooth) path $[\check{\gamma}]$ in $\mathcal{B}^\sigma_k(Y)$ which approaches $[\mathfrak{a}]$ and $[\mathfrak{b}]$ at the two ends. We can therefore decompose $M([\mathfrak{a}],[\mathfrak{b}])$ according to the relative homotopy class of the path, an element

$$z \in \pi_1(\mathcal{B}^\sigma_k(Y), [\mathfrak{a}], [\mathfrak{b}]).$$

The set of homotopy classes is an affine space on $H^1(Y;\mathbb{Z})$, the component group of the gauge group (see Lemma 9.6.1). We write

$$M([\mathfrak{a}],[\mathfrak{b}]) = \bigcup_z M_z([\mathfrak{a}],[\mathfrak{b}]).$$

13.2 Sobolev spaces on the cylinder

Although our definition of $M_z([\mathfrak{a}],[\mathfrak{b}])$ is straightforward, we also need a different description that exhibits the moduli space as a quotient of a subset of a Hilbert space by the action of a Hilbert Lie group. Before doing so, we need to discuss Sobolev spaces and multiplication theorems on the infinite cylinder $Z = \mathbb{R} \times Y$. We deduce these from the corresponding results on compact manifolds.

We begin with the embedding theorem for the usual Sobolev spaces of functions, $L^p_k(Z)$.

Theorem 13.2.1. *There is a continuous inclusion*

$$L^p_k(Z) \hookrightarrow L^q_l(Z)$$

if $k \geq l$, $p \leq q$ and $(k - n/p) \geq (l - n/q)$, except that if the last inequality is an equality, then we require $1 < p \leq q < \infty$. This inclusion is never compact.

Proof. We use the fact that the embedding holds for a finite cylinder of fixed length, say $[-1, 1] \times Y$ (though in this situation the inequality $p \leq q$ is not necessary). Let $f_n(t, y)$ be the restriction of the function $f(t + n, y)$ to $(t, y) \in [-1, 1] \times Y$. We have then

$$\|f\|_{L^q_l} = 2^{-1/q} \left(\sum_{n=-\infty}^{\infty} \|f_n\|^q_{L^q_l} \right)^{1/q}$$

(where the right-hand side should be replaced by the supremum of $\|f_n\|_{L^q_l}$ when $q = \infty$). Taking the case p and q finite, we then have, for example,

$$\begin{aligned}
\|f\|_{L^q_l} &\leq C_1 \left(\sum_{n=-\infty}^{\infty} \|f_n\|^q_{L^q_l} \right)^{1/q} \\
&\leq C_2 \left(\sum_{n=-\infty}^{\infty} \|f_n\|^p_{L^q_l} \right)^{1/p} \\
&\leq C_3 \left(\sum_{n=-\infty}^{\infty} \|f_n\|^p_{L^p_k} \right)^{1/p} \\
&\leq C\|f\|_{L^p_k}.
\end{aligned}$$

In the second inequality, we used $p \leq q$. The constant C here depends on the Sobolev embedding constant on $[-1, 1] \times Y$ and the bounds on the derivatives of the partition of unity.

This embedding is never compact since given a function we can translate this function until the norm on any compact set is as small as we desire, and so produce a weakly convergent sequence which will only have weakly convergent subsequences regardless of the norm used. □

Here is the multiplication theorem.

Theorem 13.2.2. *Suppose $k, l \geq m$ and $1/p + 1/q \geq 1/r$, with $p, q, r \in (1, \infty)$. Then the multiplication*

$$L^p_k(Z) \times L^q_l(Z) \to L^r_m(Z)$$

is continuous in any of the following three cases:

(i) (a) $(k - n/p) + (l - n/q) \geq m - n/r$, *and*
   (b) $k - n/p < 0$, *and*
   (c) $l - n/q < 0$;
   or

(ii) (a) $\min\{(k - n/p), (l - n/q)\} \geq m - n/r$, *and*
   (b) *either* $k - n/p > 0$ *or* $l - n/q > 0$;
   or

(iii) (a) $\min\{(k - n/p), (l - n/q)\} > r - n/m$, *and*
   (b) *either* $k - n/p = 0$ *or* $l - n/q = 0$.

When the map is continuous, it is a compact operator as a function of g for fixed f, provided that in addition we have $l > m$ and $l - n/q > m - n/r$.

Proof. In each of the three cases, the multiplication theorem holds on a finite cylinder; so, using the same notation as in the proof of the previous theorem, we have:

$$\|fg\|_{L^r_m} \leq C_1 \left(\sum_{n=-\infty}^{\infty} \|(fg)_n\|^r_{L^r_m} \right)^{1/r}$$
$$\leq C_2 \left(\sum_{n=-\infty}^{\infty} \|f_n\|^r_{L^p_k} \|g_n\|^r_{L^q_l} \right)^{1/r}$$

$$\leq C_3 \left(\sum_{n=-\infty}^{\infty} \|f_n\|_{L^p_k}^p \right)^{1/p} \left(\sum_{n=-\infty}^{\infty} \|g_n\|_{L^q_l}^q \right)^{1/q}$$
$$\leq C_4 \|f\|_{L^p_k} \|g\|_{L^q_l}.$$

In the penultimate line we used the fact that the multiplication $\ell^p \times \ell^q \to \ell^r$ is continuous on sequence spaces when $1/p + 1/q \geq 1/r$.

To prove the compactness assertion consider a function $f \in L^p_k(Z)$ and a bounded sequence $g_i \in L^q_l(Z)$ with bound M and suppose that the hypotheses of the assertion are satisfied. Then on any finite cylinder the multiplication $L^p_k([t_1, t_2] \times Y) \times L^q_l([t_1, t_2] \times Y) \to L^r_m([t_1, t_2] \times Y)$ is compact as a function of either variable. Thus by a diagonal argument we can pass to a subsequence so that on any $[-n, n] \times Y, fg_i$ converges strongly in L^r_m. On the other hand

$$\|fg_i\|_{L^r_m(Z \setminus [-n,n] \times Y)} \leq CM \|f\|_{L^p_k(Z \setminus [-n,n] \times Y)}.$$

The right-hand side goes to zero with n. Thus given $\epsilon > 0$ we can find n so large that $CM \|f\|_{L^p_k(Z \setminus [-n,n] \times Y)} < \epsilon/2$ and then find i_0 such that for all $i, j \geq i_0$ we have

$$\|fg_i - fg_j\|_{L^r_m([-n,n] \times Y)} < \epsilon/2.$$

Thus

$$\|fg_i - fg_j\|_{L^r_m(Z)} < \epsilon.$$

□

13.3 Statement of results

We return to the moduli spaces defined in Subsection 13.1. Let $[\mathfrak{a}]$ and $[\mathfrak{b}]$ be critical points in $\mathcal{B}^\sigma_k(Y)$, and let $\mathfrak{a}$ and $\mathfrak{b}$ be smooth lifts in $\mathcal{C}^\sigma_k(Y)$. Choose a base configuration $\gamma_0 \in \mathcal{C}^\tau_{k,\text{loc}}(Z)$ which agrees near $\pm\infty$ with the corresponding translation-invariant configurations $\gamma_{\mathfrak{a}}$ and $\gamma_{\mathfrak{b}}$. For convenience, we may choose γ_0 to be smooth also. By choosing $\mathfrak{a}$ and $\mathfrak{b}$ in appropriate components of the gauge group orbit, we can arrange that $[\check{\gamma}_0]$ belongs to any given homotopy class z. Because $\mathcal{C}^\tau_{k,\text{loc}}(Z)$ is a subset of an affine space,

$$\tilde{\mathcal{C}}^\tau_{k,\text{loc}}(Z) \subset \mathcal{A}_{k,\text{loc}}(Z) \times L^2_{k,\text{loc}}(\mathbb{R}; \mathbb{R}) \times L^2_{k,\text{loc}}(Z; S^+),$$

we can interpret the difference of two elements as an element of a vector space:

$$\gamma - \gamma_0 \in L^2_{k,\text{loc}}(Z; iT^*Z) \times L^2_{k,\text{loc}}(\mathbb{R}; \mathbb{R}) \times L^2_{k,\text{loc}}(Z; S^+).$$

Inside this vector space is the subspace of elements of finite L^2_k norm, where on the spinor component we construct the L^2_k norm using the spin^c connection A_0 belonging to the configuration $\gamma_0 = (A_0, s_0, \phi_0)$:

$$L^2_k(Z; iT^*Z) \times L^2_k(\mathbb{R}; \mathbb{R}) \times L^2_{k,A_0}(Z; S^+).$$

We introduce a configuration space

$$\mathcal{C}^\tau_k(\mathfrak{a}, \mathfrak{b}) = \\ \{\, \gamma \in \mathcal{C}^\tau_{k,\mathrm{loc}}(Z) \mid \gamma - \gamma_0 \in L^2_k(Z; iT^*Z) \times L^2_k(\mathbb{R}; \mathbb{R}) \times L^2_{k,A_0}(Z; S^+) \,\}.$$

Although γ_0 appears in the definition, this space only depends on $\mathfrak{a}$ and $\mathfrak{b}$. The larger space $\tilde{\mathcal{C}}^\tau_k(\mathfrak{a}, \mathfrak{b})$ is defined similarly as a subset of $\tilde{\mathcal{C}}^\tau_{k,\mathrm{loc}}(Z)$. We introduce a gauge group $\mathcal{G}_{k+1}(Z)$ which is the subgroup of $\mathcal{G}_{k+1,\mathrm{loc}}$ consisting of the gauge transformations which preserve $\mathcal{C}^\tau_k(\mathfrak{a}, \mathfrak{b})$. Using the Sobolev multiplication theorems, we can see:

Lemma 13.3.1. *The group $\mathcal{G}_{k+1}(Z)$ is independent of $\mathfrak{a}$ and $\mathfrak{b}$, and can be described as*

$$\mathcal{G}_{k+1}(Z) = \{\, u : Z \to S^1 \mid 1 - u \in L^2_{k+1}(Z; \mathbb{C}) \,\}.$$

Proof. It is clear that if $1 - u$ is in L^2_{k+1}, then u preserves $\mathcal{C}^\tau_k(\mathfrak{a}, \mathfrak{b})$. Conversely, suppose u in $L^2_{k+1,\mathrm{loc}}$ and that $u(\gamma_0) = \gamma$, with $\gamma \in \mathcal{C}^\tau_k(\mathfrak{a}, \mathfrak{b})$. Write $\gamma = (A, s, \phi)$.

Let u_n be the restriction of u to $[n-1, n] \times Y$. As in the proof of Proposition 9.3.1, it follows from the fact that $A - A_0$ is in L^2_k that u_n is in the identity component. It follows that we can write $u = e^{\xi}$. Write ξ_n for the restriction of ξ to $[n-1, n] \times Y$. Following the proof of Proposition 9.3.1 again, we decompose ξ_n as $\xi^0_n + \xi^\perp_n$, where ξ^0_n is the average value, and we see that $\xi^\perp_n$ defines a Cauchy sequence in $L^2_{k+1}([0,1] \times Y)$ whose limit is zero. Indeed, the same argument shows that

$$\sum \|\xi^\perp_n\|^2_{L^2_{k+1}} < \infty.$$

This gives us Sobolev bounds on the derivatives of ξ, and hence the derivatives of u and $1 - u$. All that remains is to show that $1 - u$ is square-integrable, and this now reduces to showing that $1 - u_0$ is square-integrable, where u_0 is the discontinuous function defined by the sequence $e^{\xi^0_n}$ on the cylinder. What we know is that $(1 - u_0)\phi_0$ is square-integrable. The square integrability of $(1 - u_0)$ follows easily, because ϕ_0 is non-zero and translation-invariant on the two ends of the cylinder. □

As a corollary of the proof, we obtain:

Corollary 13.3.2. *If $u \in \mathcal{G}_{k+1}(Z)$, then the path $t \mapsto u(t, y_0)$ is a continuous path from $\mathbb{R} \to S^1$ approaching 1 at both ends, and u is in the identity component if and only if the winding number of the path is zero. The component group of the gauge group is therefore $\mathbb{Z}$.* □

Definition 13.3.3. Let $[\mathfrak{a}]$, $[\mathfrak{b}]$ be gauge-equivalence classes of critical points in $\mathcal{B}_k^\sigma(Y)$, and let $z \in \pi_1(\mathcal{B}_k^\sigma(Y), [\mathfrak{a}], [\mathfrak{b}])$ be a relative homotopy class of paths. Pick lifts $\mathfrak{a}_z$ and $\mathfrak{b}_z$ such that a path joining these lifts projects to a path in the class z. We write $\mathcal{B}_{k,z}^\tau([\mathfrak{a}], [\mathfrak{b}])$ and $\tilde{\mathcal{B}}_{k,z}^\tau([\mathfrak{a}], [\mathfrak{b}])$ for the quotient spaces

$$\mathcal{B}_{k,z}^\tau([\mathfrak{a}], [\mathfrak{b}]) = \mathcal{C}_k^\tau(\mathfrak{a}_z, \mathfrak{b}_z)/\mathcal{G}_{k+1}^\tau(Z)$$

$$\tilde{\mathcal{B}}_{k,z}^\tau([\mathfrak{a}], [\mathfrak{b}]) = \tilde{\mathcal{C}}_k^\tau(\mathfrak{a}_z, \mathfrak{b}_z)/\mathcal{G}_{k+1}^\tau(Z).$$

We define

$$\mathcal{B}_k^\tau([\mathfrak{a}], [\mathfrak{b}]) = \bigcup_z \mathcal{B}_{k,z}^\tau([\mathfrak{a}], [\mathfrak{b}])$$

$$\tilde{\mathcal{B}}_k^\tau([\mathfrak{a}], [\mathfrak{b}]) = \bigcup_z \tilde{\mathcal{B}}_{k,z}^\tau([\mathfrak{a}], [\mathfrak{b}]).$$

To within a canonical identification, $\mathcal{B}_{k,z}^\tau([\mathfrak{a}], [\mathfrak{b}])$ is independent of the choice of representatives $\mathfrak{a}_z$, $\mathfrak{b}_z$. ◊

Proposition 13.3.4. *The quotient space $\mathcal{B}_k^\tau([\mathfrak{a}], [\mathfrak{b}])$ is Hausdorff.*

Proof. This result follows from combining the argument used in the case of a finite cylinder (Proposition 9.3.1) with the argument of Lemma 13.3.1. □

The following theorem tells us that we can use $\mathcal{C}_k^\tau(\mathfrak{a}, \mathfrak{b})$ in place of $\mathcal{C}_{k,\mathrm{loc}}^\tau(Z)$ in describing the moduli space $M_z([\mathfrak{a}], [\mathfrak{b}])$.

Theorem 13.3.5. *Let $\gamma \in \mathcal{C}_{k,\mathrm{loc}}^\tau(Z)$ represent an element $[\gamma] \in M_z([\mathfrak{a}], [\mathfrak{b}])$. Let $\mathfrak{a} = \mathfrak{a}_z$, $\mathfrak{b} = \mathfrak{b}_z$ be suitable lifts, as above. Then there exists a gauge transformation u in $\mathcal{G}_{k+1,\mathrm{loc}}$ such that $u(\gamma)$ belongs to $\mathcal{C}_k^\tau(\mathfrak{a}, \mathfrak{b})$. If u and u' are two such gauge transformations, then $u^{-1}u'$ belongs to $\mathcal{G}_{k+1}(Z)$. The resulting bijection is a homeomorphism,*

$$M_z([\mathfrak{a}], [\mathfrak{b}]) \to \{ [\gamma] \in \mathcal{B}_{k,z}^\tau([\mathfrak{a}], [\mathfrak{b}]) \mid \mathfrak{F}_\mathfrak{q}^\tau(\gamma) = 0 \}.$$

A similar statement is true for the larger moduli space $\tilde{M}_z([\mathfrak{a}], [\mathfrak{b}])$.

The proof of the theorem is given at the end of this section.

13.4 Near-constant solutions on finite cylinders

Let $\mathfrak{b}$ be a critical point for the perturbed equations in $\mathcal{C}_k^\tau(Y)$, and let β be its image in $\mathcal{C}_k(Y)$ under the blow-down map. Let

$$\gamma_{\mathfrak{b}} \in \mathcal{C}_k^\tau(I \times Y)$$
$$\gamma_\beta \in \mathcal{C}_k(I \times Y)$$

be the corresponding translation-invariant solutions on a *compact* cylinder $I \times Y$. We can write

$$\gamma_{\mathfrak{b}} = (A_{\mathfrak{b}}, s_{\mathfrak{b}}, \phi_{\mathfrak{b}})$$
$$\gamma_\beta = (A_{\mathfrak{b}}, \Phi_{\mathfrak{b}})$$

with $\Phi_{\mathfrak{b}} = s_{\mathfrak{b}}\phi_{\mathfrak{b}}$. We consider an element $\gamma^\tau \in \mathcal{C}_k^\tau(I \times Y)$ covering a configuration $\gamma \in \mathcal{C}_k(I \times Y)$. We write

$$\gamma^\tau = (A, s, \phi),$$

so that

$$\gamma = (A, s\phi) = (A, \Phi),$$

and we write

$$A - A_{\mathfrak{b}} = b \otimes 1 + (c \otimes 1)dt.$$

We regard b and c as a time-dependent 1-form and 0-form on Y respectively. Similarly we write $\phi = \phi_{\mathfrak{b}} + \psi$, and regard ψ as a time-dependent section of $S \to Y$, with the property that $\phi_{\mathfrak{b}} + \psi(t)$ is of unit L^2 norm for all t. The following is a basic estimate for a solution to the perturbed equations that is close to a non-degenerate critical point. The statement of the proposition refers to Sobolev norms of the difference of two configurations in the τ model, such as

$$\|\gamma^\tau - \gamma_{\mathfrak{b}}\|_{L^2_{k,A_{\mathfrak{b}}}(I \times Y)},$$

which are to be interpreted by regarding $\mathcal{C}_k^\tau(I \times Y)$ as a subset of the affine space

$$L_k^2(I \times Y; iT^*Z) \times L_k^2(I; \mathbb{R}) \times L^2_{k,A_{\mathfrak{b}}}(I \times Y; S).$$

Proposition 13.4.1. *Suppose $\mathfrak{b} \in \mathcal{C}_k^\sigma(Y)$ is a (non-degenerate) critical point for the perturbed equations. Let $I = [t_1, t_2]$ and $I' = [t_1', t_2']$ be a pair of compact intervals, with $I' \subset \mathrm{int}(I)$. Then there are constants C_1, C_2 and a gauge-invariant neighborhood U of the constant solution $\gamma_{\mathfrak{b}}$ in $\mathcal{C}_k^\tau(I \times Y)$ such that for every γ^τ which belongs to U and solves the perturbed Seiberg–Witten equations $\mathfrak{F}_{\mathfrak{q}}^\tau(\gamma^\tau) = 0$, there is a gauge transformation $u \in \mathcal{G}_{k+1}(I \times Y)$ such that:*

(i) *in the case that $\mathfrak{b}$ is irreducible, the squared norm of $u(\gamma^\tau) - \gamma_{\mathfrak{b}}$ on the smaller interval I' is bounded by the change in $\mathcal{L}$ on the larger interval $I = [t_1, t_2]$,*

$$\|u(\gamma^\tau) - \gamma_{\mathfrak{b}}\|^2_{L^2_{k+1,A_{\mathfrak{b}}}(I' \times Y)} \le C_1\big(\mathcal{L}(t_1) - \mathcal{L}(t_2)\big);$$

(ii) *in the case that $\mathfrak{b}$ is reducible, the squared norm of $u(\gamma^\tau) - \gamma_{\mathfrak{b}}$ on the smaller interval I' is bounded by the change in $\mathcal{L}$ on the larger interval and the drop in $\Lambda_{\mathfrak{q}}$:*

$$\begin{aligned}\|u(\gamma^\tau) - \gamma_{\mathfrak{b}}\|^2_{L^2_{k+1,A_{\mathfrak{b}}}(I' \times Y)} \\ \le C_2\big(\Lambda_{\mathfrak{q}}(t_1) - \Lambda_{\mathfrak{q}}(t_2) + (\mathcal{L}(t_1) - \mathcal{L}(t_2))^{1/2}\big).\end{aligned}$$

We will first establish the existence of a gauge transformation u with

$$\|u(\gamma) - \gamma_\beta\|^2_{L^2_{k+1,A_{\mathfrak{b}}}(I' \times Y)} \le C_1\big(\mathcal{L}(t_1) - \mathcal{L}(t_2)\big).$$

In the irreducible case, the projection $\mathcal{C}_k^\tau(I \times Y) \to \mathcal{C}_k(I \times Y)$ is a diffeomorphism in the neighborhood of $\gamma_{\mathfrak{b}}$, so this inequality is equivalent to the assertion of the proposition.

We write the equations $\mathfrak{F}_{\mathfrak{q}}(\gamma) = 0$ as

$$\frac{d}{dt}\check{\gamma}(t) + \mathbf{d}_{\check{\gamma}(t)}c(t) = -(\operatorname{grad} \mathcal{L})(\check{\gamma}(t)) \tag{13.3}$$

where $\mathbf{d}$ is the linearization of the gauge-group action on $\mathcal{C}_k(Y)$ (*cf.* (4.11)). For any γ in a neighborhood of γ_β, we can choose a gauge transformation u on $I \times Y$ so that $u(\gamma)$ is in the Coulomb–Neumann slice $\mathcal{S}_{k,\gamma_\beta}$ through γ_β (Definition 9.3.7). Replacing $u(\gamma)$ by γ in our notation, the Coulomb–Neumann condition, $\gamma - \gamma_\beta \in \mathcal{K}_{k,\gamma_\beta,k}$, is:

$$\begin{aligned}\frac{d}{dt}c(t) + \mathbf{d}^*_\beta(\check{\gamma}(t) - \beta) = 0 \\ c(t_1) = c(t_2) = 0.\end{aligned} \tag{13.4}$$

In the case that β is reducible, the first equation is $dc/dt = d^*b(t)$; and by integrating over Y and using the boundary conditions, it follows in this case that

$$\int_Y c(t) = 0 \quad \text{for all } t \in I. \tag{13.5}$$

The next lemma applies to any configuration γ, not necessarily a solution to the equations.

Lemma 13.4.2. *For any configuration γ on the cylinder $Z = I \times Y$ we have*

$$\int_I \left(\left\| \frac{d}{dt}\check{\gamma} + \mathbf{d}_{\check{\gamma}} c \right\|^2 + \left\| \frac{d}{dt} c + \mathbf{d}^*_\beta(\check{\gamma} - \beta) \right\|^2 + \|(\operatorname{grad} \mathcal{L})(\check{\gamma})\|^2 \right) dt$$

$$= \int_I \left(\left\| \frac{d}{dt}\check{\gamma} \right\|^2 + \|\mathbf{d}^*_\beta(\check{\gamma} - \beta)\|^2 + \left\| \frac{d}{dt} c \right\|^2 + \|\mathbf{d}_{\check{\gamma}} c\|^2 + \|(\operatorname{grad} \mathcal{L})(\check{\gamma})\|^2 \right) dt$$

$$+ \langle c(t_1), \mathbf{d}^*_\beta(\check{\gamma}(t_1) - \beta)\rangle - \langle c(t_2), \mathbf{d}^*_\beta(\check{\gamma}(t_2) - \beta)\rangle$$

$$+ \int_I \left\langle (\check{\gamma} - \beta) \sharp c, \frac{d}{dt}\check{\gamma} \right\rangle dt$$

where $\sharp$ is a bilinear operator involving only pointwise multiplication.

Proof. This follows from integration by parts and the identities

$$\frac{d}{dt}\langle c(t), \mathbf{d}^*_\beta(\check{\gamma}(t) - \beta)\rangle = \left\langle \frac{d}{dt}c(t), \mathbf{d}^*_\beta(\check{\gamma}(t) - \beta) \right\rangle + \left\langle \mathbf{d}_\beta c(t), \frac{d}{dt}\check{\gamma}(t) \right\rangle$$

and

$$\left\langle (\mathbf{d}_{\check{\gamma}(t)} - \mathbf{d}_\beta) c(t), \frac{d}{dt}\check{\gamma}(t) \right\rangle = \left\langle (\check{\gamma} - \beta) \sharp c, \frac{d}{dt}\check{\gamma} \right\rangle.$$

□

Next we exploit the non-degeneracy of the critical point $\mathfrak{b}$.

Lemma 13.4.3. *If $\mathfrak{b}$ is an irreducible, non-degenerate critical point of $(\operatorname{grad} \mathcal{L})^\sigma$ then there is a constant $C > 0$ and an L^2_1 neighborhood U^Y of $(\beta, 0) \in \mathcal{C}_1(Y) \times L^2_1(Y, i\mathbb{R})$ such that for all $(\beta + v, c)$ in U^Y we have*

$$\|(v,c)\|^2_{L^2_1(Y)} \le C\big(\|\mathbf{d}^*_\beta v\|^2 + \|\mathbf{d}_{\beta+v} c\|^2 + \|(\operatorname{grad} \mathcal{L})(\beta+v)\|^2\big).$$

In the case that $\mathfrak{b}$ is reducible and non-degenerate, the same conclusion holds for all $(\beta + v, c)$ satisfying the additional condition $\int_Y c = 0$.

Proof. Consider the map

$$\begin{aligned}\mathcal{C}_1(Y) \times L^2_1(Y; i\mathbb{R}) &\to \mathcal{T}_0 \times L^2(Y; i\mathbb{R}) \\ &= \mathcal{J}_0 \oplus \mathcal{K}_0 \oplus L^2(Y; i\mathbb{R})\end{aligned}$$

given by

$$(\beta + v, c) \mapsto (\mathbf{d}_{\beta+v} c, (\operatorname{grad} \mathcal{L})(\beta+v), \mathbf{d}^*_\beta v).$$

The continuity of this map follows from Condition (vi) in the definition of a tame perturbation, Definition 10.5.1. Using the decomposition $\mathcal{T}_{0,\beta} = \mathcal{J}_{0,\beta} \oplus \mathcal{K}_{0,\beta}$, the linearization of this map at $(\beta, 0)$ can be written

$$(v^J, v^K, c) \mapsto (\mathbf{d}_\beta c, \operatorname{Hess}_\beta v^K, \mathbf{d}^*_\beta v^J).$$

Here $\operatorname{Hess}_\beta = \operatorname{Hess}_{\mathfrak{q},\beta}$ is the Hessian, defined at (12.4) (though we now omit the perturbation $\mathfrak{q}$ from our notation). Non-degeneracy means that $\operatorname{Hess}_\beta$ is invertible as a map $\mathcal{K}_{1,\beta} \to \mathcal{K}_{0,\beta}$. The map

$$\mathbf{d}_\beta : L^2_1(Y; i\mathbb{R}) \to \mathcal{J}_{0,\beta}$$

is invertible in the irreducible case. In the reducible case, it is invertible when the domain is restricted to those c with $\int_Y c = 0$. The result follows. □

As a corollary, we have:

Lemma 13.4.4. *There is a gauge-invariant neighborhood U of γ_β in $\mathcal{C}_k(I \times Y)$ and a constant C_0, such that for any $\gamma \in U$ which solves the equations and is in the Coulomb–Neumann slice $\mathcal{S}_{k,\gamma_\beta}$, we have*

$$\|\gamma - \gamma_\beta\|^2_{L^2_1(I\times Y)} \le C_0\big(\mathcal{L}(t_1) - \mathcal{L}(t_2)\big).$$

Proof. First note that the Coulomb–Neumann condition implies the vanishing of the second term on the left-hand side in Lemma 13.4.2, as well as the two boundary terms. The remaining two terms on the left are the perturbed analytic energy $\mathcal{E}_{\mathfrak{q}}(\gamma)$, which is twice the drop in $\mathcal{L}$. That lemma therefore gives

$$\begin{aligned}
2\big(\mathcal{L}(t_1) - \mathcal{L}(t_2)\big) \\
&= \int_I \left(\left\| \frac{d}{dt}\check{\gamma}(t) \right\|^2 + \left\| \frac{d}{dt}c(t) \right\|^2 \right) dt \\
&\quad + \int_I \Big(\|\mathbf{d}^*_\beta(\check{\gamma}(t) - \beta)\|^2 + \|\mathbf{d}_{\check{\gamma}(t)}c(t)\|^2 + \|(\operatorname{grad} \mathcal{L})(\check{\gamma}(t))\|^2 \Big) dt \\
&\quad + \int_I \left\langle (\check{\gamma} - \beta) \sharp c, \frac{d}{dt}\check{\gamma} \right\rangle dt.
\end{aligned}$$

Using Lemma 13.4.3, we deduce

$$\begin{aligned}
C'\big(\mathcal{L}(t_1) - \mathcal{L}(t_2)\big) \\
&\geq \int_I \left(\left\| \frac{d}{dt}\check{\gamma}(t) \right\|^2 + \left\| \frac{d}{dt}c(t) \right\|^2 \right) dt \\
&\quad + \int_I \|(\check{\gamma}(t) - \beta, c(t))\|^2_{L^2_1(Y)} dt \\
&\quad + \int_I \left\langle (\check{\gamma} - \beta) \sharp c, \frac{d}{dt}\check{\gamma} \right\rangle dt.
\end{aligned}$$

Thus we obtain

$$\|\gamma - \gamma_\beta\|^2_{L^2_1(I \times Y)} \leq C'\big(\mathcal{L}(t_1) - \mathcal{L}(t_2)\big) + K\|\gamma - \gamma_\beta\|^3_{L^2_1(I \times Y)}.$$

If the left-hand side is small enough, this inequality can be rearranged to obtain the result. □

To proceed further we study $\frac{d}{dt}\Lambda_{\mathfrak{q}}(\gamma(t))$ when γ solves the perturbed equations.

Lemma 13.4.5. *Suppose that $\gamma^\tau = (A, s, \phi) \in \mathcal{C}^\tau_k(I \times Y)$ solves the perturbed monopole equations on the cylinder $I \times Y$. Then we have*

$$\frac{d}{dt}\Lambda_{\mathfrak{q}}(\gamma^\tau(t)) = -2\|\phi'\|^2 + \langle \phi, L'\phi \rangle,$$

where L is the symmetric, time-dependent linear operator,

$$L\psi = D_{\check{A}}\psi - i(\mathcal{D}_{(\check{A},\check{\Phi})}\mathfrak{q}^1)(0, i\psi) \tag{13.6}$$

of Equation (10.21), *ϕ' denotes the A-covariant derivative of ϕ in the t direction, and similarly $L' = [\nabla^A_{\partial/\partial t}, L]$.*

Proof. This is essentially the same as the derivation of the inequality (10.22) from Lemma 7.1.1: the formula is the same as the identity (7.4), cast in gauge-invariant form. Here, as before, the primed notation denotes the A-covariant derivative in the t direction. □

Lemma 13.4.6. *In the case that $\mathfrak{b}$ is reducible, there is a gauge-invariant neighborhood U of $\gamma_{\mathfrak{b}}$ in $\mathcal{C}^\tau_k(I \times Y)$ and a constant C_0, such that if γ^τ lies in U and is in the Coulomb–Neumann slice $\mathcal{S}^\tau_{k,\gamma_{\mathfrak{b}}} \subset \mathcal{C}^\tau_k(I \times Y)$, then*

$$\|\psi\|^2_{L^2_{1,B}(I\times Y)} \le C_0\big(\Lambda_{\mathfrak{q}}(t_1) - \Lambda_{\mathfrak{q}}(t_2) + \big(\mathcal{L}(t_1) - \mathcal{L}(t_2)\big)^{\frac{1}{2}}\big),$$

where as before $\gamma^\tau = (A, s, \phi) = (A, s, \phi_{\mathfrak{b}} + \psi)$.

Proof. Let us write $\mathfrak{b} = (B_{\mathfrak{b}}, 0, \phi_{\mathfrak{b}})$. Because $\mathfrak{b}$ is a non-degenerate, reducible critical point, the map

$$\phi_{\mathfrak{b}} + v \mapsto (\operatorname{grad} \mathcal{L})^{\sigma,1}(B_{\mathfrak{b}}, 0, \phi_{\mathfrak{b}} + v)$$

has linearization which is an isomorphism from $L^2_{1,B_{\mathfrak{b}}}(Y; S) \cap \langle\phi_{\mathfrak{b}}\rangle^\perp$ to $L^2(Y; S) \cap \langle\phi_{\mathfrak{b}}\rangle^\perp$. So there is a constant C such that for all ψ with small enough $L^2_{1,B_{\mathfrak{b}}}$ norm, we have

$$\|\psi\|^2_{L^2_{1,B_{\mathfrak{b}}}} \le C_1 \|(\operatorname{grad} \mathcal{L})^{\sigma,1}(B_{\mathfrak{b}}, 0, \phi_{\mathfrak{b}} + \psi)\|^2.$$

From the differentiability of $(\operatorname{grad} \mathcal{L})^{\sigma,1}$, we can then deduce that there is a neighborhood of $\mathfrak{b}$ in $\mathcal{C}^\sigma_1(Y)$ such that for all $(B, r, \phi_{\mathfrak{b}} + \psi)$ in this neighborhood, we have

$$\|\psi\|^2_{L^2_{1,B_{\mathfrak{b}}}} \le C_1 \|(\operatorname{grad} \mathcal{L})^{\sigma,1}(B, r, \phi_{\mathfrak{b}} + \psi)\|^2 + K\big(\|B - B_{\mathfrak{b}}\|^2_{L^2_1(Y)} + r^2\big).$$

Hence, writing $(\breve{A}(t), s(t), \phi_{\mathfrak{b}} + \psi(t))$ for the path in $C_k^\sigma(Y)$ corresponding to γ^τ, we have

$$\begin{aligned}
\|\psi\|^2_{L^2_{1,B}(I\times Y)} &= \int_I \left(\|\psi'\|^2_{L^2(Y)} + \|\psi\|^2_{L^2_{1,B}(Y)} \right) \\
&\leq \int_I \left(\|\psi'\|^2_{L^2(Y)} + C_1 \|(\operatorname{grad} \mathcal{L})^{\sigma,1}(\breve{A}, s, \phi_{\mathfrak{b}} + \psi)\|^2 \right) \\
&\quad + K \int_I \left(\|\breve{A} - B_{\mathfrak{b}}\|^2_{L^2_1(Y)} + s^2 \right) \qquad (13.7)
\end{aligned}$$

by integrating the above inequality. In the last line, the second integral is bounded by a multiple of $\|\gamma - \gamma_\beta\|^2_{L^2_1(I\times Y)}$, and hence by

$$C\big(\mathcal{L}(t_2) - \mathcal{L}(t_1)\big) \qquad (13.8)$$

using Lemma 13.4.4. The remaining term in the last line of (13.7) becomes

$$\begin{aligned}
&\int_I \|\psi'\|^2_{L^2(Y)} + C_1 \int_I \|\phi'\|^2_{L^2(Y)} \\
&\qquad = (2 + C_1) \int_I \|\phi'\|^2_{L^2(Y)} + 2 \int_I \|c\phi_{\mathfrak{b}}\|^2_{L^2(Y)}.
\end{aligned}$$

Again, the second term can be estimated by a term of type (13.8); and by integrating the result of Lemma 13.4.5, the first term can be bounded by a fixed multiple of

$$\begin{aligned}
&\Lambda_{\mathfrak{q}}(t_1) - \Lambda_{\mathfrak{q}}(t_2) + \int_I \langle \phi, L'\phi \rangle dt \\
&\quad \leq \Lambda_{\mathfrak{q}}(t_1) - \Lambda_{\mathfrak{q}}(t_2) + C_3 \int_I \left(\|\gamma'\|_{L^2(Y)} \|\phi\|^2_{L^2_1(Y)} \right) dt \\
&\quad \leq \Lambda_{\mathfrak{q}}(t_1) - \Lambda_{\mathfrak{q}}(t_2) + C_3 \|\gamma - \gamma_\beta\|_{L^2_1(I\times Y)} \left(1 + \|\psi\|^2_{L^2_1(I\times Y)} \right) \\
&\quad \leq \Lambda_{\mathfrak{q}}(t_1) - \Lambda_{\mathfrak{q}}(t_2) + C_4 \big(\mathcal{L}(t_1) - \mathcal{L}(t_2)\big)^{1/2} \left(1 + \|\psi\|^2_{L^2_1(I\times Y)} \right)
\end{aligned}$$

where we have used Lemma 13.4.4 once more to control $\|\gamma - \gamma_\beta\|_{L^2_1(I\times Y)}$. If U is chosen small enough then we can arrange the resulting inequality to get the desired result. □

Proof of Proposition 13.4.1. Lemmas 13.4.4 and 13.4.6 give both parts of the proposition, but with L^2_1 norms in place of L^2_{k+1}, and without the need to pass to a subinterval. By a bootstrapping argument we can get control over stronger norms using the equations. □

Note that, if we work in $\mathcal{C}_k(I \times Y)$ rather than $\mathcal{C}_k^\tau(I \times Y)$, then a bound like the one in Part (i) of Proposition 13.4.1 holds also in the reducible case:

Proposition 13.4.7. *Under the same hypotheses as Proposition* 13.4.1, *let* $\beta \in \mathcal{C}_k(Y)$ *be the image of* $\mathfrak{b}$. *Then there is a constant* C_1 *and a gauge-invariant neighborhood* U *of the constant solution* γ_β *in* $\mathcal{C}_k(I \times Y)$ *such that for every* $\gamma \in \mathcal{C}_k(I \times Y)$ *which belongs to* U *and solves the perturbed Seiberg–Witten equations* $\mathfrak{F}_\mathfrak{q}(\gamma) = 0$, *there is a gauge transformation* $u \in \mathcal{G}_{k+1}(I \times Y)$ *such that the squared norm of* $u(\gamma) - \gamma_\beta$ *on the smaller interval* I' *is bounded by the change in* $\mathcal{L}$ *on the larger interval* $I = [t_1, t_2]$:

$$\|u(\gamma) - \gamma_\beta\|^2_{L^2_{k+1,B}(I' \times Y)} \le C_1\big(\mathcal{L}(t_1) - \mathcal{L}(t_2)\big).$$

Proof. This follows from Lemma 13.4.4 and bootstrapping. □

Corollary 13.4.8. *Let* $\mathfrak{b} \in \mathcal{C}_k^\tau(Y)$ *be a non-degenerate critical point, and let* β *be its image in* $\mathcal{C}_k(Y)$. *Then there is a constant* C *and a gauge-invariant neighborhood* U *of the constant solution* γ_β *in* $\mathcal{C}_k([-1,1] \times Y)$ *such that for every* $\gamma \in \mathcal{C}_k([-1,1] \times Y)$ *which belongs to* U *and solves the perturbed Seiberg–Witten equations* $\mathfrak{F}_\mathfrak{q}(\gamma) = 0$, *we have*

$$(d/dt)\Lambda_\mathfrak{q}(t)|_{t=0} \le C(\mathcal{L}(-1) - \mathcal{L}(1))^{1/2}.$$

Proof. We work in a temporal gauge. From Lemma 13.4.5, we have

$$\begin{aligned}(d/dt)\Lambda_\mathfrak{q}(t)|_{t=0} &\le 2\langle \phi, L'\phi\rangle_{L^2(0\times Y)} \\ &= \langle \phi, \rho(\dot{b})\phi\rangle + \langle \phi, i\mathcal{D}^2_{\gamma(0)}\mathfrak{q}(\dot{\gamma}, (0, i\phi))\rangle \\ &\le C\big(\|\dot{b}\|_{L^2(0\times Y)} + \|\dot{\gamma}(0)\|_{L^2(0\times Y)}\big)\|\phi\|^2_{L^2_k(0\times Y)}\end{aligned}$$

where $\gamma = (B_\mathfrak{b} + b, s\phi)$, and L' is the derivative of L along the path $\check{\gamma}(t)$. The above proposition bounds both of these terms by the change in $\mathcal{L}$. □

13.5 Exponential decay

To deduce Theorem 13.3.5 from Proposition 13.4.1, we will need the following result, which says that the value of $\mathcal{L}$ on a trajectory belonging to a moduli space $M([\mathfrak{a}], [\mathfrak{b}])$ approaches its limiting value with exponential decay.

Proposition 13.5.1. *Let $\mathfrak{b} \in \mathcal{C}_k^\sigma(Y)$ be a non-degenerate critical point. Then there exists a $\delta > 0$ such that for every solution $\gamma^\tau \in \mathcal{C}_{k,\mathrm{loc}}^\tau$ of the perturbed Seiberg–Witten equations on $[0,\infty) \times Y$ with $\lim_\rightarrow [\gamma^\tau] = [\mathfrak{b}]$, there exists a t_0 such that for all $t \geq t_0$,*

$$\mathcal{L}(\gamma^\tau(t)) - \mathcal{L}(\mathfrak{b}) \leq Ce^{-\delta t}$$

where $C = \mathcal{L}(\gamma^\tau(t_0)) - \mathcal{L}(\mathfrak{b})$.

Proof. Let $\gamma = (A, \Phi)$ be the corresponding configuration in $\mathcal{C}_{k,\mathrm{loc}}([0,\infty)\times Y)$, and let $\beta = (B, \Psi)$ be the image of $\mathfrak{b}$ in $\mathcal{C}_k(Y)$. Observe that because $\mathcal{L}$ is C^2 on $\mathcal{C}_1(Y)$ with vanishing derivative at β, we have

$$|\mathcal{L}(\beta + w) - \mathcal{L}(\beta)| \leq C_1 \|w\|^2_{L^2_{1,B}}$$

for some C_1 and all $w \in T_\beta \mathcal{C}_1(Y)$ with $[\beta + w]$ in some L^2_1 neighborhood U_1 of $[\beta]$ in $\mathcal{B}_1(Y)$. Second, note that the non-degeneracy of the Hessian tells us that

$$\|(\mathrm{grad}\,\mathcal{L})(\beta + w)\|^2_{L^2} \geq C_2 \|w\|^2_{L^2_1}$$

for all w in $\mathcal{K}_{k,\beta}$ with $[\beta + w]$ in some L^2_1 neighborhood U_2 of $[\beta]$. The condition that $[\gamma^\tau]$ is asymptotic to $[\mathfrak{b}]$ means that we can assume that the path $\check{\gamma}(t)$ lies in the Coulomb slice $\mathcal{S}_{k,\beta} \subset \mathcal{C}_k(Y)$, and that for some t_0, the path lies in $U_1 \cap U_2$ for all $t \geq t_0$. For $t \geq t_0$ then, we have

$$\begin{aligned}
\frac{d}{dt}\mathcal{L} &= -\|\,\mathrm{grad}\,\mathcal{L}(\check{\gamma}(t))\|^2_{L^2} \\
&\leq -C_2 \|\check{\gamma}(t) - \beta\|^2_{L^2_1} \\
&\leq -(C_2/C_1)\big(\mathcal{L}(\check{\gamma}(t)) - \mathcal{L}(\beta)\big).
\end{aligned}$$

The function $f(t) = \mathcal{L}(\check{\gamma}(t)) - \mathcal{L}(\beta)$ thus satisfies a differential inequality $\dot{f} \leq -\delta f$ with $\delta = C_2/C_1$, for all $t \geq t_0$, from which it follows that $f(t) \leq f(t_0)e^{-\delta(t-t_0)}$ for $t \geq t_0$. □

For future use, we note that the same differential inequalities can be used to prove a related proposition.

Proposition 13.5.2. *Let $\mathfrak{b}$ be a non-degenerate critical point, as in the previous proposition, and let β be its image in $\mathcal{C}_k(Y)$. Then there exists a neighborhood*

U of $[\beta]$ and a $\delta > 0$ such that for any solution $\gamma \in \mathcal{C}_k([t_1, t_2] \times Y)$ for which the corresponding path $[\check{\gamma}(t)]$ is in U for all t in the interval, we have

$$-C_2 e^{\delta(t-t_2)} \leq \mathcal{L}(\check{\gamma}(t)) - \mathcal{L}(\beta) \leq C_1 e^{-\delta(t-t_1)} \tag{13.9}$$

where

$$C_1 = \left|\mathcal{L}(\check{\gamma}(t_1)) - \mathcal{L}(\beta)\right|$$
$$C_2 = \left|\mathcal{L}(\check{\gamma}(t_2)) - \mathcal{L}(\beta)\right|.$$

Proof. We take U to be as in the proof of the previous proposition. The inequalities there show that

$$\mathcal{L}(\check{\gamma}(t)) - \mathcal{L}(\beta) \leq C_1 e^{-\delta(t-t_1)}$$

on the maximal interval $[t_1, t']$ where the left-hand side is non-negative (possibly the empty set); on the complement of this interval, the inequality is trivially true because the left-hand side is decreasing and therefore everywhere negative. The other inequality is similar. □

The preceding proposition has as a corollary the following uniform bound which is important for the proof of compactness in the blown-up configuration space.

Corollary 13.5.3. *Given a constant η, there is a gauge-invariant neighborhood U of the constant solution γ_β in $\mathcal{C}_k([-1,1] \times Y)$ with the following property. Let $J \subset \mathbb{R}$ be any interval and $J' = J + [-1,1]$. If we have a solution $\gamma^\tau \in \mathcal{C}_k^\tau(J' \times Y)$ such that the translates $\tau_t^* \gamma^\tau$ all satisfy*

$$\pi\left(\tau_t^* \gamma^\tau\right)|_{[-1,1]\times Y} \in U,$$

then

$$\int_J \left(\frac{d\Lambda_{\mathfrak{q}}(\check{\gamma^\tau})}{dt}\right)^+ dt \leq \eta,$$

where f^+ denotes the positive part of the function f.

Proof. We may take it that $J = [-T, T]$ for some T. Given any η_0, we may choose U so that $|\mathcal{L}(\gamma(t)) - \mathcal{L}(\beta)| \leq \eta_0$ for all γ in U and all t in $[-1,1]$.

Applying the proposition above to the interval $J' = [-T-1, T+1]$ we learn that for any solution γ^τ on J',

$$\mathcal{L}\big(\gamma^\tau(t-1)\big) - \mathcal{L}\big(\gamma^\tau(t+1)\big) \leq 2\eta_0 e^{-\delta T}\cosh(\delta t)$$

for all t in J. As long as we have chosen U so that Proposition 13.4.7 is applicable, we deduce that for all t in J, there exists a gauge transformation u with

$$\int_{t-1/2}^{t+1/2} \|u\gamma(s) - \beta\|^2_{L^2_{k+1}(Y)}\, ds \leq C\eta_0 e^{-\delta T}\cosh(\delta t),$$

where $\gamma = \pi\gamma^\tau$. Because grad $\mathcal{L}$ is smooth and vanishes at β, we can pass to a smaller U again to deduce

$$\int_{t-1/2}^{t+1/2} \big\|\operatorname{grad}\mathcal{L}(\gamma(s))\big\|^2_{L^2_k(Y)}\, ds \leq C'\eta_0 e^{-\delta T}\cosh(\delta t),$$

and hence

$$\int_{t-1/2}^{t+1/2} \big\|\operatorname{grad}\mathcal{L}(\gamma(s))\big\|_{L^2_k(Y)}\, ds \leq \big(C'\eta_0 e^{-\delta T}\cosh(\delta t)\big)^{1/2}$$

by Cauchy–Schwarz. By integrating from $-T$ to T and throwing away the contribution from the two intervals of length $1/2$ at either end, we obtain

$$\int_{-T}^{T} \|\operatorname{grad}\mathcal{L}(\gamma(t))\|_{L^2_k(Y)}\, dt \leq \int_{-T}^{T} \big(C'\eta_0 e^{-\delta T}\cosh(\delta t)\big)^{1/2} dt \leq C''\eta_0/\delta.$$

The result now follows from Lemma 10.9.1, which bounds $d\Lambda_{\mathfrak{q}}/dt$ from above by a multiple of $\|\operatorname{grad}\mathcal{L}\|_{L^2_k(Y)}$. □

13.6 Asymptotics of solutions

Given the local results and exponential decay from the previous two subsections, we can deduce a global decay result which yields the proof of Theorem 13.3.5. We again write $\mathfrak{b}$ for a critical point in $\mathcal{C}^\sigma_k(Y)$, and write the corresponding translation-invariant configuration in $\mathcal{C}^\tau_k(\mathbb{R}\times Z)$ as

$$\gamma_{\mathfrak{b}} = (A_{\mathfrak{b}}, s_{\mathfrak{b}}, \phi_{\mathfrak{b}}).$$

Proposition 13.6.1. *Let γ^τ be a solution in $\mathcal{C}^\tau_{k,\mathrm{loc}}$ of the perturbed Seiberg–Witten equations on $[0,\infty) \times Y$ and suppose that γ^τ is asymptotic to the non-degenerate critical point $[\mathfrak{b}]$. Then there is a 4-dimensional gauge transformation u in $\mathcal{G}_{k+1,\mathrm{loc}}$ on $[0,\infty) \times Y$ such that*

$$u(\gamma^\tau) - \gamma_{\mathfrak{b}} \in L^2_{k,A_{\mathfrak{b}}}([0,\infty) \times Y).$$

Proof. We begin with a lemma.

Lemma 13.6.2. *The function $\left|(d/dt)\Lambda_{\mathfrak{q}}(t)\right|$ is integrable on $[0,\infty)$.*

Proof. On the one hand, $\Lambda_{\mathfrak{q}}(t)$ approaches a limit $\Lambda_{\mathfrak{q}}(\mathfrak{b})$ as $t \to \infty$. On the other hand $(d/dt)\Lambda_{\mathfrak{q}}(t)$ (without the absolute value signs) is bounded above by an integrable function, by Corollary 13.4.8 and the exponential decay of $\mathcal{L}(t) - \mathcal{L}(\mathfrak{b})$ (Proposition 13.5.1). □

To prove the proposition, consider the sequence of cylinders $[i-1, i+1] \times Y$. From Proposition 13.4.1 we see that there is an i_0 such that for all $i \geq i_0$ we can find a sequence of gauge transformations

$$u_i \in \mathcal{G}_{k+1}\left(\left[i - \frac{3}{4}, i + \frac{3}{4}\right] \times Y\right)$$

such that if we write

$$u_i(\gamma^\tau) = \gamma_{\mathfrak{b}} + w_i$$

we have

$$\|w_i\|_{L^2_{k,A_{\mathfrak{b}}}([i-\frac{3}{4}, i+\frac{3}{4}] \times Y)} \leq N_i \tag{13.10}$$

where

$$N_i = C\big((\Lambda_{\mathfrak{q}}(i-1) - \Lambda_{\mathfrak{q}}(i+1)) + (\mathcal{L}(i-1) - \mathcal{L}(i+1))^{1/2}\big).$$

The preceding lemma and the exponential decay together tell us that $N_i \to 0$ as $i \to \infty$. Consider two adjacent intervals. The gauge transformation between the representatives, $v_i = u_{i+1}u_i^{-1}$ (defined on $[i + \frac{1}{4}, i + \frac{3}{4}] \times Y$), solves the equation

$$dv_i = v_i(a_i - a_{i+1}),$$

where a_i is the 1-form component of $w_i = (a_i, r_i, \psi_i)$. Using the fact that $|v_i| = 1$ we deduce that for some constant C_1 independent of a_i, a_{i+1} we have

$$\begin{aligned}
\|dv_i\|^2_{L^2_k([i+\frac{1}{4},i+\frac{3}{4}]\times Y)} &\leq C_1\|a_i - a_{i+1}\|^2_{L^2_k([i+\frac{1}{4},i+\frac{3}{4}]\times Y)} \\
&\leq 2C_1\|a_i\|^2_{L^2_k([i+\frac{1}{4},i+\frac{3}{4}]\times Y)} + 2C_1\|a_{i+1}\|^2_{L^2_k([i+\frac{1}{4},i+\frac{3}{4}]\times Y)} \\
&\leq C_2\big(\pounds(i+1) - \pounds(i-1)\big) + C_2\big(\pounds(i+2) - \pounds(i)\big) \\
&= 2C_2\big(\pounds(i+2) - \pounds(i-1)\big)
\end{aligned}$$

using Proposition 13.4.7 and (13.10). The terms $\pounds(i+2)-\pounds(i-1)$ are approaching zero, so for i sufficiently large, we can assume that v_i is homotopic to the identity; so we can write

$$v_i = e^{\sqrt{-1}f_i}$$

for some function f_i with $\|df_i\|_{L^2_k([i+\frac{1}{4},i+\frac{3}{4}]\times Y)}$ small.

We also have

$$v_i(\phi_{\mathfrak{b}} + \psi_i) = \phi_{\mathfrak{b}} + \psi_{i+1}.$$

Hence, using Proposition 13.4.1 in a similar way to the calculation above, we obtain (in the worse of the two cases)

$$\begin{aligned}
|v_i(i+1/2, y) - 1|^2 &\leq C_3\big((\Lambda_{\mathfrak{q}}(i+2) - \Lambda_{\mathfrak{q}}(i-1)d) \\
&\qquad + (\pounds(i+2) - \pounds(i-1))^{1/2}\big) \\
&\stackrel{\text{def}}{=} C_3\tilde{N}_i.
\end{aligned}$$

We now know that dv_i (and hence df_i) is small in L^2_k norm. Thus by adjusting f_i by a suitable multiple of 2π we can arrange that

$$\|f_i\|^2_{L^2_{k+1}} \leq C_4\tilde{N}_i.$$

Let $\mu_i(t)$ be a function which is 0 for $t \leq i + \frac{3}{8}$ and 1 for $t \geq i + \frac{5}{8}$. Set

$$\tilde{u}_i = e^{i\mu_i f_i} u_i$$

on $[i - \frac{3}{4}, i + \frac{3}{4}] \times Y$. Then $\tilde{u}_i$ agrees with $\tilde{u}_{i+1}$ on the interval $[i + \frac{5}{8}, i + \frac{3}{4}] \times Y$, so we can patch together to give a global gauge transformation u. From the estimates on the f_i and (13.10) it is easy to see that

$$\|u(\gamma^\tau) - \gamma_{\mathfrak{b}}\|^2_{L^2_{k,B}([1,\infty)\times Y)} \leq C_5 \sum_{j=1}^{\infty} \tilde{N}_i,$$

which is finite as required, by Lemma 13.6.2 and Proposition 13.5.1. □

14 Local structure of moduli spaces

In the previous section, we introduced the moduli spaces of trajectories $M_z([\mathfrak{a}],[\mathfrak{b}])$ for critical points $[\mathfrak{a}]$, $[\mathfrak{b}]$ of the flow. In Theorem 13.3.5, we saw that $M_z([\mathfrak{a}],[\mathfrak{b}])$ could be regarded as a subspace either of the configuration space $\mathcal{B}^\tau_{k,\text{loc}}(Z)$ modelled on the $L^2_{k,\text{loc}}$ spaces, or of the configuration space $\mathcal{B}^\tau_k([\mathfrak{a}],[\mathfrak{b}])$ modelled on $L^2_k(Z)$. The present section is devoted to studying the local structure of these moduli spaces. As is usual with such moduli problems, we aim to describe $M_z([\mathfrak{a}],[\mathfrak{b}])$ locally as the zero set of a non-linear Fredholm map between Banach manifolds. Our first task therefore is to review the Fredholm theory for elliptic operators on an infinite cylinder.

14.1 Translationally invariant operators on cylinders

We consider again the situation in Subsection 12.2, in which L_0 is a first-order self-adjoint elliptic differential operator acting on sections of a vector bundle $E \to Y$ and h is an operator on sections of E,

$$h : C^\infty(Y;E) \to L^2(Y;E)$$

(not necessarily symmetric), which extends to a bounded map on $L^2_j(Y;E)$ for all j in a range $|j| \leq k$. We called such an operator $L_0 + h$ a k-ASAFOE operator. We now pull back E to the cylinder $Z = \mathbb{R} \times Y$ and consider there the translation-invariant operator on Z,

$$D = \frac{d}{dt} + L_0 + h. \tag{14.1}$$

This is a bounded operator $L^2_{j+1}(Z;E) \to L^2_j(Z;E)$ for all j in the same range.

Henceforth, when talking of the spectrum of an operator on a real Hilbert space, we shall always mean the spectrum of its *complexification*. With this in mind we make the following definition.

Definition 14.1.1. We say that a k-ASAFOE operator $L_0 + h$, as above, is *hyperbolic* if the spectrum of $L_0 + h$ is disjoint from the imaginary axis. ◊

In this setting, we have the following result (*cf.* [8, 68]):

Proposition 14.1.2. *If $L_0 + h$ is hyperbolic, then for all j in the range $|j| \leq k$, the operator*

$$D = \frac{d}{dt} + L_0 + h : L^2_{j+1}(Z;E) \to L^2_j(Z;E)$$

is invertible, and in particular Fredholm.

Remark. Conversely, if the spectrum of $L_0 + h$ meets the imaginary axis, the operator is not Fredholm, because its range fails to be closed, as the interested reader can check.

Proof of Proposition 14.1.2 We will prove the proposition first in the case that $j \geq 0$ and E is a complex vector bundle. Consider the Fourier transform in the t-variable:

$$\hat{u}(\xi, y) = \int_{-\infty}^{\infty} u(t,y)e^{-it\xi}\,dt.$$

Define the norm:

$$\|\hat{u}\|^2_{F_j} = \sum_{i=0}^{j} \int_{-\infty}^{\infty} |\xi|^{2(j-i)} \|\hat{u}\|^2_{L^2_i(\xi \times Y)} d\xi.$$

Define F_j to be the completion $C^\infty_c(\mathbb{R} \times Y; E)$ with respect to this norm. Parseval's theorem implies the Fourier transform in the t variable induces an isomorphism between $L^2_k(\mathbb{R} \times Y)$ and F_k. The Fourier transform takes D to the operator

$$\hat{D} = L_0 + h + i\xi$$

and thus the proposition is equivalent to the statement that

$$\hat{D} : F_j \to F_{j-1}$$

is invertible provided that $L_0 + h$ is hyperbolic.

Lemma 14.1.3. *If $L_0 + h$ is a k-ASAFOE operator, then there are positive constants $C > 0$ and $\xi_0 > 0$ such that for any ξ with $|\xi| \geq \xi_0$ and any j with $0 \leq j \leq k$, the operator*

$$L_0 + h + i\xi : L^2_j(Y;E) \to L^2_{j-1}(Y;E)$$

is invertible and the inverse satisfies

$$\|(L_0 + h + i\xi)^{-1}\|_{\mathrm{Op}(L^2_{j-1}, L^2_j)} \leq C$$

and

$$\|(L_0 + h + i\xi)^{-1}\|_{\mathrm{Op}(L^2_{j-1}, L^2_{j-1})} \leq \frac{C}{|\xi|}.$$

Proof. We start with the case that $h = 0$. Using the spectral theorem and the ellipticity we can decompose a general smooth section as an L^2-convergent sum of eigenvectors of L_0:

$$\phi = \sum_{\lambda \in \mathrm{Spec}(L_0)} \phi_\lambda.$$

By adding a real multiple of the identity operator we can assume that L_0 is invertible. Then the L^2_j norm of ϕ is equivalent to

$$\left(\sum_\lambda |\lambda|^{2j} \|\phi_\lambda\|^2_{L^2} \right)^{\frac{1}{2}}.$$

Thus if $\xi \neq 0$,

$$\begin{aligned}
\|(L_0 + i\xi)^{-1}\phi)\|_{L^2_j} &= \sum_\lambda |\lambda|^{2j} \left\| \left(\frac{1}{\lambda + i\xi} \right) \phi_\lambda \right\|^2_{L^2} \\
&\leq \sup_\lambda \left(\frac{\lambda^2}{\lambda^2 + \xi^2} \right) \sum_\lambda |\lambda|^{2j-2} \|\phi_\lambda\|^2_{L^2} \\
&\leq \sup_\lambda \left(\frac{\lambda^2}{\lambda^2 + \xi^2} \right) \|\phi\|_{L^2_{j-1}}.
\end{aligned}$$

It follows that the operator norm of $(L_0 + i\xi)^{-1}$, as an operator from L^2_{j-1} to L^2_j, is bounded above by

$$\sup_{\lambda \in \mathrm{Spec}(L_0)} \frac{|\lambda|}{\sqrt{\lambda^2 + \xi^2}} = 1.$$

Similarly we have that the L^2_{j-1} to L^2_{j-1} operator norm of $(L_0+i\xi)^{-1}$ is bounded above by

$$\sup_{\lambda \in \mathrm{Spec}(L_0)} \frac{1}{\sqrt{\lambda^2 + \xi^2}} = \frac{1}{\sqrt{\lambda_0^2 + \xi^2}}$$

where λ_0 is the eigenvalue with absolute value closest to zero.

Thus in the case that $h = 0$ we can take ξ_0 to be any positive constant, since for $\xi \neq 0$ the operator $L_0 + i\xi$ is certainly invertible and the inverse satisfies the above estimate.

In the general case, choose $\xi_0 > 0$ so that for any ξ with $|\xi| \geq \xi_0$ we have

$$\|(L_0 + i\xi)^{-1}\|_{\mathrm{Op}(L^2_j, L^2_j)} \|h\|_{\mathrm{Op}(L^2_j, L^2_j)} \leq 1/2.$$

Then the operator

$$(L_0 + i\xi)^{-1}(L_0 + h + i\xi) = 1 + (L_0 + i\xi)^{-1} h,$$

viewed as mapping $L^2_j \to L^2_j$, is invertible and the norm of its inverse is at most 2, independent of ξ. Let G_ξ be this inverse. Then

$$H_\xi = G_\xi \circ (L_0 + i\xi)^{-1}$$

is the required inverse for $(L_0 + h + i\xi)$, satisfying the same estimates as $(L_0 + i\xi)^{-1}$ for $\xi \geq |\xi_0|$, but with different constants. □

If $L_0 + h$ is hyperbolic, then Corollary 12.2.3 tells us that for all positive j and all $\xi \in \mathbb{R}$, the operator

$$L_0 + h + i\xi : L^2_j(Y; E) \to L^2_{j-1}(Y; E)$$

is invertible. Then Lemma 14.1.3 implies that there is a constant C such that

$$\|(L_0 + h + i\xi)^{-1}\|_{\mathrm{Op}(L^2_{j-1}, L^2_j)} \leq C,$$

now for all $\xi \in \mathbb{R}$, and (for example)

$$\|(L_0 + h + i\xi)^{-1}\|_{\mathrm{Op}(L^2,L^2)} \leq C(1 + |\xi|^2)^{-1/2}.$$

With these estimates in place it follows that

$$\begin{aligned}
\|\hat{D}^{-1}\hat{u}\|_{F_j} &\leq \sum_{i=0}^{j} \int_{-\infty}^{\infty} |\xi|^{2(j-i)} \|(L_0 + h + i\xi)^{-1}\hat{u}\|^2_{L^2_i} \\
&\leq C \sum_{i=1}^{j} \int_{-\infty}^{\infty} |\xi|^{2(j-i)} \|\hat{u}\|^2_{L^2_{i-1}} + \int_{-\infty}^{\infty} |\xi|^{2j} \|(L_0 + h + i\xi)^{-1}\hat{u}\|^2_{L^2} \\
&\leq C \sum_{i=0}^{j-1} \int_{-\infty}^{\infty} |\xi|^{2(j-1-i)} \|\hat{u}\|^2_{L^2_i} + C' \int_{-\infty}^{\infty} |\xi|^{2j} (1 + |\xi|^2)^{-1} \|\hat{u}\|^2_{L^2} \\
&\leq C'' \|\hat{u}\|_{F_{j-1}}.
\end{aligned}$$

So $\hat{D}^{-1}$ extends to define a bounded linear map from F_{j-1} to F_j.

If E is a real vector bundle the Fourier transform maps real sections of E to sections of $E \otimes \mathbb{C}$ which satisfy the symmetry:

$$\overline{s(t, y)} = s(-t, y).$$

We may now repeat the above proof with sections that satisfy this property.

Now suppose that j is negative. The operator

$$\frac{d}{dt} + L_0 + h : L^2_{j+1}(Z; E) \to L^2_j(Z; E)$$

is invertible if and only if its adjoint is. Its adjoint is the operator:

$$-\frac{d}{dt} + L_0 + h^* : L^2_{-j}(Z; E) \to L^2_{-j-1}(Z; E)$$

where h^* is the adjoint of h with respect to the L^2 inner product. By assumption, h and h^* have the same mapping properties; and the spectrum of $L_0 + h$ is the conjugate of that of $L_0 + h^*$, so the condition of having spectrum disjoint from the imaginary axis is also satisfied for the adjoint. Thus the proof of invertibility also applies to the adjoint. □

14.2 Spectral flow and Fredholm theory on cylinders

Using Proposition 14.1.2, we can next analyze the case that h is time-dependent. We state the result for an operator from L^2_1 to L^2. (Our main interest in this proposition is in the calculation of the index, which in practice will be insensitive to our choice of Sobolev regularity.) For the index of the operators we encounter, we need to recall the notion of *spectral flow*.

Suppose we are given a family of operators

$$L_0 + h_t, \quad t \in [0, 1],$$

where L_0 is a first-order, self-adjoint elliptic operator, and h_t is a continuous path in the space of bounded operators on L^2. Suppose also that $L_0 + h_0$ and $L_0 + h_1$ are hyperbolic. Then the spectral flow, denoted $\mathrm{sf}\,(L_0 + h_t)$, counts the net number of eigenvalues whose real parts go from *negative* to *positive*. To give a more precise definition, first deform the path h_t so that it is smooth over $(0, 1)$, and consider the set

$$\begin{aligned} S &= \{\, (t, \lambda) \mid \lambda \in \mathrm{Spec}(L_0 + h_t) \,\} \\ &\subset (0, 1) \times \mathbb{C}. \end{aligned}$$

According to Lemma 12.2.4, for each t, the spectrum is discrete and the generalized eigenspaces are finite-dimensional. Let us say that (t, λ) is a *simple* point of S if the generalized λ-eigenspace of $L_0 + h_t$ is 1-dimensional. At simple points, S is a smooth 1-manifold, on which t is a local coordinate; we use the coordinate t to orient S at such points. In the space of bounded operators on L^2, the set of those h for which the spectrum of $L_0 + h$ has a non-simple eigenvalue lying on the imaginary axis is a locally finite union of submanifolds of codimension at least 2, as follows from Lemma 12.2.4. (See the similar argument in Subsection 12.6.) The path h_t can therefore be moved so that the intersection of S with $(0, 1) \times i\mathbb{R}$ consists entirely of simple points; and any two such paths can be joined by a homotopy of paths with the same property. If h_t is so chosen, the set S has a well-defined intersection number with $(0, 1) \times i\mathbb{R}$, near which it is a smooth oriented manifold. This intersection number is the spectral flow. The definition makes clear that, in the present context, the spectral flow depends only on the endpoints of the path.

Proposition 14.2.1. *Let L_0 be a first-order, self-adjoint elliptic operator acting on sections of a vector bundle $E \to Y$, and let h_t be a time-dependent bounded operator on $L^2(Y; E)$, varying continuously in the operator norm topology and equal to a constant $h_\pm$ on the ends. Suppose $L_0 + h_\pm$ are hyperbolic. Then* *the operator*

$$Q = \frac{d}{dt} + L_0 + h_t : L^2_1(Z;E) \to L^2(Z;E)$$

is Fredholm and has index equal to the spectral flow of the path of operators $L_0 + h_t$.

Proof. The time-dependent operators $(d/dt) + L_0 + h_\pm$ are invertible on the cylinder by Proposition 14.1.2, so let

$$\begin{aligned} G_1 &: L^2(Z;E) \to L^2_1(Z;E) \\ G_2 &: L^2(Z;E) \to L^2_1(Z;E) \end{aligned} \tag{14.2}$$

be inverses for $(d/dt) + L_0 + h_-$ and $(d/dt) + L_0 + h_+$. Let $1 = \eta_1 + \eta_2$ be a partition of unity subordinate to a covering of Z by the two half-infinite cylinders $(-\infty, 1) \times Y$ and $(-1, \infty) \times Y$. Let γ_1 be a function which is 1 on the support of η_1 and vanish where $t \geq 2$, and let γ_2 be 1 on the support of η_2 and vanishes where $t \leq -2$. Define

$$P : L^2(Z;E) \to L^2_1(Z;E) \tag{14.3}$$

by

$$Pe = \sum_{i=1,2} \gamma_i G_i \eta_i e.$$

We compute

$$QPe = e + \sum_{i=1,2} \big(\dot{\gamma}_i G_i \eta_i e + \gamma_i (h_t - h_i) G_i \eta_i e\big). \tag{14.4}$$

The operator $QP - I$ is therefore compact as an operator on L^2, as follows from the compactness of the map $L^2_1(Z) \to L^2([-2,2] \times Y)$. Similarly, $PQ - I$ is compact as an operator on L^2_1. It follows that Q is Fredholm.

To compute the index, we may assume the spectrum of $L_0 + h_\pm$ is simple, by a perturbation of the path, and that the real parts of the eigenvalues are distinct. Then we can find a finite-dimensional subspace $C \subset L^2_1(Y;E)$ spanned by generalized eigenvectors of $L_0 + h_-$, and a real scalar g such that $L_0 + h_- + g\Pi_C$ is hyperbolic and the spectral flow from $L_0 + h_+$ to $L_0 + h_- + g\Pi_C$ is zero. Here Π_C is the L^2 orthogonal projection to C.

We claim there is a path L_0+k_s ($s \in [0,1]$) joining L_0+h_+ to $L_0+h_-+g\Pi_C$ such that L_0+k_s is hyperbolic for all s. To see that this is so, choose an initial path $L_0+\tilde{k}_s$ with the property that the spectrum is simple for all s. The eigenvalues $\lambda_n(s)$ of $L_0+\tilde{k}_s$ are a family of paths in $\mathbb{C}$, only finitely many of which cross the imaginary axis. The generalized eigenvectors belonging to this finite collection of eigenvalues span a continuously varying subspace $A_s \subset L^2_1(Y;E)$. Let $B_s \subset L^2_1(Y;E)$ be the orthogonal complement with respect to the L^2 inner product. We now seek the path k_s such that $k_s|_{B_s} = \tilde{k}_s|_{B_s}$ and such that L_0+k_s leaves A_s invariant. This reduces the problem to that of finding a path of finite-dimensional hyperbolic operators joining two hyperbolic operators having the same number of eigenvalues in the left half-plane.

Using the path L_0+k_s from L_0+h_+ to $L_0+h_-+g\Pi_C$, we can now construct a homotopy of the original path L_0+h_t to a new path $L_0+\bar{h}_t$ whose endpoints are L_0+h_- and $L_0+h_-+g\Pi_C$. During this homotopy, the endpoints of the path remain hyperbolic, so we have a family of Fredholm operators on the cylinder, whose index is therefore constant. At the end of the homotopy, we may then take it that the entire path has the form $L_0+h_-+g_t\Pi_C$ for some path of scalars g_t. The operator

$$(d/dt) + L_0 + h_- + g_t\Pi_C : L^2_1(Z;E) \to L^2(Z;E)$$

now has a block triangular form, in which one block is a finite-dimensional ordinary differential operator

$$(d/dt) + M + g_t : L^2_1(\mathbb{R};C) \to L^2(\mathbb{R};C),$$

and the other diagonal block is invertible. Because of the simplifying condition that M has simple spectrum, we are reduced to computing the index of a scalar operator on $\mathbb{R}$ of the form

$$u \mapsto \frac{d}{dt}u + f_t u$$

where f_t approaches a non-imaginary constant on each end. The solution of this ODE is in L^2 if and only if $\mathrm{Re}(f_t)$ is negative at $-\infty$ and positive at $+\infty$. □

There is a special situation in which the index of the operator Q on the cylinder $Z = \mathbb{R} \times Y$ can be reinterpreted in a way that is useful for calculation. Suppose that the operators L_0+h_- and L_0+h_+ are conjugate by some smooth, unitary automorphism $u : E \to E$ of the bundle $E \to Y$: that is, we suppose

$$L_0 + h_+ = u \circ (L_0 + h_-) \circ u^{-1}$$

as operators on $L^2_k(Y;E)$. Let I be a closed interval containing $[0,1]$ in its interior, so that the path h_t is constant near the two endpoints of I, and let $E_u \to S^1 \times Y$ be the bundle obtained from $I \times E$ by gluing the two ends together using u. The operator Q on the cylinder then descends to a well-defined operator

$$Q_u : L^2_1(S^1 \times Y; E_u) \to L^2(S^1 \times Y; E_u).$$

This operator is a compact perturbation of an elliptic operator, and is therefore Fredholm, because $S^1 \times Y$ is compact. The point to be made here, however, is that the *index* of Q_u coincides with the index of the operator Q on the infinite cylinder:

Proposition 14.2.2. *When $L_0 + h_-$ and $L_0 + h_+$ are conjugate by a gauge transformation u as above, then the index of the operator Q on the cylinder is equal to the index of the operator Q_u on $S^1 \times Y$.*

Proof. The proposition can be viewed as a special case of an excision theorem for the index. The strategy which was used to prove Proposition 14.2.1 (reducing the operator to a sum of 1-dimensional ODEs) is not applicable to this problem, but a suitable technique will be introduced later, in Subsection 17.1, which allows one to reinterpret the index of the operator Q on the infinite cylinder as the index of a boundary-value problem on the finite cylinder $[0,1]\times Y$, with spectral boundary conditions. Proposition 14.2.2 can then be proved by regarding the $S^1 \times Y$ as the union of two pieces: the finite cylinder $[0,1] \times Y$, and a piece on which the operator is constant. The proposition reduces to showing that the index of the operator on $S^1 \times Y$ is the sum of the indices of the spectral boundary-value problems on the two pieces. A proof of this additivity can be based on a simple homotopy argument, a version of which appears in Subsection 20.3: see in particular the remark following Definition 20.3.2. □

The usefulness of this proposition is that, unlike the spectral flow, the index of the operator Q_u is usually easy to calculate in terms of characteristic classes, by the index theorem.

14.3 Slices for the gauge-group action

We continue to write Z for the infinite cylinder $\mathbb{R} \times Y$, and take up the task of showing that the quotient configuration space $\mathcal{B}^\tau_k(\mathfrak{a}, \mathfrak{b})$ is a closed subset of a Hilbert manifold. We carry over the constructions of Subsection 9.4, in which we treated the case of the finite cylinder $I \times Y$. We write $\mathcal{T}^\tau_j$ for the L^2_j

completion of the tangent bundle of $\tilde{\mathcal{C}}_k^\tau(\mathfrak{a},\mathfrak{b})$. At $\gamma = (A_0, s_0, \phi_0)$, the fiber is

$$\begin{aligned}\mathcal{T}_{j,\gamma}^\tau &= \{\, (a,s,\phi) \mid \operatorname{Re}\langle \phi_0|_t, \phi|_t\rangle_{L^2(Y)} = 0 \text{ for all } t\,\} \\ &\subset L^2_j(Z; iT^*Z) \oplus L^2_j(\mathbb{R};\mathbb{R}) \oplus L^2_{j,A_0}(Z; S^+).\end{aligned} \tag{14.5}$$

We write

$$\mathbf{d}^\tau : \operatorname{Lie}(\mathcal{G}_{j+1}(Z)) \times \tilde{\mathcal{C}}_k^\tau(\mathfrak{a},\mathfrak{b}) \to \mathcal{T}_j^\tau$$

for the derivative of the action of the gauge-group, regarded now as a bundle map over $\tilde{\mathcal{C}}_k^\tau(\mathfrak{a},\mathfrak{b})$:

$$\mathbf{d}_\gamma^\tau \xi = (-d\xi, 0, \xi\phi_0).$$

To define slices for the gauge-group action, we adapt Definition 9.4.2. (There is no longer a boundary condition.)

Definition 14.3.1. For a configuration $\gamma = (A_0, s_0, \phi_0)$ in $\tilde{\mathcal{C}}_k^\tau(\mathfrak{a},\mathfrak{b})$, we define $\mathcal{S}_{k,\gamma}^\tau$ to be the set of triples $(A, s, \phi) \in \tilde{\mathcal{C}}_k^\tau(\mathfrak{a},\mathfrak{b})$ satisfying

$$-d^*a + iss_0 \operatorname{Re}\langle i\phi_0, \phi\rangle + i|\phi_0|^2 \operatorname{Re}\mu_Y(\langle i\phi_0, \phi\rangle) = 0, \tag{14.6}$$

where $A = A_0 + a \otimes 1$. We write $\operatorname{Coul}_\gamma^\tau$ for the smooth map defined by the left-hand side of this equation:

$$\operatorname{Coul}_\gamma^\tau : \tilde{\mathcal{C}}_k^\tau(\mathfrak{a},\mathfrak{b}) \to L^2_{k-1}(Z; i\mathbb{R}).$$

◊

At this stage, we do not know that this definition really provides a slice for the gauge-group action at γ: this will be shown shortly.

The linearization of $\operatorname{Coul}_\gamma^\tau$ at γ extends to an operator

$$\mathbf{d}_\gamma^{\tau,\dagger} : \mathcal{T}_j^\tau \to L^2_{j-1}(Z; i\mathbb{R})$$

for all j, given by

$$\mathbf{d}_\gamma^{\tau,\dagger}(a,s,\phi) = -d^*a + is_0^2 \operatorname{Re}\langle i\phi_0, \phi\rangle + i|\phi_0|^2 \operatorname{Re}\mu_Y\langle i\phi_0, \phi\rangle$$

(*cf.* (9.16)). For $\gamma \in \tilde{\mathcal{C}}_k^\tau(\mathfrak{a},\mathfrak{b})$, we define

$$\mathcal{K}_{j,\gamma}^\tau \subset \mathcal{T}_{j,\gamma}^\tau \tag{14.7}$$

to be the kernel of $\mathbf{d}_\gamma^{\tau,\dagger}$. We write

$$\mathcal{J}_{j,\gamma}^\tau \subset \mathcal{T}_{j,\gamma}^\tau$$

for the image of $\mathbf{d}_\gamma^\tau$. The next proposition is a version of Proposition 9.4.1, now in the context of the infinite cylinder Z.

Proposition 14.3.2. *The operator $\mathbf{d}_\gamma^\tau$ is injective with closed range, and $\mathbf{d}_\gamma^{\tau,\dagger}$ is surjective. The subspaces $\mathcal{J}_{j,\gamma}^\tau$ and $\mathcal{K}_{j,\gamma}^\tau$ define complementary closed subbundles of $\mathcal{T}_j^\tau \to \tilde{\mathcal{C}}_k^\tau(\mathfrak{a},\mathfrak{b})$, and we have a smooth bundle decomposition*

$$\mathcal{T}_j^\tau = \mathcal{J}_j^\tau \oplus \mathcal{K}_j^\tau.$$

Furthermore,

$$\mathcal{J}_j^\tau = \mathcal{J}_0^\tau \cap \mathcal{T}_j^\tau.$$

Remark. In particular, the case $j = k$ in this theorem tells us that

$$T\tilde{\mathcal{C}}_k^\tau(\mathfrak{a},\mathfrak{b}) = \mathcal{J}_k^\tau \oplus \mathcal{K}_k^\tau.$$

Proof of Proposition 14.3.2. The space $\mathcal{K}_{j,\gamma}^\tau$ is closed because it is defined as the kernel of the operator $\mathbf{d}_\gamma^{\tau,\dagger}$. To see that $\mathcal{J}_{j,\gamma}^\tau$ is closed, we recall first that it is defined as the image of

$$\begin{aligned}\mathbf{d}_\gamma^\tau : L^2_{j+1}(Z; i\mathbb{R}) &\to \mathcal{T}_j^\tau(Z)\\ \xi &\mapsto (-d\xi, 0, \xi\phi_0).\end{aligned}$$

We have

$$\begin{aligned}\|\mathbf{d}_\gamma^\tau \xi\|^2_{L^2_j} &= \|d\xi\|^2_{L^2_j} + \|\xi\phi_0\|^2_{L^2_j}\\ &\geq \frac{1}{2}\|d\xi\|^2_{L^2_j} + \frac{1}{2}\int_{\mathbb{R}}\left(\|d_Y\xi(t)\|^2_{L^2_j(Y)} + \|\xi(t)\phi_0(t)\|^2_{L^2_j(Y)}\right)dt\\ &\geq \frac{1}{2}\|d\xi\|^2_{L^2_j} + C\|\xi\|^2_{L^2}\\ &\geq C'\|\xi\|^2_{L^2_{j+1}},\end{aligned}$$

which shows that the image is closed.

Consider now the operator

$$\mathbf{d}_\gamma^{\tau,\dagger}\mathbf{d}_\gamma^\tau : L^2_{j+1}(Z; i\mathbb{R}) \to L^2_{j-1}(Z; i\mathbb{R}).$$

As in the proof of Proposition 9.4.1, to show that we have a direct sum, we must prove that this operator is an isomorphism. The operator is given by

$$\xi \mapsto \Delta\xi + (\mathrm{vol}(Y))^{-1}|\phi_0|^2\mu_Y(|\phi_0|^2\xi) + s_0^2|\phi_0|^2\xi,$$

and the injectivity of this operator can be deduced by a straightforward integration by parts. From this injectivity, it follows that $\mathcal{J}^\tau_{j,\gamma} \cap \mathcal{K}^\tau_{j,\gamma}$ is zero. The closedness of the image can be proved by an argument similar to the one we have just used to prove that the image of $\mathbf{d}^\tau_\gamma$ is closed. Finally, the surjectivity of this second-order operator follows from its injectivity, because the operator is symmetric and has the regularity property,

$$\mathbf{d}_\gamma^{\tau,\dagger}\mathbf{d}_\gamma^\tau\xi \in L^2_{-j+1} \implies \xi \in L^2_{-j+3}.$$

□

The above proposition tells us that $\mathcal{S}^\tau_{k,\gamma}$ is a Hilbert submanifold of $\tilde{\mathcal{C}}^\tau_k(\mathfrak{a}, \mathfrak{b})$ in a neighborhood of γ. We write $\iota : \mathcal{S}^\tau_{k,\gamma} \to \tilde{\mathcal{C}}^\tau_k(\mathfrak{a}, \mathfrak{b})$ for the inclusion, and

$$\bar{\iota} : \mathcal{S}^\tau_{k,\gamma} \to \tilde{\mathcal{B}}^\tau_k([\mathfrak{a}], [\mathfrak{b}])$$

for the restriction of the quotient map. Then we have, as in previous cases:

Proposition 14.3.3. *The quotient space $\tilde{\mathcal{B}}^\tau_k([\mathfrak{a}], [\mathfrak{b}])$ is a Hilbert manifold. For each $\gamma \in \mathcal{C}^\tau_k(\mathfrak{a}, \mathfrak{b})$, there is an open neighborhood of γ in $\mathcal{S}^\tau_{k,\gamma}$,*

$$\mathcal{U}_\gamma \subset \tilde{\mathcal{S}}^\tau_{k,\gamma},$$

such that $\bar{\iota} : \mathcal{U}_\gamma \to \tilde{\mathcal{B}}^\tau_k([\mathfrak{a}], [\mathfrak{b}])$ is a diffeomorphism onto its image, which is an open neighborhood of $[\gamma] \in \mathcal{B}^\tau_k([\mathfrak{a}], [\mathfrak{b}])$. □

14.4 The linearized equations

The local structure of the moduli spaces $M_z([\mathfrak{a}], [\mathfrak{b}])$ can be studied following standard lines. In the previous subsection, we found suitable slices for the action of the gauge-group $\mathcal{G}_{k+1}(Z)$ on $\tilde{\mathcal{C}}^\tau_k(\mathfrak{a}, \mathfrak{b})$. Now we need to verify that, restricted to the slice, the equations defining the moduli space have Fredholm linearization. We will assume that the perturbation has been chosen so that the critical points are non-degenerate.

As in (9.18), we define $\mathcal{V}_j^\tau \to \tilde{\mathcal{C}}_k^\tau(\mathfrak{a}, \mathfrak{b})$ to be the vector bundle whose fiber over $\gamma = (A_0, s_0, \phi_0)$ is the vector space

$$\begin{aligned} \mathcal{V}_{j,\gamma}^\tau &= \{ (\eta, r, \psi) \mid \mathrm{Re}\langle \check{\phi}_0(t), \check{\psi}(t) \rangle_{L^2(Y)} = 0 \text{ for all } t \} \\ &\subset L_j^2(Z; i\,\mathfrak{su}(S^+)) \oplus L_j^2(\mathbb{R}; \mathbb{R}) \oplus L_{j,A_0}^2(Z; S^-). \end{aligned} \tag{14.8}$$

In the case of a *compact* cylinder $Z = I \times Y$, we wrote the perturbed 4-dimensional Seiberg–Witten equations as

$$\mathfrak{F}_\mathfrak{q}^\tau(\gamma) = 0,$$

where $\mathfrak{F}_\mathfrak{q}^\tau$ was a smooth section of the vector bundle $\mathcal{V}_{k-1}^\tau \to \mathcal{C}_k^\tau(Z)$. The same applies to the infinite cylinder $\mathbb{R} \times Y$: we only have to check that the required estimates hold in the global Sobolev spaces on the cylinder. The following lemma illustrates this:

Lemma 14.4.1. *The perturbed Seiberg–Witten equations define a smooth section*

$$\mathfrak{F}_\mathfrak{q}^\tau = \mathfrak{F}^\tau + \hat{\mathfrak{q}}^\tau : \tilde{\mathcal{C}}_k^\tau(\mathfrak{a}, \mathfrak{b}) \to \mathcal{V}_{k-1}^\tau.$$

Proof. The fact that the perturbations have these properties for the case of a *finite* cylinder is part of the definition of "tame". The lemma can be deduced from the results for the finite cylinder rather as the Sobolev embedding and multiplication theorems were proved in Subsection 13.2.

We illustrate the argument by showing that $\mathfrak{F}_\mathfrak{q}^\tau$ is a continuous section of $\mathcal{V}_{k-1}^\tau$. Let γ belong to $\mathcal{C}_k^\tau(\mathfrak{a}, \mathfrak{b})$. Much as in the proofs of the results of Subsection 13.2, let

$$\gamma_n \in \mathcal{C}_k^\tau([-1, 1] \times Y)$$

be the restriction of the translation $\tau_n^*(\gamma)$. Let $\gamma_\mathfrak{a}$ and $\gamma_\mathfrak{b}$ temporarily denote the configurations on $[-1, 1] \times Y$ corresponding to these two critical points. The fact that γ belongs to $\mathcal{C}_k^\tau(\mathfrak{a}, \mathfrak{b})$ (rather than just the $L_{k,\mathrm{loc}}^2$ version) is equivalent to saying that the two quantities

$$\sum_{n \ge 0} \|\gamma_n - \gamma_\mathfrak{b}\|_{L_k^2([-1,1]\times Y)}^2, \qquad \sum_{n \le 0} \|\gamma_n - \gamma_\mathfrak{a}\|_{L_k^2([-1,1]\times Y)}^2$$

are finite. For n positive, because γ_n converges to $\gamma_{\mathfrak{b}}$ and $\mathfrak{F}^\tau_{\mathfrak{q}}$ is continuously differentiable in the case of the finite cylinder, there is a constant C (depending on γ) such that

$$\|\mathfrak{F}^\tau_{\mathfrak{q}}(\gamma_n) - \mathfrak{F}^\tau_{\mathfrak{q}}(\gamma_{\mathfrak{b}})\|^2_{L^2_{k-1}([-1,1]\times Y)} \leq C\|\gamma_n - \gamma_{\mathfrak{b}}\|^2_{L^2_k([-1,1]\times Y)}$$

for all $n \geq 0$, with a similar result for n negative. So the sum

$$\sum_{n\geq 0} \|\mathfrak{F}^\tau_{\mathfrak{q}}(\gamma_n)\|^2_{L^2_{k-1}([-1,1]\times Y)}$$

is finite. This, together with the corresponding result for negative n, shows that $\mathfrak{F}^\tau_{\mathfrak{q}}(\gamma)$ is in $\mathcal{V}^\tau_{k-1}$ for the infinite cylinder. The proof of continuity proceeds similarly. □

Our moduli space $M_z([\mathfrak{a}], [\mathfrak{b}])$ is the quotient of the locus $\mathfrak{F}^\tau_{\mathfrak{q}}(\gamma) = 0$ in $\mathcal{C}^\tau_k(\mathfrak{a}, \mathfrak{b})$ by the action of the gauge-group, and to understand its local structure, we need to understand the derivative of $\mathfrak{F}^\tau_{\mathfrak{q}}$. Because $\mathcal{V}^\tau_{k-1}$ is not a trivial vector bundle, the definition of the derivative as a bundle map

$$\mathcal{D}\mathfrak{F}^\tau_{\mathfrak{q}} : \mathcal{T}^\tau_k \to \mathcal{V}^\tau_{k-1}$$

involves a projection: referring to (14.8), we define a projection

$$\Pi^\tau_\gamma : L^2_j(Z; i\,\mathfrak{su}(S^+)) \oplus L^2_j(\mathbb{R}; \mathbb{R}) \oplus L^2_{j,A_0}(Z; S^-) \to \mathcal{V}^\tau_{j,\gamma}$$

by applying the L^2 projection on each slice $\{t\} \times Y$, That is,

$$\Pi^\tau_\gamma : (\eta, r, \psi) \mapsto (\eta, r, \Pi^\perp_{\phi_0(t)}\psi)$$
$$\Pi^\perp_{\phi_0(t)}\psi = \psi - \mathrm{Re}\langle\check{\phi}_0(t), \psi(t)\rangle_{L^2(Y)}\phi_0.$$

The derivative $\mathcal{D}\mathfrak{F}^\tau_{\mathfrak{q}}$ is then defined as the derivative in the ambient Hilbert space, followed by the projection Π^τ_γ. Because of Condition (iii) in the definition of a tame perturbation (Definition 10.5.1), the derivative $\mathcal{D}\mathfrak{F}^\tau_{\mathfrak{q}}$ extends to the spaces of lower regularity, defining smooth bundle maps

$$\mathcal{D}\mathfrak{F}^\tau_{\mathfrak{q}} : \mathcal{T}^\tau_j \to \mathcal{V}^\tau_{j-1} \tag{14.9}$$

for $j \in [-k, k]$.

We will examine the operator $\mathcal{D}\mathfrak{F}^\tau_{\mathfrak{q}}$ in some detail. Let γ_0 belong to $\mathcal{C}^\tau_k(\mathfrak{a}, \mathfrak{b})$ and consider the operator at γ_0:

$$\mathcal{D}\mathfrak{F}^\tau_{\mathfrak{q}} : \mathcal{T}^\tau_{j,\gamma_0} \to \mathcal{V}^\tau_{j-1,\gamma_0}. \tag{14.10}$$

Let $\check{\gamma}_0(t) = (B_0(t), r_0(t), \phi_0(t))$ be the path $\mathbb{R} \to \mathcal{C}^\sigma_{k-1}(Y)$ defined by γ_0. If we temporarily ignore questions of regularity, we can regard elements of the codomain here as sections along $\check{\gamma}_0$ of the tangent bundle $\mathcal{T}^\sigma \to \mathcal{C}^\sigma(Y)$: this is because we can use Clifford multiplication as usual to identify the endomorphism η of the spin bundle S with a 1-form on Y, so that elements (η, r, ψ) of $\mathcal{V}^\tau$ become triples (b, r, ψ), with ψ orthogonal to the spinor $\phi_0(t)$ on each t slice.

The domain of the operator (14.10) can be interpreted similarly. The domain consists of triples (a, r, ψ), where a is a 4-dimensional connection on the cylinder and r and ψ are as before. We can write a as

$$a = b + c\,dt$$

where b is in temporal gauge and c is an imaginary-valued function on the cylinder. In this way, a becomes a pair of (b, c), consisting of a path b in the space of 1-forms on Y and a path c in the space of 0-forms on Y. The triple (b, r, ψ) is the same data we saw in the codomain. So, continuing to ignore the Sobolev regularity, we can regard the domain $\mathcal{T}^\tau_{\gamma_0}$ as sections along γ_0 of the bundle

$$\mathcal{T}^\sigma(Y) \oplus L^2(Y, i\mathbb{R})$$

where (b, r, ψ) defines the path in the first summand and c defines the path in the second. We will write such a section as (V, c), where $V(t) = (b(t), r(t), \psi(t))$ defines an element of $\mathcal{T}^\sigma_{\gamma_0(t)}(Y)$, and $c(t)$ is in $L^2(Y, i\mathbb{R})$. Because $\mathcal{T}^\sigma(Y)$ is not a trivial vector bundle along the path, we should not write dV/dt for the derivative of V along the path. Instead, we again use the projection $\Pi^\perp$ to the orthogonal complement of $\phi_0(t)$ to define a covariant derivative: if $V = (b, r, \psi)$, we set

$$\frac{D}{dt}V = \left(\frac{db}{dt}, \frac{dr}{dt}, \Pi^\perp_{\phi_0(t)} \frac{d\psi}{dt} \right). \tag{14.11}$$

With this notation in place, we can write down the linear operator (14.9). Recall that we have a vector field $(\operatorname{grad} \mathcal{L})^\sigma$ on $\mathcal{C}^\sigma_k(Y)$, whose derivative defines a bundle map

$$\mathcal{D}(\operatorname{grad} \mathcal{L})^\sigma : \mathcal{T}^\sigma_j(Y) \to \mathcal{T}^\sigma_{j-1}(Y).$$

Let us write an element of the domain of (14.9) as (V, c), where $V = (b, r, \psi)$ is a section of $\mathcal{T}^\sigma(Y)$ along γ_0. Let us also suppose that γ_0 is in temporal gauge on the cylinder. Then the operator (14.9) is given by

$$(V,c) \mapsto \frac{D}{dt}V + \mathcal{D}(\operatorname{grad} \mathcal{L})^\sigma(V) + \mathbf{d}^\sigma_{\gamma_0(t)} c.$$

Here $\mathbf{d}^\sigma$ is defined by the derivative of the gauge group action on $\mathcal{C}^\sigma(Y)$, as in (9.10). It is a bundle map on $\mathcal{C}^\sigma(Y)$ from the trivial bundle $L^2_j(Y, i\mathbb{R})$ to $\mathcal{T}^\sigma_{j-1}(Y)$.

If we restrict the operator to pairs (V,c) with $c = 0$, we see just the operator

$$V \mapsto \frac{D}{dt}V + \mathcal{D}(\operatorname{grad} \mathcal{L})^\sigma(V).$$

which is familiar as the linearization of the equation for a flow line of the vector field $-(\operatorname{grad} \mathcal{L})^\sigma$ on $\mathcal{C}^\sigma(Y)$. Imposing the condition $c = 0$ is a gauge-fixing condition, but it is not the type of gauge-fixing condition which leads to an elliptic system of equations. Instead, we need to impose a Coulomb-type gauge-fixing condition, namely the linearization of the Coulomb slice condition introduced in the previous subsection. That is, we impose the condition

$$\mathbf{d}^{\tau,\dagger}_{\gamma_0}(V,c) = 0.$$

When we regard V and c as sections of Hilbert vector bundles along the path γ_0 (and continuing to suppose that γ_0 is in temporal gauge), the condition becomes

$$\frac{d}{dt}c + \mathbf{d}^{\sigma,\dagger}_{\gamma_0(t)}(V) = 0,$$

where $\mathbf{d}^{\sigma,\dagger}$ is the linearized gauge-fixing operator on Y defined at (12.10). An appropriate linear operator for the elliptic theory is thus the sum of the linearized equations (14.9) with the gauge-fixing condition $\mathbf{d}^{\tau,\dagger}_{\gamma_0} = 0$, as an operator acting on Sobolev spaces on the infinite cylinder:

$$\begin{aligned} Q_{\gamma_0} &= \mathcal{D}_{\gamma_0}(\mathfrak{F}^\tau_{\mathfrak{q}}) \oplus \mathbf{d}^{\tau,\dagger}_{\gamma_0} \\ Q_{\gamma_0} : \mathcal{T}^\tau_{j,\gamma_0} &\to \mathcal{V}^\tau_{j-1,\gamma_0} \oplus L^2_{j-1}(Z; i\mathbb{R}). \end{aligned} \tag{14.12}$$

In the path notation, this combined operator has the form

$$(V,c) \mapsto \frac{D}{dt}(V,c) + L_{\gamma_0(t)}(V,c)$$

where for each $\mathfrak{a}$ in $\mathcal{C}^\sigma_k(Y)$ we write

$$L_{\mathfrak{a}} : \mathcal{T}^\sigma_{j,\mathfrak{a}} \oplus L^2_j(Y; i\mathbb{R}) \to \mathcal{T}^\sigma_{j-1,\mathfrak{a}} \oplus L^2_{j-1}(Y; i\mathbb{R})$$

for the linear operator given by

$$L_{\mathfrak{a}} = \begin{bmatrix} \mathcal{D}_{\mathfrak{a}}(\operatorname{grad} \mathcal{L})^{\sigma} & \mathbf{d}^{\sigma}_{\mathfrak{a}} \\ \mathbf{d}^{\sigma,\dagger}_{\mathfrak{a}} & 0 \end{bmatrix}.$$

This operator has appeared before. It is precisely the *extended Hessian* on the blown-up configuration space: the operator $\widehat{\operatorname{Hess}}^{\sigma}_{\mathfrak{q},\mathfrak{a}}$ defined at (12.11).

To summarize, the linearization of the perturbed Seiberg–Witten equations on the infinite cylinder can be combined with a linearized gauge-fixing condition to obtain an equation which can be cast in the form

$$\frac{D}{dt}(V,c) + L_{\gamma(t)}(V,c) = 0.$$

We are now ready to state and prove the basic Fredholm property of the linearized equations on the infinite cylinder $Z = \mathbb{R} \times Y$.

Theorem 14.4.2. *For each pair of (non-degenerate) critical points $\mathfrak{a}$, $\mathfrak{b}$ and each γ_0 in $\mathcal{C}^{\tau}_k(\mathfrak{a},\mathfrak{b})$, the linear operator*

$$Q_{\gamma_0} = \mathcal{D}_{\gamma_0}\mathfrak{F}^{\tau}_{\mathfrak{q}} \oplus \mathbf{d}^{\tau,\dagger}_{\gamma_0} : \mathcal{T}^{\tau}_{j,\gamma_0} \to \mathcal{V}^{\tau}_{j-1,\gamma_0} \oplus L^2_{j-1}(Z; i\mathbb{R})$$

is Fredholm for all j in the range $1 \le j \le k$, and satisfies a Gårding inequality,

$$\|u\|_{L^2_j} \le C_1 \|Q_{\gamma_0} u\|_{L^2_{j-1}} + C_2 \|u\|_{L^2_{j-1}}.$$

The index of Q_{γ_0} is independent of j.

Proof. Suppose first that γ_0 is in temporal gauge and that γ_0 is translation-invariant on each of the two ends of the cylinder, so that $\gamma_0(t) = \mathfrak{a}$ for t large and negative, and $\gamma_0(t) = \mathfrak{b}$ for t large and positive. Write the path $\gamma_0(t)$ as $(B_0(t), r_0(t), \phi_0(t))$. We have seen above that the operator can be cast in the form

$$(V,c) \mapsto \frac{D}{dt}(V,c) + L_{\gamma_0(t)}(V,c). \tag{14.13}$$

This is almost exactly the form which was considered in the general setting of Proposition 14.2.1, but we have to deal with the fact that $(V(t), c(t))$ is a section along the path $\gamma_0(t)$ of a Hilbert vector bundle, and is therefore not a section of a fixed vector bundle over Y. The same point arose earlier, in Subsection 12.4, and we can use the same device here to avoid the problem: as in (12.13) we recast $V = (b, r, \psi)$ as a pair $(b, \boldsymbol{\psi})$ by setting

$$\boldsymbol{\psi}(t) = \psi(t) + r(t)\phi_0(t).$$

The operator then takes the form

$$(b, \boldsymbol{\psi}, c) \mapsto \frac{d}{dt}(b, \boldsymbol{\psi}, c) + L_0(b, \boldsymbol{\psi}, c) + \bar{h}_{\gamma_0(t)}(b, \boldsymbol{\psi}, c) \tag{14.14}$$

where L_0 is a self-adjoint elliptic differential operator acting on sections of $iT^*Y \oplus S \oplus i\mathbb{R}$ on Y, just as in (12.17). (The operator $\bar{h}$ in this formula does not quite coincide with the one that appears in (12.17), because the term $(d/dt)\boldsymbol{\psi}$ picks up also a time derivative of ϕ_0.)

We have already seen that $L_{\gamma_0(t)}$ is the extended Hessian, $\widehat{\mathrm{Hess}}^{\sigma}_{\mathfrak{q},\gamma_0(t)}$, so on the two ends of the cylinder we see the operators $\widehat{\mathrm{Hess}}^{\sigma}_{\mathfrak{q},\mathfrak{a}}$ and $\widehat{\mathrm{Hess}}^{\sigma}_{\mathfrak{q},\mathfrak{b}}$. Our examination of this operator in Lemma 12.4.3 showed that $\widehat{\mathrm{Hess}}^{\sigma}_{\mathfrak{q},\mathfrak{a}}$ and $\widehat{\mathrm{Hess}}^{\sigma}_{\mathfrak{q},\mathfrak{b}}$ are invertible with real eigenvalues when $\mathfrak{a}$ and $\mathfrak{b}$ are non-degenerate critical points. In particular, they are hyperbolic.

Proposition 14.2.1 therefore tells us that (14.13) is Fredholm as a map

$$L^2_1(Z; iT^*Y \oplus S \oplus i\mathbb{R}) \to L^2(Z; iT^*Y \oplus S \oplus i\mathbb{R}).$$

In other words, the operator Q_{γ_0} is Fredholm in the case $j = 1$. To deal with the case $j \geq 1$, we note that the operators $Q_{\gamma_{\mathfrak{a}}}$ and $Q_{\gamma_{\mathfrak{b}}}$ corresponding to the constant trajectories are invertible in the topologies $L^2_j \to L^2_{j-1}$, by Proposition14.1.2; so in the proof of Proposition 14.2.1, the Green's operators G_i map L^2_{j-1} to L^2_j, as does the parametrix P. The Fredholm property of Q_{γ_0} on L^2_j follows.

The Gårding inequality is a straightforward consequence of the mapping properties of h_γ and the elliptic estimate for $(d/dt) + L_0$ on a finite cylinder. From this follows the regularity statement, that if $u \in L^2_1$ and $Q_\gamma u \in L^2_{j-1}$, then $u \in L^2_j$. This in turn implies that the index is independent of j. This completes the proof of Theorem 14.4.2, under the additional assumption that γ_0 is in temporal gauge and constant on the ends.

If γ_0 is altered on a compact subset $I \times Y$, then Q_{γ_0} changes by a compact operator, when regarded as an operator acting on sections $(b, \boldsymbol{\psi}, c)$ of a fixed vector bundle. The subset of $\mathcal{C}^\tau_k(\mathfrak{a}, \mathfrak{b})$ consisting of elements γ which are equal to γ_0 on the ends is dense; and since Q_γ depends continuously on γ it follows that the difference $Q_{\gamma_0} - Q_\gamma$ is compact, for all γ in $\mathcal{C}^\tau_k(\mathfrak{a}, \mathfrak{b})$ (because the compact operators are a closed set in the operator norm topology). The theorem therefore holds without the additional conditions on γ_0. □

Combining Theorem 14.4.2 with Proposition 14.3.2, we deduce:

Proposition 14.4.3. *The restriction of the bundle map $\mathcal{D}\mathfrak{F}^\tau_{\mathfrak{q}}$,*

$$\mathcal{D}\mathfrak{F}^\tau_{\mathfrak{q}} : \mathcal{K}^\tau_{j,\gamma} \to \mathcal{V}^\tau_{j-1,\gamma},$$

is Fredholm, and has the same index as Q_γ.

Proof. The space $\mathcal{K}^\tau_{j,\gamma}$ is, by definition, the kernel of $\mathbf{d}^{\tau,\dagger}_\gamma$. So the kernel of $\mathcal{D}\mathfrak{F}^\tau_{\mathfrak{q}}$ restricted to $\mathcal{K}^\tau_{j,\gamma}$ is the same as the kernel of the sum $\mathbf{d}^{\tau,\dagger}_\gamma \oplus \mathcal{D}\mathfrak{F}^\tau_{\mathfrak{q}}$, which is the operator Q_γ. The operators have the same cokernel because of the surjectivity of $\mathbf{d}^{\tau,\dagger}_\gamma$, which was established in Proposition 14.3.2. □

The proof of Theorem 14.4.2 shows that the index of Q_γ depends only on the endpoints, $\mathfrak{a}$ and $\mathfrak{b}$, of γ, and is equal to the spectral flow of the family of extended Hessians $\widehat{\text{Hess}}^\sigma_{\mathfrak{q},\check{\gamma}(t)}$.

We give this important quantity a name:

Definition 14.4.4. Given critical points $\mathfrak{a}$, $\mathfrak{b}$ in $\mathcal{C}^\sigma_k(Y)$, we write $\text{gr}(\mathfrak{a}, \mathfrak{b})$ for the index of Q_γ, where γ is any chosen element of $\mathcal{C}^\tau_k(\mathfrak{a}, \mathfrak{b})$. If $[\mathfrak{a}]$, $[\mathfrak{b}]$ are the gauge orbits of $\mathfrak{a}$, $\mathfrak{b}$, and z is the relative homotopy class of the path $\pi \circ \check{\gamma}$, we may write $\text{gr}_z([\mathfrak{a}], [\mathfrak{b}])$ for $\text{gr}(\mathfrak{a}, \mathfrak{b})$. We call $\text{gr}(\mathfrak{a}, \mathfrak{b})$ the *relative grading* of $\mathfrak{a}$ and $\mathfrak{b}$. ◊

From Proposition 14.2.1 and the additivity of the spectral flow, the following proposition is immediate.

Proposition 14.4.5. *If $\mathfrak{a}$, $\mathfrak{b}$ and $\mathfrak{c}$ are three critical points, then*

$$\text{gr}(\mathfrak{a}, \mathfrak{c}) = \text{gr}(\mathfrak{a}, \mathfrak{b}) + \text{gr}(\mathfrak{b}, \mathfrak{c}).$$

□

It is natural to ask how the relative grading $\text{gr}_z([\mathfrak{a}], [\mathfrak{b}])$ depends on the choice of relative homotopy class z for the path. Because of the additivity, this question reduces to the question of computing $\text{gr}_z([\mathfrak{a}], [\mathfrak{a}])$ for a closed loop z based at $[\mathfrak{a}]$. Expressing this in terms of $\mathcal{C}^\sigma_k(Y)$ rather than the quotient space $\mathcal{B}^\sigma_k(Y)$, we can ask for a calculation of $\text{gr}(\mathfrak{a}, \mathfrak{a}')$ when $\mathfrak{a}' = u(\mathfrak{a})$ for some gauge transformation $u : Y \to S^1$; this is an equivalent question, because to each u there corresponds a closed loop z_u in the quotient space, obtained as the image of any path joining $\mathfrak{a}$ to $\mathfrak{a}'$ in $\mathcal{C}_k^\sigma(Y)$. The result of this calculation is given in the next lemma:

Lemma 14.4.6. *For the closed loop z_u based at $[\mathfrak{a}]$ in $\mathcal{B}_k^\sigma(Y)$, we have*

$$\mathrm{gr}_{z_u}([\mathfrak{a}],[\mathfrak{a}]) = ([u] \cup c_1(S))[Y],$$

where $[u]$ denotes the homotopy class of $u : Y \to S^1$, identified with an element of $H^1(Y;\mathbb{Z})$.

Proof. We can use Proposition 14.2.2 to reinterpret the index $\mathrm{gr}_{z_u}([\mathfrak{a}],[\mathfrak{a}])$ as the index of an operator on $S^1 \times Y$. The relevant operator is the linearization of the perturbed 4-dimensional Seiberg–Witten equations together with gauge fixing, and the relevant spin^c structure on $S^1 \times Y$ is the structure $\mathfrak{s}_u$ obtained by pulling back the spin^c structure $\mathfrak{s}$ from Y and using u to identify the two ends. For this spin^c structure $\mathfrak{s}_u$, we have

$$c_1(\mathfrak{s}_u) = c_1(\mathfrak{s}) + 2\eta \cup [u],$$

where η is the oriented generator of $H^1(S^1;\mathbb{Z})$. The index of the 4-dimensional Seiberg–Witten equations (with gauge fixing) on a closed 4-manifold is given in Theorem 1.4.4. In the simple case of $S^1 \times Y$, the formula becomes

$$\begin{aligned} d &= \frac{1}{4} c_1(\mathfrak{s}_u)^2 [S^1 \times Y] \\ &= ([u] \cup c_1(\mathfrak{s}))[Y] \end{aligned}$$

as claimed. □

If we think of the finite-dimensional model discussed in Section 2, then we should think of $\mathrm{gr}_z([\mathfrak{a}],[\mathfrak{b}])$ as playing the role of the difference in Morse indices for two hyperbolic critical points of a flow. The fact that this quantity depends on the homotopy class of the path z joining $[\mathfrak{a}]$ to $[\mathfrak{b}]$ in $\mathcal{B}^\sigma(Y)$ has no analogy in the finite-dimensional case, essentially because the spectral flow for a loop of finite-dimensional operators is always zero. There is another point to make, to clarify the analogy with the finite-dimensional situation. For a flow on a finite-dimensional manifold B, we can indeed interpret the difference of Morse indices between two hyperbolic critical points a and b in terms of a spectral flow, namely the spectral flow of the "Hessian" (any covariant derivative of the vector field) along a path from a to b. The analog of that Hessian in our setting is the operator

$$\mathrm{Hess}_j^\sigma : \mathcal{K}_j^\sigma \to \mathcal{K}_{j-1}^\sigma$$

acting on the subbundle $\mathcal{K}_j^\sigma$ transverse to the gauge orbit. The definition of $\mathrm{gr}(\mathfrak{a}, \mathfrak{b})$, however, relates it to the spectral flow of the *extended* Hessian (12.12). Along the path, however, the terms x and y in (12.12) are compact operators which vanish at the endpoints; so the family of extended Hessians has the same spectral flow as the family of operators

$$\begin{pmatrix} 0 & \mathbf{d}^\sigma_{\gamma(t)} \\ \mathbf{d}^{\sigma,\dagger}_{\gamma(t)} & 0 \end{pmatrix} \oplus \mathrm{Hess}^\sigma_{\mathfrak{q},\gamma(t)}.$$

The first block is invertible, as we saw in the proof of Lemma 12.4.3. So we can regard $\mathrm{gr}(\mathfrak{a}, \mathfrak{b})$ as the spectral flow of the family of operators $\mathrm{Hess}^\sigma_{\mathfrak{q},\gamma(t)}$, as expected.

14.5 Regularity and boundary-obstructed trajectories

We now turn to the moduli spaces $\tilde{M}_z([\mathfrak{a}], [\mathfrak{b}]) \supset M_z([\mathfrak{a}], [\mathfrak{b}])$, which we identify with subsets of $\tilde{\mathcal{B}}^\tau_{k,z}([\mathfrak{a}], [\mathfrak{b}])$ using Theorem 13.3.5. Recall that the former notation (with the tilde) denotes the space obtained by dropping the condition $s \geq 0$ in the construction of the blow-up. The slice result, Proposition 14.3.3, together with the Fredholm property of the linearized equations on the slice (Proposition 14.4.3), tells us that if $[\gamma]$ belongs to the moduli space $\tilde{M}_z([\mathfrak{a}], [\mathfrak{b}])$, then a neighborhood of $[\gamma]$ in $\tilde{M}_z([\mathfrak{a}], [\mathfrak{b}])$ is identified with the zero set of a map

$$\mathfrak{F}^\tau_\mathfrak{q}|_{\mathcal{U}_\gamma} : \mathcal{U}_\gamma \to \mathcal{V}^\tau_{k-1},$$

whose linearization at $\gamma \in \mathcal{U}_\gamma$ is the Fredholm operator

$$\mathcal{D}_\gamma \mathfrak{F}^\tau_\mathfrak{q} : \mathcal{K}^\tau_{k,\gamma} \to \mathcal{V}^\tau_{k-1,\gamma}$$

which was described just above in Proposition 14.4.3. From this local description, it follows now by the implicit function theorem in Banach spaces, that if $\mathcal{D}_\gamma \mathfrak{F}^\tau_\mathfrak{q}$ is *surjective*, then the moduli space $\tilde{M}_z([\mathfrak{a}], [\mathfrak{b}])$ is a smooth manifold near $[\gamma]$. In this surjective case, the dimension of the moduli space will be the dimension of the kernel of the linearization, which is the *index* of $\mathcal{D}_\gamma \mathfrak{F}^\tau_\mathfrak{q}$. This is also the index of Q_γ, by Proposition 14.4.3 again, so we can state:

Corollary 14.5.1. *If* $[\gamma]$ *belongs to* $\tilde{M}_z([\mathfrak{a}], [\mathfrak{b}])$ *and the linear operator* Q_γ *is surjective, then* $\tilde{M}_z([\mathfrak{a}], [\mathfrak{b}])$ *is a smooth manifold in a neighborhood of* $[\gamma]$*, and its dimension is given by the index of* Q_γ*: that is to say, the quantity* $\mathrm{gr}_z([\mathfrak{a}], [\mathfrak{b}])$. □

It is important to note, however, that it is not reasonable to expect that the derivative $\mathcal{D}_\gamma \mathfrak{F}_\mathfrak{q}^\tau$ will always be surjective, for any residual set of perturbations. To see why this is so, let us write $\gamma = (A, s, \phi)$ and separate two possibilities. Either $s : \mathbb{R} \to \mathbb{R}$ is identically zero, or it is everywhere non-zero; that is, the solution is either reducible or irreducible. In the latter case, if s is everywhere positive, then $[\gamma]$ belongs to $M_z([\mathfrak{a}], [\mathfrak{b}]) \subset \tilde{M}_z([\mathfrak{a}], [\mathfrak{b}])$. So we can identify $M_z([\mathfrak{a}], [\mathfrak{b}])$ as the quotient of $\tilde{M}_z([\mathfrak{a}], [\mathfrak{b}])$ by the involution

$$\mathbf{i} : [A, s, \phi] \mapsto [A, -s, \phi]$$

(see (9.7)). As in the finite-dimensional case (see Definition 2.4.1), we can distinguish two types of critical points in the reducible case:

Definition 14.5.2. If $\mathfrak{a} \in \mathcal{C}_k^\sigma(Y)$ belongs to the reducible locus, we call $\mathfrak{a}$ *boundary-stable* if $\Lambda_\mathfrak{q}(\mathfrak{a}) > 0$ and *boundary-unstable* if $\Lambda_\mathfrak{q}(\mathfrak{a}) < 0$. ◊

Remark. For now, we are only interested in applying this definition to *critical* points $\mathfrak{a}$ of the perturbed functional. Later, it will sometimes be convenient to have this concept also for general points on the reducible locus (in which case it is possible that $\Lambda_\mathfrak{q}(\mathfrak{a}) = 0$, so that $\mathfrak{a}$ is neither boundary-stable nor-unstable).

Lemma 14.5.3. *Let $\mathfrak{a}$, $\mathfrak{b}$ be critical points, and suppose $M_z([\mathfrak{a}], [\mathfrak{b}])$ contains an irreducible trajectory $[\gamma] = [A, s, \phi]$. Then $\mathfrak{a}$ is either irreducible or boundary-unstable, and $\mathfrak{b}$ is either irreducible or boundary-stable.*

Proof. The equation for s is

$$\frac{ds}{dt} = -\Lambda_\mathfrak{q}(\check{\gamma}(t))s$$

where $\Lambda_\mathfrak{q}(\check{\gamma}(t))$ approaches $\Lambda_\mathfrak{q}(\mathfrak{a})$ and $\Lambda_\mathfrak{q}(\mathfrak{b})$ as t goes to $-\infty$ and $+\infty$. If $\mathfrak{a}$ is reducible, then s approaches 0 at $-\infty$, so $\Lambda_\mathfrak{q}(\mathfrak{a})$ must be positive. Similarly $\Lambda_\mathfrak{q}(\mathfrak{b})$ must be negative if $\mathfrak{b}$ is reducible. □

If γ is a reducible trajectory (connecting, therefore, two reducible critical points), then the operator Q_γ of Theorem 14.4.2 decomposes as a sum of two operators, the "boundary" and "normal" parts,

$$Q_\gamma = Q_\gamma^\partial \oplus Q_\gamma^\nu, \tag{14.15}$$

reflecting the decomposition of the domain and codomain into the parts where the involution $\mathbf{i}$ is trivial and non-trivial. The first operator is

$$Q_\gamma^\partial = (\mathcal{D}_\gamma \mathfrak{F}_\mathfrak{q}^\tau)^\partial \oplus \mathbf{d}_\gamma^{\tau,\dagger},$$

where $(\mathcal{D}_\gamma \mathfrak{F}^\tau_\mathfrak{q})^\partial : (\mathcal{T}^\tau_{k,\gamma})^\partial \to (\mathcal{V}^\tau_{k,\gamma})^\partial$ is the part invariant under the involution, and Q^ν_γ is the operator

$$\begin{aligned} Q^\nu_\gamma &: L^2_k(\mathbb{R}; i\mathbb{R}) \to L^2_{k-1}(\mathbb{R}; i\mathbb{R}) \\ Q^\nu_\gamma(s) &= (ds/dt) + \Lambda_\mathfrak{q}(\check{\gamma})s. \end{aligned} \tag{14.16}$$

Lemma 14.5.4. *The dimensions of the kernel and cokernel of Q^ν_γ are:*

(i) 1 *and* 0 *if* $\mathfrak{a}$, $\mathfrak{b}$ *are boundary-unstable and -stable respectively;*
(ii) 0 *and* 1 *if* $\mathfrak{a}$, $\mathfrak{b}$ *are boundary-stable and -unstable respectively;*
(iii) 0 *and* 0 *in the remaining cases, that* $\mathfrak{a}$ *and* $\mathfrak{b}$ *are either both boundary-unstable or both boundary-stable.*

Proof. This is straightforward. □

From this lemma, we see that the cokernels of Q_γ and Q^∂_γ are the same for reducible trajectories γ, except in one case. We label this case:

Definition 14.5.5. We say that a moduli space $M([\mathfrak{a}], [\mathfrak{b}])$ is *boundary-obstructed* if $\mathfrak{a}$, $\mathfrak{b}$ are both reducible, $\mathfrak{a}$ is boundary-stable, and $\mathfrak{b}$ is boundary-unstable. ◊

In the boundary-obstructed case, the cokernel of Q_γ (or equivalently the cokernel of the operator $\mathcal{D}_\gamma \mathfrak{F}^\tau_\mathfrak{q}$) is at least 1-dimensional, so we cannot expect the linearized equations to be surjective. We make the following definition:

Definition 14.5.6. Let $[\gamma]$ be a solution in $M_z([\mathfrak{a}], [\mathfrak{b}])$. If the moduli space is not boundary-obstructed, we say that γ is *regular* if Q_γ is surjective. In the boundary-obstructed case, γ must be reducible (see below), and we say γ is *regular* if Q^∂_γ is surjective. We say that $M_z([\mathfrak{a}], [\mathfrak{b}])$ is regular if all its elements are regular. ◊

Note that in the boundary-obstructed case, regularity means that the cokernel of Q_γ has dimension 1. The following proposition is an expansion of Corollary 14.5.1, incorporating now all the various possibilities.

Proposition 14.5.7. *Suppose the moduli space $M_z([\mathfrak{a}], [\mathfrak{b}])$ is regular, and let $d = \mathrm{gr}_z([\mathfrak{a}], [\mathfrak{b}])$. Then the moduli space is:*

- *a smooth d-manifold consisting entirely of irreducible solutions if either $\mathfrak{a}$ or $\mathfrak{b}$ is irreducible;*
- *a smooth d-manifold with boundary if $\mathfrak{a}$, $\mathfrak{b}$ are boundary-unstable and -stable respectively;*

- *a smooth d-manifold consisting entirely of reducibles if $\mathfrak{a}$, $\mathfrak{b}$ are either both boundary-stable or both boundary-unstable;*
- *a smooth $(d+1)$-manifold consisting entirely of reducibles in the boundary-obstructed case.*

In the second case, the boundary of the moduli space consists of the reducible elements.

Proof. In all but the last case, regularity is equivalent to the surjectivity of Q_γ, which is equivalent to the surjectivity of $\mathcal{D}_\gamma \mathfrak{F}^\tau_{\mathfrak{q}}$. In these cases, the larger moduli space $\tilde{M}_z([\mathfrak{a}],[\mathfrak{b}])$ is a manifold of dimension d. The moduli space $M_z([\mathfrak{a}],[\mathfrak{b}])$ is the quotient of this by the involution $\mathbf{i}$, and the three cases correspond to the situation that the involution has no fixed points, a fixed submanifold of codimension 1, or is trivial. The fixed-point set cannot be of codimension larger than 1 because the vectors normal to the fixed manifold are the kernel of Q^ν_γ. In the last case, the moduli space is contained in the reducibles and is cut out by an equation whose linearization is Q^∂_γ, which is surjective and has index $d+1$, because Q^ν_γ has index -1 in this case. □

14.6 The simplest moduli spaces

We continue to assume that the perturbation has been chosen so that the critical points in the blown-up configuration space are all non-degenerate. Suppose now that $\mathfrak{a}_1$ and $\mathfrak{a}_2$ are two *reducible* critical points in $\mathcal{C}^\sigma_k(Y)$ whose images under the blow-down map are the same:

$$\pi(\mathfrak{a}_1) = \pi(\mathfrak{a}_2) = \alpha \in \mathcal{C}_k(Y).$$

If we write α as (B,Ψ) on Y, then according to Proposition 12.2.5 and Proposition 10.3.1, the critical points $\mathfrak{a}_1$ and $\mathfrak{a}_2$ correspond to eigenvalues of the perturbed Dirac operator $D_{\mathfrak{q},B}$, a complex-linear self-adjoint operator with simple spectrum. These eigenvalues can be recovered as the value of $\Lambda_\mathfrak{q}$ at the critical points:

$$\lambda_1 = \Lambda_\mathfrak{q}(\mathfrak{a}_1), \quad \lambda_2 = \Lambda_\mathfrak{q}(\mathfrak{a}_2).$$

Let z_0 be the homotopy class of paths joining $[\mathfrak{a}_1]$ to $[\mathfrak{a}_2]$ in $\mathcal{B}^\sigma_k(Y)$ arising from paths joining $\mathfrak{a}_1$ to $\mathfrak{a}_2$ in $\mathcal{C}^\sigma_k(Y)$. We can analyze the moduli spaces $M_{z_0}([\mathfrak{a}_1],[\mathfrak{a}_2])$ quite explicitly, guided by the finite-dimensional case of linear flows on projective spaces discussed in Subsection 2.3.

Let i be the "signed count" of the number of eigenvalues of $D_{\mathfrak{q},B}$ between λ_1 and λ_2, including one endpoint:

$$i = \begin{cases} |\{\, \mu \in \mathrm{Spec}(D_{\mathfrak{q},B}) \mid \lambda_2 < \mu \le \lambda_1 \,\}|, & \text{if } \lambda_1 \ge \lambda_2 \\ -|\{\, \mu \in \mathrm{Spec}(D_{\mathfrak{q},B}) \mid \lambda_2 > \mu \ge \lambda_1 \,\}|, & \text{if } \lambda_1 \le \lambda_2. \end{cases}$$

Proposition 14.6.1. *Let i be as above. If $\lambda_1 \ge \lambda_2$, so that $i \ge 0$, then $M_{z_0}([\mathfrak{a}_1], [\mathfrak{a}_2])$ is diffeomorphic to an open subset of a projective space $\mathbb{CP}^i$, obtained as the complement of two hyperplanes. If $\lambda_1 < \lambda_2$ then the moduli space is empty.*

Proof. Let γ be in temporal gauge and represent a point in the moduli space $M_{z_0}([\mathfrak{a}_1], [\mathfrak{a}_2])$. We have $\mathcal{L}(\mathfrak{a}_1) = \mathcal{L}(\mathfrak{a}_2)$, so the image of $\check{\gamma}$ under the blow-down map must be a constant path. After gauge transformation, this may as well be a constant path at $\alpha \in \mathcal{C}_k(Y)$. In particular, $\check{\gamma}(t)$ is reducible for all t, and we can write the path in $\mathcal{C}_k^{\sigma}(Y)$ as

$$\check{\gamma}(t) = (B, 0, \phi(t)),$$

where $\phi(t)$ is of unit L^2 norm. The remaining gauge freedom is the action of constant gauge transformations, the circle group S^1. The equation that $\phi(t)$ must satisfy appeared at (10.9), and can be written

$$\frac{d}{dt}\phi(t) = -D_{\mathfrak{q},B}\phi(t) + \lambda(t)\phi(t),$$

where $\lambda(t)$ runs from λ_1 to λ_2. The operator $D_{\mathfrak{q},B}$ is independent of t. The moduli space is the quotient by S^1 of the set of solutions having the correct limiting behavior: $\phi(t)$ must approach the S^1 orbit of ϕ_1 and ϕ_2 as t goes to the two ends. As we did in Subsection 6.3, we write $\Phi = s_0\phi$, where s_0 is a positive solution of the ODE (6.12); then Φ satisfies the 4-dimensional, translation-invariant perturbed Dirac equation

$$\frac{d}{dt}\Phi(t) = -D_{\mathfrak{q},B}\Phi(t).$$

The asymptotics of $|\Phi(t)|_{L^2(Y)}$ are the same as those of s_0, and so we have

$$\Phi(t) \sim c_1 e^{-\lambda_1 t}\phi_1, \quad \text{as } t \to -\infty$$
$$\Phi(t) \sim c_2 e^{-\lambda_2 t}\phi_2, \quad \text{as } t \to +\infty.$$

The moduli space is the quotient by $\mathbb{R}^+ \times S^1 = \mathbb{C}^*$ of the set of solutions Φ with these asymptotics. The asymptotic conditions mean that we can write

$$\Phi = \sum_{\mu} c_\mu e^{-\mu t} \phi_\mu,$$

where μ runs through the eigenvalues in the interval $[\lambda_2, \lambda_1]$, the spinors ϕ_μ are corresponding unit eigenvectors for $D_{\mathfrak{q},B}$ on Y, and c_μ are complex coefficients, with c_{λ_1} and c_{λ_2} both non-zero. (If $\lambda_1 < \lambda_2$, there are no solutions.) Thus the moduli space is identified with the complement of two hyperplanes in $\mathbb{CP}^i$, where $i + 1$ is the number of eigenvalues in the closed interval $[\lambda_2, \lambda_1]$. □

The result of this proposition is consistent with the moduli spaces being regular. We shall verify this regularity in the next section. Granted this regularity, we can deduce:

Corollary 14.6.2. *If λ_1 and λ_2 have the same sign, then*

$$\mathrm{gr}(\mathfrak{a}_1, \mathfrak{a}_2) = 2i.$$

If λ_1 is positive and λ_2 is negative, then

$$\mathrm{gr}(\mathfrak{a}_1, \mathfrak{a}_2) = 2i - 1.$$

If λ_1 is negative, and λ_2 is positive, then

$$\mathrm{gr}(\mathfrak{a}_1, \mathfrak{a}_2) = 2i + 1,$$

(which is a negative integer).

Proof. If $\lambda_1 \geq \lambda_2$, then the moduli space is non-empty, and regular, as stated. The case that λ_1 is positive and λ_2 is negative is the boundary-obstructed case. The proposition above tells us the dimension of the moduli spaces in these cases, and we can deduce the value of $\mathrm{gr}(\mathfrak{a}_1, \mathfrak{a}_2)$ from Proposition 14.5.7. In the case that $\lambda_2 \geq \lambda_1$, we can use the identity $\mathrm{gr}(\mathfrak{a}_1, \mathfrak{a}_2) = -\mathrm{gr}(\mathfrak{a}_2, \mathfrak{a}_1)$ to complete the proof. □

Remark. As an alternative proof of this corollary, one can directly compute the spectral flow of the family of operators that appears in the proof of Theorem 14.4.2.

15 Transversality for moduli spaces of trajectories

15.1 Statement and proof

Let us recall that we have a fixed Banach space of perturbations, $\mathcal{P}$, which is "large" in the sense made precise by Definition 11.6.3. The main result of this section is that there exist perturbations in $\mathcal{P}$ such that the moduli spaces of trajectories are regular (see Definition 14.5.6 and Proposition 14.5.7). More precisely:

Theorem 15.1.1. *There is a* $\mathfrak{q} \in \mathcal{P}$ *such that:*

(i) *all the critical points* $\mathfrak{a} \in \mathcal{C}_k^\sigma(Y)$ *are non-degenerate;*
(ii) *for each pair of critical points* $\mathfrak{a}$, $\mathfrak{b}$, *and each relative homotopy class* z, *the moduli space* $M_z([\mathfrak{a}], [\mathfrak{b}])$ *is regular, in the sense of Definition 14.5.6.*

Note that unless the first condition is satisfied, the second condition cannot be interpreted. To indicate their dependence on $\mathfrak{q}$, we shall sometimes write $\mathcal{L}_\mathfrak{q}$ or $M_{z,\mathfrak{q}}([\mathfrak{a}], [\mathfrak{b}])$ below, instead of our usual notation, $\mathcal{L}$ or $M_z([\mathfrak{a}], [\mathfrak{b}])$, for the perturbed functional and moduli spaces.

Proof of Theorem 15.1.1. First, Theorem 12.1.2 provides us with a perturbation $\mathfrak{q}_0$ satisfying the first condition. In the based configuration space (see (11.3)), $\mathcal{B}_k^o(Y) = \mathcal{C}_k(Y)/\mathcal{G}_{k+1}^o(Y)$, the image of the critical set for this perturbation is a finite collection of points (the reducible critical points) and circles (the irreducibles). By Proposition 11.2.1, we can choose a map

$$p_0 : \mathcal{B}_k^o(Y) \to \mathbb{R}^n \times \mathbb{T} \times \mathbb{C}^m$$

defined by coexact forms c_i and sections Υ_j as in (11.4), such that p_0 separates the orbits of the different critical points. For each critical point $[\alpha] \in \mathcal{B}_k(Y)$, let $\mathcal{O}_{[\alpha]} \subset \mathcal{B}_k^o(Y)$ be an open, S^1-invariant neighborhood of the corresponding S^1 orbit, and let these be chosen so that their images under p_0 have disjoint closures. Write

$$\mathcal{O} = \bigcup_{[\alpha]} \mathcal{O}_{[\alpha]} \subset \mathfrak{B}^o(Y).$$

The condition on disjoint closures means that any path joining $p_0([\alpha])$ to $p_0([\beta])$ must leave $\overline{p_0(\mathcal{O})}$ if $[\alpha]$ and $[\beta]$ are distinct critical orbits. We will also require the neighborhoods $\mathcal{O}$ to be small enough that no essential loop based at any $p_0([\alpha])$ is contained in $\overline{p_0(\mathcal{O})}$.

Let $\mathcal{P}_{\mathcal{O}} \subset \mathcal{P}$ be the set of perturbations

$$\mathcal{P}_{\mathcal{O}} = \{ \mathfrak{q} \in \mathcal{P} \mid \mathfrak{q}|_{\mathcal{O}} = \mathfrak{q}_0|_{\mathcal{O}} \}.$$

This is a closed linear subspace of $\mathcal{P}$, and therefore a Banach space.

Lemma 15.1.2. *There is an open neighborhood of* $\mathfrak{q}_0$ *in* $\mathcal{P}_{\mathcal{O}}$ *such that for all* $\mathfrak{q}$ *in this neighborhood, the perturbed vector field* grad $\mathcal{L}_{\mathfrak{q}}$ *has no critical points outside* $\mathcal{O}$*: the critical points of* grad $\mathcal{L}_{\mathfrak{q}}$ *and* grad $\mathcal{L}_{\mathfrak{q}_0}$ *are therefore the same, as are the critical points of* $(\text{grad}\,\mathcal{L}_{\mathfrak{q}})^\sigma$ *and* $(\text{grad}\,\mathcal{L}_{\mathfrak{q}_0})^\sigma$.

Proof. This follows from Proposition 11.6.4. □

For each perturbation $\mathfrak{q} \in \mathcal{P}_{\mathcal{O}}$, and each pair of critical points $\mathfrak{a}$, $\mathfrak{b}$ of $(\text{grad}\,\mathcal{L}_{\mathfrak{q}})^\sigma$ in $\mathcal{O}$, we have a moduli space $M_{z,\mathfrak{q}}([\mathfrak{a}],[\mathfrak{b}])$. Because of the lemma, the main theorem follows from the next proposition. □

Proposition 15.1.3. *The set of perturbations* $\mathfrak{q} \in \mathcal{P}_{\mathcal{O}}$ *satisfying Condition* (ii) *of Theorem* 15.1.1 *for all* $\mathfrak{a}$, $\mathfrak{b}$ *whose images belong to* $\mathcal{O} \subset \mathcal{B}_k(Y)$ *is a residual subset of* $\mathcal{P}_{\mathcal{O}}$.

Proof. We introduce the parametrized moduli space

$$\mathfrak{M}_z([\mathfrak{a}],[\mathfrak{b}]) \subset \mathcal{B}^\tau_{k,z}([\mathfrak{a}],[\mathfrak{b}]) \times \mathcal{P}_{\mathcal{O}}$$

which is the quotient by $\mathcal{G}^\tau_{k+1}(Z)$ of the zero set of

$$\begin{aligned} \mathfrak{W} &: \mathcal{C}^\tau_k(\mathfrak{a},\mathfrak{b}) \times \mathcal{P}_{\mathcal{O}} \to \mathcal{V}^\tau_{k-1}(Z) \\ \mathfrak{W} &: (\gamma,\mathfrak{q}) \mapsto \mathfrak{F}^\tau_{\mathfrak{q}}(\gamma). \end{aligned}$$

The derivative of $\mathfrak{W}$ at $(\gamma,\mathfrak{q})$ is a map

$$\mathcal{D}_{(\gamma,\mathfrak{q})}\mathfrak{W} : \mathcal{T}^\tau_{k,\gamma} \times T_{\mathfrak{q}}\mathcal{P}_{\mathcal{O}} \to \mathcal{V}^\tau_{k-1,\gamma}(Z)$$

which, when γ is reducible, has a summand

$$(\mathcal{D}_{(\gamma,\mathfrak{q})}\mathfrak{W})^\partial : (\mathcal{T}^\tau_{k,\gamma})^\partial \times T_{\mathfrak{q}}\mathcal{P}_{\mathcal{O}} \to (\mathcal{V}^\tau_{k-1,\gamma})^\partial(Z)$$

as in (14.15). Following a familiar strategy for these arguments, we shall show:

(i) the derivative $\mathcal{D}\mathfrak{W}$ is surjective at all irreducible points $(\gamma,\mathfrak{q})$ in $\mathfrak{W}^{-1}(0)$; and

(ii) the summand $(\mathcal{D}\mathfrak{W})^\partial$ is surjective at all reducible critical points.

As in Subsection 14.5, it then follows that $\mathfrak{M}_z([\mathfrak{a}],[\mathfrak{b}])$ is the quotient by $\mathbb{Z}/2$ of a Banach manifold $\tilde{\mathfrak{M}}_z([\mathfrak{a}],[\mathfrak{b}])$. The projection of $\tilde{\mathfrak{M}}_z([\mathfrak{a}],[\mathfrak{b}])$ to $\mathcal{P}_{\mathcal{O}}$ is Fredholm of index $\mathrm{gr}_z([\mathfrak{a}],[\mathfrak{b}])$, or $\mathrm{gr}_z([\mathfrak{a}],[\mathfrak{b}])+1$ in the case that $\mathfrak{a}$, $\mathfrak{b}$ are both reducible, and boundary-stable and -unstable respectively. So the Sard–Smale theorem provides a residual subset of regular values in $\mathcal{P}_{\mathcal{O}}$: this is the residual subset of the proposition.

Consider then an element $(\gamma,\mathfrak{q})$ in $\mathfrak{W}^{-1}(0)$. For the moment, we will assume either that the images $[\alpha]$ and $[\beta]$ of $[\mathfrak{a}]$ and $[\mathfrak{b}]$ in $\mathcal{B}_k(Y)$ are distinct, or (if $[\alpha]=[\beta]$) that the homotopy class z is non-trivial; the cases excluded by this assumption are the moduli spaces which we described in Proposition 14.6.1 in the previous section, and we will establish their regularity at the end of the present proof. Because of our assumption, the path

$$\underline{\gamma}=\pi\circ\check{\gamma}:\mathbb{R}\to\mathcal{C}_k(Y)$$

corresponding to γ has non-constant image in $\mathcal{B}_k(Y)$; and by our hypothesis concerning the closures of the $p_0(\mathcal{O}_\alpha)$, there is an open interval $J\subset\mathbb{R}$ such that

$$p_0(\underline{\gamma}(J))\cap\overline{p_0(\mathcal{O})}=\varnothing. \tag{15.1}$$

By the unique continuation result, Proposition 10.8.2, the closure $\bar{J}$ is embedded in $\mathcal{B}_k(Y)$ via $\underline{\gamma}$.

The partial derivative of $\mathfrak{W}$ in the $\mathcal{C}_k^\tau(\mathfrak{a},\mathfrak{b})$ directions is $\mathcal{D}_\gamma\mathfrak{F}_\mathfrak{q}^\tau$; so if $\mathcal{D}_\gamma\mathfrak{F}_\mathfrak{q}^\tau$ is surjective we are done. If $\mathcal{D}_\gamma\mathfrak{F}_\mathfrak{q}^\tau$ is not surjective, consider a non-zero element $V\in\mathcal{V}_{0,\gamma}^\tau(Z)$, orthogonal to the image of $\mathcal{D}_\gamma\mathfrak{F}_\mathfrak{q}^\tau$ with respect to the L^2 inner product on $\mathcal{V}_{k-1,\gamma}^\tau(Z)$. The pair $(V,0)\in\mathcal{V}_{0,\gamma}^\tau(Z)\times L^2(Z;i\mathbb{R})$ is then L^2-orthogonal to the image of the Fredholm operator

$$Q_\gamma:\mathcal{T}_{1,\gamma}^\tau\to\mathcal{V}_{0,\gamma}^\tau(Z)\times L^2(Z;i\mathbb{R}).$$

(See Theorem 14.4.2.) As before, we write $Q_\gamma=(d/t)+L_0+h$, where h is defined slicewise by a continuous path of operators $\mathbb{R}\to\mathrm{Hom}(L^2,L^2)$, and L_0 is an elliptic operator on Y. Thus $(V,0)$ is a weak solution of

$$-\frac{D}{dt}(V,0)+L_0^*(V,0)+h^*(V,0)=0.$$

Because $(d/dt)+L_0$ is elliptic, it follows that V is in L_1^2, i.e. $V\in\mathcal{V}_{1,\gamma}^\tau(Z)$. Unique continuation holds for this linear equation, by an application of Lemma 7.1.3, so the restriction of V to J is non-zero.

The usual isomorphism

$$i\,\mathfrak{su}(S^+) \oplus S^- \to T^*(Y) \oplus S$$

on Z gives rise also to an isomorphism

$$\mathcal{V}_j^\tau(Z) \oplus L_j^2(Z; i\mathbb{R}) \to \mathcal{T}_j^\tau(Z).$$

Using the first of these, the section V gives rise to $\check{V}$, an L_1^2 section of $\mathcal{T}_0^\sigma(Y)$ along the path $\check{\gamma}$ in $\mathcal{C}_k^\sigma(Y)$. Using the second isomorphism, we can regard $(0, V)$ as an element of $\mathcal{T}_{1,\gamma}^\tau(Z)$.

Lemma 15.1.4. *For all $t \in \mathbb{R}$, the element $\check{V}(t)$ in $\mathcal{T}_0^\sigma(Y)$ belongs to the orthogonal complement of the tangent space to the $\mathcal{G}_{k+1}(Y)$ orbit through $\check{\gamma}(t)$, with respect to the L^2 inner product on $\mathcal{T}_0(Y)$. In particular, the image of $\check{V}(t)$ in $[\mathcal{T}_0^\sigma]$ (the L^2 completion of $T\mathcal{B}_k^\sigma(Y)$) is non-zero for all t.*

Proof. We have an integration-by-parts formula, for any $U \in \mathcal{T}_{j,\gamma}^\tau(Z)$ and $W \in \mathcal{V}_{1,\gamma}^\tau(Z) \oplus L_1^2(Z; i\mathbb{R})$,

$$\langle Q_\gamma U, W\rangle_{L^2(\{t\}\times Y)} - \langle U, Q_\gamma^* W\rangle_{L^2(\{t\}\times Y)} = \frac{d}{dt}\langle U, W\rangle_{L^2(\{t\}\times Y)}, \tag{15.2}$$

using the standard isomorphisms above to interpret the term on the right. We apply this to our element of the cokernel having the form $W = (V, 0)$. If U is a tangent vector to the gauge orbit then it is in the kernel of $\mathcal{D}_\gamma \mathfrak{F}_\mathfrak{q}^\tau$ at the solution γ, so $Q_\gamma U$ has the form $(0, x)$. The left-hand side thus vanishes, and we see that $\langle \check{U}(t), \check{V}(t)\rangle_t$ is independent of t. On the other hand, this is an integrable function of t, because U and V belong to L^2. So the inner product is zero for all t. □

We now focus on the case that γ is irreducible. The blow-down map $\pi : \mathcal{C}_k^\sigma(Y) \to \mathcal{C}_k(Y)$ is a diffeomorphism in a neighborhood of $\check{\gamma}(J)$, and carries the path $\check{\gamma}$ to $\underline{\gamma}$. We seek to construct a cylinder function f so that the corresponding perturbation $\delta\mathfrak{q}^\sigma = (\operatorname{grad} f)^\sigma$ has

$$\left\langle \delta\mathfrak{q}^\sigma(t), \check{V}(t)\right\rangle_{\mathcal{T}^\sigma_{0,\check{\gamma}(t)}} \geq 0 \tag{15.3}$$

with strict inequality at t_0. This will contradict the supposition that V is orthogonal to the image of $\mathcal{D}\mathfrak{W}$, and so prove that $\mathcal{D}\mathfrak{W}$ is surjective at γ. The blow-down map π is not an isometry for the L^2 inner products on $\mathcal{T}$ and $\mathcal{T}^\sigma$; but over the irreducibles, there is an endomorphism κ of $\mathcal{T}$ such that

$$\langle u, v\rangle_{\mathcal{T}^\sigma_{0,\check\gamma(t)}} = \langle \pi_* u, \kappa\pi_* v\rangle_{\mathcal{T}_{0,\check\gamma(t)}}.$$

(At $(B, \Psi) \in \mathcal{C}_k(Y)$, the map κ is given by

$$(b, \psi_0 + \psi^\perp) \mapsto (b, \psi_0 + (\|\Psi\|)^{-2}\psi^\perp),$$

where ψ_0 and $\psi^\perp$ are parallel to and real-orthogonal to Ψ.) Write $\bar V(t) = \kappa\pi_*\check V(t)$, so that the desired inequality (15.3) becomes

$$\langle \delta\mathfrak{q}(t), \bar V(t)\rangle_{\mathcal{T}_{0,\check\gamma(t)}} \geq 0.$$

From the definition of $\delta\mathfrak{q}$ as a gradient with respect to this inner product, this becomes the metric-independent statement

$$(\mathcal{D}f)(\bar V(t)) \geq 0.$$

To achieve this, we now appeal to Proposition 11.2.1 to augment the collection of forms c_i and sections Υ_j defining p_0, so that the larger collection gives

$$p : \mathcal{C}_k(Y) \to \mathbb{R}^n \times \mathbb{T} \times \mathbb{C}^m$$

which now embeds $\underline{\gamma}(S)$ in $\mathbb{R}^n \times \mathbb{T} \times \mathbb{C}^m$, where $S \supset J$ is the compact subset of $\mathbb{R}$ defined as

$$S = \{\, t \in \mathbb{R} \mid p(\underline{\gamma}(t)) \notin p(\mathcal{O})^+ \,\}$$

and $p(\mathcal{O})^+$ is an open neighborhood of $\overline{p(\mathcal{O})}$ disjoint from $p(\underline{\gamma}(\bar J))$. (Recall $\underline{\gamma} = \pi \circ \check\gamma$). Furthermore, because of Lemma 15.1.4 above, we can choose p such that the image $p_*(\bar V)$ of the vector field along $\bar J$ is non-zero. Pick a $t_0 \in J$. As in Corollary 11.2.2, we can find a cylinder function $f = g \circ p$, arising from some

$$g : \mathbb{R}^n \times \mathbb{T} \times \mathbb{C}^m \to \mathbb{R},$$

such that $(\mathcal{D}f)(\bar V(t)) \geq 0$ for all t in J, with strict inequality at t_0. By multiplying g by a suitable cut-off function on $\mathbb{R}^n \times \mathbb{T} \times \mathbb{C}^m$, we then arrange that $\mathcal{D}_{\underline{\gamma}(t)}f(\bar V(t)) = 0$ for $t \in \mathbb{R} \setminus J$, because $p(\underline{\gamma}(\bar J))$ is disjoint from $p(\underline{\gamma}(\mathbb{R} \setminus \bar J))$. Replacing p and g by suitable approximations, we may safely assume that g

corresponds to a perturbation belonging to the Banach space $\mathcal{P}$. This completes the proof of surjectivity of $\mathcal{D}\mathfrak{W}$ at the irreducible solutions.

Next we consider a trajectory $\gamma \in \mathfrak{W}^{-1}(0)$ belonging to the reducibles. We need to show that the summand $(\mathcal{D}_\gamma \mathfrak{W})^\partial$ is surjective. Supposing it is not surjective, we take an element V in the cokernel, just as before, though we are now supposing $V \in (\mathcal{V}_1^\sigma)^\partial(Z)$. Lemma 15.1.4 still applies. We can write

$$\begin{aligned} \gamma &= (A, 0, \phi) \\ \check{\gamma}(t) &= (\check{A}(t), 0, \phi(t)) \\ \check{V}(t) &= (\omega(t), 0, \psi(t)) \end{aligned}$$

in the coordinates of $\mathcal{C}_k^\sigma(Y)$, where $\phi(t)$ is of unit length and orthogonal to $\psi(t)$, and $\omega(t)$ is an imaginary-valued 1-form on Y.

We write an element of $(\mathcal{T}_k^\sigma)^\partial(Z)$ as a pair $(\delta A, \delta\phi)$. If we consider a tangent vector

$$\delta = (\delta A, \delta\phi, \delta\mathfrak{q}) \in (\mathcal{T}_k^\sigma)^\partial(Z) \times T\mathcal{P}$$

at $(\gamma, \mathfrak{q})$, then the inner product

$$\langle (\mathcal{D}\mathfrak{W})^\partial(\delta), V \rangle_{L^2(Z)}$$

is the sum of terms:

$$\begin{aligned} &\langle \rho(d^+(\delta A)) + (\mathcal{D}\hat{\mathfrak{q}}^0)(\delta A, 0) + (\delta\hat{\mathfrak{q}}^0)(A, 0), \rho(\omega) \rangle \\ &+ \langle \rho(\delta A)\phi + (\mathcal{D}^2\hat{\mathfrak{q}}^1)((\delta A, 0), (0, \phi)), \psi \rangle \\ &+ \langle D_A^+(\delta\phi) + (\mathcal{D}\hat{\mathfrak{q}}^1)(0, \delta\phi), \psi \rangle \\ &+ \langle (\mathcal{D}\delta\hat{\mathfrak{q}}^1)(0, \phi), \psi \rangle. \end{aligned} \tag{15.4}$$

Our assumption is that this pairing is zero, for all δ. We will get a contradiction by showing that $\check{V} = (\omega, 0, \psi)$ is zero. Consider first vectors $\delta = (0, \delta\phi, 0)$ with $\delta A = 0$ and $\delta\mathfrak{q} = 0$. These variations show that ψ is in the cokernel of the linear Fredholm operator

$$L : \psi \mapsto D_A^+\psi + (\mathcal{D}_{(A,0)}\delta\hat{\mathfrak{q}})(0, \psi),$$

so ψ is in the kernel of the adjoint L^*. Now, when γ is in temporal gauge,

$$L = \rho(\partial/\partial t) \circ (d/dt) + D_{\check{A},\mathfrak{q}}$$

so

$$-\frac{d}{dt}\psi + D_{\check{A},\mathfrak{q}}\psi = 0. \tag{15.5}$$

(Recall that $D_{\check{A},\mathfrak{q}}$ is symmetric.) Unique continuation now implies that the restriction of ψ to $\{t\} \times Y$ is always non-zero, unless ψ is zero is everywhere. To show that ψ is in fact zero, we next consider a perturbation $\delta = (0, 0, \delta\mathfrak{q})$, where $\delta\mathfrak{q}$ arises from a cylinder function f that is identically zero on the reducibles (*cf.* Section 12.6). For such perturbations, $\delta\hat{\mathfrak{q}}^0$ is zero at γ, and the only potentially non-zero term in the sum is the last one. We will show that we can choose $\delta\mathfrak{q}$ to make it non-zero if ψ is non-zero.

Lemma 15.1.5. *For each $t \in \mathbb{R}$, the spinors $\psi(t)$ and $\phi(t)$ are orthogonal as elements of $L^2(Y; S)$ with respect to the complex-valued hermitian inner product, using the usual isomorphisms $S^{\pm}|_{\{t\}\times Y} = S$.*

Proof. The spinors ϕ and ψ on Z are solutions of two equations that are formally adjoint:

$$\frac{d}{dt}\phi + D_{\check{A},\mathfrak{q}}\phi = 0, \qquad -\frac{d}{dt}\psi + D_{\check{A},\mathfrak{q}}\psi = 0.$$

The argument is therefore the same as was used in the proof of Lemma 15.1.4. □

As in the irreducible case , we can find an interval J such that the path $p \circ \pi \circ \check{\gamma}$ in $\mathcal{C}_k(Y)$ given by $(\check{A}(t), 0)$ embeds $\bar{J}$, disjoint from $\overline{p(\mathcal{O})}$ and disjoint from $p \circ \underline{\gamma}(\mathbb{R} \setminus \bar{J})$. (To do this, we may need to augment p, as before.) We can also arrange that $p_*(\psi)$ and $p_*(\phi)$ are complex-linearly independent along $p \circ \underline{\gamma}(J)$. (They are vector fields in the direction of the $\mathbb{C}^m$ factor.) Now choose a cylinder function $f \in \mathcal{P}_{\mathcal{O}}$ that arises from an S^1-invariant function

$$g : \mathbb{R}^n \times \mathbb{T} \times \mathbb{C}^m \to \mathbb{R}$$

that is zero on the fixed-point set $\mathbb{R}^n \times T \times \{0\}$ and whose second derivative $(\nabla^2 g)(p_*(\phi), p_*(\psi))$ is positive at $\underline{\gamma}(t_0)$ and non-negative for all t. Such a g can be constructed because the second derivative in the $\mathbb{C}^m$ directions at a point on the fixed-point set can be the real part of any complex bilinear form. The resulting variation $\delta\mathfrak{q}$ makes the last term of (15.4) non-zero, as desired, so showing that ψ must be zero.

We now have $V = (\omega, 0, 0)$, and it remains to show that ω is zero. The argument is similar to the other arguments. By considering a variation first with only δA non-zero, we see that ω satisfies an equation adjoint to the equation

$$d^+(\delta A) + (\mathcal{D}\hat{\mathfrak{q}}^0)(\delta A, 0) = 0,$$

so ω has unique continuation and is therefore non-zero along J. The argument is completed by considering a tangent vector $\delta = (0, 0, \delta\mathfrak{q})$, where $\delta\mathfrak{q}$ is the gradient of a cylinder function f with derivative so chosen that

$$(\mathcal{D}_{\underline{\gamma}(t)} f)\omega(t) \geq 0,$$

with strict inequality at some $t_0 \in J$.

Finally, there remains what we excluded at the beginning of this argument: the case that $\pi[\mathfrak{a}] = \pi[\mathfrak{b}] = [\alpha]$ and z is the trivial homotopy class of paths. In this case, recall from Proposition 14.6.1 that the moduli space consists of reducibles. We shall show that this moduli space is regular, as long as the perturbation has been chosen so that the critical points are non-degenerate. For a $[\gamma]$ in this moduli space, we can follow the discussion of the reducible case above: in particular, an element of the cokernel of Q^∂_γ gives rise to

$$\check{V}(t) = (\omega(t), 0, \psi(t))$$

where $\psi(t)$ satisfies an equation

$$-\frac{d}{dt}\psi + D_{\mathfrak{q},B}\psi = 0,$$

and B is the connection on Y corresponding to the critical point α (see (15.5) and the proof of Proposition 14.6.1). The translation-invariant operator that appears in this equation is invertible, as a map from $L^2_1(Z; S)$ to $L^2(Z; S)$, by Proposition 14.1.2: the operator $D_{\mathfrak{q},B}$ is hyperbolic because 0 is not in its spectrum. This leaves us only to show that ω is zero. The equation satisfied by ω is also translation-invariant; and the relevant operator on Y is also hyperbolic, because of the non-degeneracy of the critical point α. So the same argument applies. □

Notes and references for Chapter IV

The Fredholm theory of differential operators on cylindrical manifolds, of the sort discussed in this chapter, is developed in [8] and [68]. The more general L^p

theory, and higher-order operators, are treated in [68]. Gauge theory on cylinders, and on manifolds with cylindrical ends, is developed in [32, 73, 78, 83, 99] and [112]; the related theory for manifolds with periodic ends was developed previously in [109]. We have discussed the equations on the cylinder only in the case that the critical points of the perturbed functional are non-degenerate: this has allowed us to base our constructions on the ordinary Sobolev spaces $L^2_k(Z)$, rather than weighted Sobolev spaces used (for example) in [78, 112]. More significantly, it has obviated the need for any discussion of center-manifold theory in the Kuranishi description of the moduli spaces.

Smale's infinite-dimensional version of Sard's theorem is proved in [103]. Achieving transversality for the moduli spaces of trajectories by perturbing the Chern–Simons (or in our case the Chern–Simons–Dirac) functional goes back to Floer's work in [32].

V
Compactness and gluing

In our discussion of finite-dimensional Morse theory in Chapter I, we touched on the important fact that the space of flow lines between two critical points, a and b, of a Morse function is in general not compact. There is an obvious source of non-compactness arising from the fact that a trajectory $\gamma(t)$ can be reparametrized as $\gamma(t + t_0)$; but there is an additional source of non-compactness, wherein a sequence of trajectories can converge to a "broken trajectory": a concatenation of trajectories between intermediate critical points. The space of trajectories (modulo reparametrization) can be compactified by adding such broken trajectories. Section 16 establishes essentially the same result in the context of the infinite-dimensional Morse theory of the Chern–Simons–Dirac functional. These compactness theorems for global trajectories leverage our earlier compactness results for solutions on finite cylinders, from Chapters II and III.

The remaining sections of this chapter are concerned with *gluing*. The question here is to understand the structure of the compactification of the trajectory space near the broken trajectories. The term "gluing" arises because we are investigating whether, for example, the existence of a pair of trajectories with a common endpoint implies the existence of a continuous unbroken trajectory nearby. Our treatment of this question is based on a careful examination of the space of trajectories which are defined on a finite time interval and which remain in the neighborhood of a critical point. The moduli spaces of solutions on a finite cylinder are studied in Section 17, and the solutions near a critical point are analyzed in Section 18. In Section 19, these results are applied to global trajectories and the gluing problem.

16 Compactness of trajectory spaces

The moduli spaces of trajectories, $M_z([\mathfrak{a}],[\mathfrak{b}])$, can be compactified by introducing *broken trajectories*. In this section, we define and topologize the space of broken trajectories and establish its compactness.

Throughout this section, we continue to suppose a tame perturbation $\mathfrak{q}$ has been chosen so that all critical points $[\mathfrak{a}]$ in $\mathcal{B}_k^\sigma(Y,\mathfrak{s})$ are non-degenerate (in the sense of Definition 12.1.1). We will not need to assume that the moduli spaces of trajectories are regular, at least while discussing the compactification of a single moduli space $M_z([\mathfrak{a}],[\mathfrak{b}])$. Regularity of the moduli spaces will be needed later in this section, where we will show (for example) that only finitely many of these moduli spaces can contain irreducible trajectories.

16.1 Broken trajectories and convergence

Let $\mathfrak{a}$ and $\mathfrak{b}$ be critical points, and z the corresponding homotopy class of paths joining the $[\mathfrak{a}]$ to $[\mathfrak{b}]$ in $\mathcal{B}_k^\tau(Y)$. We say that a trajectory $[\gamma]$ belonging to a moduli space $M_z([\mathfrak{a}],[\mathfrak{b}])$ is *non-trivial* if it is not invariant under the action of $\mathbb{R}$ by translations on the cylinder $Z = \mathbb{R}\times Y$. This is equivalent to saying that either $[\mathfrak{a}] \neq [\mathfrak{b}]$, or $[\mathfrak{a}] = [\mathfrak{b}]$ and z is non-trivial.

Definition 16.1.1. An *unparametrized* trajectory is an equivalence class of ***non-trivial*** trajectories in $M_z([\mathfrak{a}],[\mathfrak{b}])$ under the action of translations. We write

$$\breve{M}_z([\mathfrak{a}],[\mathfrak{b}])$$

for the space of unparametrized trajectories. ◊

Definition 16.1.2. An *unparametrized broken trajectory* joining $[\mathfrak{a}]$ to $[\mathfrak{b}]$ consists of the following data:

- an integer $n \geq 0$, the *number of components*;
- an $(n+1)$-tuple of critical points $[\mathfrak{a}_0], \ldots, [\mathfrak{a}_n]$ with $[\mathfrak{a}_0] = [\mathfrak{a}]$ and $[\mathfrak{a}_n] = [\mathfrak{b}]$, the *restpoints*;
- for each i with $1 \leq i \leq n$, an unparametrized trajectory $[\breve{\gamma}_i]$ in $\breve{M}_{z_i}([\mathfrak{a}_{i-1}],[\mathfrak{a}_i])$, the *ith component* of the broken trajectory.

The *homotopy class* of the broken trajectory is the class of the path obtained by concatenating representatives of the classes z_i, or the constant path at $[\mathfrak{a}]$ if $n=0$. We write $\breve{M}_z^+([\mathfrak{a}],[\mathfrak{b}])$ for the space of unparametrized broken trajectories in the homotopy class z, and denote a typical element by $[\breve{\boldsymbol{\gamma}}] = ([\breve{\gamma}_1],\ldots,[\breve{\gamma}_n])$, even though this notation does not accurately reflect the special case $n=0$. ◊

The case $n = 0$ is included here for later bookkeeping purposes: note that if z is the class of the constant path at $[\mathfrak{a}]$, then $\breve{M}_z([\mathfrak{a}],[\mathfrak{a}])$ is empty, while $\breve{M}_z^+([\mathfrak{a}],[\mathfrak{a}])$ is a single point, a broken trajectory with no components.

We topologize the space of unparametrized broken trajectories as follows. Let $[\breve{\boldsymbol{\gamma}}] = ([\breve{\gamma}_1],\ldots,[\breve{\gamma}_n])$ belong to $\breve{M}_z^+([\mathfrak{a}],[\mathfrak{b}])$, with $[\breve{\gamma}_i] \in \breve{M}_{z_i}([\mathfrak{a}_{i-1}],[\mathfrak{a}_i])$ being represented by a (parametrized) trajectory

$$[\gamma_i] \in M_{z_i}([\mathfrak{a}_{i-1}],[\mathfrak{a}_i]).$$

Let $U_i \subset \mathcal{B}^\tau_{k,\mathrm{loc}}(Z)$ be an open neighborhood of $[\gamma_i]$, and let $T \in \mathbb{R}^+$. We define $\Omega = \Omega(U_1,\ldots,U_n,T)$ to be the subset of $\breve{M}_z^+(\mathfrak{a},\mathfrak{b})$ consisting of unparametrized broken trajectories $[\breve{\boldsymbol{\delta}}] = ([\breve{\delta}_1],\ldots,[\breve{\delta}_m])$ satisfying the following condition: there exists a map $(\jmath,s) : \{1,\ldots,n\} \to \{1,\ldots,m\} \times \mathbb{R}$ such that

- $[\tau_{s(i)}\delta_{\jmath(i)}] \in U_i$, and
- if $1 \le i_1 < i_2 \le n$, then either $\jmath(i_1) < \jmath(i_2)$, or $\jmath(i_1) = \jmath(i_2)$ and $s(i_1)+T \le s(i_2)$.

Here $\tau_s\delta$ denotes the translate, $\tau_s\delta(t) = \delta(s+t)$. We take the sets of the form $\Omega(U_1,\ldots,U_n,T)$ to be a neighborhood base for $[\breve{\boldsymbol{\gamma}}]$ in $\breve{M}_z^+([\mathfrak{a}],[\mathfrak{b}])$.

Our main result here is:

Theorem 16.1.3. *The space of unparametrized broken trajectories $\breve{M}_z^+([\mathfrak{a}],[\mathfrak{b}])$ is compact.*

Theorem 16.1.3 follows from a slightly stronger proposition, whose proof is our main goal in this section. For an element γ in $\mathcal{C}^\tau_{k,\mathrm{loc}}(\mathbb{R}\times Y)$, we write $\mathcal{E}_{\mathfrak{q}}(\gamma)$ for the energy of $\boldsymbol{\pi}\circ\gamma$ in $\mathcal{C}_{k,\mathrm{loc}}(\mathbb{R}\times Y)$, as in (10.12). The energy may be infinite. For a solution of the perturbed equations, the energy is twice the drop in $\mathcal{L}$. We define the energy of a broken trajectory to be the sum of the energies of its components. In particular for broken trajectories in $\breve{M}_z^+([\mathfrak{a}],[\mathfrak{b}])$ the energy depends only on z and we write $\mathcal{E}_{\mathfrak{q}}(z)$ for this energy.

Proposition 16.1.4. *For any $C > 0$ and any $[\mathfrak{a}]$, $[\mathfrak{b}]$, there are only finitely many z with energy $\mathcal{E}_{\mathfrak{q}}(z) \le C$ for which $\breve{M}_z^+([\mathfrak{a}],[\mathfrak{b}])$ is non-empty. Furthermore* *each* $\breve{M}_z^+([\mathfrak{a}],[\mathfrak{b}])$ *is compact. In other words the space of broken trajectories* $[\breve{\boldsymbol{\gamma}}] \in \breve{M}^+([\mathfrak{a}],[\mathfrak{b}])$ *with energy* $\mathcal{E}_{\mathfrak{q}}(\breve{\boldsymbol{\gamma}}) \le C$ *is compact.*

The proof of this proposition will appear in Subsection 16.3 after we prove a similar result for the images of these moduli spaces under the blow-down map $\boldsymbol{\pi}$.

A basic tool is the compactness result on finite cylinders, Theorem 10.9.2. This gives us the following result concerning trajectories in $\mathcal{B}^{\tau}_{k,\mathrm{loc}}(Z)$, with Z the infinite cylinder $\mathbb{R} \times Y$.

Proposition 16.1.5. *If* $[\gamma^n]$ *is a sequence of trajectories in* $\mathcal{B}^{\tau}_{k,\mathrm{loc}}(Z)$ *(i.e. solutions of the perturbed Seiberg–Witten equations), and if the energies* $\mathcal{E}_{\mathfrak{q}}(\gamma^n)$ *are finite and uniformly bounded, and* $|\Lambda_{\mathfrak{q}}(\gamma^n(t))|$ *is uniformly bounded,*

$$|\Lambda_{\mathfrak{q}}(\gamma^n(t))| \le \Lambda_0, \quad \textit{for all } n \textit{ and all } t,$$

then there exists a subsequence converging in the topology of $\mathcal{B}^{\tau}_{k,\mathrm{loc}}(Z)$ *to a trajectory* $[\gamma]$. □

Without the hypothesis on $|\Lambda_{\mathfrak{q}}(\gamma^n(t))|$, Theorem 10.7.1 implies the following weaker version:

Proposition 16.1.6. *If* $[\gamma^n]$ *is a sequence of trajectories in* $\mathcal{B}^{\tau}_{k,\mathrm{loc}}(Z)$ *and if the energies* $\mathcal{E}_{\mathfrak{q}}(\gamma^n)$ *are finite and uniformly bounded then the trajectories* $[\boldsymbol{\pi} \circ \gamma^n]$ *have a subsequence converging in the topology of* $\mathcal{B}_{k,\mathrm{loc}}(Z)$ *to a trajectory* $[\bar{\gamma}] \in \mathcal{B}_{k,\mathrm{loc}}(Z)$. □

16.2 Proof of compactness downstairs

Before proving Proposition 16.1.4, we prove a version of the same proposition for the moduli spaces of trajectories "downstairs" in $\mathcal{B}_{k,\mathrm{loc}}(Z)$ (without the blow-up), using the second of the two "local" compactness propositions above. For critical points $[\alpha]$, $[\beta]$ in $\mathcal{B}_k(Y)$, we introduce the moduli spaces

$$\begin{aligned} N_z([\alpha],[\beta]) &\subset \mathcal{B}_{k,\mathrm{loc}}(Z) \\ N([\alpha],[\beta]) &= \bigcup_z N_z([\alpha],[\beta]) \end{aligned} \tag{16.1}$$

of solutions to the perturbed equations, asymptotic to $[\alpha]$ and $[\beta]$ at the two ends. If $\alpha = \boldsymbol{\pi}(\mathfrak{a})$ and $\beta = \boldsymbol{\pi}(\mathfrak{b})$, we have the map

$$\boldsymbol{\pi} : M_z([\mathfrak{a}],[\mathfrak{b}]) \to N_z([\alpha],[\beta]).$$

We define a trajectory in $N_z([\alpha],[\beta])$ to be non-trivial if it is not invariant under the translations $\mathbb{R}$. Note that π maps non-trivial trajectories in $M_{z_0}([\mathfrak{a}],[\mathfrak{a}'])$ to trivial trajectories in $N_{z_0}([\alpha],[\alpha])$ when z_0 is the trivial homotopy class. For these moduli spaces, we can introduce also topological spaces $\breve{N}([\alpha],[\beta])$, $\breve{N}^+([\alpha],[\beta])$, and $\breve{N}^+_z([\alpha],[\beta])$, in parallel with the definitions above. We introduce $\breve{N}^+$ for the disjoint union

$$\breve{N}^+ = \bigcup_{[\alpha],[\beta]} \breve{N}^+([\alpha],[\beta]).$$

The following proposition concerning the compactness of $\breve{N}^+$ differs from the version for $\breve{M}^+$ (Proposition 16.1.4), in that it asserts the compactness of the union over all $[\alpha]$ and $[\beta]$.

Proposition 16.2.1. *For any $C > 0$, there are only finitely many $[\alpha]$, $[\beta]$, and z with $\mathcal{E}_{\mathfrak{q}}(z) \le C$ and such that the space $\breve{N}^+_z([\alpha],[\beta])$ is non-empty. Furthermore each $\breve{N}^+_z([\alpha],[\beta])$ is compact. In other words the space of broken trajectories $[\breve{\boldsymbol{\gamma}}] \in \breve{N}^+$ with energy $\mathcal{E}_{\mathfrak{q}}(\breve{\boldsymbol{\gamma}}) \le C$ is compact.*

Proof. We begin with two lemmas. For each of the finitely many critical points $[\alpha]$ in $\mathcal{B}_k(Y)$, choose a representative α, and let γ_α be the corresponding translation-invariant configuration in temporal gauge on the cylinder. We will also regard γ_α as a configuration on a finite cylinder $[t_1,t_2] \times Y$, by restriction. Let I be any closed interval, and for each $[\alpha]$, let $U_\alpha \subset \mathcal{C}_k(I \times Y)$ be a gauge-invariant neighborhood of γ_α.

In the following lemma (and below) we introduce $\mathcal{E}_{\mathfrak{q}}^{[a,b]}(\gamma)$ as notation for the perturbed energy of γ restricted to the interval $[a,b]$:

$$\mathcal{E}_{\mathfrak{q}}^{[a,b]}(\gamma) = 2\big(\mathcal{L}(\check{\gamma}(a)) - \mathcal{L}(\check{\gamma}(b))\big).$$

Lemma 16.2.2. *Let $U_\alpha \subset \mathcal{C}_k(I \times Y)$ be neighborhoods as above, let C be any constant, and let I' be any other interval of non-zero length. Then there exists $\epsilon > 0$ such that if γ is a trajectory satisfying $\mathcal{E}_{\mathfrak{q}}(\gamma) \le C$ and $\mathcal{E}_{\mathfrak{q}}^{I'}(\gamma) \le \epsilon$, then $\gamma|_{I\times Y} \in U_\alpha$ for some critical point $[\alpha]$.*

Proof. If not then there exists a sequence γ_i with $\mathcal{E}_{\mathfrak{q}}^{I'}(\gamma_i) \to 0$, none of which belongs to any U_α. But we know that, after gauge transformation, a subsequence converges in $\mathcal{C}_{k,\mathrm{loc}}(Z)$ to a limit γ_∞, which must have energy zero on I', and which must therefore be a constant trajectory, gauge-equivalent to some γ_α on the whole of Z by unique continuation. This means that a subsequence of the $\gamma_i|_{I\times Y}$ converges to γ_α in $\mathcal{C}_k(I \times Y)$ after gauge transformation. So, in this subsequence, $\gamma_i|_{I\times Y}$ is eventually an element of U_α. This is a contradiction. □

Definition 16.2.3. Let $\mathcal{U}$ be a collection of gauge-invariant open neighborhoods $U_\alpha \subset \mathcal{C}_k(I \times Y)$ of the critical points. We say that $\mathcal{U}$ has the *separating property* for $I \times Y$ if the following holds. There should exist neighborhoods $V_{[\alpha]} \subset \mathcal{B}_{k-1}(Y)$ of the critical points $[\alpha]$ in $\mathcal{B}_{k-1}(Y)$ such that:

- the sets $V_{[\alpha]}$, $V_{[\beta]}$ are disjoint whenever $[\alpha] \neq [\beta]$;
- each $V_{[\alpha]}$ is path-connected and simply connected;
- if γ belongs to U_α, then the gauge-equivalence class of $\check{\gamma}(t)$ is in $V_{[\alpha]}$, for all t in I.

For future use (see the following subsection), we note that the same definition can be repeated verbatim in the blown-up context, with $\mathcal{C}_k^\tau(I \times Y)$ in place of $\mathcal{C}_k(I \times Y)$, and $\mathcal{B}_{k-1}^\sigma(Y)$ in place of $\mathcal{B}_{k-1}(Y)$. ◊

We will use this definition in the proof of the following basic lemma.

Lemma 16.2.4. *Let $[\gamma] \in \mathcal{B}_{k,\mathrm{loc}}(Z)$ be a solution of the equations with finite energy. Then $[\gamma]$ belongs to $N([\alpha], [\beta])$ for some critical points $[\alpha]$ and $[\beta]$.*

Proof. Fix an interval I, and let $\mathcal{U}$ be a collection of neighborhoods with the separating property for $I \times Y$. Consider the translates $\tau_t\gamma$. The finite energy condition implies $\mathcal{E}_{\mathrm{q}}^I(\tau(t)\gamma) \to 0$ as $t \to +\infty$. So from Lemma 16.2.2, we deduce that there exists t_0 such that for all $t \geq t_0$, the translate $(\tau_t\gamma)|_I$ belongs to $U_{\beta_t} \in \mathcal{U}$ for some critical point $[\beta_t]$. The gauge-equivalence class of the critical point $[\beta_t]$ must be independent of t, because of the separating property of these neighborhoods, for we have $[\check{\gamma}(t')] \in V_{[\beta_t]}$ for all t' in $t + I$. We write $[\beta]$ for this critical point.

In this argument, the initial interval I can be chosen as large as we wish, and the separating neighborhoods are arbitrary. We conclude that $[\tau_t\gamma]$ converges to $[\gamma_\beta]$ in the topology of $\mathcal{B}_{k,\mathrm{loc}}(Z)$, as $t \to +\infty$. Similarly, $[\tau_t\gamma]$ converges to some $[\gamma_\alpha]$ as $t \to -\infty$. □

Turning now to the situation addressed by the statement of the proposition, we will prove that any sequence of broken trajectories, all with energy $\mathcal{E}_{\mathrm{q}} \leq C$, has a convergent subsequence. The energies of parametrized trajectories $[\gamma] \in N([\alpha], [\beta])$ belonging to different homotopy classes have energies differing by multiples of $2\pi^2$ (see Lemma 4.1.3). Since there are only finitely many critical points in $\mathcal{B}_k(Y)$, there is a constant $\mathcal{E}_0 > 0$, such that any non-trivial trajectory $[\gamma]$ has energy at least $\mathcal{E}_0$. So the broken trajectories in our sequence all have at most $C/\mathcal{E}_0$ components. We can therefore pass to a subsequence in which each broken trajectory has the same number of components and the same restpoints. By this means, we reduce to the case that our sequence of broken trajectories is actually a sequence of (unparametrized) trajectories $[\check{\gamma}^n] \in \check{N}([\alpha],[\beta])$. We take $[\gamma^n]$ to be parametrized representatives in $N([\alpha],[\beta])$. As usual, we write $[\check{\gamma}^n]$ for the corresponding parametrized paths in $\mathcal{B}_k(Y)$.

Let I be the interval $[-1,1]$, and let $\mathcal{U}$ be a collection of neighborhoods with the separating property for $I \times Y$. Lemma 16.2.2 supplies us with an $\epsilon > 0$, such that for any trajectory with $\mathcal{E}^I_{\mathrm{q}}(\gamma) \le \epsilon$, we have that $\gamma|_I$ belongs to $U_\alpha \in \mathcal{U}$ for some $[\alpha]$. For each n, we have

$$\sum_{p \in \mathbb{Z}} \mathcal{E}^I(\tau_p \gamma^n) \le 2C.$$

It follows that, for each n, there are at most $2C/\epsilon$ integers p such that

$$\tau_p \gamma^n \notin \bigcup_{[\alpha]} U_\alpha.$$

Replacing the γ^n with a subsequence, we may suppose that, for every n, there exist exactly l integers p with this property. Let these integers be

$$p_1^n < \cdots < p_l^n.$$

We may pass to a further subsequence with the property that, for each m in the range $1 \le m \le l-1$, the difference $p_{m+1}^n - p_m^n$ *either* increases to infinity with n *or* is constant (independent of n). We can then define an equivalence relation $\sim$ on the set $\{1,\ldots,l\}$ by declaring

$$m \sim m' \iff \lim_{n\to\infty} |p_m^n - p_{m'}^n| < \infty.$$

The equivalence classes are strings of adjacent integers, and $p_m^n - p_{m'}^n$ is independent of n if $m \sim m'$. Pick representatives $m_1 < m_2 < \cdots < m_d$, one from each equivalence class.

We now appeal to the conditions that define the separating property for the neighborhoods U_α. Let I_i^n denote the interval

$$\begin{aligned} I_i^n &= [a_i^n, b_i^n] \\ a_i^n &= \min\{ p_m^n \mid m \sim m_i \} \\ b_i^n &= \max\{ p_m^n \mid m \sim m_i \}. \end{aligned}$$

The length of the interval I_i^n is independent of n; and the separations between these intervals (the lengths $a_{i+1}^n - b_i^n$, for $1 \le i \le d-1$) increase without bound. We name these lengthening intervals J:

$$J_i^n = [b_i^n, a_{i+1}^n], \quad 1 \le i \le d-1.$$

We may assume these have length at least 2. We also write

$$J_0^n = (-\infty, a_1]$$
$$J_d^n = [b_d, \infty).$$

Thus, for each n, the line $\mathbb{R}$ is decomposed as the union of the fixed-length intervals I_i^n and the lengthening or infinite intervals J_i^n.

For each n and i, the image of the interval J_i^n under $[\check{\gamma}^n]$ lies entirely in one of the neighborhoods $V_{[\tilde{\alpha}]}$. We may assume that $[\tilde{\alpha}]$ depends only on i, not on n, by passing to a subsequence. So we can write:

$$[\check{\gamma}^n(J_i^n)] \subset V_{[\alpha_i]},$$

for $0 \le i \le d$. We must have $[\alpha_0] = [\alpha]$ and $[\alpha_d] = [\beta]$. The image of I_i^n under $[\check{\gamma}^n]$ gives a path in $\mathcal{B}_{k-1}(Y)$, joining a point in $V_{[\alpha_{i-1}]}$ to a point in $V_{[\alpha_{i-1}]}$. To such a path, we can assign a well-defined homotopy class of paths z_i^n, from $[\alpha_{i-1}]$ to $[\alpha_i]$, because of the connectivity of these neighborhoods. For each n the concatenation of the homotopy classes z_i^n is the homotopy class z_n to which γ_n belongs.

Fix i with $1 \le i \le d$, and consider the translates

$$\tau_{p_{m_i}^n} \gamma^n.$$

(Any element of I_i^n would do in place of $p_{m_i}^n$.) Passing to a subsequence again, we may assume (because of Proposition 16.1.6) that for each i there exists a γ_i which is a limit of this sequence:

$$[\tau_{p_{m_i}^n} \gamma^n] \to [\gamma_i]$$

in $\mathcal{B}_{k,\mathrm{loc}}(Z)$. The limit γ_i is a non-trivial trajectory, because its restriction to I does not belong to any U_α. Because of Lemma 16.2.4, we have

$$[\gamma_i] \in N_{z_i}([\alpha_i'], [\beta_i'])$$

for some critical points $[\alpha_i']$ and $[\beta_i']$. The convergence ensures that the endpoints $([\alpha_i'], [\beta_i'])$ of $[\breve{\gamma}_i]$ are the same as the endpoints of the path $\breve{\gamma}^n|_{I_i^n}$: that is,

$$[\alpha_i'] = [\alpha_{i-1}]$$
$$[\beta_i'] = [\alpha_i].$$

For n sufficiently large, convergence also ensures that $z_i^n = z_i$. This means, in particular, that the concatenation of the homotopy classes z_i is z_n, and z_n is therefore independent of n. This verifies that the sequence of unparametrized trajectories $[\breve{\gamma}^n]$ is converging to the broken trajectory $([\breve{\gamma}_1], \ldots, [\breve{\gamma}_d])$. □

16.3 Proof of compactness upstairs

We now turn to proving Proposition 16.1.4. The proof follows the same line of argument as in the downstairs model but we need to have control over both $\Lambda_{\mathfrak{q}}$ and the energy. We take up this issue now.

For a trajectory $\gamma^\tau \in M([\mathfrak{a}], [\mathfrak{b}])$ define $K(\gamma^\tau)$ to be the total variation of $\Lambda_{\mathfrak{q}}$,

$$K(\gamma^\tau) = \int_{\mathbb{R}} \left| \frac{d\Lambda_{\mathfrak{q}}(\gamma^\tau)}{dt} \right| dt,$$

and define

$$K_+(\gamma^\tau) = \int_{\mathbb{R}} \left(\frac{d\Lambda_{\mathfrak{q}}(\gamma^\tau)}{dt} \right)^+ dt$$

where $f^+ = \max\{0, f\}$ denotes the positive part of a function f. A priori, both of these may be infinite; but if one is finite then so is the other, and we have

$$\Lambda_{\mathfrak{q}}(\mathfrak{a}) - \Lambda_{\mathfrak{q}}(\mathfrak{b}) = K(\gamma^\tau) - 2K_+(\gamma^\tau).$$

For an interval I we define $K^I(\gamma^\tau)$ and $K_+^I(\gamma^\tau)$ by replacing $\mathbb{R}$ by I as the domain of integration in the above formulae. For a broken trajectory, we define K and K_+ to be the sums of the corresponding function of its components. The following lemma depends crucially on the decay results proved in Subsection 13.5, and on Corollary 13.5.3 in particular.

Lemma 16.3.1. *For any broken trajectory $\boldsymbol{\gamma}$, $K(\boldsymbol{\gamma})$ is finite. Furthermore the function $K : \breve{M}^+([\mathfrak{a}], [\mathfrak{b}]) \to \mathbb{R}$ is bounded on any subset on which the energy is bounded.*

Proof. We can deal with K_+ instead of K. The function K_+ is identically zero on any component of a broken trajectory whose blow-down is a constant trajectory, since $\Lambda_{\mathfrak{q}}$ is strictly decreasing on such components. Proposition 16.2.1 tells us that a bound on the energy gives a bound on the number of components for which the blow-down is non-constant. Because of these two observations, we can reduce the problem to showing that for a sequence of *unbroken* trajectories $[\gamma^n] \in M([\mathfrak{a}], [\mathfrak{b}])$ with bounded energy, $K_+(\gamma^n)$ is uniformly bounded. We argue by contradiction and suppose that $K_+(\gamma^n)$ is increasing without bound.

Choose a constant η, and for each downstairs critical point $[\alpha]$ choose a gauge-invariant neighborhood $U_\alpha \subset \mathcal{C}_k([-1, 1] \times Y)$ such that the conclusion of Corollary 13.5.3 holds. From the proof of Proposition 16.2.1, after passing to a subsequence, we can find for each n a decomposition of the real line into intervals I_i^n for $i = 1, \ldots, d$ and J_i^n with $i = 0, 1, \ldots, d$ such that

- I_i^n have length independent of n, and
- for each subinterval $[t - 1, t + 1] \subset J_i^n$ of length 2, the translate $(\tau_t^* \gamma^n)|_{[-1,1]}$ lies in U_α for some α.

Let $\tilde{J}_i^n \subset J_i^n$ be the smaller interval obtained by bringing in the endpoint distance 1. Corollary 13.5.3 then implies that the contribution to K_+ from the $\tilde{J}_i^n$ is bounded by $(d + 1)\eta$. Let $\tilde{I}_i^n \supset I_i^n$ be the larger intervals obtained by moving the endpoints out distance 1 (so that, for each n, the real line is now the union of the intervals $\tilde{I}_i^n$ and the intervals $\tilde{J}_i^n$). Since these intervals $\tilde{I}_i^n$ have uniformly bounded length, Proposition 16.1.6 and Lemma 10.9.1 imply that the contribution to K_+ from the intervals $\tilde{I}_i^n$ is uniformly bounded. □

With this in place the proof of Proposition 16.1.4 proceeds as follows. Fix $[\mathfrak{a}], [\mathfrak{b}]$ and $C > 0$ as in the statement of the proposition and suppose we have a sequence $[\gamma^n] \in M([\mathfrak{a}], [\mathfrak{b}])$ with $\mathcal{E}(\gamma^n) \leq C$. Lemma 16.3.1 tells us that K is bounded on this sequence, so the values $\Lambda_{\mathfrak{q}}(\gamma^n(t))$ $(t \in \mathbb{R}, n \in \mathbb{N})$ are bounded above and below. Let I be any closed interval, and for each critical point $[\mathfrak{c}]$ in $\mathcal{B}_k^\sigma(Y)$, let $U_{\mathfrak{c}} \subset \mathcal{C}_k^\sigma(I \times Y)$ be a gauge-invariant neighborhood of $\gamma_{\mathfrak{c}}$. The lemma below is a version of Lemma 16.2.2, adapted to the blown-up situation.

Lemma 16.3.2. *Let $U_{\mathfrak{c}} \subset \mathcal{C}_k^\tau(I \times Y)$ be neighborhoods as above, let C be any constant, and let I' be any other interval of non-zero length. Then there exist*

$\epsilon_1, \epsilon_2 > 0$ *such that if* $\gamma \in M([\mathfrak{a}], [\mathfrak{b}])$ *is a trajectory satisfying*

$$\mathcal{E}_{\mathfrak{q}}(\gamma) \leq C$$
$$\mathcal{E}_{\mathfrak{q}}^{I'}(\gamma) \leq \epsilon_1$$
$$K^{I'}(\gamma) \leq \epsilon_2,$$

then $\gamma|_{I\times Y} \in U_{\mathfrak{c}}$ *for some critical point* $[\alpha]$.

Proof. If not then there exists a sequence γ_i with $\mathcal{E}_{\mathfrak{q}}(\gamma_i) \leq C$, $\mathcal{E}_{\mathfrak{q}}^{I'}(\gamma_i) \to 0$ and $K^{I'}(\gamma_i) \to 0$, none of which belongs to any $U_{\mathfrak{c}}$. The first bound implies that $\Lambda_{\mathfrak{q}}$ is uniformly bounded above and below on these trajectories, as we pointed out just above; so after gauge transformation, a subsequence converges in $\mathcal{C}_{k,\mathrm{loc}}(Z)$ to a limit γ_∞, by Proposition 16.1.5. The limit must have energy zero on I' and zero change in $\Lambda_{\mathfrak{q}}$ on I'. It must therefore be a constant trajectory, gauge-equivalent to some $\gamma_{\mathfrak{c}}$ on the whole of Z by unique continuation. This means that a subsequence of the $\gamma_i|_{I\times Y}$ converges to $\gamma_{\mathfrak{c}}$ in $\mathcal{C}_k^\sigma(I \times Y)$ after gauge transformation. So, in this subsequence, $\gamma_i|_{I\times Y}$ is eventually an element of $U_{\mathfrak{c}}$. This is a contradiction. □

Recall that in Definition 16.2.3 we have already extended the notion of *separating neighborhoods* to the blown-up context. With this understood, the following adaptation of Lemma 16.2.2 is proved in just the same way as the original version:

Lemma 16.3.3. *Let* $[\gamma] \in \mathcal{B}_{k,\mathrm{loc}}^\tau(Z)$ *be a solution of the equations with finite energy and finite* $K(\gamma)$. *Then* $[\gamma]$ *belongs to* $M([\mathfrak{a}], [\mathfrak{b}])$ *for some critical points* $[\mathfrak{a}]$ *and* $[\mathfrak{b}]$. □

With these two lemmas in hand, the remainder of the proof of Proposition 16.1.4 is essentially identical to the downstairs case, except that wherever we previously had a condition such as $\mathcal{E}_{\mathfrak{q}}^I(\gamma) < \epsilon$, we should now have both $\mathcal{E}_{\mathfrak{q}}^I(\gamma) < \epsilon_1$ and $K^I(\gamma) < \epsilon_2$.

To spell out the argument a little, consider a sequence $[\gamma^n]$ in $M([\mathfrak{a}], [\mathfrak{b}])$ with energy bounded by C. Set $I = [-1, 1]$ and choose a separating collection of neighborhoods $\mathcal{U} = \{U_{\mathfrak{c}}\}$ for the interval I in the blown-up configuration space. Let $V_{\mathfrak{c}} \subset \mathcal{B}_{k-1}^\sigma(Y)$ be the corresponding neighborhoods in the 3-dimensional configuration space. An important preliminary observation is that, although there may be infinitely many critical points in the blown-up picture (unlike the situation downstairs), only *finitely many* of these critical points are in play: we have seen that there is a uniform bound

$$\left|\Lambda_{\mathfrak{q}}(\gamma^n(t))\right| \le \Lambda_0 \quad (t \in \mathbb{R}, n \in \mathbb{N}),$$

and because the values of $\Lambda_{\mathfrak{q}}$ at the critical points are a discrete set, there will only be finitely many critical points $\mathfrak{c}$ for which $\gamma^n(t)$ can ever meet the neighborhood $V_{\mathfrak{c}}$.

Lemma 16.3.2 provides ϵ_1 and ϵ_2 such that for any trajectory with $\mathcal{E}^I_{\mathfrak{q}}(\gamma) \le \epsilon_1$ and $K^I_{\mathfrak{q}}(\gamma) \le \epsilon_2$, we have that $\gamma|_I$ belongs to $U_\alpha \in \mathcal{U}$ for some $[\alpha]$. Let K_0 be a uniform upper bound for $K(\gamma^n)$. It follows that, for each n, there are at most $(2C/\epsilon_1) + (2K_0/\epsilon_2)$ integers p such that

$$\tau_p \gamma^n \notin \bigcup_{[\mathfrak{c}]} U_{\mathfrak{c}}.$$

As in the previous proof, after passing to a subsequence, we arrive at the following picture. For each n, there is a decomposition of the line into intervals I^n_i $(1 \le i \le d)$ and J^n_i $(0 \le i \le s)$, with

$$I^n_i = [a^n_i, b^n_i]$$

and

$$\begin{aligned} J^n_0 &= (-\infty, a_1] \\ J^n_i &= [b^n_i, a^n_{i+1}], \quad 1 \le i \le d-1 \\ J^n_d &= [b_d, \infty). \end{aligned}$$

The endpoints of these intervals are integers, and the length of I^n_i is independent of n. Furthermore, there are critical points $\mathfrak{a} = \mathfrak{a}_0, \ldots, \mathfrak{a}_d = \mathfrak{b}$, such that

$$p \in \mathbb{Z} \cap \operatorname{interior}(J^n_i) \implies \tau_p \gamma^n \in U_{\mathfrak{a}_i},$$

while

$$p \in \mathbb{Z} \cap I^n_i \implies \tau_p \gamma^n \notin \bigcup_{[\mathfrak{c}]} U_{\mathfrak{c}}.$$

The proof is completed as before. □

16.4 Finiteness results

Our compactness result, Theorem 16.1.3, refers to a single moduli space, and although we have a slightly stronger result in Proposition 16.1.4, the statement of that proposition required a bound on the energy. If the moduli spaces of trajectories are not regular (in particular, if there are moduli spaces which are non-empty despite $\mathrm{gr}_z([\mathfrak{a}], [\mathfrak{b}])$ being negative), then the results we have stated cannot be improved. However, if we assume the perturbation $\mathfrak{q}$ is chosen so that the moduli spaces are regular, then we have stronger finiteness results for the set of non-empty moduli spaces. We state two such propositions, together with a corollary of the first one, before turning to their proof.

Proposition 16.4.1. *Suppose that all the moduli spaces $M_z([\mathfrak{a}], [\mathfrak{b}])$ are regular. Then for given $[\mathfrak{a}]$ and $[\mathfrak{b}]$, there are only finitely many homotopy classes z for which the moduli space $\breve{M}_z^+([\mathfrak{a}], [\mathfrak{b}])$ is non-empty.*

Corollary 16.4.2. *In a broken trajectory $[\breve{\boldsymbol{\gamma}}] = ([\breve{\gamma}_1], \ldots, [\breve{\gamma}_n])$, the restpoints $[\mathfrak{a}_0], \ldots, [\mathfrak{a}_n]$ are distinct.* □

Proposition 16.4.3. *Suppose that all the moduli spaces $M_z([\mathfrak{a}], [\mathfrak{b}])$ are regular.*

(i) *If $c_1(\mathfrak{s})$ is torsion then for a given $[\mathfrak{a}]$ and any d_0, there are only finitely many pairs $([\mathfrak{b}], z)$ for which the moduli space $\breve{M}_z^+([\mathfrak{a}], [\mathfrak{b}])$ is non-empty and of dimension at most d_0.*
(ii) *If $c_1(\mathfrak{s})$ is non-torsion, suppose also that the perturbation has been chosen so that there are no reducible solutions. Then there are only finitely many triples $([\mathfrak{a}], [\mathfrak{b}], z)$ for which the moduli space $\breve{M}_z^+([\mathfrak{a}], [\mathfrak{b}])$ is non-empty (without restriction on the dimension).*

Remark. When $c_1(\mathfrak{s})$ is non-torsion and there is no perturbation, there are no reducible solutions. By choosing the perturbation to be small, we can therefore fulfill the extra hypothesis in the second case of this proposition.

Proof of Proposition 16.4.1. This proposition follows from Proposition 16.1.4 if we can show that for given $[\mathfrak{a}]$ and $[\mathfrak{b}]$, there is a bound on the energy of solutions in non-empty moduli spaces, independent of the homotopy class z.

We recall that a reducible critical point $\mathfrak{a}$ corresponds to a pair (α, λ), where $\alpha = (B, 0)$ is the critical point $\pi(\mathfrak{a})$ in $\mathcal{C}_k(Y)$, and λ is an element of $\mathrm{Spec}(D_{\mathfrak{q},B})$. For such an $\mathfrak{a}$, we define

$$\iota(\mathfrak{a}) = \begin{cases} \left|\left(\mathrm{Spec}(D_{\mathfrak{q},B}) \cap [0, \lambda)\right)\right|, & \lambda > 0 \\ 1/2 - \left|\left(\mathrm{Spec}(D_{\mathfrak{q},B}) \cap [\lambda, 0]\right)\right|, & \lambda < 0. \end{cases} \tag{16.2}$$

For an irreducible critical point $\mathfrak{a}$, we set $\iota(\mathfrak{a}) = 0$. This definition is set up so that, if $[\mathfrak{a}]$ and $[\mathfrak{a}']$ are two critical points whose images under the projection π are equal to the same critical point $[\alpha]$ in $\mathcal{B}_k(Y)$, then

$$\mathrm{gr}_{z_0}([\mathfrak{a}], [\mathfrak{a}']) = 2(\iota(\mathfrak{a}) - \iota(\mathfrak{a}')) \tag{16.3}$$

for the trivial homotopy class z_0. This equality is a restatement of Corollary 14.6.2. The following lemma provides the z-independent energy bound required to prove Proposition 16.4.1. □

Lemma 16.4.4. *If all moduli spaces are regular, then there exists a C such that for every* $[\mathfrak{a}]$, $[\mathfrak{b}]$ *and* z, *and every broken trajectory* $[\breve{\boldsymbol{\gamma}}]$ *in* $\breve{M}_z^+([\mathfrak{a}], [\mathfrak{b}])$, *we have the energy bound*

$$\mathcal{E}_{\mathfrak{q}}(\breve{\boldsymbol{\gamma}}) \leq C + 8\pi^2(\iota([\mathfrak{a}]) - \iota([\mathfrak{b}])).$$

Proof. Let $[\breve{\boldsymbol{\gamma}}] = ([\breve{\gamma}_1], \ldots, [\breve{\gamma}_l])$ be a broken trajectory in $\breve{M}_z^+([\mathfrak{a}], [\mathfrak{b}])$, with

$$[\breve{\gamma}_i] \in \breve{M}_{z_i}([\mathfrak{a}_{i-1}], [\mathfrak{a}_i]).$$

The space $\breve{M}_{z_i}([\mathfrak{a}_{i-1}], [\mathfrak{a}_i])$ is then non-empty, and it is a manifold (possibly with boundary) of dimension 1 less than the dimension of $M_{z_i}([\mathfrak{a}_{i-1}], [\mathfrak{a}_i])$. Thus $\dim \breve{M}_{z_i}([\mathfrak{a}_{i-1}], [\mathfrak{a}_i])$ is either $\mathrm{gr}_{z_i}([\mathfrak{a}_{i-1}], [\mathfrak{a}_i]) - 1$, or $\mathrm{gr}_{z_i}([\mathfrak{a}_{i-1}], [\mathfrak{a}_i])$. (See Proposition 14.5.7.) In either case,

$$\mathrm{gr}_{z_i}([\mathfrak{a}_{i-1}], [\mathfrak{a}_i]) \geq 0,$$

and by additivity of gr, we have

$$\mathrm{gr}_z([\mathfrak{a}_0], [\mathfrak{a}_l]) \geq 0,$$

because z is the class of the composite of all the z_i. The energy $\mathcal{E}_{\mathfrak{q}}(\breve{\boldsymbol{\gamma}})$ is equal to the quantity $\mathcal{E}_{\mathfrak{q}}(z)$, defined as twice the change in $\mathcal{L}$ along any path $\tilde{\zeta}$ in $\mathcal{C}_k^\sigma(Y)$ whose image ζ in $\mathcal{B}_k^\sigma(Y)$ belongs to the class $z \in \pi_1(\mathcal{B}_k^\sigma(Y), [\mathfrak{a}], [\mathfrak{b}])$.

Consider now the quantity

$$\mathcal{E}_{\mathfrak{q}}(w) + 4\pi^2\,\mathrm{gr}_w([\mathfrak{a}], [\mathfrak{b}]), \tag{16.4}$$

for $w \in \pi_1(\mathcal{B}_k^\sigma(Y), [\mathfrak{a}], [\mathfrak{b}])$. We claim this depends only on the endpoints $[\mathfrak{a}]$ and $[\mathfrak{b}]$, not on the homotopy class w. If we compare two different homotopy classes, their difference is the class of a closed loop z_u whose lift to $\mathcal{C}_k^\sigma(Y)$ joins $\mathfrak{a}$ to $u\mathfrak{a}$, for some $u \in \mathcal{G}_{k+1}(Y)$. The difference between the values of (16.4) for the two classes is

$$\begin{aligned}\mathcal{E}_{\mathfrak{q}}(z_u) + 4\pi^2 \operatorname{gr}_{z_u}([\mathfrak{a}],[\mathfrak{a}]) &= -4\pi^2([u] \cup c_1(S))[Y] \\ &\quad + 4\pi^2([u] \cup c_1(S))[Y] \\ &= 0,\end{aligned}$$

where the first equality follows from Lemma 4.1.3 and Lemma 14.4.6.

The quantity

$$\mathcal{E}_{\mathfrak{q}}(w) + 4\pi^2\big(\operatorname{gr}_w([\mathfrak{a}],[\mathfrak{b}]) - 2\iota(\mathfrak{a}) + 2\iota(\mathfrak{b})\big) \tag{16.5}$$

can now be seen to depend only on the critical points $[\alpha] = [\boldsymbol{\pi}(\mathfrak{a})]$ and $[\beta] = [\boldsymbol{\pi}(\mathfrak{b})]$ in $\mathcal{B}_k(Y)$, using (16.3). Since there are only finitely many critical points in $\mathcal{B}_k(Y)$, there is a C such that this quantity is at most C, for all $[\mathfrak{a}]$, $[\mathfrak{b}]$ and w. Returning to our broken trajectory $[\breve{\boldsymbol{\gamma}}]$, we therefore have

$$\begin{aligned}\mathcal{E}(\breve{\boldsymbol{\gamma}}) &= \mathcal{E}(z) \\ &\le C - 4\pi^2\big(\operatorname{gr}_z([\mathfrak{a}],[\mathfrak{b}]) - 2\iota([\mathfrak{a}]) + 2\iota([\mathfrak{b}])\big) \\ &\le C + 8\pi^2(\iota([\mathfrak{a}]) - \iota([\mathfrak{b}])).\end{aligned}$$

□

Proof of Proposition 16.4.3. In the case that $c_1(\mathfrak{s})$ is torsion, $\mathcal{L}$ descends to a single-valued real function on $\mathcal{B}_k^\sigma(Y)$. Furthermore, since the function $\mathcal{L}$ on $\mathcal{B}_k^\sigma(Y)$ is pulled back from $\mathcal{B}_k(Y)$, and since the image of the critical set in $\mathcal{B}_k(Y)$ is finite, the function $\mathcal{L}$ takes only finitely many values on the critical set in $\mathcal{B}_k^\sigma(Y)$. The energy $\mathcal{E}_{\mathfrak{q}}$ of a trajectory is twice the change in $\mathcal{L}$, so there is a uniform bound on the energy of all solutions.

The expression (16.5) depends only on $\boldsymbol{\pi}([\mathfrak{a}])$ and $\boldsymbol{\pi}([\mathfrak{b}])$ and therefore takes only finitely many values. We have upper and lower bounds on $\mathcal{E}_{\mathfrak{q}}$, and assumed that the dimension is bounded and $[\mathfrak{a}]$ is fixed, so we conclude that $\iota([\mathfrak{b}])$ is bounded above and below, leaving only finitely many possibilities for $[\mathfrak{b}]$. This combined with the previous proposition implies the result for the torsion case.

In the case that $c_1(\mathfrak{s})$ is not torsion, if we choose our perturbation so that there are no reducible critical points, the critical set is finite so the result follows directly from the previous proposition. □

16.5 The compactification as a stratified space

In Section 19, we will examine the structure of the compactification $\breve{M}_z^+([\mathfrak{a}],[\mathfrak{b}])$ in detail. At present, we make only some remarks, beginning with a convenient definition. We will assume now that the perturbation is chosen so that all moduli spaces are regular.

Definition 16.5.1. A space N^d is a *d-dimensional space stratified by manifolds* if there are closed subsets

$$N^d \supset N^{d-1} \supset \cdots \supset N^0 \supset N^{-1} = \varnothing$$

such that $N^d \neq N^{d-1}$ and each space $N^e \setminus N^{e-1}$ (for $0 \le e \le d$) is either empty or homeomorphic to a manifold of dimension e. The difference $N^e \setminus N^{e-1}$ is called the *e-dimensional stratum*. We will also use the term *stratum* to refer to any union of path components of $N^e \setminus N^{e-1}$. ◊

Proposition 16.5.2. *Suppose that $M_z([\mathfrak{a}],[\mathfrak{b}])$ is non-empty and of dimension d. Then the compactification $\breve{M}_z^+([\mathfrak{a}],[\mathfrak{b}])$ is a $(d-1)$-dimensional space stratified by manifolds.*

If $M_z([\mathfrak{a}],[\mathfrak{b}])$ contains irreducible trajectories, then the $(d-1)$-dimensional stratum of $\breve{M}_z^+([\mathfrak{a}],[\mathfrak{b}])$ consists of the irreducible part of $\breve{M}_z([\mathfrak{a}],[\mathfrak{b}])$.

Proof. From its definition, $\breve{M}_z^+([\mathfrak{a}],[\mathfrak{b}])$ is a disjoint union of subspaces of the form

$$\breve{M}_{z_1}([\mathfrak{a}_0],[\mathfrak{a}_1]) \times \cdots \times \breve{M}_{z_i}([\mathfrak{a}_{i-1}],[\mathfrak{a}_i]) \times \cdots \times \breve{M}_{z_l}([\mathfrak{a}_{l-1}],[\mathfrak{a}_l]). \quad (16.6)$$

The space (16.6) is a product of spaces each of which is either a manifold or a manifold with boundary, by Proposition 14.5.7. A manifold with boundary is a simple example of a space stratified by manifolds, and the product (16.6) obtains a product stratification. In this way, $\breve{M}_z^+([\mathfrak{a}],[\mathfrak{b}])$ is eventually a union of subspaces each of which is a manifold of some dimension. To show that this compactification is actually a space stratified by manifolds, we need to understand the closure relation between the strata and their dimensions. We must verify that the closure in $\breve{M}_z^+([\mathfrak{a}],[\mathfrak{b}])$ of one of the e-dimensional manifolds M^e arising from a product of the form (16.6) is contained in the union of M^e and other such manifolds $M^{e'}$ of strictly smaller dimension. It is sufficient to treat the case of the stratum $\breve{M}_z([\mathfrak{a}],[\mathfrak{b}])$.

The statement that $M_z([\mathfrak{a}],[\mathfrak{b}])$ has dimension d means that $\mathrm{gr}_z([\mathfrak{a}],[\mathfrak{b}])$ is $d-\epsilon$, where $\epsilon = 1$ if the moduli space is boundary-obstructed and 0 otherwise (see Proposition 14.5.7). Consider a space of the form (16.6), and for brevity let us write

$$M_i = M_{z_i}([\mathfrak{a}_{i-1}], [\mathfrak{a}_i]).$$

Let the dimension of M_i be d_i and let $\mathrm{gr}_{z_i}([\mathfrak{a}_{i-1}], [\mathfrak{a}_i]) = d_i - \epsilon_i$ with $\epsilon_i = 0$ or 1 in the same way. The additivity of gr means that we have

$$d - \epsilon = \sum_{i=1}^{l} d_i - \sum_{i=1}^{l} \epsilon_i.$$

On the other hand, the dimension of (16.6) is

$$\left(\sum_i d_i\right) - l = d - \epsilon + \left(\sum_i \epsilon_i\right) - l. \tag{16.7}$$

We need a simple lemma at this point:

Lemma 16.5.3. (i) *In a broken trajectory $[\breve{\gamma}] = ([\breve{\gamma}_1], \ldots, [\breve{\gamma}_l])$ belonging to $\breve{M}_z^+([\mathfrak{a}], [\mathfrak{b}])$, two adjacent components $[\breve{\gamma}_i]$, $[\breve{\gamma}_{i+1}]$ cannot both be boundary-obstructed.*

(ii) *If $M_z([\mathfrak{a}], [\mathfrak{b}])$ contains irreducible trajectories, then neither $[\breve{\gamma}_1]$ nor $[\breve{\gamma}_l]$ can be boundary-obstructed.*

(iii) *If $M_z([\mathfrak{a}], [\mathfrak{b}])$ consists only of reducibles but is not boundary-obstructed, then at most one of $[\breve{\gamma}_1]$ and $[\breve{\gamma}_l]$ is boundary-obstructed.*

We note an elementary corollary:

Corollary 16.5.4. *The number of components $[\breve{\gamma}_i]$ of $[\breve{\gamma}]$ that are boundary-obstructed is at most:*

(i) $(l+1)/2$ *in general;*

(ii) $(l-1)/2$ *if $M_z([\mathfrak{a}], [\mathfrak{b}])$ contains irreducible trajectories;*

(iii) $l/2$ *if $M_z([\mathfrak{a}], [\mathfrak{b}])$ consists only of reducibles but is not boundary-obstructed.*

□

Proof of Lemma 16.5.3. If $[\breve{\gamma}_i]$ is boundary-obstructed, then (from the definition) the critical point $[\mathfrak{a}_{i-1}]$ is boundary-stable and $[\mathfrak{a}_i]$ is boundary-unstable. The trajectory $[\breve{\gamma}_{i+1}]$ cannot then be boundary-obstructed, because $[\mathfrak{a}_i]$ is not boundary-stable. This proves the first statement. If $M_z([\mathfrak{a}], [\mathfrak{b}])$ contains irreducible trajectories, then $[\mathfrak{a}]$ cannot be reducible and boundary-stable, nor can $[\mathfrak{b}]$ be reducible and boundary-unstable. So neither $[\breve{\gamma}_1]$ nor $[\breve{\gamma}_l]$ can be boundary-obstructed, which is the second statement. The third statement is similar. □

The corollary tells us that

$$\left(\sum_i \epsilon_i\right) \leq (l+1)/2$$

with equality only if $\epsilon = 1$. It follows in this case that the expression (16.7) is at most $d - 1/2$, and therefore at most $d - 1$ (because it is an integer). Equality occurs in two cases. The first case is the case $l = 1$, which is the case that (16.6) is just $\breve{M}_z([\mathfrak{a}], [\mathfrak{b}])$. The second case of equality is when $l = 2$, exactly one of $[\breve{\gamma}_1]$, $[\breve{\gamma}_2]$ is boundary-obstructed, and $\epsilon = 0$. Let us suppose that it is $[\breve{\gamma}_1]$ that is boundary-obstructed (the other case is similar). By the lemma, $M_z([\mathfrak{a}], [\mathfrak{b}])$ consists entirely of reducibles, and $[\mathfrak{a}_0]$, $[\mathfrak{a}_1]$, $[\mathfrak{a}_2]$ are boundary-stable, boundary-unstable and boundary-stable respectively. Because it consists of reducibles, the closure of $\breve{M}_z([\mathfrak{a}], [\mathfrak{b}])$ in $\breve{M}_z^+([\mathfrak{a}], [\mathfrak{b}])$ cannot contain all of $\breve{M}_1 \times \breve{M}_2$: it can only contain the reducible part of this product, which comprises the $(d-2)$-manifold $\breve{M}_1 \times \partial\breve{M}_2$. Thus in all cases the closure of $\breve{M}_z([\mathfrak{a}], [\mathfrak{b}])$ is contained in the union of $\breve{M}_z([\mathfrak{a}], [\mathfrak{b}])$ and manifolds of lower dimension. □

Remark. Our definition of a space stratified by manifolds allows some pathology, as the following (compact) example shows. Let C_n be the circle in $\mathbb{R}^2$ of radius $1/n$ centered at $(-1/n, 0)$. Let N^1 be the union of the circles C_n for all $n \geq 1$ and the line segment joining $(0,0)$ to $(1,0)$. Let $N^0 \subset N^1$ consist of the two points $(0,0)$ and $(1,0)$. Then $N^1 \setminus N^0$ is a 1-manifold.

At no point will we prove any result that rules out the possibility that $\breve{M}_z^+([\mathfrak{a}], [\mathfrak{b}])$ might be homeomorphic to N^1, with $\breve{M}_z([\mathfrak{a}], [\mathfrak{b}])$ corresponding to the subset $N^1 \setminus N^0$. The authors believe that a strengthening of the results of Subsection 19.4 would rule out such pathology for a suitable choice of perturbation $\mathfrak{q}$.

Let $M_z([\mathfrak{a}], [\mathfrak{b}])$ be a d-dimensional moduli space containing some irreducible trajectories. We have $\mathrm{gr}_z([\mathfrak{a}], [\mathfrak{b}]) = d$, because a moduli space containing irreducibles is not boundary-obstructed. We saw in Proposition 16.5.2 that $\breve{M}_z^+([\mathfrak{a}], [\mathfrak{b}])$ is a compact space stratified by manifolds, of dimension $d - 1$. We now wish to describe all the contributions to the codimension-1 stratum. The proposition below describes all the strata in the compactification that can have codimension 1. In the special case that the dimension d of

$M_z([\mathfrak{a}],[\mathfrak{b}])$ is 2, the proposition should be compared with Lemma 2.4.3, which is the corresponding statement for the finite-dimensional model.

Consider again the contribution

$$\breve{M}_{z_1}([\mathfrak{a}_0],[\mathfrak{a}_1]) \times \cdots \times \breve{M}_{z_l}([\mathfrak{a}_{l-1}],[\mathfrak{a}_l]) \tag{16.8}$$

to $\breve{M}_z^+([\mathfrak{a}],[\mathfrak{b}])$. Let d_i and ϵ_i be as above, so that $d_i - \epsilon_i = \mathrm{gr}_{z_i}([\mathfrak{a}_{i-1}],[\mathfrak{a}_i])$. We refer to

$$(d_1 - \epsilon_1, \ldots, d_l - \epsilon_l)$$

as the *grading vector* of (16.8) and refer to $(\epsilon_1, \ldots, \epsilon_l)$ as the *obstruction vector*.

Proposition 16.5.5. *Let $M_z([\mathfrak{a}],[\mathfrak{b}])$ be a d-dimensional moduli space containing irreducibles, so that $\breve{M}_z^+([\mathfrak{a}],[\mathfrak{b}])$ is a compact $(d-1)$-dimensional space stratified by manifolds, with top stratum the irreducible part of $\breve{M}_z([\mathfrak{a}],[\mathfrak{b}])$. Then the $(d-2)$-dimensional stratum of $\breve{M}_z^+([\mathfrak{a}],[\mathfrak{b}])$ is the union of pieces of three types:*

(i) *the top stratum of a part (16.8) with grading vector (d_1, d_2) and obstruction vector $(0,0)$;*
(ii) *the top stratum of a part (16.8) with grading vector $(d_1, d_2 - 1, d_3)$ and obstruction vector $(0,1,0)$;*
(iii) *the intersection of $\breve{M}_z([\mathfrak{a}],[\mathfrak{b}])$ with the reducibles, if $M_z([\mathfrak{a}],[\mathfrak{b}])$ contains both reducibles and irreducibles.*

The third case occurs only when $[\mathfrak{a}]$ is boundary-unstable and $[\mathfrak{b}]$ is boundary-stable. In the first case, $d_1 + d_2 = d$. In the second case, $d_1 + d_2 + d_3 = d + 1$. In all cases, the d_i are positive.

Proof. Consider the space (16.8). By Corollary 16.5.4, we have $\sum \epsilon_i \le (l-1)/2$, because the moduli space $M_z([\mathfrak{a}],[\mathfrak{b}])$ contains reducibles. The expression (16.7) for the dimension of this stratum is therefore bounded by

$$d - \epsilon - \frac{1}{2} - \frac{l}{2}.$$

We are seeking broken trajectories ($l \ge 2$) for which this dimension is $d-2$. There are two possibilities. The first is that $l = 2$, $\epsilon = 0$ and $\sum \epsilon_i = 0$. This is the first case in the proposition. The second possibility is that $l = 3$, $\epsilon = 0$ and $\sum \epsilon_i = 1$. This means that, of the three components in the broken trajectory, exactly one is boundary-obstructed. The boundary-obstructed component must be the middle one of the three: $[\mathfrak{a}]$ cannot be boundary-stable and $[\mathfrak{b}]$

cannot be boundary-unstable, because the moduli space $M_z([\mathfrak{a}],[\mathfrak{b}])$ contains irreducibles. □

16.6 Compactification of reducible trajectories

We will write $M_z^{\mathrm{red}}([\mathfrak{a}],[\mathfrak{b}])$ for the subset of $M_z([\mathfrak{a}],[\mathfrak{b}])$ consisting of the reducible trajectories. Recall that, under our transversality hypotheses, this is either empty, or all of $M_z([\mathfrak{a}],[\mathfrak{b}])$, or the boundary of $M_z([\mathfrak{a}],[\mathfrak{b}])$ in the case that $[\mathfrak{a}]$ is boundary-unstable and $[\mathfrak{b}]$ is boundary-stable. For reducible critical points, we introduce a modified relative grading, setting

$$\begin{aligned}\bar{\mathrm{gr}}_z([\mathfrak{a}],[\mathfrak{b}]) &= \dim M_z^{\mathrm{red}}([\mathfrak{a}],[\mathfrak{b}]) \\ &= \mathrm{gr}_z([\mathfrak{a}],[\mathfrak{b}]) - o[\mathfrak{a}] + o[\mathfrak{b}],\end{aligned} \tag{16.9}$$

where

$$o[\mathfrak{a}] = \begin{cases} 0, & \text{if } [\mathfrak{a}] \text{ is boundary-stable} \\ 1, & \text{if } [\mathfrak{a}] \text{ is boundary-unstable.} \end{cases} \tag{16.10}$$

Along with M^{red}, we can introduce $\breve{M}^{\mathrm{red}}$ and $\breve{M}^{\mathrm{red}+}$, as the intersections of $\breve{M}$, $\breve{M}^+$ with the reducibles.

The situation for these reducible moduli spaces is somewhat simpler. The dimension of $M_z^{\mathrm{red}}([\mathfrak{a}],[\mathfrak{b}])$ is always equal to $\bar{\mathrm{gr}}_z([\mathfrak{a}],[\mathfrak{b}])$, which is additive. (The dimension of $M_z([\mathfrak{a}],[\mathfrak{b}])$ is not additive, because of the boundary-obstructed case.) Also, $M_z^{\mathrm{red}}([\mathfrak{a}],[\mathfrak{b}])$ is always a manifold, never a manifold with boundary. We omit the proof of the following proposition, which summarizes the adaptations of the results of the previous subsection to the case of the reducible moduli spaces.

Proposition 16.6.1. *Suppose $M_z^{\mathrm{red}}([\mathfrak{a}],[\mathfrak{b}])$ is non-empty and of dimension d. Then the space of unparametrized, broken reducible trajectories, $\breve{M}_z^{\mathrm{red}+}([\mathfrak{a}],[\mathfrak{b}])$, is a compact $(d-1)$-dimensional space stratified by manifolds. The top stratum consists of $\breve{M}_z^{\mathrm{red}}([\mathfrak{a}],[\mathfrak{b}])$ alone. The $(d-l)$-dimensional stratum consists of the spaces of unparametrized broken trajectories with l factors:*

$$\breve{M}_{z_1}^{\mathrm{red}}([\mathfrak{a}_0],[\mathfrak{a}_1]) \times \cdots \times \breve{M}_{z_l}^{\mathrm{red}}([\mathfrak{a}_{l-1}],[\mathfrak{a}_l]).$$

□

17 The moduli space on a finite cylinder

In this section we will set up the theory required to handle moduli spaces on manifolds with boundary. On a manifold with boundary, moduli spaces are no longer finite dimensional but we shall see that they are Hilbert manifolds when we specify the regularity of the configurations involved. The main tool is the Fredholm theory for certain kinds of boundary-value problems. The model for these results is the work of Atiyah, Patodi and Singer [8]. The appropriate boundary conditions for our moduli problems are spectral boundary conditions analogous to the situation for the $\bar{\partial}$ operator on a disk: a holomorphic function on the disk is determined by the non-negative Fourier coefficients of its restriction to the unit circle.

17.1 Atiyah–Patodi–Singer boundary-value problems

We consider a situation in which X is a compact manifold with boundary, and we suppose in addition that X has a Riemannian metric which is cylindrical in the neighborhood of the boundary. So X contains an isometric copy of $I \times Y$ for some interval $I = (-C, 0]$, with ∂X identified with $\{0\} \times Y$. (Later, we will take X to be a finite cylinder, and Y will become $Y \cup \bar{Y}$.) We suppose we have vector bundles E, F on X, with inner product, and an operator

$$D : C^\infty(X; E) \to L^2(X; F)$$

which has the form

$$D = D_0 + K$$

where D_0 is an elliptic first-order differential operator and K is an operator which extends to a bounded operator

$$K : L^2_j(X; E) \to L^2_j(X; F)$$

for $-k+1 \le j \le k-1$. Near the boundary, we assume a particular form for D_0 of the sort we considered in Subsection 14.1. We suppose that the restrictions of E and F to $I \times Y$ are equipped with fixed isomorphisms with the pull-back of a bundle $E_0 \to Y$, and that for any section u of E, we have (with respect to those isomorphisms)

$$D_0 u|_{I \times Y} = \frac{du}{dt} + L_0 u,$$

where

$$L_0 : C^\infty(Y; E_0) \to C^\infty(Y; E_0)$$

is a first-order, self-adjoint elliptic differential operator.

With these assumptions the operator D extends as a bounded operator

$$D : L^2_{j+1}(X; E) \to L^2_j(X; F)$$

for $j \leq k - 1$. As we shall see, D has closed range with finite-dimensional cokernel; however, the kernel will be infinite-dimensional. We need boundary conditions to get control of the kernel. To gain some intuition about what are good boundary conditions we can consider the case of the half-infinite cylinder $Z = (-\infty, 0] \times Y$ and the translationally invariant situation:

$$\frac{du}{dt} + L_0 u = 0.$$

A typical solution (say smooth) can be written in terms of the eigenvectors of L_0 (see Lemma 12.2.4) as follows:

$$u = \sum_{\lambda \in \mathrm{Spec} L_0} e^{-\lambda t} \phi_\lambda.$$

We can control the behavior of solutions at $-\infty$ by imposing a decay condition. For example the elements of the kernel which are in L^2 clearly can have no components with $\phi_\lambda \neq 0$ for $\lambda \geq 0$. If we hope to get a problem with finite-dimensional kernel, appropriate boundary conditions need to control the projection to the *negative* eigenspace of L_0 of the boundary values.

We will need fractional-order Sobolev spaces to discuss the boundary-value problems precisely. For any real number s the L^2_s norm on sections of $E_0 \to Y$ can be defined in terms of L_0 as

$$\begin{aligned} \|u\|^2_{L^2_s(\{t\}\times Y)} &= \big\| |L_0|^s u \big\|^2_{L^2(\{t\}\times Y)} + \|u_0\|^2_{L^2} \\ &= \sum_{\lambda \neq 0} |\lambda|^{2s} \|u_\lambda\|^2_{L^2} + \|u_0\|^2_{L^2}, \end{aligned} \tag{17.1}$$

where u_λ is the component of u in the λ eigenspace of L_0. Note that when s is integral there is no need for the absolute value of the eigenvalue in the definition. The need for fractional-order Sobolev spaces stems from the Sobolev restriction theorem (see [53, Appendix B]):

Theorem 17.1.1. *For all* $s > 1/2$*, the restriction map* $C^\infty(X;E) \to C^\infty(Y;E_0)$ *extends to a continuous map*

$$r : L^2_s(X;E) \to L^2_{s-1/2}(Y;E_0)$$

which is surjective and has a continuous left-inverse. □

We will only need the restriction theorem when s is a positive integer. We will also need a closely related extension result.

Theorem 17.1.2. *Let* $X' \subset X$ *be a codimension-zero submanifold with boundary, contained in the interior of the manifold* X*. The restriction map* $L^2_k(X) \to L^2_k(X')$ *is a surjective map with a continuous left inverse.* □

These two theorems reflect the fact that we should consider half-integer Sobolev spaces on the 3-manifold for our boundary-value problems.

Let H_0^+ and H_0^- be the closures in $L^2_{1/2}(Y;E_0)$ of the spans of the eigenvectors belonging to positive and non-positive eigenvalues of L_0 respectively. The characterization (17.1) of the Sobolev norm means that these two provide an orthogonal direct sum decomposition of $L^2_{1/2}$. We write Π_0 for the projection

$$\Pi_0 : L^2_{1/2}(Y;E_0) \to L^2_{1/2}(Y;E_0)$$

with image H_0^- and kernel H_0^+. The projection Π_0 also maps $L^2_s(Y;E_0)$ to $L^2_s(Y;E_0)$ for all s, giving rise to a similar decomposition

$$L^2_s(Y;E_0) = (H_0^+ \cap L^2_s) \oplus (H_0^- \cap L^2_s).$$

With these definitions, we can state the main result of this section.

Theorem 17.1.3. *Suppose that* X *is a compact manifold with boundary, and let* D *and* Π_0 *be as above. Then:*

(i) *For* $1 \le j \le k$*, the operator*

$$D \oplus (\Pi_0 \circ r) : L^2_j(X;E) \to L^2_{j-1}(X;F) \oplus (H_0^- \cap L^2_{j-1/2}(Y;E_0))$$

is Fredholm. In particular, the restriction of $\Pi_0 \circ r$ *to* $\ker(D)$ *is Fredholm.*

(ii) *If u_i is a bounded sequence in $L^2_j(X;E)$ and Du_i is Cauchy in L^2_{j-1}, then $(1-\Pi_0)r(u_i)$ has a convergent subsequence in $H_0^+ \cap L^2_{j-1/2}$. In particular, the restriction of $(1-\Pi_0)\circ r$ to $\ker(D)$ is compact.*

(iii) *If u is in $L^2_j(X;E)$ for $j \leq k$ and the image of u under this operator is in $L^2_{k-1}(X;F) \oplus (H_0^- \cap L^2_{k-1/2}(Y;E_0))$, then u lies in $L^2_k(X;E)$. In particular, the kernel consists of L^2_k sections.*

The main tool in proving this result is a similar result for the half-cylinder Z, which we prove first.

Theorem 17.1.4. *Let Z be a half-cylinder $(-\infty, 0] \times Y$, and let $D_0 : C^\infty(Z;E) \to L^2(Z;E)$ be an operator having the form*

$$D_0 = \frac{d}{dt} + L_0$$

where $L_0 : C^\infty(Y;E_0) \to C^\infty(Y;E_0)$ is a self-adjoint elliptic operator on Y. Assume that zero is not in the spectrum of L_0. Then the operator

$$D_0 \oplus (\Pi_0 \circ r) : L^2_j(Z;E) \to L^2_{j-1}(Z;E) \oplus (H_0^- \cap L^2_{j-1/2}(Y;E_0))$$

is an isomorphism for all $j \geq 1$. The subspace $H_0^- \cap L^2_{j-1/2}(Y;E_0)$ is precisely the image of $\ker(D_0)$ under r.

Proof. First we consider the case $j = 1$. Since L_0 is invertible, the L^2_s norm on Y can be defined in terms of L_0 by the formula (17.1), with the u_0 term zero,

$$\|u\|^2_{L^2_s(\{t\}\times Y)} = \sum_{\lambda \in \mathrm{Spec}(L_0)} |\lambda|^{2s} \|u_\lambda\|^2_{L^2},$$

and we take

$$\|u\|^2_{L^2_1(Z)} = \int_{-\infty}^{0} \left(\left\| \frac{du}{dt} \right\|^2_{L^2(Y)} + \|L_0 u\|^2_{L^2(Y)} \right) dt$$

as the definition of the L_1^2 norm on the half-infinite cylinder Z. To see that the operator in question is injective we consider

$$\begin{aligned}
\|D_0 u\|_{L^2(Z)}^2 &= \int_{-\infty}^0 \left\| \frac{du}{dt} + L_0 u \right\|_{L^2(\{t\}\times Y)}^2 dt \\
&= \int_{-\infty}^0 \left\| \frac{du}{dt} \right\|_{L^2(\{t\}\times Y)}^2 dt + \int_{-\infty}^0 \|L_0 u\|_{L^2(\{t\}\times Y)}^2 dt \\
&\quad + 2\int_{-\infty}^0 \left\langle L_0 u, \frac{du}{dt} \right\rangle_{L^2(\{t\}\times Y)} dt \\
&= \|u\|_{L_1^2(Z)}^2 + \langle L_0 r(u), r(u)\rangle \\
&\geq \|u\|_{L_1^2(Z)}^2 + \langle L_0 \Pi_0 \circ r(u), \Pi_0 \circ r(u)\rangle \\
&\geq \|u\|_{L_1^2(Z)}^2 - \|\Pi_0 \circ r(u)\|_{L_{1/2}^2(Y)}^2.
\end{aligned} \tag{17.2}$$

To derive the last inequality we have used again the invertibility of L_0 to conclude that

$$\langle L_0 \Pi_0 \circ r(u), \Pi_0 \circ r(u)\rangle = -\|\Pi_0 \circ r(u)\|_{L_{1/2}^2(Y)}^2$$

again for some positive constant C. Thus we have

$$\|D_0 u\|_{L^2(Z)}^2 + \|\Pi_0 u(0)\|_{L_{1/2}^2(Y)}^2 \geq \|u\|_{L_1^2(Z)}^2$$

so that $D_0 \oplus (\Pi_0 \circ r)$ is injective with closed range. To prove surjectivity, it suffices to prove that $D_0 \oplus (\Pi_0 \circ r)$ has dense range. To solve the equation $(D_0 u, \Pi_0 \circ r(u)) = (v, w)$, decompose u, v and w as L^2-orthogonal sums:

$$\begin{aligned}
u(t) &= \sum_\lambda u_\lambda(t) \\
v(t) &= \sum_\lambda v_\lambda(t) \\
w &= \sum_\lambda w_\lambda.
\end{aligned}$$

To prove density of the range it suffices to consider v and w that involve only finitely many of the eigenspaces; and v_λ can be assumed to be smooth and have compact support in t. We shall show that such (v, w) are in the image of L_1^2.

The equations we are trying to solve then are, for $\lambda > 0$, the equations

$$\frac{du_\lambda}{dt} + \lambda u_\lambda = v_\lambda \text{ for } t \in (-\infty, 0]$$

(with no boundary condition at $t = 0$) and, for $\lambda < 0$, the equations

$$\frac{du_\lambda}{dt} + \lambda u_\lambda = v_\lambda \text{ for } t \in (-\infty, 0]$$
$$u_\lambda(0) = w_\lambda.$$

The solution to the first equation is

$$u_\lambda(t) = \int_{-\infty}^{t} e^{-\lambda(t-s)} v_\lambda(s) ds,$$

while that to the second is

$$u_\lambda(t) = e^{-\lambda t} w_\lambda - \int_{t}^{0} e^{-\lambda(t-s)} v_\lambda(s) ds.$$

Note that in the first case if v_λ has compact support then so does u_λ. In the second case if v_λ has compact support then u_λ is exponentially decaying, and therefore is in L^2. Because the equation tells us that the t derivative of u_λ is in L^2, it follows that u_λ is in L_1^2, so completing the proof. Of course one can easily estimate the operator norm of these inverses on the eigenspaces and prove invertibility in this manner, without first proving that the range is closed (see [8]).

It remains to verify the last claim. The boundary values of the kernel of D_0 acting on L_1^2 clearly can have no component in H^+ since such a component would be exponentially increasing and hence would have infinite norm. Clearly finite linear combinations of eigenvectors of L_0 with negative eigenvalue extend to elements of the kernel that are in L_1^2. The calculation which verified (17.2) above, applied to a u arising as such an extension, implies that

$$\|u\|_{L_1^2(Z)} = \|r(u)\|_{L_{1/2}^2}.$$

Thus the $L_{1/2}^2$ closure of H^- is precisely the space of boundary values of the kernel of D_0 acting on L_1^2.

The proof for $j > 1$ is similar: what is needed is a version of (17.2). First of all we can use

$$\sum_{n+m\leq j} \int_{-\infty}^{0} \left\| \frac{d^n L_0^m u}{dt^n} \right\|_{L^2(Y)}^2$$

as the definition of (the square of) the L_j^2 norm. Then

$$\begin{aligned} \|D_0 u\|_{L_{j-1}^2(Z)}^2 &= \sum_{n+m\leq j-1} \int_{-\infty}^{0} \left\| \frac{d^n L_0^m D_0 u}{dt^n} \right\|_{L^2(Y)}^2 \\ &= \sum_{n+m\leq j-1} \int_{-\infty}^{0} \left(\left\| \frac{d^{n+1} L_0^m u}{dt^{n+1}} \right\|_{L^2(Y)}^2 + \left\| \frac{d^n L_0^{m+1} u}{dt^n} \right\|_{L^2(Y)}^2 \right) \\ &\quad + \sum_{n+m\leq j-1} 2 \int_{-\infty}^{0} \left\langle \frac{d^{n+1} L_0^m u}{dt^{n+1}}, \frac{d^n L_0^{m+1} u}{dt^n} \right\rangle_{L^2(Y)}. \end{aligned}$$

The first integral in the final formula above is equal to

$$\sum_{n=1}^{j} \int_{-\infty}^{0} \left(\left\| \frac{d^n u}{dt^n} \right\|_{L^2(Y)}^2 + \|L_0^n u\|_{L^2(Y)}^2 \right) + 2 \sum_{\substack{n+m\leq j \\ n,m\geq 1}} \int_{-\infty}^{0} \left\| \frac{d^n L_0^m u}{dt^n} \right\|_{L^2(Y)}^2, \tag{17.3}$$

while the second integral in the same formula (the cross-term) is bounded below by

$$\begin{aligned} &- \sum_{\substack{n+m\leq j-1 \\ n\geq 1}} \left(\frac{2n}{2n+1} \left\| \frac{d^{n+1} L_0^m u}{dt^{n+1}} \right\|_{L^2(Y)}^2 + \frac{2n+1}{2n} \left\| \frac{d^n L_0^{m+1} u}{dt^n} \right\|_{L^2(Y)}^2 \right) \\ &+ \sum_{m=0}^{j-1} \langle L_0^m u(0), L_0^{m+1} u(0) \rangle \end{aligned} \tag{17.4}$$

because the terms with $n = 0$ can be integrated (and we have used Cauchy–Schwarz on the remaining terms). Changing the summation index in (17.4), we can write it as

$$
-\int_0^\infty \left(\sum_{\substack{n+m\leq j \\ n\geq 2}} \frac{2n-2}{2n-1} \left\| \frac{d^n L_0^m u}{dt^n} \right\|^2_{L^2(Y)} + \sum_{\substack{n+m\leq j \\ n,m\geq 1}} \frac{2n+1}{2n} \left\| \frac{d^n L_0^m u}{dt^n} \right\|^2_{L^2(Y)} \right)
$$

$$
+\sum_{n=0}^{j-1} \langle L_0^n u(0), L^{n+1} u(0)\rangle.
$$

Reuniting this expression with the terms (17.3), we obtain

$$
\begin{aligned}
&\sum_{n=1}^{j} \int_{-\infty}^{0} \left(\frac{1}{2n-1} \left\| \frac{d^n u}{dt^n} \right\|^2_{L^2(Y)} + \|L_0^n u\|^2_{L^2(Y)} \right) \\
&+ \sum_{\substack{n+m\leq j \\ n,m\geq 1}} \left(2 - \frac{2n-2}{2n-1} - \frac{2n+1}{2n} \right) \int_{-\infty}^{0} \left\| \frac{d^n L_0^m u}{dt^n} \right\|^2_{L^2(Y)} \\
&+ \sum_{n=0}^{j-1} \langle L_0^n u(0), L_0^{n+1} u(0)\rangle \\
&\geq C_1 \|u\|^2_{L^2_j(Z)} + \sum_{n=0}^{j-1} \langle L_0^n \Pi_0 \circ r(u), L_0^{n+1} \Pi_0 \circ r(u)\rangle \\
&\geq C_2 (\|u\|^2_{L^2_j(Z)} - \|\Pi_0 \circ r(u)\|^2_{L^2_{j-1/2}(Y)}).
\end{aligned}
$$

We have used the trivial inequality

$$
\begin{aligned}
\left(2 - \frac{2n-2}{2n-1} - \frac{2n+1}{2n} \right) &= \frac{1}{2n-1} - \frac{1}{2n} \\
&> 0
\end{aligned}
$$

above. With this estimate for $\|D_0 u\|^2_{L^2_{j-1}(Z)}$ in hand, the proof proceeds as in the case $j = 1$. □

Proof of Theorem 17.1.3. The Fredholm property can be deduced from Theorem 17.1.4 by the usual parametrix patching argument: the reader can follow the proof of Proposition 14.2.1. The same argument establishes the regularity assertion in the third statement of the theorem.

To prove the second part of the theorem, suppose that we have a bounded sequence $\{u_i\}$ in $L^2_j(X; E)$ with $\{Du_i\}$ Cauchy in L^2_{j-1}. Choose a cut-off function

β with $\beta = 1$ near ∂X and $\beta = 0$ outside a collar neighborhood of ∂X. Then we can pass to a subsequence where $\{u_i\}$ converges strongly in the L^2_{j-1} topology on the support of β. It follows that

$$\begin{aligned} D_0(\beta u_i) &= \beta D_0(u_i) + d\beta \sharp u_i \\ &= \beta D u_i - K u_i + d\beta \sharp u_i \end{aligned}$$

is Cauchy in the L^2_{j-1} norm. Viewing $D_0(\beta u_i)$ as being defined on the cylinder $Z = (-\infty, 0] \times Y$, Theorem 17.1.4 tells us that there is a unique $v_i \in L^2_j(Z; E)$ with $D_0 v_i = D_0(\beta u_i)$ and $\Pi_0 \circ r(v_i) = 0$. Clearly the v_i then are Cauchy in L^2_j, and so $r(v_i)$ are Cauchy in $L^2_{j-1/2}$. Since $D_0(\beta u_i - v_i) = 0$, the last claim of Theorem 17.1.4 implies that $(1 - \Pi_0) \circ r(u_i - v_i) = 0$, proving the claim. □

Corollary 17.1.5. *Let $K^* : L^2(X; F) \to L^2(X; E)$ be the adjoint of K, let D_0^* be the formal adjoint of D_0, and write $D^* = D_0^* + K^*$. Suppose that D^* has the property that every non-zero solution of $D^* v = 0$ has non-zero restriction to the boundary Y. Then the operator*

$$D : L^2_j(X; E) \to L^2_{j-1}(X; F)$$

is surjective, for $1 \le j \le k$.

Proof. We treat the case $j = 1$, because the other cases follow by regularity. It follows from Theorem 17.1.3 that D has closed range. If D is not surjective, there is a $v \in L^2(X; F)$ orthogonal to the range. This v is a weak solution of $D_0^* v + K^* v = 0$; so, since D_0 is elliptic, v is in L^2_1 and a solution of $D^* v = 0$. If $u_\delta \in L^2_1(X; E)$ is obtained by pulling back a section $u_0 \in L^2_1(Y; E_0)$ and multiplying by a cut-off function supported in a collar of the boundary of width δ, say

$$u_\delta = \beta(t/\delta)\pi^* u_0,$$

then from the orthogonality $\langle D u_\delta, v \rangle = 0$ in $L^2(X; F)$, we deduce that v is zero on the boundary, by letting δ go to zero. This non-zero v therefore contradicts the hypothesis. □

Before continuing, we note that Theorem 17.1.3 and its proof can be extended to cover a slightly more general situation:

Proposition 17.1.6. *Suppose instead that X is not compact but has both cylindrical ends and boundary. On the cylindrical ends, suppose that the operator*

D takes the form $(d/dt) + L_1$, *where* L_1 *is a translation-invariant hyperbolic operator in the sense considered in Proposition* 14.1.2. *Then the conclusions of Theorem* 17.1.3 *continue to hold.* □

A particular case of this is the half-cylinder $Z = (-\infty, 0] \times Y$ and an operator

$$D = \frac{d}{dt} + L_0 + h(t)$$

of the same form considered in Subsection 14.2. We suppose that

$$L_0 + h(t) = L_1$$

for all sufficiently negative t and that L_0 and L_1 are hyperbolic. We continue to write Π_0 for the spectral projection defined by the (self-adjoint, elliptic) operator L_0. Then the above proposition tells us that the operator

$$(D, \Pi_0 \circ r) : L^2_j(Z;E) \to L^2_{j-1}(Z;E) \oplus H_0^- \cap L^2_{j-1/2}(Y;E_0)$$

is Fredholm. Indeed, we can say a little more:

Proposition 17.1.7. *The Fredholm operator*

$$(D, \Pi_0 \circ r) : L^2_j(Z;E) \to L^2_{j-1}(Z;E) \oplus H_0^- \cap L^2_{j-1/2}(Y;E_0)$$

in the above setting has index equal to the spectral flow from L_1 *to* L_0.

Proof. The same line of argument can be used as was applied in the proof of the similar result, Proposition 14.2.1. □

17.2 Commensurate projections

We will need an extension of Theorem 17.1.3, in which the spectral projection Π_0 is replaced by a projection Π, not too different from Π_0, as in the following definition.

Definition 17.2.1. Let Π, $\Pi' : L^2_{1/2}(Y;E_0) \to L^2_{1/2}(Y;E_0)$ be two projections (not necessarily orthogonal) which also map $L^2_{j-1/2}(Y;E_0)$ to $L^2_{j-1/2}(Y;E_0)$ for $1 \le j \le k$. We say that Π and Π' are *k-commensurate* if the difference

$$\Pi - \Pi' : L^2_{j-1/2}(Y;E_0) \to L^2_{j-1/2}(Y;E_0)$$

is compact, for all $1 \le j \le k$. We will often write H^- for $\mathrm{im}(\Pi) \subset L^2_{1/2}(Y; E_0)$ and H^+ for $\mathrm{im}(1-\Pi) \subset L^2_{1/2}(Y; E_0)$. ◊

Examples of commensurate projections arise naturally, as we shall now see. Recall the definition of k-ASAFOE from Subsection 12.2 above. We consider such an operator $L = L_0 + h$ on Y, and we consider the corresponding translation-invariant 4-dimensional operator $(d/dt) + L$ on the negative half-cylinder Z. We name this operator D^-,

$$D^- = \frac{d}{dt} + L,$$

because we also want to introduce a companion operator

$$D^+ = \frac{d}{dt} - L.$$

We can also write $D^\pm$ as $D_0^\pm + K^\pm$, where $D_0^\pm = (d/dt) \mp L_0$ as in the previous subsection.

Lemma 17.2.2. *Let L and $D^\pm$ be as above, and suppose L is hyperbolic. Let H^- (respectively H^+) be the subspace of $L^2_{1/2}(Y; E_0)$ obtained as the boundary values of the kernel of $D^- : L^2_1(Z; E) \to L^2(Z; E)$ (respectively D^+). Then $L^2_{1/2}(Y; E_0)$ is a direct sum (of closed subspaces), $H^+ \oplus H^-$. Furthermore, for each integer j in the range $1 \le j \le k+1$, we have*

$$L^2_{j-1/2}(Y; E_0) = \left(H^+ \cap L^2_{j-1/2}\right) \oplus \left(H^- \cap L^2_{j-1/2}\right). \tag{17.5}$$

Proof. The proof is accomplished by the following trick. Consider the operator

$$\begin{aligned} T : {} & L^2_j(Z; E) \oplus L^2_j(Z; E) \\ & \to L^2_{j-1}(Z; E) \oplus L^2_{j-1}(Z; E) \oplus L^2_{j-1/2}(Y; E_0), \end{aligned} \tag{17.6}$$

given by

$$(u, v) \mapsto (D^- u, D^+ v, -r(u) + r(v)).$$

We claim that T is a Fredholm operator of index 0. To verify this, we can compare T with another operator between the same spaces, namely the operator

$$T' : (u, v) \mapsto (D^- u, D^+ v, -\Pi_0 r(u) + (1 - \Pi_0) r(v))$$

which is the direct sum of $(D^-, -\Pi_0 \circ r)$ and $(D^+, (1 - \Pi_0) \circ r)$. Here Π_0 is the spectral projection for the self-adjoint elliptic operator L_0 as before. Proposition 17.1.6 tells us that the two summands are Fredholm, so T' is Fredholm. Although the difference between T' and T is not a compact operator, its restriction to $\ker D^- \oplus \ker D^+$ is compact, by Proposition 17.1.6; from this, and the fact that $D^\pm$ are surjective, the Fredholmness of T follows. The same reasoning shows that T and T' are homotopic through Fredholm operators (using the linear homotopy), so they have the same index. The index of T is therefore the sum of the indices of $(D^-, -\Pi_0 \circ r)$ and $(D^+, (1 - \Pi_0) \circ r)$. By Proposition 17.1.7, the indices of these summands are equal to respectively the spectral flow from L to L_0 and from $-L$ to $-L_0$; so the sum is zero.

The kernel of the operator T is of course pairs the (u, v) with $D^- u = 0$, $D^+ v = 0$ and $r(u) = r(v)$. If (u, v) belongs to the kernel, consider the section $\tilde{v}$ of E on the *positive* half-cylinder $\mathbb{R}^{\geq} \times Y$ given by $\tilde{v}(t, y) = v(-t, y)$. By concatenating u and $\tilde{v}$, we obtain a section z on the doubly infinite cylinder $\mathbb{R} \times Y$. This concatenated section z is in L^2_1, because $r(u) = r(v)$; but, a priori, we do not know that the derivatives of u and $\tilde{v}$ in the $\mathbb{R}$ direction match up at $t = 0$, so we do not know that z is in L^2_j. But z is also an L^2_1 solution of the equation $D^- z = 0$ on all of $\mathbb{R} \times Y$. We know that D^- is an isomorphism from L^2_1 to L^2, because L is hyperbolic. So $z = 0$. This proves that T is injective, and hence an isomorphism, because its index is zero.

It follows that for any element w of $L^2_{j-1/2}(Y; E_0)$ there are unique $u, v \in L^2_j(Z; E)$ with $D^- u = 0$, $D^+ v = 0$ and $-r(u) + r(v) = w$. Furthermore u and v depend continuously on w. □

The subspaces H^+ and H^- defined by this lemma are invariant under the action of the hyperbolic operator L, in that the restriction of L defines isomorphisms

$$L^+ : H^+ \cap L^2_{j+1/2} \to H^+ \cap L^2_{j-1/2}$$
$$L^- : H^- \cap L^2_{j+1/2} \to H^- \cap L^2_{j-1/2}$$

for $1 \leq j \leq k$. This follows from the fact that L acts on the kernel of $D^\pm$ on the half-cylinder. We already know that the spectrum of L consists of eigenvalues, and $L^\pm$ inherit this property. It is clear from the definition that an eigenfunction of L belonging to an eigenvalue of positive (respectively negative) real part must belong to L^+ (respectively L^-). So the spectra of L^+ and L^- are just the intersection of the spectrum of L with the two half-planes separated by the imaginary axis. If the generalized eigenspaces of L have dense span in $L^2_{1/2}$, then we could define H^+ and H^- as the closure of the span of the generalized eigenspaces belonging to the corresponding part of the spectrum; but it is not clear that our conditions ensure this density. (For example, a non-symmetric compact operator on a complex Hilbert space need have no eigenvectors.)

Definition 17.2.3. We call the subspaces H^+ and H^- defined by Lemma 17.2.2 the *spectral subspaces* of the hyperbolic operator L. ◊

Proposition 17.2.4. *Let L be a k-ASAFOE operator and let H^- and H^+ be the spectral subspaces, as just defined. Let Π be the projection on $L^2_{1/2}(Y;E_0)$ with image H^- and kernel H^+. Then Π maps $L^2_{j-1/2}(Y;E_0)$ to $L^2_{j-1/2}(Y;E_0)$ for $1 \leq j \leq k$, and Π is k-commensurate with the projection Π_0 determined by L_0.*

Proof. The fact that Π maps $L^2_{j-1/2}$ to $L^2_{j-1/2}$ comes straight from the lemma. To prove commensurability, we must show $\Pi - \Pi_0$ is a compact operator on $L^2_{j-1/2}$. Given a bounded sequence in $L^2_{j-1/2}$, we may write each term as two parts, using the direct sum decomposition (17.5). Because of the symmetry between D^- and D^+ in the argument, it is sufficient to consider a bounded sequence $\{w_i\}$ in $H^- \cap L^2_{j-1/2}$. We must show that the terms

$$(\Pi - \Pi_0)w_i = (1 - \Pi_0)w_i$$

have a convergent subsequence. But this follows directly from the last part of Theorem 17.1.3, because we can write $w_i = r(u_i)$, where $\{u_i\}$ is a bounded sequence in $L^2_j(Z;E)$ and $D^-u_i = 0$. □

Having found a supply of commensurate projections, we now note that the two main results of the previous subsection (Theorems 17.1.3 and 17.1.4) continue to hold if we substitute a commensurate projection for Π_0. This is the content of the next two propositions. The first is an extension of Theorem 17.1.3:

Proposition 17.2.5. *Let D and Π_0 be as in the statement of Theorem* 17.1.3, *and let Π be a projection k-commensurate with Π_0, with image H^-. Then the conclusions of that theorem continue to hold with Π_0 replaced by Π: for $1 \leq j \leq k$, the operator*

$$D \oplus (\Pi \circ r) : L^2_j(X;E) \to L^2_{j-1}(X;F) \oplus (H^- \cap L^2_{j-1/2}(Y;E_0))$$

is Fredholm. In addition, if u_i is a bounded sequence in $L^2_j(X;E)$ and Du_i is Cauchy in L^2_{j-1}, then $(1 - \Pi)r(u_i)$ has a convergent subsequence in $L^2_{j-1/2}$. In particular, the maps $\Pi \circ r$ and $(1 - \Pi) \circ r$, restricted to the kernel of D, are respectively Fredholm and compact.

Proof. First of all

$$\Pi\Pi_0 : \mathrm{im}(\Pi_0) \cap L^2_{j-1/2} \to \mathrm{im}(\Pi) \cap L^2_{j-1/2}$$

is Fredholm. To check this recall that an operator is Fredholm if and only if it is invertible modulo compact operators. The required Fredholm inverse is $\Pi_0\Pi$. The composite $\Pi_0\Pi\Pi\Pi_0$ can be written as $\Pi_0 + \Pi_0(\Pi - \Pi_0)\Pi_0$. This exhibits the operator as a compact perturbation of the identity operator (acting on the image of Π_0), and the argument is clearly symmetric in Π_0 and Π. It follows that the operator $(D, \Pi\Pi_0)$ is Fredholm, for it is equal to the composition $(\mathrm{Id}, \Pi\Pi_0) \circ (D, \Pi_0)$ and each of these operators is Fredholm. Finally (D, Π) is Fredholm for it is equal to the sum $(D, \Pi\Pi_0) + (0, \Pi(\Pi - \Pi_0))$, the first being Fredholm and the second compact.

Finally suppose that we have a bounded sequence $\{u_i\}$ in $L^2_j(X;E)$ with $\{Du_i\}$ Cauchy in L^2_{j-1}. We already know that we can pass to a subsequence where $(1 - \Pi_0) \circ r(u_i)$ converges in $L^2_{j-1/2}$. Because $r(u_i)$ is bounded in $L^2_{j-1/2}$ and $(\Pi - \Pi_0)$ is compact, we can pass to a further subsequence where $(\Pi - \Pi_0) \circ r(u_i)$ converges in $L^2_{j-1/2}$. Then

$$(1 - \Pi) \circ r(u_i) = (1 - \Pi_0) \circ r(u_i) - (\Pi - \Pi_0) \circ r(u_i)$$

converges. □

There is a larger class of projections, for which "half" of the above proposition continues to hold.

Proposition 17.2.6. *Let D be as in the statement of Proposition* 17.2.5, *and let Π be again a projection k-commensurate with Π_0, with image H^-. Let Π_1 be any linear projection on $L^2_{j-1/2}(Y;E_0)$ whose kernel is a complement of H^-:*

$$\ker(\Pi_1) \oplus (H^- \cap L^2_{j-1/2}(Y;E_0)) = L^2_{j-1/2}(Y;E_0).$$

Let H^-_1 be the image of Π_1. Then the operator

$$D \oplus (\Pi_1 \circ r) : L^2_j(X;E) \to L^2_{j-1}(X;F) \oplus H^-_1$$

is Fredholm.

Proof. Write the domain of the operator D as $C \oplus K$, where K is the kernel and C is a complement. Then $D|_C$ is Fredholm, as is $(\Pi \circ r)|_K$, from Proposition 17.2.5.

That is, we can write $D \oplus (\Pi \circ r)$ as a block matrix whose diagonal blocks are Fredholm:

$$\begin{bmatrix} D|_C & 0 \\ x & (\Pi \circ r)|_K \end{bmatrix}.$$

To deduce that the operator $D \oplus (\Pi_1 \circ r)$ is Fredholm, we can use the same decomposition; and we see that all that is needed is to show that $(\Pi_1 \circ r)|_K$ is Fredholm. To this end, we write

$$(\Pi_1 \circ r) = \big(\Pi_1 \circ \Pi\big) \circ r + \big(\Pi_1 \circ (1 - \Pi)\big) \circ r.$$

The second term in this sum is compact on K, because $(1 - \Pi) \circ r$ is compact on K by Proposition 17.2.5. The first term is Fredholm, because $\Pi \circ r|_K$ is Fredholm and Π_1 is an isomorphism on the image of Π. □

Next we have an extension of Theorem 17.1.4 to the commensurate setting, where Π is a projection of the sort arising in Lemma 17.2.2 above.

Proposition 17.2.7. *Let Z be a half-cylinder $(-\infty, 0] \times Y$, and let $D : C^\infty(Z; E) \to L^2(Z; E)$ be an operator having the form*

$$D = \frac{d}{dt} + L$$

where L is a k-ASAFOE operator on Y. Suppose that L is hyperbolic, let H^+ and H^- be the spectral subspaces as in Definition 17.2.3, and let Π be the spectral projection, with kernel H^+ and image H^-. Then the operator

$$D \oplus (\Pi \circ r) : L^2_j(Z; E) \to L^2_{j-1}(Z; E) \oplus (H^- \cap L^2_{j-1/2}(Y; E_0)) \qquad (17.7)$$

is an isomorphism for $1 \le j \le k$. The subspace $H^- \cap L^2_{j-1/2}(Y; E_0)$ is precisely the image of $\ker(D)$ *under r.*

Proof. From Proposition 17.2.4 we know the projections Π and Π_0 are commensurate; and so, by Proposition 17.2.7, the operator (17.7) is Fredholm. It is injective, since by unique continuation elements of the kernel of D are determined by their boundary values, which lie in $H^- \cap L^2_{j-1/2}(Y; E_0)$ by definition of H^-. We already know that the operator D is surjective on the doubly infinite cylinder. The result then follows, because $v \in L^2_{j-1}(Z; E)$ is the restriction of some $\tilde{v} \in L^2_{j-1}(\mathbb{R} \times Y; E)$ by Theorem 17.1.2. □

In Lemma 17.2.4 above, we saw that $L^2_{j-1/2}(Y;E_0)$ was the direct sum of the boundary values of the kernel of the operator $D = D^-$ on the two half-infinite cylinders (positive and negative). If we replace the half-infinite cylinders by finite cylinders, a similar result holds; but we do not have a *direct* sum. The next proposition extends this result to the case that the operator is not translation-invariant.

Let Z be a cylinder $I \times Y$, where I is a compact interval containing 0 in its interior. Let I_1, I_2 be the intersections of I with the negative and positive closed half-lines, $\mathbb{R}^{\leq}$ and $\mathbb{R}^{\geq}$, and let $Z = Z_1 \cup Z_2$ be the corresponding decomposition of Z. Suppose we have an operator $D : C^\infty(Z;E) \to L^2(Z;E)$ of the form

$$D = \frac{d}{dt} + L_0 + h(t)$$

where L_0 is, as usual, a self-adjoint elliptic operator on Y. We suppose that the time-dependent operator $h(t) : C^\infty(Y;E_0) \to L^2(Y;E_0)$ provides a bounded map

$$h : L^2_j(Z;E) \to L^2_j(Z;E)$$

for $j \leq k-1$. Let D_{Z_1}, D_{Z_2} be the restrictions of these operators to the two pieces. Let

$$H_{j-1/2}(Z_i) \subset L^2_{j-1/2}(\{0\} \times Y; E_0)$$

be the image of $\ker(D_{Z_i})$ under the restriction map

$$r_i : L^2_j(Z_i;E) \to L^2_{j-1/2}(\{0\} \times Y; E_0).$$

Although the definition of $H_{j-1/2}(Z_i)$ for $i = 1, 2$ resembles that of H^- and H^+ above, the two types of subspace are not comparable, because our interval is now finite.

Proposition 17.2.8. *Suppose $D : L^2_j(Z;E) \to L^2_{j-1}(Z;E)$ is surjective (for one and hence all j in the range $1 \leq j \leq k$). Then*

$$L^2_{j-1/2}(\{0\} \times Y; E_0) = H_{j-1/2}(Z_1) + H_{j-1/2}(Z_2). \tag{17.8}$$

Conversely, suppose that (17.8) *holds, and that D_{Z_1} and D_{Z_2} are surjective. Then D is surjective.*

Remark. The surjectivity hypothesis is a mild restriction, as Corollary 17.1.5 shows.

Proof of Proposition 17.2.8. The converse direction presents fewer problems, so we deal only with the forward direction of the theorem.

For $h = 0$, the result is straightforward, because the subspaces $H_{j-1/2}(Z_1)$ and $H_{j-1/2}(Z_2)$ contain the negative and positive spectral subspaces of L_0. For non-zero h, consider first the case $j = 1$. Let $a \in L^2_{1/2}(\{0\} \times Y; E_0)$. Using the result for $h = 0$, we find u_1, u_2 in $L^2_1(Z_1)$ and $L^2_1(Z_2)$ satisfying $D_0 u_i = 0$ and

$$r_1 u_1 - r_2 u_2 = a.$$

Let $u \in L^2(Z; E)$ be the element that is equal to u_i on Z_i for $i = 1, 2$. Because D is surjective, we can find $w \in L^2_1(Z)$ with $Dw = hu$. Let w_i be the restriction of w to Z_i, and let $\tilde{u}_i = u_i - w_i$. Then

$$r_1 \tilde{u}_1 - r_2 \tilde{u}_2 = a$$

because $r_1 w_1 = r_2 w_2$; and

$$\begin{aligned}(D_{Z_i})\tilde{u}_i &= Du_i - Dw_i \\ &= hu_i - Dw_i \\ &= 0.\end{aligned}$$

So we have exhibited a as an element $r_1\tilde{u}_1 - r_2\tilde{u}_2$ of $H_{1/2}(Z_1) + H_{1/2}(Z_2)$.

Now suppose $j > 1$, and let $a \in L^2_{j-1/2}(\{0\} \times Y; E_0)$. Write

$$a = a_1 + a_2 \tag{17.9}$$

with $a_i \in H_{1/2}(Z_i)$. We shall show that $a_i \in H_{j-1/2}(Z_i)$.

Lemma 17.2.9. *Let Π_0 be the spectral projection for D_0, with image the negative spectral subspace H^- as usual. If $b \in H_{j-3/2}(Z_1)$ for some $2 \le j \le k$, then*

$$(1 - \Pi_0)b \in L^2_{j-1/2}.$$

Proof. Let v be the solution to $D_0 v = 0$ on Z_1 with $v|_{0\times Y} = \Pi_0 b$, so $v \in L^2_{j-1}(Z_1; E)$. From the definition of $H_{j-3/2}(Z_1)$, there exists $\tilde{v}$ in $L^2_{j-1}(Z_1; E)$ with $D_{Z_1}\tilde{v} = 0$ and $\tilde{v}|_{0\times Y} = b$. We have

$$\begin{aligned}D_0(v - \tilde{v}) &= h(\tilde{v}) \\ &\in L^2_{j-1}(Z_1; E_0)\end{aligned}$$

and $\Pi_0(v - \tilde{v})|_{\{0\}\times Y} = 0$. These two conditions imply that $(v - \tilde{v})$ is in L^2_j on a neighborhood of $\{0\} \times Y$ in Z_1 (for we can multiply by a cut-off function and then exploit the invertibility of (D_0, Π_0) on the half-infinite cylinder, as given in Theorem 17.1.4). Applying $(1 - \Pi_0)$ to the boundary value of $v - \tilde{v}$, we conclude that $(1 - \Pi_0)b$ is in $L^2_{j-1/2}$. □

Returning to (17.9), we can suppose as an induction hypothesis that a_i is in $H_{j-3/2}(Z_i)$ for $i = 1, 2$. We apply $(1 - \Pi_0)$ to (17.9):

$$(1 - \Pi_0)a = (1 - \Pi_0)a_1 + (1 - \Pi_0)a_2.$$

We are supposing that a is in $L^2_{j-1/2}$, and the lemma tells us that $(1 - \Pi_0)a_1$ is in $L^2_{j-1/2}$; so this equation tells us that $(1 - \Pi_0)a_2$ is in $L^2_{j-1/2}$. We can also apply the lemma with opposite signs, to see that $\Pi_0 a_2$ is in $L^2_{j-1/2}$. So a_2 is in $L^2_{j-1/2}$, which completes the proof. □

17.3 Atiyah–Patodi–Singer boundary-value problems and gauge theory

Let Z be a finite cylinder $I \times Y$. Recall that we have configuration spaces $\mathcal{C}^\tau_k(Z) \subset \tilde{\mathcal{C}}^\tau_k(Z)$ and quotient spaces

$$\mathcal{B}^\tau_k(Z) \subset \tilde{\mathcal{B}}^\tau_k(Z) = \tilde{\mathcal{C}}^\tau_k(Z)/\mathcal{G}_{k+1}(Z)$$

as in Corollary 9.4.4. The space $\tilde{\mathcal{B}}^\tau_k(Z)$ is a Hilbert manifold. Inside this space is the set of gauge-equivalence classes of solutions to the perturbed Seiberg–Witten equations,

$$M(Z) \subset \tilde{M}(Z) = \{\, [\gamma] \in \tilde{\mathcal{B}}^\tau_k(Z) \mid \mathfrak{F}^\tau_{\mathfrak{q}}(\gamma) = 0 \,\}.$$

Unlike the moduli spaces of trajectories on the infinite cylinder that we considered in Definition 13.1.1, these moduli spaces on a compact cylinder (with no boundary conditions) are infinite-dimensional. Another point to bear in mind is that the moduli space $M(Z)$, as defined, does depend on the choice of k, although we omit k from the notation: a solution in $M(Z)$ will have higher regularity in the *interior* of the cylinder (after gauge transformation) but not up to the boundary.

On a finite cylinder, the moduli spaces are always smooth manifolds, with no special transversality condition required of the perturbation, as the next theorem shows.

Theorem 17.3.1. *The subspace $\tilde{M}(Z) \subset \tilde{\mathcal{B}}^\tau_k(Z)$ is a closed Hilbert submanifold. The subset $M(Z)$ is a Hilbert submanifold with boundary: it can be identified as the quotient of $\tilde{M}(Z)$ by the involution* $\mathbf{i}$.

Proof. Certainly the subspaces $\tilde{M}(Z)$ and $M(Z)$ are closed. To see that $\tilde{M}(Z)$ is a submanifold, we may consider its inverse image in $\tilde{\mathcal{C}}^\tau_k(Z)$, the solution set of the equations, and we need to show that the derivative of the equations, $\mathcal{D}_\gamma\mathfrak{F}^\tau_\mathfrak{q}$, is surjective at every solution γ. This will follow from a stronger statement, the surjectivity of the operator

$$\mathcal{Q}_\gamma = \mathcal{D}_\gamma\mathfrak{F}^\tau_\mathfrak{q} \oplus \mathbf{d}^{\tau,\dagger}_\gamma : \mathcal{T}^\tau_{j,\gamma}(Z) \to \mathcal{V}^\tau_{j-1,\gamma}(Z) \oplus L^2_{j-1}(Z; i\mathbb{R}) \tag{17.10}$$

defined as in Theorem 14.4.2, but now using the finite cylinder $Z = I \times Y$, rather than $\mathbb{R} \times Y$ which appeared there. As in the proof of Theorem 14.4.2, this operator can be cast in the form

$$\frac{d}{dt} + L_0 + \bar{h}_\gamma$$

(see (14.14)), where L_0 is self-adjoint elliptic and $\bar{h}_\gamma$ is defined slicewise and maps L^2_j to L^2_j for $0 \leq j \leq k-1$. This operator has formal adjoint

$$-\frac{d}{dt} + L_0 + \bar{h}^*_\gamma$$

which has the unique continuation property, that any solution which vanishes on the boundary is zero. Corollary 17.1.5 therefore applies: this gives the required surjectivity.

The fact that $M(Z)$ is the quotient of $\tilde{M}(Z)$ by the involution $\mathbf{i}$ is proved as in the case of the infinite cylinder: see the discussion preceding Definition 14.5.2. □

We can apply the Atiyah–Patodi–Singer theory to get some finer information about the Hilbert manifold $M(\tilde{Z})$, beyond just the smoothness which we have proved. Let us write the boundary of Z as

$$\partial Z = \bar{Y} \sqcup Y,$$

where $\bar{Y}$ denotes Y with the opposite orientation. We have restriction maps

$$R_Y : \tilde{M}(Z) \to \mathcal{B}^\sigma_{k-1/2}(Y)$$
$$R_{\bar{Y}} : \tilde{M}(Z) \to \mathcal{B}^\sigma_{k-1/2}(\bar{Y}).$$

Let $[\gamma]$ be any element of $\tilde{M}(Z)$, and let $\mathfrak{a}$ and $\bar{\mathfrak{a}}$ be the restrictions of γ to the two boundary components. We can identify the tangent space to $\mathcal{B}^\sigma_{k-1/2}(Y)$ at $[\mathfrak{a}]$ with the linear space $\mathcal{K}^\sigma_{k-1/2,\mathfrak{a}}$ transverse to the gauge orbit through $\mathfrak{a}$ (Proposition 9.3.5), so we can regard the derivatives of R_Y and $R_{\bar{Y}}$ as defining maps

$$\begin{aligned} \mathcal{D}R_Y &: T_{[\gamma]}\tilde{M}(Z) \to \mathcal{K}^\sigma_{k-1/2,\mathfrak{a}}(Y) \\ \mathcal{D}R_{\bar{Y}} &: T_{[\gamma]}\tilde{M}(Z) \to \mathcal{K}^\sigma_{k-1/2,\bar{\mathfrak{a}}}(\bar{Y}). \end{aligned} \tag{17.11}$$

Recall that the Hilbert vector bundle $\mathcal{K}^\sigma_j \to \mathcal{C}^\sigma_k(Y)$ carries a smooth family of operators with real spectrum, namely the Hessians $\text{Hess}^\sigma_{\mathfrak{q}}$ defined at (12.8). If we are not at a critical point, the operator $\text{Hess}^\sigma_{\mathfrak{q},\mathfrak{a}}$ is not a direct summand of the extended Hessian, but we see from (12.12) that by dropping the terms x and y we can eventually exhibit $\text{Hess}^\sigma_{\mathfrak{q},\mathfrak{a}}$ as a summand of an operator which is a bounded perturbation of a self-adjoint elliptic operator: that is, a summand of an ASAFOE operator. So we have the usual conclusion, that $\text{Hess}^\sigma_{\mathfrak{q},\mathfrak{a}}$ has discrete spectrum, and finite-dimensional generalized eigenspaces. If the operator is hyperbolic (that is, if zero is not an eigenvalue), then we also have a spectral decomposition

$$\mathcal{K}^\sigma_{k-1/2,\mathfrak{a}} = \mathcal{K}^+_{\mathfrak{a}} \oplus \mathcal{K}^-_{\mathfrak{a}}. \tag{17.12}$$

In the non-hyperbolic case, we pick an ϵ sufficiently small that there are no eigenvalues in $(0, \epsilon)$, and we then define $\mathcal{K}^\pm_{\mathfrak{a}}$ using the spectral decomposition arising from the operator $\text{Hess}^\sigma_{\mathfrak{q},\mathfrak{a}} - \epsilon$. (The effect of this is to put the generalized 0-eigenspace into $\mathcal{K}^-_{\mathfrak{a}}$.) In this way, we can define the decomposition (17.12) everywhere, though it is not continuous at points $\mathfrak{a}$ where the Hessian has kernel.

Where necessary, we will write $\mathcal{K}^-_{\mathfrak{a}}(Y)$ rather than just $\mathcal{K}^-_{\mathfrak{a}}$, to emphasize which 3-manifold is involved. In particular, we note

$$\mathcal{K}^-_{\mathfrak{a}}(\bar{Y}) = \mathcal{K}^+_{\mathfrak{a}}(Y).$$

We will also use the notation

$$\begin{aligned} \mathcal{K}^\sigma_{k-1/2,(\bar{\mathfrak{a}},\mathfrak{a})}(\bar{Y} \amalg Y) &\cong \mathcal{K}^\sigma_{k-1/2,\bar{\mathfrak{a}}}(Y) \oplus \mathcal{K}^\sigma_{k-1/2,\mathfrak{a}}(Y) \\ \mathcal{K}^-_{(\bar{\mathfrak{a}},\mathfrak{a})}(\bar{Y} \amalg Y) &\cong \mathcal{K}^+_{\bar{\mathfrak{a}}}(Y) \oplus \mathcal{K}^-_{\mathfrak{a}}(Y). \end{aligned}$$

With this understood, we return to the derivatives of the restriction maps R_Y and $R_{\bar{Y}}$ above. Together these define a map

$$\mathcal{D}(R_{\bar{Y}}, R_Y) : T_{[\gamma]}\tilde{M}(Z) \to \mathcal{K}^\sigma_{k-1/2,(\bar{\mathfrak{a}},\mathfrak{a})}(\bar{Y} \amalg Y).$$

What we want to explain is that the tangent space to the moduli space $\tilde{M}(Z)$ is everywhere comparable to the spectral summand $\mathcal{K}^-_{(\bar{\mathfrak{a}},\mathfrak{a})}(\bar{Y} \amalg Y)$, in a sense made precise by the next theorem.

Theorem 17.3.2. *Let γ, $\mathfrak{a}$ and $\bar{\mathfrak{a}}$ be as above, and let π be the projection*

$$\pi : \mathcal{K}^\sigma_{k-1/2,(\bar{\mathfrak{a}},\mathfrak{a})}(\bar{Y} \amalg Y) \to \mathcal{K}^-_{(\bar{\mathfrak{a}},\mathfrak{a})}(\bar{Y} \amalg Y)$$

with kernel $\mathcal{K}^+_{(\bar{\mathfrak{a}},\mathfrak{a})}(\bar{Y} \amalg Y)$. Then the two composite maps

$$\pi \circ \mathcal{D}(R_{\bar{Y}}, R_Y) : T_{[\gamma]}\tilde{M}(Z) \to \mathcal{K}^-_{(\bar{\mathfrak{a}},\mathfrak{a})}(\bar{Y} \amalg Y)$$
$$(1-\pi) \circ \mathcal{D}(R_{\bar{Y}}, R_Y) : T_{[\gamma]}\tilde{M}(Z) \to \mathcal{K}^+_{(\bar{\mathfrak{a}},\mathfrak{a})}(\bar{Y} \amalg Y)$$

are respectively Fredholm and compact.

Proof. We start with a slightly more general setting: rather than a solution γ to the equations, let us consider a general γ in $\mathcal{B}^\tau_k(Z)$, and continue to define $\mathfrak{a}$ and $\bar{\mathfrak{a}}$ as above. Because of gauge invariance, we may suppose that γ is in temporal gauge. Let Q_γ be again the operator (17.10) obtained by combining the linearization of the perturbed Seiberg–Witten equations with the Coulomb condition. We know how to write Q_γ as

$$Q_\gamma = \frac{D}{dt} + L_{\gamma(t)},$$

where $L(t)$ is equivalent to an ASAFOE operator on Y. As we observed above (see (12.12), for example), the operator $L_{\gamma(t)}$ differs by bounded terms from an operator which has $\mathrm{Hess}^\sigma_{\mathfrak{q},\gamma(t)}$ as a summand:

$$\tilde{L}_{\gamma(t)} = \begin{bmatrix} 0 & \mathbf{d}^\sigma_{\gamma(t)} \\ \mathbf{d}^{\sigma,\dagger}_{\gamma(t)} & 0 \end{bmatrix} \oplus \mathrm{Hess}^\sigma_{\mathfrak{q},\gamma(t)}. \tag{17.13}$$

In particular, the spectral decompositions arising from $L_{\gamma(t)}$ and $\tilde{L}_{\gamma(t)}$ define comparable projections. From our general theory, it follows that there is a Fredholm operator obtained from Q_γ using the spectral projections defined by $\tilde{L}_{\gamma(t)}$ at the two boundary components:

$$Q_\gamma \oplus \tilde{\Pi} \circ r : \mathcal{T}^\tau_{j,\gamma}(Z) \to \mathcal{V}^\tau_{j-1,\gamma}(Z) \oplus L^2_{j-1}(Z; i\mathbb{R}) \oplus \tilde{H}^- \tag{17.14}$$

where $\tilde{H}^- = \tilde{H}^-_{\bar{Y}} \oplus \tilde{H}^-_Y$ is defined at each end using the spectral decomposition of $\tilde{L}_\gamma$, so that

$$\tilde{H}^-_Y \subset \mathcal{T}^\sigma_{j-1/2,\mathfrak{a}}(Y) \oplus L^2_{j-1/2}(Y; i\mathbb{R})$$
$$\tilde{H}^-_{\bar{Y}} \subset \mathcal{T}^\sigma_{j-1/2,\bar{\mathfrak{a}}}(\bar{Y}) \oplus L^2_{j-1/2}(\bar{Y}; i\mathbb{R}).$$

We should remember here that the distinction between the signs in the spectrum depends on the orientation of Y; so in the case that $\mathfrak{a} = \bar{\mathfrak{a}}$, for example, and if $L_\mathfrak{a}$ is hyperbolic, then we will have $\tilde{H}^-_{\bar{Y}} = \tilde{H}^+_Y$.

The projection $\tilde{\Pi} = \tilde{\Pi}_{\bar{Y}} \oplus \tilde{\Pi}_Y$ onto the spectral subspace is not the one that is most closely related to the geometrical situation described by the theorem, however. We need to consider, instead, a projection which is not obtained as a spectral projection. In the decomposition

$$\mathcal{T}^\sigma_{k-1/2,\mathfrak{a}}(Y) \oplus L^2_{k-1/2}(Y; i\mathbb{R}) = \mathcal{J}^\sigma_{k-1/2,\mathfrak{a}} \oplus \mathcal{K}^\sigma_{k-1/2,\mathfrak{a}} \oplus L^2_{k-1/2}(Y; i\mathbb{R}),$$

let H_Y be the subspace

$$H^-_Y = \{0\} \oplus \mathcal{K}^-_\mathfrak{a} \oplus L^2_{k-1/2}(Y; i\mathbb{R}) \tag{17.15}$$

and let

$$\Pi^-_Y : \mathcal{T}^\sigma_{k-1/2,\mathfrak{a}}(Y) \oplus L^2_{k-1/2}(Y; i\mathbb{R}) \to H^-_Y$$

be the projection with kernel

$$\ker(\Pi^-_Y) = \mathcal{J}^\sigma_{k-1/2,\mathfrak{a}} \oplus \mathcal{K}^+_\mathfrak{a} \oplus \{0\}.$$

(Even in the case that $\mathfrak{a} = \bar{\mathfrak{a}}$ is hyperbolic, the two projections $\Pi^-_{\bar{Y}}$ and Π^-_Y are no longer complementary.) Consider now the operator obtained using the projection $\Pi = \Pi^-_{\bar{Y}} \oplus \Pi^-_Y$ at the boundary: set $H = H^-_{\bar{Y}} \oplus H^-_Y$ and write

$$Q_\gamma \oplus \Pi \circ r : \mathcal{T}^\tau_{k,\gamma}(Z) \to \mathcal{V}^\tau_{k-1,\gamma}(Z) \oplus L^2_{k-1}(Z; i\mathbb{R}) \oplus H. \tag{17.16}$$

Lemma 17.3.3. *The above operator $Q_\gamma \oplus \Pi \circ r$ is Fredholm.*

Proof. We apply Proposition 17.2.6, with Π playing the role of Π_1 in that proposition, and $\tilde{\Pi}$ playing the role of Π. We need to show that $\ker \Pi$ and $\tilde{H}^-$ are complementary subspaces. Although the boundary has two components, the statement to be checked can be looked at on just the component Y; and from the definitions of the two projections, we see that it is equivalent to a statement about the block

$$\begin{bmatrix} 0 & \mathbf{d}^{\sigma}_{\mathfrak{a}} \\ \mathbf{d}^{\sigma,\dagger}_{\mathfrak{a}} & 0 \end{bmatrix}$$

acting on $\mathcal{J}^{\sigma}_{j,\mathfrak{a}} \oplus L^2_j(Y; i\mathbb{R})$. We need to check that the projection

$$h^- \to \{0\} \oplus L^2_{k-1/2}(Y; i\mathbb{R})$$

from the spectral subspace h^- for this 2-by-2 block is an isomorphism. The reason this is an isomorphism is that h^- can be described as a graph of a map

$$L^2_{k-1/2}(Y; i\mathbb{R}) \to \mathcal{J}^{\sigma}_{k-1/2,\mathfrak{a}};$$

we can write

$$h^- = \{ (-\mathbf{d}^{\sigma}_{\mathfrak{a}} (\mathbf{d}^{\sigma,\dagger}_{\mathfrak{a}} \mathbf{d}^{\sigma}_{\mathfrak{a}})^{-1/2} c, c) \mid c \in L^2_{k-1/2} \},$$

because the operator $\mathbf{d}^{\sigma,\dagger}_{\mathfrak{a}} \mathbf{d}^{\sigma}_{\mathfrak{a}}$ is positive and self-adjoint, as we saw in the proof of Lemma 12.4.3. □

We return to the proof of the theorem. If γ is a solution of the equations, then we can identify the tangent space to $\tilde{M}(Z)$ at $[\gamma]$ with the space of solutions (a, s, ϕ) to the linearized equations satisfying the Coulomb–Neumann gauge-fixing conditions

$$\mathbf{d}^{\tau,\dagger}_{\gamma}(a, s, \phi) = 0$$
$$\langle a, \vec{n} \rangle = 0 \text{ at } \partial Z$$

(see (9.16)). In other words, $T_{[\gamma]}\tilde{M}(Z)$ is the set of (a, s, ϕ) which are in the kernel of Q_γ and whose projection to the summands $L^2_{k-1/2}(\bar{Y} \amalg Y; i\mathbb{R})$ at the boundary is zero. The projection to $L^2_{k-1/2}(\bar{Y} \amalg Y; i\mathbb{R})$ is one part of the boundary condition Π that appears in the lemma; the other part is the projection to $\mathcal{K}^-_{(\bar{\mathfrak{a}},\mathfrak{a})}(\bar{Y} \amalg Y)$. So the first assertion of the theorem follows.

For the second assertion, we just have to observe that $(1-\pi)$ factors through the projection $(1-\tilde{\Pi})$ to $\tilde{H}^+$, and we already know that $(1-\tilde{\Pi}) \circ r$ is compact on the kernel of Q_γ. □

18 Stable manifolds and gluing near critical points

18.1 The motivating example

Before taking up our main task in this section, we digress to illustrate the corresponding results for a linear flow in finite dimensions. Consider an invertible, self-adjoint linear transformation L on $\mathbb{R}^n$, and the corresponding flow

$$\dot{\gamma}(t) = -L\gamma(t).$$

We shall examine the solutions γ to this equation on a finite interval $[-T, T]$. Let $M(T)$ denote the space of solutions, and let

$$r : M(T) \to \mathbb{R}^n \times \mathbb{R}^n$$

be the map that evaluates a solution at the two endpoints:

$$r(\gamma) = \bigl(\gamma(-T), \gamma(T)\bigr).$$

The image of r is a copy of $\mathbb{R}^n$ (because a solution in $M(T)$ is determined by its value at any point); but we wish to parametrize the image in such a manner that the parametrizations converge as T increases to infinity. To this end, write the codomain of r as

$$\mathbb{R}^n \times \mathbb{R}^n = (H^+ \oplus H^-) \times (H^+ \oplus H^-),$$

where $H^\pm$ are, as usual, the spectral subspaces of L for the positive and negative eigenvalues. Then the image of r can be described as the locus

$$\bigl\{\, (u_+ + e^{2TL}u_-,\ e^{-2TL}u_+ + u_-) \bigm| (u_+, u_-) \in H^+ \times H^- \,\bigr\}.$$

For given u_+ and u_-, the terms $e^{2TL}u_-$ and $e^{-2TL}u_+$ decay exponentially as T increases, and the image of r approaches the subspace

$$H^+ \times H^- \subset \mathbb{R}^n \times \mathbb{R}^n.$$

We can interpret this limiting subspace as the set of boundary values of a limiting object, which we can call $M(\infty)$: we define $M(\infty)$ as the set of solutions to the equation on the disjoint union of two closed half-lines, $\mathbb{R}^{\geq} \amalg \mathbb{R}^{\leq}$. The oriented boundary of the two half-lines is the same as that of each compact interval $[-T, T]$, and there is therefore a map

$$r : M(\infty) \to \mathbb{R}^n \times \mathbb{R}^n$$

just as there is on each $M(T)$. The image of this limiting map is precisely $H^+ \times H^-$. We can summarize the situation as follows:

Proposition 18.1.1. *For all $T > 0$, we can find linear parametrizations*

$$u(T, \cdot\,) : \mathbb{R}^n \to M(T),$$

and

$$u(\infty, \cdot\,) : \mathbb{R}^n \to M(\infty),$$

with the property that the linear maps

$$\mu_T : \mathbb{R}^n \to \mathbb{R}^n \times \mathbb{R}^n$$

given by $\mu_T(x) = r \circ u(T, x)$ converge as $T \to \infty$ to the limiting map $\mu_\infty(x) = r \circ u(\infty, x)$. □

It is instructive to examine the geometry of the union of the linear subspaces $\mu_T(\mathbb{R}^n)$ as T runs through $(0, \infty]$. Let us call this union W:

$$W = \bigcup_{T \in (0,\infty]} \mu_T(\mathbb{R}^n).$$

This space certainly has a singularity at 0, so let us consider

$$W^o = \bigcup_{T \in (0,\infty]} \mu_T(\mathbb{R}^n \setminus 0).$$

The images of μ_{T_1} and μ_{T_2} are disjoint (away from 0) when $T_1 \neq T_2$, because of the unique continuation property. So W^o is the injective image of $(0, \infty] \times (\mathbb{R}^n \setminus 0)$. As such, W^o is a C^0 manifold with boundary: the boundary is $\mu_\infty(\mathbb{R}^n \setminus 0)$. However, a simple calculation shows that W^o is not a smooth submanifold-with-boundary in $\mathbb{R}^n \times \mathbb{R}^n$. To illustrate this, consider the case $n = 2$, and let the eigenvalues of L be λ and $-\mu$, with λ and μ both positive. The image of μ_T is the linear subspace of $\mathbb{R}^4 = \mathbb{R}^2 \times \mathbb{R}^2$ given by the equations

$$x_3 = e^{-2T\lambda} x_1$$
$$x_2 = e^{-2T\mu} x_4.$$

The union of these, taken over all T in $(0, \infty]$, lies on the locus in $\mathbb{R}^4$ given by

$$|x_1|^{\mu/\lambda} x_2 - |x_3|^{\mu/\lambda} x_4 = 0.$$

Within this locus, W^o is cut out by the constraints that x_3 and x_2 be non-negative multiples of x_1 and x_4 respectively, and not all four be zero. To understand the geometry, consider the case $\mu = 2\lambda$, and take the intersection of W^o with the slice $x_4 = 1$. We arrive at the locus $W_1^o \subset \mathbb{R}^3$ described by:

$$\begin{aligned} x_1^2 x_2 - x_3^2 &= 0 \\ x_2 &\geq 0 \\ \mathrm{sign}(x_3) &= \mathrm{sign}(x_1) \text{ or } 0. \end{aligned}$$

The locus defined by the first equation and the inequality $x_2 \geq 0$ is "Whitney's umbrella", which is the image in $\mathbb{R}^3$ of a map $f : \mathbb{R}^2 \to \mathbb{R}^3$ having a single non-immersion point:

$$g(y_1, y_2) = (y_1, y_2^2, y_1 y_2).$$

With the condition on the sign of x_3, we see that W_1^o is the injective image of the half-space $y_2 \geq 0$ under the map g. It is a C^0 manifold with boundary, and the boundary is the x_1 axis in $\mathbb{R}^3$. The subspace W_1^o is a smooth manifold with boundary *except* at the point $g(0,0)$: along the boundary of W_1^o, the outward-pointing tangent vector to W_1^o is the vector $(0,0,1)$ for x_1 positive, and is $(0,0,-1)$ for x_1 negative; at $x_1 = 0$ it is undefined. See Figure 4.

Let us now explain how this picture is relevant to understanding the structure of the compactification of moduli spaces of trajectories. Let B be a smooth, compact Riemannian manifold with a Morse function f. Let K_1 and K_2 be two closed submanifolds of B. For simplicity, suppose that $f = 1$ on K_1 and $f = -1$ on K_2, and suppose in addition that there is a unique critical point a of f in the set $f^{-1}[-1, 1]$. We shall also suppose that the flow is linear in suitable local

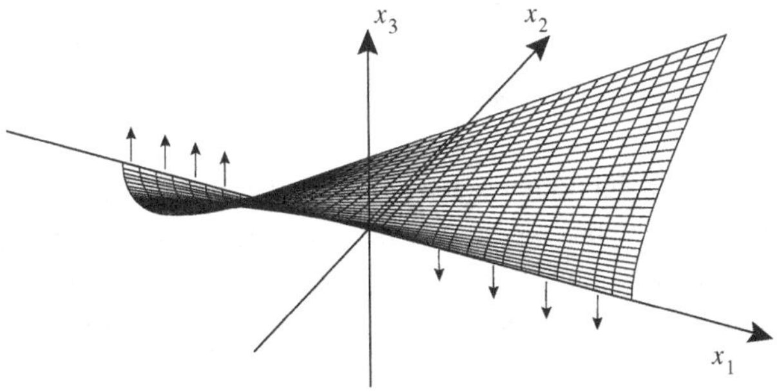

Fig. 4. Half of Whitney's umbrella.

coordinates around a. Let us consider trajectories $\gamma(t)$ of the downward gradient flow, on a finite interval, that start on K_1 and end on K_2: we can decompose these according to the length of the interval, and so we introduce the notation

$$M(K_1, K_2) = \bigcup_{S>0} M_S(K_1, K_2)$$

where

$$M_S(K_1, K_2) \\ = \{\gamma : [-S, S] \to B \mid \gamma(-S) \in K_1, \gamma(S) \in K_2, \dot{\gamma} = -\operatorname{grad}(f)\}.$$

The space $M(K_1, K_2)$ has a compactification, obtained by attaching a stratum $M_\infty(K_1, K_2)$ consisting of broken trajectories: we write

$$M_\infty(K_1, K_2) = M(K_1, a) \times M(a, K_2),$$

where

$$M(K_1, a) = \left\{\gamma : \mathbb{R}^{\geq} \to B \middle| \gamma(0) \in K_1, \lim_{t\to\infty} \gamma(t) = a\right\}$$
$$M(a, K_2) = \left\{\gamma : \mathbb{R}^{\leq} \to B \middle| \gamma(0) \in K_2, \lim_{t\to-\infty} \gamma(t) = a\right\}.$$

We suppose that the stable manifold of a meets K_1 transversely, so that $M(K_1, a)$ is a manifold; and similarly we suppose that the unstable manifold of a meets K_2 transversely. Under these circumstances, we can derive the following result:

Proposition 18.1.2. *The compactification of $M(K_1, K_2)$ obtained by adding the broken trajectories $M_\infty(K_1, K_2)$ has the structure of a C^0 manifold with boundary in a neighborhood of $M_\infty(K_1, K_2)$.*

Proof. Let Ω be a neighborhood of a on which the flow looks linear in suitable coordinates; and let Ω be chosen with the additional property that every trajectory intersects Ω in a connected set. Then an elementary argument shows that there exists an $S_1 \geq 0$ such that for all $S \geq S_1$ and all γ in $M_S(K_1, K_2)$, we have

$$\gamma\big([-S + S_1, S - S_1]\big) \subset \Omega.$$

Thus, for $S > S_1$, any γ in $M_S(K_1, K_2)$, every trajectory γ in $M_S(K_1, K_2)$ can be regarded as the concatenation of three trajectories,

$$\gamma_1 : [-S, -S + S_1] \to B$$
$$\delta : [-S + S_1, S - S_1] \to \Omega$$
$$\gamma_2 : [S - S_1, S] \to B.$$

where γ_1 has initial endpoint on K_1 and final endpoint in Ω, and γ_2 similarly runs from Ω to K_2. If we let $\phi_t : B \to B$ denote the time-t flow, and write $T = S - S_1$, then we can summarize the situation by saying that $M_S(K_1, K_2)$ is parametrized by the intersection

$$\big(\phi_{S_1}(K_1) \times \phi_{-S_1}(K_2)\big) \cap W_T \subset \Omega \times \Omega, \tag{18.1}$$

where

$$W_T = \left\{ (\delta(-T), \delta(T)) \;\middle|\; \delta : [-T, T] \to \Omega \text{ with } \dot{\delta} = -\operatorname{grad}(f) \right\}$$
$$\subset \Omega \times \Omega. \tag{18.2}$$

Because the flow is linear on Ω and trajectories meet Ω in connected sets, we have a complete understanding of W_T from our earlier analysis. We can use our parametrization μ_T, and write

$$W_T = \mu_T(\mathbb{R}^n) \cap (\Omega \times \Omega),$$

where $\mu_T : \mathbb{R}^n \to \mathbb{R}^n \times \mathbb{R}^n$ is a linear map as before. We know that, as $T \to \infty$, the linear spaces converge, so W_T approaches

$$W_\infty = \mu_\infty(\mathbb{R}^n) \cap (\Omega \times \Omega).$$

Our transversality hypotheses are equivalent to saying that the intersection (18.1) is transverse when $T = \infty$. It follows that the intersection is transverse also for finite T greater than some T_1 (i.e. for S greater than $S_1 + T_1$).

Because of this transversality, the compact manifolds $M_S(K_1, K_2)$, for $S > S_1 + T_1$, are all diffeomorphic, and we can identify the locus $S > S_1 + T_1$ in the compactification of $M(K_1, K_2)$ with a product

$$(0, \infty] \times M_\infty(K_1, K_2).$$

□

We make two remarks about this argument. First, the argument can be applied to the case that K_1 and K_2 are the intersections of level sets of f with the unstable and stable manifolds of critical points a_1 and a_2 respectively. If f has no critical values other than $f(a)$ between $f(a_1)$ and $f(a_2)$, then the proposition above effectively describes the compactification of the moduli space of unparametrized trajectories $\breve{M}(a_1, a_2)$:

Proposition 18.1.3. *Suppose that $\breve{M}(a_1, a_2), \breve{M}(a_1, a)$ and $\breve{M}(a, a_2)$ are all regular. Then the compactification, $\breve{M}^+(a_1, a_2)$, of $\breve{M}(a_1, a_2)$ is a C^0 manifold with boundary. The boundary is $\breve{M}(a_1, a) \times \breve{M}(a, a_2)$.* □

The second remark is that we should consider whether we can drop the hypothesis that the flow is linear on Ω. This we can certainly do, but we will need then to extend our analysis. What is at issue is whether the submanifolds W_T of $\Omega \times \Omega$, suitably parametrized, converge in the C^1 topology to the correct limit W_∞. This does indeed occur: we shall take up this issue in the context of the Chern–Simons–Dirac flow later. For the present, we state such a result in the finite-dimensional context, without proof.

Proposition 18.1.4. *Let $a \in B$ be a non-degenerate critical point of the flow. Then there exist a neighborhood Ω of a, a positive number T_1 and a family of maps μ_T from a fixed ball,*

$$\mu_T : B_1 \to B \times B, \qquad T \in (T_1, \infty],$$

which parametrize the loci $W_T \subset \Omega \times \Omega$ defined by (18.2), *in the sense that $W_T = \mu_T(B_1) \cap \Omega \times \Omega$. Furthermore, for finite T, the map*

$$(T, x) \mapsto \mu_T(x)$$

is a smooth function of the two variables; and as $T \to \infty$, the maps μ_T converge to μ_∞ in the C^∞ topology. □

In this section, we will prove a version of this proposition in the gauge-theory setting. The statement is formulated in Subsection 18.2 below; the proof is given first for a more abstract version in Subsection 18.3, which is applied to the gauge-theory version in Subsection 18.4. In Section 19, we will apply this result to study the structure of the compactifications of the moduli space $\breve{M}([\mathfrak{a}], [\mathfrak{b}])$.

18.2 Statement of the theorem

Let $\mathfrak{a} \in \tilde{\mathcal{C}}^\sigma_k(Y)$ be a critical point of $(\operatorname{grad} \mathcal{L})^\sigma$, and let $\gamma_\mathfrak{a}$ be the translation-invariant solution to the equations on the cylinder, in temporal gauge. We suppose our perturbation is generic, so that $\mathfrak{a}$ is non-degenerate as usual. For each $T > 0$, we can regard $\gamma_\mathfrak{a}$ as defining an element $[\gamma_\mathfrak{a}] \in \tilde{M}(Z^T)$, where $Z^T = [-T, T] \times Y$ and $\tilde{M}(Z^T)$ is defined as in the previous subsection.

We also introduce Z^∞ as the *disjoint* union of two half-infinite cylinders:

$$Z^\infty = (\mathbb{R}^\leq \times Y) \amalg (\mathbb{R}^\geq \times Y).$$

We introduce configuration spaces for Z^∞, in parallel to our previous definitions for $\mathbb{R} \times Y$ in Section 13. In particular we have $\tilde{\mathcal{C}}^\tau_{k,\mathrm{loc}}(Z^\infty)$ and its quotient space $\tilde{\mathcal{B}}^\tau_{k,\mathrm{loc}}(Z^\infty)$. Inside $\tilde{\mathcal{B}}^\tau_{k,\mathrm{loc}}(Z^\infty)$ is the moduli space $M(Z^\infty, [\mathfrak{a}])$ of solutions to the equations that are asymptotic to $[\mathfrak{a}]$ on *both* ends. As in the case of $\mathbb{R} \times Y$, we can also describe $\tilde{M}(Z^\infty, [\mathfrak{a}])$ using configuration spaces based on L^2_k rather than $L^2_{k,\mathrm{loc}}$, and so describe the moduli spaces as subsets of the Hilbert manifold

$$\tilde{\mathcal{B}}^\tau_k(Z^\infty, [\mathfrak{a}]) = \tilde{\mathcal{C}}^\tau_k(Z^\infty, \mathfrak{a})/\mathcal{G}_{k+1}(Z^\infty)$$

where

$$\begin{aligned}\tilde{\mathcal{C}}^\tau_k&(Z^\infty, \mathfrak{a}) \\ &= \{\gamma \in \tilde{\mathcal{C}}^\tau_{k,\mathrm{loc}}(Z^\infty) \mid \gamma - \gamma_\mathfrak{a} \in L^2_k(Z; iT^*Z) \oplus L^2_k(\mathbb{R}; \mathbb{R}) \oplus L^2_{k,A_0}(Z; S^+)\}\end{aligned}$$

(see Subsection 13.3). As in the case of $\tilde{M}(Z^T)$, the moduli space $\tilde{M}(Z^\infty, [\mathfrak{a}])$ is a Hilbert submanifold of $\tilde{\mathcal{B}}^\tau_k(Z^\infty, [\mathfrak{a}])$.

The manifold Z^∞, like Z^T, has boundary $Y \amalg \bar{Y}$, so we have restriction maps

$$\begin{aligned} R : \tilde{M}(Z^\infty, [\mathfrak{a}]) &\to \tilde{\mathcal{B}}^\sigma_{k-1/2}(Y \amalg \bar{Y}), \\ R : \tilde{M}(Z^T) &\to \tilde{\mathcal{B}}^\sigma_{k-1/2}(Y \amalg \bar{Y}). \end{aligned}$$

(See Theorem 17.1.1.) Let

$$\mathcal{K} = \mathcal{K}^\sigma_{k-1/2,\mathfrak{a}}(Y) \tag{18.3}$$

be the $L^2_{k-1/2}$ completion of the complement $\mathcal{K}^\sigma_{k,\mathfrak{a}}$ to the gauge-group orbit, as defined in Subsection 9.3. The aim of this section is to prove the following result, which should be compared with the finite-dimensional version above, Proposition 18.1.1.

Theorem 18.2.1. *There exists T_0 such that for all $T \geq T_0$ we can find smooth maps*

$$u(T,-) : B(\mathcal{K}) \to \tilde{M}(Z^T)$$
$$u(\infty,-) : B(\mathcal{K}) \to \tilde{M}(Z^\infty)$$

which are diffeomorphisms from a ball $B(\mathcal{K}) \subset \mathcal{K}$ onto neighborhoods of the constant solution $[\gamma_\mathfrak{a}]$. These can be chosen so that the map

$$\mu_T : B(\mathcal{K}) \to \tilde{\mathcal{B}}^\sigma_{k-1/2}(Y \amalg \bar{Y})$$

defined by composing $u(T,-)$ with the restriction maps to the boundary,

$$\mu_T(h) = R\,u(T,h),$$

is a smooth embedding of $B(\mathcal{K})$ for $T \in [T_0,\infty]$, with the following properties: as a function on $[T_0,\infty) \times B(\mathcal{K})$, the map $(T,h) \mapsto \mu_T(h)$ is smooth for finite T; and μ_T converges to μ_∞ in the C^∞_{loc} topology as $T \to \infty$,

$$\mu_T \xrightarrow{C^\infty_{\mathrm{loc}}} \mu_\infty.$$

Finally, there is an $\eta > 0$, independent of T, such that the images of the maps $u(T,-)$ can be taken to contain all solutions $[\gamma] \in M(Z^T)$ with $\|\gamma - \gamma_\mathfrak{a}\|_{L^2_k(Z^T)} \leq \eta$.

In the case that $\mathfrak{a}$ is reducible, the parametrizations u are equivariant for the $\mathbb{Z}/2$ actions arising from the standard $\mathbb{Z}/2$ action $\mathbf{i}$ on $\tilde{\mathcal{C}}^\sigma_{k-1/2}(Y)$.

The space $\tilde{M}(Z^\infty)$, of course, is a product

$$\tilde{M}(Z^\infty) = \tilde{M}(\mathbb{R}^{\geq} \times Y) \times \tilde{M}(\mathbb{R}^{\leq} \times Y).$$

The parametrization $u(\infty,-)$ provided by the theorem respects this product structure: we shall see that, in the spectral decomposition of $\mathcal{K} = \mathcal{K}^\sigma_{k-1/2,\mathfrak{a}}$ as $\mathcal{K}^+ \oplus \mathcal{K}^-$ (see (17.12)), the map $u(\infty,-)$ gives local diffeomorphisms

$$\begin{aligned} B(\mathcal{K}^+) &\to \tilde{M}(\mathbb{R}^{\geq} \times Y) \\ B(\mathcal{K}^-) &\to \tilde{M}(\mathbb{R}^{\leq} \times Y). \end{aligned} \tag{18.4}$$

Taking the boundary values, we get maps

$$\mu_\infty^+ : B(\mathcal{K}^+) \to \mathcal{B}^\sigma_{k-1/2}(\bar{Y})$$
$$\mu_\infty^- : B(\mathcal{K}^-) \to \mathcal{B}^\sigma_{k-1/2}(Y).$$

It is natural to adopt the terminology of finite-dimensional flows, and refer to the spaces parametrized by $\mu_\infty^\pm$ as the (local) stable and unstable manifolds of $[\mathfrak{a}]$.

18.3 An abstract gluing result

In order to get away from the specifics of the equations that are involved, we shall abstract the properties that are relevant to Theorem 18.2.1, and prove a version of the theorem in a more general context. Later, we shall deduce Theorem 18.2.1 from this abstract version.

On the cylinder $\mathbb{R} \times Y$, let E be the pull-back of a bundle E_0 on Y. Let D be an operator acting on sections of E, having the form

$$Du = \frac{du}{dt} + Lu,$$

where L is an operator on Y:

$$L : L^2_k(Y; E_0) \to L^2_{k-1}(Y; E_0).$$

So D is an operator $L^2_k(Z; E) \to L^2_{k-1}(Z; E)$, where Z can be taken to be either Z^T or Z^∞ (in the notation of the previous subsection).

In the case of Z^∞, the operator D acts also on *weighted* Sobolev spaces. Given a weight $\delta \in \mathbb{R}$, we consider $e^{-\delta|t|}$ as a smooth function on Z^∞, equal to $e^{-\delta t}$ on the first component and equal to $e^{\delta t}$ on the second. We define the weighted Sobolev space $L^2_{k,\delta}(Z^\infty; E)$ to be the space $e^{-\delta|t|} L^2_k(Z^\infty; E)$, so

$$s \in L^2_{k,\delta}(Z^\infty; E) \iff e^{\delta|t|} s \in L^2_k(Z^\infty; E).$$

Then D defines an operator

$$D : L^2_{k,\delta}(Z^\infty; E) \to L^2_{k-1,\delta}(Z^\infty; E).$$

We also suppose we are given a linear map

$$\Pi : L^2_{k-1/2}(Y \amalg \bar{Y}; E_0) \to H$$

for some Hilbert space H. Omitting the restriction map from our notation, we regard Π also as defining a map

$$\Pi : L^2_k(Z; E) \to H$$

for $Z = Z^T$ and $Z = Z^\infty$.

Because there is never any risk of confusion, we omit the subscripts k etc. in these norms. We also write

$$\begin{aligned} \mathcal{E}^T &= L^2_k(Z^T; E) \\ \mathcal{F}^T &= L^2_{k-1}(Z^T; E), \end{aligned}$$

and we write $\mathcal{E}^\infty_\delta$, $\mathcal{F}^\infty_\delta$ for the weighted spaces. With this notation, we impose the following hypotheses for the linear operators:

Hypothesis 18.3.1. We suppose that the linear operator

$$(D, \Pi) : \mathcal{E}^\infty \to \mathcal{F}^\infty \oplus H$$

is invertible. ◊

Lemma 18.3.2. *If* (D, Π) *satisfies Hypothesis* 18.3.1, *then the same operator acting on the weighted spaces,*

$$(D, \Pi) : \mathcal{E}^\infty_\delta \to \mathcal{F}^\infty_\delta \oplus H,$$

is also invertible, for all δ *sufficiently close to zero.*

Proof. Multiplication by $e^{\delta|t|}$ on the two components of Z^∞ gives an isometry I from $\mathcal{E}^\infty_\delta$ to $\mathcal{E}^\infty$ and from $\mathcal{F}^\infty_\delta$ to $\mathcal{F}^\infty$. The map I commutes with Π, so it is enough to consider the operator (IDI^{-1}, Π) on the unweighted spaces. We have

$$IDI^{-1} = D \pm \delta,$$

so this operator varies continuously with δ in operator norm. The result follows. □

We will fix a $\delta > 0$ satisfying the conclusion of the lemma, and we take C_0 to be a constant at least as large as the operator norm of the inverse of (D, Π) in both the unweighted and weighted spaces: that is,

$$\begin{aligned} \|u\| &\le C_0\big(\|Du\| + \|\Pi u\|\big) \\ \|u\|_\delta &\le C_0\big(\|Du\|_\delta + \|\Pi u\|\big) \end{aligned}$$

for all u. The canonical example of such a situation is the Atiyah–Patodi–Singer boundary-value problem, in which H is the $L^2_{k-1/2}$ closure of the span of eigenvectors of L belonging to negative eigenvalues on Y and the positive eigenvalues on $\bar{Y}$. In our eventual application, H will be a little different.

We shall consider a non-linear operator $D+\alpha$ having D as its linearization at $u=0$. We shall suppose that the non-linear term arises as follows. We suppose there is a continuous map

$$\alpha_0 : C^\infty(Y; E_0) \to L^2(Y; E_0)$$

and that

$$\alpha : C^\infty(\mathbb{R} \times Y; E) \to L^2_{\text{loc}}(\mathbb{R} \times Y; E)$$

is defined by restriction to slices $\{t\} \times Y$. We can also consider α over a compact interval.

Hypothesis 18.3.3. We shall suppose that α defines a smooth map

$$\alpha : L^2_k([-1, 1] \times Y; E) \to L^2_{k-1}([-1, 1] \times Y; E).$$

We suppose that $\alpha(0) = 0$ and $\mathcal{D}_0\alpha = 0$. $\diamondsuit$

As in the proof of Lemma 14.4.1, it follows that α defines a smooth map in the topologies $\mathcal{E}^T \to \mathcal{F}^T$ on the finite cylinders, and a smooth map on the infinite cylinders:

$$\alpha : \mathcal{E}^\infty \to \mathcal{F}^\infty$$
$$\alpha : \mathcal{E}^\infty_\delta \to \mathcal{F}^\infty_\delta.$$

Since α is C^1 on both $\mathcal{E}^\infty$ and $\mathcal{E}^\infty_\delta$ and has vanishing derivative at the origin, it is uniformly Lipschitz with small Lipschitz constant on small balls about 0. That is, for any $\epsilon > 0$, we can find an $\eta > 0$, such that for all u, u' in $\mathcal{E}^\infty$, we have

$$\|u\|, \|u'\| \le \eta \quad \implies \quad \|\alpha(u) - \alpha(u')\| \le \epsilon \|u - u'\|,$$

and similarly

$$\|u\|_\delta, \|u'\|_\delta \le \eta \quad \implies \quad \|\alpha(u) - \alpha(u')\|_\delta \le \epsilon \|u - u'\|_\delta$$

when u and u' are in $\mathcal{E}^\infty_\delta$.

Hypothesis 18.3.4. We will suppose $\eta_1 > 0$ is chosen so that the above Lipschitz property holds with $\epsilon = 1/(2C_0)$, where C_0 is the constant on page 326. ◊

With these hypotheses in place, we shall consider the space of zeros of the maps

$$F^T = D + \alpha : \mathcal{E}^T \to \mathcal{F}^T$$
$$F^\infty = D + \alpha : \mathcal{E}^\infty \to \mathcal{F}^\infty.$$

We write

$$M(T) = (F^T)^{-1}(0) \subset \mathcal{E}^T$$
$$M(\infty) = (F^\infty)^{-1}(0) \subset \mathcal{E}^\infty.$$

Our abstract version of the theorem is the following:

Theorem 18.3.5. *For $T \geq T_0$, the solution sets $M(T)$ and $M(\infty)$ are Hilbert submanifolds of $\mathcal{E}^T$ and $\mathcal{E}^\infty$ in a neighborhood of zero. There is an $\eta > 0$ and smooth maps from the η ball in the Hilbert space H to the solution sets,*

$$u(T, -) : B_\eta(H) \to M(T)$$
$$u(\infty, -) : B_\eta(H) \to M(\infty),$$

which are diffeomorphisms onto their image, and which satisfy $\Pi u(T,h) = \Pi u(\infty,h) = h$. Furthermore, for $T \in [T_0, \infty]$, the map

$$\mu_T : B_\eta(H) \to L^2_{k-1/2}(Y \amalg \bar{Y}; E_0) \tag{18.5}$$

defined by composing $u(T,-)$ with the restriction maps to the boundary,

$$\mu_T(h) = r\, u(T,h),$$

is a smooth embedding of $B_\eta(H)$. As a function on $[T_0, \infty) \times B_\eta(H)$, the map $(T,h) \mapsto \mu_T(h)$ is smooth for finite T; and μ_T converges to μ_∞ in the C^∞_{loc} topology as $T \to \infty$:

$$\mu_T \xrightarrow{C^\infty_{\text{loc}}} \mu_\infty.$$

Finally, there is an $\eta' > 0$ (independent of T) such that the images of the maps $u(T,-)$ contain all solutions $u \in M(T)$ with $\|u\| \leq \eta'$.

The proof uses the inverse function theorem: we state here a suitable version, without proof.

Proposition 18.3.6. *Let H_1 and H_2 be Hilbert spaces, and let $S : H_1 \to H_2$ be a continuous map with $S(0) = 0$. Suppose S has the form $Q + \beta$, where Q is linear and invertible, and β is uniformly Lipschitz on the η_1 ball $B_{\eta_1}(H_1) \subset H_1$ with Lipschitz constant ϵ:*

$$\|x\|, \|x'\| \leq \eta_1 \qquad \Longrightarrow \qquad \|\beta(x) - \beta(x')\| \leq \epsilon \|x - x'\|.$$

Suppose $\epsilon < 1/\|Q^{-1}\|$. Then S is injective on $B_{\eta_1}(H_1)$ and its image contains the ball $B_{\eta_2}(H_2)$, where

$$\eta_2 = \eta_1(1 - \epsilon\|Q^{-1}\|)/\|Q^{-1}\|.$$

In particular, for all $y \in B_{\eta_2}(H_2)$ there exists a unique $x = W(y)$ in $B_{\eta_1}(H_1)$ solving the equation

$$S(x) = y.$$

Furthermore, if β is smooth, then so is the map $W : B_{\eta_2}(H_2) \to B_{\eta_1}(H_1)$, and the norm of $\mathcal{D}_y W$ is bounded by $\|Q^{-1}\|/(1 - \epsilon\|Q^{-1}\|)$; and for all $m > 1$ and $y \in B_{\eta_2}(H_2)$, the norm of $\mathcal{D}^m_y W$ is bounded by a universal continuous function of the quantities

$$(1 - \epsilon\|Q^{-1}\|)/\|Q^{-1}\|, \ \|\mathcal{D}_{W(y)}\beta\|, \ldots, \|\mathcal{D}^m_{W(y)}\beta\|.$$

□

As a first step, we show the existence of the solution $u(\infty, h)$ for small h:

Proposition 18.3.7. *Let C_0 be the constant in Hypothesis 18.3.1, and let η_1 be as in Hypothesis 18.3.4. Then for every $h \in H$ with $\|h\| \leq \eta_1/2C_0$, there exists a unique $u = u(\infty, h)$ in $B_{\eta_1}(\mathcal{E}^\infty)$, satisfying*

$$F^\infty(u) = 0$$

$$\Pi u = h.$$

The map $u(\infty, -)$ is a smooth map from $B_{\eta_2}(H)$ to $B_{\eta_1}(\mathcal{E}^\infty)$, where $\eta_2 = \eta_1/2C_0$, and satisfies

$$\|u(\infty, h)\| \leq 2C_0\|h\|. \tag{18.6}$$

With the same constants η_1 and η_2, the map $u(\infty,-)$ is also a smooth map from $B_{\eta_2}(H)$ to $B_{\eta_1}(\mathcal{E}_\delta^\infty)$ and satisfies

$$\|u(\infty,h)\|_\delta \leq 2C_0\|h\|.$$

Proof. We apply the inverse function theorem, Proposition 18.3.6, to the map

$$(F^\infty,\Pi) : \mathcal{E}^\infty \to \mathcal{F}^\infty \oplus H.$$

(We have chosen η_1 so that α is uniformly Lipschitz on $B_{\eta_1}(\mathcal{E}^\infty)$, with Lipschitz constant ϵ, where $(1-\epsilon C_0) = 1/2$, so in the statement of the proposition above, η_2 is $\eta_1/2C_0$.) The inequality (18.6) follows from the bound on the derivative of u which is provided by Proposition 18.3.6, together with our choice of ϵ in Hypothesis 18.3.4.

The same inverse function theorem can be applied to the map on the weighted Sobolev spaces. The uniqueness result means that the solutions are the same. □

We now examine the linear operator (D,Π) acting as

$$(D,\Pi) : \mathcal{E}^T \to \mathcal{F}^T \oplus H.$$

Lemma 18.3.8. *There exists a T_0 such that for $T \geq T_0$, the operator*

$$P^T = (D,\Pi) : \mathcal{E}^T \to \mathcal{F}^T \oplus H$$

is invertible. Furthermore, as $T \to \infty$, the operator norm $\|(P^T)^{-1}\|$ is bounded by a constant independent of T.

Proof. Let the inverse of P^∞ be N^∞. On Z^T, let β_1, β_2 be a partition of unity subordinate to the cover of $[-T,T]$ by the two open sets $[-T,T/2)$ and $(-T/2,T]$. We can choose these so that the pointwise norms of their first k derivatives go to zero uniformly as $T \to \infty$. Let ϕ_1, ϕ_2 be functions with the property that $\phi_i = 1$ on the support of β_i, with $\phi_1 = 0$ on $(3T/4,T]$ and $\phi_2 = 0$ on $[-T,-3T/4)$. These can also be chosen so that their first k derivatives go to zero as T goes to infinity.

Given v in $\mathcal{F}^T$, we define $\gamma(v)$ in $\mathcal{F}^\infty$ by

$$\rho(v) = \begin{cases} \tau_{-T}^*\beta_1 v & \text{on } [0,\infty)\times Y \\ \tau_T^*\beta_2 v & \text{on } (-\infty,0]\times Y. \end{cases}$$

Similarly, given u in $\mathcal{E}^\infty$, define $\pi(u) \in \mathcal{E}^T$ by

$$\pi(u) = \phi_1 \tau_T^* u_1 + \phi_2 \tau_{-T}^* u_2,$$

where u_1 and u_2 are the parts of u on the two components $[0, \infty) \times Y$ and $(-\infty, 0] \times Y$ of Z^∞. Now define $\tilde{N}^T : \mathcal{F}^T \oplus H \to \mathcal{E}^T$ by

$$\tilde{N}^T(v, h) = \pi \circ N^\infty(\rho(v), h).$$

We find that the composites $P^T \circ \tilde{N}^T$ and $\tilde{N}^T \circ P^T$ have the form $1 + K$ and $1 + J$, where the operator norms of K and J are bounded by fixed multiples of the operator norms of the multiplication operators

$$v \mapsto (d\phi_i/dt)v$$
$$u \mapsto (d\beta_i/dt)u$$

respectively, acting on $\mathcal{F}^T$ and $\mathcal{E}^T$ (*cf.* (14.4)). Our hypotheses on the derivatives of β_i and ϕ_i ensure that these operator norms go to zero. It follows that P^T is invertible for $T \geq T_0$, and that the inverse N^T has $\|N^T - \tilde{N}^T\|$ going to zero as $T \to \infty$. The formula for $\tilde{N}^T$ and the fact that the derivatives of the cut-off functions go to zero imply that the operator norm of $\tilde{N}^T$ remains bounded as T increases. □

On account of the lemma, we can suppose that the operator norm of the inverse of P^T is bounded by the same constant C_0 as appears on page 326, for all $T \geq T_0$. We therefore deduce, just as in Proposition 18.3.7, the existence of solutions on the finite cylinder:

Corollary 18.3.9. *Let the constants C_0, η_1 and η_2 be as in Proposition* 18.3.7, *and let T_0 be as above. Then for every $h \in H$ with $\|h\| \leq \eta_2$, there exists a unique $u = u(T, h)$ in $B_{\eta_1}(\mathcal{E}^T)$, satisfying*

$$F^T(u) = 0$$
$$\Pi u = h.$$

The map $u(T, -)$ is a smooth map from $B_{\eta_2}(H)$ to $B_{\eta_1}(\mathcal{E}^T)$, and satisfies

$$\|u(T, h)\| \leq 2C_0 \|h\|.$$

□

Our next task is to compare the solution $u(T,h)$ on the finite cylinder Z^T with the solutions $u(\infty,h)$. As above, we write the two components of the latter as $u_1(h)$ and $u_2(h)$, on the two components of Z^∞. Set

$$U(T,h) = \tau_T^* u_1(h) + \tau_{-T}^* u_2(h)$$

regarded as an element of $\mathcal{E}^T$. (Here τ_T is again the map $(t,y) \mapsto (t+T,y)$ on the cylinder.) We shall see that $U(T,h)$ is close to $u(T,h)$:

Proposition 18.3.10. *For any $\eta < \eta_2$, the function*

$$\begin{aligned} \xi(T,-) &: B_\eta(H) \to \mathcal{E}^T \\ \xi(T,h) &= u(T,h) - U(T,h) \end{aligned}$$

converges to zero in the C^∞_{loc} topology as $T \to \infty$. Indeed, for each $m \geq 0$, we have

$$\left\| \mathcal{D}_h^m(\xi(T,-)) \right\| \leq K_m(h) e^{-2\delta T}, \tag{18.7}$$

where K_m is a continuous function on the ball $B_\eta(H)$.

Proof. The function $u_T = u(T,-)$ is the solution to

$$\begin{aligned} F^T u_T(h) &= 0 \\ \Pi u_T(h) &= h. \end{aligned}$$

Let us define $\zeta(h)$ and $g(h)$ by

$$\begin{aligned} F^T U_T(h) &= \zeta(h) \\ \Pi U_T(h) &= h + g(h). \end{aligned}$$

Lemma 18.3.11. *The functions $\zeta(h)$ and $g(h)$ are smooth functions of $h \in B_{\eta_2}(H)$. These two functions and their derivatives satisfy bounds of the same form as* (18.7) *(though for different continuous functions K_m).*

Proof. We begin by examining $g(h)$. Let $\Pi_1 : L^2_{k-1/2}(\bar{Y}) \to H$ and $\Pi_2 : L^2_{k-1/2}(Y) \to H$ be the two parts of the map Π. Then we have

$$\Pi U_T(h) = h + \Pi_1\big(u_2|_{\{-2T\}\times Y}\big) + \Pi_2\big(u_1|_{\{2T\}\times Y}\big).$$

By the rider to the implicit function theorem, $u(\infty, h)$ is a smooth function of h, as a map to the weighted space,

$$B_{\eta_2}(H) \to \mathcal{E}^{\infty}_{\delta};$$

and the norm of the derivative $\mathcal{D}^m_h u(\infty, -)$ is bounded by a continuous function $K'_m(h)$ on the ball. The last two terms in the above formula define a function $g(h)$ which is exhibited as the composite of the smooth function $u(\infty, -)$ and a linear map

$$\mathcal{E}^{\infty}_{\delta} \to H.$$

The linear map has norm bounded by a multiple of $e^{-2\delta T}$. The result for g follows.

To examine ζ, we see first that it involves only the non-linearity α in the map F: we have

$$\zeta(h) = \alpha(U_T(h)) - \alpha(\tau^*_T u_1(h)) - \alpha(\tau^*_{-T} u_2(h)).$$

Lemma 18.3.12. *Let $A : X_1 \times X_2 \to Z$ be a smooth map of normed vector spaces with $A(x, 0) = 0$ for all x and $A(0, y) = 0$ for all y. Let $B_\eta(H) \subset H$ be a ball in a normed space H, and let $\xi_i : B_\eta(H) \to B_{\eta_1}(X_i)$ $(i = 1, 2)$ be two smooth maps. Define continuous functions $J_{i,m}$ on $B_\eta(H)$ by*

$$J_{i,m}(h) = \sum_{n=0}^{m} \|\mathcal{D}^n_h \xi_i\|.$$

Then the composite map

$$A \circ (\xi_1, \xi_2) : B_\eta(H) \to Z$$

satisfies

$$\|A \circ (\xi_1, \xi_2)(h)\| \le K_0(h) \|\xi_1(h)\| \, \|\xi_2(h)\|,$$

while for $m \ge 1$, we have

$$\|\mathcal{D}^m_h A \circ (\xi_1, \xi_2)\| \le K_m(h) \sum_{i=1}^{m} \sum_{j=1}^{m} J_{1,m}(h)^i J_{2,m}(h)^j,$$

for some continuous functions K_m on the ball.

Proof. Let $A^{(p,q)}$ be the (p,q)-th partial derivative of A with respect to the X_1 and X_2 coordinates. On differentiating the composite l times, we obtain terms of the form

$$A^{(p,q)}_{(\xi_1,\xi_2)}(\mathcal{D}^{r_1}\xi_1,\ldots,\mathcal{D}^{r_p}\xi_1,\mathcal{D}^{s_1}\xi_2,\ldots,\mathcal{D}^{s_q}\xi_2),$$

with $\sum r_i + \sum s_j = l$. If p and q are positive, we can bound this term directly by a multiple of

$$J_{1,l}(h)^p J_{2,l}(h)^q$$

and the size of $\|\mathcal{D}^m_{(\xi_1(h),\xi_2(h))}A\|$. Suppose next that one of p and q, say p, is zero and the other, q, is positive. Then we can write

$$\begin{aligned}A^{(0,q)}_{(\xi_1,\xi_2)}&(\mathcal{D}^{s_1}\xi_2,\ldots,\mathcal{D}^{s_q}\xi_2)\\ &= \int_0^1 A^{(1,q)}_{(s\xi_1,\xi_2)}(\xi_1,\mathcal{D}^{s_1}\xi_2,\ldots,\mathcal{D}^{s_q}\xi_2)\,ds\end{aligned}$$

using the fact that $A^{(0,q)}_{(0,y)} = 0$. The integrand is bounded as the previous terms were. Finally, if p and q are both zero, we use the identity

$$A(\xi_1,\xi_2) = \int_0^1\int_0^1 \mathcal{D}^2_{(s_1\xi_1,s_2\xi_2)}A(\xi_1,\xi_2)\,ds_1 ds_2$$

and the boundedness of the second derivative. □

We apply the observation of this lemma to the function

$$A(v_1,v_2) = \alpha(v_1+v_2) - \alpha(v_1) - \alpha(v_2),$$

regarded as a map

$$L^2_k([-1,1]\times Y)\times L^2_k([-1,1]\times Y)\to L^2_{k-1}([-1,1]\times Y).$$

Let ζ_s be the map obtained from ζ by translation and restriction,

$$\begin{aligned}\zeta_s : B_{\eta_2}(H) &\to L^2_{k-1}([-1,1]\times Y)\\ h &\mapsto \big(\tau_s^*\zeta(h)\big)|_{[-1,1]\times Y}\end{aligned}$$

for $s\in[-T+1,T-1]$. The formula for ζ above shows that ζ_s is the composite of A with the two maps $v_1, v_2 : B_{\eta_2}(H)\to L^2_k([-1,1]\times Y)$ given by translates of u_1 and u_2:

$$v_1 : h \mapsto u_1(h)|_{[T+s-1,T+s+1]\times Y}$$

$$v_2 : h \mapsto u_2(h)|_{[-T+s-1,-T+s+1]\times Y}.$$

On $B_{\eta_2}(H)$, the norms of the derivatives $\mathcal{D}_h^m v_i$ are bounded continuous functions $K'_m(h)e^{-\delta(T+s)}$ and $K'_m(h)e^{-\delta(T-s)}$, because (u_1, u_2) is a smooth map to the weighted space $\mathcal{E}_\delta^\infty$. From the lemma, we therefore obtain

$$\|\mathcal{D}_h^m \zeta_s\| \le K''_m(h)e^{-2\delta T},$$

for some continuous function K''_m, independent of s. By covering Z^T with $T+1$ intervals of length 2, applying the above estimate to each of these, and using Cauchy–Schwarz, we bound the mth derivative of ζ by a function of the shape $K_m(h)e^{-2\delta T}$. This concludes the proof of Lemma 18.3.11. □

We now return to the proof of Proposition 18.3.10. Since $\eta < \eta_2$, it follows that for $T \ge T_1$ we have $\|\zeta(h)\| + \|h + g(h)\| \le \eta_2$. It follows that we can write

$$u_T(h) = W(0, h)$$

$$U_T(h) = W(\zeta(h), h + g(h))$$

where W is the inverse on $B_{\eta_2}(\mathcal{F}^T \oplus H)$ of the map $S = (F^T, \Pi)$. The final rider to the inverse function theorem provides bounds on all the derivatives of W, by continuous functions on the ball; so the proposition follows from Lemma 18.3.11 and the chain rule. □

Corollary 18.3.13. *For any $\eta < \eta_2$, the map μ_T converges to μ_∞ in the topology of $C^\infty_{\mathrm{loc}}(B_\eta(H), L^2_{k-1/2}(Y \amalg \bar{Y}; E_0))$ as T goes to infinity.*

Proof. By Proposition 18.3.10 above, we can replace

$$\mu_T(h) = ru(T, h)$$

by

$$\tilde{\mu}_T(h) = rU(T, h)$$

for $T < \infty$ without affecting the estimate to be proved. The component of $\tilde{\mu}_T(h)$ in $L^2_{k-1/2}(\bar{Y}; E_0)$ is the sum

$$\begin{aligned}\tilde{\mu}_T(h) &= u_1(h)|_{\{0\}\times Y} + u_2(h)|_{\{-2T\}\times Y}\\ &= \mu_\infty(h) + u_2(h)|_{\{-2T\}\times Y}.\end{aligned}$$

The second term converges to zero in the C^∞_{loc} topology on the ball, because u_2 factors through the inclusion $\mathcal{E}^\infty_\delta \to \mathcal{E}^\infty$. The same argument applies to the other component of μ. □

With this corollary, we have completed most of the proof of Theorem 18.3.5. What remains is to prove that, for finite T, the map $(T,h) \mapsto \mu_T(h)$ is a smooth function of the two variables. To this end, fix any $T_1 > T_0$. We have isomorphisms

$$\begin{aligned} I_T &: \mathcal{E}^{T_1} \to \mathcal{E}^T\\ I_T &: \mathcal{F}^{T_1} \to \mathcal{F}^T,\end{aligned}$$

defined by pulling back sections of E via the map between the cylinders, $(t,y) \mapsto ((T/T_1)t, y)$. Let S_T be the map

$$S_T = (F^T, \Pi) : \mathcal{E}^T \to \mathcal{F}^T \oplus H,$$

and consider

$$\begin{aligned}\tilde{S}_T &= (I_T^{-1} \oplus 1) \circ S_T \circ I_T\\ \tilde{S}_T &: \mathcal{E}^{T_1} \to \mathcal{F}^{T_1} \oplus H.\end{aligned}$$

As maps between the fixed Hilbert spaces $\mathcal{E}^{T_1}$ and $\mathcal{F}^{T_1} \oplus H$, the map $\tilde{S}_T$ varies smoothly with T, in the sense that $(T,x) \mapsto \tilde{S}_T(x)$ is a smooth function of two variables: indeed, the relevant part of this map is the term

$$I_T^{-1}(D+\alpha)I_T = \tilde{D}_T + \alpha,$$

where

$$\tilde{D}_T = (T/T_1)\frac{d}{dt} + L.$$

The same smoothness therefore applies to the inverse of $\tilde{S}_T$, and hence to the map $\tilde{u}$ defined by

$$\tilde{u}(T,h) = I_T^{-1} u(T,h),$$

which solves the equation

$$\tilde{S}_T(\tilde{u}(T,h)) = (0,h)$$

for small h. We have

$$\begin{aligned}\mu_T(h) &= r \circ u(T,h) \\ &= r \circ \tilde{u}(T,h)\end{aligned}$$

so the result follows. □

18.4 Deducing the gauge-theory version

Consider now a critical point $\mathfrak{a}$ for the perturbed flow on $\mathcal{C}^{\sigma}_k(Y)$. Let $\gamma_{\mathfrak{a}}$ denote the corresponding translation-invariant solution on the finite cylinder Z^T or Z^{∞}. Write $\gamma_{\mathfrak{a}}$ as (A_0, s_0, ϕ_0) (in temporal gauge), and let $(\check{A}_0, \check{s}_0, \check{\phi}_0)$ be notation for $\mathfrak{a}$ on the 3-manifold. The first step in deducing Theorem 18.2.1 from the abstract version is to replace neighborhoods of $[\gamma_{\mathfrak{a}}]$ in $\tilde{M}(Z^T)$ and $\tilde{M}(Z^{\infty}, [\mathfrak{a}])$ with neighborhoods of $\gamma_{\mathfrak{a}}$ in the set of solutions to the equations in the slice to the gauge-group action, a subset of $\tilde{\mathcal{C}}^{\tau}_k(Z^T)$ or $\tilde{\mathcal{C}}^{\tau}_k(Z^{\infty}, \mathfrak{a})$. In the case of Z^T, the slice

$$\mathcal{S}^{\tau}_{k,\mathfrak{a}}(Z^T) \subset \tilde{\mathcal{C}}^{\tau}_k(Z^T)$$

was defined in Definition 9.4.2. As in Subsection 14.3, we rewrite this definition as the set of triples $(A, s, \phi) \in \tilde{\mathcal{C}}^{\tau}_k(Z)$ satisfying

$$\begin{aligned}\langle a, \vec{n}\rangle &= 0 \text{ at } \partial Z \\ \mathrm{Coul}^{\tau}_{\gamma_{\mathfrak{a}}}(A, s, \phi) &= 0.\end{aligned} \tag{18.8a}$$

Here we have written $A = A_0 + a \otimes 1$. The same conditions define the slice in the case $T = \infty$:

$$\mathcal{S}^{\tau}_{k,\mathfrak{a}}(Z^{\infty}) \subset \tilde{\mathcal{C}}^{\tau}_k(Z^{\infty}, \mathfrak{a}).$$

We will study a neighborhood of $[\gamma_{\mathfrak{a}}]$ in $\tilde{M}(Z^T)$ by studying solutions of the gauge-fixing equation (18.8a) together with the perturbed Seiberg–Witten equation

$$\mathfrak{F}^{\tau}_{\mathfrak{q}}(A, s, \phi) = 0. \tag{18.8b}$$

In our model problem, we considered a non-linear equation for an element u in a *linear* space $\mathcal{E}^T$. In our present application, (A, s, ϕ) belongs to $\tilde{\mathcal{C}}_k^\tau(Z^T)$, which is not a linear space, because of the constraint that $\phi(t) = \phi|_{\{t\}\times Y}$ should lie on the unit sphere in $L^2(Y; S)$ for all t. We therefore choose a linear chart near $\gamma_{\mathfrak{a}}$, identifying a neighborhood of $\gamma_{\mathfrak{a}}$ in $\tilde{\mathcal{C}}_k^\tau(Z^T)$ with the tangent space at $\mathcal{T}_{k,\gamma_{\mathfrak{a}}}^\tau(Z^T)$ at $\gamma_{\mathfrak{a}}$, by the map

$$i : \mathcal{T}_{k,\gamma_{\mathfrak{a}}}^\tau(Z^T) \to \tilde{\mathcal{C}}_k^\tau(Z^T)$$

$$i : (a, r, \psi) \mapsto (A_0 + a \otimes 1, s_0 + r, \phi)$$

where

$$\phi(t) = \frac{\check{\phi}_0 + \psi(t)}{\sqrt{1 + \|\psi(t)\|^2}}. \tag{18.9}$$

(Because (a, r, ψ) is a tangent vector, $\psi(t)$ is L^2-orthogonal to the unit vector $\check{\phi}_0$.)

The choice of chart i enters in a more significant way when we come to setting up suitable boundary conditions for the equations. The boundary of Z^T is $\bar{Y} \amalg Y$, and we have the restriction map

$$r : \tilde{\mathcal{C}}_k^\tau(Z^T) \to \tilde{\mathcal{C}}_{k-1/2}^\sigma(\bar{Y} \amalg Y) \times L_{k-1/2}^2(\bar{Y} \amalg Y; i\mathbb{R}),$$

where the second component records the normal component of the connection A at the boundary. In addition to the Neumann-type boundary condition $\langle a, \vec{n}\rangle = 0$ at the boundary, we wish to impose a spectral boundary condition. To do so however, we need to use the chart, so as to identify the boundary data $\tilde{\mathcal{C}}_{k-1/2}^\sigma(\bar{Y} \amalg Y)$ near $(\mathfrak{a}, \mathfrak{a})$ with a linear space. To do this, we use the same formula (18.9) to define a similar chart

$$i : \mathcal{T}_{k-1/2,\mathfrak{a}}^\sigma \times L_{k-1/2}^2(Y; i\mathbb{R}) \to \tilde{\mathcal{C}}_{k-1/2}^\sigma(Y) \times L_{k-1/2}^2(Y; i\mathbb{R}).$$

We will use the decomposition:

$$\mathcal{T}_{k-1/2,\mathfrak{a}}^\sigma \oplus L_{k-1/2}^2(Y; i\mathbb{R}) = \mathcal{J}_{k-1/2,\mathfrak{a}}^\sigma(Y) \oplus \mathcal{K}_{k-1/2,\mathfrak{a}}^\sigma(Y) \oplus L_{k-1/2}^2(Y; i\mathbb{R}). \tag{18.10}$$

We have seen that the extended Hessian is ASAFOE, and it is hyperbolic because of the non-degeneracy of $\mathfrak{a}$ (Lemma 12.4.3). The restriction of this operator to

$\mathcal{K}^{\sigma}_{k-1/2,\mathfrak{a}}$ gives the direct sum decomposition of $\mathcal{K}^{\sigma}_{k-1/2,\mathfrak{a}}$ as $\mathcal{K}^{+} \oplus \mathcal{K}^{-}$, just as in (17.12). Let H^{-}_{Y} and $H^{-}_{\bar{Y}}$ be defined by

$$\begin{aligned} H^{-}_{Y} &= \{0\} \oplus \mathcal{K}^{-} \oplus L^{2}_{k-1/2}(Y; i\mathbb{R}) \\ H^{-}_{\bar{Y}} &= \{0\} \oplus \mathcal{K}^{+} \oplus L^{2}_{k-1/2}(Y; i\mathbb{R}) \end{aligned}$$

as in (17.15); let $H = H^{-}_{\bar{Y}} \oplus H^{-}_{Y}$ and $\Pi = \Pi^{-}_{\bar{Y}} \oplus \Pi^{-}_{Y}$ be as in Lemma 17.3.3. We regard Π as an operator

$$\Pi : \mathcal{T}^{\sigma}_{k-1/2,\mathfrak{a}}(\bar{Y} \amalg Y) \oplus L^{2}_{k-1/2}(\bar{Y} \amalg Y; i\mathbb{R}) \to H.$$

We will apply Theorem 18.3.5 to the following equations, for $\gamma = (A_0 + a, s, \phi)$ in a neighborhood of $\gamma_{\mathfrak{a}} = (A_0, s_0, \phi_0) \in \tilde{\mathcal{C}}^{\tau}_{k}(Z)$ on $Z = Z^{\infty}$ or Z^{T}:

$$\begin{aligned} \mathfrak{F}^{\tau}_{\mathfrak{q}}\gamma &= 0 \\ \mathrm{Coul}^{\tau}_{\mathfrak{a}}\gamma &= 0 \\ (\Pi \circ i^{-1} \circ r)\gamma &= h \end{aligned}$$

where

$$\begin{aligned} h &= (\kappa_1, c_1, \kappa_2, c_2) \\ &\in \mathcal{K}^{+} \oplus L^{2}_{k-1/2}(\bar{Y}; i\mathbb{R}) \oplus \mathcal{K}^{-} \oplus L^{2}_{k-1/2}(Y; i\mathbb{R}) \\ &= H^{-}_{\bar{Y}} \oplus H^{-}_{Y}. \end{aligned}$$

If we write $\tilde{\gamma} = i^{-1}\gamma$, so that $\tilde{\gamma}$ belongs to

$$\mathcal{T}^{\tau}_{k,\gamma_{\mathfrak{a}}},$$

then we can write the equations as

$$\begin{aligned} (Q_{\gamma_{\mathfrak{a}}} + \alpha)\tilde{\gamma} &= 0 \\ (\Pi \circ r)\tilde{\gamma} &= h, \end{aligned}$$

where $Q_{\gamma_{\mathfrak{a}}}$ is as in (14.12) and α is the remainder of the terms. The equation is not quite in the form required by Theorem 18.3.5: neither the domain nor the range is the space of sections of a finite-dimensional vector bundle. The domain can be converted to this form by the device of (12.13). The range is the vector bundle $\mathcal{V}^{\tau}_{k-1} \to \mathcal{C}^{\tau}_{k}(Z, \mathfrak{s}_Z)$ (see (9.18)) and a similar device converts it to a space of sections of a finite-dimensional vector bundle. Once we have verified the necessary hypotheses for its application, Theorem 18.3.5 will provide a solution $\gamma = u(T,h)$ or $u(\infty,h)$ to these equations when $\|h\| \leq \eta$. The boundary condition $(\Pi \circ R \circ i^{-1})\gamma = h$ includes the conditions

$$\langle a, \vec{n}\rangle = c_1 \quad \text{on } \bar{Y}$$
$$\langle a, \vec{n}\rangle = c_2 \quad \text{on } Y.$$

When c_1 and c_2 are zero, the trajectory γ is therefore in Neumann gauge, so it belongs to the slice $\mathcal{S}^{\tau}_{k,\gamma_\alpha}$. Thus, by restricting the domain of $u(T,-)$, to those h with $c_1 = c_2 = 0$, we will obtain a parametrization of the solutions γ belonging to the slice, by a ball B_η in $\mathcal{K}^- \oplus \mathcal{K}^+ = \mathcal{K}$.

We now turn to verifying the required hypotheses. The fact that the non-linear terms α satisfy Hypothesis 18.3.3 follows from Condition (i) of Definition 10.5.1, because i is a diffeomorphism. The more interesting task is to verify the invertibility of the linearized equations in the case $T = \infty$, as required by Hypothesis 18.3.1. The operator whose invertibility we must check is the sum of two operators, corresponding to the two components of Z^∞; so let us look at only one, the operator

$$(Q_{\gamma_\alpha}, \Pi^-_Y \circ r) : \mathcal{T}^{\tau}_{k,\gamma_\alpha} \to \mathcal{V}^{\tau}_{k-1,\gamma_\alpha} \oplus H^-_Y. \tag{18.11a}$$

The proof that this is an isomorphism is much the same as the proof that the similar operator in Lemma 17.3.3 is Fredholm: indeed the projection Π^-_Y is the same one that appears there. The argument uses the same device as the proof of Proposition 17.2.6. To carry out the proof, we write $Q_{\gamma_\alpha} = (d/dt) + L$, and let $H^{\pm}_L$ be the spectral subspaces in $L^2_{1/2}(Y; E_0)$. Theorem 17.2.7 tells us that the linear map

$$(Q_{\gamma_\alpha}, \Pi^-_L \circ r) : \mathcal{T}^{\tau}_{k,\gamma_\alpha} \to \mathcal{V}^{\tau}_{k-1,\gamma_\alpha} \oplus (H^-_L \cap L^2_{k-1/2}) \tag{18.11b}$$

is an isomorphism, where Π^-_L is the spectral projection with kernel H^+_L. As in the proof of Proposition 17.2.6, we decompose the domain as $C \oplus K$, where K is the kernel of Q_{γ_α}, so as to write the operator (18.11b) as

$$\begin{bmatrix} Q_{\gamma_\alpha}|_C & 0 \\ x & (\Pi^-_L \circ r)|_K \end{bmatrix},$$

where both diagonal blocks are isomorphisms. The image of the $r|_K$ is precisely H^-_L; and as in the proof of Lemma 17.3.3, the projection Π^-_Y is an isomorphism on H_L^-. So the matrix

$$\begin{bmatrix} Q_{\gamma_\mathfrak{a}}|_C & 0 \\ x & (\Pi_Y^- \circ r)|_K \end{bmatrix}$$

also defines an invertible operator. This is the operator (18.11a).

We have now verified the necessary hypotheses for Theorem 18.3.5. What we learn is that there is an $\eta_1 > 0$ and maps from the η_1 ball,

$$u(T,-) : B_{\eta_1}(\mathcal{K}) \to \mathcal{S}^\tau_{k,\gamma_\mathfrak{a}}(Z^T)$$
$$u(\infty,-) : B_{\eta_1}(\mathcal{K}) \to \mathcal{S}^\tau_{k,\gamma_\mathfrak{a}}(Z^\infty),$$

parametrizing open subsets of the set of solutions to $\mathfrak{F}^\tau_\mathfrak{q} = 0$ in the Coulomb–Neumann slices $\mathcal{S}^\tau_{k,\gamma_\mathfrak{a}}(Z^T)$. The fact that these maps are $(\mathbb{Z}/2)$-equivariant, as asserted in the last part of Theorem 18.2.1, is a simple consequence of the uniqueness of the solution. We also learn that there is an $\eta_2 > 0$ independent of T, such that the image of $u(T,-)$ contains all solutions γ in $\mathcal{S}^\tau_{k,\gamma_\mathfrak{a}}(Z^T)$ with $\|\gamma - \gamma_\mathfrak{a}\|_{L^2_k} \le \eta_2$. The statement of Theorem 18.2.1, however, refers to uniform neighborhoods in the moduli spaces $\tilde{M}(Z^T)$ defined by the inequality $\|\gamma - \gamma_\mathfrak{a}\|_{L^2_k} \le \eta$ where γ is an arbitrary gauge representative of the solution, not necessarily a representative in the slice. The following proposition addresses this point, and so completes the proof of Theorem 18.2.1.

Proposition 18.4.1. *There exists η_0 such that for all $\eta < \eta_0$, there exists η', independent of T, such that:*

(i) *the map*

$$\bar{\iota} : \{\gamma \in \mathcal{S}^\tau_{k,\gamma_\mathfrak{a}}(Z^T) \mid \|\gamma - \gamma_\mathfrak{a}\|_{L^2_k(Z)} \le \eta\} \to \tilde{\mathcal{B}}^\tau_k(Z^T)$$

is a diffeomorphism onto its image;

(ii) *the image of the above map contains all gauge-equivalence classes $[\gamma] \in \tilde{\mathcal{B}}^\tau_k(Z^T)$ represented by elements γ with*

$$\|\gamma - \gamma_\mathfrak{a}\|_{L^2_k(Z^T)} \le \eta'.$$

Proof. We must show that, for some η' independent of T and every γ with $\|\gamma - \gamma_\mathfrak{a}\|_{L^2_k} \le \eta'$, there is a gauge transformation e^ξ in $\mathcal{G}_{k+1}(Z^T)$ with

$$e^\xi \cdot \gamma \in \mathcal{S}^\tau_{k,\gamma_\mathfrak{a}}(Z^T)$$

and $\|e^{\xi}\cdot\gamma-\gamma_{\mathfrak{a}}\|_{L^2_k}\leq\eta$. The non-linear equation, to be solved for ξ, is of the form

$$\begin{aligned}\mathrm{Coul}^{\tau}_{\gamma_{\mathfrak{a}}}(e^{\xi}\cdot\gamma)&=0\\ \langle d\xi,\vec{n}\rangle&=c\quad\text{on }\partial Z^T\end{aligned}$$

which we can write as

$$\begin{aligned}\mathbf{d}^{\tau,\dagger}_{\gamma_{\mathfrak{a}}}\mathbf{d}^{\tau}_{\gamma_{\mathfrak{a}}}\xi+\beta(\xi)&=\mu\\ \langle d\xi,\vec{n}\rangle&=c\quad\text{on }\partial Z^T,\end{aligned}$$

where $\|\mu\|_{L^2_{k-1}}$ and $\|c\|_{L^2_{k-1/2}}$ are bounded by $C\eta'$. The non-linear part $\beta(\xi)$ represents a map

$$\beta:L^2_{k+1}(Z^T)\to L^2_{k-1}(Z^T)$$

which is infinitely differentiable and has $\beta(0)=0$ and $\mathcal{D}_0\beta=0$. Furthermore, it follows from the multiplication theorems that it satisfies a Lipschitz condition: for ξ_1, ξ_2 in the η ball in L^2_{k+1}, we have

$$\|\beta(\xi_1)-\beta(\xi_2)\|\leq\epsilon(\eta)\|\xi_1-\xi_2\|,$$

where $\epsilon(\eta)\to 0$ as $\eta\to 0$, and $\epsilon(\eta)$ is independent of T. Given these properties of the non-linear terms, the existence of a solution (when η' is sufficiently small) will follow from the invertibility of the linear terms; and to prove the lemma, we need to establish that the solution ξ to the linear equation

$$\begin{aligned}\mathbf{d}^{\tau,\dagger}_{\gamma_{\mathfrak{a}}}\mathbf{d}^{\tau}_{\gamma_{\mathfrak{a}}}\xi&=\mu\\ \langle d\xi,\vec{n}\rangle&=c\quad\text{on }\partial Z^T\end{aligned}\tag{18.12}$$

satisfies a bound

$$\|\xi\|_{L^2_{k+1}}\leq K\big(\|\mu\|_{L^2_{k-1}}+\|c\|_{L^2_{k-1/2}}\big)$$

for some K that is independent of T. We can view (18.12) as equations for $w=\mathbf{d}^{\tau}_{\gamma_{\mathfrak{a}}}\xi$. This w is along the gauge orbit at the solution $\gamma_{\mathfrak{a}}$, so it satisfies the linearized Seiberg–Witten equations and the projection of its boundary values to $\mathcal{K}^{\sigma}_{\mathfrak{a}}(Y)$ is zero. We can therefore apply Lemma 18.3.8 to w to obtain

$$\|w\|_{L^2_k}\leq K_1\big(\|\mu\|_{L^2_{k-1}}+\|c\|_{L^2_{k-1/2}}\big)$$

for some K_1 independent of T. Finally,

$$\|\xi\|_{L^2_{k+1}} \leq K_2 \|w\|_{L^2_k}$$

for some K_2 independent of T, by the argument of Lemma 13.3.1. □

19 Gluing trajectories

Proposition 18.1.3 described a simple model gluing theorem for finite-dimensional Morse theory. In this section we take up the task of proving such a theorem for the Chern–Simons–Dirac functional. There were two main ingredients in the finite-dimensional case and we already have the corresponding versions. First, there is the compactness theorem for the spaces of broken trajectories, Theorem 16.1.3. Second, there is Theorem 18.2.1 which is used in place of Proposition 18.1.4. Before getting into the proof there is one point which was trivial in the finite dimensional situation and needs to be addressed. We need to show that a concatenation of trajectories on adjacent finite intervals is a trajectory on the union. This is done in the first subsection.

19.1 Moduli spaces as fiber products

Consider the moduli space $\tilde{M}(I \times Y) \subset \tilde{\mathcal{B}}^{\tau}_k(I \times Y)$ of solutions to the perturbed Seiberg–Witten equations, $\mathfrak{F}^{\tau}_{\mathfrak{q}}\gamma = 0$ on $I \times Y$. We suppose that I is a compact interval, written as a union of two subintervals I_1, I_2, meeting at a single point, say 0. Restriction to the two subcylinders $I_1 \times Y$ and $I_2 \times Y$ defines a map

$$\rho : \tilde{M}(I \times Y) \to \tilde{M}(I_1 \times Y) \times \tilde{M}(I_2 \times Y).$$

We also have two restriction maps from the product

$$R_i \colon \tilde{M}(I_1 \times Y) \times \tilde{M}(I_2 \times Y) \to \mathcal{B}^{\sigma}_{k-1/2}(\{0\} \times Y), \quad i = 1, 2.$$

The image of ρ is contained in the fiber product

$$\begin{aligned} \mathrm{Fib}(R_1, R_2) &= \{\, m \mid R_1(m) = R_2(m) \,\} \\ &\subset \tilde{M}(I_1 \times Y) \times \tilde{M}(I_2 \times Y). \end{aligned}$$

Recall that the moduli spaces $\tilde{M}(I \times Y)$ etc. are Hilbert submanifolds of the corresponding configuration spaces $\tilde{\mathcal{B}}^{\tau}_k$ (Theorem 17.3.1). The map ρ, as the restriction of a smooth map on the configuration space, is a smooth map of Hilbert manifolds.

Lemma 19.1.1. *The image of* ρ *is* $\mathrm{Fib}(R_1, R_2)$*, and* $\rho : \tilde{M}(I \times Y) \to \mathrm{Fib}(R_1, R_2)$ *is a homeomorphism.*

Proof. To show surjectivity, let $\gamma_1 \in \mathcal{C}_k^\tau(I_1 \times Y)$ and $\gamma_2 \in \mathcal{C}_k^\tau(I_2 \times Y)$ be solutions to the perturbed Seiberg–Witten equations $\mathfrak{F}_\mathfrak{q}^\tau(\gamma_i) = 0$, and suppose that the restrictions $R\gamma_1$ and $R\gamma_2$ in $\mathcal{C}_{k-1/2}^\sigma(\{0\} \times Y)$ are gauge-equivalent on $\{0\} \times Y$. We must show there is a solution $\gamma \in \mathcal{C}_k^\tau(I \times Y)$ whose restrictions to $I_1 \times Y$ and $I_2 \times Y$ are gauge-equivalent to γ_1 and γ_2 respectively.

Let $v_0 \in \mathcal{G}_{k+1/2}(Y)$ be a gauge transformation that takes the restriction of γ_2 to the restriction of γ_1. We can extend v_0 to an element v of $\mathcal{G}_{k+1}(I_2 \times Y)$. (See Theorem 17.1.1.) By replacing γ_2 with $v\gamma_2$ we arrive at a situation in which the restrictions agree in $\mathcal{C}_{k-1/2}^\tau(\{0\} \times Y)$. Next, we take v_1, v_2 to be L_k^2 gauge transformations, equal to the identity on $\{0\} \times Y$, such that $\tilde{\gamma}_1 = v_1\gamma_1$ and $\tilde{\gamma}_2 = v_2\gamma_2$ are in temporal gauge on $I_1' \times Y$ and $I_2' \times Y$, where I_i' is, say, $(1/2)I_i$. We can take $\tilde{\gamma}_i$ to be the identity also in a neighborhood of the other boundary component of each $I_i \times Y$. The solutions $\tilde{\gamma}_i$ will belong to $\mathcal{C}_{k-1}^\tau$ on their respective domains, and will be of class L_k^2 outside a neighborhood of $\{0\} \times Y$.

We claim that $\tilde{\gamma}_1$ and $\tilde{\gamma}_2$ are (without further gauge transformation) the restrictions of an element $\tilde{\gamma} \in \mathcal{C}_2^\tau(I \times Y)$ to the two halves of the cylinder $I \times Y$. As in (10.9), we can regard $\tilde{\gamma}_i$ as sections of a vector bundle V pulled back from Y to the cylinder, where they satisfy an equation of the form

$$(d/dt)\tilde{\gamma}_i = T\tilde{\gamma}_i + \beta(\tilde{\gamma}_i),$$

where T is a first-order differential operator on Y and β maps L_j^2 to L_j^2 on the cylinder, and arises from a map on Y mapping L_j^2 to L_j^2 also. Our hypotheses so far mean that the restrictions of $\tilde{\gamma}_i$ to $\{0\} \times Y$ agree as elements of (in particular) $L_{3/2}^2(Y; V_0)$. From the equation above, it follows that $(d/dt)\tilde{\gamma}_i$ are elements of L_1^2 on the two parts of the cylinder with equal boundary values in $L_{1/2}^2(Y; V_0)$. These two facts together mean that $\tilde{\gamma}_i$ can be regarded as the restriction of a $\tilde{\gamma}$ of class L_2^2, as claimed.

Now we can find an L_3^2 gauge transformation v_1 such that $\gamma_1 = v_1\tilde{\gamma}$ is in Coulomb–Neumann gauge with respect to a smooth, translation-invariant trajectory γ_0. A solution to the equations, of class L_2^2 and in Coulomb gauge, is of class L_k^2 on any subcylinder. If we use the fact that $\tilde{\gamma}$ was already of class L_k^2 near $\partial I \times Y$, we can apply a cut-off function to v_1 to obtain a new gauge transformation v_2 (equal to 1 near $\partial I \times Y$), so that $\gamma = v_2\tilde{\gamma}$ is of class L^2_k on all of $I \times Y$. This establishes the surjectivity of ρ.

The injectivity of ρ means the uniqueness of the γ just constructed, up to gauge transformation. This is seen using an argument similar to the one above. Suppose γ and γ' are two elements of $\mathcal{C}^\tau_k(I \times Y)$ with the property that their restrictions to both $I_1 \times Y$ and $I_2 \times Y$ are gauge-equivalent, by gauge transformations v_1 and v_2. Then the restrictions of v_i to $\{0\} \times Y$ agree as elements of $L^2_{k+1/2}$, and a fortiori as elements of $L^2_{1/2}$. Because their $L^2_{1/2}$ boundary values agree, v_i are the two restrictions of an L^2_1 gauge transformation v. From the equation

$$v^{-1}dv = a' - a$$

(where a' and a are the 1-form components of γ, γ'), it follows by bootstrapping that v is in L^2_{k+1}.

The continuity of the inverse map to ρ follows from the continuity of the maps used to select the gauge transformations in the proof of surjectivity. □

Lemma 19.1.2. *The map*

$$(R_1, R_2):$$
$$\tilde{M}(I_1 \times Y) \times \tilde{M}(I_2 \times Y) \to \tilde{\mathcal{B}}^\sigma_{k-1/2}(\{0\} \times Y) \times \tilde{\mathcal{B}}^\sigma_{k-1/2}(\{0\} \times Y)$$

is transverse to the diagonal in the product. So $\mathrm{Fib}(R_1, R_2)$ *is a smooth Hilbert submanifold of the product.*

Proof. Let $([\gamma_1], [\gamma_2])$ belong to the fiber product. By the previous lemma, we can suppose these are the restrictions of a solution γ on $I \times Y$. The proof of Proposition 17.3.1 showed not just that $\tilde{M}(I \times Y)$ is a Hilbert submanifold of $\tilde{\mathcal{B}}^\tau_k$ at $[\gamma]$, but that the operator Q_γ is surjective. It follows from Proposition 17.2.8 that the kernels of Q_{γ_1} and Q_{γ_2} restrict to transverse linear subspaces H_1, H_2 of $T_{\gamma_0}\tilde{\mathcal{C}}^\sigma_{k-1/2}(Y) \oplus L^2_{k-1/2}(Y; i\mathbb{R})$:

$$H_1 + H_2 = T_{\gamma_0}\tilde{\mathcal{C}}^\sigma_{k-1/2}(Y) \oplus L^2_{k-1/2}(Y; i\mathbb{R}).$$

Here γ_0 is the restriction of γ as an element of $\tilde{\mathcal{C}}^\sigma_{k-1/2}(Y)$. The statement that the derivative of (R_1, R_2) is transverse at the pair $[\gamma_1], [\gamma_2]$ is a consequence, for it is equivalent to the statement that the images of H_1 and H_2 in $T_{[\gamma_0]}\tilde{\mathcal{B}}^\sigma_{k-1/2}(Y)$ are transverse. □

Because of this lemma, it makes sense to ask for a strengthening of Lemma 19.1.1:

Proposition 19.1.3. *The map ρ is a diffeomorphism from the Hilbert manifold $\tilde{M}(I \times Y)$ to its image, the fiber product* $\mathrm{Fib}(R_1, R_2)$.

Proof. We already know that ρ is a homeomorphism, and it is a smooth map. We must see that the derivative is an isomorphism at each $[\gamma]$ in the domain. This is proved in the same way as Lemma 19.1.1, but replacing the perturbed Seiberg–Witten equations by their linearization. □

Rather than a compact interval I, we can consider instead the whole line $\mathbb{R}$ and the cylinder $Z = \mathbb{R} \times Y$, decomposed as the union of manifolds with boundary, $\mathbb{R}^{\leq} \times Y$ and $\mathbb{R}^{\geq} \times Y$. On Z we consider the (finite-dimensional) moduli space $\tilde{M}([\mathfrak{a}], [\mathfrak{b}])$ for a pair of critical points $[\mathfrak{a}], [\mathfrak{b}]$. On the two half-cylinders, there are moduli spaces $\tilde{M}(\mathbb{R}^{\leq} \times Y, [\mathfrak{a}])$ and $\tilde{M}(\mathbb{R}^{\geq} \times Y, [\mathfrak{b}])$, which are Hilbert manifolds, with smooth maps R_1, R_2 to $\tilde{\mathcal{B}}^{\sigma}_{k-1/2}(\{0\} \times Y)$. The proofs above adapt readily, to show first that the moduli space $\tilde{M}([\mathfrak{a}], [\mathfrak{b}])$ is homeomorphic to the fiber product $\mathrm{Fib}(R_1, R_2)$; and second, that *if the moduli space $\tilde{M}([\mathfrak{a}], [\mathfrak{b}])$ is regular*, and if we are not in the boundary-obstructed case, then (R_1, R_2) is transverse to the diagonal at $\rho(\gamma)$, and the map ρ to the fiber product is a diffeomorphism. (Recall that the term "regular" means that Q_γ is surjective, except in the boundary-obstructed case: see Definition 14.5.6.) We will return to the boundary-obstructed case shortly.

Note that $\tilde{M}([\mathfrak{a}], [\mathfrak{b}])$ is the union of the various $\tilde{M}_z([\mathfrak{a}], [\mathfrak{b}])$ for different relative homotopy classes z; and these components may have differing dimensions. There is also a fiber product description of each $\tilde{M}_z([\mathfrak{a}], [\mathfrak{b}])$, which we can obtain using the restriction maps $\tilde{R}_i$ to the quotient of $\tilde{\mathcal{C}}^{\sigma}_{k-1/2}(Y)$ by the identity component $\mathcal{G}^{e}_{k+1/2}(Y)$ of the gauge-group.

There is also a version of this construction for more than two intervals. Consider a decomposition of $\mathbb{R}$ as a union of intervals $I_0, \ldots, I_n$, where $I_{i-1} \cap I_i = \{t_i\}$, and $t_1 \leq t_2 \leq \cdots \leq t_n$. (The intervals I_0 and I_n are non-compact.) Let $[\mathfrak{a}], [\mathfrak{b}]$ be critical points, and let

$$\tilde{M}(i) = \begin{cases} \tilde{M}(I_0 \times Y, [\mathfrak{a}]), & i = 0 \\ \tilde{M}(I_n \times Y, [\mathfrak{b}]), & i = n \\ \tilde{M}(I_i \times Y), & \text{otherwise.} \end{cases}$$

Set

$$\tilde{\mathcal{B}} = \prod_{i=1}^{n} \tilde{\mathcal{B}}^{\sigma}_{k-1/2}(\{t_i\} \times Y).$$

There are two maps

$$R_+, R_- : \prod_{i=0}^{n} \tilde{M}(i) \to \tilde{\mathcal{B}},$$

given by the restriction maps to the right-hand and left-hand endpoints of the intervals I_i respectively. (Thus R_+, for example, ignores the component in $\tilde{M}(n)$.) Define

$$\mathrm{Fib}(R_+, R_-) = \left\{ m \in \prod_{i=0}^{n} \tilde{M}(i) \,\middle|\, R_+ m = R_- m \right\}.$$

By a slight abuse of terminology we shall still refer to this as a fiber product. There is an obvious restriction map ρ from $\tilde{M}([\mathfrak{a}],[\mathfrak{b}])$ to $\mathrm{Fib}(R_+, R_-)$, and we have:

Theorem 19.1.4. *The map ρ is a homeomorphism from the moduli space $\tilde{M}([\mathfrak{a}],[\mathfrak{b}])$ to $\mathrm{Fib}(R_+, R_-)$. If we suppose, in addition, that the moduli space is regular and not boundary-obstructed, then the maps $R_\pm$ have the property that $(R_+, R_-) : \prod_{i=0}^{n} \tilde{M}(i) \to \tilde{\mathcal{B}} \times \tilde{\mathcal{B}}$ is transverse to the diagonal, so that $\mathrm{Fib}(R_+, R_-)$ is a smooth submanifold of $\prod_{i=0}^{n} \tilde{M}(i)$. In this case, the map ρ is a diffeomorphism.*

Conversely, if the moduli space is not boundary-obstructed, and $[\gamma]$ is a point in the moduli space with the property that (R_+, R_-) is transverse to the diagonal at $\rho([\gamma]) \in \mathrm{Fib}(R_+, R_-)$, then $\tilde{M}([\mathfrak{a}],[\mathfrak{b}])$ is regular in a neighborhood of $[\gamma]$.

Proof. Only the converse direction is essentially new here. Its proof is based on the converse direction in Proposition 17.2.8. □

Finally, we provide a version of this theorem in the boundary-obstructed case, where the critical points are reducible, $\mathfrak{a}$ is boundary-stable and $\mathfrak{b}$ is boundary-unstable. Let $\partial\mathcal{B}^{\sigma}_{k-1/2}(Y) \subset \tilde{\mathcal{B}}^{\sigma}_{k-1/2}(Y)$ be the codimension-1 submanifold given by $s = 0$, and let

$$\pi^{\partial} : \tilde{\mathcal{B}}^{\sigma}_{k-1/2}(Y) \to \partial\mathcal{B}^{\sigma}_{k-1/2}(Y) \tag{19.1}$$

be the projection $[B, s, \psi] \mapsto [B, 0, \psi]$. Let the intervals I_i $(i = 0, \ldots, n)$ be as before and let i_0 be some chosen index, $1 \leq i_0 \leq n$, and suppose that $t_{i_0} = 0$. Set

$$\tilde{\mathcal{B}}' = \partial \mathcal{B}^{\sigma}_{k-1/2}(\{0\} \times Y) \times \prod_{i \neq i_0} \tilde{\mathcal{B}}^{\sigma}_{k-1/2}(\{t_i\} \times Y),$$

and let

$$R'_+, R'_- : \prod_{i=0}^{n} \tilde{M}(i) \to \tilde{\mathcal{B}}'$$

be the maps obtained by composing $R_\pm$ with the projection π^∂ on the i_0 factor. Then we have:

Theorem 19.1.5. *Suppose the moduli space $M([\mathfrak{a}], [\mathfrak{b}])$ is boundary-obstructed. Then the map ρ is a homeomorphism from $M([\mathfrak{a}], [\mathfrak{b}])$ to* $\mathrm{Fib}(R'_+, R'_-)$. *If we suppose, in addition, that the moduli space is regular, then the maps $R'_\pm$ have the property that $(R'_+, R'_-) : \prod_{i=0}^{n} \tilde{M}(i) \to \tilde{\mathcal{B}}' \times \tilde{\mathcal{B}}'$ is transverse to the diagonal, so that* $\mathrm{Fib}(R'_+, R'_-)$ *is a smooth submanifold of $\prod_{i=0}^{n} \tilde{M}(i)$. In this case, the map ρ is a diffeomorphism.*

Conversely, if $[\gamma]$ is a point in the moduli space with the property that (R'_+, R'_-) is transverse to the diagonal at $\rho([\gamma])$, then $M([\mathfrak{a}], [\mathfrak{b}])$ is regular in a neighborhood of $[\gamma]$.

Proof. A priori, the space $\mathrm{Fib}(R'_+, R'_-)$ might be strictly larger than $\mathrm{Fib}(R_+, R_-)$; but in the boundary-obstructed case, the two spaces are equal because the argument which shows that the moduli space $M([\mathfrak{a}], [\mathfrak{b}])$ contains only reducibles also shows that $\mathrm{Fib}(R'_+, R'_-)$ contains only reducibles. So ρ is indeed a homeomorphism to $\mathrm{Fib}(R'_+, R'_-)$. The smoothness of $\mathrm{Fib}(R'_+, R'_-)$ and the fact that ρ is a diffeomorphism proceed as before. □

In the boundary-unobstructed case, the space $\mathrm{Fib}(R'_+, R'_-)$ is strictly larger than the image $\mathrm{Fib}(R_+, R_-)$ of the moduli space $\tilde{M}([\mathfrak{a}], [\mathfrak{b}])$ (unless both are empty). Let us take the case $n = 2$ and $t_1 = 0$, so that the two intervals I_0, I_1 are the negative and positive half-lines, and

$$\tilde{\mathcal{B}}' = \partial \mathcal{B}^{\sigma}_{k-1/2}(\{0\} \times Y).$$

We give the fiber product in this case a name:

Definition 19.1.6. For any pair of critical points $[\mathfrak{a}], [\mathfrak{b}]$ we define the *extended moduli space* $E\tilde{M}([\mathfrak{a}], [\mathfrak{b}])$ as the fiber product $\mathrm{Fib}(R'_+, R'_-)$ resulting from the decomposition of $\mathbb{R}$ into the positive and negative half-lines, as above. Note that $E\tilde{M}([\mathfrak{a}], [\mathfrak{b}])$ does not have an action of $\mathbb{R}$ by translations. ◊

To spell out the meaning of this definition, let us recall that a typical element of $\mathcal{B}^\sigma_{k-1/2}(Y)$ can be written as $[A, s, \phi]$, with $s \geq 0$ as in Subsection 6.1; we have

$$\mathcal{B}^\sigma_{k-1/2}(Y) = \mathcal{B}' \times \mathbb{R}^{\geq}$$

where $\mathbb{R}^{\geq}$ is the s coordinate. Similarly,

$$\tilde{\mathcal{B}}^\sigma_{k-1/2}(Y) = \mathcal{B}' \times \mathbb{R}.$$

An element of the extended moduli space $E\tilde{M}([\mathfrak{a}], [\mathfrak{b}])$ consists of a pair of gauge-equivalence classes of solutions to the perturbed equations, $([\gamma_-], [\gamma_+])$, defined on the negative and positive closed half-lines respectively. The restrictions of $[\gamma_-]$ and $[\gamma_+]$ to $\{0\} \times Y$ are not required to be equal: only the projections of the restrictions to $\mathcal{B}'$ are required to match. Thus we can think of an element of $E\tilde{M}([\mathfrak{a}], [\mathfrak{b}])$ as a trajectory $[\gamma]$ defined on the whole line but having a discontinuity in the s coordinate across $\{0\} \times Y$. We therefore see that $\tilde{M}([\mathfrak{a}], [\mathfrak{b}])$ arises as the fiber of a map δ,

$$\tilde{M}([\mathfrak{a}], [\mathfrak{b}]) \hookrightarrow E\tilde{M}([\mathfrak{a}], [\mathfrak{b}]) \overset{\delta}{\hookrightarrow} \mathbb{R},$$

where δ measures the size of the discontinuity in the s coordinate:

$$\delta = s([\gamma_+]|_{\{0\}\times Y}) - s([\gamma_-]|_{\{0\}\times Y}). \tag{19.2}$$

Because it determines a path in $\partial\mathcal{B}^\sigma_{k-1}(Y)$, an element of $E\tilde{M}([\mathfrak{a}], [\mathfrak{b}])$ determines a relative homotopy class, and so we can write

$$E\tilde{M}([\mathfrak{a}], [\mathfrak{b}]) = \bigcup_z E\tilde{M}_z([\mathfrak{a}], [\mathfrak{b}]).$$

We have already observed that $E\tilde{M}([\mathfrak{a}], [\mathfrak{b}])$ coincides with $\tilde{M}([\mathfrak{a}], [\mathfrak{b}])$ in the boundary-obstructed case (in other words, δ is identically zero). In the boundary-unobstructed case, if $\tilde{M}([\mathfrak{a}], [\mathfrak{b}])$ is regular, then $E\tilde{M}([\mathfrak{a}], [\mathfrak{b}])$ is regular in a neighborhood of $\tilde{M}([\mathfrak{a}],[\mathfrak{b}])$, and contains $\tilde{M}([\mathfrak{a}],[\mathfrak{b}])$ as a smooth codimension-1 submanifold, because π^∂ is a submersion. We say that $E\tilde{M}([\mathfrak{a}],[\mathfrak{b}])$ is regular at m if the map (R'_+, R'_-) is transverse to the diagonal at m. We therefore have:

Proposition 19.1.7. *If $E\tilde{M}([\mathfrak{a}],[\mathfrak{b}])$ is regular at m, then a neighborhood of m in $E\tilde{M}_z([\mathfrak{a}],[\mathfrak{b}])$ is a smooth $(d+1)$-manifold, where $d = \mathrm{gr}_z([\mathfrak{a}],[\mathfrak{b}])$. This holds in both the boundary-obstructed and -unobstructed cases.* □

Finally, we can consider the extended moduli space with a fiber product description with more than two intervals:

Theorem 19.1.8. *Theorem* 19.1.5 *continues to hold without the hypothesis that the moduli space is boundary-obstructed, if we replace $M([\mathfrak{a}],[\mathfrak{b}])$ by $E\tilde{M}([\mathfrak{a}],[\mathfrak{b}])$ in the statement.* □

19.2 Localized trajectories and centered trajectories

For an element $\gamma \in \mathcal{C}^\sigma_{k-1/2}(Y)$, we write:

$$\mathfrak{e}(\gamma) = \|(\mathrm{grad}\, \mathcal{L})^\sigma\|^2_{L^2(Y)}.$$

This is a smooth, gauge-invariant function on the blown-up configuration space, vanishing precisely at the critical points of the flow.

Lemma 19.2.1. *Given any pair of critical points $\mathfrak{a}$, $\mathfrak{b}$, there exists a constant $\epsilon > 0$ such that for any component $[\gamma_i]$ of any broken trajectory $[\breve{\gamma}] \in \breve{M}^+_z([\mathfrak{a}],[\mathfrak{b}])$, there exists a t with*

$$\mathfrak{e}(\gamma_i(t)) > \epsilon.$$

Proof. This follows from the compactness theorem for $\breve{M}^+_z([\mathfrak{a}],[\mathfrak{b}])$ and the fact that there are no non-constant trajectories (broken or unbroken) for which $\mathfrak{e}(\gamma(t))$ is identically zero. □

Let $\mathfrak{a}$ and $\mathfrak{b}$ be given, and let $\epsilon > 0$ be as in the lemma. Let β be a smooth cut-off function with $\beta(t) = 0$ for $t \le \epsilon$ and $\beta(t) > 0$ for $t > \epsilon$. If $[\gamma] \in M_z([\mathfrak{a}],[\mathfrak{b}])$, the function

$$\beta \circ \mathfrak{e}(\gamma(t))$$

is non-negative, not everywhere zero, and supported in a compact interval. We can therefore consider the *center of mass* of this distribution on the real line. For each unparametrized trajectory $[\breve{\gamma}]$ in $\breve{M}_z([\mathfrak{a}],[\mathfrak{b}])$, there is a unique parametrized representative $[\gamma]$ in $M_z([\mathfrak{a}],[\mathfrak{b}])$ for which this center of mass is zero. The idea of using $\beta \circ \mathfrak{e}(\gamma(t))$ to locate a center for a trajectory is applicable also to a solution defined on a finite interval, as in the next definition.

Definition 19.2.2. Let ϵ and β be given, and let $I = [t_1, t_2]$ be any interval of length greater than 2. A solution $[\gamma] \in M(I \times Y)$ is called *centered* if

(i) $\mathfrak{e}(\gamma(t)) < \epsilon/2$ for $t \in [t_1, t_1 + 1]$ and $t \in [t_2 - 1, t_2]$;
(ii) $\mathfrak{e}(\gamma(t)) > \epsilon$ for some t in $[t_1, t_2]$; and
(iii) the center of mass of the distribution $\beta \circ \mathfrak{e}(\gamma(t))$, namely the quantity

$$c(\gamma) = \int_I t\beta \circ \mathfrak{e}(\gamma(t))dt \Big/ \int_I \beta \circ \mathfrak{e}(\gamma(t))dt$$

is at the center of the interval $\frac{1}{2}(t_1 + t_2)$.

We write $M^{cen}(I \times Y)$ for the subset of $M(I \times Y)$ consisting of centered solutions. ◊

The point of introducing the cut-off function into the above definition is that it makes the center-of-mass function c behave quite simply: if γ satisfies the first two conditions in the definition, then so does the translate $\tau_\delta^* \gamma$ for small δ; and the center of mass of this translate is then $c(\gamma) - \delta$.

We now apply this concept to the restrictions of a trajectory $[\gamma] \in M_z([\mathfrak{a}],[\mathfrak{b}])$ to intervals $I \subset \mathbb{R}$.

Definition 19.2.3. Let $\mathfrak{a}$, $\mathfrak{b}$, ϵ and β be as above. An interval $I = [t_1, t_2]$ is a *local centering interval* for $[\gamma] \in M_z([\mathfrak{a}],[\mathfrak{b}])$ if the restriction of $[\gamma]$ to I is centered, in the sense of Definition 19.2.2.

Let $I_1, \ldots, I_n$ be a finite collection of disjoint intervals, whose centers $c_1, \ldots, c_n$ are increasing. These form a *complete collection of local centering intervals* of length L for $[\gamma]$ if each I_m is a local centering interval, the length of each I_m is L, and $\beta \circ \mathfrak{e}(\gamma(t))$ is supported in $\bigcup_m I_m$. ◊

We extend this language a little further. If $[\breve{\gamma}]$ is an unparametrized trajectory in $\breve{M}_z([\mathfrak{a}],[\mathfrak{b}])$, we can talk about a local centering interval for $[\breve{\gamma}]$ as being a local centering interval for a parametrized representative (which by translation gives a local centering interval for any other representative). We can also allow broken trajectories: $[\breve{\boldsymbol{\gamma}}]$ can be an unparametrized *broken* trajectory, with components $[\breve{\gamma}_i]$, $i = 1, \ldots, l$.

Definition 19.2.4. A *complete collection of local centering intervals* of length L for a broken trajectory

$$[\breve{\gamma}] = ([\breve{\gamma}_1], \dots, [\breve{\gamma}_l])$$

means a collection of intervals $\{I_{i,m}\}$, with $1 \le m \le n_i$, such that $(I_{i,1}, \dots, I_{i,n_i})$ is a complete collection of local centering intervals of length L for the ith component, $[\breve{\gamma}_i]$. ◊

Lemma 19.2.5. *The subset of $M(I \times Y)$ defined by the first two conditions in Definition* 19.2.2 *is open, and the centered solutions $M^{cen}(I \times Y)$ are a closed, smooth submanifold of this open set.*

Proof. The first statement is clear. The centered solutions $M^{cen}(I \times Y)$ are cut out as the level set of a smooth function

$$c : M'(I \times Y) \to \mathbb{R},$$

defined on the open subset $M'(I \times Y) \subset M(I \times Y)$ defined by the first conditions. We claim that the derivative of c is nowhere zero on $M'(I \times Y)$. Because c is gauge-invariant, we can consider instead the derivative of the function c on U, the inverse image of $M'(I \times Y)$ in $\mathcal{C}_k^\tau(I \times Y)$. For $\gamma \in U$, the differential of c is a linear function on

$$T_\gamma U = \ker\left(\mathcal{D}_\gamma \mathfrak{F}_\mathfrak{q} : \mathcal{T}_{k,\gamma}^\tau \to \mathcal{V}_{k-1,\gamma}^\tau\right)$$

which is a bounded linear functional in the L^2 norm on $T_\gamma \tilde{U}$. So dc extends to the L^2 completion.

The infinitesimal action of translations generates a "vector field", or more specifically a section of $\mathcal{T}_{k-1}^\tau$, which we write as $\dot{\gamma}$. Because the equations are translation-invariant, we have

$$\dot{\gamma} \in \ker\left(\mathcal{D}_\gamma \mathfrak{F}_\mathfrak{q} : \mathcal{T}_{k-1,\gamma}^\tau \to \mathcal{V}_{k-2,\gamma}^\tau\right).$$

The pairing $dc(\dot{\gamma})$ is -1, so dc is non-zero on $\mathcal{T}_{k-1}^\tau$. The lemma follows if we show that $T_\gamma U$ is dense in the above kernel. As in the proof of Theorem 17.3.1, we can replace $\mathcal{D}_\gamma \mathfrak{F}_\mathfrak{q}$ by the operator Q_γ: we write $\ker_j$ for the kernel of Q_γ acting on $\mathcal{T}_j^\tau$, and we claim that $\ker_k$ is dense in $\ker_1$. To verify the claim, it is enough to exhibit a right inverse P for the operator

$$Q_\gamma : \mathcal{T}_{1,\gamma}^\tau(I \times Y) \to \mathcal{V}_{0,\gamma}^\tau(I \times Y) \oplus L^2(I \times Y; i\mathbb{R})$$

having the property that P maps L^2_{j-1} to L^2_j for $1 \le j \le k$. We recall that Q_γ has the form $D_0 + h$, where D_0 is an elliptic operator of the standard form $(d/dt) + L_0$, and we observe that a suitable P is obtained by considering the unique solution u to the equation $Q_\gamma u = v$ satisfying a boundary condition $\Pi u = 0$, where Π is a finite-rank modification of the spectral projections Π_0 defined by L_0 at the two ends of the cylinder. The required properties follow from Theorem 17.1.3. □

The above definition of a complete collection of local centering intervals is constructed so as to give a description of neighborhoods of strata in the space of unparametrized broken trajectories. Let $\mathfrak{a}_0, \mathfrak{a}_1, \ldots, \mathfrak{a}_n$ be critical points, with $[\mathfrak{a}] = [\mathfrak{a}_0]$ and $[\mathfrak{b}] = [\mathfrak{a}_n]$. Let $K_m \subset M([\mathfrak{a}_{m-1}], [\mathfrak{a}_m])$ be a compact subset of the space of parametrized trajectories ($m = 1, \ldots, n$), let $\breve{K}_m$ be its image in $\breve{M}([\mathfrak{a}_{m-1}], [\mathfrak{a}_m])$, and let $\breve{\boldsymbol{K}} \subset \breve{M}^+([\mathfrak{a}_0], [\mathfrak{a}_n])$ be the corresponding subset of the space of unparametrized broken trajectories:

$$\begin{aligned} \breve{\boldsymbol{K}} &= \prod_{m=1}^{n} \breve{K}_m \\ &\subset \prod_{m=1}^{n} \breve{M}([\mathfrak{a}_{m-1}], [\mathfrak{a}_m]). \end{aligned} \tag{19.3}$$

Let ϵ be chosen as in Lemma 19.2.1, and let β be a suitable cut-off function. Use these to define the centered moduli spaces $M^{cen}(I \times Y)$. We seek to describe a neighborhood $\breve{W}$ of $\breve{\boldsymbol{K}}$ in $\breve{M}^+([\mathfrak{a}_0], [\mathfrak{a}_n])$, by presenting each nearby broken trajectory $[\breve{\boldsymbol{\gamma}}]$ in a standard form.

Lemma 19.2.6. *Given ϵ and β as above we can find L_0, depending on $\breve{\boldsymbol{K}}$, with the following property. For all $L \ge L_0$, there exists a neighborhood $\breve{W}$ of $\breve{\boldsymbol{K}}$ in $\breve{M}^+_z([\mathfrak{a}_0], [\mathfrak{a}_n])$ such that each broken trajectory*

$$\breve{\boldsymbol{\gamma}} = ([\breve{\gamma}_1], \ldots, [\breve{\gamma}_l]) \in \breve{W}$$

admits a unique complete collection of local centering intervals $\{I_{i,m}\}$ of length $2L$. This collection has n members.

Proof. For a given neighborhood $\breve{W}$, let $\breve{W}^o$ denote the intersection of $\breve{W}$ with the unbroken trajectories. It will suffice to show we can choose $\breve{W}$ so that every $[\breve{\gamma}]$ in $\breve{W}^o$ has a unique collection of local intervals I_i of length $2L$, and that the collection has n members. It will also suffice to consider the case that $\boldsymbol{K}$ is a singleton, consisting of a single broken trajectory $[\breve{\boldsymbol{\gamma}}_*]$ of n components.

Choose a collection of neighborhoods $U_{\mathfrak{c}} \subset \mathcal{B}_k^{\tau}([-1,1] \times Y)$ with the separating property, as in Definition 16.2.3. Let these be sufficiently small that the function $\mathfrak{e}$ is everywhere less than $\epsilon/2$ on the corresponding neighborhoods $V_{\mathfrak{c}} \subset \mathcal{C}_{k-1}^{\sigma}(Y)$. Any path joining $V_{\mathfrak{a}_{i-1}}$ to $V_{\mathfrak{a}_i}$ must pass through points where $\mathfrak{e} > \epsilon$.

Following the proof of the compactness theorem for the space of broken trajectories (see Subsection 16.3 in particular), we can find a neighborhood $\breve{W}$ of $[\breve{\boldsymbol{\gamma}}_*]$ and a length L_1 such that for any representative γ of an unbroken element $[\breve{\gamma}] \in \breve{W}$, we have a decomposition of $\mathbb{R}$ into intervals I_i $(1 \leq i \leq n)$ of length $2L_1$ and intervals J_i $(0 \leq m \leq n)$, such that

$$t \in J_i \implies \tau_t^* \gamma|_{[-1,1]} \in U_{\mathfrak{a}_i}. \tag{19.4}$$

Furthermore, given any T, we can choose $\breve{W}$ so that the J_i all have length at least T. (Recall that the first and last of the J_0 and J_m have infinite length, being half-lines.) The condition (19.4) ensures that $\beta \circ \mathfrak{e}$ is supported in $\bigcup_i I_i$. Furthermore, $\mathfrak{e}$ is less than $\epsilon/2$ at points distance 1 or less from the endpoints of each I_i; and $\mathfrak{e}$ must be somewhere greater than ϵ on each I_i.

At this point, the intervals I_i look like a complete collection of local centering intervals, except that we do not know that the center of mass of $\beta \circ \mathfrak{e}$ on each I_i is at the center of the interval. As long as $T > 2L_1$ however, we can translate the intervals so as to locate their centers at the appropriate centers of mass, without disturbing the other conditions.

We have now established the existence of a neighborhood $\breve{W}$ which is small enough to ensure the existence of a complete collection of local centering intervals $\{I_i\}$ of length $2L_1$ with n members. By making $\breve{W}$ smaller again, we can arrange that the lengths of the complementary intervals J_i are all at least $T > 2L_1$, and it is then clear this collection is unique. □

The lemma provides a picture of any (possibly broken) trajectory in the neighborhood $\breve{W}$ as having exactly n intervals of length $2L$ containing the support of $\beta \circ \mathfrak{e}$. Within each interval, the solution is centered. We define $n-1$ positive real numbers $S_1, \ldots, S_{n-1}$ as the distances between the centers of these intervals, taking S_m to be infinity if the intervals belong to different components. That is, let

$$\mu : \{1, \ldots, n\} \to \{\,(i,m) \mid 1 \leq i \leq l,\ 1 \leq m \leq n_i\,\}$$

be the map respecting lexicographic order, and define the mth separation S_m of the local centering intervals $\{I_{i,m}\}$ to be:

$$S_m = \begin{cases} c_{\mu(m+1)} - c_{\mu(m)}, & \text{if } \mu(m+1), \mu(m) \text{ have the same first component} \\ \infty, & \text{otherwise,} \end{cases}$$

where $c_{i,m}$ is the center of $I_{i,m}$. Because the lemma provides a neighborhood $\breve{W}$ of $\breve{\boldsymbol{K}}$ on which the centering intervals exist and are unique, we have a well-defined map

$$\boldsymbol{S} : \breve{W} \to (0, \infty]^{n-1}, \tag{19.5}$$

with components S_m. If we start with a larger compact subset $\breve{\boldsymbol{K}}$ of the stratum (19.3), then we will end up with a different neighborhood $\breve{W}$; but on the intersection of the neighborhoods the two collections of centering intervals will coincide, and the two functions $\boldsymbol{S}$ will agree. There is therefore a neighborhood $\breve{W}$ of the *entire stratum* on which we can define $\boldsymbol{S}$. The map $\boldsymbol{S}$ will provide the framework for our description of the neighborhood of the stratum.

Definition 19.2.7. Let (Q, q_0) be a topological space, let $\pi : S \to Q$ be a continuous map, and let $S_0 \subset \pi^{-1}(q_0)$. We say that π is a *topological submersion along* S_0 if for all $s_0 \in S_0$ we can find a neighborhood U of s_0 in S, a neighborhood Q' of q_0 in Q and a homeomorphism $(U \cap S_0) \times Q' \to U$ such that the diagram

$$\begin{array}{ccc} (U \cap S_0) \times Q' & \longrightarrow & U \\ \downarrow & & \downarrow \pi \\ Q' & = & Q' \end{array}$$

commutes. ◊

We are now able to state the main result about the neighborhood of the stratum

$$\prod_{i=1}^{n} \breve{M}([\mathfrak{a}_{i-1}], [\mathfrak{a}_i]) \subset \breve{M}^{+}([\mathfrak{a}_0], [\mathfrak{a}_n]) \tag{19.6}$$

in the case that none of the moduli spaces involved are boundary-obstructed. The theorem tells us that $\breve{M}^+([\mathfrak{a}_0],[\mathfrak{a}_n])$ is locally homeomorphic to a product of the stratum (19.6) and the corner of an $(n-1)$-cube.

Theorem 19.2.8. *Suppose the moduli spaces $M([\mathfrak{a}_{i-1}],[\mathfrak{a}_i])$ are boundary-unobstructed, for $i = 1,\dots,n$. Then there is a neighborhood $\breve{W}$ of the subset* (19.6) *in $\breve{M}^+([\mathfrak{a}_0],[\mathfrak{a}_n])$ and a map*

$$\boldsymbol{S} : \breve{W} \to (0,\infty]^{n-1}$$

such that $\boldsymbol{S}^{-1}(\infty,\dots,\infty)$ is the subset (19.6) *and such that $\boldsymbol{S}$ is a topological submersion along* (19.6).

19.3 Proof of the gluing theorem: the unobstructed case

Fix $L > 0$ and let $T_j \in (1,\infty]$, for $j = 1,\dots,n-1$. Write

$$\boldsymbol{T} = (T_1,\dots,T_{n-1}) \in (1,\infty]^{n-1}.$$

As in Section 18, let Z^T denote the finite cylinder $[-T,T]\times Y$ if $T < \infty$, and let Z^∞ denote the disjoint union of two half-cylinders

$$Z^\infty = Z^- \amalg Z^+.$$

Set

$$\boldsymbol{Z_T} = Z^- \amalg Z^L \amalg Z^{T_1} \amalg Z^L \amalg \cdots \amalg Z^{T_{n-1}} \amalg Z^L \amalg Z^+.$$

This disjoint union has $4n$ boundary components, $2n$ of which have the positive orientation. If we identify each positively oriented boundary component with the negatively oriented boundary component of the next part of $\boldsymbol{Z_T}$ (in the order in which they are written), we obtain l copies of the infinite cylinder $\mathbb{R}\times Y$, where $l-1$ is the number of T_j that are infinite. See Figure 5, which illustrates the case that $n = 2$ and $T_1 = \infty$. Note that we omit L from the notation for $\boldsymbol{Z_T}$.

For any cylinder $I \times Y$, let d_k denote the metric on $\mathcal{B}_k^\tau(I\times Y)$ obtained from the L^2_k distance on $\mathcal{C}_k^\tau(I\times Y)$:

$$d_k([\gamma],[\gamma']) = \inf\left\{ \|u\gamma - \gamma'\|_{L^2_{k,\mathfrak{a}}} \;\middle|\; u \in \mathcal{G}_{k+1}(I\times Y)\right\}.$$

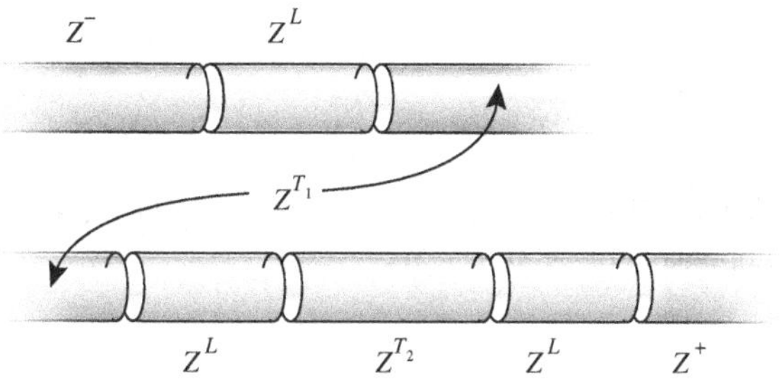

Fig. 5. Obtaining two doubly infinite cylinders by gluing the components of $\boldsymbol{Z_T}$.

Let $M_\eta(Z^\pm; [\mathfrak{a}])$ denote the moduli space of trajectories $[\gamma]$ on the half-cylinder $Z^\pm$, asymptotic to $[\mathfrak{a}]$ and having

$$d_k([\gamma], [\gamma_\mathfrak{a}]) \le \eta.$$

For $0 < T < \infty$, let $M_\eta(Z^T; [\mathfrak{a}])$ similarly denote the moduli space of all solutions $[\gamma]$ with $d_k([\gamma], [\gamma_\mathfrak{a}]) < \eta$. For $T = \infty$, we let

$$M_\eta(Z^\infty; [\mathfrak{a}]) \subset M_\eta(Z^-; [\mathfrak{a}]) \times M_\eta(Z^+; [\mathfrak{a}])$$

be the subset of pairs $([\gamma^-], [\gamma^+])$ with

$$d_k([\gamma^-], [\gamma_\mathfrak{a}])^2 + d_k([\gamma^+], [\gamma_\mathfrak{a}])^2 < \eta^2.$$

Set

$$\begin{aligned} \boldsymbol{M_T} = {} & M(Z^-; [\mathfrak{a}_0]) \times M(Z^L) \times M(Z^{T_1}) \times M(Z^L) \\ & \times \cdots \times M(Z^{T_{n-1}}) \times M(Z^L) \times M(Z^+; [\mathfrak{a}_n]). \end{aligned} \tag{19.7}$$

Inside $\boldsymbol{M_T}$ is a subset

$$\boldsymbol{M_T} \supset \boldsymbol{M}^{cen}_{\boldsymbol{T},\eta},$$

defined by

$$\begin{aligned} \boldsymbol{M}^{cen}_{\boldsymbol{T},\eta} = {} & M_\eta(Z^-; [\mathfrak{a}_0]) \times M^{cen}(Z^L) \times M_\eta(Z^{T_1}; [\mathfrak{a}_1]) \times M^{cen}(Z^L) \\ & \times \cdots \times M_\eta(Z^{T_{n-1}}; [\mathfrak{a}_{n-1}]) \times M^{cen}(Z^L) \times M_\eta(Z^+; [\mathfrak{a}_n]). \end{aligned} \tag{19.8}$$

Let $\mathrm{Fib}(R^-_{\boldsymbol{T}}, R^+_{\boldsymbol{T}}) \subset \boldsymbol{M_T}$ be the subset

$$\mathrm{Fib}(R^-_{\boldsymbol{T}}, R^+_{\boldsymbol{T}}) = \{\, m \in \boldsymbol{M_T} \mid R^-_{\boldsymbol{T}}(m) = R^+_{\boldsymbol{T}}(m) \in (\mathcal{B}^\sigma_{k-1/2}(Y))^{2n} \,\}$$

where

$$R_T^-, R_T^+ : \boldsymbol{M}_T \to (\mathcal{B}^\sigma_{k-1/2}(Y))^{2n-2}$$

are obtained from the two different restriction maps, to the negative and positive boundary components respectively. We write

$$\boldsymbol{M} = \bigcup_{\boldsymbol{T}} \boldsymbol{M}_{\boldsymbol{T}}$$

and

$$\boldsymbol{M}^{cen}_\eta = \bigcup_{\boldsymbol{T}} \boldsymbol{M}^{cen}_{\boldsymbol{T},\eta}.$$

We also write $\mathrm{Fib}(R^-, R^+) \subset \boldsymbol{M}$ for the union of the fiber products. (We only need to consider $\boldsymbol{M}, \boldsymbol{M}^{cen}_\eta$ and $\mathrm{Fib}(R^-, R^+)$ as sets, not as topological spaces.)

Let us explain what these definitions are about. If $m \in \mathrm{Fib}(R_T^-, R_T^+) \subset \boldsymbol{M}_{\boldsymbol{T}}$, then we can concatenate the component trajectories of m, using Lemma 19.1.1. Thus we have a map

$$\mathbf{c} : \mathrm{Fib}(R_T^-, R_T^+) \to \breve{M}^+([\mathfrak{a}_0], [\mathfrak{a}_n])$$

obtained by concatenation. The number of components in the broken trajectory $\mathbf{c}(m)$ is l, where again $l-1$ is the number of j with $T_j = \infty$. If η is sufficiently small and

$$m \in \mathrm{Fib}(R_T^-, R_T^+) \cap \boldsymbol{M}^{cen}_{\boldsymbol{T},\eta},$$

then the broken trajectory $\mathbf{c}(m)$ has a complete collection of n local centering intervals of length $2L$: these centering intervals are the images of the n cylinders $Z^L \subset \boldsymbol{Z}_{\boldsymbol{T}}$. We need to choose η small enough that γ is in $M_\eta(Z_T)$; then $\mathfrak{e}(\gamma(t)) \le \epsilon$ for all $t \in [-T, T]$. The separations S_i between the centers of the local centering intervals are the lengths

$$S_i(m) = \begin{cases} 2T_i + 2L, & \text{if } T_i < \infty \\ \infty, & \text{if } T_i = \infty. \end{cases} \tag{19.9}$$

Taking the union over all $\boldsymbol{T}$, we obtain a map

$$\mathbf{c} : \mathrm{Fib}(R^-, R^+) \to \breve{M}^+([\mathfrak{a}_0], [\mathfrak{a}_n]).$$

Lemma 19.2.6 provides a preferred right inverse to $\mathbf{c}$ on a subset $\breve{W} \subset \breve{M}^+([\mathfrak{a}_0], [\mathfrak{a}_n])$. The main content of the following proposition is that after making L sufficiently large and passing to a smaller $\breve{W}$ the image of this right inverse is contained in $\boldsymbol{M}_\eta^{cen}$.

Proposition 19.3.1. *Let $\breve{\boldsymbol{K}} \subset \prod \breve{M}([\mathfrak{a}_{i-1}], [\mathfrak{a}_i])$ be a compact subset. There exists $\eta_0 > 0$ such that for all $\eta < \eta_0$ and all $L > L_1(\eta)$, the image of the map*

$$\mathbf{c} : \mathrm{Fib}(R^-, R^+) \cap \boldsymbol{M}_\eta^{cen} \to \breve{M}^+([\mathfrak{a}_0], [\mathfrak{a}_n]), \tag{19.10}$$

obtained by concatenating trajectories, contains an open neighborhood $\breve{W} = \breve{W}(\eta, L)$ of $\breve{\boldsymbol{K}} \subset \prod \breve{M}([\mathfrak{a}_{i-1}], [\mathfrak{a}_i])$. Furthermore, $\mathbf{c}$ is injective on

$$\mathbf{c}^{-1}(\breve{W}) \cap \boldsymbol{M}_\eta^{cen}.$$

Proof. The injectivity asserted in the last part of the proposition follows from the uniqueness of the collection of local centering intervals. The main assertion is the first part of the proposition.

For each critical point $[\mathfrak{c}]$ in $\mathcal{B}_k^\sigma(Y)$, let $U_\mathfrak{c}$ be a gauge-invariant neighborhood of the constant trajectory $\gamma_\mathfrak{c}$ in $\mathcal{C}_k^\tau([-1,1] \times Y)$. Let these be sufficiently small that $\mathfrak{e}(\gamma(t)) < \epsilon$ for all $\gamma \in U_\mathfrak{c}$ and all $t \in [-1, 1]$. For all finite $T \geq 1$, let

$$U_\mathfrak{c}(T) \subset \mathcal{C}_k^\tau(Z^T)$$

be defined as the set of trajectories γ on $[-T, T] \times Y$, such that the translate $\tau_s^* \gamma$ belongs to $U_\mathfrak{c}$ for all s in $[-T+1, T-1]$. Extend this definition to the case $T = \infty$ in the obvious way. Let

$$\boldsymbol{M}_U^{cen} \subset \boldsymbol{M}$$

be defined just as $\boldsymbol{M}_\eta^{cen}$ is, but replacing the factors $M_\eta(Z^{T_i}; [\mathfrak{a}_i])$ in the definition (19.8) by the neighborhoods $M(Z^T) \cap (U_{\mathfrak{a}_i}(T)/\mathcal{G}_{k+1})$. The proof of Lemma 19.2.6 shows that the image of the map

$$\mathbf{c} : \mathrm{Fib}(R^-, R^+) \cap \boldsymbol{M}_U^{cen} \to \breve{M}^+([\mathfrak{a}_0], [\mathfrak{a}_n]) \tag{19.11}$$

contains a neighborhood $\breve{W}$ of the product stratum. The key point is to be able to pass from a neighborhood such as $U_{\mathfrak{a}_i}(T)$, defined by a local condition along the trajectory, to a neighborhood such as $M_\eta(Z^T; [\mathfrak{a}_i])$ defined by the L_k^2 distance d_k on the entire cylinder $[-T, T] \times Y$, uniformly in T.

We exploit our freedom to increase the length L. Suppose $[\breve{\gamma}]$ is an element of $\breve{W}$ and has preimage $m \in \mathbf{M}^{cen}_{U,L}$ under the map $\mathbf{c}_L$ (where we have now made the dependence on L explicit). If the coordinates T_i for m all exceed T_0+C, then $[\breve{\gamma}]$ has an easily described preimage m' under the map $\mathbf{c}_{L'}$ for $L' = L+C$. The T coordinates of m' are given by $T_i' = T_i - C$, and the component γ' of m' in the space $M(Z^{T_i'}; [\mathfrak{a}_i])$ is the restriction to $[-T_i + C, T_i - C]$ of the corresponding component γ of m. The proof of the proposition is therefore concluded by the following lemma. □

Lemma 19.3.2. *Let $\mathfrak{a}$ be a critical point, and let $\eta > 0$ be given. Then there exists a gauge-invariant neighborhood $U_{\mathfrak{a}}$ of $\gamma_{\mathfrak{a}}$ in $\mathcal{C}^{\tau}_k([-1,1] \times Y)$, with the following property. For all $T > 2$ and all trajectories γ in $M([-T,T] \times Y) \cap (U_{\mathfrak{a}}(T)/\mathcal{G}_{k+1})$, there is a gauge transformation u such that*

$$\|u(\gamma) - \gamma_{\mathfrak{a}}\|_{L^2_k([-T+2,T-2]\times Y)} \leq \eta. \tag{19.12}$$

(Here $U_{\mathfrak{a}}(T)$ is the neighborhood defined as in the proof of the proposition above.)

Proof. The argument is similar to the proof of Proposition 13.6.1. The patching construction in the proof of that proposition shows that, as long as $U_{\mathfrak{a}}$ is chosen so as to be able to apply Proposition 13.4.1, we can find a gauge transformation u such that the square of the left-hand side of (19.12) is bounded by the two quantities

$$\sum_{i \in [-T+2, T-2]} \big(\pounds(i+2) - \pounds(i-1)\big)^{1/2}$$

and

$$\int_{-T}^{T} \left|(d/dt)\Lambda_{\mathfrak{q}}\right| \, dt.$$

The first of these two can be made small by choosing $U_{\mathfrak{a}}$ small and using the exponential bounds provided by Proposition 13.5.2. The second can be made small similarly, using Corollary 13.5.3. □

The proposition above describes an open subset $\breve{W}$ of the moduli space $\breve{M}^+([\mathfrak{a}_0],[\mathfrak{a}_n])$ as a sort of fiber product. We have a well-defined map $\boldsymbol{S}$ on $\breve{W}$ given by the separations of the local centers, as in (19.5). The next lemma is a general result that will eventually tell us that $\boldsymbol{S}$ is a topological submersion along the fiber over infinity.

Lemma 19.3.3. *Let $\mathcal{M}$ and $\mathcal{N}$ be Hilbert manifolds. Suppose that $\mathcal{N}$ is a closed submanifold of a Hilbert space, so that we may unambiguously talk about the C^1_{loc} topology on the space of continuously differentiable maps from $\mathcal{M}$ to $\mathcal{N}$. Let (Q, q_0) be a pointed topological space and*

$$R : \mathcal{M} \times Q \to \mathcal{N}$$

a continuous map whose restriction R_q to each $\mathcal{M} \times \{q\}$ is smooth. Suppose that R_{q_0} is transverse to a smooth, closed submanifold $\Delta \subset \mathcal{N}$, and that R_q converges to R_{q_0} in the C^1_{loc} topology as $q \to q_0$. Then the projection

$$R^{-1}(\Delta) \to Q$$

is a topological submersion along $R^{-1}_{q_0}(\Delta)$.

Proof. The question is local, and therefore reduces to the case that $\mathcal{N}$ is a Hilbert space and Δ is the origin. The lemma is just a version of the implicit function theorem, with continuous dependence on a parameter q. □

We can now complete the proof of the gluing theorem for trajectories that are not boundary-obstructed .

Proof of Theorem 19.2.8. First of all, since the submersion property is local, it is enough to consider a compact subset $\boldsymbol{K}$ of the product stratum of the form (19.3): we will show $\boldsymbol{S}$ is a topological submersion along $\boldsymbol{K}$. Let $\eta \leq \eta_0$ and $L \geq L_1(\eta)$ be as in Proposition 19.3.1, and let $W = W(\eta, L)$ be the corresponding neighborhood of $\boldsymbol{K}$, as given by that proposition.

Proposition 19.3.1 identifies $\breve{W}$ with a subset of $\boldsymbol{M}^{cen}_{\eta}$. The latter is the union over all $\boldsymbol{T}$ of the spaces $\boldsymbol{M}^{cen}_{\boldsymbol{T},\eta}$. As $\boldsymbol{T}$ varies, the factors $M_\eta(Z^{T_i}; [\mathfrak{a}_i])$ in the definition (19.8) of $\boldsymbol{M}^{cen}_{\boldsymbol{T},\eta}$ vary, while the remaining factors do not.

The spaces $M_\eta(Z^{T_i}; [\mathfrak{a}_i])$ are described by Theorem 18.2.1. Recall that $M_\eta(Z^{T_i}; [\mathfrak{a}_i])$ is a Hilbert manifold with boundary (the boundary is non-empty only if $[\mathfrak{a}_i]$ is reducible), and is contained in its "double", the Hilbert manifold $\tilde{M}_\eta(Z^{T_i}; [\mathfrak{a}_i])$. For $T \geq T_0$ (including the case $T = \infty$), Theorem 18.2.1 tells us that there is a neighborhood $\tilde{U}^T_i$ in $\tilde{M}_\eta(Z^T; [\mathfrak{a}_i])$ parametrized by a diffeomorphism

$$u_i(-, T) : B_{\eta_1}(\mathcal{K}_i) \to \tilde{U}^T_i,$$

where $B_{\eta_1}(\mathcal{K}_i)$ is a ball in a Hilbert space. In the case that $\mathfrak{a}_i$ is reducible, $\tilde{U}^T_i$ meets $M_\eta(Z^T; [\mathfrak{a}_i]) \subset \tilde{M}_\eta(Z^T; [\mathfrak{a}_i])$ in a neighborhood U^T_i which we can identify with the quotient of $\tilde{U}_i^T$ by the involution $\mathbf{i} : s \mapsto -s$; the quotient is a Hilbert manifold with boundary. The parametrization is equivariant for the action of $\mathbf{i}$, and we have a parametrization

$$u_i(-, T) : b_{\eta_1}(\mathcal{K}_i) \to U_i^T,$$

where $b_{\eta_1}(\mathcal{K}_i)$ is the half-ball

$$b_{\eta_1}(\mathcal{K}_i) = B_{\eta_1}(\mathcal{K}_i) \cap \{s \geq 0\}.$$

In the irreducible case, we set $U_i^T = \tilde{U}_i^T$ and $b_{\eta_1}(\mathcal{K}_i) = B_{\eta_1}(\mathcal{K}_i)$, so as to keep the notation uniform. According to Theorem 18.2.1, the neighborhood U_i^T contains some $\tilde{M}_{\eta'}(Z^T; [\mathfrak{a}_i])$, where $\eta' \leq \eta$ is independent of T. We may assume $L \geq L_1(\eta')$ (where $L_1(\eta')$ is again as in Proposition 19.3.1).

For consistency, we also choose to parametrize a neighborhood in $M(Z^-; [\mathfrak{a}_0])$ and $M(Z^+; [\mathfrak{a}_n])$. The map (see (18.4))

$$u_0(-, \infty) : b_{\eta_1}(\mathcal{K}_0) \to M_\eta(Z^\infty; [\mathfrak{a}_0])$$

constructed by Theorem 18.2.1 restricts to a map

$$u_0(-, \infty) : b_{\eta_1}(\mathcal{K}_0^-) \to M_\eta(Z^-; [\mathfrak{a}_0])$$

where $\mathcal{K}_0 = \mathcal{K}_0^+ \oplus \mathcal{K}_0^-$ is the spectral decomposition. Note that the involution $\mathbf{i}$ is non-trivial on exactly one of the summands of the decomposition: on $\mathcal{K}_0^-$ if $\mathfrak{a}_0$ is boundary-unstable, and on $\mathcal{K}_0^+$ if $\mathfrak{a}_0$ is boundary-stable. We have a similar parametrization of a neighborhood in $M_\eta(Z^+; [\mathfrak{a}_n])$, by a ball or half-ball, $b_{\eta_1}(\mathcal{K}_n^+)$.

Combining all the u_i, we obtain, for each

$$\boldsymbol{T} = (T_1, \ldots, T_{n-1}),$$

a map

$$\mathbf{u}_{\boldsymbol{T}} : \mathcal{M}_{\eta_1}^{cen} \to \boldsymbol{M}_{\boldsymbol{T},\eta}^{cen}$$

where

$$\mathcal{M}_{\eta_1}^{cen} = \left(\prod b_{\eta_1}(\mathcal{K}_i)\right) \times b_{\eta_1}(\mathcal{K}_0^-) \times b_{\eta_1}(\mathcal{K}_n^+) \times \left(M^{cen}(Z^L)\right)^n.$$

The map $\mathbf{u}_{\boldsymbol{T}}$ is a diffeomorphism onto an open set $\boldsymbol{U}_{\boldsymbol{T}}$, and

$$\boldsymbol{M}_{\boldsymbol{T},\eta'}^{cen} \subset \boldsymbol{U}_{\boldsymbol{T}} \subset \boldsymbol{M}_{\boldsymbol{T},\eta}^{cen}. \tag{19.13}$$

Inside $\mathcal{M}^{cen}_{\eta_1}$, we have the subset

$$\mathrm{Fib}(\mathcal{R}_{\boldsymbol{T}}) = \mathrm{Fib}(R^-_{\boldsymbol{T}} \circ \mathbf{u}_{\boldsymbol{T}}, R^+_{\boldsymbol{T}} \circ \mathbf{u}_{\boldsymbol{T}}),$$

which is the inverse image of the diagonal Δ under the smooth map

$$\begin{aligned} \mathcal{R}_{\boldsymbol{T}} : \mathcal{M}^{cen}_{\eta_1} &\to \boldsymbol{\mathcal{B}} \times \boldsymbol{\mathcal{B}} \\ \mathcal{R}_{\boldsymbol{T}} &= (R^-_{\boldsymbol{T}} \circ \mathbf{u}_{\boldsymbol{T}}, R^+_{\boldsymbol{T}} \circ \mathbf{u}_{\boldsymbol{T}}), \end{aligned} \tag{19.14}$$

where

$$\boldsymbol{\mathcal{B}} = \left((\mathcal{B}^{\sigma}_{k-1/2}(Y))^{2n} \right).$$

We combine the maps $\mathcal{R}_{\boldsymbol{T}}$ for $\boldsymbol{T} \in (T_0, \infty]^{n-1}$ into a single map

$$\mathcal{R} : \mathcal{M}^{cen}_{\eta_1} \times (T_0, \infty]^{n-1} \to \boldsymbol{\mathcal{B}} \times \boldsymbol{\mathcal{B}}, \tag{19.15}$$

and we have the parametrized fiber product

$$\mathrm{Fib}(\mathcal{R}) = \mathcal{R}^{-1}(\Delta).$$

The maps $\mathbf{u}_{\boldsymbol{T}}$ combine to form a map

$$\mathbf{u} : \mathrm{Fib}(\mathcal{R}) \to \mathrm{Fib}(R^-, R^+) \cap \boldsymbol{M}^{cen}_{\eta} \subset \boldsymbol{M}.$$

Lemma 19.3.4. *The image of the composite map*

$$\mathbf{c} \circ \mathbf{u} : \mathrm{Fib}(\mathcal{R}) \to \breve{M}^+([\mathfrak{a}_0], [\mathfrak{a}_n])$$

contains the open neighborhood $\breve{W} = \breve{W}(\eta', L)$ *of the compact set* $\breve{\boldsymbol{K}}$*, given by Proposition* 19.3.1. *The composite map gives a homeomorphism*

$$(\mathbf{c} \circ \mathbf{u}) : \mathrm{Fib}(\mathcal{R}) \cap (\mathbf{c} \circ \mathbf{u})^{-1}(\breve{W}) \to \breve{W}.$$

Proof. The image of $\mathrm{Fib}(\mathcal{R})$ under the map $\mathbf{u}$ contains $\mathrm{Fib}(R^-, R^+) \cap \boldsymbol{M}^{cen}_{\eta'}$ because of the statement (19.13), and it therefore follows from Proposition 19.3.1 that the image of the composite contains $\breve{W}(\eta', L)$ as claimed. The map in the last part of the lemma is injective because of the corresponding statement in Proposition 19.3.1, and all that remains is to check the continuity of the map and its inverse.

The separations S_i of the centering intervals define continuous functions on $\breve{W}$, so the restriction of an element $[\breve{\gamma}] \in \breve{W}$ to the ith centering interval defines a continuous map from $\breve{W}$ to $M^{cen}(Z^L)$. Together, the separations and these restrictions define a continuous map

$$p : \breve{W} \to (0,\infty]^{n-1} \times \left(M^{cen}(Z^L)\right)^n.$$

The map p is a topological embedding, because of the unique continuation theorem for trajectories. The composite map $p \circ \mathbf{c} \circ \mathbf{u}$ is rather trivially continuous, because it is the identity on each $M^{cen}(Z^L)$ factor and is the map $T_i \mapsto 2(T_i + L)$ on the $(0,\infty]$ factors. So the map $\mathbf{c} \circ \mathbf{u}$ is continuous also.

For the continuity of the inverse, the remaining point is to see that, for an element

$$\left((h_1, \dots h_{n-1}, h_0^-, h_n^+), (w_1, \dots, w_n), (T_1, \dots, T_{n-1})\right)$$

of

$$\begin{aligned}
\mathrm{Fib}(\mathcal{R}) &\subset \mathcal{M}^{cen}_{\eta_1} \times (T_0, \infty]^{n-1} \\
&= \left(\prod b_{\eta_1}(\mathcal{K}_i)\right) \times b_{\eta_1}(\mathcal{K}_0^-) \times b_{\eta_1}(\mathcal{K}_n^+) \times \left(M^{cen}(Z^L)\right)^n \\
&\quad \times (T_0, \infty]^{n-1},
\end{aligned}$$

the h coordinates are continuous functions of the w_j and the T_i. Because of the definition of the fiber product, this reduces to showing that h_i can be recovered as a continuous function of T_i and the boundary values of $u_i^{T_i}(h_i)$. This becomes a statement about the map μ_T in Theorem 18.2.1: we must see that if T_n converges to $T \in (T_0, \infty]$ and $\mu_{T_n}(h_n)$ converges to $\mu_T(h)$, then h_n converges to h. This follows from the fact that μ_T depends continuously on T and for each fixed T is an embedding. □

We now return to the proof of the theorem. Under the map $\mathbf{c} \circ \mathbf{u}$ of the lemma, the two maps

$$\begin{aligned}
\boldsymbol{S} &: \breve{W} \to [0, \infty]^{n-1} \\
\boldsymbol{T} &: \mathrm{Fib}(\mathcal{R}) \to (T^0, \infty]^{n-1}
\end{aligned}$$

are related by $S_i \circ \mathbf{c} \circ \mathbf{u} = 2(T_i + L)$. The map $T_i \mapsto 2(T_i + L)$ gives a local homeomorphism of $(0, \infty]^{n-1}$ near ∞, so the proof of Theorem 19.2.8 is reduced to the following proposition. □

Proposition 19.3.5. *The map*

$$\boldsymbol{T} : \mathrm{Fib}(\mathcal{R}) \to (T^0, \infty]^{n-1}$$

is a topological submersion along the fiber over $\boldsymbol{\infty} = (\infty, \dots, \infty)$.

Proof. Some of the factors that comprise the space $\mathcal{M}^{cen}_{\eta 1}$ are half-balls. It is convenient to work with Hilbert manifolds rather than manifolds with boundary, so we introduce also the space $\tilde{\mathcal{M}}^{cen}_{\eta 1}$, obtained by replacing each $b_{\eta 1}$ with a ball $B_{\eta 1}$, and replacing each $M^{cen}(Z^L)$ by $\tilde{M}^{cen}(Z^L)$. Adding a tilde to everything, we have a map

$$\tilde{\mathcal{R}}_{\boldsymbol{T}} : \tilde{\mathcal{M}}^{cen}_{\eta 1} \to \tilde{\boldsymbol{\mathcal{B}}} \times \tilde{\boldsymbol{\mathcal{B}}}$$

where

$$\tilde{\boldsymbol{\mathcal{B}}} = \left((\tilde{\mathcal{B}}^{\sigma}_{k-1/2}(Y))^{2n} \right),$$

and we have fiber products

$$\mathrm{Fib}(\tilde{\mathcal{R}}) = \bigcup_{\boldsymbol{T}} \mathrm{Fib}(\tilde{\mathcal{R}}_{\boldsymbol{T}}). \tag{19.16}$$

Theorem 18.2.1 tells us that $\tilde{\mathcal{R}}_{\boldsymbol{T}}$ converges to $\tilde{\mathcal{R}}_{\infty}$ in the $C^{\infty}_{\mathrm{loc}}$ topology as $\boldsymbol{T} \to \boldsymbol{\infty} = (\infty, \dots, \infty)$.

Lemma 19.3.6. *The map* $\tilde{\mathcal{R}}_{\infty}$ *is transverse to the diagonal* $\Delta \subset \tilde{\boldsymbol{\mathcal{B}}} \times \tilde{\boldsymbol{\mathcal{B}}}$.

Proof. Here we use the regularity of the moduli spaces $M([\mathfrak{a}_{i-1}], [\mathfrak{a}_i])$. The map $\tilde{\mathcal{R}}_{\infty}$ is a Cartesian product of n maps, each of which is the composite with $\mathbf{u}_{\infty}$ of one of the maps

$$\begin{aligned}(R_-, R_+) : \ & \tilde{M}_{\eta}(Z^-; [\mathfrak{a}_{i-1}]) \times \tilde{M}^{cen}(Z^L) \times \tilde{M}_{\eta}(Z^+; [\mathfrak{a}_i]) \\ & \to \left(\tilde{\mathcal{B}}^{\sigma}_{k-1/2}(Y) \times \tilde{\mathcal{B}}^{\sigma}_{k-1/2}(Y)\right)^2.\end{aligned} \tag{19.17}$$

In place of $\tilde{M}^{cen}(Z^L)$, consider again the larger space $\tilde{M}'(Z^L) \subset \tilde{M}(Z^L)$ defined by the first two conditions in Definition 19.2.2 (see the proof of Lemma 19.2.5 earlier in this section); and consider the extension of (R^-, R^+) to this larger domain:

$$\begin{aligned}(R'_-, R'_+) : \ & \tilde{M}_{\eta}(Z^-; [\mathfrak{a}_{i-1}]) \times \tilde{M}'(Z^L) \times \tilde{M}_{\eta}(Z^+; [\mathfrak{a}_i]) \\ & \to \left(\tilde{\mathcal{B}}^{\sigma}_{k-1/2}(Y) \times \tilde{\mathcal{B}}^{\sigma}_{k-1/2}(Y)\right)^2.\end{aligned} \tag{19.18}$$

The fact that this map is transverse to the diagonal is an instance of Theorem 19.1.4. The lemma follows if we can show that the smooth map

$$c : \mathrm{Fib}(R'_-, R'_+) \to \mathbb{R}$$

given by taking the center of mass of the middle factor is transverse to 0. This transversality is similar to the transversality asserted in Lemma 19.2.5 (but technically easier): the action of translations defines a tangent vector to $\tilde{M}([\mathfrak{a}_{i-1}], [\mathfrak{a}_i])$ along which the derivative of c is -1. □

It follows from this lemma and Lemma 19.3.3 that the map

$$\boldsymbol{T} : \mathrm{Fib}(\tilde{\mathcal{R}}) \to (T_0, \infty]^{n-1}$$

is a topological submersion along the fiber over ∞. Let $\mathbf{i}$ be defined on $\mathrm{Fib}(\tilde{\mathcal{R}}) \subset \tilde{\mathcal{M}}^{cen}_{\eta_1}$ by simultaneously applying the involution $s \mapsto -s$ on each component. In the fiber product, we can never have an element with s positive on one component and negative on another. It follows that we can identify $\mathrm{Fib}(\mathcal{R})$ with the quotient of $\mathrm{Fib}(\tilde{\mathcal{R}})$ by the involution $\mathbf{i}$. Since $\boldsymbol{T}$ is invariant under the involution, the map

$$\boldsymbol{T} : \mathrm{Fib}(\mathcal{R}) \to (T_0, \infty]^{n-1}$$

is also a topological submersion. This completes the proof of Proposition 19.3.5, and hence Theorem 19.2.8. □

19.4 Gluing in the boundary-obstructed case

We now consider the modification that is necessary to the statement of Theorem 19.2.8 in the case that some of the moduli spaces $M([\mathfrak{a}_{i-1}], [\mathfrak{a}_i])$ are boundary-obstructed. We shall suppose that this moduli space is boundary-obstructed for

$$i \in O \subset \{1, \ldots, n\}$$

and unobstructed for i in O', the complement of O. (The unobstructed case includes the case that one or both of $\mathfrak{a}_{i-1}$, $\mathfrak{a}_i$ is irreducible.) From the previous subsection, we have the neighborhood

$$\breve{W} \supset \prod_{i=1}^{n} \breve{M}([\mathfrak{a}_{i-1}], [\mathfrak{a}_i])$$

in $\breve{M}^+([\mathfrak{a}_0], [\mathfrak{a}_1])$, on which is defined a map

$$\mathbf{S} : \breve{W} \to (0, \infty]^{n-1}$$

which describes the separation of the local centering intervals.

The following theorem gives the general structure of a suitable neighborhood $\breve{W}$ in the boundary-obstructed case. The statement of the theorem is rather elaborate: readers might first read the next subsection where some important special cases are worked out in more detail.

Theorem 19.4.1. *Suppose the moduli spaces $M([\mathfrak{a}_{i-1}], [\mathfrak{a}_i])$ are boundary-obstructed for $i \in O$ and boundary-unobstructed for $i \in O' = \{1, \ldots, n\} \setminus O$. Then there is an open set $\breve{W} \subset \breve{M}^+([\mathfrak{a}_0], [\mathfrak{a}_n])$ as above, with the following properties.*

(i) *There is a topological embedding of $\breve{W}$ in a space $E\breve{W}$ with a map $\mathbf{S}$ to $(0, \infty]^{n-1}$ such that the diagram commutes:*

$$\begin{array}{ccc} \breve{W} & \xrightarrow{\;j\;} & E\breve{W} \\ \downarrow{\scriptstyle S} & & \downarrow{\scriptstyle S} \\ (0, \infty]^{n-1} & = & (0, \infty]^{n-1}. \end{array}$$

(ii) *The map $\mathbf{S} : E\breve{W} \to (0, \infty]^{n-1}$ is a topological submersion along the fiber over ∞.*

(iii) *The image $j(\breve{W}) \subset E\breve{W}$ is the zero set of a continuous map*

$$\delta : E\breve{W} \to \mathbb{R}^O$$

which vanishes on the fiber over ∞. In particular therefore, the fiber over ∞ in both $\breve{W}$ and $E\breve{W}$ is identified with the stratum $\prod \breve{M}([\mathfrak{a}_{i-1}], [\mathfrak{a}_i])$.

(iv) *If $\breve{W}^o \subset W$ and $E\breve{W}^o \subset E\breve{W}$ are the subsets where none of the S_i is infinite, then the restriction of j to $\breve{W}^o$ is an embedding of smooth manifolds, and $\delta|_{E\breve{W}^o}$ is transverse to zero.*

(v) *Let $i_0 \in O$, and let δ_{i_0} be the corresponding component of δ. Then for all $z \in E\breve{W}$, we have:*
- *if $i_0 \geq 2$ and $S_{i_0 - 1}(z) = \infty$ then $\delta_{i_0}(z) \geq 0$;*
- *if $i_0 \leq n - 1$ and $S_{i_0}(z) = \infty$ then $\delta_{i_0}(z) \leq 0$.*

Proof. Let

$$\mathcal{R}_T : \mathcal{M}^{cen}_{\eta 1} \to \boldsymbol{\mathcal{B}} \times \boldsymbol{\mathcal{B}}$$
$$\tilde{\mathcal{R}}_T : \tilde{\mathcal{M}}^{cen}_{\eta 1} \to \tilde{\boldsymbol{\mathcal{B}}} \times \tilde{\boldsymbol{\mathcal{B}}}$$

be as in the proof of the unobstructed case, Theorem 19.2.8. Let $\mathrm{Fib}(\mathcal{R}_T)$ be the inverse images of the diagonals as before, and let $\mathrm{Fib}(\mathcal{R})$ etc. be obtained again by taking the union over $\boldsymbol{T}$. The map $\mathcal{R}_\infty$ is no longer transverse to the diagonal (compare Lemma 19.3.6).

As in Definition 19.1.6, we define a moduli space

$$E\tilde{M}(Z^L) \subset \tilde{M}([-L,0] \times Y) \times \tilde{M}([0,L] \times Y)$$

as the fiber product of the two maps

$$R'_+ : \tilde{M}([-L,0] \times Y) \to \partial \mathcal{B}^\sigma_{k-1/2}(Y)$$
$$R'_- : \tilde{M}([0,L] \times Y) \to \partial \mathcal{B}^\sigma_{k-1/2}(Y).$$

For $[\gamma] = ([\gamma_-], [\gamma_+])$ in $E\tilde{M}(Z^L)$, we can regard $\mathfrak{e}(\gamma, t)$ as defining a piecewise continuous function of t on $[-L, L]$, with a discontinuity at $t = 0$. With this understood, we define a subset

$$E\tilde{M}^{cen}(Z^L) \subset E\tilde{M}(Z^L)$$

by repeating Definition 19.2.2. For $[\gamma] \in E\tilde{M}(Z^L)$, we define $\delta([\gamma])$ to be the discontinuity in the normal coordinate s on $\tilde{\mathcal{B}}^\sigma_{k-1/2}(Y)$ at $t = 0$, as in (19.2):

$$\delta([\gamma]) = s([\gamma_+|_{\{0\}\times Y}]) - s([\gamma_-|_{\{0\}\times Y}]).$$

The moduli space $\tilde{M}^{cen}(Z^L)$ is contained in the extended moduli space as the locus $\delta = 0$:

$$\tilde{M}^{cen}(Z^L) = E\tilde{M}^{cen}(Z^L) \cap \{\delta = 0\}.$$

Just as $\tilde{M}^{cen}(Z^L)$ is a Hilbert manifold, so (a fortiori) $E\tilde{M}^{cen}(Z^L)$ is smooth also in a neighborhood of $\tilde{M}^{cen}(Z^L)$: the function δ is transverse to zero.

Define $\mathcal{E}\tilde{\mathcal{M}}^{cen}_{\eta 1}$ by taking the definition of $\tilde{\mathcal{M}}^{cen}_{\eta 1}$ and replacing the ith copy of $\tilde{M}^{cen}(Z^L)$ by $E\tilde{M}^{cen}(Z^L)$ when $i \in O$:

$$\mathcal{E}\tilde{\mathcal{M}}^{cen}_{\eta 1} = \left(\prod B_{\eta 1}(\mathcal{K}_i)\right) \times B_{\eta 1}(\mathcal{K}_0^-) \times B_{\eta 1}(\mathcal{K}_n^+) \\ \times \left(\tilde{M}^{cen}(Z^L)\right)^{O'} \times \left(E\tilde{M}^{cen}(Z^L)\right)^{O}.$$

The original $\tilde{\mathcal{M}}^{cen}_{\eta 1}$ is contained in the larger space, as the zero set of the map

$$\delta : \mathcal{E}\tilde{\mathcal{M}}^{cen}_{\eta 1} \to \mathbb{R}^O.$$

The map $\tilde{\mathcal{R}}_{\boldsymbol{T}}$ extends to the larger space,

$$\mathcal{E}\tilde{\mathcal{R}}_{\boldsymbol{T}} : \mathcal{E}\tilde{\mathcal{M}}^{cen}_{\eta 1} \to \tilde{\boldsymbol{\mathcal{B}}} \times \tilde{\boldsymbol{\mathcal{B}}},$$

and we have the inverse image of the diagonal,

$$\mathrm{Fib}(\mathcal{E}\tilde{\mathcal{R}}_{\boldsymbol{T}}) \subset \mathcal{E}\tilde{\mathcal{M}}^{cen}_{\eta 1}.$$

As in (19.15), we form a total map $\mathcal{E}\tilde{\mathcal{R}}$, and we have the fiber product

$$\mathrm{Fib}(\mathcal{E}\tilde{\mathcal{R}}) = \bigcup_{\boldsymbol{T}} \mathrm{Fib}(\mathcal{E}\tilde{\mathcal{R}}_{\boldsymbol{T}}) \subset \mathcal{E}\tilde{\mathcal{M}}^{cen}_{\eta 1} \times (T_0, \infty]^{n-1}.$$

Lemma 19.4.2. *The map $\mathcal{E}\tilde{\mathcal{R}}_\infty$ is transverse to the diagonal $\Delta \subset \tilde{\boldsymbol{\mathcal{B}}} \times \tilde{\boldsymbol{\mathcal{B}}}$.*

Proof. As in the proof of Lemma 19.3.6, the map $\mathcal{E}\tilde{\mathcal{R}}_\infty$ is a Cartesian product of n maps. Each of these n maps either is of the form (19.17), or is the composition of $\mathbf{u}_\infty$ with a map of the form

$$(R_-, R_+) : \tilde{M}_\eta(Z^-; [\mathfrak{a}_{i-1}]) \times E\tilde{M}^{cen}(Z^L) \times \tilde{M}_\eta(Z^+; [\mathfrak{a}_i]) \\ \to \left(\tilde{\mathcal{B}}^\sigma_{k-1/2}(Y) \times \tilde{\mathcal{B}}^\sigma_{k-1/2}(Y)\right)^2,$$

where $i \in O$. Elements in the inverse image of the diagonal can be concatenated to give elements of $E\tilde{M}([\mathfrak{a}_{i-1}], \mathfrak{a}_i])$. Because this moduli space is boundary-obstructed, we have $E\tilde{M}([\mathfrak{a}_{i-1}], \mathfrak{a}_i]) = \tilde{M}([\mathfrak{a}_{i-1}], [\mathfrak{a}_i])$. So δ vanishes on elements in the fiber product. The proof now proceeds as in the proof of Lemma 19.3.6, but appealing to Theorem 19.1.5 in place of Theorem 19.1.4. □

As in the proof of Theorem 19.2.8, we now conclude that the map

$$T : \mathrm{Fib}(\mathcal{E}\tilde{\mathcal{R}}) \to (0, \infty]^{n-1}$$

is a topological submersion along the fiber over ∞, as is the map

$$T : \mathrm{Fib}(\mathcal{E}\mathcal{R}) \to (0, \infty]^{n-1}$$

obtained by restricting to the set where $s \geq 0$ on each factor. The space $\mathrm{Fib}(\mathcal{E}\mathcal{R})$ contains $\mathrm{Fib}(\mathcal{R})$ as the zero locus of $\delta : \mathrm{Fib}(\mathcal{E}\mathcal{R}) \to \mathbb{R}^O$. Our definition of $\mathrm{Fib}(\mathcal{E}\mathcal{R})$ involves a choice of L, but there is an inclusion of $\mathrm{Fib}(\mathcal{E}\mathcal{R}_{T+1}(L))$ in $\mathrm{Fib}(\mathcal{E}\mathcal{R}_T(L+1))$. So we can form a union

$$\mathcal{E}\breve{\mathcal{W}} = \bigcup_{L \geq L_0} \bigcup_{T_i \geq T_0 + L} \mathrm{Fib}(\mathcal{E}\mathcal{R}_T(L)).$$

We denote by $\breve{\mathcal{W}}$ the subset of $\mathcal{E}\breve{\mathcal{W}}$ defined by $\delta = 0$. This gives the following picture, in which the map $\mathcal{E}\breve{\mathcal{W}} \to (0, \infty]^{n-1}$ is a topological submersion along the fiber over ∞:

$$\begin{array}{ccc} \breve{\mathcal{W}} & \xrightarrow{\subset} & \mathcal{E}\breve{\mathcal{W}} \\ \downarrow s & & \downarrow s \\ (0, \infty]^{n-1} & = & (0, \infty]^{n-1}. \end{array}$$

Lemma 19.3.4 identifies a neighborhood $\breve{W}$ of the stratum (19.6) with a neighborhood of the fiber over ∞ in $\breve{\mathcal{W}}$, and this establishes the first three parts of the theorem.

It remains to prove the last two parts. In Part (iv), the subset $\breve{W}^o$ is an open subset of the moduli space $\breve{M}([\mathfrak{a}_0], [\mathfrak{a}_n])$, which is a smooth manifold by our regularity assumptions. Let $\breve{\mathcal{W}}^o \subset \breve{\mathcal{W}}$ be the corresponding subset of $\breve{\mathcal{W}}$. Theorem 19.1.4 tells us that, at a point of $\breve{\mathcal{W}}^o \cap \mathcal{M}^{cen}_{\eta 1}$, the map

$$\mathcal{E}\mathcal{R} \times \delta : \mathcal{E}\mathcal{M}^{cen}_{\eta 1} \to \boldsymbol{\mathcal{B}} \times \boldsymbol{\mathcal{B}} \times \mathbb{R}^O$$

is transverse to $\Delta \times \{0\}$. This tells us that a neighborhood of $\breve{W}^o$ in $E\breve{W}^o$ is a smooth manifold, cut out transversely by δ.

Finally, for the last part of the theorem, if $i_0 \geq 2$ and $S_{i_0 - 1}(z) = \infty$, then in the i_0th factor $EM(Z^L)$ we have an element $([\gamma_-], [\gamma_+])$ whose first component is reducible, because the restriction of $[\gamma_-]$ to $-L \times Y$ agrees with an element of the unstable manifold of $[\mathfrak{a}_{i-1}]$, and $[\mathfrak{a}_{i-1}]$ is boundary-stable. It follows that the discontinuity in the s coordinate between $[\gamma_-]$ and $[\gamma_+]$ is non-negative. So δ_{i_0} is non-negative. The other case is similar. This completes the proof of Theorem 19.4.1. □

19.5 The codimension-one strata

We spell out some particular corollaries of the gluing theorems from the previous subsection. We first examine the case where no boundary-obstructed trajectories are involved, so that Theorem 19.2.8 is applicable. The simplest case occurs when $n = 2$, so that we are considering a subset

$$\breve{M}_{z_1}([\mathfrak{a}_0], [\mathfrak{a}_1]) \times \breve{M}_{z_2}([\mathfrak{a}_1], [\mathfrak{a}_2]) \subset \breve{M}_z^+([\mathfrak{a}_0], [\mathfrak{a}_2]).$$

For each of the three critical points $[\mathfrak{a}_i]$, we must consider whether the critical point is reducible or irreducible; and in the reducible case, whether it is boundary-stable or boundary-unstable. The boundary-unobstructed condition precludes having an adjacent pair of reducibles $[\mathfrak{a}_{i-1}]$, $[\mathfrak{a}_i]$ when the first is boundary-stable and the second is boundary-unstable. Recall also that a moduli space $M([\mathfrak{a}_{i-1}], [\mathfrak{a}_i])$ (and therefore also $\breve{M}([\mathfrak{a}_{i-1}], [\mathfrak{a}_i])$) may be a manifold with boundary, arising as the quotient of $\tilde{M}([\mathfrak{a}_{i-1}], [\mathfrak{a}_i])$ by the involution $\mathbf{i}$. This occurs only if $[\mathfrak{a}_{i-1}]$ is boundary-unstable and $[\mathfrak{a}_i]$ is boundary-stable. Excluding such cases, we have:

Corollary 19.5.1. *Suppose the moduli spaces $\breve{M}_{z_1}([\mathfrak{a}_0], [\mathfrak{a}_1])$ and $\breve{M}_{z_2}([\mathfrak{a}_1], [\mathfrak{a}_2])$ are both boundary-unobstructed and are manifolds (rather than manifolds with boundary, as they would be if they contained both reducible and irreducible trajectories). Then there is an open subset $\breve{W}$ of $\breve{M}_z^+([\mathfrak{a}_0], [\mathfrak{a}_2])$ which is a topological manifold with boundary, and the boundary of $\breve{W}$ consists of the broken trajectories $\breve{M}_{z_1}([\mathfrak{a}_0], [\mathfrak{a}_1]) \times \breve{M}_{z_2}([\mathfrak{a}_1], [\mathfrak{a}_2])$.* □

There are only two boundary-unobstructed cases with $n = 2$ that are not covered by this corollary. In each of these cases, the critical points are all reducible, and either the first is boundary-unstable and the other two are boundary-stable, or the first two are boundary-unstable and the last is boundary-stable. In these cases there is a neighborhood $\breve{W}$ of $\breve{M}_{z_1}([\mathfrak{a}_0], [\mathfrak{a}_1]) \times \breve{M}_{z_2}([\mathfrak{a}_1], [\mathfrak{a}_2])$ in $\breve{M}_z^+([\mathfrak{a}_0], [\mathfrak{a}_2])$ which is still locally homeomorphic to a neighborhood of $T = \infty$ in

$$\breve{M}_{z_1}([\mathfrak{a}_0], [\mathfrak{a}_1]) \times \breve{M}_{z_2}([\mathfrak{a}_1], [\mathfrak{a}_2]) \times (T_0, \infty].$$

However in this case exactly one of the two moduli spaces is a manifold with boundary so $\breve{M}^+_{z_1}([\mathfrak{a}_0],[\mathfrak{a}_2])$ is locally homeomorphic to $\mathbb{R}^{d-2} \times \mathbb{R}_{\geq 0} \times \mathbb{R}_{\geq 0}$ in such a way that $(x, 0, t)$ parametrizes the reducible trajectories and $(x, s, 0)$ parametrizes the broken trajectories.

In the case $n > 2$, the analysis of the boundary-unobstructed case is similar. Either all the factors $\breve{M}([\mathfrak{a}_{i-1}],[\mathfrak{a}_i])$ are manifolds, or exactly one of the factors is a manifold with boundary. In the latter case, all the critical points must be reducible and, for some m with $1 \leq m \leq n$, the critical points $[\mathfrak{a}_0], \ldots, [\mathfrak{a}_{m-1}]$ are boundary-unstable, while the critical points $[\mathfrak{a}_m], \ldots, [\mathfrak{a}_n]$ are boundary-stable.

We now turn to the boundary-obstructed cases. The simplest interesting case of Theorem 19.4.1 occurs when $n = 3$. Suppose we have four critical points $[\mathfrak{a}_0], \ldots, [\mathfrak{a}_3]$. Set

$$\breve{M}_i = \breve{M}([\mathfrak{a}_{i-1}],[\mathfrak{a}_i]), \quad i = 1, 2, 3.$$

Suppose that $[\mathfrak{a}_1]$ and $[\mathfrak{a}_2]$ are reducible and that $\breve{M}_2$ is boundary-obstructed. Let $\breve{M}_1^{\text{irr}} \subset \breve{M}_1$ and $\breve{M}_3^{\text{irr}} \subset \breve{M}_3$ be the subsets of irreducible trajectories, and suppose $\breve{M}_1^{\text{irr}}$, $\breve{M}_2$ and $\breve{M}_3^{\text{irr}}$ are non-empty. In this case, the theorem has the following corollary.

Corollary 19.5.2. *Suppose $\breve{M}_2$ is boundary-obstructed, as above. Then there is a space $E\breve{W}$, a topological submersion*

$$S = (S_1, S_2) : E\breve{W} \to (0, \infty] \times (0, \infty]$$

and a map

$$\delta : E\breve{W} \to \mathbb{R}$$

which vanishes identically on the fiber over (∞, ∞) and whose zero set is identified with a neighborhood $\breve{W}$ of the subset

$$\breve{M}_1^{\text{irr}} \times \breve{M}_2 \times \breve{M}_3^{\text{irr}} \subset \breve{M}^+([\mathfrak{a}_0],[\mathfrak{a}_3]).$$

Furthermore, $\delta(z)$ is strictly positive if $S_1(z) = \infty$ and $S_2(z)$ is finite; and $\delta(z)$ is strictly negative if $S_2(z) = \infty$ and $S_1(z)$ is finite. Finally, the set $E\breve{W}^o$ where S_1 and S_2 are both finite is a smooth manifold, containing $\breve{W}^o$ as the transverse zero set of the smooth function δ.

Proof. This is a restatement of the theorem, with the extra point that $\delta(z)$ is strictly positive when $S_1(z)$ is infinite and $S_2(z)$ is finite (and vice versa): the theorem only had $\delta \geq 0$. The strict inequality follows from the argument used in the proof of the theorem, with the following additional observation. If $\delta(z)$ were zero then z would represent an element $([\breve{\gamma}_1], [\breve{\gamma}'])$ in the subset $\breve{M}_1 \times \breve{M}([\mathfrak{a}_1, \mathfrak{a}_3])$. Because $[\mathfrak{a}_1]$ is boundary-stable, $[\breve{\gamma}']$ would be reducible. We choose our open set $\breve{W}$ so that it contains no such pair $([\breve{\gamma}_1], [\breve{\gamma}'])$ with $[\breve{\gamma}']$ reducible. □

This corollary prompts the following definition, which captures the structure described.

Definition 19.5.3. Let N be a d-dimensional space stratified by manifolds and $M^{d-1} \subset N$ a union of components of the $(d-1)$-dimensional stratum. We say that N has a *codimension-*1 *δ-structure* along M^{d-1} if M^{d-1} is smooth and we have the following additional data. There is an open set $W \subset N$ containing M^{d-1}, an embedding $j : W \to EW$ and a map

$$\boldsymbol{S} = (S_1, S_2) : EW \to (0, \infty]^2$$

with the following properties:

(i) the map $\boldsymbol{S}$ is a topological submersion along the fiber over (∞, ∞);
(ii) the fiber of $\boldsymbol{S}$ over (∞, ∞) is $j(M^{d-1})$;
(iii) the subset $j(W) \subset EW$ is the zero set of a map $\delta : EW \to \mathbb{R}$;
(iv) the function δ is strictly positive where $S_1 = \infty$ and S_2 is finite, and strictly negative where $S_2 = \infty$ and S_1 is finite;
(v) on the subset of EW where S_1 and S_2 are both finite, δ is smooth and transverse to zero.

◊

Thus we can restate the above corollary as saying that $\breve{M}([\mathfrak{a}_0], [\mathfrak{a}_3])$ has a codimension-1 δ-structure along the stratum $\breve{M}_1^{\text{irr}} \times \breve{M}_2 \times \breve{M}_3^{\text{irr}}$. If we recall Proposition 16.5.5, which classifies the codimension-1 strata, we see that the examples so far discussed cover all non-trivial cases in that proposition (i.e. all cases in which the trajectory is broken with more than one component). We therefore have:

Theorem 19.5.4. *Suppose that the moduli space $M_z([\mathfrak{a}], [\mathfrak{b}])$ is d-dimensional and contains irreducible trajectories, so that the moduli space $\breve{M}^+([\mathfrak{a}], [\mathfrak{b}])$ is a $(d-1)$-dimensional space stratified by manifolds. Let $M' \subset \breve{M}^+([\mathfrak{a}], [\mathfrak{b}])$ be any component of the codimension-*1 *stratum. Then, along M', the moduli space $\breve{M}^+([\mathfrak{a}], [\mathfrak{b}])$ either is a C^0 manifold with boundary, or (more generally)* *has a codimension-1 δ-structure in the sense of Definition* 19.5.3. *The latter occurs only when M' consists of 3-component broken trajectories, with the middle component boundary-obstructed.* □

19.6 Gluing reducibles

Everything we have done in this section can be redone in the more restricted context of the moduli spaces $M_z^{\text{red}}([\mathfrak{a}],[\mathfrak{b}])$ of reducible trajectories. When $[\mathfrak{a}]$ is a reducible critical point, the rider to Theorem 18.2.1 says that the parametrization of $\tilde{M}(Z^T)$ is equivariant for the involution $\mathbf{i}$. The fixed-point set in $\tilde{M}(Z^T)$ is $M^{\text{red}}(Z^T)$, and by passing to the fixed-point set throughout one obtains a version of Theorem 18.2.1 for the reducibles. With this tool, everything else is essentially unchanged, with the added simplification that there is no counterpart to the boundary-obstructed situation. Thus, for example, we arrive at the reducible version of Theorem 19.5.4 above.

Theorem 19.6.1. *Suppose that the moduli space $M_z^{\text{red}}([\mathfrak{a}],[\mathfrak{b}])$ is d-dimensional and non-empty, so that the compactified moduli space of broken reducible trajectories, $\breve{M}^{\text{red}+}([\mathfrak{a}],[\mathfrak{b}])$, is a $(d-1)$-dimensional space stratified by manifolds. Let $M' \subset \breve{M}^{\text{red}+}([\mathfrak{a}],[\mathfrak{b}])$ be any component of the codimension-1 stratum. Then, along M', the moduli space $\breve{M}^{\text{red}+}([\mathfrak{a}],[\mathfrak{b}])$ is a C^0 manifold with boundary.* □

Notes and references for Chapter V

Global compactness theorems in gauge theory were pioneered by Uhlenbeck [123, 122] and refined by Sedlacek [102], Donaldson [17] and Taubes [110]. These arguments were applied to the cylindrical case by Floer in [32].

The arguments of Sections 18 and 19 are gluing techniques of the sort first developed by Taubes [55, 108]. The approach we have followed is inspired by [112]. In particular, the treatment of moduli spaces on manifolds with boundary, and the basic concatenation results, all appear there (in the context of the anti-self-dual Yang–Mills equations), as does the emphasis on the Atiyah–Patodi–Singer spectral boundary-value problems.

VI
Floer homology

This chapter presents the construction of the Floer homology and cohomology groups of a 3-manifold, and takes the first steps in verifying the formal properties that were described in the introductory Section 3. One thing we will not do in this chapter is prove that the Floer groups are topological invariants: their definition depends on an a priori choice of Riemannian metric and a choice of perturbation for the Chern–Simons–Dirac functional. The topological invariance is an issue that is closely related to the functoriality discussed in Subsection 3.4. It will be taken up in the next chapter.

Another matter that will be postponed for later is the fact that the Floer groups of Y are naturally graded abelian groups, graded by the set of homotopy classes of oriented 2-plane fields on Y. (See Definition 3.1.2.)

20 Orienting moduli spaces

20.1 Discussion

To define Floer homology groups with coefficients that are not of characteristic 2, we will need to orient the moduli spaces of trajectories, $M([\mathfrak{a}], [\mathfrak{b}])$, for arbitrary critical points $[\mathfrak{a}]$ and $[\mathfrak{b}]$ in our configuration space $\mathcal{B}^\sigma_k(Y, \mathfrak{s})$. We will prove in this section that these moduli spaces are indeed orientable, as long as our perturbation is chosen so that the moduli spaces are regular.

Orientability alone is not enough, however. In the case of finite-dimensional Morse theory, discussed in Subsection 2.2, it was important not only that the trajectory spaces were orientable, but that the fiber of the orientation bundle of a trajectory space $M(a, b)$ could be identified with the product of the orientation bundles for the two unstable manifolds, U_a and U_b. This "factorization" of the orientation bundle for $M(a,b)$ played an important role in the construction of the Morse complex.

In the Floer homology setting, it is not immediately clear what should play the role of the orientation bundle for the unstable manifolds: unlike the moduli spaces of trajectories, the unstable manifolds are infinite-dimensional. This is one of the matters we must take up as we examine the orientability of $M([\mathfrak{a}],[\mathfrak{b}])$.

We can express the desired factorization property rather abstractly as follows. The set of orientations for a real vector space V is a 2-element set. Given two 2-element sets, say Λ_1 and Λ_2, we can form a product

$$\Lambda = \Lambda_1 \Lambda_2 \stackrel{\text{def}}{=} \Lambda_1 \times_{\mathbb{Z}/2} \Lambda_2, \tag{20.1}$$

where $\mathbb{Z}/2$ acts as the non-trivial involution on both factors. If Λ is the set of orientations for V and $V = V_1 \oplus V_2$ then the product above corresponds to the way orientations of the summands induce an orientation for V by the "first-summand-first" convention.

As before we fix a spin^c structure $\mathfrak{s}$ and we write $\mathcal{B}_k(Y)$ etc. for the configuration spaces corresponding to this spin^c structure. Given a configuration $[\mathfrak{a}] \in \mathcal{B}_k^\sigma(Y)$ and a tame perturbation $\mathfrak{q}$ we will define a 2-element set

$$\Lambda([\mathfrak{a}],\mathfrak{q}).$$

We do not assume that $\mathfrak{q}$ satisfies any transversality conditions nor do we assume that $[\mathfrak{a}]$ is a critical point. We write

$$\Lambda([\mathfrak{a}_1],\mathfrak{q}_1,[\mathfrak{a}_2],\mathfrak{q}_2) = \Lambda([\mathfrak{a}_1],\mathfrak{q}_1)\Lambda([\mathfrak{a}_2],\mathfrak{q}_2)$$

using the notation from (20.1) above; and when the perturbations are understood from the context we will abbreviate the notations to

$$\Lambda([\mathfrak{a}]) \quad \text{and} \quad \Lambda([\mathfrak{a}_1],[\mathfrak{a}_2]).$$

The 2-element sets $\Lambda([\mathfrak{a}],\mathfrak{q})$ will play the role of "orientation of the unstable manifold" (though the definition applies even when $[\mathfrak{a}]$ is not a critical point). They will be defined in such a way that if $[\mathfrak{a}_1]$, $[\mathfrak{a}_2]$ are non-degenerate critical points of $(\operatorname{grad} \mathcal{L}_\mathfrak{q})^\sigma$ and the moduli space $M([\mathfrak{a}_1],[\mathfrak{a}_2])$ is regular, then the orientation double cover of the moduli space $M([\mathfrak{a}_1],[\mathfrak{a}_2])$ is canonically identified with the product $\Lambda([\mathfrak{a}_1],[\mathfrak{a}_2]) \times M([\mathfrak{a}_1],[\mathfrak{a}_2])$. In other words $M([\mathfrak{a}_1],[\mathfrak{a}_2])$ is orientable and a choice of element from $\Lambda([\mathfrak{a}_1],[\mathfrak{a}_2])$ orients $M([\mathfrak{a}_1],[\mathfrak{a}_2])$. This is the counterpart to the way in which trajectory spaces in finite-dimensional

Morse theory are oriented by choosing orientations of the two unstable manifolds. These constructions will be carried out in Subsections 20.3 and 20.4 below.

There is another type of product rule for orientations of trajectory spaces that is relevant in the construction of the usual Morse complex. This involves the relationship between the orientations of $M(a, b)$, $M(b, c)$ and $M(a, c)$ for three critical points a, b and c. This relationship, which is used in the proof that the square of the differential on the Morse complex is zero, will be discussed in the context of our trajectory spaces in Subsection 20.5.

The whole discussion is rather more complicated than the simplest finite-dimensional model, not just because we are dealing with an infinite-dimensional situation, but because our configuration space $\mathcal{B}_k^\sigma(Y, \mathfrak{s})$ is a manifold with boundary. (Compare Subsection 2.4.) Our "regular" moduli spaces include those that are boundary-obstructed in the sense of Definition 2.4.2, and we will therefore need to consider various different versions of the relevant product laws.

20.2 Determinant lines and direct sums

We begin by recalling Quillen's construction of the determinant line bundle of a family of Fredholm operators, following Bismut and Freed [12]. Let

$$P_s : A \to B, \quad s \in S,$$

be a family of Fredholm operators between two real Hilbert spaces, parametrized by a space S, and suppose that one can choose a finite-dimensional subspace $J \subset B$ such that $P_s(A) + J = B$ for all s. (This can always be done if S is compact, and in general we can achieve this if we restrict to a sufficiently small neighborhood of a given point in S.) Then the spaces $P_s^{-1}(J)$ have constant dimension and form a vector bundle over S. The lines

$$\left(\Lambda^{\max} P_s^{-1}(J)\right) \otimes \left(\Lambda^{\max} J\right)^*$$

form a real line bundle over S. On the other hand, we can identify this line with

$$\det P_s \stackrel{\text{def}}{=} \left(\Lambda^{\max} \ker(P_s)\right) \otimes \left(\Lambda^{\max} \operatorname{coker}(P_s)\right)^*$$

using the exact sequence

$$0 \to \ker P_s \to P_s^{-1} J \to J \to \operatorname{coker} P_s \to 0,$$

because whenever we have an exact sequence of finite-dimensional vector spaces

$$0 \to V_0 \xrightarrow{a_1} V_1 \xrightarrow{a_2} \cdots \xrightarrow{a_{2k+1}} V_{2k+1} \to 0$$

there is a canonical element

$$\eta(a) \in \Lambda^{\max}(V_0) \otimes \Lambda^{\max}(V_1)^* \otimes \cdots \otimes \Lambda^{\max}(V_{2k+1})^*.$$

To define $\eta(a)$, decompose V_i as $W_i \oplus a_i(W_{i-1})$ so that, where W_i is a chosen complement of the kernel of a_{i+1}. Choose $s_i \in \Lambda^{\max} W_i$, and define

$$\eta(a) = (s_0) \otimes (a_1 s_0 \wedge s_1)^* \otimes (s_2 \wedge a_2 s_1) \otimes \cdots \otimes (a_{2k+1} s_{2k} \wedge s_{2k+1})^*.$$

(In this formula, $(a_1 s_0 \wedge s_1)^*$ denotes the basis element for the dual space of $\Lambda^{\max}(V_1)$ which evaluates to 1 on $a_1 s_0 \wedge s_1$.) This element $\eta(a)$ is independent of the choice of the s_i and the W_i. Via this canonical identification, the family of lines $\det(P_s)$ becomes a line bundle. The topology on the line bundle is independent of the choice of J. In the non-compact case, the family of lines $\det(P_s)$ acquires the structure of a line bundle over neighborhoods of every point in S, and these topologies agree on the overlaps; so the union is once again a line bundle.

It is important to note that the topology that we have defined on the family of lines $\det(P_s)$ depends on the operators P_s themselves, not just on the family of kernels and cokernels as subspaces of A and B. This issue arises if we compare the determinant line bundles of the family of operators $-P_s$ with that of P_s. For each fixed s, we have a canonical identification

$$\begin{aligned} i_s :\ & \left(\Lambda^{\max} \ker(P_s)\right) \otimes \left(\Lambda^{\max} \operatorname{coker}(P_s)\right)^* \\ &= \left(\Lambda^{\max} \ker(-P_s)\right) \otimes \left(\Lambda^{\max} \operatorname{coker}(-P_s)\right)^* \end{aligned}$$

simply because the kernels and cokernels are the same. However, the family of maps i_s is not continuous as a map of determinant line bundles. To obtain a continuous map, we can replace i_s by

$$j_s = (-1)^{\dim \operatorname{coker}(P_s)} i_s. \tag{20.2}$$

The reader can verify from the above constructions that j_s, so defined, is indeed continuous. Of course, $-j_s$ is another continuous map.

There is a product rule for determinant lines of Fredholm operators, which expresses the determinant line of a direct sum as the tensor product of the determinants. To explain the construction, we consider a family of Fredholm operators expressed as a direct sum

$$\begin{aligned} P_s &= P'_s \oplus P''_s \\ P_s &: A' \oplus A'' \to B' \oplus B''. \end{aligned}$$

After choosing J' and J'', we have identifications

$$\begin{aligned} \det P'_s &= \left(\Lambda^{\max}(P'_s)^{-1}(J')\right) \otimes \left(\Lambda^{\max} J'\right)^* \\ \det P''_s &= \left(\Lambda^{\max}(P''_s)^{-1}(J'')\right) \otimes \left(\Lambda^{\max} J''\right)^*. \end{aligned}$$

Setting $J = J' \oplus J''$ and using the isomorphisms

$$\begin{aligned} \Lambda^{\max}(P'_s)^{-1}(J') \otimes \Lambda^{\max}(P''_s)^{-1}(J'') &\to \Lambda^{\max} P_s^{-1}(J) \\ \Lambda^{\max} J' \otimes \Lambda^{\max} J'' &\to \Lambda^{\max} J \end{aligned}$$

given by $x' \otimes x'' \mapsto x' \wedge x''$, we obtain an isomorphism

$$\bar{q}(J', J'') : \det P'_s \otimes \det P''_s \to \det P_s,$$

which is a continuous isomorphism of line bundles over S. This isomorphism, however, depends on J', and not just up to homotopy: its sign depends on the dimension of J' if the index of P''_s is odd. To obtain an isomorphism that is independent of choices (up to homotopy), we set

$$\begin{aligned} q &= (-1)^r \bar{q}(J', J'') \\ r &= \dim J' \operatorname{index} P''_s. \end{aligned}$$

If α', β' are elements of $\Lambda^{\max} \ker P'_s$ and $\Lambda^{\max} \operatorname{coker} P'_s$, and if α'', β'' are similarly elements of $\Lambda^{\max} \ker P''_s$ and $\Lambda^{\max} \operatorname{coker} P''_s$, then (up to a positive scalar multiple) this q is the map

$$\begin{aligned} q &: \det P'_s \otimes \det P''_s \to \det P_s \\ \left(\alpha' \otimes (\beta')^*\right) \otimes \left(\alpha'' \otimes (\beta'')^*\right) &\mapsto (-1)^p \left((\alpha' \wedge \alpha'') \otimes (\beta' \wedge \beta'')^*\right) \end{aligned} \tag{20.3}$$

where

$$p = \dim \operatorname{coker} P'_s \times \operatorname{index} P''_s.$$

Note that in the special case that P'_s and P''_s are surjective and have positive index, this becomes the standard convention: chosen bases for $\ker P'_s$ and $\ker P''_s$ determine orientations for $\det P'_s$ and $\det P''_s$, and the corresponding orientation of $\det(P'_s \oplus P''_s)$ is obtained from the basis for $\ker(P'_s \oplus P''_s)$ obtained by taking first the basis vectors for the first summand, then the basis vectors for the second. The reader can verify the following:

Lemma 20.2.1. *The map q in* (20.3) *satisfies the natural associativity law for an operator expressed as an ordered direct sum of three terms,*

$$P = P' \oplus P'' \oplus P'''.$$

□

20.3 Construction of orientation sets

We now use the formalities of determinant line bundles to carry out the first half of construction promised in Subsection 20.1. Namely, to each pair $([\mathfrak{a}], \mathfrak{q})$, we will associate a 2-element set $\Lambda([\mathfrak{a}], \mathfrak{q})$. In Subsection 20.4 below, we will show that these have the property that, when $\mathfrak{q}_1 = \mathfrak{q}_2$ and $[\mathfrak{a}_1]$ and $[\mathfrak{a}_2]$ are critical points, the 2-element product set

$$\Lambda([\mathfrak{a}_1], \mathfrak{q}_1, [\mathfrak{a}_2], \mathfrak{q}_2) = \Lambda([\mathfrak{a}_1], \mathfrak{q}_1)\Lambda([\mathfrak{a}_2], \mathfrak{q}_2)$$

orients the corresponding moduli space of trajectories $M([\mathfrak{a}_1], [\mathfrak{a}_2])$, as long as this space is regular. The natural way to proceed is actually to define $\Lambda([\mathfrak{a}_1], \mathfrak{q}_1, [\mathfrak{a}_2], \mathfrak{q}_2)$ *first*, as the set of orientations of a suitable determinant line. The 2-element set $\Lambda([\mathfrak{a}], \mathfrak{q})$ can then be defined in terms of $\Lambda([\mathfrak{a}], \mathfrak{q}, [\mathfrak{a}_0], \mathfrak{q}_0)$ for suitable reference configurations $([\mathfrak{a}_0], \mathfrak{q}_0)$.

To proceed, let $\mathfrak{a}_1$, $\mathfrak{a}_2$ in $\mathcal{C}^\sigma_k(Y)$ be given, and let $\mathfrak{q}_1$, $\mathfrak{q}_2$ be perturbations. Let $I = [t_1, t_2]$ be an interval, and consider the space $\mathcal{C}$ of all pairs $(\gamma, \mathfrak{p})$, where

- γ is a configuration in $\mathcal{C}^\tau_k(I \times Y)$ whose restriction to the end $\{t_i\} \times Y$ is gauge-equivalent to $\mathfrak{a}_i$, for $i = 1, 2$; and
- $\mathfrak{p}$ is a continuous path in the Banach space $\mathcal{P}$ of tame perturbations, with $\mathfrak{p}(t_i) = \mathfrak{p}_i$, for $i = 1, 2$.

Associated to any $(\gamma, \mathfrak{p})$ is the operator $Q_{\gamma,\mathfrak{p}}$, the linearized Seiberg–Witten equations and gauge-fixing equation (17.10), considered here on a finite cylinder $I \times Y$ and acting on Sobolev spaces L^2_1. We write the operator as

$$Q_{\gamma,\mathfrak{p}} : \mathcal{E} \to \mathcal{F}$$

where

$$\begin{aligned}\mathcal{E} &= \mathcal{T}^{\tau}_{1,\gamma}(I \times Y) \\ \mathcal{F} &= \mathcal{V}^{\tau}_{0,\gamma}(I \times Y) \oplus L^2(I \times Y; i\mathbb{R}).\end{aligned} \tag{20.4}$$

We have made the perturbation time-dependent, but this is a minor change.

To obtain a Fredholm operator, we combine $Q_{\gamma,\mathfrak{p}}$ with a suitable boundary condition, as in Lemma 17.3.3. At the boundary components $\{t_i\} \times Y$, we again decompose

$$\mathcal{T}^{\sigma}_{1/2,\mathfrak{a}_i}(Y) \oplus L^2_{1/2}(Y; i\mathbb{R}) = \mathcal{J}^{\sigma}_{1/2,\mathfrak{a}_i} \oplus \mathcal{K}^{\sigma}_{1/2,\mathfrak{a}_i} \oplus L^2_{1/2}(Y; i\mathbb{R})$$

and we write

$$\begin{aligned}H_i^- &= \{0\} \oplus \mathcal{K}^-_{1/2,\mathfrak{a}_i} \oplus L^2_{1/2}(Y; i\mathbb{R}) \\ H_i^+ &= \{0\} \oplus \mathcal{K}^+_{1/2,\mathfrak{a}_i} \oplus L^2_{1/2}(Y; i\mathbb{R})\end{aligned} \tag{20.5}$$

where, as before, we have decomposed $\mathcal{K}^{\sigma}_{1/2,\mathfrak{a}_i}$ as

$$\mathcal{K}^{\sigma}_{1/2,\mathfrak{a}_i} = \mathcal{K}^-_{1/2,\mathfrak{a}_i} \oplus \mathcal{K}^+_{1/2,\mathfrak{a}_i}$$

using the spectral subspaces of the Hessian $\mathrm{Hess}^{\sigma}_{\mathfrak{q}_i,\mathfrak{a}_i}$, as at (17.12). As before, because we do not know that the operator is hyperbolic, it is actually the spectral subspaces of $\mathrm{Hess}^{\sigma}_{\mathfrak{q}_i,\mathfrak{a}_i} - \epsilon$ (for small positive ϵ) that we use to define this decomposition. We write $\Pi^-_{Y,\mathfrak{a}_i}$ and $\Pi^+_{Y,\mathfrak{a}_i}$ for the projections to H_i^- and H_i^+ with kernels

$$\begin{aligned}\ker(\Pi^-_{Y,\mathfrak{a}_i}) &= \mathcal{J}^{\sigma}_{1/2,\mathfrak{a}_i} \oplus \mathcal{K}^+_{1/2,\mathfrak{a}_i} \oplus \{0\} \\ \ker(\Pi^+_{Y,\mathfrak{a}_i}) &= \mathcal{J}^{\sigma}_{1/2,\mathfrak{a}_i} \oplus \mathcal{K}^-_{1/2,\mathfrak{a}_i} \oplus \{0\}.\end{aligned}$$

These are not complementary projections. Note that changing the orientation of Y interchanges the $\Pi^-_{Y,\mathfrak{a}_i}$ and $\Pi^+_{Y,\mathfrak{a}_i}$ *if* the operator $\mathrm{Hess}^{\sigma}_{\mathfrak{q}_i,\mathfrak{a}_i}$ is hyperbolic: if this operator has zero eigenvalues, then the symmetry is broken by our choice of sign for ϵ above. For convenience, we use the shorter notation $\Pi_i^{\pm}$ for the composites of these projections with the restriction maps r_i to $\{t_i\} \times Y$:

$$\begin{aligned}\Pi_i^- &= \Pi^-_{Y,\mathfrak{a}_i} \circ r_i : \mathcal{E} \to H_i^- \\ \Pi_i^+ &= \Pi^+_{Y,\mathfrak{a}_i} \circ r_i : \mathcal{E} \to H_i^+.\end{aligned}$$

Lemma 17.3.3 tells us that the operator

$$\begin{aligned} P_{\gamma,\mathfrak{p}} &= (Q_{\gamma,\mathfrak{p}}, -\Pi_1^+, \Pi_2^-) \\ \mathcal{E} &\to \mathcal{F} \oplus H_1^+ \oplus H_2^- \end{aligned} \tag{20.6}$$

is Fredholm.

We define $\Lambda(\gamma, \mathfrak{p})$ to be the set of orientations of the 1-dimensional real vector space $\det(P_{\gamma,\mathfrak{p}})$. As γ and $\mathfrak{p}$ vary, these 2-element sets define a double cover of $\mathcal{C}$: this is the orientation bundle of the determinant line bundle of this family of Fredholm operators. The gauge-group $\mathcal{G}_{k+1}(I \times Y)$ acts naturally on the determinant line bundle, and acts freely on the base $\mathcal{C}$; so the double cover descends to a double cover of the quotient $\mathcal{C}/\mathcal{G}_{k+1}$. There is fibration

$$\mathcal{C}/\mathcal{G}_{k+1} \to \mathcal{B}_k^\tau(I \times Y; [\mathfrak{a}_1], [\mathfrak{a}_2])$$

with base

$$\mathcal{B}_k^\tau(I \times Y; [\mathfrak{a}_1], [\mathfrak{a}_2]) \stackrel{\text{def}}{=} \{ [\gamma] \in \mathcal{B}_k^\tau(I \times Y) \mid R_i[\gamma] = [\mathfrak{a}_i],\ i = 1, 2 \}.$$

The fibers of this fibration are affine spaces, so the double cover of $\mathcal{C}/\mathcal{G}_{k+1}$ which we have constructed gives rise to a double cover

$$\Lambda \to \mathcal{B}_k^\tau(I \times Y; [\mathfrak{a}_1], [\mathfrak{a}_2]). \tag{20.7}$$

Definition 20.3.1. For $[\gamma]$ in $\mathcal{B}_k^\tau(I \times Y; [\mathfrak{a}_1], [\mathfrak{a}_2])$, we write

$$\Lambda_{[\gamma]}([\mathfrak{a}_1], \mathfrak{q}_1, [\mathfrak{a}_2], \mathfrak{q}_2)$$

for the 2-element set given by the fiber of the above double cover. Thus an element of the fiber $\Lambda_{[\gamma]}([\mathfrak{a}_1], \mathfrak{q}_1, [\mathfrak{a}_2], \mathfrak{q}_2)$ is the same thing as a continuous, gauge-invariant choice of orientations for the determinant lines $\det(P_{\gamma,\mathfrak{p}})$, as γ runs through all representatives of the gauge-equivalence class and $\mathfrak{p}$ runs through all perturbations with the given boundary conditions. ◊

The product law for the determinant line of a direct sum gives rise to a composition law for the double covers Λ. We consider an interval I expressed as the union of $I_1 = [t_1, t_2]$ and $I_2 = [t_2, t_3]$, and a configuration γ in $\mathcal{C}_k^\tau(I \times Y)$ whose restrictions define two configurations γ_1 and γ_2. We take also a time-dependent perturbation $\mathfrak{p}$ on I, and write $\mathfrak{p}_i$ for its restriction to I_i. Write

$$\begin{aligned} \mathcal{E}_i &= \mathcal{E}(I_i \times Y) \\ \mathcal{F}_i &= \mathcal{F}(I_i \times Y) \end{aligned}$$

for $i = 1, 2$ as in (20.4), and let $H_i^{\pm}$ be the spectral subspaces at $\{t_i\} \times Y$ for $i = 1, 2, 3$. Then we have operators

$$P_{\gamma_i, \mathfrak{p}_i} : \mathcal{E}_i \to \mathcal{F}_i \oplus H_i^+ \oplus H_{i+1}^-$$

for $i = 1, 2$, and

$$P_{\gamma, \mathfrak{p}} : \mathcal{E} \to \mathcal{F} \oplus H_1^+ \oplus H_3^-.$$

To avoid clutter, we omit the perturbations $\mathfrak{p}$ etc. from our notation of the operators P_γ and Q_γ in what follows. A solution to $Q_\gamma u = 0$ gives rise, by restriction, to solutions u_i to $Q_{\gamma_i} u_i = 0$ for $i = 1, 2$; and u_1, u_2 have the same boundary value in

$$\mathcal{H}_2 = \mathcal{T}^{\tau}_{1/2, \mathfrak{a}_2}(\{t_2\} \times Y) \oplus L^2_{1/2}(Y; i\mathbb{R}).$$

Thus an element u of the kernel of P_γ gives an element (u_1, u_2) in the kernel of

$$\begin{aligned} \tilde{P} : \mathcal{E}_1 \oplus \mathcal{E}_2 &\to \mathcal{F}_1 \oplus \mathcal{F}_2 \oplus H_1^+ \oplus H_3^- \oplus \mathcal{H}_2 \\ (u_1, u_2) &\mapsto (Q_{\gamma_1} u_1, Q_{\gamma_2} u_2, -\Pi_1^+ u_1, \Pi_3^- u_2, r_2 u_1 - r_2 u_2), \end{aligned}$$

where r_2 is the restriction to $\{t_2\} \times Y$. By the arguments of the proof of Proposition 17.2.8 in fact, the kernel and cokernel of $\tilde{P}$ are the same as those of P_γ, so providing an identification

$$\det P_\gamma = \det \tilde{P}.$$

On the other hand, the operator $P_{\gamma_1} \oplus P_{\gamma_2}$ has the same domain and codomain as $\tilde{P}$, and is given by

$$\begin{aligned} P_{\gamma_1} \oplus P_{\gamma_2} : \mathcal{E}_1 \oplus \mathcal{E}_2 &\to \mathcal{F}_1 \oplus \mathcal{F}_2 \oplus H_1^+ \oplus H_3^- \oplus H_2^- \oplus H_2^+ \\ (u_1, u_2) &\mapsto (Q_{\gamma_1} u_1, Q_{\gamma_2} u_2, -\Pi_1^+ u_1, \Pi_3^- u_2, \Pi_2^- u_1, -\Pi_2^+ u_2). \end{aligned}$$

Thus the operators $P_{\gamma_1} \oplus P_{\gamma_2}$ and $\tilde{P}$ differ only in the last components, where we see, respectively, the operators obtained by composing the restriction map (r_2, r_2) to $\mathcal{H}_2 \oplus \mathcal{H}_2$ with two different projections:

$$\begin{aligned} A_0 &: (v_1, v_2) \mapsto (\Pi^-_{Y, \mathfrak{a}_2} v_1, -\Pi^+_{Y, \mathfrak{a}_2} v_2) \\ A_1 &: (v_1, v_2) \mapsto r_2 v_1 - r_2 v_2. \end{aligned}$$

The kernels of these are

$$\begin{aligned}\ker(A_0) &= \left(\mathcal{J}^{\sigma}_{1/2,\mathfrak{a}_2} \oplus \mathcal{K}^{+}_{1/2,\mathfrak{a}_2} \oplus \{0\}\right) \oplus \left(\mathcal{J}^{\sigma}_{1/2,\mathfrak{a}_2} \oplus \mathcal{K}^{-}_{1/2,\mathfrak{a}_2} \oplus \{0\}\right)\\ &\subset \mathcal{H}_2 \oplus \mathcal{H}_2\\ \ker(A_1) &= \{\, (v_1, v_2) \mid v_1 = v_2 \,\}\\ &\subset \mathcal{H}_2 \oplus \mathcal{H}_2.\end{aligned}$$

On the other hand, we can use the operator

$$\tilde{L}_{\mathfrak{a}_2} = \begin{bmatrix} 0 & 0 & \mathbf{d}^{\sigma}_{\mathfrak{a}_2} \\ 0 & \mathrm{Hess}^{\sigma}_{\mathfrak{q},\mathfrak{a}_2} & 0 \\ \mathbf{d}^{\sigma,\dagger}_{\mathfrak{a}_2} & 0 & 0 \end{bmatrix}$$

as at (17.13) to give a spectral decomposition: an internal direct sum

$$\mathcal{H}_2 = \mathcal{H}^- + \mathcal{H}^+.$$

The kernels of both A_0 and A_1 can be described as graphs of linear maps

$$\mathcal{H}^+ \oplus \mathcal{H}^- \to \mathcal{H}^- \oplus \mathcal{H}^+.$$

In the case of A_0, the linear map is constructed as in the proof of Lemma 17.3.3, while for A_1 the map is $(x, y) \mapsto (y, x)$. Using this description, we can identify the codomain of both A_0 and A_1 with $\mathcal{H}^- \oplus \mathcal{H}^+$, and they are then homotopic through a path of projections A_s all of whose kernels are transverse to $\mathcal{H}^- \oplus \mathcal{H}^+$.

We now have a family of operators

$$P_s = (Q_{\gamma_1} \oplus Q_{\gamma_2}, -\Pi^+_1, \Pi^-_3, A_s \circ (r_2, r_2))$$

between $P_{\gamma_1} \oplus P_{\gamma_2}$ and $\tilde{P}$. All of these operators are Fredholm, by the argument of the proof of Proposition 17.2.6, because the kernels of A_s are transverse to the spectral subspace $\mathcal{H}^- \oplus \mathcal{H}^+$. This homotopy P_s gives a canonical identification

$$\det \tilde{P} = \det(P_{\gamma_1} \oplus P_{\gamma_2}),$$

up to a positive real scalar factor. We combine these last two canonical identifications with our earlier discussion concerning the determinant of a sum of operators, and we obtain an isomorphism, well-defined up to homotopy,

$$q : \det P_{\gamma_1} \otimes \det P_{\gamma_2} \to \det P_{\gamma}. \tag{20.8}$$

This q is gauge-invariant, and depends continuously on $\mathfrak{p}$, so it descends to a map on the spaces $\Lambda_{[\gamma_1]}([\mathfrak{a}_1], \mathfrak{q}_1, [\mathfrak{a}_2], \mathfrak{q}_2)$ etc., a fact which we summarize with a definition:

Definition 20.3.2. The map

$$\begin{aligned} q: \ & \Lambda_{[\gamma_1]}([\mathfrak{a}_1], \mathfrak{q}_1, [\mathfrak{a}_2], \mathfrak{q}_2)\Lambda_{[\gamma_2]}([\mathfrak{a}_2], \mathfrak{q}_2, [\mathfrak{a}_3], \mathfrak{q}_3) \\ & \to \Lambda_{[\gamma]}([\mathfrak{a}_1], \mathfrak{q}_1, [\mathfrak{a}_3], \mathfrak{q}_3) \end{aligned} \tag{20.9}$$

is defined by choosing a gauge representative for $[\gamma]$ and suitable paths of perturbations, and then applying the composition map (20.8). ◊

Remark. The above construction establishes not only the multiplicative law (20.8) for the determinants, but also an additive law for the indices of these operators, by an argument which eventually makes no reference to spectral flow. The same sort of homotopy can be used to prove additivity in the case that $S^1 \times Y$ is decomposed into two copies of $I \times Y$. In this case, the index of an operator P on the closed manifold can be expressed as the sum of the indices of the two spectral boundary-value problems. This is an argument that was referred to earlier, in the proof of Proposition 14.2.2.

We wish next to define $\Lambda([\mathfrak{a}], \mathfrak{q})$ in terms of $\Lambda_{[\gamma]}([\mathfrak{a}], \mathfrak{q}, [\mathfrak{a}_0], \mathfrak{q}_0,)$ for suitable reference configurations $([\mathfrak{a}_0], \mathfrak{q}_0)$. The obvious choice for the reference configurations is those with $\mathfrak{a}_0$ reducible and $\mathfrak{q}_0 = 0$. For this to work, we need to consider two related issues: the independence of the construction on the choice of $[\mathfrak{a}_0]$ and the choice of $[\gamma]$. This leads us to examine the situation of reducible trajectories in some detail.

So we return to the case of a single interval I, and consider now the special case that $[\gamma] \in \mathcal{B}_k^\tau(I \times Y; [\mathfrak{a}_1], [\mathfrak{a}_2])$ is reducible and $\mathfrak{q}_1 = \mathfrak{q}_2 = 0$. In this reducible case, we can define a canonical trivialization of $\Lambda_{[\gamma]}$: a map

$$\tau: \Lambda_{[\gamma]}([\mathfrak{a}_1], 0, [\mathfrak{a}_2], 0) \to \mathbb{Z}/2. \tag{20.10}$$

To define τ, we take the time-dependent perturbation $\mathfrak{p}$ to be zero. The operator Q_γ belonging to a corresponding 4-dimensional configuration on $[t_1, t_2] \times Y$ has the form

$$Q_\gamma = \frac{D}{dt} + L(t),$$

where $L(t)$ is a time-dependent operator on

$$\mathcal{T}^\sigma_{1,\gamma(t)} \oplus L^2_1(Y; i\mathbb{R}),$$

and D/dt is as in (14.11). For the case of a reducible, the shape of the operator $L(t)$ is described quite explicitly by the matrix (12.19); by scaling the off-diagonal terms x_i in the matrix (12.19) to zero, we can deform $L(t)$ to an operator which is the direct sum of the diagonal blocks (12.20). In this way, the Fredholm operator

$$P = (Q_\gamma, -\Pi_1^+, \Pi_2^-)$$

can be deformed through Fredholm operators to an operator

$$P' = (Q'_\gamma, -\Pi_1^+, \Pi_2^-),$$

where Q'_γ is the direct sum of the following pieces:

(i) the operator

$$\frac{d}{dt} + \lambda(t)$$

acting on $L^2_1(I; \mathbb{R})$;

(ii) the operator

$$\frac{d}{dt} + \begin{bmatrix} 0 & -1 \\ -1 & 0 \end{bmatrix}$$

acting on $L^2_1(I; \mathbb{R} \oplus \mathbb{R})$;

(iii) the operator

$$\frac{d}{dt} + \begin{bmatrix} 0 & -d^* \\ -d & 0 \end{bmatrix}$$

where for fixed t the matrix is interpreted as an operator on pairs $(\hat{c}, b)$, where c is an imaginary-valued function orthogonal to the constants and b is an exact 1-form;

(iv) the operator

$$\frac{d}{dt} + *d$$

where $*d$ is acting on the coclosed 1-forms on Y;

(v) the operator

$$\frac{D}{dt} + \Pi^\perp_{\mathbb{C}} D_{B(t)}$$

acting on sections along γ of the bundle of spinors L^2-orthogonal to $\phi(t)$. Here D/dt is the covariant derivative in this bundle along γ defined by L^2 projection.

Furthermore, the boundary operators Π_1^+ and Π_2^- become direct sums also; so that P' is a direct sum of five corresponding operators, $P'_1, \dots, P'_5$.

In the case of P'_2 and P'_3, the boundary projections are the projections to the first component at both ends (up to sign), and these operators are therefore invertible, as is easily verified. In the case of P'_4, the projections are complementary spectral projections at the two ends; and because the corresponding part of Q'_γ is translation-invariant, the operator P'_4 is again invertible. The operator P'_5 is a complex-linear operator. We specify the canonical complex orientation for its determinant.

Thus we see that an orientation for $\det(P)$ is canonically determined by specifying an orientation for the operator P'_1, which is the operator $d/dt + \lambda(t)$ augmented with certain boundary conditions.

We examine this 1-dimensional block in more detail. The corresponding summand of Q'_γ is the part which appears as the part "normal" to the boundary of the blown-up configuration space: it was considered in the context of the infinite cylinder in Lemma 14.5.4, where the differential operator was called Q^ν_γ. Accordingly we write

$$P^\nu_\gamma = P'_1$$

for this part of the operator P'. There are four cases, according to the sign of the real-valued function $\lambda(t)$ at $t = t_1$ and $t = t_2$. In each of the four cases, we will specify a preferred orientation of the determinant line. For $i = 1, 2$, the spectral subspace H_i^+ is either $\mathbb{R}$ or 0, according as $\lambda(t_i) > 0$ or $\lambda(t_i) \leq 0$ respectively, with the complementary rule for H_i^-. The four cases are as follows.

(i) If $\lambda(t_1)$ and $\lambda(t_2)$ are both positive, the operator is invertible, and we take the canonical orientation for the determinant.

(ii) If $\lambda(t_1)$ and $\lambda(t_2)$ are both non-positive, the operator is again invertible, and we take the canonical orientation.

(iii) In the case that $\lambda(t_1) \leq 0$ and $\lambda(t_2) > 0$, the operator is

$$P^\nu = (d/dt + \lambda) : L_1^2 \to L^2 \tag{20.11}$$

on $[t_1, t_2]$. This operator has 1-dimensional kernel and no cokernel. We orient the determinant by picking a basis vector for the kernel represented by a *negative* function $u(t)$ on $[t_1, t_2]$.

(iv) When $\lambda(t_1) > 0$ and $\lambda(t_2) \leq 0$, the operator is

$$P^{\nu} : L_1^2 \to L^2 \oplus \mathbb{R} \oplus \mathbb{R},$$
$$u \mapsto (du/dt + \lambda(t)u, -u(t_1), u(t_2)).$$

The kernel is zero, and the cokernel is 1-dimensional. The element

$$(0, 0, -1) \in L^2 \oplus \mathbb{R} \oplus \mathbb{R} \tag{20.12}$$

is not in the image of P^{ν}, so represents a basis vector for the cokernel. We use this basis element to orient the cokernel, and hence the determinant of the operator.

Having chosen preferred orientations in each case, we define the trivialization τ as in (20.10) so that it sends the preferred orientation to $0 \in \mathbb{Z}/2$. The reason for selecting these particular rules for defining τ is that we wish these trivializations of the determinant lines to be compatible with composition law q. The reader can verify that the rule for orienting the cokernel in the last of the four cases is thus forced on us by our rule for orienting the kernel in the previous case. Thus:

Lemma 20.3.3. *The chosen trivializations*

$$\tau : \Lambda_{[\gamma]}([\mathfrak{a}_1], 0, [\mathfrak{a}_2], 0) \to \mathbb{Z}/2$$

for reducible trajectories, obtained from the above choices, are consistent with the product rule for composite trajectories provided by the map q in (20.9). □

Next we prove the triviality of the double covers we have constructed:

Proposition 20.3.4. *For any choice of $\mathfrak{a}_1, \mathfrak{a}_2$ and perturbations $\mathfrak{q}_1$, $\mathfrak{q}_2$, the double cover*

$$\Lambda \to \mathcal{B}_k^{\tau}(I \times Y; [\mathfrak{a}_1], [\mathfrak{a}_2])$$

whose fibers are the sets $\Lambda_{[\gamma]}([\mathfrak{a}_1], \mathfrak{q}_1, [\mathfrak{a}_2], \mathfrak{q}_2)$ is a trivial double cover.

Proof. Any $[\gamma]$ in $\mathcal{B}_k^{\tau}(I \times Y; [\mathfrak{a}_1], [\mathfrak{a}_2])$ determines a continuous path ζ joining $[\mathfrak{a}_1]$ to $[\mathfrak{a}_2]$ in $\mathcal{B}_j^{\sigma}(Y)$, for $j \leq k-1$, and this assignment is a weak homotopy equivalence. Thus Λ is equivalent to a double cover

$$\Lambda \to \Omega\big(\mathcal{B}^{\sigma}(Y); [\mathfrak{a}_1], [\mathfrak{a}_2]\big)$$

of the space of continuous paths with fixed endpoints. To show that this double cover is trivial, it is enough to consider the case that $\mathfrak{a}_1$ and $\mathfrak{a}_2$ are both equal to a single reducible configuration $\mathfrak{a}_0$ and the perturbations are 0: for the general case, we can join $[\mathfrak{a}_1]$ and $[\mathfrak{a}_2]$ to $[\mathfrak{a}_0]$ by fixed paths, and then use the composition law q from (20.9). The inclusion of the reducibles $\partial\mathcal{B}^\sigma(Y)$ in $\mathcal{B}^\sigma(Y)$ is a homotopy equivalence, and so therefore is the inclusion

$$\Omega\big(\partial\mathcal{B}^\sigma(Y);[\mathfrak{a}_0],[\mathfrak{a}_0]\big) \hookrightarrow \Omega\big(\mathcal{B}^\sigma(Y);[\mathfrak{a}_0],[\mathfrak{a}_0]\big).$$

It is therefore enough to consider the restriction of the double cover,

$$\Lambda' \to \Omega\big(\partial\mathcal{B}^\sigma(Y);[\mathfrak{a}_0],[\mathfrak{a}_0]\big).$$

The trivialization τ shows that Λ' is trivial. □

Remark. The fundamental group of the space $\Omega(\mathcal{B}^\sigma(Y);[\mathfrak{a}_0],[\mathfrak{a}_0])$ is $\pi_2(\mathcal{B}^\sigma(Y))$, which is $\mathbb{Z}$. This space therefore has one non-trivial double cover.

Now that we know that the double cover is trivial, we can drop $[\gamma]$ from the notation, and simply write

$$\Lambda([\mathfrak{a}_1],\mathfrak{q}_1,[\mathfrak{a}_2],\mathfrak{q}_2)$$

instead of $\Lambda_{[\gamma]}([\mathfrak{a}_1],\mathfrak{q}_1,[\mathfrak{a}_2],\mathfrak{q}_2)$. We can think of an element of the former as a section of the trivial double cover Λ from the proposition. We have composition maps

$$q : \Lambda([\mathfrak{a}_1],\mathfrak{q}_1,[\mathfrak{a}_2],\mathfrak{q}_2)\Lambda([\mathfrak{a}_2],\mathfrak{q}_2,[\mathfrak{a}_3],\mathfrak{q}_3) \to \Lambda([\mathfrak{a}_1],\mathfrak{q}_1,[\mathfrak{a}_3],\mathfrak{q}_3), \quad (20.13)$$

arising from (20.9). If we take $([\mathfrak{a}_3],\mathfrak{q}_3) = ([\mathfrak{a}_1],\mathfrak{q}_1)$, we obtain from q a map

$$\rho : \Lambda([\mathfrak{a}_2],\mathfrak{q}_2,[\mathfrak{a}_1],\mathfrak{q}_1) \to \Lambda([\mathfrak{a}_1],\mathfrak{q}_1,[\mathfrak{a}_2],\mathfrak{q}_2) \quad (20.14)$$

characterized by

$$q(\lambda,\rho(\lambda)) = 1$$

where 1 is the canonical orientation in $\Lambda([\mathfrak{a}_2],\mathfrak{q}_2,[\mathfrak{a}_2],\mathfrak{q}_2)$. We also have trivializations

$$\tau : \Lambda([\mathfrak{a}_0],0,[\mathfrak{a}_0'],0) \to \mathbb{Z}/2$$

in the case that $\mathfrak{a}_0$ and $\mathfrak{a}_0'$ are reducible. Combining q and τ, we obtain isomorphisms

$$\begin{aligned}\sigma : \Lambda([\mathfrak{a}], \mathfrak{q}, [\mathfrak{a}_0], 0) &\to \Lambda([\mathfrak{a}], \mathfrak{q}, [\mathfrak{a}_0'], 0)\\ \sigma(\lambda) &= q(\lambda, \tau^{-1}(0)),\end{aligned} \tag{20.15}$$

for any $\mathfrak{a}$, whenever $\mathfrak{a}_0$ and $\mathfrak{a}_0'$ are reducible. If we let $\mathfrak{a}_0$ and $\mathfrak{a}_0'$ run through all the reducibles, the diagram of isomorphisms σ that we obtain is a commutative diagram, by Lemma 20.2.1. In other words, for each $\mathfrak{a}$, we have an equivalence relation on

$$\coprod_{[\mathfrak{a}_0] \text{ reducible}} \Lambda([\mathfrak{a}], \mathfrak{q}, [\mathfrak{a}_0], 0). \tag{20.16}$$

Definition 20.3.5. For $[\mathfrak{a}]$ in $\mathcal{B}_k^\sigma(Y)$ and any tame perturbation $\mathfrak{q}$, we define $\Lambda([\mathfrak{a}], \mathfrak{q})$ to be the 2-element set obtained as the quotient of the above disjoint union by the equivalence relation defined by the isomorphisms σ. ◇

We have now defined both $\Lambda([\mathfrak{a}], \mathfrak{q})$ and $\Lambda([\mathfrak{a}_1], \mathfrak{q}_1, [\mathfrak{a}_2], \mathfrak{q}_2)$ as promised in our introductory remarks to this section, and the composition law provides the required isomorphism

$$\Lambda([\mathfrak{a}_1], \mathfrak{q}_1)\Lambda([\mathfrak{a}_2], \mathfrak{q}_2) \to \Lambda([\mathfrak{a}_1], \mathfrak{q}_1, [\mathfrak{a}_2], \mathfrak{q}_2) \tag{20.17}$$

defined at the level of the representatives in (20.16) by

$$\lambda_1 \times \lambda_2 \mapsto q(\lambda_1, \rho(\lambda_2)).$$

(This construction is compatible with the equivalence relation arising from σ, as it needs to be, to define a map (20.17).)

Remark. Our definition of $\Lambda([\mathfrak{a}], \mathfrak{q})$ has been motivated in part by our wish to mimic as closely as we can the situation in finite-dimensional Morse theory, where one can define $\Lambda(a)$ (for a critical point a of a finite-dimensional Morse function f) to be the set of orientations of the unstable manifold U_a. One could set up finite-dimensional Morse theory differently, taking orientations of the *stable* manifolds S_a as the primary object; but the unstable manifolds are more appropriate when thinking of homology (rather than cohomology). Instead of using $\Lambda([\mathfrak{a}], \mathfrak{q}, [\mathfrak{a}_0], 0)$ as the basis of our definition, we might have used $\Lambda([\mathfrak{a}_0], 0, [\mathfrak{a}], \mathfrak{q})$, which would have more closely resembled an orientation of the stable manifold in the finite-dimensional case.

20.4 Orienting the moduli spaces

Fix a perturbation $\mathfrak{q}$ now, and write $\Lambda([\mathfrak{a}])$ in place of $\Lambda([\mathfrak{a}], \mathfrak{q})$. Our next task is to explain how an element of $\Lambda([\mathfrak{a}_1], [\mathfrak{a}_2])$ determines an orientation of the moduli space $M([\mathfrak{a}_1], [\mathfrak{a}_2])$ when the $\mathfrak{a}_i$ are non-degenerate critical points and the moduli space is regular.

Let $[\gamma]$ be an element of the moduli space $M([\mathfrak{a}_1], [\mathfrak{a}_2])$, and let ζ be the path in $\mathcal{B}_k^\sigma(Y)$ corresponding to γ. Write $Q_\gamma = (d/dt) + L$, as before. We are assuming that the critical points are non-degenerate, so there exists $T_1 \ll 0$ such that $L(t)$ is hyperbolic for all $t \leq T_1$, and there is $T_2 \gg 0$ such that $L(t)$ is hyperbolic for all $t \geq T_2$. Thus $H^-(L(t))$ varies continuously with t for $t \leq T_1$ and $t \geq T_2$. Because of this, we can identify

$$\Lambda([\mathfrak{a}_1], [\mathfrak{a}_2]) = \Lambda(\zeta(t_1), \zeta(t_2)),$$

as long as the interval $[t_1, t_2]$ contains $[T_1, T_2]$, for we have a continuous family of Fredholm operators. Choose $[t_1, t_2]$ satisfying this condition.

The boundary-unobstructed case. Suppose first that the moduli space is not boundary-obstructed. Then the determinant line of $T_{[\gamma]}M([\mathfrak{a}_1], [\mathfrak{a}_2])$ is equal to $\det P_\gamma$, where $P_\gamma = Q_\gamma$ is the operator on the *infinite* cylinder:

$$P_\gamma : L_1^2(\mathbb{R} \times Y; iT^*Y \oplus S \oplus i\mathbb{R}) \to L^2(\mathbb{R} \times Y; iT^*Y \oplus S \oplus i\mathbb{R}).$$

We regard γ as the concatenation of three pieces, γ_-, γ_0 and γ_+, where γ_0 is the restriction of γ to $[t_1, t_2]$, and the other two are the restrictions to $(-\infty, t_1]$ and $[t_2, \infty)$. On the finite interval $[t_1, t_2]$, we have the operator P_{γ_0} from (20.6) whose determinant is $\Lambda(\zeta(t_1), \zeta(t_2))$. We also have Fredholm operators

$$P_{\gamma_-} : L_1^2\big(Z^-; iT^*Y \oplus S \oplus i\mathbb{R}\big) \to L^2\big(Z^-; iT^*Y \oplus S \oplus i\mathbb{R}\big) \oplus H_1^-$$
$$P_{\gamma_+} : L_1^2\big(Z^+; iT^*Y \oplus S \oplus i\mathbb{R}\big) \to L^2\big(Z^+; iT^*Y \oplus S \oplus i\mathbb{R}\big) \oplus H_2^+,$$

where we have written $\mathbb{R} \times Y$ as the union $Z^- \cup Z^0 \cup Z^+$, with $Z^- = (-\infty, t_1] \times$ etc. These last two are invertible operators if t_1 is sufficiently negative and t_2 sufficiently positive, because they approach constant-coefficient operators in operator norm as $|t_i|$ goes to infinity. The determinants of P_{γ_-} and P_{γ_+} are therefore canonically trivial. As in the construction of the map q in (20.8) above, there is a canonical isomorphism

$$\det(P_\gamma) = \det(P_{\gamma_-}) \otimes \det(P_{\gamma_0}) \otimes \det(P_{\gamma_+}).$$

This identifies $\det T_{[\gamma]}M([\mathfrak{a}_1],[\mathfrak{a}_2])$ with $\det(P_{\gamma_0})$, and hence with $\Lambda([\mathfrak{a}_1],[\mathfrak{a}_2])$.

The boundary-obstructed case. In the boundary-obstructed case, the operator P_γ has kernel equal to the tangent space of the moduli space, but also has 1-dimensional cokernel arising from the normal part, Q^ν_γ (see (14.16)), which is the 1-dimensional block $(d/dt)+\lambda$ considered above. Our argument therefore provides an identification

$$\begin{aligned}\Lambda([\mathfrak{a}_1],[\mathfrak{a}_2]) &= \det(P_\gamma)\\ &= \det T_\gamma M \otimes \operatorname{coker}(P_{\gamma_0})^*.\end{aligned}$$

If we wish to identify $\Lambda([\mathfrak{a}_1],[\mathfrak{a}_2])$ with a set of preferred orientations for the moduli spaces, we must therefore fix an orientation for the 1-dimensional cokernel. As before we choose the element (20.12) as a basis for the cokernel, so identifying the fiber of the orientation bundle of the moduli space with $\Lambda([\mathfrak{a}_1],[\mathfrak{a}_2])$.

These particular choices of orientation conventions will be important later, but irrespective of the choices made, we have established, in all cases:

Corollary 20.4.1. *The moduli spaces $M([\mathfrak{a}],[\mathfrak{b}])$ are orientable manifolds, whenever they are regular.* □

The unparametrized moduli spaces. The unparametrized moduli space $\breve{M}([\mathfrak{a}],[\mathfrak{b}])$ is of course orientable also, being the quotient of $M([\mathfrak{a}],[\mathfrak{b}])$ by a free action of $\mathbb{R}$. To fix a specific orientation, we consider $\mathbb{R}$ acting on $M([\mathfrak{a}],[\mathfrak{b}])$ by

$$[\gamma] \mapsto [\tau_t^* \gamma],$$

where τ_t as usual is the map $(s,y) \mapsto (s+t,y)$ on the cylinder. This convention about the action means that the map from the moduli space to $\mathcal{B}^\sigma_k(Y)$ given by $[\gamma] \mapsto [\check{\gamma}(0)]$ is equivariant for the (partially defined) action of $\mathbb{R}$ on $\mathcal{B}^\sigma_k(Y)$ given by the downward gradient flow on $\mathcal{B}_k(Y)$. For a general fiber bundle

$$F \hookrightarrow E \xrightarrow{p} B$$

of smooth manifolds, we adopt the "fiber-first" convention for orientations: that is, at any $x \in E$, we choose a splitting $T_xE = T_xF \oplus T_xB$ and we declare that orientations on these three vector spaces are compatible if they agree with the "first-summand-first" convention. In this way, taking the standard orientation for $\mathbb{R}$, we obtain an orientation on $\breve{M}([\mathfrak{a}], [\mathfrak{b}])$ from an orientation on $M([\mathfrak{a}], [\mathfrak{b}])$. This is the convention we adopt whereby a choice of element from $\Lambda([\mathfrak{a}], [\mathfrak{b}])$ orients $\breve{M}([\mathfrak{a}], [\mathfrak{b}])$.

20.5 Orientations and gluing

We now turn to the question of how our chosen orientations behave under gluing. To consider the simplest case first, suppose that $M_{z_1}([\mathfrak{a}_0], [\mathfrak{a}_1])$ and $M_{z_2}([\mathfrak{a}_1], [\mathfrak{a}_2])$ are two moduli spaces neither of which is boundary-obstructed. We have an inclusion

$$\breve{M}_z^+([\mathfrak{a}_0], [\mathfrak{a}_2]) \supset \breve{M}_{z_1}([\mathfrak{a}_0], [\mathfrak{a}_1]) \times \breve{M}_{z_2}([\mathfrak{a}_1], [\mathfrak{a}_2]),$$

and according to Corollary 19.5.1, the moduli space $\breve{M}_z^+([\mathfrak{a}_0], [\mathfrak{a}_2])$ has the structure of a C^0 manifold with boundary along this stratum. Write Λ_{ij} for $\Lambda([\mathfrak{a}_i], [\mathfrak{a}_j])$ and M_{ij} similarly. Pick elements $\lambda_{01} \in \Lambda_{01}$ and $\lambda_{12} \in \Lambda_{12}$. Set

$$\lambda_{02} = q(\lambda_{01}, \lambda_{12}),$$

where q is the composition map (20.13). Using λ_{01}, λ_{12} and the product λ_{02}, we orient the moduli spaces $\breve{M}_{ij}$.

Having oriented the moduli spaces in this way, we can ask whether the product orientation of $\breve{M}_{01} \times \breve{M}_{12}$ agrees with the orientation it obtains as the boundary of $\breve{M}_{02}^+$ (using the standard outward-normal-first convention). We shall see that these orientations differ by the sign $(-1)^{d_{01}}$, where d_{01} is the dimension of M_{01}. This is one of the cases contained in Proposition 20.5.2 below.

The other case in Proposition 20.5.2 concerns a stratum in $\breve{M}_{03}^+ = \breve{M}_z^+([\mathfrak{a}_0], [\mathfrak{a}_3])$ of the form

$$\breve{M}_{01} \times \breve{M}_{12} \times \breve{M}_{23}$$

under the hypothesis that $\breve{M}_{12}$ (and only $\breve{M}_{12}$) is boundary-obstructed and that $\breve{M}_{03}$ contains irreducibles. According to Corollary 19.5.2 and Definition 19.5.3, $\breve{M}_{03}^+$ has a codimension-1 δ-structure along this stratum:

$$\breve{M}_{03}^+ \supset \breve{M}_{01} \times \breve{M}_{12} \times \breve{M}_{23}.$$

We choose orientations λ_{01}, λ_{12} and λ_{23}; and we then set

$$\lambda_{03} = q(\lambda_{01}\lambda_{12}\lambda_{23})$$

(extending the associative binary operation q to three factors). The notion of a codimension-1 δ-structure is more general than that of a manifold with boundary, but orientations can be compared in a similar way. Recall from Definition 19.5.3 that if N has a codimension-1 δ-structure along M^{d-1}, then we have an open $W \subset N$ and

$$N \supset W \overset{j}{\hookrightarrow} EW,$$

where EW is locally homeomorphic to $M^{d-1} \times (0, \infty]^2$. The image of j is the zero set of a map $\delta : EW \to \mathbb{R}$. If M^{d-1} is oriented, we can orient EW by the local homeomorphism with $M^{d-1} \times (0, \infty]^2$, and then W can be oriented as the fiber of δ (with the "fiber-coordinates-first" convention). We summarize this with a definition:

Definition 20.5.1. Let N be a d-dimensional space stratified by manifolds and $M^{d-1} \subset N$ a union of components of the $(d-1)$-dimensional stratum. Suppose that N has codimension-1 δ-structure along M^{d-1} in the sense of Definition 19.5.3. Suppose in addition that M^{d-1} and the top stratum of N are oriented. We say that M^{d-1} has the *boundary orientation* if the orientation of the top stratum of N differs from the orientation that $W \subset N$ obtains as the fiber of δ by the sign $(-1)^d$. ◊

This definition is set up so that if δ is simply

$$\delta = 1/S_2 - 1/S_1$$

(so that $W \subset EW$ is the set $S_1 = S_2$ and N has the structure of a manifold with boundary along M^{d-1}), then the boundary orientation in the above sense agrees with the standard outward-normal-first convention.

With this definition understood, we can ask whether the product orientation on the stratum

$$\breve{M}_{01} \times \breve{M}_{12} \times \breve{M}_{23}$$

is the same as the boundary orientation in the stratified space $\breve{M}^+_{03}$, oriented by λ_{03}. The answer is that the product orientation differs from the boundary orientation by the sign $(-1)^{d_{01}+1}$, where d_{01} is again the dimension of M_{01}. We state this result and the previous one as a proposition.

Proposition 20.5.2. *Suppose that $M_z([\mathfrak{a}], [\mathfrak{b}])$ contains irreducibles, so that $\breve{M}^+_z([\mathfrak{a}], [\mathfrak{b}])$ is a space stratified by manifolds with top stratum $\breve{M}_z([\mathfrak{a}], [\mathfrak{b}])$. Let M' be a codimension-1 stratum of one of the two forms*

(i) $\breve{M}_{01} \times \breve{M}_{12}$,
(ii) $\breve{M}_{01} \times \breve{M}_{12} \times \breve{M}_{23}$,

where $\breve{M}_{12}$ is boundary-obstructed in the second case. Let an element λ be chosen in $\Lambda([\mathfrak{a}],[\mathfrak{b}])$, so that the top stratum of $\breve{M}_z^+([\mathfrak{a}],[\mathfrak{b}])$ is oriented. Orient M' with the product orientation, having chosen orientations for the individual factors so that their q product is λ. Then the product orientation of M' differs from the boundary orientation by the sign

(i) $(-1)^{d_{01}}$,
(ii) $(-1)^{d_{01}+1}$

accordingly, where d_{01} is the dimension of M_{01}.

Proof. We begin with the unobstructed gluing that arises as Case (i). Let the critical points be

$$\mathfrak{a} = \mathfrak{a}_0, \mathfrak{a}_1, \mathfrak{a}_2 = \mathfrak{b}$$

as before. Let elements λ_{01} and λ_{12} be chosen in $\Lambda([\mathfrak{a}_0, \mathfrak{a}_1])$, $\Lambda([\mathfrak{a}_1],[\mathfrak{a}_2])$, and let λ_{02} be their q product. As in the statement of the proposition, use these λ_{ij} to orient M_{ij} and hence $\breve{M}_{ij}$.

The proof of the gluing theorem provides a preferred isotopy class of diffeomorphisms between an open subset of M_{02} and a subset of the product $M_{01} \times M_{12}$. More specifically, let

$$c : M_{ij} \to \mathbb{R}$$

be the map that locates the center of mass of the real function $\| \operatorname{grad} \mathcal{L} \|^2$ on the line. (Our convention about how $\mathbb{R}$ acts on M_{ij} by translations means that $-c$, not c, is equivariant.) If K_{01} and K_{12} are precompact subsets of M_{01} and M_{12} respectively, and if c is large and negative on K_{01} and positive on K_{12}, then the gluing theorem embeds $K_{01} \times K_{12}$ in M_{02}. Furthermore, this embedding is compatible, up to homotopy, with the product rule for determinant lines that is used in the definition of q. In other words, the identification of $K_{01} \times K_{12}$ with a subset of M_{02} respects our chosen orientations. Let us express this relationship by writing:

$$M_{02} \equiv M_{01} \times M_{12}.$$

The notation means that the gluing theorem provides a preferred isotopy class of diffeomorphisms between certain open sets, and that these are orientation-preserving.

As oriented manifolds, we can write

$$M_{ij} = \mathbb{R} \times \breve{M}_{ij},$$

in such a way that our preferred $\mathbb{R}$ action is the standard translation on the first factor. We can also identify $\breve{M}_{ij}$ with subset of M_{ij} satisfying $c = c_{ij}$, where $c_{ij} \in \mathbb{R}$ is any chosen constant. We therefore obtain

$$\mathbb{R} \times \breve{M}_{02} \equiv (\mathbb{R} \times \breve{M}_{01}) \times (\mathbb{R} \times \breve{M}_{12}).$$

This identification is equivariant for the $\mathbb{R}$ action which is the standard one on each $\mathbb{R}$ factor. Since $\breve{M}_{01}$ has dimension $d_{01} - 1$, we obtain

$$\mathbb{R} \times \breve{M}_{02} \equiv (-1)^{d_{01}-1}(\mathbb{R} \times \mathbb{R}) \times \breve{M}_{01} \times \breve{M}_{12}.$$

Now we divide by the $\mathbb{R}$ action on both sides, using the fiber-first convention. On the left-hand side, we obtain $\breve{M}_{02}$ as an oriented manifold. On the right, we must divide $\mathbb{R} \times \mathbb{R}$ with its standard orientation by the diagonal action of $\mathbb{R}$. We can identify the quotient $(\mathbb{R} \times \mathbb{R})/\mathbb{R}$ with the line $L \subset \mathbb{R} \times \mathbb{R}$ spanned by the vector $l = (-1, 1)$; and the quotient orientation of L is the one which l determines. As oriented manifolds, we therefore have

$$\breve{M}_{02} \equiv (-1)^{d_{01}-1}\mathbb{R}l \times \breve{M}_{01} \times \breve{M}_{12}.$$

If we fix precompact sets $\breve{K}_{01}$ and $\breve{K}_{12}$, then the trajectories in $\breve{M}_{02}$ corresponding to the points

$$tl \times \breve{K}_{01} \times \breve{K}_{12}$$

are trajectories with two local centers of mass whose separation *increases* as $t \to -\infty$. For the compactification by broken trajectories, we therefore have

$$\breve{M}_{02}^{+} \equiv (-1)^{d_{01}-1}[-\infty, \infty) \times l \times \breve{M}_{01} \times \breve{M}_{12}.$$

As oriented manifolds, we therefore have

$$\begin{aligned} \partial \breve{M}_{02}^{+} &\equiv (-1)^{d_{01}-1}\big(\partial[-\infty, \infty)\big) \times \breve{M}_{01} \times \breve{M}_{12} \\ &\equiv (-1)^{d_{01}} \breve{M}_{01} \times \breve{M}_{12}. \end{aligned}$$

This completes the proof for Case (i).

In Case (ii), we have critical points

$$\mathfrak{a} = \mathfrak{a}_0, \mathfrak{a}_1, \mathfrak{a}_2, \mathfrak{a}_3 = \mathfrak{b}$$

and we choose orientation elements λ_{ij} for $i < j$, compatible with the composition law q as before. The moduli space $M([\mathfrak{a}_1], [\mathfrak{a}_2])$ is boundary-obstructed, and we recall that in this case our orientation convention ran as follows. We have identifications

$$\det P_{\gamma|_{[t_1,t_2]}} \cong \Lambda^{\max} T_{[\gamma]} M([\mathfrak{a}_1], [\mathfrak{a}_2]) \otimes C^*,$$

where C is a 1-dimensional cokernel; we orient C using a preferred basis element, and use the fact that $\Lambda([\mathfrak{a}_1], [\mathfrak{a}_2])$ orients $\det P_{\gamma|_{[t_1,t_2]}}$ when $t_1 \ll 0$ and $t_2 \gg 0$, as in Subsection 20.4.

There is an alternative way to use $\Lambda([\mathfrak{a}_1], [\mathfrak{a}_2])$ to orient $M([\mathfrak{a}_1], [\mathfrak{a}_2])$ in the boundary-obstructed case. In the end, it gives the same result; but the alternative route fits better with the gluing arguments. This proceeds as follows. Let $[\mathfrak{a}_1]$ and $[\mathfrak{a}_2]$ be reducible configurations (not necessarily critical points) and suppose that these are boundary-stable and boundary-unstable respectively. For I a compact interval and $[\gamma] \in \mathcal{B}_k^\tau(I \times Y; [\mathfrak{a}_1], [\mathfrak{a}_2])$, we defined an operator

$$P_\gamma : \mathcal{E} \to \mathcal{F} \oplus H_1^+ \oplus H_2^-$$

at (20.6). In the present case, if γ is reducible, this operator has block P^ν which contributes a 1-dimensional piece C to the cokernel of P_γ. Let $c \in C$ be our distinguished generator of the cokernel, and consider the operator EP_γ:

$$\begin{aligned} EP_\gamma &: \mathcal{E} \oplus \mathbb{R} \to \mathcal{F} \oplus H_1^+ \oplus H_2^- \\ EP_\gamma(u, x) &= P_\gamma(u) - xc. \end{aligned}$$

Thus the domain of EP_γ is enlarged by a copy of $\mathbb{R}$ so as to kill the cokernel element c. We can use EP_γ to define a 2-element set $E\Lambda([\mathfrak{a}_1], [\mathfrak{a}_2])$, just as $\Lambda([\mathfrak{a}_1], [\mathfrak{a}_2])$ is defined using the orientations of $\det P_\gamma$. In the case that $[\mathfrak{a}_1]$ and $[\mathfrak{a}_2]$ are critical points, the kernel of EP becomes the tangent space to the boundary-obstructed moduli space, while the cokernel becomes zero; so there is a natural way in which $E\Lambda([\mathfrak{a}_1], [\mathfrak{a}_2])$ orients $M([\mathfrak{a}_1], [\mathfrak{a}_2])$. There is an obvious homotopy of operators,

$$P_\gamma \oplus \mathbf{0} \simeq EP_\gamma \tag{20.18}$$

where $\mathbf{0}$ is the zero operator, $\mathbb{R} \to \{0\}$. Using this homotopy and our convention on direct sums, we obtain a preferred isomorphism

$$e : \Lambda([\mathfrak{a}_1], [\mathfrak{a}_2]) \to E\Lambda([\mathfrak{a}_1, \mathfrak{a}_2]).$$

In this way, an element $\lambda_{12} \in \Lambda([\mathfrak{a}_1], [\mathfrak{a}_2])$ determines an element $e\lambda_{12} \in E\Lambda([\mathfrak{a}_1], [\mathfrak{a}_2])$, and hence an orientation of the boundary-obstructed moduli space $M([\mathfrak{a}_1], [\mathfrak{a}_2])$.

To summarize, an element λ_{12} in $\Lambda([\mathfrak{a}_1], [\mathfrak{a}_2])$ can be used in two different ways to orient the boundary-obstructed moduli space: either by the route described in Subsection 20.4, or via the map e to $E\Lambda([\mathfrak{a}_1], [\mathfrak{a}_2])$. The following lemma is an exercise in the definitions:

Lemma 20.5.3. *The two routes whereby an element λ_{12} in $\Lambda([\mathfrak{a}_1], [\mathfrak{a}_2])$ can be used to orient the boundary-obstructed moduli space $M([\mathfrak{a}_1], [\mathfrak{a}_2])$ lead to the same orientation.* □

The operator EP_γ is relevant also in the boundary-unobstructed case, in connection with the extended moduli space. Consider in particular the moduli space $M([\mathfrak{a}], [\mathfrak{b}])$. Let $[\gamma^\sim]$ belong to this moduli space, let $I = [t_1, t_2]$ be an interval with $t_1 \ll 0$ and $t_2 \gg 0$, and let $\gamma = \gamma^\sim|_I$. We have, up to homotopy, a preferred identification

$$\ker P_\gamma \cong T_{[\gamma^\sim]} M([\mathfrak{a}], [\mathfrak{b}])$$

and $\operatorname{coker} P_\gamma = 0$. The construction of the EP_γ makes sense even though the moduli space is not boundary-obstructed: there is no cokernel C now, but the specific generator c which we defined still leads to an element of the codomain of P_γ, and the formula for EP_γ still makes sense. With this understood, we have a similar identification

$$\ker EP_\gamma \cong T_{[\gamma^\sim]} EM([\mathfrak{a}], [\mathfrak{b}])$$

where $EM([\mathfrak{a}], [\mathfrak{b}])$ is the extended moduli space. To understand this more clearly, it is convenient to look again at our preferred generator c. In (20.12), it is defined as the element

$$c = (0, 0, -1) \in L^2 \oplus \mathbb{R} \oplus \mathbb{R}.$$

It is convenient to consider instead the element

$$c^* = (\beta, 0, 0) \in L^2 \oplus \mathbb{R} \oplus \mathbb{R},$$

where β is a positive bump-function supported near 0. Positive multiples of the elements c and c^* differ by an element of the image of P_γ, so they lead to the same orientations. If we consider EP_γ as

$$EP_\gamma(u, x) = P_\gamma u - xc^*,$$

then the interpretation of the kernel of EP_γ is that u should solve the linearized equations $P_\gamma u = 0$ except near 0; and that the behavior of the s coordinate of u near 0 is that of a (smoothed out) step discontinuity of size x. In this way, the kernel of EP_γ approximates the tangent space to the extended moduli space $EM([\mathfrak{a}], [\mathfrak{b}])$.

Now let $\lambda_{03} \in \Lambda([\mathfrak{a}], [\mathfrak{b}])$ be given. Let $e\lambda_{03}$ be the corresponding element of $E\Lambda([\mathfrak{a}], [\mathfrak{b}])$, and use these to orient $M([\mathfrak{a}], [\mathfrak{b}])$ and $EM([\mathfrak{a}], [\mathfrak{b}])$ respectively. The definition of the extended moduli space means that $M([\mathfrak{a}], [\mathfrak{b}])$ is the fiber $\delta^{-1}(0)$ of a map δ,

$$M([\mathfrak{a}], [\mathfrak{b}]) \hookrightarrow EM([\mathfrak{a}], [\mathfrak{b}]) \xrightarrow{\delta} \mathbb{R},$$

which measures the size of the discontinuity (see (19.2)). The sign of δ is positive when the value of the s coordinate increases across the discontinuity. The next lemma is another application of the definitions.

Lemma 20.5.4. *If $M([\mathfrak{a}], [\mathfrak{b}])$ and $EM([\mathfrak{a}], [\mathfrak{b}])$ are oriented using λ_{03} and $e\lambda_{03}$, then $M([\mathfrak{a}], [\mathfrak{b}])$ is the oriented fiber of the map δ, with our usual fiber-first orientation convention.* □

Before continuing with the rest of the calculation, there is one more point to consider with $E\Lambda$. As with the ordinary orientation sets $\Lambda([\mathfrak{a}_i], [\mathfrak{a}_j])$, there is a natural map corresponding to composition of paths:

$$\Lambda_{01} \times E\Lambda_{12} \times \Lambda_{23} \to E\Lambda_{03}.$$

We can therefore ask whether the following diagram commutes:

$$\begin{array}{ccc} \Lambda_{01} \times \Lambda_{12} \times \Lambda_{23} & \longrightarrow & \Lambda_{03} \\ \downarrow {\scriptstyle 1\times e\times 1} & & \downarrow {\scriptstyle e} \\ \Lambda_{01} \times E\Lambda_{12} \times \Lambda_{23} & \longrightarrow & E\Lambda_{03}. \end{array}$$

Lemma 20.5.5. *The above diagram commutes or not according to the sign $(-1)^{\dim M_{23}}$.*

Proof. Use the homotopy from EP_γ to a sum of operators, as in (20.18). After the homotopy, we are comparing the composition rules for sums

$$P_{\gamma_1} \oplus (P_{\gamma_2} \oplus \mathbf{0}) \oplus P_{\gamma_3}$$

and

$$(P_{\gamma_1} \oplus P_{\gamma_2} \oplus P_{\gamma_3}) \oplus \mathbf{0}.$$

The sign results from the interchange of the operator $\mathbf{0}$ (which has index 1) with the operator P_{γ_3}. □

In the boundary-obstructed case, the moduli spaces $M([\mathfrak{a}_1],[\mathfrak{a}_2])$ and $EM([\mathfrak{a}_1],[\mathfrak{a}_2])$ coincide; Lemma 20.5.3 says that they coincide as *oriented* manifolds, when oriented using λ_{12} and $e\lambda_{12}$ respectively. The gluing construction gives a local diffeomorphism of open sets between EM_{03} and the product $M_{01} \times EM_{12} \times M_{23}$, and Lemmas 20.5.3 and 20.5.5 together tell us that, with orientations, we have

$$EM_{03} \equiv (-1)^{d_{23}} M_{01} \times M_{12} \times M_{23},$$

where d_{ij} is the dimension of M_{ij}. Passing to the unparametrized moduli spaces, we have

$$\begin{aligned}\mathbb{R} \times E\breve{M}_{03} &\equiv (-1)^{d_{23}} (\mathbb{R} \times \breve{M}_{01}) \times (\mathbb{R} \times \breve{M}_{12}) \times (\mathbb{R} \times \breve{M}_{23}) \\ &\equiv (-1)^{d_{23}+\dim \breve{M}_{12}} (\mathbb{R} \times \mathbb{R} \times \mathbb{R}) \times \breve{M}_{01} \times \breve{M}_{12} \times \breve{M}_{23}.\end{aligned}$$

Using translations to center the middle factor, we can write

$$\begin{aligned}E\breve{M}_{03} &\equiv (-1)^{d_{23}+\dim \breve{M}_{12}+1} (\mathbb{R} \times 0 \times \mathbb{R}) \times \breve{M}_{01} \times \breve{M}_{12} \times \breve{M}_{23} \\ &\equiv (-1)^{d_{23}+d_{12}} \breve{M}_{01} \times \breve{M}_{12} \times \breve{M}_{23} \times (\mathbb{R} \times 0 \times \mathbb{R}).\end{aligned}$$

The two factors of $\mathbb{R}$ that appear at the end are not the separation coordinates that are used in Definition 20.5.1: to obtain the correct sign for the separations, we need to change the sign of the first $\mathbb{R}$ factor. So, in terms of the separation coordinates, we have

$$E\breve{M}_{03} \equiv (-1)^{d_{23}+d_{12}+1} \breve{M}_{01} \times \breve{M}_{12} \times \breve{M}_{23} \times (0, \infty]^2.$$

Using Lemma 20.5.4, we see that $\breve{M}_{03}$ is locally the oriented fiber of δ, regarded as a map

$$(-1)^{d_{23}+d_{12}+1}\breve{M}_{01} \times \breve{M}_{12} \times \breve{M}_{23} \times (0,\infty]^2 \xrightarrow{\delta} \mathbb{R}.$$

In Definition 20.5.1, the relevant quantity d is $d_{01}+d_{12}+d_{23}-2$. In the sense of that definition therefore, the orientation of $\breve{M}_{01} \times \breve{M}_{12} \times \breve{M}_{23}$ differs from the boundary orientation by

$$(-1)^{d_{23}+d_{12}+1}(-1)^{d_{01}+d_{12}+d_{23}-2},$$

which is $(-1)^{d_{01}+1}$ as the proposition asserts. □

20.6 Orienting moduli spaces of reducible trajectories

Let $[\mathfrak{a}]$ and $[\mathfrak{b}]$ be reducible critical points, and as in Subsection 16.6, let

$$M^{\text{red}}([\mathfrak{a}],[\mathfrak{b}]) \subset M([\mathfrak{a}],[\mathfrak{b}])$$

denote the moduli space of reducible trajectories. Recall that the two spaces in this inclusion are actually equal in all but one case: the special situation is when $[\mathfrak{a}]$ is boundary-unstable and $[\mathfrak{b}]$ is boundary-stable, and in this case $M([\mathfrak{a}],[\mathfrak{b}])$ is a smooth manifold with boundary and

$$M^{\text{red}}([\mathfrak{a}],[\mathfrak{b}]) = \partial M([\mathfrak{a}],[\mathfrak{b}]).$$

An element of $\Lambda([\mathfrak{a}],[\mathfrak{b}])$ orients $M([\mathfrak{a}],[\mathfrak{b}])$, by the rule specified in Subsection 20.4 above. We therefore have a rule for orienting $M^{\text{red}}([\mathfrak{a}],[\mathfrak{b}])$ also:

Definition 20.6.1. In the case that $[\mathfrak{a}]$ is boundary-unstable and $[\mathfrak{b}]$ is boundary-stable, the orientation of $M^{\text{red}}([\mathfrak{a}],[\mathfrak{b}])$ determined by an element λ of $\Lambda([\mathfrak{a}],[\mathfrak{b}])$ is the orientation it obtains as the boundary of $M([\mathfrak{a}],[\mathfrak{b}])$, with the outward-normal-first convention. In the three other cases, the orientation of $M^{\text{red}}([\mathfrak{a}],[\mathfrak{b}])$ corresponding to λ is defined by identifying $M^{\text{red}}([\mathfrak{a}],[\mathfrak{b}])$ with $M([\mathfrak{a}],[\mathfrak{b}])$.

We orient $\breve{M}^{\text{red}}([\mathfrak{a}],[\mathfrak{b}])$ as the quotient of $M^{\text{red}}([\mathfrak{a}],[\mathfrak{b}])$ by $\mathbb{R}$, using the fiber-first convention. ◊

With this convention, once an element of $\Lambda([\mathfrak{a}],[\mathfrak{b}])$ is chosen, the moduli space $M^{\text{red}}([\mathfrak{a}],[\mathfrak{b}])$ is oriented. There is a different approach, however, to orienting $M^{\text{red}}([\mathfrak{a}],[\mathfrak{b}])$. Recall that $\Lambda([\mathfrak{a}],[\mathfrak{b}])$ is defined as the set of orientations of the determinant of an operator P_γ. When γ is a reducible configuration on $I \times Y$, we have a decomposition $Q_\gamma = Q^\nu_\gamma \oplus Q^\partial_\gamma$ and a corresponding decomposition of the spectral subspaces; we can write the operator P_γ that appears in (20.6) as a sum

$$P_\gamma = P^\nu_\gamma \oplus \bar{P}_\gamma$$

where $\bar{P}_\gamma$ is the "reducible" part, and P^ν_γ is the summand corresponding to the 1-by-1 block, as it appeared in the definition of the trivializations τ in (20.10). The operator $\bar{P}_\gamma$ stands in the same relation to the moduli spaces $M^{\text{red}}([\mathfrak{a}],[\mathfrak{b}])$ as P_γ does to the moduli spaces $M([\mathfrak{a}],[\mathfrak{b}])$. We define a 2-element set $\bar{\Lambda}([\mathfrak{a}],[\mathfrak{b}])$ by repeating the definition of $\Lambda([\mathfrak{a}],[\mathfrak{b}])$ but using $\bar{P}_\gamma$ in place of P_γ: that is, $\bar{\Lambda}([\mathfrak{a}],[\mathfrak{b}])$ is the set of orientations of $\det(\bar{P}_\gamma)$, we have composition maps

$$\bar{q} : \bar{\Lambda}([\mathfrak{a}_1],[\mathfrak{a}_2]) \times \bar{\Lambda}([\mathfrak{a}_2],[\mathfrak{a}_3]) \to \bar{\Lambda}([\mathfrak{a}_1],[\mathfrak{a}_3]).$$

As in the irreducible case, the perturbations should be mentioned when the context requires: we write $\bar{\Lambda}([\mathfrak{a}],\mathfrak{q}_1,[\mathfrak{b}],\mathfrak{q}_2)$.

When the perturbation is zero, the orientation sets $\bar{\Lambda}([\mathfrak{a}_0],0,[\mathfrak{a}'_0],0)$ are canonically trivial, so we can imitate our earlier definitions and define

$$\bar{\Lambda}([\mathfrak{a}]) = \Big(\coprod_{\mathfrak{a}_0} \bar{\Lambda}([\mathfrak{a}],[\mathfrak{a}_0])\Big) \Big/\!\sim,$$

where it is understood that the zero perturbation is taken at $[\mathfrak{a}_0]$. We then have maps

$$\bar{\Lambda}([\mathfrak{a}]) \times \bar{\Lambda}([\mathfrak{b}]) \to \bar{\Lambda}([\mathfrak{a}],[\mathfrak{b}])$$

as in the irreducible case. When it comes to gluing, these "bar" orientations for reducible moduli spaces behave just as the previous orientations did for irreducible moduli spaces:

Proposition 20.6.2. *Let $[\mathfrak{a}]$ and $[\mathfrak{b}]$ be reducible critical points, so that $\breve{M}^{\text{red}+}_z([\mathfrak{a}],[\mathfrak{b}])$ is a space stratified by manifolds with top stratum $\breve{M}^{\text{red}}_z([\mathfrak{a}],[\mathfrak{b}])$. Let M' be a codimension-1 stratum of the form*

$$\breve{M}^{\text{red}}_{01} \times \breve{M}^{\text{red}}_{12}.$$

Let an element $\bar{\lambda}$ be chosen in $\bar{\Lambda}([\mathfrak{a}],[\mathfrak{b}])$, so that the top stratum of $\breve{M}^{\text{red}+}_z([\mathfrak{a}],[\mathfrak{b}])$ is oriented. Orient M' with the product orientation, having *chosen orientations for the individual factors so that their $\bar{\mathfrak{q}}$ product is $\bar{\lambda}$. Then the product orientation of M' differs from the boundary orientation by the sign $(-1)^{\bar{d}_{01}}$, where $\bar{d}_{01}$ is the dimension of M_{01}^{red}.*

In order to have a meaningful comparison between the "bar" orientations and the previous orientations, we need to fix an isomorphism between $\bar{\Lambda}([\mathfrak{a}])$ and $\Lambda([\mathfrak{a}])$. This we now do:

Definition 20.6.3. Let $[\mathfrak{a}]$ be a reducible configuration, and $\mathfrak{q}$ a perturbation. Define a map

$$r : \Lambda([\mathfrak{a}]) \to \bar{\Lambda}([\mathfrak{a}])$$

from $\Lambda([\mathfrak{a}]) = \Lambda([\mathfrak{a}], \mathfrak{q})$ to $\bar{\Lambda}([\mathfrak{a}]) = \bar{\Lambda}([\mathfrak{a}], \mathfrak{q})$ as follows. Let $[\mathfrak{a}_0]$ be a boundary-stable configuration with the zero perturbation. If $[\mathfrak{a}]$ is also boundary-stable, so that P_γ^ν is invertible, orient $\det(P_\gamma^\nu)$ with the trivial canonical orientation; and if $[\mathfrak{a}]$ is boundary-unstable, so that P_γ^ν has 1-dimensional kernel, orient $\det(P_\gamma^\nu)$ by taking a negative function as in (20.11). Then use the (ordered) direct sum decomposition $P_\gamma = P_\gamma^\nu \oplus \bar{P}_\gamma$ to identify $\det(P_\gamma)$ with $\det(\bar{P}_\gamma)$. ◊

We can now compare orientations of $M^{\mathrm{red}}([\mathfrak{a}], [\mathfrak{b}])$ and $M([\mathfrak{a}], [\mathfrak{b}])$. Fix elements $\lambda_\mathfrak{a}$ and $\lambda_\mathfrak{b}$ in $\Lambda([\mathfrak{a}])$ and $\Lambda([\mathfrak{b}])$, and set

$$\bar{\lambda}_\mathfrak{a} = r(\lambda_\mathfrak{a})$$
$$\bar{\lambda}_\mathfrak{b} = r(\lambda_\mathfrak{b}).$$

Let $\lambda_{\mathfrak{ab}}$ be the corresponding element of $\Lambda([\mathfrak{a}], [\mathfrak{b}])$:

$$\lambda_{\mathfrak{ab}} = q(\lambda_\mathfrak{a}, \rho(\lambda_\mathfrak{b})).$$

Similarly set

$$\bar{\lambda}_{\mathfrak{ab}} = \bar{q}(\bar{\lambda}_\mathfrak{a}, \bar{\rho}(\bar{\lambda}_\mathfrak{b})).$$

As an element of $\Lambda([\mathfrak{a}], [\mathfrak{b}])$, the latter orients $M^{\mathrm{red}}([\mathfrak{a}], [\mathfrak{b}])$, while the former orients $M([\mathfrak{a}], [\mathfrak{b}])$.

Lemma 20.6.4. *The oriented manifolds $M = (M([\mathfrak{a}], [\mathfrak{b}]), \lambda_{\mathfrak{ab}})$ and $\bar{M} = (M^{\mathrm{red}}([\mathfrak{a}], [\mathfrak{b}]), \bar{\lambda}_{\mathfrak{ab}})$ are related as follows:*

(i) *if $[\mathfrak{a}]$ is boundary-stable, then $M = \bar{M}$ as oriented manifolds;*

(ii) *if* $[\mathfrak{a}]$ *is boundary-unstable and* $[\mathfrak{b}]$ *is boundary-stable, then* $\bar{M}$ *is the oriented boundary of* M*;*

(iii) *if* $[\mathfrak{a}]$ *and* $[\mathfrak{b}]$ *are both boundary-unstable, then although* M *and* $\bar{M}$ *coincide as manifolds, their orientations differ by the sign* $(-1)^d$*, where* d *is the dimension of* M*.*

Proof. We illustrate the argument in the most interesting case, (iii). So let $\mathfrak{a}$ and $\mathfrak{b}$ be boundary-unstable. We have an orientation $\bar{\lambda}_{\mathfrak{b}}$ for the determinant of an operator $\bar{P} = \bar{P}(\mathfrak{b}, \mathfrak{a}_0)$, and we can write

$$\lambda_{\mathfrak{b}} = \tau \wedge \bar{\lambda}_{\mathfrak{b}}$$

where the $\wedge$ denotes the product law for orientations of the sum $P^{\nu} \oplus \bar{P}$. We wish to compare $\rho(\lambda_{\mathfrak{b}})$ with $\bar{\rho}(\bar{\lambda}_{\mathfrak{b}})$. Let τ^* be our preferred orientation of the determinant of $P^{\nu}(\mathfrak{a}_0, \mathfrak{b})$: that is, τ^* is obtained by orienting the 1-dimensional cokernel using the preferred element c from (20.12). Using the ordered direct sum decomposition

$$P(\mathfrak{a}_0, \mathfrak{b}) = P^{\nu}(\mathfrak{a}_0, \mathfrak{b}) \oplus \bar{P}(\mathfrak{a}_0, \mathfrak{b})$$

we can write down an element

$$\tau^* \wedge \bar{\rho}(\bar{\lambda}_{\mathfrak{b}}) \in \Lambda(\mathfrak{a}_0, \mathfrak{b}).$$

This element $\tilde{\lambda} = \tau^* \wedge \bar{\rho}(\bar{\lambda}_{\mathfrak{b}})$ can be compared with $\rho(\lambda_{\mathfrak{b}})$, and the sign can be determined by checking the relation which defines ρ: we have $q(\lambda_{\mathfrak{b}}, \rho(\lambda_{\mathfrak{b}}) = 1$ by definition, whereas

$$\begin{aligned} q(\lambda_{\mathfrak{b}}, \tilde{\lambda}) &= q(\tau \wedge \bar{\lambda}_{\mathfrak{b}}, \tau^* \wedge \bar{\rho}(\bar{\lambda}_{\mathfrak{b}}) \\ &= (-1)^{\text{index } \bar{P}(\mathfrak{b},\mathfrak{a}_0)} q^{\nu}(\tau, \tau^*) \wedge \bar{q}(\bar{\lambda}_{\mathfrak{b}}, \bar{\rho}(\bar{\lambda}_{\mathfrak{b}})) \\ &= (-1)^{\text{index } \bar{P}(\mathfrak{b},\mathfrak{a}_0)}. \end{aligned}$$

So

$$\rho(\lambda_{\mathfrak{b}}) = (-1)^{\text{index } \bar{P}(\mathfrak{b},\mathfrak{a}_0)} \tau^* \wedge \bar{\rho}(\bar{\lambda}_{\mathfrak{b}}).$$

We are now ready to compare $\lambda_{\mathfrak{ab}} = q(\lambda_{\mathfrak{a}}, \rho(\lambda_{\mathfrak{b}}))$ with $1 \wedge \bar{\lambda}_{\mathfrak{ab}}$, where 1 is the canonical orientation of the invertible operator $P^{\nu}(\mathfrak{a}, \mathfrak{b})$: we have

$$\begin{aligned}
\lambda_{\mathfrak{ab}} &= q(\lambda_{\mathfrak{a}}, \rho(\lambda_{\mathfrak{b}})) \\
&= (-1)^{\text{index } \bar{P}(\mathfrak{b},\mathfrak{a}_0)} q(\tau \wedge \bar{\lambda}_{\mathfrak{a}}, \tau^* \wedge \bar{\rho}(\bar{\lambda}_{\mathfrak{b}})) \\
&= (-1)^{\text{index } \bar{P}(\mathfrak{b},\mathfrak{a}_0)+\text{index } \bar{P}(\mathfrak{a},\mathfrak{a}_0)} q^{\nu}(\tau, \tau^*) \wedge \bar{q}(\bar{\lambda}_{\mathfrak{a}}, \bar{\rho}(\bar{\lambda}_{\mathfrak{b}})) \\
&= (-1)^{\text{index } \bar{P}(\mathfrak{a},\mathfrak{b})} 1 \wedge \bar{\lambda}_{\mathfrak{ab}} \\
&= (-1)^{d} 1 \wedge \bar{\lambda}_{\mathfrak{ab}}
\end{aligned}$$

as claimed. □

21 A version of Stokes' theorem

The structure of the compactification of the trajectory spaces $\breve{M}([\mathfrak{a}], [\mathfrak{b}])$ is potentially more complicated than a manifold with corners: even near the codimension-1 strata, our analysis does not rule out the sort of pathology illustrated in the example at the end of Subsection 16.5, in which the top-dimensional stratum has infinitely many path components. The purpose of this short section is to review the cohomology theory of spaces stratified by manifolds, and to prove in particular a version of Stokes' theorem in this context. In the construction of the chain complexes which define our Floer homology groups in the next section, we will need only the simplest case of 1-dimensional spaces stratified by manifolds; but later we will use the general result in developing further properties of the Floer groups, such as the module structure in Section 25.

21.1 Boundary multiplicities in stratified spaces

We start with some remarks about Čech cohomology. Recall that an open cover $\mathcal{U}$ of a space B has *covering order* $\leq d + 1$ if every $(d + 2)$-fold intersection

$$U_0 \cap U_1 \cap \cdots \cap U_{d+1}, \quad U_i \in \mathcal{U}, \ U_i \text{ distinct},$$

is empty. A metric space has *covering dimension* $\leq d$ if every open cover has a refinement with covering order $\leq d + 1$. To an open covering $\mathcal{U}$, one can associate a simplicial complex $K(\mathcal{U})$, its nerve. The Čech cohomology of B (with for example coefficients $\mathbb{Z}$) is the limit

$$\check{H}^n(B;\mathbb{Z}) = \lim_{\rightarrow} H^n_{\mathrm{Simp}}(K(\mathcal{U});\mathbb{Z})$$

as $\mathcal{U}$ runs through the open coverings of B. (Here H^n_{Simp} denotes simplicial homology.) If $\mathcal{U}' \subset \mathcal{U}$, and the open sets $U \cap B'$ ($U \in \mathcal{U}'$) cover B', then $K(\mathcal{U}'|B')$ is a subcomplex of $K(\mathcal{U})$: the notation means the nerve of this open cover of B'. The Čech cohomology of the pair (B, B') is the limit

$$\check{H}^n(B, B';\mathbb{Z}) = \lim_{\rightarrow} H^n_{\mathrm{Simp}}(K(\mathcal{U}), K(\mathcal{U}'|B');\mathbb{Z}).$$

Let $|K(\mathcal{U})|$ denote, as usual, the topological realization of the abstract simplicial complex. Via the standard coordinates on the simplices, a partition of unity $\{\phi_U\}$ subordinate to the open covering gives rise to a map

$$\phi : B \to |K(\mathcal{U})|$$

whose homotopy class depends only on the cover.

Now consider a compact d-dimensional space N^d stratified by manifolds, in the sense of Definition 16.5.1. So we have

$$N^d \supset N^{d-1} \supset \cdots \supset N^0,$$

and each stratum $M^e = N^e \setminus N^{e-1}$ is either empty or homeomorphic to a manifold of dimension e, with top stratum M^d non-empty. Such a stratified space has covering dimension at most d [84]. Suppose in addition that each M^e is oriented: we say N^d has a stratification by *oriented* manifolds. In Čech cohomology, we have

$$\check{H}^d(N^d, N^{d-1};\mathbb{Z}) = H^d_c(M^d;\mathbb{Z})$$

and (because of the orientations) this is a free abelian group with generators μ^d_α corresponding to the components M^d_α of M^d. Let

$$I_\alpha : \check{H}^d(N^d, N^{d-1};\mathbb{Z}) \to \mathbb{Z}$$

be the map which is 1 on the generator μ^d_α and zero on the others. From the long exact sequence of the triple (N^d, N^{d-1}, N^{d-2}), there is a coboundary map

$$\delta_* : H^{d-1}_c(M^{d-1};\mathbb{Z}) \to H^d_c(M^d;\mathbb{Z}), \qquad (21.1)$$

which is a map of free abelian groups,

$$\delta_* : \bigoplus_\beta H_c^{d-1}(M_\beta^{d-1};\mathbb{Z}) \to \bigoplus_\alpha H_c^d(M_\alpha^d;\mathbb{Z}).$$

In this sense, for each pair of components M_α^d, M_β^{d-1}, there is a well-defined multiplicity $\delta_{\alpha\beta}$ with which M_β^{d-1} appears in the boundary of M_α^d:

Definition 21.1.1. Let N^d be the space stratified by oriented manifolds as above. The *multiplicity* of the component M_β^{d-1} in the boundary of the component M_α^d is the integer

$$\delta_{\alpha\beta} = I_\alpha \delta_* \mu_\beta^{d-1}.$$

The *boundary multiplicity* of the component M_β^{d-1} in the stratified space N^d is the (finite) sum

$$\delta_\beta = \sum_\alpha \delta_{\alpha\beta}.$$

◊

Remark. The component M_β^{d-1} may appear in the closure of infinitely many d-dimensional components M_α^d, even though only finitely many of the $\delta_{\alpha\beta}$ will be non-zero. See the example later in this section.

As a consequence of the definition, we can write

$$I_\alpha(\delta_* x) = \sum_\beta \delta_{\alpha\beta} I_\beta x$$

for $x \in \check{H}^{d-1}(N^{d-1}, N^{d-2};\mathbb{Z}) = H_c^{d-1}(M^{d-1};\mathbb{Z})$.

21.2 Transverse open covers and Stokes' theorem

Let N^d be a compact d-dimensional space stratified by manifolds. If N^d is embedded in a metric space B, we will say that an open cover $\mathcal{U}$ of B is *transverse* to the strata if $\mathcal{U}|N^e$ has covering order $\leq e+1$ for all $e \leq d$.

Lemma 21.2.1. *Let $N_k^{d_k}$ be a countable, locally finite collection of spaces stratified by manifolds. Then every open cover $\mathcal{U}$ of B has a refinement $\mathcal{U}'$ that is transverse to the strata in every $N_k^{d_k}$.*

Proof. Let $\mathcal{U} = \{ U_\alpha \mid \alpha \in S \}$. According to [84, Theorem 12-9], there is a locally finite *closed* cover $\mathcal{G} = \{ G_\beta \mid \beta \in T \}$ which is a refinement of $\mathcal{U}$, in that there is a map $f : T \to S$ with $G_\beta \subset U_{f(\beta)}$ for all β, and which has the property that for every k and every closed stratum $N \subset N_k^{d_k}$, the intersection

$$G_{\beta_0} \cap \cdots \cap G_{\beta_{e+1}} \cap N \tag{21.2}$$

is empty whenever the β_i are distinct and $e \geq \dim N$. Now apply [84, Proposition 9-3] to the collection of closed sets $\mathcal{F}$ obtained by adding to $\mathcal{G}$ all the closed strata of the spaces $N_k^{d_k}$: this tells us that there are open sets $H_\beta \supset G_\beta$ such that

$$H_{\beta_0} \cap \cdots \cap H_{\beta_{e+1}} \cap N = \varnothing$$

whenever (21.2) is empty. If we set $U'_\beta = H_\beta \cap U_{f(\beta)}$, then we have an open cover of B which is a refinement of $\mathcal{U}$ with the required transversality to all the strata N. □

It follows from the lemma that we can compute the Čech cohomology of B as the limit

$$\check{H}^n(B; \mathbb{Z}) = \lim_{\to} H^n(K(\mathcal{U}); \mathbb{Z})$$

as $\mathcal{U}$ runs through the open coverings of B that are transverse to the stratification of a stratified space N^d. We consider a fixed open cover $\mathcal{U}$ satisfying this transversality condition. Every Čech cochain

$$u \in \check{C}^d(\mathcal{U}|N^d; \mathbb{Z}) = C^d_{\text{Simp}}(K(\mathcal{U}|N^d); \mathbb{Z})$$

is automatically coclosed, and vanishes on $K(\mathcal{U}|N^{d-1}); \mathbb{Z})$ (because there are no d-simplices in the latter). So u has a well-defined class $[u]$ in $\check{H}^d(N^d, N^{d-1}; \mathbb{Z})$. Čech cochains $u \in \check{C}^d(\mathcal{U}|N^d; \mathbb{Z})$ can therefore be *integrated* over M^d, or over any component M^d_α: we have maps

$$\begin{aligned} \langle -, [M_\alpha] \rangle &: \check{C}^d(\mathcal{U}; \mathbb{Z}) \to \mathbb{Z} \\ \langle u, [M_\alpha] \rangle &= I_\alpha[u|_{N^d}]. \end{aligned} \tag{21.3}$$

Finally, we have Stokes' theorem, with multiplicities: for $v \in \check{C}^{d-1}(\mathcal{U}; \mathbb{Z})$, we have

$$\sum_\beta \delta_{\alpha\beta} \langle v, [M_\beta^{d-1}] \rangle = \langle \delta v, [M_\alpha^d] \rangle, \tag{21.4}$$

where $\delta : \check{C}^{d-1}(\mathcal{U};\mathbb{Z}) \to \check{C}^{d}(\mathcal{U};\mathbb{Z})$ is the Čech coboundary map. In this form, Stokes' theorem is an immediate consequence of the definitions. In the simplest case, N^d is a manifold with boundary, and $M^{d-1} = N^{d-1}$ is the boundary. In this case, if the orientations are standard, the boundary multiplicities are 0 or 1 (and all 1 if N^d is connected), and the above formula becomes more recognizable:

$$\langle \delta v, [N^d] \rangle = \langle v, [\partial N^d] \rangle.$$

21.3 Computing boundary multiplicities

The next lemma allows us to compute the multiplicities of boundaries in the type of situation that occurs with the boundary-obstructed moduli spaces. (See Theorem 19.5.4 for example.)

Lemma 21.3.1. *Let N^d be a compact d-dimensional space stratified by oriented manifolds. Let M^{d-1}_β be an oriented component of M^{d-1}. Suppose that N^d has a codimension-1 δ-structure along M^{d-1}_β in the sense of Definition* 19.5.3, *and that M^{d-1}_β has the boundary orientation as defined in Definition* 20.5.1. *Then the boundary multiplicity, δ_β, of the stratum M^{d-1}_β in N^d is* 1.

Proof. Let $W \supset M^{d-1}_\beta$ and $j : W \to EW$ be as in Definition 19.5.3. We will identify W with $j(W)$, so W is the zero set of $\delta : EW \to \mathbb{R}$. The d-manifold $W \setminus M^{d-1}_\beta$ is properly embedded in $EW \setminus M^{d-1}_\beta$. Take a cochain m^{d-1}_β with compact support in M^{d-1}_β representing the generator μ^{d-1}_β of $H^{d-1}_c(M^{d-1}_\beta)$. Application of the coboundary map at the cochain level gives a compactly supported cochain δm^{d-1}_β on $EW \setminus M^{d-1}$ whose restriction to the manifold $W \setminus M^{d-1}_\beta$ we wish to integrate: the integral is the boundary multiplicity δ_β, because

$$\begin{aligned}\langle (\delta m^{d-1}_\beta), [W \setminus M^{d-1}_\beta] \rangle &= \sum_\alpha \langle (\delta m^{d-1}_\beta), [M^d_\alpha] \rangle \\ &= \sum_\alpha \delta_{\alpha\beta} \langle m^{d-1}_\beta, [M^{d-1}_\beta] \rangle \\ &= \sum_\alpha \delta_{\alpha\beta} \\ &= \delta_\beta.\end{aligned}$$

By a homotopy of δ, we can construct a (non-compact) cobordism, properly embedded in $[0,1] \times EW$, between W and W_0, where the latter is the locus

$S_1 = S_2$. The evaluation of the compactly supported cochain δm_β^{d-1} on $W_0 \setminus M_\beta^{d-1}$ is the same as on $W \setminus M_\beta^{d-1}$, and in this way we reduce to the case that W is a manifold with boundary M_β^{d-1}. Our orientation conventions mean that the *oriented* boundary of W_0 is M_β^{d-1}, and the result is now straightforward. □

Corollary 21.3.2. *Let N^1 be a compact 1-dimensional space stratified by oriented manifolds, so that $N^1 \setminus N^0$ is an oriented 1-manifold and N^0 is an oriented 0-manifold: a finite set of signed points. Suppose that N^1 has a codimension-1 δ-structure along N^0 and N^0 has the boundary orientation. Then the number of points in N^0, counted with sign, is zero.*

Proof. The number of points in N^0, counted with sign, is the integral

$$\langle 1, [N^0] \rangle.$$

The corollary follows from the lemma above and our version of Stokes' theorem (21.4). □

Example. We revisit the example which follows Proposition 16.5.2 on page 289. Let $N^1 \subset \mathbb{R}^2$ be the union of the line segment I joining $(0,0)$ to $(1,0)$ and all the circles C_n centered at $(-1/n, 0)$ and passing through $(0,0)$, for $n > 0$. Let N^0 be the 2-point set containing $(0,0)$ and $(1,0)$. Orient $I \setminus N^0$ by the vector $\partial/\partial x$, and orient the arcs $C_n \setminus (0,0)$ arbitrarily. Orient N^0 as the boundary of I. Then N^1 is a compact 1-dimensional space stratified by manifolds, and satisfies the hypotheses of the corollary above. Note that $M^1 = N^1 \setminus N^0$ has infinitely many components, though $\delta_{\alpha\beta}$ is zero for all but finitely many α, as it must be. The number of points in N^0, counted with sign, is zero, as asserted.

22 Floer homology

22.1 The basic construction

Let Y once more be a compact, connected, oriented, Riemannian 3-manifold, with a spin^c structure $\mathfrak{s}$. We may write g for the Riemannian metric. Let $\mathcal{P}$ be a large Banach space of tame perturbations, in the sense of Definition 11.6.3, and let $\mathfrak{q} \in \mathcal{P}$ be chosen so that all the critical points of $(\operatorname{grad} \mathcal{L})^\sigma$ in $\mathcal{B}_k^\sigma(Y, \mathfrak{s})$ are non-degenerate, and all moduli spaces $M([\mathfrak{a}], [\mathfrak{b}])$ are regular (Definitions 12.1.1 and 14.5.6). In addition, we suppose that, if $c_1(S)$ is not a torsion class, then the perturbation is chosen so that there are no reducible critical points (see the remark following Proposition 16.4.3). We introduce a name for such a perturbation:

Definition 22.1.1. A tame perturbation $\mathfrak{q}$ is an *admissible* perturbation for $(Y, g, \mathfrak{s})$ if the above three conditions hold: the critical points are non-degenerate, the moduli spaces are regular, and there are no reducibles unless $c_1(S)$ is torsion. ◊

All our constructions in this section depend on g, $\mathfrak{s}$ and $\mathfrak{q}$, though the metric and perturbation will not often be mentioned. Given these choices, we will define three flavors of Floer homology group,

$$\widecheck{HM}_*(Y,\mathfrak{s}), \quad \widehat{HM}_*(Y,\mathfrak{s}) \quad \overline{HM}_*(Y,\mathfrak{s}).$$

As we mentioned in the introductory remarks to this chapter, the proof that these groups do not depend on the choice of metric g and perturbation $\mathfrak{q}$ will be postponed until Chapter VII. Our construction of these three groups follows closely the Morse-theory constructions in Subsection 2.4.

Let

$$\mathfrak{C} \subset \mathcal{B}^\sigma_k(Y,\mathfrak{s})$$

be the set of critical points. We can write $\mathfrak{C}$ as a disjoint union

$$\mathfrak{C} = \mathfrak{C}^o \cup \mathfrak{C}^s \cup \mathfrak{C}^u,$$

where $\mathfrak{C}^o$ is the set of irreducible critical points, and $\mathfrak{C}^s$, $\mathfrak{C}^u$ are respectively the boundary-stable and boundary-unstable reducible critical points (Definition 14.5.2).

Given a 2-element set $\Lambda = \{x, y\}$, we write $\mathbb{Z}\Lambda$ for the infinite cyclic group

$$\mathbb{Z}\Lambda = \langle x, y \mid x = -y \rangle. \tag{22.1}$$

(See the discussion at the end of Subsection 2.2.) A choice of preferred element from Λ allows one to identify $\mathbb{Z}\Lambda$ with $\mathbb{Z}$. Set

$$\left.\begin{aligned} C^o &= \bigoplus_{[\mathfrak{a}]\in\mathfrak{C}^o} \mathbb{Z}\Lambda[\mathfrak{a}] \\ C^s &= \bigoplus_{[\mathfrak{a}]\in\mathfrak{C}^s} \mathbb{Z}\Lambda[\mathfrak{a}] \\ C^u &= \bigoplus_{[\mathfrak{a}]\in\mathfrak{C}^u} \mathbb{Z}\Lambda[\mathfrak{a}]. \end{aligned}\right\} \tag{22.2}$$

Here $\Lambda([\mathfrak{a}])$ is the 2-element set defined in Subsection 20.3. We define

$$\left.\begin{aligned} \check{C} &= C^o \oplus C^s \\ \hat{C} &= C^o \oplus C^u \\ \bar{C} &= C^s \oplus C^u. \end{aligned}\right\} \tag{22.3}$$

When necessary, we will write, for example, $\check{C}(Y, \mathfrak{s}, \mathfrak{q})$ to indicate the choices made.

On the free abelian groups $\check{C}$, $\hat{C}$ and $\bar{C}$, we are going to define differentials $\check{\partial}$, $\hat{\partial}$ and $\bar{\partial}$. In general, suppose that $[\mathfrak{a}]$ and $[\mathfrak{b}]$ are two critical points, and consider the moduli space $\breve{M}_z([\mathfrak{a}], [\mathfrak{b}])$. Our regularity assumption means that this moduli space, if non-empty, is a manifold of dimension $d - 1$ or (in the boundary-obstructed case) dimension d, where $d = \mathrm{gr}_z([\mathfrak{a}], [\mathfrak{b}])$ (see Proposition 14.5.7). Suppose that this dimension is 0, and let $[\gamma] \in \breve{M}_z([\mathfrak{a}], [\mathfrak{b}])$. As a 0-manifold, $\breve{M}_z([\mathfrak{a}], [\mathfrak{b}])$ is canonically oriented: we call the preferred orientation of a 0-manifold the *positive* orientation. On the other hand, according to our results on the orientation of moduli spaces in Subsection 20.4, the choice of an element from $\Lambda([\mathfrak{a}], [\mathfrak{b}])$ determines an orientation of $\breve{M}_z([\mathfrak{a}], [\mathfrak{b}])$. Alternatively, we can say that the product

$$\Lambda([\mathfrak{a}], [\mathfrak{b}]) \times \breve{M}_z([\mathfrak{a}], [\mathfrak{b}])$$

is canonically oriented. Comparison of this canonical orientation with the positive orientation at $[\gamma]$ gives a preferred element of $\Lambda([\mathfrak{a}], [\mathfrak{b}])$, or equivalently an isomorphism

$$\epsilon[\gamma] : \mathbb{Z}\Lambda[\mathfrak{a}] \to \mathbb{Z}\Lambda[\mathfrak{b}]. \tag{22.4}$$

The definition of $\epsilon[\gamma]$ is formulated so as to be valid whether or not $[\mathfrak{a}]$, $[\mathfrak{b}]$ are reducible, boundary-stable etc. However, in the case that $[\mathfrak{a}]$ and $[\mathfrak{b}]$ are both reducible, there is a variant of this construction. Let

$$\breve{M}_z^{\mathrm{red}}([\mathfrak{a}], [\mathfrak{b}]) \subset \breve{M}_z([\mathfrak{a}], [\mathfrak{b}])$$

be the moduli space of reducible trajectories, as before. Using the convention in Definition 20.6.1, we have a canonical orientation for

$$\Lambda([\mathfrak{a}], [\mathfrak{b}]) \times \breve{M}_z^{\mathrm{red}}([\mathfrak{a}], [\mathfrak{b}]).$$

In the case that $[\mathfrak{a}] \in \mathfrak{C}^u$ and $[\mathfrak{b}] \in \mathfrak{C}^s$, the canonical orientation is *minus* the boundary orientation of $\Lambda([\mathfrak{a}], [\mathfrak{b}]) \times \breve{M}_z([\mathfrak{a}], [\mathfrak{b}])$, because left multiplication by $\mathbb{R}$ does not commute with "taking the boundary" for oriented manifolds. Suppose that $\breve{M}_z([\mathfrak{a}],[\mathfrak{b}])$ has dimension 1, so that the boundary is zero-dimensional, and let $[\gamma] \in \breve{M}_z^{\text{red}}([\mathfrak{a}],[\mathfrak{b}])$. Comparison of the positive orientation of the zero-dimensional boundary at $[\gamma]$ with the canonical orientation determines an isomorphism,

$$\bar{\epsilon}[\gamma] : \mathbb{Z}\Lambda[\mathfrak{a}] \to \mathbb{Z}\Lambda[\mathfrak{b}]. \tag{22.5}$$

If both critical points are reducible but either $[\mathfrak{a}]$ is boundary-stable or $[\mathfrak{b}]$ is boundary-unstable, then M^{red} and M coincide, as do the canonical orientations of $\Lambda \times M^{\text{red}}$ and $\Lambda \times M$, according to Definition 20.6.1. In these cases, we set:

$$\bar{\epsilon}[\gamma] = \begin{cases} \epsilon[\gamma], & \text{if } [\mathfrak{a}] \in \mathfrak{C}^s \text{ and } [\mathfrak{b}] \in \mathfrak{C}^s \\ -\epsilon[\gamma], & \text{if } [\mathfrak{a}] \in \mathfrak{C}^u \text{ and } [\mathfrak{b}] \in \mathfrak{C}^u \\ \epsilon[\gamma], & \text{if } [\mathfrak{a}] \in \mathfrak{C}^s \text{ and } [\mathfrak{b}] \in \mathfrak{C}^u. \end{cases} \tag{22.6}$$

The explanation for the apparently anomalous minus sign in the "*uu*" case can be found in Case (iii) of Lemma 20.6.4, and will be relevant in Lemma 22.1.6 later in this section.

The simplest of the differentials to define is $\bar{\partial}$ on $\bar{C}$. We define

$$\bar{\partial} : \bar{C} \to \bar{C}$$

by

$$\bar{\partial} = \sum_{[\mathfrak{a}],[\mathfrak{b}],z} \sum_{[\gamma] \in \breve{M}_z^{\text{red}}([\mathfrak{a}],[\mathfrak{b}])} \bar{\epsilon}[\gamma].$$

Here the sum is taken over all $([\mathfrak{a}],[\mathfrak{b}])$ in $(\mathfrak{C}^s \cup \mathfrak{C}^u)^2$ and all moduli spaces $\breve{M}_z^{\text{red}}([\mathfrak{a}],[\mathfrak{b}])$ of dimension zero. Because this sum is potentially infinite, we must take care to see that it is well-defined. Fix $[\mathfrak{a}]$, and consider the sum defining the map

$$\left(\sum_{[\mathfrak{b}],z} \sum_{[\gamma] \in \breve{M}_z^{\text{red}}([\mathfrak{a}],[\mathfrak{b}])} \bar{\epsilon}[\gamma] \right) : \mathbb{Z}\Lambda[\mathfrak{a}] \to \bar{C}.$$

Only finitely many pairs $([\mathfrak{b}], z)$ enter this sum, according to Proposition 16.4.3. (This is the reason for requiring that an "admissible" perturbation should not introduce any reducible critical points in the case of non-torsion spin^c structures.) For a given pair $([\mathfrak{b}], z)$, the number of terms in the above sum is finite, by the following simple consequence of the compactness results. The differential $\bar{\partial}$ is therefore well-defined as a linear operator on $\bar{C}$.

Lemma 22.1.2. *If $\breve{M}_z([\mathfrak{a}],[\mathfrak{b}])$ has dimension zero, then it is a finite set. Similarly, if $\breve{M}_z^{\text{red}}([\mathfrak{a}],[\mathfrak{b}])$ has dimension zero, then it is finite.*

Proof. Proposition 16.5.2 tells us that the compactification $\breve{M}_z^+([\mathfrak{a}],[\mathfrak{b}])$ is a compact space stratified by manifolds, of dimension zero. The stratification is trivial when the dimension is zero, so $\breve{M}_z([\mathfrak{a}],[\mathfrak{b}])$ is already a compact 0-manifold. The case of the reducible moduli space follows similarly from Proposition 16.6.1. □

We can decompose $\bar{\partial}$ with respect to the direct sum decomposition $\bar{C} = C^s \oplus C^u$, as

$$\bar{\partial} = \begin{bmatrix} \bar{\partial}^s_s & \bar{\partial}^u_s \\ \bar{\partial}^s_u & \bar{\partial}^u_u \end{bmatrix}, \tag{22.7}$$

so for example $\bar{\partial}^s_u : C^s \to C^u$. (The superscript indicates the *domain* of the operator: the mnemonic is that we are considering the *downward* gradient flow of $\mathcal{L}$.)

In the same way, we define operators

$$\begin{aligned} \partial^o_o &: C^o \to C^o \\ \partial^o_s &: C^o \to C^s \\ \partial^u_s &: C^u \to C^s \\ \partial^u_o &: C^u \to C^o \end{aligned}$$

by

$$\left.\begin{aligned} \partial^o_o &= \sum_{[\mathfrak{a}]\in\mathfrak{C}^o} \sum_{[\mathfrak{b}]\in\mathfrak{C}^o} \sum_{[\gamma]\in\breve{M}_z([\mathfrak{a}],[\mathfrak{b}])} \epsilon[\gamma] \\ \partial^o_s &= \sum_{[\mathfrak{a}]\in\mathfrak{C}^o} \sum_{[\mathfrak{b}]\in\mathfrak{C}^s} \sum_{[\gamma]\in\breve{M}_z([\mathfrak{a}],[\mathfrak{b}])} \epsilon[\gamma] \\ \partial^u_o &= \sum_{[\mathfrak{a}]\in\mathfrak{C}^u} \sum_{[\mathfrak{b}]\in\mathfrak{C}^o} \sum_{[\gamma]\in\breve{M}_z([\mathfrak{a}],[\mathfrak{b}])} \epsilon[\gamma] \\ \partial^u_s &= \sum_{[\mathfrak{a}]\in\mathfrak{C}^u} \sum_{[\mathfrak{b}]\in\mathfrak{C}^s} \sum_{[\gamma]\in\breve{M}_z([\mathfrak{a}],[\mathfrak{b}])} \epsilon[\gamma]. \end{aligned}\right\} \tag{22.8}$$

The operators $\bar{\partial}^*_*$ whose names have a bar are defined by counting reducible trajectories in zero-dimensional moduli space $\breve{M}^{\text{red}}_z([\mathfrak{a}],[\mathfrak{b}])$, while the operators ∂^*_* without bars count irreducible trajectories in 0-dimensional moduli spaces $\breve{M}_z([\mathfrak{a}],[\mathfrak{b}])$. Note that there are two maps $C^u \to C^s$, namely $\bar{\partial}^u_s$ and ∂^u_s, but the moduli spaces which contribute to these two maps are different: if $\breve{M}^{\text{red}}_z([\mathfrak{a}],[\mathfrak{b}])$ is 0-dimensional, then $\breve{M}_z([\mathfrak{a}],[\mathfrak{b}])$ will be 1-dimensional, having $\breve{M}^{\text{red}}_z([\mathfrak{a}],[\mathfrak{b}])$ as its boundary.

There is no operator of this sort from C^o to C^u or from $C^s \to C^o$, because the corresponding moduli spaces are empty. We do however have *composite* operators

$$\bar{\partial}^s_u \partial^o_s : C^o \to C^u$$
$$\partial^u_o \bar{\partial}^s_u : C^s \to C^o.$$

Note that the operator $\bar{\partial}^s_u$ which is common to both of these composites counts trajectories belonging to boundary-obstructed moduli spaces.

Definition 22.1.3. On $\check{C} = C^o \oplus C^s$, we define an operator

$$\check{\partial} : \check{C} \to \check{C}$$
$$\check{\partial} = \begin{bmatrix} \partial^o_o & -\partial^u_o \bar{\partial}^s_u \\ \partial^o_s & \bar{\partial}^s_s - \partial^u_s \bar{\partial}^s_u \end{bmatrix}.$$

On $\hat{C} = C^o \oplus C^u$, we define

$$\hat{\partial} : \hat{C} \to \hat{C}$$
$$\hat{\partial} = \begin{bmatrix} \partial^o_o & \partial^u_o \\ -\bar{\partial}^s_u \partial^o_s & -\bar{\partial}^u_u - \bar{\partial}^s_u \partial^u_s \end{bmatrix}.$$

◊

Proposition 22.1.4. *The squares* $(\bar{\partial})^2$, $(\check{\partial})^2$ *and* $(\hat{\partial})^2$ *are zero as operators on* $\bar{C}$, $\check{C}$ *and* $\hat{C}$.

The fact that $\check{\partial}$ and $\hat{\partial}$ have square zero will follow from $\bar{\partial}^2 = 0$ and the identities in the following lemma.

Lemma 22.1.5. *We have the following identities.*

(i) $-\partial^o_o \partial^o_o + \partial^u_o \bar{\partial}^s_u \partial^o_s = 0.$
(ii) $-\partial^o_s \partial^o_o - \bar{\partial}^s_s \partial^o_s + \partial^u_s \bar{\partial}^s_u \partial^o_s = 0.$

(iii) $-\partial^o_o \partial^u_o + \partial^u_o \bar{\partial}^u_u + \partial^u_o \bar{\partial}^s_u \partial^u_s = 0$.
(iv) $-\bar{\partial}^u_s - \partial^o_s \partial^u_o - \bar{\partial}^s_s \partial^u_s + \partial^u_s \bar{\partial}^u_u + \partial^u_s \bar{\partial}^s_u \partial^u_s = 0$.

Proof. Each of the four parts is proved by considering a moduli space $\breve{M}_z([\mathfrak{a}], [\mathfrak{b}])$ of dimension 1 containing irreducible trajectories. The four operators on the left-hand sides of the identities are operators

$$A^1 : C^o \to C^o$$
$$A^2 : C^o \to C^s$$
$$A^3 : C^u \to C^o$$
$$A^4 : C^u \to C^s.$$

Each is a sum

$$A^i = \sum A^i_{[\mathfrak{a}][\mathfrak{b}]},$$

where

$$A^i_{[\mathfrak{a}][\mathfrak{b}]} : \mathbb{Z}\Lambda[\mathfrak{a}] \to \mathbb{Z}\Lambda[\mathfrak{b}]$$

and the sum is taken over all $[\mathfrak{a}]$ and $[\mathfrak{b}]$ of the appropriate type: both irreducible for A^1 and so on.

Let $\breve{M}_z([\mathfrak{a}], [\mathfrak{b}])$ be such a moduli space, so $\mathrm{gr}_z([\mathfrak{a}], [\mathfrak{b}])$ is 2. According to Theorem 19.5.4, the space $\breve{M}_z([\mathfrak{a}], [\mathfrak{b}])$ is the top stratum in a compact 1-dimensional space $\breve{M}^+([\mathfrak{a}], [\mathfrak{b}])$ stratified by manifolds, and has a codimension-1 δ-structure along the zero-dimensional stratum N^0. Choose trivializations of $\Lambda[\mathfrak{a}]$ and $\Lambda[\mathfrak{b}]$. Then both $\breve{M}_z([\mathfrak{a}], [\mathfrak{b}])$ and the zero-dimensional stratum N^0 obtain canonical orientations; Proposition 20.5.2 tells us how the canonical orientation of N^0 differs from the boundary orientation on each component. According to Corollary 21.3.2, the number of points in N^0, counted with the boundary-orientation signs, is zero. The proof of the proposition comes from equating the number of points in N^0, with these signs, to the component $A^i_{[\mathfrak{a}][\mathfrak{b}]}$, regarded now as an integer using the trivializations.

In each case, $\breve{M}^+([\mathfrak{a}], [\mathfrak{b}])$ is obtained from $\breve{M}([\mathfrak{a}], [\mathfrak{b}])$ by adjoining broken trajectories $([\breve{\gamma}_1], \ldots, [\breve{\gamma}_l])$ with $l = 2$ or 3 and grading vector

$$(\mathrm{gr}_{z_1}([\mathfrak{a}_0], [\mathfrak{a}_1]), \ldots, \mathrm{gr}_{z_l}([\mathfrak{a}_{l-1}], [\mathfrak{a}_l])) \tag{22.9}$$

equal to either

$$(1, 1) \quad \text{or} \quad (1, 0, 1), \tag{22.10}$$

because these are the only two possibilities allowed by Proposition 16.5.5. In any event, each $[\gamma_i]$ belongs to a 1-dimensional moduli space $M_{z_i}([\mathfrak{a}_{i-1}], [\mathfrak{a}_i])$.

We now consider each of the four parts of the lemma in turn. To begin with the first part, let $[\mathfrak{a}]$ and $[\mathfrak{b}]$ be irreducible critical points and suppose that $\mathrm{gr}_z([\mathfrak{a}], [\mathfrak{b}]) = 2$, so that $\breve{M}_z([\mathfrak{a}], [\mathfrak{b}])$ is 1-dimensional. There are two types of broken trajectory in the compactification, corresponding to the two types (22.10). Consider the first type, $(1, 1)$, and a broken trajectory belonging to a product

$$\breve{M}_{z_1}([\mathfrak{a}], [\mathfrak{a}_1]) \times \breve{M}_{z_2}([\mathfrak{a}_1], [\mathfrak{b}])$$

of (compact) 0-dimensional moduli spaces. Note that $[\mathfrak{a}_1]$ must be irreducible, because if it were reducible and boundary-stable, then the second moduli space would be empty, while if it were boundary-unstable then the first would be empty. According to Corollary 19.5.1, a neighborhood of this product in the compactification is a topological 1-manifold with boundary. Furthermore, Proposition 20.5.2 of Subsection 20.5 tells us that the boundary orientation of this 0-dimensional stratum is the opposite of the boundary orientation. The number of points in all the strata of this form, taken over all irreducible critical points $[\mathfrak{a}_1]$ and counted with the canonical orientations, is the component of $\partial^o_o \partial^o_o$ between $\mathbb{Z}\Lambda[\mathfrak{a}]$ and $\mathbb{Z}\Lambda[\mathfrak{b}]$. With the boundary orientations, the number of points is the component of $-\partial^o_o \partial^o_o$ and accounts for the first term in the first identity.

The second possibility allowed for in (22.10) is that the compactification contains broken trajectories with three components, belonging to a stratum

$$\breve{M}_{z_1}([\mathfrak{a}], [\mathfrak{a}_1]) \times \breve{M}_{z_2}([\mathfrak{a}_1], [\mathfrak{a}_2]) \times \breve{M}_{z_3}([\mathfrak{a}_2], [\mathfrak{b}]). \tag{22.11}$$

The middle piece should be boundary-obstructed, so $[\mathfrak{a}_1]$ is boundary stable and $[\mathfrak{a}_2]$ is boundary-unstable. This stratum is again a finite set. The space $\breve{M}^+_z([\mathfrak{a}], [\mathfrak{b}])$ has a codimension-1 δ-structure along this 0-dimensional stratum, and the canonical orientation agrees with the boundary orientation according to Proposition 20.5.2. The number of points in strata of this sort, counted with sign, is the matrix entry of $\partial^u_o \bar{\partial}^s_u \partial^o_s$ from $\mathbb{Z}\Lambda[\mathfrak{a}]$ to $\mathbb{Z}\Lambda[\mathfrak{b}]$, and accounts for the second term in the first identity.

The second and third identities are entirely similar. Take the second, for example. It is proved by considering a moduli space $\breve{M}_z([\mathfrak{a}], [\mathfrak{b}])$ of dimension 1 as above, but now with $[\mathfrak{a}]$ irreducible and $[\mathfrak{b}]$ reducible and boundary-stable. There are now two types of possible factorization of type $(1, 1)$, because the intermediate critical point $[\mathfrak{a}_1]$ can be either irreducible, or reducible and boundary-stable. These two possibilities account for the first two terms of the identity. The last term corresponds to the factorization of type $(1, 0, 1)$. The third identity has the same form: the second term has positive sign in this identity rather than the expected minus sign because the definition of $\bar{\partial}^u_u$ used the maps $\bar{\epsilon}[\gamma]$, which in this case are the negative of the standard map $\epsilon[\gamma]$.

The fourth identity is proved similarly, using a moduli space $\breve{M}([\mathfrak{a}], [\mathfrak{b}])$ of dimension 1, with $[\mathfrak{a}]$ boundary-unstable and $[\mathfrak{b}]$ boundary-stable. Even before compactifying, this moduli space is a manifold with boundary, and the boundary consists of the reducibles, which are counted by $\bar{\partial}^u_s$. This gives the first term. The minus sign is there because, as we remarked earlier, the canonical orientation of the moduli spaces $\Lambda \times \breve{M}^{\text{red}}$ that appear in the definition of $\bar{\partial}^u_s$ is minus the boundary orientation of $\Lambda \times \breve{M}$. The next three terms are the possible factorizations of type $(1, 1)$, where the intermediate critical point can be irreducible, reducible and boundary-stable, or boundary-unstable. The last term is the factorization of type $(1, 0, 1)$. The fourth term in this last identity has a plus sign for the same reason that a plus sign appeared in the second term of the third identity. □

The next lemma disposes of the case of the operator $\bar{\partial}$ in Proposition 22.1.4.

Lemma 22.1.6. *The operator $\bar{\partial}$ has square zero on $\bar{C}$. In terms of the four components of $\bar{\partial}$, we have*

$$\bar{\partial}^s_s \bar{\partial}^s_s + \bar{\partial}^u_s \bar{\partial}^s_u = 0$$
$$\bar{\partial}^s_s \bar{\partial}^u_s + \bar{\partial}^u_s \bar{\partial}^u_u = 0$$
$$\bar{\partial}^s_u \bar{\partial}^s_s + \bar{\partial}^u_u \bar{\partial}^s_u = 0$$
$$\bar{\partial}^s_u \bar{\partial}^u_s + \bar{\partial}^u_u \bar{\partial}^u_u = 0.$$

Proof. This is proved in just the same way as Lemma 22.1.5 above, but now using moduli spaces of reducible trajectories, $\breve{M}^{\text{red}}_z([\mathfrak{a}], [\mathfrak{b}])$, of dimension 1. The compactification $\breve{M}^{\text{red}+}_z([\mathfrak{a}], [\mathfrak{b}])$ is a C^0 1-manifold with boundary, and the boundary arises from broken trajectories, forming 0-dimensional strata

$$\breve{M}^{\text{red}}_{z_1}([\mathfrak{a}], [\mathfrak{a}_1]) \times \breve{M}^{\text{red}}_{z_2}([\mathfrak{a}_1], [\mathfrak{b}]).$$

Such factorizations account for all the terms in the four identities, and the only remaining thing to verify is the signs with which they enter. If a trivialization is chosen for $\Lambda([\mathfrak{a}], [\mathfrak{b}])$, so that $M^{\text{red}}_z([\mathfrak{a}], [\mathfrak{b}])$ and the product stratum above are both canonically oriented, then the canonical orientation differs from the boundary orientation by a sign that can be computed from Proposition 20.6.2

and Lemma 20.6.4. Proposition 20.6.2 tells us that there is a sign -1 if we were to orient the moduli spaces $\breve{M}^{\text{red}}$ differently, using the $\bar{\Lambda}$ orientations in place of the Λ orientations. Meanwhile, Lemma 20.6.4 compares the $\bar{\Lambda}$ orientations with the Λ orientations, and reveals an extra sign $(-1)^d = -1$ in the case that both critical points are boundary-unstable. This extra sign in the case of boundary-unstable critical points is the reason we introduced the compensating minus sign in the definition of $\bar{\epsilon}[\gamma]$ in (22.6). □

Proof of Proposition 22.1.4. We have already dealt with $\bar{\partial}^2$ in the lemma above. We compute the square of $\check{\partial}$, as a 2-by-2 matrix (Definition 22.1.3). The top left entry is

$$\partial^o_o \partial^o_o - \partial^u_o \bar{\partial}^s_u \partial^o_s,$$

which is zero by the first identity in Lemma 22.1.5. The top right entry is

$$\begin{aligned} -\partial^o_o \partial^u_o \bar{\partial}^s_u - \partial^u_o \bar{\partial}^s_u \bar{\partial}^s_s + \partial^u_o \bar{\partial}^s_u \partial^u_s \bar{\partial}^s_u & \\ &= -\partial^o_o \partial^u_o \bar{\partial}^s_u + \partial^u_o \bar{\partial}^u_u \bar{\partial}^s_u + \partial^u_o \bar{\partial}^s_u \partial^u_s \bar{\partial}^s_u \\ &= (-\partial^o_o \partial^u_o + \partial^u_o \bar{\partial}^u_u - \partial^u_o \bar{\partial}^s_u \partial^u_s) \bar{\partial}^s_u \\ &= 0. \end{aligned}$$

We used $\bar{\partial}^s_u \bar{\partial}^s_s + \bar{\partial}^u_u \bar{\partial}^s_u = 0$ in the first line (part of the statement that $\bar{\partial}^2 = 0$), and we used the third identity of Lemma 22.1.5 in the last line. The bottom left entry is

$$\partial^o_s \partial^o_o + \bar{\partial}^s_s \partial^o_s - \partial^u_s \bar{\partial}^s_u \partial^o_s$$

which vanishes because of the second identity of the lemma. The bottom right entry is

$$\begin{aligned} -\partial^o_s \partial^u_o \bar{\partial}^s_u + \bar{\partial}^s_s \bar{\partial}^s_s - \partial^u_s \bar{\partial}^s_u \bar{\partial}^s_s - \bar{\partial}^s_s \partial^u_s \bar{\partial}^s_u + \partial^u_s \bar{\partial}^s_u \partial^u_s \bar{\partial}^s_u & \\ &= -\partial^o_s \partial^u_o \bar{\partial}^s_u - \bar{\partial}^u_s \bar{\partial}^s_u + \partial^u_s \bar{\partial}^u_u \bar{\partial}^s_u - \bar{\partial}^s_s \partial^u_s \bar{\partial}^s_u + \partial^u_s \bar{\partial}^s_u \partial^u_s \bar{\partial}^s_u \\ &= (-\partial^o_s \partial^u_o - \bar{\partial}^u_s + \partial^u_s \bar{\partial}^u_u - \bar{\partial}^s_s \partial^u_s + \partial^u_s \bar{\partial}^s_u \partial^u_s) \bar{\partial}^s_u \\ &= 0 \end{aligned}$$

using the identities $\bar{\partial}^s_s \bar{\partial}^s_s = -\bar{\partial}^u_s \bar{\partial}^s_s$, $\bar{\partial}^s_u \bar{\partial}^s_s = -\bar{\partial}^u_u \bar{\partial}^s_u$ and the last identity of Lemma 22.1.5. Thus all four entries of $\check{\partial}^2$ are zero.

The proof that $\hat{\partial}^2$ is zero is entirely similar. The top left entry of the square of this operator is the same as the corresponding entry of $\check{\partial}^2$, and is zero as above.

The top right entry is

$$\partial_o^o \partial_o^u - \partial_o^u \bar{\partial}_u^u - \partial_o^u \bar{\partial}_u^s \partial_s^u,$$

which vanishes by the third part of Lemma 22.1.5. The bottom left entry is

$$\begin{aligned}
-\bar{\partial}_u^s \partial_s^o \partial_o^o &+ \bar{\partial}_u^u \bar{\partial}_u^s \partial_s^o + \bar{\partial}_u^s \partial_s^u \bar{\partial}_u^s \partial_s^o \\
&= -\bar{\partial}_u^s \partial_s^o \partial_o^o - \bar{\partial}_u^s \bar{\partial}_s^s \partial_s^o + \bar{\partial}_u^s \partial_s^u \bar{\partial}_u^s \partial_s^o \\
&= \bar{\partial}_u^s (-\partial_s^o \partial_o^o - \bar{\partial}_s^s \partial_s^o + \partial_s^u \bar{\partial}_u^s \partial_s^o),
\end{aligned}$$

which is zero by the second identity of the lemma. Finally, the bottom right entry is

$$\begin{aligned}
-\bar{\partial}_u^s \partial_s^o \partial_o^u &+ \bar{\partial}_u^u \bar{\partial}_u^u + \bar{\partial}_u^u \bar{\partial}_u^s \partial_s^u + \bar{\partial}_u^s \partial_s^u \bar{\partial}_u^u + \bar{\partial}_u^s \partial_s^u \bar{\partial}_u^s \partial_s^u \\
&= -\bar{\partial}_u^s \partial_s^o \partial_o^u - \bar{\partial}_u^s \bar{\partial}_s^u - \bar{\partial}_u^s \bar{\partial}_s^s \partial_s^u + \bar{\partial}_u^s \partial_s^u \bar{\partial}_u^u + \bar{\partial}_u^s \partial_s^u \bar{\partial}_u^s \partial_s^u \\
&= \bar{\partial}_u^s (-\partial_s^o \partial_o^u - \bar{\partial}_s^u - \bar{\partial}_s^s \partial_s^u + \partial_s^u \bar{\partial}_u^u + \partial_s^u \bar{\partial}_u^s \partial_s^u),
\end{aligned}$$

which is zero, on account of the last identity of the lemma once again. □

We can now define the *monopole Floer homology groups* of Y, in three flavors, as the homology of our three differential groups, $(\check{C}, \check{\partial})$, $(\hat{C}, \hat{\partial})$ and $(\bar{C}, \bar{\partial})$.

Definition 22.1.7. We define

$$\begin{aligned}
\widecheck{HM}_*(Y, \mathfrak{s}) &= H(\check{C}, \check{\partial}) = \ker(\check{\partial})/\mathrm{im}(\check{\partial}) \\
\widehat{HM}_*(Y, \mathfrak{s}) &= H(\hat{C}, \hat{\partial}) = \ker(\hat{\partial})/\mathrm{im}(\hat{\partial}) \\
\overline{HM}_*(Y, \mathfrak{s}) &= H(\bar{C}, \bar{\partial}) = \ker(\bar{\partial})/\mathrm{im}(\bar{\partial}).
\end{aligned}$$

When necessary, we will write $\widecheck{HM}_*(Y, \mathfrak{s}, \mathfrak{q})$ to indicate the a priori dependence on a choice of perturbation. At present, the subscript $*$ is just for decoration, as we have not defined a grading on these groups. ◊

22.2 The exact sequence of the pair

As explained in Section 3 and Subsection 2.4, we are to think of $\widecheck{HM}_*(Y, \mathfrak{s})$, $\widehat{HM}_*(Y, \mathfrak{s})$ and $\overline{HM}_*(Y, \mathfrak{s})$ as half-infinite-dimensional homology groups of $\mathcal{B}_k^\sigma(Y, \mathfrak{s})$, of the pair $(\mathcal{B}_k^\sigma(Y, \mathfrak{s}), \partial\mathcal{B}_k^\sigma(Y, \mathfrak{s}))$ and of the boundary $\partial\mathcal{B}_k^\sigma(Y, \mathfrak{s})$

respectively. In this analogy, the following exact sequence is the long exact sequence of the pair.

Proposition 22.2.1. *For any $(Y, \mathfrak{s})$, there is an exact sequence*

$$\cdots \xrightarrow{i_*} \widecheck{HM}_*(Y, \mathfrak{s}) \xrightarrow{j_*} \widehat{HM}_*(Y, \mathfrak{s}) \xrightarrow{p_*} \overline{HM}_*(Y, \mathfrak{s}) \xrightarrow{i_*} \widecheck{HM}_*(Y, \mathfrak{s}) \xrightarrow{j_*} \cdots$$

in which the maps i_*, j_* and p_* arise from the (anti-) chain maps

$$i : \bar{C} \to \check{C}, \qquad j : \check{C} \to \hat{C}, \qquad p : \hat{C} \to \bar{C},$$

given by

$$i = \begin{bmatrix} 0 & -\partial^u_o \\ 1 & -\partial^u_s \end{bmatrix}, \qquad j = \begin{bmatrix} 1 & 0 \\ 0 & -\bar{\partial}^s_u \end{bmatrix}, \qquad p = \begin{bmatrix} \partial^o_s & \partial^u_s \\ 0 & 1 \end{bmatrix}.$$

Here i and j are genuine chain maps, while p satisfies $p\hat{\partial} = -\bar{\partial}p$ (an anti-chain map).

Proof. First, it is straightforward to verify that i, j and p are (anti-) chain maps, using the identities of Lemma 22.1.5 and Lemma 22.1.6. To verify exactness, we introduce the differential group $(\check{E}, \check{e})$ that is the mapping cone of $-p$:

$$\check{E} = \hat{C} \oplus \bar{C} = (C^o \oplus C^u) \oplus (C^s \oplus C^u)$$
$$\check{e} = \begin{bmatrix} \hat{\partial} & 0 \\ p & \bar{\partial} \end{bmatrix}.$$

The fact that $\check{e}^2 = 0$ follows from $p\hat{\partial} = -\bar{\partial}p$. We have a long exact sequence of the mapping cone,

$$\cdots \xrightarrow{\bar{\imath}_*} H(\check{E}) \xrightarrow{\bar{\jmath}_*} \widehat{HM} \xrightarrow{-p_*} \overline{HM} \xrightarrow{\bar{\imath}_*} H(\check{E}) \xrightarrow{\bar{\jmath}_*} \cdots,$$

which results from the short exact sequence of differential groups

$$\bar{C} \xrightarrow{\bar{\imath}} \check{E} \xrightarrow{\bar{\jmath}} \hat{C}.$$

The boundary map in the long exact sequence is p_*, where p is as before. It remains, then, to identify $H(\check{E})$ with $\widecheck{HM}$ in such a way that the sequence above becomes the sequence in the proposition.

Define maps

$$k : \check{C} \to \check{E}, \qquad l : \check{E} \to \check{C}$$

by

$$k\begin{bmatrix} x \\ y \end{bmatrix} = \begin{bmatrix} x \\ -\bar{\partial}^s_u y \\ y \\ 0 \end{bmatrix}, \qquad l\begin{bmatrix} e \\ f \\ g \\ h \end{bmatrix} = \begin{bmatrix} e - \partial^u_o h \\ g - \partial^u_s h \end{bmatrix}.$$

The maps k and l are chain maps, as follows again from the identities in the two lemmas above. The composite $lk : \check{C} \to \check{C}$ is the identity, and the composite $kl : \check{E} \to \check{E}$ is chain-homotopic to the identity, for we have

$$kl = 1 + \check{e}K + K\check{e}$$

where

$$K\begin{bmatrix} e \\ f \\ g \\ h \end{bmatrix} = \begin{bmatrix} 0 \\ -h \\ 0 \\ 0 \end{bmatrix}.$$

It follows that k_* and l_* are mutually inverse isomorphisms between $\widecheck{HM}$ and $H(\check{E})$.

Finally, we should see that the isomorphism k_* carries the sequence of the proposition to the mapping-cone sequence. That is, we must verify that $j_* = \bar{\jmath}_* k_*$ and $\bar{\imath}_* = k_* i_*$. The first of these identities is true because it is true at the chain level: we have $j = \bar{\jmath}k$. The second identity is true because the corresponding chain maps $\bar{\imath}$ and ki are chain-homotopic: we have

$$ki - \bar{\imath} = \check{e}\bar{K} + \bar{K}\bar{\partial},$$

where $\bar{K} = K\bar{\imath}$. □

We return to a definition introduced in Subsection 3.6 (see Definition 3.6.3).

Definition 22.2.2. We define $HM_*(Y, \mathfrak{s})$ as the image of the map

$$j_* : \widecheck{HM}_*(Y, \mathfrak{s}) \to \widehat{HM}_*(Y, \mathfrak{s}).$$

This is the *reduced* Floer homology of $(Y, \mathfrak{s})$. ◊

Proposition 22.2.3. *The reduced Floer homology $HM_*(Y, \mathfrak{s})$ is of finite rank.*

Proof. If $c_1(\mathfrak{s})$ is not torsion, then the set of critical points is only $\mathfrak{C}^o$, which is finite. In this case, $\widecheck{HM}_*$, $\widehat{HM}_*$ and HM_* all coincide and are of finite rank. In the case that $c_1(\mathfrak{s})$ is torsion, we need:

Lemma 22.2.4. *The maps $\bar{\partial}^s_u$, $\bar{\partial}^u_s$ and ∂^u_s have only finitely many non-zero matrix entries in the natural bases of C^s and C^u. In particular, they have finite rank.*

Proof. These maps only arise in the case that $c_1(\mathfrak{s})$ is torsion, and in this case $\mathrm{gr}_z([\mathfrak{a}], [\mathfrak{b}])$ is independent of z. There are only finitely many critical points of $\mathrm{grad}\,\mathcal{L}$ in $\mathcal{B}_k(Y, \mathfrak{s})$ before blowing up; let $[\alpha]$ and $[\beta]$ be two of these critical points, and let $[\mathfrak{a}_i]$, $[\mathfrak{b}_j]$ be critical points in the blow-up, lying over $[\alpha]$ and $[\beta]$ and having $\iota[\mathfrak{a}_i] = i$ and $\iota[\mathfrak{b}_j] = j$. (The definition of ι is at (16.2), and the quantities i and j are each either a non-negative integer or a negative element of $\mathbb{Z} + 1/2$.)

A matrix entry of $\bar{\partial}^s_u$ may arise from $M_z([\mathfrak{a}_i], [\mathfrak{b}_j])$ only when i is positive, j is negative and $\mathrm{gr}_z([\mathfrak{a}_i], [\mathfrak{b}_j])$ is zero. But for some d depending only on $[\alpha]$ and $[\beta]$ we have

$$\mathrm{gr}_z([\mathfrak{a}_i], [\mathfrak{b}_j]) = d + 2i - 2j.$$

So there are only finitely many non-zero matrix entries. The situation with $\bar{\partial}^u_s$ and ∂^u_s is similar. □

Returning to the proof of the proposition, let us suppose that the rank of $HM_*(Y, \mathfrak{s})$ exceeds the number of points in $\mathfrak{C}^o(\mathfrak{s})$ by r. Then there is a subgroup H of $\widecheck{HM}_*(Y, \mathfrak{s})$ of rank r generated by elements having representatives in

$$0 \oplus C^s \subset C^o \oplus C^s = \check{C},$$

such that the restriction of j_* to H is injective. But if $(0, y)$ is such a representative of an element of H, then

$$j(0, y) = (0, \bar{\partial}^s_u y).$$

So the rank of H is no larger than the rank of $\bar{\partial}^s_u$. This shows that

$$\text{rank } HM_* \le \text{rank } C^o + \text{rank } \bar{\partial}^s_u,$$

which is finite on account of the lemma. □

22.3 Gradings and cohomology

In Section 3, we explained what it meant for a group G to be graded by a set $\mathbb{J}$ with an action of $\mathbb{Z}$, and we stated that the Floer groups of Y would be graded, in this sense, by the set of homotopy classes of oriented 2-plane fields on Y. This will be taken up in Section 28, but we take a first step in this direction here. For each Y, with choice of metric and spin^c structure, we define a set $\mathbb{J}(\mathfrak{s}) = \mathbb{J}(Y, \mathfrak{s})$, in such a way that the group $\widetilde{HM}_*(Y, \mathfrak{s})$ and its relatives are graded (rather tautologically) by $\mathbb{J}(\mathfrak{s})$.

We define $\mathbb{J}(Y, \mathfrak{s})$ as the quotient of $\mathcal{B}^\sigma_k(Y, \mathfrak{s}) \times \mathcal{P} \times \mathbb{Z}$ by an equivalence relation whose definition is as follows. Let $([\mathfrak{a}], \mathfrak{q}_1, m)$, $([\mathfrak{b}], \mathfrak{q}_2, n)$ be in $\mathcal{B}^\sigma_k(Y, \mathfrak{s}) \times \mathcal{P} \times \mathbb{Z}$, let ζ be a path joining $[\mathfrak{a}]$ and $[\mathfrak{b}]$, and let $\mathfrak{p}$ be a 1-parameter family of perturbations joining $\mathfrak{q}_1$ to $\mathfrak{q}_2$. Associated to ζ and $\mathfrak{p}$, we have a Fredholm operator $P_{\gamma,\mathfrak{p}}$ as in (20.6). Its index may depend on the choice of the homotopy class of the path ζ. We say that $([\mathfrak{a}], \mathfrak{q}_1, m) \sim ([\mathfrak{b}], \mathfrak{q}_2, n)$ if there exists a path ζ such that

$$\operatorname{index}(P_{\gamma,\mathfrak{p}}) = n - m.$$

Note that if $[\mathfrak{a}]$ and $[\mathfrak{b}]$ are critical points for a fixed perturbation, then index (P_γ) is equal to $\operatorname{gr}_z([\mathfrak{a}], [\mathfrak{b}])$. We define, then,

$$\mathbb{J}(\mathfrak{s}) = \big(\mathcal{B}^\sigma_k(Y, \mathfrak{s}) \times \mathcal{P} \times \mathbb{Z}\big)/\sim. \tag{22.12}$$

The map $([\mathfrak{a}], \mathfrak{q}, m) \mapsto ([\mathfrak{a}], \mathfrak{q}, m+1)$ descends to $\mathbb{J}(\mathfrak{s})$, and generates an action of $\mathbb{Z}$. As usual, we write this action additively: for $j \in \mathbb{J}(\mathfrak{s})$ and $n \in \mathbb{Z}$, we write $j + n$ for the resulting element of $\mathbb{J}(\mathfrak{s})$.

Let $\mathfrak{q}$ now be a fixed admissible perturbation. For a critical point $[\mathfrak{a}]$, either irreducible or reducible, we now define

$$\begin{aligned} \operatorname{gr}[\mathfrak{a}] &= ([\mathfrak{a}], \mathfrak{q}, 0)/\sim \\ &\in \mathbb{J}(\mathfrak{s}). \end{aligned} \tag{22.13}$$

For any path z joining $[\mathfrak{a}]$ and $[\mathfrak{b}]$, we have

$$\operatorname{gr}[\mathfrak{a}] = \operatorname{gr}[\mathfrak{b}] + \operatorname{gr}_z([\mathfrak{a}], [\mathfrak{b}]) \in \mathbb{J}(\mathfrak{s}). \tag{22.14}$$

For reducible critical points, we introduce a modified grading, defining

$$\bar{\operatorname{gr}}[\mathfrak{a}] = \begin{cases} \operatorname{gr}[\mathfrak{a}], & [\mathfrak{a}] \in \mathfrak{C}^s \\ \operatorname{gr}[\mathfrak{a}] - 1, & [\mathfrak{a}] \in \mathfrak{C}^u. \end{cases} \tag{22.15}$$

For each $j \in \mathbb{J}(\mathfrak{s})$, we set

$$\left.\begin{aligned} \check{C}_j &= \bigoplus \{ \mathbb{Z}\Lambda[\mathfrak{a}] \mid [\mathfrak{a}] \in \mathfrak{C}^o \cup \mathfrak{C}^s,\ \mathrm{gr}[\mathfrak{a}] = j \} \\ \hat{C}_j &= \bigoplus \{ \mathbb{Z}\Lambda[\mathfrak{a}] \mid [\mathfrak{a}] \in \mathfrak{C}^o \cup \mathfrak{C}^u,\ \mathrm{gr}[\mathfrak{a}] = j \} \\ \bar{C}_j &= \bigoplus \{ \mathbb{Z}\Lambda[\mathfrak{a}] \mid [\mathfrak{a}] \in \mathfrak{C}^s \cup \mathfrak{C}^u,\ \bar{\mathrm{gr}}[\mathfrak{a}] = j \}. \end{aligned}\right\} \tag{22.16}$$

In this way, each of $\check{C}$, $\hat{C}$ and $\bar{C}$ becomes a direct sum of groups indexed by $\mathbb{J}(\mathfrak{s})$. If we index each of the constituent pieces C^o, C^s and C^u using the grading gr (rather than $\bar{\mathrm{gr}}$ in the case of the latter two), then we can write

$$\left.\begin{aligned} \check{C}_j &= C^o_j \oplus C^s_j \\ \hat{C}_j &= C^o_j \oplus C^u_j \\ \bar{C}_j &= C^s_j \oplus C^u_{j+1}. \end{aligned}\right\} \tag{22.17}$$

The next lemma follows easily from the definitions:

Lemma 22.3.1. *The boundary operators* $\check{\partial} : \check{C} \to \check{C}$, $\hat{\partial} : \hat{C} \to \hat{C}$ *and* $\bar{\partial} : \bar{C} \to \bar{C}$ *all have degree* -1*: that is, we have*

$$\begin{aligned} \check{\partial}(\check{C}_j) &\subset \check{C}_{j-1} \\ \hat{\partial}(\hat{C}_j) &\subset \hat{C}_{j-1} \\ \bar{\partial}(\bar{C}_j) &\subset \bar{C}_{j-1}. \end{aligned}$$

In the same sense, the chain maps i, j and p of Proposition 22.2.1 *have degree* 0, 0 *and* -1 *respectively.* □

In particular, we have a long exact sequence

$$\xrightarrow{i_*} \widecheck{HM}_k(Y,\mathfrak{s}) \xrightarrow{j_*} \widehat{HM}_k(Y,\mathfrak{s}) \xrightarrow{p_*} \overline{HM}_{k-1}(Y,\mathfrak{s}) \xrightarrow{i_*} \widecheck{HM}_{k-1}(Y,\mathfrak{s}) \xrightarrow{j_*},$$

and we can define the reduced Floer homology group $HM_k(Y,\mathfrak{s})$ as the image of $j_* : \widecheck{HM}_k(Y,\mathfrak{s}) \to \widehat{HM}_k(Y,\mathfrak{s})$.

We now identify $\mathbb{J}(\mathfrak{s})$ up to isomorphism, as a set with $\mathbb{Z}$ action.

Lemma 22.3.2. *The action of $\mathbb{Z}$ on $\mathbb{J}(\mathfrak{s})$ is transitive. The stabilizer is the image of the map*

$$\begin{aligned} H_2(Y;\mathbb{Z}) &\to \mathbb{Z} \\ [\sigma] &\mapsto \langle c_1(\mathfrak{s}), [\sigma] \rangle. \end{aligned} \tag{22.18}$$

In particular, the action is free if and only if $c_1(\mathfrak{s})$ is torsion, and the stabilizer is always contained in $2\mathbb{Z}$.

Proof. The transitivity of the action follows from the definitions and from the fact that $\mathcal{B}_k^\sigma(Y,\mathfrak{s})$ is connected. In the same way, it follows that the stabilizer is precisely the subgroup of $\mathbb{Z}$ consisting of the quantities index (P_γ) as γ runs through all closed loops in $\mathcal{B}_k^\sigma(Y,\mathfrak{s})$. As is shown by Lemma 14.4.6, this subgroup is the image of the map (22.18). □

Lemma 22.3.3. *For each j in $\mathbb{J}(\mathfrak{s})$, the free abelian groups $\check{C}_j$, $\hat{C}_j$ and $\bar{C}_j$ are all finitely generated.*

Proof. If $c_1(\mathfrak{s})$ is not torsion, the critical-point set is finite, and there is nothing to do. In the torsion case, the action of $\mathbb{Z}$ on $\mathbb{J}(\mathfrak{s})$ is free. If $[\mathfrak{a}_i]$ and $[\mathfrak{a}_j]$ are reducible critical points whose image in $\mathcal{B}_k(Y,\mathfrak{s})$ is the same critical point $[\alpha]$, then as stated in (16.3) we have

$$\mathrm{gr}[\mathfrak{a}_i] - \mathrm{gr}[\mathfrak{a}_j] = 2\iota([\mathfrak{a}_i]) - 2\iota([\mathfrak{a}_j]).$$

So the critical points lying above $[\alpha]$ all have different gradings, and they appear in every other grading. Since there are only finitely many $[\alpha]$ to consider, the lemma follows. □

At this point we can also introduce the *cochain* complexes graded by $\mathbb{J}(\mathfrak{s})$,

$$\begin{aligned}\check{C}^j &= \mathrm{Hom}(\check{C}_j,\mathbb{Z})\\ \hat{C}^j &= \mathrm{Hom}(\hat{C}_j,\mathbb{Z})\\ \bar{C}^j &= \mathrm{Hom}(\bar{C}_j,\mathbb{Z}),\end{aligned}$$

and the corresponding Floer cohomology groups,

$$\widecheck{HM}^j(Y,\mathfrak{s}),\quad \widehat{HM}^j(Y,\mathfrak{s}),\quad \overline{HM}^j(Y,\mathfrak{s}).$$

There is a long exact sequence

$$\xleftarrow{i^*} \widecheck{HM}^k(Y,\mathfrak{s}) \xleftarrow{j^*} \widehat{HM}^k(Y,\mathfrak{s}) \xleftarrow{p^*} \overline{HM}^{k-1}(Y,\mathfrak{s}) \xleftarrow{i^*} \widecheck{HM}^{k-1}(Y,\mathfrak{s}) \xleftarrow{j^*},$$

and we define the *reduced Floer cohomology* $HM^k(Y,\mathfrak{s})$ as the image of j^*.

Definition 22.3.4. We write $\widecheck{HM}^*(Y,\mathfrak{s})$, $\widehat{HM}^*(Y,\mathfrak{s})$, $\overline{HM}^*(Y,\mathfrak{s})$ and $HM^*(Y,\mathfrak{s})$ for the total Floer cohomologies, defined as direct sums, so that, for example,

$$\widecheck{HM}^*(Y,\mathfrak{s}) = \bigoplus_{j \in \mathbb{J}(\mathfrak{s})} \widecheck{HM}^j(Y,\mathfrak{s}).$$

We may also write

$$\widecheck{HM}_*(Y,\mathfrak{s}) = \bigoplus_{j \in \mathbb{J}(\mathfrak{s})} \widecheck{HM}_j(Y,\mathfrak{s}).$$

◊

Remark. Note that we do not define $\widecheck{HM}^*(Y,\mathfrak{s})$ as a direct *product* over j in $\mathbb{J}(\mathfrak{s})$, as is sometimes done with cohomology.

22.4 The canonical mod two grading

The grading set $\mathbb{J}(Y,\mathfrak{s})$ for the groups $\widecheck{HM}^*(Y,\mathfrak{s})$ etc. is completely described by Lemma 22.3.2, which tells us that, as a set with $\mathbb{Z}$-action, it is isomorphic to $\mathbb{Z}/d$ for some *even* integer d. In general, there is no distinguished element in $\mathbb{J}(Y,\mathfrak{s})$. There is however a distinguished decomposition of $\mathbb{J}(Y,\mathfrak{s})$ into "even" and "odd" elements: that is, there is a preferred map of $\mathbb{Z}$ sets,

$$\mathbb{J}(Y,\mathfrak{s}) \to \mathbb{Z}/2.$$

To define this, pick again a reducible point $\mathfrak{a}_0$ in $\mathcal{C}_k^\sigma(Y,\mathfrak{s})$. We choose $\mathfrak{a}_0$ to be boundary-stable for the zero perturbation. (We do not mean to require here that $\mathfrak{a}_0$ is a critical point: see Definition 14.5.2 and the remark following it.) Recalling the definition of $\mathbb{J}(Y,\mathfrak{s})$ above as equivalence classes of triples $([\mathfrak{a}],\mathfrak{q},n)$, we define

$$\mathrm{gr}^{(2)}([\mathfrak{a}],\mathfrak{q},n) = \mathrm{index}\,(P_{\gamma,\mathfrak{p}})(\mathrm{mod}\,2),$$

where γ arises from any path joining $\mathfrak{a}$ to $\mathfrak{a}_0$ and $\mathfrak{p}$ is a 1-parameter family of perturbations joining $\mathfrak{q}$ to 0. This is independent of the choice of γ for the same reason that d above is even, and it descends through the equivalence relation to define a map on $\mathbb{J}(Y,\mathfrak{s})$. It is also independent of the choice of $\mathfrak{a}_0$: this is because if $\mathfrak{a}_0$ and $\mathfrak{a}_0'$ are two boundary-stable reducibles, then index $(P_{\gamma,0})$ is even for any path joining them, for the usual reason that we can take the path to lie in the reducibles, and the kernel and cokernel then arise from the complex

Dirac operator. (This is the reason for asking that $\mathfrak{a}_0$ and $\mathfrak{a}_0'$ are boundary-stable: if one of the two is boundary-unstable, then the index picks up a 1-dimensional contribution from the normal part of the operator, P^ν. See the discussion preceding Lemma 20.3.3.)

When an admissible perturbation $\mathfrak{q}$ is given, we use the abbreviation

$$\mathrm{gr}^{(2)}[\mathfrak{a}] = \mathrm{gr}^{(2)}([\mathfrak{a}], \mathfrak{q}, 0) \in \mathbb{Z}/2$$

for critical points $[\mathfrak{a}]$. Using this mod 2 grading, we can write $\widecheck{HM}_*(Y, \mathfrak{s})$ and $\widehat{HM}_*(Y, \mathfrak{s})$ as sums of even and odd parts. For $\overline{HM}_*(Y, \mathfrak{s})$, we imitate (22.15) and define

$$\bar{\mathrm{gr}}^{(2)}[\mathfrak{a}] = \begin{cases} \mathrm{gr}^{(2)}[\mathfrak{a}], & [\mathfrak{a}] \in \mathfrak{C}^s \\ \mathrm{gr}^{(2)}[\mathfrak{a}] - 1, & [\mathfrak{a}] \in \mathfrak{C}^u. \end{cases} \tag{22.19}$$

This grading decomposes $\overline{HM}_*(Y, \mathfrak{s})$ into even and odd also. In the i, j, p long exact sequence, i and j are even and p is odd. There is no additional checking needed to verify these assertions: they are consequences of the behavior of the grading by $\mathbb{J}(Y, \mathfrak{s})$ that we have already discussed, combined with the observation that $\mathrm{gr}^{(2)}$ respects the action of $\mathbb{Z}$, as a map from $\mathbb{J}(Y, \mathfrak{s})$ to $\mathbb{Z}/2$.

22.5 Duality

We now consider the effect of changing the orientation of Y. We will write $-Y$ for the oppositely oriented manifold in the present context. (We avoid the notation $\bar{Y}$ here, because we are using a similar notation to indicate the reducibles, for example in the notation $\bar{C}$.) A spin^c structure on Y gives a spin^c structure on $-Y$: we can take the same spin bundle S, and replace Clifford multiplication ρ by $-\rho$ to satisfy the orientation condition. The Dirac operator $D(-Y)$ is therefore $-D(Y)$. We can identify $C_k(Y, \mathfrak{s})$ with $C_k(-Y, \mathfrak{s})$, and the Chern–Simons–Dirac functionals are related by

$$\mathcal{L}(Y) = -\mathcal{L}(-Y).$$

If we choose a perturbation $\mathfrak{q}$ on Y, we can take $-\mathfrak{q}$ on $-Y$, so that the perturbed Chern–Simons–Dirac functionals satisfy the same relation. On $\mathcal{B}_k^\sigma(-Y, \mathfrak{s})$, the vector field $(\mathrm{grad}\, \bar{\mathcal{L}}(-Y))^\sigma$ is the negative of $(\mathrm{grad}\, \bar{\mathcal{L}}(Y))^\sigma$. The critical-point sets $\mathfrak{C}(Y, \mathfrak{s})$ and $\mathfrak{C}(-Y, \mathfrak{s})$ are the same, but for the reducible critical points, the notions of boundary-stable and boundary-unstable are switched:

$$\mathfrak{C}^s(-Y,\mathfrak{s}) = \mathfrak{C}^u(Y,\mathfrak{s})$$
$$\mathfrak{C}^u(-Y,\mathfrak{s}) = \mathfrak{C}^s(Y,\mathfrak{s}).$$

The perturbed equation on the cylinder arises formally from the equation $d\gamma/dt = -(\mathrm{grad}\,\mathcal{L})^\sigma(\gamma)$, so if $\gamma(t)$ is a trajectory for Y then $\gamma(-t)$ is a trajectory for $-Y$. This gives a canonical identification

$$M_z(Y;[\mathfrak{a}],[\mathfrak{b}]) = M_{-z}(-Y;[\mathfrak{b}],[\mathfrak{a}]). \tag{22.20}$$

It is not quite true that there is a canonical identification of, for example, the free abelian groups $C^o(-Y)$ and $C^o(Y)$. The reason lies in an asymmetry in the definition of $\Lambda([\mathfrak{a}_1],\mathfrak{q}_1,[\mathfrak{a}_2],\mathfrak{q}_2)$ in Subsection 20.3. Recall that the definition of the orientation set is in terms of the determinant of an operator $P = (Q, -\Pi_1^-, \Pi_2^+)$ whose definition in turn uses the decomposition of $\mathcal{K}^\sigma_{\mathfrak{a}_i}$ into the two spectral subspaces

$$\mathcal{K}^\sigma_{\mathfrak{a}_i} = \mathcal{K}^-_{\mathfrak{a}_i} \oplus \mathcal{K}^+_{\mathfrak{a}_i}.$$

The operator defining the spectral decomposition is $\mathrm{Hess}^\sigma_{\mathfrak{a}_i} - \epsilon$, for small ϵ, so as to put the generalized zero-eigenspace of the Hessian $\mathrm{Hess}^\sigma_{\mathfrak{a}}$ (if there is one) into $\mathcal{K}^-_{\mathfrak{a}_i}$. There is an alternative choice, which is to put this generalized eigenspace with the positive part, giving rise to a different decomposition $\mathcal{K}^\sigma = \mathcal{K}^{+'} \oplus \mathcal{K}^{-'}$. Using the primed version, we define

$$H_i^{-'} = \{0\} \oplus \mathcal{K}^{-'}_{1/2,\mathfrak{a}_i} \oplus L^2_{1/2}(Y;i\mathbb{R})$$
$$H_i^{+'} = \{0\} \oplus \mathcal{K}^{+'}_{1/2,\mathfrak{a}_i} \oplus L^2_{1/2}(Y;i\mathbb{R})$$

as in (20.5), and we have a corresponding Fredholm boundary-value problem P'_γ on $I \times Y$. (For a non-degenerate critical point, there is no zero-eigenspace for the Hessian; but our constructions exploit these operators also when the boundary values $\mathfrak{a}_i$ are not critical points.)

When we change the orientation of Y, two things happen: first, the roles of the primed and unprimed decompositions are reversed; and second, the operator changes sign (as can be seen by examining the formula (20.6)). What this means is that, if we denote by $\tilde{\gamma}$ the configuration $\gamma(-t)$ on $-Y$, then the operators $P_\gamma(Y)$ and $-P'_{\tilde{\gamma}}(-Y)$ are identified. We can define $\Lambda'([\mathfrak{a}],[\mathfrak{b}])$ in just the same manner that $\Lambda([\mathfrak{a}],[\mathfrak{b}])$ is defined, as the orientations of the determinant of P'_γ, and we then have a canonical identification

$$c : \Lambda(Y; [\mathfrak{a}], [\mathfrak{b}]) \to \Lambda'(-Y; [\mathfrak{b}], [\mathfrak{a}]), \tag{22.21}$$

using the convention (20.2) to deal with the change of sign of the operator. In the case that $\mathfrak{a}$ and $\mathfrak{b}$ are non-degenerate critical points, $\Lambda'(-Y; [\mathfrak{b}], [\mathfrak{a}])$ and $\Lambda(-Y; [\mathfrak{b}], [\mathfrak{a}])$ are the same.

Lemma 22.5.1. *Suppose that* $[\mathfrak{a}]$ *and* $[\mathfrak{b}]$ *are non-degenerate critical points. Pick* λ *in* $\Lambda(Y; [\mathfrak{a}], [\mathfrak{b}])$ *so as to orient the moduli space* $M_z([\mathfrak{a}], [\mathfrak{b}])$, *and then orient* $M_{-z}([\mathfrak{b}], [\mathfrak{a}])$ *using the element*

$$\begin{aligned}\tilde{\lambda} = c(\lambda) &\in \Lambda'(-Y; [\mathfrak{b}], [\mathfrak{a}]) \\ &= \Lambda(-Y; [\mathfrak{b}], [\mathfrak{a}]).\end{aligned}$$

Then the canonical identification of the moduli spaces in (22.20) *is orientation-preserving in the boundary-unobstructed cases, and orientation-reversing in the boundary-obstructed case.*

Proof. This follows from the formula for j_s in (20.2). The identification of the moduli spaces corresponds to the identification of the kernels of the operators; the extra minus sign in the boundary-obstructed case arises from the fact that the cokernel has dimension 1. □

The difference between P and P' is important in the definition of $\Lambda([\mathfrak{a}])$, where $[\mathfrak{a}]$ is a non-degenerate critical point for some perturbation $\mathfrak{q}$. There is no difference between the primed and unprimed decomposition at $[\mathfrak{a}]$, but $\Lambda([\mathfrak{a}])$ is defined using the 2-element set

$$\Lambda(Y; [\mathfrak{a}], \mathfrak{q}, [\mathfrak{a}_0], 0),$$

where $\mathfrak{a}_0$ is reducible and either boundary-stable or boundary-unstable (and not necessarily a critical point). Given an element λ in $\Lambda(Y; [\mathfrak{a}], \mathfrak{q}, [\mathfrak{a}_0], 0)$, we have

$$\rho(\lambda) \in \Lambda(Y; [\mathfrak{a}_0], 0, [\mathfrak{a}], \mathfrak{q})$$

where ρ is as in (20.14). Applying c, we obtain

$$c(\rho(\lambda)) \in \Lambda'(-Y; [\mathfrak{a}], -\mathfrak{q}, [\mathfrak{a}_0], 0).$$

To compare the Λ' that appears on the right here with the unprimed version, we introduce the boundary-value problem $P^\sharp_{\mathfrak{a}_0}$ on $I \times Y$ defined using the primed space $H^{+\prime}$ on the left-hand end and the unprimed space H^- on the right-hand end:

$$P^{\sharp}_{\mathfrak{a}_0} : \mathcal{E} \to \mathcal{F} \oplus H^{+\prime} \oplus H^-.$$

There is a similar operator $P^{\flat}_{\mathfrak{a}_0}$ with the roles of the primed and unprimed subspaces reversed:

$$P^{\flat}_{\mathfrak{a}_0} : \mathcal{E} \to \mathcal{F} \oplus H^{+} \oplus H^{-\prime}.$$

We wish to understand how to orient the determinants of the Fredholm operators $P^{\sharp}_{\mathfrak{a}_0}$ and $P^{\flat}_{\mathfrak{a}_0}$, and to do this we perturb each operator to a direct sum of five blocks, as we did on pages 386f. In the present case, the function λ in the first block is constant and non-zero, so the first block is invertible. The second and third blocks are also invertible, as in our previous discussion, and the fifth block is again a complex operator. The conclusion is that orienting the determinants of $P^{\flat}_{\mathfrak{a}_0}$ and $P^{\sharp}_{\mathfrak{a}_0}$ is equivalent to orienting the determinant of the fourth block, namely the operator

$$\frac{d}{dt} + *d$$

acting on coclosed, imaginary-valued 1-forms on Y with certain spectral boundary conditions. Note that the kernel of $*d$ coincides with the harmonic 1-forms, or $H^1(Y; i\mathbb{R})$. In the case of $P^{\flat}_{\mathfrak{a}_0}$, neither H^+ nor $H^{-\prime}$ contains the harmonic forms, so the kernel of the corresponding block is $H^1(Y; i\mathbb{R})$. In the case of $P^{\sharp}_{\mathfrak{a}_0}$, the same block contributes to the cokernel.

At this point we recall a definition that was introduced in Subsection 3.1.

Definition 22.5.2. We define the 2-element set $\Lambda(Y)$ to be the set of orientations of the vector space $H^1(Y; \mathbb{R})$. A choice of an element from $\Lambda(Y)$ is a *homology orientation* of Y. ◊

We fix an orientation convention about how to identify $\mathbb{R}$ with $i\mathbb{R}$ (say multiplication by i), so that $\Lambda(Y)$ becomes also the orientations of $H^1(Y; i\mathbb{R})$. The above discussion shows that a homology orientation $\mu \in \Lambda(Y)$ canonically determines orientation $\mu^{\flat}$ for the determinant of the operator $P^{\flat}_{\mathfrak{a}_0}$, by orienting the kernel of the block diagonal operator to which it is homotopic. In turn, this determines an orientation $\mu^{\sharp}$ for $\det P^{\sharp}_{\mathfrak{a}_0}$ by the rule

$$\begin{aligned} q(\mu^{\flat}, \mu^{\sharp}) &= 1 \in \Lambda(\mathfrak{a}_0, \mathfrak{a}_0) \\ &= \{1, -1\}, \end{aligned}$$

where q is the natural composition. This rule is equivalent to the alternative rule

$$\begin{aligned} q(\mu^\sharp, \mu^\flat) &= 1 \in \Lambda'(\mathfrak{a}_0, \mathfrak{a}_0) \\ &= \{1, -1\}. \end{aligned}$$

We may write $\mu^\sharp_Y$ and $\mu^\flat_Y$ to emphasize the role of Y. In particular, the first cohomology groups of Y and $-Y$ are the same, so we can also regard a cohomology orientation μ as giving rise to orientation $\mu^\sharp_{-Y}$ and $\mu^\flat_{-Y}$ for the corresponding objects on $-Y$. We have

$$q(\mu^\flat_{-Y}, \mu^\sharp_{-Y}) = 1$$

and

$$c(\mu^\flat_Y) = \mu^\flat_{-Y}$$

from which follows

$$c(\mu^\sharp_Y) = (-1)^{b_1(Y)} \mu^\sharp_{-Y}. \tag{22.22}$$

(See (22.24) below.)

Via this identification, the definition of the composition law q gives us also a map which removes the prime:

$$\begin{gathered} (x, \mu) \mapsto q(x, \mu^\sharp_{-Y}) \\ \Lambda'(-Y; [\mathfrak{a}], -\mathfrak{q}, [\mathfrak{a}_0], 0) \times \Lambda(-Y) \to \Lambda(-Y; [\mathfrak{a}], -\mathfrak{q}, [\mathfrak{a}_0], 0). \end{gathered}$$

Thus, given a homology orientation $\mu \in \Lambda(Y)$, we get a map

$$\begin{aligned} \Lambda(Y; [\mathfrak{a}], \mathfrak{q}, [\mathfrak{a}_0], 0) &\to \Lambda(-Y; [\mathfrak{a}], -\mathfrak{q}, [\mathfrak{a}_0], 0) \\ \lambda &\mapsto q\left(c(\rho(\lambda)), \mu^\sharp_{-Y}\right). \end{aligned} \tag{22.23}$$

As it stands, the map (22.23) does not define a map from $\Lambda(Y; [\mathfrak{a}])$ to $\Lambda(-Y; [\mathfrak{a}])$, because $\Lambda(Y; [\mathfrak{a}])$ is the quotient of the union

$$\coprod_{[\mathfrak{a}_0]} \Lambda(Y; [\mathfrak{a}], \mathfrak{q}, [\mathfrak{a}_0], 0)$$

by an equivalence relation; and the map (22.23) does not preserve this relation. The equivalence relation *is* respected, however, if we confine ourselves to the cases that $\mathfrak{a}_0$ is boundary-stable on Y. This leads to the following definition, in which an additional sign has been introduced in order to more closely match the signs that arise in finite-dimensional Morse theory.

Definition 22.5.3. For a critical point $[\mathfrak{a}]$ and a homology orientation μ for Y, we define

$$\omega_\mu : \Lambda(Y; [\mathfrak{a}]) \to \Lambda(-Y; [\mathfrak{a}])$$

by taking a representative $\lambda \in \Lambda(Y; [\mathfrak{a}], [\mathfrak{a}_0])$ for a boundary-stable $[\mathfrak{a}_0]$, and setting

$$\omega_\mu(\lambda) = (-1)^{\mathrm{gr}^{(2)}(\mathfrak{a})} q\left(c(\rho(\lambda)), \mu^\sharp_{-Y}\right)$$

where q is the composition operation for $-Y$ and $\mathrm{gr}^{(2)}$ is the canonical mod 2 grading. ◊

Now let $\lambda_{\mathfrak{a}}$ and $\lambda_{\mathfrak{b}}$ be orientations in $\Lambda(Y; [\mathfrak{a}])$ and $\Lambda(Y; [\mathfrak{b}])$, and write

$$\begin{aligned}\tilde{\lambda}_{\mathfrak{a}} &= \omega_\mu(\lambda_{\mathfrak{a}})\\ \tilde{\lambda}_{\mathfrak{b}} &= \omega_\mu(\lambda_{\mathfrak{b}}).\end{aligned}$$

From $\lambda_{\mathfrak{a}}$ and $\lambda_{\mathfrak{b}}$ we obtain as before an element

$$\begin{aligned}\lambda_{\mathfrak{a}\mathfrak{b}} &= q(\lambda_{\mathfrak{a}}, \rho(\lambda_{\mathfrak{b}}))\\ &\in \Lambda(Y; [\mathfrak{a}], [\mathfrak{b}]).\end{aligned}$$

Similarly, with $\tilde{q}$ and $\tilde{\rho}$ denoting the corresponding operators for $-Y$, we have an element

$$\begin{aligned}\tilde{\lambda}_{\mathfrak{a}\mathfrak{b}} &= \tilde{q}(\tilde{\lambda}_{\mathfrak{b}}, \tilde{\rho}(\tilde{\lambda}_{\mathfrak{a}}))\\ &\in \Lambda(-Y; [\mathfrak{b}], [\mathfrak{a}]).\end{aligned}$$

Recalling the map c from Lemma 22.5.1, we can compare $\tilde{\lambda}_{\mathfrak{a}\mathfrak{b}}$ with $c(\lambda_{\mathfrak{a}\mathfrak{b}})$:

Lemma 22.5.4. *We have* $\tilde{\lambda}_{\mathfrak{a}\mathfrak{b}} = (-1)^r c(\lambda_{\mathfrak{a}\mathfrak{b}})$, *where*

$$r = \mathrm{gr}^{(2)}(\mathfrak{b})(1 + \mathrm{gr}^{(2)}(\mathfrak{a})).$$

Proof. Let us write $x \bullet y$ for $q(x, y)$, and let us abbreviate $\mathrm{gr}^2(\mathfrak{a})$ and $\mathrm{gr}^2(\mathfrak{b})$ as a and b respectively. Because of the interchange of orders that is involved when passing from trajectories on Y to trajectories on $-Y$, we have a general relation

$$c(x \bullet y) = (-1)^{\mathrm{gr}(x)\,\mathrm{gr}(y)} c(y) \bullet c(x), \tag{22.24}$$

where $\mathrm{gr}(x)$ and $\mathrm{gr}(y)$ denote the indices of the operators for which x and y define orientations. So we compute as follows, writing $\mu^\flat$ and $\mu^\sharp$ for $\mu^\flat_{-Y}$ and $\mu^\sharp_{-Y}$:

$$\begin{aligned} c(\lambda_{\mathfrak{a}\mathfrak{b}}) &= c(\lambda_{\mathfrak{a}} \bullet \rho(\lambda_{\mathfrak{b}})) \\ &= (-1)^{ab} c(\rho(\lambda_{\mathfrak{b}})) \bullet c(\lambda_{\mathfrak{a}}) \\ &= (-1)^{ab} c(\rho(\lambda_{\mathfrak{b}})) \bullet (\mu^\sharp \bullet \mu^\flat) \bullet c(\lambda_{\mathfrak{a}}) \\ &= (-1)^{ab+b} \tilde{\lambda}_{\mathfrak{b}} \bullet \mu^\flat \bullet c(\lambda_{\mathfrak{a}}). \end{aligned}$$

We now claim that

$$\mu^\flat \bullet c(\lambda_{\mathfrak{a}}) = \rho(\tilde{\lambda}_{\mathfrak{a}}), \tag{22.25}$$

so that the above gives

$$\begin{aligned} c(\lambda_{\mathfrak{a}\mathfrak{b}}) &= (-1)^{ab+b} \tilde{\lambda}_{\mathfrak{b}} \bullet \rho(\tilde{\lambda}_{\mathfrak{a}}) \\ &= (-1)^{ab+b} \tilde{\lambda}_{\mathfrak{a}\mathfrak{b}} \end{aligned}$$

as the lemma asserts. To check (22.25), we turn to the defining property of ρ – namely the condition $\rho(\tilde{\lambda}_{\mathfrak{a}}) \bullet \tilde{\lambda}_{\mathfrak{a}} = 1$ – to see whether it is satisfied. To do this, we compute

$$\begin{aligned} \mu^\flat \bullet c(\lambda_{\mathfrak{a}}) \bullet \tilde{\lambda}_{\mathfrak{a}} &= (-1)^a \mu^\flat \bullet c(\lambda_{\mathfrak{a}}) \bullet c(\rho(\lambda_{\mathfrak{a}})) \bullet \mu^\sharp \\ &= \mu^\flat \bullet c\big(\rho(\lambda_{\mathfrak{a}}) \bullet \lambda_{\mathfrak{a}}\big) \bullet \mu^\sharp \\ &= \mu^\flat \bullet c(1) \bullet \mu^\sharp \\ &= \mu^\flat \bullet \mu^\sharp \\ &= 1. \end{aligned}$$

This verifies the claim and the lemma. □

Remark. The relation (22.24) and the defining property of ρ also tell us that $\rho(c(\lambda_{\mathfrak{a}}))$ and $c(\rho(\lambda_{\mathfrak{a}}))$ differ by the sign $(-1)^{\mathrm{gr}^{(2)}(\mathfrak{a})}$:

$$\rho(c(\lambda_{\mathfrak{a}})) = (-1)^{\mathrm{gr}^{(2)}(\mathfrak{a})}\, c(\rho(\lambda_{\mathfrak{a}})); \tag{22.26}$$

so the definition of ω_μ in Definition 22.5.3 can also be written

$$\omega_\mu(\lambda) = q\left(\rho(c(\lambda)), \mu^\sharp_{-Y}\right). \tag{22.27}$$

Combining the last lemma with Lemma 22.5.1, we deduce:

Corollary 22.5.5. *When the moduli spaces are equipped with the orientations* $\lambda_{\mathfrak{a}\mathfrak{b}}$ *and* $\tilde{\lambda}_{\mathfrak{a}\mathfrak{b}}$ *respectively, the natural identification of moduli spaces*

$$\breve{M}_z(Y;[\mathfrak{a}],[\mathfrak{b}]) \to \breve{M}_{-z}(-Y;[\mathfrak{b}],[\mathfrak{a}])$$

is orientation-preserving or orientation-reversing according to the sign of

$$(-1)^{\mathrm{gr}^{(2)}(\mathfrak{b})(1+\mathrm{gr}^{(2)}(\mathfrak{a}))+1}$$

in the boundary-unobstructed cases, or of

$$(-1)^{\mathrm{gr}^{(2)}(\mathfrak{b})(1+\mathrm{gr}^{(2)}(\mathfrak{a}))}$$

in the boundary-obstructed case.

Proof. The sign $\mathrm{gr}^{(2)}(\mathfrak{b})(1+\mathrm{gr}^{(2)}(\mathfrak{a}))$ comes from the preceding lemma. When passing from M to $\breve{M}$, we must divide by the action of $\mathbb{R}$; and the change of orientation of Y means an accompanying change in the direction of the $\mathbb{R}$ action on the moduli space. This results in another overall sign change. In the boundary-obstructed case, an additional minus sign comes from Lemma 22.5.1. □

We can specialize to the case of 1-dimensional moduli spaces:

Corollary 22.5.6. *In the situation of the previous corollary, suppose that* $M_z([\mathfrak{a}],[\mathfrak{b}])$ *has dimension* 1. *Then the map*

$$\breve{M}_z(Y;[\mathfrak{a}],[\mathfrak{b}]) \to \breve{M}_{-z}(-Y;[\mathfrak{b}],[\mathfrak{a}])$$

is orientation-preserving or orientation-reversing according to the sign of $(-1)^{\mathrm{gr}^{(2)}(\mathfrak{a})}$ *in the boundary-unobstructed cases. In the boundary-obstructed case, the map is orientation-preserving.* □

If we treat μ as a variable, then the maps ω_μ can be thought of as a map

$$\omega_Y : \mathbb{Z}\Lambda(Y;[\mathfrak{a}]) \to \mathbb{Z}\Lambda(-Y;[\mathfrak{a}]) \otimes \mathbb{Z}\Lambda(Y).$$

This gives preferred isomorphisms between, for example, $C^s(Y)$ and $C^u(-Y)\otimes \mathbb{Z}\Lambda(Y)$. However, it is more natural to make identifications here between dual spaces: for a free abelian group C with a canonical basis, we write $C^\dagger$ for the subspace of $\mathrm{Hom}(C,\mathbb{Z})$ consisting of homomorphisms which are zero on all but finitely many basis vectors (the *finitely supported* dual). There is a canonical isomorphism between C and $C^\dagger$. A homomorphism $e : C_1 \to C_2$ with only finitely many matrix entries in each row and column gives a homomorphism $e^\dagger : C_2^\dagger \to C_1^\dagger$. With this convention, we have identifications

$$\begin{aligned}
\omega^o &: C^o(-Y,\mathfrak{s}) \to C^o(Y,\mathfrak{s})^\dagger \otimes \mathbb{Z}\Lambda(Y)\\
\omega^s &: C^s(-Y,\mathfrak{s}) \to C^u(Y,\mathfrak{s})^\dagger \otimes \mathbb{Z}\Lambda(Y)\\
\omega^u &: C^u(-Y,\mathfrak{s}) \to C^s(Y,\mathfrak{s})^\dagger \otimes \mathbb{Z}\Lambda(Y).
\end{aligned}$$

(We emphasize that ω_Y, not ω_{-Y}, is being used here, though we have reversed the direction of the maps.) With these isomorphisms, Corollary 22.5.6 tells us:

$$\left.\begin{aligned}
\omega^o\partial^o_o(-Y) &= \eta\partial^o_o(Y)^\dagger\omega^o\\
\omega^s\partial^o_s(-Y) &= \eta\partial^u_o(Y)^\dagger\omega^o\\
\omega^o\partial^u_o(-Y) &= \eta\partial^o_s(Y)^\dagger\omega^u\\
\omega^s\partial^u_s(-Y) &= \eta\partial^u_s(Y)^\dagger\omega^u,
\end{aligned}\right\} \tag{22.28}$$

where

$$\eta : C(Y,\mathfrak{s}) \to C(Y,\mathfrak{s})$$

is the map that is equal to $(-1)^{\mathrm{gr}^{(2)}(\mathfrak{a})}$ on the summand $\mathbb{Z}\Lambda([\mathfrak{a}])$. For the components $\bar\partial^s_s$ etc. of $\bar\partial$, there is a slight twist on account of the signs in the definition of $\bar\epsilon$ when both critical points are boundary-unstable. We obtain:

$$\left.\begin{aligned}
\omega^s\bar\partial^s_s(-Y) &= -\eta\bar\partial^u_u(Y)^\dagger\omega^s\\
\omega^u\bar\partial^s_u(-Y) &= \bar\partial^s_u(Y)^\dagger\omega^s\\
\omega^s\bar\partial^u_s(-Y) &= -\bar\partial^u_s(Y)^\dagger\omega^u\\
\omega^u\bar\partial^u_u(-Y) &= -\eta\bar\partial^s_s(Y)^\dagger\omega^u.
\end{aligned}\right\} \tag{22.29}$$

The sign in the $\bar{\partial}^s_u$ case comes from Corollary 22.5.6. For the $\bar{\partial}^u_s$ case, we use the fact that the reducible moduli space $\breve{M}^{\mathrm{red}}([\mathfrak{a}],[\mathfrak{b}])$ is minus the oriented boundary of a moduli space $\breve{M}([\mathfrak{a}],[\mathfrak{b}])$, with $\mathrm{gr}^{(2)}([\mathfrak{b}]) = \mathrm{gr}^{(2)}([\mathfrak{a}])$.

We now define

$$\check{\omega} : \check{C}(-Y,\mathfrak{s}) \to \hat{C}(Y,\mathfrak{s})^{\dagger} \otimes \mathbb{Z}\Lambda(Y)$$
$$\hat{\omega} : \hat{C}(-Y,\mathfrak{s}) \to \check{C}(Y,\mathfrak{s})^{\dagger} \otimes \mathbb{Z}\Lambda(Y)$$

by the formulae

$$\check{\omega} = \begin{bmatrix} \omega^o & 0 \\ 0 & \omega^s \end{bmatrix}, \quad \hat{\omega} = \begin{bmatrix} \omega^o & 0 \\ 0 & \omega^u \end{bmatrix}.$$

The above relations then give us:

$$\begin{aligned} \check{\omega}\check{\partial}(-Y) &= \eta\hat{\partial}(Y)^{\dagger}\check{\omega} \\ \hat{\omega}\hat{\partial}(-Y) &= \eta\check{\partial}(Y)^{\dagger}\hat{\omega}. \end{aligned} \tag{22.30}$$

To make the signs work for $\bar{C}$, we introduce $\bar{\eta}$ as the $\bar{\mathrm{gr}}^{(2)}$ version of η,

$$\bar{\eta} = (-1)^{\bar{\mathrm{gr}}^{(2)}} : \bar{C}(Y) \to \bar{C}(Y),$$

so that the relations (22.29) can be rewritten

$$\begin{aligned} \omega^s\bar{\partial}^s_s(-Y) &= \bar{\eta}\bar{\partial}^u_u(Y)^{\dagger}\omega^s \\ \omega^u\bar{\partial}^s_u(-Y) &= \bar{\partial}^s_u(Y)^{\dagger}\omega^s \\ \omega^s\bar{\partial}^u_s(-Y) &= -\bar{\partial}^u_s(Y)^{\dagger}\omega^u \\ \omega^u\bar{\partial}^u_u(-Y) &= -\bar{\eta}\bar{\partial}^s_s(Y)^{\dagger}\omega^u. \end{aligned}$$

We then define

$$\bar{\omega} : \bar{C}(-Y,\mathfrak{s}) \to \bar{C}(Y,\mathfrak{s})^{\dagger} \otimes \mathbb{Z}\Lambda(Y)$$

by

$$\bar{\omega} = \begin{bmatrix} 0 & \bar{\eta}\omega^u \\ \omega^s & 0 \end{bmatrix} : C^s(-Y) \oplus C^u(-Y) \to \left(C^s(Y)^{\dagger} \oplus C^u(Y)^{\dagger}\right) \otimes \mathbb{Z}\Lambda(Y).$$

We then have

$$\bar{\omega}\bar{\partial}(-Y) = \bar{\eta}\bar{\partial}(Y)^{\dagger}\bar{\omega}. \tag{22.31}$$

At this point, we see from (22.30) and (22.31) that $\check{\omega}$, $\hat{\omega}$ and $\bar{\omega}$ respect the differentials up to sign. They therefore give well-defined maps on the (co)homology groups, and so we have constructed isomorphisms between for example $\widetilde{HM}_*(-Y,\mathfrak{s})$ and $\widehat{HM}^*(Y,\mathfrak{s}) \otimes \mathbb{Z}\Lambda(Y)$. Before proceeding, we wish to consider how the grading behaves under these duality isomorphisms. Recall the definition (22.12) of $\mathbb{J}$. Define a map

$$o : \mathcal{B}_k^\sigma(-Y,\mathfrak{s}) \times \mathbb{Z} \to \mathcal{B}_k^\sigma(Y,\mathfrak{s}) \times \mathbb{Z},$$

by

$$o([\mathfrak{a}],n) = \big([\mathfrak{a}], -n - N(\mathrm{Hess}^\sigma_{\mathfrak{q},\mathfrak{a}})\big).$$

Here $\mathrm{Hess}^\sigma_{\mathfrak{q},\mathfrak{a}}$ is again the Hessian at $\mathfrak{a}$ acting on $\mathcal{K}^\sigma_{\mathfrak{a}}$, and $N(\mathrm{Hess}^\sigma_{\mathfrak{q},\mathfrak{a}})$ is the dimension of its generalized zero-eigenspace. This is the codimension of $H^{-'}$ in H^-. Let $\sim$ be the equivalence relation on $\mathcal{B}_k^\sigma(Y,\mathfrak{s}) \times \mathbb{Z}$ defined prior to (22.12), and let $\sim'$ be the equivalence relation on $\mathcal{B}_k^\sigma(-Y,\mathfrak{s}) \times \mathbb{Z}$ defined in the same way, with $-Y$ in place of Y.

Lemma 22.5.7. *The map o satisfies*

$$([\mathfrak{a}],m) \sim' ([\mathfrak{b}],n) \quad \Longleftrightarrow \quad o([\mathfrak{a}],m) \sim o([\mathfrak{b}],n).$$

As a consequence, o gives rise to a map

$$o : \mathbb{J}(-Y,\mathfrak{s}) \to \mathbb{J}(Y,\mathfrak{s}) \tag{22.32}$$

on the equivalence classes.

Proof. Let $\gamma(t)$ be a path in $\mathcal{B}_k^\sigma(-Y,\mathfrak{s})$ joining $[\mathfrak{a}]$ to $[\mathfrak{b}]$, and let $P_\gamma(-Y)$ be the corresponding operator defined using the orientation of $-Y$. Let γ' be the opposite path, $\gamma'(t) = \gamma(-t)$. If the Hessians $\mathrm{Hess}^\sigma_{\mathfrak{a}}$ and $\mathrm{Hess}^\sigma_{\mathfrak{b}}$ are hyperbolic (that is, if they have no kernel), then $P_\gamma(-Y)$ is isomorphic to $-P_{\gamma'}(Y)$, and in this case we have

$$\mathrm{index}(P_\gamma(-Y)) = -\mathrm{index}(P_\gamma(Y)).$$

In the general case, we have

$$\mathrm{index} P_\gamma(-Y) = -\mathrm{index} P_\gamma(Y) + N(\mathrm{Hess}^\sigma_{\mathfrak{q},\mathfrak{a}}) - N(\mathrm{Hess}^\sigma_{\mathfrak{q},\mathfrak{b}}).$$

From the definition, $([\mathfrak{a}], m) \sim' ([\mathfrak{b}], n)$ if and only if there is a path γ with

$$\operatorname{index} P_\gamma(-Y) = n - m$$

or equivalently

$$\operatorname{index} P_\gamma(Y) = -n + m + N(\operatorname{Hess}^\sigma_{\mathfrak{q},\mathfrak{a}}) - N(\operatorname{Hess}^\sigma_{\mathfrak{q},\mathfrak{b}}).$$

This means that

$$\big([\mathfrak{a}], -m - N(\operatorname{Hess}^\sigma_{\mathfrak{q},\mathfrak{a}})\big) \sim \big([\mathfrak{b}], -n - N(\operatorname{Hess}^\sigma_{\mathfrak{q},\mathfrak{b}})\big).$$

□

Note that o changes the sign in the action of $\mathbb{Z}$:

$$o(j + n) = o(j) - n$$

for j in $\mathbb{J}(-Y, \mathfrak{s})$ and $n \in \mathbb{Z}$. If $[\mathfrak{a}]$ is a critical point in $\mathfrak{C}(Y) = \mathfrak{C}(-Y)$, it has a grading (22.13),

$$\operatorname{gr}_Y[\mathfrak{a}] \in \mathbb{J}(Y, \mathfrak{s}),$$

and we see that

$$o \operatorname{gr}_{-Y}[\mathfrak{a}] = \operatorname{gr}_Y[\mathfrak{a}]. \tag{22.33}$$

We also record how the canonical mod 2 grading is affected by o:

Lemma 22.5.8. *The map $o : \mathbb{J}(-Y, \mathfrak{s}) \to \mathbb{J}(Y, \mathfrak{s})$ preserves or reverses the canonical mod 2 grading according as $1 + b_1(Y)$ is even or odd respectively: that is,*

$$\operatorname{gr}^{(2)}_{-Y}(j) = (-1)^{1+b_1(Y)} \operatorname{gr}^{(2)}_Y(o(j)).$$

Proof. We consider the case of a point $\mathfrak{a}$ where the Hessian is hyperbolic. The definition of $\operatorname{gr}^{(2)}_Y(\mathfrak{a})$ is that it is the index mod 2 of $P_\gamma(Y)$, where γ runs from $(\mathfrak{a}, \mathfrak{q})$ to $(\mathfrak{a}_0, 0)$, for some boundary-stable reducible configuration $\mathfrak{a}_0$. This is equal to the index of $P'_{-\gamma}(-Y)$, where $-\gamma$ is the opposite path. This is equal mod 2 to the index of $P'_\delta(-Y)$, where δ is a path from $(\mathfrak{a}, -\mathfrak{q})$ to $(\mathfrak{a}_0, 0)$. We need to replace P' by P here, and this changes the index by $b_1(Y)$ mod 2; and we need to trade $\mathfrak{a}_0$, which is boundary-unstable on $-Y$, for some $\mathfrak{b}_0$ which is boundary-stable which will change the parity of the index again. The overall change in parity is $1 + b_1(Y)$. □

It follows now from (22.33) and the definitions that $\check{\omega}$ and $\hat{\omega}$ restrict to maps

$$\check{\omega} : \check{C}_j(-Y,\mathfrak{s}) \to \hat{C}^{o(j)}(Y,\mathfrak{s}) \otimes \mathbb{Z}\Lambda(Y)$$
$$\hat{\omega} : \hat{C}_j(-Y,\mathfrak{s}) \to \check{C}^{o(j)}(Y,\mathfrak{s}) \otimes \mathbb{Z}\Lambda(Y).$$

The map $\bar{\omega}$ on $\bar{C}$ behaves a little differently: from the definition (22.15), we see that the restriction of $\bar{\omega}$ is a map

$$\bar{\omega} : \bar{C}_j(-Y,\mathfrak{s}) \to \bar{C}^{o(j)-1}(Y,\mathfrak{s}) \otimes \mathbb{Z}\Lambda(Y).$$

Let us choose a homology orientation $\mu \in \Lambda(Y)$. This identifies $\mathbb{Z}\Lambda(Y)$ with $\mathbb{Z}$, and the maps $\check{\omega}$ etc. become isomorphisms

$$\check{\omega}_\mu : \check{C}_j(-Y,\mathfrak{s}) \to \hat{C}^{o(j)}(Y,\mathfrak{s}).$$

We now have a diagram of maps, in which the vertical maps are isomorphisms, and in which all maps are chain maps up to sign:

$$\begin{array}{ccccccccc}
\cdots & \bar{C}_k(-Y,\mathfrak{s}) & \xrightarrow{i} & \check{C}_k(-Y,\mathfrak{s}) & \xrightarrow{j} & \hat{C}_k(-Y,\mathfrak{s}) & \xrightarrow{p} & \bar{C}_{k-1}(-Y,\mathfrak{s}) & \cdots \\
 & \downarrow \bar{\omega}_\mu & & \downarrow \check{\omega}_\mu & & \downarrow \hat{\omega}_\mu & (-1)^r & \downarrow \bar{\omega}_\mu & \\
\cdots & \bar{C}^{o(k)-1}(Y,\mathfrak{s}) & \xrightarrow{p^\dagger} & \hat{C}^{o(k)}(Y,\mathfrak{s}) & \xrightarrow{j^\dagger} & \check{C}^{o(k)}(Y,\mathfrak{s}) & \xrightarrow{i^\dagger} & \bar{C}^{o(k)}(Y,\mathfrak{s}) & \cdots
\end{array}$$

Lemma 22.5.9. *In the above diagram, the left-hand and middle squares commute. The right-hand square commutes or anti-commutes according to the parity of*

$$\begin{aligned} r &= 1 + b_1(Y) + \mathrm{gr}^{(2)}_{-Y}(k) \\ &= \mathrm{gr}^{(2)}_{Y}(o(k)). \end{aligned} \tag{22.34}$$

Proof. This follows in a straightforward fashion from the formulae for i, j and p (Proposition 22.2.1) and the relations (22.28) and (22.29). □

Corollary 22.5.10. *We have the following diagram in which the first row is the long exact Floer homology sequence for $(-Y,\mathfrak{s})$, the second row is the cohomology sequence for $(Y,\mathfrak{s})$ and the vertical maps are the duality isomorphisms* *determined by a choice of homology orientation μ for Y. The first two squares commute, while the third square commutes or anti-commutes according to the parity of $r = 1 + b_1(Y) + \mathrm{gr}^{(2)}_Y(k)$.*

$$\begin{array}{ccccccc}
\cdots\ \overline{HM}_k(-Y) & \xrightarrow{i_*} & \widecheck{HM}_k(-Y) & \xrightarrow{j_*} & \widehat{HM}_k(-Y) & \xrightarrow{p_*} & \overline{HM}_{k-1}(-Y)\ \cdots \\
\downarrow \bar{\omega}_\mu & & \downarrow \check{\omega}_\mu & & \downarrow \hat{\omega}_\mu & (-1)^r & \downarrow \bar{\omega}_\mu \\
\cdots\ \overline{HM}^{o(k)-1}(Y) & \xrightarrow{p^*} & \widehat{HM}^{o(k)}(Y) & \xrightarrow{j^*} & \widecheck{HM}^{o(k)}(Y) & \xrightarrow{i^*} & \overline{HM}^{o(k)}(Y)\ \cdots
\end{array}$$

(We have omitted mention of $\mathfrak{s}$ in the diagram.) □

Corollary 22.5.11. *The map $\hat{\omega}_\mu$ corresponding to the homology orientation μ gives rise to an isomorphism from the reduced Floer homology of $-Y$ to the reduced Floer cohomology of Y:*

$$\omega_\mu : HM_k(-Y, \mathfrak{s}) \to HM^{o(k)}(Y, \mathfrak{s}).$$

□

We can also consider the sign that arises when we square the duality map. Returning for a moment to $\Lambda(Y; \mathfrak{a})$, and fixing a homology orientation μ for Y, we consider the two maps

$$\begin{aligned} \omega_{Y,\mu} &: \Lambda(Y; \mathfrak{a}) \to \Lambda(-Y; \mathfrak{a}) \\ \omega_{-Y,\mu} &: \Lambda(-Y; \mathfrak{a}) \to \Lambda(Y; \mathfrak{a}). \end{aligned}$$

Lemma 22.5.12. *The composite map*

$$\omega_{-Y,\mu}\omega_{Y,\mu} : \Lambda(Y; [\mathfrak{a}]) \to \Lambda(Y; [\mathfrak{a}])$$

is the identity if and only if

$$(\mathrm{gr}^{(2)}_Y([\mathfrak{a}]) + 1)b_1(Y)$$

is even.

Proof. Let us represent an element of $\Lambda(Y; [\mathfrak{a}])$ by

$$\lambda \in \Lambda(Y; [\mathfrak{a}], [\mathfrak{a}_0]),$$

for some boundary-stable reducible element $\mathfrak{a}_0$. Using again the bullet notation for q, and the formula (22.27) for ω_μ, we have $\omega_{Y,\mu}(\lambda)$ represented by the element

$$\begin{aligned}\omega_{Y,\mu}(\lambda) &= \rho(c(\lambda)) \bullet \mu^\sharp_{\mathfrak{a}_0} \\ &\in \Lambda(-Y;[\mathfrak{a}],[\mathfrak{a}_0]).\end{aligned}$$

The definition of ω requires us to use a representative in $\Lambda(Y;[\mathfrak{a}],[\mathfrak{a}_0])$ with $[\mathfrak{a}_0]$ boundary-*stable*; but for $-Y$, the point $[\mathfrak{a}_0]$ is boundary-*unstable*. So let us take another reducible point $\mathfrak{b}_0$ that is boundary-stable for $-Y$ (and so boundary-unstable for Y), and let τ be the canonical orientation in $\Lambda(-Y;[\mathfrak{a}_0],[\mathfrak{b}_0])$, as in (20.12). We can represent $\omega_{Y,\mu}(\lambda)$ also by

$$\rho(c(\lambda)) \bullet \mu^\sharp_{\mathfrak{a}_0} \bullet \tau \in \Lambda(-Y;[\mathfrak{a}],[\mathfrak{b}_0]).$$

We are now in a position to apply ω again. In the following calculation, we stick to the convention that $\mu^\sharp$ denotes $\mu^\sharp_{-Y}$; we can represent $\mu^\sharp_Y$ then as $(-1)^{b_1(Y)}c(\mu^\sharp)$ using (22.22). We compute

$$\begin{aligned}
\omega_{-Y,\mu}\omega_{Y,\mu}(\lambda) &= (-1)^{b_1(Y)}\rho\left(c(\rho(c(\lambda)) \bullet \mu^\sharp_{\mathfrak{a}_0} \bullet \tau)\right) \bullet c(\mu^\sharp_{\mathfrak{b}_0}) \\
&= (-1)^{b_1(Y)+\mathrm{gr}^{(2)}_Y(\mathfrak{a})(b_1(Y)+1)}\rho\left(c(\mu^\sharp_{\mathfrak{a}_0} \bullet \tau) \bullet c(\rho(c(\lambda)))\right) \bullet c(\mu^\sharp_{\mathfrak{b}_0}) \\
&= (-1)^{b_1(Y)+\mathrm{gr}^{(2)}_Y(\mathfrak{a})(b_1(Y)+1)}\rho\left(c(\rho(c(\lambda)))\right) \bullet \rho(c(\mu^\sharp_{\mathfrak{a}_0} \bullet \tau)) \bullet c(\mu^\sharp_{\mathfrak{b}_0}) \\
&= (-1)^{(\mathrm{gr}^{(2)}_Y(\mathfrak{a})+1)b_1(Y)}\rho\left(c(c(\rho(\lambda)))\right) \bullet \rho(c(\mu^\sharp_{\mathfrak{a}_0} \bullet \tau)) \bullet c(\mu^\sharp_{\mathfrak{b}_0}) \\
&= (-1)^{(\mathrm{gr}^{(2)}_Y(\mathfrak{a})+1)b_1(Y)}\,\lambda \bullet \rho(c(\mu^\sharp_{\mathfrak{a}_0} \bullet \tau)) \bullet c(\mu^\sharp_{\mathfrak{b}_0}) \\
&= (-1)^{(\mathrm{gr}^{(2)}_Y(\mathfrak{a})+1)b_1(Y)}\,\lambda \bullet \rho(\rho(c(\mu^\sharp_{\mathfrak{b}_0})) \bullet c(\mu^\sharp_{\mathfrak{a}_0} \bullet \tau)) \\
&= (-1)^{\mathrm{gr}^{(2)}_Y(\mathfrak{a})b_1(Y)}\,\lambda \bullet \rho(c(\rho(\mu^\sharp_{\mathfrak{b}_0})) \bullet c(\mu^\sharp_{\mathfrak{a}_0} \bullet \tau)) \\
&= (-1)^{\mathrm{gr}^{(2)}_Y(\mathfrak{a})b_1(Y)}\,\lambda \bullet \rho(c(\mu^\sharp_{\mathfrak{a}_0} \bullet \tau \bullet \rho(\mu^\sharp_{\mathfrak{b}_0}))) \\
&= (-1)^{(\mathrm{gr}^{(2)}_Y(\mathfrak{a})+1)b_1(Y)}\,\lambda \bullet \rho(c(\tau \bullet \mu^\sharp_{\mathfrak{b}_0} \bullet \rho(\mu^\sharp_{\mathfrak{b}_0}))) \\
&= (-1)^{(\mathrm{gr}^{(2)}_Y(\mathfrak{a})+1)b_1(Y)}\,\lambda \bullet \rho(c(\tau)) \\
&\equiv (-1)^{(\mathrm{gr}^{(2)}_Y(\mathfrak{a})+1)b_1(Y)}\,\lambda
\end{aligned}$$

where we used the relation (22.24) for $c(x \bullet y)$ in the second line, the relation $\rho(x \bullet y) = \rho(y) \bullet \rho(x)$ in the third, the relation (22.26) in the fourth, and the relation

$$\mu^\sharp_{\mathfrak{a}_0} \bullet \tau = (-1)^{b_1(Y)} \tau \bullet \mu^\sharp_{\mathfrak{b}_0}$$

in the ninth. □

We can reinterpret this lemma, for example, in relation to the maps $\check{\omega}_\mu$ and $\hat{\omega}_\mu$:

Corollary 22.5.13. *The maps*

$$\check{\omega}_{Y,\mu} : \widecheck{HM}_k(-Y, \mathfrak{s}) \to \widehat{HM}^{o(k)}(Y, \mathfrak{s})$$
$$\hat{\omega}_{-Y,\mu} : \widehat{HM}_{o(k)}(Y, \mathfrak{s}) \to \widecheck{HM}^k(-Y, \mathfrak{s})$$

satisfy

$$(\hat{\omega}_{-Y,\mu})^\dagger = (-1)^{(\mathrm{gr}^{(2)}_Y(o(k))+1)b_1(Y)} \check{\omega}_{Y,\mu},$$

where $(\hat{\omega}_{-Y,\mu})^\dagger$ denotes the map on cohomology induced from the adjoint chain map. Similarly,

$$(\check{\omega}_{-Y,\mu})^\dagger = (-1)^{(\mathrm{gr}^{(2)}_Y(o(k))+1)b_1(Y)} \hat{\omega}_{Y,\mu}.$$

□

22.6 Local coefficients

Rather than use coefficients $\mathbb{Z}$, we can construct Floer homology groups with an arbitrary abelian group Γ as coefficient group. More generally, we can take Γ to be any *local system* of abelian groups on $\mathcal{B}^\sigma_k(Y, \mathfrak{s})$, in line with our discussion of finite-dimensional Morse theory in Subsection 2.7. Thus to each point $[\mathfrak{a}]$ in $\mathcal{B}^\sigma_k(Y, \mathfrak{s})$ there is associated a group $\Gamma[\mathfrak{a}]$ and to each homotopy class z of paths from $[\mathfrak{a}]$ to $[\mathfrak{b}]$ there is associated an isomorphism $\Gamma(z) : \Gamma[\mathfrak{a}] \to \Gamma[\mathfrak{b}]$. We define

$$C^o(\Gamma) = \bigoplus_{[\mathfrak{a}] \in \mathfrak{C}^o} \mathbb{Z}\Lambda[\mathfrak{a}] \otimes \Gamma[\mathfrak{a}]$$

$$C^s(\Gamma) = \bigoplus_{[\mathfrak{a}] \in \mathfrak{C}^s} \mathbb{Z}\Lambda[\mathfrak{a}] \otimes \Gamma[\mathfrak{a}]$$

$$C^u(\Gamma) = \bigoplus_{[\mathfrak{a}] \in \mathfrak{C}^u} \mathbb{Z}\Lambda[\mathfrak{a}] \otimes \Gamma[\mathfrak{a}]$$

(compare (22.2)). We have the eight operators

$$\begin{aligned}
\partial^o_o &: C^o(\Gamma) \to C^o(\Gamma) \\
\partial^o_s &: C^o(\Gamma) \to C^s(\Gamma) \\
\partial^u_o &: C^u(\Gamma) \to C^o(\Gamma) \\
\bar{\partial}^s_s &: C^s(\Gamma) \to C^s(\Gamma) \\
\bar{\partial}^s_u &: C^s(\Gamma) \to C^u(\Gamma) \\
\bar{\partial}^u_s &: C^u(\Gamma) \to C^s(\Gamma) \\
\bar{\partial}^u_u &: C^u(\Gamma) \to C^u(\Gamma) \\
\partial^u_s &: C^u(\Gamma) \to C^s(\Gamma)
\end{aligned}$$

defined by (for example)

$$\partial^o_o = \sum_{[\mathfrak{a}]\in\mathfrak{C}^o} \sum_{[\mathfrak{b}]\in\mathfrak{C}^o} \sum_{z} \sum_{[\gamma]\in\breve{M}_z([\mathfrak{a}],[\mathfrak{b}])} \epsilon[\gamma] \otimes \Gamma(z),$$

just as in (22.8). These operators are combined to form the differentials

$$\begin{aligned}
\check{\partial} &: \check{C}(\Gamma) \to \check{C}(\Gamma) \\
\hat{\partial} &: \hat{C}(\Gamma) \to \hat{C}(\Gamma) \\
\bar{\partial} &: \bar{C}(\Gamma) \to \bar{C}(\Gamma),
\end{aligned}$$

where $\check{C}(\Gamma) = C^o(\Gamma) \oplus C^s(\Gamma)$, and so on, just as in Definition 22.1.3. These operators have square zero (the proof of Proposition 22.1.4 is essentially unchanged), and we therefore have homology groups

$$\widecheck{HM}(Y,\mathfrak{s};\Gamma), \quad \widehat{HM}(Y,\mathfrak{s};\Gamma), \quad \overline{HM}(Y,\mathfrak{s};\Gamma). \tag{22.35}$$

All our earlier constructions can be repeated in this context. The groups $\widecheck{HM}_*(Y,\mathfrak{s};\Gamma)$ are graded by the set $\mathbb{J}(\mathfrak{s})$. We can form the Floer cohomology groups in the usual way: we have groups

$$\begin{aligned}
(C^o)^*(\Gamma) &= \bigoplus_{[\mathfrak{a}]\in\mathfrak{C}^o} \mathrm{Hom}(\mathbb{Z}\Lambda[\mathfrak{a}],\mathbb{Z}) \otimes \Gamma[\mathfrak{a}] \\
(C^s)^*(\Gamma) &= \bigoplus_{[\mathfrak{a}]\in\mathfrak{C}^s} \mathrm{Hom}(\mathbb{Z}\Lambda[\mathfrak{a}],\mathbb{Z}) \otimes \Gamma[\mathfrak{a}] \\
(C^u)^*(\Gamma) &= \bigoplus_{[\mathfrak{a}]\in\mathfrak{C}^u} \mathrm{Hom}(\mathbb{Z}\Lambda[\mathfrak{a}],\mathbb{Z}) \otimes \Gamma[\mathfrak{a}],
\end{aligned}$$

with operators

$$(\partial^o_o)^* = \sum_{[\mathfrak{a}]\in\mathfrak{C}^o}\ \sum_{[\mathfrak{b}]\in\mathfrak{C}^o}\ \sum_{z}\ \sum_{[\gamma]\in\breve{M}_z([\mathfrak{a}],[\mathfrak{b}])} \epsilon[\gamma]^* \otimes \Gamma(z^{-1}),$$

from which we form the cochain complexes

$$(\check{C}^*(\Gamma), \check{\partial}^*), \quad (\hat{C}^*(\Gamma), \hat{\partial}^*), \quad (\bar{C}^*(\Gamma), \bar{\partial}^*),$$

graded by $\mathbb{J}(\mathfrak{s})$. The Floer cohomologies

$$\widecheck{HM}^*(Y, \mathfrak{s}; \Gamma), \quad \widehat{HM}^*(Y, \mathfrak{s}; \Gamma), \quad \overline{HM}^*(Y, \mathfrak{s}; \Gamma)$$

are the cohomology groups of these cochain complexes. There is a long exact sequence, as in Proposition 22.2.1; and if μ is a homology orientation of Y, there are duality isomorphisms $\check{\omega}_\mu$, $\hat{\omega}_\mu$ and $\bar{\omega}_\mu$, and a diagram

$$\begin{array}{ccccccccc}
\xrightarrow{i_*} & \widecheck{HM}_k(-Y;\Gamma) & \xrightarrow{j_*} & \widehat{HM}_k(-Y;\Gamma) & \xrightarrow{p_*} & \overline{HM}_{k-1}(-Y;\Gamma) & \xrightarrow{i_*} \\
 & \Big\downarrow \check{\omega}_\mu & & \Big\downarrow \hat{\omega}_\mu & (-1)^r & \Big\downarrow \bar{\omega}_\mu & \\
\xrightarrow{p^*} & \widehat{HM}^{o(k)}(Y;\Gamma) & \xrightarrow{j^*} & \widecheck{HM}^{o(k)}(Y;\Gamma) & \xrightarrow{i^*} & \overline{HM}^{o(k)}(Y;\Gamma) & \xrightarrow{p^*}
\end{array} \tag{22.36}$$

in which the first two squares commute and the third square commutes or anti-commutes, just as in Corollary 22.5.10.

Examples. In the introductory Subsection 3.7, we promised that to each C^∞ singular 1-cycle η on Y with real coefficients, we would assign a local system on $\mathcal{B}^\sigma(Y, \mathfrak{s})$, called Γ_η. Some of its formal properties were listed there. To construct Γ_η, we first set

$$\Gamma_\eta[\mathfrak{a}] = \mathbb{R}$$

for each $[\mathfrak{a}] \in \mathcal{B}^\sigma(Y, \mathfrak{s})$. If $\zeta : I \to \mathcal{B}^\sigma(Y, \mathfrak{s})$ is a path from $[\mathfrak{a}]$ to $[\mathfrak{b}]$ representing a homotopy class of paths z, then ζ gives rise to a gauge-equivalence class of 4-dimensional connections $[A_\zeta]$, in temporal gauge, on the cylinder $I \times Y$. Let $F_{A^t_\zeta}$ be the curvature of the connection in the associated line bundle, and let

$$f(z) = \frac{i}{2\pi} \int_{I\times\eta} F_{A^t_\zeta}.$$

This real number depends only on the endpoints and the homotopy class z. To complete the definition of the local system, we define

$$\Gamma_\eta(z) : \mathbb{R} \to \mathbb{R}$$

to be multiplication by the real number

$$\Gamma_\eta(z) = e^{f(z)}.$$

This example can be modified by replacing $\mathbb{R}$ by another field $\mathbb{K}$ and the exponential map by any group homomorphism $\exp : \mathbb{R} \to \mathbb{K}^\times$. That is, we set $\Gamma_\eta[\mathfrak{a}] = \mathbb{K}$ for all $[\mathfrak{a}]$, and define $\Gamma_\eta(z) = \exp(f(z))$. For example, one can take $\mathbb{K}$ to be the field of fractions belonging to the ring $k[\mathbb{R}]$ – the group ring of $\mathbb{R}$ with coefficients in a field k.

22.7 The Floer homology of the three-sphere

We now examine the simplest example, the 3-sphere S^3 equipped with the round metric. We have already described the Floer groups for S^3, without proof, in the introduction, Subsection 3.3. We are now in a position to verify that description from our definitions, though we have not yet dealt with the module structure or the grading by 2-plane fields: we will identify the Floer groups of S^3 only as groups at present.

The unperturbed Chern–Simons–Dirac functional $\mathcal{L}$ for the unique spin^c structure $\mathfrak{s}$ on S^3 has only one critical point, namely the unique reducible solution $[B, 0]$, where B is a spin^c connection with $F_{B^t} = 0$. This follows from a slightly more general observation.

Proposition 22.7.1. *Let Y be a Riemannian 3-manifold with non-negative scalar curvature s. Then the only critical points $[B, \Psi]$ of the unperturbed functional $\mathcal{L}$ are the reducible solutions with $\Psi = 0$.*

Proof. This is essentially the same as Proposition 4.6.1, which dealt with solutions to the 4-dimensional equations. The present 3-dimensional version can be deduced by noting that a solution to the 3-dimensional equations on Y gives a solution to the 4-dimensional equations on $S^1 \times Y$. An alternative argument is to repeat some of the calculations leading up to Corollary 4.5.3 in a purely 3-dimensional setting. □

In the blown-up configuration space $\mathcal{B}^\sigma_k(S^3, \mathfrak{s})$, the critical points of the vector field $(\mathrm{grad}\,\mathcal{L})^\sigma$ are degenerate, because the spectrum of the Dirac operator D_B

is not simple: the eigenspaces are non-trivial representations of the symmetry group Spin(4). We choose a small perturbation $\mathfrak{q} \in \mathcal{P}$ to make the critical points non-degenerate, without changing the essential features, namely that there is only one critical point of the functional downstairs in $\mathcal{B}(Y, \mathfrak{s})$ and that this critical point $[B, 0]$ is reducible. Non-degeneracy means that the spectrum of $D_{\mathfrak{q},B}$ is simple and does not contain zero (see Proposition 12.2.5). Label the eigenvalues in increasing order as λ_i, with λ_0 being the first positive eigenvalue.

The eigenvalues are in one-to-one correspondence with the gauge-equivalence classes of critical points of $(\operatorname{grad} \mathcal{L})^\sigma$, which we label accordingly as $[\mathfrak{a}_i]$ $(i \in \mathbb{Z})$. The critical points $[\mathfrak{a}_i]$ with $i \geq 0$ belong to $\mathfrak{C}^s$, and the others belong to $\mathfrak{C}^u$. The set $\mathbb{J}(S^3, \mathfrak{s})$ has a free, transitive $\mathbb{Z}$ action, and we specify an isomorphism

$$\mathbb{J}(S^3, \mathfrak{s}) \to \mathbb{Z}$$

by sending $\operatorname{gr}[\mathfrak{a}_0]$ to 0. If λ_i and λ_{i-1} have the same sign, then $\operatorname{gr}_z([\mathfrak{a}_i], [\mathfrak{a}_{i-1}]) = 2$, independent of the path z, by Corollary 14.6.2; and we have $\operatorname{gr}_z([\mathfrak{a}_0], [\mathfrak{a}_{-1}]) = 1$. Thus, under the above isomorphism, we have

$$\operatorname{gr}[\mathfrak{a}_i] = \begin{cases} 2i & (i \geq 0) \\ -2i + 1 & (i < 0). \end{cases}$$

The reducible grading $\bar{\operatorname{gr}}$ (defined at (22.15)) satisfies $\bar{\operatorname{gr}}[\mathfrak{a}_i] = 2i$ for all i.

The orientation sets $\Lambda([\mathfrak{a}_i])$ all have a preferred element, because the definition of $\Lambda([\mathfrak{a}_i])$ means that it is identified with $\Lambda([\mathfrak{a}_i], [\mathfrak{a}'])$, for any choice of reducible configuration $[\mathfrak{a}']$, and in particular for $[\mathfrak{a}'] = [\mathfrak{a}_i]$. So each $\mathbb{Z}\Lambda[\mathfrak{a}_i]$ has a preferred isomorphism with $\mathbb{Z}$. Putting it all together, we see that

- $\check{C}_j = \mathbb{Z}$ for even, non-negative j and is zero otherwise,
- $\hat{C}_j = \mathbb{Z}$ for odd, negative j and is zero otherwise, and
- $\bar{C}_j = \mathbb{Z}$ for all even j.

In all three cases, the chain groups are never non-zero in two adjacent gradings, so the differentials are all zero; and the Floer groups $\widecheck{HM}_*(S^3, \mathfrak{s})$, $\widehat{HM}_*(S^3, \mathfrak{s})$ and $\overline{HM}_*(S^3, \mathfrak{s})$ are therefore isomorphic to the chain groups. This is the result promised in Proposition 3.3.1. The canonical mod 2 grading defined in Subsection 22.4 is the one that fits with our chosen labelling: $\operatorname{gr}^{(2)}[\mathfrak{a}_i]$ is even for $i \geq 0$.

The long exact sequence relating these three is the obvious one, with $j_* = 0$, as follows easily from the definitions, because there are no irreducible solutions and the boundary maps are zero. The reduced Floer homology $HM_*(S^3, \mathfrak{s})$ is zero, because it is defined as the image of j_*.

We should note that the above description of the Floer groups of S^3 can be applied to a 3-manifold with $b_1 = 0$ and positive scalar curvature. If Y is any such 3-manifold and $\mathfrak{s}$ is any spin^c structure, then by Corollary 4.2.2 and the above proposition, there is a unique critical point $[\alpha]$ in $\mathcal{B}^\sigma(Y, \mathfrak{s})$, which is reducible. The corresponding Dirac operator D_B has no kernel, again because of the scalar curvature condition. After a small perturbation, we can take it that the spectrum of the perturbed Dirac operator is simple, and we can label the critical point in the blown-up configuration space as $[\mathfrak{a}_i]$ as before. Just as for S^3, we see that $\widecheck{HM}_j(Y, \mathfrak{s})$, for example, is $\mathbb{Z}$ in the all gradings $j = j_0 + 2i$, where $i \geq 0$ and j_0 is the grading corresponding to $\mathfrak{a}_0$.

Notes and references for Chapter VI

The orientability of a moduli space arising in gauge theory was exploited by Donaldson in his first proof [17] that a simply connected smooth 4-manifold cannot have non-standard, definite intersection form. The key feature of Donaldson's proof is recognizing the role of the determinant line bundle, which exists not just over the moduli space, but over the whole configuration space, whose topology is known. Essentially the same orientability argument was used by Floer in [32]. Our argument showing that the Seiberg–Witten moduli spaces are orientable follows the same pattern, and is by now quite standard. We have taken care to specify our conventions carefully enough to be able to deal correctly with the boundary-obstructed phenomenon and with duality.

The material on Stokes' theorem is used only briefly in the case of 1-dimensional moduli spaces in this chapter. In the following chapter, we will evaluate more general cohomology classes on moduli spaces and make use of the full result. Traditionally, evaluations of cohomology classes on moduli spaces have often been defined as intersection numbers; see [20], for example. This technique requires the construction (often ad hoc) of finite-codimension submanifolds (or subvarieties) dual to the classes being evaluated. An alternative approach, based on Čech or Alexander–Spanier cohomology, is more flexible and works well with quite general spaces stratified by manifolds. This approach was used by Frøyshov in [40].

VII

Cobordisms and invariance

Our next aim is to establish that the monopole Floer homology groups are topological invariants of a 3-manifold Y. The construction of the groups $\widecheck{HM}_*(Y, \mathfrak{s})$ etc. in the previous sections depended on a choice of Riemannian metric g and an admissible perturbation $\mathfrak{q}$ (see Definition 22.1.1). To emphasize the role that they play, we will write

$$\widecheck{HM}_*(Y, g, \mathfrak{q}, \mathfrak{s})$$

to indicate the dependence. The task of establishing that different choices of g and $\mathfrak{q}$ lead to canonically isomorphic groups is closely tied with another feature of the Floer groups, first discussed in Section 3, namely the fact that a cobordism between 3-manifolds gives rise to homomorphisms between their monopole Floer homologies. We take up both of these matters now, stating the main results before turning to the machinery needed for the proofs. By the end of this chapter, we will have verified the formal properties of the Floer groups that were discussed in Section 3, with just a few exceptions.

The main exception is the material of Subsection 3.9, concerning what happens when two manifolds with $b^+ = 0$ are joined to form a closed manifold with $b^+ \geq 2$. This and the related matter of non-exact perturbations will be discussed in the next chapter.

23 Summary of results

23.1 Metrics, perturbations and cobordisms

It will be convenient to combine all the spin^c structures on a given 3-manifold Y and form a sum of the Floer homologies. For this purpose we need a collection of perturbations, one for each spin^c structure. Let us write $\mathcal{P}(Y,\mathfrak{s})$ for a large Banach space of tame perturbations belonging to the spin^c structure $\mathfrak{s}$ on Y, and let us introduce

$$\mathcal{P}(Y) = \prod_{\mathfrak{s}} \mathcal{P}(Y,\mathfrak{s}),$$

where the product is taken over all isomorphism classes of spin^c structures. Recall that, to be tame, the perturbation must satisfy the estimates and smoothness conditions in Definition 10.5.1. When dealing with a collection of perturbations $\{\mathfrak{q}_{\mathfrak{s}}\}$, we will require some uniformity in the estimates:

Definition 23.1.1. We say that an element $\mathfrak{q} = \{\mathfrak{q}_{\mathfrak{s}}\}$ in $\mathcal{P}(Y)$ is *admissible* if all its components $\mathfrak{q}_{\mathfrak{s}}$ are admissible, and if in addition the bound in Item (iv) of Definition 10.5.1 holds with a *uniform* constant m_2, independent of $\mathfrak{s}$. ◊

The uniformity of the constant m_2 appeared in an earlier context, in Lemma 10.7.5. That lemma tells us that, for such a choice of perturbations, there can only be finitely many spin^c structures $\mathfrak{s}$ for which the corresponding perturbed functional has critical points. It follows that $\widecheck{HM}_*(Y,g,\mathfrak{q}_{\mathfrak{s}},\mathfrak{s})$ is zero, for all but finitely many $\mathfrak{s}$, as are $\widehat{HM}_*$ and $\overline{HM}_*$, as claimed in Proposition 3.1.1.

Definition 23.1.2. Given an admissible perturbation $\mathfrak{q} = \{\mathfrak{q}_{\mathfrak{s}}\}_{\mathfrak{s}}$ in $\mathcal{P}(Y)$, write $\widecheck{HM}_*(Y,g,\mathfrak{q})$ for the direct sum of the Floer homologies $\widecheck{HM}_*(Y,g,\mathfrak{q}_{\mathfrak{s}},\mathfrak{s})$ taken over all isomorphism classes of spin^c structures $\mathfrak{s}$ on Y:

$$\widecheck{HM}_*(Y,g,\mathfrak{q}) = \bigoplus_{\mathfrak{s}} \widecheck{HM}_*(Y,g,\mathfrak{q}_{\mathfrak{s}},\mathfrak{s}).$$

We make a similar definition for $\widehat{HM}_*$, $\overline{HM}_*$, and the Floer cohomology groups $\widecheck{HM}^*$, $\widehat{HM}^*$ and $\overline{HM}^*$. By the above remarks, these direct sums involve only finitely many non-zero terms. ◊

We regard $\widecheck{HM}_*(Y,g,\mathfrak{q})$ and its companions as graded abelian groups, graded by the set with $\mathbb{Z}$ action

$$\mathbb{J}(Y,g,\mathfrak{q}) = \coprod_{\mathfrak{s}} \mathbb{J}(Y,\mathfrak{s},g,\mathfrak{q}).$$

We have again changed our notation to emphasize the role of the metric and perturbation in the definition of this set. (See Subsection 22.3.) Given an element $j \in \mathbb{J}(Y,g,\mathfrak{q})$, we can write $\widecheck{HM}_j(Y,g,\mathfrak{q})$ without mention of the spin^c structure:

$\mathfrak{s}$ is determined by j. Following Definition 3.1.3, we form Floer groups $\widecheck{HM}_\bullet$ etc. by completion:

Definition 23.1.3. The Floer homology groups $\widecheck{HM}_\bullet(Y, g, \mathfrak{q})$, $\widehat{HM}_\bullet(Y, g, \mathfrak{q})$ and $\overline{HM}_\bullet(Y, g, \mathfrak{q})$ are constructed from the $*$ versions by negative completion in the sense of Definition 3.1.3. The Floer cohomology groups $\widecheck{HM}^\bullet(Y, g, \mathfrak{q})$, $\widehat{HM}^\bullet(Y, g, \mathfrak{q})$ and $\overline{HM}^\bullet(Y, g, \mathfrak{q})$ are constructed by positive completion. ◊

Recall now the category COB whose objects are compact, connected, oriented 3-manifolds and whose morphisms are isomorphism classes of connected cobordisms equipped with homology orientations, Definition 3.4.2. As a formal device, we introduce a category whose objects are 3-manifolds decorated with metrics and perturbations:

Definition 23.1.4. We write $\widetilde{\text{COB}}$ for the category in which an object is a triple $(Y, g, \mathfrak{q})$, where g is a Riemannian metric and $\mathfrak{q} \in \mathcal{P}(Y)$ an admissible perturbation. A morphism in $\widetilde{\text{COB}}$ is defined to be a COB-morphism between the underlying oriented 3-manifolds. Thus there is a forgetful functor from $\widetilde{\text{COB}}$ to COB; and two objects $(Y_0, g_0, \mathfrak{q}_0)$, $(Y_1, g_1, \mathfrak{q}_1)$ in $\widetilde{\text{COB}}$ are isomorphic if the underlying 3-manifolds are diffeomorphic. ◊

We are now in a position to state:

Theorem 23.1.5. *The construction of the Floer groups $\widecheck{HM}_\bullet(Y, g, \mathfrak{q})$ etc. can be extended to define covariant functors*

$$\widecheck{HM}_\bullet : \widetilde{\text{COB}} \to \text{GROUP}$$
$$\widehat{HM}_\bullet : \widetilde{\text{COB}} \to \text{GROUP}$$
$$\overline{HM}_\bullet : \widetilde{\text{COB}} \to \text{GROUP}$$

to the category of abelian groups. The Floer cohomology groups similarly extend to contravariant functors

$$\widecheck{HM}^\bullet : \widetilde{\text{COB}} \to \text{GROUP}$$
$$\widehat{HM}^\bullet : \widetilde{\text{COB}} \to \text{GROUP}$$
$$\overline{HM}^\bullet : \widetilde{\text{COB}} \to \text{GROUP}.$$

Remark. This theorem would not be true as stated if we were to replace $\widehat{HM}_\bullet$, for example, by the incomplete version $\widehat{HM}_*$. The reason lies in the fact that, if W is a cobordism from Y_0 to Y_1, the corresponding map

$$\widehat{HM}_\bullet(W) : \widehat{HM}_\bullet(Y_0, g_0, \mathfrak{q}_0) \to \widehat{HM}_\bullet(Y_1, g_1, \mathfrak{q}_1)$$

is a sum over all spin^c structures on W, and infinitely many of these may contribute. This is illustrated already in the case that Y_0 and Y_1 are both S^3 with the round metric and a small perturbation, as in Subsection 22.7. We saw there that

$$\widehat{HM}_*(S^3) = \sum_{\substack{j \text{ odd} \\ j<0}} \mathbb{Z}e_j,$$

where e_j stands for the generator given by the unique critical point in grading j. Thus

$$\widehat{HM}_\bullet(S^3) = \prod_{\substack{j \text{ odd} \\ j<0}} \mathbb{Z}e_j.$$

Let W be the cobordism from S^3 to S^3 obtained from $\bar{\mathbb{CP}}^2$ (that is, $\mathbb{CP}^2$ with the opposite orientation) by removing two balls, and for each odd integer m let $\mathfrak{s}_m$ be the spin^c structure on $\bar{\mathbb{CP}}^2$ whose first Chern class is m times a chosen generator. Then we can write

$$\widehat{HM}_\bullet(W) = \sum_{m \text{ odd}} \widehat{HM}_\bullet(W, \mathfrak{s}_m), \tag{23.1}$$

and we have (for a suitable choice of homology orientation on W)

$$\widehat{HM}_\bullet(W, \mathfrak{s}_m)(e_j) = e_{j-(m^2-1)/4}.$$

The sum (23.1) is infinite, and makes sense as an endomorphism of $\widehat{HM}_\bullet(S^3)$, but it is not defined on $\widehat{HM}_*(S^3)$. See Subsection 39.3.

Theorem 23.1.5 implies the topological invariance of the Floer groups. If $(g, \mathfrak{q})$ and $(g', \mathfrak{q}')$ are two choices of metric and perturbation for Y, then the objects $(Y, g, \mathfrak{q})$ and $(Y, g', \mathfrak{q}')$ are canonically isomorphic in our category $\widetilde{\text{COB}}$: the trivial cobordism $[0, 1] \times Y$ provides the isomorphism.

Corollary 23.1.6. *If $(Y, g, \mathfrak{q})$ and $(Y, g', \mathfrak{q}')$ are two objects of $\widetilde{\text{COB}}$ with the same underlying oriented 3-manifold Y, then $\widecheck{HM}_\bullet(Y, g, \mathfrak{q})$ and $\widecheck{HM}_\bullet(Y, g', \mathfrak{q}')$*

are canonically isomorphic (as are the other flavors of Floer homology and cohomology). □

Technically, although they are canonically isomorphic, $\widecheck{HM}_\bullet(Y,g,\mathfrak{q})$ and $\widecheck{HM}_\bullet(Y,g',\mathfrak{q}')$ are two different groups. If CAT is any category, we can form a new category CAT / CAN, "CAT up to canonical isomorphism". An object x of CAT / CAN consists of a set A, a family $\{x_a\}$ of objects of CAT indexed by A, and isomorphisms $\phi_{a_1a_2} : x_{a_1} \to x_{a_2}$ for all a_1 and a_2, satisfying $\phi_{a_2a_3}\phi_{a_1a_2} = \phi_{a_1a_3}$. A morphism m from $x = (A, \{x_a\}, \phi)$ to $y = (B, \{y_b\}, \psi)$ is a collection of morphisms $m_{ab} : x_a \to y_b$ with $\psi_{bb'}m_{ab}\phi_{a'a} = m_{a'b'}$. Using this terminology, we could describe the Floer groups as functors from COB to the category GROUP / CAN of groups up to canonical isomorphism. We can avoid this issue, however, because there is a functor from GROUP / CAN to GROUP: to an object $G = (A, \{G_a\}, \phi)$ in GROUP / CAN, this functor assigns the group of "cross-sections", the subgroup of $\prod G_a$ consisting of collections $\{g_a\}$ with $g_b = \phi_{ab}g_a$ for all a, b. This subgroup is isomorphic to each of the groups G_a.

With this understood, we can state the next corollary:

Corollary 23.1.7. *The Floer groups define covariant functors from the cobordism category* COB *to the category of groups,*

$$\widecheck{HM}_\bullet : \text{COB} \to \text{GROUP}$$

$$\widehat{HM}_\bullet : \text{COB} \to \text{GROUP}$$

$$\overline{HM}_\bullet : \text{COB} \to \text{GROUP},$$

and contravariant functors

$$\widecheck{HM}^\bullet : \text{COB} \to \text{GROUP}$$

$$\widehat{HM}^\bullet : \text{COB} \to \text{GROUP}$$

$$\overline{HM}^\bullet : \text{COB} \to \text{GROUP}.$$

□

Having now recorded the topological invariance of the Floer groups, we can also point out that the groups $\widecheck{HM}_\bullet(Y,\mathfrak{s})$ and its companions can be non-zero only for finitely many spin^c structures $\mathfrak{s}$ on Y, as stated earlier in Proposition 3.1.1: see Definition 23.1.2 and the remarks which precede it.

We have now recovered the most important elements of the basic structure promised in Section 3, namely groups $\widecheck{HM}_\bullet(Y)$ etc. which are topological invariants of Y and maps $\widecheck{HM}_\bullet(W)$ etc. arising from 4-dimensional cobordisms W.

At this point we can also discuss the dependence of the grading set $\mathbb{J}(Y, g, \mathfrak{q})$ on the choice of the metric and perturbation. The group homomorphisms arising from a morphism W from $(Y_0, g_0, \mathfrak{q}_0)$ to $(Y_1, g_1, \mathfrak{q}_1)$ in the category $\widetilde{\text{COB}}$ respect a *relation* between the two grading sets (see the general discussion in Subsection 3.4). When W is an isomorphism, this relation is an isomorphism between $\mathbb{J}(Y_0, g_0, \mathfrak{q}_0)$ and $\mathbb{J}(Y_1, g_1, \mathfrak{q}_1)$, respecting the $\mathbb{Z}$ action. Just as with the Floer groups themselves, we can therefore regard $\mathbb{J}(Y)$ as a topological invariant of Y: it is a functor from the category of 3-manifolds and diffeomorphisms to the category of sets with $\mathbb{Z}$ action. The following proposition ties this grading set $\mathbb{J}(Y)$ to the 2-plane fields that were described as the grading set in Subsection 3.1.

Proposition 23.1.8. *There is a natural isomorphism of sets with $\mathbb{Z}$ action, between the grading set $\mathbb{J}(Y)$ and the set of homotopy classes of oriented 2-plane fields on Y, with the $\mathbb{Z}$ action described in Definition 3.1.2.*

This proposition will be proved in Subsection 28.2.

23.2 Module structure

In the introductory Subsection 3.2, it was stated that the monopole Floer homology and cohomology groups are modules for the *ordinary* cohomology of the configuration space, via cap and cup product operations. When we combine all the spin^c structures to form the groups $\widecheck{HM}_\bullet(Y)$, the appropriate configuration space is the union of $\mathcal{B}^\sigma(Y, \mathfrak{s})$ over all isomorphism classes of spin^c structures; so we introduce

$$\boldsymbol{\mathcal{B}}^\sigma(Y) = \coprod_{\mathfrak{s}} \mathcal{B}^\sigma(Y, \mathfrak{s}). \tag{23.2}$$

It was also explained in Subsection 3.4 how this module structure could be extended in a way that meshed with the maps obtained from cobordisms. We now take up these matters in more detail.

We shall construct operations

$$\begin{aligned} \frown &: H^d(\mathcal{B}^\sigma(Y, \mathfrak{s})) \otimes \widecheck{HM}_k(Y, \mathfrak{s}) \to \widecheck{HM}_{k-d}(Y, \mathfrak{s}) \\ \smile &: H^d(\mathcal{B}^\sigma(Y, \mathfrak{s})) \otimes \widecheck{HM}^k(Y, \mathfrak{s}) \to \widecheck{HM}^{k+d}(Y, \mathfrak{s}), \end{aligned} \tag{23.3}$$

with similar operations in which $\widecheck{HM}$ is replaced by $\widehat{HM}$ or $\overline{HM}$. Strictly speaking, a choice of Riemannian metric is involved in the definition of the space

$\mathcal{B}^\sigma(Y, \mathfrak{s})$; but the cohomology ring of the space is independent of the metric, up to canonical isomorphism, so there is no real harm in our notation.

The first main properties of these "cup" and "cap" operations are the following associative laws.

Proposition 23.2.1. *The cup and cap products satisfy the associative laws*

$$\begin{aligned}(u \smile v) \smile x &= u \smile (v \smile x)\\(u \smile v) \frown \xi &= v \frown (u \frown \xi),\end{aligned} \tag{23.4}$$

for ξ *in any of* $\widecheck{HM}_*$, $\widehat{HM}_*$ *and* $\overline{HM}_*$ *and* x *in any of* $\widecheck{HM}^*$, $\widehat{HM}^*$ *and* $\overline{HM}^*$.

In particular, the groups $\widecheck{HM}^*$, $\widehat{HM}^*$ and $\overline{HM}^*$ become modules over the ring

$$H^*(\mathcal{B}^\sigma(Y, \mathfrak{s})) = \oplus_d H^d(\mathcal{B}^\sigma(Y, \mathfrak{s})),$$

while $\widecheck{HM}_*$, $\widehat{HM}_*$ and $\overline{HM}_*$ become modules for the opposite ring. (We must not take the direct *product* here.)

The ring $H^*(\mathcal{B}^\sigma(Y, \mathfrak{s}))$ was described in Proposition 9.7.1. As we explained in the introductory Section 3, we can use the description provided by Proposition 9.7.1 to rephrase the maps (23.3) as maps

$$\begin{aligned}\frown &: \mathbb{A}_\dagger(Y) \otimes \widecheck{HM}_*(Y, \mathfrak{s}) \to \widecheck{HM}_*(Y, \mathfrak{s})\\\smile &: \mathbb{A}(Y) \otimes \widecheck{HM}^*(Y, \mathfrak{s}) \to \widecheck{HM}^*(Y, \mathfrak{s}).\end{aligned}$$

We will not use that description in this chapter; we will think of the module structure primarily in terms of the maps (23.3). Taking the sum over all spin^c structures, the groups $\widecheck{HM}(Y)$ and their companions become modules over the cohomology ring of $\mathcal{B}^\sigma(Y)$, as do their completions $\widecheck{HM}_\bullet(Y)$ and so on.

For the generalization of the module structure, let W be a cobordism from Y_0 to Y_1, equipped with a Riemannian metric g_W, and let $\mathfrak{s}_W$ be a spin^c structure on W. We assume that the metric g_W is cylindrical in collar neighborhoods of the boundary components, and we write g_0, g_1 for the metrics on Y_0 and Y_1. The spin^c structure $\mathfrak{s}_W$ gives rise to spin^c structures $\mathfrak{s}_0$, $\mathfrak{s}_1$ on the boundary.

We have already introduced in Subsection 9.3 the configuration space $\mathcal{B}^\sigma(W, \mathfrak{s}_W)$. Let $[\gamma] = [A, s, \phi]$ be an element of this configuration space, so ϕ is a spinor of $L^2(W)$-norm 1 and $s \geq 0$ is a constant. If the restriction of ϕ to the boundary component Y_i is not identically zero, then $[\gamma]$ has a well-defined restriction to Y_i as an element of $\mathcal{B}^\sigma(Y_i, \mathfrak{s}_i)$. So there are partially defined restriction maps

$$r = (r_0, r_1) : \mathcal{B}^\sigma(W, \mathfrak{s}_W) \dashrightarrow \mathcal{B}^\sigma(Y_0, \mathfrak{s}_0) \times \mathcal{B}^\sigma(Y_1, \mathfrak{s}_1)$$

whose common domain of definition is the open subset

$$U = \{ [A, s, \phi] \in \mathcal{B}^\sigma(W, \mathfrak{s}_W) \mid \phi|_{Y_i} \text{ not identically zero}, i = 0, 1 \} \\ \subset \mathcal{B}^\sigma(W, \mathfrak{s}_W).$$

We again use the notation $\boldsymbol{\mathcal{B}}^\sigma(W)$ to denote the disjoint union

$$\boldsymbol{\mathcal{B}}^\sigma(W) = \coprod_{\mathfrak{s}_W} \mathcal{B}^\sigma(W, \mathfrak{s}_W)$$

over all isomorphism classes of spin^c structures on W, so that we have a restriction map

$$r = (r_0, r_1) : \boldsymbol{\mathcal{B}}^\sigma(W) \dashrightarrow \boldsymbol{\mathcal{B}}^\sigma(Y_0) \times \boldsymbol{\mathcal{B}}^\sigma(Y_1). \tag{23.5}$$

If $W = W_2 \circ W_1$, then there is a similar restriction map

$$R_i : \boldsymbol{\mathcal{B}}^\sigma(W) \dashrightarrow \boldsymbol{\mathcal{B}}^\sigma(W_i),$$

for $i = 1, 2$. Because the inclusions of subsets such as U in $\boldsymbol{\mathcal{B}}^\sigma(W)$ are weak homotopy equivalences, there are pull-back maps

$$R_i^* : H^*(\boldsymbol{\mathcal{B}}^\sigma(W_i)) \to H^*(\boldsymbol{\mathcal{B}}^\sigma(W)).$$

So given cohomology classes u_i on $\boldsymbol{\mathcal{B}}^\sigma(W_i)$ for $i = 1$ and 2, we can form a product

$$\begin{aligned} u &= u_2 u_1 \\ &= R_1^*(u_1) \smile R_2^*(u_2). \end{aligned} \tag{23.6}$$

(The reason for the reversed order in the last line is the same reason that led us to define $\mathbb{A}_\dagger$ as the opposite ring to $\mathbb{A}$ when discussing the module structure previously.)

Given a connected cobordism W from Y_0 to Y_1, with a homology orientation and a class $u \in H^*(\boldsymbol{\mathcal{B}}^\sigma(W))$, we will define maps

$$\begin{aligned} \widecheck{HM}_\bullet(u \mid W) &: \widecheck{HM}_\bullet(Y_0) \to \widecheck{HM}_\bullet(Y_1) \\ \widehat{HM}_\bullet(u \mid W) &: \widehat{HM}_\bullet(Y_0) \to \widehat{HM}_\bullet(Y_1) \\ \overline{HM}_\bullet(u \mid W) &: \overline{HM}_\bullet(Y_0) \to \overline{HM}_\bullet(Y_1). \end{aligned}$$

On Floer cohomology, we will define maps

$$\widecheck{HM}^\bullet(u \mid W) : \widecheck{HM}^\bullet(Y_1) \to \widecheck{HM}^\bullet(Y_0)$$

etc. In the special case that W is a trivial cobordism $[0,1] \times Y$, the spaces $\mathcal{B}^\sigma(W)$ and $\mathcal{B}^\sigma(Y)$ have the same weak homotopy type, and we can therefore equate their cohomology rings. In this special case, the maps $\widecheck{HM}_\bullet(u \mid W)$ and $\widecheck{HM}^\bullet(u \mid W)$ will coincide with the cap and cup products by the class u. The following proposition, proved in Section 26 below, generalizes the associative law for the cap and cup products.

Proposition 23.2.2. *Let W_1, W_2 be cobordisms with homology orientations, and let $W = W_2 \circ W_1$ be the composite. Let $u = u_2 u_1$ be a product, as in* (23.6). *Then we have*

$$\widecheck{HM}_\bullet(u \mid W) = \widecheck{HM}_\bullet(u_2 \mid W_2) \circ \widecheck{HM}_\bullet(u_1 \mid W_1),$$

along with five similar identities for $\widehat{HM}_\bullet$, $\overline{HM}_\bullet$, $\widecheck{HM}^\bullet$, $\widehat{HM}^\bullet$ and $\overline{HM}^\bullet$.

This proposition could also be formulated by introducing an appropriate category in which the morphisms were pairs $(u \mid W)$. As with $\mathcal{B}^\sigma(Y)$, there is a homomorphism

$$\mathbb{A}(W) \to H^*(\mathcal{B}^\sigma(W)),$$

and the cohomology ring of each component is generated by $\mathbb{A}(W)$ and the 2-dimensional class u_2. Based on this, one recovers the generalized module structure as we first described it in Subsection 3.4, where we used the notation $\widecheck{HM}_\bullet(a \mid W)$, for example, for the operator corresponding to an element a in $\mathbb{A}_\dagger(W)$.

The construction of the homomorphisms $\widecheck{HM}_\bullet(u \mid W)$ etc. defined by a cobordism W will be carried out in Section 25, and the proof of the composition law is contained in Section 26.

23.3 Local coefficients

Next we wish to extend these statements to cover Floer homology with local coefficients. Let (Y_0, g_0) and (Y_1, g_1) be compact, connected, oriented 3-manifolds with Riemannian metrics, and let Γ_0, Γ_1 be local systems on $\mathcal{B}^\sigma(Y_0)$ and $\mathcal{B}^\sigma(Y_1)$. Let W be a cobordism from Y_0 to Y_1, and let $r = (r_0, r_1)$ be the restriction map (23.5). The following definition is motivated by the discussion of Subsection 2.7.

Definition 23.3.1. A W-*morphism*, Γ_W, from the local system Γ_0 to Γ_1 is an isomorphism Γ_W between the pull-backs of the local systems on $\mathcal{B}^\sigma(W)$:

$$\Gamma_W : r_0^*(\Gamma_0) \to r_1^*(\Gamma_1).$$

◊

Here it is to be understood again that the restriction maps are defined only on a subset U whose inclusion in $\mathcal{B}^\sigma(W)$ is a weak homotopy equivalence. Also, although a metric is again involved in the definition of $\mathcal{B}^\sigma(W)$, the notion of a W-morphism is essentially metric-independent.

We can rephrase the definition to clarify what is involved. Fix a Riemannian metric g_W on W which is isometric to a cylinder in collar neighborhoods of the two boundary components, and let

$$[\mathfrak{a}_0] \in \mathcal{B}^\sigma(Y_0), \quad [\mathfrak{a}_1] \in \mathcal{B}^\sigma(Y_1)$$

be configurations on the two boundary components.

Definition 23.3.2. A W-*path* ξ from $[\mathfrak{a}_0]$ to $[\mathfrak{a}_1]$ is an element $[\gamma]$ in $\mathcal{B}^\sigma(W)$ whose restriction to the boundary components is the given pair: $r([\gamma]) = ([\mathfrak{a}_0], [\mathfrak{a}_1])$. Two W-paths are *homotopic* if they belong to the same path component of the fiber $r^{-1}([\mathfrak{a}_0], [\mathfrak{a}_1])$. We write

$$\pi([\mathfrak{a}_0], W, [\mathfrak{a}_1])$$

for the set of homotopy classes of W-paths. Again, the notion of a W-path is really independent of the choice of metric. ◊

If we have two cobordisms $W_1 : Y_0 \to Y_1$ and $W_2 : Y_1 \to Y_2$ and if we have classes z_1 in $\pi([\mathfrak{a}_0], W_1, [\mathfrak{a}_1])$ and z_2 in $\pi([\mathfrak{a}_1], W_2, [\mathfrak{a}_2])$, then there is a well-defined homotopy class of composite W-paths $z_2 \circ z_1$ in $\pi([\mathfrak{a}_0], W, [\mathfrak{a}_2])$. To define the composite, we can represent z_1 by a configuration γ_1 that is in temporal gauge and translation-invariant in a collar neighborhood of Y_1. We do the same with z_2, and there is then a unique way to glue the two configurations together: the gauge-group acts freely on $\mathcal{C}_k^\sigma(Y_1, \mathfrak{s}_1)$, so there is a uniquely determined identification of the spin bundles.

A special case of a homotopy class of W-paths, for W a cylinder $[0,1] \times Y$, arises from a homotopy class of paths z from $[\mathfrak{a}_0]$ to $[\mathfrak{a}_1]$ in $\mathcal{B}_k^\sigma(Y)$. Indeed, homotopy classes of $([0,1] \times Y)$-paths are the same as homotopy classes of paths. For a general cobordism $W : Y_0 \to Y_1$, we can identify the composite cobordism

$$([0,1] \times Y_1) \circ W \circ ([0,1] \times Y_0)$$

with W by a standard diffeomorphism. So the above composition law allows us to form composites

$$z_1 \circ z \circ z_0,$$

for z in $\pi([\mathfrak{a}_0], W, [\mathfrak{a}_1])$ and z_i a relative homotopy class of paths in $\mathcal{B}^\sigma(Y_i)$. When writing the composition of W-paths, we order the components (as above) in the same way that we chose for our notation for composite cobordisms and composite paths (see (2.22) for example).

We can now formulate an equivalent description of what a W-morphism is. Let Γ_0 and Γ_1 again be local systems on $\mathcal{B}^\sigma(Y_0)$ and $\mathcal{B}^\sigma(Y_1)$. Then a W-morphism assigns to each z in $\pi([\mathfrak{a}_0], W, [\mathfrak{a}_1])$ an isomorphism

$$\Gamma_W(z) : \Gamma_0[\mathfrak{a}_0] \to \Gamma_1[\mathfrak{a}_1].$$

These are required to satisfy the composition rule

$$\Gamma_W(z_1 \circ z \circ z_0) = \Gamma_1(z_1) \circ \Gamma_W(z) \circ \Gamma_0(z_0) \tag{23.7}$$

for every pair of paths z_0 in $\mathcal{B}^\sigma(Y_0)$ and z_1 in $\mathcal{B}^\sigma(Y_1)$.

We can incorporate both local coefficients and the generalized cap and cup products in an enlarged version of the category COB.

Definition 23.3.3. Given a commutative ring R, we define a category COB-LC (the LC stands for *local coefficients*) as follows. An object (Y, Γ) in COB-LC is a compact, connected, oriented 3-manifold Y together with a local system of R-modules Γ on $\mathcal{B}^\sigma(Y)$. A morphism

$$(u \mid W, \Gamma_W) : (Y_0, \Gamma_0) \to (Y_1, \Gamma_1)$$

is a connected cobordism W with a homology orientation, together with a W-morphism Γ_W from Γ_0 to Γ_1, and a cohomology class

$$u \in H^*(\mathcal{B}^\sigma(W); R).$$

Morphisms are composed using the rule (23.6) to combine the cohomology classes. ◊

Once again, an inessential choice of Riemannian metrics is implicit in the definition. To avoid this, we could perhaps replace $\mathcal{B}^\sigma(Y)$ with a larger space $\underline{\mathcal{B}}^\sigma(Y)$, the union of $\mathcal{B}^\sigma(Y)$ over all choices of metrics, and do the same for W; but the slight sloppiness in the above definition will not be a handicap. The category COB-LC is where the Floer functors are ultimately defined:

Theorem 23.3.4. *For each choice of R, the Floer homology groups with local coefficients, $\widecheck{HM}_\bullet(Y;\Gamma)$, provide covariant functors*

$$\widecheck{HM}_\bullet : \text{COB-LC} \to \text{MOD}$$
$$\widehat{HM}_\bullet : \text{COB-LC} \to \text{MOD}$$
$$\overline{HM}_\bullet : \text{COB-LC} \to \text{MOD}$$

from the category COB-LC *to the category of R-modules. The Floer cohomology groups are contravariant functors*

$$\widecheck{HM}^\bullet : \text{COB-LC} \to \text{MOD}$$
$$\widehat{HM}^\bullet : \text{COB-LC} \to \text{MOD}$$
$$\overline{HM}^\bullet : \text{COB-LC} \to \text{MOD}.$$

We illustrate W-morphisms with an example. Recall that to each real, C^∞ singular 1-cycle η on a 3-manifold Y, we associated a local system Γ_η on $\mathcal{B}^\sigma(Y)$ with fiber $\mathbb{R}$. (See Subsection 22.6.) Suppose now that we are given a relative 2-cycle ν on a cobordism $W : Y_0 \to Y_1$, and write

$$\partial\nu = \eta_1 - \eta_0$$

with $\eta_i \in Z_1(Y_i;\mathbb{R})$. We can use ν to define a W-morphism

$$\Gamma_\nu : \Gamma_{\eta_0} \to \Gamma_{\eta_1} \tag{23.8}$$

by the following recipe. Let $[\gamma] \in \mathcal{B}^\sigma(W)$ be a representative of $z \in \pi([\mathfrak{a}_0], W, [\mathfrak{a}_1])$, and write $\gamma = (A, s, \phi)$. Then define

$$\Gamma_\nu(z) : \Gamma_{\eta_0}[\mathfrak{a}_0] \to \Gamma_{\eta_1}[\mathfrak{a}_1]$$

to be the map $\mathbb{R} \to \mathbb{R}$ given by multiplication by

$$\exp\Big(\frac{i}{2\pi}\int_\nu F_{A^t}\Big).$$

The composition law (23.7) which characterizes W-morphisms is easily verified. If we replace ν by $\nu + \partial\theta$ where θ is a 3-chain, then Stokes' theorem tell us that Γ_ν is unchanged. These properties were summarized in the introduction, Subsection 3.7.

As a particular application, suppose that W is a cylinder $[0,1] \times Y$ and that η, η' are homologous real 1-cycles in Y. Let ν be a relative 2-cycle in $[0,1] \times Y$ with $\partial\nu = \{0\} \times \eta - \{1\} \times \eta'$. Then Γ_ν determines an isomorphism $\Gamma_\eta \to \Gamma_{\eta'}$. To within a boundary, ν is determined by its image in Y, which is a homology θ between η and η'. This explains the comment in Subsection 3.7, that a choice of homology ν is involved in identifying local coefficient systems on Y corresponding to homologous 1-cycles.

24 The moduli space on a manifold with boundary

Although we studied the perturbed equations on both finite and infinite cylinders in earlier chapters, we have not, until now, needed to study the more general situation that arises when the cylinders are replaced by more general manifolds with boundary. In Chapter II, we studied compactness properties of the unperturbed equations on a compact manifold with boundary, and we now need to carry over the rest of the machinery. One issue that arises is that, on a non-cylindrical space, the τ model of the blow-up is not available to us: we have to use the σ model instead. It would have been possible to use the σ model throughout, but this would have necessitated the introduction of weighted Sobolev spaces when setting up the Fredholm theory, and would have muddied the analogy with the finite-dimensional Morse-theory picture.

We start to pick up the pieces here, combining ingredients from the previous chapters to put together the results we need.

24.1 Defining the moduli space

Let X be a compact, connected, oriented Riemannian 4-manifold with non-empty boundary. As in Subsection 17.1, we suppose the Riemannian metric is cylindrical in the neighborhood of the boundary, so that X contains an isometric copy of $I \times Y$ for some interval $I = (-C, 0]$, with ∂X identified with $\{0\} \times Y$. Let $\mathfrak{s}_X$ be a spin^c structure on X, and $\mathfrak{s}$ the resulting spin^c structure on Y. We may allow Y to be disconnected, with components labelled Y^α, and we will write the configuration space as

$$\mathcal{B}_k^\sigma(Y, \mathfrak{s}) = \prod \mathcal{B}_k^\sigma(Y^\alpha, \mathfrak{s}^\alpha)$$

and similarly we write

$$\mathcal{P}(Y, \mathfrak{s}) = \prod \mathcal{P}(Y^\alpha, \mathfrak{s}^\alpha)$$

for the Banach space of tame perturbations. Later we will be most interested in the case that X has boundary $-Y_0 \cup Y_1$ (a cobordism), but at present we suppose that all components Y^α have the boundary orientation.

We recall from Section 9 the configuration space

$$\mathcal{C}^\sigma(X, \mathfrak{s}_X) = \{ (A, s, \phi) \in \mathcal{A} \times \mathbb{R} \times \Gamma(X; S^+) \mid s \geq 0, \|\phi\|_{L^2(X)} = 1 \}$$

and its Hilbert completion $\mathcal{C}^\sigma_k(X, \mathfrak{s}_X)$, with quotient $\mathcal{B}^\sigma_k(X, \mathfrak{s}_X)$, a Hilbert manifold with boundary. We have the smooth section

$$\mathfrak{F}^\sigma : \mathcal{C}^\sigma_k(X, \mathfrak{s}_X) \to \mathcal{V}^\sigma_{k-1}$$

defined by the formula from (6.5),

$$\mathfrak{F}^\sigma(A, s, \phi) = \left(\frac{1}{2} \rho_X(F^+_{A^t}) - s^2(\phi\phi^*)_0, \; D^+_A \phi \right). \tag{24.1}$$

The equations $\mathfrak{F}^\sigma = 0$ are the unperturbed Seiberg–Witten equations on X, in the σ blow-up model.

In the special case that X was a cylinder, we introduced a class of perturbations of these equations in Subsection 10.2. We now introduce a perturbation for the equations on a general X, supported in the cylindrical region $I \times Y$. Let β be a cut-off function, equal to 1 near $t = 0$ and equal to 0 near $t = -C$. Let β_0 be a bump-function with compact support in $(-C, 0)$. Let $\mathfrak{q}$ and $\mathfrak{p}_0$ be two elements of $\mathcal{P}(Y, \mathfrak{s})$. We use these to define a section

$$\hat{\mathfrak{p}} : \mathcal{C}_k(X, \mathfrak{s}_X) \to \mathcal{V}_k$$

by the formula

$$\hat{\mathfrak{p}} = \beta \hat{\mathfrak{q}} + \beta_0 \hat{\mathfrak{p}}_0. \tag{24.2}$$

Here we regard β and β_0 as functions on the collar $I \times Y \subset X$. It is understood that $\hat{\mathfrak{p}}(A, \Phi)$ depends only on the restriction of A and Φ to the collar, and that $\hat{\mathfrak{p}}(A, \Phi)$ is supported in the collar. Writing $\mathcal{V}_k = L^2_k(X; i\,\mathfrak{su}(S^+) \oplus S^-)$, we have the decomposition

$$\hat{\mathfrak{p}} = (\hat{\mathfrak{p}}^0, \hat{\mathfrak{p}}^1),$$

and we define $\hat{\mathfrak{p}}^\sigma$ on the blow-up to be the section

$$\hat{\mathfrak{p}}^\sigma : \mathcal{C}_k^\sigma(X, \mathfrak{s}_X) \to \mathcal{V}_k^\sigma$$

given by

$$\begin{aligned} \hat{\mathfrak{p}}^{0,\sigma}(A, s, \phi) &= \hat{\mathfrak{p}}^0(A, s\phi) \\ \hat{\mathfrak{p}}^{1,\sigma}(A, s, \phi) &= (1/s)\hat{\mathfrak{p}}^1(A, s\phi), \end{aligned} \tag{24.3}$$

when $s \neq 0$. We extend over $s = 0$ as in Subsection 10.2.

On X, we now have the perturbed Seiberg–Witten equation $\mathfrak{F}_{\mathfrak{p}}^\sigma = 0$, where

$$\mathfrak{F}_{\mathfrak{p}}^\sigma = \mathfrak{F}^\sigma + \hat{\mathfrak{p}}^\sigma : \mathcal{C}_k^\sigma(X, \mathfrak{s}_X) \to \mathcal{V}_{k-1}^\sigma.$$

The equation is invariant under $\mathcal{G}_{k+1}(X)$, and we have a moduli space of solutions,

$$\begin{aligned} &M(X, \mathfrak{s}_X) \subset \mathcal{B}_k^\sigma(X, \mathfrak{s}_X) \\ &M(X, \mathfrak{s}_X) = \left\{ (A, s, \phi) \;\middle|\; \mathfrak{F}_{\mathfrak{p}}^\sigma = 0 \right\} / \mathcal{G}_{k+1}(X). \end{aligned}$$

(We omit the perturbation from our notation for the moduli space at present.) We also have the larger moduli space

$$\tilde{M}(X, \mathfrak{s}_X) \subset \tilde{\mathcal{B}}_k^\sigma(X, \mathfrak{s}_X),$$

obtained by dropping the condition $s \geq 0$. As before, $M(X, \mathfrak{s}_X)$ can be identified with the quotient $\tilde{M}(X, \mathfrak{s}_X)/\mathbf{i}$, where $\mathbf{i}(A, s, \phi) = (A, -s, \phi)$.

Recall that if $X' \subset X$ is an interior domain, we have a partially defined restriction map

$$\mathcal{B}_k^\sigma(X, \mathfrak{s}_X) \dashrightarrow \mathcal{B}_k^\sigma(X', \mathfrak{s}_X).$$

In particular, there is the restriction map

$$\mathcal{B}_k^\sigma(X) \dashrightarrow \mathcal{B}_k^\sigma(I \times Y, \mathfrak{s}_X),$$

whose domain of definition is the triples (A, s, ϕ) such that the restriction of ϕ is not identically zero on the collar. Using the equivalence between the σ and τ models for the blow-up on the cylinder, there is therefore a restriction map

$$\mathcal{B}_k^\sigma(X, \mathfrak{s}_X) \dashrightarrow \mathcal{B}_k^\tau(I \times Y, \mathfrak{s}_X),$$

whose domain is now those triples such that ϕ is not zero on $t \times Y$ for any $t \in I$. Because of unique continuation, this restriction map is defined on all of $M(X, \mathfrak{s}_X)$: we have

$$M(X, \mathfrak{s}_X) \to M(I \times Y, \mathfrak{s}_X) \subset \mathcal{B}^\tau_k(I \times Y, \mathfrak{s}_X). \tag{24.4}$$

(On $I \times Y$, the τ model for the equations is implied.) There is also then a restriction map

$$R : M(X, \mathfrak{s}_X) \to \mathcal{B}^\sigma_{k-1/2}(Y, \mathfrak{s}).$$

24.2 Attaching cylindrical ends

Let X^* be the non-compact Riemannian manifold obtained by attaching a copy of the cylinder $Z = [0, \infty) \times Y$, identifying the two copies of $\{0\} \times Y$:

$$X^* = X \cup_Y Z$$
$$Z = [0, \infty) \times Y.$$

On X^*, we have the $L^2_{k,\text{loc}}$ configuration space

$$\mathcal{C}_{k,\text{loc}}(X^*, \mathfrak{s}_X) = \mathcal{A}_{k,\text{loc}} \times L^2_{k,\text{loc}}(X^*; S^+).$$

We are not now dealing with Banach spaces, but we can still construct a blown-up configuration space $\mathcal{C}^\sigma_{k,\text{loc}}(X^*, \mathfrak{s}_X)$ in the manner discussed briefly in Subsection 6.1. (See the remarks on page 115.) That is, we define the sphere $\mathbb{S}$ to be the topological quotient of $L^2_{k,\text{loc}}(X^*; S^+) \setminus 0$ by the action of $\mathbb{R}^+$, so that the real blow-up of $L^2_{k,\text{loc}}(X^*; S^+)$ at 0 can be defined as the set of pairs

$$\big\{ (\mathbb{R}^+\phi, \Phi) \bigm| \Phi \in \mathbb{R}^{\geq}\phi \big\}.$$

Thus we can define the blown-up configuration space as

$$\mathcal{C}^\sigma_{k,\text{loc}}(X^*, \mathfrak{s}_X) = \big\{ (A, \mathbb{R}^+\phi, \Phi) \bigm| \Phi \in \mathbb{R}^{\geq}\phi \big\}$$
$$\subset \mathcal{A}_{k,\text{loc}} \times \mathbb{S} \times L^2_{k,\text{loc}}(X^*; S^+).$$

The sphere $\mathbb{S}$ is a topological manifold: the charts are provided by the restriction of the quotient map to the affine subspaces

$$\{\phi \mid \xi(\phi) = 1\}$$

where ξ is a continuous linear functional on $L^2_{k,\text{loc}}(X^*; S^+)$. Because we are no longer using the unit-sphere model for $\mathbb{S}$, the bundle

$$\mathcal{V}_j^\sigma \to \mathcal{C}^\sigma_{k,\text{loc}}(X^*, \mathfrak{s}_X)$$

needs to be defined in the style of (6.6). This bundle has a continuous section

$$\mathfrak{F}^\sigma_{\mathfrak{p}} = \mathfrak{F}^\sigma + \hat{\mathfrak{p}}^\sigma.$$

The perturbing term $\hat{\mathfrak{p}}^\sigma$ is defined as before on the collar $(-C, 0] \times Y$ and is extended so as to be $\hat{\mathfrak{q}}^\sigma$ on the cylindrical-end Z. This section is invariant under the action of the gauge-group.

As in (24.4), there is a restriction map

$$\left\{ [\gamma] \in \mathcal{B}^\sigma_{k,\text{loc}}(X^*, \mathfrak{s}_X) \;\middle|\; \mathfrak{F}^\sigma_{\mathfrak{p}}(\gamma) = 0 \right\} \to \mathcal{B}^\tau_{k,\text{loc}}(Z; \mathfrak{s}).$$

The restriction of $[\gamma]$ satisfies the equation $\mathfrak{F}^\tau_{\mathfrak{q}}(\gamma) = 0$ on the cylinder. We will suppose that the perturbation $\mathfrak{q}$ on the cylinder is regular. We can now define moduli spaces on the cylindrical-end manifold.

Definition 24.2.1. For a critical point $[\mathfrak{b}]$ in $\mathcal{B}^\sigma_k(Y, \mathfrak{s})$, we define the moduli space

$$M(X^*, \mathfrak{s}_X; [\mathfrak{b}]) \subset \mathcal{B}^\sigma_{k,\text{loc}}(X^*, \mathfrak{s}_X)$$

to be the set of all $[\gamma]$ such that $\mathfrak{F}^\sigma_{\mathfrak{p}}(\gamma) = 0$ and such that the restriction of $[\gamma]$ is asymptotic to $[\mathfrak{b}]$ on the cylindrical end Z. ◊

It is also useful to form the union over all choices of spin^c structures $\mathfrak{s}_X$, to obtain the spaces $\boldsymbol{\mathcal{B}}^\sigma_k(X^*)$ and $\boldsymbol{\mathcal{B}}^\sigma_{k,\text{loc}}(X^*)$. We then define

$$\begin{aligned} M(X^*; [\mathfrak{b}]) &= \coprod_{\mathfrak{s}_X} M(X^*, \mathfrak{s}_X; [\mathfrak{b}]) \\ &\subset \boldsymbol{\mathcal{B}}^\sigma_{k,\text{loc}}(X^*), \end{aligned} \tag{24.5}$$

with the understanding that the only spin^c structures $\mathfrak{s}_X$ that make a non-empty contribution to this union are those whose restriction to Y is isomorphic to the spin^c structure $\mathfrak{s}$ to which $[\mathfrak{b}]$ belongs.

There is a fiber product description of the moduli space $M(X^*, \mathfrak{s}_X; [\mathfrak{b}])$. We have the restriction maps from $M(X^*, \mathfrak{s}_X; [\mathfrak{b}])$ to $M(X, \mathfrak{s}_X)$ and to $M(Z; [\mathfrak{b}])$.

Note that our definition of $M(X, \mathfrak{s}_X)$ uses the σ model, while our definition of $M(Z; [\mathfrak{b}])$ uses the τ model. The image of the product map

$$\rho : M(X^*, \mathfrak{s}_X; [\mathfrak{b}]) \to M(X, \mathfrak{s}_X) \times M(Z; [\mathfrak{b}])$$

is contained in the fiber product

$$\mathrm{Fib}(R_+, R_-) \subset M(X, \mathfrak{s}_X) \times M(Z; [\mathfrak{b}]),$$

where R_+ and R_- are the two restriction maps to $\mathcal{B}^\sigma_{k-1/2}(Y, \mathfrak{s})$. We have:

Lemma 24.2.2. *The map ρ is a homeomorphism from the moduli space $M(X^*, \mathfrak{s}_X; [\mathfrak{b}])$ onto its image, the fiber product* $\mathrm{Fib}(R_+, R_-)$.

Proof. Use the equivalence between the τ and σ models on the cylinder $I \times Y$, and then follow the argument in the proof of the similar result, Lemma 19.1.1. □

We let $M^{\text{red}}(X^*, \mathfrak{s}_X; [\mathfrak{b}]) \subset M(X^*, \mathfrak{s}_X; [\mathfrak{b}])$ denote the set of reducible elements: those elements having representatives (A, s, ϕ) with $s = 0$. If $[\gamma]$ belongs to $M(X^*, \mathfrak{s}_X; [\mathfrak{b}])$, and $\rho[\gamma]$ is the pair

$$([\gamma_X], [\gamma_Z]) \in M(X, \mathfrak{s}_X) \times M(Z; [\mathfrak{b}]),$$

then $[\gamma]$ is reducible if and only if either $[\gamma_X]$ or $[\gamma_Z]$ is reducible. If Y has several components, the $[\gamma]$ is reducible if and only if the restriction of $[\gamma_Z]$ to any one component is reducible.

24.3 Moduli spaces as Hilbert manifolds

Before continuing to study the moduli space on the cylindrical-end manifold X^*, we now look in more detail at the (infinite-dimensional) moduli space $M(X, \mathfrak{s}_X)$ associated with the compact manifold with boundary, X. In the case of a compact cylinder $Z = I \times Y$, we proved in Subsection 17.3 that a moduli space of this sort was a smooth Hilbert manifold: this held regardless of the choice of perturbation. We now prove the same statement for a general X.

Proposition 24.3.1. *On the manifold with boundary, X, the section $\mathfrak{F}^\sigma_\mathfrak{p}$ of $\mathcal{V}^\sigma_{k-1}$ is transverse to zero, and the subset $\tilde{M}(X, \mathfrak{s}_X) \subset \tilde{\mathcal{B}}^\sigma_k(X, \mathfrak{s}_X)$ is a smooth Hilbert submanifold. The moduli space $M(X, \mathfrak{s}_X)$ is therefore a Hilbert manifold with boundary, and can be identified with the quotient of $\tilde{M}(X, \mathfrak{s}_X))$ by the involution $s \mapsto -s$.*

Proof. Let $\gamma = (A, s, \phi)$ be a solution of $\mathfrak{F}^\sigma_\mathfrak{p} = 0$. We examine the linearization of the section $\mathfrak{F}^\sigma_\mathfrak{p}$ at γ. A tangent vector to $\mathcal{C}^\sigma_k(X, \mathfrak{s}_X)$ can be written as

$$(a, t, \psi) \in L^2_k(X; iT^*X) \oplus \mathbb{R} \oplus L^2_k(X; S^+),$$

with the constraint

$$\mathrm{Re}\langle \psi, \phi \rangle_X = 0.$$

We write $\langle \phi \rangle^\perp$ for the real-orthogonal complement of ϕ. The linearization of the term $\mathfrak{F}^\sigma$ is the operator

$$\mathcal{D}_\gamma \mathfrak{F}^\sigma : L^2_k(X; iT^*X) \times \mathbb{R} \times \langle \phi \rangle^\perp \to L^2_{k-1}(X; i\,\mathfrak{su}(S^+) \oplus S^-)$$

given by

$$\begin{aligned} \mathcal{D}_\gamma \mathfrak{F}^\sigma : \ & (a, t, \psi) \mapsto \\ & \left(\frac{1}{2} \rho_X(d^+ a) - 2ts(\phi\phi^*)_0 - s^2(\phi\psi^* + \psi\phi^*)_0, D^+_A \psi + \rho_X(a)\phi \right). \end{aligned} \tag{24.6}$$

The right-hand side of the formulae (24.3) is well-defined without the constraint that ϕ has L^2 norm 1. So the derivative of the perturbation extends from $\langle \phi \rangle^\perp$ to all of $L^2_k(X; S^+)$, where it becomes an operator

$$\mathcal{D}_\gamma \hat{\mathfrak{p}}^\sigma : L^2_k(X; iT^*X) \times \mathbb{R} \times L^2_k(X; S^+) \to L^2_{k-1}(X; i\,\mathfrak{su}(S^+) \oplus S^-).$$

With this larger domain, we write the formal adjoint of $\mathcal{D}_\gamma \hat{\mathfrak{p}}^\sigma$ as

$$(\mathfrak{r}_1, \mathfrak{r}_2, \mathfrak{r}_3) : L^2_j(X; i\,\mathfrak{su}(S^+) \oplus S^-) \to L^2_{j-1}(X; iT^*X) \times \mathbb{R} \times L^2_{j-1}(X; S^+).$$

As in the proof of Theorem 17.3.1, the surjectivity of $\mathcal{D}_\gamma \mathfrak{F}^\sigma_\mathfrak{p}$ will follow from the surjectivity of the larger operator

$$Q^\sigma_\gamma = \mathcal{D}_\gamma \mathfrak{F}^\sigma_\mathfrak{p} \oplus \mathbf{d}^{\sigma,\dagger}_\gamma, \tag{24.7}$$

where

$$\mathbf{d}^{\sigma,\dagger}_\gamma : \mathcal{T}^\sigma_{k,\gamma} \to L^2_{k-1}(X; i\mathbb{R})$$

is the operator defined by the left side of the second equation (9.12):

$$\mathbf{d}_\gamma^{\sigma,\dagger}(a,t,\psi) = -d^*a + is^2\,\mathrm{Re}\langle i\phi,\psi\rangle + i|\phi|^2\,\mathrm{Re}\,\mu_X(\langle i\phi,\psi\rangle) \qquad (24.8)$$

We wish to recast the operator Q_γ^σ in such a way as to be able to apply Corollary 17.1.5. Since Corollary 17.1.5 involves an operator whose domain is the space of sections of a bundle on X, we are led to introduce

$$\tilde{\psi} = \psi + t\phi$$

and so identify the domain of Q_γ^σ with $L_k^2(X; iT^*X \oplus S^+)$. In the coordinates $(a,\tilde{\psi})$, the operator Q_γ^σ has the form

$$Q_\gamma^\sigma = D_0 + K,$$

where K is compact and D_0 is the elliptic operator

$$(a,\tilde{\psi}) \mapsto \left(\frac{1}{2}\rho_X(d^+a), D_{A_0}^+\tilde{\psi}, -d^*a\right),$$

for some base connection A_0 that we can take to be translation-invariant in the collar. Since D_0 can now be written in the form $(d/dt) + L_0$ in the collar, we are in the general setting of Corollary 17.1.5.

What remains is to show that any solution of the formal adjoint equation

$$(Q_\gamma^\sigma)^* v = 0 \qquad (24.9)$$

which is zero on Y must be zero on all of X. We must therefore write this equation in such a way as to make clear that our unique continuation results can be applied. Write

$$v = (\eta,\pi,\xi) \in L_j^2(X; i\,\mathfrak{su}(S^+) \oplus S^- \oplus i\mathbb{R}).$$

In the original coordinates (a,t,ψ) for the domain of Q_γ^σ, the equation (24.9) becomes the following three equations for v:

$$\left.\begin{aligned} \frac{1}{2}(d^+)^*\rho_X^*\eta + \rho_X^*(\pi\phi^*) - d\xi + \mathfrak{r}_1(\eta,\pi) &= 0 \\ -2s\langle\eta,(\phi\phi^*)_0\rangle_X + \mathfrak{r}_2(\eta,\pi) &= 0 \\ \Pi\big(D_A^-\pi - 2s^2\eta(\phi) + s^2\xi\phi + (\mu_X\langle i|\phi|^2,\xi\rangle)i\phi + \mathfrak{r}_3(\eta,\pi)\big) &= 0. \end{aligned}\right\} \qquad (24.10)$$

Here $\Pi : L^2(X;S^+) \to L^2(X;S^+)$ is the projection on the real-orthogonal complement of ϕ.

We rewrite the last equation as

$$D_A^- \pi - 2s^2\eta(\phi) + s^2\xi\phi + \big(\mu_X\langle i|\phi|^2, \xi\rangle\big)i\phi + \mathfrak{r}_3(\eta, \pi) = \alpha\phi,$$

where α is real. Suppose v satisfies these equations and vanishes on the boundary. If we take the inner product of the last equation with ϕ, we obtain, after integrating by parts,

$$\begin{aligned}&\langle \pi, D_A^+\phi\rangle_X - 2s^2\langle \eta, (\phi\phi^*)_0\rangle_X \\ &\quad + \langle \eta, \mathcal{D}_\gamma\hat{\mathfrak{p}}^{0,\sigma}(0,0,\phi)\rangle_X + \langle \pi, \mathcal{D}_\gamma\hat{\mathfrak{p}}^{1,\sigma}(0,0,\phi)\rangle_X = \alpha,\end{aligned} \tag{24.11}$$

where we have again extended the domain of $\mathcal{D}_\gamma\hat{\mathfrak{p}}^\sigma$ from $\langle\phi\rangle^\perp$ to all of $L_k^2(X;S^+)$. Suppose first that s is non-zero. Then we can use the formula (24.3) to rewrite the above equation as

$$\begin{aligned}&\langle \pi, D_A^+\phi\rangle_X - 2s^2\langle \eta, (\phi\phi^*)_0\rangle_X \\ &\quad + s\langle \eta, \mathcal{D}_\gamma\hat{\mathfrak{p}}^{0}(0,\phi)\rangle_X + \langle \pi, \mathcal{D}_\gamma\hat{\mathfrak{p}}^{1}(0,\phi)\rangle_X = \alpha.\end{aligned} \tag{24.12}$$

On the other hand, the second of the three equations (24.10) is

$$-2s\langle \eta, (\phi\phi^*)_0\rangle_X + \langle \eta, \mathcal{D}_\gamma\hat{\mathfrak{p}}^{0,\sigma}(0,1,0)\rangle + \langle \pi, \mathcal{D}_\gamma\hat{\mathfrak{p}}^{1,\sigma}(0,1,0)\rangle = 0,$$

which gives

$$\begin{aligned}&-2s\langle \eta, (\phi\phi^*)_0\rangle_X + \langle \eta, \mathcal{D}_\gamma\hat{\mathfrak{p}}^{0}(0,\phi)\rangle \\ &\quad - (1/s^2)\langle \pi, \hat{\mathfrak{p}}^{1}(A, s\phi)\rangle + (1/s)\langle \pi, \mathcal{D}_\gamma\hat{\mathfrak{p}}^{1}(0,\phi)\rangle = 0.\end{aligned} \tag{24.13}$$

If we multiply this by s and subtract from the previous equation for α, we obtain

$$\langle \pi, D_A^+\phi + (1/s)\hat{\mathfrak{p}}^1(A, s\phi)\rangle_X = \alpha.$$

But the equation $\mathfrak{F}_{\mathfrak{p}}^\sigma(A, s, \phi) = 0$ means that

$$\begin{aligned}D_A^+\phi + (1/s)\hat{\mathfrak{p}}^1(A, s\phi) &= D_A^+\phi + \hat{\mathfrak{p}}^{1,\sigma}(A, s, \phi) \\ &= 0,\end{aligned}$$

so $\alpha = 0$. If we take the inner product of the third equation of (24.10) with $i\phi$, and integrate by parts again, we obtain

$$\begin{aligned}\langle \pi, iD_A^+\phi\rangle_X &+ \big(s^2 + 1/\mathrm{vol}(X)\big)\langle i|\phi|^2, \xi\rangle_X \\ &+ \langle \eta, \mathcal{D}_\gamma \hat{\mathfrak{p}}^{0,\sigma}(0,0,i\phi)\rangle_X + \langle \pi, \mathcal{D}_\gamma \hat{\mathfrak{p}}^{1,\sigma}(0,0,i\phi)\rangle_X = 0. \qquad (24.14)\end{aligned}$$

The term involving η is zero because of the S^1 invariance of the term $\hat{\mathfrak{q}}^0$, and the last term on the right can be rewritten using the S^1 equivariance of $\hat{\mathfrak{p}}^1$. This gives

$$\langle \pi, iD_A^+\phi\rangle_X + \big(s^2 + 1/\mathrm{vol}(X)\big)\langle i|\phi|^2, \xi\rangle_X + \langle \pi, i\hat{\mathfrak{p}}^{1,\sigma}(A,s,\phi)\rangle_X = 0.$$

Again, the equation $\mathfrak{F}_{\mathfrak{p}}^\sigma = 0$ tells us that the sum of the first and last terms is zero, so the term $\langle i|\phi|^2, \xi\rangle_X$ is zero also.

Having established that α and $\langle i|\phi|^2, \xi\rangle_X$ both vanish, we can recast the equations (24.10) without those terms: we have

$$\left.\begin{aligned}\frac{1}{2}(d^+)^*\rho_X^*\eta + \rho_X^*(\pi\phi^*) - d\xi + \mathfrak{r}_1(\eta,\pi) &= 0\\ -2s\langle \eta, (\phi\phi^*)_0\rangle_X + \mathfrak{r}_2(\eta,\pi) &= 0\\ D_A^-\pi - 2s^2\eta(\phi) + s^2\xi\phi + \mathfrak{r}_3(\eta,\pi) &= 0.\end{aligned}\right\} \qquad (24.15)$$

In the collar region, the equations in this last form have the shape

$$\frac{d}{dt}v + (L_0 + h(t))v = 0,$$

where L_0 is a self-adjoint elliptic operator on Y and h is a time-dependent operator on Y satisfying the conditions of the unique continuation lemma, Lemma 7.1.3. So, since v vanishes on the boundary, it vanishes on the collar too. It therefore vanishes on all of X by the argument of Proposition 7.1.4. The argument when $s = 0$ is similar. □

Having now seen that the moduli space $\tilde{M}(X, \mathfrak{s}_X) \subset \tilde{\mathcal{B}}_k^\sigma(X, \mathfrak{s}_X)$ is a smooth Hilbert manifold, we consider again the restriction map to the boundary,

$$R : \tilde{M}(X, \mathfrak{s}_X) \to \tilde{\mathcal{B}}_{k-1/2}^\sigma(Y, \mathfrak{s}).$$

In the special case of the cylinder, we earlier proved a result concerning the derivative of R, Theorem 17.3.2, which we can now extend to the case of a general 4-manifold with boundary, X. As at Theorem 17.3.2, the derivative defines a map

$$\mathcal{D}_{[\gamma]}R : T_{[\gamma]}\tilde{M}(X, \mathfrak{s}_X) \to \mathcal{K}^{\sigma}_{k-1/2,\mathfrak{a}},$$

and we have a decomposition

$$\mathcal{K}^{\sigma}_{k-1/2,\mathfrak{a}} = \mathcal{K}^{+}_{\mathfrak{a}} \oplus \mathcal{K}^{-}_{\mathfrak{a}}$$

using the spectral subspaces of the Hessian $\mathrm{Hess}^{\sigma}_{\mathfrak{a}}$, or $\mathrm{Hess}^{\sigma}_{\mathfrak{a}} - \epsilon$ in the non-hyperbolic case. (See (17.12).) Then we have:

Proposition 24.3.2. *Let X be a compact 4-manifold with boundary Y, let $[\gamma]$ belong to $\tilde{M}(X, \mathfrak{s}_X)$ and let $[\mathfrak{a}] \in \mathcal{B}^{\sigma}_{k-1/2}(Y, \mathfrak{s})$ be obtained as the restriction of $[\gamma]$ to the boundary, as above. Let π denote the projections of $\mathcal{K}^{\sigma}_{k-1/2,\mathfrak{a}}$ to $\mathcal{K}^{-}_{\mathfrak{a}}$, with kernel $\mathcal{K}^{+}_{\mathfrak{a}}$. Then the linear operators*

$$\pi \circ \mathcal{D}R : T_{[\gamma]}\tilde{M}(X, \mathfrak{s}_X) \to \mathcal{K}^{-}_{\mathfrak{a}}$$
$$(1 - \pi) \circ \mathcal{D}R : T_{[\gamma]}\tilde{M}(X, \mathfrak{s}_X) \to \mathcal{K}^{+}_{\mathfrak{a}}$$

are respectively Fredholm and compact.

Proof. The proof is essentially the same argument as in the case of a cylinder, Theorem 17.3.2. □

24.4 Transversality and cylindrical ends

We now extend our transversality results for moduli spaces of trajectories to our more general moduli spaces. We continue to suppose X is a compact manifold with (non-empty) boundary, as above. Using Lemma 24.2.2 and Proposition 24.3.1, we have a description of $M(X^*, \mathfrak{s}_X; [\mathfrak{b}])$ as the fiber product of two smooth maps on Hilbert manifolds with boundary:

$$\begin{aligned} R_+ &: M(X, \mathfrak{s}_X) \to \mathcal{B}^{\sigma}_{k-1/2}(Y, \mathfrak{s}) \\ R_- &: M(Z; [\mathfrak{b}]) \to \mathcal{B}^{\sigma}_{k-1/2}(Y, \mathfrak{s}). \end{aligned} \tag{24.16}$$

Let $([\gamma_1], [\gamma_2])$ belong to the fiber product, so

$$[\gamma_1] \in M(X, \mathfrak{s}_X)$$
$$[\gamma_2] \in M(Z; [\mathfrak{b}]).$$

Let $[\mathfrak{b}]$ be their common image in $\mathcal{B}^{\sigma}_{k-1/2}(Y, \mathfrak{s})$.

Lemma 24.4.1. *The sum of the derivatives*

$$\mathcal{D}_{[\gamma_1]}R_+ + \mathcal{D}_{[\gamma_2]}R_- : T_{[\gamma_1]}M(X, \mathfrak{s}_X) \oplus T_{[\gamma_2]}M(Z; [\mathfrak{b}]) \to T_{[\mathfrak{b}]}\mathcal{B}^\sigma_{k-1/2}(Y, \mathfrak{s})$$

is Fredholm.

Proof. According to Proposition 24.3.2, the composite maps $(1 - \pi) \circ \mathcal{D}_{[\gamma_1]}R_+$ and $\pi \circ \mathcal{D}_{[\gamma_1]}R_+$ are compact and Fredholm respectively. The same applies to $\mathcal{D}_{[\gamma_2]}R_-$, though with the roles of the two projections reversed, because the analysis that led to Proposition 24.3.2 can be extended also to the case of the half-infinite cylinder Z. The sum of the two derivatives in the lemma therefore differs by a compact operator from a direct sum of two Fredholm operators, namely the operators $\pi \circ \mathcal{D}_{[\gamma_1]}R_+$ and $(1 - \pi) \circ \mathcal{D}_{[\gamma_2]}R_-$. □

We now define regularity for our moduli spaces, in terms of the fiber product description.

Definition 24.4.2. Let $[\gamma] \in M(X^*, \mathfrak{s}_X; [\mathfrak{b}])$. If $[\gamma]$ is irreducible, we say the moduli space $M(X^*, \mathfrak{s}_X; [\mathfrak{b}])$ is *regular* at $[\gamma]$ if the maps of Hilbert manifolds

$$\begin{aligned} R_+ &: M(X, \mathfrak{s}_X) \to \mathcal{B}^\sigma_{k-1/2}(Y, \mathfrak{s}) \\ R_- &: M(Z; [\mathfrak{b}]) \to \mathcal{B}^\sigma_{k-1/2}(Y, \mathfrak{s}) \end{aligned} \tag{24.17}$$

are transverse at $\rho[\gamma]$. If $[\gamma]$ is reducible we say the moduli space is *regular* at $[\gamma]$ if the restrictions

$$\begin{aligned} R_+ &: M^{\text{red}}(X, \mathfrak{s}_X) \to \prod_\alpha \partial\mathcal{B}^\sigma_{k-1/2}(Y^\alpha, \mathfrak{s}^\alpha) \\ R_- &: M^{\text{red}}(Z; [\mathfrak{b}]) \to \prod_\alpha \partial\mathcal{B}^\sigma_{k-1/2}(Y^\alpha, \mathfrak{s}^\alpha) \end{aligned}$$

are transverse at $\rho[\gamma]$. We say the moduli space is *regular* if it is regular at all points. ◊

Proposition 24.4.3. *Let $[\mathfrak{b}]$ be a critical point, and let $[\mathfrak{b}^\alpha]$ be the restriction of $[\mathfrak{b}]$ to the αth component of Y. Suppose the moduli space $M(X^*, \mathfrak{s}_X; [\mathfrak{b}])$ is non-empty and regular. Then the moduli space is:*

(i) *a smooth manifold consisting only of irreducibles, if any $[\mathfrak{b}^\alpha]$ is irreducible;*
(ii) *a smooth manifold consisting only of reducibles, if any $[\mathfrak{b}^\alpha]$ is reducible and boundary-unstable;*

(iii) *a smooth manifold with (possibly empty) boundary if all the* $[\mathfrak{b}^\alpha]$ *are reducible and boundary-stable.*

In the last case, the boundary consists of the reducible elements of the moduli space.

Proof. Transversality means that $\tilde{M}(X^*, \mathfrak{s}_X; [\mathfrak{b}])$ is a manifold. The moduli space $M(X^*, \mathfrak{s}_X; [\mathfrak{b}])$ is the quotient of this manifold by $\mathbf{i}$, and is therefore either a manifold or a manifold with boundary. The latter occurs if and only if it contains both reducibles and irreducibles.

If $[\mathfrak{b}^\alpha]$ is irreducible, then so are all configurations in $M(Z^\alpha; [\mathfrak{b}^\alpha])$ (where Z^α is the corresponding cylindrical component), and it follows from the fiber product description that $M(X^*, \mathfrak{s}_X; [\mathfrak{b}])$ consists only of irreducibles. Similarly, if $[\mathfrak{b}^\alpha]$ is boundary-unstable, then the moduli space consists only of reducibles. □

The following definition generalizes Definition 14.5.5. (Note that in the case of the cylinder, one boundary component has the negative orientation, which changes the roles of boundary-stable and boundary-unstable.)

Definition 24.4.4. If the second case of the above proposition occurs, and more than one of the $[\mathfrak{b}^\alpha]$ is boundary-unstable, then we say that the solution $[\gamma]$ is *boundary-obstructed*. If $c + 1$ of the $[\mathfrak{b}^\alpha]$ are boundary-unstable, we say $[\gamma]$ is boundary-obstructed with *corank* c. ◊

Except in the boundary-obstructed case, if $[\gamma]$ is regular and reducible, then the maps (24.17) are transverse on the full moduli spaces. In the boundary-obstructed case, the maps are not transverse. Equivalently, the Fredholm operator of Lemma 24.4.1 is not surjective: the dimension of its cokernel is c if $[\gamma]$ is "boundary-obstructed with corank c" in the sense of the above definition.

Although it is a manifold, or a manifold with boundary, the moduli space $M(X^*, \mathfrak{s}_X; [\mathfrak{b}])$ is not equidimensional. The situation is analogous to that of $M([\mathfrak{a}], [\mathfrak{b}])$ on the cylinder, which has a decomposition into equidimensional pieces $M_z([\mathfrak{a}], [\mathfrak{b}])$, parametrized by $z \in \pi_1(\mathcal{B}^\sigma(Y, \mathfrak{s}); [\mathfrak{a}], [\mathfrak{b}])$. Let us again take the union over all spin^c structures $\mathfrak{s}_X$, and let $\mathcal{B}^\sigma(X; [\mathfrak{b}])$ be the fiber over $[\mathfrak{b}]$ of the restriction map

$$\mathcal{B}^\sigma(X) \dashrightarrow \mathcal{B}^\sigma(Y, \mathfrak{s}).$$

The analog of the homotopy class in $\pi_1(\mathcal{B}^\sigma(Y);[\mathfrak{a}],[\mathfrak{b}])$ in the manifold with cylindrical end setting is an element of

$$z \in \pi_0(\mathcal{B}^\sigma(X;[\mathfrak{b}])).$$

Any element $[\gamma] \in M(X^*;[\mathfrak{b}])$ is homotopic to a configuration which is equal to the pull-back of $[\mathfrak{b}]$ on the cylindrical end; and in this way, $[\gamma]$ determines an element of $\pi_0(\mathcal{B}^\sigma(X;[\mathfrak{b}]))$. Thus the moduli space $M(X^*;[\mathfrak{b}])$ can be written as a union

$$M(X^*;[\mathfrak{b}]) = \bigcup_z M_z(X^*;[\mathfrak{b}]), \tag{24.18}$$

and there is a similar decomposition of the configuration space:

$$\mathcal{B}^\sigma_{k,\mathrm{loc}}(X^*;[\mathfrak{b}]) = \bigcup_z \mathcal{B}^\sigma_{k,\mathrm{loc},z}(X^*;[\mathfrak{b}]).$$

This set $\pi_0(\mathcal{B}^\sigma(X;[\mathfrak{b}]))$ is a principal homogeneous space for $H^2(X,Y;\mathbb{Z})$. Each element determines, in particular, a spin^c structure $\mathfrak{s}_X$ on X, and the elements z in $\pi_0(\mathcal{B}^\sigma(X;[\mathfrak{b}]))$ belonging to a particular $\mathfrak{s}_X$ are a principal homogeneous space for $H^1(Y;\mathbb{Z})/i_Y^*(H^1(X;\mathbb{Z})$. Notice that given

$$z \in \pi_0(\mathcal{B}^\sigma(X;[\mathfrak{b}_0]))$$

and an element

$$z_1 \in \pi_1(\mathcal{B}^\sigma(Y);[\mathfrak{b}_0],[\mathfrak{b}])$$

we can form the concatenation, which we write as

$$z_1 \circ z \in \pi_0(\mathcal{B}^\sigma(X;[\mathfrak{b}])).$$

Given z in $\pi_0(\mathcal{B}^\sigma_k(X;[\mathfrak{b}]))$, let $[\gamma]$ be any element of $\mathcal{B}^\sigma_{k,z}(X;[\mathfrak{b}])$ and γ a gauge representative. Let $[\gamma_\mathfrak{b}]$ be the constant trajectory in $\mathcal{B}^\tau_k(Z,\mathfrak{s})$ corresponding to $\mathfrak{b}$. We have the operator Q^σ_γ on X, as defined in (24.7); and on Z, we have the translation-invariant operator $Q_{\gamma_\mathfrak{b}}$. There are restriction maps

$$r_+ : \ker(Q^\sigma_\gamma) \to L^2_{k-1/2}(Y; iT^*Y \oplus S \oplus i\mathbb{R})$$
$$r_- : \ker(Q_{\gamma_\mathfrak{b}}) \to L^2_{k-1/2}(Y; iT^*Y \oplus S \oplus i\mathbb{R}).$$

Definition 24.4.5. We define $\mathrm{gr}_z(X;[\mathfrak{b}])$ to be the index of

$$r_+ - r_- : \ker(Q^\sigma_\gamma) \oplus \ker(Q_{\gamma_\mathfrak{b}}) \to L^2_{k-1/2}(Y; iT^*Y \oplus S \oplus i\mathbb{R}).$$

◊

If the moduli space $M_z(X^*;[\mathfrak{b}])$ is non-empty, then we can equally well regard $\mathrm{gr}_z(X;[\mathfrak{b}])$ as being the index of the Fredholm operator of Lemma 24.4.1. The point of the definition above is that it is valid whether or not the moduli space is empty, and makes clear that $\mathrm{gr}_z(X;[\mathfrak{b}])$ depends only on $[\mathfrak{b}]$ and z. We have the simple additivity property

$$\mathrm{gr}_{z_1 \circ z}(X;[\mathfrak{b}]) = \mathrm{gr}_z(X;[\mathfrak{a}]) + \mathrm{gr}_{z_1}([\mathfrak{a}],[\mathfrak{b}]).$$

(In the case that Y has more than one component, the term $\mathrm{gr}_{z_1}([\mathfrak{a}],[\mathfrak{b}])$ is to be defined as a sum over the components.) We also have:

Proposition 24.4.6. *If the moduli space $M_z(X;[\mathfrak{b}])$ is non-empty and regular, its dimension is $\mathrm{gr}_z(X;[\mathfrak{b}])$, except in the boundary-obstructed case. If the moduli space is boundary-obstructed of corank c, then its dimension is $\mathrm{gr}_z(X;[\mathfrak{b}]) + c$.* □

We now show that regularity can be achieved by a suitable choice of perturbation.

Proposition 24.4.7. *Let $\mathfrak{q}$ be a fixed admissible perturbation for Y, let*

$$\hat{\mathfrak{p}} = \beta(t)\hat{\mathfrak{q}} + \beta_0(t)\hat{\mathfrak{p}}_0,$$

on the collar $I \times Y \subset X$, as before, and let $[\mathfrak{b}]$ be a critical point on Y. Then there is a residual subset of $\mathcal{P}(Y,\mathfrak{s})$ such that for all $\mathfrak{p}_0$ in this subset, the moduli space $M(X^*;[\mathfrak{b}])$ is regular.

Proof. To prove this proposition we deal mainly with the moduli space on the manifold with boundary X, and work with a particular spin^c structure $\mathfrak{s}_X$. We begin by introducing the appropriate parametrized moduli space for this manifold:

$$\mathfrak{M}(X,\mathfrak{s}_X) \subset \mathcal{B}^\sigma_k(X,\mathfrak{s}_X) \times \mathcal{P}(Y,\mathfrak{s})$$
$$\mathfrak{M}(X,\mathfrak{s}_X) = \{ (A,s,\phi,\mathfrak{p}_0) \mid \mathfrak{F}^\sigma_{\hat{\mathfrak{p}}} = 0 \} / \mathcal{G}_{k+1}(X),$$

which is the quotient of the zero set of

$$\mathfrak{W}^\sigma : \mathcal{C}_k^\sigma(X, \mathfrak{s}_X) \times \mathcal{P}(Y, \mathfrak{s}) \to \mathcal{V}_{k-1}^\sigma$$
$$\mathfrak{W}^\sigma : (\gamma, \mathfrak{p}) \mapsto \mathfrak{F}_{\hat{\mathfrak{p}}}^\sigma(\gamma)$$

by the gauge group. From Proposition 24.3.1 we see that the parametrized moduli space is a smooth Banach manifold. The reducible locus inside the parametrized moduli space is the set

$$\mathfrak{M}^{\text{red}}(X, \mathfrak{s}_X) = \big\{ ([A, 0, \phi], \mathfrak{p}_0) \bigm| ([A, 0, \phi], \mathfrak{p}_0) \in \mathfrak{M}^{\text{red}}(X, \mathfrak{s}_X) \big\}.$$

The differential of $\mathfrak{W}^\sigma$ at a $(\gamma, \mathfrak{p}_0)$ is a sum of the operators

$$\mathcal{D}_\gamma \mathfrak{F}_{\hat{\mathfrak{p}}}$$

and

$$\mathfrak{m} : T_{\mathfrak{p}_0}\mathcal{P}(Y, \mathfrak{s}) \to \mathcal{V}_{k-1}^\sigma$$
$$\mathfrak{m}(\mathfrak{p}) = \beta_0 \hat{\mathfrak{p}}(\gamma).$$

There is a restriction map

$$R_+ : \mathfrak{M}(X, \mathfrak{s}_X) \to \mathcal{B}_{k-1/2}^\sigma(Y, \mathfrak{s}).$$

Lemma 24.4.8. *Let* $([\gamma], \mathfrak{p}_0) \in \mathfrak{M}(X, \mathfrak{s}_X)$*, and let* $[\mathfrak{c}] = R_+([\gamma])$*. If* $[\gamma]$ *is irreducible, then the differential of the restriction map*

$$\mathcal{D}_{([\gamma], \mathfrak{p}_0)} R_+ : T_{([\gamma], \mathfrak{p}_0)} \mathfrak{M}(X, \mathfrak{s}_X) \to T_{[\mathfrak{c}]} \mathcal{B}_{k-1/2}^\sigma(Y, \mathfrak{s})$$

has range that is dense in the $L^2_{1/2}$ *topology. If* $[\gamma]$ *is reducible, then the differential of the restriction map*

$$\mathcal{D}_{([\gamma], \mathfrak{p}_0)} R_+ : T_{([\gamma], \mathfrak{p}_0)} \mathfrak{M}^{\text{red}}(X, \mathfrak{s}_X) \to \prod_\alpha T_{[\mathfrak{c}^\alpha]} \partial \mathcal{B}_{k-1/2}^\sigma(Y^\alpha, \mathfrak{s}^\alpha)$$

has range that is dense in the $L^2_{1/2}$ *topology.*

Proof. Suppose that γ is irreducible. Then the blow-down map $\pi : \mathcal{B}_k^\sigma(X, \mathfrak{s}_X) \to \mathcal{B}_k(X, \mathfrak{s}_X)$ is a diffeomorphism at γ, as is the corresponding map on Y. So in the statement of the proposition, we can replace $\mathfrak{M}$ by its image under the blow-down map, and we can replace $\mathcal{B}_{k-1/2}^\sigma(Y, \mathfrak{s})$ by $\mathcal{B}_{k-1/2}(Y, \mathfrak{s})$.

The statement in the first part of the proposition is implied by the statement that the differential at $(\pi(\gamma), \mathfrak{p}_0)$ of the map

$$(\mathfrak{W}, \tilde{R}_+) : \mathcal{C}_k(X, \mathfrak{s}_X) \times \mathcal{P}(Y) \to \mathcal{V}_{k-1} \times \mathcal{C}_{k-1/2}(Y, \mathfrak{s})$$

has range that is dense in the $L^2 \times L^2_{1/2}$ topology. Here $\tilde{R}^+$ is the restriction map on $\mathcal{C}_k(X, \mathfrak{s}_X)$. The derivative is an operator

$$\begin{aligned} &L^2_k(X; iT^*X \oplus S^+) \times \mathcal{P}(Y) \\ &\qquad \to L^2_{k-1}(X; i\,\mathfrak{su}(S^+) \oplus S^-) \oplus L^2_{k-1/2}(Y; iT^*Y \oplus S). \end{aligned}$$

Suppose the image of this operator is not dense in $L^2 \times L^2_{1/2}$. Then there is a (V, v) which at first we know to be in

$$L^2(X; i\,\mathfrak{su}(S^+) \oplus S^-) \oplus L^2_{-1/2}(Y; iT^*Y \oplus S)$$

and which is L^2-orthogonal to the image. By considering a gauge-orbit direction in $\mathcal{C}(X)$, we see that v is orthogonal to the gauge orbit through $\pi(\gamma)|_Y$ in $\mathcal{C}_{k-1/2}(Y, \mathfrak{s})$. We can add the gauge-fixing condition to the equations as in the proof of Proposition 24.3.1, and also take the restriction of the dt component of the connection form, to get an operator

$$\begin{aligned} \mathfrak{Q} \oplus r :\ & L^2_1(X; iT^*X \oplus S^+) \times \mathcal{P}(Y) \to \\ & L^2(X; i\,\mathfrak{su}(S^+) \oplus S^-) \oplus L^2(X; i\mathbb{R}) \oplus L^2_{1/2}(Y; iT^*Y \oplus S \oplus i\mathbb{R}). \end{aligned}$$

Let $W = (V, 0)$, and $w = (v, 0)$, so that (W, w) is orthogonal to the image of this operator. The operator $Q = \mathfrak{Q}(-, 0)$ is an operator of the form $D_0 + K$ where D_0 is elliptic and K is compact. By considering variations supported away from the boundary and that leave the perturbation component untouched we can conclude that W is weak solution of the equation

$$Q^* W = 0$$

and hence by our regularity results V is in L^2_1. Integration by parts and the orthogonality property give

$$\begin{aligned} \langle r(U), r(W) \rangle_{L^2(Y)} &= \langle QU, W \rangle_{L^2(X)} - \langle U, Q^* W \rangle_{L^2(X)} \\ &= -\langle r(U), w \rangle_{L^2(Y)} \end{aligned} \tag{24.19}$$

from which we deduce that $W|_{\partial X} = -w$. Arguing as in Lemma 15.1.4, and using the fact that v is orthogonal to the gauge orbit on Y, we find that the restriction of V to each $\{t\} \times Y \subset I \times Y$ in the collar is orthogonal to the gauge orbit. If V is non-zero, then, as in the proof of Proposition 15.1.3, we can find an element $\mathfrak{p}$ of $\mathcal{P}(Y, \mathfrak{s})$ such that

$$\langle \mathfrak{m}(\mathfrak{p})(\gamma), V \rangle_{L^2(X)} > 0.$$

On the other hand, the fact that (W, w) is orthogonal to the image of $(0, \mathfrak{p})$ under $(\mathfrak{Q}, r)$ means that

$$\langle \mathfrak{m}(\mathfrak{p})(\gamma), V \rangle_{L^2(X)} = 0.$$

So it must be the case that $V = 0$.

The proof in the reducible case is similar, and follows the model in Proposition 15.1.3. □

We return to the proof of Proposition 24.4.7. Let

$$\mathfrak{R}_+ : \mathfrak{M}(X, \mathfrak{s}_X) \to \mathcal{B}^\sigma_{k-1/2}(Y, \mathfrak{s})$$

be the restriction map on the parametrized moduli space. Consider the map

$$\mathfrak{R}_+ \times R_- : \mathfrak{M}(X, \mathfrak{s}_X) \times M(Z; [\mathfrak{b}]) \to \mathcal{B}^\sigma_{k-1/2}(Y, \mathfrak{s}) \times \mathcal{B}^\sigma_{k-1/2}(Y, \mathfrak{s}), \quad (24.20)$$

and let (a, b) be an element of the fiber product (the inverse image of the diagonal). From Lemma 24.4.1, the map

$$\mathcal{D}_a \mathfrak{R}_+ - \mathcal{D}_b R_- : T_a \mathfrak{M}_{L^2_1} \oplus T_b M_{L^2_1} \to [\mathcal{T}^\sigma_{1/2}]_{[\mathfrak{c}]}$$

has image which is of finite codimension. In the irreducible case, the lemma above tells us that the image is dense in the $L^2_{1/2}$ topology, and it follows that this operator is surjective. Using the regularity result, Lemma 17.2.9, we deduce that

$$\begin{aligned} \mathcal{D}_a \mathfrak{R}_+ - \mathcal{D}_b R_- : T_a \mathfrak{M}_{L^2_k} \oplus T_b M_{L^2_k} &\to [\mathcal{T}^\sigma_{k-1/2}]_{[\mathfrak{c}]} \\ &= T_{[\mathfrak{c}]} \mathcal{B}^\sigma_{k-1/2}(Y, \mathfrak{s}) \end{aligned}$$

is surjective also, using the argument in the proof of Proposition 17.2.8. In other words, the map (24.20) is transverse to the diagonal at (a, b). In the reducible case, we similarly deduce that

$$\mathcal{D}_a\mathfrak{R}_+ - \mathcal{D}_b R_- : T_a\mathfrak{M}^{\text{red}}_{L^2_k} \oplus T_b M^{\text{red}}_{L^2_k} \to \prod_\alpha T_{[\mathfrak{c}^\alpha]}\partial\mathcal{B}^\sigma_{k-1/2}(Y^\alpha, \mathfrak{s}^\alpha)$$

is surjective, so the restriction of (24.20) to the reducible loci is transverse to the diagonal.

The proposition now follows from the general transversality argument of Lemma 12.5.1. □

We can extend the transversality results to the case of a family of perturbations and metrics on X, parametrized by a manifold. Let P be a smooth (finite-dimensional) manifold, and let g^p be a smooth family of Riemannian metrics on X parametrized by $p \in P$. We want to keep the data on the cylindrical-end independent of p, so we suppose that all the metrics contain isometric copies of the collar $I \times Y$. Similarly, let $\mathfrak{p}_0^p \in \mathcal{P}(Y, \mathfrak{s}_X)$ be a smooth family of perturbations. Write

$$\mathfrak{p}^p = \beta(t)\mathfrak{q} + \beta_0(t)\mathfrak{p}_0^p$$

as usual, and let $M(X^*, \mathfrak{s}_X; [\mathfrak{b}])_p$ be the corresponding moduli space on the cylindrical-end manifold, using the metric g_p and perturbation $\mathfrak{p}^p$. We have a total space

$$\begin{aligned} M(X^*, \mathfrak{s}_X; [\mathfrak{b}])_P &= \bigcup_p \{p\} \times M(X^*, \mathfrak{s}_X; [\mathfrak{b}])_p \\ &\subset P \times \mathcal{B}^\sigma_{k,\text{loc}}(X^*, \mathfrak{s}_X). \end{aligned}$$

Remark. There is a slight abuse of notation involved in regarding $\mathcal{B}^\sigma_{k,\text{loc}}(X^*, \mathfrak{s}_X)$ as being independent of $p \in P$. The point is that a spin^c structure $\mathfrak{s}_X$ is dependent on a prior choice of metric. We really mean to take a family of spin^c structures $\mathfrak{s}_p$ and identify all the corresponding spin bundles $S^\pm_p$ with a fixed $S^\pm_{p_0}$. The metric on the spin bundle, as well as the Clifford multiplication, will vary with p.

There is a fiber product description of this moduli space, just as for the case that P is a point: it is the fiber product of the maps

$$\begin{aligned} R_+ &: M(X, \mathfrak{s}_X)_P \to \mathcal{B}^\sigma_{k-1/2}(Y, \mathfrak{s}) \\ R_- &: M(X; [\mathfrak{b}]) \to \mathcal{B}^\sigma_{k-1/2}(Y, \mathfrak{s}), \end{aligned} \tag{24.21}$$

in which $M(X, \mathfrak{s}_X)_P$ is defined as a union, parallel to the definition of $M(X^*, \mathfrak{s}_X; [\mathfrak{b}])_P$.

Definition 24.4.9. Let $(p, [\gamma]) \in M(X^*, \mathfrak{s}_X; [\mathfrak{b}])_P$, and let $\rho[\gamma] = ([\gamma_0], [\gamma_1])$. If $[\gamma]$ is irreducible, we say the moduli space $M(X^*, \mathfrak{s}_X; [\mathfrak{b}])_P$ is *regular* at $(p, [\gamma])$ if the maps of Hilbert manifolds

$$\begin{aligned} R_+ &: M(X, \mathfrak{s}_X)_P \to \mathcal{B}^\sigma_{k-1/2}(Y, \mathfrak{s}) \\ R_- &: M(Z; [\mathfrak{b}]) \to \mathcal{B}^\sigma_{k-1/2}(Y, \mathfrak{s}) \end{aligned} \tag{24.22}$$

are transverse at $\big((p, [\gamma_0]), [\gamma_1]\big)$. If $[\gamma]$ is reducible we say the moduli space is *regular* at $[\gamma]$ if the restrictions

$$R_+ : M^{\text{red}}(X, \mathfrak{s}_X)_P \to \prod_\alpha \partial \mathcal{B}^\sigma_{k-1/2}(Y^\alpha, \mathfrak{s}^\alpha)$$

$$R_- : M^{\text{red}}(Z; [\mathfrak{b}]) \to \prod_\alpha \partial \mathcal{B}^\sigma_{k-1/2}(Y^\alpha, \mathfrak{s}^\alpha)$$

are transverse at $\big((p, [\gamma_0]), [\gamma_1]\big)$. We say the moduli space is *regular* if it is regular at all points $(p, [\gamma])$. ◊

Proposition 24.4.10. *Let $\mathfrak{q}$ be a fixed admissible perturbation for Y, let g^p be a smooth family of Riemannian metrics parametrized by $p \in P$, all containing an isometric copy of the collar $I \times Y$, and let $\hat{\mathfrak{p}}^p$ be a family of perturbations having the form*

$$\mathfrak{p}^p = \beta(t)\mathfrak{q} + \beta_0(t)\mathfrak{p}^p_0,$$

as before. Let $P_0 \subset P$ be a closed subset and suppose that the parametrized moduli space $M(X^*; [\mathfrak{b}])_P$ is regular at all points $(p_0, [\gamma])$ with $p_0 \in P_0$. Then there is a new family of perturbations $\tilde{\mathfrak{p}}^p$, with

$$\tilde{\mathfrak{p}}^p = \mathfrak{p}^p \quad \textit{for all } p \in P_0,$$

such that the corresponding parametrized moduli space $M(X^*; [\mathfrak{b}])_P$ is everywhere regular.

Proof. In the case that P is a single point (and P_0 is empty), this statement becomes our earlier Proposition 24.4.7, for which the organizing principle was the general transversality result, Lemma 12.5.1. Referring to the proof of that lemma, we see that the key point is the fact that a smooth Fredholm map $Q : \mathcal{Z} \to \mathcal{P}$ between separable Banach manifolds admits a regular value. For the case of general P, we just apply the extension of this result: namely, if $Q : \mathcal{Z} \to \mathcal{P}$ is Fredholm and a map $\gamma : P \to \mathcal{P}$ is given, then we can find

$\tilde{\gamma} : P \to \mathcal{P}$ that is transverse to Q, and we can arrange furthermore that $\tilde{\gamma} = \gamma$ on P_0 if $\gamma|_{P_0}$ is already transverse to Q. □

24.5 Compactness: local results

In Section 4, we formulated a basic compactness result for convergence on interior domains, Theorem 5.1.1, for solutions to the (unperturbed) Seiberg–Witten equations on a 4-manifold with boundary. In the special case of the cylinder $X = I \times Y$, we extended the result to the perturbed equations (Theorem 10.7.1) and then to the blown-up equations (Theorem 10.9.2). The proofs of the two extensions can be adapted to our present situation: the case of a general 4-manifold with boundary, with a perturbation $\hat{\mathfrak{p}}$ supported in a collar of the boundary. We state the appropriate version here.

We begin with a lemma concerning solutions to the equations on the collar $[-C, 0] \times Y$, where the perturbation $\hat{\mathfrak{q}}$ is time-dependent.

Lemma 24.5.1. *Consider the perturbed Seiberg–Witten equations* $\mathfrak{F}^{\sigma}_{\mathfrak{p}} = 0$, *where the perturbation* $\hat{\mathfrak{p}}(t)$ *has the form* (24.2) *as before, so that* $\mathfrak{p}(t) = \mathfrak{q}$ *at* $t = 0$ *and* $\mathfrak{p}(t) = 0$ *at* $t = -C$. *Then there are constants* C_1 *and* C_2 *(depending on* C *as well as on the perturbation) such that for all solutions* $\gamma = (A, \Phi)$, *we have*

$$\begin{aligned} &2\big(\mathcal{L}(\gamma(-C)) - \mathcal{L}_{\mathfrak{q}}(\gamma(0))\big) \\ &\quad \geq \frac{1}{2}\int_{[-C,0]\times Y} \big(\tfrac{1}{8}|F_{A^t}|^2 + |\nabla_A \Phi|^2 + \tfrac{1}{4}(|\Phi|^2 - C_1)^2\big) - C_2. \end{aligned}$$

Remark. The left-hand side, of course, depends an a particular choice of perturbing function f with formal gradient $\mathfrak{q}$; and so the constant C_2 will depend on this choice. We will see in the course of the proof that if f is normalized by the condition $f(A_0, 0) = 0$ for some connection A_0 with harmonic curvature, then the constants C_1 and C_2 here depend only on the constant m_2 in Condition (iv) from Definition 10.5.1.

Proof of Lemma 24.5.1. This is a modification of Corollary 10.6.2 for the case of a time-dependent perturbation. Corollary 10.6.2 was based on two points: first, the calculation leading to Lemma 10.6.1, valid for any configuration γ; and second, the fact that twice the change in $\mathcal{L}$ is equal to the analytic energy, for solutions of the equations.

The first point needs no modification in the case of a time-dependent perturbation. For the second point, there is a change. We have a time-dependent perturbation $\mathfrak{p}(t)$ for the vector field on $\mathcal{C}(Y, \mathfrak{s})$. Let us write f_t for a gauge-invariant function on $\mathcal{C}(Y, \mathfrak{s})$ whose formal gradient is $\hat{\mathfrak{p}}$, so the perturbed Chern–Simons–Dirac functional $\mathcal{L}_t$ on the slice $\{t\} \times Y$ is $\mathcal{L} + f_t$. Then we have, for solutions of the perturbed equations,

$$2\big(\mathcal{L}(\gamma(-C)) - \mathcal{L}_{\mathfrak{q}}(\gamma(0))\big) = \int_{-C}^{0} \left(\|\dot{\gamma}(t)\|^2 + \|(\operatorname{grad} \mathcal{L}) + \mathfrak{p}(t)\|^2 \right) \\ - 2 \int_{-C}^{0} \left(\frac{d}{dt} f_t \right)(\gamma(t))\, dt. \tag{24.23}$$

The term $\mathfrak{p}(t)$ in the above formula is absorbed, just as in the proof of Lemma 10.6.1, so that we have

$$2\big(\mathcal{L}(\gamma(-C)) - \mathcal{L}_{\mathfrak{q}}(\gamma(0))\big) \\ \geq \frac{1}{2} \int_{[-C,0] \times Y} \left(\tfrac{1}{4} |F_{A^t}|^2 + |\nabla_A \Phi|^2 + \tfrac{1}{4}(|\Phi|^2 - \tilde{C}_1)^2 \right) - \tilde{C}_2 \\ - 2 \int_{-C}^{0} \left(\frac{d}{dt} f_t \right)(\gamma(t))\, dt. \tag{24.24}$$

Because the perturbation $\mathfrak{p}$ has the special form dictated by (24.2), we can take the primitive f_t for $\mathfrak{p}(t)$ to have a similar form:

$$f_t(t) = \beta(t) f + \beta_0(t) g_0,$$

where f and g_0 for $\mathfrak{q}$ and $\mathfrak{p}_0$. So the second integral in (24.23) above can be bounded by

$$K \int_{-C}^{0} \left(|f(\gamma_t)| + |g_0(\gamma_t)| \right) dt$$

where K depends on the derivatives of the cut-off functions β and β_0. We can bound a term like $|f(\gamma)|$, because we have a bound on the formal L^2 gradient $\mathfrak{q}$, of the form given in Condition (iv) of Definition 10.5.1. We thus have, for $\gamma = (B, \Psi)$ in $\mathcal{C}(Y, \mathfrak{s})$,

$$|f(\gamma)| \leq |f(\gamma_0)| + m_2(1 + \|\Psi\|_{L^2}) \|v(\gamma) - \gamma_0\|_{L^2}$$

for any reference configuration γ_0 and any gauge transformation v (because f is gauge-invariant). Let us take γ_0 to be the configuration $(B_0, 0)$, where B_0 is a connection chosen so that B_0^t has harmonic curvature. We can choose v so that

B is in Coulomb gauge with respect to B^0 and so that the periods of the 1-form $(B-B_0)^{\mathrm{t}}$ on a fixed set of generators for $H_1(Y)$ are all less than 2π in magnitude. With this v, we can then bound $\|v(\gamma)-\gamma_0\|$ by $c_1\|F_{B^{\mathrm{t}}}\|+\|\Psi\|+c_2$ for constants c_1 and c_2 depending only on Y. (See Lemma 5.1.2 for a similar point. The constant c_2 can be omitted if $b_1(Y) = 0$.) Using a Peter–Paul inequality, we have a bound of the form

$$|f(\gamma)| \le |f(\gamma_0)| + m_2\Big((\epsilon c_3)\|F_{B^{\mathrm{t}}}\|^2 + (c_4/\epsilon)\|\Psi\|^2\Big),$$

for any $\epsilon > 0$, with constants c_3 and c_4 depending on the geometry of Y. There is a similar bound for $|g_0(\gamma)|$. For small ϵ, we can absorb the curvature term here using the positive curvature term in (24.24); and the quadratic term in $\|\Psi\|$ can be absorbed by the quartic term, leading to an inequality of the sort claimed in the lemma. □

In the theorem below, X is, as usual, a compact, Riemannian 4-manifold with boundary Y, containing a collar $(-C,0]\times Y$. We write $X_\epsilon \subset X$ for a slightly smaller manifold of the form

$$X_\epsilon = X \setminus \big((-\epsilon,0]\times Y\big).$$

We take X' to be a slightly smaller manifold still:

$$X' \Subset X_\epsilon.$$

We shall suppose that the cut-off function β used in the construction of $\mathfrak{p}$ has $\dot\beta$ supported in X_ϵ, so that the perturbation is constant in the ϵ neighborhood of ∂X.

We introduce a perturbed version of the topological energy for a configuration on a manifold X with boundary Y: if $\mathcal{L}_{\mathfrak{q}} = \mathcal{L} + f$ at ∂X, then we define the perturbed topological energy as

$$\mathcal{E}^{\mathrm{top}}_{\mathfrak{q}}(\gamma) = \mathcal{E}^{\mathrm{top}}(\gamma) - 2f(\gamma). \qquad (24.25)$$

Theorem 24.5.2. *Let $\gamma_n \in \mathcal{C}^\sigma_k(X,\mathfrak{s}_X)$ be a sequence of solutions of the perturbed equations, $\mathfrak{F}^\sigma_{\mathfrak{p}}(\gamma) = 0$, where $\mathfrak{p}$ is a perturbation supported in the collar, as before. Suppose that there is a uniform bound on the perturbed topological energy,*

$$\mathcal{E}^{\mathrm{top}}_{\mathfrak{q}}(\gamma_n) \le C_1,$$

and that for each component Y^α of Y there is a uniform upper bound on $-\Lambda_\mathfrak{q}$ on the boundary component $\{-\epsilon\} \times Y^\alpha$ of X_ϵ:

$$-\Lambda_\mathfrak{q}(\gamma_n|_{\{-\epsilon\}\times Y^\alpha}) \le C_2. \tag{24.26}$$

Then there is a sequence of gauge transformations, $u_n \in \mathcal{G}_{k+1}(X)$, such that, after passing to a subsequence, the restrictions, $u_n(\gamma_n)|_{X'}$, of the transformed solutions converge in the topology of $\mathcal{C}^\sigma_k(X')$ to a solution $\gamma \in \mathcal{C}^\sigma_k(X')$ of the equations $\mathfrak{F}^\sigma_\mathfrak{p}(\gamma) = 0$. Without the hypothesis (24.26), *we obtain a weaker conclusion: convergence of $u_n(\pi\gamma_n)|_{X'}$ in $\mathcal{C}_k(X')$.* □

Proof. Write $\gamma_n = (A_n, s_n, \phi_n)$, so that $\pi(\gamma_n) = (A_n, \Phi_n)$ with $\Phi_n = s_n\phi_n$ on X.

We first prove the statement about convergence of $u_n(A_n, \Phi_n)$ in $\mathcal{C}_k(X')$, without the hypothesis (24.26). This is the counterpart of the Theorem 5.1.1, which dealt with the unperturbed case. The bound on the perturbed topological energy on X implies a bound on the unperturbed topological energy on the complement of the collar, by the lemma. So we can assume that we have convergence on the complement of the collar, by the unperturbed compactness result, Theorem 5.1.1. The lemma then also gives us a bound on the energy terms (the terms on the right in the lemma) in the collar region. These terms (the L^2 norms of F_{A^t} and $\nabla_A\Phi$, and the L^4 norm of Φ) are therefore bounded on X, and the compactness argument can proceed as before.

We next have to prove that, with the bound (24.26), we can obtain convergence in the blown-up configuration space $\mathcal{C}^\sigma(X, \mathfrak{s}_X)$. The argument is essentially the same as the cylindrical case, Theorem 10.9.2. (Once one knows that (A_n, Φ_n) is converging, the definition of tame ensures that a bound of $-\Lambda$ is equivalent to a bound on $-\Lambda_\mathfrak{q}$.) □

There is also a finiteness result connected with the compactness theorem above. This proposition, like the similar Lemma 10.7.5, requires an additional hypothesis on the components $\mathfrak{q}_\mathfrak{s}$ of the perturbation $\mathfrak{q}$ belonging to different spin^c structure $\mathfrak{s}$:

Proposition 24.5.3. *In the situation of Theorem* 24.5.2, *suppose that the perturbation $\mathfrak{q}$ has the additional property that there is a constant m_2, independent of the choice of spin^c structure $\mathfrak{s}$ on Y, such that*

$$\|\mathfrak{q}_\mathfrak{s}(B, \Psi)\|_{L^2} \le m_2\big(\|\Psi\|_{L^2} + 1\big).$$

(Compare Property (iv) *in Definition* 10.5.1.*) Let the perturbation f be normalized on $\mathcal{C}(Y, \mathfrak{s})$, for each s, so that f vanishes at a chosen configuration $(B_0, 0)$,*

where B_0 is a reference connection for which B_0^{t} has harmonic curvature. Then for a given C_1, there are only finitely many spin^c structures $\mathfrak{s}_X$ on X for which there exist solutions γ satisfying the bound

$$\mathcal{E}_{\mathfrak{q}}^{\mathrm{top}}(\gamma) \le C_1.$$

Proof. This finiteness result on the spin^c structures $\mathfrak{s}_X$ follows just as in Theorem 5.1.1 and Lemma 10.7.5. The only additional observation needed is that the values of the constants C_1 and C_2 in Lemma 24.5.1 depend only on m_2, as long as f is chosen to vanish at $(B_0, 0)$, as the proof of that lemma shows. □

24.6 Compactness: broken trajectories

Our next task is to introduce a compactification of the moduli space $M(X^*, \mathfrak{s}_X; [\mathfrak{b}])$ on the cylindrical-end manifold.

Definition 24.6.1. Let $[\mathfrak{b}]$ be a critical point, and $[\mathfrak{b}^\alpha]$ its restriction to the component $Y^\alpha \subset Y$. A *broken X-trajectory* asymptotic to $[\mathfrak{b}]$ consists of the following data:

- an element $[\gamma_0]$ in a moduli space $M_{z_0}(X^*, [\mathfrak{b}_0])$;
- for each component Y^α of Y, an unparametrized broken trajectory $[\breve{\boldsymbol{\gamma}}^\alpha]$ in a moduli space $\breve{M}^+_{z^\alpha}([\mathfrak{b}_0^\alpha], [\mathfrak{b}^\alpha])$, where $[\mathfrak{b}_0^\alpha]$ is the restriction of $[\mathfrak{b}_0]$ to Y^α.

If z_1 is the homotopy class of paths from $[\mathfrak{b}_0]$ to $[\mathfrak{b}]$ whose αth component is z^α, then the *homotopy class* of the broken X-trajectory is the element

$$z = z_1 \circ z_0 \in \pi_0(\mathcal{B}^\sigma(X, [\mathfrak{b}])).$$

We write $M_z^+(X^*, [\mathfrak{b}])$ for the space of broken X-trajectories in the homotopy class z. This contains $M_z(X^*, [\mathfrak{b}])$ as the special case when each of the broken trajectories $[\breve{\boldsymbol{\gamma}}^\alpha]$ has zero components. We write a typical element of $M_z^+(X^*, [\mathfrak{b}])$ as $([\gamma_0], [\breve{\boldsymbol{\gamma}}])$, where $[\breve{\boldsymbol{\gamma}}]$ stands for a (possibly empty) collection $[\breve{\gamma}_i^\alpha]$ of unparametrized trajectories on the components Y^α, $1 \le i \le n^\alpha$. Sometimes we will simply write $[\boldsymbol{\gamma}]$. ◊

We give $M_z^+(X^*, [\mathfrak{b}])$ a topology as follows. (See the similar definition of the topology on $\breve{M}_z^+([\mathfrak{a}], [\mathfrak{b}])$ in Subsection 16.1.) Let $([\gamma_0], [\breve{\boldsymbol{\gamma}}])$ belong to $M_z^+(X^*, [\mathfrak{b}])$, and let the components of $[\breve{\boldsymbol{\gamma}}]$ be represented by (parametrized) trajectories

$$\gamma_i^\alpha \in M_{z_i^\alpha}([\mathfrak{b}_{i-1}^\alpha], [\mathfrak{b}_i^\alpha]),$$

for $0 \leq i \leq n^\alpha$. Let $U_0 \subset \mathcal{B}^\sigma_{k,\mathrm{loc}}(X^*)$ be an open neighborhood of $[\gamma_0]$. Let $I^\alpha_i \subset \mathbb{R}$ be a compact interval, and let $U^\alpha_i \subset \mathcal{B}^\tau_k(I^\alpha_i \times Y^\alpha)$ be an open neighborhood of $[\gamma^\alpha_i|_{I^\alpha_i \times Y^\alpha}]$. Let $T \in \mathbb{R}^+$ be such that the translate $I^\alpha_i + T$ is contained in the positive half-line. We define $\Omega = \Omega(U_0, \{U^\alpha_n\}, T)$ to be the subset of $M^+_z(X^*, [\mathfrak{b}])$ consisting of broken X-trajectories $[\boldsymbol{\delta}] = ([\delta_0], [\breve{\delta}^\alpha_j])$, $1 \leq j \leq m^\alpha$, satisfying the following conditions.

(i) We have $[\delta_0]$ in U_0.
(ii) For each α with $n^\alpha \neq 0$, there exists a map $(\jmath^\alpha, s^\alpha) : \{1, \ldots, n^\alpha\} \to \{0, \ldots, m^\alpha\} \times \mathbb{R}$ such that, for all $i \in \{1, \ldots, n^\alpha\}$, we have:
  (a) if $\jmath^\alpha(i) = 0$, then $[\tau_{s^\alpha(i)}\delta^\alpha_0]_{I^\alpha_i \times Y^\alpha} \in U^\alpha_i$, where δ^α_0 denotes the restriction of δ_0 to the cylindrical end $\mathbb{R}^{\geq} \times Y^\alpha$, regarded there as an element of $\mathcal{B}^\tau_{k,\mathrm{loc}}(\mathbb{R}^{\geq} \times Y^\alpha, \mathfrak{s}^\alpha)$;
  (b) if $\jmath^\alpha(i) \neq 0$, then $[\tau_{s^\alpha(i)}\delta^\alpha_{\jmath(i)}]^\alpha_{I^\alpha_i \times Y} \in U^\alpha_i$;
  (c) if $\jmath^\alpha(1) = 0$, then $s^\alpha(1) > T$;
  (d) if $1 \leq i_1 < i_2 \leq n^\alpha$, then either $\jmath^\alpha(i_1) < \jmath^\alpha(i_2)$, or $\jmath^\alpha(i_1) = \jmath^\alpha(i_2)$ and $s^\alpha(i_1) + T \leq s^\alpha(i_2)$.

Here again $\tau_s\delta$ denotes the translate, $\tau_s\delta(t) = \delta(s + t)$. We take the sets of the form $\Omega(U_0, \{U^\alpha_n\}, T)$ to be a neighborhood base for $([\gamma_0], [\breve{\boldsymbol{\gamma}}])$ in $M^+_z(X^*, [\mathfrak{b}])$.

The proof of the next theorem runs closely parallel to the proof of Theorem 16.1.3.

Theorem 24.6.2. *Let $\mathfrak{p}_0$ be chosen so that the moduli spaces of X-trajectories, $M(X^*, [\mathfrak{b}])$, are regular for all critical points $[\mathfrak{b}]$ on Y. Then each moduli space of broken X-trajectories $M^+_z(X^*, [\mathfrak{b}])$ is compact.*

This theorem follows from a slightly stronger version: the analog of Proposition 16.1.4. To formulate this proposition (stated as Proposition 24.6.4 below) we need to define $\mathcal{E}^{\mathrm{top}}_{\mathfrak{q}}(\boldsymbol{\gamma})$, the topological energy of a broken X-trajectory, in a manner appropriate for the perturbed equations: we extend the definition (24.25) which we made in the case of the compact manifold.

Definition 24.6.3. For a configuration $\gamma \in \mathcal{B}^\sigma_{k,\mathrm{loc}}(X^*, [\mathfrak{b}])$, and $\mathcal{L} = \mathcal{L} + f$, we define the *perturbed* topological energy to be the quantity

$$\mathcal{E}^{\mathrm{top}}_{\mathfrak{q}}(\gamma) = \mathcal{E}^{\mathrm{top}}(\gamma) - 2f(\mathfrak{b}).$$

For a broken X-trajectory, the energy is the sum of the energies of the components. ◇

Note that the perturbed topological energy is gauge-invariant, and depends only on $[\mathfrak{b}]$, the homotopy class z and the perturbation. The definition coincides with our previous one in the case that X is a finite cylinder: the topological energy is then just twice the drop in $\mathcal{L}$. We can now state the strengthened version of the above theorem:

Proposition 24.6.4. *For any $C > 0$ and any $[\mathfrak{b}]$, the space of broken X-trajectories $[\breve{\boldsymbol{\gamma}}] \in \bigcup_z M_z^+(X^*, [\mathfrak{b}])$ with energy $\mathcal{E}_{\mathfrak{q}}^{\mathrm{top}}(\breve{\boldsymbol{\gamma}}) \le C$ is compact.*

Remark. Since $\mathcal{E}_{\mathfrak{q}}^{\mathrm{top}}(\breve{\boldsymbol{\gamma}})$ and $\mathcal{E}^{\mathrm{top}}(\breve{\boldsymbol{\gamma}})$ differ by a constant depending only on $[\mathfrak{b}]$, the proposition could be stated in terms of the unperturbed energy. It is stated this way to match the version we stated for the cylindrical case, Proposition 16.1.4.

Proof of Proposition 24.6.4. We deal first with a fixed $\mathfrak{s}_X$. As in the proof of Proposition 16.1.4, we can reduce to the case of showing that a sequence $[\gamma_n]$ of *unbroken* solutions,

$$[\gamma_n] \in \bigcup_z M_z(X^*, [\mathfrak{b}]),$$

has a convergent subsequence in the space of broken trajectories, as long as the energies are bounded.

The next observation is that the perturbed topological energy on the compact piece X is no larger than the topological energy on X^*:

$$\mathcal{E}_{\mathfrak{q}}^{\mathrm{top}}(\gamma_n|_X) \le \mathcal{E}_{\mathfrak{q}}^{\mathrm{top}}(\gamma_n)$$

because the contribution from the cylindrical end is the total drop in $\mathcal{L}_{\mathfrak{q}}$ along the cylindrical pieces. We can therefore use Theorem 24.5.2, which allow us to assume that the configurations $\pi(\gamma_n|_X)$ are converging in $\mathcal{C}_k(X)$. (We can replace X by a slightly larger piece in applying this theorem, so as not to have to pass to $X' \Subset X$.) On the cylindrical end part, $X^* \setminus X$, we now have a sequence of solutions to the familiar, translation-invariant perturbed equations, with a bound in the change of $\mathcal{L}_{\mathfrak{q}}$. We can apply our earlier broken-trajectory compactness argument to the images of these under the blow-down map π to obtain convergence on $X^* \setminus X$, or a slightly larger cylinder $X^* \setminus X_\epsilon$. Putting the two pieces together, we see that, after gauge transformation and passing to a subsequence, the blown-down configurations $\pi(\gamma_n)$ converge in the sense of X-trajectories.

To obtain convergence in the blown-up configuration space, we also need to control the quantities

$$\Lambda_{\mathfrak{q}}^{\alpha}(n,t) = \Lambda_{\mathfrak{q}}(\gamma_n|_{t\times Y^\alpha}).$$

Because this function is fully gauge-invariant, and because each $[\gamma_n]$ is asymptotic to $[\mathfrak{b}]$, we have

$$\lim_{t\to\infty} \Lambda_{\mathfrak{q}}^{\alpha}(n,t) \geq -L^{\alpha}$$

for all n. The boundedness of the quantity K in Lemma 16.3.1 implies as before a lower bound on $\Lambda_{\mathfrak{q}}^{\alpha}$ near $t = 0$: say

$$\Lambda_{\mathfrak{q}}^{\alpha}(n,t) \geq -L_1^{\alpha}$$

for $t \in [-\epsilon, \epsilon]$. We can then apply Theorem 24.5.2 to find gauge transformations u_n on X such that, after passing to a subsequence, the configurations $u_n(\gamma_n)$ converge in the blown-up configuration space $\mathcal{C}_k^{\sigma}(X')$. From this convergence, it now follows that we have an upper bound,

$$\Lambda_{\mathfrak{q}}^{\alpha}(n,t) \leq L_2^{\alpha},$$

for $t = -\epsilon$. Another application of Lemma 16.3.1 now provides a bound on $\Lambda_{\mathfrak{q}}^{\alpha}(n,t)$ on the entire cylindrical end $[-\epsilon, \infty) \times Y^{\alpha}$, and the proof can proceed as before. □

We also have the analogs of the two finiteness results, Proposition 16.4.1 and Proposition 16.4.3:

Proposition 24.6.5. *Suppose that the perturbation* $\mathfrak{q}$ *is admissible in the sense of Definition* 22.1.1 *and the perturbation* $\mathfrak{p}$ *is chosen so that all the moduli spaces* $M_z(X^*, [\mathfrak{b}])$ *are regular. Then for given* $[\mathfrak{b}]$, *there are only finitely many homotopy classes* z *for which the moduli space* $M_z^+(X^*, [\mathfrak{b}])$ *is non-empty.*

Proposition 24.6.6. *Suppose that the perturbation* $\mathfrak{q}$ *is admissible in the sense of Definition* 22.1.1 *and the perturbation* $\mathfrak{p}$ *is chosen so that all the moduli spaces* $M_z(X^*, [\mathfrak{b}])$ *are regular. Let* $Y^{\alpha_1}, \ldots, Y^{\alpha_n}$ *be the components of* Y *and write* $[\mathfrak{b}] = ([\mathfrak{b}^{\alpha_1}], \ldots, [\mathfrak{b}^{\alpha_n}])$. *Let* $([\mathfrak{b}^{\alpha_2}], \ldots, [\mathfrak{b}^{\alpha_n}])$ *be given. Then for any* d_0 *and* i_0, *there are only finitely many pairs* $([\mathfrak{b}^{\alpha_1}], z)$ *for which:*

(i) *the moduli space* $M_z^+(X^*, [\mathfrak{b}])$ *is non-empty;*
(ii) *the dimension of the moduli space is at most* d_0; *and*
(iii) $\iota([\mathfrak{b}^{\alpha_1}]) \geq i_0$, *where* ι *is as defined in* (16.2).

Remark. Proposition 24.6.6 has an extra hypothesis (the bound on $\iota([\mathfrak{b}^{\alpha_1}])$) which has no counterpart in Proposition 16.4.3. Unlike the case of the cylinder, there may be infinitely many different spin^c structures on a general X, all having the same restriction to the boundary. This necessitates the stronger hypothesis. This is ultimately the reason why we need to pass from $\widehat{HM}_*$ to the completion $\widehat{HM}_\bullet$.

Proof of Proposition 24.6.5. As with Proposition 16.4.1 this follows from Proposition 24.6.4 and a uniform energy bound analogous to Lemma 16.4.4. This analogous bound states that there exists a C such that for every $[\mathfrak{b}]$ and z, and every broken X-trajectory $[\boldsymbol{\gamma}]$ in $M_z^+(X^*, [\mathfrak{b}])$, we have

$$\mathcal{E}_{\mathfrak{q}}^{\mathrm{top}}(\boldsymbol{\gamma}) \leq C - 8\pi^2\iota([\mathfrak{b}]).$$

In the proof of this inequality the counterpart to (16.4) is the statement that, for $[\gamma] \in M_z(X^*, [\mathfrak{b}])$, the quantity

$$\mathcal{E}_{\mathfrak{q}}(\gamma) + 4\pi^2 \operatorname{gr}_z(X, [\mathfrak{b}])$$

does not depend on z. The remainder of the proof is essentially unaltered. □

Proof of Proposition 24.6.6. As in the proof of Proposition 16.4.3, we observe that

$$\mathcal{E}_{\mathfrak{q}}^{\mathrm{top}}(\gamma) + 4\pi^2 \left(\operatorname{gr}_z(X, \mathfrak{s}_X; [\mathfrak{b}]) - 2\iota([\mathfrak{b}])\right)$$

is a function of $\boldsymbol{\pi}([\mathfrak{b}])$ alone, and so takes only finitely many values. (Here $\iota([\mathfrak{b}])$ denotes the sum of $\iota([\mathfrak{b}^{\alpha_i}])$ over all components.) A bound on the dimension and on $-\iota([\mathfrak{b}])$ therefore provides a bound on the energy. □

We will also need a compactness result for a family of moduli spaces. Let P again be a smooth manifold parametrizing a family of metrics and perturbations on X, all of which are equal in a neighborhood of the boundary, and let $M(X^*, [\mathfrak{b}])_P$ denote the family over P. We can form a space $M^+(X^*, [\mathfrak{b}])_P$ as the disjoint union

$$M^+(X^*, [\mathfrak{b}])_P = \bigcup_p \{p\} \times M^+(X^*, [\mathfrak{b}])_p.$$

Since the metric and perturbation on the cylindrical part are independent of p, there is no difficulty in extending our definition of the topology on the fibers so as to define a topology on the total space. The small adjustments to the proof of Theorem 24.6.2 which are needed to prove the following version can safely be omitted.

Theorem 24.6.7. *Suppose that the families $M(X^*, [\mathfrak{b}])_P$ are regular for all $[\mathfrak{b}]$. Then for each $[\mathfrak{b}]$ the family of moduli spaces is proper over P: that is, the map*

$$M_z^+(X^*, [\mathfrak{b}])_P \to P$$

is proper. For fixed $[\mathfrak{b}]$, this family of moduli spaces is non-empty for only finitely many components $z \in \pi_0(\mathcal{B}_k^\sigma(X, [\mathfrak{b}]))$. □

The space $M_z^+(X^*, [\mathfrak{b}])_P$ is a disjoint union of subspaces of the form

$$M_{z_0}(X^*, [\mathfrak{b}_0])_P \times \prod_\alpha \breve{M}_{z^\alpha}^+([\mathfrak{b}_0^\alpha], [\mathfrak{b}^\alpha]), \tag{24.27}$$

in which each of the factors is a space stratified by manifolds. In this way, $M_z^+(X^*, [\mathfrak{b}])_P$ is eventually a disjoint union of manifolds. As in Proposition 16.5.2, some straightforward bookkeeping shows that the frontier of any of these manifold pieces is contained in a union of lower-dimensional pieces. The following proposition elaborates on this statement, and is the counterpart to Proposition 16.5.2:

Proposition 24.6.8. *Suppose that $M_z(X^*, [\mathfrak{b}])_P$ is non-empty, and let d be its dimension. Then the space $M_z^+(X^*, [\mathfrak{b}])_P$ is a d-dimensional space stratified by manifolds, proper over P. For $e \geq 0$, the e-dimensional stratum is the union of the e-dimensional strata of all the pieces of the form* (24.27).

If $M_z(X^*, [\mathfrak{b}])_P$ contains irreducible trajectories, then the top stratum consists of the irreducible part of $M_z(X^*, [\mathfrak{b}])_P$. □

As well as the compactification $M_z^+(X^*, [\mathfrak{b}])$, there is a smaller compactification $\bar{M}_z(X^*, [\mathfrak{b}])$ which we shall use:

Definition 24.6.9. The space $\bar{M}_z(X^*, [\mathfrak{b}])$ is the image of $M_z^+(X^*, [\mathfrak{b}])$ under the map

$$\begin{aligned} r : M_z^+(X^*, [\mathfrak{b}]) &\to \mathcal{B}_{k,\mathrm{loc}}^\sigma(X^*) \\ ([\gamma_0], [\breve{\boldsymbol{\gamma}}]) &\mapsto [\gamma_0]. \end{aligned}$$

Similarly, if P is a manifold parametrizing a family of metrics and perturbations on X, then $\bar{M}_z(X^*, [\mathfrak{b}])_P$ is defined as the image of $M_z^+(X^*, [\mathfrak{b}])_P$ under the map

$$\begin{aligned} M_z^+(X^*, [\mathfrak{b}])_P &\to P \times \mathcal{B}_{k,\mathrm{loc}}^\sigma(X^*) \\ (p, ([\gamma_0], [\breve{\boldsymbol{\gamma}}])) &\mapsto (p, [\gamma_0]). \end{aligned}$$

◇

As with M^+, the smaller $\bar{M}_z(X^*, [\mathfrak{b}])_P$ is a space stratified by manifolds, and the map to P is proper. To help understand the relationship between M^+ and $\bar{M}$, it is helpful to consider the case that X is just a finite cylinder so that X^* is the infinite cylinder. In this case (reverting to our notation used in the cylindrical case) we can consider $M_z^+([\mathfrak{a}], [\mathfrak{b}])$ and $\bar{M}_z([\mathfrak{a}], [\mathfrak{b}])$ as two different compactifications of the space of trajectories $M([\mathfrak{a}], [\mathfrak{b}])$. This is different from our earlier construction of $\breve{M}_z^+([\mathfrak{a}], [\mathfrak{b}])$, which was a compactification of the space of *unparametrized* trajectories $\breve{M}_z([\mathfrak{a}], [\mathfrak{b}])$. The same objects can be considered also for a finite-dimensional flow for an ordinary Morse function on a compact manifold B. In this case, $\bar{M}(a, b)$ is a compactification of the space of trajectories from a to b, which can be described quite simply: we can identify the trajectory space $M(a, b)$ with a subset of B, by evaluating a trajectory at $t = 0$, and the space $\bar{M}(a, b)$ is then simply the closure of $M(a, b)$ in B. For example, if B is the torus T^n with a standard Morse function given by

$$f = \sum_{i=1}^{n} \cos(\theta_i)$$

then each $\bar{M}(a, b)$ is a torus, stratified by its intersection with the coordinate hyperplanes $\theta_i = 0$ and $\theta_i = \pi$. The compactification $M^+(a, b)$ is larger, and contains one codimension-1 face for each stratum of broken trajectories of the form $\breve{M}(a, c) \times M(c, b)$ or $M(a, c) \times \breve{M}(c, b)$ (allowing that $c = b$ in the first case or $c = a$ in the second). In the case that $\bar{M}(a, b)$ is a 3-torus, the stratified space $M^+(a, b)$ is eight copies of a solid truncated octahedron, stratified by the faces, edges and vertices, as shown in Figure 6.

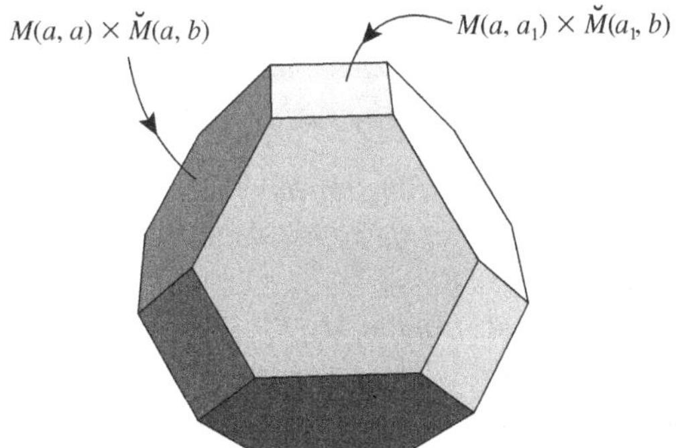

Fig. 6. A picture of the combinatorics of $M^+(a, b)$ in the case that $\bar{M}(a, b)$ is a 3-torus carrying a standard Morse function. Two of the codimension-1 faces are labelled. The critical point a_1 has index 1 less than a.

The next proposition classifies the codimension-1 strata in the space $\bar{M}_z(X^*, [\mathfrak{b}])$, in the case that the moduli space contains irreducibles. Recall that for a typical element

$$([\gamma_0], [\breve{\gamma}]) \in M_{z_0}(X^*, [\mathfrak{b}_0])_P \times \prod_\alpha \breve{M}^+_{z^\alpha}([\mathfrak{b}_0^\alpha], [\mathfrak{b}^\alpha]), \tag{24.28}$$

we write n^α for the number of components of $[\breve{\gamma}^\alpha]$. We write the ith component of $[\breve{\gamma}^\alpha]$ as

$$[\breve{\gamma}_i^\alpha] \in \breve{M}_{z_i^\alpha}([\mathfrak{b}_{i-1}^\alpha], [\mathfrak{b}_i^\alpha]).$$

Proposition 24.6.10. *Suppose $M_z(X^*, [\mathfrak{b}])_P$ contains irreducible solutions and has dimension d. Then both $M_z^+(X^*, [\mathfrak{b}])_P$ and $\bar{M}_z(X^*, [\mathfrak{b}])_P$ are d-dimensional spaces stratified by manifolds, with top stratum the irreducible part of $M_z(X^*, [\mathfrak{b}])_P$. The $(d-1)$-dimensional stratum in $M_z^+(X^*, [\mathfrak{b}])_P$ consists of elements of the following types.*

(i) *The elements (24.28) with $n^\alpha = 1$ for exactly one component $\alpha = \alpha_*$ and all other n^α zero. In this case, neither $[\breve{\gamma}_1^{\alpha_*}]$ nor $[\gamma_0]$ can be boundary-obstructed.*
(ii) *The elements (24.28) with $n^\alpha = 2$ for exactly one component $\alpha = \alpha_*$ and all other n^α zero. In this case, $[\breve{\gamma}_1^{\alpha_*}]$ must be boundary-obstructed while $[\breve{\gamma}_2^{\alpha_*}]$ and $[\gamma_0]$ are not.*
(iii) *The elements (24.28) for which $[\gamma_0]$ is boundary-obstructed of corank c. In this case $n^{\alpha_*} = 1$ for exactly $c+1$ components (say for $\alpha_* \in A$ where $|A| = c+1$), and all other n^α are zero. For α_* in A, the trajectory $[\breve{\gamma}_1^{\alpha_*}]$ cannot be boundary-obstructed.*
(iv) *The unbroken reducible solutions (all n^α are zero), in the case that $M_z(X^*, [\mathfrak{b}])_P$ contains both irreducibles and reducibles.*
(v) *The unbroken irreducible solutions lying over ∂P, if P has boundary.*

In the first three cases above, if any of the moduli spaces involved contain both reducibles and irreducibles then only the irreducibles contribute to the $(d-1)$-dimensional stratum.

The $(d-1)$-dimensional strata in $\bar{M}_z(X^*, [\mathfrak{b}])_P$ are the image under r of the strata described with the additional constraint that, in the first three cases, each of the $[\gamma_i^{\alpha_*}]$ belongs to a 1-dimensional moduli space. □

Note that in all cases of this proposition, because the moduli space $M_z(X^*, [\mathfrak{b}])_P$ contains irreducibles, each component $[\mathfrak{b}^\alpha]$ of $[\mathfrak{b}]$ must be either irreducible, or reducible and boundary-stable. In the fourth case of the proposition, all the $[\mathfrak{b}^\alpha]$ must be reducible and boundary-stable. In the third case, $[\mathfrak{b}^\alpha]$ must be reducible and boundary-stable for α not in A.

24.7 Gluing

We shall extend the results of Section 19 to the case of a manifold with cylindrical ends. One difference from the case of a cylinder is that a regular solution can be boundary-obstructed with corank greater than 1 (if the manifold has more than one end). In the cylindrical case, we saw that (under mild hypotheses) the codimension-1 strata in the compactification M^+ had neighborhoods that either were manifolds with boundary or had a codimension-1 δ-structure (Definition 19.5.3 and Theorem 19.5.4). In the case of a general X, we have a similar result; but at codimension-1 strata involving a boundary-obstructed solution of larger corank, we will see a more general structure, which we now define.

Definition 24.7.1. Let N be a d-dimensional space stratified by manifolds and $M^{d-1} \subset N$ a union of components of the $(d-1)$-dimensional stratum. We say that N has a *codimension-c δ-structure* along M^{d-1} if M^{d-1} is smooth and we have the following additional data. There is an open set $W \subset N$ containing M^{d-1}, an embedding $j : W \to EW$, and a map

$$\mathbf{S} = (S_1, \dots, S_{c+1}) : EW \to (0, \infty]^{c+1}$$

with the following properties:

(i) the map $\mathbf{S}$ is a topological submersion along the fiber over $\infty = (\infty, \dots, \infty)$;
(ii) the fiber of $\mathbf{S}$ over ∞ is $j(M^{d-1})$;
(iii) the subset $j(W) \subset EW$ is the zero set of a map $\delta : EW \to \Pi^c$, where $\Pi^c \subset \mathbb{R}^{c+1}$ is the hyperplane $\Pi^c = \{\delta \in \mathbb{R}^{c+1} \mid \sum \delta_i = 0\}$;
(iv) if $e \in EW$ has $S_{i_0}(e) = \infty$ for some i_0, then $\delta_{i_0}(e) \le 0$, with equality only if $S_i(e) = \infty$ for all i;
(v) on the subset of EW where all S_i are finite, δ is smooth and transverse to zero.

◊

Example. Let $EW \subset \mathbb{R}^{c+1}$ be the set $\{x \mid x_i \geq 0 \text{ for all } i\}$, let $S_i = 1/x_i$ as a function from EW to $(0, \infty]$, and let

$$\delta_i = cx_i - \sum_{j \neq i} x_j.$$

The zero locus of δ is the half-line $W \subset EW$ where all x_i are equal.

With this definition in place, we state our theorem concerning the structure of the compactification at codimension-1 strata.

Theorem 24.7.2. *Suppose that the moduli space $M_z(X^*, [\mathfrak{b}])_P$ is d-dimensional and contains irreducible trajectories, so that $M^+(X, [\mathfrak{b}])_P$ is a d-dimensional space stratified by manifolds having $M_z(X^*, [\mathfrak{b}])_P$ as its top stratum. Let $M' \subset M^+(X^*, [\mathfrak{b}])_P$ be any component of the codimension-1 stratum. Then, along M', the moduli space $M^+(X^*, [\mathfrak{b}])_P$ either is a C^0 manifold with boundary, or (more generally) has a codimension-c δ-structure in the sense of Definition* 24.7.1.

To flesh out the statement of this theorem, consider the five cases itemized in Proposition 24.6.10. The relevant stratum of $M^+(X^*, [\mathfrak{b}])_P$ has a neighborhood which is a manifold with boundary in the first, fourth and fifth cases of the proposition. In the second case, there is a codimension-1 δ-structure, while in the third case there is a codimension-c δ-structure. We shall indicate how to adapt the material of Section 19 to prove these assertions, focusing on the situation that arises in the third case of Proposition 24.6.10.

Proof of Theorem 24.7.2. To show how to adapt our earlier arguments, we begin by examining a particular case that does not involve any boundary-obstructed solutions. We will treat only the case that the parametrizing space P is a point, as the general parametrized case is really no more difficult.

Consider a stratum in $M_z^+(X^*, [\mathfrak{b}])$ consisting of elements of the form (24.28), with each n^α either zero or 1, and let $A = \{\alpha \mid n^\alpha = 1\}$. Such an element belongs to a stratum

$$M' = M_{z_0}(X^*, [\mathfrak{b}_0])_P \times \prod_{\alpha \in A} \breve{M}_{z^\alpha}([\mathfrak{b}_0^\alpha], [\mathfrak{b}^\alpha]), \tag{24.29}$$

and $\mathfrak{b}_0^\alpha = \mathfrak{b}^\alpha$ for $\alpha \notin A$. We suppose that none of the factors in this product are boundary-obstructed moduli spaces. This stratum has codimension $|A|$, and the case $|A| = 1$ is the first case of Proposition 24.6.10.

Let $K \subset M'$ be a compact subset. We will describe a neighborhood W of K in $M_z^+(X^*, [\mathfrak{b}])$ as a fiber product. This is the same plan that we carried through in Subsection 19.3. The main difference here is that we use a different decomposition of the 4-manifold. Fix $L > 0$, and let

$$X^L = X \cup [0, L] \times Y \qquad (24.30)$$

be the compact manifold obtained by attaching the finite cylinder of length L at the common boundary $\{0\} \times Y$. Let Z_α^T be the cylinder

$$Z_\alpha^T = [-T, T] \times Y^\alpha$$

for finite positive T. For $T = \infty$, as in Section 19, we again adopt the notation

$$\begin{aligned} Z_\alpha^\infty &= Z_\alpha^+ \amalg Z_\alpha^- \\ &= \big([0,\infty) \times Y^\alpha\big) \amalg \big((-\infty, 0] \times Y^\alpha\big). \end{aligned}$$

Fix positive real numbers T_α for each α in A, and write $\boldsymbol{T} = (T_\alpha)_{\alpha \in A}$. For any choice of $\boldsymbol{T}$, we can consider the cylindrical-end manifold X^* as obtained from the disjoint union

$$X_{\boldsymbol{T}} = X^L \amalg \left(\bigcup_{\alpha \in A} Z_\alpha^{T_\alpha} \amalg Z_\alpha^L \amalg Z_\alpha^+ \right) \amalg \left(\bigcup_{\alpha \notin A} Z_\alpha^+ \right), \qquad (24.31)$$

by identifying boundary components in pairs. (See Figure 7.) If we allow some of the components T_α to be infinite, then the identification space becomes a union of X^* and some cylinders $\mathbb{R} \times Y^\alpha$. Associated with this disjoint union

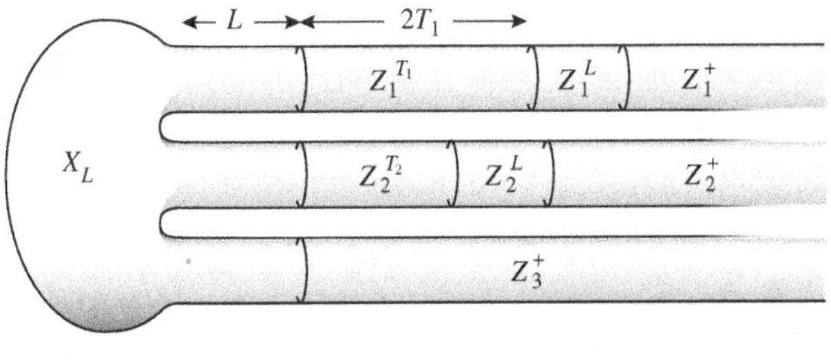

Fig. 7. Decomposing the cylindrical-end manifold X^* into the pieces (24.31).

$\boldsymbol{X_T}$ is a product of moduli spaces,

$$\boldsymbol{M_T} = M(X^L)_P \times \left(\prod_{\alpha \in A} M(Z_\alpha^{T_\alpha}) \times M(Z_\alpha^L) \times M(Z_\alpha^+; [\mathfrak{b}^\alpha]) \right)$$
$$\times \left(\prod_{\alpha \notin A} M(Z_\alpha^+; [\mathfrak{b}^\alpha]) \right).$$

(Compare this with (19.7).) Here and below we have omitted the spin^c structure from our notation. Inside $\boldsymbol{M_T}$ is a subset,

$$\boldsymbol{M_T} \supset \boldsymbol{M}_{\boldsymbol{T},\eta}^{cen},$$

obtained by replacing each $M(Z_\alpha^L)$ above by M^{cen} and replacing each $M(Z_\alpha^{T_\alpha})$ or $M(Z_\alpha^+)$ with $M_\eta(Z_\alpha^{T_\alpha})$ or $M_\eta(Z_\alpha^+)$ respectively. (See Equation (19.8).) We again define the spaces

$$\boldsymbol{M} \supset\supset \boldsymbol{M}_\eta^{cen},$$

by taking a disjoint union over $\boldsymbol{T}$ (including those where some T_α are infinite).

When we form X^* by identifying boundary components of $\boldsymbol{X_T}$, the identifications are made along three copies of each Y^α for α in A, and one copy of each Y^α for α not in A. Set

$$\mathcal{B} = \left(\prod_{\alpha \in A} \mathcal{B}_{k-1/2}^\sigma (Y^\alpha)^3 \right) \times \left(\prod_{\alpha \notin A} \mathcal{B}_{k-1/2}^\sigma (Y^\alpha) \right),$$

and let

$$R_+ : \boldsymbol{M} \to \mathcal{B} \qquad (24.32)$$

$$R_- : \boldsymbol{M} \to \mathcal{B} \qquad (24.33)$$

be obtained from restrictions to the positively and negatively oriented moduli spaces respectively. Let $\mathrm{Fib}(R_+, R_-)$ be the fiber product (i.e. the inverse image of the diagonal in $\mathcal{B} \times \mathcal{B}$ under the map (R_+, R_-)), and $\mathbf{c}$ the concatenation map,

$$\mathbf{c} : \mathrm{Fib}(R_+, R_-) : M^+(X^*, [\mathfrak{b}])_P.$$

If all T_α are finite, the image of $\mathrm{Fib}(R_+, R_-) \cap \boldsymbol{M_T}$ is contained in the top stratum, $M(X^*, [\mathfrak{b}])$. In general, if some T_α are infinite, the result of concatenation is a configuration that has $n^\alpha = 1$ if α is in A and T_α is infinite; all other n^α are zero. The proof of the following proposition, which we omit, is little different from that of the earlier version, Proposition 19.3.1.

Proposition 24.7.3. *There exists $\eta_0 > 0$ such that for all $\eta < \eta_0$ and all $L > L_1(\eta)$, the image of the map*

$$\mathbf{c} : \mathrm{Fib}(R_-, R_+) \cap \boldsymbol{M}^{cen}_\eta \to M^+(X^*, [\mathfrak{b}])_P, \tag{24.34}$$

obtained by concatenating trajectories, contains an open neighborhood $W(\eta, L)$ of the given compact piece $K \subset M'$. Furthermore, $\mathbf{c}$ is injective on

$$\mathbf{c}^{-1}(W(\eta, L)) \cap \boldsymbol{M}^{cen}_\eta.$$

□

As in Subsection 19.3, we write $S_\alpha = 2T_\alpha + 2L$ for α in A. Via the map $\mathbf{c}^{-1}$ in the above proposition, we regard S_α as defined on $W(\eta, L)$, so providing a function

$$\boldsymbol{S} : W(\eta, L) \to (0, \infty]^A.$$

The union of $W(\eta, L)$, over all $L > L_1(\eta)$ is an open neighborhood W of the entire stratum M', and $\boldsymbol{S}$ is well-defined on W. We have the analog of Theorem 19.2.8 for the present situation: the main ingredient is again the parametrization of $M_\eta(Z^T)$ provided by the material of Section 18.

Proposition 24.7.4. *Suppose the moduli spaces appearing in the product* (24.29) *are all boundary-unobstructed. Let W be the open neighborhood of M' in $M^+(X^*, [\mathfrak{b}])$ just described. Then the map*

$$\boldsymbol{S} : W \to (0, \infty]^A$$

is a topological submersion along $M' = \boldsymbol{S}^{-1}(\infty)$. □

When $|A| = 1$, this theorem states that the compactification is a C^0 manifold with boundary along M', and establishes Theorem 24.7.2 in the situation covered by the first case in Proposition 24.6.10.

We now turn to the situation that arises in the third case of Proposition 24.6.10. We are again considering a product such as (24.29) in $M^+_z(X^*, [\mathfrak{b}])$. However, we now suppose that the moduli space $M(X^*, [\mathfrak{b}_0])$ is boundary-obstructed of corank c. This means that $c + 1$ of the components $[\mathfrak{b}^\alpha_0]$ are boundary-unstable. Since all $[\mathfrak{b}^\alpha]$ are irreducible or boundary-stable (because the moduli space contains irreducibles), it must be that $[\mathfrak{b}_0^\alpha]$ is boundary-unstable precisely for $\alpha \in A$. Note that for $\alpha \in A$ the moduli space $M_{z^\alpha}([\mathfrak{b}_0^\alpha], [\mathfrak{b}^\alpha])$ contains irreducibles; it may contain reducibles as well if $\mathfrak{b}^\alpha$ is reducible. As in the last remark in Proposition 24.6.10 the codimension-1 stratum contained in M' is

$$M'' = M_{z_0}(X^*, [\mathfrak{b}_0]) \times \prod_{\alpha \in A} \breve{M}_{z^\alpha}^{\text{irr}}([\mathfrak{b}_0^\alpha], [\mathfrak{b}^\alpha]). \tag{24.35}$$

The space M'' is contained in the open set, U, where for each α either $\gamma_0|_{[0,\infty)\times Y^\alpha}$ is irreducible or some component of $[\breve{\boldsymbol{\gamma}}^\alpha]$ is irreducible.

As above we can construct a neighborhood W of M'' in $M_z^+(X^*, [\mathfrak{b}])$ and a map $\mathbf{S} : W \to (0, \infty]^A$; however, it is no longer the case that $\mathbf{S}$ is a topological submersion along M''. What is missing is the transversality at ∞. As in the cylindrical case, we will exhibit $M_z(X^*, [\mathfrak{b}])$ as a subspace of a larger moduli space $EM_z(X^*, [\mathfrak{b}])$ where this transversality can be achieved.

Let X^L be as above, (24.30). The moduli space $M(X^L)$, which is a Hilbert manifold with boundary, has a fiber product description, corresponding to the decomposition in (24.30): we can write it as

$$\text{Fib}(R_+, R_-) \subset M(X) \times M([0, L] \times Y),$$

where

$$R_+ : M(X) \to \mathcal{B}_{k-1/2}^\sigma(Y)$$
$$R_- : M([0, L] \times Y) \to \mathcal{B}_{k-1/2}^\sigma(Y).$$

Let s^α be the s coordinate on the αth factor $\mathcal{B}_{k-1/2}^\sigma(Y^\alpha)$ of the product space

$$\mathcal{B}_{k-1/2}^\sigma(Y) = \prod_\alpha \mathcal{B}_{k-1/2}^\sigma(Y^\alpha).$$

Thus

$$s^\alpha : \mathcal{B}_{k-1/2}^\sigma(Y) \to [0, \infty).$$

Define

$$\mathcal{B}'(Y) = [0, \infty) \times \left(\prod_{\alpha \in A} \partial \mathcal{B}_{k-1/2}^\sigma(Y^\alpha) \right) \times \left(\prod_{\alpha \notin A} \mathcal{B}_{k-1/2}^\sigma(Y^\alpha) \right).$$

We have a map

$$\pi : \mathcal{B}^{\sigma}_{k-1/2}(Y) \to \mathcal{B}'(Y)$$

whose first coordinate is $\sum_{\alpha\in A} s^{\alpha}$ and whose remaining components are either the projection

$$\pi^{\partial} : \mathcal{B}^{\sigma}_{k-1/2}(Y^{\alpha}) \to \partial\mathcal{B}^{\sigma}_{k-1/2}(Y^{\alpha})$$

for α in A, or the identity for α not in A. (See (19.1).) Let

$$R'_{+} = \pi \circ R_{+} : M(X) \to \mathcal{B}'(Y)$$
$$R'_{-} = \pi \circ R_{-} : M([0,L]\times Y) \to \mathcal{B}'(Y),$$

and define $EM(X^L)$ as the fiber product $\mathrm{Fib}(R'_{+}, R'_{-})$.

We can think of an element of $EM(X^L)$ as being a solution on X^L whose s coordinate is allowed to have a discontinuity at $\{0\}\times Y^{\alpha}$, for α in A, but which is constrained by the condition that the sum of the discontinuities is zero. For a configuration $([\gamma_0],[\gamma_\alpha]_{\alpha\in A})$ in $EM(X^L)$, we define δ^{α} to be the discontinuity in s^{α}:

$$\delta^{\alpha} = s(R_{-}[\gamma_\alpha]) - s^{\alpha}(R_{+}[\gamma_0]).$$

Thus we have $\delta : EM(X^L) \to \Pi^c$, where $\Pi^c \subset \mathbb{R}^A$ is the hyperplane $\sum \delta_\alpha = 0$. The moduli space $M(X^L)$ is the zero set of δ.

We define $EM(X^*,[\mathfrak{b}])$ in just the same way, except that we use the moduli space $M([0,\infty)\times Y;[\mathfrak{b}])$ on the half-infinite cylinder, in place of $M([0,L]\times Y)$ in the fiber product,

$$R'_{+} : M(X) \to \mathcal{B}'(Y)$$
$$R'_{-} : M([0,\infty)\times Y;[\mathfrak{b}]) \to \mathcal{B}'(Y).$$

Lemma 24.7.5. *Suppose $M(X^*,[\mathfrak{b}])$ is regular and not boundary-obstructed. Then the maps $R'_{\pm}$ above are transverse along $M(X^*,[\mathfrak{b}])$ in the fiber product $EM(X^*,[\mathfrak{b}])$. In the case that $M(X^*,[\mathfrak{b}])$ is boundary-obstructed and $[\mathfrak{b}^{\alpha}]$ is boundary-unstable precisely for α in A, then the corresponding moduli spaces M and EM are equal; and in this case, the maps $R'_{\pm}$ are transverse everywhere (though the maps $R_{\pm}$ are not).*

Proof. We deal with the second case in this lemma. Because $[\mathfrak{b}^{\alpha}]$ is boundary-unstable for α in A, an element in the fiber product defining EM has $[\gamma_\alpha]$

reducible for all α in A, from which it follows that each δ_α is non-positive. Since the sum is zero, each δ_α is zero, and so the configuration belongs to M.

To see that $R'_\pm$ are transverse, note that, because at least one $[\mathfrak{b}^\alpha]$ is boundary-unstable, the moduli space M consists only of reducibles. Since we are working at a reducible configuration the domains of differentials of $R'_\pm$ decompose as direct sums along the reducible configurations and the normal directions. Note that for $\alpha \notin A$, $M^{\text{red}}([0,\infty)\times Y^\alpha;[\mathfrak{b}^\alpha])$ is the boundary of $M([0,\infty)\times Y^\alpha;[\mathfrak{b}^\alpha])$ while for $\alpha \in A$ the two moduli spaces coincide. The regularity hypothesis means that

$$R_+^{\text{red}} : M^{\text{red}}(X) \to \mathcal{B}^{\text{red}}(Y)$$
$$R_-^{\text{red}} : M^{\text{red}}([0,\infty)\times Y;[\mathfrak{b}]) \to \mathcal{B}^{\text{red}}(Y)$$

are transverse. The remaining directions to hit are the initial $\mathbb{R}$ factor and $\prod_{\alpha\notin A}\mathbb{R}$ accounting for the difference between $\mathcal{B}^{\text{red}}_{k-1/2}(Y^\alpha)$ and $\mathcal{B}_{k-1/2}(Y^\alpha)$ for $\alpha \in A$. We can hit the $\mathbb{R}$ factor by variations in the normal direction to $M^{\text{red}}(X)$. The $\prod_{\alpha\notin A}\mathbb{R}$ factors are hit by variations in the normal direction to M^{red} on cylinders Z_α^+ for $\alpha \notin A$. □

As above, we can view X^* as obtained as an identification space of $\boldsymbol{X_T}$. This leads to a fiber product description of the moduli space $EM(X^*,[\mathfrak{b}])$. Under our present hypothesis that $\mathfrak{b}_0^\alpha$ is boundary-unstable precisely for $\alpha \in A$, the above lemma tells us that the fiber product description of $EM(X^*,[\mathfrak{b}_0])$ is transverse. This leads to the following analog of Theorem 19.4.1.

Proposition 24.7.6. *There is a topological embedding $j : W \to EW$ and a map $\boldsymbol{S} : EW \to (0,\infty]^A$ such that the following diagram commutes:*

$$\begin{array}{ccc} W & \xrightarrow{\;j\;} & EW \\ \downarrow{\scriptstyle S} & & \downarrow{\scriptstyle S} \\ (0,\infty]^A & = & (0,\infty]^A. \end{array}$$

Further, the map $\boldsymbol{S} : EW \to (0,\infty]^{n-1}$ is a topological submersion along the fiber over ∞, and $j(W) \subset EW$ is the zero set of a continuous map

$$\delta : EW \to \Pi^c \subset \mathbb{R}^A.$$

□

To conclude the proof of Theorem 24.7.2 in the present case, it remains to show that we can achieve the fourth condition of the definition of codimension-c δ-structure. (See Definition 24.7.1.) We can regard EW as parametrizing pairs $([\gamma_0], [\breve{\boldsymbol{\gamma}}])$ where $[\gamma_0] \in EM(X^*, [\mathfrak{b}_1])$ and $[\breve{\boldsymbol{\gamma}}^\alpha] \in \breve{M}^+_{z^\alpha}([\mathfrak{b}_1^\alpha], [\mathfrak{b}^\alpha])$ with $n_\alpha = 0$ unless $\alpha \in A$ and $S_\alpha = \infty$. In the latter case $n^\alpha = 1$ and $[\mathfrak{b}_0^\alpha] = [\mathfrak{b}_1^\alpha]$, so $[\mathfrak{b}_1^\alpha]$ is boundary-unstable. Thus the restriction of γ_0 to the end $[0, \infty) \times Y^\alpha$ is reducible whenever $S_\alpha = \infty$, from which it follows that $\delta_\alpha \leq 0$ in this case. If $\delta_{\alpha_*} = 0$ for one α_* with $S_{\alpha_*} = \infty$, then γ_0 is reducible on the compact piece X and so the components δ^α are non-negative for all α in A. Since the sum of the δ^α is zero, δ vanishes and we conclude that $([\gamma_0], [\breve{\boldsymbol{\gamma}}]) \in W$. We now see that we need to choose W to be contained in the open set U discussed after equation (24.35) so that the only configurations in W with $[\gamma_0]$ reducible have all $S_\alpha = \infty$.

We have verified the conclusions of Theorem 24.7.2 in the situation covered by Cases (i) and (iii) of Proposition 24.6.10. The proof in Case (ii) is similar to the corresponding situation in the cylindrical case. Case (iv) is the statement that the moduli space (rather than its compactification) is a manifold with boundary. □

24.8 Orienting moduli spaces

The definition of a homology orientation for a 4-manifold X with boundary was given in Definition 3.4.1. We will write $\Lambda(X)$ for the set of homology orientations X.

We wish to obtain an analytic reinterpretation of $\Lambda(X)$. Let $C \subset L^2_{k-1/2}(Y; T^*Y)$ be the coclosed 1-forms, and let

$$C = C^- \oplus C^0 \oplus C^+$$

be the spectral decomposition provided by the Fredholm operator $*d : C \to C$: the kernel, C^0, is the space of harmonic 1-forms, and $C^\pm$ are the strictly positive and negative spectral subspaces. Consider the Fredholm operator

$$B : \{ a \in L^2_k(X; T^*X) \mid d^*a = 0, d^+a = 0, \langle a_Y, \nu \rangle = 0 \} \to C^- \qquad (24.36)$$

given by restricting the 1-form a to the boundary and projecting onto C^-.

Lemma 24.8.1. *Then the kernel of B is isomorphic to $H^1(X; \mathbb{R})$. The cokernel of B is isomorphic to a maximal positive-definite subspace for the (non-degenerate) quadratic form on $I^2(X)$.*

Proof. We identify the kernel of this operator B. For a real-valued 1-form, we have

$$\int_X da \wedge da = \int_X \left(|d^+a|^2 - |d^-a|^2\right).$$

So for a in the kernel of B we have (writing b for the 1-form on Y obtained by restriction)

$$\begin{aligned} -\int_X |d^-a|^2 &= \int_X da \wedge da \\ &= \int_Y b \wedge db \\ &= \int_Y \langle b, *_Y db \rangle. \end{aligned} \tag{24.37}$$

Because $B(a) = 0$, the boundary value b satisfies

$$b - d\xi \in C^+ \oplus C^0$$

for some function ξ on Y. So the term on the last line of (24.37) is non-negative. It follows that d^-a is zero; so a is closed, and represents an element of $H^1(X;\mathbb{R})$. If $a \in \ker(B)$ is exact, then $a = df$, where f is harmonic and satisfies Neumann boundary conditions. It follows that a is zero (such an f is constant), so the map from $\ker(B)$ to $H^1(X;\mathbb{R})$ is injective. To see that it is surjective also, we recall that any 1-dimensional cohomology class is represented by a coclosed 1-form satisfying $\langle a, \nu \rangle = 0$ on the boundary. Such a 1-form is in the kernel of B, because the projection of $a|_Y$ to C is closed and therefore lies in C^0. So the kernel of B and $H^1(X;\mathbb{R})$ are isomorphic.

Next we identify the cokernel. Let

$$\begin{aligned} \mathcal{E} &= \{ a \in L^2_k(X; T^*X) \mid \langle a, \nu \rangle = 0 \text{ at } \partial X \} \\ \mathcal{F} &= \{ (f, \omega) \in L^2_{k-1}(X; \mathbb{R} \oplus \Lambda^+(X)) \mid \textstyle\int_X f = 0 \}. \end{aligned}$$

The domain of B is the kernel of the map

$$D = d^* \oplus d^+ : \mathcal{E} \to \mathcal{F};$$

and the operator D is surjective (as follows from Corollary 17.1.5). So the cokernel of B is isomorphic to the cokernel of

$$(D, B) : \mathcal{E} \to \mathcal{F} \oplus C^{-}.$$

If (f, ω, h) is in the cokernel of this last operator, then by considering variations of $a \in \mathcal{E}$ supported in the interior, we see that

$$*d\omega + df = 0. \tag{24.38}$$

By considering a variation a that is non-zero on the boundary, we then see that

$$\int_Y a \wedge \omega + \int_Y a \wedge *h = 0$$

for all a, and hence $h = - * (\omega|_Y)$. In particular, $*(\omega|_Y)$ belongs to C^{-}, and is orthogonal to the exact 1-forms on Y. Stokes' theorem now tells us that $*d\omega$ and df are L^2 orthogonal on X; and so it follows from (24.38) that $d\omega = 0$ and $df = 0$. From the definition of $\mathcal{E}$, we have $f = 0$. Thus we have identified $\text{coker}(B)$ as

$$\text{coker}(B) \cong \{\omega \in \Omega^{+}(X) \mid d\omega = 0, *(\omega|_Y) \in C^{-}\}.$$

Consider now the two subspaces

$$\begin{aligned} Z^{+} &= \{\omega \in \Omega^{+}(X) \mid d\omega = 0,\ *(\omega|_Y) \in C^{-}\} \\ Z^{-} &= \{\omega \in \Omega^{-}(X) \mid d\omega = 0,\ *(\omega|_Y) \in C^{+}\} \end{aligned}$$

and let $Z = Z^{+} \oplus Z^{-}$. There is a map

$$\begin{aligned} Z &\to H^2(X; \mathbb{R}) \\ (\omega^{+}, \omega^{-}) &\mapsto [\omega^{+} + \omega^{-}], \end{aligned} \tag{24.39}$$

which we shall show is injective. So suppose

$$\omega^{+} + \omega^{-} = da.$$

Then, much as above,

$$\begin{aligned}\int_X |da|^2 &= \int_X \omega^+ \wedge \omega^+ - \int_X \omega^- \wedge \omega^- \\ &= \int_Y a \wedge \omega^+ - \int_Y a \wedge \omega^- \\ &= \int_Y \langle a, *\omega^+ \rangle - \int_Y \langle a, *\omega^- \rangle \\ &= \int_Y \big\langle \Pi_- a - \Pi_+ a, *da \big\rangle \\ &\leq 0,\end{aligned}$$

where $*$ is the Hodge star on Y, and $\Pi_\pm$ are the projections to $C^\pm$. So $\omega^+ + \omega^-$ is zero, and $\omega^\pm$ are individually zero because they are pointwise orthogonal.

To understand the image of Z in $H^2(X;\mathbb{R})$, consider first a 2-form ω on a cylinder $\mathbb{R} \times Y$, decomposed into its self-dual and anti-self-dual parts, $\omega = \omega^+ + \omega^-$. Let $\eta^\pm(t)$ be the time-dependent 1-forms on Y obtained as $*(\omega^\pm|_{t \times Y})$. Then the equations $d\omega = d^*\omega = 0$ are equivalent to the conditions that $\eta^\pm(t)$ belong to C for all t and satisfy the equations

$$(d/dt)\eta^+ = *d\eta^+, \quad (d/dt)\eta^- = -*d\eta^-.$$

Now suppose ω is a closed and coclosed 2-form on the cylindrical-end manifold $X^* = X \cup [0,\infty) \times Y$, and let $\eta^\pm(t)$ be defined as above, for $t \geq 0$. If ω is exponentially decaying on the cylindrical end, it follows from the above observation that $\eta^+(0)$ and $\eta^-(0)$ belong to the negative and positive eigenspaces of $*d$: that is,

$$\begin{aligned}\eta^+(0) &\in C^- \\ \eta^-(0) &\in C^+.\end{aligned}$$

The converse is also true, so the image of Z is the cohomology classes in $H^2(X^*;\mathbb{R})$ represented by L^2 harmonic forms ω. The space of L^2 harmonic 2-forms is isomorphic to $I^2(X)$ in such a way that the cup product pairing Q corresponds to the pairing

$$(\omega_1, \omega_2) \mapsto \int_{X^*} \omega_1 \wedge \omega_2.$$

(See [8].) This form is strictly positive and negative on the images of Z^+ and Z^- respectively, because these are the subspaces represented by the self-dual and anti-self-dual L^2 harmonic forms.

Thus we have identified the cokernel of B with Z^+ whose (injective) image in $H^2(X;\mathbb{R})$ is a maximal positive subspace for the quadratic form Q on $I^2(X) \subset H^2(X;\mathbb{R})$. □

The lemma allows us to identify $\det(B)$ in topological terms. After a modification of this operator, we obtain an operator whose determinant line can be identified with $\Lambda(X)$, as this corollary explains:

Corollary 24.8.2. *The 2-element set $\Lambda(X)$ can be canonically identified with the set of orientations of the determinant line of the operator*

$$\tilde{B} : \{ a \in L^2_k(X; T^*X) \mid d^*a = 0, d^+a = 0, \langle a_Y, \nu \rangle = 0 \} \to C^- \oplus C^0 \tag{24.40}$$

given by restricting the 1-form a to the boundary and projecting onto $C^- \oplus C^0$.

Proof. The operator $\tilde{B}$ is homotopic to the operator $B \oplus 0$, where 0 denotes the operator $\{0\} \to C^0$ whose determinant line is $\Lambda^{\max} H^1(Y;\mathbb{R})$. The lemma identifies the determinant of B, and we can fix a particular isomorphism by using our standard convention for the orientation of a sum of operators. □

The relevance of $\Lambda(X)$ to orienting the moduli spaces on X^* is contained in the next proposition. Before stating it, we need to point out that the definition of the 2-element orientation set $\Lambda([\mathfrak{b}])$ for a non-degenerate critical point in $\mathcal{B}^\sigma(Y)$ (see Definition 20.3.5) can be extended in a straightforward way to the case that Y has more than one component: we simply join each $\mathfrak{b}^\alpha$ to a reducible configuration (with the zero perturbation) and define $\Lambda([\mathfrak{b}])$ as the set of orientations of the determinant of the resulting operator. If we have a chosen ordering of the components, then we can identify $\Lambda([\mathfrak{b}])$ with the tensor product of the $\Lambda([\mathfrak{b}^\alpha])$; but in general $\Lambda([\mathfrak{b}])$ does not depend on an ordering.

Proposition 24.8.3. *For any critical point $[\mathfrak{b}]$, the moduli space $M(X^*, [\mathfrak{b}])$ is orientable.*

(i) *If the moduli space is not boundary-obstructed, the orientation double cover is canonically identified with the product*

$$\big(\Lambda(X)\Lambda([\mathfrak{b}])\big) \times M(X^*, [\mathfrak{b}])$$

(so that an element of $\Lambda(X)\Lambda([\mathfrak{b}])$ determines an orientation of the moduli space).

(ii) *In the boundary-obstructed case, let $Y^u \subset Y$ be the union of those components Y^α for which $\mathfrak{b}^\alpha$ is boundary-unstable, and let $\Lambda H^0(Y^u)$ be the set of orientations of $H^0(Y^u;\mathbb{R})$. Then the orientation double cover of the moduli space is canonically identified with*

$$\big(\Lambda(X)\Lambda([\mathfrak{b}])\Lambda H^0(Y^u)\big) \times M(X^*, [\mathfrak{b}]).$$

Proof. As we saw in Proposition 24.3.1 the differential of the Seiberg–Witten equations on the compact manifold with boundary, X, is always a surjective operator, hence its kernel defines a gauge-invariant vector bundle over the configuration space. This bundle descends to define a vector bundle $[\ker(\mathcal{D}\mathfrak{F}^\sigma_\mathfrak{p})]$ on $\mathcal{B}^\sigma_k(X, \mathfrak{s}_X)$, a subbundle of the tangent bundle. Furthermore, restricted to the moduli space $M(X, \mathfrak{s}_X)$, $[\ker(\mathcal{D}\mathfrak{F}^\sigma_\mathfrak{p})]$ is the tangent bundle of the moduli space.

Let $[\mathfrak{b}_0]$ be a point in $\mathcal{B}^\sigma_{k-1/2}(Y, \mathfrak{s})$ (not necessarily a critical point). The tangent space $T_{[\mathfrak{b}_0]}\mathcal{B}^\sigma_{k-1/2}(Y, \mathfrak{s})$ is identified with $\mathcal{K}^\sigma_{k-1/2,\mathfrak{b}_0}$. Much as in Subsection 18.4 the operator $\mathrm{Hess}^\sigma_{\mathfrak{b}_0}$ gives a spectral decomposition

$$\mathcal{K}^\sigma_{k-1/2,\mathfrak{b}_0} = \mathcal{K}^+(Y) \oplus \mathcal{K}^-(Y)$$

where the generalized eigenspaces with purely imaginary eigenvalue go in the second summand. Choose a configuration γ_0 on X extending $\mathfrak{b}_0$. The composition

$$\Pi^- \circ \mathcal{D}_{\gamma_0} r : \ker(\mathcal{D}_\gamma \mathfrak{F}^\sigma_\mathfrak{p}) \to \mathcal{K}^-(Y) \tag{24.41}$$

is a Fredholm operator by Proposition 24.3.2.

Definition 24.8.4. Define $\Lambda_{\gamma_0}(X, [\mathfrak{b}_0], \mathfrak{q})$ to be the set of orientations of the determinant line of the operator (24.41). ◊

Because the space of perturbations is contractible, $\Lambda_{\gamma_0}(X, [\mathfrak{b}_0], \mathfrak{q})$ is well-defined: it does not depend on $\mathfrak{p}$, only on $\mathfrak{q}$. Furthermore if γ_s is a continuous family of configurations with boundary value $\mathfrak{b}_0$, parametrized by a space S, then the sets $\Lambda_{\gamma_s}(X, \mathfrak{b}_0, \mathfrak{q})$ form a double cover of S: their union has a natural topology (see Subsection 20.2). The proof of the following lemma is similar to the case of the cylinder and is omitted.

Lemma 24.8.5. *Let γ be a configuration on $X^\sim = X \cup [0,1] \times Y$,*

$$\gamma \in \mathcal{C}^\sigma(X \cup [0,1] \times Y, \mathfrak{s}_X),$$

and let $\mathfrak{b}_0$ and $\mathfrak{b}_1$ be its restrictions to $\{0\} \times Y$ and $\{1\} \times Y$. Let γ_0 be the restriction of γ to X, and let γ_1 be the path arising from the restriction of γ to $[0, 1] \times Y$. Then we have the following product rule:

$$\Lambda_\gamma(X^\sim, [\mathfrak{b}_1], \mathfrak{q}_1) = \Lambda_{\gamma_0}(X, [\mathfrak{b}_0], \mathfrak{q}_0)\Lambda_{\gamma_1}([\mathfrak{b}_0], \mathfrak{q}_0, [\mathfrak{b}_1], \mathfrak{q}_1)$$

where $\Lambda_{\gamma_1}([\mathfrak{b}_0], \mathfrak{q}_0, [\mathfrak{b}_1], \mathfrak{q}_1)$ is as defined in Definition 20.3.1. The product depends continuously on γ and is compatible with the product rule (20.9), in that it satisfies the natural associativity law in this setting. □

This $\Lambda_\gamma(X, [\mathfrak{b}], \mathfrak{q})$ is related to $\Lambda(X)$, via Corollary 24.8.2:

Lemma 24.8.6. *If both $\mathfrak{b}_0$ and γ_0 are reducible and $\mathfrak{q} = 0$ then there is a canonical isomorphism*

$$\Lambda_{[\gamma_0]}(X, [\mathfrak{b}_0], 0) \cong \Lambda(X), \tag{24.42}$$

which is continuous in $[\gamma_0]$.

Proof. After a standard homotopy the operator (24.41) becomes a direct sum of a complex-linear operator and the operator $\tilde{B}$ of equation (24.40) thought of as acting on purely imaginary rather than real forms. So the result follows from Corollary 24.8.2. □

Corollary 24.8.7. *For any $[\mathfrak{b}]$ and $\mathfrak{q}$, the double cover of $\mathcal{B}_k^\sigma(X; [\mathfrak{b}])$ obtained from $\Lambda_{[\gamma]}(X; [\mathfrak{b}], \mathfrak{q})$ is trivial, and is isomorphic to the product*

$$\big(\Lambda(X)\Lambda([\mathfrak{b}], \mathfrak{q})\big) \times \mathcal{B}_k^\sigma(X; [\mathfrak{b}]).$$

The isomorphism is compatible with composition.

Proof. In the case that $[\mathfrak{b}_0]$ is reducible and $\mathfrak{q} = 0$, we obtain from the preceding lemma an isomorphism of the double cover obtained from $\Lambda_{[\gamma]}(X; [\mathfrak{b}_0], 0)$ with the product

$$\Lambda(X) \times \mathcal{B}_k^\sigma(X; [\mathfrak{b}_0]),$$

because $\mathcal{B}_k^\sigma(X; [\mathfrak{b}_0])$ deformation-retracts onto the reducible locus $s = 0$. The general case then follows from the composition law and the definition of $\Lambda([\mathfrak{b}], \mathfrak{q})$ in Definition 20.3.5. □

We can now complete the proof of Proposition 24.8.3, which is similar to the proof of Corollary 20.4.1. The device is to replace X by $X^\sim = X \cup ([0, T] \times Y)$

for large T, and regard the moduli space on X^* as a fiber product of the moduli spaces on $X^{\sim}$ and $[T,\infty)\times Y$. Given $[\gamma]$ in the moduli space $M(X^*;[\mathfrak{b}])$, let $([\gamma_0],[\gamma_1])$ be the corresponding elements of the fiber product, and $[\mathfrak{b}']$ their common boundary value. If T is sufficiently large, then $[\gamma_1]$ is approximately constant, and the tangent space to the moduli space $M([0,T]\times Y;[\mathfrak{b}])$ maps to a subspace approximating $\mathcal{K}^+(Y)$ at $\mathfrak{b}'$. When the fiber product is transverse (i.e. when the moduli space is not boundary-obstructed), this identifies the kernel of the (surjective) operator (24.41) with the tangent space to the moduli space, and so identifies the two orientation sets as required.

In the boundary-obstructed case, we can decompose the domain and codomain of the operator (24.41) into the reducible part and the (finite-dimensional) normal part. The operator on the reducible part is surjective, with kernel approximating the tangent space to the moduli space. The normal part of the operator is the operator

$$\begin{aligned}\mathbb{R} &\to \bigoplus_{\alpha}\mathbb{R}\\ 1 &\mapsto (1,\ldots,1),\end{aligned}\tag{24.43}$$

where the sum is over all α for which $\mathfrak{b}^\alpha$ is boundary-unstable. To be explicit, we take the element $1\in\mathbb{R}$ on the left-hand side to be $-\partial/\partial s$, the outward normal to the boundary of $\mathcal{B}^\sigma_k(X)$ at the reducibles. (Recall here that, unlike Y, the 4-manifold X is supposed to be connected.) We take "1" in the αth component on the right to be the outward normal in $\mathcal{B}^\sigma_{k-1/2}(Y^\alpha)$. To orient $\Lambda(X)\Lambda([\mathfrak{b}])\times M(X^*;[\mathfrak{b}])$, we need to choose an orientation for the cokernel of the map (24.43). The codomain of this operator is $H^0(Y^u)$. An orientation for $H^0(Y^u)$ determines an orientation for the quotient $H^0(Y^u)/\mathbb{R}$ by our usual "fiber-first" convention. □

25 Maps from cobordisms

25.1 Moduli spaces on cobordisms

At this point, we switch our attention back to the situation in which W is an oriented cobordism between a pair of non-empty, connected 3-manifolds. Departing from our earlier terminology, we shall call the incoming and outgoing boundaries Y_- and Y_+ respectively. We continue to suppose that a Riemannian metric on W is given, and that the metric is cylindrical in collar neighborhoods of the boundary components. Because the boundary of W is now $-Y_-\amalg Y_+$, we have to make some adjustments to the notation of the previous section, in which

$Y = \coprod Y^\alpha$ was the oriented boundary of W. We recap some of the definitions and results, with the appropriate alterations to our notation.

We suppose that admissible perturbations $\mathfrak{q}_\pm$ are given for $Y_\pm$. We write the manifold W^* with two cylindrical ends as

$$W^* = (-\infty, 0] \times Y_- \cup W \cup [0, \infty) \times Y_+.$$

We again adopt the convention that

$$\mathcal{B}^\sigma_{k-1/2}(Y_-) = \bigcup_{\mathfrak{s}_-} \mathcal{B}^\sigma_{k-1/2}(Y_-, \mathfrak{s}_-)$$

$$\mathcal{B}^\sigma_{k-1/2}(Y_+) = \bigcup_{\mathfrak{s}_+} \mathcal{B}^\sigma_{k-1/2}(Y_+, \mathfrak{s}_+).$$

Given critical points $[\mathfrak{a}] \in \mathcal{B}^\sigma_{k-1/2}(Y_-)$ and $[\mathfrak{b}] \in \mathcal{B}^\sigma_{k-1/2}(Y_+)$ for the perturbed Chern–Simons–Dirac functionals, we have a moduli space

$$\begin{aligned} M([\mathfrak{a}], W^*, [\mathfrak{b}]) &\subset \mathcal{B}^\sigma_{k,\mathrm{loc}}(W^*) \\ &= \bigcup_{\mathfrak{s}_W} \mathcal{B}^\sigma_{k,\mathrm{loc}}(W^*, \mathfrak{s}_W). \end{aligned}$$

This moduli space decomposes as a union of pieces corresponding to the different homotopy classes of W-paths, and we write

$$\bigcup_{\mathfrak{s}_W} M([\mathfrak{a}], W^*, \mathfrak{s}_W, [\mathfrak{b}]) = \bigcup_{z} M_z([\mathfrak{a}], W^*, [\mathfrak{b}]),$$

where z runs through $\pi([\mathfrak{a}], W, [\mathfrak{b}])$ (see Definition 23.3.2). Adapting Definition 24.4.4 to account for the altered orientation of Y_-, we say that the moduli space is boundary-obstructed (with corank 1) if $[\mathfrak{a}]$ is boundary-stable and $[\mathfrak{b}]$ is boundary-unstable. This is just as in the cylindrical case. We assume that a perturbation $\mathfrak{p} = (\mathfrak{p}_-, \mathfrak{p}_+)$ is chosen in the collars of $Y_\pm$ so that the moduli spaces are regular.

We adopt the notation $\mathrm{gr}_z([\mathfrak{a}], W, [\mathfrak{b}])$ in place of $\mathrm{gr}_z(W, [\mathfrak{b}])$ now. This agrees with the dimension of the moduli space, except in the boundary-obstructed case, when the dimension of the moduli space is larger by 1.

There is a compactification of $M([\mathfrak{a}], W^*, [\mathfrak{b}])$ using broken trajectories: it is the specialization of the moduli space $M^+(X^*, [\mathfrak{b}])$ to the case of a cobordism, and we denote it by $M^+([\mathfrak{a}], W^*, [\mathfrak{b}])$. It consists of triples

$([\breve{\gamma}_-], [\gamma_0], [\breve{\gamma}_+])$, where

$$[\breve{\gamma}_-] \in \breve{M}^+([\mathfrak{a}], [\mathfrak{a}_0])$$
$$[\breve{\gamma}_+] \in \breve{M}^+([\mathfrak{b}_0], [\mathfrak{b}])$$
$$[\gamma_0] \in M([\mathfrak{a}_0], W^*, [\mathfrak{b}_0]).$$

(So $[\breve{\gamma}_\pm]$ are both possibly broken trajectories, each with any number of components, including zero.) There is also the smaller compactification, $\bar{M}([\mathfrak{a}], W^*, [\mathfrak{b}])$, which is the image of $M^+([\mathfrak{a}], W^*, [\mathfrak{b}])$ in $\mathcal{B}^\sigma_{k,\mathrm{loc}}(W^*)$ (see Definition 24.6.9).

We spell out the statement of Proposition 24.6.10 in this case, to classify the codimension-1 strata in $M_z^+([\mathfrak{a}], W^*, [\mathfrak{b}])$, in the case that the moduli space contains irreducibles.

Proposition 25.1.1. *Suppose $M_z([\mathfrak{a}], W^*, [\mathfrak{b}])$ contains irreducible solutions and has dimension d. Then both $M_z^+([\mathfrak{a}], W^*, [\mathfrak{b}])$ and $\bar{M}_z([\mathfrak{a}], W^*, [\mathfrak{b}])$ are d-dimensional spaces stratified by manifolds, with top stratum the irreducible part of $M_z([\mathfrak{a}], W^*, [\mathfrak{b}])$. The $(d-1)$-dimensional stratum in $M_z^+([\mathfrak{a}], W^*, [\mathfrak{b}])$ consists of elements of the types*

$$\breve{M}_{-1} \times M_0 \tag{25.1a}$$
$$M_0 \times \breve{M}_1 \tag{25.1b}$$
$$\breve{M}_{-2} \times \breve{M}_{-1} \times M_0 \tag{25.1c}$$
$$\breve{M}_{-1} \times M_0 \times \breve{M}_1 \tag{25.1d}$$
$$M_0 \times \breve{M}_1 \times \breve{M}_2, \tag{25.1e}$$

and finally

$$M_z^{\mathrm{red}}([\mathfrak{a}], W^*, [\mathfrak{b}]) \tag{25.1f}$$

in the case that the moduli space contains both reducibles and irreducibles. We have used M_0 to denote a typical moduli space on W^*, and $\breve{M}_{-n}$ and $\breve{M}_n$ $(n > 0)$ to denote typical unparametrized moduli spaces on Y_- and Y_+ (which change from line to line). In the strata with three factors, the middle factor is boundary-obstructed.

The image of one of the above strata has codimension 1 in the smaller compactification $\bar{M}([\mathfrak{a}], W^*, [\mathfrak{b}])$ when the unparametrized moduli spaces on the cylinder, $\breve{M}_{-1}$, $\breve{M}_1$ etc., are all 1-dimensional. □

Remark. To clarify the description of the strata, $\breve{M}_{-1} \times M_0$ in the first of the five cases is an abbreviation for a stratum of the form

$$\breve{M}_{z_{-1}}([\mathfrak{a}], [\mathfrak{a}_0]) \times M_{z_0}([\mathfrak{a}_0], W^*, [\mathfrak{b}]).$$

In the first five cases, if any of the moduli spaces contain both reducibles and irreducibles, only the top stratum is really codimension-1.

If we take W to be a cylinder $I \times Y$, so that W^* is $\mathbb{R} \times Y$, then the space $M^+([\mathfrak{a}], W^*, [\mathfrak{b}])$ is a compactification of the ordinary moduli space of trajectories $M([\mathfrak{a}], [\mathfrak{b}])$ on the cylinder, a compactification we can call $M^+([\mathfrak{a}], [\mathfrak{b}])$. This is distinct from the compactification $\breve{M}^+([\mathfrak{a}], [\mathfrak{b}])$ of the moduli space of *unparametrized* trajectories that we previously studied in Subsection 16.1. In the cylindrical case, the classification (25.1) is essentially the same as the classification of codimension-1 strata given in Proposition 16.5.5, for the space $\breve{M}^+([\mathfrak{a}], [\mathfrak{b}])$. In both cases, the statement is that the strictly broken trajectories which contribute to the codimension-1 strata either have two components, or have three components of which the middle one is boundary-obstructed. In the case of $M^+([\mathfrak{a}], [\mathfrak{b}])$, one factor in each product is a parametrized moduli space and the others are unparametrized.

25.2 Orientations and cobordisms

Because the orientation sets $\Lambda(Y; [\mathfrak{a}])$ and $\Lambda(-Y; [\mathfrak{a}])$ are not canonically isomorphic (see Subsection 22.5), the orientation for moduli spaces on cobordisms requires a few words. For critical points $[\mathfrak{a}]$ and $[\mathfrak{b}]$ on Y_- and Y_+, we denote by $\Lambda([\mathfrak{a}])$ and $\Lambda([\mathfrak{b}])$ their usual orientation sets; we put the manifold into our notation for emphasis:

$$\Lambda([\mathfrak{a}]) = \Lambda(Y_-; [\mathfrak{a}])$$
$$\Lambda([\mathfrak{b}]) = \Lambda(Y_+; [\mathfrak{b}]).$$

We can regard W either as a cobordism from Y_- to Y_+, or as simply a manifold with boundary – a cobordism from the empty set to $-Y_- \amalg Y_+$; and according to Definition 3.4.1, the appropriate notion of homology orientation is different in the two cases. To temporarily distinguish the two cases, let us write W'' for W regarded in the latter sense, with two "outgoing" boundary components. Then Definition 3.4.1 gives the relationship:

$$\Lambda(W) = \Lambda(Y_-)\Lambda(W''). \tag{25.2}$$

Because the boundary of W'' is $-Y_- \amalg Y_+$, Proposition 24.8.3 tells us that the moduli space $M([\mathfrak{a}], (W'')^*, [\mathfrak{b}])$ is canonically oriented by a choice of element from the orientation set

$$\Lambda(W'')\Lambda(-Y_- \amalg Y_+; [\mathfrak{a}] \amalg [\mathfrak{b}]), \tag{25.3}$$

which we can identify with

$$\Lambda(-Y_-; [\mathfrak{a}])\Lambda(W'')\Lambda(Y_+; [\mathfrak{b}])$$

by writing the operator on $(-Y_- \amalg Y_+) \times \mathbb{R}$ as a direct sum, using the given ordering of the two components. (In the boundary-obstructed case, Proposition 24.8.3 tells us that we also need to orient $H^0(Y)$, or equivalently choose an ordering of the two components. We choose the order with Y_- first.) Using the isomorphism of Definition 22.5.3 and the relation (25.2) above, we can write this as

$$\Lambda([\mathfrak{a}])\Lambda(W)\Lambda([\mathfrak{b}]).$$

We therefore have:

Proposition 25.2.1. *A regular moduli space $M([\mathfrak{a}], W^*, [\mathfrak{b}])$ on the cobordisms W with cylindrical ends is canonically oriented by a choice of element λ from the 2-element set $\Lambda([\mathfrak{a}])\Lambda(W)\Lambda([\mathfrak{b}])$, where $\Lambda(W)$ is the set of homology orientations of the cobordism, in the sense of Definition* 3.4.1, *and $\Lambda([\mathfrak{a}])$, $\Lambda([\mathfrak{b}])$ are as defined in Definition* 20.3.5. □

Suppose now that we choose trivializations of $\Lambda(\mathfrak{a})$, $\Lambda(\mathfrak{b})$ and $\Lambda(W)$, so as to orient the moduli space $M_z([\mathfrak{a}], W^*, [\mathfrak{b}])$ and also (using the product rule) the lower strata such as (25.1) in the compactification $M_z^+([\mathfrak{a}], W^*, [\mathfrak{b}])$. As long as the moduli space contains irreducibles, the compactification either is a manifold with boundary or has a codimension-1 δ-structure along the strata (25.1) by Theorem 24.7.2. We can therefore ask whether the orientations of these codimension-1 strata are the boundary orientation. The answer is given by the next proposition, whose proof follows the same lines as that of Proposition 20.5.2.

Proposition 25.2.2. *Suppose that $M_z([\mathfrak{a}], W^*, [\mathfrak{b}])$ contains irreducibles, so that $M_z^+([\mathfrak{a}], W^*, [\mathfrak{b}])$ is a space stratified by manifolds with top stratum $M_z([\mathfrak{a}], [\mathfrak{b}])$. Let M' be a codimension-1 stratum of $M_z^+([\mathfrak{a}], W^*, [\mathfrak{b}])$ of one of the five forms given in the first five cases of* (25.1). *Then the canonical orientation of M' differs from the boundary orientation by the sign*

(i) 1,
(ii) $(-1)^{\dim(M_0)+1}$,
(iii) $(-1)^{\dim(M_{-1})}$,
(iv) (-1),
(v) $(-1)^{\dim(M_0)}$

respectively. □

As in Subsection 20.6, which dealt with the moduli spaces of reducible solutions on the cylinder, we can also consider the compactification $M_z^{\text{red}+}([\mathfrak{a}], W^*, [\mathfrak{b}])$ of the moduli space $M_z^{\text{red}}([\mathfrak{a}], W^*, [\mathfrak{b}])$ of *reducible* solutions on a general cobordism. This is a space stratified by manifolds, and the codimension-1 strata are of one of the two forms

$$M_0^{\text{red}} \times \breve{M}_1^{\text{red}}$$
$$\breve{M}_{-1}^{\text{red}} \times M_0^{\text{red}},$$

in the same notation as above. Once again, we can compare the canonical orientations of these strata with their boundary orientations, as we did in the cylindrical case in Subsection 20.6. In particular, with straightforward modifications, Lemma 20.6.4 still holds.

25.3 Chain maps

Let W again be a cobordism from Y_- to Y_+, as above. In order to use the moduli spaces $M_z([\mathfrak{a}], W^*, [\mathfrak{b}])$ to define chain maps between the Floer complexes of Y_- and Y_+, we need a finiteness result, along the lines of Proposition 16.4.3. This lemma is merely a reformulation of Proposition 24.6.6, adapted to the case of a cobordism between connected manifolds:

Lemma 25.3.1. *For any* $[\mathfrak{a}]$ *and* d_0 *and* i_0, *there are only finitely many pairs* $(z, [\mathfrak{b}])$ *such that*

(i) *the moduli space* $M_z^+([\mathfrak{a}], W^*, [\mathfrak{b}])$ *is non-empty;*
(ii) *the dimension of the moduli space is at most* d_0; *and*
(iii) $\iota([\mathfrak{b}]) \geq i_0$, *where* ι *is as defined in* (16.2).

□

As pointed out in the remark following Proposition 24.6.6, the need for a bound on ι is also the reason that we need to form a completion of the Floer groups before proceeding further. Recall that $\widehat{HM}_\bullet(Y)$ (for example) is formed from $\widehat{HM}_*(Y)$ by negative completion, in the sense of Definition 3.1.3. As in that definition, let us choose an element $j_{\mathfrak{s}}$ in $\mathbb{J}(Y, \mathfrak{s})$ for each $\mathfrak{s}$ with $c_1(\mathfrak{s})$ torsion, and let us filter $\mathbb{J}(Y)$ by the decreasing sequence of subsets

$$\mathbb{J}(Y)[n] = \{j_{\mathfrak{s}} - m \mid \mathfrak{s} \text{ a spin}^c \text{ structure and } m \geq n\}.$$

Then the lemma above has the equivalent formulation:

Lemma 25.3.2. *For any $[\mathfrak{a}]$ and d_0 and n, there are only finitely many pairs $(z, [\mathfrak{b}])$ such that*

(i) *the moduli space $M_z^+([\mathfrak{a}], W^*, [\mathfrak{b}])$ is non-empty;*
(ii) *the dimension of the moduli space is at most d_0; and*
(iii) *the grading of $[\mathfrak{b}]$ does not belong to $\mathbb{J}(Y)[n]$.*

□

We can use negative completion at the chain level also, and indicate it always by the $\bullet$ notation. Thus we introduce the chain group $\hat{C}_\bullet(Y)$ and so on. We will also understand that a sum over all spin^c structures is implied. Thus:

$$\hat{C}_\bullet(Y) = \bigoplus_{\mathfrak{s}} \hat{C}_\bullet(Y, \mathfrak{s}).$$

With these remarks out of the way, let us recall that our aim is to define maps

$$\begin{aligned}
\widecheck{HM}_\bullet(u \mid W; \Gamma_W) &: \widecheck{HM}_\bullet(Y_-; \Gamma_-) \to \widecheck{HM}_\bullet(Y_+; \Gamma_+) \\
\widehat{HM}_\bullet(u \mid W; \Gamma_W) &: \widehat{HM}_\bullet(Y_-; \Gamma_-) \to \widehat{HM}_\bullet(Y_+; \Gamma_+) \\
\overline{HM}_\bullet(u \mid W; \Gamma_W) &: \overline{HM}_\bullet(Y_-; \Gamma_-) \to \overline{HM}_\bullet(Y_+; \Gamma_+)
\end{aligned}$$

associated to a triple (u, W, Γ_W), where W is a cobordism with a homology orientation, u is a cohomology class on $\mathcal{B}_k^\sigma(W)$, and Γ_W is a W-morphism of local systems. (See Subsections 23.2 and 23.3.) Using these "generalized cap products", we can then define the promised cap products

$$\begin{aligned}
\frown &: H^*(\mathcal{B}^\sigma(Y); R) \otimes \widecheck{HM}_\bullet(Y; \Gamma) \to \widecheck{HM}_\bullet(Y; \Gamma) \\
\frown &: H^*(\mathcal{B}^\sigma(Y); R) \otimes \widehat{HM}_\bullet(Y; \Gamma) \to \widehat{HM}_\bullet(Y; \Gamma) \\
\frown &: H^*(\mathcal{B}^\sigma(Y); R) \otimes \overline{HM}_\bullet(Y; \Gamma) \to \overline{HM}_\bullet(Y; \Gamma)
\end{aligned}$$

by taking W to be the cylinder $I \times Y$ with its canonical homology orientation, and using the formulae

$$u \frown \xi = \widecheck{HM}_\bullet(u \mid I \times Y; 1)(\xi)$$

etc.

To construct these maps, and the corresponding generalized cup products on Floer cohomology, we make use of the compact moduli space $\bar{M}_z([\mathfrak{a}], W^*, [\mathfrak{b}])$. From its definition (see Definition 24.6.9 and the end of Subsection 25.1), we have inclusions

$$M_z([\mathfrak{a}], W^*, [\mathfrak{b}]) \subset \bar{M}_z([\mathfrak{a}], W^*, [\mathfrak{b}]) \subset \mathcal{B}^\sigma_{k,\mathrm{loc}}(W^*).$$

The ambient space $\mathcal{B}^\sigma_{k,\mathrm{loc}}(W^*)$ has the same cohomology ring as $\mathcal{B}^\sigma(W)$ (and the same weak homotopy type), so we will regard u as a cohomology class on $\mathcal{B}^\sigma_{k,\mathrm{loc}}(W^*)$.

Fix a positive integer d_0, and consider all triples $(z, [\mathfrak{a}], [\mathfrak{b}])$ for which the moduli space $M_z([\mathfrak{a}], W^*, [\mathfrak{b}])$ or $M^{\mathrm{red}}_z([\mathfrak{a}], W^*, [\mathfrak{b}])$ has dimension d_0 or less. The compactifications $\bar{M}_z([\mathfrak{a}], W^*, [\mathfrak{b}])$ and $\bar{M}^{\mathrm{red}}_z([\mathfrak{a}], W^*, [\mathfrak{b}])$ form a locally finite collection of closed subsets of $\mathcal{B}^\sigma_{k,\mathrm{loc}}(W^*)$, so we can apply Lemma 21.2.1: every open cover of $\mathcal{B}^\sigma_{k,\mathrm{loc}}(W^*)$ has a refinement transverse to all strata in all compactified moduli spaces $\bar{M}$ and $\bar{M}^{\mathrm{red}}$ of dimension at most d_0. We can therefore compute the cohomology of $\mathcal{B}^\sigma_{k,\mathrm{loc}}(W^*)$ using Čech cochains carried by open covers $\mathcal{U}$ that are transverse to all these moduli spaces. Let $\mathcal{U}$ be such an open cover of $\mathcal{B}^\sigma_{k,\mathrm{loc}}(W^*)$, and let

$$u \in C^d(\mathcal{U}; R)$$

be a Čech cochain, with $d \leq d_0$. Here R is a commutative ring. If $M_z([\mathfrak{a}], W^*, [\mathfrak{b}])$ has dimension d, then there is a well-defined evaluation as in (21.3),

$$\big\langle u, [M_z([\mathfrak{a}], W^*, [\mathfrak{b}])] \big\rangle \in \mathrm{Hom}(R\Lambda[\mathfrak{a}], R\Lambda(W) \otimes R\Lambda[\mathfrak{b}]),$$

where $\otimes$ denotes tensor product over R and Hom means homomorphisms of R-modules. (Recall here that the moduli space is canonically oriented by a choice of element in $\Lambda[\mathfrak{a}]\Lambda(W)\Lambda[\mathfrak{b}]$.) Fix a homology orientation μ_W for the cobordism, so that we can regard the above evaluation as

$$\big\langle u, [M_z([\mathfrak{a}], W^*, [\mathfrak{b}])] \big\rangle \in \mathrm{Hom}(R\Lambda[\mathfrak{a}], R\Lambda[\mathfrak{b}]).$$

As usual, we define the above evaluation to be zero if the dimension of the moduli space is not d. Now let Γ_- and Γ_+ be local systems of R-modules on $\mathcal{B}^\sigma(Y_-)$ and $\mathcal{B}^\sigma(Y_-)$, and let Γ_W be a W-morphism between them (see Definition 23.3.1 and the remarks following). We define operators, for $d \le d_0$,

$$\begin{aligned}
m_o^o &: C^d(\mathcal{U};R) \otimes C_\bullet^o(Y_-;\Gamma_-) \to C_\bullet^o(Y_+;\Gamma_+)\\
m_s^o &: C^d(\mathcal{U};R) \otimes C_\bullet^o(Y_-;\Gamma_-) \to C_\bullet^s(Y_+;\Gamma_+)\\
m_o^u &: C^d(\mathcal{U};R) \otimes C_\bullet^u(Y_-;\Gamma_-) \to C_\bullet^o(Y_+;\Gamma_+)\\
m_s^u &: C^d(\mathcal{U};R) \otimes C_\bullet^u(Y_-;\Gamma_-) \to C_\bullet^s(Y_+;\Gamma_+)
\end{aligned}$$

by

$$\left.\begin{aligned}
m_o^o(u\otimes -) &= \sum_{\substack{[\mathfrak{a}]\in\mathfrak{C}^o(Y_-)\\ [\mathfrak{b}]\in\mathfrak{C}^o(Y_+)}} \sum_z \big\langle u, [M_z([\mathfrak{a}], W^*, [\mathfrak{b}])]\big\rangle \Gamma_W(z)\\
m_s^o(u\otimes -) &= \sum_{\substack{[\mathfrak{a}]\in\mathfrak{C}^o(Y_-)\\ [\mathfrak{b}]\in\mathfrak{C}^s(Y_+)}} \sum_z \big\langle u, [M_z([\mathfrak{a}], W^*, [\mathfrak{b}])]\big\rangle \Gamma_W(z)\\
m_o^u(u\otimes -) &= \sum_{\substack{[\mathfrak{a}]\in\mathfrak{C}^u(Y_-)\\ [\mathfrak{b}]\in\mathfrak{C}^o(Y_+)}} \sum_z \big\langle u, [M_z([\mathfrak{a}], W^*, [\mathfrak{b}])]\big\rangle \Gamma_W(z)\\
m_s^u(u\otimes -) &= \sum_{\substack{[\mathfrak{a}]\in\mathfrak{C}^u(Y_-)\\ [\mathfrak{b}]\in\mathfrak{C}^s(Y_+)}} \sum_z \big\langle u, [M_z([\mathfrak{a}], W^*, [\mathfrak{b}])]\big\rangle \Gamma_W(z),
\end{aligned}\right\} \tag{25.4}$$

where in each case the sum extends over all z in $\pi([\mathfrak{a}], W^*, [\mathfrak{b}])$. Lemma 25.3.2 assures us that these sums, although potentially infinite, are well-defined on the completions.

We similarly define operators on the reducible part of the Floer complexes: we have an operator

$$\bar{m} : C^d(\mathcal{U};R) \otimes \bar{C}_\bullet(Y_-;\Gamma_-) \to \bar{C}_\bullet(Y_+;\Gamma_+)$$

$$\bar{m} = \begin{bmatrix} \bar{m}_s^s & \bar{m}_s^u \\ \bar{m}_u^s & \bar{m}_u^u \end{bmatrix}, \tag{25.5}$$

where

$$\left.\begin{aligned}
\bar{m}^s_s(u \otimes -) &= \sum_{\substack{[\mathfrak{a}]\in\mathfrak{C}^s(Y_-)\\ [\mathfrak{b}]\in\mathfrak{C}^s(Y_+)}} \sum_z \big\langle u, [M^{\text{red}}_z([\mathfrak{a}], W^*, [\mathfrak{b}])] \big\rangle \Gamma_W(z) \\
\bar{m}^u_u(u \otimes -) &= \sum_{\substack{[\mathfrak{a}]\in\mathfrak{C}^u(Y_-)\\ [\mathfrak{b}]\in\mathfrak{C}^u(Y_+)}} \sum_z (-1)^d \big\langle u, [M^{\text{red}}_z([\mathfrak{a}], W^*, [\mathfrak{b}])] \big\rangle \Gamma_W(z) \\
\bar{m}^s_u(u \otimes -) &= \sum_{\substack{[\mathfrak{a}]\in\mathfrak{C}^s(Y_-)\\ [\mathfrak{b}]\in\mathfrak{C}^u(Y_+)}} \sum_z \big\langle u, [M^{\text{red}}_z([\mathfrak{a}], W^*, [\mathfrak{b}])] \big\rangle \Gamma_W(z) \\
\bar{m}^u_s(u \otimes -) &= \sum_{\substack{[\mathfrak{a}]\in\mathfrak{C}^u(Y_-)\\ [\mathfrak{b}]\in\mathfrak{C}^s(Y_+)}} \sum_z \big\langle u, [M^{\text{red}}_z([\mathfrak{a}], W^*, [\mathfrak{b}])] \big\rangle \Gamma_W(z).
\end{aligned}\right\} \tag{25.6}$$

Remarks. Compare these with the formulae (22.6), (22.7), and note that, as with $\bar{\partial}^u_s$ and ∂^u_s, the two maps $\bar{m}^u_s$ and m^u_s have the same domain and range but are defined using different moduli spaces. For $\bar{m}^u_s$, we have oriented $M^{\text{red}}_z([\mathfrak{a}], W^*, [\mathfrak{b}])$ as the boundary of $M_z([\mathfrak{a}], W^*, [\mathfrak{b}])$, as in Definition 20.6.1, and the non-zero contributions to the sum come from moduli spaces $M_z([\mathfrak{a}], W^*, [\mathfrak{b}])$ of dimension $d+1$: compare with the definition of $\bar{\epsilon}[\gamma]$ in the similar case. Note that $\bar{m}^u_s$ is a map

$$\bar{m}^u_s : C^d(\mathcal{U}; R)(Y_-; \Gamma_-) \otimes C^u_{k+1}(Y_-; \Gamma_-) \to C^s_{k-d}(Y_+; \Gamma_+),$$

which is as it should be: see (22.17). In the case of $\bar{m}^u_u$, the sign $(-1)^d$ is explained by Item (iii) of Lemma 20.6.4. (See also the remark at the end of Subsection 25.2.)

We now put together these various pieces to define operators on the complexes $\check{C}$ and $\hat{C}$.

Definition 25.3.3. On $\check{C}_\bullet = C^o_\bullet \oplus C^s_\bullet$, we define

$$\check{m} : C^d(\mathcal{U}; R) \otimes \check{C}_\bullet(Y_-; \Gamma_-) \to \check{C}_\bullet(Y_+; \Gamma_+),$$

for $d \leq d_0$, by the formula

$$\check{m} = \begin{bmatrix} m^o_o & -m^u_o \bar{\partial}^s_u(Y_-) - \partial^u_o(Y_+)\bar{m}^s_u \\ m^o_s & \bar{m}^s_s - m^u_s \bar{\partial}^s_u(Y_-) - \partial^u_s(Y_+)\bar{m}^s_u \end{bmatrix},$$

in which $\partial^u_o(Y_+)$, for example, denotes the operator ∂^u_o on Y_+. On $\hat{C}_\bullet = C^o_\bullet \oplus C^u_\bullet$, we define

$$\hat{m} : C^d(\mathcal{U}; R) \otimes \hat{C}_\bullet(Y_-; \Gamma_-) \to \hat{C}_\bullet(Y_+; \Gamma_+)$$

by the formula

$$\hat{m} = \begin{bmatrix} m^o_o & m^u_o \\ \bar{m}^s_u \partial^o_s(Y_-)\sigma - \bar{\partial}^s_u(Y_+) m^o_s & \bar{m}^u_u \sigma + \bar{m}^s_u \partial^u_s(Y_-)\sigma - \bar{\partial}^s_u(Y_+) m^u_s \end{bmatrix},$$

where σ is the sign operator $(-1)^d$ on $C^d(\mathcal{U}; R) \otimes C_\bullet$. ◊

These operators, together with $\bar{m}$, will define the generalized cup products. Note the similarity of the formulae to the definitions of the differentials $\hat{\partial}$ and $\check{\partial}$ in Definition 22.1.3. The main task of this subsection is to establish the following proposition.

Proposition 25.3.4. *The operators $\check{m}$, $\hat{m}$ and $\bar{m}$ satisfy the identities:*

$$\left.\begin{aligned} (-1)^d \check{\partial}(Y_+)\check{m}\left(u \otimes \check{\xi}\right) &= -\check{m}\left(\delta u \otimes \check{\xi}\right) + \check{m}\left(u \otimes \check{\partial}(Y_-)\check{\xi}\right) \\ (-1)^d \hat{\partial}(Y_+)\hat{m}\left(u \otimes \hat{\xi}\right) &= -\hat{m}\left(\delta u \otimes \hat{\xi}\right) + \hat{m}\left(u \otimes \hat{\partial}(Y_-)\hat{\xi}\right) \\ (-1)^d \bar{\partial}(Y_+)\bar{m}\left(u \otimes \bar{\xi}\right) &= -\bar{m}\left(\delta u \otimes \bar{\xi}\right) + \bar{m}\left(u \otimes \bar{\partial}(Y_-)\bar{\xi}\right) \end{aligned}\right\} \quad (25.7)$$

for $u \in C^d(\mathcal{U}; R)$, $\check{\xi} \in \check{C}_\bullet(Y; \Gamma_-)$ etc. and $d \le d_0 - 1$. Thus they give rise to operators

$$\left.\begin{aligned} \check{m} &: \check{H}^d(\mathcal{U}; R) \otimes \widecheck{HM}_j(Y_-; \Gamma_-) \to \widecheck{HM}_{k-d}(Y_+; \Gamma_+) \\ \hat{m} &: \check{H}^d(\mathcal{U}; R) \otimes \widehat{HM}_j(Y_-; \Gamma_-) \to \widehat{HM}_{k-d}(Y_+; \Gamma_+) \\ \bar{m} &: \check{H}^d(\mathcal{U}; R) \otimes \overline{HM}_j(Y_-; \Gamma_-) \to \overline{HM}_{k-d}(Y_+; \Gamma_+) \end{aligned}\right\} \quad (25.8)$$

for any open cover $\mathcal{U}$ of $\mathcal{B}^\tau_{k,\mathrm{loc}}(W^*)$ transverse to all the moduli spaces of dimension less than or equal to d_0.

The operators $\check{m}$, $\hat{m}$ and $\bar{m}$ at the level of homology commute with the refinement maps when we refine an open cover. We can therefore take a limit: we obtain operators

$$\begin{aligned} \check{m} &: H^d(\mathcal{B}^\sigma(W); R) \otimes \widecheck{HM}_j(Y_-; \Gamma_-) \to \widecheck{HM}_{j-d}(Y_+; \Gamma_+) \\ \hat{m} &: H^d(\mathcal{B}^\sigma(W); R) \otimes \widehat{HM}_j(Y_-; \Gamma_-) \to \widehat{HM}_{j-d}(Y_+; \Gamma_+) \\ \bar{m} &: H^d(\mathcal{B}^\sigma(W); R) \otimes \overline{HM}_j(Y_-; \Gamma_-) \to \overline{HM}_{j-d}(Y_+; \Gamma_+) \end{aligned}$$

by taking the limit of the operators (25.8) over all open covers of $\mathcal{B}^{\sigma}_{k,\mathrm{loc}}(W^*)$ transverse to the moduli spaces, and identifying the Čech cohomology $\check{H}^d(\mathcal{B}^{\sigma}_{k,\mathrm{loc}}(W^*);R)$ with $H^d(\mathcal{B}^{\sigma}(W);R)$. To obtain the generalized *cup* products, we need to dualize these definitions, so creating operators

$$\check{m}^*(u \otimes -) : \check{H}^d(\mathcal{U};R) \otimes \widecheck{HM}^\bullet(Y_+;\Gamma_+) \to \widecheck{HM}^\bullet(Y_-;\Gamma_-)$$

together with $\hat{m}^*$ and $\bar{m}^*$. At the chain level, $\check{m}^*$ for example is defined by a 2-by-2 matrix whose top left entry is the operator

$$(m^o_o)^* : C^d(\mathcal{U};R) \otimes (C^o)^\bullet(Y_+;\Gamma_+) \to (C^o)^\bullet(Y_-;\Gamma_-)$$

defined by

$$(m^o_o)^*(u \otimes -) = \sum_{\substack{[\mathfrak{a}]\in\mathfrak{C}^o(Y_-)\\ [\mathfrak{b}]\in\mathfrak{C}^o(Y_+)}} \sum_z \big\langle u, [M_z([\mathfrak{a}], W^*, [\mathfrak{b}])] \big\rangle \Gamma_W(z^{-1}).$$

The definitions of the generalized cap products $\widecheck{HM}_\bullet(u \mid W;\Gamma_W)$ and its relatives are obtained directly from those of the operators $\check{m}$ etc. with only a change of notation:

Definition 25.3.5. Let W be a cobordism from Y_- to Y_+, equipped with a Riemannian metric and a perturbation $\mathfrak{p}$, chosen to make the moduli spaces regular, as above. Let u be an element of $H^d(\mathcal{B}^\sigma(W);R)$, and let μ_W be a homology orientation of W. Let $\Gamma_W : \Gamma_- \to \Gamma_+$ be a morphism of local coefficient systems. Then we define

$$\widecheck{HM}_\bullet(u \mid W;\Gamma_W) : \widecheck{HM}_\bullet(Y_-;\Gamma_-) \to \widecheck{HM}_\bullet(Y_+;\Gamma_+)$$
$$\widehat{HM}_\bullet(u \mid W;\Gamma_W) : \widehat{HM}_\bullet(Y_-;\Gamma_-) \to \widehat{HM}_\bullet(Y_+;\Gamma_+)$$
$$\overline{HM}_\bullet(u \mid W;\Gamma_W) : \overline{HM}_\bullet(Y_-;\Gamma_-) \to \overline{HM}_\bullet(Y_+;\Gamma_+)$$

as the operators

$$\check{m}(u \otimes -)$$
$$\hat{m}(u \otimes -)$$
$$\bar{m}(u \otimes -)$$

respectively. The dual map $\check{m}^*(u \otimes -)$ defines a homomorphism between the Floer cohomology groups in the same way,

$$\begin{aligned}
\widecheck{HM}^\bullet(u \mid W; \Gamma_W) &: \widecheck{HM}^\bullet(Y_+; \Gamma_+) \to \widecheck{HM}^\bullet(Y_-; \Gamma_-) \\
\widehat{HM}^\bullet(u \mid W; \Gamma_W) &: \widehat{HM}^\bullet(Y_+; \Gamma_+) \to \widehat{HM}^\bullet(Y_-; \Gamma_-) \\
\overline{HM}^\bullet(u \mid W; \Gamma_W) &: \overline{HM}^\bullet(Y_+; \Gamma_+) \to \overline{HM}^\bullet(Y_-; \Gamma_-),
\end{aligned}$$

as the limit of the operators

$$\begin{gathered}
\check{m}^*(u \otimes -) \\
\hat{m}^*(u \otimes -) \\
\bar{m}^*(u \otimes -)
\end{gathered}$$

over open covers. ◊

The proof of Proposition 25.3.4 proceeds rather like the proof that $\check{\partial}^2$, $\hat{\partial}^2$ and $\bar{\partial}^2$ are zero. The first lemma is similar to Lemma 22.1.5. In the formulae below, we write ∂^o_o for example, without indicating any more whether this is the operator for Y_- or Y_+, since the formulae never allow any ambiguity.

Lemma 25.3.6. *We have the following identities for any open cover $\mathcal{U}$ transverse to the moduli spaces of dimension d_0 or less. In these formulae, σ is again the sign operator $(-1)^d$ on $C^d(\mathcal{U}) \otimes C_\bullet$. On the right-hand sides, δ is the Čech coboundary operator, $\delta : C^d(\mathcal{U}; \mathbb{Z}) \to C^{d+1}(\mathcal{U}; \mathbb{Z})$.*

(i) $m^o_o \partial^o_o - \partial^o_o m^o_o \sigma - m^u_o \bar{\partial}^s_u \partial^o_s - \partial^u_o \bar{m}^s_u \partial^o_s + \partial^u_o \bar{\partial}^s_u m^o_s \sigma = m^o_o(\delta \otimes 1)$.

(ii) $m^o_s \partial^o_o - \partial^o_s m^o_o \sigma + \bar{m}^s_s \partial^o_s - \bar{\partial}^s_s m^o_s \sigma - m^u_s \bar{\partial}^s_u \partial^o_s - \partial^u_s \bar{m}^s_u \partial^o_s + \partial^u_s \bar{\partial}^s_u m^o_s \sigma = m^o_s(\delta \otimes 1)$.

(iii) $m^o_o \partial^u_o - \partial^o_o m^u_o \sigma - m^u_o \bar{\partial}^u_u - \partial^u_o \bar{m}^u_u - m^u_o \bar{\partial}^s_u \partial^u_s - \partial^u_o \bar{m}^s_u \partial^u_s + \partial^u_o \bar{\partial}^s_u m^u_s \sigma = m^u_o(\delta \otimes 1)$.

(iv) $\bar{m}^u_s + m^o_s \partial^u_o - \partial^o_s m^u_o \sigma + \bar{m}^s_s \partial^u_s - \bar{\partial}^s_s m^u_s \sigma - m^u_s \bar{\partial}^u_u - \partial^u_s \bar{m}^u_u - m^u_s \bar{\partial}^s_u \partial^u_s - \partial^u_s \bar{m}^s_u \partial^u_s + \partial^u_s \bar{\partial}^s_u m^u_s \sigma = m^u_s(\delta \otimes 1)$.

Proof. The four parts are each proved by considering a moduli space $M_z([\mathfrak{a}], W^*, [\mathfrak{b}])$ of dimension $d + 1$ whose interior consists of irreducible trajectories. As in Lemma 22.1.5, the four parts correspond to the four possibilities for the type of the critical points $[\mathfrak{a}]$ and $[\mathfrak{b}]$. In each case, the identity is the result of applying our version of Stokes' theorem (21.4) to the compactification $\bar{M}_z([\mathfrak{a}], W^*, [\mathfrak{b}])$ in $\mathcal{B}^\sigma_{k,\mathrm{loc}}(W^*)$. As in the previous proof, when u is fixed, the four operators on the left-hand sides of the identities are operators

$$B^1 : C^o(Y_-;\Gamma_-) \to C^o(Y_+;\Gamma_+)$$
$$B^2 : C^o(Y_-;\Gamma_-) \to C^s(Y_+;\Gamma_+)$$
$$B^3 : C^u(Y_-;\Gamma_-) \to C^o(Y_+;\Gamma_+)$$
$$B^4 : C^u(Y_-;\Gamma_-) \to C^s(Y_+;\Gamma_+),$$

and each is a sum

$$B^i = \sum B^i_{[\mathfrak{a}],[\mathfrak{b}],z} \otimes \Gamma_W(z),$$

where

$$B^i_{[\mathfrak{a}],[\mathfrak{b}],z} : R\Lambda[\mathfrak{a}] \otimes \Gamma_-[\mathfrak{a}] \to R\Lambda[\mathfrak{b}] \otimes \Gamma_+[\mathfrak{b}].$$

The sum is taken over all $[\mathfrak{a}]$, $[\mathfrak{b}]$ and z of the appropriate type.

For the first of the four identities, we consider the case that $[\mathfrak{a}]$, $[\mathfrak{b}]$ are both irreducible. Let $M_z([\mathfrak{a}], W^*, [\mathfrak{b}])$ be such a $(d+1)$-dimensional moduli space, and let $\bar{M}_z([\mathfrak{a}], W^*, [\mathfrak{b}])$ be the compactification in $\mathcal{B}^\sigma_{k,\text{loc}}(W^*)$. It is convenient again to choose trivializations of $\Lambda[\mathfrak{a}]$ and $\Lambda[\mathfrak{b}]$, so that the top stratum $M_z([\mathfrak{a}], W^*, [\mathfrak{b}])$ is oriented. Because $d \leq d_0 - 1$, our transversality condition for the open cover means that there is a well-defined evaluation

$$\big\langle \delta u, M_z([\mathfrak{a}], W^*, [\mathfrak{b}]) \big\rangle \in R. \tag{25.9}$$

The right-hand side of the identity in the first part of the lemma is the sum

$$\big\langle \delta u, M_z([\mathfrak{a}], W^*, [\mathfrak{b}]) \big\rangle \Gamma_W(z).$$

Using Stokes' theorem (21.4), we can express (25.9) as

$$\sum_\beta \delta_\beta \big\langle u, N^d_\beta \big\rangle \tag{25.10}$$

where the sum is over all components of the d-dimensional stratum, and δ_β is the boundary multiplicity (Definition 21.1.1). The lemma is proved by identifying the sum (25.10) with the component $B^i_{[\mathfrak{a}],[\mathfrak{b}],z}$ of B^i, regarded now as an element of R using the fixed orientations.

To deal correctly with the boundary multiplicities it is simplest to pull back the calculation to the larger compactification $M^+_z([\mathfrak{a}], W^*, [\mathfrak{b}])$, using the quotient map

$$v : M_z^+([\mathfrak{a}], W^*, [\mathfrak{b}]) \to \bar{M}_z([\mathfrak{a}], W^*, [\mathfrak{b}]).$$

The image under v of any stratum of $M_z^+([\mathfrak{a}], W^*, [\mathfrak{b}])$ is contained in strata of equal or lower dimension in $\bar{M}_z([\mathfrak{a}], W^*, [\mathfrak{b}])$; so the pull-back of the open cover $\mathcal{U}$ is an open cover $\mathcal{U}^+$ of $M_z^+([\mathfrak{a}], W^*, [\mathfrak{b}])$ that is transverse to all strata. We denote the pull-back cochain by u^+. We have Stokes' theorem on $M_z^+([\mathfrak{a}], W^*, [\mathfrak{b}])$, in the form

$$\begin{aligned} \sum_\beta \delta_\beta^+ \langle u^+, (N^+)_\beta^d \rangle &= \langle \delta u^+, M_z([\mathfrak{a}], W^*, [\mathfrak{b}]) \rangle \\ &= \langle \delta u, M_z([\mathfrak{a}], W^*, [\mathfrak{b}]) \rangle \end{aligned} \tag{25.11}$$

where $(N^+)_\beta^d$ runs through those components of the d-dimensional stratum of $M_z^+([\mathfrak{a}], W^*, [\mathfrak{b}])$ *whose image in* $\bar{M}_z([\mathfrak{a}], W^*, [\mathfrak{b}])$ *is also d-dimensional*: the contribution from the other d-dimensional components of $M_z^+([\mathfrak{a}], W^*, [\mathfrak{b}])$ is zero, because $\mathcal{U}$ is transverse to the stratification of $\bar{M}_z([\mathfrak{a}], W^*, [\mathfrak{b}])$. The boundary multiplicities δ_β^+ will all be ± 1.

The components of the codimension-1 stratum N^d are already described by Proposition 25.1.1, (25.1a)–(25.1e). The case (25.1f) does not occur because $[\mathfrak{a}]$, $[\mathfrak{b}]$ are not reducible. Note that the d that appears in Proposition 25.1.1 is now $d+1$. Consider a stratum of the type (25.1a) in $M_z^+([\mathfrak{a}], W^*, [\mathfrak{b}])$. This is a stratum of the form

$$(N^+)^d = \breve{M}_{z_{-1}}([\mathfrak{a}], [\mathfrak{a}_0]) \times M_{z_0}([\mathfrak{a}_0], W^*, [\mathfrak{b}]), \tag{25.12}$$

where the second factor is d-dimensional and the first (unparametrized) moduli space is 0-dimensional – a finite set. The image of (25.12) in $\bar{M}_z([\mathfrak{a}], W^*[\mathfrak{b}])$ is the second factor. Give (25.12) its canonical orientation, determined by the chosen trivializations of $\Lambda[\mathfrak{a}]$ and $\Lambda[\mathfrak{b}]$ and $\Lambda(W)$. According to Theorem 24.7.2 and Proposition 25.2.2, the space $M_z^+([\mathfrak{a}], W^*, [\mathfrak{b}])$ is a C^0 manifold with boundary along (25.12), and the canonical orientation agrees with the boundary orientation. Thus the boundary multiplicity of each connected component of (25.12) in the stratified space $M_z^+([\mathfrak{a}], W^*, [\mathfrak{b}])$ is 1. The critical point $[\mathfrak{a}_0]$ must be irreducible; and as $[\mathfrak{a}_0]$ runs through all irreducible critical points, the total contribution to the left side of (25.11) is

$$\sum_{[\mathfrak{a}_0], z_{-1}} \epsilon([\mathfrak{a}], [\mathfrak{a}_0]) \langle u, M_{z_0}([\mathfrak{a}_0], [\mathfrak{b}]) \rangle$$

which is the component of $m_o^o \partial_o^o$ from $R\Lambda[\mathfrak{a}]$ to $R\Lambda[\mathfrak{b}]$. This gives the first term in the first identity. The remaining four terms on the left-hand side of the first identity arise from the four remaining types of codimension-1 strata, (25.1b)–(25.1e).

The other three identities in the lemma are all treated the same way. In the case of the fourth identity, the moduli space $M_z([\mathfrak{a}], W^*, [\mathfrak{b}])$, with $[\mathfrak{a}]$ boundary-unstable and $[\mathfrak{b}]$ boundary-stable, contains both irreducibles and reducibles. There is therefore an extra term to consider, corresponding to the case (25.1f) in Proposition 25.1.1: this extra term is the first term in the fourth identity. □

Lemma 25.3.7. *The identity of Proposition* 25.3.4 *holds in the case of $\bar{m}$: in terms of the four components of $\bar{m}$, we have*

$$
\begin{aligned}
-\bar{m}_s^s\bar{\partial}_s^s + \bar{\partial}_s^s\bar{m}_s^s\sigma - \bar{m}_s^u\bar{\partial}_u^s + \bar{\partial}_s^u\bar{m}_u^s\sigma &= -\bar{m}_s^s(\delta\otimes 1)\\
-\bar{m}_s^s\bar{\partial}_s^u + \bar{\partial}_s^s\bar{m}_s^u\sigma - \bar{m}_s^u\bar{\partial}_u^u + \bar{\partial}_s^u\bar{m}_u^u\sigma &= -\bar{m}_s^u(\delta\otimes 1)\\
-\bar{m}_u^s\bar{\partial}_s^s + \bar{\partial}_u^s\bar{m}_s^s\sigma - \bar{m}_u^u\bar{\partial}_u^s + \bar{\partial}_u^u\bar{m}_u^s\sigma &= -\bar{m}_u^s(\delta\otimes 1)\\
-\bar{m}_u^s\bar{\partial}_s^u + \bar{\partial}_u^s\bar{m}_s^u\sigma - \bar{m}_u^u\bar{\partial}_u^u + \bar{\partial}_u^u\bar{m}_u^u\sigma &= -\bar{m}_u^u(\delta\otimes 1).
\end{aligned}
$$

Proof. As with Lemma 22.1.6, this lemma is parallel to the previous one, but involves the reducible moduli spaces. □

Proof of Proposition 25.3.4. We have just verified the identity for $\bar{m}$ in the preceding lemma; and the identities for $\check{m}$ and $\hat{m}$ proceed quite formally from the case of $\bar{m}$ and the identities of Lemma 25.3.6. Let us examine the identity to be proved for $\check{m}$: we can write it as

$$
-\check{m}\check{\partial} + \check{\partial}\check{m}\sigma + \check{m}(\delta\otimes 1) = 0.
$$

We again use the decomposition $\check{C} = C^o \oplus C^s$ to write the left side in 2-by-2 matrix form. For illustration, let us show that the top right matrix entry is zero, as asserted. From the definitions of $\check{m}$ and $\check{\partial}$, we calculate the top right entry as

$$
\begin{aligned}
&m_o^o\partial_o^u\bar{\partial}_u^s + m_o^u\bar{\partial}_u^s\bar{\partial}_s^s + \partial_o^u\bar{m}_u^s\bar{\partial}_s^s - m_o^u\bar{\partial}_u^s\partial_s^u\bar{\partial}_u^s - \partial_o^u\bar{m}_u^s\partial_s^u\bar{\partial}_u^s\\
&- \partial_o^o m_o^u\bar{\partial}_u^s\sigma - \partial_o^o\partial_o^u\bar{m}_u^s\sigma - \partial_o^u\bar{\partial}_u^s\bar{m}_s^s\sigma + \partial_o^u\bar{\partial}_u^s m_s^u\bar{\partial}_u^s\sigma + \partial_o^u\bar{\partial}_u^s\partial_s^u\bar{m}_u^s\sigma\\
&- m_o^u\bar{\partial}_u^s(\delta\otimes 1) - \partial_o^u\bar{m}_u^s(\delta\otimes 1).
\end{aligned}
$$

Using the identity $\bar{\partial}^u_u \bar{\partial}^s_u = -\bar{\partial}^s_u \bar{\partial}^s_s$, and adding and subtracting a term $\partial^u_o \bar{m}^u_u \bar{\partial}^s_u$, we rewrite this as

$$\begin{aligned}
&(m^o_o \partial^u_o - m^u_o \bar{\partial}^u_u - m^u_o \bar{\partial}^s_u \partial^u_s - \partial^u_o \bar{m}^s_u \partial^u_s - \partial^o_o m^u_o \sigma + \partial^u_o \bar{\partial}^s_u m^u_s \sigma - \partial^u_o \bar{m}^u_u)\bar{\partial}^s_u \\
&+ \partial^u_o (\bar{m}^u_u \bar{\partial}^s_u + \bar{m}^s_u \bar{\partial}^s_s - \bar{\partial}^s_u \bar{m}^s_s \sigma) + \partial^u_o \bar{\partial}^s_u \partial^u_s \bar{m}^s_u \sigma - \partial^o_o \partial^u_o \bar{m}^s_u \sigma \\
&- m^u_o (\delta \otimes 1)\bar{\partial}^s_u - \partial^u_o \bar{m}^s_u (\delta \otimes 1).
\end{aligned}$$

The terms on the first line cancel with the first term on the last line, because of the third identity in Lemma 25.3.6. Adding and subtracting a term $\partial^u_o \bar{\partial}^u_u \bar{m}^s_u \sigma$ to and from the remainder, we obtain

$$\begin{aligned}
&\partial^u_o (\bar{m}^u_u \bar{\partial}^s_u + \bar{m}^s_u \bar{\partial}^s_s - \bar{\partial}^s_u \bar{m}^s_s \sigma - \bar{\partial}^u_u \bar{m}^s_u \sigma) \\
&+ (\partial^u_o \bar{\partial}^u_u - \partial^o_o \partial^u_o + \partial^u_o \bar{\partial}^s_u \partial^u_s)\bar{m}^s_u \sigma \\
&- \partial^u_o \bar{m}^s_u (\delta \otimes 1).
\end{aligned}$$

The terms on the second line vanish, thanks to the third identity in Lemma 22.1.5. The terms on the first line cancel with the last term, on account of the third part of Lemma 25.3.7. So this expression is indeed zero. Verification of the other components of the identities for $\check{m}$ and $\hat{m}$ is very similar. □

At this point, it has not been made clear whether the homomorphisms such as $\widetilde{HM}_\bullet(u \mid W; \Gamma_W)$ depend on the choice of metric g_W and perturbation $\mathfrak{p}$. (We still understand that the metrics and perturbations on $Y_\pm$ are fixed.) The next proposition states that they do not.

Proposition 25.3.8. *Let $g(0)$ and $g(1)$ be two metrics on W, isometric in a collar of the boundary to the same cylindrical metric. Let $\mathfrak{p}(0)$ and $\mathfrak{p}(1)$ be two perturbations on W, again constructed using the same perturbations on $Y_\pm$. Assume that the corresponding moduli spaces on W^* are regular in both cases. Let u and Γ_W be as above, and let $\check{m}(0)$ and $\check{m}(1)$ be defined by the formulae of Definition 25.3.3, using the moduli spaces obtained from $(g(0), \mathfrak{p}(0))$ and $(g(1), \mathfrak{p}(1))$ respectively. Then there is an operator*

$$\check{K} : C^d(\mathcal{U}; \mathbb{Z}) \otimes \check{C}_\bullet(Y_-; \Gamma_-) \to \check{C}_\bullet(Y_+; \Gamma_+)$$

for $d \leq d_0$, satisfying the chain-homotopy identity

$$\begin{aligned}
(-1)^d \check{\partial} \check{K}(u \otimes \check{\xi}) = &+\check{K}(\delta u \otimes \check{\xi}) - \check{K}(u \otimes \check{\partial}\check{\xi}) \\
&+ (-1)^d \check{m}(0)(u \otimes \check{\xi}) - (-1)^d \check{m}(1)(u \otimes \check{\xi})
\end{aligned}$$

for $u \in C^d(\mathcal{U};\mathbb{Z})$ $(d < d_0)$ *and* $\check{\xi} \in \check{C}_\bullet(Y_-;\Gamma_-)$. *There are similar chain homotopies for* $\hat{C}_\bullet$ *and* $\bar{C}_\bullet$.

Corollary 25.3.9. *The homomorphisms* $\widecheck{HM}_\bullet(u \mid W;\Gamma_W)$ *etc. do not depend on the metric* g_W *or the perturbation* $\mathfrak{p}$ *on the interior of* W. □

Proof of Proposition 25.3.8. The chain-homotopy will be constructed using the moduli spaces $M([\mathfrak{a}], W^*, [\mathfrak{b}])_P$ for a family of metrics and perturbations parametrized by $P = [0,1]$.

Consider more generally a compact, oriented manifold P as the parameter space, and suppose first that P *does not have boundary*. Suppose that the family of metrics and perturbations parametrized by P is such that the moduli spaces $M_z([\mathfrak{a}], W^*, [\mathfrak{b}])_P$ are regular (Definition 24.4.9). An element of $\Lambda(\mathfrak{a})\Lambda(W)\Lambda(\mathfrak{b})$ determines an orientation of the parametrized moduli space, using the "fiber-first" convention and regarding P as the "base". Using open covers transverse to the parametrized moduli spaces, we define maps $\check{m}(P)$ etc. by exactly the same formulae as $\check{m}$ etc. in Definition 25.3.3, but using the moduli spaces $M_z([\mathfrak{a}], W^*, [\mathfrak{b}])_P$ in the definition of the components m^o_o etc. The "fiber-first" convention means that the introduction of P has no effect on the orientations of the lower strata, so the identities (25.7) continue to hold with $\check{m}(P)$ etc. in place of $\check{m}$.

Suppose now that P has oriented boundary Q, and that the moduli spaces $M_z([\mathfrak{a}], W^*, [\mathfrak{b}])_Q$ are also regular. The compactification $M^+_z([\mathfrak{a}], W^*, [\mathfrak{b}])_P$ is still a space stratified by manifolds; its codimension-1 strata now include the top stratum of $M_z([\mathfrak{a}], W^*, [\mathfrak{b}])_Q$, in addition to the strata we previously classified. In the identities corresponding to those of Lemma 25.3.6, there will therefore be one extra term: we will have, for example,

$$\begin{aligned} &m^o_o(P)\partial^o_o - (-1)^{\dim P}\partial^o_o m^o_o(P)\sigma - m^u_o(P)\bar{\partial}^s_u\partial^o_s - \partial^u_o\bar{m}^s_u(P)\partial^o_s \\ &\quad + (-1)^{\dim P}\partial^u_o\bar{\partial}^s_u m^o_s(P)\sigma + (-1)^{\dim P} m^o_o(Q)\sigma = m^o_o(P)(\delta\otimes 1). \end{aligned}$$

The sign $(-1)^{\dim P}\sigma = (-1)^{d-\dim P}$ accompanies the term $m^o_o(Q)$ because the orientation of $M_z([\mathfrak{a}], W^*, [\mathfrak{b}])_Q$ differs from the boundary orientation of $M_z([\mathfrak{a}], W^*, [\mathfrak{b}])_P$ by this sign, on account of the fiber-first convention: the number $d - \dim P$ is the dimension of the fiber $M_z([\mathfrak{a}], W^*, [\mathfrak{b}])_p$ for the contributing moduli spaces. The result is that, when P has boundary, $\check{m}(P)$ satisfies an identity like those of (25.7), but with one extra term:

$$\begin{aligned} &(-1)^{d-\dim P}\check{\partial}\check{m}(P)(u\otimes\check{\xi}) \\ &\quad = -\check{m}(P)(\delta u\otimes\check{\xi}) + \check{m}(P)(u\otimes\check{\partial}\check{\xi}) + (-1)^{d-\dim P}\check{m}(Q)(u\otimes\check{\xi}). \end{aligned}$$

Taking P to be the interval $[0, 1]$ and $\check{K}$ to be $\check{m}(P)$ gives the result, as $\check{m}(Q) = \check{m}(1) - \check{m}(0)$. □

25.4 Mod two gradings and cobordisms

If W is a cobordism from Y_- to Y_+, then the maps on the Floer groups $\widecheck{HM}_\bullet(Y_-) \to \widecheck{HM}_\bullet(Y_+)$ etc. obtained from a pair $(u \mid W)$ are not homogeneous in general: they do not have a well-defined degree, because the map is defined as a sum over all spin^c structures on W. However, the degree of the map *mod* 2 is well-defined. To formalize this a little, recall that we have a canonical mod 2 grading of the Floer groups (Subsection 22.4). We can ask whether the maps $\widecheck{HM}_\bullet(u \mid W; \Gamma_W)$ preserve the mod 2 grading. The answer depends on a quantity that appears in the following definition.

Definition 25.4.1. Let W be an oriented cobordism between connected (non-empty) 3-manifolds Y_- and Y_+. For such a cobordism, we define

$$\iota(W) = \frac{1}{2}\big(\chi(W) + \sigma(W) + b_1(Y_+) - b_1(Y_-)\big),$$

where χ is the Euler number and σ is the signature of the quadratic form on $I^2(W)$. ◊

We state two important properties of $\iota(W)$ as a lemma.

Lemma 25.4.2.

(i) *For any cobordism W, the quantity $\iota(W)$ is an integer (rather than a half-integer).*
(ii) *If $W = W_2 \circ W_1$ is a composite cobordism, then $\iota(W) = \iota(W_1) + \iota(W_2)$.*

Proof. The second part holds because both χ and σ are additive (the former because 3-manifolds have Euler number zero). The first part can be verified directly, using the fact that σ is equal to the dimension of I^2 mod 2, and exploiting the long exact sequence of the pair $(W, \partial W)$.

A more illuminating explanation of both parts is to reinterpret $-\iota(W)$ as the index of the operator $d^* \oplus d^+$ acting on weighted Sobolev spaces on the cylindrical-end manifold W^*,

$$d^* \oplus d^+ : e^{\delta w} L^2_1(W^*; \Lambda^1) \to e^{\delta w} L^2(W^*; \Lambda^0 \oplus \Lambda^+), \qquad (25.13)$$

where w is equal to the coordinate function t on both ends and δ is a small positive weight. □

The operator (25.13) appears as one block in the linearization of the Seiberg–Witten equations at a reducible configuration; the other block is the complex Dirac operator. So if $\mathfrak{a}_0$ and $\mathfrak{b}_0$ are reducible configurations on Y_- and Y_+ respectively, and if $\gamma \in \mathcal{C}_k^\sigma(W)$ is a configuration on the compact manifold W whose restriction to the two boundary components is $\mathfrak{a}_0$, $\mathfrak{b}_0$, then we have

$$\iota(W) = \text{index}\,(P_\gamma) \pmod 2,$$

where P_γ is the linearized operator with gauge fixing and our usual Atiyah–Patodi–Singer boundary conditions. As in the cylindrical case, we then obtain

$$\dim M_z([\mathfrak{a}], W^*, [\mathfrak{b}]) = \iota(W) + \mathrm{gr}^{(2)}(\mathfrak{a}) + \mathrm{gr}^{(2)}(\mathfrak{b}) \pmod 2$$

for arbitrary critical points $\mathfrak{a}$, $\mathfrak{b}$ and any W-path, z. When defining the map $\widecheck{HM}_\bullet(u \mid W; \Gamma_W)$ for u a cohomology class of degree d, the non-zero matrix entries are contributed by moduli spaces of dimension d. We therefore have:

Proposition 25.4.3. *For u a class in $H^d(\mathcal{B}^\sigma(W))$, and Γ_W an arbitrary W-morphism of local coefficient systems, the maps*

$$\widecheck{HM}_\bullet(u \mid W; \Gamma_W)$$
$$\widehat{HM}_\bullet(u \mid W; \Gamma_W)$$
$$\overline{HM}_\bullet(u \mid W; \Gamma_W)$$

are either even or odd with resect to the canonical mod 2 grading, according to the parity of

$$\iota(W) + d.$$

□

25.5 Exact sequences, duality, and conjugate spin^c structures

We have now completed the construction of the homomorphisms such as $\widecheck{HM}_\bullet(u \mid W; \Gamma_W)$, associated to a cobordism W between 3-manifolds, and we turn to briefly address a few formal properties. We postpone treatment of the most important property – the composition law for a composite cobordism – until the next section.

The first proposition deals with the maps i_*, j_* and p_* in the long exact sequence of Proposition 22.2.1.

Proposition 25.5.1. *Let $(u \mid W, \Gamma_W)$ be a morphism in the category* COB-LC, *from (Y_0, Γ_0) to (Y_1, Γ_1), and suppose the cohomology class u has degree d. Then in the diagram*

$$\begin{array}{ccccccccc}
\longrightarrow & \overline{HM}_\bullet(Y_0) & \xrightarrow{i_*} & \widecheck{HM}_\bullet(Y_0) & \xrightarrow{j_*} & \widehat{HM}_\bullet(Y_0) & \xrightarrow{p_*} & \overline{HM}_\bullet(Y_0) & \longrightarrow \\
& \downarrow{\scriptstyle \overline{HM}_\bullet(u \mid W)} & & \downarrow{\scriptstyle \widecheck{HM}_\bullet(u \mid W)} & & \downarrow{\scriptstyle \widehat{HM}_\bullet(u \mid W)} & & \downarrow{\scriptstyle \overline{HM}_\bullet(u \mid W)} & \\
\longrightarrow & \overline{HM}_\bullet(Y_1) & \xrightarrow{i_*} & \widecheck{HM}_\bullet(Y_1) & \xrightarrow{j_*} & \widehat{HM}_\bullet(Y_1) & \xrightarrow{p_*} & \overline{HM}_\bullet(Y_1) & \longrightarrow
\end{array}$$

the first two squares commute, while the third square commutes or anti-commutes according to the sign $(-1)^d$. (We have omitted the local coefficients from the diagram, for compactness.) The same signs apply to the dual diagram

$$\begin{array}{ccccccccc}
\longleftarrow & \overline{HM}^\bullet(Y_0) & \xleftarrow{i^*} & \widecheck{HM}^\bullet(Y_0) & \xleftarrow{j^*} & \widehat{HM}^\bullet(Y_0) & \xleftarrow{p^*} & \overline{HM}^\bullet(Y_0) & \longleftarrow \\
& \uparrow{\scriptstyle \overline{HM}^\bullet(u \mid W)} & & \uparrow{\scriptstyle \widecheck{HM}^\bullet(u \mid W)} & & \uparrow{\scriptstyle \widehat{HM}^\bullet(u \mid W)} & & \uparrow{\scriptstyle \overline{HM}^\bullet(u \mid W)} & \\
\longleftarrow & \overline{HM}^\bullet(Y_1) & \xleftarrow{i^*} & \widecheck{HM}^\bullet(Y_1) & \xleftarrow{j^*} & \widehat{HM}^\bullet(Y_1) & \xleftarrow{p^*} & \overline{HM}^\bullet(Y_1) & \longleftarrow .
\end{array}$$

Proof. The proof is a straightforward manipulation of the definitions: the equalities hold at the chain level. □

If we take W to be a cylinder $[0, 1] \times Y$, we obtain a result about the cap and cup products.

Corollary 25.5.2. *The maps*

$$i_* : \overline{HM}_\bullet(Y) \to \widecheck{HM}_\bullet(Y)$$
$$j_* : \widecheck{HM}_\bullet(Y) \to \widehat{HM}_\bullet(Y)$$
$$p_* : \widehat{HM}_\bullet(Y) \to \overline{HM}_\bullet(Y)$$

are homomorphisms of $H^*(\mathcal{B}^\sigma(Y))$-modules, up to sign: they satisfy

$$i_*(u \frown \bar{\xi}) = u \frown i_* \bar{\xi}$$
$$j_*(u \frown \check{\xi}) = u \frown j_* \check{\xi}$$
$$p_*(u \frown \hat{\xi}) = (-1)^d u \frown p_* \hat{\xi}$$

for u in $H^d(\mathcal{B}^\sigma(Y))$. *Similarly, the maps* i^*, j^* *on the Floer cohomology groups satisfy*

$$i^*(u \cup \bar{x}) = u \cup i^*\bar{x}$$
$$j^*(u \cup \check{x}) = u \cup j^*\check{x}$$
$$p^*(u \cup \hat{x}) = (-1)^d u \cup p^*\hat{x}.$$

□

Next we consider the interaction of the maps obtained from cobordisms and the duality maps ω_μ. For the special case $u = 1$, the following proposition was stated in the introductory Subsection 3.4.

Proposition 25.5.3. *Let* W *be a morphism in* COB, *from* Y_0 *to* Y_1. *Let* μ_0 *and* μ_1 *be homology orientations for* Y_0 *and* Y_1. *Let* $W^\dagger$ *be the same oriented 4-manifold, regarded as a cobordism from* $-Y_1$ *to* $-Y_0$, *and give* $W^\dagger$ *the homology orientation determined by* μ_0, μ_1 *and the given homology orientation of* W. *Then the duality maps*

$$\check{\omega}_{\mu_i} : \widecheck{HM}_\bullet(-Y_i) \to \widehat{HM}^\bullet(Y_i)$$
$$\hat{\omega}_{\mu_i} : \widehat{HM}_\bullet(-Y_i) \to \widecheck{HM}^\bullet(Y_i)$$
$$\bar{\omega}_{\mu_i} : \overline{HM}_\bullet(-Y_i) \to \overline{HM}^\bullet(Y_i)$$

satisfy

$$\check{\omega}_{\mu_0}\widecheck{HM}_\bullet(u \mid W^\dagger)\check{\omega}_{\mu_1}^{-1}(\hat{x}) = (-1)^s\widehat{HM}^\bullet(u \mid W)(\hat{x})$$
$$\hat{\omega}_{\mu_0}\widehat{HM}_\bullet(u \mid W^\dagger)\hat{\omega}_{\mu_1}^{-1}(\check{x}) = (-1)^s\widecheck{HM}^\bullet(u \mid W)(\check{x})$$
$$\bar{\omega}_{\mu_0}\overline{HM}_\bullet(u \mid W^\dagger)\bar{\omega}_{\mu_1}^{-1}(\bar{x}) = (-1)^t\overline{HM}^\bullet(u \mid W)(\bar{x}),$$

where

$$s = (\mathrm{gr}^{(2)}(x) + \iota(W))(d + \iota(W))$$
$$t = \bar{\mathrm{gr}}^{(2)}(x)(d + \iota(W)) + d\iota(W),$$

for u in $H^d(\mathcal{B}^\sigma(W)) = H^d(\mathcal{B}^\sigma(W^\dagger))$.

Proof. In each case, the proposition comes down to comparing two different orientations on a moduli space associated to the cylindrical-end manifold W^*. In the case of the cylinder, the relevant comparison is the same one that was established in Lemma 22.5.4. For a general cobordism W, the computation can be done in essentially the same way. □

Again, there is a corollary to be drawn in the case that W is a cylinder.

Corollary 25.5.4. *If μ is a homology orientation of Y, the duality maps*

$$\check{\omega}_\mu : \widecheck{HM}_\bullet(-Y) \to \widehat{HM}^\bullet(Y)$$
$$\hat{\omega}_\mu : \widehat{HM}_\bullet(-Y) \to \widecheck{HM}^\bullet(Y)$$
$$\bar{\omega}_\mu : \overline{HM}_\bullet(-Y) \to \overline{HM}^\bullet(Y)$$

satisfy

$$\check{\omega}_\mu(u \cap \check{\xi}) = (-1)^s u \cup \check{\omega}_\mu \check{\xi}$$
$$\hat{\omega}_\mu(u \cap \hat{\xi}) = (-1)^s u \cup \hat{\omega}_\mu \hat{\xi}$$
$$\bar{\omega}_\mu(u \cap \bar{\xi}) = (-1)^s u \cup \bar{\omega}_\mu \bar{\xi}$$

where

$$s = \mathrm{gr}^{(2)}(o(j))d,$$

for u in $H^d(\mathcal{B}^\sigma(Y))$ and $\check{\xi}$, $\hat{\xi}$, $\bar{\xi}$ in grading j. □

There is a different type of duality present in Floer homology. If $\mathfrak{s}$ is a spin^c structure on Y, there is a complex conjugate spin^c structure $\mathfrak{s}^*$: the corresponding spin bundle $\bar{S}$ is the conjugate of S, while Clifford multiplication and the Dirac operator are unchanged as real-linear maps. We can identify $\mathcal{B}_k^\sigma(Y, \mathfrak{s})$ with $\mathcal{B}_k^\sigma(Y, \mathfrak{s}^*)$: the critical points and trajectories are the same for $\mathfrak{s}$ and its conjugate.

There is just one point at which our construction of the Floer groups distinguishes between $\mathfrak{s}$ and $\mathfrak{s}^*$: we used the complex orientation of the kernel and cokernel of the Dirac operator to define the 2-element orientation set $\Lambda(\mathfrak{a})$, for $[\mathfrak{a}] \in \mathcal{B}_k^\sigma(Y, \mathfrak{s})$. (The complex orientations of a complex vector space V and its conjugate $\bar{V}$ are the same only if the dimension of V is even.)

Let us denote by Λ the orientation sets associated with $\mathfrak{s}$ as before, and use Λ^* for the orientation sets arising from $\mathfrak{s}^*$. For any path ζ from $[\mathfrak{a}_1]$ to $[\mathfrak{a}_2]$ in $\mathcal{B}_k^\sigma(Y, \mathfrak{s}) = \mathcal{B}_k^\sigma(Y, \mathfrak{s}^*)$, there is a canonical identification

$$\Lambda(\mathfrak{a}_1, \mathfrak{q}_1, \mathfrak{a}_2, \mathfrak{q}_2, \zeta) = \Lambda^*(\mathfrak{a}_1, \mathfrak{q}_1, \mathfrak{a}_2, \mathfrak{q}_2, \zeta). \tag{25.14}$$

This is because both sides are defined as the set of orientations of the determinant line of a single operator P_γ. We can recast this statement as saying that a path ζ joining $\mathfrak{a}_1$ to $\mathfrak{a}_2$ determines an isomorphism

$$\Lambda(\mathfrak{a}_1)\Lambda^*(\mathfrak{a}_1) \to \Lambda(\mathfrak{a}_2)\Lambda^*(\mathfrak{a}_2).$$

(We have omitted the perturbations now from the notation.) The isomorphism may depend on ζ. We therefore have a local system of groups $\Delta_{\mathfrak{s}}$, with

$$\Delta_{\mathfrak{s}}[\mathfrak{a}] = \mathbb{Z}\Lambda[\mathfrak{a}] \otimes \mathbb{Z}\Lambda^*[\mathfrak{a}].$$

There are canonical isomorphisms of chain complexes, such as

$$\check{C}(Y, \mathfrak{s}^*; \Gamma) = \check{C}(Y, \mathfrak{s}; \Gamma \otimes \Delta_{\mathfrak{s}})$$

for any system of local coefficients Γ. Hence:

Proposition 25.5.5. *For any local system* Γ, *we have canonical isomorphisms*

$$\widecheck{HM}_*(Y, \mathfrak{s}^*; \Gamma) \to \widecheck{HM}_*(Y, \mathfrak{s}; \Gamma \otimes \Delta_{\mathfrak{s}})$$
$$\widehat{HM}_*(Y, \mathfrak{s}^*; \Gamma) \to \widehat{HM}_*(Y, \mathfrak{s}; \Gamma \otimes \Delta_{\mathfrak{s}})$$
$$\overline{HM}_*(Y, \mathfrak{s}^*; \Gamma) \to \overline{HM}_*(Y, \mathfrak{s}; \Gamma \otimes \Delta_{\mathfrak{s}}).$$

These commute with the maps i_*, j_* *and* p_* *of Proposition* 22.2.1, *as well as with the cap and cup products.* □

The local system $\Delta_{\mathfrak{s}}$, which has fiber $\mathbb{Z}$, can be characterized as follows. Let $[\mathfrak{a}_0]$ be a basepoint, choose an identification of the fiber there with $\mathbb{Z}$, and let z be a homotopy class of loops based at $[\mathfrak{a}_0]$. The automorphism

$$\Delta_{\mathfrak{s}}(z) : \mathbb{Z} \to \mathbb{Z}$$

is ± 1 and is 1 if and only if the rules defined prior to Lemma 20.3.3 specify the same standard trivialization for the 2-element sets

$$\Lambda(\mathfrak{a}_0, 0, \mathfrak{a}_0, 0, z) = \Lambda^*(\mathfrak{a}_0, 0, \mathfrak{a}_0, 0, z).$$

This 2-element set is, by definition, the set of orientations of the determinant line of an operator P_γ, which has a complex Dirac operator as a diagonal block. The *real* index of this Dirac operator is given by the formula in the first part of Lemma 14.4.6: it is

$$-([u] \cup c_1(S))[Y],$$

where u is the element of $H^1(Y;\mathbb{Z})$ corresponding to the class z. The complex index is half of this. (The class $c_1(S)$ is always divisible by 2.) The complex orientation of the real determinant line is the same for $\mathfrak{s}$ and $\mathfrak{s}^*$ if and only if the complex index is even. We therefore obtain:

Lemma 25.5.6. *The local system $\Delta_{\mathfrak{s}}$ is trivial if and only if the first Chern class, $c_1(S)$, of the corresponding spin bundle is divisible by 4 (rather than just by 2) in $H^2(Y;\mathbb{Z})/\text{torsion}$.* □

Even when the local system $\Delta_{\mathfrak{s}}$ is non-trivial, the Floer homology groups for the spin^c structures $\mathfrak{s}$ and $\mathfrak{s}^*$ are actually isomorphic, though the isomorphism is not canonical:

Proposition 25.5.7. *For any local system Γ, we have isomorphisms*

$$\widecheck{HM}_*(Y,\mathfrak{s};\Gamma) \to \widecheck{HM}_*(Y,\mathfrak{s};\Gamma\otimes\Delta_{\mathfrak{s}})$$
$$\widehat{HM}_*(Y,\mathfrak{s};\Gamma) \to \widehat{HM}_*(Y,\mathfrak{s};\Gamma\otimes\Delta_{\mathfrak{s}})$$
$$\overline{HM}_*(Y,\mathfrak{s};\Gamma) \to \overline{HM}_*(Y,\mathfrak{s};\Gamma\otimes\Delta_{\mathfrak{s}}),$$

which commute with the maps i_*, j_* and p_* of Proposition 22.2.1, *but not in general with the cap product. So we have isomorphisms*

$$\widecheck{HM}_*(Y,\mathfrak{s}^*;\Gamma) \to \widecheck{HM}_*(Y,\mathfrak{s};\Gamma),$$

etc., where $\mathfrak{s}^*$ is the conjugate spin^c structure.

Proof. Consider more generally a local system Γ carrying an automorphism ϕ. Suppose that $c_1(S)$ is not torsion, so we have a non-trivial homomorphism defined by the spectral flow of the Hessian operator (see (12.4) for example):

$$\text{sf} : \pi_1(\mathcal{B}) \to \mathbb{Z}.$$

Here we write $\mathcal{B}$ for $\mathcal{B}^\sigma(Y,\mathfrak{s})$. Let $p : \mathcal{B}^1 \to \mathcal{B}$ be the infinite cyclic cover corresponding to the subgroup of $\pi_1(\mathcal{B})$ that is the kernel of sf. From Γ, ϕ and p, we can build a new local system Γ_ϕ on $\mathcal{B}$ as follows. The fiber $\Gamma_\phi[\mathfrak{a}]$ is the quotient of

$$\Gamma[\mathfrak{a}] \times p^{-1}[\mathfrak{a}]$$

by the equivalence relation generated by

$$(\gamma,x) \sim (\phi_{[\mathfrak{a}]}\gamma, \tau^{-1}x).$$

Here $\tau : \mathcal{B}^1 \to \mathcal{B}^1$ is the covering transformation corresponding to the positive generator of the image of sf. If ζ is a path from $[\mathfrak{a}]$ to $[\mathfrak{b}]$, let

$$\tau_\zeta : p^{-1}[\mathfrak{a}] \to p^{-1}[\mathfrak{b}]$$

be defined by path lifting. The map

$$\Gamma(\zeta) \times \tau_\zeta : \Gamma[\mathfrak{a}] \times p^{-1}[\mathfrak{a}] \to \Gamma[\mathfrak{b}] \times p^{-1}[\mathfrak{b}]$$

respects the equivalence relation $\sim$, and so descends to the quotient, where it defines

$$\Gamma_\phi(\zeta) : \Gamma_\phi[\mathfrak{a}] \to \Gamma_\phi[\mathfrak{b}].$$

In the special case that ϕ is -1, we recover the tensor product $\Gamma \otimes \Delta_{\mathfrak{s}}$ this way (up to isomorphism).

Let d_0 be the positive generator of the image of sf. Let $[\mathfrak{a}_0]$ be the chosen basepoint, and let x_0 be a chosen basepoint in $\mathcal{B}^1$ lying over $[\mathfrak{a}_0]$. For each critical point $[\mathfrak{a}]$, we can choose a path ζ from $[\mathfrak{a}_0]$ to $[\mathfrak{a}]$ so that the Fredholm operator P_ζ satisfies

$$\text{index } P_\zeta \in [0, d_0 - 1].$$

In this way, we obtain a preferred element $x = \tau_\zeta(x_0)$ in $p^{-1}[\mathfrak{a}]$ for each critical point $[\mathfrak{a}]$, and hence a preferred isomorphism

$$\theta : \Gamma[\mathfrak{a}] \to \Gamma_\phi[\mathfrak{a}],$$

sending γ to the equivalence class of (γ, x). Thus our choice of basepoint determines an isomorphism of groups,

$$\theta : \check{C}_j(Y, \mathfrak{s}; \Gamma) \to \check{C}_j(Y, \mathfrak{s}; \Gamma_\phi)$$

for all $j \in \mathbb{J}(Y, \mathfrak{s})$.

The map θ is not a chain map. If $\check{\partial}_j$ denotes the component of the differential

$$\check{\partial}_j : \check{C}_j \to \check{C}_{j-1},$$

then we have

$$\theta^{-1}\check{\partial}_j\theta = \begin{cases} \check{\partial}_j\phi^{-1}, & j = j_0 \\ \check{\partial}_j, & j \neq j_0. \end{cases}$$

Here $j_0 = \text{gr}[\mathfrak{a}_0] \in \mathbb{J}(Y, \mathfrak{s})$. (See the definition of $\mathbb{J}(Y, \mathfrak{s})$ above at (22.12).) Because ϕ is an automorphism of the local system Γ, the differential $\check{\partial}$ commutes with ϕ. So both the kernel and image of $\check{\partial}_j$ are the same as those of $\check{\partial}\phi^{-1}$. So although θ is not a chain map, it induces isomorphisms

$$\theta : \widecheck{HM}_j(Y, \mathfrak{s}; \Gamma) \to \widecheck{HM}_j(Y, \mathfrak{s}; \Gamma_\phi)$$

for all j.

The isomorphism θ does not, in general, respect cap products. If u is in $H^d(\mathcal{B})$ and ξ is in $\check{C}_j(Y, \mathfrak{s}; \Gamma)$, then we have

$$\theta^{-1}(u \frown \theta\xi) = \phi^{-r}(u \frown \xi),$$

where

$$r = \left|\{ e \mid 0 \le e \le d-1,\ j - e = j_0 \in \mathbb{J}(Y, \mathfrak{s}) \}\right|.$$

□

25.6 The module structure for S^3

We earlier identified the Floer groups of S^3, as groups: the description of these groups was first given in Proposition 3.3.1, and proved in Subsection 22.7. However, we have not yet identified the module structure of the Floer groups of S^3, though the result was stated without proof in Subsection 3.3: see (3.10). We shall now verify that the module structure is as stated.

Note that, because of the short exact sequence (3.9), it is enough to identify $\overline{HM}_\bullet(S^3)$ as a module over $H^*(\mathcal{B}^\sigma(S^3))$. In Subsection 3.3, the result was stated as

$$\overline{HM}_\bullet(S^3) \cong \mathbb{Z}[U_\dagger^{-1}, U_\dagger]].$$

Recall from Subsection 22.7 that with a small admissible perturbation we can arrange that the critical points in $\mathcal{B}^\sigma(S^3)$ all lie over a single point in $\mathcal{B}(S^3)$. We again label these $[\mathfrak{a}_i]$ for $i \in \mathbb{Z}$, and we identify $\mathbb{J}(S^3)$ with $\mathbb{Z}$ again in such a way that $\text{gr}[\mathfrak{a}_i] = 2i$ for $i \ge 0$. In the complex $\bar{C}(S^3)$, we then have

$$\bar{\text{gr}}[\mathfrak{a}_i] = 2i$$

for *all* i positive and negative. These are the generators of the complex, and the differential is zero.

The assertion to be proved is that

$$u_2 \frown [\mathfrak{a}_i] = [\mathfrak{a}_{i-1}] \tag{25.15}$$

for all i, where u_2 is the standard generator for the cohomology of $\mathcal{B}^\sigma(S^3) \cong \mathbb{CP}^\infty$. To compute this, we need to understand the moduli space $M([\mathfrak{a}_i], W^*, [\mathfrak{a}_{i-1}])$, where W is the cylindrical cobordism from S^3 to S^3, and W^* is obtained from W by adding cylindrical ends. Of course, W^* is just the usual cylinder $\mathbb{R} \times S^3$. There is a slight difference however. When viewing W as a cobordism, we used perturbations supported in collars of the boundary components, and extended these to be translation-invariant on the two ends; whereas the usual setup on a cylinder $\mathbb{R} \times Y$ is to consider translation-invariant perturbations. Nevertheless, it can be easily verified that there is no harm in using a translation-invariant perturbation on the entire cylinder in the case that W is cylindrical. We can therefore safely regard the cap product as being defined using the moduli space $M([\mathfrak{a}_i], [\mathfrak{a}_{i-1}])$ associated with the cylinder.

The space $M([\mathfrak{a}_i], [\mathfrak{a}_{i-1}])$ is identified in a slightly more general setting in Proposition 14.6.1, which tells us that this moduli space is $\mathbb{CP}^1 \setminus \{0, \infty\}$. More explicitly, we can consider the image of $M([\mathfrak{a}_i], [\mathfrak{a}_{i-1}])$ under the restriction map to $\mathcal{B}^\sigma(\{0\} \times S^3)$, where its image lies in the copy of $\mathbb{CP}^\infty$ lying over the unique critical point in $\mathcal{B}(S^3)$. This image is the set

$$\{\, \phi \mid \phi = a\phi_i + b\phi_{i-1},\ a, b \text{ both non-zero and } |a|^2 + |b|^2 = 1 \,\}/S^1,$$

where ϕ_{i-1} and ϕ_i are unit eigenvectors corresponding to the two eigenvalues. The compactification $\bar{M}([\mathfrak{a}_i], [\mathfrak{a}_{i-1}])$ is the copy of $\mathbb{CP}^1$ obtained by dropping the constraints that a and b are non-zero. We chose a canonical orientation of each $\Lambda([\mathfrak{a}_i])$ in Subsection 22.7, and the resulting orientation of the moduli space is the complex orientation of $\mathbb{CP}^1$. The class u_2 evaluates to 1 on $\mathbb{CP}^1$, so we have $\bar{m}(u_2 \otimes [\mathfrak{a}_i]) = [\mathfrak{a}_{i-1}]$. We therefore have

$$u_2 \frown [\mathfrak{a}_i] = [\mathfrak{a}_{i-1}]$$

for all i. In the Floer *cohomology* group we have $u_2 \smile [\mathfrak{a}_{i-1}]^\dagger = [\mathfrak{a}_i]^\dagger$.

26 Composing cobordisms

26.1 Stretching a composite cobordism

To show that the groups and homomorphisms that we have constructed define a functor from the category COB-LC (Definition 23.3.3) as asserted in

Theorem 23.3.4, it remains to check two properties. The first of these properties is elementary:

Proposition 26.1.1. *If W is the trivial, cylindrical cobordism from $(Y, g, \mathfrak{q})$ to itself and Γ_W is the identity map from Γ_Y to Γ_Y, then $\widetilde{HM}_\bullet(1 \mid W; \Gamma_W)$ etc. are the identity maps on the appropriate Floer (co)homology groups.*

Proof. Because u is 1, the homomorphism is defined using 0-dimensional moduli spaces. The 0-dimensional moduli spaces $M_z([\mathfrak{a}], W^*, [\mathfrak{b}])$ must consist of translation-invariant solutions on the infinite cylinder W^*: non-constant solutions otherwise belong to moduli spaces of dimension at least 1. In particular, the moduli space is 0-dimensional only when $[\mathfrak{a}] = [\mathfrak{b}]$ and z is trivial: it then consists of a single point, oriented with the sign $+1$ by our conventions, provided we give the cylinder its canonical homology orientation. □

The second property is much less trivial: it is the composition law, Proposition 23.2.2. We restate the proposition here, with local coefficients.

Proposition 26.1.2. *Let Y_0, Y_1 and Y_2 be 3-manifolds with metrics and admissible perturbations, and let Γ_i be a local system of R-modules on Y_i, R a commutative ring. Let W_{01}, W_{12} be cobordisms with homology orientations, from Y_0 to Y_1 and from Y_1 to Y_2, and let Γ_{01} and Γ_{12} be homomorphisms of local coefficient systems. Let cohomology classes*

$$u_{01} \in H^{d_{01}}(\mathcal{B}^\sigma(W_{01}); R)$$
$$u_{12} \in H^{d_{12}}(\mathcal{B}^\sigma(W_{12}); R)$$

be given, and let $u = u_{12}u_{01}$ be the product in $H^d(\mathcal{B}^\sigma(W))$, as defined in (23.6), where $W = W_{12} \circ W_{01}$ is the composite cobordism with homology orientation. Then we have

$$\widetilde{HM}_\bullet(u \mid W; \Gamma_W) = \widetilde{HM}_\bullet(u_{12} \mid W_{12}; \Gamma_{W_{12}}) \circ \widetilde{HM}_\bullet(u_{01} \mid W_{01}; \Gamma_{W_{01}})$$
$$\widetilde{HM}^\bullet(u \mid W; \Gamma_W) = \widetilde{HM}^\bullet(u_{01} \mid W_{01}; \Gamma_{W_{01}}) \circ \widetilde{HM}^\bullet(u_{12}, W_{12}; \Gamma_{W_{12}}),$$

with parallel formulae for $\widehat{HM}$ and $\overline{HM}$.

Remarks. In the case that W_{01} and W_{12} are both cylindrical and $Y_0 = Y_1 = Y_2$, this proposition is the associative law for the cap product, Proposition 23.2.1. For a general composite cobordism $W = W_{12} \circ W_{01}$, the statement of the proposition uses the fact that homology orientations of the two factors can be "composed" to give a homology orientation of W, as stated following Definition 3.4.1. To define this composition, we can reinterpret $\Lambda(W)$ in the same way

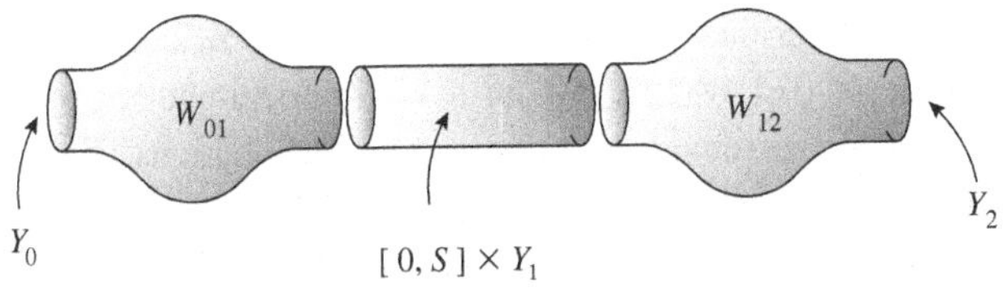

Fig. 8. The composite cobordism $W(S)$.

as we did for a manifold X with boundary in Corollary 24.8.2. The appropriate modification to state is that, for a cobordism W from Y_0 to Y_2, we can interpret $\Lambda(W)$ as the set of orientations of the determinant of an operator

$$\begin{aligned}\tilde{B} : \{\, a \in L^2_k(W; T^*W) \mid d^*a = 0, d^+a = 0, \langle a|_{\partial W}, \nu\rangle = 0 \,\}\\ \to C^+(Y_0) \oplus \big(C^- \oplus C^0\big)(Y_2) \qquad (26.1)\end{aligned}$$

given by restricting the 1-form a to the boundary and projecting onto the given spectral subspaces at the two ends. With this reinterpretation, the composition rule can be defined in the same way as for the similar situation in Definition 20.3.2.

The proof of Proposition 26.1.2 will be based on an application of Stokes' theorem to a suitable moduli space. Let $W(S)$ be the composite cobordism, with a cylinder of length S inserted in the middle (see Figure 8):

$$W(S) = W_{01} \cup \big([0, S] \times Y_1\big) \cup W_{12}. \qquad (26.2)$$

This is a cobordism diffeomorphic to W, from Y_0 to Y_2, but the metric varies with S. As S increases, the neck stretches.

Let $\mathfrak{q}^0, \mathfrak{q}^1, \mathfrak{q}^2$ be admissible tame perturbations for the three 3-manifolds Y_0, Y_1 and Y_2. Extend these to t-dependent perturbations $\hat{\mathfrak{p}}$ supported in the collars of the four components of the boundaries of W_{01} and W_{12}, as is done in (24.2). When W_{01} and W_{12} are joined together to form the composite cobordism W, the perturbations near the two copies of Y_1 match, so there is a well-defined perturbation of the equations on W, compatible with the restriction maps to the two components. Similarly on $W(S)$, we have a perturbation $\hat{\mathfrak{p}}$ which is equal to the perturbation $\hat{\mathfrak{q}}^1$ on the cylindrical piece $[0, S] \times Y_1$.

Remark. Note that this sort of perturbation on W or $W(S)$ is a little different from those that arose before, because it is no longer supported in a collar neighborhood of the boundary.

Let $[\mathfrak{a}]$ and $[\mathfrak{b}]$ be critical points for two 3-manifolds Y_0 and Y_2, and consider a moduli space $M_z([\mathfrak{a}], W(S)^*, [\mathfrak{b}])$ on the cylindrical-end manifold $W(S)^*$. As pointed out in the remark just above, our perturbations are now a little different; but we maintain the notation $M_z([\mathfrak{a}], W(S)^*, [\mathfrak{b}])$. As S varies, these form a parametrized moduli space, parametrized by $S \in [0, \infty)$:

$$\mathcal{M}_z([\mathfrak{a}], [\mathfrak{b}]) = \bigcup_{S \in [0,\infty)} \{S\} \times M_z([\mathfrak{a}], W(S)^*, [\mathfrak{b}]). \tag{26.3}$$

We can choose the perturbations $\hat{\mathfrak{p}}$ on W_{01} and W_{12} (independent of S) so as to make the parametrized moduli space regular, for all $[\mathfrak{a}]$, $[\mathfrak{b}]$ and z. This is because we can regard the total space as a fiber product, in which one of the factors is the union

$$\bigcup_{S \in [0,\infty)} \{S\} \times M([0, S] \times Y_1);$$

so we can argue as in the proof of Proposition 24.4.7. Thus:

Proposition 26.1.3. *For suitable choice of perturbation, all the moduli spaces* $\mathcal{M}_z([\mathfrak{a}], [\mathfrak{b}])$ *are smooth manifolds with boundary. The boundary is the fiber over* $S = 0$*, which is the moduli space* $M_z([\mathfrak{a}], W, [\mathfrak{b}])$. □

We can also form a union of the compactifications,

$$\bigcup_{S \geq 0} \{S\} \times M_z^+([\mathfrak{a}], W(S)^*, [\mathfrak{b}]).$$

The map to $[0, \infty)$ is proper by Theorem 24.6.7, but the total space is not compact because of the non-compactness of the base. To form a compactification, we add a fiber over $S = \infty$, which we denote by $M_z^+([\mathfrak{a}], W(\infty)^*, [\mathfrak{b}])$ and which we define as follows. An element of this space is a quintuple

$$([\breve{\gamma}_0], [\gamma_{01}], [\breve{\gamma}_1], [\gamma_{12}], [\breve{\gamma}_2]), \tag{26.4}$$

where

$$[\breve{\gamma}_0] \in \breve{M}^+([\mathfrak{a}_0], [\mathfrak{b}_0])$$
$$[\gamma_{01}] \in M([\mathfrak{b}_0], W_{01}^*, [\mathfrak{a}_1])$$
$$[\breve{\gamma}_1] \in \breve{M}^+([\mathfrak{a}_1], [\mathfrak{b}_1])$$

$$[\gamma_{12}] \in M([\mathfrak{b}_1], W_{12}^*, [\mathfrak{a}_2])$$
$$[\breve{\boldsymbol{\gamma}}_2] \in \breve{M}^+([\mathfrak{a}_2], [\mathfrak{b}_2])$$
$$[\mathfrak{a}_0] = [\mathfrak{a}]$$
$$[\mathfrak{b}_2] = [\mathfrak{b}]$$

and the homotopy classes of these elements compose to give z. The union

$$\mathcal{M}_z^+([\mathfrak{a}], [\mathfrak{b}]) = \bigcup_{S \in [0,\infty]} \{S\} \times M_z^+([\mathfrak{a}], W(S)^*, [\mathfrak{b}])$$

has a topology whose definition is similar to the topology we defined on $M_z^+(X, [\mathfrak{b}])$ in Subsection 24.6. For simplicity of exposition, we explain only what it means for a sequence

$$[\gamma_n] \in M_z([\mathfrak{a}], W(S_n)^*, [\mathfrak{b}])$$

to converge to a quintuple (26.4) in the case that $[\breve{\boldsymbol{\gamma}}_i]$ is the trivial trajectory with zero components for $i = 1, 2, 3$. So the quintuple is really just a pair

$$[\gamma_{01}] \in M_{z_1}([\mathfrak{a}], W_{01}^*, [\mathfrak{a}_1])$$
$$[\gamma_{12}] \in M_{z_2}([\mathfrak{a}_1], W_{12}^*, [\mathfrak{b}])$$

with $z_2 \circ z_1 = z$. If V is a precompact open subset of the cylindrical-end manifold W_{01}^*, then there is a canonical isometric copy of V' of V inside $W(S)^*$ for all sufficiently large S. Because of this, it makes sense to ask that $[\gamma_n]$ converges to $[\gamma_{01}]$ on compact subsets of W_{01}^*: it means that for all such V, the restriction $[\gamma_n|_{V'}]$ converges to $[\gamma_{01}|_V]$. If $[\gamma_n]$ converges on compact subsets of W_{01}^* to $[\gamma_{01}]$ and on compact subsets of W_{12}^* to $[\gamma_{12}]$, then we say that the sequence converges in $\mathcal{M}_z^+([\mathfrak{a}], [\mathfrak{b}])$.

Proposition 26.1.4. *For any* $[\mathfrak{a}]$, $[\mathfrak{b}]$ *and* z, *the moduli space* $\mathcal{M}_z^+([\mathfrak{a}], [\mathfrak{b}])$ *is compact.*

Proof. The new part of this proposition is the assertion that a sequence in $\mathcal{M}_z^+([\mathfrak{a}], [\mathfrak{b}])$ with the S coordinate going to infinity has a convergent subsequence. As usual, we may suppose that each element of this sequence is unbroken, so we have

$$[\gamma_n] \in M_z([\mathfrak{a}], W(S_n)^*, [\mathfrak{b}]),$$

with S_n increasing. As in the basic cylindrical case, the proof proceeds by first showing that, after passing to a subsequence, we have convergence of the images $\pi[\gamma_n]$, where π again is the map from the blown-up configuration space $\mathcal{B}_k^\sigma$ to $\mathcal{B}_k$. This is the counterpart of Proposition 16.2.1, and the proof is little changed from the corresponding step in the proof of the similar Proposition 24.6.4: the essential points are the additivity of the topological energy and the local compactness result, Theorem 24.5.2. The remaining step is to obtain uniform bounds on the function Λ along the trajectory.

Write $W(S_n)^*$ as the union

$$W(S_n)^* = \big((-\infty, 0] \times Y_0\big) \cup W_{01} \cup \big([0, S_n] \times Y_1\big) \cup W_{12} \cup \big([0, \infty) \times Y_2\big). \tag{26.5}$$

Let $\Lambda_0(n), \ldots, \Lambda_5(n)$ be the values of the function $\Lambda_{\mathfrak{q}}$ at various junctions in this decomposition:

$$\begin{aligned}
\Lambda_0(n) &= \Lambda_{\mathfrak{q}}(\mathfrak{a}) \\
\Lambda_1(n) &= \Lambda_{\mathfrak{q}}(\gamma_n|_{0\times Y_0}) \\
\Lambda_2(n) &= \Lambda_{\mathfrak{q}}(\gamma_n|_{0\times Y_1}) \\
\Lambda_3(n) &= \Lambda_{\mathfrak{q}}(\gamma_n|_{S_n\times Y_1}) \\
\Lambda_4(n) &= \Lambda_{\mathfrak{q}}(\gamma_n|_{0\times Y_2}) \\
\Lambda_5(n) &= \Lambda_{\mathfrak{q}}(\mathfrak{b}).
\end{aligned}$$

The differences between consecutive values,

$$\Lambda_0(n) - \Lambda_1(n), \quad \ldots \quad , \Lambda_4(n) - \Lambda_5(n),$$

are the drops in $\Lambda_{\mathfrak{q}}$ across the five pieces in the decomposition (26.5). On the cylindrical pieces, we have lower bounds

$$\left.\begin{aligned}
\Lambda_0(n) - \Lambda_1(n) &\geq -C \\
\Lambda_2(n) - \Lambda_3(n) &\geq -C \\
\Lambda_4(n) - \Lambda_5(n) &\geq -C
\end{aligned}\right\} \tag{26.6}$$

by the argument of Lemma 16.3.1. It is also the case that an upper bound on Λ "propagates" across the cobordism W_{01}: that is, if there exists C_1 such that $\Lambda_1(n) \leq C_1$ for all n, then there exists C_2 such that $\Lambda_2(n) \leq C_2$ for all n. To see this, suppose the contrary: $\Lambda_1(n) \leq C_1$ for all n while $\Lambda_2(n)$

increases without bound. Then $\Lambda_1(n)$ and $-\Lambda_2(n)$ are both bounded above, by a constant independent of n. The values of $\Lambda_{\mathfrak{q}}$ and $-\Lambda_{\mathfrak{q}}$ are then also bounded above, independent of n on collar neighborhoods of the respective boundary components in W_{01}. It then follows from Theorem 24.5.2 that we have convergence on an interior submanifold W'_{01} after passing to a subsequence. In particular $\Lambda_{\mathfrak{q}}$ is bounded above, independent of n in the subsequence, on the positive boundary component of W'_{01}, and hence also on the positive boundary component of W_{01}. This contradicts the assumption that $\Lambda_2(n)$ was increasing without bound.

The quantity $\Lambda_0(n)$ is independent of n, so using the three inequalities (26.6) and the propagation of upper bounds across the cobordisms, we obtain an upper bound on $\Lambda_i(n)$ for all n and i. Similarly we obtain a lower bound, using the fact that $\Lambda_5(n)$ is fixed too. □

There is a finiteness statement that accompanies the compactness proposition above, and which is proved in the same way: it is the analog of Lemma 25.3.1. (See also Proposition 24.6.6.) The statement is:

Lemma 26.1.5. *For any* $[\mathfrak{a}]$ *and* d_0 *and* i_0, *there are only finitely many pairs* $(z, [\mathfrak{b}])$ *such that*

(i) *the moduli space* $\mathcal{M}^+_z([\mathfrak{a}], [\mathfrak{b}])$ *is non-empty;*
(ii) *the dimension of the moduli space is at most* d_0*; and*
(iii) $\iota([\mathfrak{b}]) \geq i_0$, *where* ι *is as defined in* (16.2).

□

In addition to $\mathcal{M}^+_z([\mathfrak{a}], [\mathfrak{b}])$, we have a smaller compactification of the parametrized moduli space $\mathcal{M}_z([\mathfrak{a}], [\mathfrak{b}])$ on the cylinder, similar in spirit to the compactification $\bar{M}_z([\mathfrak{a}], [\mathfrak{b}])$ of $M_z([\mathfrak{a}], [\mathfrak{b}])$. The latter compactification was defined as a subset of $\mathcal{B}^\sigma_{k,\mathrm{loc}}(W^*)$, but could equally well have been defined as a subset of $\mathcal{B}^\sigma_k(W)$, because of the unique continuation results. For the parametrized moduli space $\mathcal{M}_z([\mathfrak{a}], [\mathfrak{b}])$, we define the compactification

$$\bar{\mathcal{M}}_z([\mathfrak{a}], [\mathfrak{b}]) \subset [0, \infty] \times \mathcal{B}^\sigma_k(W_{01}) \times \mathcal{B}^\sigma_k(W_{12}) \tag{26.7}$$

to be the image of $\mathcal{M}^+_z([\mathfrak{a}], [\mathfrak{b}])$ via a map

$$r : \mathcal{M}^+_z([\mathfrak{a}], [\mathfrak{b}]) \to [0, \infty] \times \mathcal{B}^\sigma_k(W_{01}) \times \mathcal{B}^\sigma_k(W_{12})$$

defined as follows. For finite S, an element of $\mathcal{M}^+_z$ can be written as a triple $([\breve{\boldsymbol{\gamma}}_0], [\gamma_{02}], [\breve{\boldsymbol{\gamma}}_2])$, where $[\gamma_{02}]$ is a solution on the cylindrical-end manifold

$W(S)^*$. For finite S, we define

$$r : \big(S, [\check{\gamma}_0], [\gamma_{02}], [\check{\gamma}_2]\big) \mapsto \big(S, [\gamma_{02}]|_{W_{01}}, [\gamma_{02}]|_{W_{12}}\big).$$

When S is ∞, we write a typical element as a quintuple, as in (26.4), and we define

$$r : \big(\infty, [\check{\gamma}_0], [\gamma_{01}], [\check{\gamma}_1], [\gamma_{12}], [\check{\gamma}_2]\big) \mapsto \big(\infty, [\gamma_{01}]|_{W_{01}}, [\gamma_{12}]|_{W_{12}}\big).$$

The following two propositions are adaptations of our earlier results. See, in particular, the classification of codimension-1 strata of $M_z^+([\mathfrak{a}], W^*, [\mathfrak{b}])$ given by (25.1), and the result concerning boundary orientations in Proposition 25.2.2. Our indexing of the moduli spaces has changed to suit our new circumstances, but otherwise the proofs need little modification and are again omitted.

Proposition 26.1.6. *Suppose the parametrized moduli space $\mathcal{M}_z([\mathfrak{a}], [\mathfrak{b}])$ defined at* (26.3) *contains irreducibles and has dimension $d + 1$. Then the compactification $\mathcal{M}_z^+([\mathfrak{a}], [\mathfrak{b}])$ is a space stratified by manifolds and has $\mathcal{M}_z([\mathfrak{a}], [\mathfrak{b}])$ as its top stratum. In the fiber over $S = \infty$, the strata of dimension d are the top strata of the following pieces of $M_z^+([\mathfrak{a}], W(\infty)^*, [\mathfrak{b}])$:*

(i) $M_{01} \times M_{12}$,
(ii) $\check{M}_0 \times M_{01} \times M_{12}$,
(iii) $M_{01} \times \check{M}_1 \times M_{12}$,
(iv) $M_{01} \times M_{12} \times \check{M}_2$,

where M_{ij} denotes a typical moduli space on W_{ij} and $\check{M}_i$ denotes a typical unparametrized moduli space on $\mathbb{R} \times Y_i$. In the last three cases, the middle moduli space of the three is boundary-obstructed . The moduli space has a codimension-1 δ-structure along the strata in the last three cases, and is a manifold with boundary along the stratum in the first case. □

Choose trivializations of $\Lambda(W)$, $\Lambda([\mathfrak{a}])$ and $\Lambda([\mathfrak{b}])$. This orients $M_z([\mathfrak{a}], W(S)^*, [\mathfrak{b}])$ for all S, including $S = \infty$. Contrary to our usual convention, we orient the parametrized moduli space $\mathcal{M}_z([\mathfrak{a}], [\mathfrak{b}])$ by putting the S coordinate *first*.

Proposition 26.1.7. *The canonical orientations of the four types of strata in the previous proposition differ from their boundary orientations by the signs*

(i) 1,
(ii) $(-1)^{d_0+d_{01}+1}$,

(iii) $(-1)^{d_1}$,
(iv) (-1)

accordingly, where d_0 *and* d_1 *are the dimensions of the parametrized moduli spaces* M_0 *and* M_1*, and* d_{01} *is the dimension of* M_{01}. □

26.2 Proof of the composition law

We now turn to the proof of Proposition 26.1.2. Because of the unique continuation property, each moduli space $M_z([\mathfrak{a}], W_{01}^*, [\mathfrak{b}])$ is embedded in $\mathcal{B}_k^\sigma(W_{01})$ by the restriction map. Fix d_0, and let $\mathcal{U}_{01}$ be an open cover of $\mathcal{B}_k^\sigma(W_{01})$, transverse to all moduli spaces $M_z([\mathfrak{a}], W_{01}, [\mathfrak{b}])$ of dimension at most d_0. Let $\mathcal{U}_{12}$ be a similar open cover for $\mathcal{B}_k^\sigma(W_{12})$. We may suppose that the cohomology classes in the proposition are given by Čech cocycles, also called u_{01} and u_{12}, carried by these open covers:

$$u_{01} \in C^{d_{01}}(\mathcal{U}_{01}; R)$$
$$u_{12} \in C^{d_{12}}(\mathcal{U}_{12}; R).$$

Let $\mathcal{U}_{01} \times \mathcal{U}_{12}$ be the product open cover on the product space, and let $\mathcal{V}$ be an open cover of

$$[0, \infty] \times \mathcal{B}_k^\sigma(W_{01}) \times \mathcal{B}_k^\sigma(W_{12}) \tag{26.8}$$

that refines the open cover $[0, \infty] \times \mathcal{U}_{01} \times \mathcal{U}_{12}$ and is transverse to all strata in the subset $\bar{\mathcal{M}}_z([\mathfrak{a}], [\mathfrak{b}])$ defined in (26.7), whenever this moduli space has dimension $d_0 + 1$ or less. Let

$$\mathcal{B}_k^\sigma(W)^\circ \subset \mathcal{B}_k^\sigma(W)$$

be the domain on which the restriction maps to both W_{01} and W_{12} are defined. Let $\mathcal{U}_{02}^\circ$ be the pull-back of $\mathcal{V}$ to this domain via the map

$$\mathcal{B}_k^\sigma(W)^\circ \to \{0\} \times \mathcal{B}_k^\sigma(W_{01}) \times \mathcal{B}_k^\sigma(W_{12}),$$

and let

$$u_{02}^\circ \in C^{d_{01}+d_{12}}(\mathcal{U}_{02}^\circ; R)$$

be the resulting representative for the pull-back of the cocycle $u_{01} \times u_{12}$. At the level of cohomology, u_{02}° represents the class $u = u_{12}u_{01}$ as defined in the proposition. What this construction has provided is an "external" product

$$\begin{aligned} C^{d_{01}}(\mathcal{U}_{01}; R) \otimes C^{d_{12}}(\mathcal{U}_{12}; R) &\to C^{d_{01}+d_{12}}(\mathcal{V}; R) \\ u_{01} \otimes u_{12} &\mapsto u_{01} \times u_{12} \end{aligned}$$

and an "internal" product

$$\begin{aligned} c : C^{d_{01}}(\mathcal{U}_{01}; R) \otimes C^{d_{12}}(\mathcal{U}_{12}; R) &\to C^{d_{01}+d_{12}}(\mathcal{U}^{\circ}_{02}; R) \\ u_{01} \otimes u_{12} &\mapsto u^{\circ}_{02}. \end{aligned} \tag{26.9}$$

Note that the moduli spaces $M_z([\mathfrak{a}], W^*, [\mathfrak{b}])$ are embedded in $\mathcal{B}^{\sigma}_k(W)^{\circ}$. When $S = 0$, the space $W(0)$ is just W, and the fiber of $\bar{\mathcal{M}}_z([\mathfrak{a}], [\mathfrak{b}])$ over $S = 0$ is a copy of $\bar{M}_z([\mathfrak{a}], W^*, [\mathfrak{b}])$, embedded in the product space (26.8). (See the remark on page 537, however, concerning the perturbations.) The transversality hypothesis on $\mathcal{V}$ therefore implies that the open cover $\mathcal{U}^{\circ}_{02}$ is transverse to the moduli spaces $M_z([\mathfrak{a}], W^*, [\mathfrak{b}])$.

Proof of Proposition 26.1.2. We regard $\mathcal{M}_z([\mathfrak{a}], [\mathfrak{b}])$ as a subset of the product space

$$[0, \infty] \times \mathcal{B}^{\sigma}_k(W_{01}) \times \mathcal{B}^{\sigma}_k(W_{12})$$

using the inclusion (26.7). When $d = d_{01} + d_{12}$ is less than or equal to d_0, we have a well-defined evaluation

$$\big\langle u_{01} \times u_{12}, \mathcal{M}_z([\mathfrak{a}], [\mathfrak{b}]) \big\rangle \in \mathrm{Hom}(R\Lambda[\mathfrak{a}], R\Lambda[\mathfrak{b}])$$

whenever $\dim M_z([\mathfrak{a}], W^*, [\mathfrak{b}])$ is $d - 1$. Using this, we define maps

$$\left.\begin{aligned} K^o_o &: C^{d_{01}}(\mathcal{U}_{01}) \otimes C^{d_{12}}(\mathcal{U}_{12}) \otimes C^o_\bullet(Y_0; \Gamma_0) \to C^o_\bullet(Y_2; \Gamma_2) \\ K^o_s &: C^{d_{01}}(\mathcal{U}_{01}) \otimes C^{d_{12}}(\mathcal{U}_{12}) \otimes C^o_\bullet(Y_0; \Gamma_0) \to C^s_\bullet(Y_2; \Gamma_2) \\ K^u_o &: C^{d_{01}}(\mathcal{U}_{01}) \otimes C^{d_{12}}(\mathcal{U}_{12}) \otimes C^u_\bullet(Y_0; \Gamma_0) \to C^o_\bullet(Y_2; \Gamma_2) \\ K^u_s &: C^{d_{01}}(\mathcal{U}_{01}) \otimes C^{d_{12}}(\mathcal{U}_{12}) \otimes C^u_\bullet(Y_0; \Gamma_0) \to C^s_\bullet(Y_2; \Gamma_2) \end{aligned}\right\} \tag{26.10}$$

for $d_{01} + d_{12} \le d_0$, by

$$\left.\begin{aligned}
K^o_o(u_{01} \otimes u_{12} \otimes -) &= \sum_{\substack{[\mathfrak{a}]\in\mathfrak{C}^o(Y_0)\\ [\mathfrak{b}]\in\mathfrak{C}^o(Y_2)}} \sum_z \bigl\langle u_{01} \times u_{12}, \mathcal{M}_z([\mathfrak{a}],[\mathfrak{b}])\bigr\rangle \Gamma_{02}(z),\\
K^o_s(u_{01} \otimes u_{12} \otimes -) &= \sum_{\substack{[\mathfrak{a}]\in\mathfrak{C}^o(Y_0)\\ [\mathfrak{b}]\in\mathfrak{C}^s(Y_2)}} \sum_z \bigl\langle u_{01} \times u_{12}, \mathcal{M}_z([\mathfrak{a}],[\mathfrak{b}])\bigr\rangle \Gamma_{02}(z),\\
K^u_o(u_{01} \otimes u_{12} \otimes -) &= \sum_{\substack{[\mathfrak{a}]\in\mathfrak{C}^u(Y_0)\\ [\mathfrak{b}]\in\mathfrak{C}^o(Y_2)}} \sum_z \bigl\langle u_{01} \times u_{12}, \mathcal{M}_z([\mathfrak{a}],[\mathfrak{b}])\bigr\rangle \Gamma_{02}(z),\\
K^u_s(u_{01} \otimes u_{12} \otimes -) &= \sum_{\substack{[\mathfrak{a}]\in\mathfrak{C}^u(Y_0)\\ [\mathfrak{b}]\in\mathfrak{C}^s(Y_2)}} \sum_z \bigl\langle u_{01} \times u_{12}, \mathcal{M}_z([\mathfrak{a}],[\mathfrak{b}])\bigr\rangle \Gamma_{02}(z).
\end{aligned}\right\} \tag{26.11}$$

(The ring of coefficients is understood to be R.) The non-zero contributions to these sums come from moduli spaces $\mathcal{M}_z([\mathfrak{a}],[\mathfrak{b}])$ of dimension $d = d_{01}+d_{12}$. We also have an operator

$$\bar{K} : C^{d_{01}}(\mathcal{U}_{01}) \otimes C^{d_{12}}(\mathcal{U}_{12}) \otimes \bar{C}_\bullet(Y_0;\Gamma_0) \to \bar{C}_\bullet(Y_2;\Gamma_2)$$
$$\bar{K} = \begin{bmatrix} \bar{K}^s_s & \bar{K}^u_s \\ \bar{K}^s_u & \bar{K}^u_u \end{bmatrix},$$

where

$$\bar{K}^s_s(u_{01} \otimes u_{12} \otimes -) = \sum_{\substack{[\mathfrak{a}]\in\mathfrak{C}^s(Y_0)\\ [\mathfrak{b}]\in\mathfrak{C}^s(Y_2)}} \sum_z \bigl\langle u_{01} \times u_{12}, \mathcal{M}^{\text{red}}_z([\mathfrak{a}],[\mathfrak{b}])\bigr\rangle \Gamma_{02}(z)$$

$$\begin{aligned}\bar{K}^u_u(u_{01} \otimes u_{12} \otimes -) &= (-1)^{d-1}\\
&\quad \times \sum_{\substack{[\mathfrak{a}]\in\mathfrak{C}^u(Y_0)\\ [\mathfrak{b}]\in\mathfrak{C}^u(Y_2)}} \sum_z \bigl\langle u_{01} \times u_{12}, \mathcal{M}^{\text{red}}_z([\mathfrak{a}],[\mathfrak{b}])\bigr\rangle \Gamma_{02}(z)\end{aligned}$$

$$\bar{K}^s_u(u_{01} \otimes u_{12} \otimes -) = \sum_{\substack{[\mathfrak{a}]\in\mathfrak{C}^s(Y_0)\\ [\mathfrak{b}]\in\mathfrak{C}^u(Y_2)}} \sum_z \bigl\langle u_{01} \times u_{12}, \mathcal{M}^{\text{red}}_z([\mathfrak{a}],[\mathfrak{b}])\bigr\rangle \Gamma_{02}(z)$$

$$\bar{K}^u_s(u_{01} \otimes u_{12} \otimes -) = \sum_{\substack{[\mathfrak{a}]\in\mathfrak{C}^u(Y_0)\\ [\mathfrak{b}]\in\mathfrak{C}^s(Y_2)}} \sum_z \bigl\langle u_{01} \times u_{12}, \mathcal{M}^{\text{red}}_z([\mathfrak{a}],[\mathfrak{b}])\bigr\rangle \Gamma_{02}(z).$$

Definition 26.2.1. Let $\tau : C^a(\mathcal{U}_{01}) \otimes C^b(\mathcal{U}_{12}) \to C^b(\mathcal{U}_{12}) \otimes C^a(\mathcal{U}_{01})$ be the operator interchanging the two factors. Let σ_{01}, σ_{12} and σ be the sign operators on $C^*(\mathcal{U}_{01}) \otimes C^*(\mathcal{U}_{12})$ given by

$$\begin{aligned}\sigma_{01} &= (-1)^a\\ \sigma_{12} &= (-1)^b\\ \sigma &= (-1)^{a+b}\end{aligned}$$

on $C^a(\mathcal{U}_{01}) \otimes C^b(\mathcal{U}_{12})$. We define an operator

$$\check{K} : C^{d_{01}}(\mathcal{U}_{01}) \otimes C^{d_{12}}(\mathcal{U}_{12}) \otimes \check{C}_\bullet(Y_0; \Gamma_0) \to \check{C}_\bullet(Y_2; \Gamma_2)$$
$$\check{K} = \begin{bmatrix} K^o_o & -K^u_o\bar{\partial}^s_u - m^u_o\bar{m}^s_u\tau\sigma_{01} - \partial^u_o\bar{K}^s_u \\ K^o_s & \bar{K}^s_s - K^u_s\bar{\partial}^s_u - m^u_s\bar{m}^s_u\tau\sigma_{01} - \partial^u_s\bar{K}^s_u \end{bmatrix}. \tag{26.12}$$

In this formula, the term $K^u_o\bar{\partial}^s_u$ (for example) stands for $K^u_o \circ (1 \otimes 1 \otimes \bar{\partial}^s_u)$, where the operator $\bar{\partial}^s_u$ belongs to Y_0; and $m^u_o\bar{m}^s_u$ stands for $m^u_o \circ (1 \otimes \bar{m}^s_u)$, where $\bar{m}^s_u$ and m^u_o are the operators of Subsection 25.3, defined using the open covers $\mathcal{U}_{01}$ and $\mathcal{U}_{12}$ and the cobordisms W_{01} and W_{12} respectively. ◊

Remark. The need for τ in the above definition is related again to our wish to have operators acting on the left. The term $\bar{m}^s_u m^o_s \tau$, for example acts as

$$\bar{m}^s_u m^o_s \tau(u_{01} \otimes u_{12} \otimes \xi) = \bar{m}^s_u(u_{12} \otimes m^o_s(u_{01} \otimes \xi)),$$

where the inner m^o_s belongs to the "first" factor W_{01} of the composite cobordism and the outer $\bar{m}^s_u$ belongs to W_{12}.

With suitable signs, the operator $\check{K}$ provides a chain-homotopy which will establish the composition laws for $\widetilde{HM}_\bullet(Y)$. The next lemma expresses the chain-homotopy formula. In the statement, the operators $\check{m}_{01}$, $\check{m}_{12}$ and $\check{m}_{02}$ are the operators of Definition 25.3.3, corresponding to the cobordisms W_{01}, W_{12} and W, using the open covers $\mathcal{U}_{01}$, $\mathcal{U}_{12}$ and $\mathcal{U}^\circ_{02}$. (The fact that the last of these open covers only covers the subset $\mathcal{B}^\sigma_k(W)^\circ$ is of no consequence.) There is a slight modification, however, because in the case of $\check{m}_{02}$ we have in mind an operator defined by the moduli spaces on the composite cobordism W^*, equipped with perturbations as described in the remark on page 537. This is an inessential matter, because the resulting map $\check{m}_{02}$ will be chain-homotopic to the usual chain map, defined using perturbations supported near the boundary components in W_{02}.

Lemma 26.2.2. *The operator $\check{K}$ satisfies the identity*

$$\begin{aligned} &-\check{K}(\underline{\delta}\otimes 1)-\check{K}(1\otimes 1\otimes\check{\partial})-\check{\partial}\check{K}(\sigma\otimes 1)\\ &-\check{m}_{02}(c\otimes 1)+\check{m}_{12}(1\otimes\check{m}_{01})(\tau\otimes 1)=0,\end{aligned}$$

in which $\underline{\delta} : C^*(\mathcal{U}_{01})\otimes C^*(\mathcal{U}_{12})\to C^*(\mathcal{U}_{01})\otimes C^*(\mathcal{U}_{12})$ is the coboundary on the tensor product complex,

$$\begin{aligned}\underline{\delta}&=(\delta\otimes 1)+(1\otimes\delta)\sigma_{01}\\ &=\delta_{01}+\delta_{12}\sigma_{01},\end{aligned}$$

and the operator c is the product operator (26.9).

Proof. In the lemma, we have written out the identity in full, writing $(1\otimes\check{m}_{01})$, for example, rather than just $\check{m}$. In more abbreviated style, consistent with Definition 26.2.1, we can write the identity as

$$-\check{K}\underline{\delta}-\check{K}\check{\partial}-\check{\partial}\check{K}\sigma-\check{m}c+\check{m}\check{m}\tau=0, \tag{26.13}$$

there being no potential ambiguity about which cobordism each $\check{m}$ belongs to. For example, writing the left-hand side out as a 2-by-2 matrix, the top right entry is, in full,

$$\begin{aligned} &K^u_o\underline{\delta}\bar{\partial}^s_u-m^u_o\bar{m}^s_u\tau\delta_{01}\sigma_{01}+m^u_o\bar{m}^s_u\tau\delta_{12}+\partial^u_o\bar{K}^s_u\underline{\delta}\\ &+K^o_o\partial^u_o\bar{\partial}^s_u+K^u_o\bar{\partial}^s_u\bar{\partial}^s_s+m^u_o\bar{m}^s_u\tau\bar{\partial}^s_s\sigma_{01}+\partial^u_o\bar{K}^s_u\bar{\partial}^s_s\\ &-K^u_o\bar{\partial}^s_s\partial^u_s\bar{\partial}^s_u-m^u_o\bar{m}^s_u\tau\partial^u_s\bar{\partial}^s_u\sigma_{01}-\partial^u_o\bar{K}^s_u\partial^u_s\bar{\partial}^s_u\\ &+\partial^o_oK^u_o\bar{\partial}^s_u\sigma+\partial^o_om^u_o\bar{m}^s_u\tau\sigma_{12}+\partial^o_o\partial^u_o\bar{K}^s_u\sigma\\ &+\partial^u_o\bar{\partial}^s_u\bar{K}^s_s\sigma-\partial^u_o\bar{\partial}^s_uK^u_s\bar{\partial}^s_u\sigma-\partial^u_o\bar{\partial}^s_um^u_s\bar{m}^s_u\tau\sigma_{12}-\partial^u_o\bar{\partial}^s_u\partial^u_s\bar{K}^s_u\sigma\\ &+m^u_oc\bar{\partial}^s_u+\partial^u_o\bar{m}^s_uc\\ &-m^o_om^u_o\tau\bar{\partial}^s_u-m^o_o\partial^u_o\bar{m}^s_u\tau-m^u_o\bar{\partial}^s_u\bar{m}^s_s\tau-\partial^u_o\bar{m}^s_u\bar{m}^s_s\tau\\ &+m^u_o\bar{\partial}^s_um^u_s\bar{\partial}^s_u\tau+\partial^u_o\bar{m}^s_um^u_s\bar{\partial}^s_u\tau+m^u_o\bar{\partial}^s_u\partial^u_s\bar{m}^s_u\tau+\partial^u_o\bar{m}^s_u\partial^u_s\bar{m}^s_u\tau.\end{aligned}$$

We have used the fact that σ_{01} and δ_{01} anti-commute. The fact that this long expression vanishes (along with its companions in the other three corners of the 2-by-2 matrix) is deduced from the following two lemmas, by the same sort of manipulations as Proposition 25.3.4 is deduced from Lemma 25.3.6 and Lemma 25.3.7. Since these manipulations are both lengthy and straightforward, we omit them, and content ourselves with indicating the proof of the first of the two lemmas below. □

Lemma 26.2.3. *We have the following identities:*

(i) $-K^o_o\partial^o_o - \partial^o_o K^o_o\sigma + K^u_o\bar{\partial}^s_u\partial^o_s + \partial^u_o\bar{K}^s_u\partial^o_s + \partial^u_o\bar{\partial}^s_u K^o_s\sigma - m^o_o c + m^o_o m^o_o\tau + m^u_o\bar{m}^s_u\partial^o_s\tau\sigma_{01} - m^u_o\bar{\partial}^s_u m^o_s\tau - \partial^u_o\bar{m}^s_u m^o_s\tau = K^o_o(\underline{\delta}\otimes 1)$.

(ii) $-K^o_s\partial^o_o - \partial^o_s K^o_o\sigma - \bar{K}^s_s\partial^o_s - \bar{\partial}^s_s K^o_s\sigma + K^u_s\bar{\partial}^s_u\partial^o_s + \partial^u_s\bar{K}^s_u\partial^o_s + \partial^u_s\bar{\partial}^s_u K^o_s\sigma - m^o_s c + m^o_s m^o_o\tau + \bar{m}^s_s m^o_s\tau + m^u_s\bar{m}^s_u\partial^o_s\tau\sigma_{01} - m^u_s\bar{\partial}^s_u m^o_s\tau - \partial^u_s\bar{m}^s_u m^o_s\tau = K^o_s(\underline{\delta}\otimes 1)$.

(iii) $-K^o_o\partial^u_o - \partial^o_o K^u_o\sigma + K^u_o\bar{\partial}^u_u + \partial^u_o\bar{K}^u_u + K^u_o\bar{\partial}^s_u\partial^u_s + \partial^u_o\bar{K}^s_u\partial^u_s + \partial^u_o\bar{\partial}^s_u K^u_s\sigma - m^u_o c + m^o_o m^u_o\tau + m^u_o\bar{m}^u_u\tau\sigma_{01} + m^u_o\bar{m}^s_u\partial^u_s\tau\sigma_{01} - m^u_o\bar{\partial}^s_u m^u_s\tau - \partial^u_o\bar{m}^s_u m^u_s\tau = K^u_o(\underline{\delta}\otimes 1)$.

(iv) $-\bar{K}^u_s - K^o_s\partial^u_o - \partial^o_s K^u_o\sigma - \bar{K}^s_s\partial^u_s - \bar{\partial}^s_s K^u_s\sigma + K^u_s\bar{\partial}^u_u + \partial^u_s\bar{K}^u_u + K^u_s\bar{\partial}^s_u\partial^u_s + \partial^u_s\bar{K}^s_u\partial^u_s + \partial^u_s\bar{\partial}^s_u K^u_s\sigma - m^u_s c + m^o_s m^u_o\tau + \bar{m}^s_s m^u_s\tau + m^u_s\bar{m}^u_u\tau\sigma_{01} + m^u_s\bar{m}^s_u\partial^u_s\tau\sigma_{01} - m^u_s\bar{\partial}^s_u m^u_s\tau - \partial^u_s\bar{m}^s_u m^u_s\tau = K^u_s(\underline{\delta}\otimes 1)$.

Lemma 26.2.4. *The analog of* (26.13) *holds for* $\bar{K}$. *We have:*

$$-\bar{K}\underline{\delta} - \bar{K}\bar{\partial} - \bar{\partial}\bar{K}\sigma - \bar{m}c + \bar{m}\bar{m}\tau = 0.$$

□

Proof of Lemma 26.2.3. Each of these identities is proved by applying Stokes' theorem to the evaluation of $\delta(u_{01}\times u_{12})$ on moduli spaces $\bar{\mathcal{M}}_z([\mathfrak{a}],[\mathfrak{b}])$ containing irreducibles. As in Lemmas 22.1.5 and 25.3.6, the four parts correspond to the four possibilities for $[\mathfrak{a}]$ and $[\mathfrak{b}]$.

Let us just examine the first identity, which is an equality between two operators

$$C^{d_{01}}(\mathcal{U}_{01};R)\otimes C^{d_{12}}(\mathcal{U}_{12};R)\otimes C^o(Y_0;\Gamma_0)\to C^o(Y_2;\Gamma_2).$$

Let $[\mathfrak{a}]$, $[\mathfrak{b}]$ and z be such that $\mathrm{gr}_z([\mathfrak{a}],W,[\mathfrak{b}])$ is $d = d_{01}+d_{12}$. For each such $[\mathfrak{a}]$, $[\mathfrak{b}]$ and z, the two operators in the identity have a contribution

$$C^{d_{01}}(\mathcal{U}_{01};R)\otimes C^{d_{12}}(\mathcal{U}_{12};R)\to \mathrm{Hom}(R\Lambda[\mathfrak{a}]\otimes\Gamma_0[\mathfrak{a}], R\Lambda[\mathfrak{b}]\otimes\Gamma_2[\mathfrak{b}]),$$

or simply

$$C^{d_{01}}(\mathcal{U}_{01};R)\otimes C^{d_{12}}(\mathcal{U}_{12};R)\to \mathrm{Hom}(\Gamma_0[\mathfrak{a}],\Gamma_2[\mathfrak{b}]),$$

if we trivialize $\Lambda[\mathfrak{a}]$ and $\Lambda[\mathfrak{b}]$. Because evaluation commutes with refinement, this component of the right-hand side of the first identity is the map

$$u_{01}\otimes u_{12}\mapsto \big\langle \delta(u_{01}\times u_{12}), \mathcal{M}_z([\mathfrak{a}],[\mathfrak{b}])\big\rangle\, \Gamma_{02}(z),$$

from the definition of K^o_o. We use Stokes' theorem (21.4) to re-express this as

$$u_{01} \otimes u_{12} \mapsto \sum_{\beta} \delta_\beta \langle u_1 \times u_2, N^d_\beta \rangle \Gamma_{02}(z),$$

where the sum is over all components of the codimension-1 stratum of $\bar{\mathcal{M}}_z([\mathfrak{a}], [\mathfrak{b}])$, and δ_β are the boundary multiplicities. As in the proof of Lemma 25.3.6, we can pull back to the larger compactification, in this case to $\mathcal{M}^+_z([\mathfrak{a}], [\mathfrak{b}])$, and sum over components $(N^d)^+_\beta$ of the codimension-1 stratum of $\mathcal{M}^+_z([\mathfrak{a}], [\mathfrak{b}])$ whose image is d-dimensional in $\bar{\mathcal{M}}_z([\mathfrak{a}], [\mathfrak{b}])$. Thus we can rewrite the expression as

$$u_{01} \otimes u_{12} \mapsto \sum_{\beta} \delta^+_\beta \langle u_1 \times u_2, (N^d)^+_\beta \rangle \Gamma_{02}(z).$$

We will identify this operator with the corresponding component of the operator on the left-hand side of the identity.

The next task is to enumerate the relevant contributions to the codimension-1 stratum $(N^d)^+$. First, there is the fiber over $S = 0$, which is a copy of the moduli space associated to the composite cobordism $W = W(0)$:

$$\{0\} \times M_z([\mathfrak{a}], W^*, [\mathfrak{b}]). \tag{26.14}$$

Next there are contributions lying over $0 < S < \infty$, which take the form of parametrized moduli spaces

$$\bigcup_{S \in R} \{S\} \times N^{d-1}(S), \tag{26.15}$$

where $N^{d-1}(S) \subset M^+_z([\mathfrak{a}], W(S)^*, [\mathfrak{b}])$ is a codimension-1 stratum of one of the first five types enumerated in Proposition 25.1.1. Finally, there are the contributions from the strata over $S = \infty$, which are described in Proposition 26.1.6.

We now explain how each of these types of stratum contributes to the left-hand side of the first identity in the lemma, and the sign with which it contributes. The stratum (26.14) occurs with multiplicity -1, because $\{0\}$ is minus the boundary of $[0, \infty)$; and this stratum contributes the term $m^o_o c$ to the identity. The five possibilities listed in (25.1) for the types of components that arise in (26.15) contribute the first five terms of the identity. The sign with which the stratum (26.15) appears in the boundary of $\mathcal{M}^+_z([\mathfrak{a}], [\mathfrak{b}])$ is the opposite sign to that with which the corresponding stratum N^{d-1} appears in $M^+_z([\mathfrak{a}], W^*, [\mathfrak{b}])$; these signs were the ones that appeared in Lemma 25.3.6.

The remaining terms in the first identity are the four terms that have two ms in them. These terms arise from codimension-1 strata in $\mathcal{M}_z^+([\mathfrak{a}],[\mathfrak{b}])$ over $S=\infty$, which are (in the notation of Proposition 26.1.6)

$$\{\infty\} \times M_{01} \times M_{12}$$

$$\{\infty\} \times \breve{M}_0 \times M_{01} \times M_{12}$$

$$\{\infty\} \times M_{01} \times \breve{M}_1 \times M_{12}$$

$$\{\infty\} \times M_{01} \times M_{12} \times \breve{M}_2.$$

Such strata have dimension d in the smaller compactification $\bar{\mathcal{M}}_z([\mathfrak{a}],[\mathfrak{b}])$ only if the unparametrized moduli spaces $\breve{M}_i$ on the cylindrical pieces are zero-dimensional. The signs with which these occur in the boundary are given in Proposition 26.1.7. The quantities d_0 and d_1 there are now 1, so that the signs are respectively 1, $(-1)^{d_{01}}$, -1, -1, which accounts for the signs of the remaining four terms. □

We can now complete the proof of Proposition 26.1.2. Let u_{01} and u_{12} be Čech cocycles and $\breve{\xi}$ a cycle in the complex $\check{C}_\bullet(Y_0;\Gamma_0)$. Let

$$u_{02} = c(u_{01} \otimes u_{12})$$

as in (26.9). Then the chain-homotopy formula of Lemma 26.2.2 tells us

$$-\breve{m}_{02}(u_{02} \otimes \breve{\xi}) + \breve{m}_{12}(u_{12} \otimes \breve{m}_{01}(u_{01} \otimes \breve{\xi})) = (-1)^d \breve{\partial}\breve{K}(u_{01} \otimes u_{12} \otimes \breve{\xi}),$$

so the two terms on the left are homologous in $\check{C}_\bullet(Y_2;\Gamma_2)$. At the level of homology then, we have

$$\breve{m}_{02}(u_{02} \otimes \breve{\xi}) = \breve{m}_{12}(u_{12} \otimes \breve{m}_{01}(u_{01} \otimes \breve{\xi})).$$

The cocycle u_{02} represents the class of $(-1)^{d_{01}d_{12}}u$, where $u = u_{12}u_{01}$ is the product class defined in (23.6). Recalling also the signs in Definition 25.3.5, we can write the last equality as

$$\begin{aligned}&\widecheck{HM}_\bullet(u \mid W_{02};\Gamma_{02})\\&\quad = \widecheck{HM}_\bullet(u_{12} \mid W_{12};\Gamma_{12})\widecheck{HM}_\bullet(u_{01} \mid W_{01};\Gamma_{01}).\end{aligned}$$

This establishes the composition law for $\widecheck{HM}_\bullet$. The composition law for $\widehat{HM}_\bullet$ can be proved by constructing a similar chain-homotopy $\hat{K}$, or it can be deduced from the long exact sequence of Proposition 22.2.1, the naturality given in Proposition 25.5.1 and the five-lemma. The composition laws for the operators on the Floer *co*-homology follow by duality. □

27 Closed four-manifolds

27.1 Invariants of closed four-manifolds

We now revisit the monopole invariants $\mathfrak{m}(X, \mathfrak{s}_X)$ discussed in Section 1, with the aim of establishing the results of Subsection 3.6. If X is a closed, oriented, connected 4-manifold equipped with a Riemannian metric and spin^c structure $\mathfrak{s}_X$, we used $N(X, \mathfrak{s}_X)$ in Section 1 to denote the moduli space of solutions of the (possibly perturbed) Seiberg–Witten equations:

$$N(X, \mathfrak{s}_X) \subset \mathcal{B}(X, \mathfrak{s}_X).$$

Although it is not essential to the definition of $\mathfrak{m}(X, \mathfrak{s}_X)$, it is more consistent with our present approach to instead consider the moduli space $M(X, \mathfrak{s}_X)$ of solutions to the Seiberg–Witten equations in the blown-up configuration space, defined as

$$M(X, \mathfrak{s}_X) = \{ [\gamma] \in \mathcal{B}^\sigma_k(X, \mathfrak{s}_X) \mid \mathfrak{F}^\sigma(\gamma) = 0 \},$$

where

$$\mathfrak{F}^\sigma(A, s, \phi) = \left(\frac{1}{2}\rho(F^+_{A^t}) - s^2(\phi\phi^*)_0,\ D^+_A\phi \right), \tag{27.1}$$

as in (6.5). (Compare Definition 24.2.1.) Recall here that ϕ is a section of $S^+ \to X$ of unit L^2 norm and $s \in [0, \infty)$. Given an imaginary-valued 2-form ω on X, we can also consider the perturbed equation $\mathfrak{F}^\sigma_\omega(A, s, \phi) = 0$, where

$$\mathfrak{F}^\sigma_\omega(A, s, \phi) = \left(\frac{1}{2}\rho(F^+_{A^t} - 4\omega^+) - s^2(\phi\phi^*)_0,\ D^+_A\phi \right). \tag{27.2}$$

(Compare with (1.10) in Section 1.) As we have done previously, we shall omit the perturbation from our notation and continue to write $M(X, \mathfrak{s}_X)$ for the moduli space of solutions to the perturbed equations. We also have the larger moduli space

$$\tilde{M}(X, \mathfrak{s}_X) \subset \tilde{\mathcal{B}}^\sigma_k(X, \mathfrak{s}_X)$$

obtained by dropping the condition $s \geq 0$.

The following is the counterpart of the regularity theorem for $N(X, \mathfrak{s}_X)$ which we stated (without proof) as Proposition 1.4.2. (We are now working with Sobolev completions rather than spaces of smooth sections.)

Lemma 27.1.1. *For a residual set of perturbing 2-forms $\omega \in L^2_{k-1}(X; i\Lambda^2(X))$, the section $\mathfrak{F}^\sigma_\omega$ is transverse to zero and the corresponding moduli space $\tilde{M}(X, \mathfrak{s}_X)$ is a smooth, compact manifold of dimension*

$$d = \frac{1}{4}\left(c_1^2(S_X^+)[X] - 2\chi(X) - 3\sigma(X)\right).$$

For such perturbations, the moduli space $M(X, \mathfrak{s}_X)$ is a smooth manifold with (possibly empty) boundary, and can be identified with the quotient of $\tilde{M}(X, \mathfrak{s}_X)$ by the involution $s \mapsto -s$.

Proof. Consider first the locus of solutions of $D^+_A\phi = 0$, as a subset of $\tilde{\mathcal{B}}^\sigma_k(X, \mathfrak{s}_X)$. This subset $\tilde{\mathcal{Z}} \subset \tilde{\mathcal{B}}^\sigma_k(X, \mathfrak{s}_X)$ is a Hilbert submanifold. To verify this, consider the linearization of the equation $D^+_A\phi = 0$ at a solution in $\mathcal{C}^\sigma_k(X, \mathfrak{s}_X)$, which yields the operator

$$Q : (b, \psi) \mapsto \rho(b)\phi + D^+_A\psi.$$

If $\eta \in L^2_{k-1}(S^-)$ is L^2-orthogonal to the image of Q, then by considering just variations $(0, \psi)$, we see that $D^-_A\eta = 0$. If η is not zero, then the unique continuation results imply that η cannot vanish on an open set. The same observation applies to ψ, so there is an open set on which both η and ψ are non-zero. We can then find on this open set a form b such that $\rho(b)\psi$ is proportional pointwise to η. This shows that η cannot be orthogonal to the image of Q unless it is zero. Elliptic theory then tells us that Q is surjective, and that $\tilde{\mathcal{Z}}$ is a submanifold, as claimed.

Next, we can describe $\tilde{M}(X, \mathfrak{s}_X)$ as the fiber over $2\rho(\omega^+)$ of the map

$$\varpi : \tilde{\mathcal{Z}} \to L^2_{k-1}(X; i\,\mathfrak{su}(S^+))$$

given by

$$(A, s, \phi) \mapsto \frac{1}{2}\rho(F^+_{A^t}) - s^2(\phi\phi^*)_0.$$

This map is Fredholm, so for a residual set of ω^+ the fiber $\tilde{M}(X, \mathfrak{s}_X)$ is regular and its dimension is the index of the Fredholm map.

To calculate the index, we note that the tangent space to $\tilde{\mathcal{Z}}$ at (A, s, ϕ) can be identified with

$$\{ (b, t, \psi) \mid D_A^+ \psi + \rho(b)\phi = 0, \langle \phi, \psi \rangle_{L^2} = 0, -d^* b + ist \operatorname{Re}\langle i\phi, \psi \rangle = 0 \},$$

where the first condition is the linearization of the Dirac equation, and the remaining conditions define the tangent space of $\mathcal{B}_k^\sigma(X)$, via the Coulomb slice conditions of Definition 9.3.7. Dropping the zeroth-order terms, we see that the index of the linearization of ϖ is equal to the index of the Fredholm operator

$$\varpi' : \{ (b, t, \psi) \mid \langle \phi, \psi \rangle_{L^2} = 0 \} \to L^2_{k-1}(X; S^- \oplus i\mathbb{R} \oplus i\Lambda^+)$$

given by

$$\varpi'(b, t, \psi) = (D_A^+ \psi, d^* b, d^+ b).$$

This index in turn is $i_1 + i_2$, where i_1 and i_2 are the (real) indices of the operators

$$\begin{aligned} D_A^+ &: L^2_k(X; S^+) \to L^2_{k-1}(X; S^-) \\ d^* \oplus d^+ &: L^2_k(X; iT^*X) \to L^2_{k-1}(X; i\mathbb{R} \oplus i\Lambda^+). \end{aligned} \tag{27.3}$$

The complex index of the Dirac operator is given in (1.7), and i_1 is twice this. The index i_2 is

$$b_1(X) - b_0(X) - b^+(X) = \frac{1}{2}\big(\chi(X) + \sigma(X)\big).$$

The sum of these two gives the result stated.

The fact that $M(X, \mathfrak{s}_X)$ is a quotient of $\tilde{M}(X, \mathfrak{s}_X)$ by the involution is an immediate consequence of the fact that the equations are invariant under $s \mapsto -s$. □

The statement and proof of the lemma above are valid for any X. But the lemma hides the fact that the boundary of the moduli space may be empty for topological reasons:

Lemma 27.1.2. *If $b^+(X)$ is positive, then there is a residual set of perturbing 2-forms ω satisfying the conditions of Lemma* 27.1.1 *and such that, in addition, the moduli space $M(X, \mathfrak{s}_X)$ has empty boundary.*

Proof. This is a restatement of Lemma 1.4.3: the equations have no solutions with $s = 0$ unless ω satisfies

$$4\int_X \omega \wedge \kappa = -(2\pi i c_1(S^+) \cup [\kappa])[X]$$

for all self-dual harmonic forms $\kappa \in \mathcal{H}^+$. The residual set of the lemma is the intersection of the residual set of the previous lemma with the complement of an affine linear subspace defined by this last condition. □

Remark. When the boundary is empty (that is, when there are no reducible solutions), the moduli spaces $N(X, \mathfrak{s}_X)$ and $M(X, \mathfrak{s}_X)$ are diffeomorphic, via the blow-down map $\pi : \mathcal{B}^\sigma_k(X, \mathfrak{s}_X) \to \mathcal{B}_k(X, \mathfrak{s}_X)$.

When transversality holds, we also have a finiteness result, the specialization of Proposition 24.6.5 to the case of a closed manifold:

Lemma 27.1.3. *Suppose the perturbing 2-form ω is chosen so that the moduli spaces $M(X, \mathfrak{s}_X)$ are regular for all $\mathfrak{s}_X$. Then there are only finitely many spin^c structures $\mathfrak{s}_X$ for which the moduli space $M(X, \mathfrak{s}_X)$ is non-empty.* □

As in Subsection 24.4, we can also consider a family of Riemannian metrics g^p on X, parametrized by a smooth manifold P, perhaps with non-empty boundary $\partial P = Q$. Given a family of forms ω_p for p in P, we form the $M(X, \mathfrak{s}_X)_P$ as the union over P of the corresponding moduli spaces. With the same slight abuse of notation as was explained in the remark preceding Definition 24.4.9, we can write

$$M(X, \mathfrak{s}_X)_P \subset P \times \mathcal{B}^\sigma_k(X, \mathfrak{s}_X).$$

Lemma 27.1.4. *Suppose $b^+(X) > \dim P$ and let g^p be a family of metrics parametrized by a smooth manifold P with boundary Q. Suppose ω_q is a family of perturbing 2-forms, defined for $q \in Q$, and suppose that the parametrized moduli space $M(X, \mathfrak{s}_X)_Q$ is regular and contains no reducible solutions. Then there is a family of forms ω_p, extending the given family to all of P, such that $M(X, \mathfrak{s}_X)_P$ is again regular and contains no reducible solutions. In this case $M(X, \mathfrak{s}_X)_P$ is a manifold with boundary $M(X, \mathfrak{s}_X)_Q$.*

Proof. This generalization of the previous lemmas is proved in the same manner as Proposition 24.4.10. □

As a particular case, if $b^+(X) > 1$, we can take P to be an interval. The conclusion of the lemma then provides a cobordism between the moduli spaces corresponding to the two points of the boundary Q.

Corollary 27.1.5. *Suppose $b^+(X) > 1$. Let g_0 and g_1 be two Riemannian metrics on X, and let ω_0 and ω_1 be imaginary-valued 2-forms such that the corresponding moduli spaces $M(X, \mathfrak{s}_X)_0$ and $M(X, \mathfrak{s}_X)_1$ are regular and have no reducibles. Then these moduli spaces are cobordant in $\mathcal{B}^\sigma_k(X, \mathfrak{s}_X)$.* □

A choice of homology orientation μ_X for X determines an orientation for the moduli space $M(X, \mathfrak{s}_X)$, using the same conventions as were laid out in the proof of Proposition 24.8.3. The cobordism in the corollary above is an oriented cobordism.

Once the moduli space is oriented and without boundary, it has a fundamental class; and given a cohomology class $u \in H^d(\mathcal{B}^\sigma(X, \mathfrak{s}_X))$, there is a pairing

$$\langle u, [M(X, \mathfrak{s}_X)] \rangle \in \mathbb{Z}. \tag{27.4}$$

Because of the corollary, this integer is independent of the choice of metric and perturbation if $b^+(X) > 1$. The following definition generalizes our previous definition of $\mathfrak{m}(u_2^e \mid X, \mathfrak{s}_X)$ slightly (see Definition 1.5.4), because we now replace u_2^e by an arbitrary cohomology class in $\boldsymbol{\mathcal{B}}^\sigma(X)$

Definition 27.1.6. Let X be a closed, connected, oriented 4-manifold X with $b^+(X) > 1$, equipped with a homology orientation. The *monopole invariant* of X for the spin^c structure $\mathfrak{s}_X$ is the map

$$\mathfrak{m}(- \mid X, \mathfrak{s}_X) : H^*(\mathcal{B}^\sigma(X, \mathfrak{s}_X)) \to \mathbb{Z}$$

defined by

$$\mathfrak{m}(u \mid X, \mathfrak{s}_X) = \langle u, [M(X, \mathfrak{s}_X)] \rangle.$$

◊

For a given X, this function of u is zero for all but finitely many $\mathfrak{s}_X$, by Lemma 27.1.3. We again write $\boldsymbol{\mathcal{B}}^\sigma(X)$ for the union $\bigcup_{\mathfrak{s}_X} \mathcal{B}^\sigma(X, \mathfrak{s}_X)$.

Definition 27.1.7. Let R be a commutative ring R. Then given any $R^\times$-valued map Γ_X on the set of spin^c structures, and any $u \in H^*(\boldsymbol{\mathcal{B}}^\sigma(X); R)$, we write

$$\mathfrak{m}(u \mid X, \Gamma_X) = \sum_{\mathfrak{s}_X} \Gamma_X(\mathfrak{s}_X) \mathfrak{m}(u \mid X, \mathfrak{s}_X).$$

As a special case, when $\Gamma_X = 1$, we write

$$\mathfrak{m}(u \mid X) = \sum_{\mathfrak{s}_X} \mathfrak{m}(u \mid X, \mathfrak{s}_X).$$

◊

Remark. Note that an $R^\times$-valued function Γ_X on the set of spin^c structures is essentially the same thing as a W-morphism Γ_W, where W is the cobordism from S^3 to S^3 obtained by removing two balls from X and we take the trivial rank-1 local coefficient system on both copies of $\mathcal{B}^\sigma(S^3)$. The example to have in mind corresponds to the W-morphism Γ_ν corresponding to a closed 2-form ν, as in (23.8). Given a real 2-cycle ν on X, or simply a real homology class $h = [\nu]$, we can define a function Γ_ν on spin^c structures on X by

$$\Gamma_\nu(\mathfrak{s}_X) = \exp\langle c_1(\mathfrak{s}_X), h\rangle. \tag{27.5}$$

In this case, the above definition becomes

$$\mathfrak{m}(u \,|\, X, \Gamma_\nu) = \sum_{\mathfrak{s}_X} \mathfrak{m}(u \,|\, X, \mathfrak{s}_X) \exp\langle c_1(\mathfrak{s}_X), h\rangle.$$

This is the function of $h \in H_2(X; \mathbb{R})$ that we called $\mathfrak{m}(u \,|\, X, h)$ in Subsection 3.8: see (3.25) and (3.28).

27.2 Reducible solutions on cobordisms

A key point in the definition of $\mathfrak{m}(X \mid \mathfrak{s}_X)$ is the absence of reducible solutions (for a generic perturbation) when $b^+(X) \geq 1$, and the absence of reducible solutions in 1-parameter families of perturbations when $b^+(X) \geq 2$. We now turn to consider the analog of these statements for solutions on cobordisms.

Let W be an oriented cobordism between connected, oriented 3-manifolds Y_- and Y_+. As usual we equip W with a Riemannian metric which is cylindrical in a collar neighborhood of both boundary components, and we write W^* for the cylindrical-end manifold obtained by attaching $(-\infty, 0] \times Y_-$ and $[0, \infty) \times Y_+$ to W.

Let L be a line bundle on W^* whose restriction to the ends either is trivial, or more generally has torsion first Chern class. We examine the equation $F_A^+ = \omega^+$ for a connection A in L with square-integrable curvature. The condition on the first Chern class means that $c_1(L)$ belongs to the subspace $I^2(W^*) \subset H^2(W^*; \mathbb{R})$, the image of the compactly supported cohomology.

Lemma 27.2.1. *Let L be a complex line bundle on W^* whose Chern class restricts to a torsion class on the two ends. Let A be a connection on L with square-integrable curvature. Let κ be a square-integrable closed, self-dual 2-form on W^*. Then*

$$\int_{W^*} F_A^+ \wedge \kappa = -2\pi i \langle c_1(L), [\kappa] \rangle$$

where the pairing on the right is the non-degenerate cup pairing on $I^2(W^*)$.

Proof. For convenience, we can replace L by L^k for some k, so that L is topologically trivial on the ends. If the connection A is a trivial connection on the ends of the manifold, with respect to some trivialization of the bundle, then $(i/2\pi)F_A$ is a compactly supported 2-form representing $c_1(L)$ as an element of $I^2(W^*) \subset H^2(W^*;\mathbb{R})$, and the result is clear.

Consider next the case that A is not trivial but flat on the ends. The curvature 2-form $(i/2\pi)F_A$ is still compactly supported, and its image in $I^2(W^*)$ still represents $c_1(L)$: as the flat connection on the ends is changed by the addition of a closed 1-form θ, the class of F_A in $H^2_c(W^*; i\mathbb{R}) \cong H^2(W, \partial W; i\mathbb{R})$ changes by the image $[\theta] \in H^1(\partial W; i\mathbb{R})$ under the coboundary map.

Finally, if F_A is merely L^2 but not compactly supported, the result can be proved by making suitable cut-offs. For any $\epsilon > 0$ we can find cylinders $[0,1] \times Y_\pm$ contained in the ends and flat connections $A_\pm$ on these cylinders with $\|A - A_\pm\|_{L^2_1} \le \epsilon$. We can then use a standard cut-off function to replace A by a connection A' that is flat on the ends. The difference $F_A - F_{A'}$ will converge to zero in L^2 norm as ϵ goes to zero. □

Remark. The space $I^2(W^*)$, the image of the compactly supported cohomology in the ordinary cohomology, appeared in the guise of $I^2(W)$, the image of $H^2(W;\partial W;\mathbb{R})$ in $H^2(W;\mathbb{R})$ in Subsection 24.8.

Lemma 27.2.2. *The closed, self-dual square-integrable 2-forms on W^* represent a maximal positive-definite subspace $I^+(W^*) \subset I^2(W^*)$ for the non-degenerate cup-pairing.*

Proof. This follows from the related result, that the square-integrable harmonic forms represent $I^2(W^*)$. A proof is contained in [8]. The statement can be compared with the related result, Lemma 24.8.1. □

Recall that we write $b^+(W)$ for the dimension of a maximal positive-definite subspace of $I^2(W)$ (or equivalently $I^2(W^*)$).

Corollary 27.2.3. *If $b^+(W)$ is non-zero, then exists a compactly supported imaginary-valued 2-form ω on W^* such that there exists no connection A with L^2 curvature in any line bundle $L \to W^*$ satisfying the equation $F_A^+ = \omega^+$.* □

We now return to the Seiberg–Witten equations. Fix metrics again on $Y_\pm$ and W, and let $\mathfrak{q}_\pm$ be admissible perturbations for $Y_\pm$. On the cylindrical-end manifold W^* we take a perturbation $\hat{\mathfrak{p}}$ defined as in (24.2), and we then have the perturbed equations $\mathfrak{F}^\sigma_{\hat{\mathfrak{p}}}(\gamma) = 0$ defining the moduli space $M([\mathfrak{a}], W^*, \mathfrak{s}_W, [\mathfrak{b}])$ for critical points $[\mathfrak{a}]$, $[\mathfrak{b}]$ on $Y_\pm$, as in Definition 24.2.1. In the interior of the compact part W, away from the collar regions, the equations are the equations $\mathfrak{F}^\sigma(A, s, \phi) = 0$, where $\mathfrak{F}^\sigma$ is as in Equation (27.1). At this point we observe

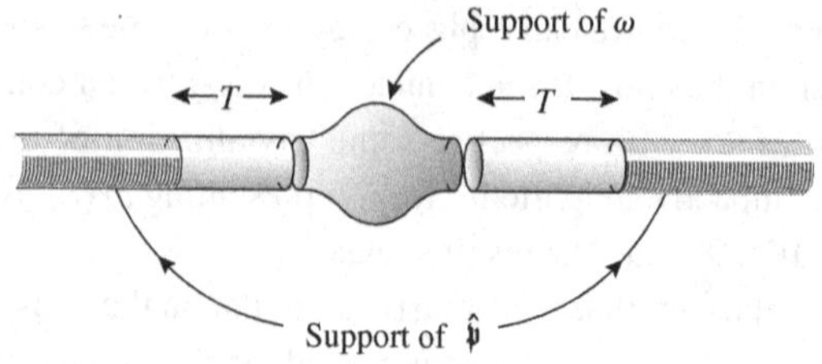

Fig. 9. Supports of ω and $\hat{\mathfrak{p}}$ separated by cylinders of length T in $W(T)$.

that we can perturb the equations $\mathfrak{F}^{\sigma}_{\hat{\mathfrak{p}}} = 0$ further by introducing a 2-form ω supported in the interior of W and replacing the equations over W with the equation from (27.2), as we previously did in the case of a closed manifold. We write the modified equations on W^* as $\mathfrak{F}^{\sigma}_{\hat{\mathfrak{p}},\omega}(\gamma) = 0$, and we continue to denote the space of solutions by $M([\mathfrak{a}], W^*, \mathfrak{s}_W, [\mathfrak{b}])$.

In W, we take it that there are cylindrical regions $I \times Y_{\pm}$ disjoint from the support of both ω and $\hat{\mathfrak{p}}$. Let g_T be a Riemannian metric on W obtained from the original metric by increasing the length of both of these cylinders by a quantity $T > 0$. Let $W(T)$ denote the manifold W equipped with this metric, and $W(T)^*$ the corresponding cylindrical-end manifold. Of course, $W(T)^*$ is isometric to W^*; but on $W(T)^*$ the support of ω and the support of $\hat{\mathfrak{p}}$ are separated by two long cylinders. (See Figure 9.) We examine the equations $\mathfrak{F}^{\sigma}_{\hat{\mathfrak{p}},\omega}(\gamma) = 0$ for the metric g_T with T large, and the corresponding moduli space $M_z([\mathfrak{a}], W(T)^*, \mathfrak{s}_W, [\mathfrak{b}])$ belonging to a particular component z.

Proposition 27.2.4. *Suppose that $b^+(W) > 0$, and choose ω so that*

$$4 \int_{W^*} \omega \wedge \kappa \neq (-2\pi i)\langle c_1(\mathfrak{s}_X), [\kappa]\rangle$$

for at least one closed, self-dual, square-integrable 2-form κ on W^*. Then there exists T_0 such that for all $T \geq T_0$, there are no reducible solutions in the moduli space $M_z([\mathfrak{a}], W(T)^*, \mathfrak{s}_W, [\mathfrak{b}])$ defined by the equations $\mathfrak{F}^{\sigma}_{\hat{\mathfrak{p}},\omega} = 0$.

Proof. Suppose the moduli space contains reducibles for all sufficiently large T, and let $(A(T), 0, \phi(T))$ represent a reducible element. Identify all the manifolds $W(T)^*$ isometrically with W^*. After passing to a subsequence and applying gauge transformations, the connections $A(T)$ will converge on compact subsets of W^* to a connection A with $F_A^+ = 4\omega^+$. The curvature F_A will be square-integrable, by the same sort of energy arguments used in Proposition 24.6.4 and Proposition 26.1.4, and we obtain a contradiction to Lemma 27.2.1. □

Remark. If the moduli spaces are regular, we can choose T_0 to be independent of $\mathfrak{s}_W, [\mathfrak{a}], [\mathfrak{b}]$ and z. This can be seen from the finiteness result, Proposition 24.6.5, combined with the observation that the existence of reducible trajectories is an issue which does not involve the blow-up picture, $\mathcal{B}^\sigma(Y)$: before blowing up, there are only finitely many critical points $[\alpha]$ in $\mathcal{B}(Y)$.

We can also deal with a family of Riemannian metrics and perturbations. Let P be a compact manifold with boundary, and let g^P be a family of Riemannian metrics on W parametrized by P, all of which are isometric to each other in cylindrical regions near $Y_\pm$. We obtain a family of metrics on the cylindrical-end manifold W^*, and we denote by $\mathcal{H}^+(W^*, g^P)$ the space of closed, self-dual, square-integrable, real-valued 2-forms κ on W^* for the given metric. Simple transversality gives us:

Lemma 27.2.5. *Suppose $b^+(W) > \dim P$. Then there exists a family of compactly supported 2-forms ω_p on W with the following transversality condition: for all $p \in P$ there exists κ_p in $\mathcal{H}^+(W^*, g^p)$ such that*

$$4 \int_{W^*} \omega_p \wedge \kappa_p \neq (-2\pi i)\langle c_1(\mathfrak{s}_X), [\kappa_p] \rangle.$$

If ω_q is previously specified for q in $Q = \partial P$, then ω_p can be chosen so as to agree with the given family on the boundary. □

We can now state a version of Proposition 27.2.4 for the family of metrics g^P. Let $g^P(T)$ be obtained by increasing the neck length, as before, and let $M([\mathfrak{a}], W^*(T), \mathfrak{s}_W, [\mathfrak{b}])_P$ denote the corresponding parametrized moduli space.

Proposition 27.2.6. *Suppose that ω_p is chosen for all $p \in P$ so that the conclusion of the above lemma holds. Then there exists T_0 such that for all $T \geq T_0$, there are no reducible solutions in the parametrized moduli space $M_z([\mathfrak{a}], W(T)^*, \mathfrak{s}_W, [\mathfrak{b}])_P$.* □

27.3 Definition of $\overrightarrow{HM}$

Let X be a closed, connected, oriented 4-manifold, and let W be the cobordism from S^3 to S^3 obtained by removing two balls from X. Note that $b^+(W) =$

$b^+(X)$. The cobordism W induces maps

$$\widecheck{HM}_\bullet(u \mid W) : \widecheck{HM}_\bullet(S^3, \mathfrak{s}) \to \widecheck{HM}_\bullet(S^3, \mathfrak{s})$$
$$\widehat{HM}_\bullet(u \mid W) : \widehat{HM}_\bullet(S^3, \mathfrak{s}) \to \widehat{HM}_\bullet(S^3, \mathfrak{s})$$
$$\overline{HM}_\bullet(u \mid W) : \overline{HM}_\bullet(S^3, \mathfrak{s}) \to \overline{HM}_\bullet(S^3, \mathfrak{s})$$

as given in Subsection 25.3 and we might hope that these maps determine $\mathfrak{m}(u \mid X)$. This is not the case. To examine these maps, we take the round metric on S^3 and recall from Subsection 22.7 that there are then no irreducible critical points for S^3, that the reducible critical points in $\mathcal{B}^\sigma(S^3, \mathfrak{s})$ all lie over a unique critical point in $\mathcal{B}(S^3, \mathfrak{s})$, and that the differentials involved in the construction of the Floer homology are all zero. From Definition 25.3.3 we can then see that in this case the map $\check{m}$ is given by the matrix

$$\begin{bmatrix} 0 & 0 \\ 0 & \bar{m}^s_s \end{bmatrix}$$

so the map only involves reducible solutions on the cobordism W. As we saw in the previous subsection, if $b^+(X) > 0$ then we can arrange that there are no reducible solutions, so the map $\widecheck{HM}_\bullet(u \mid W)$ is zero. Similar remarks apply to $\widehat{HM}$ and $\overline{HM}$.

The invariant $\mathfrak{m}(u \mid X)$ involves the irreducible solutions of the Seiberg–Witten equations on X. We should expect that these are related to the irreducible solutions on the cylindrical-end manifold W^*; and because there are no irreducible critical points for S^3, these irreducible solutions should be asymptotic to critical points in $\mathfrak{C}^u$ on the incoming end of W^* and $\mathfrak{C}^s$ on the outgoing end of W^*. In Equations (25.4) we showed how a cobordism defines a map $m^u_s(u \otimes -) : C^u(S^3) \to C^s(S^3)$, using just these irreducible solutions on W^*. We shall see that when $b^+(X) \geq 2$ the map on homology induced by m^u_s is independent of both the metric and perturbation on W and that the induced map determines $\mathfrak{m}(u \mid X)$ in a simple way. Indeed, the map we shall obtain from m^u_s is the map

$$\overrightarrow{HM}_\bullet(u \mid W) : \widehat{HM}_\bullet(S^3) \to \widecheck{HM}_\bullet(S^3)$$

promised in Subsection 3.5; and its relationship to $\mathfrak{m}$ is stated in the equation (3.18). We shall now construct $\overrightarrow{HM}_\bullet(u \mid W)$ for cobordisms between general 3-manifolds.

Let

$$(u, W, \Gamma_W) : (Y_-, \Gamma_-) \to (Y_+, \Gamma_+)$$

be a morphism in the category COB-LC, and suppose that $b^+(W) \geq 1$. From the results above, we know that we can choose a 2-form ω, compactly supported in the interior of W, and a suitable Riemannian metric, so that the moduli spaces $M([\mathfrak{a}], W^*, [\mathfrak{b}])$ defined using the equations $\mathfrak{F}^{\sigma}_{\hat{\mathfrak{p}},\omega}$ contain no reducible solutions. The four components $\bar{m}^s_s(u \otimes -)$ etc. of the map $\bar{m}$, defined at (25.5), are all zero, because they involve only the reducible parts of the moduli spaces.

In the same spirit as Definition 25.3.3, we make the following definition. It is understood here that $\mathcal{U}$ is an open cover transverse to all the moduli spaces of dimension $d \leq d_0$.

Definition 27.3.1. We define

$$\vec{m} : C^d(\mathcal{U}; R) \otimes \hat{C}_\bullet(Y_-; \Gamma_-) \to \check{C}_\bullet(Y_+; \Gamma_+),$$

for $d \leq d_0$, by the formula

$$\vec{m} = \begin{bmatrix} m^o_o & m^u_o \\ m^o_s & m^u_s \end{bmatrix}.$$

◊

Proposition 27.3.2. *We have the identity*

$$(-1)^d \check{\partial}(Y_+)\vec{m}(u \otimes \hat{\xi}) = -\vec{m}(\delta u \otimes \hat{\xi}) + \vec{m}(u \otimes \hat{\partial}(Y_-)\hat{\xi}).$$

Hence the map $\vec{m}$ on chains descends to the level of homology to define a map

$$\vec{m} : H^d(\mathcal{U}; R) \otimes \widehat{HM}_\bullet(Y_-; \Gamma_-) \to \widecheck{HM}_\bullet(Y_+; \Gamma_+).$$

Proof. The four matrix components of this identity are (after an overall change of sign, and writing σ again for the operator $(-1)^d$ on $C^d(\mathcal{U}; R)$):

(i) $m^o_o \partial^o_o - \partial^o_o m^o_o \sigma - m^u_o \bar{\partial}^s_u \partial^o_s + \partial^u_o \bar{\partial}^s_u m^o_s \sigma = m^o_o(\delta \otimes 1)$;
(ii) $m^o_s \partial^o_o - \partial^o_s m^o_o \sigma - \bar{\partial}^s_s m^o_s \sigma - m^u_s \bar{\partial}^s_u \partial^o_s + \partial^u_s \bar{\partial}^s_u m^o_s \sigma = m^o_s(\delta \otimes 1)$;
(iii) $m^o_o \partial^u_o - \partial^o_o m^u_o \sigma - m^u_o \bar{\partial}^u_u - m^u_o \bar{\partial}^s_u \partial^u_s + \partial^u_o \bar{\partial}^s_u m^u_s \sigma = m^u_o(\delta \otimes 1)$;
(iv) $m^o_s \partial^u_o - \partial^o_s m^u_o \sigma - \bar{\partial}^s_s m^u_s \sigma - m^u_s \bar{\partial}^u_u - m^u_s \bar{\partial}^s_u \partial^u_s + \partial^u_s \bar{\partial}^s_u m^u_s \sigma = m^u_s(\delta \otimes 1)$.

These are the same identities as appear in Lemma 25.3.6, without the terms involving $\bar{m}^s_s$, $\bar{m}^u_s$, $\bar{m}^s_u$ and $\bar{m}^u_u$, all of which are zero in the present situation. □

As in Subsection 25.3, we can take a limit over transverse open covers to obtain a map

$$\vec{m} : H^d(\mathcal{B}^\sigma(W); R) \otimes \widehat{HM}_\bullet(Y_-;\Gamma_-) \to \widecheck{HM}_\bullet(Y_+;\Gamma_+).$$

To within a sign change, this is the map $\overrightarrow{HM}$ that we seek to define.

Definition 27.3.3. Suppose $b^+(W) \geq 2$. Let u be an element of $H^d(\mathcal{B}^\sigma(W); R)$, and let μ_W be a homology orientation of W. Let $\Gamma_W : \Gamma_- \to \Gamma_+$ be a morphism of local coefficient systems. Choose a Riemannian metric g and perturbing 2-form satisfying the hypothesis of Proposition 27.2.4, and let T_0 be as in the conclusion of that proposition. Then we define

$$\overrightarrow{HM}_\bullet(u \mid W; \Gamma_W)_{(g,\omega)} : \widehat{HM}_\bullet(Y_-;\Gamma_-) \to \widecheck{HM}_\bullet(Y_+;\Gamma_+)$$

as the operator

$$\vec{m}(u \otimes -),$$

where $\vec{m}$ is calculated using a metric $g(T)$ with $T \geq T_0$, so that there are no reducible solutions on W. The dual map $\check{m}^*(u \otimes -)$ defines a homomorphism between the Floer cohomology groups in the same way:

$$\overrightarrow{HM}^\bullet(u \mid W; \Gamma_W)_{(g,\omega)} : \widecheck{HM}^\bullet(Y_+;\Gamma_+) \to \widehat{HM}^\bullet(Y_-;\Gamma_-).$$

◊

The maps $\overrightarrow{HM}^\bullet$ and $\overrightarrow{HM}_\bullet$ are related by duality, along the same lines as Proposition 25.5.3. Thus, if $W^\dagger$ denotes the same oriented manifold as W, but now regarded as a cobordism from $-Y_-$ to $-Y_+$, then

$$\check{\omega}_{\mu_0} \overrightarrow{HM}_\bullet(u \mid W^\dagger) \hat{\omega}_{\mu_1}^{-1}(\hat{x}) = (-1)^s \overrightarrow{HM}^\bullet(u \mid W)(\hat{x}),$$

for $\hat{x}$ in $\widehat{HM}^\bullet(Y_+)$. The sign $(-1)^s$ is the same as appears in Proposition 25.5.3.

When $b^+(W) = 1$, the map $\overrightarrow{HM}_\bullet(u \mid W; \Gamma_W)_{(g,\omega)}$ may depend on g and ω. But when $b^+ \geq 2$, they do not:

Proposition 27.3.4. *If $b^+(W) \geq 2$, the map $\overrightarrow{HM}^\bullet(u \mid W; \Gamma_W)_{(g,\omega)}$ does not depend on g and ω.*

Proof. We suppose given (g_0, ω_0) and (g_1, ω_1) satisfying the hypothesis of Proposition 27.2.4. By Lemma 27.2.5, we can find a family of metrics g_t and forms ω_t parametrized by the interval $[0, 1]$, so that by Proposition 27.2.6 we can then find T_0 such that the moduli spaces on W defined by the metrics $g_t(T)$ with perturbing 2-forms ω_t contain no reducibles (for all $t \in [0, 1]$ and $T \geq T_0$).

We now follow the proof of Proposition 25.3.8. The parametrized moduli spaces $M([\mathfrak{a}], W, [\mathfrak{b}])_{[0,1]}$, defined using the metric $g(T)$ for large T, contain no irreducibles, and define operators $m^o_o([0, 1])$, $m^o_s([0, 1])$ etc. (The absence of reducibles in the family means that the operators $\bar{m}^s_s([0, 1])$ etc. are zero.) These we combine to define an operator

$$\vec{K} : C^d(\mathcal{U}; R) \otimes \hat{C}_\bullet(Y_-; \Gamma_-) \to \check{C}_\bullet(Y_+; \Gamma_+)$$

by the same recipe that defined $\vec{m}$, but using the parametrized moduli spaces:

$$\vec{K} = \begin{bmatrix} m^o_o([0, 1]) & m^u_o([0, 1]) \\ m^o_s([0, 1]) & m^u_s([0, 1]) \end{bmatrix}.$$

This operator satisfies an identity of the same shape as the identity satisfied by $\vec{m}$, but with an additional pair of terms arising from the boundary of the $P = [0, 1]$:

$$\begin{aligned}(-1)^d \check{\partial} \vec{K}(u \otimes \hat{\xi}) = &-\vec{K}(\delta u \otimes \hat{\xi}) + \vec{K}(u \otimes \hat{\partial} \hat{\xi}) \\ &+ (-1)^d \vec{m}(0)(u \otimes \hat{\xi}) - (-1)^d \vec{m}(1)(u \otimes \hat{\xi}).\end{aligned}$$

Thus $\vec{K}$ provides a chain-homotopy, and the maps obtained from $\vec{m}(0)$ and $\vec{m}(1)$ are the same at the level of homology. □

Having defined $\overrightarrow{HM}_\bullet(u \mid W; \Gamma_W)$, we now turn to establishing the properties that were discussed in the introductory subsection, Subsection 3.5. We begin with the composition laws described in Theorem 3.5.3. As we proceed, we note that the statement and proof of this theorem are readily adaptable to the case of local coefficients.

Proof of Theorem 3.5.3. The above proposition tells us that, when $b^+(W) \geq 2$, we have a well-defined map

$$\overrightarrow{HM}_\bullet(u \mid W; \Gamma_W) : \widehat{HM}_\bullet(Y_0; \Gamma_0) \to \widecheck{HM}_\bullet(Y_1; \Gamma_1)$$

associated to a morphism

$$(u, W, \Gamma_W) : (Y_0, \Gamma_0) \to (Y_1, \Gamma_1)$$

in our cobordism category. The commutative diagram in Theorem 3.5.3 asserts that we have the relations:

$$j_* \circ \overrightarrow{HM}_\bullet(u \mid W; \Gamma_W) = \widehat{HM}_\bullet(u \mid W; \Gamma_W)$$
$$\overrightarrow{HM}_\bullet(u \mid W; \Gamma_W) \circ j_* = \widecheck{HM}_\bullet(u \mid W; \Gamma_W).$$

These can be directly verified at the chain level by multiplying the 2-by-2 matrices (Definition 27.3.1 and Proposition 22.2.1) which define $\vec{m}$ and j, and comparing with the definition of $\check{m}$ and $\hat{m}$ in Definition 25.3.3, under the hypothesis that $\bar{m}_u^s$ etc. are zero.

It remains to prove the composition laws (3.14). We will indicate the proof only of the first of the two composition laws,

$$\overrightarrow{HM}_\bullet(W) = \widecheck{HM}_\bullet(W_{12}) \circ \overrightarrow{HM}_\bullet(W_{01}) \tag{27.6}$$

as the second can be deduced from the first using duality. (We have renamed W_1 and W_2 from (3.14), to fit with the terminology of the present section.) The proof has the same setup as the proof of the composition laws in Subsection 26.2. We use the notation of that subsection here. In place of the operator (26.12), we require an operator

$$\vec{K} : C^{d_{01}}(\mathcal{U}_{01}; R) \otimes C^{d_{12}}(\mathcal{U}_{12}; R) \otimes \hat{C}_\bullet(Y_0; \Gamma_0) \to \check{C}_\bullet(Y_2; \Gamma_2)$$

satisfying an identity:

$$\begin{aligned} &- \vec{K}(\underline{\delta} \otimes 1) - \vec{K}(1 \otimes 1 \otimes \hat{\partial}) - \check{\partial} \vec{K} \sigma \\ &- \vec{m}_{02}(c \otimes 1) + \check{m}_{12}(1 \otimes \vec{m}_{01})(\tau \otimes 1) = 0. \end{aligned} \tag{27.7}$$

(Compare Lemma 26.2.2.) Because $b^+(W_{01})$ is positive, we can choose a metric g_{01} on W_1 and perturbing 2-form ω_{01} satisfying the condition of Proposition 27.2.4, so that the moduli spaces on $W_{01}(T)^*$ contain no reducibles, for $T \geq T_0$, where $W_{01}(T)$ is defined using the metric $g_{01}(T)$ as in that proposition. Now let $W(S, T)$ be the composite cobordism, defined as in (26.2), with a neck of length S in the middle and using the metric $g_{01}(T)$ on W_{01}.

Lemma 27.3.5. *There exists T_0, such that for all $T \geq T_0$, the moduli spaces on $W(S, T)$ contain no reducible solutions.*

Proof. This is proved in the same way as Proposition 27.2.4. □

Now fix $T \geq T_0$, and form the parametrized moduli space

$$\mathcal{M}_z([\mathfrak{a}_0],[\mathfrak{b}_2]) = \bigcup_{S\in[0,\infty)} \{S\} \times M_z([\mathfrak{a}_0], W(S,T)^*, [\mathfrak{b}_2]), \tag{27.8}$$

as in (26.3), and its compactification $\mathcal{M}_z^+([\mathfrak{a}_0],[\mathfrak{b}_2])$. Once again, we use a perturbation which is non-zero on the middle neck, $[0,S] \times Y_1$: see the remark on page 536. Using these moduli spaces, we can define maps K_o^o,K_s^o, K_o^u and K_s^u, exactly as in (26.11). The absence of reducibles means that the maps $\bar{K}_s^s$, $\bar{K}_u^s$, $\bar{K}_s^u$ and $\bar{K}_u^u$ are all zero. The required map $\vec{K}$ is then defined by the matrix

$$\vec{K} = \begin{bmatrix} K_o^o & K_o^u \\ K_s^o & K_s^u \end{bmatrix}.$$

The four matrix components of the required identity (27.7) coincide with the identities in Lemma 26.2.3, once these are simplified using the vanishing of $\bar{K}_s^s$ and its companions and the vanishing of $\bar{m}_s^s$ etc. for the cobordism W_1. For example, in the first identity of Lemma 26.2.3, the terms $\partial_o^u \bar{K}_u^s \partial_s^o$ and $m_o^u \bar{m}_u^s \partial_s^o \tau\sigma_1$ vanish, leaving an identity

$$- K_o^o \partial_o^o - \partial_o^o K_o^o \sigma_{12} + K_o^u \bar{\partial}_u^s \partial_s^o + \partial_o^u \bar{\partial}_u^s K_s^o \sigma_{12} - \underline{m_o^o c}$$
$$+ m_o^o m_o^o \tau - m_o^u \bar{\partial}_u^s m_s^o \tau - \partial_o^u \bar{m}_u^s m_s^o \tau = K_o^o(\underline{\delta} \otimes 1).$$

This is the top left component of the identity (27.7). □

We now check that when a cobordism W is factored as the composition of two cobordisms W_1, W_2, both with $b^+ \geq 1$, the map $\overrightarrow{HM}_\bullet(W)$ coincides with the map $Z(W_1, W_2)$ defined by the diagram chase as in (3.15). Note that the proof of the composition law (27.6) which we have just given did not use the condition $b^+(W_1) \geq 2$ in an essential way, but only the condition $b^+(W_1) \geq 1$. So if $b^+(W_1)$ and $b^+(W_2)$ are both positive, so that $b^+(W) \geq 2$, then we can write

$$\overrightarrow{HM}_\bullet(W) = \widecheck{HM}_\bullet(W_2) \circ \overrightarrow{HM}_\bullet(W_1)_{(g_1,\omega_1)}$$

for any appropriate choice of g_1 and ω_1. The composite on the right calculates $Z(W_1, W_2)$.

27.4 Closed four-manifolds revisited

We shall now show that the invariants of a closed 4-manifold X given by Definition 27.1.7 can be recovered from $\overrightarrow{HM}_\bullet(W)$, where W is the cobordism from

S^3 to S^3 obtained by removing two balls from X, as promised in Subsection 3.6. (See (3.19) and Proposition 3.6.1.) We restate the result here.

Proposition 27.4.1. *Let X be a closed, oriented 4-manifold with $b^+(X) \geq 2$, equipped with a homology orientation, and let W be the cobordism from S^3 to S^3 obtained by removing two balls from X. Let $\mathfrak{m}(u \mid X)$ be the Seiberg–Witten invariants of the closed manifold, as in Definition 27.1.7. Let 1 be the standard generator of $\widehat{HM}_\bullet(S^3)$ and $\check{1} \in \widecheck{HM}^\bullet(S^3) \cong \mathbb{Z}[[U]]$ the generator for the Floer cohomology group. Then*

$$\mathfrak{m}(u \mid X) = \langle \overrightarrow{HM}_\bullet(u \mid W)(1), \check{1} \rangle,$$

for u a class of degree d, where the angle brackets denote the $\mathbb{Z}$-valued pairing between $\widehat{HM}_\bullet(S^3)$ and $\widecheck{HM}^\bullet(S^3)$.

Proof. The proof will be another stretching argument. Consider the 4-ball B^4 equipped with a metric with positive scalar curvature containing a collar region $[0, 1] \times S^3$ in which the metric is cylindrical. Let the metric on S^3 be the round metric. Let $(B^4)^*$ be obtained from B^4 by attaching a cylindrical end $[0, \infty) \times S^3$. Let $[B_0, 0] \in \mathcal{B}(S^3, \mathfrak{s}_0)$ denote the unique critical point for the unperturbed functional $\mathcal{L}$ (so B_0 is the connection with $F_{B_0^t} = 0$). Choose a small perturbation $\mathfrak{q} \in \mathcal{P}(S^3)$ as in Subsection 22.7, so that there is still a unique critical point $[B, 0]$ and so that the perturbed Dirac equation $D_{\mathfrak{q},B}$ has simple spectrum. A family of self-adjoint operators on S^3 joining D_{B_0} to $D_{\mathfrak{q},B}$ will have no spectral flow. As in Subsection 22.7, we can label the eigenvalues of $D_{\mathfrak{q},B}$ in increasing order as λ_i, with λ_0 the first positive eigenvalue, and label the corresponding critical points in $\mathcal{B}_k^\sigma(Y, \mathfrak{s}_0)$ as $\mathfrak{a}_i$. Choose a perturbation $\hat{\mathfrak{p}}$ on $(B^4)^*$, equal to $\mathfrak{q}$ on the end as usual, so that the corresponding moduli spaces $M((B^4)^*, [\mathfrak{a}_i])$ are regular.

Lemma 27.4.2. *If $\mathfrak{q}$ and $\hat{\mathfrak{p}}$ are sufficiently small, then there is a unique gauge-equivalence class $[A_0]$ of connections on $(B^4)^*$ such that $M((B^4)^*, [\mathfrak{a}_i])$ consists only of solutions $[A, s, \phi]$ with $s = 0$ and $[A] = [A_0]$.*

The moduli space $M((B^4)^*, [\mathfrak{a}_i])$ is empty if $i \geq 0$. The moduli space $M((B^4)^*, [\mathfrak{a}_{-i}])$ has dimension $2i - 2$ for $i \geq 1$ and can be identified with

$$\mathbb{CP}^{i-1} \setminus \mathbb{CP}^{i-2} = \mathbb{P}(V_i) \setminus \mathbb{P}(V_{i-1}),$$

where V_i is the space of solutions ϕ to the perturbed Dirac equation on $(B^4)^*$ with growth $|\phi(t)| \leq Ce^{-\lambda_{-i}t}$ on the cylindrical end. In particular, $M((B^4)^*, [\mathfrak{a}_{-1}])$ is a single point.

Proof. Examine first the unperturbed equation $\mathfrak{F}(A, \Phi) = 0$ on $(B^4)^*$, without blowing up. Because the scalar curvature is positive, there is no solution with Φ non-zero and $|\Phi(t)| \to 0$ on the end, by the same argument as Proposition 4.6.1. There are no square-integrable, closed, anti-self-dual 2-forms on the manifold, so if $(A, 0)$ is a solution of the equations $F^+_{A^t} = 0$, asymptotic to the constant solution on the cylindrical end, then F_{A^t} must be zero. Up to gauge equivalence, there is a unique such connection A_0. This solution to the unperturbed equations is regular, so persists under sufficiently small perturbations.

The description of $M((B^4)^*, [\mathfrak{a}_{-i}])$ as $\mathbb{P}(V_i) \setminus \mathbb{P}(V_{i-1})$ now follows from the definition of the moduli space, much as in Proposition 14.6.1; and the only additional point to verify is that the complex dimension of V_i is $2i$. It is enough to check that the L^2 index of the unperturbed Dirac operator $D^+_{A_0}$ is zero on $(B^4)^*$, and this follows again from the positivity of the scalar curvature. □

We have a similar result for the oppositely oriented ball $-B^4$, with an "incoming end", $(-\infty, 0] \times S^3$, attached. The cylindrical-end manifold $(-B^4)^*$ is isometric to $(B^4)^*$ by an orientation-preserving map, but because of our (possibly asymmetric) perturbation and our asymmetric labelling of the eigenvalues (λ_0 is positive), the final result looks slightly different. We have

$$M([\mathfrak{a}_i], (-B^4)^*) \cong \mathbb{CP}^i \setminus \mathbb{CP}^{i-1}$$

for $i \geq 0$, so that $M([\mathfrak{a}_0], (-B^4)^*)$ is a point. These moduli spaces are empty for negative i.

We return to the proof of the proposition. Equip W with a metric g and 2-form ω, so that for $T \geq T_0$ there are no reducible solutions on the cylindrical-end manifold $W(T)^*$, as in the statement of Proposition 27.2.4. Let $X(S, T)$ be the closed manifold (diffeomorphic to X) obtained by attaching two cylinders $[0, S] \times S^3$ and the balls B^4 and $-B^4$ (see Figure 10):

$$X(S, T) = B^4 \cup \big([0, S] \times S^3\big) \cup W(T) \cup \big([0, S] \times S^3\big) \cup (-B^4).$$

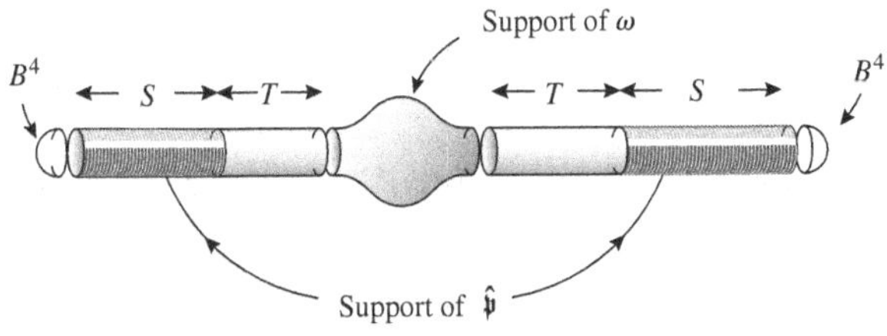

Fig. 10. The manifold $X(S, T)$ obtained by adding cylinders and two 4-balls to W.

The equations on X carry the perturbation $\hat{\mathfrak{p}}$ on both copies of $[0,S] \times S^3$ and the 2-form ω in the interior of W. Consider the moduli space $M(X(S,T),\mathfrak{s}_X)$ for the manifold equipped with this metric and perturbation, and a fixed spin^c structure $\mathfrak{s}_X$. Form the union

$$\mathcal{M}(X,\mathfrak{s}_X) = \bigcup_{S\in[0,\infty)} \{S\} \times M(X(S,T),\mathfrak{s}_X),$$

as in (26.3). This has a compactification formed by attaching a fiber at $S=\infty$: as in Subsection 26.1, we define $M(X(\infty,T),\mathfrak{s}_X)$ to be the set of quintuples

$$([\gamma_0],[\breve{\gamma}_1],[\gamma_2],[\breve{\gamma}_3],[\gamma_4])$$

where

$$\begin{aligned}
&[\gamma_0] \in M((B^4)^*,[\mathfrak{a}_{i_1}])\\
&[\breve{\gamma}_1] \in \breve{M}^+([\mathfrak{a}_{i_1}],[\mathfrak{a}_{i_2}])\\
&[\gamma_2] \in M([\mathfrak{a}_{i_2}],W(T)^*,\mathfrak{s}_W,[\mathfrak{a}_{i_3}])\\
&[\breve{\gamma}_3] \in \breve{M}^+([\mathfrak{a}_{i_3}],[\mathfrak{a}_{i_4}])\\
&[\gamma_4] \in M([\mathfrak{a}_{i_4}],(-B^4)^*),
\end{aligned}$$

and we form

$$\mathcal{M}^+(X,\mathfrak{s}_X) = \bigcup_{S\in[0,\infty]} \{S\} \times M(X(S,T),\mathfrak{s}_X).$$

This is a space stratified by manifolds. Its codimension-1 strata consist of the fiber over 0 and those strata over $S=\infty$ with $\mathfrak{a}_{i_1}=\mathfrak{a}_{i_2}$ and $\mathfrak{a}_{i_3}=\mathfrak{a}_{i_4}$, so that $[\breve{\gamma}_1]$ and $[\breve{\gamma}_3]$ belong to point moduli spaces. The latter strata are of the form

$$M((B^4)^*,[\mathfrak{a}]) \times M([\mathfrak{a}],W(T)^*,\mathfrak{s}_W,[\mathfrak{b}]) \times M([\mathfrak{b}],(\bar{B}^4)^*),$$

where each of $[\mathfrak{a}]$ and $[\mathfrak{b}]$ is a critical point on S^3. Among these, there is one stratum

$$M((B^4)^*,[\mathfrak{a}_{-1}]) \times M([\mathfrak{a}_{-1}],W(T)^*,\mathfrak{s}_W,[\mathfrak{a}_0]) \times M([\mathfrak{a}_0],(\bar{B}^4)^*) \tag{27.9}$$

which is diffeomorphic to $M([\mathfrak{a}_{-1}],W(T)^*,\mathfrak{s}_W,[\mathfrak{a}_0])$ by the lemma. We have a continuous map

$$r : \mathcal{M}^+(X,\mathfrak{s}_X) \to [0,\infty] \times \mathcal{B}^\sigma(W(T),\mathfrak{s}_W)$$

whose image is again a space stratified by manifolds. The only codimension-1 strata in $r(\mathcal{M}^+(X, \mathfrak{s}_X))$ are the moduli space $M(X(0,T), \mathfrak{s}_X)$ over $S = 0$, and the moduli space $M([\mathfrak{a}_{-1}], W(T)^*, \mathfrak{s}_W, [\mathfrak{a}_0])$ over $S = \infty$. The latter is the image of the stratum (27.9). The proposition now follows by the usual application of Stokes' theorem, because the evaluation

$$\big\langle u, [M([\mathfrak{a}_{-1}], W(T)^*, \mathfrak{s}_W, [\mathfrak{a}_0])]\big\rangle$$

calculates the contribution from the spin^c structure $\mathfrak{s}_W$ to the matrix entry of $\overrightarrow{HM}_\bullet(u \mid W)$ from the generator 1 in $\widehat{HM}_\bullet(S^3)$ corresponding to $[\mathfrak{a}_{-1}]$ and the generator of $\widecheck{HM}_\bullet(S^3)$ corresponding to $[\mathfrak{a}_0]$, which is the generator that pairs with $\check{1} \in \widecheck{HM}^\bullet(S^3)$. □

The proof of the proposition above establishes an equality for each spin^c structure separately; so we can also formulate a version with "local coefficients" with no change in the proof. Let R be a ring, let Γ_{S^3} be the trivial local coefficient system on $\mathcal{B}^\sigma(S^3)$ with fiber R, and let $\Gamma_W : \Gamma_{S^3} \to \Gamma_{S^3}$ be a W-morphism, where W is still the complement of two balls in X. As we remarked earlier, such a Γ_W is equivalent to an $R^\times$-valued function Γ_X on the set of spin^c structures $\mathfrak{s}_X$. Then we have:

Proposition 27.4.3. *Let X and W be as above, and let Γ_W be a W-morphism corresponding to an $R^\times$-valued function Γ_X on the set of spin^c structures on X. Then*

$$\mathfrak{m}(u \mid X; \Gamma_X) = \big\langle \overrightarrow{HM}_\bullet(u \mid W; \Gamma_W)(1), \check{1}\big\rangle,$$

where the angle brackets denote the R-valued pairing between $\widecheck{HM}_\bullet(S^3; R)$ and $\widecheck{HM}^\bullet(S^3; R)$. □

Taking Γ_X to be as in Equation (27.5) we recover Proposition 3.8.1. The related formulae (3.27) and (3.29) in Subsection 3.8 are formal consequences of this proposition and the composition laws.

27.5 The wall-crossing formula

So far, we have only considered the monopole invariants $\mathfrak{m}(X, [\nu])$ for closed manifolds X with $b^+(X) \geq 2$. Recall that when $b^+ = 1$, the difficulty which arises is that, although the associated moduli spaces $M(X, \mathfrak{s}_X)$ are closed manifolds for generic choice of metric and perturbing 2-form, it is not the case in general that the pairings $\langle [u_2^d], M(X, \mathfrak{s}_X)\rangle$ are independent of the choices made.

This issue was raised earlier in Subsection 1.5 (see the discussion following Theorem 1.5.2). We return to it now, to investigate more carefully how the pairing depends on the metric and perturbation.

Suppose then that X is an oriented closed manifold with $b^+(X) = 1$. Our starting points are Lemma 27.1.1 and Lemma 27.1.2, which tell us that for any metric g on Y there is a dense open set of forms ω in $L^2_{k-1}(X; i\Lambda^2(X))$ for which the corresponding moduli space $M(X, \mathfrak{s}_X)$ is smooth, compact and without boundary. The second condition – that the moduli space has no boundary – arises by ensuring that there are no reducible solutions to the equations; and this condition is equivalent to requiring

$$4\int_X \omega \wedge \kappa \neq -(2\pi i c_1(S^+) \smile [\kappa])[X] \tag{27.10}$$

where κ is any non-zero self-dual harmonic form for the metric g on X. (There is only one such form, up to scalar multiple, because b^+ is 1.) A picture of what is going on can be obtained from the proof of Lemma 27.1.1. Inside the configuration space $\tilde{\mathcal{B}}^\sigma_k(X, \mathfrak{s}_X)$ is the Hilbert submanifold

$$\tilde{\mathcal{Z}} \subset \tilde{\mathcal{B}}^\sigma_k(X, \mathfrak{s}_X)$$

defined as the gauge-equivalence classes of triples (A, s, ϕ) with $D^+_A\phi = 0$. The configuration space $\mathcal{B}^\sigma_k(X, \mathfrak{s}_X)$ is the subset of $\tilde{\mathcal{B}}^\sigma_k(X, \mathfrak{s}_X)$ where $s \geq 0$, and we have equally a Hilbert submanifold with boundary

$$\mathcal{Z} \subset \mathcal{B}^\sigma_k(X, \mathfrak{s}_X).$$

(The s coordinate is a product factor in all these cases, and the boundary of $\mathcal{Z}$ occurs at $s = 0$.) As in the proof of the lemma, we have a map

$$\varpi : \mathcal{Z} \to L^2_{k-1}(X; i\,\mathfrak{su}(S^+))$$

given by

$$(A, s, \phi) \mapsto \frac{1}{2}\rho(F^+_{A^t}) - s^2(\phi\phi^*)_0,$$

and the moduli space $M(X, \mathfrak{s}_X)$ for a given perturbing 2-form ω is the fiber of ϖ over $2\rho(\omega^+)$. The restriction of ϖ to $\partial\mathcal{Z}$ is a map

$$\partial\varpi : \partial\mathcal{Z} \to \mathcal{W}_{\mathfrak{s}_X}$$

where $\mathcal{W}_{\mathfrak{s}_X} \subset L^2_{k-1}(X; i\,\mathfrak{su}(S^+))$ is the affine linear subspace of codimension 1 consisting of all elements $2\rho(\omega^+)$ where ω runs through forms with

$$4\int_X \omega \wedge \kappa = -(2\pi i c_1(S^+) \cup [\kappa])[X].$$

The domain and codomain of $\partial\varpi$ are codimension-1 submanifolds of the domain and codomain of ϖ; so, like ϖ, the map $\partial\varpi$ is a proper Fredholm map, and its index is given by the same formula as appears in Lemma 27.1.1. In particular, the generic fiber of $\partial\varpi$ is a smooth compact manifold whose dimension is the same as that of $M(X, \mathfrak{s}_X)$. The normal direction to $\mathcal{W}_{\mathfrak{s}_X}$ is oriented by a choice of a non-zero self-dual harmonic form, while the normal direction to $\partial\mathcal{Z}$ in $\mathcal{Z}$ is canonically oriented. The difference between the determinant lines of $\mathcal{D}\varpi$ and $\mathcal{D}\partial\varpi$ is therefore the line of self-dual harmonic forms, and it follows that the fibers of $\partial\varpi$ can be canonically oriented by a choice of orientation for $H^1(X;\mathbb{R})$. A regular fiber $(\partial\varpi)^{-1}(w) \subset \mathcal{B}^\sigma_k(X, \mathfrak{s}_X)$, being a closed oriented manifold, has a homology class,

$$[(\partial\varpi)^{-1}(w)] \in H_d(\mathcal{B}^\sigma(X, \mathfrak{s}_X).$$

The overall sign of the class depends on orienting $H^1(X;\mathbb{R})$, but the homology class is otherwise independent of the choice of w in $\mathcal{W}_{\mathfrak{s}_X}$.

Now let ω_0 and ω_1 be two choices of perturbing 2-form for the Seiberg–Witten equations on X, and let $M(X, \mathfrak{s}_X)_0$ and $M(X, \mathfrak{s}_X)_1$ be the corresponding moduli spaces. Suppose that $2\rho(\omega_0^+)$ and $2\rho(\omega_1^+)$ are both regular values of ϖ, so that the moduli spaces are smooth, and suppose also that both perturbing 2-forms satisfy the condition (27.10), so that the moduli spaces are closed. Fix a homology orientation of X, to orient both moduli spaces, so that we have classes $[M(X, \mathfrak{s}_X)_0]$ and $[M(X, \mathfrak{s}_X)_1]$ in $H_d(\mathcal{B}^\sigma_k(X, \mathfrak{s}_X))$. We can consider two cases. First, it may be that $2\rho(\omega_0^+)$ and $2\rho(\omega_1^+)$ lie on the same side of the affine hyperplane $\mathcal{W}_{\mathfrak{s}_X}$ in $L^2_{k-1}(X; i\,\mathfrak{su}(S^+))$. In this case, we can join ω_0 and ω_1 by a path of forms ω_t all of which satisfy the inequality (27.10). If the intermediate forms are chosen so that $2\rho(\omega_t^+)$ is transverse to ϖ, then we obtain a cobordism between the moduli space, showing that

$$[M(X, \mathfrak{s}_X)_0] = [M(X, \mathfrak{s}_X)_1]$$

as before. However, if $2\rho(\omega_0^+)$ and $2\rho(\omega_1^+)$ lie on different sides of $\mathcal{W}_{\mathfrak{s}_X}$, the classes may not be equal. We can choose ω_t so that the path $P : [0, 1] \to L^2_{k-1}(X; i\,\mathfrak{su}(S^+))$ given by $P(t) = 2\rho(\omega_t^+)$ is transverse to both ϖ and $\partial\varpi$, and crosses $\mathcal{W}_{\mathfrak{s}_X}$ at one point w. In this case the parametrized moduli space $M(X,\mathfrak{s}_X)_P$ is a $(d+1)$-manifold with three boundary components:

$$\partial M(X,\mathfrak{s}_X)_P = M(X,\mathfrak{s}_X)_0 \cup M(X,\mathfrak{s}_X)_1 \cup (\partial\varpi)^{-1}(w).$$

Thus

$$[M(X,\mathfrak{s}_X)_0] = [M(X,\mathfrak{s}_X)_1] \pm [(\partial\varpi)^{-1}(w)].$$

The fundamental class of the moduli $M(X,\mathfrak{s}_X)$ therefore depends only on *which side of the wall*, $\mathcal{W}_{\mathfrak{s}_X}$, the element $2\rho(\omega^+)$ lies on.

To say something more, and to pin down the sign here, consider the case that $b_1(X) = 0$, so that the fiber $(\partial\varpi)^{-1}(w)$ can be canonically oriented. Indeed, the fiber is something we can describe quite explicitly: it is the quotient by the gauge-group of the space of solutions (A,ϕ) to the equations

$$\begin{aligned} D_A^+\phi &= 0 \\ F_{A^t}^+ &= 4\omega^+ \end{aligned} \tag{27.11}$$

(where w is again written as $2\rho(\omega^+)$). In the case that $b_1 = 0$ and w lies on the wall, the second equation determines the connection A uniquely up to gauge equivalence, so $(\partial\varpi)^{-1}(w)$ is the (complex) projectivization of the kernel of the Dirac operator D_A^+. If w is a regular value of $\partial\varpi$, then this Dirac operator is surjective, and so

$$(\partial\varpi)^{-1}(w) = \mathbb{CP}^{d/2}$$

where d is the (real) dimension of the moduli space $M(X,\mathfrak{s}_X)$. This fiber is indeed canonically oriented, for we can give it the complex orientation. We can therefore write

$$[M(X,\mathfrak{s}_X)_0] = [M(X,\mathfrak{s}_X)_1] \pm [\mathbb{CP}^{d/2}]. \tag{27.12}$$

To understand the sign, let us first specify a homology orientation by picking a non-zero, real, self-dual harmonic form κ, so as to orient $H^+(X)$. (We still suppose that $H^1(X;\mathbb{R})$ is zero.) If ω is a form for which the inequality (27.10) holds, let us say that $\omega \in \Omega^2(X; i\mathbb{R})$ lies on the *positive* side of the wall if

$$4i\int_X \omega\wedge\kappa > (2\pi c_1(S^+)\smile[\kappa])[X].$$

Let us choose ω_0 and ω_1 to lie on the negative and positive sides respectively. The question of the sign in (27.12) is now well-defined, and with our conventions, the sign is negative:

$$[M(X, \mathfrak{s}_X)_0] = [M(X, \mathfrak{s}_X)_1] - [\mathbb{CP}^{d/2}]. \tag{27.13}$$

When we evaluate the class $u_2^{d/2}$ against these homology classes, the extra $\mathbb{CP}^{d/2}$ yields 1 in the formula. Rather than having a single, well-defined monopole invariant

$$\mathfrak{m}(u_2^{d/2} \mid X, \mathfrak{s}_X) = \langle u_2^{d/2}, [M(X, \mathfrak{s}_X)] \rangle,$$

we have two versions,

$$\mathfrak{m}_+(u_2^{d/2} \mid X, \mathfrak{s}_X) = \langle u_2^{d/2}, [M(X, \mathfrak{s}_X)_1] \rangle$$
$$\mathfrak{m}_-(u_2^{d/2} \mid X, \mathfrak{s}_X) = \langle u_2^{d/2}, [M(X, \mathfrak{s}_X)_0] \rangle,$$

obtained using the moduli spaces from the positive and negative sides of the wall. They are related by

$$\mathfrak{m}_+(u_2^{d/2} \mid X, \mathfrak{s}_X) = \mathfrak{m}_-(u_2^{d/2} \mid X, \mathfrak{s}_X) + 1. \tag{27.14}$$

This is the *wall-crossing formula* for the case that $b_1(X) = 0$.

The wall-crossing formula can be extended also to the case of manifolds with b_1 positive. In the general case, the fiber of $\partial\varpi$ is still described as the moduli space of solutions to the pair of equations (27.11), but now the solutions of the second equation, $F^+_{A^t} = 4\omega^+$, mod gauge, form a torus $\mathbb{T}$ of dimension $b_1(X)$. Let us consider the case that the dimension d of the moduli space (and of the generic fiber of $\partial\varpi$) is zero. In this case, the torus $\mathbb{T}$ parametrizes a family of Dirac operators D^+_A of complex index $1 - b_1/2$. The index of the family defines an element ζ in the K-theory of the torus, $K(\mathbb{T})$, and generically the number of points in $\mathbb{T}$ where D^+_A has kernel, counted with sign, is given by

$$c_{b_1/2}(-\zeta)[\mathbb{T}].$$

In the case $b_1 = 2$, for example, we are evaluating c_1 of the index of a family of operators parametrized by a 2-torus $\mathbb{T}$, and an application of the index theorem for families leads to an answer

$$\frac{1}{2}\big(c_1(\mathfrak{s}_X) \cup a_1 \cup a_2\big)[X], \tag{27.15}$$

where a_1 and a_2 form a basis for $H^1(X;\mathbb{Z}) = \mathbb{Z}^2$. (The overall sign depends on an orientation of H^1.) The calculation for general b_1 appears in [65].

Example. A particularly simple case is the manifold $X = \mathbb{CP}^2$. If we choose the homology orientation given by the symplectic form, then the zero perturbation $\omega = 0$ lies on the positive or negative side of the wall, according as $c_1(\mathfrak{s})$ is respectively a negative or positive multiple of the Kähler class h. Furthermore, because the manifold has positive scalar curvature, the unperturbed equations admit no irreducible solutions, and so the corresponding monopole invariant is zero. Thus, writing $d(\mathfrak{s})$ for the dimension of the moduli space,

$$\mathfrak{m}_-(u_2^{d(\mathfrak{s})/2} \mid \mathbb{CP}^2, \mathfrak{s}) = 0, \text{if } c_1(\mathfrak{s}) \cup h > 0$$
$$\mathfrak{m}_+(u_2^{d(\mathfrak{s})/2} \mid \mathbb{CP}^2, \mathfrak{s}) = 0, \text{if } c_1(\mathfrak{s}) \cup h < 0.$$

(The case $c_1 = 0$ does not arise, because $c_1(\mathfrak{s})$ is an odd multiple of the generator.) Using the wall-crossing formula, we obtain

$$\mathfrak{m}_+(u_2^{d(\mathfrak{s})/2} \mid \mathbb{CP}^2, \mathfrak{s}) = \begin{cases} 0, & \text{if } c_1(\mathfrak{s}) \cup h < 0 \\ 1, & \text{if } c_1(\mathfrak{s}) \cup h > 0, \end{cases}$$

while similarly

$$\mathfrak{m}_-(u_2^{d(\mathfrak{s})/2} \mid \mathbb{CP}^2, \mathfrak{s}) = \begin{cases} 0, & \text{if } c_1(\mathfrak{s}) \cup h > 0 \\ -1, & \text{if } c_1(\mathfrak{s}) \cup h < 0. \end{cases}$$

We can calculate invariants for the manifold $X = S^2 \times T^2$ in a similar way, at least for spin^c structures $\mathfrak{s}_X$ with $c_1(\mathfrak{s}_X) = 0$ (so that the moduli space is zero-dimensional). Again, there is a metric with positive scalar curvature, so the invariants are zero for small perturbations ω. Each non-zero invariant is related to a zero invariant by the wall-crossing formula (27.15). In that formula, the term $a_1 \cup a_2$ is the class Poincaré dual to $[S^2]$ in X. So the non-zero invariants are given by a formula

$$\pm(c_1(\mathfrak{s}_X) \cdot [S^2])/2. \tag{27.16}$$

The proof that $\mathfrak{m}_\pm$ are independent also of the choice of Riemannian metric on X proceeds in the same way as the case of manifolds with $b_+ \geq 2$. The affine subspace $\mathcal{W}_{\mathfrak{s}_X} = \mathcal{W}_{\mathfrak{s}_X}$ depends on the metric g through the self-dual harmonic form $\kappa = \kappa_g$; but if we choose an orientation for $H^+(X)$, we can still make sense of the "positive" and "negative" sides as g varies, and so talk consistently about $\mathfrak{m}_+$ and $\mathfrak{m}_-$. It makes sense, in particular, to ask whether the zero perturbation $\omega = 0$ lies on the positive or negative side of the wall $\mathcal{W}_{\mathfrak{s}_X,g}$ for a given g, or whether it lies *on* the wall. This is the question of the sign of the pairing

$$(c_1(S^+) \cup [\kappa_g])[X]$$

for a self-dual g-harmonic form κ_g of the chosen orientation. Specifically, if this pairing is *negative*, then the moduli space $[M(X, \mathfrak{s}_X]$ for small perturbations ω computes the invariant $\mathfrak{m}_+$, while if the pairing is *positive*, then it computes $\mathfrak{m}_-$. Geometrically, if the Chern class is not torsion, the spin^c structure $\mathfrak{s}_X$ determines a hyperplane in $H^2(X; \mathbb{R})$, namely the space orthogonal to $c_1(S^+)$ with respect to the cup product; and from this point of view, the question is, on which side of this hyperplane does the ray spanned by the harmonic form $[\kappa_g]$ lie?

There is an addition observation to be made here. When $b^+ = 1$ and $b_1 = 0$, the formula for the dimension d of the moduli space $M(X, \mathfrak{s}_X)$ can be written

$$d = (1/4)(c_1^2(S^+)[X] - 9 + b^-(X)).$$

A necessary condition to obtain a non-zero invariant $\mathfrak{m}$ is that d is non-negative. So in the case that $b^-(X) \leq 8$, we are only concerned with classes $c_1(S^+)$ with *positive square*. Since κ_g also defines a class in the positive cone, it is not possible for $c_1(S^+)$ and $[\kappa_g]$ to be orthogonal when $b^- \leq 8$. We still need a homology orientation to fix the overall sign of the invariant; but if we choose the perturbing 2-form ω always to be small, then it will never lie on the wall, and we obtain in this way a well-defined invariant, independent of g. In the borderline case, when $b^- = 9$, the same applies if $c_1(S^+)$ is not torsion, because a non-zero null vector cannot be orthogonal to a positive vector when $b^+ = 1$. Figure 11 shows the situation when $b^- \geq 10$. In this range, the class $c_1(S^+)$ may have negative square, so that the complementary hyperplane intersects the positive cone. As the metric g varies, the positive ray spanned by $[\kappa_g]$ may cross this hyperplane; and when this happens, the invariant defined by the moduli space $M(X, \mathfrak{s}_X)$ for small perturbations ω will change by 1.

Gluing formulae when $b^+ = 1$. Finally, we mention the gluing formula, Proposition 3.9.3, and its extension to the case of manifolds with $b^+ = 1$. Recall the setting of Proposition 3.9.3. We have a closed manifold X with a real 2-cycle ν, and a decomposition of X as $X_1 \cup X_2$ along a manifold Y, which also decomposes ν as $\nu_1 + \nu_2$. We write η for the 1-cycle $\partial \nu_1$ (equivalently $-\partial \nu_2$) in Y, and we recall the important hypothesis that the class $[\eta]$ is non-zero in $H_1(Y; \mathbb{R})$.

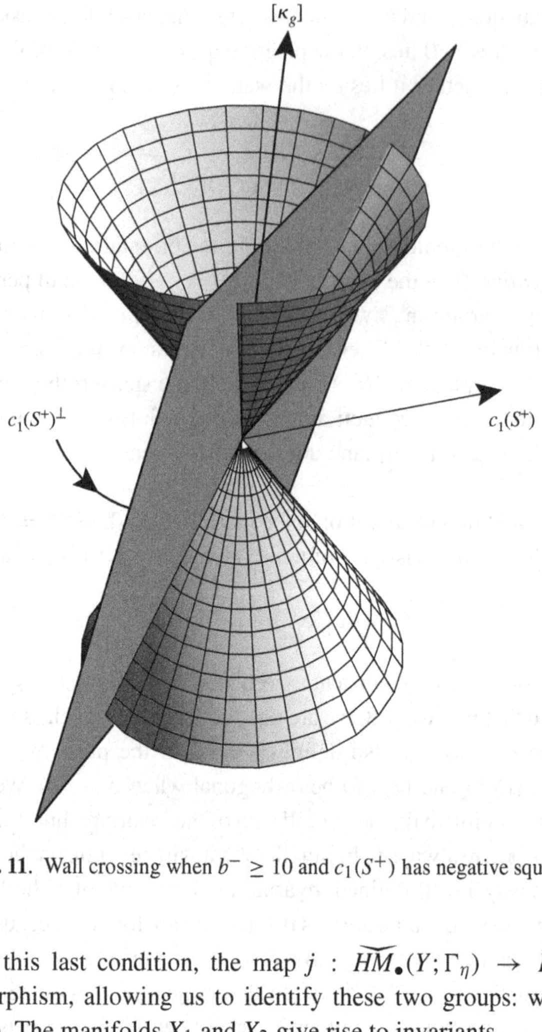

Fig. 11. Wall crossing when $b^- \geq 10$ and $c_1(S^+)$ has negative square.

Because of this last condition, the map $j : \widecheck{HM}_\bullet(Y;\Gamma_\eta) \to \widehat{HM}_\bullet(Y;\Gamma_\eta)$ is an isomorphism, allowing us to identify these two groups: we just write $HM_\bullet(Y;\Gamma_\eta)$. The manifolds X_1 and X_2 give rise to invariants

$$\begin{aligned}\psi_{(X_1,\nu_1)} &\in HM_\bullet(Y;\Gamma_\eta)\\ \psi_{(X_2,\nu_2)} &\in HM_\bullet(-Y;\Gamma_{-\eta})\end{aligned}$$

and with appropriate homology orientations, Proposition 3.9.3 expresses the invariant of the closed manifold X as a pairing,

$$\mathfrak{m}(X,[\nu]) = \left\langle \psi_{(X_1,\nu_1)}, \psi_{(X_2,\nu_2)} \right\rangle_{\omega_\mu},$$

as long as $b^+(X) \geq 2$, so that the left-hand side is defined. The slightly paradoxical observation that we made earlier was that the topological setup only guarantees that $b^+(X) \geq 1$, so the left-hand side may *not* be well-defined as it stands, even though the pairing on the right is unambiguous. We will not prove the pairing formula until later in the book (Section 32), but we do now have the means to resolve the paradox by explaining what the pairing on the right actually computes in this situation.

The decomposition of X along Y allows us to define a family of metrics g_T on X (for $T \geq 0$) by starting, as usual, with a metric g_0 which is cylindrical near Y and then lengthening the cylinder by T while keeping the metric constant outside the cylindrical region. Let us suppose that $b^+(X) = 1$. This means that each X_i has $b^+ = 0$, and the image of $H_2(Y;\mathbb{R})$ in $H_2(X;\mathbb{R})$ is 1-dimensional. (See Lemma 3.9.2.) Let $k \in H^2(X;\mathbb{R})$ be the Poincaré dual of a non-zero element in the image of this map. The element k is a null vector for the cup square, so it lies on the boundary of the positive cone: it therefore picks out one component of the cone, or equivalently an orientation of $H^+(X)$. Having chosen k therefore, we can talk unambiguously about $\mathfrak{m}_+(X,\mathfrak{s}_X)$ and $\mathfrak{m}_-(X,\mathfrak{s}_X)$ (although the overall sign of these invariants depends additionally on an orientation of $H^1(X;\mathbb{R})$ if b_1 is non-zero).

For each Riemannian metric g_T, let κ_T be a self-dual harmonic form, lying in the distinguished component of the positive cone. As $T \to \infty$, the metric degenerates; but the rays spanned by the classes $[\kappa_T]$ in $H^2(X;\mathbb{R})$ have a well-defined limit. If they are suitably normalized, we have

$$[\kappa_T] \to k$$

as $T \to \infty$. For each spin^c structure $\mathfrak{s}_X$, let us define an invariant $\mathfrak{m}_k(X,\mathfrak{s}_X)$ by specifying:

$$\mathfrak{m}_k(X,\mathfrak{s}_X) = \begin{cases} \mathfrak{m}_+(X,\mathfrak{s}_X), & \text{if } (c_1(\mathfrak{s}_X) \cup k)[X] < 0 \\ \mathfrak{m}_-(X,\mathfrak{s}_X), & \text{if } (c_1(\mathfrak{s}_X) \cup k)[X] > 0. \end{cases}$$

In the borderline case when $(c_1(\mathfrak{s}_X) \cup k)[X] = 0$ we resolve the situation using the sign of the pairing $k[\nu]$, between the limiting cohomology class of the self-dual forms and the homology class ν: we set

$$\mathfrak{m}_k(X,\mathfrak{s}_X) = \begin{cases} \mathfrak{m}_+(X,\mathfrak{s}_X), & \text{if } (c_1(\mathfrak{s}_X) \cup k)[X] = 0 \text{ and } k[\nu] < 0 \\ \mathfrak{m}_-(X,\mathfrak{s}_X), & \text{if } (c_1(\mathfrak{s}_X) \cup k)[X] = 0 \text{ and } k[\nu] > 0. \end{cases}$$

We combine these into a generating function, much as $\mathfrak{m}(X,[\nu])$ is defined in the case $b^+ \geq 2$: that is,

$$\mathfrak{m}_k(X,[\nu]) = \sum_{\mathfrak{s}_X} \mathfrak{m}_k(X,\mathfrak{s}_X)\exp\langle c_1(\mathfrak{s}_X),[\nu]\rangle.$$

(This is not a finite sum, though all but finitely many of the non-zero terms arise from spin^c structures $\mathfrak{s}_X$ with $c_1(\mathfrak{s}_X) \cup k = 0$. Although there will be infinitely many terms of the latter sort, the sum converges, as we will see in the example below.) With this notation, we have the following extension of Proposition 3.9.3.

Proposition 27.5.1. *Suppose X has $b^+(X) = 1$, and let Y separate X into manifolds X_1, X_2 each with $b^+ = 0$. Let ν and η be as above, and let $\psi_{(X_i,\nu_i)}$ be the invariants of the two manifolds with boundary. Let k be Poincaré dual to a non-zero element of the image of $H_2(Y)$ in $H_2(X)$, and define $\mathfrak{m}_k(X,[\nu])$ as above. Then we have*

$$\mathfrak{m}_k(X,[\nu]) = \langle \psi_{(X_1,\nu_1)}, \psi_{(X_2,\nu_2)} \rangle_{\omega_\mu},$$

for appropriate homology orientations, as in Proposition 3.9.3.

To see an example of the calculation of $\mathfrak{m}_k(X,[\nu])$, from the definition, consider the manifold

$$X = \mathbb{CP}^2 \# 9\bar{\mathbb{CP}}^2,$$

which has $b^+ = 1$ and $b^- = 9$. The dimension formula tells us that the moduli spaces $M(X,\mathfrak{s}_X)$ have dimension zero only when $c_1(\mathfrak{s}_X)^2 = 0$. The manifold has odd intersection form, so the invariants $\mathfrak{m}_\pm(X,\mathfrak{s}_X)$ are potentially non-zero for spin^c structures whose first Chern class is a non-zero vector on the null cone. Pick one component, C^+, of the cone C, so as to define a homology orientation.

As discussed above, we are in a situation where the orthogonal space to each class $c_1(\mathfrak{s}_X)$ does not meet the interior of the positive cone. For perturbations ω which are small, we therefore have an invariant which does not depend on the metric g, because $c_1(\mathfrak{s}_X)$ and $[\kappa_g]$ cannot be orthogonal. On the other hand, the manifold X admits a metric of positive scalar curvature, so the invariant for small perturbations and any metric g is always zero. This gives:

$$\begin{aligned} \mathfrak{m}_-(X,\mathfrak{s}_X) &= 0, \quad \text{if } c_1(\mathfrak{s}_X) \in \partial C^+ \\ \mathfrak{m}_+(X,\mathfrak{s}_X) &= 0, \quad \text{if } c_1(\mathfrak{s}_X) \in \partial C^- . \end{aligned} \tag{27.17}$$

The manifold contains a smooth 2-torus F with zero self-intersection number: in algebro-geometric terms, it can be obtained by starting with a cubic curve in $\mathbb{CP}^2$, blowing up $\mathbb{CP}^2$ at nine points along this cubic, and taking F to be the proper transform of the curve. Let X_1 be a closed tubular neighborhood of F, and X_2 the closure of the complement. This decomposes X into two pieces with common a boundary a 3-torus $Y = \partial X_1$. The image of $H_2(Y;\mathbb{Z})$ in $H_2(X;\mathbb{Z})$ is generated by $[F]$. Let k be the class in X Poincaré dual to $[F]$. We suppose orientations are chosen so that k lies in the boundary of C^+.

If a non-zero vector c in $H^2(X;\mathbb{R})$ lies on the boundary of the positive cone C, then $c \cup k$ is non-zero unless c is proportional to k. If they are not proportional, the sign of the pairing is positive if c is in ∂C^+ and negative otherwise. From the definition of $\mathfrak{m}_k(X, \mathfrak{s}_X)$ and the vanishing results (27.17), we therefore see that

$$\mathfrak{m}_k(X, \mathfrak{s}_X) = 0, \quad \text{if } c_1(\mathfrak{s}_X) \notin \operatorname{span}(k).$$

In the special case that $c_1(\mathfrak{s}_X)$ and k are proportional (and hence orthogonal), the value of $\mathfrak{m}_k(X, \mathfrak{s}_X)$ depends on ν. Let us pick a homology class h with $k(h) > 0$, and choose ν with

$$[\nu] = \lambda h$$

for some non-zero λ. For spin^c structures whose first Chern class is orthogonal to k, we then have

$$\mathfrak{m}_k(X, \mathfrak{s}_X) = \begin{cases} \mathfrak{m}_+(X, \mathfrak{s}_X), & \text{if } \lambda < 0 \\ \mathfrak{m}_-(X, \mathfrak{s}_X), & \text{if } \lambda > 0. \end{cases}$$

Because k is a primitive class, the first Chern classes of spin^c structures that are proportional to k have the form

$$c_1(\mathfrak{s}_X) = lk,$$

for l an odd integer. For this spin^c structure, we know that $\mathfrak{m}_+(X, \mathfrak{s}_X) = 0$ if $l < 0$ and $\mathfrak{m}_-(X, \mathfrak{s}_X) = 0$ if $l > 0$, by (27.17). From the wall-crossing formula, we have $\mathfrak{m}_-(X, \mathfrak{s}_X) = -1$ if $l < 0$ and $\mathfrak{m}_+(X, \mathfrak{s}_X) = 1$ if $l > 0$. Using the definition of $\mathfrak{m}_k(X, [\nu])$, we see that if λ is *negative*

$$\mathfrak{m}_k(X, \lambda h) = \sum_{\substack{l \equiv 1 \ (2) \\ l > 0}} \exp(\lambda l k \cdot h)$$

while if λ is *positive* then

$$\mathfrak{m}_k(X, \lambda h) = \sum_{\substack{l \equiv 1 \ (2) \\ l < 0}} - \exp(\lambda l k \cdot h).$$

The sum is convergent in either case; and conveniently, both of these cases can be summarized in a single formula, valid for all $[\nu]$ with $k \cdot [\nu]$ non-zero:

$$\mathfrak{m}_k(X, [\nu]) = \frac{1}{2\sinh\bigl(k \cdot [\nu]\bigr)}. \tag{27.18}$$

A similar calculation can be done for the manifold $X = S^2 \times T^2$. Let $k = \text{P.D.}[F]$ be the class dual to the torus T^2 again, and decompose X along the 3-torus into two copies of $D^2 \times T^2$. For a spin^c structure $\mathfrak{s}_X$ with $c_1(\mathfrak{s}_X) = mk$ (with m now an even integer), the wall-crossing formula (27.15) evaluates to $m/2$. We again choose a homology class h having positive pairing with k, and set $[\nu] = \lambda h$. Then, as above, if λ is negative, we have

$$\mathfrak{m}_k(S^2 \times T^2, \lambda h) = \sum_{\substack{m \equiv 0 \ (2) \\ m > 0}} \frac{m}{2} \exp(\lambda m k \cdot h),$$

while if λ is positive, we have

$$\mathfrak{m}_k(S^2 \times T^2, \lambda h) = \sum_{\substack{m \equiv 0 \ (2) \\ m < 0}} -\frac{m}{2} \exp(\lambda m k \cdot h).$$

(There is an overall sign ambiguity here, depending on an orientation of $H^1(X)$.) The formulae can again be combined into a single formula,

$$\mathfrak{m}_k(S^2 \times T^2, [\nu]) = \left(\frac{1}{2\sinh\bigl(k \cdot [\nu]\bigr)}\right)^2, \tag{27.19}$$

valid whenever $k \cdot [\nu]$ is non-zero.

We will return to these calculations and Proposition 27.5.1 in Section 38, where we will use the proposition to extend these calculations to other 4-manifolds.

28 Canonical gradings

The Floer homology groups $\widecheck{HM}_\bullet(Y)$ and their companions are "graded" abelian groups, but the grading set, let us recall, is not $\mathbb{Z}$. In Section 3, we promised to describe a grading by the set $\pi_0(\Xi(Y))$, the set of homotopy classes of oriented 2-plane fields, or equivalently the homotopy classes of nowhere-zero vector fields. Thus far, we have graded these groups by a set with a rather different definition: a set we denoted $\mathbb{J}(Y)$, defined in Subsection 22.3. This space was a union over spin^c structures,

$$\mathbb{J}(Y) = \coprod_{\mathfrak{s}} \mathbb{J}(Y, \mathfrak{s}).$$

In this section we will study $\mathbb{J}(Y)$ as an object naturally associated to Y. We will show how $\mathbb{J}(Y)$ can be identified with $\pi_0(\Xi(Y))$, and discuss also a "$\mathbb{Q}$-grading" of the Floer groups: a $\mathbb{Z}$-equivariant map from $\mathbb{J}(Y, \mathfrak{s})$ to $\mathbb{Q}$ which can be defined whenever $c_1(\mathfrak{s})$ is a torsion class.

28.1 Spin^c structures, two-plane fields and complex structures

We will write ξ for a typical oriented 2-plane field on Y: an oriented, rank-2 subbundle of the tangent bundle. The following lemma provides a relationship between 2-plane fields and spin^c structures.

Lemma 28.1.1. *On an oriented Riemannian 3-manifold Y, there is a one-to-one correspondence between*

(i) *oriented 2-plane fields ξ;*
(ii) *1-forms θ of length 1; and*
(iii) *isomorphism classes of pairs $(\mathfrak{s}, \Phi)$ consisting of a spin^c structure and a unit-length spinor Φ.*

Proof. Given an oriented 2-plane field ξ, there is a unique unit-length 1-form θ which annihilates ξ and is positive on the positively-oriented normal field to ξ. This gives a bijection between (i) and (ii). If a pair $(\mathfrak{s}, \Phi)$ is given, there is a unique 1-form θ such that the i and $-i$ eigenspaces of $\rho(\theta)$ on the corresponding spin bundle S are $\mathbb{C}\Phi$ and $\Phi^\perp$ respectively.

To recover the pair $(\mathfrak{s}, \Phi)$ from θ and ξ, we proceed as follows. We can define the spin bundle S to be the rank-2 complex vector bundle $\underline{\mathbb{C}} \oplus \xi$. We define Φ to be the section 1 of the first summand. Clifford multiplication by θ is defined to be i on the first summand and $-i$ on the second. Clifford multiplication by a

1-form α orthogonal to θ is defined so that $\rho(\alpha)\Phi = (0, \alpha^\dagger)$, where $\alpha^\dagger$ is the vector in ξ which is dual to α using the Riemannian metric. □

An oriented 2-plane field on an oriented Riemannian 3-manifold Y gives an almost-complex structure on the cylinder $\mathbb{R} \times Y$: we define J so that (i) the 2-planes ξ are J-invariant and have their complex orientation, and (ii) the complex orientation of the tangent bundle to $\mathbb{R} \times Y$ is the product orientation. The above construction which assigns a spin^c structure to a 2-plane field on Y has a more familiar 4-dimensional counterpart involving complex structures. Given an almost-complex structure J, acting isometrically on the tangent bundle of a Riemannian 4-manifold X, we can construct a spin^c structure from the Dolbeault spaces of forms $\Lambda^{p,q}$. We set

$$S^+ = \Lambda^{0,0} \oplus \Lambda^{0,2}$$
$$S^- = \Lambda^{0,1},$$

and we define Clifford multiplication

$$\rho : T^*X \otimes S^+ \to S^-$$

by the symbol of $\sqrt{2}(\bar\partial + \bar\partial^*)$. Slightly more concretely, ρ can be characterized by two properties: first, for any unit-length element e in T^*X, we have

$$\rho(e)(1) = \frac{1}{\sqrt{2}}(e + ie \circ J) \in \Lambda^{0,1}$$

for the element 1 in $\Lambda^{0,0}$; and second, the determinant of $\rho(e)$, as a map from $\Lambda^2 S^+$ to $\Lambda^2 S^-$, coincides with the canonical identification of $\Lambda^{0,2}$ with $\Lambda^2(\Lambda^{0,1})$.

In the case that X is a cylinder $\mathbb{R} \times Y$ and J is such that it preserves a 2-plane field ξ coming from Y, this construction coincides with the one in the lemma above. On a general X, we do not recover all spin^c structures this way, only those spin^c structures admitting a non-vanishing section of S^+.

These two constructions are useful in combination, to prove the following proposition.

Proposition 28.1.2. *Let Y be an oriented Riemannian 3-manifold equipped with a spin^c structure $\mathfrak{s}$. Then there exists an oriented 4-manifold X with oriented boundary Y, carrying a spin^c structure $\mathfrak{s}_X$ whose restriction to the boundary is isomorphic to $\mathfrak{s}$.*

Proof. We use the fact that every 3-manifold is an oriented boundary to obtain a manifold X_1 with boundary Y. Let ξ be a 2-plane field corresponding to $\mathfrak{s}$ on Y, and let J be the corresponding almost-complex structure on a collar of Y in X_1. An almost-complex structure on a 4-manifold is the same thing as a unit section of Λ^+; and from this we see that, although J may not extend to all of X_1, it will extend to $X_1 \setminus Z$, where Z is a union of smoothly embedded circles (the zero set of a section of Λ^+ transverse to zero). In this way, the proof of the proposition is reduced to the case that Y is $S^1 \times S^2$ (the boundary of the regular neighborhood of a component of Z). For this manifold, the result is easy to verify. □

28.2 Gradings and two-plane fields

Returning to dimension 3, we note that the spin^c structure which is associated to a given 2-plane field ξ by Lemma 28.1.1 is not enough to recover the homotopy class of ξ in general: the extra data provided by the section Φ is needed also. The next lemma quantifies this statement. In the statement of the lemma, the *divisibility* of a class $\epsilon \in H^2(Y;\mathbb{Z})$ is defined to be 0 if ϵ is torsion, and is defined to be the divisibility of the image of ϵ in the free abelian group $H^2(Y;\mathbb{Z})/\text{torsion}$ otherwise.

Lemma 28.2.1. *Let Y and a spin^c structure $\mathfrak{s}_0$ be given. The pairs $(\Phi, \mathfrak{s})$ consisting of a spin^c structure $\mathfrak{s}$ isomorphic to $\mathfrak{s}_0$ and a nowhere-zero section Φ of the associated bundle $S \to Y$ are classified up to homotopy by $\mathbb{Z}/(d\mathbb{Z})$, where d is the divisibility of $c_1(\mathfrak{s}_0)$.*

Proof. Let S_0 be the spin bundle of $\mathfrak{s}_0$. The unit sphere bundle in S_0 is a topologically trivial 3-sphere bundle $P \to Y$, and if we choose a trivialization of P we obtain a bijection between the homotopy classes of sections of P and the set $[Y, S^3]$, which we can identify with $\mathbb{Z}$ using the degree of the map. Another way to say this is that given two sections Φ_0 and Φ_1, there is a well-defined difference element $\delta(\Phi_0, \Phi_1) \in \mathbb{Z}$ which determines the homotopy class of Φ_1 once Φ_0 is known. We can construct δ directly as the Euler class of the pull-back of S to the cylinder $I \times Y$, relative to the sections Φ_0, Φ_1 at the boundary:

$$\delta(\Phi_0, \Phi_1) = e(I \times S_0, \Phi_0 \amalg \Phi_1)[I \times Y, \partial I \times Y].$$

The spin bundle S_0 is acted on by the automorphism group $\mathcal{G} = \text{Map}(Y, S^1)$. Thus $\mathcal{G}$ acts on sections of P; and the component group $[Y, S^1] = H^1(Y)$

therefore acts on $[Y, S^3] = \mathbb{Z}$. To compute this action, we examine

$$e(I \times S_0, \Phi_0 \amalg u\Phi_0)[I \times Y, \partial I \times Y]$$

for $u : Y \to S^1$. This computes the Euler number, or equivalently the second Chern number, of the bundle on $S^1 \times Y$ obtained from S_0 using u as a clutching function. As a complex vector bundle, S_0 is isomorphic to $\underline{\mathbb{C}} \oplus L$, where L is a line bundle with the same first Chern class; and with this observation we can calculate

$$\delta(\Phi_0, u\Phi_0) = \big([u] \cup c_1(\mathfrak{s})\big)[Y].$$

The homotopy classes of pairs (S, Φ) are therefore classified by the quotient of $\mathbb{Z}$ by the image of the map $H^1(Y) \to \mathbb{Z}$ given by pairing with $c_1(\mathfrak{s})$. That image is $d\mathbb{Z}$. □

Let Y and $\mathfrak{s}$ be given, equipped with a metric and admissible perturbation. We will now explain how to associate to each critical point $[\mathfrak{a}]$ a homotopy class of non-vanishing sections Φ_0 of the bundle $S \to Y$. By Proposition 28.1.2 we can find a manifold X with oriented boundary Y carrying a spin^c structure $\mathfrak{s}_X$ extending $\mathfrak{s}$. Write S^+ for the associated spin bundle. Recall that the configuration space $\mathcal{B}^\sigma(X^*, \mathfrak{s}_X; [\mathfrak{a}])$ of configurations asymptotic to $[\mathfrak{a}]$ on the cylindrical-end manifold falls into different connected components in general. Pick a component z and consider the corresponding configuration space $\mathcal{B}^\sigma_z(X^*, \mathfrak{s}_X; [\mathfrak{a}])$ and the index $\operatorname{gr}_z(X, \mathfrak{s}_X; [\mathfrak{a}]) \in \mathbb{Z}$ (which is the dimension of $M_z(X^*, [\mathfrak{a}])$ if this moduli space is regular and non-empty). We can choose a section Φ_0 of $S = S^+|_Y$ such that the relative Euler class satisfies

$$e(S^+, \Phi_0)[X, \partial X] = \operatorname{gr}_z(X, \mathfrak{s}_X; [\mathfrak{a}]). \tag{28.1}$$

Note that isomorphism class of (S, Φ_0), up to homotopy, is independent of z by the lemma, because if we change z then gr_z changes by a multiple of d.

Proposition 28.2.2. *Let Φ_0 be a section of $S \to Y$ determined as above by the condition* (28.1). *Then the isomorphism class of (S, Φ_0), up to homotopy of Φ_0, is independent of the choice of bounding 4-manifold X, and depends only on Y, $\mathfrak{s}$ and* $[\mathfrak{a}]$.

Proof. We can find a manifold X^o with oriented boundary $-Y$ and a spin^c structure extending $\mathfrak{s}$, so that the union $X \cup X^o$ is closed. Using the additivity of gr_z and the additivity of the relative Euler classes, we reduce to the case that X is a closed manifold. That is, we need to know that for a closed manifold X

with spin^c structure the difference of the dimension of the moduli space on the closed manifold and the Euler number of S^+ is independent of X. Indeed this difference is zero, as the next lemma states. (See Lemma 27.1.1 for the formula for the dimension of the moduli space on a closed manifold.) □

Lemma 28.2.3. *For a closed 4-manifold with spin^c structure, we have*

$$\left(c_2(S^+) - \frac{1}{4}c_1(S^+)^2\right)[X] = -\frac{1}{4}\big(2\chi(X) + 3\sigma(X)\big).$$

Proof. Both sides are equal to $(1/4)p_1(\Lambda^+ X)[X]$. □

Combining the above proposition with Lemma 28.1.1, we have a well-defined way of associating to each $(\mathfrak{s}, [\mathfrak{a}])$ on Y a homotopy class of 2-plane fields ξ. We denote this association by gr^π (where π stands to remind us of "2-plane"):

Definition 28.2.4. We write $\Xi(Y)$ for the space of all oriented 2-plane fields on Y. Given a critical point $[\mathfrak{a}]$ belonging to a spin^c structure $\mathfrak{s}$ on Y with an admissible tame perturbation $\mathfrak{q}$, we write $\mathrm{gr}^\pi([\mathfrak{a}])$ for the homotopy class of the 2-plane field ξ on Y corresponding to (S, Φ_0) under the equivalences in Lemma 28.1.1. Here Φ_0 is a nowhere-zero section of $S \to Y$ chosen according to the prescription (28.1). ◊

We can view gr^π as defining an isomorphism of sets with $\mathbb{Z}$-action,

$$\mathrm{gr}^\pi : \mathbb{J}(Y) \to \pi_0(\Xi(Y)).$$

To be explicit about the $\mathbb{Z}$ action on $\pi_0(\Xi(Y))$, and in particular its sign, we observe first that if $j \in \mathbb{J}(Y)$ and $\mathrm{gr}^\pi(j)$ corresponds to a pair (S, Φ), then $\mathrm{gr}^\pi(j+n)$ corresponds to a pair (S, Φ'), where the degrees of $\Phi : Y \to S^3$ and $\Phi' : Y \to S^3$ are related by

$$\deg(\Phi') = \deg(\Phi) - n.$$

(The minus sign is there because the grading of $[\mathfrak{a}]$ reflects minus the dimension of $M_z(X^*, [\mathfrak{a}])$.) So we should make the homotopy classes of sections of the sphere bundle $P \to Y$ into a $\mathbb{Z}$-set using *minus* the degree, and give $\pi_0(\Xi(Y))$ the inherited action.

When dealing with $\overline{HM}_\bullet(Y)$ we should again use a modified grading. If $[\mathfrak{a}]$ is a reducible critical point, we define

$$\bar{\mathrm{gr}}^\pi([\mathfrak{a}]) = \begin{cases} \mathrm{gr}^\pi([\mathfrak{a}]), & [\mathfrak{a}] \text{ is boundary-stable} \\ \mathrm{gr}^\pi([\mathfrak{a}]) - 1, & [\mathfrak{a}] \text{ is boundary-unstable,} \end{cases}$$

where the notation $\mathrm{gr}^{\pi}([\mathfrak{a}])-1$ refers to the $\mathbb{Z}$ action defined above. In this way, we make $\widecheck{HM}_\bullet(Y)$, $\widehat{HM}_\bullet(Y)$ and $\overline{HM}_\bullet(Y)$ into abelian groups graded $\pi_0(\Xi)$.

As an illustration, we can describe a 2-plane field on S^3 whose class is $\mathrm{gr}^{\pi}([\mathfrak{a}_0])$, where $[\mathfrak{a}_0]$ is the standard generator of $\widecheck{HM}_\bullet(S^3)$. The 3-sphere is the oriented boundary of the ball in $\mathbb{R}^4$, and we can calculate the index $\mathrm{gr}_z(B^4,[\mathfrak{a}_0])$ from Lemma 27.4.2. (There is only one component z because the sphere has no H_2.) That lemma tells us that $\mathrm{gr}_z(B^4,[\mathfrak{a}_{-i}]) = 2i-2$ for positive i; and for $i=0$ we therefore have $\mathrm{gr}_z(B^4,[\mathfrak{a}_0]) = -1$ by an application of Corollary 14.6.2. Let $S^{\pm}$ be the spin bundles of $\mathbb{R}^4$, let Ψ_- be a constant section of S^-, and let Φ be the section of S^+ on $\mathbb{R}^4$ given by

$$\Phi(x) = \rho(x)\Psi_-, \quad x \in \mathbb{R}^4.$$

Then Φ provides a non-vanishing section of $S = S^+|_{S^3}$ which extends to the ball to have a single zero, of degree -1, at the origin. So we have the correct relation (28.1), and the 2-plane field ξ corresponding to this Φ is the one we seek. This ξ can be characterized, up to homotopy, as a 2-plane field invariant under the subgroup $SU(2)_+$ in $\mathrm{Spin}(4) = SU(2)_+ \times SU(2)_-$ (the subgroup that acts trivially on S^-). This is the vector field we called ξ_- in the introductory Subsection 3.3.

If we recall that the generators for $\widecheck{HM}_\bullet(S^3)$ are the critical points $[\mathfrak{a}_i]$ for $i \geq 0$ (see Subsection 22.7), then we see that $\widecheck{HM}_\bullet(S^3)$ is non-zero in degrees $[\xi_-]+n$ for even, positive integers n, as stated in Subsection 3.3.

28.3 Torsion spin^c structures

If c is an integral 2-dimensional cohomology class on a cobordism W such that $c|_{\partial W}$ is a torsion class, then there is a well-defined square $\langle c, c\rangle$ in $\mathbb{Q}$, obtained as

$$\langle c, c\rangle = (\tilde{c} \cup \tilde{c})[W, \partial W],$$

where $\tilde{c}$ is any class in $H^2(W,\partial W;\mathbb{Q})$ whose image in $H^2(W;\mathbb{Q})$ is the same as the image of c. We can use this cup square to define a $\mathbb{Q}$-valued grading as follows.

Given a spin^c structure on Y, we again choose 4-manifold X with boundary Y over which the spin^c structure extends (see Proposition 28.1.2). We let W be the cobordism from S^3 to Y obtained by removing a ball from X. In the following definition, an admissible tame perturbation on Y is understood as always, and $[\mathfrak{a}_0]$ is again the critical point on S^3 which gives the lowest-degree generator for $\widecheck{HM}_\bullet(S^3)$.

Definition 28.3.1. Let $\mathfrak{s}$ be a spin^c structure on Y with $c_1(\mathfrak{s})$ torsion, and $[\mathfrak{a}]$ be a corresponding critical point. Let W be any cobordism from S^3 to Y over which $\mathfrak{s}$ extends, as above, and let z be any W-path from $[\mathfrak{a}_0]$ to $[\mathfrak{a}]$. We define a rational number $\mathrm{gr}^{\mathbb{Q}}([\mathfrak{a}])$ by the formula

$$\mathrm{gr}^{\mathbb{Q}}([\mathfrak{a}]) = -\mathrm{gr}_z\big([\mathfrak{a}_0], W, [\mathfrak{a}]\big) + \frac{1}{4}\big\langle c_1(S^+), c_1(S^+)\big\rangle - \iota(W) - \frac{1}{4}\sigma(W)$$

where $\iota(W)$ is the characteristic number defined in Definition 25.4.1 and S^+ is the spin bundle for the corresponding spin^c structure corresponding to z. For reducible critical points, we also define a modified grading,

$$\bar{\mathrm{gr}}^{\mathbb{Q}}([\mathfrak{a}]) = \begin{cases} \mathrm{gr}^{\mathbb{Q}}([\mathfrak{a}]), & [\mathfrak{a}] \text{ is boundary-stable} \\ \mathrm{gr}^{\mathbb{Q}}([\mathfrak{a}]) - 1, & [\mathfrak{a}] \text{ is boundary-unstable.} \end{cases}$$

◊

The above definition makes sense, because $\mathrm{gr}^{\mathbb{Q}}([\mathfrak{a}])$ as defined is independent of the choice of W and z. To see this, note first that $\iota(W)$, $\sigma(W)$ and $\langle c_1, c_1\rangle$ are all additive under the operation of composing cobordisms and W-paths, including the obvious extension to the case that we use a "cobordism" with one or both boundary components empty. Because gr_z is additive also, we can reduce to showing that the above expression is zero in the case of a closed manifold (replacing $\mathrm{gr}_z([\mathfrak{a}_0], W, [\mathfrak{a}])$ by the formal dimension d of the moduli space. We can confirm the vanishing of this expression by examining the definition of $\iota(W)$ and the formula for d in Lemma 27.1.1.

Definition 28.3.1 gives a $\mathbb{Z}$-equivariant map from $\mathbb{J}(Y, \mathfrak{s})$ to $\mathbb{Q}$ for each $\mathfrak{s}$ with torsion first Chern class. This identifies each such $\mathbb{J}(Y, \mathfrak{s})$ with a coset of $\mathbb{Z}$ in $\mathbb{Q}$. The definition is normalized so that the standard generator $[\mathfrak{a}_0]$ for $\widetilde{HM}_\bullet(S^3)$ has $\mathbb{Q}$-grading zero.

The definition of $\iota(W)$ already involves the signature $\sigma(W)$, so the reader may wonder why the formula in Definition 28.3.1 is written as it is. The point is just that $\iota(W)$ is an integer; so the formula makes clear that the fraction part of $\mathrm{gr}^{\mathbb{Q}}([\mathfrak{a}])$ is equal to the fractional part of $(\langle c_1, c_1\rangle - \sigma(W))/4$. If $[\mathfrak{a}]$ is a reducible critical point, we can say a little more:

Lemma 28.3.2. *If* $[\mathfrak{a}]$ *is a boundary-stable, reducible critical point and* $b_1(Y) = 0$, *then*

$$\mathrm{gr}^{\mathbb{Q}}([\mathfrak{a}]) = \frac{1}{4}\big\langle c_1(S^+), c_1(S^+)\big\rangle - \frac{1}{4}\sigma(W) \pmod{2\mathbb{Z}}$$

where $S = S^+ \oplus S^-$ *is any extension of the spin*^{c} *structure over a cobordism* W *from* S^3 *to* Y.

Proof. We can compute $\mathrm{gr}_z([\mathfrak{a}_0], [\mathfrak{a}])$ in Definition 28.3.1 by using a reducible configuration on W, in which case the index of the operator that defines gr_z is the sum of two terms: one term is the (real) index of a perturbed Dirac operators, which is even; and the other term is equal to $\iota(W)$. The remaining terms in the formula for $\mathrm{gr}^{\mathbb{Q}}([\mathfrak{a}])$ are the terms in the lemma. □

Corollary 28.3.3. *If Y is an integral homology 3-sphere and $[\mathfrak{a}]$ is a boundary-stable reducible critical point, then $\mathrm{gr}^{\mathbb{Q}}([\mathfrak{a}])$ is an even integer.*

Proof. When Y has no homology, the intersection form of W is unimodular, and the square of any characteristic vector is then equal to the signature mod 8. □

Remark. The last lemma above highlights the fact that the quantity

$$\frac{1}{4}\langle c_1(S^+), c_1(S^+)\rangle - \frac{1}{4}\sigma(W) \quad (\mathrm{mod}\ 2\mathbb{Z}) \tag{28.2}$$

depends only on the restriction of the spin^c structure to Y, and is otherwise independent of both W and the extension of the spin^c structure. It is worth noting that knowing only $c_1(\mathfrak{s})$ on Y (rather than the restriction of $\mathfrak{s}$ itself) is not enough to determine (28.2). A good example is the manifold $Y = \mathbb{RP}^3$, which we can regard as the boundary of a disk bundle Z of degree 2 over a 2-sphere Σ. We can take W to be the complement of a ball in Z, and we can consider the spin^c structures $\mathfrak{s}_0$ and $\mathfrak{s}_1$ on W which are uniquely characterized by specifying that $c_1(\mathfrak{s}_0) = 0$ and $c_1(\mathfrak{s}_1) \cdot [\Sigma] = 2$. As rational classes on W, these first Chern classes have square 0 and 2 respectively, so the corresponding quantities (28.2) differ. It follows that $\mathfrak{s}_0$ and $\mathfrak{s}_1$ have different restrictions as spin^c structures on Y, although both have the same first Chern class on Y.

If we have a cobordism W between two general 3-manifolds Y_0 and Y_1, and if $\mathfrak{s}$ is a spin^c structure on W, then we have an associated map

$$\widecheck{HM}_*(W, \mathfrak{s}) : \widecheck{HM}_*(Y_0, \mathfrak{s}_0) \to \widecheck{HM}_*(Y_1, \mathfrak{s}_1)$$

where $\mathfrak{s}_i$ is the restriction of $\mathfrak{s}$ to Y_i. If both $\mathfrak{s}_i$ have torsion first Chern class, then the two groups above are both graded by cosets of $\mathbb{Z}$, via the map $\mathrm{gr}^{\mathbb{Q}} : \mathbb{J}(Y_i, \mathfrak{s}_i) \to \mathbb{Q}$. With respect to these $\mathbb{Q}$-gradings, the map $\widecheck{HM}_*(W, \mathfrak{s})$ has a well-defined degree, a formula for which can be read off from the definition of $\mathrm{gr}^{\mathbb{Q}}$ using the additivity of that expression: the degree of the map is given by the rational number

$$\frac{1}{4}\langle c_1(S^+), c_1(S^+)\rangle - \iota(W) - \frac{1}{4}\sigma(W). \tag{28.3}$$

It is also worth recording how the canonical $\mathbb{Q}$-grading behaves under the duality isomorphisms:

Proposition 28.3.4. *When $c_1(\mathfrak{s})$ is torsion, the duality isomorphism*

$$\check{\omega}_\mu : \widetilde{HM}_\bullet(-Y) \to \widehat{HM}^\bullet(Y)$$

maps elements in $\mathbb{Q}$-grading j to elements in $\mathbb{Q}$-grading $-1 - b_1(Y) - j$. □

Notes and references for Chapter VII

The fact that the instanton Floer homology groups of 3-manifolds extend to a "functor" from a cobordism category to the category of groups was understood by Floer, and is treated in [32], where the composition law is proved by a stretching argument, much as we have done. (This proof is sometimes called the "continuation method".) The functoriality was used by Floer to prove the topological invariance of his instanton homology groups. The resulting "pairing formulae" for the instanton invariants of closed 4-manifolds were pointed out by Donaldson. From the beginning, it was understood that the behavior of reducible solutions (particularly on manifolds with $b^+ = 0$) meant that this functorial picture did not fit into the axiomatic framework of a "topological quantum field theory" in the sense of [4]. The structure we have described, involving the three flavors of Floer homology and the maps $\overrightarrow{HM}_\bullet(W)$ defined by cobordisms W with $b^+ \geq 2$, is modelled on the similar story for Heegaard Floer groups, developed in [97].

The $\mathbb{Q}$-grading of the Floer groups in the case of rational homology 3-spheres was observed by Frøyshov [41], and appears also in [97].

The exceptional place of 4-manifolds with $b^+ = 1$, and associated wall-crossing formulae, are features also of Donaldson's instanton invariants. Indeed, the very first examples of simply connected 4-manifolds that were homeomorphic but not diffeomorphic were discovered using an invariant that was special to this case, see [18].

VIII

Non-exact perturbations

One of the differences between the Morse theory of the Chern–Simons–Dirac functional and the standard picture from finite-dimensional Morse theory is that, on the quotient configuration space $\mathcal{B}(Y, \mathfrak{s})$, our functional is not a single-valued real function: its derivative has non-zero periods around loops in $\mathcal{B}(Y, \mathfrak{s})$. The perturbations that we have introduced have been single-valued: they have not changed the periods of the functional. It is natural to consider a larger class of perturbations, allowing the periods of the functional to change. We call such perturbations "non-exact", and they are the subject of this chapter.

In general, when we make a non-exact perturbation, the Morse homology groups of the corresponding gradient flow will change. Indeed, with ordinary coefficients, the Morse complex may cease to be defined. Nevertheless, there are isomorphisms in certain cases, between the various groups that arise this way.

One application of non-exact perturbations is to the proof of a gluing result, Proposition 3.9.3 (although the proposition itself makes no reference to non-exact perturbations). The argument is presented in Section 32.

29 Closed two-forms as perturbations

29.1 Perturbations revisited

The class of tame perturbations that we defined in Section 10 consists of sections $\mathfrak{q}$ of the L^2 tangent bundle $\mathcal{T}_0 \to \mathcal{C}(Y, \mathfrak{s})$ that are formal gradients of functions,

$$f : \mathcal{C}(Y, \mathfrak{s}) \to \mathbb{R},$$

invariant under the gauge-group $\mathcal{G}(Y)$. Rather than asking that f is invariant under $\mathcal{G}(Y)$, we can instead ask that its gradient is invariant, and that f itself is invariant only under the identity component $\mathcal{G}^e(Y)$. We refer to such sections of $\mathcal{T}_0$ as *non-exact* perturbations.

A non-exact perturbation determines a homomorphism

$$f_* : \mathcal{G}(Y)/\mathcal{G}^e(Y) \to \mathbb{R}$$

by

$$f_*(u) = f(u\mathfrak{a}) - f(\mathfrak{a}),$$

where $\mathfrak{a} \in \mathcal{C}(Y, \mathfrak{s})$ is any chosen point and $u : Y \to S^1$ is a chosen representative in $\mathcal{G}(Y)$. (The right-hand side is independent of the choice of $\mathfrak{a}$ because $\mathcal{C}(Y, \mathfrak{s})$ is connected.) Such homomorphisms f_* are parametrized by the real second cohomology of Y: there is a unique element c in $H^2(Y; \mathbb{R})$ such that

$$f_*(u) = (c \cup [u])[Y],$$

where $[u]$ denotes the cohomology class corresponding to u, the de Rham class of the form $u^{-1}du/(2\pi i)$. We call c the *period class* for the non-exact perturbation on $\mathcal{C}(Y, \mathfrak{s})$. The notion of *tameness* for a perturbation, as defined in Definition 10.5.1, extends straightforwardly to non-exact perturbations, because tameness is a condition on the gradient of f. We can therefore talk about tame perturbations with period class c.

An example of a tame perturbation with period class c is the formal gradient of the function $f : \mathcal{C}(Y, \mathfrak{s}) \to \mathbb{R}$ defined by

$$f_\omega(B, \Psi) = \int_Y (B - B_0)^{\mathrm{t}} \wedge \omega, \qquad (29.1)$$

where ω is a closed, imaginary-valued 2-form belonging to the class $(i/4\pi)c$. The *general* tame perturbation with class c is obtained from the above example by adding an exact perturbation $\mathfrak{q}$: we write such a perturbation as

$$\mathrm{grad}(f_\omega) + \mathfrak{q}.$$

If $\mathcal{P}(Y, \mathfrak{s})$ is a large Banach space of (exact) perturbations, as described in Definition 11.6.3, then we define an affine space of non-exact perturbations $\mathcal{P}_\omega(Y, \mathfrak{s})$ by

$$\mathcal{P}_\omega(Y, \mathfrak{s}) = \mathrm{grad}(f_\omega) + \mathcal{P}(Y, \mathfrak{s}).$$

For future reference, we give the perturbed functional corresponding to the perturbation (29.1) a name: we write

$$\mathcal{L}_\omega(B, \Psi) = \mathcal{L}(B, \Psi) + \int_Y (B - B_0)^{\mathrm{t}} \wedge \omega.$$

For this functional, the equations for a critical point (B, s, ψ) in the blown-up model $C^\sigma(Y, \mathfrak{s})$ are

$$\begin{aligned} \frac{1}{2}\rho(F_{B^{\mathrm{t}}} - 4\omega) - s^2(\psi\psi^*)_0 &= 0 \\ D_B\psi &= 0. \end{aligned} \tag{29.2}$$

The corresponding perturbed 4-dimensional equations for $(A, s, \phi) \in C^\tau(\mathbb{R}\times Y)$ on the cylinder are

$$\left.\begin{aligned} \frac{1}{2}\rho(F^+_{A^{\mathrm{t}}} - 4\omega^+) - s^2(\phi\phi^*)_0 &= 0 \\ \frac{d}{dt}s + \Lambda(A, s, \phi)s &= 0 \\ D^+_A\phi - \Lambda(A, s, \phi)\phi &= 0 \end{aligned}\right\} \tag{29.3}$$

(*cf.* Equations (6.11) and (6.10)). Here we have pulled ω back to the cylinder before taking its self-dual part. In the σ model for the blow-up, the equations are the same 4-dimensional equations we saw in (27.2): on a general 4-manifold equipped with a closed 2-form ω, we studied the equations

$$\begin{aligned} \frac{1}{2}\rho(F^+_{A^{\mathrm{t}}} - 4\omega^+) - s^2(\phi\phi^*)_0 &= 0 \\ D^+_A\phi &= 0, \end{aligned} \tag{29.4}$$

for (A, s, ϕ) in $C^\sigma(X, \mathfrak{s}_X)$. In the blown-down coordinates (A, Φ), the equations are

$$\begin{aligned} \frac{1}{2}\rho(F^+_{A^{\mathrm{t}}} - 4\omega^+) - (\Phi\Phi^*)_0 &= 0 \\ D^+_A\Phi &= 0. \end{aligned} \tag{29.5}$$

For these modified equations (29.5), we can write down appropriate modifications of the integration-by-parts formulae from Subsection 4.5, and in particular the relationship (4.16) between the topological and analytic energy. Given a general 4-manifold X with compact boundary Y, and a *closed* 2-form ω, we define the *topological* energy $\mathcal{E}_\omega^{\text{top}}$ as

$$\mathcal{E}_\omega^{\text{top}}(A,\Phi) = \frac{1}{4}\int_X (F_{A^t} - 4\omega)\wedge(F_{A^t} - 4\omega) - \int_Y \langle \Phi|_Y, D_B\Phi|_Y\rangle.$$

(The fact that ω is closed means that this expression is invariant under deformations of (A,Φ) supported in the interior of X.) We then have an identity

$$\mathcal{E}_\omega^{\text{an}}(A,\Phi) = \mathcal{E}_\omega^{\text{top}}(A,\Phi) + \|\mathfrak{F}_\omega(A,\Phi)\|^2, \tag{29.6}$$

where the *analytic* energy $\mathcal{E}_\omega^{\text{an}}$ is defined as

$$\begin{aligned}\mathcal{E}_\omega^{\text{an}}(A,\Phi) = \frac{1}{4}\int_X |F_{A^t} - 4\omega|^2 + \int_X |\nabla_A\Phi|^2 + \frac{1}{4}\int_X \left(|\Phi|^2 + (s/2)\right)^2 \\ - \int_X \frac{s^2}{16} + 2\int_X \langle \Phi, \rho(\omega)\Phi\rangle - \int_Y (H/2)|\Phi|^2,\end{aligned}$$

and $\mathfrak{F}_\omega(A,\Phi)$ is the left-hand side of the two equations (29.5). (As before, B denotes the spin^c connection obtained by restriction to the boundary and H is the second fundamental form.) If we expand it out, the equation (29.6) becomes the following identity, which is valid whether or not ω is closed, and which can be verified by a straightforward modification of the earlier argument:

$$\begin{aligned}&\int_X \left|\tfrac{1}{2}\rho(F_{A^t}^+ - 4\omega^+) - (\Phi\Phi^*)_0\right|^2 + \int_X |D_A^+\Phi|^2 \\ &= \frac{1}{4}\int_X |F_{A^t} - 4\omega|^2 + \int_X |\nabla_A\Phi|^2 + \frac{1}{4}\int_X \left(|\Phi|^2 + (s/2)\right)^2 - \int_X \frac{s^2}{16} \\ &\quad + 2\int_X \langle \Phi, \rho(\omega)\Phi\rangle - \frac{1}{4}\int_X (F_{A^t} - 4\omega)\wedge(F_{A^t} - 4\omega) \\ &\quad + \int_Y \langle \Phi|_Y, D_B\Phi|_Y\rangle - \int_Y (H/2)|\Phi|^2.\end{aligned}$$

In the cylindrical case, when ω is the pull-back of a form on Y, the topological energy is twice the drop in the functional $\mathcal{L}_\omega$ between the two ends of the cylinder. Note that, although the last term in the analytic energy is not positive, the integrand is quadratic in Φ, and so can be absorbed by the positive quartic term in $|\Phi|$.

For a general non-exact perturbation $\operatorname{grad}(f_\omega) + \mathfrak{q}$ on Y, we can write the perturbed functional $\not{\mathcal{L}}$ as

$$\not{\mathcal{L}} = \mathcal{L}_\omega + g,$$

where g is fully gauge-invariant and has gradient $\mathfrak{q}$, and we define the corresponding topological energy $\mathcal{E}^{\text{top}}_{\mathfrak{q}}$ on X as

$$\mathcal{E}^{\text{top}}_{\omega,\mathfrak{q}}(A, \Phi) = \mathcal{E}^{\text{top}}_{\omega}(A, \Phi) + 2g(A, \Phi).$$

The term $\mathcal{E}^{\text{top}}_{\omega}(A, \Phi)$ depends on the extension of ω to a closed 2-form on X. In the case of a cobordism, the term $g(A, \Phi)$ becomes the difference between the value of g at the two ends. In particular, in the cylindrical case $cE^{\text{top}}_{\omega,\mathfrak{q}}(A, \Phi)$ is twice the drop in the perturbed functional along the cylinder.

The Chern–Simons–Dirac function $\mathcal{L}$ is itself a function on $\mathcal{C}(Y, \mathfrak{s})$ which is invariant under $\mathcal{G}^e(Y)$ and whose gradient is invariant under $\mathcal{G}(Y)$. It too has a period class therefore, which we calculated in Lemma 4.1.3. The period class for $\mathcal{L}$ is $2\pi^2 c_1(\mathfrak{s})$. So if a non-exact perturbation has period class c, then the period class of the corresponding perturbed functional $\not{\mathcal{L}} = \mathcal{L} + f$ is $2\pi^2 c_1(\mathfrak{s}) + c$, the sum of the two.

Definition 29.1.1. Let $\not{\mathcal{L}} = \mathcal{L}_\omega + g$ be a non-exact perturbation of the Chern–Simons–Dirac functional, corresponding to a closed 2-form ω. Let c be the period class of f_ω, so that $2\pi c_1(\mathfrak{s}) + c$ is the period class of $\not{\mathcal{L}}$. We say that $\not{\mathcal{L}}$ is a *balanced* perturbation of $\mathcal{L}$ if $\not{\mathcal{L}}$ is fully gauge-invariant, or equivalently if the period class $2\pi^2 c_1(\mathfrak{s}) + c$ of $\not{\mathcal{L}}$ is zero. We say that $\not{\mathcal{L}}$ is *monotone* if the period class of $\not{\mathcal{L}}$ is a real multiple of $c_1(\mathfrak{s})$. It is *positively* or *negatively* monotone if

$$2\pi^2 c_1(\mathfrak{s}) + c = t 2\pi^2 c_1(\mathfrak{s})$$

and t is positive or negative respectively. ◊

In the balanced case, $\not{\mathcal{L}}$ descends to a single-valued function on $\mathcal{B}(Y, \mathfrak{s})$. This is the only case in which reducible solutions need concern us, as the following lemma states.

Lemma 29.1.2. *If $\mathcal{L}_\omega = \mathcal{L} + f_\omega$ is not balanced, then there are no reducible critical points for $\mathcal{L}_\omega$ in $\mathcal{B}(Y, \mathfrak{s})$.*

Proof. The equations for a reducible critical point $(B, 0, \psi)$ of the perturbed equations in the blown-up picture are:

$$F_{B^t} - 4\omega = 0$$
$$D_B\psi = 0.$$

The first equation implies that $c_1(\mathfrak{s})$ is equal to $(2i/\pi)[\omega]$, or $-(1/2\pi^2)c$, so there are no solutions except in the balanced case. □

29.2 Compactness and finiteness

Most of the theory we have developed needs very little modification in the case of a non-exact perturbation. Let us fix a spin^c structure $\mathfrak{s}$ and a closed 2-form ω on Y, with period class c. With no change to the previous proofs, one shows that there exist admissible perturbations $\text{grad}(f_\omega) + \mathfrak{q}$ in $\mathcal{P}_\omega(Y, \mathfrak{s})$ such that the critical points $[\mathfrak{a}]$ in $\mathcal{B}_k^\sigma(Y, \mathfrak{s})$ are non-degenerate and such that the moduli spaces (which we continue to denote by $M([\mathfrak{a}], [\mathfrak{b}])$) are regular. If the perturbation is not balanced, we can also assume that there are no reducible critical points, by the preceding lemma.

The space of broken trajectories $\breve{M}_z^+([\mathfrak{a}], [\mathfrak{b}])$ can be defined as before, along with its various cousins, and this space is still compact, for any $[\mathfrak{a}]$, $[\mathfrak{b}]$ and z, just as in Theorem 16.1.3. The essential point in this is to verify that Theorem 10.7.1, and Lemma 10.6.1 on which it depends, still hold. The principal raw ingredient of the argument is the identity (29.6) relating the analytic and topological energy in our new, perturbed setting: with $\mathcal{E}_\omega^{\text{an}}$ in place of $\mathcal{E}^{\text{an}}$, we can repeat the previous arguments without difficulty.

Despite these similarities, there is one important point at which the non-exact theory departs from the exact case. This is in the finiteness results of Subsection 16.4, and in particular at Proposition 16.4.1 and Proposition 16.4.3. We state a version of Proposition 16.4.3 appropriate for the new situation.

Proposition 29.2.1. *Suppose that all the moduli spaces $M_z([\mathfrak{a}], [\mathfrak{b}])$ for the non-exact perturbation $\mathfrak{q}$ are regular.*

(i) *Suppose $\mathfrak{L}$ is balanced. Then for a given $[\mathfrak{a}]$ and any d_0, there are only finitely many pairs $([\mathfrak{b}], z)$ for which the moduli space $\breve{M}_z^+([\mathfrak{a}], [\mathfrak{b}])$ is non-empty and of dimension at most d_0.*
(ii) *Suppose $\mathfrak{L}$ is negatively monotone, and that the perturbation has been chosen so that there are no reducible solutions. Then the same conclusion holds as in the balanced case above.*

(iii) *Suppose* $\mathcal{L}$ *is positively monotone, and that the perturbation has again been chosen so that there are no reducible solutions. Then for a given* $[\mathfrak{a}]$, *there are only finitely many pairs* $([\mathfrak{b}], z)$ *for which the moduli space* $\breve{M}_z^+([\mathfrak{a}], [\mathfrak{b}])$ *is non-empty (without restriction on the dimension).*

Proof. The proof of Proposition 16.4.3 in the non-torsion case hinged on the fact that, for paths w joining two given critical points $[\mathfrak{a}]$ and $[\mathfrak{b}]$, the energy $\mathcal{E}^{\text{top}}_{\omega,\mathfrak{q}}(w)$ (i.e. twice the drop in $\mathcal{L}$) is negatively proportional to the relative grading $\text{gr}_w([\mathfrak{a}], [\mathfrak{b}])$. This proportionality is expressed as the statement that the quantity in (16.4) is independent of w. In the positively monotone case, the same is true: there is a positive constant t such that

$$\mathcal{E}^{\text{top}}_{\omega,\mathfrak{q}}(w) + t\,\text{gr}_w([\mathfrak{a}], [\mathfrak{b}]) \tag{29.7}$$

is independent of w. With this in hand, the proof of the third case of the present proposition is the same as the proof of the non-torsion case of Proposition 16.4.3.

Now consider the negatively monotone case. There is now a *negative* t such that the quantity (29.7) is independent of w. To say that the dimension of $M_w([\mathfrak{a}], [\mathfrak{b}])$ is at most d_0 is to bound $\text{gr}_w([\mathfrak{a}], [\mathfrak{b}])$ from above; and since t is negative this bounds $\mathcal{E}^{\text{top}}_{\omega,\mathfrak{q}}(w)$ from above. Proposition 16.1.4 still holds for non-exact perturbations; so the energy bound implies that only finitely many of these moduli spaces can be non-empty. There are only finitely many critical points in the absence of reducibles, so the result follows.

Finally we do the balanced case. Suppose that the moduli spaces $M_{z_i}([\mathfrak{a}], [\mathfrak{b}_i])$ are non-empty. Because the image of the critical set under π is finite, we may as well assume that $\pi[\mathfrak{b}_i] = [\beta]$ for all i. In the balanced case, $\mathcal{L}$ descends to a single-valued function on $\mathcal{B}_k(Y, \mathfrak{s})$, so the energy of trajectories in all these moduli spaces is uniformly bounded, as in the torsion case of Proposition 16.4.3. Proposition 16.2.1 now tells us that there are only finitely many possibilities for the homotopy class of the path $\pi(z_i)$ in $\mathcal{B}_k(Y, \mathfrak{s})$. A bound on the dimension (above by d_0 and below by zero) now bounds $\iota([\mathfrak{b}_i])$ above and below, so only finitely many $[\mathfrak{b}_i]$ occur. □

In the light of the proposition above, it is natural to define a class of *admissible* non-exact perturbations, extending the definition made for the exact case in Definition 22.1.1. We say that a tame non-exact perturbation is regular if the corresponding critical points are all non-degenerate, the moduli spaces of trajectories are regular, and there are no reducible critical points unless the perturbed functional is balanced.

30 Floer groups and non-exact perturbations

30.1 Floer homology in the monotone and balanced cases

In the monotone and balanced cases, the construction of Floer homology groups proceeds with essentially no modification from the exact case. We have learned that, if the perturbation is admissible and if there are no reducibles in the non-exact case, then for any given $[\mathfrak{a}]$ and d_0, there are only finitely many $[\mathfrak{b}]$ and z for which the moduli space $M_z([\mathfrak{a}], [\mathfrak{b}])$ can be non-empty. This allows us to construct the Floer complexes $\check{C}_*$, $\hat{C}_*$ and $\bar{C}_*$ as before. We give names to the resulting groups:

Definition 30.1.1. Let Γ be a local system of abelian groups on $\mathcal{B}^\sigma(Y, \mathfrak{s})$. If the perturbation $\operatorname{grad}(f_\omega) + \mathfrak{q} \in \mathcal{P}_\omega(Y, \mathfrak{s})$ has period class c and is balanced or monotone, we write

$$\widecheck{HM}_*(Y, \mathfrak{s}, c; \Gamma), \quad \widehat{HM}_*(Y, \mathfrak{s}, c; \Gamma), \quad \overline{HM}_*(Y, \mathfrak{s}, c; \Gamma)$$

for the resulting Floer homology groups, where c is the period class of f_ω. In particular, $\widecheck{HM}_*(Y, \mathfrak{s}, 0; \Gamma)$ for example coincides with our original $\widecheck{HM}_*(Y, \mathfrak{s}; \Gamma)$. ◊

These groups depend only on the period class c of the perturbation and are independent of the metric on Y: this can be proved, as in the exact case, by considering maps defined by a cylindrical cobordism $W = [a, b] \times Y$, on which we use a t-dependent perturbation in the class $\mathcal{P}_\omega(Y, \mathfrak{s})$ and a varying metric. (We will study cobordisms in more detail in Section 31 below.) The results of Section 22, meanwhile, carry over. We have, for example, the i–j–p exact sequence, a grading of the Floer homology group by $\mathbb{J}(Y, \mathfrak{s})$, and duality isomorphisms as in (22.36), such as

$$\omega_\mu : \widecheck{HM}_*(-Y, \mathfrak{s}, c; \Gamma) \to \widehat{HM}^*(Y, \mathfrak{s}, c; \Gamma).$$

(Note that c does not change sign.) There are also cup and cap products, making these groups into modules over $H^*(\mathcal{B}^\sigma(Y, \mathfrak{s}))^{\mathrm{opp}}$.

There is an additional point that arises in the balanced case. Recall that in the exact case, in addition to the usual chain complexes $\check{C}_*(Y, \mathfrak{s}; \Gamma)$, $\hat{C}_*(Y, \mathfrak{s}; \Gamma)$ and $\bar{C}_*(Y, \mathfrak{s}; \Gamma)$, we introduced their *negative completions*, $\check{C}_\bullet(Y, \mathfrak{s}; \Gamma)$, $\hat{C}_\bullet(Y, \mathfrak{s}; \Gamma)$ and $\bar{C}_\bullet(Y, \mathfrak{s}; \Gamma)$. (See page 514.) These are different from the $*$ versions only in the case that $c_1(\mathfrak{s})$ is torsion (which is the case that the exact perturbations are balanced). In the general case that $c_1(\mathfrak{s})$ is not torsion and the perturbed functional $\mathcal{L}$ is balanced, we can again introduce completions $\hat{C}_\bullet(Y, \mathfrak{s}, c; \Gamma)$

etc., but their definition needs to be phrased differently, because there is no $\mathbb{Z}$ grading on $\hat{C}_*(Y, \mathfrak{s}, c; \Gamma)$ if $c_1(\mathfrak{s})$ is not torsion. Rather than use a $\mathbb{Z}$ grading, we can define a filtration of $\hat{C}_*(Y, \mathfrak{s}, c; \Gamma)$ as follows. Label the reducible critical points for the perturbed functional on $\mathcal{B}_k(Y, \mathfrak{s})$ as $[\alpha^1]$, ..., $[\alpha^p]$, and label the corresponding critical point in the blow-up as $[\mathfrak{a}_i^r]$ with $i \in \mathbb{Z}$ and $r = 1, \ldots, p$, so that, as usual, the points $[\mathfrak{a}_i^r]$ correspond to the eigenvalues λ_i^r of the perturbed Dirac operator at $[\alpha^r]$, arranged in increasing order, with λ_0^r the first positive eigenvalue. Then define a subgroup

$$\hat{C}_*(Y, \mathfrak{s}, c; \Gamma)_m \subset \hat{C}_*(Y, \mathfrak{s}, c; \Gamma)$$

to be the sum of the contributions from the critical points $[\mathfrak{a}_i^r]$ with $i \leq -m$. We define

$$\hat{C}_\bullet(Y, \mathfrak{s}, c; \Gamma) \supset \hat{C}_*(Y, \mathfrak{s}, c; \Gamma) \qquad (30.1)$$

to be the completion of this filtered group. We define $\bar{C}_\bullet(Y, \mathfrak{s}, c; \Gamma)$ in the same manner. For $\check{C}_*$, there is no completion to be done; but as in the exact case, we use $\check{C}_\bullet(Y, \mathfrak{s}, c; \Gamma)$ as a synonym for $\check{C}_*(Y, \mathfrak{s}, c; \Gamma)$. We write

$$\widecheck{HM}_\bullet(Y, \mathfrak{s}, c; \Gamma), \quad \widehat{HM}_\bullet(Y, \mathfrak{s}, c; \Gamma), \quad \overline{HM}_\bullet(Y, \mathfrak{s}, c; \Gamma).$$

Remark. In the exact case, when $c_1(\mathfrak{s})$ is torsion, we can form the completion at the chain level, or at the level of homology: the homology of the completion is the completion of the homology. In the non-exact case, where there is no $\mathbb{Z}$ grading, we need to form the completion at the chain level. As we shall see, there are cases where $\widehat{HM}_\bullet(Y, \mathfrak{s}, c; \Gamma)$ is zero, though $\widehat{HM}_*(Y, \mathfrak{s}, c; \Gamma)$ is non-zero. An example of a related phenomenon in ordinary homology is the 2-step complex

$$\mathbb{Z}[u] \xrightarrow{d} \mathbb{Z}[u]$$

where d is multiplication by $1 - u$. This complex has non-trivial first homology group. However, if we consider the same d but replace $\mathbb{Z}[u]$ by the completion $\mathbb{Z}[[u]]$, then the homology becomes trivial, because $1 - u$ has an inverse in the ring of power series. When the filtration is defined using the homological degree, as in the exact case, completion commutes with taking homology.

30.2 Floer homology in the general non-exact case

If the period class of the perturbed functional $\mathcal{L} = \mathcal{L} + f$ is not proportional to $c_1(\mathfrak{s})$ (i.e. if we are not in the monotone case), then there will always be a closed loop w in $\mathcal{B}^\sigma(Y,\mathfrak{s})$ on which the spectral flow is zero and on which the drop in $\mathcal{L}$ is positive. If $[\mathfrak{a}]$ and $[\mathfrak{b}]$ are two critical points, and z is a relative homotopy class of paths joining them, with $\mathrm{gr}_z([\mathfrak{a}],[\mathfrak{b}]) = d$, then the homotopy classes

$$z_k = z + kw \in \pi_1(\mathcal{B}^\sigma_k(Y,\mathfrak{s}); [\mathfrak{a}], [\mathfrak{b}])$$

have the property that $\mathrm{gr}_{z_k}([\mathfrak{a}],[\mathfrak{b}]) = d$ for all k, while the energy $\mathcal{E}^{\mathrm{top}}_{\omega,\mathfrak{q}}(z_k)$ increases without bound. It may well be that infinitely many of the moduli spaces $M_{z_k}([\mathfrak{a}],[\mathfrak{b}])$ are non-empty. Any analog of Proposition 29.2.1 therefore fails.

Let us consider how this affects our construction of the Morse complex $\check{C}_*(Y,\mathfrak{s},c;\Gamma)$. Note first that, because we are now concerned with a situation in which c is not balanced, we may assume there are no reducible critical points. This means that the set of all critical points will be finite, and it also means that there will be no difference between $\check{C}_*$ and $\hat{C}_*$. We simply write C_* for this object, which we define as

$$C_*(Y,\mathfrak{s},c;\Gamma) = \bigoplus_{[\mathfrak{a}]} \mathbb{Z}\Lambda[\mathfrak{a}] \otimes \Gamma[\mathfrak{a}].$$

When we try to define the differential, we encounter a problem. We wish to define, as before,

$$\partial = \sum_{[\mathfrak{a}]} \sum_{[\mathfrak{b}]} \sum_{z} \sum_{[\gamma] \in \breve{M}_z([\mathfrak{a}],[\mathfrak{b}])} \epsilon[\gamma] \otimes \Gamma(z) \tag{30.2}$$

where the sum is over all moduli spaces for which $M_z([\mathfrak{a}],[\mathfrak{b}])$ is 1-dimensional. But the contribution for a fixed $[\mathfrak{a}]$ and $[\mathfrak{b}]$ is potentially an infinite sum, because infinitely many classes z may contribute. After choosing trivializations for $\Lambda[\mathfrak{a}]$ and $\Lambda[\mathfrak{b}]$, the contribution for a given pair of critical points takes the form

$$\sum_z n_z \Gamma(z), \tag{30.3a}$$

where z runs through all relative homotopy classes satisfying the conditions

$$\mathrm{gr}_z([\mathfrak{a}],[\mathfrak{b}]) = 1 \tag{30.3b}$$

and $[z]$ counts the trajectories γ with appropriate sign. Although this sum may have infinitely many non-zero terms, the *support*

$$\mathrm{supp}(n) = \{\, z \mid n_z \neq 0 \,\}$$

is constrained, on account of Proposition 16.1.4: for all C, the intersection

$$\mathrm{supp}(n) \cap \{ z \mid \mathcal{E}^{\mathrm{top}}_{\omega,\mathfrak{q}}(z) \le C \}$$

is finite. The support of n therefore satisfies a finiteness condition, which we now define:

Definition 30.2.1. Let c be the period class of f_ω, and let $\mathcal{E}^{\mathrm{top}}_{\omega,\mathfrak{q}}$ be a corresponding perturbation of the topological energy. We say that a subset

$$S \subset \pi_1(\mathcal{B}^\sigma_k(Y,\mathfrak{s}), [\mathfrak{a}], [\mathfrak{b}])$$

is *c-finite* if the following two conditions hold:

(i) for all C, the intersection

$$S \cap \{ z \mid \mathcal{E}^{\mathrm{top}}_{\omega,\mathfrak{q}}(z) \le C \}$$

is finite;

(ii) there exists d such that $|\,\mathrm{gr}_z([\mathfrak{a}],[\mathfrak{b}])| \le d$ for all z in S.

(This condition depends on the period class c, but not otherwise on ω or $\mathfrak{q}$.) ◊

Remarks. In the monotone case, including the balanced case, a set S has this property only if it is finite.

What we want, then, is a local system of abelian groups in which a sum of the form (30.3a) makes sense whenever the support of $n \mapsto n_z$ is c-finite.

We consider a local system of complete topological abelian groups Γ on $\mathcal{B}^\sigma(Y,\mathfrak{s})$. This means that each $\Gamma[\mathfrak{a}]$ is a complete topological group, and the homomorphisms $\Gamma(z) : \Gamma[\mathfrak{a}] \to \Gamma[\mathfrak{b}]$ are continuous. We suppose that $0 \in \Gamma[\mathfrak{a}]$ has a neighborhood basis (not necessarily countable) consisting of subgroups: so $\Gamma[\mathfrak{a}]$ is a complete *filtered group*, filtered by the open subgroups. We write $\mathrm{Hom}(\Gamma[\mathfrak{a}],\Gamma[\mathfrak{b}])$ for the group of continuous homomorphisms, which we equip with the compact–open topology. In this way, $\mathrm{Hom}(\Gamma[\mathfrak{a}],\Gamma[\mathfrak{b}])$ is again a topological group, with a topology obtained from a filtration by open subgroups. A neighborhood basis for 0 in $\mathrm{Hom}(\Gamma[\mathfrak{a}],\Gamma[\mathfrak{b}])$ consists of the subgroups

$$\begin{aligned}\Omega(N,V) &= \{ k : \Gamma[\mathfrak{a}] \to \Gamma[\mathfrak{b}] \mid k(N) \subset V \} \\ &\subset \mathrm{Hom}(\Gamma[\mathfrak{a}],\Gamma[\mathfrak{b}])\end{aligned}$$

where N runs over compact subsets of $\Gamma[\mathfrak{a}]$ and V runs through open subgroups of $\Gamma[\mathfrak{b}]$. Because an open subgroup V is also closed, we can equivalently define the neighborhood basis by letting N run over *precompact* subsets, rather than compact subsets. To recognize a precompact set, we note that a subset $N \subset \Gamma[\mathfrak{a}]$ is precompact if and only if $(N+U)/U$ is finite for all open subgroups U of $\Gamma[\mathfrak{a}]$.

With this topology, a sufficient condition for a countable series $\sum_{k\in K} k$ in $\mathrm{Hom}(\Gamma[\mathfrak{a}], \Gamma[\mathfrak{b}])$ to converge is that the terms converge to zero and that the terms are *equicontinuous*: that is, for each open subgroup U in $\Gamma[\mathfrak{b}]$, there exists an open subgroup V in $\Gamma[\mathfrak{a}]$ such that $k(V) \subset U$ for all $k \in K$. In many simple cases, $\Gamma[\mathfrak{a}]$ contains a compact set which generates an open subgroup, and in such cases, the equicontinuity is automatic. The reason for considering equicontinuous series is that this condition allows us to rearrange sums: if we have countable sets $K \subset \mathrm{Hom}(\Gamma[\mathfrak{a}], \Gamma[\mathfrak{b}])$ and $H \subset \mathrm{Hom}(\Gamma[\mathfrak{b}], \Gamma[\mathfrak{c}])$, both of which are equicontinuous and have terms going to zero, then we have

$$\left(\sum_{h\in H} h\right) \circ \left(\sum_{k\in K} k\right) = \sum_{H\times K} h \circ k.$$

We now impose a condition on our local systems so as to ensure that the infinite sums we encounter are summable and equicontinuous:

Definition 30.2.2. Let ω be an imaginary-valued closed 2-form on Y, and let c be the corresponding period class. We say that a local system of complete, filtered abelian groups Γ is *c-complete* if it satisfies the following two conditions, for every pair of points $[\mathfrak{a}]$, $[\mathfrak{b}]$:

(i) For any c-finite set $S \subset \pi_1(\mathcal{B}^\sigma_k, [\mathfrak{a}], [\mathfrak{b}])$, the set $\{\Gamma(z) \mid z \in S\} \subset \mathrm{Hom}(\Gamma[\mathfrak{a}], \Gamma[\mathfrak{b}])$ is equicontinuous.

(ii) For any c-finite set $S \subset \pi_1(\mathcal{B}^\sigma_k, [\mathfrak{a}], [\mathfrak{b}])$, the homomorphisms $\Gamma(z)$ converge to zero in the compact–open topology as z runs through S.

◊

These two conditions on Γ imply that a series such as

$$\sum_z n_z \Gamma(z)$$

is convergent in the compact–open topology to a continuous limit, provided that the support of $z \mapsto n_z$ is c-finite. Furthermore, the equicontinuity condition means that we can rearrange sums, as noted above. Thus the sum defining the operator ∂ in (30.2) is convergent, and the proof that $\partial\partial = 0$ goes through. In the monotone case, any local system of groups is c-complete, with the discrete topology. We summarize this discussion with a definition.

Definition 30.2.3. Let $\text{grad}(f_\omega) + \mathfrak{q}$ be an admissible, non-exact perturbation. Assume that we are not in the balanced case, and that the perturbation is chosen so that there are no reducible critical points. Let c be the period class of f_ω and let Γ be a c-complete local system of filtered abelian groups. We define the complex $C_*(Y, \mathfrak{s}, c; \Gamma)$ to be the (finite) direct sum

$$C_*(Y, \mathfrak{s}, c; \Gamma) = \bigoplus_{[\mathfrak{a}]} \mathbb{Z}\Lambda[\mathfrak{a}] \otimes \Gamma[\mathfrak{a}],$$

and define $\partial : C_* \to C_*$ by the convergent series (30.2). We define

$$HM_*(Y, \mathfrak{s}, c; \Gamma)$$

to be the homology of the complex (C_*, ∂). For consistency, we also introduce the notations

$$\widecheck{HM}_*(Y, \mathfrak{s}, c; \Gamma), \quad \widehat{HM}_*(Y, \mathfrak{s}, c; \Gamma), \quad \overline{HM}_*(Y, \mathfrak{s}, c; \Gamma),$$

using the first two as synonyms for HM_* and defining the third to be zero. $\diamond$

If we wish to construct Floer *cohomology* groups $HM^*(Y, \mathfrak{s}, c; \Gamma)$, we need the coefficients to satisfy a condition dual to c-completeness: whenever $S \subset \pi_1(\mathcal{B}^\sigma_k, [\mathfrak{a}], [\mathfrak{b}])$ is c-finite, we require

(i) the set $\{\, \Gamma(z^{-1}) \mid z \in S \,\} \subset \text{Hom}(\Gamma[\mathfrak{b}], \Gamma[\mathfrak{a}])$ is equicontinuous; and
(ii) the homomorphisms $\Gamma(z^{-1})$ converge to zero in the compact–open topology as z runs through S.

We call such a local system *c-dual-complete*. If Γ is a c-complete local system for Y, then it is a c-dual-complete system for the manifold $-Y$ with reversed orientation. The duality isomorphism ω_μ is still defined in this context.

Examples. We consider examples of c-complete local systems, following Novikov [87]. (The situation with Floer homology is slightly different from that in [87], because there may be non-trivial spectral flow on closed loops. In a finite-dimensional situation, there is no need for a counterpart to the second clause in our definition of the finiteness condition, Definition 30.2.1.)

The first case is when $b_1(Y) = 1$. If $c_1(\mathfrak{s})$ is non-zero as a real cohomology class, or if both $c_1(\mathfrak{s})$ and c are zero, then the perturbation is monotone, and all local systems are c-complete. So consider the case that $c_1(\mathfrak{s}) = 0$ and c is non-zero.

The fundamental group of $\mathcal{B}^\sigma(Y, \mathfrak{s})$ is $\mathbb{Z}$, and we can describe a local system up to isomorphism by giving the fiber at a basepoint $[\mathfrak{a}_0]$ and specifying the automorphism $\Gamma(z)$, where z is a closed loop at the basepoint representing a generator. Let z_0 be the generator for which $\mathcal{E}^{\text{top}}_\omega(z_0)$ is *positive*. Let R be a commutative ring, and let $R[t, t^{-1}]$ be the ring of finite Laurent series in a variable t, with coefficients in R. Let $U_{-k} \subset R[t, t^{-1}]$ be the R-submodule spanned by the generators t^j for $j \leq -k$. Topologize $R[t, t^{-1}]$ so that the subgroups U_{-k} are a neighborhood base at 0, and let $R[t, t^{-1}]^-$ be the completion. This is the ring of formal Laurent series, infinite in the negative direction:

$$\sum_{i=-\infty}^{k_0} r_i t^i.$$

We specify a local system by taking the fiber at $[\mathfrak{a}_0]$ to be $R[t, t^{-1}]^-$ and taking $\Gamma(z_0)$ to be multiplication by t^{-1}. A subset S of π_1 can be written in terms of the generator z_0 as

$$\{ k_s z_0 \mid s \in S \};$$

the set S satisfies the c-finiteness condition when the coefficients $k_s \in \mathbb{Z}$ are bounded below. This coefficient system is c-complete, and a series such as

$$\sum_{s \in S} n_s \Gamma(k_s z_0)$$

converges to the continuous homomorphism given by multiplication in $R[t, t^{-1}]^-$ by the element

$$\sum_{s \in S} n_s t^{-k_s}.$$

Consider now the general case. Let $I \subset \mathbb{R}$ be the set of periods of the perturbed functional, viewed as a discrete group: it is the image of the homomorphism

$$\begin{aligned} \mathcal{E}^{\text{top}}_\omega : \pi_1(\mathcal{B}^\sigma_k(Y, \mathfrak{s})) &\to \mathbb{R} \\ z &\mapsto \mathcal{E}(z). \end{aligned}$$

Let $R[I]$ be the group ring of I. For $\kappa \in \mathbb{R}$, let $U_{-\kappa}$ be the R-submodule spanned by the generators i in I satisfying $i \leq -\kappa$. Use these as open neighborhoods, and form the completion $R[I]^-$, which is a topological ring. We define a local system of topological groups by taking the fiber at $[\mathfrak{a}_0]$ to be $R[I]^-$, and specifying that for each closed loop z based at $[\mathfrak{a}_0]$, the automorphism $\Gamma(z)$ should be multiplication by the group ring element $-\mathcal{E}^{\text{top}}_{\omega}(z) \in I$. This defines a c-complete local system in the general case. One can write elements of $R[I]^-$ in a Laurent-series notation, by introducing a formal variable t and allowing the exponent i in t^i to be an element of I. In this way, a typical element is

$$\sum_{i \in I} r_i t^i,$$

where the support of $i \mapsto r_i$ is required to have finite intersection with every positive half-line $[C, \infty)$. If R is a field, then so is $R[I]^-$.

Although the construction just described is applicable in all cases, there is a modification of this construction that is possible in the case that $c_1(\mathfrak{s})$ is non-torsion. Let $\text{Ann}(c_1(\mathfrak{s})) \subset \pi_1(\mathcal{B}^\sigma(Y, \mathfrak{s}))$ be the annihilator of $c_1(\mathfrak{s})$, or equivalently the set of loops on which the spectral flow is trivial. Let $I' \subset I$ be the image of $\mathcal{E}^{\text{top}}_{\omega}$ restricted to this annihilator. Unless there is a loop with $\mathcal{E}^{\text{top}}_{\omega} = 0$ and non-trivial spectral flow, the subgroup I' is a proper subgroup of I, and the quotient is $\mathbb{Z}$. In any event, we can choose a projection

$$p : I \to I'.$$

Now construct a local system whose fiber at $[\mathfrak{a}_0]$ is $R[I']^-$ and in which the loops z based at $[\mathfrak{a}_0]$ act by

$$\Gamma(z) = \left(\text{multiplication by } p(-\mathcal{E}^{\text{top}}_{\omega}(z)) \in I'\right).$$

This is a c-complete local system. Unlike the case of $R[I]^-$, this modified version makes use of the second clause in the definition of the c-finiteness condition, Definition 30.2.1. In any monotone case, the group I' is trivial, and the local system Γ constructed this way is the trivial local system R.
In [87], Novikov introduced a completeness condition which is equivalent to c-complete in the case that the spectral flow around all loops is trivial, but which is in general stronger. We call this condition *strongly* c-complete:

Definition 30.2.4. Let c be the period of f_ω, and let $\mathcal{E}^{\text{top}}_{\omega,\mathfrak{q}}$ be a corresponding perturbed energy function. We say that a local system of complete, filtered abelian groups Γ is *strongly c-complete* if it satisfies the following two conditions for every pair of points $[\mathfrak{a}]$, $[\mathfrak{b}]$:

(i) for all subsets S of $\pi_1(\mathcal{B}_k^\sigma, [\mathfrak{a}], [\mathfrak{b}])$ such that

$$S \cap \{ z \mid \mathcal{E}_{\omega,\mathfrak{q}}^{\text{top}}(z) \leq C \} \tag{30.4}$$

is finite for all C, the set of homomorphisms $\{ \Gamma(z) \mid z \in S \} \subset \operatorname{Hom}(\Gamma[\mathfrak{a}], \Gamma[\mathfrak{b}])$ is equicontinuous;

(ii) for the same class of subsets S, the homomorphisms $\Gamma(z)$ converge to zero in the compact–open topology as z runs through S.

In other words, the notion of strongly c-complete is obtained from that of c-complete (Definition 30.2.2) by replacing the c-finiteness condition on S by the weaker finiteness condition on (30.4). ◊

In the example above, the local system with fiber $R[I']^-$ defined above is always c-complete, while $R[I]^-$ is strongly c-complete. In the case of first Betti number 1 and a non-torsion spin^c structure, the trivial local system with fiber $\mathbb{Z}$ is c-complete, while a strongly c-complete system has fiber $\mathbb{Z}[I]^-$, isomorphic to the ring of negative Laurent series with integer coefficients.

31 Some isomorphisms

31.1 Statement of results

We now begin to address the question of the extent to which the variants of Floer homology which we obtain by non-exact perturbations differ from the earlier versions. We begin with a statement about the balanced case. Recall that, in the balanced case, we have two versions of the groups, the second of which was obtained by completing the complex: see (30.1). In the balanced case, there is no need to consider any completion of the *coefficients* Γ: we may take Γ to be any local system of abelian groups, without topology.

Theorem 31.1.1. *For any Y, any spin^c structure $\mathfrak{s}$ on Y and any coefficients Γ, we have isomorphisms*

$$\begin{aligned}
\widecheck{HM}_\bullet(Y, \mathfrak{s}, c_b; \Gamma) &\cong \widecheck{HM}_\bullet(Y, \mathfrak{s}; \Gamma) \\
\widehat{HM}_\bullet(Y, \mathfrak{s}, c_b; \Gamma) &\cong \widehat{HM}_\bullet(Y, \mathfrak{s}; \Gamma) \\
\overline{HM}_\bullet(Y, \mathfrak{s}, c_b; \Gamma) &\cong \overline{HM}_\bullet(Y, \mathfrak{s}; \Gamma),
\end{aligned}$$

where $c_b = -2\pi c_1(\mathfrak{s})$ denotes the period class corresponding to a balanced perturbation $\mathcal{L}$ of $\mathcal{L}$.

Let us expand on the statement of this theorem. First, we can recall that $\widecheck{HM}_\bullet$ is always isomorphic to $\widecheck{HM}_*$ in both the exact and non-exact cases, as negative completion does not affect the complex $\check{C}_*$. Second, we note that if $c_1(\mathfrak{s})$ is torsion, then there is no content to this theorem, because the balanced perturbations are exact in this case. Third, we observe that if $c_1(\mathfrak{s})$ is *not* torsion, then the groups on the right-hand sides coincide with the "$*$" versions, because the negative completion has no effect when there are no reducibles. In this third case additionally, $\overline{HM}_\bullet(Y, \mathfrak{s}; \Gamma)$ is zero by default and $\widecheck{HM}_\bullet(Y, \mathfrak{s}; \Gamma)$ is isomorphic to $\widehat{HM}_\bullet(Y, \mathfrak{s}; \Gamma)$, so the theorem implies that these statements hold also for the balanced versions.

Next, we have an extension of the above result to the positively monotone case:

Theorem 31.1.2. *Let $\mathfrak{s}$ be a spin^c structure with $c_1(\mathfrak{s})$ non-torsion, and let Γ be any system of coefficients. Then we have an isomorphism*

$$HM_\bullet(Y, \mathfrak{s}, c; \Gamma) \cong HM_\bullet(Y, \mathfrak{s}; \Gamma)$$

for any period class c corresponding to a positively monotone perturbation, $\not{\mathcal{L}}$, of the Chern–Simons–Dirac functional $\mathcal{L}$.

Neither of the two theorems above says anything about the case that $c_1(\mathfrak{s})$ is torsion. In the torsion case, strictly non-exact perturbations of the functional are never monotone, and it is therefore necessary to use a c-complete local coefficient system Γ. In these circumstances, we again have an isomorphism:

Theorem 31.1.3. *Suppose that $c_1(\mathfrak{s})$ is torsion, let c be any non-zero period class, and let Γ be a c-complete local system of groups on $\mathcal{B}^\sigma(Y, \mathfrak{s})$. Then we have isomorphisms*

$$\begin{aligned}
\widecheck{HM}_*(Y, \mathfrak{s}; \Gamma) &\cong \widecheck{HM}_*(Y, \mathfrak{s}, c; \Gamma) \\
\widehat{HM}_*(Y, \mathfrak{s}; \Gamma) &\cong \widehat{HM}_*(Y, \mathfrak{s}, c; \Gamma) \\
\overline{HM}_*(Y, \mathfrak{s}; \Gamma) &\cong \overline{HM}_*(Y, \mathfrak{s}, c; \Gamma);
\end{aligned}$$

consequently, we have $\overline{HM}_*(Y, \mathfrak{s}; \Gamma) = 0$ and $\widecheck{HM}_*(Y, \mathfrak{s}; \Gamma) \cong \widehat{HM}_*(Y, \mathfrak{s}; \Gamma)$, because these properties hold (by definition) for the groups on the right.

The isomorphisms in all the above theorems respect the cup and cap products by elements of $H^*(\mathcal{B}^\sigma(Y, \mathfrak{s}))$, and behave as expected with respect to Poincaré duality. We are not in a position to discuss their naturality with respect to the maps obtained from cobordisms, because such maps have not been defined for the Floer groups with non-exact perturbations.

31.2 Proof of Theorem 31.1.1

We begin by proving Theorem 31.1.1. The technique will later be adapted to prove the other theorems. As noted above, we may as well assume that $c_1(\mathfrak{s})$ is not torsion. Fix a closed 2-form ω_b on Y so that $\mathcal{L} + f_{\omega_b}$ is balanced. Let $U = [0,1] \times Y$ be a cylindrical cobordism from Y to Y, equipped with 2-form $\tilde{\omega}$ which is equal to ω_b near $\{0\} \times Y$ and which is zero near $\{1\} \times Y$. (This form cannot be closed.) Extend this form to the cylindrical-end manifold $U^* = \mathbb{R} \times Y$ so that it is equal to ω_b on the incoming end and zero on the outgoing end. On the 4-dimensional cylinder U^*, we can now consider the equations

$$\mathfrak{F}_{\tilde{\omega}}(A, \Phi) = 0,$$

on $\mathcal{B}_{k,\mathrm{loc}}(U^*, \mathfrak{s})$, together with the corresponding equations on the blown-up configuration spaces. We choose (exact) tame perturbations $\mathfrak{q}_0 = \operatorname{grad} f_0$ and $\mathfrak{q}_1 = \operatorname{grad} f_1$, so that the two perturbed functionals

$$\begin{aligned} \pounds_0 &= \mathcal{L} + f_{\omega_b} + f_0 \\ \pounds_1 &= \mathcal{L} + f_1 \end{aligned}$$

both have non-degenerate critical points and regular moduli spaces of trajectories on $\mathbb{R} \times Y$ as usual. We choose a corresponding time-dependent perturbation $\mathfrak{p}$ on the cylinder, equal to $\mathfrak{q}_0$ on the incoming end and $\mathfrak{q}_1$ on the outgoing end, so that we can consider the equations $\mathfrak{F}_{\tilde{\omega},\mathfrak{p}}(A, \Phi)$ on U^*. Given critical points $\mathfrak{a}$ and $\mathfrak{b}$ for $\pounds_0$ and $\pounds_1$, we have moduli spaces of solutions of the blown-up equations, $M([\mathfrak{a}], U^*, [\mathfrak{b}])$. Each of these is a union

$$M([\mathfrak{a}], U^*, [\mathfrak{b}]) = \bigcup_z M_z([\mathfrak{a}], U^*, [\mathfrak{b}]).$$

Here z runs over the homotopy classes of paths from $[\mathfrak{a}]$ to $[\mathfrak{b}]$ in $\mathcal{B}_k^\sigma(Y, \mathfrak{s})$. We will use these moduli spaces to define chain maps, giving rise to homomorphisms

$$\begin{aligned} \check{\imath} &: \widecheck{HM}_\bullet(Y, \mathfrak{s}, c_b) \to \widecheck{HM}_\bullet(Y, \mathfrak{s}) \\ \hat{\imath} &: \widehat{HM}_\bullet(Y, \mathfrak{s}, c_b) \to \widehat{HM}_\bullet(Y, \mathfrak{s}), \end{aligned} \tag{31.1}$$

commuting with j_*. (On the right, the two groups are isomorphic, indeed identical, and j_* is the identity.) Later, we will construct inverse homomorphisms, to show that the homomorphisms are isomorphisms. We use $\mathbb{Z}$ coefficients here, and will continue to do so: the extension to arbitrary coefficients Γ is straightforward.

We give U^* its standard homology orientation, so that to each trajectory γ in a zero-dimensional moduli space $M_z([\mathfrak{a}], U^*, [\mathfrak{b}])$ we can associate a homomorphism

$$\epsilon([\gamma]) : \mathbb{Z}\Lambda([\mathfrak{a}]) \to \mathbb{Z}\Lambda([\mathfrak{b}])$$

by comparing the positive orientation of the point $[\gamma]$ with the orientation of the moduli space determined by given elements of $\Lambda([\mathfrak{a}])$, $\Lambda([\mathfrak{b}])$. We define homomorphisms

$$\begin{aligned} m_o^o &: C^o(Y, \mathfrak{s}, c_b) \to C^o(Y, \mathfrak{s}) \\ m_o^u &: C^u(Y, \mathfrak{s}, c_b) \to C^o(Y, \mathfrak{s}) \end{aligned}$$

by counting trajectories with signs:

$$\begin{aligned} m_o^o &= \sum_{[\mathfrak{a}]\in\mathfrak{C}^o(Y,\mathfrak{s},c_b)} \sum_{[\mathfrak{b}]\in\mathfrak{C}^o(Y,\mathfrak{s})} \sum_{[\gamma]\in M_z([\mathfrak{a}],U^*,[\mathfrak{b}])} \epsilon[\gamma] \\ m_o^u &= \sum_{[\mathfrak{a}]\in\mathfrak{C}^u(Y,\mathfrak{s},c_b)} \sum_{[\mathfrak{b}]\in\mathfrak{C}^o(Y,\mathfrak{s})} \sum_{[\gamma]\in M_z([\mathfrak{a}],U^*,[\mathfrak{b}])} \epsilon[\gamma]. \end{aligned} \tag{31.2}$$

We then define

$$\begin{aligned} \check{m} &: \check{C}_\bullet(Y, \mathfrak{s}, c_b) \to \check{C}_\bullet(Y, \mathfrak{s}) \\ \hat{m} &: \hat{C}_\bullet(Y, \mathfrak{s}, c_b) \to \hat{C}_\bullet(Y, \mathfrak{s}) \end{aligned}$$

by

$$\begin{aligned} \check{m} &= \begin{bmatrix} m_o^o & 0 \end{bmatrix} \\ \hat{m} &= \begin{bmatrix} m_o^o & m_o^u \end{bmatrix}, \end{aligned} \tag{31.3}$$

using the descriptions

$$\begin{aligned} \check{C}_\bullet(Y, \mathfrak{s}, c_b) &= C_\bullet^o(Y, \mathfrak{s}, c_b) \oplus C_\bullet^u(Y, \mathfrak{s}, c_b) \\ \check{C}_\bullet(Y, \mathfrak{s}) = \hat{C}_\bullet(Y, \mathfrak{s}) &= C_\bullet^o(Y, \mathfrak{s}). \end{aligned}$$

We must check that these homomorphisms are well-defined, using our compactness and finiteness results. To make this work, we use the topological energy $\mathcal{E}_0^{\mathrm{top}}$ corresponding to the *balanced* functional $\mathcal{L}_0$ as a book-keeping device. Before proceeding, let us consider what possibilities we are trying to rule out. Fix $[\mathfrak{a}]$ and $[\mathfrak{b}]$, and consider $[\gamma]$ in $M([\mathfrak{a}], U^*, [\mathfrak{b}])$. Since $\mathcal{L}_0$ is fully gauge-invariant, the corresponding topological energy

$$\mathcal{E}_0^{\mathrm{top}} = 2\big(\mathcal{L}_0([\mathfrak{a}]) - \mathcal{L}_0([\mathfrak{b}])\big)$$

is a fixed quantity, not depending on a choice of homotopy class z. The topological energy of γ on the half-cylinder $(-\infty, 0] \times Y$ is non-negative, because γ corresponds to a gradient trajectory of $\mathcal{L}_0$ there. The potential problem is that $\mathcal{L}_0$ may be *increasing* along the other end, $[1, \infty) \times Y$; and unless we can bound this increase, we do not have an energy bound on either end. The following lemma remedies the situation: it tells us that, for a gradient trajectory of $\mathcal{L}_1$, the value of $\mathcal{L}_0$ cannot increase too much.

Lemma 31.2.1. *Let I be an interval and consider the equations $\mathfrak{F}_{\mathfrak{q}_1}(A, \Phi) = 0$ on $I \times Y$, corresponding to the gradient-flow of $\mathcal{L}_1$. Then there exists a constant K, depending only on the perturbations and not on I, such that for each solution γ and each $t \in I$, we have either*

(i) $\frac{d}{dt}\mathcal{L}_0(t) \leq 0$, *or*

(ii) $\big|\mathcal{L}_0(t)\big| \leq K$.

Proof. We have

$$\frac{d}{dt}\mathcal{L}_0(t) = -\big\langle \operatorname{grad} \mathcal{L}_0, \operatorname{grad} \mathcal{L}_0 \big\rangle - \big\langle \mathfrak{r}, \operatorname{grad} \mathcal{L}_0 \big\rangle,$$

where $\mathfrak{r}$ is the difference in the perturbations:

$$\mathfrak{r} = \mathfrak{q}_1 - \mathfrak{q}_0 - \operatorname{grad} f_{\omega_b}.$$

So for each t in I, either $\frac{d}{dt}\mathcal{L}(t) \leq 0$, or

$$\| \operatorname{grad} \mathcal{L}_0 \|_{L^2(\{t\} \times Y)} \leq \| \mathfrak{r} \|_{L^2(\{t\} \times Y)}.$$

Suppose that the latter holds, for some t in I, and write $\check{\gamma}(t) = (A, \Phi)$. The hypotheses on the perturbations mean that the L^2 norm of $\mathfrak{r}$ is bounded by a multiple of $\|\Phi\|_{L^2}$. So from this bound on $\operatorname{grad} \mathcal{L}$ we obtain, via the usual inequalities, bounds on the L^2_1 norms of $\tilde{A} - A_0$ and $\tilde{\Phi}$, where $(\tilde{A}, \tilde{\Phi})$ is gauge-equivalent to (A, Φ). The functional $\mathcal{L}_0$, being balanced, is fully gauge-invariant; and $\mathcal{L}_0$ is controlled by the L^2_1 norm (or indeed the $L^2_{1/2}$ norm) of (A, Φ). □

Corollary 31.2.2. *There is a constant K_1 such that for all $[\gamma]$ belonging to one of the moduli spaces $M_z([\mathfrak{a}], U^*, [\mathfrak{b}])$, the topological energy $\mathcal{E}_0^{\text{top}}$ corresponding to the balanced functional $\mathcal{L}_0$ is bounded below on $[1, \infty) \times Y$ by $-K_1$:*

$$\mathcal{E}_0^{\text{top}}(\gamma|_{[1,\infty)\times Y}) \geq -K_1,$$

or equivalently,

$$\mathcal{L}_0(\gamma|_{\{1\}\times Y}) - \mathcal{L}_0(\mathfrak{b}) \geq -K_1/2.$$

□

If we examine the proof of Lemma 31.2.1, we see that it can also be applied when the term $\mathfrak{r}$ is time-dependent. We can therefore apply the same argument to the compact part of the cylinder, U, to obtain:

Corollary 31.2.3. *There is a constant K_2 such that for all $[\gamma]$ belonging to one of the moduli spaces $M_z([\mathfrak{a}], U^*, [\mathfrak{b}])$, the topological energy $\mathcal{E}_0^{\text{top}}$ corresponding to the balanced functional $\mathcal{L}_0$ is bounded below on $U = [0, 1] \times Y$ by $-K_2$:*

$$\mathcal{E}_0^{\text{top}}(\gamma|_U) \geq -K_1,$$

or equivalently,

$$\mathcal{L}_0(\gamma|_{\{0\}\times Y}) - \mathcal{L}_0(\gamma|_{\{1\}\times Y}) \geq -K_2/2.$$

□

From the two corollaries above, we have lower bounds on the topological energy $\mathcal{E}_0^{\text{top}}$ on the compact part of U^* and on the outgoing end. On the incoming end, $\mathcal{E}_0^{\text{top}}$ is non-negative. On the set of critical points of $\mathcal{L}_0$ or $\mathcal{L}_1$, the balanced functional $\mathcal{L}_0$ takes only finitely many values. Using the additivity of the topological energy, we now see that there is a constant K_3 such that, for any pair of critical points $[\mathfrak{a}], [\mathfrak{b}]$, any trajectory $[\gamma]$ in $M([\mathfrak{a}], U^*, [\mathfrak{b}])$, and any $t \in \mathbb{R}$, we have bounds

$$-K_3 \leq \mathcal{L}_0(\gamma|_{\{t\}\times Y}) \leq K_3. \tag{31.4}$$

From these bounds, we can obtain convergence on compact sets. For example, it follows that if $[\gamma_n]$ is any sequence belonging to moduli spaces $M_{z_n}([\mathfrak{a}_n], U^*, [\mathfrak{b}_n])$, then the blown-down configurations $[\pi\gamma_n]$ converge in $\mathcal{B}_{k,\mathrm{loc}}(U^*)$ after passing to a subsequence.

Having obtained convergence on compact sets, we need to investigate global convergence in the sense of broken trajectories, in the downstairs moduli spaces $N([\alpha], U^*, [\beta])$, as we did in (16.1). Because there is no blow-up involved, there are only finitely many critical points $[\alpha]$, $[\beta]$. The next proposition tells us that only finitely many moduli spaces are non-empty:

Proposition 31.2.4. *For each $[\alpha]$ and $[\beta]$, there are only finitely many homotopy classes z such that the moduli space $N_z([\alpha], U^*, [\beta])$ is non-empty. The corresponding moduli spaces $N_z^+([\alpha], U^*, [\beta])$ of broken trajectories are compact.*

Proof. Let $[\gamma_n]$ belong to $N_{z_n}([\alpha], U^*, [\beta])$. The bound (31.4), and the fact that γ_n corresponds to a gradient trajectory of $\mathcal{L}_0$ on the incoming end, means as usual that we have convergence of a subsequence, in the sense of broken trajectories, on the incoming end $(-\infty, 0] \times Y$. The fact that we have convergence on compact sets means that the same is true for the slightly larger piece $U^- = (-\infty, 1] \times Y$. It follows that the topological energy $\mathcal{E}_1^{\mathrm{top}}$ corresponding to the functional $\mathcal{L}_1$ is uniformly bounded on this part of U^*: there is a constant E such that

$$\left|\mathcal{E}_1^{\mathrm{top}}(\gamma_n|_{U^-})\right| \le E$$

for all n. On the remainder of U^*, namely the outgoing end $U^+ = [1, \infty) \times Y$, the energy $\mathcal{E}_1^{\mathrm{top}}$ is non-negative, because on this piece we have the gradient-flow equations of $\mathcal{L}_1$. Putting these together, we have

$$\mathcal{E}_1^{\mathrm{top}}(\gamma_n) \ge -E$$

on U^*. Because $\mathcal{L}_1$ is an exact perturbation, the quantity

$$\mathcal{E}_1^{\mathrm{top}}(\gamma_n) + 4\pi^2 \dim N_{z_n}([\alpha], U^*, [\beta]) \tag{31.5}$$

is independent of the homotopy class z_n (see (16.5), for example); so we also have an upper bound for $\mathcal{E}_1^{\mathrm{top}}(\gamma_n)$. Thus we obtain

$$\left|\mathcal{E}_1^{\mathrm{top}}(\gamma_n|_{U^+})\right| \le E'$$

for some E' independent of n. This allows us to extract a subsequence that converges in the sense of broken trajectories on U^+. We already have convergence on U^-, so we now have convergence on U^*. This shows that only finitely many homotopy classes can arise among the z_n, and that each moduli space of broken trajectories is compact. □

Having dealt with the downstairs moduli spaces, the corresponding result for the upstairs moduli spaces follows quite quickly. It is only on the incoming end that there are any reducible critical points, so the only question is whether there can be infinitely many non-empty moduli spaces of the form

$$M_{z_n}([\mathfrak{a}_n], U^*, [\mathfrak{b}]),$$

where $[\mathfrak{b}]$ is the preimage of some $[\beta]$ and all $[\mathfrak{a}_n]$ belong to the preimage of some $[\alpha]$. Because $\mathfrak{b}$ must be irreducible, these moduli spaces consist of irreducible solutions only, and this forces all the $\mathfrak{a}_n$ to be boundary-unstable. By the preceding proposition, we may take it that all the paths z_n project to the same homotopy class of paths from $[\alpha]$ to $[\beta]$ downstairs; and in this case, if $[\mathfrak{a}_n]$ corresponds to the eigenvalue λ_n of the perturbed Dirac operator at α, and if λ_n decreases without bound, then the dimension of the above moduli spaces becomes negative (see (16.5) again). Thus only finitely many $[\mathfrak{a}_n]$ can occur, and we have:

Corollary 31.2.5. *There are only finitely many triples* $([\mathfrak{a}], [\mathfrak{b}], z)$ *for which the moduli space* $M_z([\mathfrak{a}], U^*, [\mathfrak{b}])$ *is non-empty; and for each such triple the corresponding space of broken trajectories* $M_z^+([\mathfrak{a}], U^*, [\mathfrak{b}])$ *is compact.* □

This compactness theorem validates our definition (31.2), and hence the definition of $\check{m}$ and $\hat{m}$. With compactness out of the way, the arguments previously used in the context of exact perturbations adapt without difficulty: the fact that $\check{m}$ and $\hat{m}$ are chain maps follows from the usual considerations, and so we have constructed the desired maps (31.1) as

$$\check{\imath} = \check{m}_*$$
$$\hat{\imath} = \hat{m}_*.$$

It is a straightforward matter to verify that these maps commute with j_*.

To show that these maps are invertible, we construct maps in the opposite direction, which we will then show are inverses of $\check{\imath}$ and $\hat{\imath}$. To construct these, we replace the cylindrical cobordism U with another cylindrical cobordism $V = [0, 1] \times Y$, this time carrying a 2-form which is equal to ω_b at the end

$\{1\} \times Y$ and zero at the end $\{0\} \times Y$. The corresponding cylindrical-end manifold V^*, equipped with a corresponding 2-form, will define maps

$$n_o^o : C^o(Y, \mathfrak{s}) \to C^o(Y, \mathfrak{s}, c_b)$$
$$n_s^o : C^o(Y, \mathfrak{s}) \to C^s(Y, \mathfrak{s}, c_b)$$

and hence maps

$$\check{n} : \check{C}_\bullet(Y, \mathfrak{s}) \to \check{C}_\bullet(Y, \mathfrak{s}, c_b)$$
$$\hat{n} : \hat{C}_\bullet(Y, \mathfrak{s}) \to \hat{C}_\bullet(Y, \mathfrak{s}, c_b)$$

by

$$\check{n} = \begin{bmatrix} n_o^o \\ n_s^o \end{bmatrix}$$

and

$$\hat{n} = \begin{bmatrix} n_o^o \\ 0 \end{bmatrix}.$$

Corollary 31.2.5 has a mirror-image version applicable to U^*, so these maps are well-defined. Once again, $\check{n}$ and $\hat{n}$ are chain maps, and they define maps on the homology groups

$$\begin{aligned} \check{\jmath} &: \widecheck{HM}_\bullet(Y, \mathfrak{s}) \to \widecheck{HM}_\bullet(Y, \mathfrak{s}, c_b) \\ \hat{\jmath} &: \widehat{HM}_\bullet(Y, \mathfrak{s}) \to \widehat{HM}_\bullet(Y, \mathfrak{s}, c_b). \end{aligned} \tag{31.6}$$

To show that the maps $\check{\jmath}$ and $\hat{\jmath}$ are inverse to $\check{\imath}$ and $\hat{\imath}$, we examine the composites. The composites

$$\begin{aligned} \check{\jmath} \circ \check{\imath} &: \widecheck{HM}_\bullet(Y, \mathfrak{s}, c_b) \to \widecheck{HM}_\bullet(Y, \mathfrak{s}, c_b) \\ \hat{\jmath} \circ \hat{\imath} &: \widehat{HM}_\bullet(Y, \mathfrak{s}, c_b) \to \widehat{HM}_\bullet(Y, \mathfrak{s}, c_b) \end{aligned} \tag{31.7}$$

present a more interesting story than $\check{\imath} \circ \check{\jmath}$ and $\hat{\imath} \circ \hat{\jmath}$, so we will discuss only the former two.

The maps $\check{\imath}$ etc. are defined via the chain maps $\check{m}$ etc. on the negative completions of the complexes. The reader should notice that the maps could equally well have been defined without completion, on the $*$ version of the complexes. However, if we had used the $*$ versions, the composites (31.7) would *not* have been the identity (though the proof that the other composites, $\check{\imath} \circ \check{\jmath}$ and $\hat{\imath} \circ \hat{\jmath}$, are 1 would still have held).

To examine the composites (31.7), we form the composite cobordism $W = V \circ U$. Recall that with our conventional notation for composition of cobordisms, this means that the outgoing end of U is attached to the incoming end of V. We also form $W(S)$ by inserting a cylinder of length S between the two parts, as in (26.2), and we then attach two cylindrical ends to form $W(S)^*$. The cylindrical cobordism $W(S)^*$ carries a 2-form $\tilde{\omega}(S)$, which is equal to ω_b on the two ends and is zero on the interior neck of length S. For each pair of critical points $[\mathfrak{a}]$, $[\mathfrak{b}]$ of $(\operatorname{grad} \mathcal{L}_0)^\sigma$, and each homotopy class of paths z, we have a moduli space $M_z([\mathfrak{a}], W(S)^*, [\mathfrak{b}])$. As in (26.3), we take the union over S to form a parametrized moduli space

$$\mathcal{M}_z([\mathfrak{a}], [\mathfrak{b}]) = \bigcup_{S \in [0,\infty)} \{S\} \times M_z([\mathfrak{a}], W(S)^*, [\mathfrak{b}]). \tag{31.8}$$

A parametrized moduli space of this sort was the basis of the proof of the composition law for cobordisms in Section 26: the moduli spaces (31.8) were used to define a chain-homotopy. The same constructions will carry over to the non-exact case, as long as we can establish the necessary finiteness results. The next proposition addresses the finiteness issue.

Proposition 31.2.6. *Suppose that, for some fixed* $[\mathfrak{a}]$, *there are infinitely many distinct pairs* $([\mathfrak{b}_n], [z_n])$ *such that the moduli space* $\mathcal{M}_{z_n}([\mathfrak{a}], [\mathfrak{b}_n])$ *is non-empty. Then* $\iota(\mathfrak{b}_n) \to -\infty$ *as* $n \to \infty$.

Proof. Let $[\gamma_n]$ belong to the nth moduli space $M_{z_n}([\mathfrak{a}], W(S_n)^*, [\mathfrak{b}_n])$. As before, we obtain bounds (31.4) on the functional $\mathcal{L}_0$ and hence on the topological energy $\mathcal{E}_0^{\text{top}}$, leading to convergence of subsequences on compact sets. If we write $W(S_n)$ as a union

$$W(S_n) = W^- \cup [0, S_n] \times Y \cup W^+$$

then we also have convergence in the sense of broken trajectories, downstairs, on both W^- and W^+, and hence bounds on the other topological energy:

$$\left|\mathcal{E}_1^{\text{top}}(\gamma_n|_{W^-})\right| \le E$$
$$\left|\mathcal{E}_1^{\text{top}}(\gamma_n|_{W^+})\right| \le E.$$

On $[0, S_n] \times Y$, the energy $\mathcal{E}_1^{\text{top}}$ is non-negative, because there we have the gradient-flow equation for $\mathcal{L}_1$. If γ_n is irreducible for all n, then

$M_{z_n}([\mathfrak{a}], W(S_n)^*, [\mathfrak{b}_n])$ is a subset of some $N_{z_n}([\alpha], W(S_n)^*, [\beta_n])$, and we can apply the same type of dimension argument that we used for Proposition 31.2.4 to obtain an upper bound on $\mathcal{E}_1^{\text{top}}$, which leads to a finiteness result. So if the pairs $([\mathfrak{b}_n], z_n)$ are distinct, it must be that infinitely many of the $[\gamma_n]$ are reducible. We may then assume that all $[\mathfrak{b}_n]$ lie over the same downstairs critical point $[\beta]$, and that all $[\gamma_n]$ are reducible. We may also assume that $\mathcal{E}_1^{\text{top}}(\gamma_n) \to \infty$ as n increases.

At this point, we again use a dimension argument. As in (16.5), the quantity

$$\mathcal{E}_1^{\text{top}}(\gamma_n) + 4\pi^2\big(\text{gr}_{z_n}([\mathfrak{a}], [\mathfrak{b}_n]) - 2\iota(\mathfrak{a}) + 2\iota(\mathfrak{b}_n)\big) \tag{31.9}$$

is independent of n. The non-emptiness of the parametrized moduli spaces means that $\text{gr}_{z_n}([\mathfrak{a}], [\mathfrak{b}_n]) \geq -1$; so as $\mathcal{E}_1^{\text{top}}(\gamma_n)$ goes to $+\infty$, it follows that $\iota(\mathfrak{b}_n)$ must go to $-\infty$. □

The proposition tells us that we can use the moduli spaces $\mathcal{M}_z([\mathfrak{a}], [\mathfrak{b}])$ in the definition of maps between chain complexes, provided that we use negative completion when dealing with $C^u(Y, \mathfrak{s}, c_b)$. The proofs that the compositions (31.7) are the identity can now proceed, following the general outline of Section 26. This completes the proof of Theorem 31.1.1. □

31.3 Proof of the remaining two isomorphisms

Having proved Theorem 31.1.1, we can now turn to the proof of its two companion theorems. Theorem 31.1.2 can be deduced very quickly. Because of the previous theorem, we need only establish isomorphisms

$$\begin{aligned}
\widecheck{HM}_\bullet(Y, \mathfrak{s}, c_b; \Gamma) &\cong \widecheck{HM}_\bullet(Y, \mathfrak{s}, c; \Gamma)\\
\widehat{HM}_\bullet(Y, \mathfrak{s}, c_b; \Gamma) &\cong \widehat{HM}_\bullet(Y, \mathfrak{s}, c; \Gamma)\\
\overline{HM}_\bullet(Y, \mathfrak{s}, c_b; \Gamma) &\cong \overline{HM}_\bullet(Y, \mathfrak{s}, c; \Gamma),
\end{aligned}$$

between the balanced and positively monotone groups. We can write $\mathcal{E}_0^{\text{top}}$ again for the topological energy corresponding to the balanced functional, and $\mathcal{E}_1^{\text{top}}$ for the topological energy corresponding to the perturbation $\mathcal{L}_1$ whose period class $2\pi c_1(\mathfrak{s}) + c$ is positively proportional to $c_1(\mathfrak{s})$. The construction of the maps $\check{\imath}$ etc. can be repeated almost verbatim from the proof of Theorem 31.1.1. The only minor change is that the quantity (31.5) is no longer independent of the path z_n: instead, we simply have another positive constant τ such that

$$\mathcal{E}_1^{\text{top}}(\gamma_n) + \tau \dim N_{z_n}([\alpha], U^*, [\beta]) \tag{31.10}$$

is independent of n. This allows us to bound the energy $\mathcal{E}_1^{\text{top}}(\gamma_n)$ as before, and deduce the necessary finiteness. In proving that the composites $\check{j} \circ \check{\imath}$ and $\hat{j} \circ \hat{\imath}$ are the identity, there is a corresponding change at Formula (31.9), but the argument is otherwise identical. □

The last of the three is Theorem 31.1.3. Because this theorem is concerned only with torsion spin^c structures, the exact perturbations are also the perturbations which make the functional $\mathcal{L}$ balanced. The groups on the left-hand sides of the isomorphisms in Theorem 31.1.3 are therefore the balanced groups, and we are once again looking to construct isomorphisms

$$\begin{aligned}\check{\imath} &: \widecheck{HM}_*(Y, \mathfrak{s}, c_b; \Gamma) \to \widecheck{HM}_*(Y, \mathfrak{s}, c; \Gamma)\\ \hat{\imath} &: \widehat{HM}_*(Y, \mathfrak{s}, c_b; \Gamma) \to \widehat{HM}_*(Y, \mathfrak{s}, c; \Gamma),\end{aligned}$$

as we were in the immediately previous case, Theorem 31.1.2. Now, however, the coefficients Γ are assumed to be c-complete; and we are not taking the negative completion that was previously implied by the $\bullet$ notation. We again write $\mathcal{L}_0$ and $\mathcal{L}_1$ for the perturbed functionals that are used in the construction of the groups on the left- and right-hand sides respectively, and $\mathcal{E}_0^{\text{top}}$ and $\mathcal{E}_1^{\text{top}}$ for the corresponding notions of topological energy. The situation is again that $\mathcal{L}_0$ will have reducible critical points, but $\mathcal{L}_1$ will not.

We define chain maps

$$\begin{aligned}m_o^o &: C^o(Y, \mathfrak{s}, c_b; \Gamma) \to C^o(Y, \mathfrak{s}; \Gamma)\\ m_o^u &: C^u(Y, \mathfrak{s}, c_b; \Gamma) \to C^o(Y, \mathfrak{s}; \Gamma)\end{aligned}$$

by counting trajectories with signs and using the coefficient system Γ. That is,

$$\begin{aligned}m_o^o &= \sum_{[\mathfrak{a}] \in \mathfrak{C}^o(Y, \mathfrak{s}, c_b)} \sum_{[\mathfrak{b}] \in \mathfrak{C}^o(Y, \mathfrak{s})} \sum_{z} \sum_{[\gamma] \in M_z([\mathfrak{a}], U^*, [\mathfrak{b}])} \epsilon[\gamma] \otimes \Gamma(z)\\ m_o^u &= \sum_{[\mathfrak{a}] \in \mathfrak{C}^u(Y, \mathfrak{s}, c_b)} \sum_{[\mathfrak{b}] \in \mathfrak{C}^o(Y, \mathfrak{s})} \sum_{z} \sum_{[\gamma] \in M_z([\mathfrak{a}], U^*, [\mathfrak{b}])} \epsilon[\gamma] \otimes \Gamma(z).\end{aligned} \tag{31.11}$$

As usual, the sum is over zero-dimensional moduli spaces; and the summand is to be interpreted as a homomorphism

$$\epsilon[\gamma] \otimes \Gamma(z) : \mathbb{Z}[\mathfrak{a}] \otimes \Gamma[\mathfrak{a}] \to \mathbb{Z}[\mathfrak{b}] \otimes \Gamma[\mathfrak{b}].$$

These sums, however, may have infinitely many non-zero terms; and in order to show that the sums converge, we need to establish that, for each $[\mathfrak{a}]$, the set of homotopy classes z that contribute to the above sums is *energy-finite* for the energy function $\mathcal{E}_1^{\text{top}}$ (see Definitions 30.2.2 and 30.2.1). Because $c_1(\mathfrak{s})$ is torsion, the relative grading $\text{gr}_z([\mathfrak{a}], [\mathfrak{b}])$ depends only on the endpoints of z, and not on its relative homotopy class; so the energy-finite condition simplifies. What we need to prove is therefore that, for fixed $[\mathfrak{a}]$ and any constant C, there are only finitely many solutions $[\gamma]$ which contribute to the above sums and which satisfy in addition the energy bound

$$\mathcal{E}_1^{\text{top}}(\gamma) \leq C. \tag{31.12}$$

Let $[\gamma_n]$ be a sequence of solutions contributing to the sum and let z_n denote the corresponding sequence of homotopy classes. In the earlier proof of Theorem 31.1.1, we used the fact that the quantity (31.5) was independent of the homotopy class z_n in order to obtain an upper bound on the energy $\mathcal{E}_1^{\text{top}}(\gamma)$. There is no longer any invariant quantity of this sort, but the upper bound on the energy has now been added as a hypothesis (31.12). So the proof of finiteness continues as before. Indeed, we have slightly more: for given C, there can be only finitely many triples $([\mathfrak{a}_n], [\mathfrak{b}_n], z_n)$ such that $\mathcal{E}_1^{\text{top}}(z_n) \leq C$ and such that the corresponding moduli space is non-empty. (In other words, we did not need to fix $[\mathfrak{a}]$.)

This establishes that the homomorphisms (31.11) are well-defined. The chain maps $\check{m}$ and $\hat{m}$ are defined as before, (31.3), and give rise to the desired maps $\check{\imath}$ and $\hat{\imath}$ on cohomology.

The maps $\check{\jmath}$ and $\hat{\jmath}$ are constructed in the same way, as in the proof of Theorem 31.1.1. The proof that these are the inverses of $\check{\imath}$ and $\hat{\imath}$ respectively uses the moduli spaces $\mathcal{M}_z([\mathfrak{a}], [\mathfrak{b}])$ defined as in (31.8), for critical points $[\mathfrak{a}]$, $[\mathfrak{b}]$ of $(\text{grad}\, \mathcal{L}_0)^\sigma$ in $\mathcal{B}_k^\sigma(Y, \mathfrak{s})$. Once again, the question is whether these moduli spaces have the necessary finiteness properties to define maps on the groups $C_*^o(Y, \mathfrak{s}, c_b; \Gamma)$ and the companions C^s and C^u.

Because $c_1(\mathfrak{s})$ is torsion, the critical points have a grading by $\mathbb{J}(Y, \mathfrak{s}) \cong \mathbb{Z}$; so for given $[\mathfrak{a}]$, there are only finitely many $[\mathfrak{b}]$ for which $\mathcal{M}_z([\mathfrak{a}], [\mathfrak{b}])$ can be zero-dimensional; and in the other direction, for each $[\mathfrak{b}]$, there are only finitely many $[\mathfrak{a}]$. There is therefore no need to use negative completion in order for the maps defined by $\mathcal{M}_z([\mathfrak{a}], [\mathfrak{b}])$ to be well-defined: the issue is only whether, for given $[\mathfrak{a}]$ and $[\mathfrak{b}]$, the set of homotopy classes of paths z for which $\mathcal{M}_z([\mathfrak{a}], [\mathfrak{b}])$ is zero-dimensional is energy-finite for the energy function $\mathcal{E}_1^{\text{top}}$. In other words, we must show that given any $[\mathfrak{a}]$, $[\mathfrak{b}]$ and C, there are only finitely many non-empty, zero-dimensional moduli spaces $\mathcal{M}_z([\mathfrak{a}], [\mathfrak{b}])$ with energy $\mathcal{E}_1^{\text{top}} \leq C$. The proof runs as before, but uses the given upper bound of $\mathcal{E}_1^{\text{top}}$ instead of deducing one via (31.9). □

31.4 Proportional period classes

The isomorphism theorems that we have proved in this section leave open some questions on how the Floer homology groups might vary as we vary the period class of a non-exact perturbation. There is one further result that we can prove along these lines however. To motivate this, we can look again at Theorem 31.1.2, which tells us that the Floer groups do not change as we vary the period class c along a particular open ray emanating from the class of the balanced perturbation. In Theorem 31.1.2, the relevant ray is the set of period classes of positively monotone perturbations. But in fact a similar result holds along all rays from the balanced perturbation.

If $\mathcal{L}_1$ and $\mathcal{L}_2$ are two non-exact perturbations of the Chern–Simons–Dirac functional $\mathcal{L}$, then to say that their period classes lie on the same ray or half-line is to say that the periods of $\mathcal{L}_1$ and $\mathcal{L}_2$ are non-trivial and positively proportional. In other words, there are positive constants μ_1 and μ_2 such that the difference

$$\mu_1 \mathcal{L}_1 - \mu_2 \mathcal{L}_2$$

is invariant under the full gauge-group and so descends a single-valued function on $\mathcal{B}(Y, \mathfrak{s})$. As long as the period classes of $\mathcal{L}_1$ and $\mathcal{L}_2$ are not actually equal, then the above difference is proportional to some balanced perturbation $\mathcal{L}_0$ of $\mathcal{L}$: after scaling the μ_i and possibly interchanging the roles of $\mathcal{L}_1$ and $\mathcal{L}_2$ to get the correct sign, we may assume that

$$\mu_1 \mathcal{L}_1 - \mu_2 \mathcal{L}_2 = \mathcal{L}_0, \tag{31.13}$$

where $\mathcal{L}_0$ is balanced. If we write the period classes of the perturbations as c_1 and c_2, then this proportionality also means that the notions of c_1-complete and c_2-complete for a local system Γ are the same. We can therefore fix a local system Γ that is c-complete for both period classes, and consider the corresponding Floer groups $HM_\bullet(Y, \mathfrak{s}, c_i; \Gamma)$, for $i = 1, 2$. (Note that, because the c_i are not balanced, there are no reducibles and there is only one flavor of Floer group.) We then have:

Theorem 31.4.1. *Suppose, as above, that the period classes c_1 and c_2 are such that the corresponding perturbed functionals $\mathcal{L}_1$ and $\mathcal{L}_2$ are not balanced but satisfy the proportionality condition* (31.13) *for positive constants μ_i. Then for any local system Γ that is c_i-complete (for one and hence both c_i), the corresponding Floer groups are isomorphic:*

$$HM_\bullet(Y, \mathfrak{s}, c_1; \Gamma) \cong HM_\bullet(Y, \mathfrak{s}, c_2; \Gamma).$$

The theorem tells us that, up to isomorphism, $HM_\bullet(Y, c_1; \Gamma)$ is constant as c varies along a ray emanating from the balanced class c_b. The key point in the proof of this theorem, as in the previous isomorphism theorems, is to obtain uniform bounds on the value of the functionals $\mathcal{L}_i$ along trajectories belonging to moduli spaces on a cylindrical manifold. We take closed 2-forms ω_1 and ω_2 such that the corresponding perturbations f_{ω_i} of the Chern–Simons–Dirac functional $\mathcal{L}$ on Y have period class c_1 and c_2 respectively. (So the de Rham class of ω_j is $4\pi c_j/i$.) We again write U for the cylinder $[0, 1] \times Y$, equipped with a form $\tilde{\omega}$ which now interpolates between the form ω_1 near $\{0\} \times Y$ and ω_2 near $\{1\} \times Y$. The manifold

$$U^* = U^- \cup U \cup U^+$$

is the infinite cylinder, obtained by attaching cylindrical ends to U, equipped with a perturbation of the Seiberg–Witten equations, so that solutions on U^* give rise to gradient trajectories of $\mathcal{L}_1$ on the negative end U^- and of $\mathcal{L}_2$ on the positive end U^+. We will again make use of a *balanced* non-exact perturbation of $\mathcal{L}$, which we call $\mathcal{L}_0$. This functional, which is invariant under the full gauge-group, will be used as an auxiliary to the argument.

The strategy again is to use moduli spaces on U^* to define a map from $HM_\bullet(Y, \mathfrak{s}, c_1; \Gamma)$ to $HM_\bullet(Y, \mathfrak{s}, c_2; \Gamma)$. A map in the opposite direction can be constructed in the same manner, and one must then show the maps are mutually inverse. Let $\mathfrak{a}$ and $\mathfrak{b}$ be critical points for $\mathcal{L}_1$ and $\mathcal{L}_2$, and consider a moduli space

$$M_z([\mathfrak{a}], U^*, [\mathfrak{b}])$$

of solutions to these perturbed equations. The first point is that we have uniform bounds, above and below, independent of $[\mathfrak{a}]$, $[\mathfrak{b}]$ and z, for the value of $\mathcal{L}_0$ at $\gamma(t)$ for any γ in such a moduli space. This is what we obtained at (31.4), using Lemma 31.2.1 and Corollary 31.2.3 in the previous proofs; and the same argument applies. From this we again conclude that if we have a sequence

$$[\gamma_n] \in M_{z_n}([\mathfrak{a}_n], U^*, [\mathfrak{b}_n])$$

then we can pass to a subsequence that converges on every compact set in U^*, up to gauge transformation.

When Γ is c_i-complete, we can use these moduli spaces to define a map on the Floer groups provided we have a compactness theorem for the spaces $M_z^+([\mathfrak{a}], U^*, [\mathfrak{b}])$ and a *finiteness* theorem: specifically, we need to know that given any constant C, there are only finitely many homotopy classes of paths

z_n for which the moduli space above is non-empty and for which, in addition, the total drop in $\mathcal{L}_1$ along z_n is bounded by C.

On U^-, each $[\gamma_n]$ defines a gradient trajectory $\check{\gamma}_n$ of $\mathcal{L}_1$; so $\mathcal{L}_1(\check{\gamma}_n(t))$ is decreasing for t in $(-\infty, 0]$. To proceed further, we need a bound on the magnitude of the decrease, assuming only a uniform bound on the total change of $\mathcal{L}_1$ along the paths z_n. Convergence on compact sets already tells us that, on the compact piece $[0, 1] \times U$, the change in $\mathcal{L}_1$ is bounded above and below. What remains, therefore, is to bound the possible *increase* in $\mathcal{L}_1$ along the positive end U^+: we want a bound

$$\mathcal{L}_1(\mathfrak{b}_n) - \mathcal{L}_1(\gamma_n(1)) \leq K, \tag{31.14}$$

for some K independent of n. Along U^+, it is the function $\mathcal{L}_2$ which is decreasing: we have

$$\mathcal{L}_2(\mathfrak{b}_n) - \mathcal{L}_2(\gamma_n(1)) \leq 0.$$

On the other hand, the proportionality condition says that

$$\mathcal{L}_1 = (\mu_2/\mu_1)\mathcal{L}_2 + (1/\mu_1)\mathcal{L}_0,$$

and because we already know that $\mathcal{L}_0$ is uniformly bounded above and below, the monotonicity of $\mathcal{L}_2$ implies the desired inequality (31.14). We now know enough to conclude that the functional $\mathcal{L}_1$ is uniformly bounded above and below along any sequence of trajectories γ_n belonging to moduli spaces $M_{z_n}([\mathfrak{a}_n], U^*, [\mathfrak{b}_n])$, provided only that the total drop in $\mathcal{L}_1$ (or equivalently $\mathcal{L}_2$) along the paths z_n is uniformly bounded above by a constant C. There is a similar bound on $\mathcal{L}_2$, and we can proceed without further difficulty to obtain maps

$$HM_\bullet(Y, \mathfrak{s}, c_1; \Gamma) \to HM_\bullet(Y, \mathfrak{s}, c_2; \Gamma)$$

and

$$HM_\bullet(Y, \mathfrak{s}, c_2; \Gamma) \to HM_\bullet(Y, \mathfrak{s}, c_1; \Gamma).$$

The bounds necessary to prove that these two maps are mutually inverse are established using the same ideas. □

31.5 An exact sequence

We have seen in Theorem 31.4.1 that as the period class c of the perturbation moves along a straight line passing through c_b, the corresponding Floer groups remain unchanged except as c passes c_b. If we look at a particular line and choose period classes c_- and c_+ which lie on this line and are separated by c_b, then the groups we can consider are

$$HM_*(Y,\mathfrak{s},c_-,\Gamma_-)$$
$$\widehat{HM}_*(Y,\mathfrak{s},c_b),\quad \overline{HM}_*(Y,\mathfrak{s},c_b),\quad \widecheck{HM}_*(Y,\mathfrak{s},c_b)$$
$$HM_*(Y,\mathfrak{s},c_+,\Gamma_+)$$

for some coefficient systems Γ_- and Γ_+ which are c_--complete and c_+-complete respectively. It is natural to ask what further relations there are amongst these. In particular, we will focus here on the case that $c_1(\mathfrak{s})$ is non-torsion and where c_+ and c_- are positively and negatively monotone. (So c_+ may as well be zero.) In this special case, we can take $\Gamma_\pm$ both to be $\mathbb{Z}$, because the completeness condition becomes trivial. In the following theorem, the first isomorphism follows from Theorem 31.1.1, after we recall that there is no difference between $\widecheck{HM}_*$ and $\widecheck{HM}_\bullet$ (unlike $\widehat{HM}_*$ and $\widehat{HM}_\bullet$, which are different in general). The second isomorphism is proved in a very similar manner, and we omit the argument.

Theorem 31.5.1. *Suppose $c_1(\mathfrak{s})$ is not torsion, let c_+ and c_- be period classes for positively and negatively monotone perturbations respectively, and let c_b be the class of the balanced perturbation. Then we have isomorphisms (with $\mathbb{Z}$ coefficients)*

$$HM_*(Y,\mathfrak{s},c_+) \cong \widecheck{HM}_*(Y,\mathfrak{s},c_b)$$
$$HM_*(Y,\mathfrak{s},c_-) \cong \widehat{HM}_*(Y,\mathfrak{s},c_b).$$

□

Corollary 31.5.2. *There is a long exact sequence*

$$\cdots \xrightarrow{i_*} HM_*(Y,\mathfrak{s},c_+) \xrightarrow{j_*} HM_*(Y,\mathfrak{s},c_-) \xrightarrow{p_*} \overline{HM}_*(Y,\mathfrak{s},c_b)$$
$$\xrightarrow{i_*} HM_*(Y,\mathfrak{s},c_+) \xrightarrow{j_*} \cdots$$

Proof. The sequence is the usual exact sequence (Proposition 22.2.1) relating $\widecheck{HM}_*$, $\widehat{HM}_*$ and $\overline{HM}_*$ for the manifold $(Y,\mathfrak{s},c_b)$, reinterpreted using the isomorphisms of the theorem. □

To summarize, when we consider a non-torsion spin^c structure and non-exact perturbations on the monotone line, then (with $\mathbb{Z}$ coefficients and no completion), we encounter only three distinct groups; and these three are related by a long exact sequence.

32 Applications to gluing

32.1 A generating function and a gluing formula

In this section, we shall prove a gluing result, along the lines of Proposition 3.9.3. The latter proposition requires, for its statement, that $\overline{HM}_\bullet(Y;\Gamma_\eta)$ is zero, something that we will not establish in this chapter. The version of the gluing proposition that we will prove here will use instead a suitable c-complete local coefficient system for which the vanishing of $\overline{HM}_\bullet$ can be deduced from Theorem 31.1.3. The original version of Proposition 3.9.3 will be proved afterwards, in Subsection 32.3.

Fix a closed imaginary-valued 2-form ω on Y, and let $\mathfrak{s}$ be a fixed spin^c structure with $c_1(\mathfrak{s})$ torsion. We introduce a local coefficient system Π_ω that is a variant of the examples introduced in Subsection 30.2. For the fiber of Π_ω we take everywhere the ring $R[\mathbb{R}]^-$, the completed group ring of $\mathbb{R}$ with coefficients in a commutative ring with 1. We again introduce a formal symbol t and write elements of $R[\mathbb{R}]^-$ as sums $\sum r_i t^i$, where i runs through the reals, and the coefficients r_i satisfy the finiteness constraint, that the support of $i \mapsto r_i$ meets every positive half-line in a finite set. If z is a homotopy class of paths in $\mathcal{B}^\sigma(Y)$ joining $[\mathfrak{a}]$ to $[\mathfrak{b}]$, then $\Pi_\omega(z) : R[\mathbb{R}]^- \to R[\mathbb{R}]^-$ is defined to be multiplication by t^{-x}, where

$$x = \mathcal{E}_\omega^{\text{top}}(z)$$

is the ω-perturbed topological energy of a corresponding configuration on the cylinder. This local system is c-complete, where c is the cohomology class of $(i/4\pi)\omega$ (which is also the period class of $\mathcal{L}_\omega$ because $c_1(\mathfrak{s})$ is torsion). Suppose now that c is *non-zero*. From Theorem 31.1.3, we then obtain

$$\overline{HM}_*(Y,\mathfrak{s};\Pi_\omega) = 0,$$

and j is an isomorphism,

$$j : \widecheck{HM}_*(Y,\mathfrak{s};\Pi_\omega) \xrightarrow{\cong} \widehat{HM}_*(Y,\mathfrak{s};\Pi_\omega).$$

Let X be a closed 4-manifold that contains Y as a separating hypersurface, so that $X = X_1 \cup X_2$, with $\partial X_1 = Y$ and $\partial X_2 = -Y$. We suppose that ω extends to a closed form ω_X on X. We write ω_i for the restriction of ω_X to X_i. If $b^+(X)$ is 2 or more, and if a homology orientation is chosen, then we have well-defined integer invariants $\mathfrak{m}(X, \mathfrak{s}_X)$ for each spin^c structure $\mathfrak{s}_X$ on X. From these, we can form a generating function

$$\sum_{\mathfrak{s}_X} t^{-\mathcal{E}^{\text{top}}_{\omega}(\mathfrak{s}_X)} \mathfrak{m}(X, \mathfrak{s}_X),$$

regarded as an element of $R[\mathbb{R}]^-$. Here $\mathcal{E}^{\text{top}}_{\omega}(\mathfrak{s}_X)$ denotes the topological energy of any configuration belonging to the spin^c structure, which is just the quantity

$$\frac{1}{4}\int_X (F_{A^t} - 4\omega) \wedge (F_{A^t} - 4\omega).$$

We can also restrict the sum to those spin^c structures $\mathfrak{s}_X$ whose restriction to Y is $\mathfrak{s}$: we denote this restricted sum by

$$\mathfrak{m}_{\mathfrak{s}} = \sum_{\mathfrak{s}_X|_Y = \mathfrak{s}} t^{-\mathcal{E}^{\text{top}}_{\omega}(\mathfrak{s}_X)} \mathfrak{m}(X, \mathfrak{s}_X). \tag{32.1}$$

We shall write down a gluing formula for this invariant $\mathfrak{m}_{\mathfrak{s}}$. Let W_1 denote the cobordism $X_1 \setminus B^4$, from S^3 to Y. We can regard Γ_ω as defining a local coefficient system Γ_0 (with no monodromy) also on $\mathcal{B}^\sigma(S^3)$; and we have a W_1-morphism of local coefficient systems,

$$\Gamma_{\omega_1} : \Gamma_0 \to \Gamma_\omega$$

defined using the perturbed topological energy of a configuration on W_1:

$$\Gamma_{\omega_1}(z) = t^{-\mathcal{E}^{\text{top}}_{\omega_1}(z)}.$$

Using this cobordism and the canonical element 1 in $\widehat{HM}_\bullet(S^3; \Gamma_0)$, we obtain in the usual way an element

$$\psi_1 = \widehat{HM}_\bullet(W_1; \Gamma_{\omega_1})(1) \in \widehat{HM}_\bullet(Y; \Gamma_\omega).$$

In this construction, we can again restrict the spin^c structures on W_1 to those whose restriction to Y is $\mathfrak{s}$; and in this way we obtain a restricted version of the above invariant, $\psi_{1,\mathfrak{s}} \in \widehat{HM}_\bullet(Y, \mathfrak{s}; \Gamma_\omega)$. Because j is an isomorphism, we can also regard $\psi_{1,\mathfrak{s}}$ as lying in the reduced group

$$\psi_{1,\mathfrak{s}} \in HM_\bullet(Y, \mathfrak{s}; \Gamma_\omega).$$

In a similar way, we obtain an invariant

$$\psi_{2,\mathfrak{s}} \in HM_\bullet(-Y,\mathfrak{s};\Pi^*_\omega),$$

where Π^*_ω is the coefficient system whose definition mimics that of Π_ω but which uses the orientation $-Y$ in place of Y. We can describe Π^*_ω as a coefficient system on $\mathcal{B}^\sigma(Y;\mathfrak{s})$, using the original orientation of Y, by saying that its fiber at each point is $R[\mathbb{R}]^-$ (the same negative completion), and that for a path z from $[\mathfrak{a}]$ to $[\mathfrak{b}]$, we have

$$\Pi^*_\omega(z) = t^{+\mathcal{E}^{\text{top}}_\omega(z)}.$$

(This is the inverse of $\Pi_\omega(z)$.) When a homology orientation μ for Y is given, we have a duality isomorphism

$$\omega_\mu : HM_\bullet(-Y,\mathfrak{s};\Pi^*_\omega) \to HM^\bullet(Y,\mathfrak{s};\Pi^*_\omega). \tag{32.2}$$

There is also a pairing, resulting from the obvious multiplication pairing at the chain level,

$$HM_\bullet(Y,\mathfrak{s};\Pi_\omega) \otimes HM^\bullet(Y,\mathfrak{s};\Pi^*_\omega) \to R[\mathbb{R}]^-.$$

Combining this pairing with (32.2), we obtain a pairing

$$\langle -,-\rangle_{\omega_\mu} : HM_\bullet(Y,\mathfrak{s};\Pi_\omega) \otimes HM_\bullet(-Y,\mathfrak{s};\Pi^*_\omega) \to R[\mathbb{R}]^-;$$

and using this we can form the element $\langle \psi_1, \psi_2\rangle_{\omega_\mu}$ in $R[\mathbb{R}]^-$.

Proposition 32.1.1. *If $b^+(X) \geq 2$ and the cohomology class $c = (i/4\pi)[\omega]$ is non-zero, then we have the following pairing formula for the restricted invariant $\mathfrak{m}_\mathfrak{s}$ in $R[\mathbb{R}]^-$:*

$$\mathfrak{m}_\mathfrak{s} = \langle \psi_1, \psi_2\rangle_{\omega_\mu},$$

whether or not either of the X_j has $b^+ = 0$.

Proof. As a preliminary step, we can reformulate the result without referring to the duality isomorphism ω_μ. We have maps

$$\begin{aligned}
\widehat{HM}_\bullet(W_1;\Pi_{\omega_1})_\mathfrak{s} &: \widehat{HM}_\bullet(S^3;\Pi_0) \to \widehat{HM}_\bullet(Y,\mathfrak{s};\Pi_\omega)\\
j^{-1} &: \widehat{HM}_\bullet(Y,\mathfrak{s};\Pi_\omega) \to \widecheck{HM}_\bullet(Y,\mathfrak{s};\Pi_\omega)\\
\widecheck{HM}_\bullet(W_2;\Pi_{\omega_2})_\mathfrak{s} &: \widecheck{HM}_\bullet(Y,\mathfrak{s};\Pi_\omega) \to \widecheck{HM}_\bullet(S^3;\Pi_0),
\end{aligned}$$

where, in the first and third lines, the subscript $\mathfrak{s}$ in the notation for the map indicates that we use only the contribution of spin^c structures on the cobordism that restrict to $\mathfrak{s}$ on Y. The composition of these three gives a map

$$\widecheck{HM}_\bullet(W_2; \Pi_{\omega_2})_\mathfrak{s} \circ j^{-1} \circ \widehat{HM}_\bullet(W_1; \Pi_{\omega_1})_\mathfrak{s} : \widehat{HM}(S^3; \Pi_0) \to \widecheck{HM}(S^3; \Pi_0).$$

On the other hand, the cobordism $W : S^3 \to S^3$ obtained by removing two balls from X (the composite $W_2 \circ W_1$) has $b^+ \geq 2$ and gives rise to a map with the same domain and codomain:

$$\overrightarrow{HM}_\bullet(W; \Pi_{\omega X}) : \widehat{HM}(S^3; \Pi_0) \to \widecheck{HM}(S^3; \Pi_0).$$

(See Subsection 27.3 for the definition of this map.) Here $\Pi_{\omega X}$ is a W-morphism defined in the same manner as Π_{ω_1} etc., using the perturbed topological energy $\mathcal{E}^{\text{top}}_{\omega X}$. There is also a version of this map that counts only the contribution from spin^c structures $\mathfrak{s}_X$ on X whose restriction to Y is $\mathfrak{s}$. We denote this map by $\overrightarrow{HM}_\bullet(W; \Pi_{\omega X})_\mathfrak{s}$. The real content of the proposition is a composition law, expressing the equality of these two:

$$\widecheck{HM}_\bullet(W_2; \Pi_{\omega_2})_\mathfrak{s} \circ j^{-1} \circ \widehat{HM}_\bullet(W_1; \Pi_{\omega_1})_\mathfrak{s} = \overrightarrow{HM}_\bullet(W; \Pi_{\omega X})_\mathfrak{s}. \tag{32.3}$$

Indeed, the proposition follows from this equality because we have (for formal reasons)

$$\langle \psi_1, \psi_2 \rangle_{\omega_\mu} = \left\langle \widecheck{HM}_\bullet(W_2; \Pi_{\omega_2})_\mathfrak{s} \circ j^{-1} \circ \widehat{HM}_\bullet(W_1; \Pi_{\omega_1})_\mathfrak{s}(1), \check{1} \right\rangle$$

where 1 and $\check{1}$ are the canonical generators for $\widehat{HM}_\bullet(S^3; \Pi_0)$ and $\widecheck{HM}^\bullet(S^3; \Pi_0^*)$, and the angle brackets denote the $R[\mathbb{R}]^-$-valued pairing between homology and cohomology. (See also Proposition 3.8.1.)

In order to prove the composition law (32.3), we need to get a handle on j^{-1}. Note that j is an isomorphism between homology groups, but the chain complexes giving rise to these homology groups are not isomorphic: so j and its inverse are non-trivial maps. We can construct an explicit j^{-1} at the chain level using the isomorphisms of Theorem 31.1.3. The proof of Theorem 31.1.1 provides isomorphisms

$$\hat{\imath} : \widehat{HM}_\bullet(Y, \mathfrak{s}; \Pi_\omega) \to \widehat{HM}_\bullet(Y, \mathfrak{s}, c; \Pi_\omega)$$

and

$$\check{\jmath} : \widecheck{HM}_\bullet(Y, \mathfrak{s}, c; \Pi_\omega) \to \widecheck{HM}_\bullet(Y, \mathfrak{s}; \Pi_\omega),$$

both of which are obtained from solutions to the suitably perturbed equations on a cylindrical cobordism. The groups $\widehat{HM}_\bullet(Y,\mathfrak{s},c;\Pi_\omega)$ and $\widecheck{HM}_\bullet(Y,\mathfrak{s},c;\Pi_\omega)$ are not just isomorphic, but actually equal, because there are no reducible critical points for the non-exact perturbation. So we have a composite isomorphism,

$$\check{\jmath}\circ\hat{\imath}:\widehat{HM}_\bullet(Y,\mathfrak{s};\Pi_\omega)\to\widecheck{HM}_\bullet(Y,\mathfrak{s};\Pi_\omega). \tag{32.4}$$

Theorem 31.1.3 tells us that this composite is j^{-1}.

The merit of the formula (32.4) is that it expresses j^{-1} (and hence everything on the left-hand side of (32.3)) in terms of solutions to various versions of the Seiberg–Witten equations on cobordisms. We can therefore regard (32.3) as simply another composition law for maps resulting from cobordisms.

When we substitute $\check{\jmath}\circ\hat{\imath}$ for j^{-1} in (32.3), we have a composite of four maps. If we split these four maps into two composites of two maps each, we can write the left-hand side in (32.3) as the composite of the two maps

$$\begin{aligned}\hat{\imath}\circ\widehat{HM}_\bullet(W_1;\Pi_{\omega_1})_\mathfrak{s}&:\widehat{HM}_\bullet(S^3;\Pi_0)\to HM_\bullet(Y,\mathfrak{s},c;\Pi_\omega)\\ \widecheck{HM}_\bullet(W_2;\Pi_{\omega_2})_\mathfrak{s}\circ\check{\jmath}&:HM_\bullet(Y,\mathfrak{s},c;\Pi_\omega)\to\widecheck{HM}_\bullet(S^3;\Pi_0).\end{aligned} \tag{32.5}$$

(We have simply written $HM_\bullet(Y,\mathfrak{s},c;\Pi_\omega)$ for the two equal groups $\widecheck{HM}_\bullet$ and $\widehat{HM}_\bullet$.) Each of the two composites in (32.5) has a more direct interpretation, as we now explain.

Consider first the composite $\hat{\imath}\circ\widehat{HM}_\bullet(W_1;\Pi_{\omega_1})_\mathfrak{s}$. Let W_1^* denote the cylindrical-end manifold obtained from the cobordism W_1, and consider the form ω_1 extended to W_1^* so as to be equal to the translation-invariant form ω on the cylindrical end $\mathbb{R}^{\geq}\times Y$ and zero on the cylindrical end $\mathbb{R}^{\leq}\times S^3$. After choosing an auxiliary (exact) tame perturbation $\mathfrak{p}$ to achieve transversality, we have the 4-dimensional equations $\mathfrak{F}^\sigma_{\omega_1,\mathfrak{p}}=0$ on W_1^*, and corresponding moduli spaces

$$M_z([\mathfrak{a}],W_1,\omega_1,[\mathfrak{b}])\subset\mathcal{B}^\sigma_{k,\mathrm{loc}}(W_1^*).$$

Here $[\mathfrak{a}]$ is a critical point in $\mathcal{B}^\sigma(S^3)$, and $[\mathfrak{b}]$ is a critical point for the non-exact perturbation $(\operatorname{grad}\mathcal{L}_{\mathfrak{q},\omega})^\sigma$ in $\mathcal{B}^\sigma(Y,\mathfrak{s})$. Using these moduli spaces, we can define a map (for example)

$$m^u_o:C^u(S^3;\Pi_0)\to C^o(Y,\mathfrak{s},c;\Pi_\omega)$$

by the recipe

$$\sum_{[\mathfrak{a}]}\sum_{[\mathfrak{b}]}\sum_{z}\sum_{[\gamma]\in M_z([\mathfrak{a}],W_1^*,\omega_1,[\mathfrak{b}])} \epsilon[\gamma]\otimes\Pi_{\omega_1}(z)$$

(compare (31.11), for example), in which $\Pi_{\omega_1}(z) : R[\mathbb{R}]^- \to R[\mathbb{R}]^-$ is multiplication by $t^{-\mathcal{E}^{\text{top}}_{\omega_1(z)}}$. For fixed $[\mathfrak{a}]$ and $[\mathfrak{b}]$, there may be infinitely many z which contribute to this sum; but for any constant C, only finitely many z with energy less than C will contribute. So, because Π_ω is c-complete, the map m^u_o is well-defined. As long as the perturbation on S^3 is chosen small, so that $\widehat{HM}_\bullet(S^3;\Pi_0)$ is simply the homology of the complex $(C^u, \bar\partial^u_u)$, then m^u_o is a chain map, and gives rise to a map on homology,

$$(m^u_o)_* : \widehat{HM}_\bullet(S^3;\Pi_0) \to HM_\bullet(Y,\mathfrak{s},c;\Pi_\omega).$$

(Again, there being no reducibles for the non-exact perturbation, the homology group on the right is just the homology of (C^o, ∂^o_o).) In a similar manner, we obtain a map

$$(m^o_s)_* : HM_\bullet(Y,\mathfrak{s},c;\Pi_\omega) \to \widecheck{HM}_\bullet(S^3;\Pi_0)$$

using moduli spaces on W_2^*.

The equality (32.3) is now a consequence of three separate composition laws: first a composition law involving W_1,

$$(m^u_o)_* = \hat\imath \circ \widehat{HM}_\bullet(W_1;\Pi_{\omega_1})_{\mathfrak{s}};$$

second a similar composition law involving W_2,

$$(m^o_s)_* = \widecheck{HM}_\bullet(W_2;\Pi_{\omega_2})_{\mathfrak{s}} \circ \check{\jmath};$$

and finally a natural composition law for the composite cobordism $W = W_2 \circ W_1$,

$$(m^o_s)_* \circ (m^u_o)_* = \overrightarrow{HM}_\bullet(W;\Pi_{\omega_X})_{\mathfrak{s}}.$$

With each of these three, we are on familiar territory. The three identities above all follow in the usual way from chain-homotopies obtained from parametrized moduli spaces, the parameter being the length of the neck. Because we are dealing with non-exact perturbations, we only have to check in addition that the chain-homotopies (which will involve infinite sums) are really well-defined.

We have seen a model for such arguments in the proofs of Theorems 31.1.1–31.1.3; and in the present situation, because the definition of the coefficient system Π_ω makes direct use of the energy $\mathcal{E}_\omega^{\text{top}}$, checking the convergence of the sums that arise is straightforward. □

32.2 Remarks on the generating function

It is worth expanding a little the generating function $\mathfrak{m}_{\mathfrak{s}}$ that is defined in the previous subsection, (32.1), and the more general sum that precedes it (without the restriction to spin^c structures $\mathfrak{s}_X$ that restrict to $\mathfrak{s}$ on Y):

$$\sum_{\mathfrak{s}_X} t^{-\mathcal{E}_{\omega X}^{\text{top}}(\mathfrak{s}_X)} \mathfrak{m}(X, \mathfrak{s}_X). \tag{32.6}$$

As we noted above, on the closed manifold, we have

$$\mathcal{E}_\omega^{\text{top}} = \frac{1}{4}\int_X (F_{A^t} - 4\omega) \wedge (F_{A^t} - 4\omega).$$

The 2-form F_{A^t} represents $(2\pi/i)c_1(\mathfrak{s}_X)$. If we introduce the class

$$k = -\frac{2i}{\pi}[\omega_X]$$

in $H^2(X;\mathbb{R})$, then we can write

$$\mathcal{E}_{\omega X}^{\text{top}}(\mathfrak{s}_X) = -\pi^2 \int_X (c_1(\mathfrak{s}_X) + k)^2,$$

so that

$$t^{-\mathcal{E}_{\omega X}^{\text{top}}(\mathfrak{s}_X)} = t^{\pi^2 (c_1(\mathfrak{s}_X)^2 + 2c_1(\mathfrak{s}_X) \smile k + k^2)[X]}.$$

We can make our expression more closely resemble the generating function $\mathfrak{m}(X, h)$ from (3.25) by writing h for the Poincaré dual of k, and formally substituting

$$t^{2\pi^2} = e.$$

We can also use the fact that the spin^c structures which contribute to the sum (32.6) are those for which the moduli space on X is zero-dimensional, which means that

$$c_1(\mathfrak{s}_X)^2[X] = 2\chi(X) + 3\sigma(X).$$

Thus the term $t^{-\mathcal{E}^{\text{top}}_{\omega X}(\mathfrak{s}_X)}$ becomes

$$\exp\left(\frac{h\cdot h}{2}+\frac{2\chi+3\sigma}{2}+\langle c_1(\mathfrak{s}_X),h\rangle\right).$$

With these same substitutions, the generating function (32.6) becomes

$$\exp\left(\frac{h\cdot h}{2}+\frac{2\chi+3\sigma}{2}\right)\sum_{\mathfrak{s}_X}\exp(\langle c_1(\mathfrak{s}_X),h\rangle),$$

or equivalently

$$\exp\left(\frac{h\cdot h}{2}+\frac{2\chi+3\sigma}{2}\right)\mathfrak{m}(X,h),$$

where $\mathfrak{m}(X,h)$ is the original version of the generating function from (3.25).

32.3 Proof of Proposition 3.9.3

We can deduce the original version of Proposition 3.9.3 by a largely formal argument, starting from Proposition 32.1.1. One part of the argument is not formal: we need to use the vanishing of $\overline{HM}_\bullet(Y;\Gamma_\eta)$ for non-zero classes $[\eta]$, as asserted in Proposition 3.9.1. We use, in fact, a slight amplification of Proposition 3.9.1 which will be proved in Subsection 35.2 in the next chapter.

To keep the notation under control, let us write $\mathcal{R}$ for the group ring of $\mathbb{R}$ with real coefficients,

$$\mathcal{R}=\mathbb{R}[\mathbb{R}],$$

and $\mathcal{R}^-$ for the completion $\mathbb{R}[\mathbb{R}]^-$ that we introduced previously. The latter is a field, and we have inclusions

$$\mathcal{R}\subset\mathcal{Q}\subset\mathcal{R}^-,$$

where $\mathcal{Q}$ is the field of fractions of $\mathcal{R}$. We can write an element of $\mathcal{R}$ again as an expression $\sum r_i t^i$, with real exponents i and real coefficients, in a formal variable t.

Fix once more a spin^c structure $\mathfrak{s}$ on Y, and an imaginary-valued closed 2-form ω in a *non-zero* cohomology class. We introduced in Subsection 32.1 above the local system Π_ω on $\mathcal{B}^\sigma(Y,\mathfrak{s})$. We now introduce a slightly modified version of the construction, which is a little closer to our construction of Γ_η. Fix a real 1-cycle η in the class dual to $-(i/4\pi)\omega$: we will dispense with ω and use η in its place. Define a local system $\mathcal{R}_\eta$ on $\mathcal{B}^\sigma(Y,\mathfrak{s})$ as follows. The fiber of $\mathcal{R}_\eta$ at every point is $\mathcal{R}$. And for a relative homotopy class of paths z, represented by a particular path $\zeta : I \to \mathcal{B}^\sigma(Y,\mathfrak{s})$, we define $\mathcal{R}_\eta(z)$ to be multiplication by $t^{f(z)}$, where $f(z)$ is the real number

$$f(z) = \frac{i}{2\pi}\int_{I\times\eta} F_{A^t_\zeta}, \tag{32.7}$$

and A is the connection on the cylinder $I \times Y$ obtained from the path ζ. In other words, $\mathcal{R}_\eta$ is defined in exactly the same manner as we defined Γ_η on page 445, but replacing the e in the exponential function with the formal symbol t. We can apply the same construction with $\mathcal{Q}$ and $\mathcal{R}^-$ in place of $\mathcal{R}$, and we obtain local coefficient systems

$$\mathcal{R}_\eta \subset \mathcal{Q}_\eta \subset \mathcal{R}^-_\eta.$$

The local system $\mathcal{R}^-_\eta$ is isomorphic to the system Π_ω considered earlier. It is c-complete for the non-zero class $c = (i/4\pi)[\omega]$; and we therefore have a vanishing result,

$$\overline{HM}_\bullet(Y,\mathfrak{s};\mathcal{R}^-_\eta) = 0.$$

The inclusion $\mathcal{Q} \subset \mathcal{R}^-$ is a field extension, and we have the general relationship

$$\overline{HM}_\bullet(Y,\mathfrak{s};\mathcal{Q}_\eta) \otimes_{\mathcal{Q}} \mathcal{R}^- = \overline{HM}_\bullet(Y,\mathfrak{s};\mathcal{R}^-_\eta).$$

So we also have a vanishing result for the smaller field:

$$\overline{HM}_\bullet(Y,\mathfrak{s};\mathcal{Q}_\eta) = 0.$$

For the coefficient system $\mathcal{R}_\eta$, we do not have vanishing: the previous line tells us

$$\overline{HM}_\bullet(Y,\mathfrak{s};\mathcal{R}_\eta) \otimes_{\mathcal{R}} \mathcal{Q} = 0,$$

which says that every element of the $\mathcal{R}$-module $\overline{HM}_\bullet(Y,\mathfrak{s};\mathcal{R}_\eta)$ is torsion. We can extract much more explicit information about the torsion, from the calculations in the next chapter (Subsection 35.2): we state a slightly simplified version of the result here, without proof.

Proposition 32.3.1. *If $p \in \mathbb{R}$ is any period of η (that is, the pairing of η with an integer cohomology class), then $\overline{HM}_\bullet(Y, \mathfrak{s}; \mathcal{R}_\eta)$ is annihilated by some power of the element $a = 1 - t^{2p}$ in $\mathcal{R}$.*

Fix a non-zero p as in the proposition, and set $a = 1 - t^{2p}$. Consider the ring $\mathcal{S} = \mathcal{R}[a^{-1}]$ obtained from $\mathcal{R}$ by inverting a. We have

$$\mathcal{R} \subset \mathcal{S} \subset \mathcal{Q}.$$

Because $\mathcal{S}$ is a torsion-free $\mathcal{R}$-module, we have

$$\overline{HM}_\bullet(Y, \mathfrak{s}; \mathcal{S}_\eta) = \overline{HM}_\bullet(Y, \mathfrak{s}; \mathcal{R}_\eta) \otimes_{\mathcal{R}} \mathcal{S};$$

and the tensor product on the right is zero, because some power of the unit $a \in \mathcal{S}$ annihilates $\overline{HM}_\bullet(Y, \mathfrak{s}; \mathcal{R}_\eta)$. Thus

$$\overline{HM}_\bullet(Y, \mathfrak{s}; \mathcal{S}_\eta) = 0. \tag{32.8}$$

Consider now the ring homomorphism $\tau : \mathcal{R} \to \mathbb{R}$ given by $t \mapsto e$. Because e is not a zero of $1 - t^{2p}$, this ring homomorphism extends to a homomorphism $\tau : \mathcal{S} \to \mathbb{R}$, and there is a corresponding homomorphism of local systems

$$\tau : \mathcal{S}_\eta \to \Gamma_\eta.$$

From (32.8) and the universal coefficient theorem, we obtain (as claimed in Proposition 3.9.1)

$$\overline{HM}_\bullet(Y, \mathfrak{s}; \Gamma_\eta) = 0.$$

Consider again the situation in which Y separates a closed manifold X as $X_1 \cup X_2$; and as in Proposition 3.9.3, suppose that there are 2-chains ν_1 and ν_2 in X_1 and X_2, with boundary η and $-\eta$ respectively. Write $\nu = \nu_1 + \nu_2$. Let W_1 be the cobordism from S^3 to Y obtained by removing a ball from X_1, and similarly let W_2 be the cobordism from Y to S^3. The vanishing of $\overline{HM}_\bullet(Y, \mathfrak{s}; \Gamma_\eta)$ means that

$$j : \widecheck{HM}_\bullet(Y, \mathfrak{s}; \Gamma_\eta) \to \widehat{HM}_\bullet(Y, \mathfrak{s}; \Gamma_\eta)$$

is an isomorphism. We also have homomorphisms arising from the pairs (W_i, ν_i):

$$\widehat{HM}_\bullet(W_1; \Gamma_{\nu_1})_{\mathfrak{s}} : \widehat{HM}_\bullet(S^3; \mathbb{R}) \to \widehat{HM}_\bullet(Y, \mathfrak{s}; \Gamma_\eta)$$
$$\widecheck{HM}_\bullet(W_2; \Gamma_{\nu_2})_{\mathfrak{s}} : \widecheck{HM}_\bullet(Y, \mathfrak{s}; \Gamma_\eta) \to \widecheck{HM}_\bullet(S^3; \mathbb{R}).$$

The subscript $\mathfrak{s}$ again indicates that we count only the contributions from spin^c structures which restrict to $\mathfrak{s}$ on Y. The key point we have to verify is that a composition law holds, exactly parallel to (32.3). This is the content of the following lemma.

Lemma 32.3.2. *We have*

$$\widecheck{HM}_\bullet(W_2;\Gamma_{\nu_2})_{\mathfrak{s}} \circ j^{-1} \circ \widehat{HM}_\bullet(W_1;\Gamma_{\nu_1})_{\mathfrak{s}} = \overrightarrow{HM}_\bullet(W;\Gamma_\nu)_{\mathfrak{s}}$$

as homomorphisms from $\widehat{HM}_\bullet(S^3;\mathbb{R})$ *to* $\widecheck{HM}_\bullet(S^3;\mathbb{R})$.

Proof. In addition to Γ_η and the W_i-morphisms Γ_{ν_i}, we have the local coefficient system $\mathcal{R}_\eta$, obtained by replacing e with the formal symbol t; and there are W_i-morphisms $\mathcal{R}_{\nu_i}$ constructed analogously. Thus we have, for example,

$$\widehat{HM}_\bullet(W_1;\Gamma_{\nu_1})_{\mathfrak{s}} : \widehat{HM}_\bullet(S^3;\mathcal{R}) \to \widehat{HM}_\bullet(Y,\mathfrak{s};\mathcal{R}_\eta).$$

We can apply the same construction with $\mathcal{S}$, $\mathcal{Q}$ or $\mathcal{R}^-$ in place of $\mathcal{R}$. In the case of $\mathcal{R}^-$, we already know that the counterpart of the equality in the lemma holds,

$$\widecheck{HM}_\bullet(W_2;\mathcal{R}^-_{\nu_2})_{\mathfrak{s}} \circ j^{-1} \circ \widehat{HM}_\bullet(W_1;\mathcal{R}^-_{\nu_1})_{\mathfrak{s}} = \overrightarrow{HM}_\bullet(W;\mathcal{R}^-_\nu)_{\mathfrak{s}}, \qquad (32.9)$$

because $\mathcal{R}^-_\eta$ is c-complete and the arguments of Subsection 32.1 can be applied. Indeed, the remarks in Subsection 32.2 show that $\mathcal{R}^-_\nu$ differs from the earlier $\Pi_{\omega X}$ by inconsequential factors when the dimension of the moduli space is fixed.

The homomorphisms of local systems

$$\begin{array}{ccc} & & \mathcal{R}^-_\eta \\ & \overset{i}{\nearrow} & \\ \mathcal{S}_\eta & & \\ & \underset{\tau}{\searrow} & \\ & & \Gamma_\eta \end{array}$$

give rise to natural transformations of the corresponding Floer groups, respecting j. For each of $\mathcal{R}^-_\eta$, $\mathcal{S}_\eta$ and Γ_η, the map j is invertible. So from (32.9), we deduce that the corresponding equality holds for the coefficients $\mathcal{S}_\eta$, and also for Γ_η as required. □

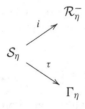

The lemma above can be applied to any spin^c structure $\mathfrak{s}$ on Y with $c_1(\mathfrak{s})$ torsion. In the non-torsion case, the equality still holds, but is more elementary, because j is the identity. Taking the sum over all $\mathfrak{s}$, we therefore obtain

$$\widecheck{HM}_\bullet(W_2; \Gamma_{\nu_2}) \circ j^{-1} \circ \widehat{HM}_\bullet(W_1; \Gamma_{\nu_1}) = \overrightarrow{HM}_\bullet(W; \Gamma_\nu).$$

Proposition 3.9.3 follows formally from this equality, just as Proposition 32.1.1 followed from (32.3). □

Notes and references for Chapter VIII

The fact that one can perturb the 4-dimensional Seiberg–Witten equations by adding a 2-form was exploited by Witten in [125], where a holomorphic 2-form was used to simplify the calculation of the invariants for Kähler surfaces. Taubes [114] perturbed the equations by a large multiple of the symplectic form to prove a fundamental result about the invariants of symplectic 4-manifolds. The Floer homology groups arising from non-exact perturbations of the Chern–Simons–Dirac functional arise naturally from such perturbations of the 4-dimensional equations.

IX
Calculations

Until now, we have calculated the Floer homology groups $\widecheck{HM}_\bullet$, $\widehat{HM}_\bullet$ and $\overline{HM}_\bullet$ only for the simplest case of the 3-sphere. In this chapter, we shall add to this small supply of calculations. First, we shall examine $\overline{HM}_\bullet(Y)$, which can be calculated for quite general 3-manifolds, at least in the sense that the resulting Floer groups have filtrations for which the graded quotients can be explicitly identified. In the case of a product 3-manifold, $S^1 \times \Sigma$, we will describe the differentials explicitly, to calculate $\overline{HM}_\bullet$.

The groups $\widecheck{HM}_\bullet$ and $\widehat{HM}_\bullet$ are usually hard to calculate. When the homomorphism $j : \widecheck{HM}_\bullet \to \widehat{HM}_\bullet$ is zero, one can determine $\widecheck{HM}_\bullet$ and $\widehat{HM}_\bullet$ completely once one knows $\overline{HM}_\bullet$ and some additional information concerning gradings. We will illustrate this in the case of $S^1 \times S^2$. We will then examine the case of T^3, which is the simplest manifold for which j is non-zero (when using twisted local coefficients). The calculation for T^3 extends to the similar case of a general flat 3-manifold.

An understanding of the Floer groups of T^3 allows us to calculate the monopole invariants of 4-manifolds which decompose along 3-tori into standard pieces. In particular, it leads to a calculation of the invariants of elliptic surfaces, which we take up in Section 38.

33 Coupled Morse theory

The group $\overline{HM}_*(Y, \mathfrak{s})$ is calculated from a complex that does not involve the irreducible solutions to the (perturbed) Seiberg–Witten equations, but only the reducible solutions. Before perturbation, the reducible critical points for the functional $\mathcal{L}$ on $\mathcal{C}(Y, \mathfrak{s})$ are pairs (A, Φ), where the spin^c connection A on Y has $F_{A^t} = 0$, and $\Phi = 0$. When $c_1(\mathfrak{s})$ is torsion, the gauge-equivalence classes of these pairs comprise a torus whose dimension is $b_1(Y)$, as we explained in Subsection 4.2. This torus parametrizes a family of self-adjoint Fredholm operators, namely the Dirac operators D_A on Y. (Essentially the same situation arises for a balanced perturbation by a closed 2-form on Y, in the case that $c_1(\mathfrak{s})$ is not torsion.)

In the next two sections of this chapter, we shall see how to compute the group $\overline{HM}_*(Y,\mathfrak{s})$ starting only from the family of Dirac operators parametrized by the torus. To understand the situation, we first generalize the picture: we replace the torus by a general compact manifold Q, and instead of the family of Dirac operators, we consider a family L of self-adjoint Fredholm operators parametrized by Q (of a type we make precise below). Associated to Q and L will be a group $\bar{H}_*(Q,L)$, which we shall define using a Morse-theory construction analogous to the construction of $\overline{HM}(Y,\mathfrak{s})$. We call $\bar{H}_*(Q,L)$ the homology of Q *coupled* to the family of operators L, or simply the *coupled* homology of Q.

A family of (complex) self-adjoint Fredholm operators on Q is classified by a map

$$u : Q \to U(\infty)$$

to the infinite unitary group, or equivalently, by an element of $K^1(Q)$. Up to isomorphism, the coupled group $\bar{H}(Q,L)$ depends only on the homotopy class of the classifying map. We shall explain how to calculate it in a case sufficiently general for our application to the Floer groups.

33.1 Self-adjoint operators and the infinite unitary group

Let H be a separable complex Hilbert space, let $K : H \to H$ be a compact self-adjoint operator without kernel, and let $H_1 = K(H)$. The subspace $H_1 \subset H$ is dense in H and is itself a Hilbert space, with norm

$$\|h\|_1 = \|K^{-1}h\|.$$

Let $K = K^+ \oplus K^-$ be the decomposition of K into positive and negative parts, let H_1^+ and H_1^- be the images of these, and let H^+ and H^- be their closures in H. We shall suppose that all these are infinite-dimensional.

The model we have in mind is the case when H arises as $L^2(Y;E)$, the L^2 sections of a vector bundle E on a compact manifold Y, and the subspace H_1 is the Sobolev space $L^2_1(Y;E)$. We need to impose some restriction on K (and hence on H_1) in order to rule out potential pathology.

Definition 33.1.1. Let J be a compact, positive self-adjoint operator on a Hilbert space, having no kernel. Let the eigenvalues of J be written as

$$\mu_i = \lambda_i^{-1},$$

with $0 < \lambda_0 \leq \lambda_1 \leq \cdots$. We say that the spectrum of J is *mild* if there exists a constant C such that, for all N,

$$\frac{\lambda_{2N}}{\lambda_N} \leq C.$$

◊

Let $B(H : H_1)$ be the Banach algebra of bounded operators x on H with the additional property that

$$x(H_1) \subset H_1$$
$$x^*(H_1) \subset H_1$$

and x, x^* have finite H_1-operator norm. (Here x^* is the Hilbert space adjoint on H.) The norm on $B(H : H_1)$ is

$$\max\left\{\|x\|_H, \|x\|_{H_1}, \|x^*\|_{H_1}\right\}.$$

Let $U(H : H_1) \subset B(H : H_1)$ be the group

$$\{x \mid x^*x = 1\}.$$

It is a Banach Lie group with the inherited topology. Let $S(H : H_1)$ denote the space of operators

$$S(H : H_1) = \{L : H_1 \to H \mid L \text{ is Fredholm of index zero}$$
$$\text{and } \langle L\phi, \psi\rangle = \langle \phi, L\psi\rangle \text{ for all } \phi, \psi\}.$$

Such an L is an unbounded self-adjoint operator on H. In our model case, $S(H : H_1)$ includes the first-order self-adjoint differential operators acting on sections of E.

Lemma 33.1.2. *If L belongs to $S(H : H_1)$, then there is a complete orthonormal system e_i for H, such that $e_i \in H_1$ for all i, and*

$$Le_i = \lambda_i e_i.$$

Proof. An elementary argument shows that $L+\epsilon$ has no kernel for sufficiently small ϵ; and the fact that $L+\epsilon$ has index zero then tells us that it is an isomorphism. Now regard the inverse as defining a compact operator on H, and apply the standard theory of compact operators. The eigenvectors of this compact operator lie in H_1, because the eigenvalues are non-zero and H_1 is the image of the compact operator. □

The eigenvalues λ_i have no accumulation point, and there are three possible cases: (i) the eigenvalues are bounded below; (ii) the eigenvalues are bounded above; or (iii) we can order the eigenvalues so that $\lambda_i \leq \lambda_{i+1}$, $-\infty < i < \infty$, and

$$\lambda_i \to \pm\infty, \quad \text{as } i \to \pm\infty.$$

Definition 33.1.3. Given H, H_1 and K, we write

$$S_*(H : H_1) \subset S(H : H_1)$$

for the operators $L \in S(H : H_1)$ of the third of the above types, satisfying the following additional condition. Let $H_1^{\pm}(L)$ and $H^{\pm}(L)$ be the closures in H_1 and H respectively of the span of the eigenvectors belonging to non-negative and negative eigenvalues of L. Then we require that there is an element $u \in U(H : H_1)$ with

$$u(H^{\pm}) = H^{\pm}(L)$$
$$u(H_1^{\pm}) = H_1^{\pm}(L),$$

where the spaces $H^{\pm}$ that appear on the left are defined, as before, in terms of the eigenvectors of K. ◊

Suppose Q is a compact Riemannian manifold, and suppose we are given a smooth principal bundle over Q with structure group $U(H : H_1)$: that is, the principle bundle is defined in terms of transition functions on Q that are smooth maps to $U(H : H_1)$. Associated to the principal bundle are a pair of Hilbert bundles $\mathcal{H}_1 \subset \mathcal{H}$ over Q.

Definition 33.1.4. By a *family of self-adjoint operators of type* $S_*(H : H_1)$ over Q, we mean a principal $U(H : H_1)$ bundle, as just described, together with a bundle map

$$L : \mathcal{H}_1 \to \mathcal{H}$$

which is a smooth section of the associated bundle $S_*(\mathcal{H}:\mathcal{H}_1) \to Q$ with fiber $S_*(H:H_1)$. ◊

According to Kuiper's theorem [64], the unitary group $U(H)$ is contractible. We have the following variant:

Proposition 33.1.5. *If the spectrum of* $|K| = K^+ - K^-$ *is mild, then the group*

$$U(H:H_1) = B(H:H_1) \cap U(H)$$

is contractible.

Proof. We may as well assume that K is positive, and so write K instead of $|K|$. Let e_n be a complete orthonormal system in H, indexed by the non-negative integers, with

$$K^{-1}e_n = \lambda_n e_n$$

and $\lambda_{2N}/\lambda_N \leq C$ for all N. This is the condition that the spectrum is "mild". If H' is a closed subspace of H and $H'_1 = H' \cap H_1$, then we shall say that $(H':H'_1)$ is equivalent to $(H:H_1)$ if there is a unitary isomorphism ψ from H to H' such that both ψ and its inverse are bounded with respect to the H_1 norms on both sides. The mild condition makes it easy to construct such subspaces $H' \subset H$. For example, given any sequence of integers $n_1, n_2, \ldots$, such that n_i/i is bounded above and below, we can take H' to be the closed span of the e_{n_i}, and the map $H \to H'$ given by $e_i \mapsto e_{n_i}$ restricts to a bounded map $H_1 \to H'_1$ with bounded inverse. By the same token, there is a unitary isomorphism

$$\Phi : H \to \hat{\bigoplus}_{i=1}^{\infty} H \qquad (33.1)$$

such that both Φ and its inverse are bounded with respect to the H_1 norms on both sides. (The symbol $\hat{\bigoplus}$ means we are to take the Hilbert-space sum.)

We now follow Kuiper's argument, from [64], with appropriate modifications. We begin by showing that the larger group $GL(H:H_1)$ of invertible elements in $B(H:H_1)$ is contractible. It will be enough to show the homotopy groups vanish, so we consider a map from a sphere, $f_0 : S^k \to GL(H:H_1)$. The map f_0 is a homotopic to map $f_1 : S^k \to GL(H:H_1)$ with the additional property that f_1 is linear on each simplex of some triangulation of S^k. Let $W \subset B(H:H_1)$ be the finite-dimensional space spanned by the image of f_1, and let N be its dimension.

We will inductively construct a sequence of elements a_i and a_i^o in H_1, as follows. Take $a_1 = e_1$ in H_1, and consider the space

$$\mathbb{C}a_1 + W(a_1) \subset H_1. \tag{33.2}$$

Let us say that elements of H_1 are "doubly orthogonal" if they are orthogonal for both the H and H_1 inner products. The linear space (33.2) has dimension at most $N + 1$; and to be doubly orthogonal to this space imposes $2N + 2$ linear conditions at most. We can therefore find an element a_1^o, of unit H norm, which is doubly orthogonal to the above subspace and which lies in the span of the vectors

$$e_1, \ldots, e_{2N+3}.$$

Let $A_1 \subset H_1$ be the linear space

$$A_1 = \mathbb{C}a_1 + \mathbb{C}a_1^o + W(a_1).$$

Having defined a_1 and a_1^o, we choose an a_2 with the property that

$$\mathbb{C}a_2 \oplus W(a_2)$$

is doubly orthogonal to A_1. This constraint imposes at most $2(N + 1)(N + 2)$ linear conditions on a_2, so we can select a_2 of unit H norm lying in the span of the vectors

$$e_2, \ldots, e_{2(N+1)(N+2)+2}.$$

Next choose a_2^o to be doubly orthogonal to both A_1 and $\mathbb{C}a_2 \oplus W(a_2)$. This a_2 can be chosen to lie in the span of the vectors

$$e_2, \ldots, e_{4N+8}.$$

Define A_2 to be the space

$$A_2 = \mathbb{C}a_2 + \mathbb{C}a_2^o + W(a_2).$$

Proceeding in this way, we construct a sequence of subspaces

$$A_1, A_2, A_3, \ldots$$

which are mutually doubly orthogonal. Each has dimension at most $N+2$ and has the form

$$A_i = \mathbb{C}a_i + \mathbb{C}a_i^o + W(a_i),$$

with a_i^o doubly orthogonal to the other two summands. Furthermore, we may arrange that

$$a_i, a_i^o \subset \text{span}\{e_i, \ldots, e_{Mi}\} \tag{33.3}$$

where M is an integer depending only on N.

Let $H' \subset H$ be the H-closure of the span of the doubly orthogonal sequence vectors a_i, and let $H_1' \subset H_1$ be the closure in the H_1 topology. The mild condition on the spectrum, together with the constraint imposed on the a_i by (33.3), means that the pair $(H' : H_1')$ is equivalent to $(H : H_1)$. More specifically, we have inequalities

$$C_1\lambda_i \le \|a_i\|_{H_1} \le C_2\lambda_i$$
$$C_1\lambda_i \le \|a_i^o\|_{H_1} \le C_2\lambda_i$$

for some constants C_1 and C_2.

The next step is to homotope f_1 in stages to a new map f_4 with the property that $f_4(s)|_{H'} = 1$ for all $s \in S^k$. Given s, set $w = f_1(s) \in W$. Because the image of S^k is a compact subset of $GL(H : H_1)$, there are positive constants C_j independent of $s \in S^k$ and independent of i such that

$$C_3 \le \|w(a_i)\|_H \le C_4$$
$$C_5\lambda_i \le \|w(a_i)\|_{H_1} \le C_6\lambda_i.$$

These bounds, and the mutual double orthogonality of a_i^o and $w(a_i)$, give us a uniform bound on the H_1 operator norm of the rotation expressed by

$$\rho_{i,t}^s = \begin{bmatrix} \cos(\pi t/2) & -\sin(\pi t/2) \\ \sin(\pi t/2) & \cos(\pi t/2) \end{bmatrix}$$

with respect to the H-orthonormal basis

$$\frac{w(a_i)}{\|w(a_i)\|}, \quad a_i^o$$

of the 2-dimensional space spanned by these vectors. We can therefore define a continuous path $\rho_t^s \in U(H : H_1)$ to be the direct sum of these 2-by-2 blocks $\rho_{i,t}^s$.

The transformation ρ_1 carries $w(a_i)$ to $\|w(a_i)\|a_i^o$, for all i. Define a homotopy from f_1 to a new map f_2 by the formula

$$f_{1+t}(s) = \rho_t^s f_1(s).$$

The endpoint of this homotopy has

$$f_2(s)(a_i) = c_{s,i}a_i^o$$

where the constants $c_{s,i}$ all satisfy

$$C_3 \leq c_{s,i} \leq C_4.$$

We now construct a similar homotopy from f_2 to a new f_3, using rotations $\sigma_{i,t}$ in the planes spanned by a_i and a_i^o: we define

$$f_{2+t}(s) = \sigma_t f_2(s)$$

in such a way that

$$f_3(s)(a_i) = c_{s,i}a_i.$$

Because of the bounds on the $c_{s,i}$, we can define a homotopy

$$f_{3+t}(s) = \tau_t^s f_3(s),$$

by specifying that

$$\tau_t^s(a_i) = (1-t)c_{s,i} + t$$

and arranging that $\tau_t^s = 1$ on the H-orthogonal complement of H'. At the end of this homotopy, we have an f_4, with

$$f_4(s)(a_i) = a_i$$

for all i. That is, $f_4(s)$ is the identity on H'.

Now write $H = H' \oplus H''$, where as before H' is the span of the a_i and H'' is its H-orthogonal complement. We have a corresponding decomposition of H_1 as $H_1' \oplus H_1''$, where $H_1'' = H'' \cap H_1$. The pairs $(H' : H_1')$ and $(H'' : H_1'')$ are both equivalent to $(H : H_1)$. With respect to this decomposition, we have

$$f_4(s) = \begin{bmatrix} 1 & X(s) \\ 0 & B(s) \end{bmatrix}.$$

Define a homotopy to a new f_5 by

$$f_{4+t}(s) = \begin{bmatrix} 1 & (1-t)X(s) \\ 0 & B(s) \end{bmatrix}$$

so that $f_5(s)$ is block diagonal:

$$f_5(s) = \begin{bmatrix} 1 & 0 \\ 0 & B(s) \end{bmatrix}.$$

The final stages of the homotopy use the isomorphism (33.1). This isomorphism allows us to view $f_5(s)$ as an infinite block diagonal matrix

$$f_5(s) = \operatorname{diag}(B(s), 1, 1, \dots)$$

acting on the the Hilbert sum. The remaining device is to use a homotopy from $f_5(s)$ to

$$f_6(s) = \operatorname{diag}(B(s), B^{-1}(s), B(s), \dots)$$

and thence to the constant map

$$f_7(s) = \operatorname{diag}(1, 1, 1, \dots).$$

These last two homotopies both exploit a standard homotopy between $\operatorname{diag}(1, 1)$ and $\operatorname{diag}(B^{-1}, B)$, applied to adjacent diagonal blocks.

These homotopies establish the vanishing of the homotopy groups of $GL(H : H_1)$. In the usual case of $U(H)$, one then appeals to "polar decomposition" to show that $U(H)$ and $GL(H)$ have the same homotopy type. In the present context, it is not clear that we have a counterpart to the usual polar decomposition theorem for operators on Hilbert space; so it is not a priori clear that the homotopy type of $U(H : H_1)$ is the same as that of the group of invertible elements, $GL(H : H_1)$. However, we shall obtain a "local version" of polar decomposition for a tubular neighborhood of $U(H : H_1)$, and apply that instead.

Let $S(H) \subset B(H)$ be the symmetric bounded operators on H, and consider the map

$$\pi : GL(H : H_1) \to B(H : H_1) \cap S(H)$$

given by

$$A \mapsto A^*A.$$

This map is a submersion, and its image is an open set in the codomain. Let V_ϵ be the metric ball of radius ϵ in $B(H : H_1) \cap S(H)$, centered on the point 1. We can choose ϵ small enough that V_ϵ is contained in the image of π. Let $\Omega_\epsilon \subset GL(H : H_1)$ be $\pi^{-1}(V_\epsilon)$. The map

$$\pi : \Omega_\epsilon \to V_\epsilon$$

is a fibration: its fibers are orbits of $U(H : H_1)$ acting by left multiplication on $GL(H : H_1)$. Because the base is contractible, Ω_ϵ has the same homotopy type as $U(H : H_1)$. We can think of Ω_ϵ as a tubular neighborhood of the unitary subgroup $U(H : H_1)$ in $GL(H : H_1)$.

To show that $U(H : H_1)$ is contractible, we will show that any map $f_0 : S^k \to U(H : H_1)$ can be contracted to a constant map in the larger space Ω_ϵ. To do this, we will reexamine the sequence of homotopies from f_0 to f_7 described above, and check that they can be chosen so as to remain in Ω_ϵ.

We begin by choosing an ϵ_0 less than ϵ: further requirements will be placed on ϵ_0 shortly. There is no difficulty about making the piecewise linear approximation f_1, homotopic to f_0 in Ω_{ϵ_0}, as this step is applicable to an open set in any Banach space. The subsequent homotopies from f_1 to f_5 all have the form

$$f_{n+t}(s) = Z_t(s) f_n(s). \tag{33.4}$$

In the cases $n = 1$ and $n = 2$, the transformation Z_t is a map from S^k into the unitary group; so these homotopies remain inside Ω_{ϵ_0} because Ω_{ϵ_0} is invariant under left multiplication by $U(H : H_1)$. In the case $n = 3$, we can look at the homotopy f_{3+t} and see from its shape that the norm of $f(s)^* f(s)$ is decreasing along the homotopy; so we remain withing Ω_{ϵ_0}.

The homotopy from f_4 to f_5 has the shape (33.4) also; but for this case the transformations $Z_t(s)$ are not unitary. We have

$$f_{4+t}(s)^* f_{4+t}(s) = \begin{bmatrix} 1 & (1-t)X(s) \\ (1-t)X(s)^* & (1-t)^2 X(s)^* X(s) + B(s)^* B(s) \end{bmatrix}.$$

Since $f_4(s)$ lies in Ω_{ϵ_0}, the term $X(s)$ has norm at most ϵ_0 in $B(H : H_1)$. So the distance between $f_{4+t}(s)$ and $f_4(s)$ is $O(\epsilon_0)$. We can therefore choose $\epsilon_0 < \epsilon$ small enough to ensure that $f_{4+t}(S^k)$ is contained in Ω_ϵ.

At this point, it is convenient to introduce an additional intermediate step in the homotopy. In Ω_ϵ, we can deform f_4 to a homotopic map

$$\tilde{f}_4 = \begin{bmatrix} 1 & 0 \\ 0 & \tilde{B}(s) \end{bmatrix}$$

with $\tilde{B}(s)$ now taking values in $U(H : H_1)$. After this, we proceed with the remaining two steps of the homotopy as before: they have the form (33.4), so they remain inside Ω_ϵ. This completes the proof. □

Just as the above proposition is a variant of Kuiper's theorem, adapted to our present setting, so the next proposition is a modification of a result from [10].

Proposition 33.1.6. *If the positive operators K^+ and $-K^-$ have mild spectrum, then the space $S_*(H : H_1)$ has the homotopy type of the infinite unitary group, $U(\infty) = \lim U(n)$.*

Proof. Let $\beta : \mathbb{R} \to [-1, 1]$ be a monotonic function with $\beta(\lambda) = 1$ for $\lambda \geq \Lambda_+$ and $\beta(\lambda) = -1$ for $\lambda \leq \Lambda_-$. The map $L \mapsto \beta(L)$ defines a continuous map from $S_*(H : H_1)$ to

$$S^f(H : H_1) = \{A \in B(H : H_1) \cap S(H) \mid$$
$$A^2 - 1 \text{ has finite rank, } \operatorname{im}(A^2 - 1) \subset H_1 \text{ and } \|A\| = 1\}.$$

The image is the subset

$$S^f_*(H : H_1) \subset S^f(H : H_1)$$

consisting of operators satisfying the additional condition that there exist unitary isomorphisms

$$\ker(A - 1) \to H^+$$
$$\ker(A + 1) \to H^-$$

which restrict to bounded isomorphisms

$$\ker(A - 1) \cap H_1 \to H_1^+$$
$$\ker(A + 1) \cap H_1 \to H_1^-.$$

The fiber of the resulting map

$$\beta : S_*(H : H_1) \to S^f_*(H : H_1)$$

over an element A can be identified with the space of pairs (L^+, L^-), where

$$L^+ : \ker(A - 1) \cap H_1 \to \ker(A - 1)$$

is a symmetric Fredholm operator of index zero having spectrum contained in $[\Lambda_+, \infty)$, with a similar condition on L^-. The fiber is contractible, and by an argument similar to that in [10, Lemma 3.7 ff.], it follows that β is a homotopy equivalence.

Next we have the exponential map

$$\epsilon : S_*^f(H : H_1) \to U^f(H : H_1)$$
$$A \mapsto -\exp(\pi i A),$$

whose image $U^f(H : H_1)$ is the subset of $U(H : H_1)$ consisting of elements u with $u - 1$ of finite rank. The fibers of this map are homeomorphic to copies of

$$U(H : H_1)/\big(U(H^+; H_1^+) \times U(H^-; H_1^-)\big),$$

a space which parametrizes the orbit of the decomposition $H = H^+ \oplus H^-$ under the action of $U(H : H_1)$. These fibers are contractible by Proposition 33.1.5; so by the same arguments from [10], the map ϵ is also a homotopy equivalence.

Let e_i $(i \geq 0)$ be an orthonormal basis of H consisting of eigenvectors of K, arranged in decreasing order of the absolute value of the eigenvalue. Let $E_n \subset H_1$ be the span of the first n of these eigenvectors, and let $U(n) \subset U(H : H_1)$ consist of the elements u such that $u - 1$ is supported in E_n. An approximation argument shows that the direct limit of these inclusions,

$$U(\infty) \to U^f(H : H_1),$$

is a weak homotopy equivalence, and hence a homotopy equivalence as these spaces have the homotopy type of CW complexes. □

Corollary 33.1.7. *If the operators K^+ and $-K^-$ have mild spectrum, then families of self-adjoint operators of type $S_*(H : H_1)$ over Q are classified by*

$$[Q, U(\infty)].$$

Proof. Proposition 33.1.5 provides a unique homotopy class of trivializations of any $U(H : H_1)$ bundle, so the corollary follows from Proposition 33.1.6. □

33.2 A model for the universal family

Having learnt that families of self-adjoint operators (of type $S_*(H : H_1)$) are classified by maps to $U(\infty)$, it is natural to look for a universal family of such self-adjoint operators, parametrized by the unitary group itself. More precisely, the classification result, Corollary 33.1.7, assures us that there is a family of self-adjoint operators over the compact unitary group $U(N)$ for which the classifying map is the inclusion $U(N) \to U(\infty)$. We will identify such a map. We begin with a lemma concerning the closely related unitary group $U^f(H)$, defined as the unitary transformations u such that $u - 1$ has finite rank. The lemma gives a criterion for when two maps to $U^f(H)$ are homotopic.

Lemma 33.2.1. *Let Q be any space, and let u_1, u_2 be two maps from Q to $U^f(H)$, the group of unitary transformations $u : H \to H$ such that $u - 1$ has finite rank. Suppose there exists a continuous map θ to the space of bounded operators of finite rank,*

$$\theta : Q \to B^f(H),$$

such that $\theta u_1 = u_2\theta$ and such that, for all $q \in Q$, the restriction

$$\theta|_{\ker(u_1+1)} : \ker(u_1 + 1) \to \ker(u_2 + 1)$$

is an isomorphism. Then u_1 and u_2 are homotopic.

Proof. We will demonstrate the equivalent statement, that the map

$$u = \begin{bmatrix} u_1 & 0 \\ 0 & u_2^{-1} \end{bmatrix}$$

from Q to $U^f(H \oplus H)$ is null-homotopic. Replace θ by a smaller multiple if necessary so that $\|\theta(q)\| < \pi$ for all q. Let $\tilde{u}$ be the map

$$\begin{aligned} \tilde{u} &= \exp \begin{bmatrix} 0 & -\theta^* \\ \theta & 0 \end{bmatrix} \begin{bmatrix} u_1 & 0 \\ 0 & u_2^{-1} \end{bmatrix} \\ &= \begin{bmatrix} c(\theta)u_1 & -s(\theta^*)u_2^{-1} \\ s(\theta)u_1 & c(\theta^*)u_2^{-1} \end{bmatrix}, \end{aligned}$$

where

$$\begin{aligned} c(\theta) &= \cos(\sqrt{\theta^*\theta}) \\ s(\theta) &= \theta \frac{\sin(\sqrt{\theta^*\theta})}{\sqrt{\theta^*\theta}}. \end{aligned}$$

The maps u and $\tilde{u}$ are homotopic maps to $U^f(H \oplus H)$. The operators $c(\theta)$ and $c(\theta^*)$ commute with u_1 and u_2 respectively, and we have the relation $s(\theta)u_1 = u_2 s(\theta)$.

We now show that -1 is not an eigenvalue of $\tilde{u}(q)$, for any q in Q. Suppose on the contrary that $h = (h_1, h_2)$ is a unit vector in $H \oplus H$ with $\tilde{u}(q)h = -h$. We then have (omitting q),

$$\begin{aligned} -1 &= \mathrm{Re}\langle h, \tilde{u}(q)h\rangle \\ &= \mathrm{Re}\langle h_1, c(\theta)u_1 h_1\rangle + \mathrm{Re}\langle h_2, c(\theta^*)u_2^{-1}h_2\rangle \end{aligned} \tag{33.5}$$

because the cross-terms cancel. The operator $c(\theta)$ is self-adjoint, and our assumption that $\|\theta\| < \pi$ means that its spectrum is contained in $(-1, 1]$. We therefore have

$$\|c(\theta)u_1 h_1\| \leq \|h_1\|$$

with equality only if $u_1 h_1$ is in the kernel of θ. A similar remark applies to the h_2 term. So the only way that (33.5) can hold is if $u_1 h_1$ and $u_2^{-1} h_2$ belong to the kernels of θ and θ^* respectively; and we then see that we must have $u_1 h_1 = -h_1$ and $u_2^{-1} h_2 = -h_2$. But the hypothesis of the lemma states that θ is injective on the kernel of $u_1 + 1$, and the hypothesis implies also that θ^* is injective on the kernel of $u_2 + 1$. So we have a contradiction.

Having learned that -1 is never an eigenvalue of $\tilde{u}(q)$, we conclude that $\tilde{u} : Q \to U^f(H \oplus H)$ is null-homotopic. Indeed, the subspace of $U^f(H \oplus H)$ consisting of operators for which -1 is not in the spectrum is contractible: if

$$g_t : S^1 \setminus \{-1\} \to S^1 \setminus \{-1\}$$

is a homotopy that retracts the punctured circle to the point $\{1\}$, then $u \mapsto g_t(u)$ is a contraction of the corresponding subset of $U^f(H \oplus H)$ to the identity element. □

For $z \in U(N)$, let $C^\infty(S^1; z)$ denote the space of smooth functions

$$h : \mathbb{R} \to \mathbb{C}^N$$

satisfying

$$h(t+1) = zh(t).$$

Let $\|h\|$ and $\|h\|_1$ denote the L^2 norm and L^2_1 norm:

$$\|h\|^2 = \int_0^1 |h(t)|^2\, dt$$

$$\|h\|_1^2 = \int_0^1 \left(|h(t)|^2 + |(d/dt)h(t)|^2\right) dt.$$

Let $H(z)$ and $H_1(z)$ denote the completions of $C^\infty(S^1; z)$ in these two norms. Write H and H_1 for the ordinary $\mathbb{C}^N$-valued Sobolev spaces $H(1)$ and $H_1(1)$.

As z varies, these spaces form a bundle over $U(N)$ with structure group $U(H : H_1)$. Let

$$L(z) : H_1(z) \to H(z)$$

be the Fredholm operator

$$h \mapsto -i\frac{d}{dt}h.$$

The operators $L(z)$ form a family of self-adjoint Fredholm operators of type $S_*(H : H_1)$ parametrized by $z \in U(N)$. According to Corollary 33.1.7, the family $L(z)$ is classified by a map

$$\psi : U(N) \to U(\infty).$$

Proposition 33.2.2. *The classifying map $\psi : U(N) \to U(\infty)$ is homotopic to the inclusion.*

Proof. We first examine the operator $L(z)$ and its spectrum. Let ζ be a self-adjoint transformation of $\mathbb{C}^N$ with $\exp(i\zeta) = z$. We have

$$\mathrm{spec}(L(z)) = \mathrm{spec}(\zeta) + 2\pi\mathbb{Z},$$

for the following reason. If $v \in \mathbb{C}^N$ is an eigenvector of ζ with eigenvalue λ, then

$$h(t) = e^{i(\lambda+2\pi k)t}v$$

is an eigenvector of $L(z)$ with eigenvalue $\lambda + 2\pi k$, for any k in $\mathbb{Z}$; and these eigenvectors form a complete orthonormal system in $H(z)$ if we also let v run through an orthonormal basis of eigenvectors for z in $\mathbb{C}^N$.

Let $\beta : \mathbb{R} \to [-1, 1]$ satisfy

$$\beta(\lambda) = \lambda/\pi, \quad \text{for } -\pi \le \lambda \le \pi,$$

and let $\beta(\lambda) = 1$ for $\lambda \ge \pi$ and -1 for $\lambda \le -\pi$. Let $A(z)$ be the bounded self-adjoint operator $\beta(L(z))$, and let

$$\begin{aligned} u(z) &= \epsilon(A(z)) \\ &= -\exp(i\pi A(z)) \in U^f(H(z)). \end{aligned}$$

The spectrum of $A(z)$ is

$$\operatorname{spec}(A(z)) = \left(\frac{1}{\pi}\left(\operatorname{spec}(\zeta) \cap (-\pi, \pi)\right)\right) \cup \{-1, 1\},$$

and the spectrum of $u(z)$ is

$$\operatorname{spec}(u(z)) = -\operatorname{spec}(z) \cup \{1\}.$$

With the exception of 1, all points in $\operatorname{spec}(u(z))$ are eigenvalues whose (finite) multiplicities are the same as their multiplicities as eigenvalues of z. Moreover, there is a densely defined map

$$\begin{aligned} \operatorname{ev}(z) : H(z) &\to \mathbb{C}^N \\ h &\mapsto h(0) \end{aligned}$$

and for each $\lambda \ne 1$ in the spectrum, the restriction

$$\operatorname{ev}(z) : \ker(u(z) - \lambda) \to \ker(z - \lambda)$$

is an isomorphism. The evaluation map ev is not defined on the whole L^2 space $H(z)$, but if we choose a continuous map

$$f : \mathbb{R} \to \mathbb{R}$$

with compact support and equal to 1 on $[-\pi, \pi]$, then we can set

$$\tilde{\operatorname{ev}}(z) = \operatorname{ev}(z) \circ f(L(z)).$$

This is a bounded map

$$\tilde{\operatorname{ev}}(z) : H(z) \to \mathbb{C}^N$$

depending continuously on z and satisfying

$$\tilde{\mathrm{ev}}(z)u(z) = z\,\tilde{\mathrm{ev}}(z).$$

It restricts to an isomorphism on (in particular) the -1 eigenspaces. After applying Kuiper's theorem to trivialize the bundle of Hilbert spaces $H(z)$, and embedding $\mathbb{C}^N$ in H as the span of the first N basis vectors, we arrive at the situation described in the lemma above. □

Remark. The proposition above can also be deduced from the index theorem for families of self-adjoint operators, see [9, section 3].

33.3 Coupled homology in the absence of spectral flow

We are now ready to define "coupled Morse homology". Let Q be a smooth compact manifold, and $L : \mathcal{H}_1 \to \mathcal{H}$ be a family of self-adjoint Fredholm operators of type $S_*(H : H_1)$ as in Definition 33.1.4. Let a Riemannian metric on Q be given and let

$$f : Q \to \mathbb{R}$$

be a Morse function satisfying the Morse–Smale transversality condition. We also suppose that we are given a smooth connection ∇ for the principal bundle with structure group $U(H : H_1)$. In local trivializations, ∇ is defined by a smooth connection 1-form taking values in the Banach space that is the Lie algebra of $U(H : H_1)$. For each critical point q of f, we suppose that the spectrum of $L(q) : H_1(q) \to H(q)$ is simple and does not contain zero. (Compare Lemma 2.5.5 or Proposition 12.2.5.)

From this data, we shall now form a chain complex. We mimic the construction of the complex $\bar{C}_*$ from Subsection 2.5, replacing the finite-dimensional vector bundle N that appeared there with the Hilbert bundle $\mathcal{H}$. To keep the comparison with the finite-dimensional case close by, we shall suppose that the family of operators L has no spectral flow around loops in Q. This is equivalent to saying that the classifying map for the family is a map from Q to $SU(\infty) \subset U(\infty)$.

Because there is no spectral flow around loops, we can label the eigenvalues of $L(q)$ at each critical point q as

$$\cdots < \lambda_{-1}(q) < \lambda_0(q) < \lambda_1(q) < \cdots$$

in such a way that for any path $q(t)$ joining critical points q_1 and q_2, the spectral flow along $q(t)$ satisfies

$$\mathrm{sf} = s(q_2) - s(q_1)$$

where

$$s(q) = |\{i \mid i < 0 \text{ and } \lambda_i(q) > 0\}|. \tag{33.6}$$

Having so labelled the eigenvalues, we take $\phi_i(q)$ to be a unit eigenvector in $H(q)$ belonging to the eigenvalues $\lambda_i(q)$. We define $\bar{C}_* = \bar{C}_*(Q, L)$ to be a free abelian group with one generator for each pair (q, i) (*cf.* Subsection 2.5); but as we are working with an oriented theory, we now set

$$\bar{C}_* = \bigoplus_q \bigoplus_i \mathbb{Z}\Lambda_q \tag{33.7}$$

where Λ_q is the 2-element set of orientations of the unstable manifold of $-\operatorname{grad}(f)$ at q. See (22.1). This is a graded abelian group: we define

$$\bar{C}_n = \bigoplus_{\substack{q,i \\ \text{index }(q)+2i=n}} \mathbb{Z}\Lambda_q. \tag{33.8}$$

To define the boundary map on the complex, we write down the same equations as appeared in (2.21):

$$\frac{d}{dt}\gamma + (\operatorname{grad} f)_{\gamma(t)} = 0 \tag{33.9a}$$

$$\gamma^*(\nabla)\phi + (L(\gamma(t))\phi)dt = 0. \tag{33.9b}$$

We can interpret these as equations for a smooth path $\gamma : \mathbb{R} \to Q$ and a section ϕ of $\gamma^*(\mathcal{H})$, i.e. a section of $\mathcal{H}$ along the path. More precisely, let

$$\mathbb{H}_{\mathrm{loc}}(\gamma) = L^2_{\mathrm{loc}}(\mathbb{R}, \gamma^*(\mathcal{H}))$$

be the L^2_{loc} sections of $\gamma^*(\mathcal{H})$ along the path, and let

$$\mathbb{H}_{1,\mathrm{loc}}(\gamma) = L^2_{\mathrm{loc}}(\mathbb{R}, \gamma^*(\mathcal{H}_1)) \cap L^2_{1,\mathrm{loc}}(\mathbb{R}, \gamma^*(\mathcal{H})).$$

Then in the above equations we can regard ϕ as an element of $\mathbb{H}_{1,\mathrm{loc}}(\gamma)$.

Definition 33.3.1. Given critical points q_0 and q_1 of f in Q, and integers i_0 and i_1, we define $M(q_0, i, q_1, j)$ to be the quotient by $\mathbb{C}^*$ of the space of solutions (γ, ϕ) to the equations (33.9), with the following properties:

(i) the flow line $\gamma(t)$ joins q_0 to q_1,
(ii) $\phi(t) \sim c_0 e^{-\lambda_i t}\phi_i(q_0)$, as $t \to -\infty$,
(iii) $\phi(t) \sim c_1 e^{-\lambda_j t}\phi_j(q_1)$, as $t \to +\infty$,

for some non-zero constants c_0 and c_1. (The $\mathbb{C}^*$ action as usual scales ϕ.) ◊

We can set up suitable Sobolev spaces so that the operator that appears in (33.9b) is Fredholm. For each trajectory γ joining two critical points in q_0 and q_1 in Q, let

$$\mathbb{H}(\gamma) = L^2(\mathbb{R}, \gamma^*(\mathcal{H}))$$

and let

$$\mathbb{H}_1(\gamma) = L^2(\mathbb{R}, \gamma^*(\mathcal{H}_1)) \cap L^2_1(\mathbb{R}, \gamma^*(\mathcal{H})).$$

For each i and each critical point q, choose $\bar{\lambda}_i$ with

$$\lambda_i(q) < \bar{\lambda}_i(q) < \lambda_{i+1}(q).$$

Given i and j, let $w_{ij} : \mathbb{R} \to \mathbb{R}$ be a function with

$$w_{ij}(t) = \begin{cases} \bar{\lambda}_{i-1}(q_1)t & \text{for } t > 1 \\ \bar{\lambda}_j(q_0)t & \text{for } t < -1. \end{cases}$$

We can define weighted versions of these spaces. We set

$$\mathbb{H}(\gamma; w_{ij}) = e^{-w(t)}\mathbb{H}(\gamma)$$

and

$$\mathbb{H}_1(\gamma; w_{ij}) = e^{-w(t)}\mathbb{H}_1(\gamma).$$

Let $M(q_0, q_1)$ denote the moduli space of trajectories γ for the downward gradient flow of f. Recall that we are assuming that this flow satisfies the Morse–Smale transversality conditions so that this is a manifold.

Proposition 33.3.2. *For fixed $\gamma \in M(q_0, q_1)$, the operator acting on ϕ on the right-hand side of Equation* (33.9b) *is Fredholm, viewed as an operator*

$$P_{\gamma,i,j} : \mathbb{H}_1(\gamma; w_{ij}) \to \mathbb{H}(\gamma; w_{ij}).$$

The index of this operator is given by

$$\text{index}_{\mathbb{C}}(P_{\gamma,i,j}) = i - j + 1.$$

The operator varies smoothly as a function of γ.

Proof. The proof can be modelled on the standard "freezing coefficients" argument, that is used to show that an elliptic operator is Fredholm (but see also Proposition 14.2.1). First one shows that if γ is a constant trajectory, then the corresponding translationally invariant operator is invertible. One needs also to check the map $h \mapsto \beta(t)h$ is a compact operator, thought of as a map $\mathbb{H}_1(\gamma) \hookrightarrow \mathbb{H}(\gamma)$, whenever the support of β is compact. □

The moduli space $M(q_0, i, q_1, j)$ can now be described as

$$\bigcup_{\gamma \in M(q_0,q_1)} \mathbb{P}\big(\ker(P_{\gamma,i,j})\big)$$

(with the understanding that $\mathbb{P}(\{0\}) = \varnothing$). We describe a condition which ensures that the moduli space is smooth. Let $(\gamma, [\phi])$ be an element of $M(q_0, i, q_1, j)$. We say that this element is *regular* if the map

$$\begin{aligned} T_\gamma M(q_0, q_1) &\to \text{coker}(P_{\gamma,i,j}) \\ \dot{\gamma} &\mapsto \mathcal{D}_\gamma P_{-,i,j}(\dot{\gamma})\phi + \text{im}\, P_{\gamma,i,j} \end{aligned}$$

is surjective. We say that the moduli space is regular if it is regular at every $(\gamma, [\phi])$.

Remark. There are two alternative viewpoints on the moduli space $M(q_0, i, q_1, j)$, corresponding to our two versions of "blowing up" when we studied the Seiberg–Witten equations on a cylinder (Section 6): the description we have given above corresponds to the σ model for the blow-up, though we are now using weighted Sobolev spaces, as is appropriate in the non-compact case. There is also a "τ-model" description of these moduli spaces (see Subsection 6.3 and the discussion in Subsection 2.5). To describe this, let $\mathbb{P}(\mathcal{H})$ be the projectivization of the Hilbert bundle $\mathcal{H}$ over Q. If (γ, ϕ) solves the equations (33.9), then $\phi(t)$ is non-zero for all t and determines a section of $\mathbb{P}(\mathcal{H})$ along the path γ. As in Subsection 2.5, we can therefore interpret the moduli space $M(q_0, i, q_1, j)$ as a space of paths joining the points $[\phi_i(q_0)]$ and $[\phi_j(q_1)]$. We can interpret these paths, formally, as the flow lines of a vector field V^σ on the total space of $\mathbb{P}(\mathcal{H})$. (Compare (2.20).) The paths can be lifted to paths in the unit sphere bundle of $\mathcal{H}$, satisfying the equations

$$\frac{d}{dt}q + (\operatorname{grad} f)_{q(t)} = 0 \tag{33.10a}$$

$$q^*(\nabla)\phi + \big((L(\gamma(t)) - \Lambda(\phi))\phi\big)dt = 0, \tag{33.10b}$$

where the function

$$\Lambda : \mathbb{P}(\mathcal{H}_1) \to \mathbb{R}$$

is defined by

$$\Lambda(\phi) = \langle L\phi, \phi\rangle_H \big/ \langle \phi, \phi\rangle_H.$$

The regularity condition above then corresponds to the Morse–Smale condition for the vector field V^σ.

The next lemma justifies our use of the word "regular", and shows that regularity can be achieved by deforming L.

Lemma 33.3.3. *If the moduli space $M(q_0, i, q_1, j)$ is regular, then it is a smooth manifold of dimension*

$$\text{index}\,(q_0) - \text{index}\,(q_1) + 2(i - j).$$

Given any family of self-adjoint Fredholm operators L of type $S_*(H : H_1)$ over Q, there exists a homotopic family $\tilde{L}$, equal to L in a neighborhood of the critical set of f, such that $M(q_0, i, q_1, j)$ is regular for all (q_0, i, q_1, j).

Proof. The formula for the dimension is the real index of the operator that is the linearization of the pair of equations (33.9), less 2 for the action of $\mathbb{C}^*$. The proof of the second part can be modelled on the proof for the case of the Seiberg–Witten equations, given in Section 15. □

Hypothesis 33.3.4. We assume from now on that L is chosen so that all moduli spaces are regular. ◇

For a moduli space $M(q_0, i, q_1, j)$ of non-constant trajectories, we write $\breve{M}(q_0, i, q_1, j)$ for the moduli space of unparametrized trajectories

$$\breve{M}(q_0, i, q_1, j) = M(q_0, i, q_1, j)/\mathbb{R}.$$

This is contained in a larger space of unparametrized *broken trajectories*, $\breve{M}^+(q_0, i, q_1, j)$, as in Subsection 2.1 (see also Definition 16.1.2). The topology on $\breve{M}^+(q_0, i, q_1, j)$ can be defined by imitating the definition in Subsection 16.1. We have:

Proposition 33.3.5. *The moduli spaces of unparametrized broken trajectories $\breve{M}^+(q_0, i, q_1, j)$ are compact.*

Proof. The proof for the Seiberg–Witten equations provides a model. The two ingredients are first the standard Morse-theory argument on the compact manifold Q, and second a differential inequality for the function $\Lambda(\phi)$ defined above, which we now explain. The issue is the following. Let γ_n be a sequence of gradient trajectories from q_0 to q_1, converging to a broken trajectory, in the usual sense of Morse theory for the Morse function f on Q. For each n, let ϕ_n be an element of the kernel of $P_{\gamma_n, i, j}$. To establish compactness, following our earlier model, we need a uniform bound on the total variation of the real functions $\Lambda(\phi_n)$: see Lemma 16.3.1, which is the corresponding statement in the Seiberg–Witten case.

To obtain this bound, we note that, for a fixed trajectory $\gamma_n(t)$, the corresponding 1-parameter family of operators $L(t) = L(\gamma_n(t))$ satisfies an inequality of the form (7.6b); and the constants in that inequality are uniform in n. We can therefore follow the proof of the Lemma 7.1.3 to obtain an inequality of the same type as (7.9), which (recalling that $\dot{l}$ in (7.9) corresponds to $-\Lambda$) we can write:

$$\dot{\Lambda}(\phi_n) - C_3|\Lambda(\phi_n)| - C_5 \leq 0.$$

As it stands, this inequality is not sufficient (it allows Λ to grow exponentially). We can do better by noting that the constants C_1 and C_2 in (7.6b) can be replaced by functions of t, of the form $c_i|\dot{\gamma}_n|$. This provides a differential inequality for Λ which, when combined with the exponential decay results for the converging trajectories γ_n, is sufficient to bound the total variation as required. □

Because the operator $P_{\gamma, i, j}$ is complex, we can identify the orientation bundle of the manifold $M(q_0, i, q_1, j)$ with the pull-back of the orientation bundle of $M(q_0, q_1)$. In the case that $\breve{M}(q_0, i, q_1, j)$ is zero-dimensional, and hence canonically oriented, each element $(\gamma,[\phi])$ therefore determines an element

$$\epsilon(\gamma,[\phi]) \in \mathrm{Hom}(\mathbb{Z}\Lambda_{q_0}, \mathbb{Z}\Lambda_{q_1}).$$

As in the usual Morse complex, we define

$$\bar{\partial} : \bar{C}_* \to \bar{C}_*$$

by the formula

$$\bar{\partial} = \sum \sum_{(\gamma,[\phi])} \epsilon(\gamma,[\phi]),$$

in which the first sum is over all unparametrized moduli spaces $\breve{M}(q_0,i,q_1,j)$ of dimension zero. The operator $\bar{\partial}$ has degree -1, and we have:

Proposition 33.3.6. *The operator $\bar{\partial} : \bar{C}_* \to \bar{C}_*$ has square zero.*

Remarks on the proof. The proposition requires proving the gluing theorem for the moduli spaces of trajectories $M(q_0,i,q_1,j)$, leading to a result such as Corollary 19.5.1. To set this proof up in a manner similar to our proof for the Seiberg–Witten equations, one can work with the alternative model (33.10) for the trajectory spaces, so that the picture resembles the standard one in which trajectories are glued near a hyperbolic critical point of a flow. The model argument in Subsection 18.3 can then be applied. □

Definition 33.3.7. We write $\bar{H}_*(Q,L)$ for the cohomology of the complex $(\bar{C}_*,\bar{\partial})$ defined above. ◊

Proposition 33.3.8. *Let L_0 and L_1 be homotopic families of self-adjoint Fredholm operators of type $S_*(H : H_1)$ over Q. Assume that all the trajectory moduli spaces $M(q_0,i,q_1,j)$ for both of these families are regular. Then the homology groups $\bar{H}_*(Q,L_0)$ and $\bar{H}_*(Q,L_1)$ are isomorphic, by an isomorphism that preserves degree.*

Proof. Let $L(s,q)$ be a family of self-adjoint Fredholm operators of type $S_*(H : H_1)$ over $\mathbb{R} \times Q$, chosen so that $L(s,-) = L_0$ for $s \le -1$ and $L(s,-) = L_1$ for $s \ge 1$. One constructs a chain map

$$m : \bar{C}_*(Q,L_0) \to \bar{C}_*(Q,L_1)$$

using the moduli spaces of solutions $(\gamma, [\phi])$ to the non-autonomous equations

$$\frac{d}{dt}\gamma + (\operatorname{grad} f)_{\gamma(t)} = 0$$
$$\gamma^*(\nabla)\phi + (L(t, \gamma(t))\phi)dt = 0.$$

That this *is* a chain map, and that it is invertible, are statements that have proofs along the same lines as the arguments in Subsection 25.3.

There is one additional twist however. The chain map m may not have degree zero. Indeed, the degree of m depends on the choice of homotopy $L(s, -)$ $(-1 \leq s \leq 1)$ between L_0 and L_1. If we choose a basepoint q_0 in Q and examine the operators $L(s, q_0)$, we obtain a path from $L_0(q_0)$ to $L_1(q_0)$ in the space $S_*(H : H_1)$, a space with the homotopy type of $U(\infty)$, whose fundamental group is $\mathbb{Z}$. Given two homotopies $L(s, q)$ and $L'(s, q)$, there is therefore a difference element, $\delta = L - L'$ in $\mathbb{Z}$. The degrees of the corresponding chain maps m and m' differ by 2δ. The mod 2 grading is canonical, being just the index mod 2 of the corresponding critical point in Q. So we can always choose the homotopy so that m has degree zero. □

Remark. The isomorphism constructed in the proof of this proposition may depend on the choice of homotopy.

The group $\bar{H}_*(Q, L)$ has the structure of a module over the ordinary cohomology ring of $\mathbb{P}(\mathcal{H})$, by a construction like that of Subsection 25.3. We have continuous embeddings of each moduli space $M(q_0, i, q_1, j)$ in $\mathbb{P}(\mathcal{H})$ (by evaluation of γ and ϕ at $t = 0$), and the module maps are defined using Čech representatives of classes in $H^*(\mathbb{P}(\mathcal{H}))$, defined with respect to open covers $\mathcal{U}$ transverse to the moduli spaces.

In particular, $\bar{H}_*(Q, L)$ is a module over the ordinary cohomology of the *base* Q; and there is in addition a distinguished 2-dimensional class u_2 on $\mathbb{P}(\mathcal{H})$, minus the first Chern class of the tautological bundle. Because Q is finite-dimensional, we can choose a nowhere-vanishing smooth section s of the dual bundle of $\mathcal{H}$, whose zero set $s^{-1}(0)$ will be a smooth codimension-2 submanifold. We can choose a Čech representative $\tilde{u}_2$ for u_2 supported in a neighborhood of $s^{-1}(0)$, and the transversality conditions will hold if $s^{-1}(0)$ is transverse to the moduli spaces $M(q_0, i, q_1, j)$. The resulting map

$$(u_2 \frown -) : \bar{H}_*(Q, L) \to \bar{H}_{*-2}(Q, L) \tag{33.11}$$

arises from a chain map

$$(\tilde{u}_2 \frown -) : \bar{C}_*(Q, L) \to \bar{C}_{*-2}(Q, L)$$

whose matrix entry from (q_0, i) to (q_1, j) counts (with sign) the elements (γ, ϕ) in the 2-dimensional moduli space $M(q_0, i, q_1, j)$ satisfying the constraint

$$s(\phi(0)) = 0.$$

The map u_2 has a property not shared by the operators arising from other cohomology classes:

Lemma 33.3.9. *The map $(\tilde{u}_2 \frown -) : \bar{C}_k(Q, L) \to \bar{C}_{k-2}(Q, L)$ is invertible, for all k.*

Proof. The complex $\bar{C}_*(Q, L)$ has a decreasing filtration

$$\cdots \supset \mathcal{F}_k\bar{C}_*(Q, L) \supset \mathcal{F}_{k-1}\bar{C}_*(Q, L) \supset \cdots \tag{33.12}$$

where $\mathcal{F}_k\bar{C}_*(Q, L)$ is spanned by the generators corresponding to pairs (q, i') where $q \in Q$ is a critical point of f of ordinary Morse index at most k. If $e_{q,i} \in \mathbb{Z}\Lambda_q$ is a generator corresponding to the pair (q, i), and if q has Morse index k, then at the chain level we have

$$\tilde{u}_2 \frown e_{q,i} = e_{q,i-1} \pmod{\mathcal{F}_{k-2}},$$

just as in (25.15), because the moduli space $M(q, i, q, i-1)$ can be identified with the copy of $\mathbb{CP}^1 \setminus \{0, \infty\}$ spanned by $\phi_i(q)$ and $\phi_{i-1}(q)$, which will intersect $s^{-1}(0)$ in one point. So the map is an isomorphism on the graded complex associated to the filtration. The result follows. □

34 Calculation of coupled homology

34.1 The standard family over S^3

We take Q to be the 3-sphere, $SU(2)$, and take on Q a standard Morse function with one maximum and one minimum. We take these to be the points

$$q_+ = \begin{bmatrix} i & 0 \\ 0 & -i \end{bmatrix}$$

$$q_- = \begin{bmatrix} -i & 0 \\ 0 & i \end{bmatrix}.$$

In Subsection 33.2, we described a family of operators L over $U(N)$ of type $S_*(H : H_1)$, where H_1 was the space $L^2_1(S^1; \mathbb{C}^N) \subset L^2(S^1; \mathbb{C}^N)$. Taking $N = 2$ and restricting to the special unitary group, we obtain a family of self-adjoint Fredholm operators $L^{SU(2)}$ over $SU(2)$. The operator $L^{SU(2)}(z)$ is the operator $-id/dt$ acting on $H_1(z)$, as described earlier.

Take the standard orientation on the 3-sphere. The complex $\bar{C}_*$ in this case then becomes a free abelian group with basis elements indexed by pairs (q, i), where $q \in \{q_+, q_-\}$ and $i \in \mathbb{Z}$. Let us denote these generators by

$$e_i^+, e_i^- \in \bar{C}_*.$$

Proposition 34.1.1. *In the model complex on $SU(2)$, the boundary map $\bar{\partial} : \bar{C}_* \to \bar{C}_*$ is given by*

$$\begin{aligned} \bar{\partial} e_i^+ &= e_{i+1}^- \\ \bar{\partial} e_i^- &= 0 \end{aligned}$$

up to an overall sign.

Remark. It follows that the homology of the complex is trivial; but it is the complex itself that interests us here.

Proof of Proposition 34.1.1. Because $\bar{\partial}$ has degree -1, the only non-zero matrix entries of $\bar{\partial}$ are from e_i^+ to e_{i+1}^- (see (33.8)). The unparametrized trajectories of $-\operatorname{grad}(f)$ from q_+ to q_- comprise a 2-sphere $\breve{M}(q_+, q_-)$. If γ is one of these trajectories, then there exists a trajectory $(\gamma, [\phi])$ in $M(q_+, i, q_-, i+1)$ if and only if there is a non-zero element ϕ in the kernel of the operator

$$P_{\gamma,i,i+1} : \mathbb{H}_1(\gamma; w_{i,i+1}) \to \mathbb{H}(\gamma; w_{i,i+1})$$

that appears in Proposition 33.3.2. We must count, with sign, the number of points in the 2-sphere for which this Fredholm operator (of index zero) has kernel; and this is the same as calculating

$$\langle c_1(\operatorname{index}(P_{-,i,i+1})), [S^2] \rangle \tag{34.1}$$

up to an overall sign. Our description of the operator $L(z)$ can be recast as a Dirac operator on the circle; and the operators $P_{\gamma,i,i+1}$ are Cauchy–Riemann operators on $\mathbb{R} \times S^1$, acting on weighted Sobolev spaces of sections of a rank-2 bundle.

By an excision argument, we can replace $\mathbb{R} \times S^1$ with a torus $S^1 \times S^1$. This involves replacing the paths γ from q_+ to q_- with loops δ based at q_+ by composing with a fixed path from q_- to q_+. We obtain a family of Cauchy–Riemann operators on $S^1 \times S^1$ parametrized by a family of loops $\delta_s : S^1 \to SU(2)$ parametrized by $s \in S^2$.

The family of Cauchy–Riemann operators on $S^1 \times S^1$ is carried by a vector bundle

$$E \to S^1 \times S^1 \times S^2$$

which is constructed as follows: take the trivial bundle

$$\mathbb{C}^2 \times S^1 \times [0,1] \times S^2$$

and make the identifications

$$(v, \theta, 1, s) \sim (\delta_s(\theta), \theta, 0, s).$$

By the index theorem for families, the pairing (34.1) is equal to minus the second Chern number of the bundle E. This in turn is equal to the degree of the map

$$(\theta, s) \mapsto \delta_s(\theta)$$

from $S^1 \times S^2$ to $SU(2)$. This is equal to ± 1, because there is a unique unparametrized trajectory $\breve{\gamma}$ in $\breve{M}(q_+, q_-)$ passing through any (non-critical) point. □

34.2 Families pulled back from the three-sphere

When Q carries a family of operators L which is the pull-back of the standard family L^{S^3} by a map $Q \to S^3$, then we can obtain a description of the coupled Morse homology $\bar{H}_*(Q, L)$ that eventually makes no explicit reference to the Fredholm operators. The main step in this direction is the following proposition. See also Lemma 34.3.2 below.

Proposition 34.2.1. *Let Q be a compact manifold, let L be a family of self-adjoint Fredholm operators of type $S_*(H : H_1)$ over Q, and suppose that the classifying map for this family factors through the inclusion $SU(2) \hookrightarrow U(\infty)$. Let*

$$\xi \in H^3(Q; \mathbb{Z})$$

be the pull-back of the generator of $H^3(U(\infty); \mathbb{Z})$ via the classifying map.

Then there is a Riemannian metric on Q, a Morse function f satisfying the Morse–Smale condition, and a homotopic family of operators $\tilde{L}$, such that the corresponding complex $\bar{C}_*(Q,\tilde{L})$ can be described as follows. As a graded abelian group,

$$\bar{C}_*(Q,\tilde{L}) = C_*(Q,f) \otimes \mathbb{Z}[T^{-1},T], \tag{34.2}$$

where $C_*(Q,f)$ is the ordinary Morse complex and T has degree 2. The differential has the form

$$\bar{\partial}x = \partial x + T(\tilde{\xi} \cap x)$$

where ∂ is the Morse differential on $C_*(Q,f)$, and

$$\tilde{\xi} \in C^3(\mathcal{U})$$

is a certain Čech representative for the class ξ with respect to an open cover $\mathcal{U}$ that is transverse to the trajectory spaces $M(a,b) \subset Q$.

Remark. The cap product

$$\cap : C^d(\mathcal{U}) \otimes C_i(Q,f) \to C_{i-d}(Q,f)$$

is defined as in Subsection 25.3. Note that the Čech representative $\tilde{\xi}$ in the proposition is special (and is constructed in the proof). In particular, for this representative, the map $x \mapsto \tilde{\xi} \cap x$ has square zero at the chain level, for otherwise the formula for $\bar{\partial}$ does not define a differential.

Corollary 34.2.2. *There is a homology spectral sequence abutting to $\bar{H}_*(Q,L)$ whose E^2 and E^3 terms are given by*

$$E^2_{s,2j} = E^3_{s,2j} = T^j H_s(Q)$$

and whose differential $d^3_{s,2j}$ is the map

$$\begin{aligned} d^3_{s,2j} : E^3_{s,2j} &\to E^3_{s-3,2j+2} \\ T^j[x] &\mapsto T^{j+1}\xi \cap [x]. \end{aligned}$$

Proof. This is the spectral sequence arising from the filtration $\mathcal{F}_k$ of $\bar{C}_*(Q,L)$ defined at (33.12). □

The higher differentials d^5, d^7 etc. in the spectral sequence are obtained from a construction resembling that of Massey products such as $\{\xi, \ldots, \xi, [x]\}$. Later in this section, we will describe in more detail the relationship between this construction and the "twisted de Rham cohomology" introduced by Atiyah and Segal in [5].

Corollary 34.2.3. *If the higher differentials d^{2l+1} $(l \geq 2)$ are all zero, then $\bar{H}_*(Q, L)$ has a decreasing filtration*

$$\cdots \supset \mathcal{F}_s\bar{H}_*(Q,L) \supset \mathcal{F}_{s-1}\bar{H}_*(Q,L) \supset \cdots$$

such that the associated graded groups satisfy

$$\frac{\mathcal{F}_s\bar{H}_*(Q,L)}{\mathcal{F}_{s-1}\bar{H}_*(Q,L)} \cong \frac{\ker(\beta_s)}{\mathrm{im}(\beta_{s+3})} \otimes \mathbb{Z}[T^{-1}, T]$$

where

$$\beta_s : H_s(Q) \to H_{s-3}(Q)$$

is given by the cap product with ξ. This isomorphism respects the homological degree, when we interpret T as having degree 2. □

Proof of Proposition 34.2.1. Let L^{S^3} be a family of operators on S^3 homotopic to the standard family described in Subsection 34.1 above, but arrange that the family is constant outside of a small 3-ball $W \subset S^3$ centered at a point w on the equator. Let

$$v : Q \to S^3$$

be a map homotopic to the classifying map for the family L on Q, so that we can identify L with $v^*(L^{S^3})$. We can arrange that v is transverse to w and that $v^{-1}(W)$ is diffeomorphic to the product

$$v^{-1}(W) \cong v^{-1}(w) \times W$$

in such a way that the map v coincides with projection to the second factor. Choose a Riemannian metric on Q which is a product metric on $v^{-1}(W)$, and choose a Morse function f that is equal to the composite $y \circ v$ on $v^{-1}(W)$, where $y : S^3 \to [-1, 1]$ is a linear coordinate vanishing at w.

The effect of these choices is that if $\gamma : \mathbb{R} \to Q$ is a trajectory for the gradient of the function f, then the path $v \circ \gamma : \mathbb{R} \to S^3$ passes through $W \cap \{y = 0\}$

at most once; and if $v \circ \gamma$ *does* pass through $W \cap \{y = 0\}$, then the part of this path that lies in W is a trajectory of the gradient flow of the linear function y.

We can also arrange that f is Morse–Smale, and that for every trajectory space $M(a, b) \subset Q$, the restriction of v to $M(a, b)$ is transverse to w. After replacing W by a smaller neighborhood if necessary, we can arrange that for all the 3-dimensional moduli spaces $M(a, b)$, the intersection $v^{-1}(W) \cap M(a, b)$ has a product structure,

$$v^{-1}(W) \cap M(a, b) \cong \left(v^{-1}(w) \cap M(a, b) \right) \times W, \qquad (34.3)$$

where $v^{-1}(w) \cap M(a, b)$ is a finite set. We can also arrange that the $v^{-1}(W) \cap M(a, b)$ is empty if the dimension of $M(a, b)$ is less than 3.

Now fix a and b in Q, critical points with index difference 3. Suppose that $f(a) > 0$ and $f(b) < 0$. Let $\check{M}(a, b)$ be the moduli space of unparametrized trajectories, which we can identify with the trajectories $\gamma : \mathbb{R} \to Q$ with $f(0) = 0$. This moduli space parametrizes a family of Fredholm operators $P_{\gamma, i, j}$, as described earlier. If γ does not pass through $v^{-1}(W)$, then the operator $P_{\gamma, i, j}$ is translation-invariant because L is constant outside $v^{-1}(W)$. In this case $P_{\gamma, i, j}$ has no kernel if $j > i$. If γ does pass through $v^{-1}(W)$, then the operator $P_{\gamma, i, j}$ is the same as the operator $P_{\delta, i, j}$, where

$$\delta : \mathbb{R} \to S^3$$

is the unique trajectory of the standard Morse function y on S^3 whose intersection with the equator $y = 0$ is the point $(v \circ \gamma)(0)$.

Specialize now to the case that $j = i + 1$, so that the operator $P_{\gamma, i, i+1}$ has index zero. We wish to find finitely many elements γ in $\check{M}(a, b)$ for which the operator has kernel, and count these elements with sign. By our calculation for S^3 in Subsection 34.1 above, there are finitely many trajectories δ for the Morse function y such that $P_{\delta, i, i+1}$ has kernel, and the number of such trajectories counted with sign is 1. All these trajectories pass through W. Because of the product structure (34.3), the number of γ in $\check{M}(a, b)$ for which $P_{\gamma, i, i+1}$ has kernel, counted with sign, is the number of points in $v^{-1}(w) \cap M(a, b)$ counted with sign, once an orientation of $M(a, b)$ is given.

Now choose an open cover of S^3 and a Čech representative for the generator of $H^3(S^3)$ whose support is contained in W. Let $\tilde{\xi}$ be the pull-back of this representative to Q, carried by the pulled-back open cover $\mathcal{U}$. Then the number of points in $v^{-1}(w) \cap M(a, b)$ coincides with the evaluation of the compactly supported Čech cocycle,

$$\left\langle \tilde{\xi}, [M(a, b)] \right\rangle$$

in the notation of Subsection 25.3. According to our definition of $\bar{\partial}$, this quantity is therefore the matrix entry of $\bar{\partial}$,

$$\bar{\partial}(a, i, b, i+1) : (\mathbb{Z}\Lambda_a)_i \to (\mathbb{Z}\Lambda_b)_{i+1}$$

in the notation of (33.7).

As alternative notation, we can identify $\bar{C}_*(Q, L)$ with $C_*(Q,f) \otimes \mathbb{Z}[T^{-1}, T]$ in the obvious way, and write

$$\bar{\partial} = \partial_1 + T\partial_{3,i} + T^2\partial_{5,i} + \cdots , \tag{34.4}$$

where ∂_1 is the ordinary Morse differential and

$$\partial_{m,i} : C_i(Q,f) \to C_{i-m}(Q,f).$$

In these terms, we have calculated $\partial_{3,i}$ and shown it to be the chain-level cap product with $\tilde{\xi}$, independent of i.

The maps $\partial_{m,i}$ for $m > 3$ are zero, because the operators $P_{\gamma,i,j}$ have no kernel when $j > i+1$. This is because these operators are again isomorphic to operators $P_{\delta,i,j}$ for suitable trajectories δ of the Morse function y on S^3, none of which have kernel for $j > i+1$. □

We can also identify the action of the class u_2 in Proposition 34.2.1.

Proposition 34.2.4. *Under the isomorphism* (34.2) *arising in Proposition* 34.2.1, *the action of $\tilde{u}_2$ on $\bar{C}_*(Q, L)$ becomes the map*

$$C_*(Q,f) \otimes \mathbb{Z}[T^{-1}, T] \to C_*(Q,f) \otimes \mathbb{Z}[T^{-1}, T]$$

given by multiplication by T^{-1}, for a suitable chain representative of the cohomology class u_2.

Proof. We are continuing to suppose that $\mathcal{H}$ on Q, and the operator L, are both pulled back from S^3 by the map v. To define a suitable representative of u_2, we take a section s of the dual bundle of $\mathcal{H}$ which is also pulled back from S^3 and take $\tilde{u}_2$ to be supported in the neighborhood of $s^{-1}(0)$. We choose the section on S^3 so that it is constant outside the neighborhood of w.

We saw in the proof of Lemma 33.3.9 that the cap product with $\tilde{u}_2$ has the property that, if $x \in \mathcal{F}_k$, then

$$\tilde{u}_2 \frown x = T^{-1}x + y$$

for some $y \in \mathcal{F}_{k-2}$. We wish to show that $\tilde{u}_2 \cap x = T^{-1}x$. To investigate y, we examine the matrix entry of $(\tilde{u}_2 \cap -)$ from $(\mathbb{Z}\Lambda_a)_i \to (\mathbb{Z}\Lambda_b)_{i'}$ for $i' \geq i$.

In the case $i' = i$, the moduli space $M(a,b) \subset Q$ is 2-dimensional. The corresponding matrix entry for $\tilde{u}_2$ counts pairs $(\gamma, [\phi])$, where $\gamma \in M(a,b)$ and ϕ is a solution of the Fredholm equation $P_{\gamma,i,i}\phi = 0$ satisfying the additional constraint $s(\phi(0)) = 0$. Our assumptions ensure that the image of $M(a,b)$ under the map $v : Q \to S^3$ is disjoint from the neighborhood of w, so the operator $P_{\gamma,i,i}$ has the form $(d/dt) + L$ for some L independent of t. The solution ϕ must be the eigensolution $e^{-\lambda_i t}\phi_i(a)$, independent of γ; and this will not satisfy the constraint $s(\phi(0)) = 0$ if s is chosen generically. The matrix entry is therefore zero in this case.

In the case $i' > i$, the moduli space $M(a,b)$ has dimension 4 or more. In this case, the vanishing of the matrix entry can be seen by a dimension argument, based on the fact that the dimension of $M(a,b)$ exceeds the dimension of S^3. □

34.3 Relation with twisted cohomology

In [5], Atiyah and Segal introduce the "twisted cohomology" of a compact smooth manifold Q equipped with a closed 3-form ζ, as

$$H^*_\zeta(Q) = H^*(\Omega^*(Q), d + \zeta).$$

That is, $H^*_\zeta(Q)$ is the cohomology of the differential

$$x \mapsto dx + \zeta \wedge x$$

acting on real forms. This cohomology is $(\mathbb{Z}/2)$-graded.

There is a slight formal difference between this twisted de Rham cohomology and the coupled Morse theory, in that the latter involves a formal variable T of degree 2, whose inclusion makes the theory $\mathbb{Z}$-graded rather than just $(\mathbb{Z}/2)$-graded. Another formal difference is that, for coupled Morse theory, we have been discussing homology rather than cohomology. To dispense with these two points first, let us modify the Atiyah–Segal construction by introducing a formal variable T and defining $\bar{H}^*_\zeta(Q)$ as the $\mathbb{Z}$-graded cohomology theory given by

$$\bar{H}^*_\zeta(Q) = H^*\big(\Omega^*(Q) \otimes \mathbb{R}[T^{-1}, T],\ d + T\zeta\big);$$

so the differential is the map

$$x \mapsto dx + T\zeta \wedge x \tag{34.5}$$

acting on differential forms with values in Laurent polynomials in T. The construction of the coupled Morse homology $\bar{H}_*(Q, L)$ leads as usual to a corresponding cohomology theory, $\bar{H}^*(Q, L)$; and after passing to real coefficients this is what we should compare with the de Rham version. The following theorem confirms that these two coincide:

Theorem 34.3.1. *Let L be a family of operators of type $S_*(H : H_1)$ over the manifold Q, whose classifying map factors through the 3-sphere as in Proposition 34.2.1. Let ζ be a closed 3-form on Q whose de Rham class represents the pull-back of the 3-dimensional generator of the cohomology of S^3. Then the coupled Morse cohomology $\bar{H}^*(Q, L) \otimes \mathbb{R}$ is isomorphic to the twisted de Rham cohomology $\bar{H}^*_\zeta(Q)$ defined by the differential (34.5) above.*

We begin the proof with a discussion of the issues involved. A cohomology version of Proposition 34.2.1 tells us that the coupled Morse homology arises from a differential

$$x \mapsto dx + T\zeta_1 \cup x \tag{34.6}$$

acting on the Morse complex with values in $\mathbb{R}[T^{-1}, T]$. Here d is the ordinary Morse coboundary map, ζ_1 is a Čech representative for the 3-dimensional class, and the operation $\cup$ is the Morse-theoretic cup product at the chain level. The structure of this differential is so close to the de Rham version (34.5) that the theorem would seem inevitable; but some work is required to pass from Morse homology to de Rham cohomology. In particular, in (34.6), it is necessary to impose the condition that the map $x \mapsto \zeta_1 \cup (\zeta_1 \cup x)$ is zero at the chain level, or we do not have a differential at all, and this condition has no parallel on the de Rham side. The fact that $\zeta_1 \cup \zeta_1$ need not be zero reflects the fact that the anti-commutativity of the cup product holds only at the cohomology level, not at the chain level, for simplicial or Čech cohomology.

Our first step is to pass from Morse homology to simplicial homology. Let K be a simplicial complex with a homeomorphism $\phi : |K| \to Q$ which is smooth on each simplex. Let Δ^n be the standard closed n-simplex, regarded as the intersection of the plane $\sigma x_i = 1$ with the positive "octant". Let $\tilde{\Delta}^n$ be the corresponding locus in the unit sphere S^n obtained by radial projection. On $\tilde{\Delta}^n$, we consider a standard Morse function

$$g = -\sum x_i^4$$

and its gradient vector field V. Via radial projection, V becomes a vector field on Δ^n. The stationary points are precisely the barycenters of Δ^n and all its proper "facets" (subsimplices). The index of the critical point at the barycenter of a k-dimensional facet is k. Furthermore, there is a trajectory of V from an index-k critical point to an index-$(k-1)$ critical point precisely when the corresponding $(k-1)$-simplex is in the boundary of the k-simplex. If we equip each top-dimensional simplex of $|K|$ with such a vector field and transfer these to Q via the homeomorphism ϕ, we obtain a piecewise smooth, C^0 vector field V on Q.

Although it is neither smooth nor a gradient, the vector field V on Q is quite adequate for the constructions of Morse theory, including the construction of the coupled Morse complex which defines $\bar{H}^*(Q,L)$. This is because, for any pair of critical points a and b, the moduli space of trajectories $M(a,b)$ can be non-empty only if a is the barycenter of some Δ^k and b is the barycenter of some facet of Δ^k; in this case, the moduli space of trajectories $M(a,b)$ and its compactification by broken trajectories coincide with the corresponding objects defined by a smooth gradient flow of a function such as g above on a standard smooth k-simplex in the k-sphere.

What we have done here is described a flow on a triangulated smooth manifold with the property that its Morse complex coincides with the simplicial chain complex of the triangulation. To work with orientations, we can fix an ordering of the vertices of the complex, and define the boundary maps as usual in simplicial homology. That is, the simplicial chain complex is defined as having generators the simplices

$$[e_0, e_1, \ldots, e_k]$$

with the e_i strictly increasing; and the boundary map is

$$\partial[e_0, e_1, \ldots, e_k] = \sum_{i=0}^{k}(-1)^i[e_0, \ldots, \widehat{e_i}, \ldots, e_k].$$

Now triangulate S^3 and let $v : |K| = Q \to S^3$ be a simplicial representative of the map by which the family L is pulled back from S^3. Order the vertices in both complexes so that the map on vertices is monotone increasing. As in the proof of Proposition 34.2.1, we suppose that the family L^{S^3} on S^3 is constant outside a small ball $W \subset S^3$. We now impose the additional constraint that W is contained entirely in the interior of one 3-simplex, say $\Delta^3_0 \subset S^3$. Furthermore, we ask that W is contained in a small neighborhood of the highest-numbered vertex e_{top} of Δ^3_0; what we require is that the only trajectories of V on S^3 which meet W are trajectories belonging to $M(a, e_{\text{top}})$, where a is the critical point at the barycenter of Δ^3_0.

Following the arguments in Proposition 34.2.1, we obtain the following picture. Let w be the center of the ball W as before, and let Ξ be the piecewise linear submanifold of Q arising as $v^{-1}(w)$. The choices we have made mean that the contribution ∂_3 to the differential in the coupled Morse complex (see (34.4)) counts unparametrized trajectories of V that meet Ξ. Let ζ_1 denote the simplicial 3-cocycle obtained as the pull-back by the simplicial map v of the cocycle on S^3 which assigns 1 to the distinguished 3-simplex Δ^3_0. This ζ_1 assigns 1 to every 3-simplex in Q which meets Ξ and zero to all others. Then ∂_3 becomes simply the simplicial cap product with ζ_1. Passing to cohomology, we can summarize this part of the discussion:

Lemma 34.3.2. *There is a triangulation of Q as a simplicial complex $|K|$ and a simplicial 3-cocycle ζ_1 representing the pull-back of the generator from S^3, such that the coupled Morse cohomology can be computed as the cohomology of the differential*

$$x \mapsto \delta x + T(\zeta_1 \cup x)$$

acting on the simplicial cochain complex $C^*_\Delta(K) \otimes \mathbb{R}[T^{-1}, T]$. Here δ is the usual simplicial coboundary map and $\cup$ is the standard simplicial cup product, so that

$$(\alpha \cup \beta)[e_0, \dots, e_{p+q}] = \big(\alpha[e_0, \dots, e_p]\big)\big(\beta[e_p, \dots, e_{p+q}]\big)$$

for p- and q-cocycles α and β. Note that the representative ζ_1 has been chosen so that $\zeta_1 \cup \zeta_1 = 0$. □

Proof of Theorem 34.3.1. The lemma above has taken us from coupled Morse cohomology to the more familiar simplicial cohomology, with a twisted differential. The remaining step is to pass from simplicial cohomology to de Rham cohomology. We state a version of the necessary result, without the variable T:

Lemma 34.3.3. *Let Q be a compact smooth manifold, with a C^∞ triangulation $\phi : |K| \to Q$. Let ζ_1 be a 3-cocycle on K arising as the pull-back of a cocycle on S^3 supported on a single simplex, via an order-preserving simplicial map. Let ζ_2 be a smooth 3-form representing the same class in de Rham cohomology. Then the $(\mathbb{Z}/2)$-graded cohomology of the differential*

$$x \mapsto \delta x + \zeta_1 \cup x \tag{34.7}$$

acting on $C^*_\Delta(K;\mathbb{R})$ *is isomorphic to the twisted de Rham cohomology arising from the differential*

$$\chi \mapsto d\chi + \zeta_2 \wedge \chi$$

acting on differential forms.

Proof. For each vertex of K, there is a standard open neighborhood in Q consisting of the interiors of all simplices which contain the vertex. The nerve of the resulting open cover $\mathcal{U}$ coincides with K; so the Čech complex $C^*(\mathcal{U};\mathbb{R})$ with coefficients in the constant sheaf $\mathbb{R}$ matches the simplicial cochain complex. We set up our conventions for Čech cohomology to match our conventions for simplicial cohomology, making use of the given ordering of the vertices of K (which is the indexing set for $\mathcal{U}$). The coboundary map (34.7) can then be viewed as a differential on $C^*(\mathcal{U};\mathbb{R})$.

Both the Čech complex $C^*(\mathcal{U};\mathbb{R})$ and the de Rham complex are subcomplexes of the Čech–de Rham complex, $C\Omega^*(\mathcal{U})$ (see [14] for example). This is defined by

$$C\Omega^p(\mathcal{U}) = \bigoplus_{r+a=p} C^r(\mathcal{U};\Omega^a),$$

where $C^r(\mathcal{U};\Omega^a)$ is the usual Čech group with coefficients in the sheaf Ω^a:

$$C^r(\mathcal{U};\Omega^a) = \sum_{i_0<\cdots<i_r} \Omega^a(U_{i_0} \cap \cdots \cap U_{i_r}).$$

The differential on the Čech–de Rham complex,

$$D : C\Omega^p(\mathcal{U}) \to C\Omega^{p+1}(\mathcal{U}),$$

is given by the formula

$$D\phi = \delta\phi + (-1)^r d\phi$$

for ϕ in $C^r(\mathcal{U};\Omega^a)$, in which δ and d are the Čech and de Rham derivatives respectively. The cup product $\bar{\cup}$ on $C\Omega^*(\mathcal{U})$ is defined by

$$\phi \,\bar{\cup}\, \psi = (-1)^{as} \phi \cup \psi$$

for

$$\phi \in C^r(\mathcal{U};\Omega^a)$$

$$\psi \in C^s(\mathcal{U};\Omega^b).$$

On the right of the above formula, $\cup$ denotes the naive cup product

$$(\phi \cup \psi)_{i_0,\ldots,i_{r+s}} = \phi_{i_0,\ldots,i_r} \wedge \psi_{i_r,\ldots,i_{r+s}}.$$

With this convention, we have the usual sign in the Leibniz rule:

$$D(\phi \mathbin{\bar\cup} \psi) = D\phi \mathbin{\bar\cup} \psi + (-1)^p \phi \mathbin{\bar\cup} D\psi$$

for ϕ in $C\Omega^p(\mathcal{U})$. We have inclusions

$$C^*(\mathcal{U};\mathbb{R}) \subset C\Omega^*(\mathcal{U})$$
$$\Omega^*(Q) \subset C\Omega^*(\mathcal{U}),$$

the former from the inclusion of the constant sheaf $\mathbb{R}$ in Ω^0 and the latter as the kernel of δ on the Čech 0-cochains $C^0(\mathcal{U};\Omega^*)$. In this way, both ζ_1 and ζ_2 are 3-cocycles in $(C\Omega^*(\mathcal{U}), D)$, where they are cohomologous: for some $\eta \in C\Omega^2(\mathcal{U})$, we have

$$\zeta_1 + D\eta = \zeta_2.$$

Both have $\bar\cup$ square zero, so we can consider the differentials

$$D_{\zeta_1} : C\Omega^*(\mathcal{U}) \to C\Omega^*(\mathcal{U}) D_{\zeta_2} : C\Omega^*(\mathcal{U}) \to C\Omega^*(\mathcal{U})$$

given by

$$D_{\zeta_n}\phi = D\phi + \zeta_n \mathbin{\bar\cup} \phi.$$

By the usual argument [14], the two cohomologies which appear in the statement of the lemma are isomorphic to the cohomologies

$$H^*(C\Omega^*(\mathcal{U}), D_{\zeta_1})$$
$$H^*(C\Omega^*(\mathcal{U}), D_{\zeta_2})$$

respectively. So we are left to show that these last two groups are isomorphic. To do this, we will seek an element

$$z = 1 + z_1 + z_2 + \cdots$$

in $C\Omega^*(\mathcal{U})$, with $z_n \in C\Omega^{2n}(\mathcal{U})$, such that cupping with z intertwines the two derivatives:

$$D_{\zeta_1}(z \mathbin{\bar\cup} \phi) = z \mathbin{\bar\cup} D_{\zeta_2}\phi, \quad \forall \phi.$$

In terms of z_n, this says

$$Dz_{n+1} = -\zeta_1 \bar{\cup}\, z_n + z_n \,\bar{\cup}\, \zeta_2. \tag{34.8}$$

The awkwardness in constructing z stems from the fact that $\bar{\cup}$ is not graded-commutative: like the cup product on simplicial or Čech cohomology, it is graded-commutative only at the level of cohomology, not at the level of cochains.

In the commutative case, one should define z as the exponential of η, so that $z_n = (1/n!)\eta^n$. The reason that this does not work in our case is that η and $D\eta$ do not commute:

$$\eta \,\bar{\cup}\, D\eta \neq D\eta \,\bar{\cup}\, \eta. \tag{34.9}$$

The non-commutativity at the cochain level in simplicial homology gives rise to a hierarchy of products, Steenrod's higher products $\cup_i$ defined in [104]. In the simplicial case (or by the same formulae in the Čech case), these satisfy $\cup_0 = \cup$ and

$$\begin{aligned}\delta(u \cup_i v) = {} & (-1)^{p+q-i} u \cup_{i-1} v + (-1)^{pq+p+q} v \cup_{i-1} u \\ & + \delta u \cup_i v + (-1)^p u \cup_i \delta v\end{aligned} \tag{34.10}$$

for u, v in $C^p(\mathcal{U};\mathbb{R})$ and $C^q(\mathcal{U};\mathbb{R})$ respectively. Just as we defined $\bar{\cup}$ in terms of $\cup$, so we can define $\bar{\cup}_i$ on the Čech–de Rham complex in terms of the Čech product $\cup_i$: the appropriate sign for ϕ in $C^r(\mathcal{U};\Omega^a)$ and ψ in $C^s(\mathcal{U};\Omega^b)$ is

$$\phi \,\bar{\cup}_i\, \psi (-1)^{as+i(a+b)} \phi \cup_i \psi$$

with which the reader can verify that we again have

$$\begin{aligned}D(\phi \,\bar{\cup}_i\, \psi) = {} & (-1)^{p+q-i} \phi \,\bar{\cup}_{i-1}\, \psi + (-1)^{pq+p+q} \psi \,\bar{\cup}_{i-1}\, \phi \\ & + D\phi \,\bar{\cup}_i\, \psi + (-1)^p \phi \,\bar{\cup}_i\, D\psi\end{aligned} \tag{34.11}$$

for ϕ and ψ in $C\Omega^p(\mathcal{U})$ and $C\Omega^q(\mathcal{U})$, just as in Steenrod's formula (34.10).

A correct formula for z_n involves $\bar{\cup}_1$. We make use of the fact that ζ_1 and ζ_2 are both special: the latter is in the graded center of $C\Omega^*(\mathcal{U})$; while ζ_1 satisfies

$\zeta_1 \bar{\cup} \zeta_1 = 0$. These properties pass to the higher products, so that we have, for example

$$\zeta_2 \bar{\cup}_1 \phi = 0$$

for all ϕ, and

$$\zeta_1 \bar{\cup}_1 \zeta_1 = 0$$

because ζ_1 is pulled back from S^3. We also have, for example,

$$\zeta_1 \bar{\cup} \phi \bar{\cup} (\zeta_1 \bar{\cup}_1 \psi) = 0$$

for any ϕ and ψ. Consider now the 4-cochain w defined by

$$w = (D\eta \bar{\cup}_1 \eta).$$

Using the above observations and (34.11), we can verify that Dw is the commutator of $D\eta$ and η:

$$\begin{aligned} Dw &= D(D\eta \bar{\cup}_1 \eta) \\ &= D\eta \bar{\cup} \eta - \eta \bar{\cup} D\eta + DD\eta \bar{\cup}_1 \eta - D\eta \bar{\cup}_1 D\eta \\ &= D\eta \bar{\cup} \eta - \eta \bar{\cup} D\eta + 0 - (\zeta_2 - \zeta_1) \bar{\cup}_1 (\zeta_2 - \zeta_1) \\ &= D\eta \bar{\cup} \eta - \eta \bar{\cup} D\eta. \end{aligned} \tag{34.12}$$

Once we have this, a routine calculation shows that a working formula for a z_n which solves the relations (34.8) is

$$z_n = \frac{1}{n!}\left(\eta^n + \sum_{i=1}^{n-1}(n-i)\eta^{i-1} \bar{\cup}\, w \,\bar{\cup}\, \eta^{n-1-i}\right).$$

To verify this, note first that, because ζ_2 is central, we can rewrite the desired relation (34.8) as

$$Dz_{n+1} = D\eta \bar{\cup} z_n.$$

Now we compute, using the fact that any product involving both w and Dw is zero,

$$
\begin{aligned}
(n+1)!Dz_{n+1} &= D\left(\eta^{n+1} + \sum_{i=1}^{n}(n+1-i)\eta^{i-1}w\eta^{n-i}\right) \\
&= \sum_{j=1}^{n+1}\eta^{j-1}(D\eta)\eta^{n-j+1} + \sum_{i=1}^{n}(n+1-i)\eta^{i-1}(Dw)\eta^{n-i} \\
&\quad + \sum_{i=2}^{n}(n+1-i)(i-1)D\eta\eta^{i-2}w\eta^{n-i} \\
&\quad + \sum_{i=1}^{n-1}(n+1-i)(n-i)D\eta\eta^{i-1}w\eta^{n-i-1} \\
&= (n+1)(D\eta)\eta^{n} + (n+1)\sum_{i=1}^{n-1}(n-i)D\eta\eta^{i-1}w\eta^{n-i-1} \\
&= (n+1)!D\eta z_n.
\end{aligned}
$$

This completes the proof of the lemma. □

Theorem 34.3.1 now follows from the above lemma (with the variable T reinstated) and the previous one. □

Remark. If we work over the integers, and consider $C^*_\Delta(K)$ equipped with the differential

$$x \mapsto \delta x + \zeta_1 \cup x$$

where ζ_1 is pulled back from S^3 by a map $v : Q \to S^3$ as in Lemma 34.3.3, then we know that the isomorphism class of the resulting cohomology depends only on the homotopy class of the composite map

$$v : Q \to S^3 \hookrightarrow U(\infty),$$

i.e. on the resulting element of $K^1(Q)$. This follows from Lemma 34.3.2 to the fact $K^1(Q)$ classifies the families of self-adjoint operators L over Q of type $S_*(H : H_1)$. Over the reals or rationals, the isomorphism class of this cohomology group depends only on the cohomology class $\chi = [\zeta_1]$; but our results do not make clear whether this is true over $\mathbb{Z}$. There remains the possibility that two maps $Q \to S^3$, which determine different elements of $K^1(Q)$ but the same element of $H^3(Q;\mathbb{Z})$, might give rise to different cohomology groups. The authors do not have an example to show whether this occurs.

34.4 Families with spectral flow

We now suppose that the family L of self-adjoint Fredholm operators has spectral flow. Let $l \in \mathbb{Z}$ generate the image of the spectral flow, regarded as a map $\mathrm{sf} : H_1(Q) \to \mathbb{Z}$. Because our operators are complex, we will take this to be the complex spectral flow.

When the spectral flow is non-trivial, we are not able to label the eigenvalues of L at the critical points $q \in Q$ so as to satisfy the conditions (33.6). Instead, we label the eigenvalues $\lambda_i(q)$, still in increasing order, but with $\lambda_0(q)$ the first positive eigenvalue of $L(q)$.

A complex $\bar{C}_*(Q,L)$ is still defined in this situation, but it is not graded by $\mathbb{Z}$. As a group, we still take

$$\bar{C}_*(Q,L) = \bigoplus_q \bigoplus_i \mathbb{Z}\Lambda_q,$$

the sum being over all critical points in Q. The boundary map is defined as before by counting, with sign, the points of the zero-dimensional components of the moduli spaces $\breve{M}(q_0,i,q_1,j) = M(q_0,i,q_1,j)/\mathbb{R}$ defined just as in Definition 33.3.1. Now, however, the dimension of a component depends on the homotopy class of the trajectory γ from q_0 to q_1. Write z for a typical homotopy class, and let $M_z(q_0,i,q_1,j)$ be the corresponding subset of $M(q_0,i,q_1,j)$. When the necessary transversality holds, we have

$$\dim M_z(q_0,i,q_1,j) = \text{index}\,(q_0) - \text{index}\,(q_1) - 2j + 2i + 2\,\mathrm{sf}\,(L,z),$$

where the last term is the spectral flow of the family L along a path in the component z.

As a group, we can again identify $\bar{C}_*(Q,L)$ as

$$\bar{C}_*(Q,L) = C(Q,f) \otimes \mathbb{Z}[T^{-1},T]$$

in the obvious way. We write the boundary map $\bar{\partial}$ as a sum

$$\bar{\partial} = \bar{\partial}_1 + \bar{\partial}_3 + \cdots,$$

where

$$\bar{\partial}_m : C_i(Q,f) \otimes \mathbb{Z}[T^{-1},T] \to C_{i-m}(Q,f) \otimes \mathbb{Z}[T^{-1},T].$$

The differential $\bar{\partial}_1$ can be regarded as the ordinary Morse differential on $C(Q,f)$ with coefficients in a non-trivial local system with fiber $\mathbb{Z}[T^{-1},T]$. Namely, for each critical point q, let $\Gamma(q)$ be a copy of $\mathbb{Z}[T^{-1},T]$, and for each path z between critical points let

$$\Gamma(z) : \Gamma(q_0) \to \Gamma(q_1)$$

be multiplication by T^s, where $s = \mathrm{sf}(L,z)$. The ordinary Morse complex of (Q,f) with coefficients in this local system is

$$C_*(Q,f;\Gamma) = \bigoplus_q \mathbb{Z}\Lambda_q \otimes \Gamma(q)$$

with differential

$$\sum_\gamma \epsilon(\gamma) \otimes \Gamma([\gamma]), \tag{34.13}$$

where the sum is over trajectories in zero-dimensional moduli spaces $\breve{M}(q_0,q_1)$. This is exactly the differential $\bar{\partial}_1$.

If $\mathcal{U}$ is an open cover transverse to the trajectory spaces $M(q_0,q_1) \subset Q$, then we have a cap product

$$\cap : C^d(\mathcal{U}) \otimes C_i(Q,f;\Gamma) \to C_{i-d}(Q,f;\Gamma). \tag{34.14}$$

With this understood, we can state:

Proposition 34.4.1. *Let Q and L be as before, and suppose that the classifying map for this family factors through the inclusion $U(2) \hookrightarrow U(\infty)$. Let*

$$\xi_1 \in H^1(Q;\mathbb{Z}) \tag{34.15}$$

$$\xi_3 \in H^3(Q;\mathbb{Z}) \tag{34.16}$$

be the pull-backs of the two generators of $H^*(U(2);\mathbb{Z})$ via the classifying map.

Then there is a Riemannian metric on Q, a Morse function f satisfying the Morse–Smale condition, and a homotopic family of operators $\tilde{L}$, such that the corresponding complex $\bar{C}_*(Q,\tilde{L})$ can be described as follows. As an abelian group,

$$\bar{C}_*(Q,\tilde{L}) = C_*(Q,f;\Gamma_{\xi_1})$$

(the ordinary Morse complex of Q with coefficients in a local system). The differential has the form

$$\bar{\partial} = \bar{\partial}_1 + \bar{\partial}_3$$

where $\bar{\partial}_1$ is the Morse differential on $C_*(Q,f;\Gamma_{\xi_1})$ for the local system, and

$$\bar{\partial}_3 x = T(\tilde{\xi}_3 \frown x).$$

Here

$$\tilde{\xi}_3 \in C^3(\mathcal{U})$$

is a certain Čech representative for the class ξ_3 carried by a transverse open cover, and the cap product is the map referred to at (34.14). *The multiplication by T in the formula for $\bar{\partial}_3$ refers to the structure of Γ_{ξ_1} as a local system of $\mathbb{Z}[T^{-1},T]$-modules.*

As in the case of zero spectral flow treated in Proposition 34.2.1, there is a spectral sequence arising from this description of the differential. We still have the filtration of $\bar{C}_*(Q,L)$; but $\bar{C}_*(Q,L)$ is now only a differential group (it is not a complex with a $\mathbb{Z}$ grading), and the filtration is by differential subgroups $\mathcal{F}_k\bar{C}_*(Q,L)$.

Corollary 34.4.2. *There is a homology spectral sequence abutting to $\bar{H}_*(Q,L)$ whose E^2 and E^3 terms are given by*

$$E^2_s = E^3_s = H_s(Q;\Gamma_{\xi_1})$$

and whose differential d^3_s is the map

$$\begin{aligned} d^3_s : E^3_s &\to E^3_{s-3} \\ [x] &\mapsto T(\xi_3 \frown [x]). \end{aligned}$$

□

Corollary 34.4.3. *If the higher differentials d^{2l+1} $(l \geq 2)$ all vanish, then $\bar{H}_*(Q,L)$ has a decreasing filtration*

$$\cdots \supset \mathcal{F}_s\bar{H}_*(Q,L) \supset \mathcal{F}_{s-1}\bar{H}_*(Q,L) \supset \cdots$$

such that the associated graded groups satisfy

$$\frac{\mathcal{F}_s\bar{H}_*(Q,L)}{\mathcal{F}_{s-1}\bar{H}_*(Q,L)} \cong \frac{\ker(\beta_s)}{\operatorname{im}(\beta_{s+3})}$$

where

$$\beta_s : H_s(Q;\Gamma_{\xi_1}) \to H_{s-3}(Q;\Gamma_{\xi_1})$$

is given by the cap product with ξ_3. □

Proof of Proposition 34.4.1. We can choose a diffeomorphism of $U(2)$ with $S^1 \times S^3$ so that the first coordinate is the determinant. Let W be a small neighborhood of a point w in S^3, and let $y : S^3 \to [-1, 1]$ be a linear function vanishing at w. We can homotope the standard family of operators L parametrized by $U(2) \cong S^1 \times S^3$ so that, outside $S^1 \times W$, it coincides with the family $L^{S^1} \oplus L_0$, where L^{S^1} is the standard family parametrized by $U(1) \cong S^1$, and L_0 is the constant family given by the Dirac operator $i\partial/\partial\theta$ on $L^2_1(S^1;\mathbb{C})$.

Choose $v : Q \to S^1 \times S^3$ in the homotopy class of the classifying map of the given family of operators on Q, and arrange that v is transverse to $S^1 \times w$. As before, we can arrange that $v^{-1}(S^1 \times W)$ has a product structure,

$$v^{-1}(S^1 \times W) \cong v^{-1}(S^1 \times \{w\}) \times W,$$

and has a product metric. We choose our Morse function f so that it coincides with y in $v^{-1}(S^1 \times W)$. For $\tilde{L}$ we take the pull-back by v of the family on $S^1 \times S^3$.

To examine the matrix entries of $\bar{\partial}_3$, let a and b be critical points of f, let z be a homotopy class of paths joining them, and consider the operators $P_{\gamma,i,j}$ for $\gamma \in M_z(a,b)$ and $j = i + 1 + \operatorname{sf}(L,z)$. This operator has index zero. Let $\delta = v \circ \gamma$ be the corresponding path in $S^1 \times S^3$, so that the operators $P_{\gamma,i,j}$ and $P_{\delta,i,j}$ are isomorphic. Our choices mean that $P_{\delta,i,j}$ can have kernel only if δ passes through $S^1 \times W$; and whether or not it has kernel is determined entirely by the intersection of the path with $S^1 \times W$. From this point on, the proof proceeds as in the case of no spectral flow. For ξ_3 we take the pull-back by v of a Čech cocycle on $S^1 \times S^3$ representing the 3-dimensional generator and supported in $S^1 \times W$. □

35 Application to the Floer groups $\overline{HM}$

In the previous sections, we have defined a group $\bar{H}_*(Q,L)$ associated to a compact manifold Q carrying a suitable family of self-adjoint operators L, and we have shown how to calculate this group when L has a particularly restricted form. We now return to Floer homology, where we can realize our Floer group $\overline{HM}_*(Y,\mathfrak{s})$ as an example of such a "coupled homology" group.

With our present framework, there is very little difference between the cases that $c_1(\mathfrak{s})$ is torsion and non-torsion, if we use a non-exact perturbation in the "balanced" class for the non-torsion case; but we still choose to describe them separately. After stating theorems describing the general structure of the Floer group, we shall give some examples.

35.1 A general calculation of $\overline{HM}$

Let Y be an oriented 3-manifold, and let $\mathfrak{s}$ be a spin^c structure with $c_1(\mathfrak{s})$ torsion. The reducible critical points of the unperturbed Chern–Simons–Dirac functional $\mathcal{L}$ in $\mathcal{B}(Y,\mathfrak{s})$ comprise the torus $\mathbb{T}$ of gauge-equivalence classes of spin^c connections A in S such that $\mathrm{tr}(A)$ is flat. This torus $\mathbb{T}$ is isomorphic to $H^1(Y;\mathbb{R})/H^1(Y;\mathbb{Z})$. If we introduce $A_*(Y)$ as the exterior algebra generated by the free abelian group $A_1(Y) = H^1(Y;\mathbb{Z})$, then we can identify $A_k(Y)$ with the homology group of $\mathbb{T}$ in dimension k:

$$A_k(Y) = H_k(\mathbb{T};\mathbb{Z}).$$

There is a distinguished map

$$\beta : A_k(Y) \to A_{k-3}(Y)$$

defined using the cup product on Y and the fundamental class:

$$\alpha_1\alpha_2\ldots\alpha_k \mapsto$$
$$\sum_{i_1<i_2<i_3} (-1)^{i_1+i_2+i_3}\langle \alpha_{i_1}\cup\alpha_{i_2}\cup\alpha_{i_3},[Y]\rangle\, \alpha_1\ldots\hat{\alpha}_{i_1}\ldots\hat{\alpha}_{i_2}\ldots\hat{\alpha}_{i_3}\ldots\alpha_k.$$

Here juxtaposition denotes the product in $A_*(Y)$. The following theorem describes $\overline{HM}_*(Y,\mathfrak{s})$ over the rationals when $c_1(\mathfrak{s})$ is torsion.

Theorem 35.1.1. *Let Y be a closed, connected, oriented 3-manifold and $\mathfrak{s}$ a spin^c structure with $c_1(\mathfrak{s})$ torsion. Then $\overline{HM}_*(Y,\mathfrak{s};\mathbb{Q})$ has a decreasing* *filtration*

$$\cdots \supset \mathcal{F}_s\overline{HM}_*(Y,\mathfrak{s};\mathbb{Q}) \supset \mathcal{F}_{s-1}\overline{HM}_*(Y,\mathfrak{s};\mathbb{Q}) \supset \cdots$$

such that the associated graded groups satisfy

$$\frac{\mathcal{F}_s\overline{HM}_*(Y,\mathfrak{s};\mathbb{Q})}{\mathcal{F}_{s-1}\overline{HM}_*(Y,\mathfrak{s};\mathbb{Q})} \cong \frac{\ker(\beta_s)}{\operatorname{im}(\beta_{s+3})} \otimes \mathbb{Q}[T^{-1},T]$$

where

$$\beta_s : A_s(Y) \to A_{s-3}(Y)$$

is as above. If $b_1(Y)$ is even, this isomorphism respects the canonical mod 2 grading, when T is assigned even degree and $A_s(Y)$ is assigned grading s mod 2. If $b_1(Y)$ is odd, then these isomorphisms have degree 1 mod 2.

Proof. There is a gauge-invariant retraction

$$p : \mathcal{B}(Y,\mathfrak{s}) \to \mathbb{T}$$

arising from a linear map on the affine space of connections. Let $f_1 : \mathcal{B}(Y,\mathfrak{s}) \to \mathbb{R}$ be the composite $f \circ p$, where f is a Morse function on $\mathbb{T}$, and let $\rlap{\,-}\mathcal{L} = \mathcal{L} + f_1$. This is a perturbation belonging to our allowed class $\mathcal{P}$.

The reducible critical points of $\rlap{\,-}\mathcal{L}$ are exactly the critical points $[\alpha]$ of f in $\mathbb{T} \subset \mathcal{B}(Y,\mathfrak{s})$. The torus $\mathbb{T}$ is invariant under the gradient flow of $-\operatorname{grad}(\rlap{\,-}\mathcal{L})$, and the trajectories there are the gradient trajectories for the Morse function f.

In the blow-up $\mathcal{B}^\sigma(Y,\mathfrak{s})$, the reducible critical points $[\mathfrak{a}]$ are the gauge orbits of configurations $(\alpha,[\phi])$, where $[\alpha] \in \mathbb{T}$ is a critical point of f and $\phi \in L^2(Y;S)$ is an eigenvector of the corresponding Dirac operator D_α. If we add a further perturbation from the class $\mathcal{P}$, chosen so that its gradient is zero on the reducible locus in $\mathcal{C}(Y,\mathfrak{s})$, then the picture is unchanged except that the Dirac operator D_α becomes a more general self-adjoint Fredholm operator $D_{\mathfrak{q},\alpha}$. (See Subsection 12.2.) We can choose this perturbation so that the reducible critical points $[\mathfrak{a}]$ are non-degenerate in the blow-up and so that the perturbed Seiberg–Witten moduli spaces $M_z^{\text{red}}([\mathfrak{a}],[\mathfrak{b}])$ are regular.

At this point, the complex that computes $\overline{HM}(Y,\mathfrak{s})$ is of the type $\bar{C}_*(Q;L)$ we have been considering in this chapter: for Q we have the torus $\mathbb{T}$, and for L we have the family of perturbed Dirac operators $D_{\mathfrak{q},\alpha}$, for $[\alpha] \in \mathbb{T}$.

Lemma 35.1.2. *The classifying map $v : \mathbb{T} \to U(\infty)$ for the family of self-adjoint Dirac operators D_α on Y factors through the inclusion $SU(2) \hookrightarrow$*

$U(\infty)$. The pull-back of the 3-dimensional generator is the class

$$\begin{aligned}\xi_3 &\in H^3(\mathbb{T};\mathbb{Z})\\ &\cong \Lambda^3 H^1(\mathbb{T};\mathbb{Z})\\ &\cong \Lambda^3 H^1(Y;\mathbb{Z})^*\end{aligned}$$

given by the triple cup product,

$$(a_1, a_2, a_3) \mapsto \langle a_1 \cup a_2 \cup a_3, [Y]\rangle.$$

Proof. The homotopy classes $[\mathbb{T}, U(\infty)]$ classify $K^{-1}(\mathbb{T})$, which is torsion-free. Two elements of $K^{-1}(\mathbb{T})$ are therefore equal if they have the same Chern character in $H^{\mathrm{odd}}(\mathbb{T};\mathbb{Q})$; and two maps $v : \mathbb{T} \to U(\infty)$ are homotopic if they induce the same map on rational cohomology.

The Chern character of the element $\zeta \in K^{-1}(\mathbb{T})$ arising from the family of Dirac operators can be computed using the index theorem in [9, section 3]. Let $\mathcal{S} \to \mathbb{T} \times Y$ be the vector bundle that carries the family of spin^c connections: that is, $\mathcal{S}$ is a rank-2 bundle with unitary connection, and the restriction of $\mathcal{S}$ to $[\alpha] \times Y$ is isomorphic to (S, α) as a bundle with connection. We can normalize $\mathcal{S}$ by making it trivial on $\mathbb{T} \times \{y_0\}$ for some basepoint y_0. There is a line bundle $\mathcal{P} \to \mathbb{T} \times Y$, the Poincaré line bundle, with a connection that is flat on each $[\alpha] \times Y$, such that

$$\mathcal{S} = \mathcal{P} \otimes p_2^*(S)$$

where p_2 is the projection to Y. The first Chern class of $\mathcal{P}$ is the class

$$c_1(\mathcal{P}) = \sum_i p_1^*(a_i^*) \cup p_2^*(a_i),$$

where the classes a_i are a basis for $H^1(Y;\mathbb{Z})$ and a_i^* are the dual basis in

$$H^1(\mathbb{T};\mathbb{Z}) \cong \mathrm{Hom}(H^1(Y;\mathbb{Z}), \mathbb{Z}).$$

Then the index theorem tells us in this case that

$$\begin{aligned}\mathrm{ch}(\zeta) &= \int_Y \mathrm{ch}(\mathcal{P})\\ &= \frac{1}{6}\int_Y c_1(\mathcal{P})^3\\ &= \sum_{i_1<i_2<i_3} \langle a_{i_1} a_{i_2} a_{i_3}, [Y]\rangle\, a_{i_1}^* \cup a_{i_2}^* \cup a_{i_3}^*. \qquad (35.1)\end{aligned}$$

(Note that $c_1(\mathcal{P})^r$ is zero for $r \geq 4$.) Under the isomorphisms in the statement of the lemma, this element of $H^3(\mathbb{T};\mathbb{Q})$ coincides with ξ_3.

The restriction of the universal family to $SU(2)$ has Chern character equal to the 3-dimensional generator. Since every 3-dimensional cohomology class on the torus is the pull-back of the generator on S^3 by some map, it follows that there is a map $v : \mathbb{T} \to SU(2)$ such that the pull-back of the universal family of self-adjoint Fredholm operators by v has the same Chern character as the family of Dirac operators. By our opening remarks, it follows that v is the classifying map for the family of Dirac operators. □

When we identify $H_*(\mathbb{T};\mathbb{Z})$ with $A_*(Y)$, then the operation of cap product with ξ_3 becomes the map β described in the statement of the theorem. So the theorem would now follow from Proposition 34.2.1 and Corollary 34.2.3 if we could verify that the higher differentials in the spectral sequence (that is, d^5 and beyond) are all zero. Indeed, there would be no need to pass to rational coefficients, if one could prove that the higher differentials in the spectral sequence vanished over $\mathbb{Z}$.

While we do not have an argument that works over $\mathbb{Z}$, we *can* see that the higher differentials in the spectral sequence vanish with rational (or equivalently real) coefficients by exploiting the de Rham interpretation detailed in Subsection 34.3 above. The results of that section mean that it suffices to consider the de Rham model, where we consider the differential

$$(d + T\zeta) : \Omega^*(\mathbb{T}) \otimes \mathbb{R}[T^{-1}, T] \to \Omega^*(\mathbb{T}) \otimes \mathbb{R}[T^{-1}, T]$$

in which ζ is a closed 3-form representing the class of ξ_3. In the corresponding spectral sequence, we again see, at the E^3 term, the differential

$$[x] \mapsto T[\zeta \wedge x]$$

$$H^*(\mathbb{T}) \otimes \mathbb{R}[T^{-1}, T] \to H^*(\mathbb{T}) \otimes \mathbb{R}[T^{-1}, T]$$

and the question again is whether the higher differentials are non-zero. In this de Rham model, however, the vanishing of the higher differentials is a straightforward consequence of the fact that, on a flat torus, the cohomology is represented by the parallel forms, which are closed under cup product. This is the same mechanism that leads to the vanishing of the Massey products on a torus.

This completes the calculation of the group $\overline{HM}_\bullet(Y)$. The statement about the canonical mod 2 grading follows from the original definition of $\mathrm{gr}^{(2)}$ in Subsection 22.4. □

As a corollary of this theorem, we have:

Corollary 35.1.3. *For any 3-manifold Y and any spin^c structure $\mathfrak{s}$ with $c_1(\mathfrak{s})$ torsion, the group $\overline{HM}_\bullet(Y,\mathfrak{s})$ is non-zero (and has infinite rank as a $\mathbb{Z}$-module).*

Proof. Because of Theorem 35.1.1, the statement reduces to showing that at least one of the groups

$$\frac{\ker(\beta_s)}{\operatorname{im}(\beta_{s+3})}$$

has non-zero rank, where

$$\beta : A_s(Y) \to A_{s-3}(Y)$$

is as in the statement of the theorem. Let $\zeta = e^{2\pi i/6}$, and consider the quantity

$$\sum_s \operatorname{rank}\left(\frac{\ker(\beta_s)}{\operatorname{im}(\beta_s)}\right)\zeta^s. \tag{35.2}$$

Just as the Euler characteristic of a bounded complex is equal to the Euler characteristic of its homology, the above sum is also equal to

$$\sum_s \operatorname{rank} A_s(Y)\zeta^s \tag{35.3}$$

because β drops degree by 3 and we can regard $A_*(Y)$ therefore as a sum of three complexes, each having β as differential. The sum (35.3) is just the evaluation at ζ of the Poincaré polynomial of the torus $\mathbb{T}$, which is

$$(1-\zeta)^{b_1(Y)}.$$

This quantity is non-zero, whatever $b_1(Y)$, so the result follows. □

We also have:

Corollary 35.1.4. *For any 3-manifold Y and any spin^c structure $\mathfrak{s}$ with $c_1(\mathfrak{s})$ torsion, the group $\widecheck{HM}_*(Y,\mathfrak{s})$ is non-zero in infinitely many gradings, as is $\widehat{HM}_*(Y,\mathfrak{s})$.*

Proof. This follows from the previous corollary, and the fact that $\widecheck{HM}_*(Y,\mathfrak{s})$ and its siblings are graded by $\mathbb{J}(Y,\mathfrak{s})$ which is a copy of $\mathbb{Z}$ when $c_1(\mathfrak{s})$ is torsion. More specifically, it follows from the definitions that the set of gradings in which

$\widecheck{HM}_*(Y,\mathfrak{s})$ is non-zero is bounded below, and the set of gradings in which $\widehat{HM}_*(Y,\mathfrak{s})$ is non-zero is bounded above. The previous corollary establishes that $\overline{HM}_*(Y,\mathfrak{s})$ is non-zero in gradings extending infinitely in both directions. So the long exact sequence relating the three flavors reveals that the gradings in which $\widecheck{HM}_*(Y,\mathfrak{s})$ and $\widehat{HM}_*(Y,\mathfrak{s})$ are non-zero are unbounded above and unbounded below respectively. □

We can also describe the completed version of this Floer group, in similar terms. The description of $\overline{HM}_\bullet(Y,\mathfrak{s})$ can be obtained from that of $\overline{HM}_*(Y,\mathfrak{s})$ by replacing the ring of finite Laurent series $\mathbb{Z}[T,T^{-1}]$ by the completion $\mathbb{Z}[T,T^{-1}]]$ which allows Laurent series which extend infinitely in the direction of negative powers of T.

We should also address the question of the module structure of the group $\overline{HM}_\bullet(Y,\mathfrak{s})$. Recall (from Subsection 3.2, for example) that $\overline{HM}_\bullet(Y,\mathfrak{s})$ is a module for the ring $\mathbb{A}_\dagger(Y)$, which is the exterior algebra on $H_1(Y;\mathbb{Z})/\text{torsion}$ tensored by a polynomial algebra $\mathbb{Z}[U_\dagger]$. More invariantly, $\mathbb{A}_\dagger(Y)$ is the cohomology ring of $\mathcal{B}^\sigma(Y,\mathfrak{s})$, with the negative of the standard grading and the opposite multiplication. The proof of Theorem 35.1.1 identifies $\overline{HM}_\bullet(Y,\mathfrak{s})$ with $\bar{H}_*(Q,L)$, where Q is the torus $\mathbb{T}$, and L is the family of Dirac operators. In general, as we explained in Subsection 33.3, the group $\bar{H}_*(Q,L)$ is a module over the cohomology of $\mathbb{P}(\mathcal{H})$, where $\mathcal{H}$ is the Hilbert bundle over Q. In the case of the torus $\mathbb{T}$, the cohomology rings of $\mathbb{P}(\mathcal{H})$ and $\mathcal{B}^\sigma(Y,\mathfrak{s})$ are canonically identified. It is easy to verify that the two module structures coincide.

The proof of the theorem also shows that, once a spin^c structure $\mathfrak{s}$ on Y is chosen, the structure of $\overline{HM}_\bullet(Y,\mathfrak{s})$ is entirely determined by the cohomology $H^1(Y;\mathbb{Z})$ and the triple cup product $\Lambda^3 H^1(Y;\mathbb{Z}) \to \mathbb{Z}$, for this data determines the family of Dirac operators on the torus $\mathbb{T}$, up to homotopy in the class of operators of type $S_*(H : H_1)$. We therefore have:

Proposition 35.1.5. *Let $(Y,\mathfrak{s})$ and $(Y',\mathfrak{s}')$ be two 3-manifolds carrying spin^c structures $\mathfrak{s}$ and $\mathfrak{s}'$ each of which has torsion first Chern class. Suppose that $H^1(Y;\mathbb{Z})$ and $H^1(Y';\mathbb{Z})$ are isomorphic by an isomorphism h that respects the triple cup product, regarded as a 3-form on the first cohomology. Then $\overline{HM}_\bullet(Y,\mathfrak{s})$ and $\overline{HM}_\bullet(Y',\mathfrak{s}')$ are isomorphic. This isomorphism respects the canonical mod 2 grading, as well as the module structure, once we identify $\mathbb{A}_\dagger(Y)$ with $\mathbb{A}_\dagger(Y')$ using h.* □

Note that the above proposition holds over $\mathbb{Z}$, not just over $\mathbb{Q}$, because in the course of the proof of Theorem 35.1.1, we saw that the cohomology ring of Y determines the family of operators L on $\mathbb{T}$ up to homotopy. This proposition verifies the claim made earlier, in Proposition 3.3.2. □

Suppose now that $c_1(\mathfrak{s})$ is not torsion. Let $\mathbb{T}$ be once more the torus $H^1(Y;\mathbb{R})/H^1(Y;\mathbb{Z})$ and let

$$\begin{aligned}\xi_1 &\in H^1(\mathbb{T};\mathbb{Z}) \\ &\cong \mathrm{Hom}(H^1(Y;\mathbb{Z}),\mathbb{Z})\end{aligned}$$

be the element corresponding to the homomorphism

$$a \mapsto \frac{1}{2}(a \cup c_1(\mathfrak{s}))[Y].$$

Let $\xi_3 \in H^3(\mathbb{T};\mathbb{Z})$ be the element arising from the triple cup product on $H^1(Y;\mathbb{Z})$ as in the proof of the preceding theorem. Let Γ_{ξ_1} be the local coefficient system on $\mathbb{T}$ with fiber $\mathbb{Z}[T^{-1},T]$ and whose holonomy around a loop γ in $\mathbb{T}$ is multiplication by T^k, where $k = \xi_1[\gamma]$.

Let $c_b \in H^2(Y;\mathbb{R})$ be the class of the balanced perturbation (Definition 29.1.1), and consider the Floer group $\overline{HM}_*(Y,\mathfrak{s},c_b)$ defined with respect to a perturbation with period class c_b (and integer coefficients).

Theorem 35.1.6. *Let Y be a closed, connected, oriented 3-manifold and $\mathfrak{s}$ a spin^c structure with $c_1(\mathfrak{s})$ non-torsion, as above. Then for the class of the balanced perturbation, $\overline{HM}_*(Y,\mathfrak{s},c_b)$ is isomorphic to the homology of a complex $(C,\bar\partial_1+\bar\partial_3)$, where*

$$C = C_*(\mathbb{T},f;\Gamma_{\xi_1})$$

is the Morse complex of the torus with local coefficients Γ_{ξ_1}, for a suitable Morse function f,

$$\bar\partial_1 : C_*(\mathbb{T},f;\Gamma_{\xi_1}) \to C_*(\mathbb{T},f;\Gamma_{\xi_1})$$

is the Morse differential for the local coefficient system, and

$$\bar\partial_3 : C_*(\mathbb{T},f;\Gamma_{\xi_1}) \to C_*(\mathbb{T},f;\Gamma_{\xi_1})$$

is given by the chain-level cap product $x \mapsto \tilde\xi_3 \cap x$, where

$$\tilde\xi_3 \in C^3(\mathcal{U})$$

is a Čech representative of ξ_3 with respect to an open cover $\mathcal{U}$ of $\mathbb{T}$ transverse to the Morse stratification.

Corollary 35.1.7. *The group $\overline{HM}_*(Y;c_b,\mathfrak{s})$ has a decreasing filtration*

$$\cdots \supset \mathcal{F}_s\overline{HM}_*(Y;c_b,\mathfrak{s}) \supset \mathcal{F}_{s-1}\overline{HM}_*(Y;c_b,\mathfrak{s}) \supset \cdots$$

such that the associated graded groups satisfy

$$\frac{\mathcal{F}_s\overline{HM}_*(Y;c_b,\mathfrak{s})}{\mathcal{F}_{s-1}\overline{HM}_*(Y;c_b,\mathfrak{s})} \otimes \mathbb{Q} \cong \frac{\ker(\beta_s)}{\operatorname{im}(\beta_{s+3})} \otimes \mathbb{Q}$$

where

$$\beta_s : H_s(\mathbb{T};\Gamma_{\xi_1}) \to H_{s-3}(\mathbb{T};\Gamma_{\xi_1})$$

is given by the cap product with ξ_3. □

Proof of Theorem 35.1.6 Let ω be a harmonic 2-form on Y belonging to the class $(4\pi/i)c_b$. A perturbation of the correct period class is the formal gradient on $\mathcal{C}(Y,\mathfrak{s})$ of the function (29.1). With this perturbation, the critical points with $\Phi = 0$ in $\mathcal{B}(Y,\mathfrak{s})$ comprise the torus $\mathbb{T}$ of gauge-equivalence class connections $[A]$ with $d(A - A_0)^t$ harmonic. After this initial step, the proof proceeds as with Theorem 35.1.1, but appealing to Proposition 34.4.1. □

As in the torsion case, we can also consider the completed version $\overline{HM}_\bullet(Y,\mathfrak{s},c_b)$, which we can again obtain from the above description by passing to the ring of infinite Laurent series $\mathbb{Z}[T,T^{-1}]]$. When ξ_1 is non-zero, this Floer group is trivial however: this follows from the theorem above, because the ordinary homology of $\mathbb{T}$ with local coefficients Γ_{ξ_1} is already trivial. The vanishing of the twisted homology of the torus follows from the fact that $1 - T^d$ is invertible, together with the following general lemma.

Lemma 35.1.8. *Let Γ be a local system of free R-modules on a torus $\mathbb{T}$. Let z be a primitive loop in $\mathbb{T}$, and suppose that the corresponding automorphism of Γ at the basepoint is given by multiplication by the unit $r \in R$. If $(1 - r)$ is not a zero-divisor, then the R-module $H_*(\mathbb{T};\Gamma)$ is annihilated by the element $(1 - r)$. In particular, if $(1 - r)$ is a unit, then the homology is zero.*

Proof. If the torus is simply a circle, then the zeroth and first homology groups are the kernel and cokernel of multiplication by $1 - r$ on Γ at the basepoint. So if $1 - r$ is not a zero-divisor, then only the first homology is non-zero, and it is annihilated by $1 - r$. In the higher-dimensional case, we can regard the n-torus as a fiber bundle over the $(n-1)$-torus, with fiber a circle, and calculate accordingly. □

35.2 Local coefficients

The theorems above describe the Floer groups $\overline{HM}_*$ and $\overline{HM}_\bullet$ with $\mathbb{Z}$ or $\mathbb{Q}$ coefficients. We are also interested in the Floer groups with values in a local coefficient system, such as the system Γ_η mentioned in Subsection 3.7 and defined in more detail in Subsection 22.6. We treat the case that $c_1(\mathfrak{s})$ is *torsion*.

If Γ is a local system on $\mathcal{B}(Y,\mathfrak{s})$, then Theorem 35.1.1 continues to be applicable, with minor modification. As before, we need to either tensor with $\mathbb{Q}$ or suppose from the start that Γ is a local system of vector spaces over a field of characteristic zero. Under these conditions, the group $\overline{HM}_*(Y,\mathfrak{s};\Gamma)$ has a filtration for which the associated graded is again of the form

$$\frac{\ker(\beta_s)}{\mathrm{im}(\beta_s)} \otimes \mathbb{Z}[T^{-1},T]$$

where now β_s is the map

$$\beta_s : H_s(\mathbb{T};\Gamma) \to H_{s-3}(\mathbb{T};\Gamma)$$

given by cap product with the integer class ξ corresponding to the triple cup product. (Note that the torus sits inside $\mathcal{B}(Y,\mathfrak{s})$, and Γ therefore defines a local system on $\mathbb{T}$.)

Lemma 35.1.8 is useful again here. The description of $\overline{HM}_*(Y,\mathfrak{s};\Gamma)$ shows that it is annihilated by anything that annihilates $H_*(\mathbb{T};\Gamma)$. Consider, for example, the local system $\mathcal{R}_\eta$ corresponding to a real 1-cycle η in Y, with fiber $\mathcal{R} = \mathbb{R}[\mathbb{R}]$, as defined in Subsection 32.3. Let $S \subset \mathbb{T}$ be a primitive loop. The monodromy of the local system $\mathcal{R}_\eta$ around S has the form t^f, where f is the first Chern class of a spin^c bundle on $S^1 \times Y$, evaluated on the real cycle $S^1 \times \eta$. (See the definition of $f(z)$ in (32.7).) This quantity f may take the value $2p$, where p is any period of η. Lemma 35.1.8 therefore tells us that $\overline{HM}_*(Y,\mathfrak{s};\mathcal{R}_\eta)$ has a filtration for which all the terms in the associated graded $\mathcal{R}$-module are annihilated by $1 - t^{2p}$. This means that $1 - t^{2p}$ strictly lowers the filtration degree in $\overline{HM}_*(Y,\mathfrak{s};\mathcal{R}_\eta)$; and since there are only finitely many steps in the filtration, we conclude that some power of $(1-t^{2p})$ annihilates $\overline{HM}_*(Y,\mathfrak{s};\mathcal{R}_\eta)$. This proves Proposition 32.3.1. □

For the local system Γ_η, which has fiber $\mathbb{R}$, we similarly deduce that $\overline{HM}_*(Y,\mathfrak{s};\Gamma_\eta)$ is zero if η belongs to a non-trivial real cohomology class, for this group is annihilated by a power of the non-zero real number $1 - e^{2p}$, for any non-zero period p. This completes the proof of Proposition 3.9.1. □

35.3 Examples

Rational-homology spheres. If Y is a rational homology sphere, then every spin^c structure $\mathfrak{s}$ has torsion first Chern class. For each $\mathfrak{s}$, $\overline{HM}_*(Y,\mathfrak{s})$ is isomorphic to $\mathbb{Z}[T,T^{-1}]$, and the action of the class $U_\dagger$ in $\mathbb{A}_\dagger(Y)$ (corresponding to the 2-dimensional class u_2) is multiplication by T^{-1}. Thus, as a $\mathbb{Z}[U_\dagger^{-1},U_\dagger]$ module, $\overline{HM}_*(Y,\mathfrak{s})$ is free of rank 1. The non-zero terms are all in even degree for the canonical mod 2 grading. These statements can be deduced directly from Theorem 35.1.1, or they can be seen as an instance of Proposition 35.1.5. But the direct explanation is more straightforward. Before blowing up the configuration space, there is exactly one reducible critical point $\alpha_\mathfrak{s}$ in $\mathcal{B}(Y,\mathfrak{s})$ for each spin^c structure $\mathfrak{s}$ (as well as possible irreducible critical points): this critical point corresponds to the spin^c connection A for which A^{tr} is flat. After blowing up and perturbing to achieve non-degeneracy, the critical points correspond to pairs $(\alpha_\mathfrak{s},\lambda)$, where λ runs through the eigenvalues of the perturbed Dirac operator. These generators correspond to the generators of $\mathbb{Z}[U_\dagger^{-1},U_\dagger]$ as an abelian group. This picture is the same as the one presented earlier for the special case of S^3 in Subsection 22.7. The difference is that we now do not know anything about the irreducible critical points, so the result is only for $\overline{HM}$. We state the result as a proposition.

Proposition 35.3.1. *Let Y be a rational homology sphere, and let $\mathfrak{s}$ be a spin^c structure on Y. Then*

$$\overline{HM}_*(Y,\mathfrak{s}) \cong \mathbb{Z}[U_\dagger^{-1},U_\dagger]$$

as modules over $\mathbb{Z}[U_\dagger]$, with all generators appearing in even grading for the canonical mod 2 grading. Similarly

$$\overline{HM}_\bullet(Y,\mathfrak{s}) \cong \mathbb{Z}[U_\dagger^{-1},U_\dagger]]$$

as modules over $\mathbb{Z}[[U_\dagger]]$. □

Manifolds with vanishing triple cup product. Let Y be a 3-manifold for which the triple cup product vanishes, as a skew trilinear form on $H^1(Y)$. If $b_1 = 1$ or 2, the vanishing of this form is inevitable; but there are 3-manifolds with arbitrary first Betti number satisfying this condition. Simple examples are connected sums in which the summands are copies of $S^1 \times S^2$ or any other manifold with $b_1 < 3$. For such manifolds, the class ξ_3 on $\mathbb{T}$ is zero; and the proofs of Theorems 35.1.1 and 35.1.6 show that the cochain representative $\tilde{\xi}_3$ can be taken to be zero also.

In the case that $c_1(\mathfrak{s})$ is torsion, the proof of Theorem 35.1.1 then tells us that $\overline{HM}_*(Y, \mathfrak{s})$ is isomorphic to $A_*(Y) \otimes \mathbb{Z}[T^{-1}, T]$, or equivalently $H_*(\mathbb{T}; \mathbb{Z}) \otimes \mathbb{Z}[T^{-1}, T]$. There is no need here to pass to the graded object associated to the filtration, nor is it necessary to tensor with $\mathbb{Q}$, for we have complete information about the differential when $\tilde{\xi}$ is zero at the cochain level. The isomorphism is again an isomorphism of graded abelian groups, up to an overall degree shift. In particular, the rank of $\overline{HM}_j(Y, \mathfrak{s})$ is $2^{b_1(Y)-1}$ for all j.

If $c_1(\mathfrak{s})$ is not torsion, let ξ_1 be the corresponding element of $H^1(\mathbb{T})$ as before. If the divisibility of $c_1(\mathfrak{s})$ in $H^2(Y; \mathbb{Z})/\text{torsion}$ is $2d$ with d positive, then we can choose a basis γ_i $(i = 1, \ldots, b_1(Y))$ for $\mathbb{T}$ such that $\xi_1[\gamma_1] = d$ and $\xi_1[\gamma_i] = 0$ for $i \neq 1$. Because there is no ξ_3, the group $\overline{HM}_*(Y, \mathfrak{s}, c_b)$ described by Theorem 35.1.6 is the homology of $\mathbb{T}$ with coefficients Γ_{ξ_1}. Write

$$\mathbb{T} = \mathbb{T}_1 \otimes \mathbb{T}',$$

where the first factor is the circle with homology generated by γ_1, and the homology of the second is generated by the γ_i with $i > 1$. Then

$$H_*(\mathbb{T}; \Gamma_{\xi_1}) = H_*(\mathbb{T}_1; \Gamma_{\xi_1}) \otimes H_*(\mathbb{T}'; \mathbb{Z}).$$

In $\mathbb{T}_1$, there are two trajectories from the maximum to the minimum of the standard Morse function; and then we can normalize the coefficient system so that the corresponding differential in the Morse complex is the map

$$\bar{\partial}_1 : \mathbb{Z}[T^{-1}, T] \to \mathbb{Z}[T^{-1}, T]$$

given by

$$p(T) \mapsto (T^d - 1)p(T).$$

The differential is injective, and its cokernel is the rank-d group $\mathbb{Z}[T]/(T^d - 1)$. Thus

$$\overline{HM}_*(Y, \mathfrak{s}, c_b) \cong \Big(\mathbb{Z}[T]/(T^d - 1)\Big) \otimes H_*(\mathbb{T}'; \mathbb{Z}), \tag{35.4}$$

where $\mathbb{T}'$ is a torus of dimension $b_1(Y) - 1$. As we saw earlier, the version $\overline{HM}_\bullet(Y, \mathfrak{s}, c_b)$ formed from the completed complex is zero, because $(T^d - 1)$ is a unit in the ring of formal Laurent series.

Manifolds with $b_1 = 3$. The triple cup product may be non-zero on a 3-manifold with $b_1 = 3$. The simplest example is the 3-torus, in which the product of the three standard generators of H^1 is the generator for H^3. In general, if $b_1(Y) = 3$ and a_1, a_2, a_3 is a basis for $H^1(Y;\mathbb{Z})$, then we can write

$$a_1 \cup a_2 \cup a_3 = m\nu,$$

where ν is the oriented generator of $H^3(Y;\mathbb{Z})$. We can arrange that $m \geq 0$ (so that the 3-torus has $m = 1$). Manifolds with $b_1 = 3$ and arbitrary m can be constructed by integer surgeries on the Borromean rings [100].

There is a further simplification that we can make in the case $b_1(Y) = 3$. In the proofs of Theorems 35.1.1 and 35.1.6 above, the relevant Morse complex is the Morse complex of a Morse function f on the torus $\mathbb{T}$ which we cannot choose arbitrarily: in the proof of Proposition 34.2.1, we required f to have a particular form in the neighborhood of $v^{-1}(w)$, where $w : \mathbb{T} \to S^3$ was a classifying map for the family. In the present case, v is a map of degree m between 3-manifolds, and $v^{-1}(w)$ is a finite set: the restriction on the Morse function does not now prevent us from taking f to be a small perturbation of the standard Morse function on the torus, with $v^{-1}(w)$ lying in some level set.

Suppose that $\mathfrak{s}$ has torsion first Chern class. For the standard Morse function, the differential in the Morse complex of the torus $\mathbb{T}$ is zero, and the Morse complex $C_*(\mathbb{T}, f)$ can be identified with $A_*(Y)$. There is no spectral sequence in this case: the complex that computes $\overline{HM}_*(Y, \mathfrak{s})$ is $A_*(Y) \otimes \mathbb{Z}[T^{-1}, T]$, with boundary map the map $T(\xi_3 \cap -)$. The only non-zero component of the differential is the component

$$A_3(Y) \otimes \mathbb{Z}[T^{-1}, T] \to A_0(Y) \otimes \mathbb{Z}[T^{-1}, T]$$

given by multiplication by mT (the groups $A_3(Y)$ and $A_0(Y)$ are both $\mathbb{Z}$).

The homology of this complex is isomorphic to $\mathbb{Z}^3 \oplus \mathbb{Z}/m\mathbb{Z}$ in every even dimension, and $\mathbb{Z}^3$ in every odd dimension. The $\mathbb{Z}/m\mathbb{Z}$ summand arises from the cokernel of the non-zero component of the differential above, and the $\mathbb{Z}^3$ summands arise from $A_1(Y)$ and $A_2(Y)$. Taking account of the change in the canonical mod 2 grading that is detailed in Theorem 35.1.1, we obtain:

Proposition 35.3.2. *Suppose Y has $b_1 = 3$ and let m, as above, be the integer determined by the triple cup product. Then for each spin^c structure $\mathfrak{s}$ with $c_1(\mathfrak{s})$ torsion, the group $\overline{HM}_j(Y, \mathfrak{s})$ is isomorphic to $\mathbb{Z}^3$ if $\mathrm{gr}^{(2)}(j)$ is even and is isomorphic to $\mathbb{Z}^3 \oplus \mathbb{Z}/m\mathbb{Z}$ if $\mathrm{gr}^{(2)}(j)$ is odd.* □

When $c_1(\mathfrak{s})$ is non-torsion, we again write $\mathbb{T}$ as a product $\mathbb{T}_1 \times \mathbb{T}'$, where the second factor is now a 2-torus. Once again, we take the Morse function f to be standard. The Morse complex $C_*(\mathbb{T},f;\Gamma_{\xi_1})$ can be written as

$$(C_1 \oplus C_0) \otimes H_*(\mathbb{T}';\mathbb{Z}),$$

where $(C_1 \oplus C_0) = \mathbb{Z}[T^{-1},T] \oplus \mathbb{Z}[T^{-1},T]$ is the Morse complex of the circle $\mathbb{T}_1$, in which the differential has one non-zero component

$$\mathbb{Z}[T^{-1},T] \xrightarrow{\bar{\partial}_1} \mathbb{Z}[T^{-1},T]$$

given by $p(T) \mapsto (T^d - 1)p(T)$ as before. There will be m Morse trajectories on $\mathbb{T}$ that pass through points of $v^{-1}(w)$; and the spectral flow along each of these will be either d or 0, with our normalization. Let a and b be the number of trajectories of each type, so $a+b=m$. The only non-zero component of the operator $\bar{\partial}_3$ is the map

$$C_1 \otimes H_*(\mathbb{T}';\mathbb{Z}) \to C_0 \otimes H_*(\mathbb{T}';\mathbb{Z})$$

given by

$$p(T) \otimes \delta \mapsto (aT^{d+1} + bT)p(T)(\nu_2 \frown \delta),$$

where ν_2 is the 2-dimensional generator of $H^*(\mathbb{T}';\mathbb{Z})$. Let us write $H^*(\mathbb{T}') = \mathbb{Z}^4$, in such a way that the cap product $(\nu_2 \frown -)$ maps the fourth generator to the first. Then our complex has the form

$$(\mathbb{Z}^4 \oplus \mathbb{Z}^4) \otimes \mathbb{Z}[T^{-1},T],$$

and the differential maps the first summand to the second by a map given by the 4-by-4 matrix

$$\begin{bmatrix} T^d-1 & 0 & 0 & mT \\ 0 & T^d-1 & 0 & 0 \\ 0 & 0 & T^d-1 & 0 \\ 0 & 0 & 0 & T^d-1 \end{bmatrix}. \tag{35.5}$$

This map is injective, and the cokernel is a free abelian group of finite rank, $4d$. The group $\overline{HM}_*(Y,\mathfrak{s})$ is graded by the set $\mathbb{J}(\mathfrak{s}) \cong \mathbb{Z}/(2d)$, and its rank is 2 in each grading.

The manifold $S^1 \times \Sigma_g$. We now examine the case that Y is the product $S^1 \times \Sigma_g$ of a circle and a closed 2-manifold of genus g. In this case, we can write

$$\mathbb{T} = \mathbb{T}_1 \times (\mathbb{T}_2)^g,$$

where the first factor is dual to the circle S^1. The 3-dimensional class ξ_3 has the form

$$\xi_3 = \eta \cup \omega$$

where $\omega = \sum \omega_i$ is the sum of the pull-backs of the 2-dimensional generators of the 2-dimensional torus factors, and η is dual to the circle.

We treat the case that $c_1(\mathfrak{s})$ is the pull-back to $S^1 \times \Sigma_g$ of a class in $H^2(\Sigma_g)$ equal to $2d$ times the generator; we denote this spin^c structure by $\mathfrak{s}_d$. On $\mathbb{T}$, the 1-dimensional class ξ_1 which determines the spectral flow is now

$$\xi_1 = d \times \eta.$$

We can improve on the general result of Theorem 35.1.6 by choosing a particular Morse function and a particular representative of ξ_3. On the circle, let f_1 be the Morse function $f_1(\theta) = \cos(g\theta)$. Let f_2 be a standard Morse function on $\mathbb{T}_{2g}$ and set

$$f = f_1 + \epsilon\beta(\theta) f_2,$$

where $\beta(\theta) = 0$ in the neighborhood of the $2g$ zeros of f_1 on the circle and $\beta(\theta) = 1$ in the neighborhood of the $2g$ critical points of f_1. Let

$$\mathbb{T}^{(i)}_{2g-2} \subset \mathbb{T}_{2g}$$

be the subtorus

$$\mathbb{T}^{(i)}_{2g-2} = \mathbb{T}_2 \times \cdots \times \{q\} \times \cdots \times \mathbb{T}_2,$$

where q is a basepoint in the ith copy of $\mathbb{T}_2$, and let

$$S = \bigcup_{i=1}^{g} \{\theta_i\} \times \mathbb{T}^{(i)}_{2g-2} \subset \mathbb{T},$$

where $\theta_i = 2\pi(4i - 3)/(4g)$ is the ith of the g zeros of f_1 at which $df_1/d\theta$ is negative. The submanifold S is dual to ξ_3; and using the standard framing of S in $\mathbb{T}$, we can construct a map $v : \mathbb{T} \to S^3$ such that $v^{-1}(w) = S$, as a framed submanifold. We can do this in such a way that f_1 coincides with $z \circ v$ on a tubular neighborhood of S.

The critical points of f are pairs $(\theta, p) \in \mathbb{T}_1 \times \mathbb{T}_{2g}$, where θ is a multiple of π/g and p is a critical point of f_2. We can normalize the spectral flow, so that for a trajectory of f, the (complex) spectral flow is d if the trajectory joins a critical point of the form $(0, p)$ to some $((2g-1)\pi/g, p')$, and is zero otherwise.

For such representatives, the Morse complex can be described explicitly. We can write

$$C_*(\mathbb{T}, f; \Gamma_{\xi_1}) = H_*(\mathbb{T}_{2g}; R)^g \oplus H_*(\mathbb{T}_{2g}; R)^g \tag{35.6}$$

where $R = \mathbb{Z}[T^{-1}, T]$. The first summand comes from the critical points in the g copies of $\mathbb{T}_{2g}$ lying over the points $\theta = 2n\pi/g$ in $\mathbb{T}_1$ $(0 \le n < g)$, and the second summand comes from the copies lying over the points $\theta = (2n+1)\pi/g$. The grading of the first summand is shifted by 1. The differential $\bar{\partial}_1$ for the Morse complex (twisted by T^d in the case that d is non-zero) has non-zero components only from the first summand to the second. We write this component as

$$\bar{\partial}_1 : H_*(\mathbb{T}_{2g}; R)^g \to H_*(\mathbb{T}_{2g}; R)^g.$$

It has the block form

$$\begin{bmatrix} 1 & -1 & 0 & \cdots & \cdots & 0 \\ 0 & 1 & -1 & \cdots & \cdots & 0 \\ \vdots & & \ddots & \ddots & & \vdots \\ & & & & & \\ 0 & 0 & 0 & \cdots & 1 & -1 \\ -T^d & 0 & 0 & \cdots & 0 & 1 \end{bmatrix}.$$

The map $\bar{\partial}_3$ counts trajectories which intersect S and can be expressed in the same block form as

$$\begin{bmatrix} T\Lambda_1 & & & \\ & T\Lambda_2 & & \\ & & \ddots & \\ & & & T\Lambda_g \end{bmatrix},$$

where $\Lambda_i : H_*(\mathbb{T}_{2g}) \to H_*(\mathbb{T}_{2g})$ denotes the cap product with the class ω_i (defined above). The sum of the two matrices describes the component of $\bar{\partial} = \bar{\partial}_1 + \bar{\partial}_3$ from the first summand to the second (the only non-zero component). So the sum of the kernel and cokernel of this matrix is isomorphic to $\overline{HM}_*(Y, \mathfrak{s}, c_b)$.

We can use the fact that $(1 + T\Lambda_i)$ is an invertible operator on $H_*(\mathbb{T}_{2g}; R)$, with inverse $(1 - T\Lambda_i)$, to reduce the matrix by row and column operations to the form

$$\begin{bmatrix} 1 & & & \\ & 1 & & \\ & & \ddots & \\ & & & A \end{bmatrix},$$

where A is the operator

$$A = \prod_{i=1}^{g} (1 + T\Lambda_i) - T^d,$$

as a map

$$A : H_*(\mathbb{T}_{2g}; R) \to H_*(\mathbb{T}_{2g}; R).$$

The fact that $\Lambda_i^2 = 0$ means that we can write $(1 + T\Lambda_i)$ as $\exp(T\Lambda_i)$; so the formula for A can be more succinctly written as

$$A = \exp(T\Lambda) - T^d,$$

where $\Lambda = \sum \Lambda_i$ is the cap product with ω.

Returning to the description of the Morse complex (35.6), we see that the Floer homology can be described as:

$$\overline{HM}_*(S^1 \times \Sigma, \mathfrak{s}_d, c_b) = (\ker A)\langle 1 \rangle \oplus (\operatorname{coker} A), \tag{35.7}$$

where the notation $\langle 1 \rangle$ indicates that the grading has been shifted up by 1.

If we work over the rational field, then we can analyze the situation a little further. Let

$$P_j \subset H_j(T_{2g}; \mathbb{Q}) \quad (j \leq g)$$

be the kernel of $\Lambda : H_j(T_{2g}; \mathbb{Q}) \to H_{j-2}(T_{2g}; \mathbb{Q})$ (the primitive part of the homology). The dimension of P_j is given by

$$\dim P_j = \binom{2g}{j} - \binom{2g}{j-2}.$$

The Jordan canonical form of the nilpotent operator Λ has one Jordan block of size $g - j + 1$ for each vector in P_j. Each Jordan block of size $k + 1$ gives a diagonal block in A of the form

$$\left.\begin{bmatrix} 1-T^d & T & T^2/2 & T^3/6 & \dots \\ & 1-T^d & T & T^2/2 & \dots \\ & & \ddots & & \\ & & & & T \\ & & & & 1-T^d \end{bmatrix}\right\} (k+1). \tag{35.8}$$

For $d \neq 0$, the operator A is injective, so

$$\overline{HM}_*(Y, \mathfrak{s}, c_b) \cong H_*(\mathbb{T}_{2g}; R)/\mathrm{im}(A).$$

By row and column operations over $\mathbb{Q}[T^{-1}, T]$, such a block can be reduced to

$$\begin{bmatrix} 1 & & & \\ & 1 & & \\ & & \ddots & \\ & & & (1-T^d)^{k+1} \end{bmatrix}.$$

This block contributes a summand $\mathbb{Q}[T^{-1}, T]/(1-T^d)^{k+1}$ to the cokernel of A over $\mathbb{Q}$. Thus we obtain:

Proposition 35.3.3. *Let d be non-zero and let $\mathfrak{s}_d$ be the spin^c structure whose first Chern class is the pull-back of d times the generator on Σ. Then for c_b the class of a balanced perturbation, we have*

$$\begin{aligned} \overline{HM}_*(S^1 \times \Sigma, \mathfrak{s}_d, c_b) \otimes \mathbb{Q} &\cong \bigoplus_{k=0}^{g} \left(\frac{\mathbb{Q}[T^{-1}, T]}{(1-T^d)^{k+1}} \right) \otimes_{\mathbb{Q}} P_{g-k} \\ &\cong \bigoplus_{k=0}^{g} \left(\frac{\mathbb{Q}[T^{-1}, T]}{(1-T^d)^{k+1}} \right)^{m_k}, \end{aligned}$$

where

$$m_k = \binom{2g}{g-k} - \binom{2g}{g-k-2} = \dim P_{g-k}.$$

□

When $d = 0$, the analysis of the cokernel of A is essentially unchanged: each Jordan block now contributes a summand $\mathbb{Q}[T^{-1}, T]$. In this case, however, A is not injective: its kernel and cokernel are isomorphic. Because the first summand in (35.6) has degree shifted by 1, the end result is:

Proposition 35.3.4. *For the trivial spin^c structure $\mathfrak{s}_0$ on $S^1 \times \Sigma$, we have*

$$\overline{HM}_*(S^1 \times \Sigma, \mathfrak{s}_0) \otimes \mathbb{Q} \cong (P_*\langle 1\rangle \oplus P_*) \otimes \mathbb{Q}[T^{-1}, T],$$

where $P_* \subset H_*(\mathbb{T}_{2g}; \mathbb{Q})$ is the primitive part of the homology (the kernel of Λ) and $P_*\langle 1\rangle$ is the same group with grading shifted by 1. In particular, for all k, we have

$$\begin{aligned}\dim_{\mathbb{Q}} \overline{HM}_k(S^1 \times \Sigma, \mathfrak{s}_0) \otimes \mathbb{Q} &= \dim P_* \\ &= \binom{2g}{g} + \binom{2g}{g-1}.\end{aligned}$$

□

36 The manifold $S^1 \times S^2$

36.1 Positive scalar curvature

The calculation of $\overline{HM}_\bullet(Y)$ in the previous sections was possible because $\overline{HM}_\bullet$ involves only the reducible solutions of the monopole equations. In general, we cannot calculate $\widecheck{HM}_\bullet(Y)$ or $\widehat{HM}_\bullet(Y)$ without information about the irreducible solutions also; and for most 3-manifolds, a direct understanding of these solutions is not possible.

There is, however, a particularly simple situation if Y has a metric of positive scalar curvature, for in such cases all solutions are reducible. We have already exploited this observation in our discussion of homology spheres in Subsection 22.7. We can consider manifolds with non-zero Betti number and positive scalar curvature in a similar manner.

This class of manifolds is not large. The results of Schoen and Yau [101] tell us that an orientable 3-manifold with positive scalar curvature can always be obtained from a manifold with $b_1 = 0$ by making a connected sum of a number of copies of $S^1 \times S^2$. Thurston's geometrization conjecture would further imply that the remaining manifold with $b_1 = 0$ is a sum of spherical space forms (see [51] for a survey).

Suppose then that Y has positive scalar curvature, and let $\mathfrak{s}$ be a spin^c structure with $c_1(\mathfrak{s})$ torsion. Let $\mathbb{T}$ again be the torus $H^1(Y;\mathbb{R})/H^1(Y;\mathbb{Z})$ which parametrizes reducible solutions to the unperturbed equations, let f be a Morse function on $\mathbb{T}$ and let $f_1 = f \circ p$ be the corresponding function on $\mathcal{B}(Y, \mathfrak{s})$. (See the proof of Theorem 35.1.1 for this notation.)

The connections $[A]$ in $\mathbb{T}$ are those with A^t flat, and it follows from the Weitzenböck formula that the corresponding Dirac operator D_A has no kernel. This means, in particular, that if $[A(t)]$ is any path in $\mathbb{T}$, then the corresponding family of 3-dimensional Dirac operators has no spectral flow. If $[\alpha]$ and $[\beta]$ are critical points of Morse function on $\mathbb{T}$, then we can label the corresponding critical points in the blow-up as $[\mathfrak{a}_i]$ and $[\mathfrak{b}_j]$ in increasing order of the index, with $[\mathfrak{a}_0]$ and $[\mathfrak{b}_0]$ corresponding to the first positive eigenvalues of the Dirac operator at α and β respectively. Thus the points $[\mathfrak{a}_i]$ and $[\mathfrak{b}_i]$ are boundary-stable for $i \geq 0$ and boundary-unstable for $i < 0$. The absence of spectral flow means that the relative grading in the complex $\bar{C}_*$ is given by

$$\bar{\mathrm{gr}}([\mathfrak{a}_i], [\mathfrak{b}_j]) = \mathrm{index}_f[\alpha] - \mathrm{index}_f[\beta] + 2i - 2j, \tag{36.1}$$

where the first two terms are the ordinary Morse indices of these two critical points of the function f on $\mathbb{T}$. Our first lemma is a straightforward consequence:

Lemma 36.1.1. *The component $\bar{\partial}^s_u$ of the boundary map $\bar{\partial}$ in the complex $\bar{C}_*(Y, \mathfrak{s})$ is zero.*

Proof. The differential $\bar{\partial}$ counts trajectories in the blow-up $\mathcal{B}^\sigma(Y, \mathfrak{s})$ which lie over ordinary Morse trajectories of f on $\mathbb{T} \subset \mathcal{B}(Y, \mathfrak{s})$. For any such trajectory, from $[\mathfrak{a}_i]$ to $[\mathfrak{b}_j]$ say, the difference $\mathrm{index}_f[\alpha] - \mathrm{index}_f[\beta]$ in the formula above is non-negative. If the trajectory contributes to $\bar{\partial}^s_u$ then $i \geq 0$ and $j < 0$, from which it follows that

$$\bar{\mathrm{gr}}([\mathfrak{a}_i], [\mathfrak{b}_j]) \geq 2.$$

The contributions to $\bar{\partial}$, however, come only from trajectories satisfying $\bar{\mathrm{gr}}([\mathfrak{a}_i], [\mathfrak{b}_j]) = 1$. □

The next lemma makes use of positive scalar curvature in a slightly more delicate way.

Lemma 36.1.2. *If we replace the Morse function f by a smaller positive multiple, ϵf, then for small enough ϵ, the component $\bar{\partial}^u_s$ of the boundary map $\bar{\partial}$ in the complex $\bar{C}_*(Y, \mathfrak{s})$ is zero.*

Proof. A gradient trajectory of f gives rise to a family $A^t(t)$ of flat connections on Y. The corresponding 4-dimensional connection on the cylinder has curvature whose size pointwise is controlled by dA^t/dt, and hence by the size of the gradient of f on the torus $\mathbb{T}$. If we replace f by ϵf, then the trajectories are reparametrized but otherwise unchanged, and the size of the curvature of the 4-dimensional connection scales by ϵ.

The contributions to $\bar{\partial}^u_s$ arise from solutions (A, ϕ) to the coupled equations on the 4-dimensional cylinder, with ϕ a non-zero solution of the Dirac equation $D^+_A \phi = 0$ having exponential decay at both ends. The exponential decay allows us to apply the Weitzenböck formula (4.14) and integrate by parts. The positive contribution of the term involving the scalar curvature dominates the term involving $F^+_{A^t}$ pointwise once ϵ is sufficiently small, and the usual contradiction establishes that there is no solution ϕ of this sort. □

Now let us examine the complexes $\check{C}_*$ and $\hat{C}_*$ which compute $\widecheck{HM}_*(Y, \mathfrak{s})$ and $\widehat{HM}_*(Y, \mathfrak{s})$. Recall that, in general, the underlying group of the complex $\check{C}_*$ is the sum $C^o_* \oplus C^s_*$. The first summand is generated by the irreducible critical points, of which there are none, so

$$\check{C}_* = C^s_*$$

and similarly

$$\hat{C}_* = C^u_*.$$

The formula for the differential $\check{\partial}$ for $\check{C}_*$ is given in Definition 22.1.3, and when C^o_* is absent this formula becomes

$$\check{\partial} = \bar{\partial}^s_s - \partial^u_s \bar{\partial}^s_u.$$

When $\bar{\partial}^s_u$ is zero also (as in our present case), we simply have $\check{\partial} = \bar{\partial}^s_s$. Similarly (after examining the sign in the definition of $\hat{\partial}$), we see that $\hat{\partial} = -\bar{\partial}^u_u$.

It follows from this discussion that when the parameter ϵ is small enough to make $\bar{\partial}^u_s = 0$, we simply have

$$\bar{\partial} = \begin{bmatrix} \check{\partial} & 0 \\ 0 & -\hat{\partial}. \end{bmatrix}$$

So, as differential groups, we have (to within an irrelevant sign)

$$\bar{C}_* = \check{C}_* \oplus \hat{C}_*[-1],$$

where the notation $[-1]$ tells us that a boundary-unstable critical point contributes a generator in $\bar{C}_*$ whose degree is 1 lower than the corresponding generator of $\hat{C}_*$.

On a 3-manifold with positive scalar curvature, the triple cup product is zero. This one can see, for example, by using the families index theorem in the form (35.1). So our general structure theorem for $\overline{HM}_*$ gives us

$$\overline{HM}_*(Y,\mathfrak{s}) = H_*(\mathbb{T}) \otimes \mathbb{Z}[U_\dagger^{-1}, U_\dagger].$$

Taking account of the information about gradings in (36.1), we obtain the following picture.

Proposition 36.1.3. *If Y has strictly positive scalar curvature, and $\mathfrak{s}$ is a spin^c structure on Y with $c_1(\mathfrak{s})$ torsion, then the map j in the long exact sequence relating $\widecheck{HM}_*$, $\widehat{HM}_*$ and $\overline{HM}_*$ is zero, and we have*

$$\overline{HM}_*(Y,\mathfrak{s}) = \widecheck{HM}_*(Y,\mathfrak{s}) \oplus \widehat{HM}_*(Y,\mathfrak{s})[-1]$$

as groups graded by $\mathbb{J}(Y,\mathfrak{s}) \cong \mathbb{Z}$. Furthermore, we can choose the identification of $\mathbb{J}(Y,\mathfrak{s})$ with $\mathbb{Z}$ so that, as $\mathbb{Z}$-graded modules over $\mathbb{Z}[U_\dagger]$, we have

$$\overline{HM}_*(Y,\mathfrak{s}) = H_*(\mathbb{T}) \otimes \mathbb{Z}[U_\dagger^{-1}, U_\dagger]$$
$$\widehat{HM}_*(Y,\mathfrak{s}) = H_*(\mathbb{T}) \otimes \mathbb{Z}[U_\dagger],$$

and

$$\widecheck{HM}_*(Y,\mathfrak{s}) = H_*(\mathbb{T}) \otimes \left(\mathbb{Z}[U_\dagger^{-1}, U_\dagger]/\mathbb{Z}[U_\dagger]\right).$$

If $c_1(\mathfrak{s})$ is not torsion, then the Floer groups are zero. □

We can apply this result to $Y = S^1 \times S^2$. In this case, $\mathbb{T}$ is a circle, and its homology has two generators in degrees differing by 1. We recall that $U_\dagger$ has degree -2, and we learn from the above theorem that if $\mathfrak{s}_0$ is the spin^c structure with $c_1 = 0$, then $\overline{HM}_k(S^1 \times S^2, \mathfrak{s}_0)$ is isomorphic to $\mathbb{Z}$ for all k in $\mathbb{J}(Y,\mathfrak{s}_0)$. Furthermore, there exists a k_0 such that $\widecheck{HM}_k$ is non-zero precisely in degrees $k \geq k_0$ while $\widehat{HM}_k$ is non-zero precisely in degrees $k \leq k_0$. (Remember the shift of grading by 1 when reading the last sentence.)

When we identify $\mathbb{J}(Y)$ with the set of homotopy classes of 2-plane fields on Y, then we can find the class of 2-plane fields which must correspond to k_0 using a symmetry argument, exploiting the duality for the Floer groups. The manifold $Y = S^1 \times S^2$ admits an orientation-reversing self-homeomorphism, so $\widecheck{HM}_*(Y)$ is isomorphic to $\widehat{HM}^*(Y)$. Since both of these groups are non-zero only in grading k_0, it must be that k_0 is invariant under the orientation-reversing self-homeomorphism. There is only one class of oriented 2-plane fields on $S^1 \times S^2$

which is invariant under the orientation-reversing map $(\theta, x) \mapsto (-\theta, x)$. This class can also be characterized as the one represented by any 2-plane field ξ_0 which is invariant under rotation of the S^1 factor and has Euler class zero. Thus we recover the statement of Proposition 3.10.3 of Subsection 3.10.

When the scalar curvature is positive, the conclusion that $\overline{HM}_*$ is the sum of $\widecheck{HM}_*$ and $\widehat{HM}_*$ as groups remains valid if we replace $\mathbb{Z}$ by any local coefficient system. Since we already know that $\overline{HM}_*(Y; \Gamma_\eta)$ is zero if the class $[\eta]$ is non-zero, we can deduce the vanishing of $\widecheck{HM}_*(Y; \Gamma_\eta)$ and $\widehat{HM}_*(Y; \Gamma_\eta)$ also. In the case of $S^1 \times S^2$, this result is Proposition 3.10.4.

37 The three-torus

37.1 Perturbations of the Chern–Simons–Dirac functional

Although we have calculated $\overline{HM}_\bullet$ for the 3-torus, we have not yet calculated $\widecheck{HM}_\bullet$ or $\widehat{HM}_\bullet$. That task will be taken up here. The line of argument that we use is applicable to any flat 3-manifold (i.e. to any quotient of a flat torus).

Let Y be the 3-torus T^3 equipped with a flat metric, and let $S = \mathbb{C}^2 \times T^3$ be the spin^c bundle for the unique spin^c structure $\mathfrak{s}_0$ with $c_1 = 0$. Because T^3 has zero scalar curvature, Proposition 22.7.1 tells us that all solutions of the unperturbed Seiberg–Witten equations on T^3 are reducible. However, to examine the Floer groups, we must perturb the Chern–Simons–Dirac functional to achieve transversality; and in so doing, we may introduce irreducible critical points. With a careful choice of perturbation, this potential difficulty can be avoided. We will describe a perturbation of the Chern–Simons–Dirac functional $\mathcal{L}$ for which we can explicitly describe all the critical points (A, Φ): all critical points will be non-degenerate, and all will lie on the reducible locus. This, with a small amount of extra input, will allow a straightforward calculation of the Floer groups.

Let $\mathbb{T}$ denote the dual 3-torus which parametrizes the flat spin^c connections in S up to gauge equivalence, and let $[A_0] \in \mathbb{T}$ denote the connection with trivial holonomy. The point $[A_0]$ is the only class of flat connections in $\mathbb{T}$ for which the corresponding Dirac operator D_{A_0} has kernel: the kernel is 2-dimensional and consists of parallel sections of S. There is a gauge-invariant projection $p : \mathcal{C}(T^3, \mathfrak{s}) \to \mathbb{T}$ which sends (A, Φ) to $[A']$, where A' is the unique flat connection such that $A - A'$ is orthogonal to the harmonic 1-forms and $A_0 - A'$ is harmonic. (See also Section 11.) We consider two types of tame perturbations of $\mathcal{L}$. For the first, we fix a Morse function f on $\mathbb{T}$, and pull it back to $\mathcal{C}(T^3, S)$ using p, as we have done before. We call the resulting function f still. The second perturbation we consider is the function $\|\Phi\|^2$, the L^2 norm of the spinor. We consider an arbitrary combination of these, with small coefficients ϵ and δ: we set

$$\mathcal{\tilde{L}} = \mathcal{L} - (\delta/2)\|\Phi\|^2 + \epsilon f. \tag{37.1}$$

Proposition 37.1.1. *Let f be chosen so that it has a local minimum at $[A_0] \in \mathbb{T}$. Then there exist $\delta_0 > 0$ and a positive function $\epsilon_0 : (0, \delta_0) \to \mathbb{R}$, such that whenever δ and ϵ satisfy $0 < \delta \leq \delta_0$ and $0 < \epsilon \leq \epsilon_0(\delta)$, the perturbed functional $\mathcal{\tilde{L}}$ above has only reducible critical points in $\mathcal{C}(T^3, \mathfrak{s})$.*

Before proving the proposition, we make a few remarks. The signs are important here: if we change the sign of one of ϵ and δ, then the proposition fails. This dependence on this sign can be seen by looking at a finite-dimensional model which captures some of the behavior of $\mathcal{L}$ near the point $(A_0, 0) \in \mathcal{C}(T^3, \mathfrak{s})$. (This model is essentially the restriction of $\mathcal{L}$ to the kernel of the Hessian of $\mathcal{L}$ in the Coulomb slice through $(A_0, 0)$.) Take V to be a 2-dimensional hermitian vector space, and consider the function L on $i\,\mathfrak{su}(V) \times V$ defined by

$$L(\alpha, \Phi) = \frac{1}{2}\langle \Phi, \alpha\Phi \rangle.$$

This has a degenerate critical point at the origin. We perturb it, and write

$$\tilde{L}(\alpha, \Phi) = L(\alpha, \Phi) - (\delta/2)|\Phi|^2 + \epsilon|\alpha|^2.$$

The equations for a critical point of $\tilde{L}$ are:

$$\alpha\Phi = \delta\Phi$$
$$(\Phi\Phi^*)_0 + 2\epsilon\alpha = 0.$$

Applying the second equation to Φ we have

$$(\Phi\Phi^*)_0\Phi + 2\epsilon\alpha\Phi = 0,$$

and using the first equation this becomes

$$(1/2)|\Phi|^2\Phi + 2\epsilon\delta\Phi = 0.$$

When $\epsilon\delta$ is positive, this equation tells us that Φ is zero, and the second equation then gives $\alpha = 0$. So in this case the only critical point is the origin, which is now non-degenerate. If $\epsilon\delta$ is negative, then in addition to the origin, there is a

3-sphere of critical points, consisting of all pairs (α, Φ) with $|\Phi|^2 = -4\epsilon\delta$ and $\alpha = -(\Phi\Phi^*)_0/(4\epsilon)$. Like the bona fide Chern–Simons–Dirac functional, this finite-dimensional model has a circle symmetry, and the 3-sphere of critical points which arises when $\epsilon\delta$ descends to a 2-sphere in the quotient space. If we replaced $|\alpha|^2$ above with a different positive-definite quadratic form, with distinct eigenvalues, then the 2-sphere would be replaced by eight points corresponding to the eight eigenvectors of the quadratic form lying on the sphere.

Proof of Proposition 37.1.1. Consider first the perturbed functional (37.1) in the case $\epsilon = 0$. The equations for a critical point (A, Φ) are then:

$$\begin{aligned} D_A\Phi &= \delta\Phi \\ \tfrac{1}{2}\rho(F_{A^t}) &= (\Phi\Phi^*)_0. \end{aligned} \tag{37.2}$$

Suppose that there exist a sequence δ_i converging to zero, and corresponding *irreducible* solutions (A_i, Φ_i) (so Φ_i is non-zero). We will obtain a contradiction.

In the blown-up configuration space $\mathcal{C}^\sigma(T^3, \mathfrak{s})$, we have corresponding points (A_i, s_i, ϕ_i), with $\|\phi_i\| = 1$ in L^2 norm; and after passing to a subsequence and applying gauge transformations, these will converge to a solution (A, s, ϕ) of the unperturbed equations in $\mathcal{C}^\sigma(T^3, \mathfrak{s})$. The only solutions for the unperturbed equations are the reducible ones; so $s = 0$ and $[A]$ belongs to $\mathbb{T}$. Furthermore, by taking the limit of the first equation as $\delta_i \to 0$, we see that $D_A\phi = 0$. This implies that $A = A_0$ up to gauge transformation, for otherwise D_A has no kernel. We can therefore take it that $A = A_0$ and ϕ is a parallel section of $S = \mathbb{C}^2 \times T^3$. Because $c_1(S) = 0$, the curvature F_{A^t} is exact. So the second equation gives

$$\int_{T^3} \rho^{-1}(\phi_i\phi_i^*)_0 \wedge \eta = 0$$

for all closed 1-forms η, and in particular for the parallel 1-forms. The limiting ϕ satisfies the same condition, but since ϕ and η are parallel we obtain

$$\rho^{-1}(\phi\phi^*)_0 \wedge \eta = 0$$

pointwise. This tells us that $(\phi\phi^*)_0 = 0$, and hence $\phi = 0$, contradicting the fact that ϕ has unit L^2 norm.

From this contradiction, we conclude that there is a $\delta_1 > 0$ such that for $|\delta| \leq \delta_1$ the only solutions of the above equations are reducible. Let us now examine the Dirac equation

$$D_A\phi = \delta\phi \tag{37.3}$$

for $[A]$ in the torus $\mathbb{T}$ of flat connections and δ small. After gauge transformation, we may suppose that $A = A_0 + b \otimes 1_S$ for an imaginary-valued parallel 1-form b. The operator D_A then preserves the A_0-parallel spinors (the kernel of D_{A_0}), and it follows that for δ sufficiently small, the only solutions to the above equation have ϕ parallel with respect to A_0. The equation is then

$$\rho(b)\phi = \delta\phi. \tag{37.4}$$

This tells us that the parallel 1-form b has pointwise length δ. For each such b, the δ-eigenspace is 1-dimensional. We conclude that, for some $\delta_0 > 0$ and any positive $\delta < \delta_0$, the flat connections $[A] \in \mathbb{T}$ for which the equation (37.3) has a solution comprise a round sphere Σ_δ surrounding $[A_0]$: in the metric coming from the pointwise norm of the parallel 1-forms b, this sphere has radius δ. By choosing δ_0 perhaps smaller, we can arrange that the gradient of the Morse function $f : \mathbb{T} \to \mathbb{R}$ has positive inner product with the radial direction at all points of Σ_δ, because $[A_0]$ is a minimum for f.

Fix δ in the interval $0 < \delta \leq \delta_0$, and consider next the perturbed functional (37.1) with ϵ non-zero. Let us suppose that there is a sequence of positive numbers ϵ_i converging to zero and corresponding irreducible solutions (A_i, Φ_i). As before, we can suppose that the corresponding points (A_i, s_i, ϕ_i) in the blown-up configuration space are converging to a solution (A, s, ϕ) of the equations (37.2). By our choice of δ, this limiting solution must have $s = 0$. Furthermore, ϕ satisfies the equation (37.3), so $[A]$ lies on the small sphere Σ_δ. We again write $A = A_0 + b \otimes 1_S$, so that ϕ is parallel with respect to A_0 and satisfies (37.4). From this last equation, we obtain

$$\begin{aligned}\langle \rho(b), (\phi\phi^*)_0 \rangle &= \delta|\phi|^2/2 \\ &> 0\end{aligned}$$

pointwise on T^3. On the other hand, the perturbed equation for (A_i, s_i, ϕ_i) has the form

$$\tfrac{1}{2} * F_{A_i^t} + s_i^2 \rho^{-1}(\phi_i\phi_i^*)_0 = -\epsilon_i \eta_i$$

where η_i is a parallel imaginary-valued 1-form arising from the gradient of the perturbation f. More precisely, if the harmonic projection of A_i is gauge-equivalent to $A_i^h = A_0 + b_i \otimes 1_S$ in the torus $\mathbb{T}$, then η_i is the gradient of f, regarded as a periodic function on the space of imaginary-valued parallel 1-forms. We have $b_i \to b$, and so $\eta_i \to \eta$, where by our choice of δ_0 we have

$\langle b, \eta \rangle > 0$. For large i we therefore have (using again the exactness of $F_{A_i^t}$)

$$\int_{T^3} \langle \rho^{-1}(\phi_i \phi_i^*)_0, b \rangle < 0,$$

and so the limiting A_0-parallel spinor ϕ satisfies

$$\langle \rho^{-1}(\phi \phi^*)_0, b \rangle \leq 0$$

pointwise. This contradicts the earlier inequality above, which went the other way. The contradiction tells us that for $\delta \leq \delta_0$ and all sufficiently small ϵ (depending on δ), there can be no irreducible solutions. □

37.2 The Floer groups with $\mathbb{Z}$ coefficients

Let ϵ and δ be chosen small, as in the above proposition and its proof. For f, we choose a Morse function of a standard type, with eight critical points: a minimum at $[A_0]$, a single maximum and three each of index 1 and 2. The Morse differential for $(\mathbb{T}, f)$ is zero. As we saw, the Dirac operator D_A for $[A]$ in $\mathbb{T}$ has kernel only when $[A]$ lies on the small 2-sphere Σ_δ. We arrange that the critical points other than $[A_0]$ are separated from $[A_0]$ by this sphere. The (complex) spectral flow for the family of Dirac operators corresponding to a path which starts outside Σ_δ and ends at $[A_0]$ is -1 with our choice of sign for δ.

Let us label the critical points of f as

$$\begin{array}{ccc} & w & \\ z^1 & z^2 & z^3 \\ y^1 & y^2 & y^3 \\ & x & \end{array}$$

so that w is the maximum, x is the minimum and the y^α and z^α have index 1 and 2 respectively. In $\mathcal{B}^\sigma(T^3, \mathfrak{s})$, we label the critical points

$$\begin{array}{ccc} & w_i & \\ z_i^1 & z_i^2 & z_i^3 \\ y_i^1 & y_i^2 & y_i^3 \\ & x_i & \end{array} \tag{37.5}$$

with i running through $\mathbb{Z}$. The indexing is chosen so that $i = 0$ corresponds to the first positive eigenvalue of the Dirac operator at the critical point, and

$i = -1$ corresponds to the first negative eigenvector. The complex $\check{C}_*(T^3, \mathfrak{s})$ therefore has generators corresponding to Entries (37.5) with $i \geq 0$.

We can calculate the canonical $\mathbb{Q}$ grading of these critical points, using the definition in Definition 28.3.1. To make the calculation, we use the cobordism W from S^3 to T^3 obtained by removing a ball from $D^2 \times T^2$, and make use of the vanishing of the L^2 index of the Dirac operator on the cylindrical-end version of this cobordism. From this we obtain

$$\left.\begin{aligned} \mathrm{gr}^{\mathbb{Q}}(w_i) &= 2i \\ \mathrm{gr}^{\mathbb{Q}}(z_i^\alpha) &= 2i - 1 \\ \mathrm{gr}^{\mathbb{Q}}(y_i^\alpha) &= 2i - 2 \\ \mathrm{gr}^{\mathbb{Q}}(x_i) &= 2i - 1 \end{aligned}\right\} \tag{37.6a}$$

for $i \geq 0$, while

$$\left.\begin{aligned} \mathrm{gr}^{\mathbb{Q}}(w_i) &= 2i + 1 \\ \mathrm{gr}^{\mathbb{Q}}(z_i^\alpha) &= 2i \\ \mathrm{gr}^{\mathbb{Q}}(y_i^\alpha) &= 2i - 1 \\ \mathrm{gr}^{\mathbb{Q}}(x_i) &= 2i \end{aligned}\right\} \tag{37.6b}$$

for $i < 0$. In these formulae, note that the grading of x_i is shifted by 2 from what one would expect if there were no spectral flow for the Dirac operators. When viewed as generators of the complex that computes $\overline{HM}_*$, we must recall that the gradings of the boundary-unstable critical points are shifted by 1: the "bar" grading $\bar{\mathrm{gr}}^{\mathbb{Q}}$ is given by the formulae (37.6a) for all i, positive or negative.

As there are no irreducible critical points, the complex that computes $\widecheck{HM}_*(T^3, \mathfrak{s})$ is the group C^s generated by the boundary-stable critical points equipped with the differential

$$\bar{\partial}^s_s - \partial^u_s \bar{\partial}^s_u.$$

From our study of the structure of $\overline{HM}_*$ in general, we have some information about $\bar{\partial}$ and its four components $\bar{\partial}^s_s$, $\bar{\partial}^s_u$, $\bar{\partial}^u_s$ and $\bar{\partial}^u_u$. First, each trajectory γ that contributes to $\bar{\partial}$ lies over a trajectory γ_f of the Morse function f on the finite-dimensional manifold $\mathbb{T}$, and the trajectory γ_f must have index 1 or 3. For each trajectory γ_f of index 1, there is a unique trajectory γ which contributes to $\bar{\partial}$, which gives us trajectories of the type

$$\left.\begin{aligned} w_i &\to z_i^\alpha \\ z_i^\alpha &\to y_i^\beta \\ y_i^\beta &\to x_{i-1}. \end{aligned}\right\} \qquad (37.7)$$

(We see x_{i-1} rather than x_i because of the spectral flow.) The only trajectories γ_f of index 3 run from w to x, giving us, for each i, a possible contribution

$$w_i \to x_i. \qquad (37.8)$$

With $\mathbb{Z}$ coefficients, the differential in the finite-dimensional Morse complex of $(\mathbb{T}, f)$ is zero, and accordingly the components of (37.7) are zero also. The only non-zero contribution to $\bar{\partial}$ therefore comes from trajectories from w_i to x_i, and is a map (one for each i)

$$\mathbb{Z}\Lambda[w_i] \to \mathbb{Z}\Lambda[x_i], \qquad (37.9)$$

where $\Lambda[\mathfrak{a}]$ denotes the 2-element set of orientations at $[\mathfrak{a}]$ as before. Since we already know that $\overline{HM}_*(T^3)$ is equal to $\mathbb{Z}^3$ in each degree, the map (37.9) must be an isomorphism for each i. (Referring back to our earlier calculation, we see that this reflects the fact that the map $\beta : A_3(T^3) \to A_0(T^3)$ defined by the triple cup product is an isomorphism: see Subsection 35.1.)

The maps (37.9) with $i \geq 0$ contribute to $\bar{\partial}^s_s$, while those with $i < 0$ contribute to $\bar{\partial}^u_u$. The maps $\bar{\partial}^u_s$ and $\bar{\partial}^s_u$ are zero. We therefore have

$$\bar{\partial} = \begin{bmatrix} \bar{\partial}^s_s & 0 \\ 0 & \bar{\partial}^u_u \end{bmatrix}$$

and accordingly

$$\overline{HM}_k(T^3, \mathfrak{s}_0) = \widecheck{HM}_k(T^3, \mathfrak{s}_0) \oplus \widehat{HM}_{k+1}(T^3, \mathfrak{s}_0).$$

Looking at the degrees of the generators of the two complexes we see:

$$\widecheck{HM}_k(T^3, \mathfrak{s}_0) = \begin{cases} \mathbb{Z}^3, & \mathrm{gr}^{\mathbb{Q}}(k) \geq -2 \\ 0, & \text{otherwise,} \end{cases}$$

and

$$\widehat{HM}_k(T^3, \mathfrak{s}_0) = \begin{cases} \mathbb{Z}^3, & \mathrm{gr}^{\mathbb{Q}}(k) \leq -2 \\ 0, & \text{otherwise.} \end{cases}$$

The map $j : \widetilde{HM}_k \to \widehat{HM}_k$ is zero, because $\bar{\partial}^s_u$ is zero. (See the definition of j in Proposition 22.2.1.)

We have described these groups using the $\mathbb{Z}$-valued canonical $\mathbb{Q}$ grading, but we can also describe the gradings using the corresponding 2-plane fields. The element of $\mathbb{J}(T^3)$ with $c_1 = 0$ and $\mathrm{gr}^{\mathbb{Q}} = -2$ is the unique element which is invariant under an the orientation-reversing self-diffeomorphism $t \mapsto -t$ of T^3. The corresponding homotopy class of oriented 2-plane fields must therefore be the class represented by the parallel 2-plane fields. Denoting by ξ_0 such a 2-plane field, we see for example that $\widetilde{HM}_*(T^3, \mathfrak{s}_0)$ is non-zero in the gradings $[\xi_0] + j$ for $j \geq 0$. Thus we recover the statement of Proposition 3.10.1 from Chapter I.

37.3 Local coefficients on the three-torus

We now replace $\mathbb{Z}$ coefficients by the local system Γ_η on $\mathcal{C}(T^3, \mathfrak{s}_0)$. We take η to be a real 1-cycle with non-zero homology class in T^3, so that $\overline{HM}_*(T^3, \mathfrak{s}_0; \Gamma_\eta)$ is zero by Proposition 3.9.1.

Each critical point w_i etc. now contributes a summand $\Gamma_\eta[w_i] = \mathbb{R}$ to the complex $\check{C}$, and the differential $\bar{\partial}$ again has components of two types: first, the contribution from index-1 trajectories of the Morse function f on $\mathbb{T}$ as in (37.7); second, from the index-3 trajectories as in (37.8). The first kind give us maps

$$\Gamma_\eta[w_i] \to \sum_{\alpha=1}^{3} \Gamma_\eta[z_i^\alpha]$$

$$\sum_{\alpha=1}^{3} \Gamma_\eta[z_i^\alpha] \to \sum_{\alpha=1}^{3} \Gamma_\eta[y_i^\beta]$$

$$\sum_{\alpha=1}^{3} \Gamma_\eta[y_i^\beta] \to \Gamma_\eta[x_{i-1}]$$

which coincide with the differential in the finite-dimensional Morse complex of $\mathbb{T}$. The second give us maps

$$\Gamma_\eta[w_i] \to \Gamma_\eta[x_i].$$

Let us examine the complex $\hat{C}_*(T^3, \mathfrak{s}_0)$ which calculates $\widehat{HM}_*(T^3, \mathfrak{s}_0)$. The generators of this complex, arranged by their canonical $\mathbb{Q}$ grading, begin with:

$$\begin{array}{lcccc} \text{index } -1: & & w_{-1} & & \\ \text{index } -2: & z^1_{-1} & z^2_{-1} & z^3_{-1} & \boldsymbol{x}_{-1} \\ \text{index } -3: & y^1_{-1} & y^2_{-1} & y^3_{-1} & w_{-2}. \end{array}$$

Excluding for a moment the generators in boldface, the differentials among the remaining generators here coincide with the ordinary Morse differentials, arising from trajectories of the first kind. The only other contributions to the differential $\hat{\partial} = -\bar{\partial}^u_u - \bar{\partial}^s_u \partial^u_s$ which appear between these generators are possible maps

$$\Gamma_\eta[w_{-1}] \to \Gamma_\eta[x_{-1}].$$

(Potentially, there are two contributions here: one from the reducible trajectories contributing to $\bar{\partial}^u_u$ and one from $\bar{\partial}^s_u \partial^u_s$, which would involve irreducible trajectories also.) Without calculating this last contribution, we already have enough information here to conclude that $\widehat{HM}_k(T^3, \mathfrak{s}_0; \Gamma_\eta)$ is 1-dimensional in canonical grading $\mathrm{gr}^{\mathbb{Q}}(k) = -2$, and that a generator can be taken to be the image of the chain

$$1 \in \Gamma_\eta[x_{-1}] \cong \mathbb{R}.$$

In all lower gradings, $\widehat{HM}_*$ and $\overline{HM}_*$ coincide; so this is the only non-zero contribution to $\widehat{HM}_*$.

Because $\overline{HM}_*$ is zero, we know that j is an isomorphism; so $\widecheck{HM}_*$ is also 1-dimensional. We can see this directly, by looking at the generators as we just did for $\widehat{HM}_*$. In degrees -1 and -2, we have generators

$$\begin{array}{lcccc} \text{index } -1: & z^1_0 & z^2_0 & z^3_0 & \boldsymbol{x}_0 \\ \text{index } -2: & y^1_0 & y^2_0 & y^3_0, & \end{array}$$

and there are no generators in lower degree. The differential from the zs to the ys in this picture coincides with the corresponding Morse differential with local coefficients on $\mathbb{T}$; so this map has rank 2 and a 1-dimensional cokernel. There are no possible trajectories from x_0 to the ys. So we confirm again that $\widecheck{HM}_k(T^3, \mathfrak{s}_0; \Gamma_\eta)$ is $\mathbb{R}$ in canonical grading -2. We can see the non-zero map j

arise explicitly at the chain level, from the Morse trajectories

$$y_0^\alpha \to x_{-1} \qquad (\alpha = 1, 2, 3)$$

which contribute to $\bar{\partial}^s_u$.

This completes the proof of Proposition 3.10.2, with the exception of the statement that the group of self-diffeomorphisms of T^3 which preserve the class $[\eta]$ acts trivially on the group $\widecheck{HM}_*(T^3;\Gamma_\eta) \cong \mathbb{R}$. This is true by default if $[\eta]$ is proportional to a rational class, because in this case the component group G of this group of diffeomorphisms is an extension

$$\mathbb{Z}^2 \to G \to SL_2(\mathbb{Z}).$$

The latter is the group of special unitary automorphisms of $\mathbb{Z}^3$ preserving the first basis vector: it is a perfect group, and therefore cannot act non-trivially on $\mathbb{R}$. For general η, the result follows by continuity. □

37.4 Other flat three-manifolds

In addition to the 3-torus, there are five other compact, orientable 3-manifolds which admit flat metrics. Four of these five Euclidean space forms are bundles over S^1 with fiber T^2: the monodromies of these fiber bundles are automorphisms of T^2 fixing a point and having order 2, 3, 4 or 6. The last of the five is the Hantzsche–Wendt manifold, which is the quotient of T^3 by an action of the Klein 4-group. The first four all have $b_1 = 1$, while the Hantzsche–Wendt manifold has $b_1 = 0$ and $H_1 = \mathbb{Z}/4 \oplus \mathbb{Z}/4$.

Of the four non-trivial flat T^2 bundles over S^1, only one admits an orientation-reversing diffeomorphism; so as oriented manifolds, we should count these as seven examples. All can be dealt with together in a uniform manner, leading to a result very similar to the one we obtained for the 3-torus. In particular, if $[\eta] \in H_1(Y;\mathbb{R})$ is a non-zero class in any of these seven oriented manifolds, then

$$\widecheck{HM}_*(Y;\Gamma_\eta) \cong \widehat{HM}_*(Y;\Gamma_\eta) \cong \mathbb{R}. \qquad (37.10)$$

The grading in which these groups are non-zero corresponds to the class of the 2-plane field ξ_0 which is tangent to the fibers of the fibration over S^1.

We sketch briefly why (37.10) holds for one of these manifolds Y. Because Y is flat, the only solutions to the unperturbed 3-dimensional Seiberg–Witten equations on Y are the reducible solutions $(A, 0)$, with A a flat connection. For each torsion spin^c structure $\mathfrak{s}$, there is a circle $\mathbb{T}(\mathfrak{s})$ pararametrizing the flat spin^c connections $[A]$ up to gauge (because $b_1(Y) = 1$). If D_A has no kernel for any $[A]$ in $\mathbb{T}(\mathfrak{s})$, then the spin^c structure $\mathfrak{s}$ will make no contribution to the Floer groups with twisted local coefficients. So we must look for spin^c structures admitting spin^c connections with parallel spinors.

Lemma 37.4.1. *If Y is one of the non-trivial flat torus bundles over S^1, then the only spin^c structure for which there are flat connections admitting parallel spinors is the spin^c structure $\mathfrak{s}_0$ corresponding to the 2-plane field ξ_0 tangent to the fibers.*

For this spin^c structure $\mathfrak{s}_0$, there are exactly two points $[A_0]$, $[A_1]$ on the circle $\mathbb{T}(\mathfrak{s}_0)$ for which the corresponding Dirac operator D_A has kernel.

Proof. This proposition is quite easy to verify. If Φ is a parallel spinor for a flat spin^c connection A, then the quadratic $\rho^{-1}(\Phi\Phi^*)_0$ is a parallel 1-form. For these manifolds, all parallel 1-forms annihilate the tangents to the fibers: up to rescaling by positive reals, there are two possibilities, differing by sign. □

To determine the contribution of $\mathfrak{s}_0$ to $\widehat{HM}_*$ and $\widecheck{HM}_*$, we can use the same type of perturbation (37.1) as before. For the Morse function f, we take a function on the circle $\mathbb{T}(\mathfrak{s}_0)$ having two local minima, at $[A_0]$ and $[A_1]$. Because Y is covered by a flat 3-torus, our previous analysis tells us that there are no irreducible solutions. The reducible solutions in $\mathcal{B}(Y, \mathfrak{s}_0)$ correspond to the four critical points x^0, x^1, y^0, y^1 on $\mathbb{T}(\mathfrak{s}_0)$, where the x^i are the two local minima. We can label the two maxima as y^0 and y^1 so that, on paths joining y^0 to either of the local minima, there is no spectral flow for the family of Dirac operators $D_A - \delta$; while on paths joining y^1 to the local minima, the complex spectral flow is -1. We label the critical points in the blow-up as x_i^α and y_i^α as before, with $i = 0$ corresponding to the first positive eigenvalue.

We make a table of the generators that contribute to the complex $\check{C}(Y, \mathfrak{s}_0)$ that computes $\widecheck{HM}_*(Y, \mathfrak{s}_0)$, grouping generators with the same grading in rows:

$$\begin{array}{ccc} \cdots & & \cdots \\ y_1^0 & & y_2^1 \\ x_1^0 & & x_1^1 \\ y_0^0 & & y_1^1 \\ x_0^0 & & x_0^1 \\ & y_0^1. & \end{array} \tag{37.11}$$

The only differentials that arise are those corresponding to differentials in the ordinary Morse complex of the circle $\mathbb{T}(\mathfrak{s}_0)$ with the Morse function f. These differentials run from the ys to the xs. From the second row up in the table above, the complex is just a number of copies of the Morse complex of the circle. The generator corresponding to y_0^1 is not hit by any differential. We conclude that, with twisted local coefficients, the homology $\widecheck{HM}_*(Y, \mathfrak{s}_0; \Gamma_\eta)$ is $\mathbb{R}$, and that a generator is provided by the element 1 in $\Gamma_\eta[y_0^1]$. Because $\overline{HM}_*$ is zero for twisted coefficients, $\widehat{HM}_*(Y, \mathfrak{s}_0; \Gamma_\eta)$ is isomorphic to $\widecheck{HM}_*(Y, \mathfrak{s}_0; \Gamma_\eta)$.

With $\mathbb{Z}$ coefficients, the description of $\check{C}(Y, \mathfrak{s}_0)$ still applies and tells us that $\widecheck{HM}_k(Y, \mathfrak{s}_0)$ is $\mathbb{Z}$ for all $k = k_0 + n$, where k is the grading of y_0^1 and $n \geq 0$. We also see from this description that the map

$$i : \overline{HM}_*(Y, \mathfrak{s}_0) \to \widecheck{HM}_*(Y, \mathfrak{s}_0)$$

is onto, so the map j in the long exact sequence is zero and $\widehat{HM}_k(Y, \mathfrak{s}_0)$ is $\mathbb{Z}$ for all k of the form $k_0 + n$ with $n < 0$.

There remains the question of why the grading of the generator y_0^1 corresponds to the 2-plane field ξ_0. In the case of T^3, we argued using the existence of an orientation-reversing self-diffeomorphism of the manifold; but, for the flat torus bundles with monodromy of orders 3, 4 and 6, this argument is not available to us. Instead, we can argue that the corresponding 2-plane field must be ξ_0 using a later result, Theorem 41.4.1, which tells us (amongst other things) that if a 3-manifold fibers over the circle with fibers of positive genus, then the Floer group $\widecheck{HM}_k(Y; \Gamma_\eta)$ must be non-zero in the grading corresponding to the 2-plane field tangent to the fibers. (Of course, one could also work more directly, from the definitions.)

Finally, we mention again the remaining flat 3-manifold, the Hantzsche–Wendt manifold. Since b_1 is zero in this case, non-trivial local coefficient systems such as Γ_η do not arise. With $\mathbb{Z}$ coefficients, we can verify that the Floer groups for each spin^c structure $\mathfrak{s}$ follow the same pattern as for manifolds of positive scalar curvature, as outlined in Subsection 22.7. In particular, j is zero. The point that underlies this calculation is that, for each flat spin^c connection A, the Dirac operator D_A has no kernel. Indeed, if D_A had kernel, then as above there would be a non-zero parallel 1-form obtained as $\rho^{-1}(\Phi\Phi^*)_0$; but there is no such form, because $b_1 = 0$. Once we know that D_A has no kernel, we are in the same situation as when Y has positive scalar curvature: we can perturb to achieve transversality without introducing irreducible critical points. Note that, although the Hantzsche–Wendt manifold is known as the unique flat 3-manifold with $b_1 = 0$, the argument we have just used did not depend on any particular properties of this example.

38 Elliptic surfaces

38.1 The smooth classification of elliptic surfaces

Elliptic surfaces are smooth, connected complex surfaces E admitting a holomorphic map to a Riemann surface Σ, $\pi : E \to \Sigma$, such that there is a regular value of π with fiber an elliptic curve. In any elliptic surface, all but finitely many fibers will be smooth elliptic curves; the possibilities for the singular fibers were classified by Kodaira [56, 57]. We shall restrict ourselves to the simply connected case in which case $\Sigma = \mathbb{CP}^1$. We will also restrict ourselves to *relatively minimal* elliptic surfaces. This means that there are no exceptional curves in the fibers of π. We will describe the C^∞ classification of the underlying 4-manifolds of simply connected elliptic surfaces, following Kodaira.

The simplest such elliptic surface is obtained by taking a generic pair of homogenous cubic polynomials q_0 and q_1 in three variables. They give rise to a rational map

$$r : \mathbb{CP}^2 \dashrightarrow \mathbb{CP}^1, \tag{38.1}$$

and after blowing up $\mathbb{CP}^2$ at nine points we get an honest map from

$$\pi : E(1) \to \mathbb{CP}^1.$$

The fiber $\pi^{-1}([z_0 : z_1])$ is the projectivization of the zero set of the homogenous cubic $z_0 q_0 - z_1 q_1$ and hence (unless the q_i are particularly special) the generic fiber is an elliptic curve. From the C^∞ point of view, the regular fibers of π are embedded 2-dimensional tori with trivialized normal bundle (with the trivialization provided by the derivative of the projection π). As a smooth 4-manifold, $E(1)$ is independent of the choice of the two general cubics. In addition, the isotopy classes of both the fiber and the trivialization of the fiber's normal bundle are independent of the choices made. As a smooth manifold $E(1)$ is diffeomorphic to the connected sum

$$\mathbb{CP}^2 \# 9(-\mathbb{CP}^2).$$

This manifold appeared earlier as an example in Subsection 27.5. The homology class of the fibers is the same class $[F]$ as we discussed in Subsection 27.5.

We get further elliptic surfaces by the process sometimes called the *fiber sum*. Consider a pair of 4-manifolds X and X' each containing tori F and F' with trivialized normal bundle. Let $\hat{X}$ and $\hat{X}'$ be the manifolds with boundary obtained by removing tubular neighborhoods of F and F'. Thanks to the trivialization of the normal bundle of the fiber we can write the boundaries as products

$$\partial\hat{X} = S^1 \times F$$
$$\partial\hat{X}' = S^1 \times F'$$

respecting orientation. Choose an orientation-reversing self-diffeomorphism of S^1 and an orientation-preserving diffeomorphism $\phi : F \to F'$. (The first choice is essentially unique, the second is not.) Together, these give an orientation-reversing diffeomorphism $\partial\hat{X} \to \partial\hat{X}'$, and we can form a new oriented manifold

$$X\#_{F,\phi}X'$$

by identifying the boundaries. If F and F' arise as fibers of elliptic fibrations π and π', then $X\#_{F,\phi}X'$ is also elliptically fibered.

We can apply this construction to two copies of $E(1)$: we set

$$E(2) = E(1)\#_{F,1}E(1).$$

Although we have specified the identity map 1 as the diffeomorphism from F to F here, it does not matter which map we choose. One way to understand this is as follows. If γ is a loop in S^2 so that the map π is a submersion over γ, then we obtain a torus bundle over the circle, by pulling back π over γ. This torus bundle has a monodromy, well-defined up to isotopy, $\psi : F \to F$. The construction makes clear that

$$E(1)\#_{F,\psi}E(1) \cong E(1)\#_{F,1}E(1)$$

as smooth 4-manifolds. On the other hand, in the case of $E(1)$, we can obtain any element of the mapping class group of the torus by suitable choice of γ. In fact, in the case of $E(1)$, rather more is true; we record the following:

Proposition 38.1.1. *Let $\widehat{E(1)}$ denote the manifold with T^3 as its boundary obtained by removing an open tubular neighborhood of the fiber F from $E(1)$. Then any orientation-preserving diffeomorphism of the boundary T^3 extends to a diffeomorphism of $E(1)$.* □

So the same manifold $E(2)$ can be obtained by gluing two copies of $\widehat{E(1)}$ using *any* orientation-reversing diffeomorphism of the boundaries.

We can repeat the construction, defining $E(n)$ inductively:

$$E(n) = E(n-1)\#_F E(1).$$

Once again, the choice of ϕ does not affect the diffeomorphism type, so we omit it from our notation. The fiber sum construction as we have described it is not a holomorphic operation. Nevertheless, it can be realized in a holomorphic manner. In particular, each of the 4-manifolds $E(n)$ arises as the total space of a genuine elliptic surface. It is natural to write $E(0)$ for the trivial elliptic surface $S^2 \times F$; we then have

$$E(n_1)\#_F E(n_2) = E(n_1 + n_2)$$

all $n_i \geq 0$. For $n \geq 1$, the manifold $E(n)$ is simply connected and has $b^+ = 2n + 1$; its signature is $-8n$, and the manifold is spin if n is even. From the point of view of complex surfaces, the integer n records the holomorphic Euler characteristic, $h^{0,0} - h^{0,1} + h^{0,2}$.

The other construction we need is that of adding a multiple fiber. Consider the circle group S^1 acting on the unit sphere S^3 in $\mathbb{C}^2$ by the action $(z_1, z_2) \mapsto (\lambda z_1, \lambda^m z_2)$ for some fixed $m \geq 2$. All orbits are free except the orbit of $(0, 1)$, which has stabilizer of order m. The quotient is topologically a 2-sphere, and the quotient map $\pi_m : S^3 \to S^2$ exhibits S^3 as a Seifert-fibered space over S^2, with one multiple fiber of multiplicity m. If we take the product of this picture with an extra circle, we obtain a C^∞ elliptic surface $\pi : Q_m \to S^2$. All fibers in Q_m are regular except one, the *multiple* fiber. We write F as usual for a regular fiber. The multiple fiber is also an embedded 2-torus, but the derivative of π_m is singular along the whole multiple fiber. If we write $\hat{Q}_m$ for the complement of an open tubular neighborhood of a fiber, then $\hat{Q}_m$ is simply a copy of $D^2 \times T^2$ as a smooth manifold, because it is a regular neighborhood of the multiple fiber. But again, as an elliptic surface, $\hat{Q}_m$ is not the trivial product example.

We write $E(n)_m$ for the fiber sum

$$E(n)_m = E(n)\#_F Q_m$$

defined using any diffeomorphism ϕ of the generic fibers F. As long as $n \geq 1$, the result is independent of ϕ, for the same reason as before. The elliptic surface $E(n)_m$ fibers over S^2, with one multiple fiber of multiplicity m. Again, we have described the construction in C^∞ terms, but there is a complex-analytic version of the construction. Two caveats are in order. First, although the construction can be performed in the complex-analytic context, introducing a multiple fiber in this way can turn an algebraic surface into a complex surface that is not algebraic and not even Kähler. For example if we start with a 4-torus $T^4 = T^2 \times F$, then the fiber sum $T^4\#_F Q_m$ has odd first Betti number; this manifold therefore has no Kähler structure. The second remark is that introducing a multiple fiber of "multiplicity 1" – that is, forming a fiber sum with Q_1 – is in general a non-trivial operation, although the effect on $E(n)$ is trivial in the C^∞ setting. We can also form a fiber sum with several Q_m of different multiplicities, so forming elliptic surfaces

$$E(n)_{m_1,\dots,m_r}$$

with r multiple fibers.

Kodaira's classification establishes that the underlying smooth 4-manifold of any simply connected elliptic complex surface is diffeomorphic to one of

$$\left.\begin{array}{ll} E(n) & (n \geq 1) \\ E(n)_p & (n \geq 2, p \geq 2) \\ E(n)_{p,q} & (n \geq 1, p > q \geq 2, (p,q) = 1) \end{array}\right\} \tag{38.2}$$

or to a blow-up of one of these at any number of points.

The reason for excluding $E(n)_p$ from the list when $n = 1$ can be seen in Proposition 38.1.1; the manifold $\hat{Q}_m$ is just a standard $D^2 \times T^2$, so that proposition tells us that $E(1)_p$ and $E(1)$ are diffeomorphic manifolds, even though they have different structures as elliptic fibrations. Indeed, a diffeomorphism between these manifolds cannot preserve even the homology class of the fiber, because our description of $E(1)$ shows that the fiber class in $E(1)$ is primitive in integral homology, whereas in $E(1)_p$ the regular fiber is divisible by p, for it is homologous to p times the multiple fiber. The following proposition clarifies the point.

Proposition 38.1.2. *Let $E(1)$ be the standard elliptic surface, as above, and let $E(1)_m$ be obtained as $E(1)\#_F Q_m$, introducing a multiple fiber of multiplicity m. Let F and F' be regular fibers in the elliptic fibrations $E(1)$ and $E(1)_m$ respectively. Then there is a diffeomorphism $\psi : E(1)_m \to E(1)$ with $\psi_*[F'] = m[F]$.* □

38.2 Gluing along three-tori

Our computation of the Floer homology groups of the 3-torus allows us to understand how the monopole invariants of 4-manifolds behave when gluing 4-manifolds with 3-torus boundary. We take the point of view that we are interested in determining the generating function $\mathfrak{m}(X, [\nu])$ on a given homology class $[\nu]$. We discuss the simplest case, when the Poincaré dual of $[\nu]$ has non-trivial restriction to the splitting 3-torus. Our calculations are based on the framework of the following theorem.

Theorem 38.2.1. *Let X_{12} and X_{34} be oriented 4-manifolds which are each decomposed as a union of manifolds with boundary,*

$$X_{12} = X_1 \cup X_2 \quad \text{and} \quad X_{34} = X_3 \cup X_4.$$

We assume that the boundary of each of the pieces in either decomposition is diffeomorphic to T^3. Suppose further that these 4-manifolds contain 2-cycles $\nu_{12} \subset X_{12}$ and $\nu_{34} \subset X_{34}$ so that ν_{12} meets ∂X_1 transversally in the 1-cycle η and ν_{34} meets ∂X_3 transversally in η'. We assume that $[\eta] \in H_1(\partial X_1)$ is non-zero. Finally suppose that we are given an orientation-preserving diffeomorphism $f : \partial X_1 \to \partial X_3$ which takes η to η'. Then we can form the 4-manifolds

$$X_{14} = X_1 \cup_f X_4 \quad \text{and} \quad X_{32} = X_3 \cup_{f^{-1}} X_2.$$

Writing $\nu_i = X_i \cap \nu_{12}$ for $i = 1, 2$ and $\nu_i = X_i \cap \nu_{34}$ for $i = 2, 4$, we can construct corresponding 2-cycles in X_{14} and X_{32};

$$\nu_{14} = \nu_1 \cup_f \nu_4 \quad \text{and} \quad \nu_{32} = \nu_3 \cup_{f^{-1}} \nu_2.$$

If each closed 4-manifold has $b^+ \geq 2$, then the generating functions for the monopole invariants of these manifolds are related by

$$\mathfrak{m}(X_{12}, [\nu_{12}])\mathfrak{m}(X_{34}, [\nu_{34}]) = \mathfrak{m}(X_{14}, [\nu_{14}])\mathfrak{m}(X_{32}, [\nu_{32}]).$$

In the case that one or more X_{ij} has $b^+ = 1$, then the same result holds if we replace the invariant $\mathfrak{m}$ by $\mathfrak{m}_k$ defined in Subsection 27.5, where $k \in H^2(X_{ij})$ is Poincaré dual to a non-zero element in the image of $H_2(\partial X_i)$ in $H_2(\partial X_{ij})$.

Proof. From Proposition 3.9.3 we can write the invariants of the four 4-manifolds as pairings in the appropriate Floer groups: if $b^+ \geq 2$ for all four closed manifolds, then we have

$$\begin{aligned}
\mathfrak{m}(X_{12}, [\nu_{12}]) &= \langle \psi_{X_1,\nu_1}, \psi_{X_2,\nu_2} \rangle_{\omega_\mu} \\
\mathfrak{m}(X_{34}, [\nu_{34}]) &= \langle \psi_{X_3,\nu_3}, \psi_{X_4,\nu_4} \rangle_{\omega_\mu} \\
\mathfrak{m}(X_{14}, [\nu_{14}]) &= \langle \psi_{X_1,\nu_1}, \psi_{X_4,\nu_4} \rangle_{\omega_\mu} \\
\mathfrak{m}(X_{32}, [\nu_{32}]) &= \langle \psi_{X_3,\nu_3}, \psi_{X_2,\nu_2} \rangle_{\omega_\mu}.
\end{aligned}$$

If any of the manifolds have $b^+ = 1$, then we must use $\mathfrak{m}_k$ in place of $\mathfrak{m}$, and appeal to Proposition 27.5.1 for the appropriate version of the gluing formula. The pairings on the right-hand sides of these formulae all take place in the vector space $HM_\bullet(T^3;\Gamma_\eta)$ and its dual. The result now follows from Proposition 3.10.2, which states

$$HM_\bullet(T^3;\Gamma_\eta) \cong \mathbb{R};$$

the equality of the products is simply a formal consequence of the fact that this space is 1-dimensional. □

We can apply this theorem to obtain a formula for the monopole invariants of fiber sums. Let X and X' again be closed 4-manifolds containing embedded 2-tori F and F' with trivialized normal bundles. Form as before the fiber sum $X\#_{F,\phi}X'$ using any orientation-preserving diffeomorphism $\phi : F \to F'$. Let ν and ν' be real 2-cycles in the two manifolds, having non-zero pairing with F and F'. We can suppose that ν and ν' meet the tubular neighborhoods of F and F' each in a standard 2-disk with a non-zero real multiplicity. If $[\nu]\cdot F$ is equal to $[\nu']\cdot F'$, then we are in the situation described in the theorem: we can write

$$X = N(F) \cup \hat{X}$$
$$X' = N(F') \cup \hat{X}'$$

where $N(F)$ and $N(F')$ are tubular neighborhoods. Cutting and re-gluing as specified in the theorem, we obtain new manifolds

$$S^2 \times T^2 = N(F) \cup N(F')$$

and

$$X\#_{F,\phi}X' = \hat{X} \cup \hat{X}'.$$

The cycles ν and ν' give rise to cycles in $S^2 \times T^2$ and the fiber sum. In $S^2 \times T^2$, the cycle we obtain is a copy of the 2-sphere with multiplicity $[\nu]\cdot F$. In the fiber sum, it is natural to call the 2-cycle $\nu\#\nu'$. The theorem then gives

$$\begin{aligned}&\mathfrak{m}(X,[\nu])\mathfrak{m}(X',[\nu'])\\ &\quad = \mathfrak{m}_k\big(S^2\times T^2, ([\nu]\cdot[F])[S^2]\big)\mathfrak{m}\big(X\#_{F,\phi}X',[\nu\#\nu']\big).\end{aligned} \tag{38.3}$$

The invariant of $S^2 \times T^2$ was computed in Subsection 27.5, at (27.19), using the wall-crossing formula. So we can rewrite the formula as

$$\mathfrak{m}(X,[\nu])\mathfrak{m}(X',[\nu']) = \left(\frac{1}{2\sinh\big([\nu]\cdot[F]\big)}\right)^2 \mathfrak{m}\big(X\#_{F,\phi}X',[\nu\#\nu']\big). \tag{38.4}$$

The above formula for the invariant of a fiber sum allows us to calculate inductively the invariants of the elliptic surfaces $E(n)$:

Proposition 38.2.2. *The invariants of the elliptic surface $E(n)$ for $n \geq 2$ are given by*

$$\mathfrak{m}(E(n), h) = \big(2\sinh(h \cdot F)\big)^{n-2}$$

for any $h \in H_2(E(n); \mathbb{R})$, where F is the fiber. In the case $n = 1$, the above is a valid formula for $\mathfrak{m}_k(E(1), h)$, where k is the dual of the fiber class, as long as $h \cdot F$ is non-zero.

Proof. The formula in the case of $E(1) = \mathbb{CP}^2 \# 9\overline{\mathbb{CP}}^2$ was already established in (27.18). For larger n, apply the gluing theorem with $X = E(1)$ and $X' = E(n)$, with 2-cycles ν and ν' having $[\nu] \cdot F = [\nu'] \cdot F'$. Then $X \#_F X' = E(n+1)$, and we obtain

$$\begin{aligned}\left(\frac{1}{2\sinh\big([\nu]\cdot[F]\big)}\right)^2 & \mathfrak{m}\big(E(n+1), [\nu \# \nu']\big) \\ &= \mathfrak{m}_k\big(E(1), [\nu]\big)\mathfrak{m}\big(E(n), [\nu']\big) \\ &= \left(\frac{1}{2\sinh\big([\nu]\cdot[F]\big)}\right)\mathfrak{m}\big(E(n), [\nu']\big).\end{aligned}$$

Any homology class h arises as $[\nu \# \nu']$ on $E(n+1)$ in this way, as long as it has non-zero pairing with the fiber; so the result follows by induction on n, as long as $h \cdot F$ is non-zero. When $n \geq 2$, the result for $h \cdot F = 0$ follows by continuity, because when $b^+ \geq 2$ the invariant is a continuous function of the homology class. □

The 4-manifold $E(2)$ is better known as the *K3* surface. In this special case, the formula becomes

$$\mathfrak{m}(K3, h) = 1$$

for all h. In terms of the individual spin^c structures, this means that $\mathfrak{m}(X, \mathfrak{s}_X)$ is 1 for the spin^c with $c_1(\mathfrak{s}_X) = 0$ and is 0 for all other spin^c structures. This result can be derived from a different starting point, making use of the fact that the *K3* surface has a Ricci-flat Kähler metric.

We can also compute the invariants of elliptic surfaces with multiple fibers. We are helped by the fact that, in the case $n = 1$, the 4-manifolds $E(1)_m$ and

$E(1)$ are diffeomorphic; which means that, in a sense, we already know the invariants of $E(1)_m$. More specifically, Proposition 38.1.1 tells us that there is a diffeomorphism carrying the fiber class $[F']$ in $E(1)_m$ to $[mF]$, and taking therefore the ray spanned by the dual $k' = \text{P.D.}[F']$ to the ray spanned by $k = \text{P.D.}[F]$. Our calculation for $E(1)$ therefore gives

$$\mathfrak{m}_{k'}(E(1)_m, h) = \left(2\sinh\left(\frac{F' \cdot h}{m}\right)\right)^{-1}$$

where F' is the regular fiber. Once we have this, we can apply Theorem 38.2.1 to obtain the result for a general elliptic surface with multiple fibers:

Theorem 38.2.3. *Let $X = E(n)_{m_1,\ldots,m_r}$ be the elliptic surface with holomorphic Euler characteristic $n \geq 2$ and multiple fibers of multiplicity $m_1, \ldots, m_r$. Let $[F]$ be the class of a regular fiber. Then the monopole invariants of X are given by*

$$\mathfrak{m}(X, h) = 2^{n-2} \frac{\sinh(F \cdot h)^{n-2+r}}{\prod_{s=1}^{r} \sinh\big((F \cdot h)/m_s\big)}.$$

In the case $n = 1$, the same formula holds as an expression of $\mathfrak{m}_k(X, h)$, where k is the dual of the fiber class. □

When $n \geq 2$, the formula in this theorem can be expressed as a polynomial function of $\exp(fh)$, where $f = F/\prod m_i$:

$$\mathfrak{m}(X, h) = P_{n,m_1,\ldots,m_r}(e^{f \cdot h}).$$

It follows quite easily from this formula that, in the list of elliptic surfaces (38.2), no two are diffeomorphic, at least for $n \geq 2$, essentially because the polynomials P_n, $P_{n,p}$ and $P_{n,p,q}$ are all distinct. In the case $n = 1$, one can pass to the "small-perturbation" invariant $\mathfrak{m}(X, h)$, to obtain the same result. By contrast, for each n, the list of elliptic surfaces (38.2) contains either one homeomorphism type (if n is odd), or two homeomorphism types (if n is even) distinguished by whether or not the manifold is spin. (For even n, the manifold $E(n)_{p,q}$ is spin if and only if both p and q are odd.) See the references at the end of this chapter.

An alternative way to summarize these calculations is as follows. Let η be a 1-cycle in a torus T^3, with non-zero homology class. We know that $\widehat{HM}_\bullet(T^3, \Gamma_\eta)$ is $\mathbb{R}$, but we can now specify a canonical generator, up to sign, for this copy of $\mathbb{R}$. To do this, realize T^3 as the boundary of $\widehat{E(1)}$, and extend η to a relative

2-cycle ν with $\partial \nu = \eta$. Regarding T^3 as the oriented boundary of $\widehat{E(1)}$, we obtain an invariant

$$\psi_{\widehat{E(1)},\nu} \in \widehat{HM}_{\bullet}(T^3, \Gamma_\eta).$$

Our calculation of the invariant of $E(1)$ and Proposition 38.1.1 imply that this element is independent of the choice of extension: it is a preferred basis element

$$\psi(1) = \widehat{HM}_{\bullet}(T^3, \Gamma_\eta).$$

If we us this basis element to identify the Floer homology group with $\mathbb{R}$, then given any other manifold $\hat{X}$ with boundary T^3, and given a 2-chain ν in $\hat{X}$ with boundary η representing a non-zero class, we can regard the associated invariant $\psi_{\hat{X},\nu}$ as a real number. For example, in these terms, we have

$$\psi_{(D^2 \times T^2, \nu)} = \frac{1}{2\sinh(T^2 \cdot \nu)}.$$

Notes and references for Chapter IX

The fact that the Floer homology of the 3-torus with twisted coefficients is 1-dimensional means that a suitable invariant of a compact 4-manifold with 3-torus boundary can be interpreted as a single real number. This observation, and the consequent gluing formulae for manifolds cut along 3-tori, have been pursued and developed by several authors. In particular, Taubes' paper [118] develops gluing results which correspond to the case of twisted local coefficients on the torus (in our exposition), leading to the same formalism which we used in our discussion of elliptic surfaces in this chapter. Other gluing results for decompositions along 3-tori are obtained in [80], where formulae are obtained corresponding to the case of ordinary coefficients, where the Floer group is $\mathbb{Z}^3$. A similar phenomenon arises in instanton Floer theory, where it was used, for example, to understand the effect of "logarithmic transforms" on Donaldson's polynomial invariants in [49].

Because of the conjectured isomorphism between the Seiberg–Witten Floer homology groups and the Heegaard Floer groups of Ozsváth and Szabó (see Subsection 3.12), it is interesting to compare the calculations here with what is known in the case of Heegaard Floer homology. The Heegaard homology group $HF^\infty(Y)$ is completely understood (at the time of writing) only for small values of $b_1(Y)$ and in some examples with higher Betti number. Nevertheless, it is known that the Heegaard group $HF^\infty(Y)$ has a description of the same sort as we obtained in Corollary 34.4.2 for $\overline{HM}_\bullet(Y)$; it is only the vanishing of the higher differentials which is not known to hold in general. See [90]. The result of Proposition 35.3.3 and Proposition 35.3.4 for $S^1 \times \Sigma$ matches a corresponding calculation in Heegaard Floer homology by Jabuka and Mark [54].

The fact that the Floer groups are non-trivial, as stated in Corollary 35.1.4, plays a role in Taubes' proof of the Weinstein conjecture on the existence of closed orbits for the Reeb flow on a contact 3-manifold [107].

The calculation of the invariants of elliptic surfaces has a long history. A calculation of a rather special instanton invariant for $E(1)$ and $E(1)_{2,3}$ in [18] gave the first example of an h-cobordant pair of simply connected 4-manifolds that could be shown not to be diffeomorphic. Further contributions to the calculation of instanton invariants for elliptic surfaces, and to the proof that the elliptic surfaces listed in (38.2) represent distinct diffeomorphism types, include [37, 38, 81, 79, 11, 67, 36, 105, 60, 30]. A detailed study of the topology of elliptic surfaces is given in [39].

X
Further developments

In this final chapter, we give a taste of some further results concerning Floer homology and its topological applications. We do not always give complete proofs, nor do we aim to give a comprehensive survey.

Section 39 discusses Frøyshov's invariant and its application to negative-definite cobordisms. The topological results of this section, originating from Frøyshov's paper [41], can be seen as a natural extension of Donaldson's theorem, that the intersection form of a smooth, closed 4-manifold must be the standard diagonal form if it is positive- or negative-definite.

If a 3-manifold has positive scalar curvature, then the map j from $\widecheck{HM}_\bullet$ to $\widehat{HM}_\bullet$ is identically zero, which in particular means that the Floer groups $HM_\bullet(Y, \mathfrak{s})$ must vanish for all spin^c structures with non-torsion first Chern class. Section 40 explores an extension of this idea, an inequality between the genus of an embedded surface $\Sigma \subset Y$ and the value of $c_1(\mathfrak{s})$ on Σ which holds whenever $HM_\bullet(Y, \mathfrak{s})$ is non-zero. While Section 40 exploits scalar curvature to bound the genus from above, a rather deeper result shows that these bounds are essentially sharp. More precisely, the Thurston norm (which encodes the genus of embedded surfaces as a function of their homology class) can be recovered from the Floer groups of a 3-manifold. The proof of this result is in Section 41. It uses a circle of ideas involving symplectic structures, contact structures and foliations. The central result of Section 41 is a non-vanishing theorem for Floer homology, Theorem 41.4.1.

Many of the topological applications of Floer homology rest on a long exact sequence relating the Floer groups of three 3-manifolds obtained by different Dehn fillings of a given 3-manifold with torus boundary. We present the exact sequence (without proof), and explain some topological applications. Such topological applications were pioneered in the context of Heegaard homology by Ozsváth and Szabó. Further references are given, as usual, at the end of the chapter.

39 Homology spheres and negative-definite cobordisms

39.1 Frøyshov's invariant

For homology 3-spheres, there is an integer-valued invariant that one can extract from the Floer homology groups, using the canonical grading from Subsection 28.3. An invariant of this type was first constructed by Frøyshov in [42] in the context of instanton Floer homology. The Seiberg–Witten version was introduced in [43], and a similar construction was used by Ozsváth and Szabó for their Heegaard Floer homology [89]. The definition can be extended to *rational* homology 3-spheres equipped with a spin^c structure, but we keep to the simplest case here.

So let Y be a homology 3-sphere, and let $\mathfrak{s}$ be its unique spin^c structure. In $\mathcal{B}(Y)$, with a small perturbation of $\mathcal{L}$, there is exactly one reducible critical point, contributing a sequence of generators to $\overline{HM}_{\bullet}(Y)$, just as in the case of S^3 (see also Subsection 35.3). We work with real coefficients, so that

$$\overline{HM}_{\bullet}(Y) \cong \mathbb{R}[U_{\dagger}^{-1}, U_{\dagger}]],$$

as modules over the ring $\mathbb{R}[[U_{\dagger}]]$. The construction of Subsection 28.3 gives $\overline{HM}_{*}(Y)$ a canonical $\mathbb{Z}$ grading; and in the case of a homology sphere, $\overline{HM}_{k}(Y)$ is non-zero for even integers k, as stated in Corollary 28.3.3. In the above isomorphism, we can regard the element $1 \in \mathbb{R}[U_{\dagger}^{-1}, U_{\dagger}]]$ as having degree zero and $U_{\dagger}$ as having degree -2; and in this case the isomorphism can be taken to *preserve the canonical grading*. As an abbreviation, let us write

$$L = \mathbb{R}[U_{\dagger}^{-1}, U_{\dagger}]],$$

so that $\overline{HM}_{\bullet}(Y) \cong L$ as graded modules over the ring

$$S = \mathbb{R}[[U_{\dagger}]].$$

The only non-trivial proper S-submodules of L are the submodules $L_h \subset L$, defined as

$$L_h = U_{\dagger}^{h+1} S$$

for h in $\mathbb{Z}$. This is the closure of the span of all graded components of L in degree less than $-2h$. (In particular, our notation implies we are regarding S itself as a submodule of L, consisting of the elements involving non-negative powers of $U_{\dagger}$, i.e. the elements of non-positive degree.)

From the 3-manifold Y, we obtain a preferred submodule of L, from the image of the map

$$p : \widehat{HM}_\bullet(Y) \to \overline{HM}_\bullet(Y)$$

or equivalently the kernel of

$$i : \overline{HM}_\bullet(Y) \to \widecheck{HM}_\bullet(Y).$$

This submodule is proper and non-zero, so it coincides with L_h for some h. In this way, one obtains an integer invariant $h(Y)$ associated to an oriented homology 3-sphere Y. The definition is normalized so that h is zero for the standard 3-sphere:

Definition 39.1.1. If Y is an oriented integral homology 3-sphere, the Frøyshov invariant $h(Y)$ is defined as the unique integer h such that the image of $p : \widehat{HM}_\bullet(Y) \to \overline{HM}_\bullet(Y) \cong L$ coincides with the S-submodule L_h. $\diamond$

Because i defines an injective map on $\overline{HM}_\bullet(Y)/\mathrm{im}(p)$, an alternative way to think about h is in terms of the image of i: the definition says that the lowest grading of any element of the image of i in $\widecheck{HM}_\bullet(Y)$ is $-2h$.

A notable property of h, which motivated Frøyshov's definition, is its monotonicity with respect to negative-definite cobordisms. To describe this, suppose we have a cobordism W between homology 3-spheres Y_0 and Y_1. Suppose that W has zero integral first homology and negative-definite intersection form. The cobordism gives a commutative diagram

$$\begin{array}{ccc} \widehat{HM}_\bullet(Y_0) & \longrightarrow & \widehat{HM}_\bullet(Y_1) \\ \downarrow p & & \downarrow p \\ \overline{HM}_\bullet(Y_0) & \xrightarrow{\overline{HM}_\bullet(W)} & \overline{HM}_\bullet(Y_1), \end{array} \tag{39.1}$$

and hence a commutative diagram

$$\begin{array}{ccc} L_{h_0} & \longrightarrow & L_{h_1} \\ \downarrow & & \downarrow \\ L & \xrightarrow{x} & L, \end{array} \tag{39.2}$$

in which the two vertical maps are the inclusions, and h_i denotes $h(Y_i)$. The map x is what becomes of $\overline{HM}_\bullet(W)$ under the identification of the two groups

$\overline{HM}_\bullet(Y_i)$ with L, and this map can be described quite precisely, as we now explain.

For each choice of spin^c structure $\mathfrak{s}_W$ on W, there is a unique spin^c connection A_0 (up to gauge equivalence) on the cylindrical-end manifold W^+ having $F^+_{A_0^t} = 0$, because of the topological assumptions on Y. Let us label the reducible critical point in $\mathcal{B}^\sigma(Y_0)$ with grading $2k$ as $\mathfrak{a}_k$, and write $\mathfrak{b}_k$ similarly on Y_1. For each k and k', the compactified moduli space $M_z^{\text{red}}([\mathfrak{a}_k], W, [\mathfrak{b}_{k'}])$ is either empty (if the formal dimension is negative) or a complex projective space, whose dimension depends on W, $\mathfrak{s}_W$ and the difference $k - k'$. In particular, looking at the contribution of the zero-dimensional moduli spaces, we see that the contribution of the spin^c structure $\mathfrak{s}_W$ to the map

$$x : L \to L$$

has a simple form, given by multiplication by $U_\dagger^{-r}$. We can read off r from the formula for the canonical $\mathbb{Q}$ grading, Definition 28.3.1, as we did in (28.3): we have

$$2r = \frac{1}{4}\langle c_1(S^+), c_1(S^+)\rangle - \iota(W) - \frac{1}{4}\sigma(W),$$

which becomes

$$r = \frac{1}{8}\left(\langle c_1(S^+), c_1(S^+)\rangle + b_2(W)\right).$$

Note that $c_1(S^+)$ is a *characteristic vector* in the negative-definite, unimodular lattice $H^2(W;\mathbb{Z})$, which is to say that

$$\langle c, a\rangle = \langle a, a\rangle \pmod 2$$

for all integer classes a. For a characteristic vector, the square $\langle c, c\rangle$ is equal to minus the rank mod 8, which confirms that r is an integer.

When we sum over all spin^c structure, we obtain a formula for x:

Proposition 39.1.2. *Let W be a negative-definite cobordism with $b_1 = 0$, between homology spheres Y_0 and Y_1, and let $x : L \to L$ be the map arising from*

$$\overline{HM}_\bullet(W) : \overline{HM}_\bullet(Y_0) \to \overline{HM}_\bullet(Y_1)$$

after identifying each $\overline{HM}_\bullet(Y_i)$ *with L, preserving the canonical grading. Then x is given by multiplication by the power series*

$$x = \sum_c U_\dagger^{-(c\cdot c + b_2(W))/8},$$

where the sum is over all characteristic vectors. □

Note that this power series extends infinitely in the direction of positive powers of $U_\dagger$, and depends only on the intersection form Q_W (as a quadratic form over $\mathbb{Z}$). The series contains at least one non-zero term involving a non-positive power of $U_\dagger$, according to the following result of Elkies concerning unimodular forms:

Proposition 39.1.3 ([27]). *Let Q be a unimodular quadratic form over* $\mathbb{Z}$, *either positive- or negative-definite. Then there is a characteristic vector c for Q satisfying*

$$|Q(c)| \le \text{rank}(Q).$$

If Q is not the standard diagonal form, then there is a characteristic vector c which achieves strict inequality here. □

Note that for *even* forms, the result is clear, for in that case 0 is a characteristic vector. For the diagonal form $Q = \text{diag}(1, \ldots, 1)$, the characteristic vector $c = (1, \ldots, 1)$ achieves equality, but no characteristic vector achieves strict inequality. We introduce the notation

$$\rho(Q) = \frac{1}{8}\Big(\text{rank}(Q) - \inf_c |Q(c)|\Big),$$

and interpret Elkies' result as saying that $\rho(Q)$ is non-negative, and is strictly positive if Q is not standard. The leading term in the power series x above is a non-zero multiple of $U_\dagger^{-\rho(Q_W)}$:

$$x = U_\dagger^{-\rho(Q_W)}\big(a_0 + a_1 U_\dagger + \cdots\big).$$

From this, we can now deduce:

Theorem 39.1.4 (Frøyshov [43]). *Let* Y_0 *and* Y_1 *be integral homology 3-spheres, and W an oriented cobordism from* Y_0 *to* Y_1 *with* $H_1(W;\mathbb{Z}) = 0$

and Q_W negative-definite. Then the invariants $h_0 = h(Y_0)$ and $h_1 = h(Y_1)$ satisfy

$$h_0 \geq h_1 + \rho(Q_W).$$

In particular, $h_0 \geq h_1$, with strict inequality if Q_W is not the standard diagonal form.

Proof. The result follows from the commutative diagram (39.2). Our description of the power series x and the definition of L_h together tell us that

$$xL_{h_0} = L_{h_0-\rho}$$

where $\rho = \rho(Q_W)$. The diagram therefore tells us that

$$L_{h_0-\rho} \subset L_{h_1},$$

which means that

$$h_0 - \rho \geq h_1.$$

□

Remark. We have considered only the case that $H_1(W;\mathbb{Z}) = 0$ so as to simplify the exposition. If $H_1(W;\mathbb{Z})$ is a torsion group, very little changes, except that for each characteristic vector c in $H^2(W;\mathbb{Z})/\text{torsion}$ there will be more than one spin^c structure: there will be $|\text{Tor}|$ spin^c structures for each c, where Tor is the torsion subgroup. The only effect of this is that the formula for x given in Proposition 39.1.2 needs to be corrected by including a factor of $|\text{Tor}|$. In addition, even if W has non-zero first Betti number, it can be altered by surgery to make $b_1 = 0$ without affecting the intersection form; so in the above theorem the hypothesis on $H_1(W;\mathbb{Z})$ can be entirely removed.

The strict monotonicity property in Frøyshov's theorem, i.e. the statement that $h_0 > h_1$ if Q_W is non-standard, already implies that if W is a negative-definite cobordism from S^3 to S^3, then Q_W is standard. This provides a proof of the theorem of Donaldson, first proved using Yang–Mills theory, that a closed negative-definite 4-manifold must have standard intersection form. Another application is to the case of even intersection forms, where $\rho(W)$ is $\text{rank}(Q_W)/8$:

Corollary 39.1.5. *If Y is a homology sphere arising as the oriented boundary of a 4-manifold X with even, negative-definite intersection form, then*

$$b_2(X)/8 \leq -h(Y).$$

□

As an example, Frøyshov computes the invariant h for the Poincaré homology sphere Y, oriented as the boundary of the negative-definite E_8 plumbing: the result, $h(Y) = -1$, implies that Y cannot be the boundary of a negative-definite 4-manifold with even intersection form if the form has rank larger than 8.

39.2 Frøyshov's invariant with mod two coefficients

One can define a Frøyshov-type invariant using any field in place of $\mathbb{R}$. We use $\bar{h}(Y)$ to denote the invariant obtained by using a field of characteristic 2. (In [43], the notation h_p is used for the invariant obtained by using coefficients with characteristic p.) Theorem 39.1.4 continues to hold for this modified version of the invariant. In the proof, it might at first appear that we have made use of the fact that no cancellation can occur in the sum that defines the power series x, something that would cease to hold if the coefficients had finite characteristic. But one can simply consider the spin^c structures on a cobordism W one by one, rather than using the map $\overline{HM}_\bullet(W)$ which is defined using them all. Equivalently, one can use a W-morphism to separate the spin^c structures.

To illustrate this, consider the field $\mathbb{K}$ of characteristic 2 obtained as the field of fractions of the ring $\mathbb{F}_2[\mathbb{R}]$, the group ring of $\mathbb{R}$ over the field $\mathbb{F}_2$. This field comes equipped with a group homomorphism

$$\exp : \mathbb{R} \to \mathbb{K}^\times \tag{39.3}$$

arising from the tautological inclusion of $\mathbb{R}$ in $\mathbb{F}_2[\mathbb{R}]^\times$. Associated to a real 1-cycle η in a 3-manifold is a local coefficient system Γ_η on $\mathcal{B}^\sigma(Y)$, as described in Subsections 3.7 and 22.6. As defined, Γ_η has fiber $\mathbb{R}$. However, as pointed out in Subsection 22.6, we can modify the definition of Γ_η by replacing $\mathbb{R}$ with any other field, such as $\mathbb{K}$, as long as it is equipped with an exponential map such as (39.3). In this way, given a real 1-cycle η, we obtain a local coefficient system on $\mathcal{B}^\sigma(Y)$ which we will denote by $\mathbb{K}_\eta$. Its fibers are all canonically isomorphic to $\mathbb{K}$, which is a field of characteristic 2. As with Γ_η, if W is a cobordism between 3-manifolds and ν is a 2-chain in W with

$$\partial\nu = -\eta_0 + \eta_1,$$

then ν gives rise to a W-morphism of local coefficient systems,

$$\mathbb{K}_\nu : \mathbb{K}_{\eta_0} \to \mathbb{K}_{\eta_1},$$

and hence a map

$$\overline{HM}_\bullet(W; \mathbb{K}_\nu) : \overline{HM}_\bullet(Y_0; \mathbb{K}) \to \overline{HM}_\bullet(Y_1; \mathbb{K}).$$

We now consider the case that Y_0 and Y_1 are homology 3-spheres and ν_0 and ν_1 are zero. The 2-chain ν is therefore a cycle in W. We can again identify

$$\overline{HM}_\bullet(Y_i; \mathbb{K}) \cong \bar{L},$$

where $\bar{L}$ is now the ring of Laurent series

$$\bar{L} = \mathbb{K}[U_\dagger^{-1}, U_\dagger]].$$

This identification is again chosen to preserve the canonical grading. If W is negative-definite and $H_1(W; \mathbb{Z}) = 0$, then we have a formula for $\overline{HM}_\bullet(W; \mathbb{K}_\nu)$ as multiplication by the Laurent series

$$\bar{x} = \sum_c \exp(c \cdot \nu) U_\dagger^{-(c \cdot c + b_2(W))/8} \tag{39.4}$$

where the sum is again over all characteristic vectors, and the exponential map that appears is again the tautological map $\mathbb{R} \to \mathbb{K}^\times$. (There is nothing special about our choice of $\mathbb{K}$, of course, as far as this formula is concerned: the same formula would hold for $\mathbb{K} = \mathbb{R}$, with exp being the usual exponential map.) If ν is chosen so that it has non-zero pairing with every non-zero integer class, then there will be no chance of cancellation occurring in this sum, and $\bar{x}$ will therefore have the form

$$\bar{x} = U_\dagger^{-\rho(Q_W)} \big(\bar{a}_0 + \bar{a}_1 U_\dagger + \cdots\big),$$

with non-zero leading coefficient $\bar{a}_0 \in \mathbb{K}$. In this way, we obtain a convenient framework in which to carry over our previous arguments to the case of a field of characteristic 2. We shall make further use of the field $\mathbb{K}$ later in this chapter.

39.3 The blow-up formula

The *blow-up* of a smooth 4-manifold means the connected sum

$$\tilde{X} = X \# \bar{\mathbb{CP}}^2.$$

(The terminology comes from algebraic geometry.) In Subsection 3.11, we deduced a general vanishing result for the monopole invariants of connected sums, Corollary 3.11.3. That result, however, is applicable only when both summands have b^+ non-zero: it does not apply to the case of $\tilde{X}$, since the intersection form of $\bar{\mathbb{CP}}^2$ is negative. Indeed, if X has non-zero monopole invariants, then so does $\tilde{X}$, and there is a straightforward formula relating the two. This is the *blow-up formula*, which we now describe.

Rather than consider the case of $\bar{\mathbb{CP}}^2$ alone, we can consider a general closed 4-manifold N with negative-definite intersection form Q_N and trivial integral first homology. (One can also treat the case that $H_1(N;\mathbb{Z})$ is a non-zero torsion group in much the same way: see the remark following the proof of Theorem 39.1.4.) We set

$$\tilde{X} = X \# N.$$

A homology orientation for X determines one also for $\tilde{X}$, and we suppose that these have been fixed. We can then consider the monopole invariants

$$\mathfrak{m}(\tilde{u} \mid \tilde{X}) = \sum_{\tilde{\mathfrak{s}}} \langle \tilde{u}, M(\tilde{X}, \tilde{\mathfrak{s}}) \rangle$$

where the sum is over all spin^c structures on $\tilde{X}$ and $\tilde{u}$ is a cohomology class on $\boldsymbol{\mathcal{B}}^\sigma(\tilde{X})$.

Note that each component $\mathcal{B}^\sigma(\tilde{X}, \tilde{\mathfrak{s}})$ of $\boldsymbol{\mathcal{B}}^\sigma(\tilde{X})$ has the same cohomology ring as the components of $\mathcal{B}^\sigma(X)$, namely the algebra

$$\mathbb{A}(X) = \Lambda\,(H_1(X)/\text{torsion}) \otimes \mathbb{Z}[U].$$

We can therefore regard $\mathfrak{m}(- \mid \tilde{X})$ as a function on $\mathbb{A}(X)$. The following theorem expresses this invariant of $\tilde{X}$ in terms of the invariant of X. As a technical point in the statement, we note that the expression $\mathfrak{m}(a \mid X)$ makes sense not just for $a \in \mathbb{A}(X)$, but also for a in the completion

$$\mathbb{A}_\bullet(X) = \Lambda\,(H_1(X)/\text{torsion}) \otimes \mathbb{Z}[[U]].$$

The reason is that only finitely many moduli spaces $M(X, \mathfrak{s})$ are non-empty, so only finitely many terms of a make any contribution. With this understood, we can state the blow-up formula:

Theorem 39.3.1. *For any $a \in \mathbb{A}_\bullet(X)$, we have*

$$\mathfrak{m}(a \mid \tilde{X}) = \mathfrak{m}(xa \mid X)$$

where $x \in \mathbb{A}_\bullet(X)$ denotes the expression

$$x = \sum_c U^{-(c\cdot c + b_2(N))/8},$$

and the sum is over all characteristic vectors in $H^2(N;\mathbb{Z})$.

Some remarks are in order. First, Donaldson's theorem [17] tells us that the intersection form of N is the standard, diagonal form. So the series x actually only depends on the Betti number $b_2(N)$, and it is simply the b_2'th power of the series that arises from the quadratic form (-1):

$$x = \left(\sum_{n \text{ odd}} U^{(n^2-1)/8}\right)^{b_2(N)}.$$

Second, the simple-type conjecture (Conjecture 1.6.2, slightly generalized), would say that $\mathfrak{m}(U^e a \mid X) = 0$, for e positive. If X satisfies this condition (which holds in all known cases at the time of writing), then the only odd integers n which contribute are $n = \pm 1$. In such cases, the blow-up formula says more simply

$$\mathfrak{m}(a \mid \tilde{X}) = 2^{b_2(N)}\mathfrak{m}(a \mid X).$$

A third remark is that, if we drop the condition that $H_1(N;\mathbb{Z})$ is zero and instead ask only that $b_1(N)$ is zero, then we have essentially the same formula with an additional factor of the order of the torsion group. (See the similar remark after the proof of Theorem 39.1.4 above.)

In addition to the invariant $\mathfrak{m}(u \mid X)$, there is the generalization

$$\mathfrak{m}(u \mid X, h) = \sum_{\mathfrak{s}} \langle u, M(X,\mathfrak{s})\rangle \exp\langle c_1(\mathfrak{s}), h\rangle$$

for h in $H^2(X;\mathbb{R})$. The blow-up formula extends to this more general invariant. We write $H_2(\tilde{X};\mathbb{R})$ as a direct sum, and so write a typical element

$$\tilde{h} = (h, l),$$

where $l \in H_2(N;\mathbb{R})$. Then we have:

Theorem 39.3.2. *For any $a \in \mathbb{A}_{\bullet}(X)$ and $\tilde{h} = (h, l)$ in $H_2(\tilde{X}; \mathbb{R})$, we have*

$$\mathfrak{m}(a \mid \tilde{X}, \tilde{h}) = \mathfrak{m}(x_l a \mid X, h)$$

where $x_l \in \mathbb{A}_{\bullet}(X)$ denotes the expression

$$x_l = \sum_c U^{-(c \cdot c + b_2(N))/8} \exp\langle c, l \rangle,$$

and the sum is over all characteristic vectors in $H^2(N; \mathbb{Z})$.

The formula for x_l that appears can be expressed in terms of a ϑ-function. In the case that $b_2(N)$ is 1, for example, we pick a generator E for $H_2(N)$ and write $l = \lambda E$ for $\mathbb{R}$; we can then write x_l as

$$\sum_m U^{m(m+1)/2} e^{-(2m+1)\lambda}.$$

This can be written in terms of the Jacobi ϑ-function

$$\vartheta_2(z, q) = \sum_m q^{(m+1/2)^2} e^{(2m+1)iz}$$

as

$$x_l = U^{-1/8} \vartheta_2(U^{1/2}, i\lambda).$$

In the case of larger b_2, after diagonalizing the quadratic form, and writing $l = (\lambda_1, \lambda_2, \ldots, \lambda_{b_2(N)})$, we obtain

$$x_l = U^{-b_2(N)/8} \prod_{k=1}^{b_2(N)} \vartheta_2(U^{1/2}, i\lambda_k).$$

When X has simple type, all terms involving higher powers of U disappear. So for $N = \bar{\mathbb{CP}}^2$, if we write $\tilde{h} = (h, \lambda E)$ where E is a generator of $H_2(\bar{\mathbb{CP}}^2; \mathbb{Z})$, then the formula becomes

$$\mathfrak{m}(\tilde{X}, h + \lambda E) = 2\cosh(\lambda)\, \mathfrak{m}(X, h). \tag{39.5}$$

If we go back to considering individual spin^c structures on X rather than the generating function, then this last formula can be expressed more directly. Let

$\mathfrak{s}$ be a spin^c structure on X, and let $\tilde{\mathfrak{s}}$ be a spin^c structure on the blow-up $\tilde{X}$ whose first Chern class is given by

$$c_1(\tilde{\mathfrak{s}}) = c_1(\mathfrak{s}) \pm \text{P.D.}[E]$$

with respect to the direct sum description of the cohomology. Then the result is simply:

$$\mathfrak{m}(\tilde{X}, \tilde{\mathfrak{s}}) = \mathfrak{m}(X, \mathfrak{s}). \tag{39.6}$$

Proof of Theorem 39.3.2 The proof of the blow-up formula in the general form of Theorem 39.3.2 is a straightforward adaptation of our earlier result about negative-definite cobordisms, and in particular the formula (39.4) in the case of the real field. To understand this, recall that the invariant $\mathfrak{m}(u \mid X, h)$ of a closed manifold X can be expressed in terms of the map on Floer groups resulting from the cobordism $W_X : S^3 \to S^3$ obtained by removing two balls from X. The formula relating $\mathfrak{m}(u \mid X, h)$ to the map

$$\overrightarrow{HM}_\bullet(u \mid W_X; \Gamma_\nu) : \widehat{HM}_\bullet(S^3) \to \widecheck{HM}_\bullet(S^3)$$

is given in Proposition 27.4.3. Here ν is a closed cycle in W_X representing the class h. When we replace X by $\tilde{X} = X \# N$, then we can write the corresponding cobordism $W_{\tilde{X}}$ as a composite,

$$W_{\tilde{X}} = W_X \circ W_N,$$

where W_N is obtained from N by removing two balls. We also have a composition law from Theorem 3.5.3 (see also Subsection 27.3),

$$\overrightarrow{HM}_\bullet(u \mid W_{\tilde{X}}; \Gamma_{\tilde{\nu}}) = \overrightarrow{HM}_\bullet(u \mid W_X; \Gamma_\nu) \circ \widehat{HM}_\bullet(u \mid W_N; \Gamma_\pi) \tag{39.7}$$

where $\tilde{\nu} = \nu + \pi$ is a cycle representing the class $\tilde{h} = h + l$ in the connected sum. The factor

$$\widehat{HM}_\bullet(u \mid W_N; \Gamma_\pi) : \widehat{HM}(S^3) \to \widehat{HM}(S^3)$$

that appears in (39.7) is known to us. Indeed, $\widehat{HM}(S^3)$ is a submodule of $\overline{HM}(S^3)$, and this map is just the restriction of the map

$$\overline{HM}_\bullet(u \mid W_N; \Gamma_\pi) : \overline{HM}(S^3) \to \overline{HM}(S^3)$$

which we calculated above: see Equation (39.4), which is the formula for the case of a general negative-definite cobordism W. Thus (39.7) expresses $\overrightarrow{HM}_\bullet(u \mid W_{\tilde{X}}; \Gamma_{\tilde{\nu}})$ in terms of $\overrightarrow{HM}_\bullet(u \mid W_X; \Gamma_\nu)$ and a known power series (39.4). Using the relationship between the latter and $\mathfrak{m}(u \mid X, h)$, we obtain the blow-up formula in the form given in the theorem. □

40 Genus bounds and scalar curvature

40.1 Dimension three

Recall that when we decompose $\widecheck{HM}_\bullet(Y)$ and its companions according to spin^c structures,

$$\widecheck{HM}_\bullet(Y) = \bigoplus_{\mathfrak{s}} \widecheck{HM}_\bullet(Y, \mathfrak{s}),$$

there are only finitely many spin^c structures which contribute. This was stated as Proposition 3.1.1, and the proof was based on an L^2 bound on the curvature, (10.17). If we take a closer look at this bound, and exploit the topological invariance of the Floer groups by choosing very particular Riemannian metrics, we can get a much sharper result than just the finiteness of the set of contributing spin^c structures.

The result we shall state involves the spin^c structures $\mathfrak{s}$ with $c_1\mathfrak{s}$ non-torsion. For such spin^c structures, $\widecheck{HM}_\bullet(Y, \mathfrak{s})$ and $\widehat{HM}_\bullet(Y, \mathfrak{s})$ are isomorphic; so, as we have done before, we simply write $HM_\bullet(Y, \mathfrak{s})$.

Proposition 40.1.1. *Let Σ be a smoothly embedded, oriented connected 2-manifold in Y of genus at least 1, and let $\mathfrak{s}$ be a spin^c structure. If*

$$\left|\langle c_1(\mathfrak{s}), [\Sigma]\rangle\right| > 2\,\mathrm{genus}(\Sigma) - 2$$

then there is a Riemannian metric on Y for which the unperturbed Seiberg–Witten equations admit no solution belonging to the spin^c structure $\mathfrak{s}$.

Corollary 40.1.2. *Under the hypotheses of the proposition above, if*

$$\left|\langle c_1(\mathfrak{s}), [\Sigma]\rangle\right| > 2\,\mathrm{genus}(\Sigma) - 2 \geq 0$$

then the group $HM_\bullet(Y, \mathfrak{s})$ is zero. □

To state the corollary the other way around, the spin^c structures $\mathfrak{s}$ for which $HM_\bullet(Y,\mathfrak{s})$ is non-zero are constrained by the inequality

$$\left|\langle c_1(\mathfrak{s}),[\Sigma]\rangle\right| \leq 2\,\text{genus}(\Sigma) - 2.$$

If we represent a basis for $H_2(Y,\mathbb{Z})$ by embedded surfaces Σ_i, then the corresponding inequalities bound the coefficients of $c_1(\mathfrak{s})$ when expressed in terms of the dual basis for $H^2(Y,\mathbb{Z})/\text{torsion}$. In this way, the image of $c_1(\mathfrak{s})$ in $H^2(Y,\mathbb{R})$ is constrained to lie in a compact polytope. We shall return to this in more detail when we discuss the Thurston norm later in this chapter.

Proof of Proposition 40.1.1. To prove the proposition, we construct a Riemannian metric on Y which contains a long cylindrical piece, $[-T,T]\times\Sigma$. To be more specific, we first fix a Riemannian metric h_Σ on Σ with constant sectional curvature and total area 1. We then consider a 1-parameter family of metrics h_T on Y, with the property that (Y,h_T) contains an isometric copy of the cylinder $[-T,T]\times Y$ with the product metric

$$dt^2 + h_\Sigma.$$

We further require that the geometry of (Y,h_T) in the complement of the cylindrical piece is independent of T.

On the cylindrical part of (Y,h_T), the scalar curvature s is a constant, and by the Gauss–Bonnet theorem this constant is $-8\pi(g-1)$, where g is the genus of Σ. Since the geometry is fixed outside the cylindrical piece, we therefore have

$$\int_Y s^2\,d\text{vol} = T(8\pi)^2(g-1)^2 + C$$

for this metric, where C is a constant independent of T.

Now let A be any spin^c connection for the spin^c structure $\mathfrak{s}$. By the Chern–Weil theorem, we have

$$\int_\Sigma iF_{A^t} = 2\pi\langle c_1(\mathfrak{s}),[\Sigma]\rangle.$$

So for each $t\in[-T,T]$, we have

$$\int_{\{t\}\times\Sigma} \left|F_{A^t}|_{t\times\Sigma}\right|^2 d\text{vol}_\Sigma \geq 4\pi^2\langle c_1(\mathfrak{s}),[\Sigma]\rangle^2,$$

because Σ has area 1. Integrating with respect to t, we find

$$\int_Y |F_{A^t}|^2 \, d\mathrm{vol}_Y \geq 4\pi^2 T \langle c_1(\mathfrak{s}), [\Sigma]\rangle^2.$$

Having estimated the L^2 norms of both s and F_{A^t}, we now use the basic identity of Corollary 4.5.3: in its 4-dimensional version, this identity states that on a closed 4-manifold X, solutions (A, Φ) to the Seiberg–Witten equations are characterized by

$$\begin{aligned} 0 = \frac{1}{4}\int_X |F_{A^t}|^2 + \int_X |\nabla_A \Phi|^2 + \frac{1}{4}\int_X \left(|\Phi|^2 + (s/2)\right)^2 - \int_X \frac{s^2}{16} \\ - \frac{1}{4}\int_X F_{A^t} \wedge F_{A^t}. \end{aligned} \tag{40.1}$$

The 3-dimensional version of this identity can be obtained by considering $X = S^1 \times Y$ and pulling back a solution from the 3-manifold. In this case, the term $F_{A^t} \wedge F_{A^t}$ is zero, and we have

$$0 \;=\; \frac{1}{4}\int_Y |F_{A^t}|^2 \;+\; \int_Y |\nabla_A \Phi|^2 \;+\; \frac{1}{4}\int_Y \left(|\Phi|^2 \;+\; (s/2)\right)^2 \;-\; \int_Y \frac{s^2}{16}$$

for any solution on Y. In particular,

$$\int_Y |F_{A^t}|^2 \leq \frac{1}{4}\int_Y s^2.$$

If a solution exists on (Y, h_T), then the above estimates give us

$$4\pi^2 T \langle c_1(\mathfrak{s}), [\Sigma]\rangle^2 \leq (1/4) T (8\pi)^2 (g-1)^2 + C,$$

or more simply

$$\langle c_1(\mathfrak{s}), [\Sigma]\rangle^2 \leq (2g-2)^2 + C/(4\pi^2 T).$$

If a solution (A_T, Φ_T) exists for *all* metrics in the family h_T, then it follows that

$$\left|\langle c_1(\mathfrak{s}), [\Sigma]\rangle\right| \leq 2g - 2.$$

This proves the proposition. □

A simple illustration of Corollary 40.1.2 is the case $Y = S^1 \times \Sigma$. A basis for the second homology in this case is provided by the surface $\{p\} \times \Sigma$ and the tori $S^1 \times \gamma_i$, for loops γ_i in Σ forming a basis for $H_1(\Sigma)$. The corollary tells us that, for $HM_\bullet(Y, \mathfrak{s})$ to be non-zero, it is necessary that $c_1(\mathfrak{s})$ evaluate to zero on the tori $S^1 \times \gamma_i$, which means that the spin^c is pulled back from the Σ. Furthermore, the remaining pairing $\langle c_1(\mathfrak{s}), [\Sigma]\rangle$ is constrained to lie in the interval from $-(2g-2)$ to $(2g-2)$. (In this simple case, much more can be said, in fact: the solutions of the equations can be parametrized explicitly.)

Corollary 40.1.2 says nothing about the case that Σ is a 2-sphere. There is, however, a vanishing theorem for the Floer homology groups, with a non-trivial local coefficient system, for 3-manifolds containing a 2-sphere in a non-trivial homology class.

Proposition 40.1.3. *Suppose that Y contains an embedded 2-sphere Σ in a non-trivial homology class. Let η be a real 1-cycle such that the Poincaré dual of the class $[\eta]$ has non-zero pairing with $[\Sigma]$. Then*

$$HM_\bullet(Y; \Gamma_\eta) = 0.$$

Proof. Fix a spin^c structure $\mathfrak{s}$ and consider the contribution $HM_\bullet(Y, \mathfrak{s}; \Gamma_\eta)$. As in the proof of the previous proposition, we consider a metric h_T on Y containing an isometric copy of a cylinder $[-T, T] \times \Sigma$, equipping Σ with a round metric of radius 1.

Suppose $HM_\bullet(Y, \mathfrak{s}; \Gamma_\eta)$ is non-zero. Then for all T there must exist solutions to the Seiberg–Witten equations on (Y, h_T). The inequalities which we used previously no longer lead to a contradiction for any finite T. But if we take the limit as T increases to infinity, then a straightforward compactness argument shows that there will be a solution (B, Ψ) to the 3-dimensional Seiberg–Witten equations on the infinite cylinder $\mathbb{R} \times \Sigma$. Furthermore, the spinor Ψ will satisfy a pointwise bound.

It is not hard to see, in fact, that the limiting spinor Ψ must be zero on the cylinder. One way to demonstrate this is to use the differential inequality (4.22), which gives

$$\Delta|\Psi|^2 \leq -|\Psi|^2.$$

Because we know that Ψ is bounded, this differential inequality forces Ψ to be zero, as one can see, for example, by replacing $|\Psi|^2$ by its average value over the 2-spheres, reducing the question to a differential inequality in one variable.

Knowing that Ψ is zero, we see that $F_{B^t} = 0$, so $c_1(\mathfrak{s})$ is zero on the 2-sphere Σ. To finish the argument, we exploit the fact that we have non-trivial local coefficients, and use Theorem 31.1.3. We perturb the equations by a closed, non-exact 2-form ω of period class $c = (4\pi/i)[\omega]$. We arrange that c is non-zero on the class $[\Sigma]$. If ω is smaller than a fixed multiple of the scalar curvature, then the compactness argument can be repeated: the conclusion is that, if solutions of the perturbed equations exist on (Y, h_T) for all T, then there exists a solution (B, Ψ) on the cylinder, with $\Psi = 0$ and

$$F_{B^t} = 4\omega.$$

But there is no such B because $c_1(\mathfrak{s})$ is zero on Σ and $[\omega]$ is non-zero. With the non-exact perturbation, there are therefore no solutions when T is large. From Theorem 31.1.3, we deduce that

$$HM_\bullet(Y, \mathfrak{s}; \Gamma) = 0$$

for any c-complete local system Γ. The local system Γ_η is not c-complete; but if we choose c to be a small multiple of the Poincaré dual of η, then we can use the universal coefficient theorem, as in Subsection 32.3, to deduce the vanishing of $HM_\bullet(Y, \mathfrak{s}; \Gamma_\eta)$ also. □

40.2 Dimension four

The results in the previous subsection can be extended from dimension 3 to dimension 4. We consider a closed 4-manifold X (oriented as always) and an embedded, oriented, connected surface $\Sigma \subset X$. To begin with, we suppose that Σ has *trivial normal bundle* in X. In this situation, we have the following adaptation of Proposition 40.1.1.

Proposition 40.2.1. *Let Σ be a smoothly embedded, oriented connected 2-manifold in X of genus at least 1 with trivial normal bundle and let $\mathfrak{s}$ be a spin^c structure. If*

$$\left|\langle c_1(\mathfrak{s}), [\Sigma]\rangle\right| > 2\,\text{genus}(\Sigma) - 2$$

then there is a Riemannian metric on X for which the unperturbed Seiberg–Witten equations admit no solution belonging to the spin^c structure $\mathfrak{s}$.

Proof. Let $Y \subset X$ be the copy of $S^1 \times \Sigma$ that arises as the boundary of a tubular neighborhood of the surface. Fix metric h_Σ on Σ with constant curvature and area 1 as before, and let h_Y be the product metric on $S^1 \times \Sigma$ in which the S^1 factor has length 1. The total volume of (Y, h_Y) is thus 1, and its scalar curvature is the constant $8\pi(1-g)$. As we previously did in dimension 3, we consider on X a family of metrics h_T containing a cylindrical piece $[-T, T] \times Y$ with the metric $dt^2 + h_Y$, and whose geometry is independent of T on the complement. Much as before, we have

$$\int_X s^2 \, d\mathrm{vol} = T(8\pi)^2(g-1)^2 + C$$

for the metric h_T.

Now let A be a spin^c connection for the spin^c structure $\mathfrak{s}$. On each copy of Σ of the form

$$\{t\} \times \{\theta\} \times \Sigma \subset [-T, T] \times Y,$$

we can apply the Chern–Weil formula as before to obtain

$$\int_{\{t\}\times\{\theta\}\times\Sigma} \left|F_{A^t}|_\Sigma\right|^2 d\mathrm{vol}_\Sigma \geq 4\pi^2 \langle c_1(\mathfrak{s}), [\Sigma]\rangle^2,$$

so

$$\int_X |F_{A^t}|^2 \, d\mathrm{vol}_X \geq 4\pi^2 T \langle c_1(\mathfrak{s}), [\Sigma]\rangle^2,$$

just as in dimension 3.

If (A, Φ) is a solution of the Seiberg–Witten equations, we use the identity (40.1) again. There is now an additional term, which was absent in the 3-dimensional case, namely the term

$$\frac{1}{4}\int_X F_{A^t} \wedge F_{A^t}.$$

But this term is a topological quantity, independent of the metric and of A: it is equal to $-\pi^2 c_1^2(\mathfrak{s})[X]$. The identity therefore yields

$$\int_X |F_{A^t}|^2 \leq -4\pi^2 c_1^2(\mathfrak{s})[X] + \frac{1}{4}\int_X s^2 \, d\mathrm{vol}.$$

The additional topological constant does not affect the remainder of the argument: if a solution (A_T, Φ_T) exists for the metric h_T for all T, then our estimates again yield

$$\left|\langle c_1(\mathfrak{s}), [\Sigma]\rangle\right| \leq 2\,\mathrm{genus}(\Sigma) - 2$$

as in the 3-dimensional case. □

When $b^+(X) > 1$, we can draw a corollary phrased in terms of the invariants $\mathfrak{m}(X, \mathfrak{s})$. We recall from Subsection 1.5 that a class c in $H^2(X;\mathbb{Z})$ is a *basic class* if it arises as $c_1(\mathfrak{s})$ for some spin^c structure $\mathfrak{s}$ with $\mathfrak{m}(X, \mathfrak{s})$ non-zero. We also take the opportunity to point out that, in dimension 4, an oriented embedded surface $\Sigma \subset X$ has trivial normal bundle if and only if the self-intersection number $\Sigma \cdot \Sigma$ is zero. We can then state:

Corollary 40.2.2. *Let X be a closed, oriented smooth 4-manifold with $b^+(X) > 1$. Let Σ be a connected, oriented surface embedded in X, of genus at least 1, with $\Sigma \cdot \Sigma = 0$. Then*

$$2\,\text{genus}(\Sigma) - 2 \geq \big|\langle c, [\Sigma]\rangle\big|$$

for all basic classes $c \in H^2(X;\mathbb{Z})$. □

This result can be extended to surfaces with positive normal bundle if we incorporate a term $\Sigma \cdot \Sigma$ into the inequality. This is an application of the blow-up formula. We state the result:

Theorem 40.2.3. *Let X be a closed, oriented smooth 4-manifold with $b^+(X) > 1$. Let Σ be a connected, oriented surface embedded in X, of genus at least 1, with $\Sigma \cdot \Sigma \geq 0$. Then*

$$2\,\text{genus}(\Sigma) - 2 \geq \big|\langle c, [\Sigma]\rangle\big| + \Sigma \cdot \Sigma$$

for all basic classes $c \in H^2(X;\mathbb{Z})$. □

Proof. We have already proved the result for the case $\Sigma \cdot \Sigma = 0$. If $\Sigma \cdot \Sigma$ is positive, we can proceed by induction on $\Sigma \cdot \Sigma$. Let $\tilde{X}$ denote the connected sum $X \# \bar{\mathbb{CP}}^2$, and let $E \in H_2(\bar{\mathbb{CP}}^2;\mathbb{Z})$ be a generator, represented by a sphere S. Inside $\tilde{X}$, we can form the connected sum of Σ and S, to obtain an embedded surface

$$\begin{aligned}\tilde{\Sigma} &= \Sigma \# S \\ &\subset \tilde{X}.\end{aligned}$$

This surface has the same genus as Σ. It represents the class $[\Sigma] + E$, so

$$\tilde{\Sigma} \cdot \tilde{\Sigma} = \Sigma \cdot \Sigma - 1.$$

If c is a basic class for X, then the blow-up formula, in the form (39.6), tells us that the class

$$\tilde{c} = c \pm \text{P.D.}[E]$$

is a basic class for $\tilde{X}$. Because $\tilde{\Sigma}$ has smaller self-intersection number, our induction hypothesis tells us

$$2\,\text{genus}(\tilde{\Sigma}) - 2 \geq \left|\langle \tilde{c}, [\tilde{\Sigma}]\rangle\right| + \tilde{\Sigma} \cdot \tilde{\Sigma},$$

which becomes

$$2\,\text{genus}(\Sigma) - 2 \geq \left|\langle c, [\Sigma]\rangle \mp 1\right| + \Sigma \cdot \Sigma - 1.$$

Choosing the favorable sign, we obtain

$$2\,\text{genus}(\Sigma) - 2 \geq \left|\langle c, [\Sigma]\rangle\right| + \Sigma \cdot \Sigma$$

as required. □

The inequality for the genus that appears in the theorem above is usually called the *adjunction inequality*. This terminology arises from algebraic geometry, where the *adjunction formula* expresses the genus of a smooth algebraic curve C in an algebraic surface X in terms of its homology class: in algebro-geometric notation, the formula is written

$$2\,\text{genus}(C) - 2 = C \cdot C + K_X \cdot C.$$

Here K_X is the canonical class (in cohomology, minus the first Chern class of X); so we can also write this formula

$$2\,\text{genus}(C) - 2 = C \cdot C - \langle c_1(X), [C]\rangle.$$

Of course, when we are dealing with C^∞ embedded surfaces rather than algebraic curves, the homology class of the surface can no longer determine the genus. The theorem, however, tells us that the equality that holds in the algebro-geometric case becomes an inequality in the C^∞ case. To apply the theorem to the case of an algebraic surface X, we need to know the additional important fact that on any smooth algebraic surface X with $b^+(X) > 1$, the classes $\pm c_1(X)$ are basic classes. (This result is discussed below in Subsection 41.2.) Thus:

Corollary 40.2.4. *Let X be a smooth Kähler surface with $b^+(X) > 1$, and let $C \subset X$ be a smooth, connected algebraic curve with $C \cdot C \geq 0$. Let Σ be a C^∞ embedded surface, representing the same homology class as C. Then the genus of Σ is not less than the genus of C.* □

In fact, the hypotheses $b^+(X) > 1$ and $C \cdot C \geq 0$ can both be removed from this corollary. The case that $b^+(X) = 1$ includes the interesting case of the projective plane, in which case the corollary is traditionally known as the Thom conjecture: it was proved in [59] and [82]. The case that $C \cdot C < 0$ needs additional techniques, and was proved by Ozsváth and Szabó in [88]. All these results extend also to symplectic 4-manifolds X, when we replace the notion of a smooth algebraic curve by a *symplectic submanifold* $C \subset X$, i.e. a smooth 2-dimensional submanifold on which the symplectic form is everywhere positive.

41 Foliations and non-vanishing theorems

41.1 Taut foliations

Let Y be a smooth 3-manifold, and let ξ be a field of 2-planes. The 2-plane field defines a *foliation* $\mathcal{F}$ of Y with smooth leaves if, for each $y \in Y$, there is a 2-manifold Σ and a smooth embedding $\Sigma \to Y$ whose image passes through y and is everywhere tangent to ξ. In this case, for each y, there is a maximal connected Σ, whose image is the *leaf* through y. The leaves of a foliation $\mathcal{F}$ need not be closed, but they partition Y.

The definition of a foliation just given does not require that the 2-plane field ξ be smooth, though the smoothness of the leaves imposes some restriction. We do however require continuity of ξ for our purposes: thus we will be working with foliations with *smooth leaves and C^0 tangent planes*. Locally, ξ can always be defined as the kernel of a non-vanishing 1-form α. This can be done globally if ξ is *coorientable*. An example of a non-smooth foliation of $\mathbb{R}^3$, with smooth leaves, is provided by the continuous 1-form

$$\alpha = \begin{cases} dz + z dx, & z > 0 \\ dz, & z \leq 0. \end{cases}$$

In the case that α is C^1, the corresponding 2-plane field defines a foliation if and only if the Frobenius integrability condition holds:

$$\alpha \wedge d\alpha = 0.$$

Our foliations will always be cooriented: equivalently, since Y will be oriented, we will have a chosen orientation for the 2-plane field.

Suppose now that Y is closed. A foliation $\mathcal{F}$ (oriented, with smooth leaves and C^0 tangent planes) is said to be *taut* if there is a smooth closed 2-form ω whose restriction to the leaves is everywhere positive. This condition implies, in particular, that there is no closed leaf which separates Y, as one sees from Stokes' theorem. A more topological condition, which implies the existence of such a 2-form, is that there is a smooth closed curve δ in Y which is transverse to the leaves and meets every leaf. Given such a δ, one can construct a 2-form ω positive on the leaves as follows. Start by constructing a form ω_δ, supported in the tubular neighborhood, which represents the Poincaré dual class and is non-negative on the leaves. Then observe that given any y in Y, we can construct a new δ' passing through y, by "pushing δ along the leaves". Using compactness, we can construct a collection of forms ω_{δ_i} whose supports cover Y and whose sum will be a form ω which is everywhere positive on the leaves. At least when $\mathcal{F}$ is smooth, one can also go in the other direction, as stated in the following lemma from [106, p. 244, Remark].

Lemma 41.1.1. *If $\mathcal{F}$ is a smooth, oriented foliation of a closed oriented 3-manifold Y, and if there is a closed 2-form ω that is positive on the leaves, then there is a closed curve δ that is everywhere transverse to the leaves and meets every leaf.*

Proof. It will be sufficient to show that, given any leaf L_0, there is a curve $\delta : [a, b] \to Y$, not necessarily closed, that is transverse to the leaves and meets L_0 twice: one can then form a single closed curve by pushing along the leaves and using a compactness argument.

The existence of the closed form means that, for any volume form on Y, there is a nowhere-zero, volume-preserving vector field on Y which is transverse to the leaves. We write ϕ_t for the neighborhood flow. Given any leaf L_0, let $\Omega_0 \subset L_0$ be an open set and ϵ a positive real, chosen so that the flow ϕ_t for $t \in (-\epsilon, \epsilon)$ embeds $(-\epsilon, \epsilon) \times \Omega_0$ as an open set Ω in Y. By the Poincaré recurrence theorem, there is a point x in Ω and a $t_1 > 2\epsilon$ such that $\phi_t(x)$ also lies in Ω. The path $t \mapsto \phi_t(x)$ then meets L_0 at least twice: once each within distance ϵ of $t = 0$ and $t = t_1$. □

We will not make use of the above lemma, as we will always take the existence of the closed 2-form as our characterization of tautness. (The advantage of the alternative formulation – using a curve transverse to the leaves – is that it makes sense even when the foliation has no smoothness. If the foliation has smooth leaves and C^0 tangent planes, it is also easy to see that a C^0 curve transverse to the leaves can be approximated by a smooth one.)

A general existence theorem for taut foliations of 3-manifolds with non-zero Betti number was proved by Gabai in [44]. Before stating the result, we recall that the Thurston norm of a closed orientable surface Σ with connected components $\Sigma_1, \ldots, \Sigma_n$ is defined as

$$|\Sigma| = \sum_{i=1}^{n} \max\{0, 2g(\Sigma_i) - 2\}, \tag{41.1}$$

where g denotes the genus. Every integral 2-dimensional homology class in a closed 3-manifold is represented by a closed embedded surface, and therefore also by a closed embedded surface of smallest possible norm. Indeed given a class h we can find a representative Σ satisfying the following three conditions:

(i) Σ achieves the smallest possible Thurston norm among surfaces representing the class h;
(ii) no component of Σ is a sphere;
(iii) each genus-1 component of Σ represents a non-trivial homology class (i.e. does not separate the 3-manifold.)

Remarks. To achieve (ii), one can add a handle to any spherical component, thus turning it into a torus, without increasing the norm. In (iii), only the genus-1 components need be mentioned, because the higher-genus components are automatically non-trivial in homology by Item (i).

The following is a somewhat restricted statement of Gabai's theorem. Recall that a closed orientable 3-manifold Y is *irreducible* if every embedded 2-sphere bounds a ball. This condition means that Y is not a connected sum and is not $S^1 \times S^2$.

Theorem 41.1.2 (Gabai [44]). *Let Y be a closed, oriented, irreducible 3-manifold and Σ a closed, oriented embedded surface representing a non-zero homology class. Suppose that Σ satisfies the three conditions above. Then there exists a taut, oriented foliation $\mathcal{F}$ of Y having Σ as a union of closed leaves. This foliation can be taken to be C^∞, except possibly along genus-1 components of Σ.* □

Remarks. (i) The fact that $\mathcal{F}$ is smooth away from the genus-1 closed leaves allows us to arrange also that $\mathcal{F}$ has smooth leaves and C^0 tangent planes. This is because the structure of $\mathcal{F}$ in the neighborhood of a genus-1 closed leaf is determined by a pair (f, g) of commuting germs of self-homeomorphisms of $(\mathbb{R}, 0)$, both of which are smooth away from 0. Given (f, g), it is straightforward to construct a corresponding foliation of a neighborhood of T^2 in such a way that the tangent planes are continuous.

(ii) The condition that no component of Σ separates Y can always be achieved by discarding any components that do. This will not increase the norm. Furthermore, when this condition is satisfied and Y is irreducible, no component of Σ can be a sphere (because all spheres bound balls).

Corollary 41.1.3. *If Y is a closed, oriented, irreducible 3-manifold with non-zero Betti number, then Y admits a taut foliation with smooth leaves and C^0 tangent planes.* □

To give a little context for Gabai's theorem, we point out that, without the tautness condition, there is a quite general existence theorem for smooth foliations of manifolds of any dimension [126, 119, 120]. In particular, any 2-plane field on a 3-manifold is homotopic to one that defines a foliation. Taut foliations of 3-manifolds, by contrast, are constrained: on a given 3-manifold, only finitely many homotopy classes of 2-plane fields are realized [62].

We recall also the closely related concept of an (oriented) *contact structure*. On an oriented 3-manifold Y, an oriented 2-plane field ξ is a contact structure compatible with the orientation of Y if it is defined as the kernel of a 1-form α which satisfies

$$\alpha \wedge d\alpha > 0.$$

Thus a contact structure is a 2-plane field which is "maximally non-integrable". The sign here makes reference to the given orientation of Y. Note that we can change the orientation of ξ, or the sign of α, without consequence. As with foliations, there is a general existence theorem for contact structures homotopic to any given 2-plane field [69, 72].

41.2 Taubes' theorem

The non-vanishing theorem whose proof we will shortly give is based on a rather simpler result of Taubes which tells us that the monopole invariants of symplectic 4-manifolds are not zero.

Recall from Subsection 28.1 that an almost complex structure J on a 4-manifold X determines a spin^c structure $\mathfrak{s}_J$. A symplectic structure ω determines a preferred homotopy class of almost complex structures and so also a preferred spin^c structure $\mathfrak{s}_\omega$. When the 4-manifold is X closed, the formal dimension of the Seiberg–Witten moduli space for the spin^c structure $\mathfrak{s}_\omega$ is zero, so there is a well-defined monopole invariant $\mathfrak{m}(X, \mathfrak{s}_\omega)$, as long as $b^+(X) > 1$. With this notation in place we can state a version of Taubes' theorem.

Theorem 41.2.1 (Taubes [114]). *Let X be a smooth closed 4-manifold with $b^+(X) > 1$ admitting a symplectic form ω. Then*

$$\mathfrak{m}(X, \mathfrak{s}_\omega) = \pm 1$$

and

$$0 \geq \langle c_1(\mathfrak{s}_\omega) \cup [\omega], [X] \rangle.$$

Furthermore for any other spin^c structure $\mathfrak{s}$ with $\mathfrak{m}(X, \mathfrak{s}) \neq 0$, we have

$$\langle c_1(\mathfrak{s}) \cup [\omega], [X] \rangle \geq \langle c_1(\mathfrak{s}_\omega) \cup [\omega], [X] \rangle,$$

with equality if and only if $\mathfrak{s} = \mathfrak{s}_\omega$. □

Remarks. By $c_1(\mathfrak{s})$ we mean as usual the first Chern class of either of the spin bundles $S^\pm$. In the case of $\mathfrak{s}_\omega$, this coincides with the first Chern class of the complex vector bundle (TX, J), for a compatible complex structure J.

One can expand on the statement of the theorem by using the fact that

$$\mathfrak{m}(X, \mathfrak{s}) = \pm \mathfrak{m}(X, \bar{\mathfrak{s}})$$

when $\mathfrak{s}$ and $\bar{\mathfrak{s}}$ are complex conjugate spin^c structures. We obtain

$$\Big| \langle c_1(\mathfrak{s}) \cup [\omega], [X] \rangle \Big| \leq \Big| \langle c_1(\mathfrak{s}_\omega) \cup [\omega], [X] \rangle \Big|,$$

whenever $\mathfrak{m}(X, \mathfrak{s})$ is non-zero, with equality if and only if $\mathfrak{s}$ is $\mathfrak{s}_\omega$ or $\bar{\mathfrak{s}}_\omega$.

In the language of "basic classes" (see Subsection 1.5), the non-vanishing of $\mathfrak{m}(X, \mathfrak{s}_\omega)$ and $\mathfrak{m}(X, \bar{\mathfrak{s}}_\omega)$ means that both $c_1(X)$ and $-c_1(X)$ are basic classes for a symplectic manifold with $b^+(X) > 1$. This fact, and also the inequality of Taubes' theorem, can be seen as generalizations of previously known results for Kähler manifolds, due to Witten [125].

41.3 Embedding three-manifolds in symplectic four-manifolds

It is interesting to try to characterize those 3-manifolds which embed in closed symplectic 4-manifolds, or even into closed Kähler 4-manifolds. We can ask the more specific question: given a 3-manifold with a closed nowhere-zero 2-form ω, when does the 3-manifold embed in a symplectic 4-manifold (X, ω_X) in such a way that ω is the restriction of ω_X? (Every 2-dimensional cohomology class on any 3-manifold can be represented by a nowhere-vanishing form; and one may even specify the homotopy class: this follows from the existence theorem for contact structures, which supplies us with nowhere-vanishing exact 2-forms.) In this subsection, by pulling together many different results, we will prove a fairly general result for the symplectic case:

Theorem 41.3.1. *Let Y be a closed oriented 3-manifold. Suppose that Y has a taut foliation $\mathcal{F}$ which either* (a) *is smooth, or* (b) *has smooth leaves and C^0 tangent planes. In Case* (b), *suppose also that $\mathcal{F}$ is smooth outside the closed leaves and has holonomy. Let ω be a closed 2-form on Y which is positive on the leaves of $\mathcal{F}$. Then there is a closed symplectic 4-manifold (X, ω_X) containing Y as a separating submanifold, such that the restriction of ω_X to Y is ω.*

Furthermore, if Y is not $S^1 \times S^2$, then we can arrange that the map $H^2(X;\mathbb{Z}) \to H^2(Y;\mathbb{Z})$ is surjective, and that the two components X_1, X_2 into which Y divides X both have $b^+ > 0$.

Remark. The foliations which arise from Gabai's theorem 41.1.2 satisfy either (a) or (b). In Case (b), the condition that the foliation has holonomy excludes the case that $\mathcal{F}$ is a C^0 fibration with smooth fibers and C^0 tangent planes. In this latter case, there is a new smooth structure on Y which makes the fibration smooth (for the fibration is determined by an element of the mapping class group of the fiber, and the mapping class group is insensitive to the choice of C^∞ versus C^0). For the new smooth structure $\tilde{Y}$, there will be a closed 2-form $\tilde{\omega}$ positive on the leaves, but $(\tilde{Y}, \tilde{\omega})$ is not the same as the original (Y, ω), considered as a smooth manifold equipped with a 2-form. The theorem provides an embedding of $(\tilde{Y}, \tilde{\omega})$, not of (Y, ω).

Although the statement has been repackaged here, the above theorem is essentially due to Eliashberg [25] and independently Etnyre [28], drawing on earlier results of Eliashberg and Thurston [26], Giroux [48], and Etnyre and Honda [29].

We shall explain some of the steps in the proof of the theorem. We immediately dispose of the exceptional case of $S^1 \times S^2$, on which both the foliation and the form ω can be taken to be standard and which can be embedded in the symplectic manifold $S^2 \times S^2$. We turn to the construction for the case of a manifold other than $S^1 \times S^2$.

Given a closed, oriented 3-manifold Y carrying a smooth non-vanishing closed 2-form ω, we can explicitly construct a symplectic form on the manifold with boundary, $[-1, 1] \times Y$. To see this let α be a 1-form such that $\omega \wedge \alpha \neq 0$. Then for ϵ sufficiently small the 2-form

$$\Omega = \omega + \epsilon d(t\alpha) \tag{41.2}$$

is symplectic on $[-1, 1] \times Y$, and its restriction as a 2-form to $\{0\} \times Y$ is ω. Thus the question of which 3-manifolds carrying closed nowhere-zero 2-forms embed into closed symplectic 4-manifolds fits into the more general question of which symplectic 4-manifolds with boundary embed into closed symplectic 4-manifolds.

In complex geometry there is the notion of pseudo-convexity for a bounded domain with smooth boundary in $\mathbb{C}^n$. In the symplectic case, one can also define notions of "convexity" for a boundary; but there is more than one possible formulation. We say that a boundary component Y of a symplectic $2n$-manifold (X, ω) is *weakly convex* if we can equip Y with a contact structure ξ, compatible with the boundary orientation of Y, such that the restriction of ω to the $(2n-2)$-planes ξ_y ($y \in Y$) is everywhere non-degenerate (or equivalently, the 1-dimensional null-space of $\omega|_Y$ is everywhere transverse to ξ). We say that Y is *strongly convex* (sometimes called *contact type*) if, in some collar neighborhood of Y, there is a 1-form α satisfying $d\alpha = \omega$ whose restriction to Y defines a contact structure compatible with the orientation.

Note that the weakly convex condition does not require even that ω be exact near the boundary, unlike the strong condition. When the boundary is strongly convex and a primitive α is given, one can define a vector field V near the boundary by the condition

$$\iota_V \omega = \alpha$$

and from the Cartan formula for the Lie derivative we then have

$$\mathcal{L}_V \omega = \omega.$$

Thus the flow ϕ_t generated by V has the property that $\phi_t^*(\omega) = e^t \omega$. We can therefore choose a collar neighborhood $(-1, 0] \times Y$ of the boundary on which ω can be expressed as

$$\omega = d(e^t \alpha).$$

In the case of a domain $X \subset \mathbb{C}^n$ with smooth, pseudo-convex boundary Y, one can choose a pluri-subharmonic function ϕ having Y as a level set, in which case the symplectic form

$$\omega_\phi = \frac{i}{4}\partial\bar{\partial}\phi$$

makes Y strongly convex, with the primitive given by

$$\alpha_\phi = \frac{i}{8}(\bar{\partial} - \partial)\phi.$$

For the differential form Ω on $[-1,1]\times Y$ defined by (41.2), there is no reason in general to expect that the boundary components $\{-1\} \times Y$ and $\{1\} \times Y$ are even weakly convex (and they cannot be strongly convex if ω is not exact). However, one can say more if ω arises from a taut foliation:

Theorem 41.3.2 (Eliashberg–Thurston [26]). *Let Y be a closed oriented 3-manifold other than $S^1 \times S^2$. Suppose that Y has a foliation $\mathcal{F}$ which either* (a) *is smooth, or* (b) *has smooth leaves and C^0 tangent planes. In Case* (b), *suppose also that $\mathcal{F}$ is smooth outside the closed leaves and has holonomy. Let ω be a smooth closed 2-form on Y which is positive on the leaves of $\mathcal{F}$. Then there exist a smooth 1-form α on Y, with $\alpha \wedge \omega > 0$, and an $\epsilon_0 > 0$ such that for all $\epsilon < \epsilon_0$:*

- *the 2-form Ω on $[-1,1] \times Y$ defined by* (41.2) *is symplectic; and*
- *the boundary components $\{-1\} \times Y$ and $\{1\} \times Y$ are both weakly convex.*

□

The main point of the theorem above is the construction of the contact structures ξ_- and ξ_+ on Y which establish the weak convexity of $\{-1\} \times Y$ and $\{1\} \times Y$. Eliashberg and Thurston show that, given any foliation $\mathcal{F}$ with tangent plane field ξ_0 satisfying the given hypotheses, one can construct smooth contact structures ξ_+ and ξ_- which are C^0-close to ξ_0 and which are compatible with the two different orientations of Y. This is proved first in [26, Theorem 2.4.1] for any *smooth* foliation that is not the trivial foliation of $S^1 \times S^2$ by 2-spheres. The case of a non-smooth foliation with holonomy and C^0 tangent planes is Proposition 2.9.4 of [26]. We can, in particular, assume that $\xi_\pm$ are close enough to ξ_0 that ω is positive on both these 2-plane fields. For α, we take a smooth approximation to the continuous 1-form α_0 that defines ξ_0, so chosen that $\alpha \wedge \omega$ is positive. We have already observed that Ω is a symplectic form when ϵ is small; and by the same token, it is positive on $\xi_\pm$ when ϵ is small, because ω is. This gives us the weak convexity of the boundary components.

We have now seen that, given Y, $\mathcal{F}$ and ω as in the statement of Theorem 41.3.1, we can embed Y in a symplectic manifold with boundary, $([-1,1] \times Y, \Omega)$, so that the restriction of Ω to $\{0\} \times Y$ is ω, and so that the boundary components are weakly convex. The remaining big step in the proof of Theorem 41.3.1 is the following result:

Theorem 41.3.3. *Let (W, ω_W) be a compact symplectic 4-manifold with weakly convex boundary. Then there is a closed symplectic 4-manifold (X, ω_X) containing W as a submanifold, such that the restriction of ω_X to W is ω_W.* □

The above statement is from [25], and is proved independently in [28]. Both proofs depend on Giroux's work [48] on open-book decompositions of contact 3-manifolds. The case that (W, ω_W) is a regular sub-level set of a pluri-subharmonic function on a Stein manifold was treated earlier by Lisca and Matić, and with a different construction by Akbulut and Ozbagci [3]. The case of a strongly convex boundary is proved in [29].

The constructions in [25] and [28] which supply the proof of Theorem 41.3.3 allow considerable freedom in construction of X. In particular, we can take it that the connected components of $X \setminus W$ correspond one-to-one with the components of ∂W, and that each component has large b^+. One can also arrange that $H_1(X)$ is zero, and that the restriction map $H^2(X) \to H^2(W)$ is surjective. By combining this theorem with Theorem 41.3.2, one obtains Theorem 41.3.1: the given 3-manifold Y, with its 2-form ω, is embedded first in the cylinder $W = [-1,1] \times Y$ with weakly convex boundary using the first theorem, and then in a closed symplectic manifold using the second theorem.

41.4 A non-vanishing theorem for Floer homology

We are now ready to state and prove a general non-vanishing theorem for Floer homology groups of 3-manifolds with taut foliations. Recall that $\widecheck{HM}_\bullet(Y)$ and $\widehat{HM}_\bullet(Y)$ are always non-zero if we use ordinary coefficients: indeed, they have infinite rank. To be non-trivial, a non-vanishing theorem must assert the non-vanishing of the *reduced* Floer homology, i.e. the image of the map

$$j_* : \widecheck{HM}_\bullet(Y) \to \widehat{HM}_\bullet(Y).$$

Our theorem involves not just the trivial coefficient system, but also the local coefficient system Γ_η defined on page 445 (though other variations are possible).

Let us also recall that if we fix a grading element

$$k \in \mathbb{J}(Y) = \bigcup_{\mathfrak{s}} \mathbb{J}(Y, \mathfrak{s})$$

then the map

$$j_* : \widetilde{HM}_k(Y; \Gamma_\eta) \to \widehat{HM}_k(Y; \Gamma_\eta)$$

is automatically an isomorphism in either of two cases: the first case is when the spin^c structure $\mathfrak{s}$ corresponding to k has non-torsion first Chern class, for in this case there are no reducible solutions; and the second case is when the homology class $[\eta] \in H_1(Y; \mathbb{R})$ is non-zero, by Proposition 3.9.1. In any event, we always write

$$HM_k(Y; \Gamma_\eta) = \operatorname{im}\Big(j_* : \widetilde{HM}_k(Y; \Gamma_\eta) \to \widehat{HM}_k(Y; \Gamma_\eta)\Big),$$

with the understanding that all three groups in this formula are isomorphic if either of the above two conditions holds.

Our non-vanishing theorem, then, concerns the group $HM_\bullet(Y; \Gamma_\eta)$. We should note, however, that this group *is* zero in some important cases. For example, we know that $HM_\bullet(Y; \Gamma_\eta)$ is zero for all η if Y admits a metric of positive scalar curvature (see Subsection 36.1). This applies to S^3 and $S^1 \times S^2$, whose groups we earlier described in detail. When $\eta = 0$, we also know that $HM_\bullet(Y; \mathbb{R})$ is zero in the case that Y is a 3-torus. These examples limit the possible generality of a non-vanishing theorem. The result which we state asserts the non-vanishing of $HM_k(Y; \Gamma_\eta)$ for specific k and η arising from a taut foliation of Y.

Theorem 41.4.1. *Let Y be a closed, oriented* 3*-manifold other than $S^1 \times S^2$, carrying a taut foliation $\mathcal{F}$ which either* (a) *is smooth or* (b) *has smooth leaves and C^0 tangent planes. In Case* (b)*, suppose also that $\mathcal{F}$ is smooth outside the closed leaves and has holonomy. Let ω be a closed* 2*-form on Y which is positive on the leaves. Then*

$$HM_k(Y; \Gamma_\eta) \neq 0$$

when $[\eta] = \mathrm{P.D.}[\omega]$ and $k \in \mathbb{J}(Y)$ is the grading element corresponding to the homotopy class of the 2*-plane field $[\xi]$ given by the tangent planes to $\mathcal{F}$ (as in Subsection* 28.2*).*

Because the theorem above only states that the reduced homology group is non-zero for certain η, we turn briefly to the question of how this group might vary with η. In line with our previous remarks, we observe that the conclusion of the theorem is equivalent to the non-vanishing of the group $\widetilde{HM}_k(Y;\Gamma_\eta)$ if either:

- the Euler class $e(\mathcal{F})$ (by which we mean the Euler class of the tangent plane-field ξ on Y) is not a torsion class; or
- the homology class $[\eta]$ is non-zero in $H_1(Y;\mathbb{R})$.

(For the first case, we recall that $e(\mathcal{F})$ is also the first Chern class of the corresponding spin^c structure.) Unlike $HM_k(Y;\Gamma_\eta)$ in general, the group $\widetilde{HM}_k(Y;\Gamma_\eta)$ is simply the k'th homology of a complex $(C_*, \check{\partial}_*)$ (graded by $\mathbb{J}(Y)$), in which only the differential $\check{\partial}_*$ depends on η: the vector space C_* does not change. Because the Floer groups depend only on the homology class $[\eta]$ (up to isomorphism), we may choose a real vector space $\mathcal{H}_1$ inside the space of real 1-cycles which represents the homology $H_1(Y;\mathbb{R})$, and we can restrict η to lie in $\mathcal{H}_1$. The matrix entries of $\check{\partial}_\eta$ are then analytic functions of η on the finite-dimensional vector space $\mathcal{H}_1$. The complex $\check{C}_*$ is finite-dimensional in each grading; so the semi-continuity of homology groups tells us that the rank of the homology group $H_k(\check{C}_*, \check{\partial}_\eta)$ achieves its minimum on an open dense subset of $\mathcal{H}_1$ (the complement of an analytic set).

The condition that a closed 2-form be positive on the leaves is an open condition; so Theorem 41.4.1 tells us that $HM_k(Y;\Gamma_\eta)$, and hence also $\widetilde{HM}_k(Y;\Gamma_\eta)$, is non-zero for all η in a non-empty open subset of $\mathcal{H}_1$. The semi-continuity then gives:

Corollary 41.4.2. *Let Y and $\mathcal{F}$ be as in Theorem* 41.4.1, *and let $k \in \mathbb{J}(Y)$ again be the grading element corresponding to the tangent* 2*-plane field. Then the group $\widetilde{HM}_k(Y;\Gamma_\eta)$ is non-zero for all cycles η.*

In particular, if either $e(\mathcal{F})$ is non-torsion or $[\eta]$ is non-zero, then the reduced group $HM_k(Y;\Gamma_\eta)$ is non-zero, for it is isomorphic to $\widetilde{HM}_k(Y;\Gamma_\eta)$ in these cases. □

Remark. There is an important case of Theorem 41.4.1 which is not covered by the corollary above, namely the case that $e(\mathcal{F})$ is zero and there is an exact 2-form ω positive on the leaves. In this case, the theorem tells us that $HM_k(Y;\mathbb{R})$ is non-zero. The above corollary makes no assertion about $HM_k(Y;\mathbb{R})$ in this case, because j_* is not an isomorphism.

Gabai's theorem supplies us with taut foliations, so we can draw an important corollary:

Corollary 41.4.3. *Suppose that Y is a closed, oriented irreducible 3-manifold with non-zero Betti number. Then for all real 1-cycles η with non-zero cohomology class, the group $HM_\bullet(Y;\Gamma_\eta)$ is non-zero. If we suppose in addition that there is at least one class in $H_2(Y;\mathbb{Z})$ that cannot be represented by a union of embedded tori, then the reduced group*

$$HM_\bullet(Y) = \mathrm{im}(j_*)$$

is non-zero also for the trivial coefficient system $\mathbb{R}$.

Proof. Let Σ be a norm-minimizing representative of any non-zero class in $H_2(Y;\mathbb{Z})$, satisfying the three conditions on page 743. According to Theorem 41.1.2, there is a taut foliation having Σ as a union of closed leaves. As stated in the remark following Theorem 41.3.1, this foliation satisfies the hypotheses of Theorem 41.4.1, so we have non-vanishing when $[\eta]$ is non-zero, by the previous corollary.

If there is a class that cannot be represented by tori, then we may arrange that some connected component $\Sigma' \subset \Sigma$ is a surface of negative Euler number, and it follows that $e(\mathcal{F})$ is not a torsion class, because Σ' is a leaf of the foliation and $\langle e(\mathcal{F}), [\Sigma'] \rangle$ is the Euler number. □

Using a refinement of the existence theorem for foliations, Gabai proves in [45] that if K is a non-trivial knot in the 3-sphere, then the 3-manifold Y obtained by longitudinal surgery (0-surgery) on K is irreducible, and in particular is not $S^1 \times S^2$. Furthermore, the genus of a norm-minimizing representative for the generator of $H_2(Y;\mathbb{Z})$ is the same as the classical genus of the knot (the smallest possible genus for a Seifert surface). We therefore have:

Corollary 41.4.4. *Suppose that Y is obtained by longitudinal surgery on a non-trivial knot K in S^3. Then for all real 1-cycles η with non-zero cohomology class, the group $HM_\bullet(Y;\Gamma_\eta)$ is non-zero. If the genus of the knot is greater than 1, then the reduced group $HM_\bullet(Y)$ is non-zero also for the trivial coefficient system.* □

Note that the trivial knot (the "unknot") is a genuine exception. The manifold obtained by 0-surgery on the unknot is just $S^1 \times S^2$, and the reduced group $HM_\bullet(S^1 \times S^2;\Gamma_\eta)$ (the image of j_*) is zero, for all η, as stated in Propositions 3.10.3 and 3.10.4. We may therefore amplify the preceding corollary:

Corollary 41.4.5. *Let Y be obtained by longitudinal surgery on a knot K in S^3, and let η represent a generator of homology in Y. Then $HM_\bullet(Y;\Gamma_\eta)$ is zero if and only if K is the unknot.* □

In this sense, via Gabai's theorem, Floer homology detects the unknot, distinguishing it from other knots. We shall see in the following subsection that we can strengthen this statement: Floer homology detects also the genus of a knot.

We turn to the proof of Theorem 41.4.1. Let Y and η be as in the statement of the theorem. According to Theorem 41.3.1, we can embed Y as a submanifold in a symplectic manifold (X, ω_X), separating X into two pieces X_1, X_2 both with $b^+ > 1$. We can also ensure that the restriction of ω_X to Y is ω, and that the restriction of $c_1(X, \omega_X)$ to Y is $e(\xi)$. (Here, $c_1(X, \omega_X)$ refers as usual to the first Chern class of an almost-complex structure compatible with the symplectic form ω_X.)

Let $\mathfrak{s}_\omega$ denote the canonical spin^c structure on X corresponding to the symplectic form ω_X, and let

$$h = \text{P.D.}[\omega_X] \in H_2(X;\mathbb{R}).$$

The restriction of the dual of h to Y is $[\eta]$, so we can represent h by a 2-cycle ν meeting Y in the 1-cycle

$$\eta = \partial\nu_1 = -\partial\nu_2.$$

Recall from (3.27) the pairing formula

$$\mathfrak{m}(X, h) = \langle \psi_{(X_1,\nu_1)}, \psi_{(X_2,\nu_2)} \rangle_{\omega_\mu}, \tag{41.3}$$

where

$$\psi_{(X_1,\nu_1)} \in HM_\bullet(Y;\Gamma_\eta)$$
$$\psi_{(X_2,\nu_2)} \in HM_\bullet(-Y;\Gamma_\eta).$$

(The applicability of this pairing formula depends on the fact that $b^+(X_1)$ and $b^+(X_2)$ are both positive.)

If we knew that the invariant $\mathfrak{m}(X, h)$ were non-zero, then this pairing formula would tell us at once that $HM_\bullet(Y;\Gamma_\eta)$ is non-zero. Although we shall not establish that $\mathfrak{m}(X, h)$ is non-zero, we have the following result:

Lemma 41.4.6. *There exists $\lambda_0 > 0$ such that $\mathfrak{m}(X, \lambda h) \neq 0$ for all λ with $|\lambda| > \lambda_0$.*

Proof. We have

$$\mathfrak{m}(X, \lambda h) = \sum_{\mathfrak{s}} \mathfrak{m}(X, \mathfrak{s}) \exp\bigl(\lambda \langle c_1(\mathfrak{s}), h \rangle\bigr),$$

from the definition. In this sum, $\mathfrak{s}$ runs through all isomorphism classes of spin^{c}structures, but the non-zero contributions come from the finitely many spin^c structures with $\mathfrak{m}(X, \mathfrak{s}) \neq 0$. According to Theorem 41.2.1, there is a non-zero contribution from $\mathfrak{s}_\omega$, and therefore also from the conjugate spin^c structure, $\bar{\mathfrak{s}}_\omega$. In addition, that theorem tells us that all other spin^c structures $\mathfrak{s}$ with non-zero invariant satisfy

$$\bigl|\langle c_1(\mathfrak{s}), h \rangle\bigr| < \bigl|\langle c_1(\mathfrak{s}_\omega), h \rangle\bigr|.$$

It also tells us that $\mathfrak{s}_\omega$ and $\bar{\mathfrak{s}}_\omega$ must be equal if the right-hand side is zero. As a function of λ then, $\mathfrak{m}(X, \lambda h)$ is a sum of exponentials with a non-zero leading term: the sum is dominated by the term arising from $\bar{\mathfrak{s}}_\omega$ or $\mathfrak{s}_\omega$ as $\lambda \to +\infty$ or $-\infty$ respectively. □

The pairing formula (41.3) and the lemma above together tell us that

$$HM_\bullet(Y; \Gamma_{\lambda\eta}) \neq 0$$

for all sufficiently large λ. If the class $[\eta]$ is non-zero, then $HM_\bullet = \widetilde{HM}_\bullet$ here, and the semi-continuity argument tells us that $HM_\bullet(Y; \Gamma_\eta)$ is non-zero (the case $\lambda = 1$). On the other hand, if $[\eta]$ is zero, then the factor of λ is irrelevant, and we have the same conclusion.

We have now almost proved the theorem, but our conclusion so far is only that $HM_\bullet(Y; \Gamma_\eta)$ is non-zero, without yet knowing in which degree it might be non-zero. To deal with this point, consider the following modified version of $\mathfrak{m}$: we set

$$\mathfrak{m}'(X, h) = \sum_{\mathfrak{s}}{}' \, \mathfrak{m}(X, \mathfrak{s}) \exp\langle c_1(\mathfrak{s}), h \rangle,$$

where the sum now runs *only* over isomorphism classes of spin^c structures $\mathfrak{s}$ satisfying

$$\mathfrak{s}|_{X_i} = \mathfrak{s}_\omega|_{X_i}, \quad i = 1, 2.$$

For this modified invariant, we can give a pairing formula of the same shape as (41.3). To do this, recall that the definition of $\psi_{(X_i,\nu_i)}$ expresses it as a sum over all isomorphism classes of spin^c structures on X_i; we then define

$$\psi'_{(X_1,\nu_1)} \in HM_k(Y;\Gamma_\eta)$$
$$\psi'_{(X_2,\nu_2)} \in HM_k(-Y;\Gamma_\eta)$$

to be the contribution to $\psi_{(X_1,\nu_1)}$ and $\psi_{(X_2,\nu_2)}$ arising from the spin^c structures $\mathfrak{s}_1$ and $\mathfrak{s}_2$ respectively. In both cases, because ψ' is defined in terms of zero-dimensional moduli spaces, the grading $k \in \mathbb{J}(Y, \mathfrak{s}_\omega|_Y)$ is completely determined by knowing the topology of one of the X_i and the corresponding spin^c structure $\mathfrak{s}_i$. According to our canonical identification of $\mathbb{J}(Y)$ with the space of homotopy classes of 2-plane fields $\pi_0(\Xi)$ described in Subsection 28.2, the grading element k is the one corresponding to the 2-plane field determined by the almost-complex structure, i.e. the 2-plane field ξ defined by the foliation $\mathcal{F}$. The pairing formula then reads:

$$\mathfrak{m}'(X, h) = \langle \psi'_{(X_1,\nu_1)}, \psi'_{(X_2,\nu_2)} \rangle_{\omega_\mu}. \tag{41.4}$$

The proof of the lemma above shows equally that $\mathfrak{m}'(X, \lambda h)$ is non-zero for large λ, so we conclude in the same way that $HM_k(Y;\Gamma_\eta)$ is non-zero as desired. This completes the proof of Theorem 41.4.1. □

41.5 The unit ball of the dual Thurston norm

Recall the definition of the norm (41.1) for an embedded surface $\Sigma \subset Y$. From this notion, Thurston [121] defined a semi-norm on the homology group $H_2(Y;\mathbb{R})$, also called the Thurston norm. For an *integer* class h in $H_2(Y;\mathbb{Z})$, the Thurston norm is defined as

$$|h| = \inf\{\, |\Sigma| \mid \Sigma \subset Y, [\Sigma] = h \,\}.$$

It is shown in [121] that this function on the integer classes extends to a semi-norm on the real homology group. This semi-norm fails to be a norm if there are non-zero homology classes represented by embedded tori in Y, for the Thurston norm of these classes is zero.

There is a dual norm on the second *cohomology*, though because there may be homology classes of norm zero, we must allow the dual norm to take the value ∞, or we must restrict its domain. We adopt the latter policy, by writing $\mathcal{T} \subset H_2(Y;\mathbb{R})$ to be the span of the classes of Thurston norm zero and defining a dual norm on the annihilator $\mathrm{Ann}(\mathcal{T}) \subset H^2(Y;\mathbb{R})$ by the usual recipe: for $\alpha \in \mathrm{Ann}(\mathcal{T})$,

$$|\alpha| = \inf\{\, C \geq 0 \mid \alpha(h) \leq C|h|,\ \forall h \in H_2(Y;\mathbb{R})\,\}.$$

In [121], it is shown that if $\mathcal{F}$ is a taut foliation on Y, then the Euler class $e(\mathcal{F})$ has dual Thurston norm at most 1 (and in particular, this class annihilates $\mathcal{T}$). Thus the convex hull of the classes $e(\mathcal{F})$, as $\mathcal{F}$ runs through taut foliations, is contained in the unit ball of the dual norm, a compact subset of $H^2(Y;\mathbb{R})$. The reverse inclusion also holds, if Y is irreducible. This is a consequence of Gabai's theorem 41.1.2. One therefore has:

Theorem 41.5.1 (Gabai [44]). *If Y is a closed, orientable irreducible 3-manifold, then the unit ball of the dual Thurston norm in* $\mathrm{Ann}(\mathcal{T}) \subset H^2(Y;\mathbb{R})$ *is the convex hull of the classes $e(\mathcal{F})$, as $\mathcal{F}$ runs through all taut foliations on Y (with smooth leaves and C^0 tangent planes).* □

Combining this result with the already established relationship between foliations and the non-vanishing of the Floer groups, we shall derive:

Theorem 41.5.2. *If Y is a closed, orientable irreducible 3-manifold, then the unit ball of the dual Thurston norm in* $\mathrm{Ann}(\mathcal{T}) \subset H^2(Y;\mathbb{R})$ *is the convex hull of the classes $c_1(\mathfrak{s})$, as $\mathfrak{s}$ runs through all spin^c structures on Y for which the Floer group $\widecheck{HM}_\bullet(Y,\mathfrak{s};\mathbb{R})$ is non-zero.*

Proof. From Corollary 41.4.2, we already know that that the classes $c_1(\mathfrak{s})$ with $\widecheck{HM}_\bullet(Y,\mathfrak{s};\mathbb{R}) \neq 0$ include the Euler classes of all taut foliations on Y. So in the present theorem, we do know that the convex hull of these classes $c_1(\mathfrak{s})$ contains the unit ball, because of Theorem 41.5.1. What remains is to show that

$$|c_1(\mathfrak{s})| \leq 1, \qquad (41.5)$$

whenever $\widecheck{HM}_\bullet(Y,\mathfrak{s};\mathbb{R}) \neq 0$, where the norm denotes the dual Thurston norm. Since the inequality is not in question when $c_1(\mathfrak{s})$ is zero or torsion, we may as well assume that $c_1(\mathfrak{s})$ is non-torsion and replace $\widecheck{HM}_\bullet$ here by $HM_\bullet$.

This inequality for the norm has a more down-to-earth interpretation: it is the statement that, for any connected, oriented embedded surface $\Sigma \subset Y$ of genus 1 or more, and any $\mathfrak{s}$ with $HM_\bullet(Y,\mathfrak{s};\mathbb{R}) \neq 0$, we have

$$\langle c_1(\mathfrak{s}), [\Sigma] \rangle \leq 2g(\Sigma) - 2.$$

This is the adjunction inequality, Corollary 40.1.2 which we discussed in Section 40, so the proof of Theorem 41.5.2 is complete. □

A particular case of Theorem 41.5.2 occurs when Y is obtained by longitudinal surgery on a knot $K \subset S^3$. As we remarked earlier in the context of Corollary 41.4.4, the results of [45] tell us that the unit ball of the dual Thurston norm is the interval $[-(2g-2), (2g-2)]$, where g denotes the classical genus of K and the cohomology $H^2(Y;\mathbb{R})$ has been identified with $\mathbb{R}$ by the choice of an integral generator σ. If we write $\mathfrak{s}_n$ for the spin^c structures on Y with $c_1(\mathfrak{s}_n) = 2n\sigma$, then we can restate the conclusion of the theorem in this case as:

Corollary 41.5.3. *Let Y be obtained by longitudinal surgery on a non-trivial knot $K \subset S^3$ of genus g. Then*

$$g - 1 = \sup\{\, n \mid \widecheck{HM}_{\bullet}(Y, \mathfrak{s}_n) \neq 0 \,\}.$$

□

This corollary tells us that the Floer homology groups detect the genus of a knot. This result is of significance, since the genus is an invariant with a geometric–topological nature and is not accessible through, for example, the Alexander or Jones polynomials of the knot. Although the proof of the corollary leverages Gabai's results, which tell us that taut foliations also "detect" the genus of a knot, the Floer groups have functorial properties which have no counterpart on the foliations side. In particular, as we have seen, cobordisms between 3-manifolds give maps between their Floer groups. This functoriality is a strong tool when combined with an additional property of Floer homology, which we discuss in the next section, namely a long exact sequence relating the Floer groups of 3-manifolds obtained by different surgeries on a knot.

42 Surgery and exact triangles

42.1 Surgery and elementary cobordisms

By a *knot* in a smooth 3-manifold Y we simply mean a smooth submanifold diffeomorphic to the circle, and by a *framing* ϕ of a knot K we mean a trivialization of its normal bundle. If K is oriented and a framing is given, then a tubular neighborhood of K can be identified with $S^1 \times D^2$, canonically up to isotopy.

If X is an oriented 4-manifold having Y as a boundary component, and if an oriented framed knot $K \subset Y$ is given, then we can form a new 4-manifold X' by attaching a 2-handle $D^2 \times D^2$ to X in such a way that the portion $\partial D^2 \times D^2$ is attached to the tubular neighborhood of K using the identification supplied by the framing. The resulting manifold does not depend on the orientation of K, but does depend on the framing.

If we are given for reference a 2-chain $\Delta \subset X$ with boundary K, then the various homotopy classes of framings can be indexed by the integers, as follows. The union of Δ and the "core" 2-disk $D^2 \times \{0\}$ in the 2-handle is a closed 2-cycle Σ. We say that the framing is the l-framing (relative to the chosen Δ) if the self-intersection number of $\Sigma \subset X$ is l. This notion is again insensitive to the orientation of K.

In the special case that X has the form of a cylinder $[0, 1] \times Y_0$, and when (K_0, ϕ_0) is a framed knot in $\{1\} \times Y_0$, the resulting 4-manifold is a cobordism from Y_0 to a new manifold Y_1: we call a cobordism an *elementary cobordism* if it arises in this way, by the addition of a single 2-handle.

Given an elementary cobordism W_0 from Y_0 to Y_1 as above, the manifold Y_1 itself comes equipped with a new knot $K_1 \subset Y_1$, namely the curve $\{0\} \times S^1$ on the 2-handle: this is the boundary of the "co core" $\{0\} \times D^2$. We equip K_1 with the framing -1 relative to the co core, and call this framing ϕ_1. In this way, given Y_0 with a framed knot (K_0, ϕ_0) we have produced a new 3-manifold Y_1 with a framed knot (K_1, ϕ_1). Repeating the procedure, we obtain an elementary cobordism W_1 from Y_1 to a new manifold Y_2, together with a framed knot (K_2, ϕ_2).

The reason for choosing the framing to be -1 here, rather than any other integer, is two-fold. The interesting topological consequence of choosing ± 1 for the framing is that the resulting sequence of 3-manifolds has a periodicity of order 3, as stated in the next lemma. The reason for choosing -1 rather than $+1$ appears later, when we bring Floer homology into the picture.

Lemma 42.1.1. *If 3-manifolds Y_n are produced by the procedure above, then there is an orientation-preserving diffeomorphism*

$$Y_3 \xrightarrow{\cong} Y_0$$

which carries the oriented framed knot (K_3, ϕ_3) to (K_0, ϕ_0).

Proof. The periodicity can be understood by first observing that the knot complements are the same, for all n. More specifically, let Z_n denote the manifold with torus boundary obtained by removing an open tubular neighborhood of K_n. Let μ_n and λ_n be the meridional and longitudinal curves on the torus boundary, as determined by the framing ϕ_n: that is, μ_n and λ_n are the oriented curves on ∂Z_n which the framing identifies with $\{\text{point}\} \times S^1$ and $S^1 \times \{\text{point}\}$. If ∂Z_n is oriented as the boundary of Z_n, then the intersection number of the two curves is

$$\mu_n \cdot \lambda_n = -1.$$

The point then is that we can naturally identify Z_{n+1} with Z_n in such a way that the isotopy classes of these curves satisfy

$$\mu_{n+1} = \lambda_n$$
$$\lambda_{n+1} = -\mu_n - \lambda_n.$$

In the homology of T^2 therefore, the classes of the new curves are related to the classes of the old via the matrix

$$\begin{bmatrix} 0 & -1 \\ 1 & -1 \end{bmatrix}.$$

Because this matrix has order 3, the curves μ_{n+3}, λ_{n+3} coincide with μ_n, λ_n up to isotopy. ☐

Because all the manifolds Z_n are the same, an alternative starting point for this construction is a 3-manifold Z with torus boundary, from which all the 3-manifolds Y_n can be obtained by gluing a solid torus $S^1 \times D^2$ to ∂Z. The data which we need to describe how the attaching is done is given by three simple closed curves

$$\mu_0, \mu_1, \mu_2$$

on ∂Z, with intersection numbers

$$\mu_0 \cdot \mu_1 = \mu_1 \cdot \mu_2 = \mu_2 \cdot \mu_0 = -1, \qquad (42.1)$$

a condition that also ensures

$$\mu_0 + \mu_1 + \mu_2 = 0.$$

Given these curves, we obtain 3-manifolds Y_0, Y_1, Y_2 by attaching $S^1 \times D^2$ to Z in such a way as to identify {point} $\times S^1$ with μ_0, μ_1 or μ_2 respectively. Thus these manifolds are the *Dehn fillings* of Z along the three curves. Note that the orientation of the curves is not relevant to the Dehn filling, but only serves to describe the relationship of the three. We may change the orientation of all three of the μ_i simultaneously while preserving the algebraic constraints above.

Consider, in particular, the case that $Z \subset S^3$ is a knot complement: that is, Z is obtained from S^3 by removal of an open tubular neighborhood of a knot K. In this case, once we orient K, there are a canonical pair of curves on ∂Z (up to isotopy), namely the classical meridian and longitude, m and l. The latter is a curve on ∂Z which is null-homologous in Z, while the former bounds a disk in the tubular neighborhood of K. They are oriented so that $m \cdot l = -1$ in ∂Z. An arbitrary oriented simple closed curve μ can be described, up to isotopy, by its homology class, which is a linear combination

$$[\mu] = p[m] + q[l]$$

for some integers (p, q) with no common factor. If we forget the orientation of μ, then its isotopy class is described by the ratio

$$r = p/q \in \mathbb{Q} \cup \{\infty\}.$$

For a knot K in S^3, we denote by $S^3_r(K)$ the manifold obtained by Dehn filling of the knot complement K along the curve μ whose slope is r. In the case $r = \infty$, the resulting manifold is S^3. If $r = p/q$ in lowest terms, with $p \geq 0$, then the first homology of $S^3_r(K)$ is $\mathbb{Z}/p\mathbb{Z}$. In the special case $r = 0$, the first homology is $\mathbb{Z}$: this is the case that we earlier called longitudinal surgery, because the corresponding curve μ is the longitude of the knot.

To obtain curves μ_n satisfying the conditions (42.1), we require pairs

$$(p_0, q_0), (p_1, q_1), (p_2, q_2)$$

with

$$-p_n q_{n+1} + p_{n+1} q_n = -1,$$

for all n, where the subscripts are interpreted mod 3. As a particular case, we can take it that the corresponding slopes $r_n = p_n/q_n$ are given by

$$r_0 = 0, \quad r_1 = 1/(q+1), \quad r_2 = 1/q$$

corresponding to the pairs $(0, 1)$, $(-1, -q-1)$ and $(1, q)$. Another special case occurs when

$$r_0 = p, \quad r_1 = p+1, \quad r_2 = \infty,$$

corresponding to the pairs $(p, 1)$, $(-p-1, -1)$, $(1, 0)$. In the general case, we can isolate two possibilities: either

(i) one of the slopes r_n is zero, in which case we may take it that $r_2 = 0$ after cyclic relabelling; or
(ii) after relabelling the indices by a cyclic permutation, p_0 and p_2 have the same sign, and p_1 has the opposite sign.

In the second case, we can change the overall sign of all (p_n, q_n) so that p_0 and p_2 are positive, and write the three rational numbers as

$$r_0 = \frac{p}{q}, \quad r_1 = \frac{p+p'}{q+q'}, \quad r_2 = \frac{p'}{q'} \quad (p, p' > 0). \tag{42.2}$$

In each of these cases, we have 3-manifolds Y_n obtained by different Dehn fillings of the knot complement, repeating with period 3 and each related to the next by an elementary cobordism.

42.2 The exact sequence

We continue to suppose, as above, that a sequence of manifolds and framed knots (Y_n, K_n, ϕ_n) is generated by the above construction, together with elementary cobordisms W_n, so that

$$\partial W_n = (-Y_n) \amalg (Y_{n+1}).$$

Theorem 42.2.1. *Let $\mathbb{F}_2$ be the field of two elements. Then the sequence*

$$\longrightarrow \widetilde{HM}_\bullet(Y_{n-1}; \mathbb{F}_2) \xrightarrow{\widetilde{HM}_\bullet(W_{n-1})} \widetilde{HM}_\bullet(Y_n; \mathbb{F}_2) \xrightarrow{\widetilde{HM}_\bullet(W_n)} \widetilde{HM}_\bullet(Y_{n+1}; \mathbb{F}_2) \longrightarrow$$

is exact, as are the corresponding sequences with $\widehat{HM}$ or $\overline{HM}$ in place of $\widetilde{HM}$. □

Because the manifolds and cobordisms repeat with period 3, the exact sequence of the theorem is a long exact sequence relating the Floer groups of the 3-manifolds Y_0, Y_1 and Y_2. As is usual in a homology exact sequence, the maps induced by two of the cobordisms W_n have even degree, and one has odd degree (with respect to the canonical mod 2 grading, Subsection 22.4). The question of which map has odd degree is taken up in the subsection below.

Exact sequences of the sort appearing in this theorem were pioneered by Floer [33, 34, 15], though his results for instanton homology were not so general. A general surgery exact sequence was established by Ozsváth and Szabó for their Heegaard Floer homology groups [94]. A second proof for the Heegaard Floer groups was given in [96], and the latter proof was adapted to the Seiberg–Witten case in [63].

As is often the case for an exact sequence, the proof that the composite of two consecutive maps in the sequence is zero is relatively straightforward. As in Floer's original argument (see [15]), the key point is that the cobordism $W_{n+1} \circ W_n$ contains an embedded 2-sphere Σ of self-intersection number -1. When one has such an embedded sphere, there is a self-diffeomorphism ψ of the 4-manifold, mapping $[\Sigma]$ to $-[\Sigma]$, acting on the homology by reflection in $[\Sigma]$. The spin^c structures on $W_{n+1} \circ W_n$ can therefore be placed in pairs $\mathfrak{s}$, $\mathfrak{s}'$, with $\mathfrak{s}' = \psi^*(\mathfrak{s})$. The contributions of $\mathfrak{s}$ and $\mathfrak{s}'$ to the map

$$\widetilde{HM}_\bullet(W_{n+1}) \circ \widetilde{HM}_\bullet(W_n)$$

are equal, because of diffeomorphism invariance. Since we are working with mod 2 coefficients, the contributions of the spin^c structures cancel in pairs.

The above argument seems to hinge on the use of $\mathbb{F}_2$ as the coefficients. But in fact, by using suitably twisted coefficients, the contributions of $\mathfrak{s}$ and $\mathfrak{s}'$ can be made to cancel also over $\mathbb{Z}$. It is expected that, with such coefficients, the exact sequence remains valid more generally, but the details have not been checked.

42.3 Mod two gradings in the exact sequence

To determine which of the maps $\widetilde{HM}_\bullet(W_n)$ have even and which have odd degree, we need to look a little more closely at the topology of the cobordisms. The degree of $\widetilde{HM}_\bullet(W_n)$ depends on the Euler number and signature of W_n. The Euler number is always 1, but the signature is more subtle. Recall that the manifolds Y_n can all be obtained from a single manifold Z with torus boundary. On ∂Z, there is a distinguished curve σ, namely the curve representing a primitive class in the kernel of the map $H_1(\partial Z) \to H_1(Z)$. (The orientation of σ is not determined.) In W_n, there is a closed 2-cycle S, formed as the union of three pieces: a surface $T \subset Z$ with boundary σ, the core of the 2-handle, and a piece in the solid torus $\partial D^2 \times D^2$. To determine the signature of W_n, we need to know the sign of the self-intersection number $S \cdot S$.

To understand $S \cdot S$, recall that, in forming Y_{n+1}, we attach a solid torus $D^2 \times S^1$, so picking out curves μ_n and λ_n on ∂Z, as the images of $\{\text{point}\} \times S^1$ and $S^1 \times \{\text{point}\}$. More invariantly, these are the curves μ_n and μ_{n+1} on ∂Z which bound meridional disks in the fillings Y_n and Y_{n+1}. The orientations of these two satisfy

$$\mu_n \cdot \mu_{n+1} = -1$$

in ∂Z. It is the relationship between the three curves σ, μ_n and μ_{n+1} in ∂Z that determines the sign of $S \cdot S$:

Lemma 42.3.1. *The self-intersection number $S \cdot S$ is positive if*

$$\mu_n \cdot \sigma, \quad \mu_{n+1} \cdot \sigma$$

have the same (non-zero) sign, and is negative if they have opposite signs. If either of $\mu_n \cdot \sigma$ or $\mu_{n+1} \cdot \sigma$ is zero, then $S \cdot S$ is zero. □

In the case that both $\mu_n \cdot \sigma$ and $\mu_{n+1} \cdot \sigma$ are non-zero, then the manifolds Y_n and Y_{n+1} have the same first Betti number, and the space $I^2(W_n)$ (the image of relative in absolute cohomology) has rank 1; the signature of W_n is then ± 1, according to the sign of $S \cdot S$. Referring to Definition 25.4.1, we see that the quantity $\iota(W_n)$, which determines the effect on the mod 2 grading, is given by

$$\iota(W_n) = \begin{cases} 0, & \text{if } \operatorname{sign}(\mu_n \cdot \sigma) = -\operatorname{sign}(\mu_{n+1} \cdot \sigma) \\ 1, & \text{if } \operatorname{sign}(\mu_n \cdot \sigma) = \operatorname{sign}(\mu_{n+1} \cdot \sigma). \end{cases}$$

If either $\mu_n \cdot \sigma$ or $\mu_{n+1} \cdot \sigma$ is zero, then W_n has signature zero, and the Betti numbers of the two 3-manifolds differ by 1: the manifold Y_n has the larger Betti number if $\mu_n \cdot \sigma = 0$, so that σ is proportional to μ_n; otherwise, Y_{n+1} has larger Betti number. We therefore have the remaining cases,

$$\iota(W_n) = \begin{cases} 0, & \text{if } \mu_n \cdot \sigma = 0 \\ 1, & \text{if } \mu_{n+1} \cdot \sigma = 0. \end{cases}$$

Let us illustrate this by seeing what happens in the case that Z is a classical knot complement, in which case σ is the longitudinal curve l on ∂Z. We take Y_0, Y_1 and Y_2 to be obtained by Dehn filling with slopes r_0, r_1 and r_2, where the r_n are given by (42.2), with both p and p' positive. In this case, we can write

$$[\mu_n] = p_n[m] + q_n[l]$$

where

$$\begin{aligned}
(p_0, q_0) &= (p, q) \\
(p_1, q_1) &= (-p - p', -q - q') \\
(p_2, q_2) &= (p', q'),
\end{aligned}$$

and $\sigma \cdot [\mu_n] = p_n$. We therefore see that $\widetilde{HM}_\bullet(W_n)$ is a map of even degree for $n = 0$ and 1 (mod 3), and is of odd degree for $n = 2$ (mod 3). A particular case is the sequence relating the surgeries

$$S^3_p(K), \quad S^3_{p+1}(K), \quad S^3_\infty(K).$$

When $p \geq 0$, the map induced by the cobordism from $S^3_\infty(K)$ to $S^3_p(K)$ is the one with odd degree. When $p \leq -2$, it is the cobordism from $S^3_{p+1}(K)$ to $S^3_\infty(K)$ that is odd. When $p = -1$, it is the cobordism from $S^3_p(K)$ to $S^3_{p+1}(K)$ which is odd.

Another special case, where one of the manifolds has larger Betti number, is the exact sequence relating the Floer groups of the Dehn fillings

$$S^3_0(K), \quad S^3_{1/(q+1)}(K), \quad S^3_{1/q}(K) \tag{42.3}$$

in which the map

$$\widetilde{HM}_\bullet(S^3_{1/q}(K)) \to \widetilde{HM}_\bullet(S^3_0(K))$$

has odd degree, while the other two maps have even degree.

42.4 The exact sequence and local coefficients

The surgery exact sequence becomes more useful in an elaborated version using a local coefficient system.

We observed that the knot complements Z_n can all be identified. The product $I \times Z_n$ is also a submanifold (with corners) of W_n. In particular, if we have a real C^∞ singular 1-cycle η in Z_0, then we can regard it also as a 1-cycle η_n in Y_n for each n; and there are 2-chains

$$\nu_n = I \times \eta_n$$

in W_n, with $\partial \nu_n = -\eta_n + \eta_{n+1}$.

The stated theorem for the surgery exact sequence requires, in its present form, a coefficient system of characteristic 2, and this prevents us from using the local systems Γ_{η_n}, which have fiber $\mathbb{R}$ (see Subsections 3.7 and 22.6). As in Subsection 39.2, we can modify the definition of Γ_η by using the field $\mathbb{K}$, the field of fractions of the ring $\mathbb{F}_2[\mathbb{R}]$. We again denote the resulting local system on $\mathcal{B}^\sigma(Y)$ by $\mathbb{K}_\eta$. Its fibers are all canonically isomorphic to $\mathbb{K}$, which is a field of characteristic 2.

For each n, we have Floer groups with local coefficients,

$$\widecheck{HM}_\bullet(Y_n; \mathbb{K}_{\eta_n}),$$

and the 2-chains ν_n in the cobordisms W_n give homomorphisms

$$\widecheck{HM}_\bullet(W_n; \mathbb{K}_{\nu_n}) : \widecheck{HM}_\bullet(Y_n; \mathbb{K}_{\eta_n}) \to \widecheck{HM}_\bullet(Y_{n+1}; \mathbb{K}_{\eta_{n+1}}).$$

The following version of the surgery exact sequence reduces to Theorem 42.2.1 in the case that η is zero.

Theorem 42.4.1. *With η_n and ν_n as above, the sequence*

$$\to \widecheck{HM}_\bullet(Y_{n-1}; \mathbb{K}_{\eta_{n-1}}) \to \widecheck{HM}_\bullet(Y_n; \mathbb{K}_{\eta_n}) \to \widecheck{HM}_\bullet(Y_{n+1}; \mathbb{K}_{\eta_{n+1}}) \to,$$

in which the maps are provided by the pairs (W_n, ν_n), is exact, as are the corresponding sequences for $\widehat{HM}_\bullet$ and $\overline{HM}_\bullet$.

The proof of this version of the theorem is, not surprisingly, very little different from the proof of Theorem 42.2.1. In particular, the part of the proof that we sketched above, showing that the composite of consecutive maps in the sequence is zero, remains valid. (This argument now uses the fact that the union of the 2-cycles ν_n does not meet the embedded spheres of self-intersection -1, so that the spin^c structures $\mathfrak{s}$ and $\mathfrak{s}^*$ continue to contribute cancelling terms.)

42.5 Floer homology and the Alexander polynomial

When a short exact sequence of chain complexes gives rise to a long exact sequence for the corresponding homology groups, then one also has a linear relation among the Euler characteristics of the homology groups (as long as these are defined). In particular, when Y_0, Y_1 and Y_2 are 3-manifolds obtained by three different surgeries on a knot, and when the surgery coefficients are such that one has a long exact sequence of their Floer groups, as in Theorem 42.4.1, then one should expect that an alternating sum of Euler characteristics will be zero.

The study of the Euler characteristic of the Floer homology, and the development of the corresponding surgery formulae, both predate the construction of the Floer groups themselves. One of the important antecedents is Casson's surgery formula for the Casson invariant [2]. We will state a result, first proved by Meng and Taubes [74], for the Seiberg–Witten Floer groups of a 3-manifold Y obtained by 0-surgery on a knot K in S^3. The result relates the Euler characteristic of the Floer groups of Y to the *Alexander polynomial* of the knot K.

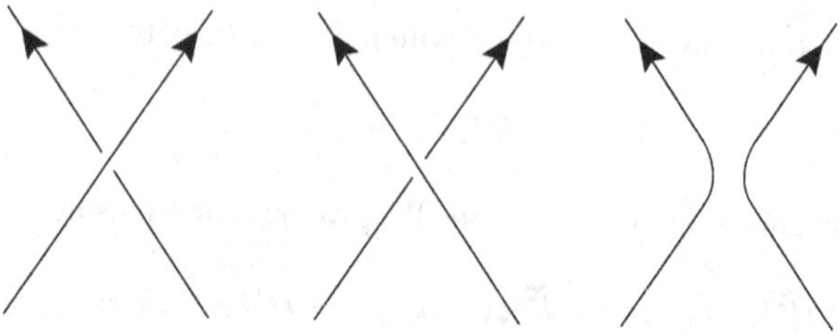

Fig. 12. Three skein-related links, L_+, L_- and L_0.

We recall here that the Alexander polynomial $\Delta_L(t)$ of a knot or link L in S^3 is a Laurent polynomial in $t^{1/2}$ which can be characterized by Conway's skein relation

$$\Delta_{L_+}(t) - \Delta_{L_-}(t) = (t^{1/2} - t^{-1/2})\Delta_{L_0}(t)$$

with the normalization that the Alexander polynomial of the unknot is 1. In the skein relation, L_+, L_- and L_0 denote three links whose projections differ at a single crossing, as shown in Figure 12. (Despite the appearance of $t^{1/2}$, the Alexander polynomial lies in $\mathbb{Z}[t, t^{-1}]$ in the case of a knot, or a link with an odd number of components.) This version of the Alexander polynomial is often called the *symmetrized* Alexander polynomial, because if we write it as

$$\sum a_k t^k$$

then $a_{-k} = \pm a_k$, with the sign depending on whether k is an integer or a half-integer.

To obtain a clean formulation of the relationship between the Floer groups and the Alexander polynomial, it is convenient to study the Floer homology groups using a *non-exact* perturbation of the sort discussed in Section 29. For now, let Y be any 3-manifold with $H^2(Y;\mathbb{Z}) = \mathbb{Z}$, and let h be a generator. For each integer k, there is a unique spin^c structure $\mathfrak{s}_k$ with

$$c_1(\mathfrak{s}_k) = 2kh.$$

We will use $\mathbb{Z}/2$ coefficients for our Floer group (without further comment in this subsection), and we consider the groups

$$HM_\bullet(Y, \mathfrak{s}_k, c)$$

where $c \in H^2(Y;\mathbb{R})$ is the period class of the non-exact perturbation. Recall that the perturbation is *balanced* if $2\pi c_1(\mathfrak{s}_k) + c = 0$. In general, if we write $c = \lambda h$, so that

$$2\pi c_1(\mathfrak{s}_k) + c = (4\pi k + \lambda)h,$$

then the group $HM_\bullet(Y, \mathfrak{s}_k, c)$ depends only on k and the *sign* of $4\pi k + \lambda$ (positive, negative or zero): this is the content of Theorem 31.4.1. All these groups have a canonical $\mathbb{Z}/2$ grading, so we can consider their Euler characteristics.

Definition 42.5.1. After choosing a generator h for the group $H^2(Y;\mathbb{Z}) = \mathbb{Z}$ and an integer $k \neq 0$, we write

$$\chi(Y, \mathfrak{s}_k)_b$$

for the Euler characteristic of the $(\mathbb{Z}/2)$-graded homology group $HM_\bullet(Y, \mathfrak{s}_k)$. We write

$$\chi(Y, \mathfrak{s}_k)_+$$

for the Euler characteristic of the group $HM_\bullet(Y, \mathfrak{s}_k, c)$, where $c = \lambda h$ is any period class with $4\pi k + \lambda > 0$. We similarly write

$$\chi(Y, \mathfrak{s}_k)_-$$

for the Euler characteristic of this homology group in the case that $4\pi k + \lambda < 0$. Note that $\chi(Y, \mathfrak{s}_k)_b$ coincides with $\chi(Y, \mathfrak{s}_k)_+$ in the case that k is positive and with $\chi(Y, \mathfrak{s}_k)_-$ in the case that k is negative.

In the case $k = 0$, we write

$$\chi(Y, \mathfrak{s}_0) = \chi(Y, \mathfrak{s}_-)_+ = \chi(Y, \mathfrak{s}_0)_-$$

for the Euler characteristic of the group

$$HM_\bullet(Y, \mathfrak{s}_0, \Gamma) \cong HM_\bullet(Y, \mathfrak{s}_0, c, \Gamma) \tag{42.4}$$

where c is any non-zero period class and Γ is any c-complete local system of fields. ◊

Remark. Some clarifying remarks are needed in the case that k is zero. The two groups in (42.4) are isomorphic by Theorem 31.1.3. The group $HM_\bullet(Y, \mathfrak{s}_0, c, \Gamma)$ is the homology of a bounded complex of finite-dimensional vector spaces, with generators corresponding to the critical points of the non-exact perturbed functional; so one can see that the Euler characteristic is independent of Γ. Also, when k is zero, one can use the symmetry provided by complex conjugation to prove that c and $-c$ lead to the same Euler characteristic.

The next lemma is an example of a "wall-crossing" formula, related to the results of Subsection 27.5:

Lemma 42.5.2. *We have*

$$\chi(Y, \mathfrak{s}_k)_+ = \chi(Y, \mathfrak{s}_k)_- - k.$$

Proof. We consider the case that k is non-zero, because the result holds for $k = 0$ by the definition and the remark above. The homology groups $HM_*(Y, \mathfrak{s}_k, \lambda h)$ for the λh either side of the balanced case are related by a long exact sequence, as stated in Corollary 31.5.2. The third term in the long exact sequence is $\overline{HM}_*(Y, \mathfrak{s}_k, c_b)$, where c_b is the balanced class, and the map p_* in the sequence has odd degree. In the statement of Corollary 31.5.2, the class that is referred to as c_+ refers to the case that $4\pi k + \lambda$ has the same sign as k. So when k is positive, we have

$$\chi(Y, \mathfrak{s}_k)_+ = \chi(Y, \mathfrak{s}_k)_- + \chi\big(\overline{HM}_*(Y, \mathfrak{s}_k, c_b)\big).$$

When k is negative, the roles of c_+ and c_- are switched, so that

$$\chi(Y, \mathfrak{s}_k)_+ = \chi(Y, \mathfrak{s}_k)_- - \chi\big(\overline{HM}_*(Y, \mathfrak{s}_k, c_b)\big).$$

The group $\overline{HM}_*(Y, \mathfrak{s}_k, c_b)$ is calculated in Subsection 35.3, at Equation (35.4). This formula tells us that the group has rank $|k|$. Furthermore, in the case that b_1 is 1, the non-zero part of this group occurs in odd degree. (See Theorem 35.1.1.) So

$$\chi\big(\overline{HM}_*(Y, \mathfrak{s}_k, c_b)\big) = -|k|.$$

This gives the result stated in the lemma. □

We can combine the Euler numbers $\chi(Y, \mathfrak{s}_k)$ into generating functions

$$\left.\begin{aligned}\chi(Y)_b(t) &= \sum_{k\in\mathbb{Z}} t^k \chi(Y, \mathfrak{s}_k)_b \\ \chi(Y)_+(t) &= \sum_{k\in\mathbb{Z}} t^k \chi(Y, \mathfrak{s}_k)_+ \\ \chi(Y)_-(t) &= \sum_{k\in\mathbb{Z}} t^k \chi(Y, \mathfrak{s}_k)_- .\end{aligned}\right\} \qquad (42.5)$$

Without the non-exact perturbation, finitely many of the Floer groups $HM_\bullet(Y, \mathfrak{s}_k)$ are non-zero, so $\chi(Y)_b(t)$ is a polynomial in t. This and the above wall-crossing formula give us

$$\begin{aligned}\chi(Y)_+(t) &= \chi(Y)_b(t) - \sum_{k<0} kt^k \\ \chi(Y)_-(t) &= \chi(Y)_b(t) + \sum_{k>0} kt^k\end{aligned}$$

so that $\chi(Y)_+$ and $\chi(Y)_-$ are Laurent series which extend infinitely in the negative and positive directions respectively.

As a particular example, if Y is $S^1 \times S^2$, then $\chi(Y)_b = 0$, and so

$$\begin{aligned}\chi(S^1 \times S^2)_- &= \sum_{k>0} kt^k \\ &= t/(1-t)^2\end{aligned}$$

where it is understood that the rational function is to be expanded as a positive power series in t. As another example, if Y is obtained by 0-surgery on the trefoil knot, then Y_0 is one of the flat torus-bundles over the circle discussed in Subsection 37.4. From the results of that subsection, we know that $\chi(Y, \mathfrak{s}_k)_b$ is 1 for $k = 0$ and zero for all other k. So in this example,

$$\begin{aligned}\chi(Y)_- &= 1 + \sum_{k>0} kt^k \\ &= (1 - t + t^2)/(1-t)^2.\end{aligned}$$

These two calculations are special cases of the following general result:

Theorem 42.5.3 (Meng–Taubes [74]). *If Y is obtained by 0-surgery on a knot K in S^3, then*

$$\chi(Y)_- = \left(\sum_{k>0} kt^k\right)\Delta_K(t)$$
$$= \frac{t}{(1-t)^2}\Delta_K(t)$$

where Δ_K is the symmetrized Alexander polynomial of the knot K, normalized so that $\Delta_K(1) = 1$.

Remark. In the statement of the theorem, the rational function $t/(1-t)^2$ is to be expanded as a positive power series in t, as shown. We can write this rational function as $1/(t^{-1/2} - t^{1/2})^2$ to exhibit its symmetry with respect replacing t by t^{-1}. The same rational function has a negative expansion,

$$-\sum_{k<0} kt^k.$$

If we use this expansion in negative powers, we obtain an expression for $\chi(Y)_+$:

$$\chi(Y)_+ = \left(-\sum_{k<0} kt^k\right)\Delta_K(t).$$

The theorem can be deduced using an argument, based on the surgery exact sequence for the Floer groups and the skein relation satisfied by $\Delta_K(t)$. Such an argument is given by Ozsváth and Szabó in the context of Heegaard Floer homology, in [92, section 2]. An earlier related argument is in [31].

42.6 Applications to surgery

Applications of Floer homology to classical questions about Dehn surgery on knots were given in [63]. In this subsection, we give a flavor of those results, by indicating how the surgery exact sequence can be used to prove the following illustrative theorem:

Theorem 42.6.1. *Let K be a knot in S^3, and let $S^3_{1/q}(K)$ be obtained by Dehn surgery, with $q \in \mathbb{Z}$. If q is non-zero and*

$$S^3_{1/q}(K) = S^3,$$

then K is the unknot.

This result was proved for $|q| \geq 2$ by Culler, Gordon, Luecke and Shalen in [16], as a corollary of their cyclic surgery theorem. For $q = \pm 1$, it was proved by Gordon and Luecke in [50]. Since Dehn surgery with coefficient p/q yields a manifold with non-trivial homology if $p \neq \pm 1$, the theorem can be restated as saying that non-trivial Dehn surgery on a non-trivial knot cannot yield S^3. The argument using Floer homology in [63] is independent of [16] and [50], except in that it draws on Gabai's work. The more general result of [63] states that, if K_u is the unknot and

$$S^3_r(K) = S^3_r(K_u)$$

as oriented manifolds for some rational number r, then K is also unknotted.

To prove the above theorem, we apply the surgery exact sequence with local coefficients. Let η be a closed 1-cycle in the knot complement $Z \subset S^3$, representing a generator of the real homology. To make the notation more compact in the commutative diagrams that follow, let us introduce the abbreviation

$$S_r = S^3_r(K)$$

for the manifold obtained by Dehn surgery. Let us also write

$$\begin{aligned}\check{S}_r &= \widecheck{HM}_\bullet(S_r; \mathbb{K}_\eta)\\ \hat{S}_r &= \widehat{HM}_\bullet(S_r; \mathbb{K}_\eta)\\ \bar{S}_r &= \overline{HM}_\bullet(S_r; \mathbb{K}_\eta),\end{aligned}$$

where $\mathbb{K}_\eta$ is the local system used in Subsection 42.4. The Floer groups of the manifolds S_0, $S_{1/(q+1)}$, $S_{1/q}$ are related by a long exact sequence: see the example (42.3). There is one such sequence for each flavor of Floer homology, and the three sequences are the rows of the following commutative diagram, in which the columns are also exact, being the exact sequences of Proposition 22.2.1.

$$
\begin{array}{ccccccccc}
\longrightarrow & \hat{S}_0 & \longrightarrow & \hat{S}_{1/(q+1)} & \longrightarrow & \hat{S}_{1/q} & \longrightarrow & \hat{S}_0 & \longrightarrow \\
 & \downarrow & & \downarrow & & \downarrow p & & \downarrow & \\
\longrightarrow & \bar{S}_0 & \longrightarrow & \bar{S}_{1/(q+1)} & \xrightarrow{\cong} & \bar{S}_{1/q} & \longrightarrow & \bar{S}_0 & \longrightarrow \\
 & \downarrow & & \downarrow & & \downarrow i & & \downarrow & \\
\longrightarrow & \check{S}_0 & \xrightarrow{B_q} & \check{S}_{1/(q+1)} & \xrightarrow{C_q} & \check{S}_{1/q} & \xrightarrow{A_q} & \check{S}_0 & \longrightarrow \\
 & \downarrow & & \downarrow & & \downarrow j & & \downarrow & \\
\longrightarrow & \hat{S}_0 & \longrightarrow & \hat{S}_{1/(q+1)} & \longrightarrow & \hat{S}_{1/q} & \longrightarrow & \hat{S}_0 & \longrightarrow
\end{array}
$$

We have one such diagram for each q. An important feature is that the manifold S_0 has non-zero Betti number; and because η represents a non-zero class in its homology, we have

$$\bar{S}_0 = 0$$

from Proposition 3.9.1. As we have indicated in the diagram, it follows that the map $\bar{S}_{1/(q+1)} \to \bar{S}_{1/q}$ is an isomorphism.

There is one other topological input to the argument, which is contained in the next lemma.

Lemma 42.6.2 ([63]). *The composite $B_{q-1} \circ A_q : \check{S}_{1/q} \to \check{S}_{1/q}$ is zero.*

Proof. As explained in [63], this lemma follows from a borderline case of the adjunction inequality. The composite cobordism which induces the map $B_{q-1} \circ A_q$ contains an embedded 2-sphere with trivial normal bundle, representing a non-zero homology class, and this forces the corresponding map to vanish. □

Corollary 42.6.3. *The map A_q is zero, for all $q \geq 0$, so we have a short exact sequence relating $\widecheck{HM}_\bullet$ of S_0, $S_{1/(q+1)}$ and $S_{1/q}$:*

$$0 \to \check{S}_0 \to \check{S}_{1/(q+1)} \to \check{S}_{1/q} \to 0.$$

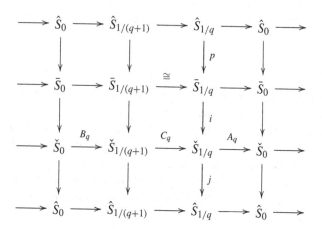

Proof. The proof uses the lemma, and is by induction on q. For $q = 0$, the manifold $S_{1/0}$ is S^3, for which we know that the map $j : \check{S}_{1/0} \to \hat{S}_{1/0}$ is zero and the map $i : \bar{S}_{1/0} \to \check{S}_{1/0}$ is surjective. From the commutativity of the diagram, it follows that C_0 is surjective also, and hence that A_0 is zero.

For the induction step, suppose A_{q-1} is zero. Then B_{q-1} is injective, from the exactness of the rows. The composite $B_{q-1} \circ A_q$ is zero by the lemma, and it follows that A_q is zero. □

Lemma 42.6.4. *If $j : \check{S}_{1/(q+1)} \to \hat{S}_{1/(q+1)}$ is zero for some $q \geq 0$, then so too is $j : \check{S}_{1/q} \to \hat{S}_{1/q}$.*

Proof. The preceding corollary tells us that C_q is surjective. The hypothesis of the lemma and the commutativity of the diagram tell us that $j \circ C_q$ is zero as a map $\check{S}_{1/(q+1)} \to \hat{S}_{1/q}$; so it follows that j is zero also. □

The lemma and the preceding corollary together give us the following simplification of part of the commutative diagram, whenever $j : \check{S}_{1/(q+1)} \to \hat{S}_{1/q}$ is zero (the rows and columns are exact):

$$
\begin{array}{ccccccccc}
 & & & & \hat{S}_{1/(q+1)} & \longrightarrow & \hat{S}_{1/(q)} & & \\
 & & & & \big\downarrow & & \big\downarrow p & & \\
 & & & & \bar{S}_{1/(q+1)} & \xrightarrow{\cong} & \bar{S}_{1/q} & & \\
 & & & & \big\downarrow & & \big\downarrow i & & \\
0 & \longrightarrow & \check{S}_0 & \longrightarrow & \check{S}_{1/(q+1)} & \xrightarrow{C_q} & \check{S}_{1/q} & \longrightarrow & 0 \\
 & & & & \big\downarrow & & \big\downarrow & & \\
 & & & & 0 & & 0. & &
\end{array}
$$

By the definition of Frøyshov's invariant, this diagram can be described in the following way, where $\bar{L}$ is again $\mathbb{K}[U_\dagger^{-1}, U_\dagger]]$ and $h(q+1)$ and $h(q)$ are the

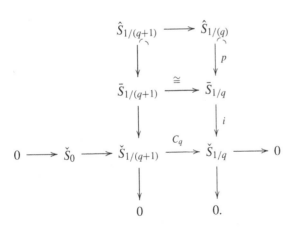

Frøyshov invariants $\bar{h}(S_{q+1})$ and $\bar{h}(S_q)$ using coefficients of characteristic 2:

$$\begin{array}{ccccccccc}
 & & & & \bar{L}_{h(q+1)} & \longrightarrow & \bar{L}_{h(q)} & & \\
 & & & & \big\downarrow & & \big\downarrow & & \\
 & & & & \bar{L} & \xrightarrow{\ x\ } & \bar{L} & & \\
 & & & & \big\downarrow & & \big\downarrow{\scriptstyle i} & & \\
0 & \longrightarrow & \check{S}_0 & \longrightarrow & \bar{L}/\bar{L}_{h(q+1)} & \xrightarrow{\ C_q\ } & \bar{L}/\bar{L}_{h(q)} & \longrightarrow & 0 \\
 & & & & \big\downarrow & & \big\downarrow & & \\
 & & & & 0 & & 0. & &
\end{array}$$

From this we read off

$$\dim_{\mathbb{K}} \check{S}_0 = \dim_{\mathbb{K}}(\bar{L}_{h(q)}/\bar{L}_{h(q+1)}),$$

or in other words

$$\bar{h}(S_{1/q}) - \bar{h}(S_{1/(q+1)}) = \dim_{\mathbb{K}} HM_{\bullet}(S_0; \mathbb{K}_\eta). \tag{42.6}$$

Lemma 42.6.4 gives us exactly the same relationship also for a smaller q, and we therefore obtain the following proposition.

Proposition 42.6.5. *Suppose that for some $q_0 \geq 0$ the homology 3-sphere S_{1/q_0} has $j = 0$ for its Floer groups with coefficients $\mathbb{K}$. Then the characteristic-2 Frøyshov invariant $\bar{h}(S_{1/q_0})$ is given by*

$$\bar{h}(S_{1/q_0}) = q_0 \dim_{\mathbb{K}} HM_{\bullet}(S_0; \mathbb{K}_\eta),$$

where S_0 is the manifold obtained by longitudinal surgery and the coefficient system $\mathbb{K}_\eta$ is the non-trivial local system described in Subsection 42.4 above.

Proof. Apply (42.6) to the successive differences, starting with the fact that $S_{1/0}$ is S^3, which has $\bar{h} = 0$. □

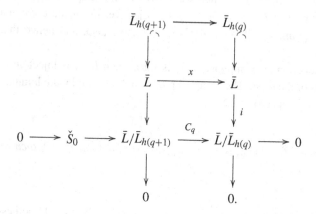

We can now deduce the theorem. By replacing the knot with its mirror image if necessary, we can reduce our considerations to the case $q > 0$. Suppose then that $S_{1/q}$ is S^3 when $q = q_0 \geq 1$. Being S^3, the manifold S_{1/q_0} has $j = 0$, so the above proposition is applicable. Also, $\bar{h}(S^3)$ is zero; so the proposition tells us

$$HM_\bullet(S_0; \mathbb{K}_\eta) = 0.$$

From Corollary 41.4.5, it follows that K is the unknot. □

The statement and proof of Proposition 42.6.5 can be extended in the following way:

Proposition 42.6.6. *Let Y be an integral homology 3-sphere, let K be a knot in Y, and let $Y_{1/q}$ denote the integral homology 3-sphere obtained by $1/q$ Dehn surgery on K. Suppose that the homomorphism $j : \widetilde{HM}_\bullet \to \widehat{HM}_\bullet$ is zero for both the manifolds $Y_{1/p}$ and $Y_{1/q}$, where p and q are integers with $p < q$. Then j is zero also for $Y_{1/q'}$ for integers q' in the range $p \leq q' \leq q$. Furthermore,*

$$\bar{h}(Y_{1/q}) = \bar{h}(Y_{1/p}) + (q - p) \dim HM_\bullet(Y_0; \mathbb{K}_\eta),$$

where Y_0 is the manifold obtained by longitudinal surgery. □

This proposition highlights the class of 3-manifolds for which $j = 0$. Ozsváth and Szabó introduced the term *L-space* for a rational homology sphere which has the corresponding property in their Heegaard Floer homology theory. For monopole Floer homology, we already know that a rational homology sphere is an L-space if it has positive scalar curvature. Other examples of L-spaces arise from applications of the surgery exact triangle, as the above proposition shows.

A particular example is the Poincaré homology sphere Y_P, the oriented boundary of the negative-definite E_8 plumbing. This manifold is obtained by $1/q$ surgery on a trefoil knot, with $q = -1$. Longitudinal surgery on the same trefoil knot K gives a manifold $S^3_0(K)$ with $b_1 = 1$ which is a non-trivial flat torus bundle over the circle. By the results of Subsection 37.4, we know that $HM_\bullet(Y_0; \mathbb{K}_\eta)$ has rank 1, so the above proposition tells us that the $\bar{h}$-invariant of the Poincaré homology sphere Y_P is -1, a result obtained by Frøyshov (*cf.* [41]). Having obtained the $\bar{h}$ invariant, we can turn Proposition 42.6.6 around: it tells us that if Y_P is obtained by $(1/q)$-surgery on *any* knot K in S^3, then q must be -1 and the rank of $HM_\bullet(Y_0(K); \mathbb{K}_\eta)$ must be 1.

One can take this result a little further. If $HM_\bullet(Y_0(K); \mathbb{K}_\eta)$ has rank 1, then the Euler characteristic defined in (42.5) is $\chi(Y_0)_b(t) = \pm 1$ (with the sign depending on whether the generator is in even or odd grading); and hence

$$\chi(Y_0)_-(t) = \begin{cases} (1 - t + t^2)/(1 - t)^2, & \text{or} \\ (-1 + 3t - t^2)/(1 - t)^2. \end{cases}$$

From the relationship between the Alexander polynomial of K and $\chi(Y_0)_-$ we deduce that

$$\Delta_K(t) = \begin{cases} t^{-1} - 1 + t, & \text{or} \\ -t^{-1} + 3 - t. \end{cases} \tag{42.7}$$

In the case that $(1/q)$-surgery on K yields the Poincaré homology sphere, the single generator of $HM_\bullet(Y_0(K); \mathbb{K}_\eta)$ lies in even degree, and

$$\Delta_K(t) = t^{-1} - 1 + t$$

is the only possibility. This argument was used in [95] (using Heegaard Floer homology), and it proves:

Proposition 42.6.7 (Ozsváth–Szabó [95]). *If $(1/q)$-surgery on a knot K yields the Poincaré sphere, oriented as the boundary of the negative-definite E_8 plumbing, then q must be -1 and K must have the same Alexander polynomial as the trefoil.* □

Theorem 42.6.1 has the following extension, as an application of Proposition 42.6.6.

Theorem 42.6.8. *Let Y be an integral homology 3-sphere, and let p and q be distinct integers. If*

$$Y_{1/p}(K) = Y_{1/q}(K)$$

as oriented 3-manifolds, and if $j : \widecheck{HM}_\bullet(Y_{1/p}(K)) \to \widehat{HM}_\bullet(Y_{1/p}(K))$ is zero, then $Y_0(K)$ is reducible. □

42.7 Fibered knots

We have just seen that if the manifold $Y_0(K)$ obtained by zero-surgery on a knot K has Floer homology $HM_\bullet(Y_0(K); \mathbb{K}_\eta)$ of rank 1, then the Alexander polynomial of K is one of the two possibilities in (42.7) above. These two polynomials are the Alexander polynomials of the trefoil and figure-eight knot respectively: the two *fibered* knots of genus 1. A striking strengthening of this observation concerning the Alexander polynomials was proved by Ghiggini:

Theorem 42.7.1 ([47]). *Let K be a knot in S^3 and let $Y_0(K)$ be obtained by longitudinal surgery. Let η be a 1-cycle generating $H_1(Y_0(K); \mathbb{R})$. Then the*

Floer group $HM_\bullet(Y_0(K); \mathbb{K}_\eta)$ has rank 1 if and only if K is a fibered knot of genus 1: that is, a trefoil or figure-eight knot.

Combined with the previous arguments, this result shows that the trefoil is indeed the only knot which can give rise to the Poincaré homology sphere by surgery. The proof of the theorem has two parts. We already know that K has genus 1, so the generator of $H_2(Y_0(K); \mathbb{Z})$ is represented by an embedded 2-torus Σ. The first part of the argument is then to show that, if we cut along the torus Σ and re-glue by any element ϕ of $SL(2, \mathbb{Z})$ to obtain a new 3-manifold Y_ϕ, then the two groups

$$HM_\bullet(Y_\phi; \mathbb{K}_\eta), \quad HM_\bullet(Y_0(K); \mathbb{K}_\eta)$$

have the same rank. For this, it is enough to show that if ϕ and ϕ' differ by a single Dehn twist, then the corresponding Floer groups are the same. In this case, the manifolds Y_ϕ and $Y_{\phi'}$ differ by a Dehn surgery along a curve γ lying on Σ, and their Floer groups are related by a surgery exact sequence. The third manifold that appears in the long exact sequence contains a sphere representing a generator of homology (obtained by ordinary surgery on Σ along γ), and the Floer groups of this third manifold with twisted coefficients are zero, by Proposition 40.1.3. The groups $HM_\bullet(Y_\phi; \mathbb{K}_\eta)$ and $HM_\bullet(Y_{\phi'}; \mathbb{K}_\eta)$ are therefore isomorphic as desired. The second part of Ghiggini's argument leverages Gabai's existence theorems for taut foliations, to show that, if K is not fibered, then for suitable ϕ the manifold Y_ϕ admits two distinct taut foliations that can be distinguished by the Euler classes of the corresponding 2-plane fields. Because of the fundamental non-vanishing theorem, Theorem 41.4.1, this shows that, if K is not fibered, then there are two different spin^c structures which make non-zero contributions to $HM_\bullet(Y_\phi; \mathbb{K}_\eta)$. The rank of this group is therefore at least 2 if K is not fibered, as is their rank of $HM_\bullet(Y_0(K); \mathbb{K}_\eta)$, by the first part of the argument.

In the setting of Heegaard homology, Theorem 42.7.1 has been extended by Ni [85], who shows that fibered knots of genus g are characterized amongst all genus-g knots by the condition

$$\dim_{\mathbb{K}} HM_*(Y_0(K), \mathfrak{s}_c; \mathbb{K}_\eta) = 1,$$

where $\mathfrak{s}_c$ is the spin^c structure whose first Chern class is $2g - 2$ times the generator in $H^2(Y_0(K); \mathbb{Z})$. This result was first conjectured by Ozsváth and Szabó in [91].

Notes and references for Chapter X

For a survey of applications of foliations to topology from 1990, see [46]. Taubes' non-vanishing theorem for symplectic manifolds was first proved in [114], though the proof was simplified shortly afterwards (see [116, 21, 58]). A similar statement for Kähler manifolds was observed earlier, by Witten [125].

One precursor of Theorem 41.3.3 is the theorem of Etnyre and Honda [29] that a closed contact 3-manifold can be realized as a strongly *concave* boundary of a symplectic 4-manifold. The question of which contact 3-manifolds arose as concave boundaries appears to have received little attention before [29], perhaps because there were no techniques to address the question prior to Giroux's work [48]. The question of which contact 3-manifolds arise as (weakly) *convex* boundaries arose earlier in Eliashberg's work: see [24] for a survey of this and other developments in contact topology.

An earlier approach to the non-vanishing for Floer homology appeared in [62] (though the result there was formulated in terms of the non-emptiness of the set of critical points, rather than in terms of the corresponding Floer groups). The argument in [62] made use of the symplectic structure with weakly convex boundary on the cylinder $[-1, 1] \times Y$, as supplied by Theorem 41.3.2, but did not use the embedding theorem, Theorem 41.3.3, or Giroux's results (which were not known at the time that [62] was written). The approach in [62] was to define monopole invariants of symplectic 4-manifolds with contact boundary, essentially using the contact structure to supply a boundary condition for the Seiberg–Witten equations. Taubes' non-vanishing theorem extends to this context, when the 4-manifold admits a symplectic form which is positive on the contact planes at the boundary. This non-vanishing result can be applied to the cylinder $[-1, 1] \times Y$ to obtain the non-vanishing of the Floer homology of Y.

After Floer's treatment of a surgery exact sequence for the instanton homology of homology spheres, the first proof of a surgery exact triangle with the generality of Theorem 42.2.1 was proved for Heegaard Floer groups by Ozsváth and Szabó [94]. The non-vanishing theorem for symplectic manifolds also holds for the Heegaard invariants for 4-manifolds, so topological applications such as those of Subsection 42.6 can be made using either monopole Floer homology or Heegaard Floer homology, using the same arguments.

Additional properties of the invariant $h(Y)$ are proved by Frøyshov in [43], in the more general setting of rational homology spheres. In particular, Frøyshov establishes the additivity of this invariant under connected sums.

References

[1] S. Agmon and L. Nirenberg. Lower bounds and uniqueness theorems for solutions of differential equations in a Hilbert space. *Comm. Pure Appl. Math.*, **20** (1967), 207–229.

[2] S. Akbulut and J. D. McCarthy. *Casson's Invariant for Oriented Homology 3-Spheres*, vol. 36 of *Mathematical Notes*. Princeton University Press (1990).

[3] S. Akbulut and B. Ozbagci. On the topology of compact Stein surfaces. *Internat. Math. Res. Notices* **15** (2002), 769–782.

[4] M. Atiyah. Topological quantum field theories. *Inst. Hautes Etudes Sci. Publ. Math.*, **68** (1988), 175–186 (1989).

[5] M. Atiyah and G. Segal. Twisted K-theory and cohomology. arXiv:math.KT/0510674.

[6] M. F. Atiyah and R. Bott. The Yang–Mills equations over Riemann surfaces. *Philos. Trans. Roy. Soc. London Ser. A*, **308** (1983), 523–615.

[7] M. F. Atiyah, N. J. Hitchin and I. M. Singer. Self-duality in four-dimensional Riemannian geometry. *Proc. Roy. Soc. London Ser. A*, **362** (1978), 425–461.

[8] M. F. Atiyah, V. K. Patodi and I. M. Singer. Spectral asymmetry and Riemannian geometry. I. *Math. Proc. Cambridge Philos. Soc.*, **77** (1975), 43–69.

[9] M. F. Atiyah, V. K. Patodi and I. M. Singer. Spectral asymmetry and Riemannian geometry. III. *Math. Proc. Cambridge Philos. Soc.*, **79** (1976), 71–99.

[10] M. F. Atiyah and I. M. Singer. Index theory for skew-adjoint Fredholm operators. *Inst. Hautes Etudes Sci. Publ. Math.*, **37** (1969), 5–26.

[11] S. Bauer. Diffeomorphism types of elliptic surfaces with $p_g = 1$. *J. Reine Angew. Math.*, **451** (1994), 89–148.

[12] J.-M. Bismut and D. S. Freed. The analysis of elliptic families. I. Metrics and connections on determinant bundles. *Comm. Math. Phys.*, **106** (1986), 159–176.

[13] R. Bott. Morse theory indomitable. *Inst. Hautes Etudes Sci. Publ. Math.*, **68** (1988), 99–114.

[14] R. Bott and L. W. Tu. *Differential Forms in Algebraic Topology*, vol. 82 of *Graduate Texts in Mathematics*. Springer-Verlag (1982).

[15] P. J. Braam and S. K. Donaldson. Floer's work on instanton homology, knots and surgery. In *The Floer Memorial Volume*, vol. 133 of *Progress in Mathematics*. Birkhäuser (1995), 195–256.

[16] M. Culler, C. M. Gordon, J. Luecke and P. B. Shalen. Dehn surgery on knots. *Ann. of Math. (2)*, **125** (1987), 237–300.
[17] S. K. Donaldson. An application of gauge theory to four-dimensional topology. *J. Differential Geom.*, **18** (1983), 279–315.
[18] S. K. Donaldson. Irrationality and the h-cobordism conjecture. *J. Differential Geom.*, **26** (1987), 141–168.
[19] S. K. Donaldson. The orientation of Yang-Mills moduli spaces and 4-manifold topology. *J. Differential Geom.*, **26** (1987), 397–428.
[20] S. K. Donaldson. Polynomial invariants for smooth four-manifolds. *Topology*, **29** (1990), 257–315.
[21] S. K. Donaldson. The Seiberg–Witten equations and 4-manifold topology. *Bull. Amer. Math. Soc. (N.S.)*, **33** (1996), 45–70.
[22] S. K. Donaldson. *Floer Homology Groups in Yang-Mills Theory*, vol. 147 of *Cambridge Tracts in Mathematics*. Cambridge University Press (2002). With the assistance of M. Furuta and D. Kotschick.
[23] S. K. Donaldson and P. B. Kronheimer. *The Geometry of Four-Manifolds*. Oxford University Press (1990).
[24] Y. Eliashberg. Contact 3-manifolds twenty years since J. Martinet's work. *Ann. Inst. Fourier (Grenoble)*, **42** (1992), 165–192.
[25] Y. Eliashberg. A few remarks about symplectic filling. *Geom. Topol.*, **8** (2004), 277–293 (electronic).
[26] Y. M. Eliashberg and W. P. Thurston. *Confoliations*, vol. 13 of *University Lecture Series*. American Mathematical Society (1998).
[27] N. D. Elkies. A characterization of the $\mathbf{Z}^n$ lattice. *Math. Res. Lett.*, **2** (1995), 321–326.
[28] J. B. Etnyre. On symplectic fillings. *Algebr. Geom. Topol.*, **4** (2004), 73–80 (electronic).
[29] J. B. Etnyre and K. Honda. On symplectic cobordisms. *Math. Ann.*, **323** (2002), 31–39.
[30] R. Fintushel and R. J. Stern. Rational blowdowns of smooth 4-manifolds. *J. Differential Geom.*, **46** (1997), 181–235.
[31] R. Fintushel and R. J. Stern. Knots, links, and 4-manifolds. *Invent. Math.*, **134** (1998), 363–400.
[32] A. Floer. An instanton-invariant for 3-manifolds. *Comm. Math. Phys.*, **118** (1988), 215–240.
[33] A. Floer. Instanton homology, surgery, and knots. In *Geometry of Low-Dimensional Manifolds, 1 (Durham, 1989)*, vol. 150 of *London Math. Soc. Lecture Note Ser.* Cambridge University Press (1990), 97–114.
[34] A. Floer. Instanton homology and Dehn surgery. In *The Floer Memorial Volume*, vol. 133 of *Progr. Math.* Birkhäuser (1995), 77–97.
[35] D. S. Freed and K. K. Uhlenbeck. *Instantons and Four-Manifolds*, vol. 1 of *Mathematical Sciences Research Institute Publications*. Springer-Verlag, 2nd edn. (1991).
[36] R. Friedman. Vector bundles and SO(3)-invariants for elliptic surfaces. *J. Amer. Math. Soc.*, **8** (1995), 29–139.
[37] R. Friedman and J. W. Morgan. On the diffeomorphism types of certain algebraic surfaces. I. *J. Differential Geom.*, **27** (1988), 297–369.

[38] R. Friedman and J. W. Morgan. On the diffeomorphism types of certain algebraic surfaces. II. *J. Differential Geom.*, **27** (1988), 371–398.
[39] R. Friedman and J. W. Morgan. *Smooth Four-Manifolds and Complex Surfaces*, vol. 27 of *Ergebnisse der Mathematik und ihrer Grenzgebiete (3)*. Springer-Verlag (1994).
[40] K. A. Frøyshov. On Floer homology and four-manifolds with boundary. Ph.D. thesis, Oxford University (1995).
[41] K. A. Frøyshov. The Seiberg–Witten equations and four-manifolds with boundary. *Math. Res. Lett.*, **3** (1996), 373–390.
[42] K. A. Frøyshov. Equivariant aspects of Yang-Mills Floer theory. *Topology*, **41** (2002), 525–552.
[43] K. A. Frøyshov. Monopole floer homology for rational homology 3-spheres (2004). Preprint.
[44] D. Gabai. Foliations and the topology of 3-manifolds. *J. Differential Geom.*, **18** (1983), 445–503.
[45] D. Gabai. Foliations and the topology of 3-manifolds III. *J. Differential Geom.*, **26** (1987), 479–536.
[46] D. Gabai. Foliations and 3-manifolds. In *Proceedings of the International Congress of Mathematicians (Kyoto, 1990)*. Mathematical Society of Japan (1991), 609–619.
[47] P. Ghiggini. Knot Floer homology detects genus-one fibred links. arXiv:math.GT/0603445.
[48] E. Giroux. Géométrie de contact: de la dimension trois vers les dimensions supérieures. In *Proceedings of the International Congress of Mathematicians (Beijing, 2002)*. Higher Ed. Press (2002), 405–414.
[49] R. E. Gompf and T. S. Mrowka. Irreducible 4-manifolds need not be complex. *Ann. of Math. (2)*, **138** (1993), 61–111.
[50] C. M. Gordon and J. Luecke. Knots are determined by their complements. *Bull. Amer. Math. Soc. (N.S.)*, **20** (1989), 83–87.
[51] M. Gromov and H. B. Lawson, Jr. Positive scalar curvature and the Dirac operator on complete Riemannian manifolds. *Inst. Hautes Etudes Sci. Publ. Math.*, **58** (1983), 83–196 (1984).
[52] F. Hirzebruch and H. Hopf. Felder von Flächenelementen in 4-dimensionalen Mannigfaltigkeiten. *Math. Ann.*, **136** (1958), 156–172.
[53] L. Hörmander. *The Analysis of Linear Partial Differential Operators. III*, vol. 274 of *Grundlehren der Mathematischen Wissenschaften*. Springer-Verlag (1985).
[54] S. Jabuka and T. Mark. The Heegaard Floer homology of a surface times a circle. arXiv:math.GT/0502328.
[55] A. Jaffe and C. Taubes. *Vortices and Monopoles*, vol. 2 of *Progress in Physics*. Birkhäuser (1980).
[56] K. Kodaira. On compact analytic surfaces. II. *Ann. of Math. (2)*, **77** (1963), 563–626.
[57] K. Kodaira. On compact analytic surfaces. III. *Ann. of Math. (2)*, **78** (1963), 1–40.
[58] D. Kotschick. The Seiberg–Witten invariants of symplectic four-manifolds (after C. H. Taubes). Séminaire Bourbaki, Vol. 1995/96. *Astérisque* (1997), Exp. No. 812, 4, 195–220.

[59] P. B. Kronheimer and T. S. Mrowka. The genus of embedded surfaces in the projective plane. *Math. Res. Lett.*, **1** (1994), 797–808.

[60] P. B. Kronheimer and T. S. Mrowka. Recurrence relations and asymptotics for four-manifold invariants. *Bull. Amer. Math. Soc. (N.S.)*, **30** (1994), 215–221.

[61] P. B. Kronheimer and T. S. Mrowka. Embedded surfaces and the structure of Donaldson's polynomial invariants. *J. Differential Geom.*, **41** (1995), 573–734.

[62] P. B. Kronheimer and T. S. Mrowka. Monopoles and contact structures. *Invent. Math.*, **130** (1997), 209–255.

[63] P. B. Kronheimer, T. S. Mrowka, P. Ozsváth and Z. Szabó. Monopoles and lens space surgeries. *Ann. of Math. (2)*, **165** (2007), 457–546.

[64] N. H. Kuiper. The homotopy type of the unitary group of Hilbert space. *Topology*, **3** (1965), 19–30.

[65] T. J. Li and A. Liu. General wall crossing formula. *Math. Res. Lett.*, **2** (1995), 797–810.

[66] A. Lichnerowicz. Spineurs harmoniques. *C. R. Acad. Sci. Paris*, **257** (1963), 7–9.

[67] P. Lisca. On the Donaldson polynomials of elliptic surfaces. *Math. Ann.*, **299** (1994), 629–639.

[68] R. B. Lockhart and R. C. McOwen. Elliptic differential operators on noncompact manifolds. *Ann. Scuola Norm. Sup. Pisa Cl. Sci. (4)*, **12** (1985), 409–447.

[69] R. Lutz. Structures de contact sur les fibrés principaux en cercles de dimension trois. *Ann. Inst. Fourier (Grenoble)*, **27** (1977), ix, 1–15.

[70] M. Marcolli. *Seiberg–Witten Gauge Theory*, vol. 17 of *Texts and Readings in Mathematics*. Hindustan Book Agency (1999). With an appendix by the author and Erion J. Clark.

[71] M. Marcolli and B.-L. Wang. Equivariant Seiberg–Witten Floer homology. *Comm. Anal. Geom.*, **9** (2001), 451–639.

[72] J. Martinet. Formes de contact sur les variétés de dimension 3. In *Proceedings of Liverpool Singularities Symposium, II (1969/1970)*, vol. 209 of *Lecture Notes in Mathematics*. Springer–Verlag (1971), 142–163.

[73] G. Matić. SO(3)-connections and rational homology cobordisms. *J. Differential Geom.*, **28** (1988), 277–307.

[74] G. Meng and C. H. Taubes. $\underline{\mathrm{SW}}$ = Milnor torsion. *Math. Res. Lett.*, **3** (1996), 661–674.

[75] J. Milnor. *Morse Theory*. Based on lecture notes by M. Spivak and R. Wells. Vol. 51 of *Annals of Mathematics Studies*. Princeton University Press (1963).

[76] J. D. Moore. *Lectures on Seiberg–Witten Invariants*, vol. 1629 of *Lecture Notes in Mathematics*. Springer-Verlag, 2nd edn. (2001).

[77] J. W. Morgan. *The Seiberg–Witten Equations and Applications to the Topology of Smooth Four-Manifolds*, vol. 44 of *Mathematical Notes*. Princeton University Press (1996).

[78] J. W. Morgan, T. Mrowka and D. Ruberman. *The L^2-Moduli Space and a Vanishing Theorem for Donaldson Polynomial Invariants*, vol II of *Monographs in Geometry and Topology*. International Press (1994).

[79] J. W. Morgan and T. S. Mrowka. On the diffeomorphism classification of regular elliptic surfaces. *Internat. Math. Res. Notices*, (1993), 183–184.

[80] J. W. Morgan, T. S. Mrowka and Z. Szabó. Product formulas along T^3 for Seiberg–Witten invariants. *Math. Res. Lett.*, **4** (1997), 915–929.

[81] J. W. Morgan and K. G. O'Grady. *Differential Topology of Complex Surfaces*, vol. 1545 of *Lecture Notes in Mathematics*. Springer-Verlag (1993).

[82] J. W. Morgan, Z. Szabó and C. H. Taubes. A product formula for the Seiberg–Witten invariants and the generalized Thom conjecture. *J. Differential Geom.*, **44** (1996), 706–788.

[83] T. S. Mrowka. A local Mayer-Vietoris principle for Yang-Mills moduli spaces. Ph.D. thesis, University of California, Berkeley (1989).

[84] K. Nagami. *Dimension theory*, vol 37 of *Pure and Applied Mathematics*. Academic Press (1970). With an appendix by Yukihiro Kodama.

[85] Y. Ni. Knot Floer homology detects fibred knots. arXiv:math.GT/0607156.

[86] L. I. Nicolaescu. *Notes on Seiberg–Witten Theory*, vol. 28 of *Graduate Studies in Mathematics*. American Mathematical Society (2000).

[87] S. P. Novikov. Multivalued functions and functionals. An analogue of the Morse theory. *Dokl. Akad. Nauk SSSR*, **260** (1981), 31–35.

[88] P. Ozsváth and Z. Szabó. The symplectic Thom conjecture. *Ann. of Math. (2)*, **151** (2000), 93–124.

[89] P. Ozsváth and Z. Szabó. Absolutely graded Floer homologies and intersection forms for four-manifolds with boundary. *Adv. Math.*, **173** (2003), 179–261.

[90] P. Ozsváth and Z. Szabó. On the Floer homology of plumbed three-manifolds. *Geom. Topol.*, **7** (2003), 185–224 (electronic).

[91] P. Ozsváth and Z. Szabó. Heegaard diagrams and holomorphic disks. In *Different Faces of Geometry*, in *International Mathematics Series (New York)*. Kluwer/Plenum (2004), 301–348.

[92] P. Ozsváth and Z. Szabó. Holomorphic disks and knot invariants. *Adv. Math.*, **186** (2004), 58–116.

[93] P. Ozsváth and Z. Szabó. Holomorphic disks and topological invariants for closed three-manifolds. *Ann. of Math. (2)*, **159** (2004), 1027–1158.

[94] P. Ozsváth and Z. Szabó. Holomorphic disks and three-manifold invariants: properties and applications. *Ann. of Math. (2)*, **159** (2004), 1159–1245.

[95] P. Ozsváth and Z. Szabó. On Heegaard Floer homology and Seifert fibered surgeries. In *Proceedings of the Casson Fest*, vol. 7 of *Geometry and Topology Monographs*. Geometry and Topology Publications (2004), 181–203 (electronic).

[96] P. Ozsváth and Z. Szabó. On the Heegaard Floer homology of branched double-covers. *Adv. Math.*, **194** (2005), 1–33.

[97] P. Ozsváth and Z. Szabó. Holomorphic triangles and invariants for smooth four-manifolds. *Adv. Math.*, **202** (2006), 326–400.

[98] R. S. Palais. *Foundations of Global Non-Linear Analysis*. W. A. Benjamin, Inc. (1968).

[99] D. Ruberman. Rational homology cobordisms of rational space forms. *Topology*, **27** (1988), 401–414.

[100] D. Ruberman and S. Strle. Mod 2 Seiberg–Witten invariants of homology tori. *Math. Res. Lett.*, **7** (2000), 789–799.

[101] R. Schoen and S. T. Yau. Existence of incompressible minimal surfaces and the topology of three-dimensional manifolds with nonnegative scalar curvature. *Ann. of Math. (2)*, **110** (1979), 127–142.

[102] S. Sedlacek. A direct method for minimizing the Yang-Mills functional over 4-manifolds. *Comm. Math. Phys.*, **86** (1982), 515–527.

[103] S. Smale. An infinite dimensional version of Sard's theorem. *Amer. J. Math.*, **87** (1965), 861–866.

[104] N. E. Steenrod. Products of cocycles and extensions of mappings. *Ann. of Math. (2)*, **48** (1947), 290–320.

[105] A. Stipsicz and Z. Szabó. The smooth classification of elliptic surfaces with $b^+ > 1$. *Duke Math. J.*, **75** (1994), 1–50.

[106] D. Sullivan. Cycles for the dynamical study of foliated manifolds and complex manifolds. *Invent. Math.*, **36** (1976), 225–255.

[107] C. H. Taubes. The Seiberg–Witten equations and the Weinstein conjecture (2006). arXiv:math.SG/0611007.

[108] C. H. Taubes. Self-dual Yang-Mills connections on non-self-dual 4-manifolds. *J. Differential Geom.*, **17** (1982), 139–170.

[109] C. H. Taubes. Gauge theory on asymptotically periodic 4-manifolds. *J. Differential Geom.*, **25** (1987), 363–430.

[110] C. H. Taubes. A framework for Morse theory for the Yang-Mills functional. *Invent. Math.*, **94** (1988), 327–402.

[111] C. H. Taubes. Casson's invariant and gauge theory. *J. Differential Geom.*, **31** (1990), 547–599.

[112] C. H. Taubes. *L^2 Moduli Spaces on 4-Manifolds with Cylindrical Ends*, vol I in *Monographs in Geometry and Topology*. International Press (1993).

[113] C. H. Taubes. The role of reducibles in Donaldson–Floer theory. In *Topology, Geometry and Field Theory*. World Scientific Publishing (1994), 171–191.

[114] C. H. Taubes. The Seiberg–Witten invariants and symplectic forms. *Math. Res. Lett.*, **1** (1994), 809–822.

[115] C. H. Taubes. Unique continuation theorems in gauge theories. *Comm. Anal. Geom.*, **2** (1994), 35–52.

[116] C. H. Taubes. The Seiberg–Witten and Gromov invariants. *Math. Res. Lett.*, **2** (1995), 221–238.

[117] C. H. Taubes. *The Seiberg–Witten Invariants*. Selected Lectures in Mathematics. American Mathematical Society (1995). Lectures presented in San Francisco, California, January 1995.

[118] C. H. Taubes. The Seiberg–Witten invariants and 4-manifolds with essential tori. *Geom. Topol.*, **5** (2001), 441–519 (electronic).

[119] W. P. Thurston. The theory of foliations of codimension greater than one. *Comment. Math. Helv.*, **49** (1974), 214–231.

[120] W. P. Thurston. Existence of codimension-one foliations. *Ann. of Math. (2)*, **104** (1976), 249–268.

[121] W. P. Thurston. A norm for the homology of 3-manifolds. *Mem. Amer. Math. Soc.*, **59** (1986), 99–130.

[122] K. K. Uhlenbeck. Removable singularities in Yang-Mills fields. *Comm. Math. Phys.*, **83** (1982), 11–29.

[123] K. K. Uhlenbeck. Connections with L^p bounds on curvature. *Comm. Math. Phys.*, **83** (1982), 31–42.

[124] E. Witten. Supersymmetry and Morse theory. *J. Differential Geom.*, **17** (1982), 661–692.

[125] E. Witten. Monopoles and four-manifolds. *Math. Res. Lett.*, **1** (1994), 769–796.

[126] J. W. Wood. Foliations of codimension one. *Bull. Amer. Math. Soc.*, **76** (1970), 1107–1111.

Glossary of notation

A	Used to denote a typical spin^c connection, usually on a 4-manifold, page 5.
$\mathbb{A}(Y)$	The graded ring $\Lambda(H_1(Y)/\text{torsion}) \otimes \mathbb{Z}[U]$, over which the Floer groups of Y are modules, page 56.
$\mathbb{A}_\dagger(Y)$	The ring $\mathbb{A}(Y)$ equipped with the opposite multiplication and a reversed grading, page 56.
$\mathfrak{a}, \mathfrak{b}$	Used to denote solutions of the Seiberg–Witten equations in the blown-up configuration space, $\mathcal{C}^\sigma_k(Y, \mathfrak{s})$. Also referred to as critical points, page 196.
$[\mathfrak{a}], [\mathfrak{b}]$	Used to denote the gauge equivalence classes represented by solutions $\mathfrak{a}$, $\mathfrak{b}$ of the Seiberg–Witten equations in the blown-up configuration space, $\mathcal{B}^\sigma_k(Y, \mathfrak{s})$. Also referred to as critical points, page 197.
$\check{A}$	If A is a connection on bundle over a 4-dimensional cylinder, $\mathbb{R} \times Y$, then for each $t \in \mathbb{R}$ we obtain by restriction a connection $\check{A}(t)$ on $\{t\} \times Y$; thus $\check{A}$ is a path of connections on Y, page 97.
B	Used to denote a typical spin^c connection, usually on a 3-manifold, page 6.
$b^+(X)$	For a closed oriented 4-manifold X, the second betti number can be written $b_2 = b^+ + b^-$, where the difference $b^+ - b^-$ is the signature. Thus b^+ is the dimension of a maximal positive-definite subspace for the quadratic form on $H^2(X; \mathbb{R})$.
$b^+(W)$	For an oriented 4-dimensional cobordism, the dimension of a maximal positive-definite subspace for the non-degenerate quadratic form on the image of $H^2(W, \partial W; \mathbb{R})$ in $H^2(W; \mathbb{R})$.
$\mathcal{B}(X, \mathfrak{s})$	The quotient of $\mathcal{C}(X, \mathfrak{s})$ by the gauge group $\mathcal{G}(X)$, page 8.
$\mathcal{B}^\sigma(X, \mathfrak{s})$	The quotient of $\mathcal{C}^\sigma(X, \mathfrak{s})$ by the gauge group $\mathcal{G}(X)$, page 138.
$\mathcal{B}^\sigma_k(X, \mathfrak{s})$	The quotient of $\mathcal{C}^\sigma_k(X, \mathfrak{s})$ by the action of the group $\mathcal{G}_{k+1}(X)$, page 138.
$\mathcal{B}^*(X, \mathfrak{s})$	The quotient of $\mathcal{C}^*(X, \mathfrak{s})$ by the gauge group $\mathcal{G}(X)$, page 11.
$\mathcal{B}(Y, \mathfrak{s})$	The quotient of $\mathcal{C}(Y, \mathfrak{s})$ by the gauge group $\mathcal{G}(Y)$, page 138.
$\mathcal{B}^\sigma(Y, \mathfrak{s})$	The quotient of $\mathcal{C}^\sigma(Y, \mathfrak{s})$ by the gauge group $\mathcal{G}(Y)$, page 138.

$\mathcal{B}^\sigma_k(Y,\mathfrak{s})$ The quotient of $\mathcal{C}^\sigma_k(Y,\mathfrak{s})$ by the action of the group $\mathcal{G}_{k+1}(Y)$, page 138.

$\boldsymbol{\mathcal{B}}^\sigma(Y)$ The union of $\mathcal{B}^\sigma(Y,\mathfrak{s})$ taken over all spin^c structures $\mathfrak{s}$ on Y, page 454.

$\mathcal{B}^*(Y,\mathfrak{s})$ The quotient of $\mathcal{C}^*(Y,\mathfrak{s})$ by the gauge group $\mathcal{G}(Y)$, page 140.

$\tilde{\mathcal{B}}^\sigma(Y,\mathfrak{s})$ The quotient of $\tilde{\mathcal{C}}^\sigma(Y,\mathfrak{s})$ by the action of the gauge-group $\mathcal{G}(Y)$, page 323.

$\mathcal{B}^\tau_{k,\mathrm{loc}}(Z)$ On an infinite cylinder $Z=\mathbb{R}\times Y$, the quotient of $\mathcal{C}^\tau_{k,\mathrm{loc}}(Z)$ by the action of the gauge group, page 219.

$\tilde{\mathcal{B}}^\tau_k(Z)$ On a finite cylinder $Z=[a,b]\times Y$, the quotient of $\tilde{\mathcal{C}}^\tau_k(Z)$ by the action of the gauge group, page 138.

$c_1(\mathfrak{s})$ For spin^c structure $\mathfrak{s}$, the first Chern class of either the associated bundle S (in the 3-dimensional case) or S^+ (in the 4-dimensional case).

c_n The nth Chern class.

$\mathcal{C}(X,\mathfrak{s})$ The configuration space of pairs (A,Φ) consisting of a smooth spin^c connection A and spinor Φ, on a 4-manifold X, possibly with boundary, page 91.

$\mathcal{C}^\sigma(X,\mathfrak{s})$ The blow-up of $\mathcal{C}(X,\mathfrak{s})$ along the locus $\Phi=0$. A typical element is represented as a triple (A,s,ϕ), where A is a spin^c connection, s is a non-negative real number and ϕ is a spinor on X with L^2 norm 1, page 113.

$\mathcal{C}^\sigma_k(X,\mathfrak{s})$ The L^2_k Sobolev completion of $\mathcal{C}^\sigma(X,\mathfrak{s})$, page 135.

$\mathcal{C}^*(X,\mathfrak{s})$ The irreducible part of $\mathcal{C}(X,\mathfrak{s})$: the configurations (A,Φ) with Φ non-zero, page 140.

$\mathcal{C}(Y,\mathfrak{s})$ The configuration space of pairs (B,Ψ) consisting of a smooth spin^c connection B and spinor Ψ, on a closed 3-manifold Y, page 85.

$\mathcal{C}^\sigma(Y,\mathfrak{s})$ The blow-up of $\mathcal{C}(Y,\mathfrak{s})$ along the locus $\Psi=0$. A typical element is represented as a triple (B,r,ψ), where B is a spin^c connection, r is a non-negative real number and ψ is a spinor on Y with L^2 norm 1, page 115.

$\mathcal{C}^\sigma_k(Y,\mathfrak{s})$ The L^2_k Sobolev completion of $\mathcal{C}^\sigma(Y,\mathfrak{s})$, page 135.

$\mathcal{C}^*(Y,\mathfrak{s})$ The irreducible part of $\mathcal{C}(Y,\mathfrak{s})$: the configurations (B,Ψ) with Ψ non-zero, page 113.

$\tilde{\mathcal{C}}^\sigma(Y,\mathfrak{s})$ The double of $\mathcal{C}^\sigma(Y,\mathfrak{s})$. A typical element is represented as a triple (B,r,ψ), where B is a spin^c connection, r is a real number and ψ is a spinor on Y with L^2 norm 1. It contains $\mathcal{C}^\sigma(Y,\mathfrak{s})$ as the subset where $r\geq 0$, page 137.

$\mathcal{C}^\tau_{k,\mathrm{loc}}(Z)$ On an infinite cylinder $Z=\mathbb{R}\times Y$, a space of triples (A,s,ϕ), with s a non-negative real-valued function of the first coordinate, lying in an $L^2_{k,\mathrm{loc}}$ Sobolev space, page 218.

$\tilde{\mathcal{C}}^\tau_k(Z)$ On a finite cylinder $Z=[a,b]\times Y$, a space of triples (A,s,ϕ), with s a real-valued function of the first coordinate, lying in an L^2_k Sobolev space. This space contains $\mathcal{C}^\tau_k(Z)$ as the locus where s is everywhere non-negative, page 137.

$\check{C}_*$, $\hat{C}_*$, $\bar{C}_*$ The Floer homology groups $\widehat{HM}_*(Y,\mathfrak{s})$ etc. are the homology of differentials $\check{\partial}:\check{C}_*\to\check{C}_*$ etc., page 412.

C^s, C^u, C^o — The summands of the groups $\check{C}_*$ etc. formed from critical points that are boundary-stable, boundary-unstable, and irreducible respectively, page 412.

$\mathcal{D}$ — In the context of a Banach manifold, $\mathcal{D}$ is used to denote a derivative.

$\check{\partial}, \hat{\partial}, \bar{\partial}$ — The differentials on $\check{C}_*$, $\hat{C}_*$, $\bar{C}_*$ (*qq.v.*), page 415.

∂^u_s etc. — The individual maps $C^u \to C^s$ etc. from which the differentials $\check{\partial}$ and $\hat{\partial}$ are constructed, page 415.

$\bar{\partial}^u_s$ etc. — The maps $C^u \to C^s$ etc. from which the differential $\bar{\partial}$ is constructed. These are defined in terms of the reducible solutions to the Seiberg–Witten equations, page 414.

$\mathfrak{F}$ — On a 4-manifold X, the Seiberg–Witten equations for a pair (A, Φ) in $\mathcal{C}(X, \mathfrak{s})$ are written $\mathfrak{F}(A, \Phi) = 0$, page 8.

$\mathfrak{F}_{\mathfrak{q}}$ — The Seiberg–Witten equations on a cylinder, perturbed by $\hat{\mathfrak{q}}$, page 154.

$\mathfrak{F}^\sigma$ — On a 4-manifold X, the Seiberg–Witten equations for a triple (A, s, ϕ) in the blown-up configuration space $\mathcal{C}^\sigma(X, \mathfrak{s})$ are written $\mathfrak{F}^\sigma(A, s, \phi) = 0$, page 114.

$\mathfrak{F}^\tau$ — On a cylinder $Z = [a, b] \times Y$, the Seiberg–Witten equations for a triple (A, s, ϕ) in the τ model of the blown-up configuration space $\mathcal{C}^\tau(Z, \mathfrak{s})$ are written $\mathfrak{F}^\tau(A, s, \phi) = 0$, page 120.

$\mathrm{gr}([\mathfrak{a}])$ — For a critical point $\mathfrak{a}$, the grading in which the corresponding generator of the Floer complex $\check{C}_*$ or $\hat{C}_*$ appears. An element of $\mathbb{J}(Y)$, page 424.

$\mathrm{gr}^{(2)}([\mathfrak{a}])$ — The image of the grading element $\mathrm{gr}([\mathfrak{a}])$ under the canonical map $\mathbb{J}(Y) \to \mathbb{Z}/2$, page 427.

$\bar{\mathrm{gr}}^{(2)}([\mathfrak{a}])$ — For a reducible critical point $\mathfrak{a}$, the image of the grading element $\bar{\mathrm{gr}}([\mathfrak{a}])$ (*q.v.*) under the canonical map $\mathbb{J}(Y) \to \mathbb{Z}/2$, page 428.

$\bar{\mathrm{gr}}([\mathfrak{a}])$ — For a reducible critical point $\mathfrak{a}$, the grading in which the corresponding generator of the Floer complex $\bar{C}_*$ appears. An element of $\mathbb{J}(Y)$, equal to $\mathrm{gr}([\mathfrak{a}]) - 1$ if $\mathfrak{a}$ is boundary-unstable, page 293.

$\mathrm{gr}^{\mathbb{Q}}([\mathfrak{a}])$ — The image of the grading element $\mathrm{gr}([\mathfrak{a}])$ under the canonical map $\mathbb{J}(Y, \mathfrak{s}) \to \mathbb{Q}$ in the case that $c_1(\mathfrak{s})$ is torsion, page 587.

$\mathrm{grad}\, \mathcal{L}$ — The gradient of the Chern–Simons–Dirac functional on $\mathcal{C}(Y, \mathfrak{s})$, page 86.

$(\mathrm{grad}\, \mathcal{L})^\sigma$ — Formally, the vector field on the blown-up configuration space, $\mathcal{C}^\sigma(Y, \mathfrak{s})$ obtained from $\mathrm{grad}\, \mathcal{L}$ on $\mathcal{C}(Y, \mathfrak{s})$, page 117.

$\mathcal{G}(X)$ — The gauge group, consisting of all smooth maps $u : X \to S^1$, page 5.

$\mathcal{G}_{k+1}(X)$ — The L^2_{k+1} Sobolev completion of the gauge group $\mathcal{G}(X)$, page 135.

$\mathcal{G}_{k+1}(Y)$ — The L^2_{k+1} Sobolev completion of the gauge group $\mathcal{G}(Y)$, page 135.

Γ — A system of local coefficients on $\mathcal{B}^\sigma(Y, \mathfrak{s})$ or on $\boldsymbol{\mathcal{B}}^\sigma(Y)$, page 443.

Γ_η — The local system on $\boldsymbol{\mathcal{B}}^\sigma(Y)$ with fiber $\mathbb{R}$, corresponding to a choice of C^∞ real 1-cycle η in Y, page 445.

$\widehat{HM}, \widecheck{HM}$ — Two flavors of Floer homology, page 420.

$\overline{HM}$ — The third flavor of Floer homology, constructed using the reducible critical points, which appears with $\widehat{HM}$ and $\widecheck{HM}$ in a long exact sequence, page 420.

$\widehat{HM}_*(Y)$	One flavor of the Floer homology of a 3-manifold Y, page 420.
$\widehat{HM}^*(Y)$	One flavor of the Floer cohomology of a 3-manifold Y, page 426.
$\widehat{HM}_j(Y)$	One summand of $\widehat{HM}_*(Y)$ in grading j. The group $\overline{HM}_*(Y)$ is graded by the set of homotopy classes of oriented 2-plane fields on Y, page 425.
$\widehat{HM}_*(Y,\mathfrak{s})$	The summand of $\widehat{HM}_*(Y)$ corresponding to the spin^c structure $\mathfrak{s}$ on Y, page 420.
$\widehat{HM}_\bullet(Y,\mathfrak{s})$	The completion of $\widehat{HM}_*(Y,\mathfrak{s})$, page 513.
$HM_\bullet(Y)$	The image of $\widecheck{HM}_\bullet(Y)$ in $\widehat{HM}_\bullet(Y)$ under the map j in the long exact sequence; a finitely generated abelian group. Often identified with both $\widecheck{HM}_\bullet(Y)$ and $\widehat{HM}_\bullet(Y)$ when j is an isomorphism, page 68.
$\widehat{HM}_\bullet(Y;\Gamma)$	The Floer homology of type $\widehat{HM}$ with coefficients in a local system Γ, page 444.
$\widehat{HM}_\bullet(Y,\mathfrak{s},c)$	The Floer homology obtained from a non-exact perturbation of $\mathcal{L}$, with period class c, page 597.
$\widehat{HM}_\bullet(W)$	The map from $\widehat{HM}_\bullet(Y_1)$ to $\widehat{HM}_\bullet(Y_2)$ arising from a cobordism W equipped with a homology orientation, page 451.
$\mathbf{i}$	The involution $r \mapsto -r$ on $\tilde{\mathcal{C}}^\sigma(Y,\mathfrak{s})$ (*q.v.*). The quotient space can be identified with $\mathcal{C}^\sigma(Y,\mathfrak{s})$. Also used for the corresponding involution on $M([\mathfrak{a}],[\mathfrak{b}])$ for example, page 137.
$I^2(X)$	For a 4-manifold with boundary, the image of $H^2(X,\partial X;\mathbb{R})$ in $H^2(X;\mathbb{R})$. Also used for a cobordism W, page 60.
$\iota(W)$	A characteristic number of a cobordism, page 526.
$\mathbb{J}(Y)$	The set with Z-action by which the Floer groups of Y are graded. This set can be identified with $\pi_0(\Xi(Y))$ (*q.v.*), page 424.
$\mathcal{J},\mathcal{K}$	On the irreducible part $\mathcal{C}^*_k(Y,\mathfrak{s}) \subset \mathcal{C}_k(Y,\mathfrak{s})$ of the configuration space, the L^2 completion of the tangent bundle has an orthogonal decomposition as $\mathcal{J}\oplus\mathcal{K}$, where the $\mathcal{J}$ is tangent to the gauge orbits, page 140.
$\mathcal{J}^\sigma,\mathcal{K}^\sigma$	The L^2 completion of the tangent bundle to $\mathcal{C}^\sigma_k(Y,\mathfrak{s})$ decomposes as $\mathcal{J}^\sigma \oplus \mathcal{K}^\sigma$. Away from the reducibles, this decomposition corresponds to the decomposition $\mathcal{J}\oplus\mathcal{K}$ via the blow-down map $\pi : \mathcal{C}^\sigma_k(Y,\mathfrak{s}) \to \mathcal{C}_k(Y,\mathfrak{s})$, page 142.
$\mathcal{K}^+,\mathcal{K}^-$	The spectral decomposition of $\mathcal{K}^\sigma$ determined by the Hessian of the perturbed Chern–Simons–Dirac functional, page 313.
$L^2_k(M)$	For $k \geq 0$, the space of distributions on the Riemannian manifold M having k derivatives in L^2. On a closed manifold, the space L^2_{-k} is the dual to L^2_k. We also use fractional Sobolev spaces, in particular $L^2_{k-1/2}(Y)$; see Subsection 17.1.
$L^2_{j,k}(\mathbb{R}\times Y)$	An anisotropic Sobolev space, Subsection 11.4.
$\mathcal{L}$	The Chern–Simons–Dirac functional, as a function on $\mathcal{C}(Y,\mathfrak{s})$ or $\mathcal{C}^\sigma(Y,\mathfrak{s})$, page 85.
$\pounds$	The perturbed Chern–Simons–Dirac functional, page 152.
$\pounds_{\mathfrak{q}}$	The perturbed Chern–Simons–Dirac functional given by $\mathcal{L}+f$, where f is a function with gradient $\mathfrak{q}$, page 265.

$\Lambda([\mathfrak{a}])$	A 2-element set canonically associated with a critical point $[\mathfrak{a}]$, used in orienting the moduli spaces $M([\mathfrak{a}],[\mathfrak{b}])$, page 376.
$\Lambda(B,r,\psi)$	The gauge-invariant function of (B,r,ψ) on $\mathcal{C}^\sigma(Y)$ defined by $\langle\psi, D_B\psi\rangle_{L^2}$ for a spinor ψ of L^2 norm 1, page 117.
$\Lambda_\mathfrak{q}(B,r,\psi)$	The perturbation of the function $\Lambda(B,r,\psi)$ defined by $\langle\psi, D_{\mathfrak{q},B}\psi\rangle_{L^2}$, page 157.
$M([\mathfrak{a}],[\mathfrak{b}])$	On the infinite cylinder $Z=\mathbb{R}\times Y$, the space of solutions to the perturbed Seiberg–Witten equations in $\mathcal{B}^\tau_{k,\mathrm{loc}}(Z)$ which are asymptotic to the critical points $[\mathfrak{a}]$ and $[\mathfrak{b}]$ at the two ends, page 220.
$\breve{M}([\mathfrak{a}],[\mathfrak{b}])$	The moduli space of unparametrized trajectories: the quotient of $M([\mathfrak{a}],[\mathfrak{b}])$ by translations, excluding the constant trajectory if $[\mathfrak{a}]=[\mathfrak{b}]$, page 275.
$\breve{M}^+([\mathfrak{a}],[\mathfrak{b}])$	The compactification of $\breve{M}([\mathfrak{a}],[\mathfrak{b}])$ (*q.v.*) by broken trajectories, page 276.
$\tilde{M}([\mathfrak{a}],[\mathfrak{b}])$	On the infinite cylinder $Z=\mathbb{R}\times Y$, the space of solutions to the perturbed Seiberg–Witten equations in $\tilde{\mathcal{B}}^\tau_{k,\mathrm{loc}}(Z)$ which are asymptotic the critical points $[\mathfrak{a}]$ and $[\mathfrak{b}]$ at the two ends. This contains $M([\mathfrak{a}],[\mathfrak{b}])$ just as $\tilde{\mathcal{B}}^\tau_{k,\mathrm{loc}}(Z)$ contains $\mathcal{B}^\tau_{k,\mathrm{loc}}(Z)$, page 220.
$M_z([\mathfrak{a}],[\mathfrak{b}])$	The moduli space $M([\mathfrak{a}],[\mathfrak{b}]$ has a decomposition into parts $M_z([\mathfrak{a}],[\mathfrak{b}])$ according to the homotopy classes z for paths from $[\mathfrak{a}]$ to $[\mathfrak{b}]$ in $\mathcal{B}^\sigma_k(Y,\mathfrak{s})$, page 220.
$M([\mathfrak{a}],W^*,[\mathfrak{b}])$	For the manifold with cylindrical ends, W^* (*q.v.*), obtained by attaching half-infinite cylinders to a cobordism W, the moduli space of solutions to the perturbed Seiberg–Witten equations, asymptotic to $[\mathfrak{a}]$ and $[\mathfrak{b}]$ on the two ends. Like $M(X^*,\mathfrak{b})$ (*q.v.*), it is defined using a fiber product description, page 509.
$\bar{M}([\mathfrak{a}],W^*,[\mathfrak{b}])$	A compactification of $M([\mathfrak{a}],W^*,[\mathfrak{b}])$ which is smaller than $M^+([\mathfrak{a}],W,[\mathfrak{b}])$, page 510.
$M^+([\mathfrak{a}],W^*,[\mathfrak{b}])$	The compactification of $M([\mathfrak{a}],W^*,[\mathfrak{b}])$ (*q.v.*) by broken trajectories, page 509.
$M(X,\mathfrak{s})$	The space of solutions to the (possibly perturbed) Seiberg–Witten equations on a compact 4-manifold X, as a subset of the blown-up configuration space $\mathcal{B}^\sigma_k(X,\mathfrak{s})$. This space is infinite-dimensional if X has non-empty boundary, page 463.
$M(X^*,\mathfrak{s},[\mathfrak{b}])$	The space of solutions to the (possibly perturbed) Seiberg–Witten equations on a 4-manifold X^* (*q.v.*) with a cylindrical end $\mathbb{R}^{\geq}\times Y$, asymptotic to $[\mathfrak{b}]$ on the end, defined using a fiber product description. The 3-manifold Y may have more than one component, page 465.
$\bar{M}(X^*,[\mathfrak{b}])$	A compactification of $M(X^*,[\mathfrak{b}])$ which is smaller than $M^+(X,[\mathfrak{b}])$, page 491.
$M^+(X^*,[\mathfrak{b}])$	The compactification of $M(X^*,[\mathfrak{b}])$ (*q.v.*) by broken trajectories, page 485.
$M(X,\mathfrak{s})_P$	The space of solutions to the Seiberg–Witten equations for a family of metrics and perturbations, parametrized by a manifold P, page 479.
$\tilde{M}(X,\mathfrak{s})$	The double of $M(X,\mathfrak{s})$, as a subset of $\tilde{\mathcal{B}}^\sigma_k(X,\mathfrak{s})$, page 463.

$\mathcal{P}$	A Banach space of tame perturbations for the Seiberg–Witten equations on Y, page 191.
π	A blow-down map, such as the map from $\mathcal{C}^\sigma(Y,\mathfrak{s})$ to $\mathcal{C}(Y,\mathfrak{s})$, page 113.
$\mathfrak{q}$	A typical tame perturbation of the gradient of the Chern–Simons–Dirac functional on Y. Formally, $\mathfrak{q}$ is the gradient of a function f on $\mathcal{C}(Y,\mathfrak{s})$ with respect to the L^2 inner product, page 153.
Q_γ	A linear operator associated to a trajectory γ, constructed from the linearized Seiberg–Witten equations and a gauge-fixing condition, page 254.
$\hat{\mathfrak{q}}$	The perturbation of the 4-dimensional Seiberg–Witten equations on a cylinder $Z = [a,b] \times Y$ resulting from a perturbation $\mathfrak{q}$ of the gradient of the Chern–Simons–Dirac functional, page 153.
ρ	Clifford multiplication on the 3-manifold Y: a bundle map $\rho : TY \to \mathrm{Hom}(S,S)$, whose domain is also extended to the bundles of forms, $\Lambda^i Y$, page 2.
ρ_X	Clifford multiplication on the 4-manifold X, page 4.
$\mathfrak{s}$	A typical spin^c structure, usually on a 3-manifold Y.
$S^+ \oplus S^-$	The decomposition of the spin^c bundle on a 4-manifold into its two half-spin bundles.
$\mathfrak{s}_X$	A typical spin^c structure on a 4-manifold X.
$\mathbb{S}(V)$	The unit sphere in V: if V is a topological vector space equipped with a continuous norm, then $\mathbb{S}(V)$ denotes the set of elements of norm 1. Without a norm, one can define $\mathbb{S}(V)$ as the quotient of $V \setminus 0$ by the action of the positive reals, page 113.
σ, τ	The two different ways to construct a blown-up version of the configuration space are referred to as the σ model and the τ model. The former is used both on a 3-manifold and on a general 4-manifold. The latter is applicable only to a cylindrical manifold. In finite-dimensional Morse theory on a manifold B, the τ model corresponds to the path space of a blow-up of B. See Section 6.
$\mathbb{T}$	The torus $H^1(Y;\mathbb{R})/H^1(Y;\mathbb{Z})$ which parametrizes flat $U(1)$ connections in the trivial bundle on Y, page 55.
$\mathcal{T}_j$	The L^2_j completion of the L^2_k tangent bundle of $\mathcal{C}_k(Y,\mathfrak{s})$, page 137.
$\mathcal{T}^\sigma_j$	The L^2_j completion of the L^2_k tangent bundle of $\mathcal{C}\sigma_k(Y,\mathfrak{s})$, page 136.
W	Usually used to denote a cobordism between connected 3-manifolds
w_2	The second Stiefel–Whitney class.
W^*	The complete manifold with cylindrical ends obtained by attaching two half-infinite cylinders to the boundary components of a 4-dimensional cobordism W, page 509.
X	Usually used to denote a compact, oriented 4-manifold, possibly with boundary.
X^*	The complete manifold with cylindrical ends obtained by attaching half-infinite cylinders to the boundary components of a 4-manifold X, page 464.

$\Xi(Y)$	The space of all oriented 2-plane fields on Y. Its path-components can be identified with $\mathbb{J}(Y)$ (*q.v.*), the set with Z-action by which the Floer groups are graded, page 51.
Y	Usually used to denote a compact, oriented 3-manifold, usually connected and without boundary.
Z	Usually used to denote a cylinder such as $[a,b] \times Y$ or $\mathbb{R} \times Y$.
$\mathbb{Z}\Lambda$	If Λ is a 2-element set, then $\mathbb{Z}\Lambda$ is an infinite cyclic group constructed so that a choice of element from Λ corresponds to a choice of generator for $\mathbb{Z}\Lambda$, page 20.
Z^T	The cylinder $[-T, T] \times Y$, page 323.
Z^∞	In Subsection 18.2, the disjoint union of two half-infinite cylinders $(\mathbb{R}^{\geq} \times Y) \amalg (\mathbb{R}^{\leq} \times Y)$, regarded as a limit of Z^T as $T \to \infty$, page 323.
z	Often used to denote a homotopy class of paths between critical points $[\mathfrak{a}], [\mathfrak{b}]$ in $\mathcal{B}^\sigma_k(Y, \mathfrak{s})$, or more generally an X-path or a W-path. The last letter of the alphabet, page 220.

Index

Index entries in boldface refer to a page where a term is defined.

Printed in the United States
By Bookmasters